Approaches for Enhancing Abiotic Stress Tolerance in Plants

Approaches for Enhancing Abiotic Stress Tolerance in Plants

Edited by
Mirza Hasanuzzaman, Kamrun Nahar, Masayuki Fujita,
Hirosuke Oku, and M. Tofazzal Islam

CRC Press
Taylor & Francis Group
Boca Raton London New York

CRC Press is an imprint of the
Taylor & Francis Group, an **informa** business

CRC Press
Taylor & Francis Group
6000 Broken Sound Parkway NW, Suite 300
Boca Raton, FL 33487-2742

First issued in paperback 2021

© 2019 by Taylor & Francis Group, LLC
CRC Press is an imprint of Taylor & Francis Group, an Informa business

No claim to original U.S. Government works

ISBN 13: 978-1-03-209396-3 (pbk)
ISBN 13: 978-0-8153-4642-5 (hbk)

Library of Congress Cataloging-in-Publication Data

Names: Hasanuzzaman, Mirza, editor.
Title: Approaches for enhancing abiotic stress tolerance in plants / editors:
Mirza Hasanuzzaman, Kamrun Nahar, Masayuki Fujita, Hirosuke Oku, Tofazzal
Islam.
Description: Boca Raton, FL : CRC Press, Taylor & Francis Group, 2019.
Identifiers: LCCN 2018032176| ISBN 9780815346425 (hardback : alk. paper) |
ISBN 9781351104715 (adobe) | ISBN 9781351104708 (epub) | ISBN
9781351104692 (mobi/kindle)
Subjects: LCSH: Plants--Effect of stress on.
Classification: LCC QK754 .A67 2019 | DDC 581.7--dc23
LC record available at https://lccn.loc.gov/2018032176

Visit the Taylor & Francis Web site at
http://www.taylorandfrancis.com

and the CRC Press Web site at
http://www.crcpress.com

Contents

Preface

In an open environment, plant growth and productivity are governed by several environmental factors that can be biotic or abiotic. Such environmental factors sometimes become very harsh to the growth and development of the plants. Global climate change is predicted to increase the frequency and intensity of environmental stresses such as salinity, drought, metal/metalloid toxicity, heat, chilling/freezing, flooding and atmospheric pollutants to plants. In the era of industrial development, some anthropogenic activities are also causing serious threats to the productivity of crops. Yield loss of crops due to abiotic stresses is higher than the loss caused by pests, diseases and weeds. Furthermore, environmental stresses to the crop plants are beyond the control of farmers. Yield loss of staple food crops can be up to 70% due to the detrimental effects of abiotic stresses.

The increasing world population, the subsequent higher demand for food and the increasing occurrence of abiotic stresses have made agriculture challenging in the 21st century. However, attempts to improve crop yield under stressful environments by improvement of plants through classical breeding have been largely unsuccessful mainly due to the multi-genic origin of the adaptive responses. Due to the physiological and genetic complexity of the stress tolerance traits, the real progress in crop breeding for stress tolerance may be achieved only via a painfully slow pyramiding of essential physiological traits. Therefore, a well-focused approach combining the molecular, physiological, biochemical and metabolic aspects of salt tolerance is essential to develop stress-tolerant crop varieties. Numerous studies indicated the factors governing the defense system in plants and the necessity to generate tolerant varieties which can acclimatize and adapt to the stressful environments without having any adverse impacts on their productivity. However, the molecular responses of plants to a combination of abiotic stresses are unique and cannot be directly extrapolated from the responses of plants alone. A large body of the literature suggested that, though with a certain degree of overlap, each stress causes a unique mechanism of response, tailored to the specific needs of the plant and that each combination of two or more different stresses may also have a specific response. The recent progress in molecular biology and genomics studies on many major crop plants are generating a wealth of information for the improvement of crop plants to abiotic stress. However, plant improvement for any abiotic stress is not merely a number of genes put together.

Further progress in the field may be achieved when various omics tools are intrinsically interspersed with the precise understanding of plant function and put into an environmental context.

This book compiles the recent updates of our understanding of various approaches in conferring abiotic stress tolerance. It includes 27 chapters contributed by 110 leading experts, spanning from the diverse areas of the field of plant physiology, environmental sciences, crop science, molecular biology and biotechnology. The first chapter presents the general outline of various abiotic stresses. The impacts of climate change on crop production in the world as well as in South Asia is described in Chapter 2. Plant responses to the salt stress are described in Chapter 3. Plant responses and signaling to drought stress tolerance are reviewed in Chapters 4 and 5, respectively. Variability in physiological, biochemical and molecular mechanisms of chickpea varieties to water stress is the subject matter of Chapter 6. Chapters 7 and 8 discuss the plant responses and tolerance to cold stress and cold stress tolerance in rice, respectively. Heavy metals in soils exert toxicity to plants and remarkably reduce the yield of crops. Chapter 9 updates heavy metal toxicity in plants and its mitigation. Boron is an essential nutrient, but higher levels of boron are toxic to plants. Responses of plants to nutrient deficiency and toxicity due to higher levels of boron are focused in Chapter 10. Adaptation of plants to elevated levels of ozone is described in Chapter 11. Various hydrocarbons are known as soil contaminants. Chapter 12 reviews hydrocarbon contamination and its amelioration in soils. An overview of abiotic stress-induced oxidative damage in plants is the subject matter of Chapter 13. The roles of transcription factors in the antioxidant responses of plants under abiotic stresses are reviewed in Chapter 14. Chapter 15 discusses modern approaches to enhance antioxidant defense systems in plants. Coordination and auto-propagation of ROS signals in plants are covered in Chapter 16. Synthesis of various osmolytes in plants under abiotic stresses is critical for plant tolerance. Regulation of osmolyte syntheses and improvement of abiotic stress tolerance in plants are discussed in Chapter 17. Plasma membrane plays an important role in protecting the cell from dehydration. Chapter 18 updates the roles of plasma membrane proteins in tolerance of dehydration in the plant cell.

Trehalose metabolism and signaling is an area of emerging significance. In less than a decade, our views on the importance of trehalose metabolism and its role in plants have gone through something of a revolution. Chapter 19 focuses on trehalose metabolism in plants under abiotic stressful environments. Proline is a small molecule biosynthesized in plants which plays a significant role in plants' tolerance to salinity and drought. The proline metabolism behavior of the durum wheat dehydrin transgenic *Arabidopsis thaliana* to salt stress is discussed in Chapter 20. Nitric oxide and polyamines have significant roles in plant tolerance to abiotic stresses. Molecular mechanisms of nitric oxide- and polyamines-induced plant tolerance to abiotic stresses are focused in Chapters 21 and 22, respectively. Molecular biological and genomic approaches broadened our understanding about the plants' responses to the abiotic stresses at molecular and genomic levels. Chapters 23 and 24 update current molecular and genomic knowledge about plants' tolerance and responses to the abiotic stresses. The hallmark attributes of plant transcription factors and the potential of WRKY, MYB and NAC in abiotic stresses are discussed in Chapter 25. Clustered Regularly Interspaced Short Palindromic Repeats (CRISPR)/Cas9 (CRISPR-associated enzyme 9) technology is becoming a faster, cheaper and precise tool for editing the genome of living organisms, including plants. It will revolutionize the engineering of crop plants, including for the enhancement of tolerance to the abiotic stresses. Chapter 26 updates current status and future prospects of CRISPR/Cas9 for engineering crop plants for abiotic stress tolerance. Plants harbor taxonomically diverse microorganisms. Plant-associated beneficial microorganisms (such as plant probiotic bacteria) play important roles in promoting abiotic stress tolerance in plants. Chapter 27 focused on the effects of various beneficial microorganisms in protecting plants from abiotic stresses. The wealth of information compiled in this volume updates our understanding the effects, mechanisms and interrelationships between and among various stresses, the defense strategies of plants to adapt to harsh environments, and the expression of genes involved in the synthesis of regulatory proteins. This fundamental knowledge and understanding is needed for the development of stress-tolerant plant varieties.

We, the editors, sincerely thank the authors for their outstanding works and timely contributions to publish this unique volume of book. We are highly thankful to Dr. Mahbub Alam, Lecturer, Department of Agriculture, Noakhali Science and Technology University, for his valuable help in formatting and incorporating editorial changes in the manuscripts. The Senior Editor (Biological Science), CRC Press Randy Brehm deserves our sincere thanks for prompt responses during the acquisition of this book. We are also thankful to other editorial staffs of CRC for precious help in formatting and incorporating editorial changes in the manuscripts. The editors and contributing authors hope that this book will be the reference of researchers for updating knowledge about the mechanisms and approaches for environmental stress tolerance.

Mirza Hasanuzzaman, Kamrun Nahar, Masayuki Fujita, Hirosuke Oku, and M. Tofazzal Islam

Editors

Mirza Hasanuzzaman is a Professor of Agronomy at Sher-e-Bangla Agricultural University, Dhaka, Bangladesh. In 2012, he received his PhD on 'Plant Stress Physiology and Antioxidant Metabolism' from the United Graduate School of Agricultural Sciences, Ehime University, Japan with the Japanese Government (MEXT) Scholarship. Later, he completed his postdoctoral research in Center of Molecular Biosciences (COMB), University of the Ryukyus, Okinawa, Japan with the Japan Society for the Promotion of Science (JSPS) postdoctoral fellowship. Subsequently, he joined as Adjunct Senior Researcher at the University of Tasmania with the Australian Government's Endeavour Research Fellowship. He joined as a Lecturer in the Department of Agronomy, Sher-e-Bangla Agricultural University in June 2006. He was promoted to Assistant Professor, Associate Professor and Professor in June 2008, June 2013 and June 2017, respectively. Prof. Hasanuzzaman has been devoting himself in research in the field of crop science, especially focused on Environmental Stress Physiology since 2004. He has been performing as team leader/principal investigator of different projects funded by World Bank, FAO, University Grants Commission of Bangladesh, Ministry of Science and Technology (Bangladesh) and so on.

Prof. Hasanuzzaman published over 100 articles in peer-reviewed journals and books. He has edited two books and written 35 book chapters on important aspects of plant physiology, plant stress responses and environmental problems in relation to plant species. These books were published by internationally renowned publishers (Springer, Elsevier, CRC Press, Wiley, etc.). His publications got over 2000 citations with h-index: 23 (according to Scopus). Prof. Hasanuzzaman is a Research supervisor of undergraduate and graduate students and supervised 20 M.S. students so far. He is the Editor and Reviewer of more than 50 peer-reviewed international journals and the recipient of Publons' Peer Review Award 2017. Hasanuzzaman is an active member of about 40 professional societies and acting as Publication Secretary of the Bangladesh Society of Agronomy. He has been honored by different authorities due to his outstanding performance in different fields like research and education. He received the World Academy of Science (TWAS) Young Scientist Award 2014. He has attended and presented 25 papers and posters in national and international conferences in different countries (USA, UK, Germany, Australia, Japan, Austria, Sweden, Russia, etc.).

Kamrun Nahar is an Associate Professor, Department of Agricultural Botany at Sher-e-Bangla Agricultural University, Dhaka, Bangladesh. She received her PhD on 'Environmental Stress Physiology of Plants' in 2016 from the United Graduate School of Agricultural Sciences, Ehime University, Japan with the Japanese Government (MEXT) Scholarship. Dr. Nahar has been involved in research with field crops, emphasizing stress physiology, since 2006. She has completed several research works and is also continuing a research project funded by Sher-e-Bangla Agricultural University Research System and the Ministry of Science and Technology (Bangladesh). She is supervising M.S. students. Dr. Nahar has published a number of articles in peer-reviewed journals and books with reputed publishers. She has published 50 articles and chapters related to plant physiology and environmental stresses with Springer, Elsevier, CRC Press, Wiley, etc. Her publications reached about 2000 citations with h-index: 22 (according to Scopus). She is involved in editorial activities and reviewer of international journals. She is an active member of about 20 professional societies. Dr. Nahar has attended different international conferences and presented ten papers and posters in national and international conferences in different countries (the United States, Australia, Japan, Austria, Russia, China, etc.).

Masayuki Fujita is a Professor in the Laboratory of Plant Stress Responses, Faculty of Agriculture, Kagawa University, Kagawa, Japan. He received his B.Sc. in Chemistry from Shizuoka University, Shizuoka, Japan and his M.Agr. and PhD in Plant Biochemistry from Nagoya University, Nagoya, Japan. His research interests include physiological, biochemical and molecular biological responses based on secondary metabolism in plants under various abiotic and biotic stresses; phytoalexin, cytochrome P450, glutathione *S*-transferase and phytochelatin; and redox reaction and antioxidants. In the last decade, his works were focused on oxidative stress and antioxidant defense in plants under environmental stress. His group investigates the role of different exogenous protectants in enhancing antioxidant defense and methylglyoxal detoxification systems in plants. He has supervised four M.S. students and 13 PhD students as main supervisor. He has about 150 publications in journals and books and has edited four books.

Hirosuke Oku is a Professor in the Center of Molecular Biosciences at the Tropical Biosphere Research Center in University of the Ryukyus, Okinawa, Japan. He obtained his Bachelor of Science in Agriculture from University of the Ryukyus in 1980. He received his PhD in Biochemistry from Kyushu University, Japan in 1985. In the same year, he started his career as Assistant Professor in the Faculty of Agriculture, University of the Ryukyus. He became Professor in 2009. He received several prestigious awards and medals including the Encouragement Award of Okinawa Research (1993) and Encouragement Award of Japanese Society of Nutrition and Food Science (1996). Prof. Oku is the group leader of the Molecular Biotechnology Group of the Center of Molecular Biosciences at University of the Ryukyus, His research works focused on lipid biochemistry; molecular aspects of phytomedicine; secondary metabolites biosynthesis and abiotic stress tolerance of tropical forest trees. He has about ten PhD students and over 20 M.S. students. Prof. Oku has over 50 peer-reviewed publications in his record.

M. Tofazzal Islam is a Professor of the Department of Biotechnology of Bangabandhu Sheikh Mujibur Rahman Agricultural University in Bangladesh. He did his M.S. and PhD in Applied Biosciences at Hokkaido University in Japan. Dr. Islam received postdoctoral research experiences at Hokkaido University, University of Goettingen, University of Nottingham and West Virginia University under the JSPS, Alexander von Humboldt, Commonwealth and Fulbright Fellowships, respectively. He published articles in many international journals and book series (>200 peer-reviewed articles, total citation 1664, h-index 22, i10-index 48; RG score 39.06). Dr. Islam was awarded many prizes and medals including the Bangladesh Academy of Science Gold Medal in 2011, University Grants Commission Bangladesh Awards in 2004 and 2008 and Best Young Scientist Award 2003 from the JSBBA. Prof. Islam is the Chief Editor of a book series, *Bacillus and Agrobiotechnology*, published by Springer. His research interests include genomics, genome editing, plant probiotics and novel biologicals, and bioactive natural products.

Contributors

Srivani S Adimulam
International Crops Research Institute for the Semi-
Arid Tropics (ICRISAT)
Telangana, India

Madhoolika Agrawal
Laboratory of Air Pollution and Global Climate Change
Department of Botany, Institute of Science, Banaras
Hindu University
Varanasi, India

S.B. Agrawal
Laboratory of Air Pollution and Global Climate Change
Department of Botany, Institute of Science, Banaras
Hindu University
Varanasi, India

Christianne M. Aikins
Climate Analytics Team, Climate Services and
Research Department
APEC Climate Center
Busan, Republic of Korea

Khalid Farooq Akbar
University of Lahore
Sargodha, Pakistan

Aamir Ali
Department of Botany
University of Sargodha
Sargodha, Pakistan

Imtiaz Ali Khan
Department of Agriculture
University of Swabi
Khyber Pakhtunkhwa, Pakistan

Qasim Ali
Department of Botany
Government College University Faisalabad
Faisalabad, Pakistan

Shafaqat Ali
Department of Environmental Sciences and
Engineering
Government College University Faisalabad
Faisalabad, Pakistan

Shahzad Ali Shahid
Department of Chemistry
Government College University Faisalabad
Faisalabad, Pakistan

Himanshu Bariya
Department of Life Sciences
Hemchandracharya North Gujarat University
Gujarat, India

Nicolle Louise Ferreira Barros
Instituto de Ciências Biológicas
Universidade Federal do Pará
Belém, Brazil

Pragya Barua
National Institute of Plant Genome Research
Jawaharlal Nehru University Campus
New Delhi, India

Supratim Basu
New Mexico Consortium
Los Alamos, New Mexico

Pooja Bhatnagar-Mathur
International Crops Research Institute for the Semi-
Arid Tropics (ICRISAT)
Telangana, India

Muhammad Faraz Bhatti
Atta-ur-Rahman School of Applied Biosciences
National University of Sciences and Technology
Islamabad, Pakistan

Pankaj Bhowmik
National Research Council of Canada
Saskatoon, Saskatchewan, Canada

Hassiba Bouazzi
Biotechnology and Plant Improvement Laboratory
Center of Biotechnology of Sfax, University of Sfax
Sfax, Tunisia

Faical Brini
Biotechnology and Plant Improvement Laboratory
Center of Biotechnology of Sfax, University of Sfax
Sfax, Tunisia

Nataša Čerekovič
Institute of Sciences of Food Productions
National Research Council (ISPA-CNR)
Lecce, Italy

Niranjan Chakraborty
National Institute of Plant Genome Research
Jawaharlal Nehru University Campus
New Delhi, India

Subhra Chakraborty
National Institute of Plant Genome Research
Jawaharlal Nehru University Campus
New Delhi, India

Antra Chatterjee
Laboratory of Algal Biology, Molecular Biology
    Section
Center of Advanced Study in Botany, Institute of
    Science, Banaras Hindu University
Varanasi, India

Jong Ahn Chun
Climate Analytics Team, Climate
    Services and Research
Department
APEC Climate Center
Busan, Republic of Korea

Shah Fahad
Department of Agriculture
University of Swabi
Khyber Pakhtunkhwa, Pakistan

and

College of Life Science,
    Linyi University, Linyi
    Shandong, China

Adeeb Fatima
Laboratory of Air Pollution and Global
    Climate Change
Department of Botany
Institute of Science
Banaras Hindu University
Varanasi, India

Nadia Fatnassi
Institute of Sciences of Food Productions
National Research Council (ISPA-CNR)
Lecce, Italy

Kaouthar Feki
Biotechnology and Plant Improvement Laboratory
Center of Biotechnology of Sfax
University of Sfax
Sfax, Tunisia

Dipak Gayen
National Institute of Plant Genome Research
Jawaharlal Nehru University Campus
New Delhi, India

Annesha Ghosh
Laboratory of Air Pollution and Global Climate Change
Department of Botany
Institute of Science
Banaras Hindu University
Varanasi, India

Alvina Gul
Atta-ur-Rahman School of Applied Biosciences (ASAB)
National University of Sciences and Technology
    (NUST)
Islamabad, Pakistan

Dipinte Gupta
Plant Biotechnology Lab
Department of Botany
Faculty of Science
Dayalbagh Educational Institute (Deemed University)
Dayalbagh, India

Noman Habib
Department of Botany
Government College University Faisalabad
Faisalabad, Pakistan

Muhammad Zulqurnain Haider
Department of Botany
Government College University Faisalabad
Faisalabad, Pakistan

Abdul Hamid
Laboratory of Air Pollution and Global Climate Change
Department of Botany
Institute of Science
Banaras Hindu University
Varanasi, India

John T. Hancock
Center for Research in Biosciences
University of the West of England (UWE)
Bristol, UK

Md. Mahmudul Hassan
Department of Genetics and Plant Breeding
Patuakhali Science and Technology University
Patuakhali, Bangladesh

Saeid Hazrati
Department of Agronomy and Medicinal Plants
    Production
Faculty of Agriculture
Azarbaijan Shahid Madani University
Tabriz, Iran

Abdullah Ijaz Hussain
Department of Chemistry
Government College University Faisalabad
Faisalabad, Pakistan

Syed Makhdoom Hussain
Department of Zoology
Government College University Faisalabad
Faisalabad, Pakistan

Noshin Ilyas
Department of Botany
PMAS Arid Agriculture University
Rawalpindi, Pakistan

Naeem Iqbal
Department of Botany
Government College University Faisalabad
Faisalabad, Pakistan

M. Tofazzal Islam
Davis College of Agriculture, Natural Resources and
    Design
West Virginia University
Morgantown, West Virginia

and

Department of Biotechnology
Bangabandhu Sheikh Mujibur Rahman Agricultural
    University
Gazipur, Bangladesh

Muhammad Jamil
Department of Botany
University of Sargodha
Sargodha, Pakistan

Muhammad Jamil
Department of Biotechnology & Genetic Engineering
Kohat University of Science and Technology
Kohat, Pakistan

Sami Ullah Jan
Department of Plant Biochemistry and Molecular
    Biology School of Life Sciences
University of Science and Technology of China
Hefei, People's Republic of China

Sumit Jangra
Department of Molecular Biology, Biotechnology
and Bioinformatics
CCS Haryana Agricultural University
Hisar, India

Mohsen Janmohammadi
Department of Plant Production and Genetics
Faculty of Agriculture
University of Maragheh
Maragheh, Iran

Muhammad Tariq Javed
Department of Botany
Government College University Faisalabad
Faisalabad, Pakistan

Ambuj Bhushan Jha
Crop Development Center Department of Plant
    Sciences
University of Saskatchewan
Saskatoon, Saskatchewan Canada

Disha Kamboj
Department of Molecular Biology, Biotechnology
and Bioinformatics
CCS Haryana Agricultural University
Hisar, India

Vigya Kesari
Molecular Biology Section, Laboratory of Algal Biology
Center of Advanced Study in Botany
Institute of Science, Banaras Hindu University
Varanasi, India

Daeha Kim
Climate Analytics Team, Climate Services and Research
Department
APEC Climate Center
Busan, Republic of Korea

Nicholas E. Korres
Department of Crop, Soil, and Environmental
    Sciences
University of Arkansas
Fayetteville, Arkansas

Nilesh Vikram Lande
National Institute of Plant Genome Research
Jawaharlal Nehru University Campus
New Delhi, India

Eun-Jeong Lee
Climate Analytics Team, Climate Services and Research
Department
APEC Climate Center
Busan, Republic of Korea

Wooseop Lee
Climate Analytics Team, Climate Services and Research
Department
APEC Climate Center
Busan, Republic of Korea

Sanai Li
Climate Analytics Team, Climate Services and
    Research
Department
APEC Climate Center
Busan, Republic of Korea

Eduardo C. Machado
Laboratory of Plant Physiology
    "Coaracy M. Franco"
Center for R&D in Ecophysiology and
    Biophysics
Agronomic Institute (IAC)
Campinas, Brazil

Deyvid Novaes Marques
Instituto de Ciências Biológicas
Universidade Federal do Pará
Belém, Brazil

Roomina Mazhar
Department of Botany
PMAS Arid Agriculture University
Rawalpindi, Pakistan

Aakash Mishra
Department of Plant Sciences
University of California
Davis, California

Sushma Mishra
Plant Biotechnology Lab
Department of Botany
Faculty of Science
Dayalbagh Educational Institute
    (Deemed University)
Dayalbagh, India

Hamid Mohammadi
Department of Agronomy and Medicinal Plants
    Production
Faculty of Agriculture
Azarbaijan Shahid Madani University
Tabriz, Iran

Kutubuddin Molla
Pennsylvania State University
University Park, Pennsylvania

and

National Rice Research Institute
Cuttack, India

Joseph Msanne
New Mexico Consortium
Los Alamos, New Mexico

A. Mujeeb-Kazi
Texas A&M University
College Station, Texas

Durgesh Nandini
Department of Biotechnology
Shri A. N. Patel Postgraduate Institute
Gujarat, India

Abdul Aziz Napa
Department of Plant Science
Quaid-i-Azam University
Islamabad, Pakistan

Zahra Noreen
Department of Botany
University of Education
Lahore, Pakistan

Lymperopoulos Panagiotis
New Mexico Consortium
Los Alamos, New Mexico

Ashutosh K. Pandey
Laboratory of Air Pollution and
    Global Climate Change
Department of Botany
Institute of Science
Banaras Hindu University
Varanasi, India

Santisree Parankusam
International Crops Research Institute for the Semi-
    Arid Tropics (ICRISAT)
Telangana, India

Amit Kumar Patel
Molecular Biology Section
Laboratory of Algal Biology
Center of Advanced Study in Botany
Institute of Science
Banaras Hindu University
Varanasi, India

Ashish Patel
Department of Life Sciences
Hemchandracharya North
Gujarat University
Gujarat, India

Palmiro Poltronieri
Institute of Sciences of Food Productions
National Research Council (ISPA-CNR)
Lecce, Italy

Priti
Department of Molecular Biology
Biotechnology and Bioinformatics
CCS Haryana Agricultural University
Hisar, India

Roel C. Rabara
New Mexico Consortium
Los Alamos, New Mexico

Mahfuzur Rahman
Davis College of Agriculture
Natural Resources and Design
West Virginia University
Morgantown, West Virginia

Amit Kumar Rai
Center for Genetic Disorders
Banaras Hindu University
Varanasi, India

Kshama Rai
Laboratory of Air Pollution and Global Climate Change
Department of Botany
Institute of Science
Banaras Hindu University
Varanasi, India

L.C. Rai
Molecular Biology Section
Laboratory of Algal Biology
Center of Advanced Study in Botany
Institute of Science
Banaras Hindu University
Varanasi, India

Richa Rai
Department of Botany
Banaras Hindu University
Varanasi, India

Ruchi Rai
Molecular Biology Section
Laboratory of Algal Biology
Center of Advanced Study in Botany
Institute of Science
Banaras Hindu University
Varanasi, India

Shweta Rai
Molecular Biology Section
Laboratory of Algal Biology
Center of Advanced Study in Botany
Institute of Science
Banaras Hindu University
Varanasi, India

Rajiv Ranjan
Plant Biotechnology Lab
Department of Botany
Dayalbagh Educational Institute
(Deemed University)
Dayalbagh, India

Sávio Pinho dos Reis
Instituto de Ciências Biológicas
Universidade Federal do Pará
Belém, Brazil

and

Centro de Ciências Biológicas e da Saúde
Universidade do Estado do Pará
Marabá, Brazil

Rafael V. Ribeiro
Laboratory of Plant Physiology "Coaracy M. Franco"
Center for R&D in Ecophysiology and Biophysics
Agronomic Institute (IAC)
Campinas, Brazil

and

Laboratory of Crop Physiology
Department of Plant Biology
Institute of Biology
University of Campinas (UNICAMP)
Campinas, Brazil

Maimona Saeed
Department of Botany
PMAS Arid Agriculture University
Rawalpindi, Pakistan

Walid Saibi
Biotechnology and Plant Improvement
    Laboratory
Center of Biotechnology of Sfax
University of Sfax
Sfax, Tunisia

Angelo Santino
Institute of Sciences of Food Productions
National Research Council (ISPA-CNR)
Lecce, Italy

Shah Saud
College of Horticulture
Northeast Agricultural University Harbin
Heilongjiang, China

Amedea B. Seabra
Center for Natural and Human Sciences Federal
    University of ABC
Santo André, Brazil

Sumreena Shahid
Department of Botany
Government College University Faisalabad
Faisalabad, Pakistan

Neidiquele M. Silveira
Laboratory of Plant Physiology
    "Coaracy M. Franco"
Center R&D in Ecophysiology and Biophysics
Agronomic Institute (IAC)
Campinas, Brazil

Babar Shahzad
School of Land and Food
University of Tasmania
Hobart, Australia

Alka Shankar
Molecular Biology Section
Laboratory of Algal Biology
Center of Advanced Study in
Botany Institute of Science
Banaras Hindu University
Varanasi, India

Kiran K. Sharma
International Crops Research Institute for the Semi-
    Arid Tropics (ICRISAT)
Telangana, India

Pallavi Sharma
Center for Life Sciences
Central University of Jharkhand
Brambe, Ranchi, Jharkhand, India

Shilpi Singh
Molecular Biology Section
Laboratory of Algal Biology
Center of Advanced Study in Botany
Institute of Science
Banaras Hindu University
Varanasi, India

Suruchi Singh
Laboratory of Air Pollution and Global Climate Change
Department of Botany
Banaras Hindu University
Varanasi, India

Cláudia Regina Batista de Souza
Instituto de Ciências Biológicas
Universidade Federal do Pará
Belém, Brazil

Ágnes Szepesi
Department of Plant Biology
Institute of Biology
University of Szeged
Szeged, Hungary

Mohsin Tanveer
School of Land and Food
University of Tasmania
Hobart, Australia

Liliane de Souza Conceição Tavares
Instituto de Ciências Biológicas
Universidade Federal do Pará
Belém, Brazil

Aruna V. Varanasi
Department of Horticulture
University of Arkansas
Fayetteville, Arkansas

Vijay K. Varanasi
Department of Crop, Soil, and Environmental Sciences
University of Arkansas
Fayetteville, Arkansas

Neelam R. Yadav
Department of Molecular Biology
Biotechnology and Bioinformatics
CCS Haryana Agricultural
University
Hisar, India

Ram C. Yadav
Center for Plant Biotechnology
CCS Haryana Agricultural
University
Hisar, India

1 Abiotic Stress in Plants
A General Outline

Ashutosh K. Pandey, Annesha Ghosh, Kshama Rai, Adeeb Fatima, Madhoolika Agrawal, and S.B. Agrawal

CONTENTS

1.1 INTRODUCTION

Plants encounter several abiotic stresses which are the major constraints to plant growth and productivity, causing extensive crop losses linked to food security across the world (Mittler, 2006). Plants have to exploit their immediate environment to maximum effect to cope up with the challenging adverse abiotic stresses. Plants have developed unique characteristics to defend themselves via modulating their phenotypes with physiological, biochemical, molecular changes, thereby strengthening their stress tolerance mechanisms for survival (Mohanta et al., 2017). Apparently, the most widely studied linear signaling pathways are actually a part of a more complicated signaling web network comprising overlapping between its web branches, resulting in induction of genes by more than one particular abiotic stress.

World population is rising at an alarming rate and is likely to attain about 9.8 billion by the end of the year 2050. On the contrary, agricultural productivity is not increasing at a requisite rate to keep up the pace with increasing food demand. The major abiotic stress factors posing serious threats to the agriculture and environment are drought, flooding, radiation (UV-B) soil salinity, extreme temperatures, heavy metal toxicity and numerous air pollutants that are pumped into the atmosphere at an alarming rate since the preindustrial times. Due to rising concentrations of CO_2 and other atmospheric trace gases, global temperatures have increased by about 1°C over the course of the last century and will likely rise even more in coming decades as reported by the Intergovernmental Panel on Climate Change (IPCC, 2014). Increased drought and salinization of arable land are expected to have devastating global effects on crop production (Wang et al., 2003).

The anthropogenically induced depletion of the stratospheric ozone layer is adding two more potential abiotic stresses, i.e., UV-B radiation and high light. Various model studies have suggested that the continuous rate of stratospheric ozone layer depletion would increase the penetration of UV-B by 9% (McKenzie et al., 2007). In 2015, a report from NASA suggested that due to the successful implementation of the Montreal Protocol (1987), the stratospheric ozone hole is recovering, but it will take a large time up to 2075 for its full recovery (NASA, 2015). The equatorial region of the globe is more prone to high temperature and intense UV-B perceptions due to a thinner layer of ozone in the stratosphere. Similarly, the major driving factor behind most of the abiotic stresses is the anthropogenic misuse of natural resources, heavy industrial activities, intense vehicular traffic and so on, and these factors are loading the atmosphere with tons of air pollutants (primary and secondary) ranging from oxides of sulfur (Sox), oxides of nitrogen (Nox), tropospheric ozone (O$_3$) and volatile organic compounds (VOCs) .

Here, we review plant responses to different abiotic stresses at cellular, morphological, physiological, biochemical and molecular levels. We have emphasized different antioxidative mechanisms and signaling transduction cascades generated in response to a particular stimulus to develop a better understanding of adaptive mechanisms operating in the plants. In the light of abiotic-stress-induced phytotoxicity, this compilation could be useful to the plant breeders in screening the cultivars and also in developing tolerant varieties depending on different stress-response relationships.

1.2 SOIL SALINITY STRESS

Soil salinity has been a hazard to agriculture in some parts of the world for over 3000 years; in recent times, the threat has further pronounced to a larger area (Flowers, 2006). Soil salinization is a major factor contributing to the loss of productivity of cultivated soils. The area of salinized soils is increasing, and this phenomenon is particularly severe in irrigated soils. It was estimated that about 20% (45 million ha) of irrigated land, producing one-third of the world's food, is salt-affected (Shrivastava and Kumar, 2015). Globally, soil salinization is more common in arid and semi-arid regions than in humid regions. The amount of world agricultural land destroyed by salt accumulation is estimated to be 10 million ha (Pimentel et al., 2004) each year. This rate can be accelerated by climate change, excessive use of groundwater (mainly coastal area), increasing use of low-quality water in irrigation and massive introduction of irrigation associated with intensive farming and poor drainage. The use of low-quality water for irrigation can lead to the accumulation of salts in the soil, since the leaching fraction is reduced and the salts contained in the irrigation water are not leached enough. It is estimated that, by 2050, 50% of the world's arable land will be affected by salinity (Bartels and Sunkar, 2005).

1.2.1 Effect of Soil Salinity on Growth and Yield

Soil salinity reduces the productivity of many crops which have a low tolerance to soil salinity. The first morphological symptoms of salt stress are wilting, yellowing of leaves and stunted growth. In a second phase, the damage manifests as chlorosis of green parts, leaf tip burning, necrosis of leaves, and the oldest leaves display scorching (Shannon and Grieve, 1998). Although plant species differ in their sensitivity or tolerance to salts (Marschner, 1995), high soil salinity has detrimental effects on seed germination and plant growth (Taiz and Zeiger, 2006) and in due course kills growing plants (Garg and Gupta, 1997).

The salt-induced water deficit and many nutrient interactions in salt-stressed plants are the major constraints for plant growth in saline soils. Indeed, salinity effects on plant growth reduction are a time-dependent process, and Munns et al. (1995) proposed a two-phase model to depict the response of plant growth to salinity. The first phase is very rapid, and growth reduction is ascribed to the development of a water deficit. The second phase is due to the accumulation of salts in the shoot at toxic levels and is very slow. The presence of high concentrations of sodium, calcium and magnesium causes damage to soil structure and is accompanied by an increase in the compactness of soils with a decrease in filterability, hydraulic conductivity and the oxygen availability in the root zone. Another effect of a high concentration of sodium is an increase of alkalization. Excess sodium (Na$^+$) in the soil competes with Ca^{2+}, K$^+$ and other cations to reduce their availability to crops. Therefore, soils with high levels of exchangeable sodium (Na$^+$) may impact plant growth by dispersion of soil particles, nutrient deficiencies or imbalances and specific toxicity to sodium sensitive plants. Salinity delays germination and emergence; the young salt-stressed seedlings may be more susceptible to hypocotyl and cotyledon injury (Esechie et al., 2002) or attack by pathogens. A decrease in plant biomass, leaf area and growth has been observed in different vegetable crops under salt stress (Giuffrida et al., 2013; Zribi et al., 2009). It was found that shoot growth is often suppressed more than the root growth by soil salinity (Ramoliya et al., 2006). Salt toxicity primarily occurs in the older leaves where Na and Cl build up in the transpiring leaves over a long period of time, resulting in high salt concentration and thus causing leaf death. Investigators found that crops are most sensitive during vegetative and early reproductive stages, less sensitive during flowering and least sensitive during the seed filling stage. Salinity often reduces the number of florets per ear, increases sterility and affects the time of flowering and maturity in both wheat (Maas and Poss, 1989) and rice (Khatun and Flowers, 1995).

Parameters such as yield and its components, including seed/fruit mass, spikelets per spike (for cereals), spike length, fertility rates in the spikes and 1000-grain mass, have also been shown to be affected by salinity stress (Gholizadeh et al., 2014). Salt stress decreases marketable yield due to decreased productivity and an increased yield of unmarketable fruits, roots, tubers and leaves having low commercial value.

1.2.2 Effect of Soil Salinity on Plant Physiology

Salinity affects photosynthesis by decreasing CO_2 availability as a result of diffusion limitations (Flexas et al., 2007) and a reduction of the contents of photosynthetic pigments (Ashraf and Harris, 2013). Salt accumulation in spinach inhibits photosynthesis (Di Martino et al., 1999), primarily by decreasing stomatal and mesophyll conductance to CO_2 (Delfine et al., 1998) and reducing chlorophyll content, which can affect light absorbance (Alvino et al., 2000). Salinity lowers the total photosynthetic capacity of the plant through decreased leaf growth and inhibited photosynthesis, limiting its ability to grow (Yeo and Flowers, 2008). Salt stress does not directly inhibit primary reactions of photosynthesis but inhibits product export and fine-tuning of primary reactions by interfering with an optimal arrangement of proteins and membranes (Huchzermeyer, 2000). The accumulation of injurious ions may inhibit photosynthesis and protein synthesis, inactivate enzymes and damage chloroplasts and other organelles (Taiz and Zeiger, 2002). Maximal photochemical efficiency, indicated by high F_v/F_m values of chlorophyll fluorescence, remains high under mild salt stress, while the growth rate was reduced under the same salinity level (Lee et al., 2004). In radish, about 80% of the growth reduction at high salinity could be credited to the reduction of leaf area expansion and hence to a reduction of light interception. The remaining 20% of the salinity effect on growth was most likely explained by a decrease in stomatal conductance (Marcelis and Van Hooijdonk, 1999).

Salt stress results in severe damage to plants when its inhibitory effects occur in the presence of high light intensity. Then, photosystem II (PSII) activity will result in high oxygen concentrations, especially if stomata are closed under stress. Concomitantly, chlorophyll will remain in its active state for a prolonged period, as the electron transport rate is reduced by inhibited off the flow of products. Thus, the probability of a transfer of electrons from activated chlorophyll to molecular oxygen to form superoxide will increase. Superoxide will rapidly dismutate to yield O_2 and the less reactive H_2O_2. But in the presence of some cations such as Cu and Fe, highly reactive $\bullet OH-$ may be formed. Under salt stress, there is a general decline in Calvin cycle intermediates during the phase of stress adaptation. Decreased RuBP concentration might be caused by the general rundown of the Calvin cycle and a decrease in assimilation rate (Tezara et al., 1999).

In various plant species, salt stress often leads to dilations of the thylakoid membranes (Zhen et al., 2011) and is likely to contribute to impaired photosynthetic rates under saline conditions. It was found that grana become destabilized if the ratio of monovalent and divalent cations is impaired (Hesse et al., 1976). Stomatal closure and subsequent inhibition of gas exchange is a secondary effect of salt stress.

1.2.3 Antioxidants Activity under Soil Salinity

Salt stress imposes a water deficit because of osmotic effects on a wide variety of metabolic activities such as enhanced generation of reactive oxygen species (ROS) (Ahmad and Umar, 2011). These ROS are highly reactive and can alter normal cellular metabolism. To prevent damage by ROS, plants have developed a complex antioxidant defense system. The primary components of this system include carotenoids, ascorbate, glutathione and tocopherols, in addition to enzymes such as superoxide dismutase (SOD), catalase (CAT), glutathione peroxidase (GPX), peroxidases (POD) and the enzymes such as ascorbate peroxidase (APX) and glutathione reductase (GR) (Ahmad and Umar, 2011). Many components of this antioxidant defense system can be found in various subcellular compartments (Ahmad and Umar, 2011; Hernandez et al., 2000).

An association between salt tolerance and increased activation of antioxidant enzymes has been confirmed in pea (Ahmad et al., 2008), rice (Dionisio-Sese and Tobita, 1998), tomato, soybean, maize (de Azevedo Neto et al., 2006) and mustard (Ahmad et al., 2012). Under salt stress, increases in activity of SOD, APX, GR, CAT and POX as well as higher antioxidant activity in tolerant species/varieties have been reported by various workers (Azooz et al., 2011; Koyro et al., 2012).

Overexpression of mitochondrial Mn-SOD and chloroplastic Cu/Zn-SOD can provide enhanced tolerance to salt stress (Badawi et al., 2004; Wang et al., 2004). Similar results have been found in *Triticum aestivum* (Sairam et al., 2002), *Pisum sativum* (Ahmad et al., 2008) and *Brassica juncea* (Ahmad et al., 2012). Earlier studies suggested that the increased SOD activity enables the plants to resist the potential oxidative damage caused by NaCl salinity (Ahmad and Umar, 2011). APX activity had a key role in response to salt stress in the comparison of the activities of other antioxidant enzymes in salt-sensitive and salt-tolerant cultivars (Ahmad et al., 2012). There are reports showing that GR activity increased in NaCl-tolerant pea variety as compared to NaCl-sensitive pea (Hernandez et al., 2000). Catalase activity has been found to increase under salt stress in soybean (Comba et al., 1998), and mustard (Ahmad et al., 2012). De Azevedo Neto et al. (2006) also found higher CAT activity in two maize cultivars differing in salt tolerance. Ascorbic acid (AsA) is an important antioxidant that reacts not only with H_2O_2 but also with $\bullet O^{2-}$, $\bullet OH$ and lipid hydro-peroxidases (Ahmad and Umar, 2011). Several studies have revealed that AsA plays an important role in improving plant tolerance to abiotic stress (Ahmad and Umar, 2011; Ahmad and Umar, 2011). Glutathione is involved in the ascorbate/glutathione cycle and the regulation of the protein thiol–disulfide redox status of plants in response to abiotic and biotic stress (Yousuf et al., 2012). Ruiz and Blumwald (2002) reported that glutathione content in wild canola (*Brassica napus* L) plants increased under salt stress; this suggests a possible protective mechanism against salt-induced oxidative damage.

1.2.4 Signaling and Its Significance in Alleviating Salt Stress in Plants

High salt concentration lay down plant stressors such as ion toxicity, as a consequence of ion penetration in excess and disorder of nutrient balances, as usually seen in the displacement of monovalent ions (K^+ by Na^+). During the steps forward of salinity within the plant, all the key biochemical reactions, such as photosynthesis, protein synthesis, energy and lipid metabolisms, are disturbed (Bazihizina et al., 2012). Salt stress leads to the leak of the membrane, ion imbalance, support lipid peroxidation and formation of ROS, which are scavenged by enzymatic and non-enzymatic reactions (Roychoudhury et al., 2008). Salinity adversely affects plant growth and development, hindering seed germination, seedling growth, enzyme activity, DNA,

RNA and protein synthesis (Seckin et al., 2009; Tabur and Demir, 2010).

Plant hormones are classified as specific organic substances that at very low concentration act on target tissues as regulators of growth and development. Plant hormones (abscisic acid (ABA), jasmonic acid (JA), salicylic acid (SA) and ethylene (ET)) adjust the physiological reaction of plants exposed to salt stress (Kaya et al., 2009). Among the physiological responses to salinity stress, the plant hormone ABA, a sesquiterpenoid, plays an important role. The accumulation of ABA in response to salt stress is a cell signaling process, including initial stress signal perception, cellular signal transduction and regulation of expression of genes encoding key enzymes in ABA biosynthesis and catabolism (Zhang et al., 2006). This phytohormone plays a dual role in its physiological regulation. It shows inhibitive functions when it is accumulated in large amounts under stress to help plant survival through inhibition of stomatal opening and plant size expansion (Rai et al., 2011a). In addition, ABA is involved in the restriction of ET synthesis, which is a growth inhibitor under salt stress. Abscisic acid is ubiquitous in lower and higher plants and participates in the complex processes during the life cycle of plants (Javid et al., 2011). Under salinity stress, different signaling pathways are activated including the Mitogen-Activated Protein Kinase (MAPK) cascade. MAPK molecules are a set of proteins that can perform different functions in plants, including cell cycle, plant growth and development, plant response to stress. Adjusting plant behavior, especially Na^+ cellular concentration, under salt stress is an important issue in ensuring plant survival (Lee et al., 2009). MAPK signaling can importantly affect such an event by affecting the activity of proton pumps, Na^+ localization into vacuoles and the regulation of microtubules arrangement (Bögre et al., 1999). Ca^{2+} act as an important secondary messenger in regulating the plant growth and development under normal as well as in stress conditions. It is also known that externally supplied Ca^{2+} alleviates the adverse effects of salinity in many plant species (Sarwat et al., 2013). At the molecular front, a large number of calcium transporters, sensor/decoder elements and calcium-dependent transcription factors are known to be regulated by Ca^{2+} at different levels either by direct binding of Ca^{2+}, or calmodulin or by other kinases/phosphatases (Srivastava et al., 2013).

1.3 DROUGHT

According to the report of the IPCC (2014), the emission pattern of greenhouse gases will increase global warming further and is likely to enhance extreme climatic events in the future scenario. Along with the growing impact of climate change on natural water resources, the changing pattern of precipitation globally (Dai, 2013) and the intensification of agricultural activities affecting the dynamic of soil moisture (Zhang et al., 2017a) have enhanced the frequency and severity of droughts worldwide. Drought is usually a recurring climatic event resulting from a natural reduction in the amount of precipitation received over an extended period (Wilhite, 2000). Although other climatic variables such as high temperatures, high winds and low relative humidity determine the severity of the event in many parts of the world, it is still unpredictable as it depends on many other environmental factors such as occurrence and distribution of rainfall, evaporative demands and moisture storage capacity of soils (Wery et al., 1994).

Traditionally, drought can be classified into following the categories: (a) meteorological drought, which is usually defined as lack of sufficient precipitation over a region for a particular period; (b) agricultural drought is mainly concerned with the impact of inadequate soil moisture, affecting the crop growth cycle and developmental processes; (c) hydrological drought emphasizes the impact of dry period on the subsurface hydrology; (d) socioeconomic drought is usually related to the demand and supply of an economic entity (water) (Mishra and Singh, 2010; Shi et al., 2018). Here, we mainly emphasized agricultural drought as it is one of the most important environmental factors determining the crop productivity. Interaction with such factors and plant response may require critical and comprehensive understanding in assessing its impact on an array of plant processes such as physiological, morphological, antioxidative and signal transduction and, finally, crop yield and its quality.

Drought-affected areas have increased globally since 1980 due to enhanced global warming phenomenon, with an 8% increase of drought-affected area by the first decade of the century, and were observed to be maximum over northern mid-high latitudes. However, reduction in precipitation over Africa, eastern Australia, Southeast Asia and southern Europe is the major reason for the enhancement of dryness and aridity in these regions (Dai, 2013). The semi-arid region of northern China has frequently been affected by drought owing to variable interannual and decadal precipitation and temperature (Zhang et al., 2017b). Furthermore, Wang et al. (2014) have observed an increase of about 14.3% drought-prone areas worldwide during the period 1902–1949 and 1950–2008.

Water deficit conditions proved to be a limiting factor at the early developmental growth stages, which is generally dominated by phases of cell division, cell differentiation and cell elongation, and such phases are affected under drought stress owing to a reduction in cell turgor pressure and other physiological events (Taiz and Zeiger, 2010). Nonami (1998) confirmed that cell elongation is primarily correlated with the water absorption capacity of the elongating cells under water deficient conditions. Drought-induced changes at the cellular level have been documented by Harb et al. (2010), showing acclamatory responses through cell wall expansion in *Arabidopsis thaliana*, as an early avoidance strategy under water limiting condition. Liu and Stützel (2004) reported a decrease in specific leaf area in drought treated plants, determining the reduction in cell expansion and hence resulting in thinner leaves.

The allocation of biomass to different plant organs, especially the reproductive sink, determine the crop yield and productivity. A common observation reported from drought studies is the enhancement of root–shoot ratio, determining the reduction of more shoot biomass as compared to root biomass under water deficient condition (Blum, 1996). Similar observations have been reported by Erice et al. (2010) in alfalfa, maintaining higher root–shoot ratio under water stress conditions. This could be due to the fact that the accumulation of solutes in the root tip under water deficient condition may result in potential differences between the surrounding soil and the root hairs, which in turn attract more water to these tips and consequently are able to maintain root turgor pressure and growth (Liu and Stützel, 2004). Likewise, reduced leaf biomass was also observed in *Jatropha curcas* L. seedling by 28% of the total produced leaf biomass under drought stress (Achten et al., 2010).

1.3.1 Morphological Responses to Drought Stress

The adverse impacts of drought stress on plant growth and development have been studied for different crop plants such as rice (Lafitte et al., 2007; Manickavelu et al., 2006; Pantuwan et al., 2002; Tripathy et al., 2000), maize (Kamara et al., 2003; Monneveux et al., 2006), soybean (Samarah et al., 2006; Specht et al., 2001), barley (Samarah, 2005), cowpea (Turk et al., 1980), *Amaranthus* spp. (Liu and Stützel, 2004) and wheat (Loutfy et al., 2012).

Some of the common morphological alterations due to drought stress are as follows:

- Impaired germination and weakened stand establishment (Harris et al., 2002).
- Reduction in seed vigor index by 85.8% and germination percentage by 63.3% in different wheat

cultivars due to osmotic stress induced using polyethylene glycol (PEG-6000) as reported by Dhanda et al. (2004).
- Reduction in germination stresses tolerance index, plant height stress index and dry matter stress index while root length stress index showed increment under PEG-induced drought stress in sunflower seedlings (Ahmad et al., 2009).
- Reduction in plant height in pea and wheat seedlings by 11.7% and 14.5%, respectively, under drought treatment (Alexieva et al., 2001).
- Furthermore, tiller abortion and change in the rooting pattern has also been observed in three upland rice cultivars CG14 (*Oryza glaberrima*), WAB56-104 (*O. sativa* tropical *japonica*, improved) and WAB450-24-3-2-P18-HB (CG14 × WAB56-104 hybrid) under drought stress conditions as reported by (Asch et al., 2005).

1.3.2 Physiological Responses of Plants

Severe drought conditions lead to cell contraction and ultimately result in reduction of cellular volume, which, in turn, induces the enhancement of cell viscosity (Farooq et al., 2009). Such increase in cell viscosity due to the high concentration of solute accumulation may prove to be detrimental to normal plant functioning and photosynthetic machinery (Hoekstra et al., 2001). Stomatal limitation has been documented under drought stress in different crop species such as maize (Cochard, 2002), wheat (Khan and Soja, 2003), soybean (Liu et al., 2003; Ohashi et al., 2006), kidney bean (Miyashita et al., 2005) and rice (Praba et al., 2009). Besides stomatal closure, reduction of stomatal size was also reported under moderate drought conditions (Farooq et al., 2012). Klamkowski and Treder (2008) reported CO_2 deficit and stomatal closure contributed to a reduction in photosynthesis in strawberry cultivars under moderate drought conditions. Such finding was supported by the work of Miyashita et al. (2005), who observed reductions in photosynthesis and transpiration rate in kidney beans (*Phaseolus vulgaris* L.) due to stomatal limitation.

Besides the reduction of stomatal conductance, stomatal limitation is also taken into account as a major key factor determining the detrimental impact on the carbon assimilating process in plants under drought treatment. Flexas and Medrano (2002) reported downregulation of different metabolic activities due to stomatal limitation under water stress conditions, which impaired Ribulose-bisphosphate (RuBP) regeneration and adenosine triphosphate (ATP) synthesis, which ultimately induces the events of photoinhibition and disruption of normal photochemistry. Drought-induced photorespiration in plants has been reported by Massacci et al. (2008) in cotton plants, which can be an acclimation strategy to counterbalance the over-excitation in the PSII under such stress. On the contrary, drought-induced photorespiration can offset carbon fixation and the assimilation process, leading to the generation of ROS in the photosynthetic tissues of the plants (Farooq et al., 2012).

1.3.3 ANTIOXIDANT METABOLISM UNDER DROUGHT STRESS

Higher plants often face ROS toxicity owing to a reduction in the CO_2/O_2 ratio in photosynthetic tissues and enhancement of photorespiration under water deficit condition. Uncontrolled generation of ROS may result in membrane leakiness and lipid peroxidation and ultimately lead to malondialdehyde (MDA) production and impairment of functional macromolecule such as DNA, protein, lipid, nucleic acid and chlorophyll pigments (Moussa and Abdel-Aziz, 2008). Drought inducing free radical burst inside the cellular and subcellular components promote production of enzymatic antioxidants such as SOD, CAT, GR, APX, dehydroascorbate reductase (DHAR), POD and non-enzymatic antioxidants like AsA, flavonoids, anthocyanins, carotenoids and α-tocopherol, imparting resistance against such environmental abiotic stress at different growth stages (Reddy et al., 2004). A significant increase of antioxidative enzymes viz., APX, MDHAR, DHAR and GR, was reported by Sharma and Dubey (2005) owing to imposed drought treatment in order to control oxidative damage in rice seedlings.

Accumulation of proline is one of the important adaptive plant responses to drought stress condition in plants. Bandurska et al. (2017) reported an increase in proline concentration in the leaves and roots of the barley genotypes Syrian breeding line Cam/B1/CI and the German cultivar Maresi. The accumulation of osmolytes like amino acid, protein and sugar is a common scenario directly correlated to improved drought tolerance mechanism owing to its capacity to cope with osmotic stress and maintenance of nutrient homeostasis (Iqbal et al., 2014). Zhang et al. (2017c) observed an increased accumulation of free proline content accompanied by free amino acids, as well as soluble protein under pre-flowering drought stage in the leaves of peanut cultivars, contributing to osmotic regulation and improving drought tolerance mechanism in the cultivars. Likewise, enhancement of glycinebetaine (GB) and free proline in maize plants highlights the safeguarding role of these non-enzymatic antioxidant molecules against oxidative injury under drought stress (Moussa and Abdel-Aziz, 2008).

1.3.4 YIELD ATTRIBUTES UNDER DROUGHT CONDITION

Recently, agricultural drought has been a serious concern for yield losses in different crop plants, such as wheat (Zhao et al., 2017) maize, (Kamara et al., 2003), barley (Samarah, 2005), rice (Lafitte et al., 2007; Pantuwan et al., 2002) and chickpea (Mafakheri et al., 2010), which is directly correlated with the stringency and duration of the stress period. At post-anthesis, drought has been proved to be detrimental to grain yield reduction, irrespective of stress severity in barley, as reported by Samarah (2005). The study also documented the shortening of the time span of grain filling processes in barley under drought stress as compared to well-watered plants. Moreover, drought-induced maturity acceleration associated with faster rate of grain filling has been reported in common beans (*Phaseolus vulgaris* L.) which displayed a positive correlation with the seed yield, determining the drought adaptation strategy in the resistant cultivars (Rosales-Serna et al., 2004; Table 1.1).

1.3.5 SIGNALING AND DROUGHT STRESS

Chemical signaling events which induces stress tolerance mechanisms includes the involvement of ROS, calcium, calcium-regulated proteins and plant hormones via signal transduction pathways and also activates cell programming at the genetic level are shown in Figure 1.1. Plant hormone signaling is known to play a significant role in establishing stress tolerance mechanism through controlling stomatal movement under drought conditions (Sarwat and Tuteja, 2017). The study highlighted the positive role of ABA and JA influencing stomatal closure mechanism in plants under water-stressed conditions. ABA act in association with protein kinases and phosphatases, resulting in activation and inactivation of an ion channel in the guard cell plasma membrane under stressful conditions (Kumar et al., 2013a). Moreover, guard cell accumulation of ABA promotes stomatal closure as an adaptation strategy under water-stressed conditions, thereby minimizing water loss (Miura and Tada, 2014). Drought-induced stomatal closure processes involve activation of Ca^{2+} permeable channel in the plasma membrane in presence of 1,4,5-triphosphate (IP_3), a secondary messenger molecule, mediating ABA signal, causing stimulation of cytoplasmic Ca^{2+} influx

TABLE 1.1

Different Abiotic Stress Induced Percentage Changes in Yield Parameters of Different Crop Species

Abiotic Stress	Crop Studied	Experimental Level/Dose/Concentration		Yield Parameters	Changes (%)	References
		Control	Stress			
Salinity stress	*H. vulgare* L.	EC–1.5 dS m⁻¹	EC–13.3 dS m⁻¹	Grain yield (kg ha⁻¹)	(−)45.4	Hammami et al. (2017)
	Z. mays L. Longping–206	EC–0.78 dS m⁻¹	EC–3.75, 6.25 dS m⁻¹	Grain yield (kg ha⁻¹)	(−)13.2, 20.7	Feng et al. (2017)
	O. sativa L.	0 mM NaCl	150mM NaCl	Grain yield	(−)36.17–50	Hasanuzzaman et al. (2009)
	Foeniculum vulgare Mill.	EC–0.25 dS m⁻¹	EC–12 dS m⁻¹	Seed yield (g plant⁻¹)	(−)94.4	Semiz et al. (2012)
	V. radiate L. Wilczekcv. Pusaratna	0 mM NaCl	50, 75mM NaCl	Grain yield plant⁻¹ (g)	(−)90.8, 100	Sehrawat et al. (2015)
	P. vulgaris cv. Contender	0.5mM NaCl	10mM NaCl	Fresh pod yield	(−)22.8	Kontopoulou et al. (2015)
	H. vulgare L.	EC–3 dS m⁻¹	EC–14 dS m⁻¹	Grain yield (kg ha⁻¹)	(−)87.2	Jamshidi and Javanmard (2017)
	P. vulgaris L. cv. Bronco	Deionized water	150mM NaCl	Pod yield	(−)41.5	Rady (2011)
	O. sativa L. cv. Aychade Fidji	1311 uS cm⁻¹	4184 uS cm⁻¹	Yield (t ha⁻¹)	(−)45 (−)12.1	Gay et al. (2010)
	O. sativa L.	EC< 3 dS m⁻¹	3 ≤EC< 5 dS m⁻¹	Grain yield (kg ha⁻¹)	(−)20	Clermont-Dauphin et al. (2010)
Drought Stress	*O. sativa*	Well watered	Moderate drought	Grain yield (kg ha⁻¹)	(−)44	Torres and Henry (2018)
	Phaseolus vulgaris cv. DBS 360	Irrigated to FC	Withholding irrigation from 49 days after planting for 24 days	Grain yield (t ha⁻¹)	(−)42.5	Mathobo et al. (2017)
	G. max cv. MJiHuang13	60–70% RWC	30–40% RWC	Seed yield (g plant⁻¹)	(−)38.8	Li et al. (2013a)
	Hordeum vulgare cv. Valfajr Reihan03 Nosrat Yusuf	70 mm evaporation from the pan	High Drought—110 mm evaporation from the pan	Grain yield (kg ha⁻¹)	(−)54.1 (−)50.5 (−)64.1 (−)38.5	Soleymani (2017)
	T. turgidum spp. *Durum* cv. Banamichi Samayoa Jupare Aconchi Yavaros Cocorit Rio Colorado Mohawk CMH83.2578	Optimum irrigation:1 pre–planting+4 auxiliary	Irrigation: 1 pre–planting + 1 auxiliary 30 DAP	Grain yield (t ha⁻¹)	(−)48.9 (−)44.8 (−)42.2 (−)45.7 (−)53 (−)43.4 (−)51.2 (−)51.6 (−)53.8	Li et al. (2013b)

(Continued)

TABLE 1.1 (CONTINUED)
Different Abiotic Stress Induced Percentage Changes in Yield Parameters of Different Crop Species

Abiotic Stress	Crop Studied	Experimental Level/Dose/Concentration		Yield Parameters	Changes (%)	References
		Control	Stress			
	Zea mays L.	Full irrigation throughout	Irrigation for only 28 d	Grain yield (kg ha⁻¹)	(−)70	Adebayo and Menkir (2014)
	O. sativa L.	Flood condition	Terminal drought	Grain yield (g m⁻²)	(−)52–55	Monkham et al. (2015)
	Brassica juncea L.	Irrigation done twice—40 and 80 DAS	Rainfed condition receiving 0.5–0.9mm rainfall	Seed yield (kg ha⁻¹)	(−)4.3 to 60	Chauhan et al. (2007)
	P. vulgaris L.	Irrigation was done every 7 days	Irrigation was done every 12 days	Seed yield (g)	(−)61	Habibi (2011)
	Capsicum annuum	1.43 times of crop water requirement	0.50 times of crop water requirement	Fruit yield (g)	(−)28.8	Kurunc et al. (2011)
	G. max L. Merril.	Well watered—75% FC	35% FC	Grain yield plant⁻¹ (g)	(−)21.9	Anjum et al. (2011)
	T. aestivum L.	Full irrigation	Drought stress at the start of stem elongation stage	Grain yield (kg ha⁻¹)	(−)46.1	Keyvan (2010)
	Panicum miliaceum cv. Birjand Sarbisheh	Well watered—874 mm depth of water	Drought stress at ear emergence–764.2 mm depth of water	Seed yield (t ha⁻¹)	(−)40.7 (−)43.3	Seghatoleslami et al. (2008)
	T. aestivum cv. GK Elet Emese Plainsman V. Cappelle D.	60% of total SWHC	25% of total SWHC	Total grain yield (g)	(−)78 (−)60.5 (−)9.41 (−)55.1	Guóth et al. (2009)
	H. vulgare L. Rum ACSAD176 Athroh Yarmouk	Well watered—75% FC	Severe drought—25% FC	Grain yield (g plant⁻¹)	(−)84.6 (−)84.8 (−)84.3 (−)86.7	Samarah et al. (2009)
	Zea mays cv. JK628 ND93	Well watered	Moderate drought	Grain yield	(−)33.7 (−)62.3	Wei et al. (2010)
	P. vulgaris L.	Watered every week	Watered every three week	100 seed wt. (g)	(−)8	Martínez et al. (2007)
	Z. mays L. cv. Run Nong 35 Dong Dan 80 Wan Dan 13	Well watered—100% FC	Severe drought—40% FC	Grain yield plant⁻¹ (g)	(−)43.1 (−)13.5 (−)22.5	Anjum et al. (2017)
Heavy metal	*B. vulgaris* var. Allgreen H–1	Unamended soil	20% sewage sludge (w/w) 40% sewage sludge (w/w)	Yield (g pot⁻¹)	(−)6.8 (−)21.8	Singh and Agrawal (2007)
	T. aestivum L. PBW34 WH542	0 mg Cd kg⁻¹ soil	100 mg Cd kg⁻¹ soil	Grain Yield (g plant⁻¹)	(−)25.8 (−)52.7	Khan et al. (2007)

(Continued)

TABLE 1.1 (CONTINUED)

Different Abiotic Stress Induced Percentage Changes in Yield Parameters of Different Crop Species

Abiotic Stress	Crop Studied	Experimental Level/Dose/Concentration		Yield Parameters	Changes (%)	References
		Control	Stress			
	P. sativum L.	334.5 mg Copper kg^{-1} soil	1,338 mg Cu kg^{-1} soil	Seed yield (g plant^{-1})	(−)15	Wani et al. (2008)
	P. vulgaris L. cv. Bronco	Deionized water	1 mM Cd^{+2}	Pods yield pot^{-1} (g)	(−)53.8	Rady (2011)
	O. sativa L. cv. Pusasugandha 3	0 kg m^{-2} Sewage sludge	3, 4, 5, 6 and 9 kg m^{-2} Sewage sludge (Zn, Cu, Mn, Cd, Pb, Ni and Cr)	Yield (g m^{-2})	(+) 60, 111, 125, 134 and 137	Singh and Agrawal (2010a)
	V. radiata L. cv. Malviya janpriya (HUM 6)	0 kg m^{-2} Sewage sludge	6,9 and 12 kg m^{-2} Sewage sludge (Zn, Cu, Mn, Cd, Pb, Ni and Cr)	Yield (g m^{-2})	(+) 39, 76 and 60	Singh and Agrawal (2010b)
	O. sativa. L. var. MSE–9	0 uM Copper	10, 50 and 100 uM Cu	Shoot f. w. (g)	(−)8.5, 10.5 and 35.3	Thounaojam et al. (2012)
	C. arientinum L.	Unamended soil	5.75 and 11.5 mg Cd kg−1 soil 97.5 and 195 mg Pb kg−1 soil	Seed yield	(−)14 and 19 (−)12.3 and 8.8	Wani et al. (2007)
Flooding Stress	*T. aestivum* cv. Chuanmai 104	*Without waterlogging*	Waterlogging at tillering	Grain yield (kg ha−1)	(−)12.07	Wu et al. (2018)
	Soft red winter wheat	*Without waterlogging*	Waterlogging at late tillering	Grain yield (g m−2)	(−)34	Arguello et al. (2016)
	Brassica napus L.	*Without waterlogging*	Waterlogging at early flowering stage; water level—1–2 cm above soil surface	Seed yield	(−)79.2	Xu et al. (2015)
	T. aestivum cv. Nishikazekomugi Iwainodaichi UNICULM	*Without waterlogging*	Waterlogging—From jointing stage till maturity	Grain yield (g m−2)	(−)64–73 (−)37–73 (−)4–58	Hayashi et al. (2013)
	T. aestivum L.	*Without waterlogging*	Waterlogging for 44 d at 93 DAS, 58 d at 64 DAS	Grain yield (t ha−1)	(−)20, 24	Dickin and Wright (2008)
	O. sativa L.	*Normal water depth(0–10 cm)*	Stagnant Flooding and partial submergence–water level 50 cm depth	Grain yield (g m−2)	(−)1–43	Kuanar et al. (2017)
	G. max L.	*Saturated soil culture—Water depth 20 cm under soil surface*	Temporary flooding—5 cm upper soil surface	Average productivity (t ha−1)	(−)28.3	Ghulamahdi et al. (2016)
	G. max L.	*Non flooded*	Flooding at pod fill reproductive stage	Yield (kg ha−1)	(−)20–39	Rhine et al. (2010)
	V. radiate L. Wilczek	*Non flooded*	Waterlogging at vegetative stage for 3, 6, 9 d	Grain yield (g plant−1)	(−)20, 34, 52	Kumar et al. (2013b)

(Continued)

TABLE 1.1 (CONTINUED)

Different Abiotic Stress Induced Percentage Changes in Yield Parameters of Different Crop Species

Abiotic Stress	Crop Studied	Experimental Level/Dose/Concentration		Yield Parameters	Changes (%)	References
		Control	Stress			
	Sorghum bicolor L. Moench	*Control pots—Free drainage*	Waterlogging at vegetative stage	Stalk yield	(–)22	Promkhambut et al. (2010)
	G. max L.	*Optimum soil moisture*	Excess water at early vegetative stage; water level—15–20 cm	Seed yield (g m–2)	(–)25–31	Bajgain et al. (2015)
UV–B radiation	*Vigna unguiculata* L. CB–27, CB–25	0 KJ	0 + 15 KJ	Seed wt. (g plant^{-1})	(–)56, 41	Surabhi et al. (2009)
	Solanum tuberosum L. var Kufri Badshah	9.6 KJ/m²/day	9.6 + 7.2 KJ/m²/day	No. of tubers and fresh wt. of pods	(–)53.4 and 16.4	Singh et al. (2010)
	Pisum sativum L. Var Arkel	4.8 KJ/m²/day	4.8 + 7.2 KJ/m²/day	No. of tubers and wt. of pod	(–)13.9 and 28.7	Singh et al. (2015)
	Glycine max L.	5.2 KJ/m²/day	5.2 + 7.2 KJ/m²/day	No. of seeds pod^{-1}, No. of pods plant^{-1}, wt. of seeds pod^{-1} and pod wt. plant^{-1}	(–)18–29, 21–34, 22–36 and 28–38	Choudhary and Agrawal (2014)
	Linum usitatissium L.	5.2 KJ/m²/day	5.2 + 7.2 KJ/m²/day	Seed yield	(–)56.9	Tripathi and Agrawal (2013)
	Brassica campestris L.	5.2 KJ/m²/day	5.2 + 7.2 KJ/m²/day	Pod wt. plant^{-1} and Seed wt. plant^{-1}	(–)42.2, 59.7	Tripathi and Agrawal (2012)
Heat Stress	*Triticum aestivum* L. cv. Batis	Control	1200 (STT, ºC min) 1900 (STT, ºC min)	Grain yield (g)	(–)24 (–)16	Rezaei et al. (2018)
	Oryza sativa L.	Ambient night time temp.– 27°C	High night time temp. 32°C	Grain yield	(–) 95	Mohammed and Tarpley (2009)
	T. aestivum L. cv. Golia Sever *T. durum* L. cv. Acalou TE 9306	Mean day/ night—25/14°C	Elevated day/ night—31/21°C	Yield spike^{-1} (g)	(–)6.1 (–)17.5 (–)21 (–)8.6	Dias and Lidon (2009)
	T. aestivum L.	Mean day/ night—24/14°C	Elevated day/ night—31/18°C	Grain yield (g plant^{-1})	(–)39	Prasad et al. (2011)
	T. aestivum L. cv. Excalibur Gladius Janz Krichauff	Unheated control	35°C Single day heating at Green anther stage and Early grain stage 7–10 d after anthesis	Grain yield (g m^{-2})	(–)17.7 and 18.8 (–)21.5 and 25.8 (–)35 and 30.2 (–)19.2 and 22	Talukder et al. (2014)
Cold Stress	*Cicer arietinum* L. cv. GPF2 L550	Avg. day/ night—28/17°C	Avg. day/ night—11.7/2.3°C	Avg. seed wt. (mg)	(–)40 (–)50	Nayyar et al. (2007)

(Continued)

TABLE 1.1 (CONTINUED)
Different Abiotic Stress Induced Percentage Changes in Yield Parameters of Different Crop Species

Abiotic Stress	Crop Studied	Experimental Level/Dose/Concentration		Yield Parameters	Changes (%)	References
		Control	Stress			
	O. sativa L. cv. WYJ24 NJ46 NJ 5055 WYJ27	Avg. day/ night—23/15°C	Avg. day/ night—17/15°C	1000—grain wt. (g)	(−)13.2 (−)5.2 (−)7.5 (−)7.1	Zhu et al. (2017)
	O. sativa L.	32°C	13°C	Grain yield (g hill^{-1})	(−)24.7	Ghadirnezhad and Fallah (2014)
	C. arietinum L. cv. GPF2	Avg. max/ min—28/14°C	Avg. max/ min—8.3–9.6/2.8–5.3°C	Seed yield plant^{-1} (g)	(−)45.6	Kaur et al. (2011)
O₃ stress	B. vulgaris L. Patriot	34–39 ppb, 8 h mean	62 ppb, 8 h mean	Root yield (t f.w. ha^{-1})	(−)14	De Temmerman et al. (2007)
	O. sativa L. SY63 WYJ3	Ambient— 13.8–74.2 ppb, 7 h mean	Elevated— Ambient × 1.5	Grain yield (g hills^{-5})	(−)20.7 (−)6.3	Pang et al. (2009)
	Gossypium hirsutum L. Romanos Allegria	< 4 ppb, 7 h day–1	100 ppb, 7 h day–1	Seed wt. (g)	(−)57.3 (−)50	Zouzoulas et al. (2009)
	T. aestivum L.	~3.85 ppb	40.1 ppb, 8 h mean	Yield plant^{-1}(g)	(−)20.7	Rai et al. (2007)
	O. sativa L. Saurabh 950 NDR 97	2.93–4.36 ppb	30.5–45.4 ppb, 8 h mean	Yield (g plant^{-1})	(−)11.5 (−)16	Rai and Agrawal (2008)
	B. campestris L. Kranti	3.08–4.01 ppb	41.6–54.2 ppb	Yield (g plant^{-1})	(−)16.4	Singh et al. (2009)
	G. max L. PK472 Bragg	5.6 ppb 8 ppb	4 h mean 70 ppb 100 ppb	Yield (g plant^{-1})	(−)20 (−)33.6	Singh et al. (2010)
	T. aestivum L. cv. HUW 510	4.7 ppb	12 h mean 45.3 ppb 50.4 ppb 55.6 ppb	Grain yield (g m^{-2})	(−)20.0 (−)37.0 (−)46.0	Sarkar and Agrawal (2010)
	T. aestivum L. cv. Sonalika	4.7 ppb	12 h mean 45.3 ppb 50.4 ppb 55.6 ppb	Grain yield (g m^{-2})	(−)11.0 (−)25.0 (−)38.5	Sarkar and Agrawal (2010)
	B. campestris L. Sanjukta Vardan	49.4 ppb	Ambient + 10 ppb (3 h)	Test wt.	(−)12.5 (−)47	Tripathi and Agrawal (2012)
	L. usitatissimum L. Padmini T–397	50.3 ppb	Ambient + 10 ppb (3 h)	wt. of seeds plant^{-1}	(−)40.5 (−)46.7	Tripathi and Agrawal (2013)
	T. aestivum L. HUW–37 K–9107	48.4 ppb, 8 h mean	55.2 ppb, 8h mean	Yield (g plant^{-1})	(−)39 (−)12.4	Mishra et al. (2013)

(Continued)

TABLE 1.1 (CONTINUED)
Different Abiotic Stress Induced Percentage Changes in Yield Parameters of Different Crop Species

Abiotic Stress	Crop Studied	Experimental Level/Dose/Concentration		Yield Parameters	Changes (%)	References
		Control	Stress			
	Trifolium alexandrium L. Bundel	49.6 ppb, 6 h mean	57.5 ppb, 6 h mean	Total biomass (g plant^{-1})	(–)13.5	Chaudhary and Agrawal (2013)
	Wardan				(–)18.2	
	JHB–146				(–)9.1	
	Fahli				(–)6.5	
	Saidi				(–)4.4	
	Mescavi					
	Z. mays L. HQPM1	Ambient—55.6 ppb	Ambient+15 ppb (5 h)	wt. of kernels plant^{-1}	(–)7.2	Singh et al. (2014)
	DHM117		Ambient+30 ppb (5 h)		(–)10.1	
			Ambient+15 ppb (5 h)		(–)9.5	
			Ambient+30 ppb (5 h)		(–)13.8	
	G. max L.	30–35 ppb	60–70 ppb	Grain yield (kg m^{-2})	(–)47	Bou Jaoude et al. (2008)
	V. radiata L. M–28	CF	41–73 ppb	Seed yield plant^{-1} (g)	(–)47.1	Ahmed (2009)
	M–6601				(–)51.1	
	S. tuberosum L. Desiree	12.4, 12 h mean	12 h mean, 25.8 ppb	Commercial tuber production	(–)53	Calvo et al. (2009)
			42.5 ppb		(–)65	
	P. vulgaris L. S156,	0 ppb	60 ppb	Seed yield (g plant^{-1})	(–)77	Flowers et al. (2007)
	R 123				(–)19	
	R 331				(–)35	
	O. sativa L.	CF	62 ppb	Seed yield (g)	(–)14	Ainsworth (2008)
	O. sativa L	27–28 ppb	73–77 ppb	Seed wt (g m^{-2})	(–)14	Reid and Fiscus (2008)
	O. sativa L. cv. SY 63	69.4–81.4 ppb, 7 h mean	56–59 ppb, 7 h mean	Relative yield	(–)82.5	Shi et al. (2009)
	LYPJ				(–)85	
	T. aestivum L. cv. Sufi	CF	CF+60 ppb	Relative yield	(–)88.5–55.5	Akhtar et al. (2010)
	Bijoy		CF+ 100 ppb for 7 h for 3 months		(–)66.8–54.4	
	G. max L.	20–24.2 ppb	8 h mean 33.2–45.4 ppb	Average seed yield (g dw plant^{-1})	(–)8.7	Zhang et al. (2017d)
			73.2–85.4 ppb		(–)37.7	
	T. aestivum L. Feng Kang 13	4–28 ppb	96–108 ppb 145–160 ppb for 8 h for 67 d	Grain wt. spike^{-1} (g)	(–)18.5	Zhu et al. (2015)
					(–)35.7	
	P. vulgaris L. cv. HUM–1	Ambient— 36.6–85.4 ppb, 8 h mean	Elevated— 41.2–95.2 ppb	Seed yield (g m^{-2})	(–)15.4	Chaudhary and Agrawal (2015)
	HUM–2				(–)13.8	
	HUM–6				(–)13.5	
	HUM–23				(–)9.7	
	HUM–24				(–)12.0	
	HUM–26				(–)10.3	

(Continued)

TABLE 1.1 (CONTINUED)

Different Abiotic Stress Induced Percentage Changes in Yield Parameters of Different Crop Species

Abiotic Stress	Crop Studied	Experimental Level/Dose/Concentration		Yield Parameters	Changes (%)	References
		Control	Stress			
	V. unguiculata L. Blackeye Asontem	13 ppb, 24 h mean	39 ppb, 24 h mean	Yield (ton ha^{-1})	(–)57.2 (–)37.9	Tetteh et al. (2015)
	S. tuberosum L. cv. Kufri chandramukhi	Ambient—49.6 ppb, 8 h mean	Elevated— Ambient + 20 ppb	Total tuber f.w. 60 DAE 90 DAE	(–)16.5 (–)47.6	Kumari and Agrawal (2014)
	T. durum Desf. Colombo Sculptur	AOT 40–826 ppb.h.	AOT 40–21, 121 ppb.h.	Grain wt. (g)	(–)10.5 (–)16	Monga et al. (2015)
Sulphur dioxide	*T. aestivum* L.	Filtered chamber (2.82 ppb)	Non filtered chamber (11.1 ppb)	Grain yield	(–)20.7	Rai et al. (2007)
	P. miliaceum	Ambient	Ambient +0.1 ppm	wt. of grains plant^{-1}	(–)71.6	Agrawal et al. (1983)
	O. sativa L.	Filtered chamber (1.26ppb)	Non filtered chamber (6.99ppb)	wt. of grains plant^{-1}	(–)>10	Rai and Agrawal (2008)
Oxides of nitrogen	*T. aestivum* L.	Filtered chamber (6.59ppb)	Non filtered chamber (43.1ppb)	Grain yield	(–)84.7	Rai et al. (2007)
	V. radiata L.	11.73 nl l–1	26.91 nl l–1	Seed wt. plant^{-1} No. of pods plant^{-1}	(–)51.12,(–) 20	Ahmed (2009)

• FC—Field capacity; RWC—Relative Water Content; SWHC—Soil Water Holding Capacity; DAS—Days after Sowing; EC—Electrical Conductivity; CF— Charcoal-Filtered Chambers

and subsequent depolarization of plasma membrane (Harrison, 2012). In addition to this, Suh et al. (2016) reported hydrogen peroxide (H_2O_2) mediated activation of Ca^{2+} permeable channel in the plasma membrane, thereby suggesting an indirect role of H_2O_2 in controlling stomatal aperture under water stress situation.

Transmembrane proteins like aquaporins have been found out to be an important group of proteins conferring drought tolerance in crop plants (Zargar et al., 2017). Among different aquaporins, plasma intrinsic aquaporins (PIP) localized in plasma membrane and tonoplast intrinsic aquaporins (TIP) localized in the vacuolar membrane are two of the most widely studied aquaporins due to their role in water and solute transportation under scarce water conditions (Hove and Bhave, 2011). Although ABA is considered to be one of the major hormones determining drought stress signaling and regulating several PIPs activities, the responses of *PIP* genes and ABA are different under water deficit conditions, suggesting evidence of both ABA-dependent and ABA-independent PIP genes

regulating pathways under drought stress conditions (Lian et al., 2006; Zargar et al., 2017).

Moreover, several phospholipid systems are evidenced to be significantly involved in generating an array of messenger molecules in plant cells when exposed to osmotic stress (Farooq et al., 2009). Recently, Kumar et al. (2015) evidenced a proportionate increase in the level of β-sitosterol, a major phytosterol, in rice seedlings with the severity of drought stress in drought-tolerant rice cultivar N22, inferring the importance of phytosterols in inducing drought tolerance signaling pathway. The study also reported an enhancement in the activity of HMG-CoA reductase, a key enzyme responsible for the biosynthesis of phytosterols, under osmotic stress in the rice seedling. Besides this, many studies have reported the significant role of phospholipase Dα (PLDα), a membrane phospholipid, in the abscisic acid signal transduction pathway, resulting in stomatal closure under osmotic stress (Sang et al., 2001; Zhang et al., 2004).

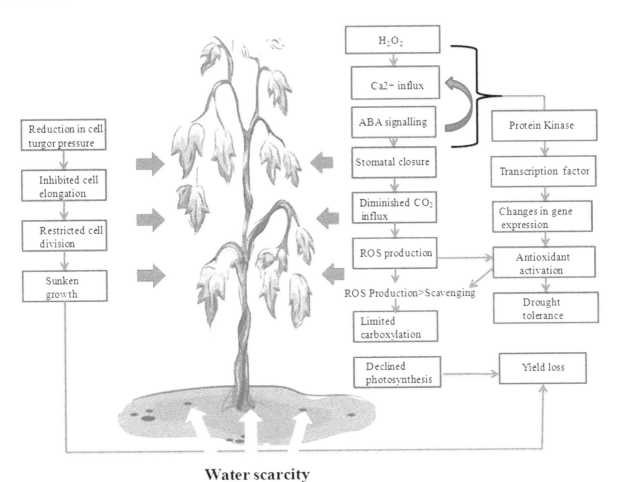

Water scarcity

FIGURE 1.1 Schematic diagram showing morphological, physiological and antioxidative responses under drought stress. Signaling transduction has also been shown in response to water scarce condition. Drought stress causes disturbances in the balance between the production of reactive oxygen species and the antioxidants, causing reduction in the scavenging capacity of reactive oxygen species by antioxidants, which induces oxidative stress. Under water limiting condition, stomatal closure occurs plausibly via ABA signaling and indirectly by H_2O_2 accumulation, which reduces the CO_2 influx. Reduction in CO_2 influx not only restricts carboxylation directly but also stimulates the formation of more reactive oxygen species, ultimately affecting plant productivity. Drought tolerant species exhibit a better ABA signaling pathway along with Ca^{2+} influx and H_2O_2 accumulation which activates specific protein kinases synthesis and regulates the downstream responses such as changes in gene expression and activation of antioxidative molecules.

1.4   METAL TOXICITY

Plants require mineral nutrient for normal growth and metabolism, but an excess of these metals can cause major production losses in agriculture as well as pose serious human health issues. Approximately 60 elements in different plant tissues have been identified biochemically, among which 17 have been considered as essential elements for the normal growth and metabolism. (Table 1.2).

The toxic level of metal concentration inside the plant's cells is the reflection of localized concentration of metal ions. Plant dry weight can indicate the metal ions present in plant tissues, in addition, is also dependent on

various other factors, i.e., chemical constituents of soil, pH, organic matter, cation exchange capacity, climatic conditions and plant age (Shah et al., 2010).

Some of the metals, such as zinc, iron, nickel, and copper, are essential micronutrients needed for normal structural, physiological and biochemical processes in a plant cell, which in turn are dependent on the metal-dependent enzyme and proteins. Whereas, metals like arsenic, mercury, cadmium, aluminum and lead are non-essential and potentially highly toxic even under low concentrations. However, even micronutrients could pose toxicity in plants above the levels required by the plants (Singh et al., 2016) (Table 1.2)

TABLE 1.2
Showing Available Forms, Phytotoxic Limits, Functional Role, and Toxic Effects of Different Metals on Plants

Elements	Available Forms	Range	Critical Limits	Functional Role	Toxic Effects	References
Cobalt (Co)	Co^{2+} and Co^{3+}	0.1–10 µg g^{-1}DW	30–40 µg g^{-1} DW	Coenzymes, Vitamin B12. for example, (cobalamin)	Growth↓ PSII activity↓ DNA damage↑ Metal displacement↑ Inhibits karyokinesis and cytokinesis	Jaleel et al. (2008); El-Sheekh et al. (2003); Bracher et al. (2015)
Copper (Cu)	Cu^+ and Cu^{2+}	2–250 µg g^{-1} DW	20–30 µg g^{-1}DW	Cofactor for enzymes in photosynthetic electron flow and Cu/Zn-SOD enzymes	Chlorosis, necrosis Inhibit root elongation, Photosynthetic competence Quantum efficiency of PSII↓ MDA content↑ ROS production↑	Hong et al. (2012); Yu et al. (2008); Tsay et al. (1995) Lombardi and Sebastiani (2005); Yurekli and Porgali (2006); Azooz et al. (2012); Sharma (2010)
Iron (Fe)	Fe^{2+} and Fe^{3+}	140 µg g^{-1}DW		Role in photosynthetic, respiratory, nitrogen fixation system. Hormone production and component of ferrodoxin and cytochrome.	Growth↓ Photosynthesis↓ Chlorophyll↓ ROS production↑ Damage lipids, proteins andnucleic acid	Møller et al. (2007): Estevez et al. (2001); Chatterjee et al. (2006); Marschner (1995); Schutzendubel and Polle (2002)
Manganese (Mn)	Mn^{2+}	15–100 µg g^{-1}DW		Biosynthesis of enzymes such as, malic enzyme, isocitrate dehydrogenase, and nitrate reductase. Involved in carbohydrate and nitrogen metabolism, synthesis of fatty acid, acyl lipids, and carotenoid Protection of PSII from photo damage. Involved in modulating carbohydrate and lipid metabolism, nucleic acid biosynthesis.	Chlorophyll↓ Leaf cupping, distortion and disintegration in Chloroplast. Photosynthesis, CO_2↓ assimilation and stomatal conductance ROS production↑ Alter chromosome and mitosis as well as % chromosomal aberrations.	Adriano (2004); Kitao et al. (2001); Fiskesjö (1988)
Molybdenum (MO)	MoO_4^{2-}	1–10 µg g^{-1} DW	135 mg Kg^{-1} (in *Hordeum vulgare* L.)	Involved in nitrogen metabolism. As cofactor (molybdopterin) of proteins and enzymes Promotes growth.	Interferes with various metabolic processes, thus causing physiological damage.	Mendel (2013); McGrath et al. (2010); Rout and Das (2002); Davis et al. (1978).

(Continued)

TABLE 1.2 (CONTINUED)
Showing Available Forms, Phytotoxic Limits, Functional Role, and Toxic Effects of Different Metals on Plants

Elements	Available Forms	Range	Critical Limits	Functional Role	Toxic Effects	References
Nickel (Ni)	Ni^{2+}	25–246 $\mu gg-1$ DW		Part of metalloenzyme, for example, urease, methyl coenzyme, hydrogenase, Promote seed germination and iron uptake.	Chlorophyll content↓ Interruption of electron transport. Inhibits germination, root meristem mitotic activity. Induces chromosomal aberrations Increases Fe deficiency↑ Crop yield↓	López and Magnitski (2011); Negi et al. (2014); Seregin and Kozhevnikova (2006); Iyaka (2011); Poonkothai and Vijayavathi (2012)
Zinc (Zn)	Zn^{2+}	8–100 $\mu g\,g^{-1}$ DW	Concentration >300 $mgKg^{-1}$	DNA and RNA regulation and stabilization. Involved in protein synthesis gene expression. As cofactor of various enzymes (such as SOD) Constituent of various enzymes (oxidoreductases, transferases and hydrolases) as well as ribosome formation of carbohydrates and chlorophyll	Alters leaf morphology and anatomy, Water potential↑ Impairs PSII activity, nutrient uptake, Disrupts electron transport chain in chloroplast Photosynthesis↓	Alloway (2004); Barber (1995); Gupta et al. (2012); Yadav (2010); Mishra and Dubey (2005); Mousavi et al. (2013); Sensi et al. (2003).
Chromium (Cr)	Cr^{3+} and Cr^{6+}	0.2–1 $\mu g\,g^{-1}$DW		Decreased shoot growth	Alters soil pH. Seed dry matter reduced, adversely affecting stem, leaves. Inhibits cell division, elongation of plant root. Decreased shoot growth	Mortvedt and Giordano et al. (1975); Fargašová (2012)
Aluminum (Al)	Al^{3+}		2–3 μgg^{-1}(pH below ≤.5)	Non nutrient element. At low concentration may stimulate plant growth	Curling or rolling of young leaves. Purple discoloration on stems, leaves and leaf veins. Veins yellowing and dead leaf tips. Stomata aperture size and photosynthesis rate↓ Altered absorption and transportation of major nutrients such as P, K, Ca and Mg.	Ma et al. (2001); Silva (2012); Mossor-Pietraszewska (2001); Rout et al. (2001); Vardar and Ünal (2007)

Toxicity consequences of heavy metal include cellular damage, enhanced production of ROS leading to oxidative stress and reduced transpiration rate, photosynthesis and altered carbohydrate metabolism ultimately resulting in adverse effects on plant growth and development processes in the plant. Chemical properties of the heavy metals play an important role in deciding the detoxification fate of metal ion. For example, metal ions at high concentration may either interfere with the functional site in protein or can displace some of the essential ions resulting in loss of functions or may lead to increased ROS production. Plants possess several efficient systems to detoxify the heavy metals through various homeostatic mechanisms that serve to control the uptake, accumulation, trafficking and detoxification of metals.

Heavy metals can strongly bind to oxygen, nitrogen and sulfur atoms; therefore, they can easily bind to functional SH-group in enzyme or can even replace the functional elements in the prosthetic group of enzyme. For example, divalent cations such as CO^{2+}, Ni^{2+} and Zn^{2+} can displace Mg^{2+} ion in ribulaose1, 5-bisphosphate-carboxylase/oxygenase, resulting in the activity loss (Nieboer and Richardson, 1980).

1.4.1 DEFENSE STRATEGIES ADOPTED BY THE PLANT IN RESPONSE TO HEAVY METAL STRESS

The first line of defense in the plant includes thick cuticle, cell wall mycorrhizal symbiosis and presence of trichomes. Trichomes can even serve as storage sites as well as secrete various secondary metabolites against metal toxicity (Emamverdian et al., 2015).

Formation of (a) siderophores: these are large chelator molecules capable of binding with the heavy metals to form a large complex, thereby preventing the entry inside the cell cytoplasm of the plants; (b) molecules like carboxylic acid, malate, citrate and oxaloacetate: these molecules can bind with the heavy metal in order to prevent the uptake of the heavy metals (Alford et al., 2010); (c) free histidines, glutathione, nicotianamine and polyamines are some of the other molecules which can bind with the heavy metals at their SH-group to prevent entry into the cell cytoplasm; (d) metallothioneins and phytochelatins: these are cysteine-rich molecules capable of binding heavy metal ions due to the presence of the SH-group.

Plants possess several classes of transport proteins, such as CPx-ATPases for Cu and Cd, ABC transporter for CD transport into the vacuole, ZIP transporter (ZRT, IRT related proteins) for Fe and Zn, Nramp transporter for several of heavy metals and cation diffusion facilitator (CDF) family (Hall, 2002).

1.4.2 ANTIOXIDANTS ACTIVITY UNDER METAL TOXICITY

Plants, upon exposure to heavy metal stress, could excessively produce ROS leading to upset in the equilibrium of cellular redox systems, and in order to maintain a normal homeostasis, plants employ a highly efficient defense system called antioxidant defense to mitigate the harmful effects of the ROS. The antioxidant defense system comprises both non-enzymatic and enzymatic constituents. Such enzymatic antioxidants (i.e., SOD, CAT, APX, GR) and non-enzymatic antioxidant (AsA and reduced glutathione (GSH)), carotenoids, alkaloids, tocopherols, proline and phenolics serve to be ROS scavengers under heavy metal stress in several studies (Emamverdian et al., 2015).

Several studies have also reported the role of antioxidant defense system under heavy metal toxicity. For example, in root mitochondria of pea plants, SOD activity increased by 20% at 20 µM Cr content, and it substantially reduced under 200 µM SOD activities (Dixit at al., 2002). Jain and Ali (2000) reported decreased activity of CAT in sugarcane, at 20–80 ppm of Cr dose. At excess metal concentrations, CAT activity was reduced and SOD activity remained unaffected (Gwóźdź et al., 1997; Shah et al., 2010). In a study on spinach (*Spinacea oleracea* L.), exposure to excess (500 µM) concentration of Co, Ni, Cu, Zn and Cd showed interesting pattern of antioxidant defense, as SOD activity increased whereas CAT activity declined with Co, Ni, Cu and Cd, whereas no significant changes were noticed under Zn treatments. APX activity significantly increased under high levels of Cd and Zn treatment and slightly increased under Ni treatment. GR activity was enhanced under CO, Ni, Cu, Zn and Cd treatment. The most remarkable increase was reported with Co and Cd treatments (Pandey et al., 2009).

1.4.3 ZN METAL TRANSPORTERS

Plants possess several metal transporters for the uptake of micronutrients from the soil. ZIP (ZRT-like zinc-regulated transporter gene) has been identified as a unique metal transporter family, found in bacteria, fungi, mammals as well as in plants. The proteins encoded by the ZIP have a high affinity for Zn^{2+} as well as other divalent cations such as Cd and Cu. ZIP 1, ZIP 2 and ZIP 3, identified from the yeast cells, play a significant role as Zn transporters in plants (Guerinot, 2000).

1.4.4 Fe Transporter

Plants uptake Fe mainly by two major strategies from the soil; in the non-grasses family, under Fe limited conditions plants activate a process that leads to both acidification of the soil and reduction of Fe^{3+}, whereas in the grass species, the chelation-based strategy is involved.

Fe is mainly found in non-available form, mainly as Fe^{3+} oxidation state in the soil. Non-grasses secrete protons into the rhizosphere with the activation of H+ATPase present in the plasma membrane, resulting in lowering of pH, thereby increasing the solubility of the Fe^{3+}. Fe deficiency acts as a trigger for the uptake of the Fe^{2+} from the soil into the roots. An inducible plasma membrane, ferric reductase oxidase (FRO), catalyzes the reduction of Fe^{3+}, which is encoded by the gene *FRO2*.

Fe^{2+} is transported to the root cells by IRT1 (iron-regulated transporter 1), a member of ZIP metal transporter family. It is able to transport other divalent metals such as Zn, Mn and Cd.

Grasses adopt different strategies, such as the formation of phytosiderophores (PSs) that are related to mugineic acid which can directly chelate with the Fe^{3+} in the rhizosphere, thus rendering the uptake by the roots. PSs are able to chelate with many other divalent cations such as Zn^{2+}, Cu^{2+}, Mn^{2+}, Ni^{2+} and Co^{2+}(Rusell et al., 2012).

1.4.5 Al Transporters

Aluminum (Al) being the non-nutrient element is of greatest agricultural importance because Al creates an acidic environment (< 5 pH) resulting in significant loss in crop production. Al^{3+} forms a complex with the organic acids, inorganic phosphate and sulfate affecting the plants, inducing damage at the cellular and molecular levels.

Al resistance mechanism can be classified into: (a) symplasmic tolerance, in which plants continue to uptake Al from soil without affecting the normal cell functioning; (b) Al exclusion strategy of the root apex, in which plants protect them against Al toxicity by excluding organic anion which can directly bind with the Al^{3+} ion in order to restrict their entry across the plasma membrane of the root cells.

Genes involved in the Al-exclusion type encode membrane proteins of the AL^{3+}-activated malate transporter (ALMT) and multi-drug and toxin extrusion (MATE) families (Rusell et al., 2012). In many of the plant species, such as *Arabidopsis, Brassica* and *Sorghum* spp., genes for organic transporters are activated by Al^{3+} via specific receptors or non-specific stress response. Plants which are highly tolerant to elements are good excluders, restricting the accumulation of the toxic metals either by limiting uptake or root-to-shoot translocation.

1.5 FLOODING STRESS

With growing population, inappropriate land use planning and the rising of the likelihood of extreme rainfall events, urban flooding has been seriously affecting the rainfed low-lying areas of South and Southeast Asia, including Bangladesh, India, China and Nepal (Ismail et al., 2013). Likewise, the report presented by IPCC (2007) predicted the occurrence of the flooding would be more frequent in South and Southeast Asia in the near future. According to the estimate of the World Bank, flood-prone areas have increased by 21 million hectares within one decade in India (Sarkar, 2016). The flooded condition is usually triggered by high and unpredictable rainfall, which has enhanced globally since the 1950s due to climate change (Oh et al., 2014). It is reported that the constraints of flooding stress have affected about 10% global agricultural areas and yield loss ranges between 15% and 80% among varying crops, depending on the biotic and abiotic factors such as plant species, soil type and duration of the stress exposure (Patel et al., 2014).

1.5.1 Morphological and Anatomical Responses

Soil flooding induced hypoxia in the submerged organs and plant parts, causing detrimental effect on the growth and productivity in varying crops such as rice (Kuanar et al., 2017), soybean (Tewari and Arora, 2016), maize (Abiko et al., 2012), wheat (Malik et al., 2002) and sorghum (Promkhambut et al., 2010). Formation of adventitious rooting as an effective tactic to provide efficient anchorage and transportation of water and nutrient to the above-ground plant parts (Ezin et al., 2010; Promkhambut et al., 2010; Zaidi et al., 2003). Development of lysigenous aerenchyma in the root not only helps in the formation of internal gas spaces but also reduces oxygen-consuming cell number, resulting in maintenance of oxygen balance under such stressful conditions (Sauter, 2013). Increased aerenchymal area along with connected longitudinal gas spaces per tiller has been reported in rice varieties under stagnant flooding condition, which could be a possible adaptive mechanism to provide a rapid means of oxygen diffusion over long distances within the plant, essential for plant survival under such hypoxic conditions (Kuanar et al., 2017). Formation of a barrier to radial oxygen loss (ROL) in the outer cell layer of the root has been reported in wild variety of maize species (*Zea nicaraguensis*) to check oxygen leakage from the root tip to the surrounding rhizospheric zone under waterlogged conditions (Abiko et al., 2012).

Waterlogging-induced hypoxic condition generally resulted in root injury with concomitant shoot damage, which adversely affects nutrient allocation and normal functioning of the plants (Jackson and Ricard, 2003). Tewari and Arora (2016) have reported the negative impact of flooding stress on the shoot length in the soybean crop.

Biomass accumulation in soybean cultivars has been reported under flooding stress condition at flowering initiation growth phase of the plant (Youn et al., 2008). Reductions in dry leaf matter by 48% in rice cultivars AC1996 and FR13A were affected under the flooded soil as compared to control (Kuanar et al., 2017). The adverse effect of flooding treatment on leaf length and number of leaves in tomato genotypes LA 1421 and LA 1579 with maximum leaf length reduction in genotype LA 1579 was reported by Ezin et al. (2010) as compared to control under flooding stress. Similarly, flooding stress was also reported to cause about 71% reduction in shoot growth in all the cultivars of sorghum (*Sorghum bicolor* (L.) Moench), owing to a decrease in plant height and leaf area and hence negatively affecting the leaf dry matter allocation in all the cultivars (Promkhambut et al., 2010) (Table 1.1).

1.5.2 Physiological Adaptation

Flooding-induced disruption in physiological functioning has resulted in yield and productivity reduction in different crop species. Reduced stomatal conductance, photosynthetic rate and enhanced transpiration are common responses that can occur in hours or days, conditional to the tolerance capacity of each species and cultivar to waterlogging (Promkhambut et al., 2010).

Flooding-induced stomatal closure generally corresponds to an adaptive reaction of the plant, mainly leading to a reduction of root water permeability and the restricted water losses by transpiration, ultimately helping plants to escape wilting situation (Soldatini et al., 1990). Promkhambut et al. (2010) observed an increase in intercellular CO_2 in the flooded plants, which might be due to inefficient diffusion of internal CO_2 from sub-stomatal cavities to the site of carboxylation which further caused a detrimental impact on the assimilation process in the sweet sorghum cultivars. In addition to this, Mutava et al. (2015) observed a reduction in the phloem transportation rate to the roots, resulting in starch accumulation which ultimately caused a decrease in the net photosynthesis rate in soybean plants under flooding stress. Another study revealed initial enhancement of leaf respiration in susceptible cultivars of mung beans (Pusa Baisakhi and MH-1K-24), which was later maintained at a normal rate under prolonged flooding conditions, indicating an acceleration of maintained respiration during adaptation period and its concomitant energy consumption for homeostasis maintenance (Kumar et al., 2013a).

1.5.3 Antioxidative Response and Signaling Cascades under Flooding Condition

Secondary oxidative stress development both in roots and in shoots is inevitable under prolonged flooded condition (Sairam et al., 2008; Simova-Stoilova et al., 2012). The prevention of ROS formation and the countering of oxidative damage is a highly relevant defense mechanism both during short and long-duration waterlogging stress (Colmer and Voesenek, 2009). Hypoxic tissues exhibited enhanced superoxide radical production from mitochondria owing to the donation of accumulated electrons at Complex III enzyme (ubiquinone: cytochrome c reductase) of the electron transport chain to O_2 (Sairam et al., 2008).

Enhancement in the level of ROS, especially H_2O_2 accumulation, has been well documented in *Trifolium* genotypes in response to soil flooding (Simova-Stoilova et al., 2012). In response to such oxidative outburst, different studies have reported an increase in the level of antioxidants (AsA and glutathione) and antioxidative enzyme (APX, SOD, CAT, POD and GR) to counteract the oxidative stress generated under soil flooding (Hossain et al., 2009; Simova-Stoilova et al., 2012; Yan et al., 1996; Yordanova et al., 2004). Yordanova et al. (2004) have observed a substantial increase in the total endogenous peroxide concentration regardless of the significant increase in the activities of POD, CAT and APX under flooding conditions which further accelerated the photo-oxidative injury due to oxygen deficiency in barley roots. Furthermore, in rice plants, root oxidase activity has been reported to decline faster under stagnant soil flooding conditions, indicating reduced oxygen is releasing efficiency of roots, ultimately leading to more oxidative damage due to hypoxic condition (Kaunar et al., 2017).

Apparently, like all other stresses, flooding is no exception in generating an array of signaling messengers characterized by several hormonal, ionic and other signaling molecules. Ethylene is one of the most widely studied plant growth regulators, playing a significant role in the adaptation mechanism in hypoxic root and shoot under flooding (Dat et al., 2004). The occurrence of ET accumulation in the plants subjected to flooding is a common observation owing to its slower rate of diffusion in the water and promotion of higher ET synthesis in the hypoxic root and aerobic shoot in the flooded soil (Sairam et al., 2008). Ethylene synthesis occurs in the presence of sufficient oxygen, thereby causing the transportation of immediate precursor of ET, i.e., 1-amino cyclopropane-1-carboxylic acid (ACC) towards the more aerobic plant parts (Sairam et al., 2008).

Hypoxic tissue exhibits development of aerenchyma, which is well associated with ET synthesis (Sauter, 2013). However, a molecular mechanism related to aerenchymal formation is still unclear. Peschke and Sachs (1994) reported the induction of *XET1*gene encoding a cell wall loosening enzyme, xyloglucan endotransglycosylase, which was found associated with aerenchyma development. This gene was observed to be specifically induced under the hypoxic/anoxic condition since other stresses were found to be incapable of inducing this gene (Peschke and Sachs, 1994; Sairam et al., 2008). Besides ET, ABA (Komatsu et al., 2013), gibberellic acid (GA) (Komatsu et al., 2013), indole acetic acid (IAA) (Dat et al., 2004) and cytokinin (CK) (Dat et al., 2004) are other growth regulators known to play an indirect role in the signaling events under flooded conditions.

Flooding stress has also been reported to induce ionic-dependent signaling pathway in different plants such as soybean (Wang and Komatsu, 2017), maize (Subbaiah et al., 1994; Subbaiah and Sachs, 2001), rice (Tsuji et al., 2000) and *Arabidopsis* (Sedbrook et al., 1996). Yin et al. (2014) has analyzed the temporal profiles of flooding-responsive proteins during the initial stages of flooding stress, using cluster analysis in soybean plants, and observed that most of the proteins' alteration involved calcium-related signal transduction. Moreover, the cytosolic Ca^{2+} level has been observed to increase transiently in response to a few minutes of flooding treatment in maize crops (Subbaiah and Sachs, 2001). In addition to this, strong evidence from different literatures exhibited the initiation of a signaling pathway by cytosolic Ca^{2+} through biosynthesis of Ca^{2+}/Calmodulin dependent glutamate decarboxylase (GAD) and α-aminobutyric acid (GABA) in flooded soybean (Komatsu et al., 2011; Wang and Komatsu, 2017). Furthermore, enhancement in the level of GABA by several folds has been observed by Komatsu et al. (2011) in response to flooding.

1.6 RADIATION—A POTENT ABIOTIC STRESS

Radiation is one of the most important driving factors since the dawn of the living world, and the sun is the source of all radiation either harmful or beneficial. We say 'harmful' to all such things which can be fatal to human society; alike, we say 'beneficial' to those which have some vital role in the survival of the human race, but most radiation is beyond human perception. The electromagnetic radiation of the sun comprised of several radiations which played a pivotal role in the evolution of life on the terrestrial domain; sunlight is necessary for the growth and development of plants, but sometimes excess of essential factors can be damaging too.

1.6.1 ULTRAVIOLET RADIATION

Photosynthesis and respiration in plants are the vital processes which require sunlight either directly or indirectly, and under such circumstances, plants being immobile, they inevitably come under the exposure of other types of radiation. The widely studied portion of the solar spectrum which causes the major deleterious effects is the intense light and ultraviolet rays. Ultraviolet radiation lies beyond the realm of the violet portion of visible light; hence, it is called 'Ultraviolet'. The ultraviolet radiation is broadly classified into three major wavebands ranging from 100–280 nm termed as UV-C, highly energetic and very deleterious but fully absorbed by stratospheric O_3 layer; 280–320 nm termed as UV-B, also very energetic and fatal, but approximately 95% portion is absorbed by O_3 layer; 320–400 nm termed as UV-A, not harmful and is sometimes regulatory in nature (Takshak and Agrawal, 2014). After the Industrial Revolution, enormous anthropogenic activities and industrialization processes took place which caused the incidence of heavy air pollution leading to the depletion of the stratospheric O_3 layer, which was first observed by Molina and Sherwood Rowland in 1974, and this was later confirmed by Farman and his coworkers in 1985. The stratospheric ozone layer is a thick covering of O_3 molecules around the earth; its thickness is measured in Dobson unit (DU) and one DU is equal to 0.01mm. The average thickness of the stratospheric ozone layer is 220 DU; concentration below this is termed as depletion. The thickness of the stratospheric ozone layer varies with the places on the earth surface, i.e., the stratospheric ozone concentration is minimum at equator and maximum at poles, but there are several other atmospheric factors which also govern the amount of perception of UV-B at a definite place, like cloud cover, position of sun, time of the day, latitude and altitude. Anthropogenic activities cause enormous emission of CFCs, which results in the depletion of the O_3 layer and consequently the formation of the stratospheric ozone hole. The depleting stratospheric ozone layer can no longer restrict UV-B, and the incidence of more penetration of UV-B has been observed. Evidence reported that such perception of intense light and higher

UV-B radiation resulted in various cellular damage and also caused impairment of various sensitive cell organelles functioning in the plants (Takshak and Agrawal, 2014). The major cellular damage alters the whole physiology of the plants leading to changes in the biochemistry, morphology and genetic makeup of the plants.

1.6.2 MORPHOLOGICAL, PHYSIOLOGICAL AND BIOCHEMICAL CHANGES

In plants, the radiation is first perceived by the leaves, and under such conditions, various symptoms can be observed, including the changes in coloration of the leaves. Bronzing or browning of leaves due to the accumulation of secondary metabolites like anthocyanins and flavonoids (Takshak and Agrawal, 2016) takes place, which later proceeds into chlorosis and necrosis because of pheophytinization (Demkura et al., 2010). On continuous exposure, cupping or curling of leaves due to partial degradation of IAA was reported in several studies (Cechin et al., 2012; Takshak and Agrawal, 2015; Zuk-Golaszewska et al., 2011). The elevated level of UV-B is known to alter the anatomical structure of the leaves, which affects the physiology of the plant. Some studies have reported an increase in leaf thickness due to increase in spongy mesophyll cell layers (Nagel et al., 1998; Takshak and Agrawal, 2015) while Kakani et al. (2003) have observed a reduction in leaf thickness due to increased palisade layer under higher dose of UV-B radiation. Caasi-Lit et al. (1997) reported rupturing of the granal stacks and chloroplast envelope in the rice crop, and such disturbed leaf anatomy negatively affected the assimilation rate in plants. Furthermore, degradation of IAA under UV-B stress has resulted in the shortening of internodes, reduction in leaf area, plant height and canopy layer (Musil et al., 2002). Delayed flower emergence and seed setting have also been observed under elevated dose of UV-B radiation.,

At the physiological level, the photosynthetic rate varies with climatic conditions, plant species, cultivars, photosynthetically active radiation (PAR) and UV-B. Earlier reports have clearly presented the damage induced with the exposure of UV-B radiation on the photosynthetic machinery, thylakoid membrane, light harvesting complexes and both the photosystems (I and II), causing deleterious impact on the assimilative performance of the plants. Photosystem II is one of the most sensitive components of photophosphorylation and is responsible for the splitting of water in the presence of light (Correia et al., 1999; Kakani et al., 2003; Savitch et al., 2001). Correia et al. (1999) have reported significant reduction of photosynthesis by 25–46% in wheat crop under elevated doses of UV-B compared to ambient conditions. Photosynthesis in plants is also regulated by an important factor, stomatal conductance; several studies have demonstrated that elevated UV-B leads to the reduction of stomatal conductance and, hence, is responsible for CO_2 assimilation to some extent (Zhao et al., 2003). Jansen and Van Den Noort (2000) summarized that stomata exposed to elevated UV-B lose their ability to readjust and regulate their original functioning, leading to the partial opening of stomata.

1.6.3 BIOCHEMICAL RESPONSE AGAINST UV-B

Lower doses of UV-B can be regulatory and provide early adaptation prior to extreme conditions. However, higher doses of UV-B induces the activation of higher synthesis of phenolics, specific flavonol glycosides and anthocyanins to cope with the severe distress of massively produced ROS (Emiliani et al., 2013; Hectors et al.,2014; Hideg et al., 2013). Furthermore, pigments accumulations, increased wax deposition, lignin production, secondary metabolites formations, mutations and genetic material impairment, etc. are some of the common biochemical responses due to increased production of ROS in response to such abiotic stresses. Various antioxidative metabolism (enzymatic or non-enzymatic), gets activated to scavenge ROS. Flavonoid mutants showed highly sensitive behavior towards higher UV-B, proving that flavonoids have a direct role in plant defense mechanisms and pathways. Biosynthetic pathway of flavonoids is mainly regulated by a group of genes encoding biosynthetic enzymes, including *chalcone synthase (CHS)*, *chalcone isomerase (CHI)*, *flavanone 3-hydroxylase* (F3H), *flavonol synthase (FLS)*, *dihydroflavonol 4-reductase (DFR)* and *leucoanthocyanidin dioxygenase (LDOX)* (Tilbrook et al., 2013), which also help in scavenging excessively produced ROS under such radiation stress.

1.6.4 ANTIOXIDATIVE DEFENSE IN RESPONSE TO UV-B

It is well known that ROS is produced as a by-product of various metabolic pathways, but its balance between production and scavenging can be disturbed by several environmental factors (biotic or abiotic) (Gill and Tuteja, 2010). In response to pathogen attack, the defense is termed as oxidative burst, which leads to the production of ROS. However, in the presence of transition metal ions, H_2O_2 is reduced to hydroxyl radical by superoxide, and these hydroxyl radicals are more reactive than superoxides and hydrogen peroxides, so the evolutionary process has evolved some sophisticated strategies to control the level of ROS, such as non-enzymatic and enzymatic ROS scavenging mechanisms (Apel and Heribert, 2004). ROS generated in a cell can trigger several signaling pathways by changing the gene expression and modifying the activity of transcription factors (Apel and Heribert, 2004).

1.6.5 SIGNAL TRANSDUCTION

Plants use sunlight as one of the major limiting environmental factors to regulate a large range of processes leading from germination, growth, stomatal regulations, circadian rhythm, flowering to development, but, in particular, high light and UV-B exposure induces stressful conditions in the normal life cycle of the plants, and the extent of such stress is determined by factors like fluence rate, exposure time and exposed organs of the plants. The UV-B induced stress includes disturbances of various cellular processes such as impairment of electron transport chain caused by the increased activity of NADPH-oxidase and peroxidases (Hideg et al., 2013; Jenkins, 2009; Müller-Xing et al., 2014), disturbed plant–pathogen interaction (Jensen, 2009; Müller-Xing et al., 2014), increased ROS production and genetic damage, etc.

According to several studies on *Arabidopsis* mutants, perception of light and UV-B involves specific photoreceptors including phytochromes (phy A-E), cryptochromes (cry1 and cry 2), phototropins (phot 1 and phot 2), zeitlupe gene (ZTL, FKF1 and LKP2) families and recently discovered UV-B receptor, UV Resistance Locus 8 (UVR8) (Fankhauser and Ulm, 2011; Favory et al., 2009; Jenkins, 2009; Kliebenstein et al., 2002; Müller-Xing et al., 2014; Rizzini et al., 2011; Schäfer and Nagy, 2006).

UVR8 gene was discovered by characterizing an *Arabidopsis* mutant which is hypersensitive to UV-B (Kliebenstein et al., 2002). Lower doses of UV-B for longer time affects the morphology, metabolism and acclimation in plants by regulating the UVR8, COP1 and HY5/HY5 HOMOLOG (HYH) genes to control the formation of ROS (Bandurska et al., 2013; Brown and Jenkins, 2008; Hideg et al., 2013; Müller-Xing et al., 2014). However, higher doses of UV-B drives ROS, leading to the increased synthesis of phytohormones (JA, ET, SA), which further activates the defense processes against UV-B stress (Brosché and Strid, 2003; Nawkar et al., 2013; Figure 1.2.)

Transcriptome analysis by Brown et al. (2005) concludes that UVR8 receptor regulates the genes involved in the oxidative stress response, photooxidative damage, DNA repair, flavonoid biosynthesis genes (CHS, CHI, and FLS1) in *Arabidopsis* mutant (Brown and Jenkins, 2008). Generally, UVR8 exists as a homodimer, but after UV-B exposure UVR8 undergoes monomerization and starts interacting with COP1 to proceed the signaling event; here, COP1 further undergoes ubiquitination and degradation by E3-ubiquitin ligase (Heijde and Ulm, 2013; Rizzini et al., 2011). It is also evident that not only UVR8 pathway but other signaling cascades are also involved under high UV-B exposure; one of the most well-known is MAPK cascade (González Besteiro et al., 2011; Ulm et al., 2002).

1.6.6 YIELD RESPONSES TO UV-B RADIATION

Elevated UV-B alters the timing of various regulatory processes like flowering, dormancy and senescence. Even delay in flowering leads to lower available pollinators, resulting in lesser seed setting and dispersal, which adversely affects the agricultural produce. In *Populus* species, a significant reduction in the height, diameter and biomass was reported (Schumaker et al., 1997; Singh et al., 2006). Elevated UV-B reduces the rate of photosynthesis (P_n), stomatal conductance (g_s) and chlorophyll florescence (F_v/F_m), leading to declining in photosynthates production and allocation. A major portion of the carbon is allocated towards the defense system of the plants, ultimately declining the yield of the plant.

Choudhary and Agrawal (2014) reported a reduction of yield in terms of number of seeds per pod (29–18%), number of seeds per plant (34–21%), weight of seeds per pod (36–22%), weight of seeds per plant (38–28%), harvest index (24–15%) and test weight (24–13%) in four different cultivars of *Glycine max* L. This study also documented the deterioration of seed quality in all the four cultivars in terms of Quality Response Index (QRI), which is illustrated as the integration of overall effects of any stress on soluble protein (SP), total soluble sugar (TSS), total free amino acids (TFAA) and starch content (SC). The QRI depicts the sensitivity of seed quality against imposed stress, and in the case of UV-B, the QRI hierarchy of *Glycine max* L. was JS-335 (-98.3)>PK-416 (-88.5)>JS 97-52 (-62.8)>PS-1042 (-53.2) (Choudhary and Agrawal, 2014) (Table 1.1).

1.7 TEMPERATURE STRESS

1.7.1 PLANT RESPONSES TO HEAT STRESS

Global temperature is estimated to increase by 1.4 to 5°C by the year 2100 (IPCC, 2007). Estimates suggest that there could be a 17% decrease in the crop yield for each degree Celsius increase in the average growing season (Rao et al., 2006). Plants have learned to adapt to fluctuations in temperature by developing

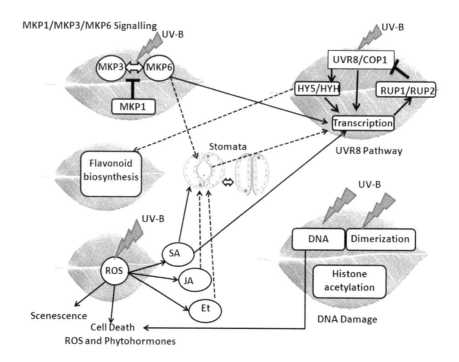

FIGURE 1.2 Schematic diagram representing the different signaling pathways induced by UV-B radiation (modified from Müller-Xing et al., 2014).The MPK3 and MPK6 (MAP kinase) signaling pathway is activated by UV stress, and MKP1 acts as a suppressor. In UVR8 signaling, UVR8 with COP1 (UV photoreceptor) undergoes dimerization, which upregulates the expression of HY5 and HYH genes, which further activates UV response genes. As a result, RUP1 and RUP2 express a negative feedback loop by binding directly to the protein. HY5 upregulate the expression of enzymes involved in flavonoids' biosynthesis. UV-B promotes the formation of ROS, which induces the production of phytohormones including salicylic acid (SA), jasmonic acid (JA) and ethylene (ET). High ROS production causes senescence and cell death. UV radiation can cause dimerization, which breaks the DNA double strand, which triggers cell death. DNA damage repair mechanism comprises of chromatin modifications, including histone acetylation and remodeling, to correct DNA lesions (modified from Müller-Xing et al., 2014).

different stress-tolerance strategies during the course of their evolution.

Heat stress adversely affects the growth and morphology of various plants. Although, plant responses are highly dependent on degree and duration of stress, plant species, the age of the plants, and different growth stage. Heat stress inhibits seed germination, plant emergence and promotes abnormal seedlings, poor seedling vigor, reduced radicle and plumule growth of germinated seedlings, reduction in plant growth and development, alteration in photosynthesis, dry matter partitioning and reduction in net assimilation rate (Kumar et al., 2011; Wahid et al., 2007). Normal physiological processes in plants are also disrupted due to high temperature, resulting in a reduction of PSII activity, photosynthetic pigments, enhanced ROS production, altered starch and sucrose synthesis, reduced ADP-glucose pyrophosphorylase and invertase activities (Rodriguez et al., 2005).

Physiological changes in the plants are the result of biochemical alterations in response to environmental stress. Plasma membrane acts as a primary target for heat stress, inducing more fluidity of lipid bilayer,

causing induction of calcium influx and cytoskeletal rearrangement, leading to upregulation of MAPKs and calcium-dependent protein kinase (CDPK). These proteins, in turn, mediate the activation of various tolerance responses, including the production of antioxidant defense against ROS produced or production of osmolytes produced (Hasanuzzaman et al., 2013). Heat acclimation triggers heat responsive molecular mechanisms, particularly the accumulation of heat shock proteins (HSPs) (Hua, 2009; Kotak et al., 2007). These proteins, in turn, accelerate transcription and translation of other proteins such as dehydrin, LEA, Pir proteins and ubiquitin (Bray, 2000) and also activate phytohormones such as ABA and other protective molecules such as proline, sugars, sugar alcohols (polyols), tertiary sulfonium compounds and tertiary and quaternary ammonium (Hasanuzzaman et al., 2013; Singh and Grover, 2008).

1.7.2 HSPs as Molecular Chaperons

HSPs function as molecular chaperones, correcting misfolding in protein and refolding of denatured proteins.

HSPs play an important role against various stress responsive mechanism(s) in plants such as, (a) interaction with antioxidant and osmolytes, (b) stress signal transduction, gene activation and cellular redox state, (c) physiological processes such as membrane stability, efficiency of assimilate partitioning, photosynthesis and water and nutrient utilization. Photosynthesis is an important factor determining growth and yield, but it can be severely affected due to heat stress. Rubisco Activase is a thermolabile protein needed for proper functionality of Rubisco (Al-Whaibi and Mohamed, 2011; Kotak et al., 2007).

1.7.3 Cold Stress

Cold stress can be considered as another important abiotic stress resulting in growth suppression and inducing several cellular abnormalities (Balestrasse et al., 2010). Several crop plants, such as soybean, cotton, maize, wheat and rice, from tropical and subtropical regions are known to be more susceptible to lower temperature, whereas herbaceous and woody plants can withstand freezing temperatures ranging up to −30°C. Cold stress induces various types of membrane injuries, including cellular lysis, cellular dehydration, negatively affecting the membrane lipid composition and causing fracture lesion depending on the temperature and the degree of freezing (Salinas, 2002). Moreover, genetic analyses indicated that plants' acclimatization to cold stress involves an array of reprogramming of gene expression, resulting in metabolic–structural changes through signal transduction pathway (Heidarvand and Amiri, 2010).

Based on the degree of temperature, cold stress can be categorized into chilling and freezing stress. Plants' exposure to a temperature ranging from 0–10°C induces chilling stress, whereas temperature below zero degrees induces freezing stress. Effects of potential chilling symptoms are evident at the cellular level and include the formation of a surface lesion, the appearance of water-soaked areas on the leaf surfaces, discoloration of tissues, ET production, early breakage of tissues, desiccation and accelerated senescence (Sharma et al., 2005). Furthermore, chilling stress reduces the fluidity of the cellular membrane, owing to the formation of unsaturation of fatty acid in membrane lipids, and causes alteration of lipid composition in the cell membrane (Wang et al., 2006). Chilling temperature also induces dehydration, owing to a reduction in water uptake by roots, and encourages stomatal closure. Freezing temperature also resulted in membrane damage due to severe cellular dehydration associated with ice formation in the intracellular spaces, causing the physical disruption of cells and tissues.

1.7.4 Antioxidative Response and Signaling Cascade under Cold Stress

Low-temperature-induced oxidative stress resulted in the generation of ROS, which disequilibrates the electron transfer reactions and alters the accompanied biochemical reactions. Thus, the generation of ROS leads to cellular damage, resulting in programmed cellular death of plant due to damage of PSII reaction center and cellular lipids (Suzuki and Mittler, 2006). Plants that are susceptible to such stress cannot withstand such deleterious conditions; however, the tolerant ones are able to cope with such stressed situations by triggering an array of signaling cascades leading to cold acclimatization responses.

Cold stress signal transduction initiates with the perception of cold temperature. Nordin Henriksson and Trewavas (2003) reported that such onset of signaling reactions occurs with a very brief exposure to low temperature; however, little is known about the cold sensors in the plants to date. Cooling rate plays an important role in determining the temperature sensing systems in plants. For example, in *Arabidopsis thaliana*, roots were reported to be very sensitive to cooling rates below 0.01°C/s dT/dt (Plieth et al., 1999). Moreover, Murata and Los (1997) indicated that as a result of temperature shift, an apparent sensor protein plays an important role in identifying the phase transitions process in microcellular spaces of the plasma membrane, resulting in both qualitative as well as quantitative alteration of lipid composition under low-temperature stress (Wang et al., 2006). Cold acclimatization responses trigger the degree of fatty acid unsaturation and the content of phospholipids in the membrane, causing membrane rigidification, which plays an important role in cold perception (Mikami and Murata, 2003). Vaultier et al. (2006) also indicated the role of membrane rigidification in cold perception and signal transduction process and analyzed the activation of diacylglycerol kinase (DAGK) as a very early event occurring within seconds in the plants due to cold exposure.

Örvar et al. (2000) and Sangwan et al. (2002) reported that cold stress-induced rearrangement of the actin cytoskeleton coupled with alteration of membrane fluidity and upstream of Ca^{2+} influx triggers cold stress responses in plants at a temperature below 25°C. For example, in cell suspension culture of alfalfa, an actin microfilament stabilizer jasplakinolide prevented the induction of cold acclimation gene (cas30) and Ca^{2+} influx at 4°C, but such events were induced at ~25°C in the presence of a membrane rigidifier, dimethyl sulfoxide (Örvar et al., 2000; Sangwan et al., 2002). This suggests the major role of the actin cytoskeleton in cold signaling transduction pathway and also emphasizes the importance of such cytoskeleton for several other physiological functions involved in cold acclimation processes (Örvar et al., 2000).

1.8 STRESS CAUSED BY VARIOUS AIR POLLUTANTS

1.8.1 Tropospheric Ozone

1.8.1.1 Formation of O_3

Formation of tropospheric O_3 in the atmosphere involves several reactions between primary air pollutants such as NOx, methane and non-methanic VOCs and hydrocarbon particles, etc., and it is not directly emitted into the atmosphere (Finlayson-Pitts and Pitts, 1997) (Figure 1.3). O_3 concentration in the atmosphere is determined by a complete set of photochemical reactions. The high temperature in the combustion chamber of the engines generally favors the formation of nitrogen dioxide (NO_2). Ultraviolet elements of solar radiation thus facilitate the dissociation of NO_2 in the atmosphere, resulting in nitrogen monoxide (NO) and singlet oxygen ($^1O_2$). The $^1O_2$ formed rapidly reacts with molecular oxygen naturally present in the atmosphere and give rise to O_3. But the

persistence of O_3 concentration may not be stable in the atmosphere as it again reacts with nitric oxide to produce NO_2 and O_2. Thus, equilibrium is formed between O_3 destruction and formation (Singh et al., 2015). Moreover, when non-methanic hydrocarbons react with NO, toxic organic substances such as peroxyacetyl nitrate (PAN— CH_3C (O) $OONO_2$) are formed.

1.8.1.2 Growth and Yield Attributes due to O_3 Stress

Symptoms of O_3 toxicity usually occur between the veins on an adaxial portion of the older and middle-aged leaves but may also involve both abaxial and adaxial leaf surfaces in some species (Cho et al., 2011). O_3-induced visible foliar injury such as chlorotic stippling or interveinal yellowing have been reported by Sarkar and Agrawal (2010) in the leaves of mature rice plants, under both ambient and elevated O_3 concentration in open-top chambers (OTCs), and observed that the magnitude of injury depends on the cumulative effect of both duration and concentration of O_3 exposure. Ahmad et al. (2013) have observed the development of O_3-induced visible foliar injuries on potato, onion and cotton plants when mean monthly O_3 concentrations exceeded 45 ppb in north-west Pakistan.

Responses to O_3 concentration have been well pronounced on growth and biomass of plants from several studies reported to date, and such responses exhibited

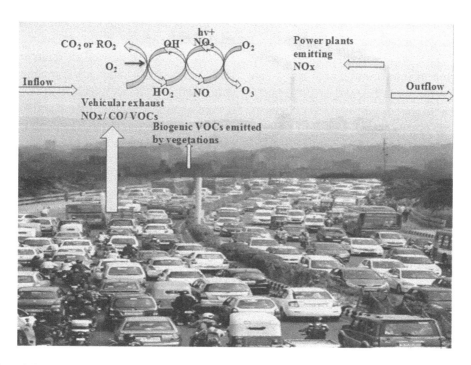

FIGURE 1.3 Pictorial presentation showing emission of different O_3 precursors and resulting photochemical formation of tropospheric O_3 (modified from Volz-Thomas and Mihelic, 1990).

variability owing to species or cultivar differences and different developmental stages of the plant (Guidi et al., 2009; Morgan et al., 2003). Sarkar and Agrawal (2010) reported adverse impact of ambient and elevated O_3 (ambient + 10ppb and ambient + 20 ppb) dose on two wheat cultivars (HUW 510 and Sonalika) and observed that cultivar HUW 510 displayed higher degree of O_3 damage in its vegetative parts than cultivar Sonalika, whereas the damage to reproductive structures (pollen viability and viable pollen floret per plants) was more in Sonalika as compared to HUW 510. Likewise, variability in responses was used to discriminate O_3 sensitivity of ten different wheat cultivars exposed to eight O_3 regimes by Saitanis et al. (2014). O_3-induced reduction in plant height, number of leaves, number of tillers and total leaf area have been reported in different crop species such as wheat (Mishra et al., 2013; Pleijel et al., 2018), maize (Singh et al., 2014), mung bean (Chaudhary and Agrawal, 2015) and soybeans (Singh et al., 2010).

Previous studies have reported that O_3-induced biomass reduction in below ground plant parts is often associated with a reduction in root–shoot ratio (Anderson, 2003). Furthermore, some studies have reported an increased carbon allocation in leaves due to O_3 exposure, which is mainly attributed to a reduction in phloem loading and transportation to meet the higher carbon demand to repair the O_3-induced foliar tissue damage (Cho et al., 2011). A meta-analysis study performed by Morgan et al. (2003) for investigating the response of soybean (*Glycine max* (L.) Merr.) to an O_3 exposure of 60 ppb exhibited about a 21% reduction in shoot and root dry biomass. Moreover, Saitanis et al. (2014) also observed a reduction in total biomass in different wheat cultivars due to O_3 fumigation with a maximum reduction of about 24.9% in cultivar Akbar followed by 20.2% in cultivar Sufi.

Several controlled environment and field studies have observed that current background O_3 concentrations are adversely affecting the yields of different crops species such as wheat (Mishra et al., 2013; Rai et al., 2007; Sarkar and Agrawal, 2010; Wahid, 2006a), rice (Sarkar et., 2015; Maggs and Ashmore, 1998; Wahid et al., 1995), soybean (Jaoudé et al., 2008; McGrath et al., 2015; Singh et al., 2010), maize (Singh et al., 2014), barley (Pleijel et al., 1992; Wahid, 2006b), mustard (Singh et al., 2009) and mung bean (Chaudhary et al., 2015). A regression analysis study carried out by McGrath et al. (2015) showed that ambient O_3 concentration in the United States had resulted in about 5% and 10% yield loss in soybean and maize respectively during the period 1980–2011 in the rainfed

field conditions. Feng and Kobayashi (2009) reported yield losses of more than 10% for soybean, wheat and rice and more than 20% for beans under the projected O_3 concentration of 51–75 ppb, thereby indicating that future peaking of O_3 concentration poses a serious threat to the food security worldwide. Furthermore, previous studies have observed higher O_3 sensitivity of leguminous crops followed by crops of Poaceae family such as wheat, rice and barley (Feng and Kobayashi, 2009; Mills et al., 2007; Sarkar and Agrawal, 2010) (Table 1.1).

1.8.1.3 Physiological Response due to O_3 Stress

The assessment of physiological processes is a more reliable measure to evaluate the intrinsic O_3 injuries to plants, especially because physiological damage can occur earlier and even at lower O_3 concentration before the appearance of visible injury. Different experimental studies have been conducted to evaluate the effects of ambient and elevated O_3 exposure on physiological processes which have exhibited detrimental impact on the assimilation rate of plants. Lowering of stomatal conductance (g_s), depression in net photosynthetic capacity and carboxylation efficiency are some of the common O_3-induced phytotoxic impacts on plant physiological processes (Cho et al., 2011; Morgan et al., 2003; Rai et al., 2011b). Impaired activity of the mesophyll cells and structural damage to the cellular membrane as evidenced by enhanced intercellular CO_2 concentrations and lipid peroxidation are associated with the O_3-induced photosynthetic loss, as reported by Rai et al. (2011b). O_3-induced alteration of photosynthetic electron transport rate in plants via a decrease in efficiency of excitation capture has been reported by Guidi et al. (2001). Reduction of F_v/F_m ratio indicates photo inhibition to PSII complexes, making plants more sensitive to light exposure (Rai et al., 2011b).

1.8.1.4 Antioxidative and ROS Signaling During O_3 Stress

O_3 acts as a gaseous signaling molecule and elicits cellular responses which can result in drastic changes in gene expression (Cho et al., 2011). Since O_3 causes oxidation of unsaturated lipid in the plasma membrane, certain membrane lipid may serve as receptor molecules and promote O_3-induced lipid signaling chains (Baier et al., 2005). Studies have reported that several plasma membrane lipid-based signaling molecules like jasmonates and lipid hydroperoxides play an important role in controlling a variety of downstream processes (Kangasjärvi et al., 2005).

O_3 flux in the apoplastic region is determined by uptake largely via stomatal aperture present on the leaf surfaces. Although the mechanism through which O_3 influences stomatal conductance is still not clear, researches have postulated activation of O_3-induced abscisic acid signaling pathway and outburst of ROS directly in guard cells (Kangasjärvi et al., 2005). Due to the short residence time of O_3 in the apoplastic region, it rapidly gets degraded to form ROS and/or reacts with cellular biomolecules such as protein, lipid, DNA or apoplastic fluid present there (Mishra et al., 2013). ROS produces cellular membrane damage and causes detrimental effects on the normal functioning of cells. Plants have developed several mechanisms to make use of non-enzymatic and enzymatic antioxidants present in different cellular compartments to cope up with an oxidative injury caused by O_3 stress (Singh et al., 2015).

Generation of ROS in the apoplastic region causes activation of Ca^{2+} channels, which in turns triggers the Ca^{2+} influx in the cytosolic region, thereby resulting in activation of MAPK cascades in the cytosol, which plays a significant role in signal transduction pathway (Cho et al., 2011). Such activation of MAPK, in turn, appears to be involved in the upregulation of ET, SA and JA signaling pathways, which participates to bring about changes in gene expression. Castagna and Ranieri (2009) observed that ET and SA together enhance lesion development in plants and ultimately cause cell death. Cellular death causes the production of certain products from lipid peroxidation and can serve as substrates for synthesis of JA, which further decreases production of ET-dependent ROS and, ultimately, the spread of cell death (Kangasjärvi et al., 2005).

The pattern of gene expression induced due to O_3 exposure has been an important topic of discussion in the recent studies. Frei et al. (2010) evaluated the potential tolerance mechanism in rice crops by investigating the alteration in gene expressions induced due to O_3 exposure. The study documented strong expression of gene encoding antioxidant enzymes (CAT and POX) in the O_3-sensitive genotype SL15. On the contrary, another gene encoding ascorbate oxidase was identified with lower expression in O_3-tolerant genotype SL41, suggesting its involvement in enhancing AsA status under O_3 exposure. Another study reported that 300 ppb O_3 dose induced enhancement of mRNA levels for glutathione S-transferase (GST), phenylalanine ammonia-lyase (PAL), a neutral peroxidase, and a cytosolic Cu/Zn SOD, with GST mRNA showing 26-fold inductions 3 h after the initiation of O3 dose, as compared to

the control plants receiving ambient air in *Arabidopsis thaliana* (Sharma and Davis, 1994). However, Ahmad et al. (2015) observed O_3-induced upregulation of CC14 gene by three to four-fold and overexpression of cysteine proteases in maize leaves after 20 d treatment, ultimately resulting in causing protein degradation and programmed cellular death.

1.8.2 SULFUR DIOXIDE (SO_2)

Sulfur dioxide is a toxic gas, acidic in nature, with an irritating, pungent odor. Naturally, SO_2 is released by the volcanic emissions, weathering processes, hot springs, sea sprays, microbial activities and so on, but anthropogenic activities like the burning of fossil fuels such as coal and petroleum, biomass burning, vehicular and industrial emissions have added extra SO_2 into the atmosphere, which has ultimately resulted in various changes in all domains of the ecosystem. The degree of phytotoxicity of SO_2 to plants is determined by the concentration of SO_2, exposure duration, environmental condition, soil status and the genetic makeup of the plants (De Kok, 1990).

After the Industrial Revolution, the atmospheric concentration of SO_2 inclined up to a very serious level, and it was documented in the form of the Great Smog of London in 1952, which drew the attention of the scientific societies, and several acts and amendments were made and strictly implemented. China, in particular, from 1996 to 2000, showed a relative decrease in SO_2 by 13% from 24.3 Tg to 21.2 Tg (Lu et al., 2011). It was also evident that this reduction was due to the decline in economic growth and coal use and the implementation of air pollution control rules. But the economic rise again dramatically increases (2000–2006) and decreases (2006–2010) the levels of atmospheric SO_2 between 2000 and 2010.

The problem of SO_2 is not confined only to urban or industrialized areas, but it reaches to the rural areas, and it also has the potential to damage vegetation and crop production (Agrawal, 2000). Pollutants rarely occur singly, and their interactive effects are more damaging. Air pollutants are mainly perceived by the leaves through the stomatal opening. Agrawal et al. (1987) suggested that upon entry through stomata, SO_2 reacts with O_2 to form sulfite and bisulfate ions, which further photo-oxidize to less toxic sulfate ion, which leads to the reaction of free radicals with the formation of cytotoxic superoxide radical ($\bullet O^-_2$), and more accumulation of such free radicals are injurious for plant health. Productions of ROS due to SO_2 absorption within the cellular space (Bartosz, 1997)

$$HSO_3^- + {}^{\bullet}O_2^- + 2H^+ \rightarrow HSO_3^{\bullet} \text{ (bisulphate radical)} + 2\,{}^{\bullet}OH$$

$$HSO_3^- + {}^{\bullet}OH + H^+ \rightarrow HSO_3^{\bullet} + H_2O$$

$$HSO_3^{\bullet} + O_2 \rightarrow SO_3^{2-} + O_2^{\bullet -} + H^+$$

$$HSO_3^{\bullet} + {}^{\bullet}OH \rightarrow SO_3^{2-} + H_2O$$

$$2HSO_3^{\bullet} \rightarrow SO_3^{2-} + HSO_3^- + H^+$$

$$SO_3 + H_2O \rightarrow SO_4^{2-} + 2H^+$$

$${}^{\bullet}O_2^- + {}^{\bullet}O_2^- + 2H^+ \rightarrow O_2 + H_2O_2$$

In sensitive plants, the defense system of the plants are weak and have low ascorbate due to its peroxidase-catalyzed oxidation (Nandi et al., 1984); the •O$^-_2$-induced reactions predominates, leading to the oxidation of chlorophyll and membrane lipids (Malhotra, 1984), which stimulates the production of isoflavonoids (Rubin et al., 1983), resulting in colored visible injury formation, ultimately causing the reduction of photosynthetic potential and dry matter production in plants (Agrawal et al., 1987). SO_2 reacts with water and gets converted into acid, leading to the development of injuries with regular margins, resulting in the marginal and bifacial chlorosis followed by necrosis. Degradation in leaf area causes a reduction in photosynthesis, leading to low photosynthates accumulation and allocation towards sinks. In various studies, it has been found that the exposure of SO_2 may lead to the pheophytinization of chlorophyll (it is a process in which the central Mg ion present in chlorophyll is replaced by H$^+$) or a reversible swelling of thylakoids within the chloroplasts (Rao and LeBlanc, 1966; Wellburn et al., 1972). In an integrated study of SO_2 and CO_2, the response of plants towards the perception of SO_2 gets modified (Agrawal and Deepak, 2003).

Agrawal and Deepak (2003) suggested that plants exposed to higher SO_2 doses (0.5 ppm) lead to the production of increased phenolics, MDA, SOD, POD (Rai and Agrawal, 2008) but the reduction in total chlorophyll content, AsA, foliar nitrogen and foliar starch. Physiological studies of *Oryza sativa* showed a significant reduction in the photosynthetic rate and stomatal conductance (Rai and Agrawal, 2008). In tolerant plants, the antioxidant defense system plays an important role in the detoxification of free radicals upon exposure to high SO_2 concentrations. Agrawal et al. (1987) suggested a similar increase in SOD activity and AsA content in response to high SO_2 concentration in *Oryza sativa* (Table 1.1).

1.8.3 OXIDES OF NITROGEN (NOX)

Increased pollutant concentration in the atmosphere due to anthropogenic activities have greatly altered the biogeochemical cycles of the earth. The enormous extraction, purification and combustion of fossil fuels are the major factors causing the emission of huge quantity of air pollutants. But the incidences of acid rain due to emissions of sulfur from coal, gasoline and oil combustion draws the attention of the world towards the emission of nitrogen compounds too, because it is related to the quality of not only air but also water, soil and the ecosystem (Hoegberg et al., 2006). The NOx generates several problems related to the environment and are commonly termed as NOx; these are nitric oxide (NO_2), nitrous oxide, (N_2O), ammonia (NH_3), nitrogen dioxide (NO_2), nitrate (NO^-_3), nitrite (NO^-_2), etc. and they contribute in the cycling of nitrogen and reactive capacity of atmosphere (Morin et al., 2008). Oxides of nitrogen play a crucial role in the formation of O_3. Increased emissions of NOx alters the atmospheric chemistry. It has been reported that the total amount of NOx emitted of about 28 529 metric tons was from vehicular sources (Lal and Patil, 2001; Larssen et al., 1994). Sahai et al. (2011), suggested that biomass burning contributes maximally to total NOx emissions (mainly NO_2) and has increased from 72.0 Gg to 140.6 Gg between 1980 and 2010 in India. They also reported that burning of agricultural residues (from rice, wheat, sugarcane) have combinedly increased from 58.9 Gg to 117.4 Gg of NOx between 1980 and 2010 (Oksanen et al., 2013).

$$NO_2 \xrightarrow{\text{sunlight}} NO + O$$

$$O_2 + O \rightarrow O_3$$

Among NOx, N_2O is a greenhouse gas, mainly produced from the denitrification under anaerobic conditions, excess use of fertilizers for better yield, leading to N_2O emission from these unutilized fertilizers (Mosier et al., 1998). However, N_2O is less reactive in the lower atmosphere, but its retention time is high, and slowly it moves towards the higher atmosphere where its photolysis results in the formation of nitric oxide, which further reacts with O_3 and causes the destruction of the O_3 layer. NO, however, is produced by the combustion of nitrogen and oxygen at high temperatures and such condition is formed in all vehicles; so, the main source of NO is vehicular emission and fertilizer industries. NO reacts with carbon and others to form CO, CO_2, N_2, NO_2 and so on, which ultimately causes the formation of acids like nitric acid (HNO_3) which get deposited in the form of precipitation on the surface of plants, soil, monuments, etc. Even NOx leads to the formation of PAN, which is very reactive and is a secondary air pollutant.

The impacts of NOx have been extensively surveyed on the forest ecosystem, and it was found that the humid temperate regions (Dise and Wright, 1995) of the world are facing the problem of increased nitrogen deposition (Adams et al., 2004), resulting in a situation termed as nitrogen saturation. (It is the availability of ammonium and nitrate in excess of total combined plants and microbial nutritional demand.) (Hoegberg et al., 2006).

The NOx can enter from the cuticular region and stomata of the plants into the mesophyll where NOx reacts with water, and the formation of small chlorotic and necrotic patches takes place; the bleaching of tips and margins are also observed. The chronic dry or wet nitrogen deposition leads to frost damage and physiological disturbances, which reduces the productivity in coniferous forests. In some recent studies, it has been reported that due to anthropogenically induced acidic precipitations, the soil is gradually turning acidic in nature, which changes the physicochemical properties of the soil, which alters the cation exchange capacities (Redling et al., 2013) and leads to increase in pool of mobile metal pools, which ultimately causes the disturbances in forest ecosystem. According to Rai et al. (2007), *Triticum aestivum* grown in filtered chambers showed increased number of tillers (33.6% at 60 DAG), root, shoot, leaf biomass (31.9% at 80 DAG), photosynthetic rate (27.1%), stomatal conductance and transpiration rate, F_v/F_m (5.4% at 60 DAG); however, plants grown in non-filtered chambers showed increment in MDA content (41.6% at 60DAG), total phenolics (38.55 at 60 DAG), peroxidase (23.4–38.1%), AsA (6.6–11.2%) and proline (36–46.2%). Both the results indicate that plants exposed to NOx are under high stress, so they produce more secondary metabolites and hence the yield is compromised (Table 1.1).

1.8.4 VOLATILE ORGANIC COMPOUNDS (VOCs)

VOCs are organic compound released from plants which have sufficiently high vapor pressure to be vaporized into the atmosphere under normal circumstances (Yuan et al., 2009). About 10% of the carbon fixed by photosynthesis is released back into the atmosphere as VOCs (Kesselmeier and Staudt, 1999; Peñuelas and Llusià, 2005). The major pathway for the synthesis of VOCs is the isoprenoid, the lipoxygenase or the shikimic acid pathway (Laothawornkitkul et al., 2009)

The most common VOCs consist of alkanes, alkenes, carbonyls, alcohols, esters and acids (Kesselmeier and Staudt, 1999; Peñuelas and Llusià, 2005). The total current annual global VOCs emission from the vegetation has been estimated to be approximately $700–1000 \times 1012$ g C y-1, while in future, a predicted 2–3°C rise in the mean global temperature could further increase emission by 30–45% (Laothawornkitkul et al., 2009; Peñuelas and Llusià, 2005).

1.8.4.1 Role of VOCs in Plant Metabolism and Defense

VOCs play an important role in plant metabolism, defense and communications in plants. Plants emit many of these VOCs. VOCs are formed from various sources such as evaporation from fuels, incomplete combustion of fuel, industrial processes and biomass burning (Holopainen, 2004). Estimates suggest that at a global scale, natural VOCs (biogenic VOCs (BVOCs)) emission from vegetation greatly exceed the anthropogenic sources in the atmosphere (Guenther et al., 2000; Hallquist et al., 2009; Peñuelas and Llusià, 2005). Annual global VOCs emission from the vegetation has been estimated to be approximately $700–1000 \times 10^{12}$ g C y^{-1}, and this can further be expected to increase by 30–45% under the present climate change scenario (Laothawornkitkul et al., 2009; Peñuelas and Llusià, 2005).

Plants can emit a considerable amount of carbon reserve in the form of VOCs such as isoprene and mono- and sesquiterpenes. VOCs have a significant role in the growth and development, reproduction and communication within and between plants. In addition, plants may also have a protective role in the abiotic stress (Laothawornkitkul et al., 2009). Some of the common responses in plants to VOCs in plants are epinasty, chlorosis, reduction in the dry weight, decreased flowering per plant, reduced shoot weight and leaf area (Cape, 2003).

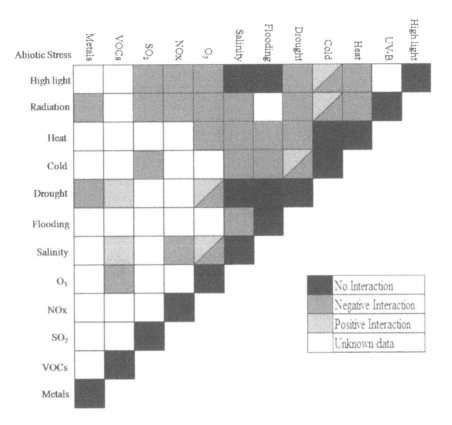

FIGURE 1.4 Stress matrix. Different combinations of abiotic stresses are shown in the form of a matrix to highlight the important interactions existing among stresses. The color-coded matrix is used to indicate the combination of stresses that were studied in different crops in terms of its growth and yield parameters (modified from Suzuki et al., 2014).

1.9 CONCLUSIONS

Plants have to face several environmentally induced abiotic stresses due to their sessile nature, but they slowly monitor the changes in their prevailing environment and also try to cope up with these environmental challenges. Abiotic stresses remain as one of the major challenges to the natural environment and crop production (Figure 1.4). Although plants have evolved from the extremes of environmental conditions such as fluctuations in temperature, harsh sunlight and damaging radiations, the domestication of the crops around the globe mainly indicates that the selection of many of the crop species and/or cultivars were mostly based on the production rather than the suitability under various environmental stress factors, leading to productivity losses.

Numerous studies indicate the factors governing the defense system in plants and the necessity to generate tolerant varieties which can acclimatize and adapt to stressful conditions without having any adverse impacts on the productivity. Since it is the need and necessity of every nation to meet the demand for food security by increasing the agricultural production, currently, the major issues are to produce more in less area under the prevailing environmental constraints. This compilation emphasizes in brief on the major defense strategies and/or pathways involved in crop plants under different abiotic stresses. The information provided can be used to understand the effects, mechanisms and interrelation between various stresses, the strategies of defense adopted by plants and the signaling of genes involved in the synthesis of regulatory proteins for the development of stress-tolerant plant varieties.

ACKNOWLEDGMENTS

The authors are grateful to the Head of Department of Botany, Banaras Hindu University, Varanasi, India for providing the necessary facilities to carry out the work. Ashutosh Kumar Pandey wishes to express his gratitude to the University Grant Commission, New Delhi for providing the D.S. Kothari Postdoctoral Fellowship. Annesha Ghosh, Kshama Rai and Adeeb Fatima are thankful to the University Grant Commission and Centre for Advanced Studies (CAS), Department of Botany, Banaras Hindu University and DST-PURSE (Promotion of University Research and Scientific Excellence).

REFERENCES

Abiko, T., Kotula, L., Shiono, K., Malik, A.I., Colmer, T.D., Nakazono, M. (2012) Enhanced formation of aerenchyma and induction of a barrier to radial oxygen loss in adventitious roots of *Zea nicaraguensis* contribute to its waterlogging tolerance as compared with maize (*Zea mays* ssp. mays). *Plant, Cell and Environment* 35, no. 9: 1618–1630.

Achten, W.M.J., Maes, W.H., Reubens, B., Mathijs, E., Singh, V.P., Verchot, L., Muys, B. (2010) Biomass production and allocation in *Jatropha curcas* L. seedlings under different levels of drought stress. *Biomass and Bioenergy* 34, no. 5: 667–676.

Adams, M., Ineson, P., Binkley, D., Cadisch, G., Tokuchi, N., Scholes, M., Hicks, K. (2004) Soil functional responses to excess nitrogen inputs at global scale. *AMBIO: A Journal of the Human Environment* 33, no. 8: 530–536.

Adebayo, M.A., Menkir, A. (2014) Assessment of hybrids of drought tolerant maize (*Zea mays* L.) inbred lines for grain yield and other traits under stress managed conditions. *Nigerian Journal of Genetics* 28, no. 2: 19–23.

Adriano, D.C., Wenzel, W.W., Vangronsveld, J., Bolan, N.S. (2004) Role of assisted natural remediation in environmental cleanup. *Geoderma* 122, no. 2–4: 121–142.

Agrawal, M., Nandi, P. K., Rao, D. N. (1983) Ozone and sulphur dioxide effects on Panicum miliaceum plants. Bulletin of the *Torrey Botanical Society*, 435–441.

Agrawal, M. (2000) Researches on air pollution impact on vegetation in India. A review. *The Botanica* 50: 76–85.

Agrawal, M., Deepak, S.S. (2003) Physiological and biochemical responses of two cultivars of wheat to elevated levels of CO_2 and SO_2, singly and in combination. *Environmental Pollution* 121, no. 2: 189–197.

Agrawal, S.B., Agrawal, M., Nandi, P.K., Rao, D.N. (1987) Effects of sulphur dioxide and quinalphos singly and in combination on the metabolic function and growth of *Oryza sativa* L. *Proceedings of the Indian National Science Academy* 53: 72.

Ahmad, P., Umar, S. (2011) *Oxidative stress: Role of antioxidants in plants*, 19–53. Studium Press: New Delhi, India.

Ahmad, P., Sarwat, M., Sharma, S. (2008) Reactive oxygen species, antioxidants and signaling in plants. *Journal of Plant Biology* 51, no. 3: 167–173.

Ahmad, S., Ahmad, R., Ashraf, M.Y., Ashraf, M., Waraich, E.A. (2009) Sunflower (*Helianthus annuus* L.) response to drought stress at germination and seedling growth stages. *Pakistan Journal of Botany* 41, no. 2: 647–654.

Ahmad, P., Kumar, A., Ashraf, M., Akram, N.A. (2012) Salt-induced changes in photosynthetic activity and oxidative defense system of three cultivars of mustard (*Brassica juncea* L.). *African Journal of Biotechnology* 11, no. 11: 2694.

Ahmad, M.N., Büker, P., Khalid, S., Van Den Berg, L., Shah, H.U., Wahid, A., Emberson, L., et al. (2013) Effects of ozone on crops in north-west Pakistan. *Environmental Pollution* 174: 244–249.

Ahmad, R., Zuily-Fodil, Y., Passaquet, C., Khan, S.A., Repellin, A. (2015) Molecular cloning, characterization and differential expression of novel phytocystatin gene during tropospheric ozone stress in maize (*Zea mays*) leaves. *Comptes Rendus Biologies* 338, no. 3: 141–148.

Ahmed, S. (2009) Effects of air pollution on yield of mungbean in Lahore, Pakistan. *Pakistan Journal of Botany* 41, no. 3: 1013–1021.

Ainsworth, E.A. (2008) Rice production in a changing climate: A meta-analysis of responses to elevated carbon dioxide and elevated ozone concentration. *Global Change Biology* 14, no. 7: 1642–1650.

Akhtar, N., Yamaguchi, M., Inada, H., Hoshino, D., Kondo, T., Izuta, T. (2010) Effects of ozone on growth, yield and leaf gas exchange rates of two Bangladeshi cultivars of wheat (*Triticum aestivum* L.). *Environmental Pollution* 158, no. 5: 1763–1767.

Alexieva, V., Sergiev, I., Mapelli, S., Karanov, E. (2001) The effect of drought and ultraviolet radiation on growth and stress markers in pea and wheat. *Plant, Cell and Environment* 24, no. 12: 1337–1344.

Alford, É.R., Pilon-Smits, E.A.H., Paschke, M.W. (2010) Metallophytes-a view from the rhizosphere. *Plant and Soil* 337, no. 1–2: 33–50.

Alloway, B.J. (2004) *Zinc in soils and crop nutrition*, 1–116. International Zinc Association, IZA Publications: Brussels, Belgium.

Alvino, A., D'Andria, R., Delfine, S., Lavini, A., Zanetti, P. (2000) Effect of water and salinity stress on radiation absorption and efficiency in sunflower. *Italian Journal of Agronomy* 4, no. 2: 53–60.

Al-Whaibi, Mohamed, H. (2011) Plant heat-shock proteins: A mini review. *Journal of King Saud University—Science* 23, no. 2: 139–150.

Anderson, C.P. (2003) Source-sink balance and carbon allocation belowground in plants exposed to ozone. *New Phytologist* 157, no. 2: 213–228.

Anjum, S.A., Wang, L., Farooq, M., Khan, I., Xue, L. (2011) Methyl jasmonate induced alteration in lipid peroxidation, antioxidative defense system and yield in soybean under drought. *Journal of Agronomy and Crop Science* 197, no. 4: 296–301.

Anjum, S.A., Ashraf, U., Tanveer, M., Khan, I., Hussain, S., Shahzad, B., Zohaib, A., et al. (2017) Drought induced changes in growth, osmolyte accumulation and antioxidant metabolism of three maize hybrids. *Frontiers in Plant Science* 8: 69.

Apel, K., Heribert, H. (2004) Reactive oxygen species: Metabolism, oxidative stress, and signal transduction. *Annual Review of Plant Biology* 55: 373–399.

Arguello, M.N., Mason, R.E., Roberts, T.L., Subramanian, N., Acuña, A., Addison, C.K., Lozada, D.N., et al. (2016) Performance of soft red winter wheat subjected to field soil waterlogging: Grain yield and yield components. *Field Crops Research* 194: 57–64.

Asch, F. Dingkuhn, M., Sow, A., Audebert, A. (2005) Drought-induced changes in rooting patterns and assimilate partitioning between root and shoot in upland rice. *Field Crops Research* 93, no. 2–3: 223–236.

Ashraf, M., Harris, P.J.C. (2013) Photosynthesis under stressful environments: An overview. *Photosynthetica* 51, no. 2: 163–190.

Azooz, M.M., Abou-Elhamd, M.F., Al-Fredan, M.A. (2011) Evaluation of salicylic acid (SA) application on growth, osmotic solutes and antioxidant enzyme activities on broad bean seedlings grown under diluted seawater. *International Journal of Plant Physiology and Biochemistry* 3, no. 14: 253–264.

Azooz, M.M., Abou-Elhamd, M.F., Al-Fredan, M.A. (2012) Biphasic effect of copper on growth, proline, lipid peroxidation and antioxidant enzyme activities of wheat (*Triticum aestivum* cv. Hasaawi) at early growing stage. *Australian Journal of Crop Science* 6, no. 4: 688–694.

Badawi, G.H., Yamauchi, Y., Shimada, E., Sasaki, R., Kawano, N., Tanaka, K., Tanaka, K. (2004) Enhanced tolerance to salt stress and water deficit by overexpressing superoxide dismutase in tobacco (*Nicotiana tabacum*) chloroplasts. *Plant Science* 166, no. 4: 919–928.

Baier, M., Kandlbinder, A., Golldack, D., Dietz, K.J. (2005) Oxidative stress and ozone: Perception, signaling and response. *Plant, Cell and Environment* 28, no. 8: 1012–1020.

Bajgain, R., Kawasaki, Y., Akamatsu, Y., Tanaka, Y., Kawamura, H., Katsura, K., Shiraiwa, T. (2015) Biomass production and yield of soybean grown under converted paddy fields with excess water during the early growth stage. *Field Crops Research* 180: 221–227.

Balestrasse, K.B., Tomaro, M.L., Batlle, A., Noriega, G.O. (2010) The role of 5-aminolevulinic acid in the response to cold stress in soybean plants. *Phytochemistry* 71, no. 17–18: 2038–2045.

Bandurska, H., Niedziela, J., Chadzinikolau, T. (2013) Separate and combined responses to water deficit and UV-B radiation. *Plant Science* 213: 98–105.

Bandurska, H., Niedziela, J., Pietrowska-Borek, M., Nuc, K., Chadzinikolau, T., Radzikowska, D. (2017) Regulation of proline biosynthesis and resistance to drought stress in two barley (*Hordeum vulgare* L.) genotypes of different origin. *Plant Physiology and Biochemistry* 118: 427–437.

Barber, S. A. (1995) *Soil nutrient bioavailability: a mechanistic approach*. John Wiley & Sons: Hoboken.

Bartels, D., Sunkar, R. (2005) Drought and salt tolerance in plants. *Critical Reviews in Plant Sciences* 24, no. 1: 23–58.

Bartosz, G. (1997) Oxidative stress in plants. *Acta Physiologiae Plantarum* 19, no. 1: 47–64.

Bazihizina, N., Barrett-Lennard, E.G., Colmer, T.D. (2012) Plant growth and physiology under heterogeneous salinity. *Plant and Soil* 354, no. 1–2: 1–19.

Blum, A. (1996) Crop responses to drought and the interpretation of adaptation. *Plant Growth Regulation* 20, no. 2: 135–148.

Bögre, L., Calderini, O., Binarova, P., Mattauch, M., Till, S., Kiegerl, S., Jonak, C., et al. (1999) A MAP kinase is activated late in plant mitosis and becomes localized to the plane of cell division. *The Plant Cell* 11, no. 1: 101–113.

Bracher, A., Hauser, T., Liu, C., Hartl, F.U., Hayer-Hartl, M. (2015) Structural analysis of the rubisco-assembly chaperone RbcX-II from *Chlamydomonas reinhardtii*. *PLOS ONE* 10, no. 8: e0135448.

Bray, E.A. (2000) Responses to abiotic stresses. In *Biochemistry and molecular biology of plants*, 1158–1203. American Society of Plant Physiologists: Rockville, MD.

Brosché, M., Strid, A. (2003) Molecular events following perception of ultraviolet-B radiation by plants. *Physiologia Plantarum* 117, no. 1: 1–10.

Brown, B.A., Cloix, C., Jiang, G.H., Kaiserli, E., Herzyk, P., Kliebenstein, D.J., Jenkins, G.I. (2005) A UV-B specific signaling component orchestrates plant UV protection. *Proceedings of National Academy of Sciences* 102, no. 50: 18225–18230.

Brown, B.A., Jenkins, G.I. (2008) UV-B signaling pathways with different fluence-rate response profiles are distinguished in mature *Arabidopsis* leaf tissue by requirement for UVR8, HY5, and HYH. *Plant Physiology* 146, no. 2: 576–588.

Caasi-Lit, M., Whitecross, M.I., Nayudu, M., Tanner, G.J. (1997) UV-B irradiation induces differential leaf damage, ultrastructural changes and accumulation of specific phenolic compounds in rice cultivars. *Australian Journal of Plant Physiology* 24, no. 3: 261–274.

Calvo, E., Calvo, I., Jimenez, A., Porcuna, J.L., Sanz, M.J. (2009) Using manure to compensate ozone-induced yield loss in potato plants cultivated in the east of Spain. *Agriculture, Ecosystems and Environment* 131, no. 3–4: 185–192.

Cape, J.N. (2003) Effects of airborne volatile organic compounds on plants. *Environmental Pollution* 122, no. 1: 145–157.

Castagna, A., Ranieri, A. (2009) Detoxification and repair process of ozone injury: from O_3 uptake to gene expression adjustment. *Environmental Pollution* 157, no. 5: 1461–1469.

Cechin, I., Rocha, V.d.J., Fumis, T.d.F. (2012) Sensitivity of yellow passion fruit to ultraviolet-B radiation. *Pesquisa Agropecuária Brasileira* 47, no. 10: 1422–1427.

Chatterjee, C., Gopal, R., & Dube, B. K. (2006) Impact of iron stress on biomass, yield, metabolism and quality of potato (Solanum tuberosum L.). *Scientia Horticulturae* 108, no. 1: 1–6.

Chaudhary, N., Agrawal, S.B. (2013) Intraspecific responses of six Indian clover cultivars under ambient and elevated levels of ozone. *Environmental Science and Pollution Research International* 20, no. 8: 5318–5329.

Chaudhary, N., Agrawal, S.B. (2015) The role of elevated ozone on growth, yield and seed quality amongst six cultivars of mung bean. *Ecotoxicology and Environmental Safety* 111: 286–294.

Chauhan, J.S., Tyagi, M.K., Kumar, A., Nashaat, N.I., Singh, M., Singh, N.B., Jakhar, M.L., Welham, S.J. (2007) Drought effects on yield and its components in Indian mustard (*Brassica juncea* L.). *Plant Breeding* 126, no. 4: 399–402.

Cho, K., Tiwari, S., Agrawal, S.B., Torres, N.L., Agrawal, M., Sarkar, A., Shibato, J., et al. (2011) Tropospheric ozone and plants: absorption, responses, and consequences. *Reviews of Environmental Contamination and Toxicology* 212: 61–111.

Choudhary, K.K., Agrawal, S.B. (2014) Cultivar specificity of tropical mung bean (*Vigna radiata* L.) to elevated ultraviolet-B: Changes in antioxidative defense system, nitrogen metabolism and accumulation of jasmonic and salicylic acids. *Environmental and Experimental Botany* 99: 122–132.

Clermont-Dauphin, C., Suwannang, N., Grünberger, O., Hammecker, C., Maeght, J.L. (2010) Yield of rice under water and soil salinity risks in farmers' fields in northeast Thailand. *Field Crops Research* 118, no. 3: 289–296.

Cochard, H. (2002) Xylem embolism and drought-induced stomatal closure in maize. *Planta* 215, no. 3: 466–471.

Colmer, T.D., Voesenek, L.A.C.J. (2009) Flooding tolerance: Suites of plant traits in variable environments. *Functional Plant Biology* 36, no. 8: 665–681.

Comba, M.E., Benavides, M.P., Tomaro, M.L. (1998) Effect of salt stress on antioxidant defense system in soybean root nodules. *Australian Journal of Plant Physiology* 25, no. 6: 665–671.

Correia, C.M., Torres-Pereira, M.S., Torres-Pereira, J.M.G. (1999) Growth, photosynthesis and UV-B absorbing compounds of Portuguese Barbela wheat exposed to ultraviolet-B radiation. *Environmental Pollution* 104, no. 3: 383–388.

Dai, A. (2013) Increasing drought under global warming in observations and models. *Nature Climate Change* 3, no. 1: 52–58.

Dat, J.F., Capelli, N., Folzer, H., Bourgeade, P., Badot, P.M. (2004) Sensing and signaling during plant flooding. *Plant Physiology and Biochemistry* 42, no. 4: 273–282.

Davis, R.D., Beckett, P.H.T., Wollan, E. (1978) Critical levels of twenty potentially toxic elements in young spring barley. *Plant and Soil* 49, no. 2: 395–408.

de Azevedo Neto, A.D., Prisco, J.T., Enéas-Filho, J., de Abreu, C.E.Bd, Gomes-Filho, E. (2006) Effect of salt stress on antioxidative enzymes and lipid peroxidation in leaves and roots of salt-tolerant and salt-sensitive maize genotypes. *Environmental and Experimental Botany* 56, no. 1: 87–94.

De Kok, L.J. (1990) Sulfur metabolism in plants exposed to atmospheric sulfur. In *Higher plants*, 111–130. Backhuys Publishers: Leiden, the Netherlands.

De Temmerman, L., Legrand, G., Vandermeiren, K. (2007) Effects of ozone on sugar beet grown in open-top chambers. *European Journal of Agronomy* 26, no. 1: 1–9.

Delfine, S., Alvino, A., Zacchini, M., Loreto, F. (1998) Consequences of salt stress on conductance to CO_2 diffusion, Rubisco characteristics and anatomy of spinach leaves. *Australian Journal of Plant Physiology* 25, no. 3: 395–402.

Demkura, P.V., Abdala, G., Baldwin, I.T., Ballaré, C.L. (2010) Jasmonate-dependent and-independent pathways mediate specific effects of solar ultraviolet B radiation on leaf phenolics and anti-herbivore defense. *Plant Physiology* 152, no. 2: 1084–1095.

Dhanda, S.S., Sethi, G.S., Behl, R.K. (2004) Indices of drought tolerance in wheat genotypes at early stages of plant growth. *Journal of Agronomy and Crop Science* 190, no. 1: 6–12.

Di Martino, C., Delfine, S., Alvino, A., Loreto, F. (1999) Photorespiration rate in spinach leaves under moderate NaCl stress. *Photosynthetica* 36, no. 1–2: 233–242.

Dias, A.S., Lidon, F.C. (2009) Evaluation of grain filling rate and duration in bread and durum wheat, under heat stress after anthesis. *Journal of Agronomy and Crop Science* 195, no. 2: 137–147.

Dickin, E., Wright, D. (2008) The effects of winter waterlogging and summer drought on the growth and yield of winter wheat (*Triticum aestivum* L.). *European Journal of Agronomy* 28, no. 3: 234–244.

Dionisio-Sese, M.L., Tobita, S. (1998) Antioxidant responses of rice seedlings to salinity stress. *Plant Science* 135, no. 1: 1–9.

Dise, N.B., Wright, R.F. (1995) Nitrogen leaching from European forests in relation to nitrogen deposition. *Forest Ecology and Management* 71, no. 1–2: 153–161.

Dixit, V., Pandey, V., Shyam, R. (2002) Chromium ions inactivate electron transport and enhance superoxide generation in vivo in pea (*Pisum sativum* L. cv. Azad) root mitochondria. *Plant, Cell and Environment* 25, no. 5: 687–693.

El-Sheekh, M.M., El-Naggar, A.H., Osman, M.E.H., El-Mazaly, E. (2003) Effect of cobalt on growth, pigments and the photosynthetic electron transport in *Monoraphidium minutum* and *Nitzchia perminuta*. *Brazilian Journal of Plant Physiology* 15, no. 3: 159–166.

Emamverdian, A., Ding, Y., Mokhberdoran, F., Xie, Y. (2015) Heavy metal stress and some mechanisms of plant defense response. *The Scientific World Journal* 2015: 756120.

Emiliani, J., Grotewold, E., Ferreyra, M.L.F., Casati, P. (2013) Flavonols protect *Arabidopsis* plants against UV-B deleterious effects. *Molecular Plant* 6, no. 4: 1376–1379.

Erice, G., Louahlia, S., Irigoyen, J.J., Sanchez-Diaz, M., Avice, J.C. (2010) Biomass partitioning, morphology and water status of four alfalfa genotypes submitted to progressive drought and subsequent recovery. *Journal of Plant Physiology* 167, no. 2: 114–120.

Esechie, H.A., Al-Saidi, A., Al-Khanjari, S. (2002) Effect of sodium chloride salinity on seedling emergence in chickpea. *Journal of Agronomy and Crop Science* 188, no. 3: 155–160.

Estevez, M.S., Malanga, G., Puntarulo, S. (2001) Iron-dependent oxidative stress in *Chlorella vulgaris*. *Plant Science* 161, no. 1: 9–17.

Ezin, V., Pena, R.D.L., Ahanchede, A. (2010) Flooding tolerance of tomato genotypes during vegetative and reproductive stages. *Brazilian Journal of Plant Physiology* 22, no. 2: 131–142.

Fankhauser, C., Ulm, R. (2011) Light-regulated interactions with SPA proteins underlie cryptochrome-mediated gene expression. *Genes and Development* 25, no. 10: 1004–1009.

Fargašová, A. (2012) Plants as models for chromium and nickel risk assessment. *Ecotoxicology* 21, no. 5: 1476–1483.

Farman, J.C., Gardiner, B.G., Shanklin, J.D. (1985) Large losses of total ozone in Antarctica reveal seasonal ClOx/NOx interaction. *Nature* 315, no. 6016: 207–210.

Farooq, M., Wahid, A., Kobayashi, N., Fujita, D., Basra, S.M.A. (2009) Plant drought stress: Effects, mechanisms and management. *Agronomy for Sustainable Development* 29, no. 1: 185–212.

Farooq, M., Hussain, M., Wahid, A., Siddique, K.H.M. (2012) Drought stress in plants: An overview. In *Plant responses to drought stress*, 1–33. Springer: Berlin–Heidelberg, Germany.

Favory, J.J., Stec, A., Gruber, H., Rizzini, L., Oravecz, A., Funk, M., Albert, A., et al. (2009) Interaction of COP1 and UVR8 regulates UV-B induced photomorphogenesis and stress acclimation in *Arabidopsis. The EMBO Journal* 28, no. 5: 591–601.

Feng, Z., Kobayashi, K. (2009) Assessing the impacts of current and future concentrations of surface ozone on crop yield with meta-analysis. *Atmospheric Environment* 43, no. 8: 1510–1519.

Feng, G., Zhang, Z., Wan, C., Lu, P., Bakour, A. (2017) Effects of saline water irrigation on soil salinity and yield of summer maize (*Zea mays* L.) in subsurface drainage system. *Agricultural Water Management* 193: 205–213.

Finlayson-Pitts, B.J., Pitts, J.N. (1997) Tropospheric air pollution: Ozone, airborne toxics, polycyclic aromatic hydrocarbons, and particles. *Science* 276, no. 5315: 1045–1052.

Fiskesjö, G. (1988) The Allium test—An alternative in environmental studies: The relative toxicity of metal ions. *Mutation Research/Fundamental and Molecular Mechanisms of Mutagenesis* 197, no. 2: 243–260.

Flexas, J., Medrano, H. (2002) Drought-inhibition of photosynthesis in C3 plants: Stomatal and non-stomatal limitations revisited. *Annals of Botany* 89, no. 2: 183–189.

Flexas, J., Diaz-Espejo, A., Galmes, J., Kaldenhoff, R., Medrano, H., Ribas-Carbo, M. (2007) Rapid variations of mesophyll conductance in response to changes in CO2 concentration around leaves. *Plant, Cell and Environment* 30, no. 10: 1284–1298.

Flowers, T. (2006) Preface. *Journal of Experimental Botany* 57, no. 5: iv–iv.

Flowers, M.D., Fiscus, E.L., Burkey, K.O., Booker, F.L., Dubois, J.B. (2007) Photosynthesis, chlorophyll fluorescence, and yield of snap bean (*Phaseolus vulgaris* L.) genotypes differing in sensitivity to ozone. *Environmental and Experimental Botany* 61, no. 2: 190–198.

Frei, M., Tanaka, J.P., Chen, C.P., Wissuwa, M. (2010) Mechanisms of ozone tolerance in rice: Characterization of two QTLs affecting leaf bronzing by gene expression profiling and biochemical analyses. *Journal of Experimental Botany* 61, no. 5: 1405–1417.

Garg, B.K., Gupta, I.C. (1997) *Saline wastelands environment and plant growth*. Scientific Publishers: Rajasthan, India.

Gay, F., Maraval, I., Roques, S., Gunata, Z., Boulanger, R., Audebert, A., Mestres, C. (2010) Effect of salinity on yield and 2-acetyl-1-pyrroline content in the grains of three fragrant rice cultivars (*Oryza sativa* L.) in Camargue (France). *Field Crops Research* 117, no. 1: 154–160.

Ghadirnezhad, R., Fallah, A. (2014) Temperature effect on yield and yield components of different rice cultivars in flowering stage. *International Journal of Agronomy* 2014: 1–4.

Gholizadeh, A., Dehghani, H., Dvorak, Y. (2014) Determination of the most effective traits on wheat yield under saline stress. *Agricultural Advances* 3: 103–110.

Ghulamahdi, M., Chaerunisa, S.R., Lubis, I., Taylor, P. (2016) Response of five soybean varieties under saturated soil culture and temporary flooding on tidal swamp. *Procedia Environmental Sciences* 33: 87–93.

Gill, S.S., Tuteja, N. (2010) Reactive oxygen species and antioxidant machinery in abiotic stress tolerance in crop plants. *Plant Physiology and Biochemistry* 48, no. 12: 909–930.

Giuffrida, F., Scuderi, D., Giurato, R., Leonardi, C. (2013) Physiological response of broccoli and cauliflower as affected by NaCl salinity. In *VI International Symposium on Brassicas and XVIII Crucifer Genetics Workshop* 1005: 435–441.

González Besteiro, M.A., Bartels, S., Albert, A., Ulm, R. (2011) Arabidopsis MAP kinase phosphatase 1 and its target MAP kinases 3 and 6 antagonistically determine UV-B stress tolerance, independent of the UVR8 photoreceptor pathway. *The Plant Journal* 68, no. 4: 727–737.

Guenther, A., Geron, C., Pierce, T., Lamb, B., Harley, P., Fall, R. (2000) Natural emissions of non-methane volatile organic compounds, carbon monoxide, and oxides of nitrogen from North America. *Atmospheric Environment* 34, no. 12–14: 2205–2230.

Guerinot, M.L. (2000) The ZIP family of metal transporters. *Biochimica et Biophysica Acta-Biomembranes* 1465, no. 1–2: 190–198.

Guidi, L., Nali, C., Lorenzini, G., Filippi, F., Soldatini, G.F. (2001) Effect of chronic ozone fumigation on the photosynthetic process of poplar clones showing different sensitivity. *Environmental Pollution* 113, no. 3: 245–254.

Guidi, L., Degl'Innocenti, E., Martinelli, F., Piras, M. (2009) Ozone effects on carbon metabolism in sensitive and insensitive Phaseolus cultivars. *Environmental and Experimental Botany* 66, no. 1: 117–125.

Guóth, A., Tari, I., Gallé, Á., Csiszár, J., Pécsváradi, A., Cseuz, L., Erdei, L. (2009) Comparison of the drought stress responses of tolerant and sensitive wheat cultivars during grain filling: Changes in flag leaf photosynthetic activity, ABA levels, and grain yield. *Journal of Plant Growth Regulation* 28, no. 2: 167–176.

Gupta, S.K., Rai, A.K., Kanwar, S.S., Sharma, T.R. (2012) Comparative analysis of zinc finger proteins involved in plant disease resistance. *PLOS ONE* 7, no. 8: e42578.

Gwóźdź, E.A., Przymusiński, R., Rucińska, R., Deckert, J. (1997) Plant cell responses to heavy metals: Molecular and physiological aspects. *Acta Physiologiae Plantarum* 19, no. 4: 459–465.

Habibi, G. (2011) Influence of drought on yield and yield components in white bean. *World Academy of Science, Engineering and Technology* 55: 244–253.

Hall. J.L. (2002) Cellular mechanisms for heavy metal detoxification and tolerance. *Journal of Experimental Botany* 53, no. 366: 1–11.

Hallquist, M., Wenger, J.C., Baltensperger, U., Rudich, Y., Simpson, D., Claeys, M., Dommen, J., et al. (2009) The formation, properties and impact of secondary organic aerosol: Current and emerging issues. *Atmospheric Chemistry and Physics* 9, no. 14: 5155–5236.

Hammami, Z., Gauffreteau, A., BelhajFraj, M., Sahli, A., Jeuffroy, M.-H., Rezgui, S., Bergaoui, K., et al. (2017) Predicting yield reduction in improved barley (Hordeum vulgare L.) varieties and landraces under salinity using selected tolerance traits. *Field Crops Research* 211: 10–18.

Harb, A., Krishnan, A., Ambavaram, M.M., Pereira, A. (2010) Molecular and physiological analysis of drought stress in Arabidopsis reveals early responses leading to acclimation in plant growth. *Plant Physiology* 154, no. 3: 1254–1271.

Harris, D., Tripathi, R.S., Joshi, A. (2002) On-farm seed priming to improve crop establishment and yield in dry direct-seeded rice. In *Direct seeding: Research strategies and opportunities*, 231–240. International Research Institute, Manila, Philippines.

Harrison, M.A. (2012) Cross-talk between phytohormone signaling pathways under both optimal and stressful environmental conditions. In *Phytohormones and abiotic stress tolerance in plants*, 49–76. Springer, Berlin–Heidelberg, Germany.

Hasanuzzaman, M., Fujita, M., Islam, M.N., Ahamed, K.U., Nahar, K. (2009) Performance of four irrigated rice varieties under different levels of salinity stress. *International Journal of Integrative Biology* 6: 85–90.

Hasanuzzaman, M., Nahar, K., Alam, M.M., Roychowdhury, R., Fujita, M. (2013) Physiological, biochemical, and molecular mechanisms of heat stress tolerance in plants. *International Journal of Molecular Sciences* 14, no. 5: 9643–9684.

Hayashi, T., Yoshida, T., Fujii, K., Mitsuya, S., Tsuji, T., Okada, Y., Hayashi, E., et al. (2013) Maintained root length density contributes to the waterlogging tolerance in common wheat (*Triticum aestivum* L.). *Field Crops Research* 152: 27–35.

Hectors, K., Van Oevelen, S., Geuns, J., Guisez, Y., Jansen, M.A., Prinsen, E. (2014) Dynamic changes in plant secondary metabolites during UV acclimation in Arabidopsis thaliana. *Physiologia Plantarum* 152, no. 2: 219–230.

Heidarvand, L., Amiri, R.M. (2010) What happens in plant molecular responses to cold stress? *Acta Physiologiae Plantarum* 32, no. 3: 419–431.

Heijde, M., Ulm, R. (2013) Reversion of the Arabidopsis UV-B photoreceptor UVR8 to the homodimeric ground state. *Proceedings of the National Academy of Sciences of the United States of America* 110, no. 3: 1113–1118.

Hernandez, J.A., Jimenez, A., Mullineaux, P., Sevilia, F. (2000) Tolerance of pea (Pisum sativum L.) to long-term salt stress is associated with induction of antioxidant defenses. *Plant, Cell and Environment* 23, no. 8: 853–862.

Hesse, H., Jank-Ladwig, R., Strotmann, H. (1976) On the reconstitution of photophosphorylation in CF1-extracted chloroplasts. *Zeitschrift für Naturforschung C* 31, no. 7–8: 445–451.

Hideg, E., Jansen, M.A., Strid, A. (2013) UV-B exposure, ROS, and stress: Inseparable companions or loosely linked associates? *Trends in Plant Science* 18, no. 2: 107–115.

Hoegberg, P., Fan, H., Quist, M., Binkley, D.A.N., Tamm, C.O. (2006) Tree growth and soil acidification in response to 30 years of experimental nitrogen loading on boreal forest. *Global Change Biology* 12, no. 3: 489–499.

Hoekstra, F.A., Golovina, E.A., Buitink, J. (2001) Mechanisms of plant desiccation tolerance. *Trends in Plant Science* 6, no. 9: 431–438.

Holopainen, J.K. (2004) Multiple functions of inducible plant volatiles. *Trends in Plant Science* 9, no. 11: 529–533.

Hong, R., Kang, T.Y., Michels, C.A., Gadura, N. (2012) Membrane lipid peroxidation in copper alloy-mediated contact killing of Escherichia coli. *Applied and Environmental Microbiology* 78, no. 6: 1776–1784.

Hossain, Z., López-Climent, M.F., Arbona, V., Pérez-Clemente, R.M., Gómez-Cadenas, A. (2009) Modulation of the antioxidant system in citrus under waterlogging and subsequent drainage. *Journal of Plant Physiology* 166, no. 13: 1391–1404.

Hove, R.M., Bhave, M. (2011) Plant aquaporins with non-aqua functions: Deciphering the signature sequences. *Plant Molecular Biology* 75, no. 4–5: 413–430.

Hua, J. (2009) From freezing to scorching, transcriptional responses to temperature variations in plants. *Current Opinion in Plant Biology* 12, no. 5: 568–573.

Huchzermeyer, B. (2000) *Biochemical principles of salt tolerance. Sustainable halophyte utilization in the Mediterranean and subtropical dry regions*, 130–133. University of Osnabrück Publication: Germany.

Intergovernmental Panel on Climate Change (2014) *Climate change 2014—Impacts, adaptation and vulnerability: Regional aspects.* Cambridge University Press: UK.

Iqbal, N., Umar, S., Khan, N.A., Khan, M.I.R. (2014) A new perspective of phytohormones in salinity tolerance: Regulation of proline metabolism. *Environmental and Experimental Botany* 100: 34–42.

Ismail, A.M., Singh, U.S., Singh, S., Dar, M.H., Mackill, D.J. (2013) The contribution of submergence-tolerant (Sub1) rice varieties to food security in flood-prone rainfed lowland areas in Asia. *Field Crops Research* 152: 83–93.

Iyaka, Y.A. (2011) Nickel in soils: a review of its distribution and impacts. *Scientific Research and Essays* 6, no. 33: 6774–6777.

Jackson, M.B., Ricard, B. (2003) Physiology, biochemistry and molecular biology of plant root systems subjected to flooding of the soil. In *Root ecology*, 193–213. Springer, Berlin–Heidelberg, Germany.

Jain, C.K., Ali, I. (2000) Arsenic: Occurrence, toxicity and speciation techniques. *Water Research* 34, no. 17: 4304–4312.

Jaleel, C.A., Changxing, Z., Jayakumar, K., Iqbal, M. (2008) Low concentration of cobalt increases growth, biochemical constituents, mineral status and yield in Zea mays. *Journal of Scientific Research* 1, no. 1: 128–137.

Jamshidi, A., Javanmard, H.R. (2017) Evaluation of barley (*Hordeum vulgare* L.) genotypes for salinity tolerance under field conditions using the stress indices. *Ain Shams Engineering Journal* 8, no. 1: 1–102.

Jansen, M.A.K., Van Den Noort, R.E. (2000) Ultraviolet-B radiation induces complex alterations in stomatal behaviour. *Physiologia Plantarum* 110, no. 2: 189–194.

Jaoudé, M.B., Katerji, N., Mastrorilli, M., Rana, G. (2008) Analysis of the ozone effect on soybean in the Mediterranean region: II. The consequences on growth, yield and water use efficiency. *European Journal of Agronomy* 28, no. 4: 519–525.

Javid, M.G., Sorooshzadeh, A., Moradi, F., Modarres Sanavy, S.A.M., Allahdadi, I. (2011) The role of phytohormones in alleviating salt stress in crop plants. *Australian Journal of Crop Science* 5, no. 4: 726–734.

Jenkins, G.I. (2009) Signal transduction in responses to UV-B radiation. *Annual Review of Plant Biology* 60: 407–431.

Jensen, M.K., Hagedorn, P.H., De Torres-Zabala, M., Grant, M.R., Rung, J.H., Collinge, D.B., Lyngkjaer, M.F. (2008) Transcriptional regulation by an NAC (NAM-ATAF1,2-CUC2) transcription factor attenuates ABA signaling for efficient basal defence towards Blumeria graminis f.sp. hordei in Arabidopsis. *The Plant Journal* 56, no. 6: 867–880.

Kakani, V.G., Reddy, K.R., Zhao, D., Sailaja, K. (2003) Field crop responses to ultraviolet-B radiation: A review. *Agricultural and Forest Meteorology* 120, no. 1–4: 191–218.

Kamara, A.Y., Menkir, A., Badu-Apraku, B., Ibikunle, O. (2003) The influence of drought stress on growth, yield and yield components of selected maize genotypes. *The Journal of Agricultural Science* 141, no. 1: 43–50.

Kangasjärvi, J., Jaspers, P., Kollist, H. (2005) Signaling and cell death in ozone-exposed plants. *Plant, Cell and Environment* 28, no. 8: 1021–1036.

Kaur, G., Kumar, S., Thakur, P., Malik, J.A., Bhandhari, K., Sharma, K.D., Nayyar, H. (2011) Involvement of proline in response of chickpea (*Cicer arietinum* L.) to chilling stress at reproductive stage. *Scientia Horticulturae* 128, no. 3: 174–181.

Kaya, C., Ashraf, M., Sonmez, O., Aydemir, S., Tuna, A.L., Cullu, M.A. (2009) The influence of arbuscular mycorrhizal colonization on key growth parameters and fruit yield of pepper plants grown at high salinity. *Scientia Horticulturae* 121, no. 1: 1–6.

Kesselmeier, J., Staudt, M. (1999) Biogenic volatile organic compounds (VOC): An overview on emission, physiology and ecology. *Journal of Atmospheric Chemistry* 33, no. 1: 23–88.

Keyvan, S. (2010) The effects of drought stress on yield, relative water content, proline, soluble carbohydrates and chlorophyll of bread wheat cultivars. *Journal of Animal and Plant Sciences* 8, no. 3: 1051–1060.

Khan, S. Soja, G. (2003) Yield responses of wheat to ozone exposure as modified by drought-induced differences in ozone uptake. *Water, Air, and Soil Pollution* 147, no. 1–4: 299–315.

Khan, N.A., Singh, S., Nazar, R. (2007) Activities of antioxidative enzymes, sulfur assimilation, photosynthetic activity and growth of wheat (*Triticum aestivum*) cultivars differing in yield potential under cadmium stress. *Journal of Agronomy and Crop Science* 193, no. 6: 435–444.

Khatun, S., Flowers, T.J. (1995) Effects of salinity on seed set in rice. *Plant, Cell and Environment* 18, no. 1: 61–67.

Kitao, M., Lei, T.T., Nakamura, T., Koike, T. (2001) Manganese toxicity as indicated by visible foliar symptoms of Japanese white birch (*Betula platyphylla* var. japonica). *Environmental Pollution* 111, no. 1: 89–94.

Klamkowski, K., Treder, W. (2008) Response to drought stress of three strawberry cultivars grown under greenhouse conditions. *Journal of Fruit and Ornamental Plant Research* 16: 179–188.

Kliebenstein, D.J., Lim, J.E., Landry, L.G., Last, R.L. (2002) *Arabidopsis* UVR8 regulates ultraviolet-B signal transduction and tolerance and contains sequence similarity to human regulator of chromatin condensation 1. *Plant Physiology* 130, no. 1: 234–243.

Komatsu, S., Yamamoto, A., Nakamura, T., Nouri, M.Z., Nanjo, Y., Nishizawa, K., Furukawa, K. (2011) Comprehensive analysis of mitochondria in roots and hypocotyls of soybean under flooding stress using proteomics and metabolomics techniques. *Journal of Proteome Research* 10, no. 9: 3993–4004.

Komatsu, S., Han, C., Nanjo, Y., Altaf-Un-Nahar, M., Wang, K., He, D., Yang, P. (2013) Label-free quantitative proteomic analysis of abscisic acid effect in early-stage soybean under flooding. *Journal of Proteome Research* 12, no. 11: 4769–4784.

Kontopoulou, C.K., Bilalis, D., Pappa, V.A., Rees, R.M., Savvas, D. (2015) Effects of organic farming practices and salinity on yield and greenhouse gas emissions from a common bean crop. *Scientia Horticulturae* 183: 48–57.

Kotak, S., Larkindale, J., Lee, U., von Koskull-Döring, P., Vierling, E., Scharf, K.D. (2007) Complexity of the heat stress response in plants. *Current Opinion in Plant Biology* 10, no. 3: 310–316.

Koyro, H.W., Ahmad, P., Geissler, N. (2012) Abiotic stress responses in plants: An overview. In *Environmental adaptations and stress tolerance of plants in the era of climate change*, 1–28. Springer: New York, NY.

Kuanar, S.R., Ray, A., Sethi, S.K., Chattopadhyay, K., Sarkar, R.K. (2017) Physiological basis of stagnant flooding tolerance in rice. *Rice Science* 24, no. 2: 73–84.

Kumar, S., Kaur, R., Kaur, N., Bhandhari, K., Kaushal, N., Gupta, K., Bains, T.S., et al. (2011) Heat-stress induced inhibition in growth and chlorosis in mungbean (*Phaseolus aureus* Roxb.) is partly mitigated by ascorbic acid application and is related to reduction in oxidative stress. *Acta Physiologiae Plantarum* 33, no. 6: 2091–2101.

Kumar, M.N., Jane, W.N., Verslues, P.E. (2013a) Role of the putative osmosensor *Arabidopsis* histidine kinase1 in dehydration avoidance and low-water-potential response. *Plant Physiology* 161, no. 2: 942–953.

Kumar, P., Pal, M., Joshi, R., Sairam, R.K. (2013b) Yield, growth and physiological responses of mung bean (*Vigna radiata* (L.) Wilczek) genotypes to waterlogging at vegetative stage. *Physiology and Molecular Biology of Plants* 19, no. 2: 209–220.

Kumar, M.S., Ali, K., Dahuja, A., Tyagi, A. (2015) Role of phytosterols in drought stress tolerance in rice. *Plant Physiology and Biochemistry* 96: 83–89.

Kumar, A., Nayak, A.K., Pani, D.R., Das, B.S. (2017) Physiological and morphological responses of four different rice cultivars to soil water potential based deficit irrigation management strategies. *Field Crops Research* 205: 78–94.

Kumari, S., Agrawal, M. (2014) Growth, yield and quality attributes of a tropical potato variety (*Solanum tuberosum* L. Cv. Kufri chandramukhi) under ambient and elevated carbon dioxide and ozone and their interactions. *Ecotoxicology and Environmental Safety* 101: 146–156.

Kurunc, A., Unlukara, A., Cemek, B. (2011) Salinity and drought affect yield response of bell pepper similarly. *Acta Agriculturae Scandinavica, Section B—Soil and Plant Science* 61, no. 6: 514–522.

Lafitte, H.R., Yongsheng, G., Yan, S., Li, Z.K. (2007) Whole plant responses, key processes, and adaptation to drought stress: The case of rice. *Journal of Experimental Botany* 58, no. 2: 169–175.

Lal, S., Patil, R.S. (2001) Monitoring of atmospheric behavior of NOx from vehicular traffic. *Environmental Monitoring and Assessment* 68, no. 1: 37–50.

Laothawornkitkul, J., Taylor, J.E., Paul, N.D., Hewitt, C.N. (2009) Biogenic volatile organic compounds in the Earth system. *The New Phytologist* 183, no. 1: 27–51.

Larssen, S., Gram, F., Hagen, L.O., Jansen, H., Olsthoorn, X. (1994) *URBAIR-Bombay city specific report*. Maharashtra Pollution Control Board: Mumbai, India.

Lee, G., Carrow, R.N., Duncan, R.R. (2004) Photosynthetic responses to salinity stress of halophytic seashore paspalum ecotypes. *Plant Science* 166, no. 6: 1417–1425.

Li, D., Liu, H., Qiao, Y., Wang, Y., Cai, Z., Dong, B., Shi, C., et al. (2013a) Effects of elevated CO2 on the growth, seed yield, and water use efficiency of soybean (*Glycine max* (L.) Merr.) under drought stress. *Agricultural Water Management* 129: 105–112.

Li, Y.F., Wu, Y., Hernandez-Espinosa, N., Peña, R.J. (2013b) Heat and drought stress on durum wheat: Responses of genotypes, yield, and quality parameters. *Journal of Cereal Science* 57, no. 3: 398–404.

Lian, H.L., Yu, X., Lane, D., Sun, W.N., Tang, Z.C., Su, W.A. (2006) Upland rice and lowland rice exhibited different PIP expression under water deficit and ABA treatment. *Cell Research* 16, no. 7: 651–660.

Liu, F., Stützel, H. (2004) Biomass partitioning, specific leaf area, and water use efficiency of vegetable amaranth (*Amaranthus* spp.) in response to drought stress. *Scientia Horticulturae* 102, no. 1: 15–27.

Liu, F., Jensen, C.R., Andersen, M.N. (2003) Hydraulic and chemical signals in the control of leaf expansion and stomatal conductance in soybean exposed to drought stress. *Functional Plant Biology* 30, no. 1: 65–73.

Lombardi, L., Sebastiani, L. (2005) Copper toxicity in *Prunus cerasifera*: Growth and antioxidant enzymes responses of in vitro grown plants. *Plant Science* 168, no. 3: 797–802.

López, M.Á., Magnitski, S. (2011) Nickel: The last of the essential micronutrients. *Agronomía Colombiana* 29: 49–56.

Loutfy, N., El-Tayeb, M.A., Hassanen, A.M., Moustafa, M.F., Sakuma, Y., Inouhe, M. (2012) Changes in the water status and osmotic solute contents in response to drought and salicylic acid treatments in four different cultivars of wheat (*Triticum aestivum*). *Journal of Plant Research* 125, no. 1: 173–184.

Lu, F., Cui, X., Zhang, S., Jenuwein, T., Cao, X. (2011) *Arabidopsis* REF6 is a histone H3 lysine 27 demethylase. *Nature Genetics* 43, no. 7: 715–719.

Ma, J.F., Ryan, P.R., Delhaize, E. (2001) Aluminum tolerance in plants and the complexing role of organic acids. *Trends in Plant Science* 6, no. 6: 273–278.

Maas, E.V., Poss, J.A. (1989) Salt sensitivity of wheat at various growth stages. *Irrigation Science* 10, no. 1: 29–40.

Mafakheri, A., Siosemardeh, A., Bahramnejad, B., Struik, P.C., Sohrabi, Y. (2010) Effect of drought stress on yield, proline and chlorophyll contents in three chickpea cultivars. *Australian Journal of Crop Science* 4, no. 8: 580–585.

Maggs, R., Ashmore, M.R. (1998) Growth and yield responses of Pakistan rice (Oryza sativa L.) cultivars to O3 and NO2. *Environmental Pollution* 103, no. 2–3: 159–170.

Malhotra, S.S. (1984) Biochemical and physiological impact of major pollutants. In *Air pollution and plant life*: 113–157. John Wiley & Sons: West Sussex, UK.

Malik, A.I., Colmer, T.D., Lambers, H., Setter, T.L., Schortemeyer, M. (2002) Short-term waterlogging has long-term effects on the growth and physiology of wheat. *New Phytologist* 153, no. 2: 225–236.

Manickavelu, A., Nadarajan, N., Ganesh, S.K., Gnanamalar, R.P., Babu, R.C. (2006) Drought tolerance in rice: Morphological and molecular genetic consideration. *Plant Growth Regulation* 50, no. 2–3: 121–138.

Marcelis, L.F.M., Van Hooijdonk, J. (1999) Effect of salinity on growth, water use and nutrient use in radish (*Raphanus sativus* L.). *Plant and Soil* 215, no. 1: 57–64.

Marschner, H. (1995) *Mineral nutrition of higher plants*. Academic Press, London, UK.

Marschner, H. (2011) *Marschner's mineral nutrition of higher plants*. Academic Press: Amsterdam, the Netherlands.

Martínez, J.P., Silva, H.F.L.J., Ledent, J.F., Pinto, M. (2007) Effect of drought stress on the osmotic adjustment, cell wall elasticity and cell volume of six cultivars of common beans (*Phaseolus vulgaris* L.). *European Journal of Agronomy* 26, no. 1: 30–38.

Massacci, A., Nabiev, S.M., Pietrosanti, L., Nematov, S.K., Chernikova, T.N., Thor, K., Leipner, J. (2008) Response of the photosynthetic apparatus of cotton (*Gossypium hirsutum*) to the onset of drought stress under field conditions studied by gas-exchange analysis and chlorophyll fluorescence imaging. *Plant Physiology and Biochemistry* 46, no. 2: 189–195.

Mathobo, R., Marais, D., Steyn, J.M. (2017) The effect of drought stress on yield, leaf gaseous exchange and chlorophyll fluorescence of dry beans (*Phaseolus vulgaris* L.). *Agricultural Water Management* 180: 118–125.

McGrath, S.P., Micó, C., Curdy, R., Zhao, F.J. (2010) Predicting molybdenum toxicity to higher plants: Influence of soil properties. *Environmental Pollution* 158, no. 10: 3095–3102.

McGrath, J.M., Betzelberger, A.M., Wang, S., Shook, E., Zhu, X.G., Long, S.P., Ainsworth, E.A. (2015) An analysis of ozone damage to historical maize and soybean yields in the United States. *Proceedings of the National Academy of Sciences* 112, no. 46: 14390–14395.

McKenzie, R.L., Aucamp, P.J., Bais, A.F., Björn, L.O., Ilyas, M. (2007) Changes in biologically-active ultraviolet radiation reaching the Earth's surface. *Photochemical and Photobiological Sciences* 6, no. 3: 218–231.

Mendel, R.R. (2013) The molybdenum cofactor. *The Journal of Biological Chemistry* 288, no. 19: 13165–13172.

Mikami, K., Murata, N. (2003) Membrane fluidity and the perception of environmental signals in cyanobacteria and plants. *Progress in Lipid Research* 42, no. 6: 527–543.

Mills, G., Buse, A., Gimeno, B., Bermejo, V., Holland, M., Emberson, L., Pleijel, H. (2007) A synthesis of AOT40-based response functions and critical levels of ozone for agricultural and horticultural crops. *Atmospheric Environment* 41, no. 12: 2630–2643.

Mishra, A.K., Singh, V.P. (2010) A review of drought concepts. *Journal of Hydrology* 391, no. 1–2: 202–216.

Mishra, A.K., Rai, R., Agrawal, S.B. (2013) Differential response of dwarf and tall tropical wheat cultivars to elevated ozone with and without carbon dioxide enrichment: Growth, yield and grain quality. *Field Crops Research* 145: 21–32.

Mittler, R. (2006) Abiotic stress, the field environment and stress combination. *Trends in Plant Science* 11, no. 1: 15–19.

Miura, K., Tada, Y. (2014) Regulation of water, salinity, and cold stress responses by salicylic acid. Frontiers in Plant Science 5: 4.

Miyashita, K.,Tanakamaru, S., Maitani, T., Kimura, K. (2005) Recovery responses of photosynthesis, transpiration, and stomatal conductance in kidney bean following drought stress. *Environmental and Experimental Botany* 53, no. 2: 205–214.

Mohammed, A.R., Tarpley, L. (2009) Impact of high night time temperature on respiration, membrane stability, antioxidant capacity, and yield of rice plants. *Crop Science* 49, no. 1: 313–322.

Mohanta, T.K., Bashir, T., Hashem, A., Abd-Allah, E. F. (2017) Systems biology approach in plant abiotic stresses. *Plant Physiology and Biochemistry* 121: 58–73.

Molina, M.J., Sherwood Rowland, F.S. (1974) Stratospheric sink for chlorofluoromethanes: Chlorine atom-catalyzed destruction of ozone. *Nature* 249, no. 5460: 810–812.

Møller, I.M., Jensen, P.E., Hansson, A. (2007) Oxidative modifications to cellular components in plants. *Annual Review of Plant Biology* 58: 459–481.

Monga, R., Marzuoli, R., Alonso, R., Bermejo, V., González-Fernández, I., Faoro, F., Gerosa, G. (2015) Varietal screening of ozone sensitivity in Mediterranean durum wheat (*Triticum durum*, Desf.). *Atmospheric Environment* 110: 18–26.

Monkham, T., Jongdee, B., Pantuwan, G., Sanitchon, J., Mitchell, J.H., Fukai, S. (2015) Genotypic variation in grain yield and flowering pattern in terminal and intermittent drought screening methods in rainfed lowland rice. *Field Crops Research* 175: 26–36.

Monneveux, P., Sanchez, C., Beck, D., Edmeades, G.O. (2006) Drought tolerance improvement in tropical maize source populations. *Crop Science* 46, no. 1: 180–191.

Morgan, P.B., Ainsworth, E.A., Long, S.P. (2003) How does elevated ozone impact soybean? A meta-analysis of photosynthesis, growth and yield. *Plant, Cell and Environment* 26, no. 8: 1317–1328.

Morin, S., Savarino, J., Frey, M.M., Yan, N.,Bekki, S., Bottenheim, J.W., Martins, J.M. (2008) Tracing the origin and fate of NOx in the Arctic atmosphere using stable isotopes in nitrate. *Science* 322, no. 5902: 730–732.

Mortvedt, J.J., Giordano, P.M. (1975) Response of corn to zinc and chromium in municipal wastes applied to soil 1. *Journal of Environment Quality* 4, no. 2: 170–174.

Mosier, A., Kroeze, C., Nevison, C., Oenema, O., Seitzinger, S., Van Cleemput, O. (1998) Closing the global atmospheric N2O budget: Nitrous oxide emissions through the agricultural nitrogen cycle: OECD/IPCC/IEA phase II development of IPCC guidelines for national greenhouse gas inventory methodology. *Nutrient Cycling in Agroecosystems* 52, no. 2–3: 225–248.

Mossor-Pietraszewska, T. (2001) Effect of aluminum on plant growth and metabolism. *Acta Biochimica Polonica—English Edition* 48, no. 3: 673–686.

Mousavi, S.R., Galavi, M., Rezaei, M. (2013) Zinc (Zn) importance for crop production-a review. *International Journal of Agronomy and Plant Production* 4, no. 1: 64–68.

Moussa, H.R., Abdel-Aziz, S.M. (2008) Comparative response of drought tolerant and drought sensitive maize genotypes to water stress. *Australian Journal of Crop Science* 1, no. 1: 31–36.

Müller-Xing, R., Xing, Q., Goodrich, J. (2014) Footprints of the sun: Memory of UV and light stress in plants. *Frontiers in Plant Science* 5: 474.

Munns, R., Schachtman, D.P., Condon, A.G. (1995) The significance of a two-phase growth response to salinity in wheat and barley. *Australian Journal of Plant Physiology* 22, no. 4: 561–569.

Murata, N., Los, D.A. (1997) Membrane fluidity and temperature perception. *Plant Physiology* 115, no. 3: 875–879.

Musil, C.F., Björn, L.O., Scourfield, M.W.J., Bodeker, G.E. (2002) How substantial are ultraviolet-B supplementation inaccuracies in experimental square-wave delivery systems? *Environmental and Experimental Botany* 47, no. 1: 25–38.

Mutava, R.N., Prince, S.J.K., Syed, N.H., Song, L., Valliyodan, B., Chen, W., Nguyen, H.T. (2015) Understanding abiotic stress tolerance mechanisms in soybean: A comparative evaluation of soybean response to drought and flooding stress. *Plant Physiology and Biochemistry* 86: 109–120.

Nagel, L.M., Bassman, J.H., Edwards, G.E., Robberecht, R., Franceshi, V.R. (1998) Leaf anatomical changes in *Populus trichocarpa, Quercus rubra, Pseudotsuga menziesii* and *Pinus ponderosa* exposed to enhanced ultraviolet-B radiation. *Physiologia Plantarum* 104, no. 3: 385–396.

Nandi, P.K., Agrawal, M., Rao, D.N. (1984) SO_2-induced enzymatic changes and ascorbic acid oxidation in *Oryza sativa. Water, Air, and Soil Pollution* 21, no. 1–4: 25–32.

NASA (2015) The antartic ozone hole will recover. Retrieved from: http://svs.gsfc.nasa.gov/cgi-bin/details.cgi?aid=3 0602 (accessed on 9th Feb. 2016).

Nawkar, G.M., Maibam, P., Park, J.H., Sahi, V.P., Lee, S.Y., Kang, C.H. (2013) UV-induced cell death in plants. *International Journal of Molecular Sciences* 14, no. 1: 1608–1628.

Nayyar, H., Kaur, G., Kumar, S., Upadhyaya, H.D. (2007) Low temperature effects during seed filling on chickpea genotypes (*Cicer arietinum* L.): Probing mechanisms affecting seed reserves and yield. *Journal of Agronomy and Crop Science* 193, no. 5: 336–344.

Negi, A., Singh, H.P., Batish, D.R., Kohli, R.K. (2014) Ni+ 2-inhibited radicle growth in germinating wheat seeds involves alterations in sugar metabolism. *Acta Physiologiae Plantarum* 36, no. 4: 923–929.

Nieboer, E., Richardson, D.H.S. (1980) The replacement of the nondescript term 'heavy metals' by a biologically and chemically significant classification of metal ions. *Environmental Pollution Series B, Chemical and Physical* 1, no. 1: 3–26.

Nonami, H. (1998) Plant water relations and control of cell elongation at low water potentials. *Journal of Plant Research* 111, no. 3: 373–382.

Nordin Henriksson, K., Trewavas, A.J. (2003) The effect of short-term low temperature treatments on gene expression in *Arabidopsis* correlates with changes in intracellular Ca2+ levels. *Plant, Cell and Environment* 26, no. 4: 485–496.

Oh, M., Nanjo, Y., Komatsu, S. (2014) Gel-free proteomic analysis of soybean root proteins affected by calcium under flooding stress. *Frontiers in Plant Science* 5: 559.

Ohashi, Y., Nakayama, N., Saneoka, H., Fujita, K. (2006) Effects of drought stress on photosynthetic gas exchange, chlorophyll fluorescence and stem diameter of soybean plants. *Biologia Plantarum* 50, no. 1: 138–141.

Oksanen, E., Pandey, V., Pandey, A.K., Keski-Saari, S., Kontunen-Soppela, S., Sharma, C. (2013) Impacts of increasing ozone on Indian plants. *Environmental Pollution* 177: 189–200.

Örvar, B.L., Sangwan, V., Omann, F., Dhindsa, R.S. (2000) Early steps in cold sensing by plant cells: The role of actin cytoskeleton and membrane fluidity. *The Plant Journal* 23, no. 6: 785–794.

Pandey, N., Pathak, G.C., Pandey, D.K., Pandey, R. (2009) Heavy metals, Co, Ni, Cu, Zn and Cd, produce oxidative damage and evoke differential antioxidant responses in spinach. *Brazilian Journal of Plant Physiology* 21, no. 2: 103–111.

Pang, J. Kobayashi, K., Zhu, J. (2009) Yield and photosynthetic characteristics of flag leaves in Chinese rice (*Oryza sativa* L.) varieties subjected to free-air release of ozone. *Agriculture, Ecosystems and Environment* 132, no. 3–4: 203–211.

Pantuwan, G., Fukai, S., Cooper, M., Rajatasereekul, S., O'toole, J.C. (2002) Yield response of rice (*Oryza sativa* L.) genotypes to drought under rainfed lowland: 3. Plant factors contributing to drought resistance. *Field Crops Research* 73, no. 2–3: 181–200.

Patel, P.K., Singh, A.K., Tripathi, N., Yadav, D., Hemantaranjan, A. (2014) Flooding: Abiotic constraint limiting vegetable productivity. *Advances in Plants and Agriculture Research* 1, no. 3: 96–103.

Peñuelas, J., Filella, I., Stefanescu, C., Llusià, J. (2005) Caterpillars of *Euphydryas aurinia* (Lepidoptera: Nymphalidae) feeding on *Succisa pratensis* leaves induce large foliar emissions of methanol. *The New Phytologist* 167, no. 3: 851–857.

Peschke, V.M., Sachs, M.M. (1994) Characterization and expression of transcripts induced by oxygen deprivation in maize (*Zea mays* L.). *Plant Physiology* 104, no. 2: 387–394.

Pimentel, D., Berger, B., Filiberto, D., Newton, M., Wolfe, B., Karabinakis, E., Clark, S., et al. (2004) Water resources: Agricultural and environmental issues. *BioScience* 54, no. 10: 909–918.

Pleijel, H., Skärby, L., Ojanperä, K., Selldén, G. (1992) Yield and quality of spring barley, Hordeum vulgare L., exposed to different concentrations of ozone in open-top chambers. *Agriculture, Ecosystems & Environment* 38, no. 1–2: 21–29.

Pleijel, H., Broberg, M.C., Uddling, J., Mills, G. (2018) Current surface ozone concentrations significantly decrease wheat growth, yield and quality. *The Science of the Total Environment* 613–614: 687–692.

Plieth, C., Hansen, U.P., Knight, H., Knight, M.R. (1999) Temperature sensing by plants: The primary characteristics of signal perception and calcium response. *The Plant Journal* 18, no. 5: 491–497.

Poonkothai, M.V.B.S., Vijayavathi, B.S. (2012) Nickel as an essential element and a toxicant. *International Journal of Environmental Science* 1, no. 4: 285–288.

Praba, M.L., Cairns, J.E., Babu, R.C., Lafitte, H.R. (2009) Identification of physiological traits underlying cultivar differences in drought tolerance in rice and wheat. *Journal of Agronomy and Crop Science* 195, no. 1: 30–46.

Prasad, P.V.V., Pisipati, S.R., Momčilović, I., Ristic, Z. (2011) Independent and combined effects of high temperature and drought stress during grain filling on plant yield and chloroplast expression in spring wheat. *Journal of Agronomy and Crop Science* 197, no. 6: 430–441.

Promkhambut, A., Younger, A., Polthanee, A., Akkasaeng, C. (2010) Morphological and physiological responses of sorghum (*Sorghum bicolor* L. Moench) to waterlogging. *Asian Journal of Plant Sciences* 9, no. 4: 183–193.

Rady, M.M. (2011) Effect of 24-epibrassinolide on growth, yield, antioxidant system and cadmium content of bean (*Phaseolus vulgaris* L.) plants under salinity and cadmium stress. *Scientia Horticulturae* 129, no. 2: 232–237.

Rai, R., Agrawal, M. (2008) Evaluation of physiological and biochemical responses of two rice (*Oryza sativa* L.) cultivars to ambient air pollution using open top chambers at a rural site in India. *The Science of the Total Environment* 407, no. 1: 679–691.

Rai, R., Agrawal, M., Agrawal, S.B. (2007) Assessment of yield losses in tropical wheat using open top chambers. *Atmospheric Environment* 41, no. 40: 9543–9554.

Rai, M.K., Shekhawat, N.S., Harish, Gupta, A.K., Phulwaria, M., Ram, K., Jaiswal, U. (2011a) The role of abscisic acid in plant tissue culture: A review of recent progress. *Plant Cell, Tissue and Organ Culture* 106, no. 2: 179–190.

Rai, R., Agrawal,M., Agrawal, S.B. (2011b) Effects of ambient O_3 on wheat during reproductive development: Gas exchange, photosynthetic pigments, chlorophyll fluorescence, and carbohydrates. *Photosynthetica* 49, no. 2: 285–294.

Ramoliya, P.J., Patel, H.M., Joshi, J.B., Pandey, A.N. (2006) Effect of salinization of soil on growth and nutrient accumulation in seedlings of *Prosopis cineraria*. *Journal of Plant Nutrition* 29, no. 2: 283–303.

Rao, D.N., LeBlanc, F. (1966) Effects of sulfur dioxide on the lichen alga, with special reference to chlorophyll. *Bryologist* 69, no. 1: 69–75.

Rao, K.V.M., Raghavendra, A.S., Reddy, J. eds. (2006) *Physiology and molecular biology of stress tolerance in plants*. Springer: Dordrecht, the Netherlands.

Reddy, A.R., Chaitanya, K.V., Vivekanandan, M. (2004) Drought-induced responses of photosynthesis and antioxidant metabolism in higher plants. *Journal of Plant Physiology* 161, no. 11: 1189–1202.

Redling, K., Elliott, E., Bain, D., Sherwell, J. (2013) Highway contributions to reactive nitrogen deposition: Tracing the fate of vehicular NOx using stable isotopes and plant biomonitors. *Biogeochemistry* 116, no. 1–3: 261–274.

Reid, C.D., Fiscus, E.L. (2008) Ozone and density affect the response of biomass and seed yield to elevated CO_2 in rice. *Global Change Biology* 14: 60–76.

Rezaei, E.E., Siebert, S., Manderscheid, R., Müller, J., Mahrookashani, A., Ehrenpfordt, B., Haensch, J., et al. (2018) Quantifying the response of wheat yields to heat stress: The role of the experimental setup. *Field Crops Research* 217: 93–103.

Rhine, M.D., Stevens, G., Shannon, G., Wrather, A., Sleper, D. (2010) Yield and nutritional responses to waterlogging of soybean cultivars. *Irrigation Science* 28, no. 2: 135–142.

Rizzini, L., Favory, J.J., Cloix, C., Faggionato, D., O'Hara, A., Kaiserli, E., Baumeister, R., et al. (2011) Perception of UV-B by the *Arabidopsis* UVR8 protein. *Science* 332, no. 6025: 103–106.

Rodriguez, M., Canales, A., Borras-Hidalgo, O. (2005). Molecular aspects of abiotic stress in plants. *Biotecnologia Aplicada* 22, no. 1: 1–10.

Rosales-Serna, R., Kohashi-Shibata, J., Acosta-Gallegos, J.A., Trejo-López, C., Ortiz-Cereceres, J., Kelly, J.D. (2004) Biomass distribution, maturity acceleration and yield in drought-stressed common bean cultivars. *Field Crops Research* 85, no. 2–3: 203–211.

Rout, G.R., Das, P. (2002) Rapid hydroponic screening for molybdenum tolerance in rice through morphological and biochemical analysis. *Rostlinna Vyroba* 48, no. 11: 505–512.

Rout, G.R., Samantaray, S., Das, P. (2001) Aluminum toxicity in plants: A review. *Agronomie* 21, no. 1: 3–21.

Roychoudhury, A., Basu, S., Sarkar, S.N., Sengupta, D.N. (2008) Comparative physiological and molecular responses of a common aromatic indica rice cultivar to high salinity with non-aromatic indica rice cultivars. *Plant Cell Reports* 27, no. 8: 1395–1410.

Rubin, B., Penner, D., Saettler, A.W. (1983) Induction of isoflavonoid production in *Phaseolus vulgaris* L. leaves by ozone, sulfur dioxide and herbicide stress. *Environmental Toxicology and Chemistry* 2, no. 3: 295–306.

Ruiz, J.M., Blumwald, E. (2002) Salinity-induced glutathione synthesis in *Brassica napus*. *Planta* 214, no. 6: 965–969.

Rusell, J., Ougham, H., Thomas, H., Waaland, S. (2012) *Molecular life of plants*. Wiley-Blackwell: West Sessex, UK.

Sahai, S., Sharma, C., Singh, S.K., Gupta, P.K. (2011) Assessment of trace gases, carbon and nitrogen emissions from field burning of agricultural residues in India. *Nutrient Cycling in Agroecosystems* 89, no. 2: 143–157.

Sairam, R.K., Rao, K.V., Srivastava, G.C. (2002) Differential response of wheat genotypes to long term salinity stress in relation to oxidative stress, antioxidant activity and osmolyte concentration. *Plant Science* 163, no. 5: 1037–1046.

Sairam, R.K., Kumutha, D., Ezhilmathi, K., Deshmukh, P.S., Srivastava, G.C. (2008) Physiology and biochemistry of waterlogging tolerance in plants. *Biologia Plantarum* 52, no. 3: 401–412.

Saitanis, C.J., Bari, S.M., Burkey, K.O., Stamatelopoulos, D., Agathokleous, E. (2014) Screening of Bangladeshi winter wheat (*Triticum aestivum* L.) cultivars for sensitivity to ozone. *Environmental Science and Pollution Research International* 21, no. 23: 13560–13571.

Salinas, J.U.L.I.O. (2002) Molecular mechanisms of signal transduction in cold acclimation. *Plant Signal Transduction* 38: 116.

Samarah, N.H. (2005) Effects of drought stress on growth and yield of barley. *Agronomy for Sustainable Development* 25, no. 1: 145–149.

Samarah, N.H., Mullen, R.E., Cianzio, S.R., Scott, P. (2006) Dehydrin-like proteins in soybean seeds in response to drought stress during seed filling. *Crop Science* 46, no. 5: 2141–2150.

Samarah, N.H., Alqudah, A.M., Amayreh, J.A., McAndrews, G.M. (2009) The effect of late-terminal drought stress on yield components of four barley cultivars. *Journal of Agronomy and Crop Science* 195, no. 6: 427–441.

Sang, Y., Zheng, S., Li, W., Huang, B., Wang, X. (2001) Regulation of plant water loss by manipulating the expression of phospholipase Dα. *The Plant Journal* 28, no. 2: 135–144.

Sangwan, V., Örvar, B.L., Beyerly, J., Hirt, H., Dhindsa, R.S. (2002) Opposite changes in membrane fluidity mimic cold and heat stress activation of distinct plant MAP kinase pathways. *The Plant Journal* 31, no. 5: 629–638.

Sarkar, A., Agrawal, S.B. (2010) Elevated ozone and two modern wheat cultivars: An assessment of dose dependent sensitivity with respect to growth, reproductive and yield parameters. *Environmental and Experimental Botany* 69, no. 3: 328–337.

Sarkar, R.K. (2016) Stagnant flooding tolerance in rice: Endeavours and Achievements. *NRRI Research Bulletin*, no. 11. ICAR-National Rice Research Institute, Cuttack, Odisha. India, 48.

Sarwat, M., Tuteja, N. (2017) Hormonal signaling to control stomatal movement during drought stress. *Plant Gene* 11: 143–153.

Sarwat, M., Ahmad, P., Nabi, G., Hu, X. (2013) Ca2+ signals: The versatile decoders of environmental cues. *Critical Reviews in Biotechnology* 33, no. 1: 97–109.

Sauter, M. (2013) Root responses to flooding. *Current Opinion in Plant Biology* 16, no. 3: 282–286.

Savitch, L.V., Pocock, T., Krol, M., Wilson, K.E., Greenberg, B.M., Huner, N.P. (2001) Effects of growth under UVA radiation on CO2 assimilation, carbon partitioning, PSII photochemistry and resistance to UVB radiation in *Brassica napus* cv. Topas. *Functional Plant Biology* 28, no. 3: 203–212.

Schäfer, E., Nagy, F. eds. (2006) *Photomorphogenesis in plants and bacteria: Function and signal transduction mechanisms.* Springer: Dordrecht, the Netherlands.

Schumaker, M.A., Bassman, J.H., Robberecht, R., Radamaker, G.K. (1997) Growth, leaf anatomy, and physiology of Populus clones in response to solar ultraviolet-B radiation. *Tree Physiology* 17, no. 10: 617–626.

Schutzendubel, A., Polle, A. (2002) Plant responses to abiotic stresses: Heavy metal-induced oxidative stress and protection by mycorrhization. *Journal of Experimental Botany* 53, no. 372: 1351–1365.

Seckin, B., Sekmen, A.H., Türkan, İ (2009) An enhancing effect of exogenous mannitol on the antioxidant enzyme activities in roots of wheat under salt stress. *Journal of Plant Growth Regulation* 28, no. 1: 12–20.

Sedbrook, J.C., Kronebusch, P.J., Borisy, G.G., Trewavas, A.J., Masson, P.H. (1996) Transgenic AEQUORIN reveals organ-specific cytosolic Ca2+ responses to anoxia in Arabidopsis thaliana seedlings. *Plant Physiology* 111, no. 1: 243–257.

Seghatoleslami, M.J., Kafi, M., Majidi, E. (2008) Effect of drought stress at different growth stages on yield and water use efficiency of five proso millet (*Panicum miliaceum* L.) genotypes. *Pakistan Journal of Botany* 40, no. 4: 1427–1432.

Sehrawat, N., Yadav, M., Bhat, K.V., Sairam, R.K., Jaiwal, P.K. (2015) Effect of salinity stress on mung bean (Vigna radiata (L.) wilczek) during consecutive summer

and spring seasons. *Journal of Agricultural Sciences, Belgrade* 60, no. 1: 23–32.

Semiz, G.D., Ünlükara, A., Yurtseven, E., Suarez, D.L., Telci, I. (2012) Salinity impact on yield, water use, mineral and essential oil content of fennel (Foeniculum vulgare Mill.). *Journal of Agricultural Science* 18: 177–186.

Sensi, S.L., Ton-That, D., Sullivan, P.G., Jonas, E.A., Gee, K.R., Kaczmarek, L.K., Weiss, J.H. (2003) Modulation of mitochondrial function by endogenous Zn2+ pools. *Proceedings of the National Academy of Sciences of the United States of America* 100, no. 10: 6157–6162.

Seregin, I.V., Kozhevnikova, A.D. (2006) Physiological role of nickel and its toxic effects on higher plants. *Russian Journal of Plant Physiology* 53, no. 2: 257–277.

Shah, F.U.R., Nasir, A., Khan, R.M., Jose, R.P.V. (2010) Heavy metal toxicity in plants. In *Plant adaptation and phytoremediation*, 71–97. Springer: Dordrecht, the Netherlands.

Shannon, M.C., Grieve, C.M. (1998) Tolerance of vegetable crops to salinity. *Scientia Horticulturae* 78, no. 1–4: 5–38.

Sharma, Y.K., Davis, K.R. (1994) Ozone-induced expression of stress-related genes in Arabidopsis thaliana. *Plant Physiology* 105, no. 4: 1089–1096.

Sharma, P., Dubey, R.S. (2005) Drought induces oxidative stress and enhances the activities of antioxidant enzymes in growing rice seedlings. *Plant Growth Regulation* 46, no. 3: 209–221.

Sharma, P., Sharma, N., Deswal, R. (2005) The molecular biology of the low-temperature response in plants. *BioEssays* 27, no. 10: 1048–1059.

Sharma, R.K., Devi, S., dan Dhyani, P.P., (2010). Comparative assessment of the toxic effects of copper and cypermethrin using seeds of *Spinacia Oleracea* L. plants. *Tropical Ecology* 51: 375–387.

Shi, G., Yang, L., Wang, Y., Kobayashi, K., Zhu, J., Tang, H., Pan, S., *et al.* (2009) Impact of elevated ozone concentration on yield of four Chinese rice cultivars under fully open-air field conditions. *Agriculture, Ecosystems and Environment* 131, no. 3–4: 178–184.

Shi, H., Chen, J., Wang, K., Niu, J. (2018) A new method and a new index for identifying socioeconomic drought events under climate change: A case study of the East River basin in China. *The Science of the Total Environment* 616–617: 363–375.

Shrivastava, P., Kumar, R. (2015) Soil salinity: A serious environmental issue and plant growth promoting bacteria as one of the tools for its alleviation. *Saudi Journal of Biological Sciences* 22, no. 2: 123–131.

Silva, S. (2012) Aluminum toxicity targets in plants. *Journal of Botany* 2012: 1–8.

Simova-Stoilova, L., Demirevska, K., Kingston-Smith, A., Feller, U. (2012) Involvement of the leaf antioxidant system in the response to soil flooding in two *Trifolium* genotypes differing in their tolerance to waterlogging. *Plant Science* 183: 43–49.

Singh, R.P., Agrawal, M. (2007) Effects of sewage sludge amendment on heavy metal accumulation and consequent responses of Beta vulgaris plants. *Chemosphere* 67, no. 11: 2229–2240.

Singh, R.P., Agrawal, M. (2010a) Effect of different sewage sludge applications on growth and yield of Vigna radiata L. field crop: Metal uptake by plant. *Ecological Engineering* 36, no. 7: 969–972.

Singh, R.P., Agrawal, M. (2010b) Variations in heavy metal accumulation, growth and yield of rice plants grown at different sewage sludge amendment rates. *Ecotoxicology and Environmental Safety* 73, no. 4: 632–641.

Singh, A., Grover, A. (2008) Genetic engineering for heat tolerance in plants. *Physiology and Molecular Biology of Plants* 14, no. 1–2: 155–166.

Singh, S.S., Kumar, P., Rai, A.K. (2006) Ultraviolet radiation stress: Molecular and physiological adaptations in trees. In *Abiotic stress tolerance in plants*, 91–110. Springer, Dordrecht, the Netherlands.

Singh, P., Agrawal, M., Agrawal, S.B. (2009) Evaluation of physiological, growth and yield responses of a tropical oil crop (*Brassica campestris* L. var. Kranti) under ambient ozone pollution at varying NPK levels. *Environmental Pollution* 157, no. 3: 871–880.

Singh, E., Tiwari, S., Agrawal, M. (2010) Variability in antioxidant and metabolite levels, growth and yield of two soybean varieties: An assessment of anticipated yield losses under projected elevation of ozone. *Agriculture, Ecosystems and Environment* 135, no. 3: 168–177.

Singh, A.A., Agrawal, S.B., Shahi, J.P., Agrawal, M. (2014) Investigating the response of tropical maize (*Zea mays* L.) cultivars against elevated levels of O3 at two developmental stages. *Ecotoxicology* 23, no. 8: 1447–1463.

Singh, A.A., Singh, S., Agrawal, M., Agrawal, S.B. (2015) Assessment of ethylene diurea-induced protection in plants against ozone phytotoxicity. *Reviews of Environmental Contamination and Toxicology* 233: 129–184. .

Singh, A., Parihar, P., Singh, R., Prasad, S.M. (2016) An assessment to show toxic nature of beneficial trace metals: Too much of good thing can be bad. *International Journal of Current Multidisciplinary Studies* 2: 141–144.

Soldatini, G.F., Ranieri, A., Gerini, O. (1990) Water balance and photosynthesis in Zea mays L. seedlings exposed to drought and flooding stress. *Biochemie und Physiologie der Pflanzen* 186, no. 2: 145–152.

Soleymani, A. (2017) Light response of barley (*Hordeum vulgare* L.) and corn (*Zea mays* L.) as affected by drought stress, plant genotype and N fertilization. *Biocatalysis and Agricultural Biotechnology* 11: 1–8.

Specht, J.E., Chase, K., Macrander, M., Graef, G.L., Chung, J., Markwell, J.P., Germann, M., Orf, J.H., Lark, K.G. (2001) Soybean response to water. *Crop Science* 41, no. 2: 493–509.

Srivastava, A.K., Rai, A.N., Patade, V.Y., Suprasanna, P. (2013) Calcium signaling and its significance in alleviating salt stress in plants. In *Salt stress in plants*, 197–218. Springer, New York, NY.

Subbaiah, C.C., Sachs, M.M. (2001) Altered patterns of sucrose synthase phosphorylation and localization precede callose induction and root tip death in anoxic maize seedlings. *Plant Physiology* 125, no. 2: 585–594.

Subbaiah, C.C., Zhang, J., Sachs, M.M. (1994) Involvement of intracellular calcium in anaerobic gene expression and survival of maize seedlings. *Plant Physiology* 105, no. 1: 369–376.

Suh, J.Y., Kim, S.J., Oh, T.R., Cho, S.K., Yang, S.W., Kim, W.T. (2016) Arabidopsis Tóxicos en Levadura 78 (AtATL78) mediates ABA-dependent ROS signaling in response to drought stress. *Biochemical and Biophysical Research Communications* 469, no. 1: 8–14.

Surabhi, G.K., Reddy, K.R., Singh, S.K. (2009) Photosynthesis, fluorescence, shoot biomass and seed weight responses of three cowpea (Vigna unguiculata (L.) Walp.) cultivars with contrasting sensitivity to UV-B radiation. *Environmental and Experimental Botany* 66, no. 2: 160–171.

Suzuki, N., Mittler, R. (2006) Reactive oxygen species and temperature stresses: A delicate balance between signaling and destruction. *Physiologia Plantarum* 126, no. 1: 45–51.

Suzuki, N., Rivero, R.M., Shulaev, V., Blumwald, E., Mittler, R. (2014) Abiotic and biotic stress combinations. *New Phytologist* 203, no. 1: 32–43.

Tabur, S., Demir, K. (2010) Role of some growth regulators on cytogenetic activity of barley under salt stress. *Plant Growth Regulation* 60, no. 2: 99–104.

Taiz, L., Zeiger, E. (2002) *Plant physiology*. Sinauer Associates: Sunderland, Massachusetts, U.S.A.

Taiz, L., Zeiger, E. (2006) *Plant Physiology*, 4th edn. Sinauer Associates Inc., Sunderland, Massachusetts, U.S.A.

Taiz, L., Zeiger, E. (2010) *Plant physiology*, 5th edn. Sinauer Associates, Sunderland, Massachusetts, U.S.A.

Takshak, S., Agrawal, S.B. (2014) Secondary metabolites and phenylpropanoid pathway enzymes as influenced under supplemental ultraviolet-B radiation in *Withania somnifera* Dunal, an indigenous medicinal plant. *Journal of Photochemistry and Photobiology B: Biology* 140: 332–343.

Takshak, S., Agrawal, S.B. (2015) Defense strategies adopted by the medicinal plant *Coleus forskohlii* against supplemental ultraviolet-B radiation: Augmentation of secondary metabolites and antioxidants. *Plant Physiology and Biochemistry* 97: 124–138.

Takshak, S., Agrawal, S.B. (2016) Ultraviolet-b radiation: A potent elicitor of phenylpropanoid pathway compounds. *Journal of Scientific Research* 60: 79–96.

Talukder, A.S.M.H.M., McDonald, G.K., Gill, G.S. (2014) Effect of short-term heat stress prior to flowering and early grain set on the grain yield of wheat. *Field Crops Research* 160: 54–63.

Tetteh, R., Yamaguchi, M., Wada, Y., Funada, R., Izuta, T. (2015) Effects of ozone on growth, net photosynthesis and yield of two African varieties of *Vigna unguiculata*. *Environmental Pollution* 196: 230–238.

Tewari, S., Arora, N.K. (2016) Soybean production under flooding stress and its mitigation using plant growth-promoting microbes. In *Environmental stresses in soybean production*, 23–40. Academic Press: London, UK.

Tezara, W., Mitchell, V.J., Driscoll, S.D., Lawlor, D.W. (1999) Water stress inhibits plant photosynthesis by decreasing coupling factor and ATP. *Nature* 401, no. 6756: 914–917.

Thounaojam, T.C., Panda, P., Mazumdar, P., Kumar, D., Sharma, G.D., Sahoo, L., Panda, S.K. (2012) Excess copper induced oxidative stress and response of antioxidants in rice. *Plant Physiology and Biochemistry* 53: 33–39.

Tilbrook, K., Arongaus, A.B., Binkert, M., Heijde, M., Yin, R., Ulm, R. (2013) The UVR8 UV-B photoreceptor: Perception, signaling and response. *The Arabidopsis Book* 11: e0164.

Torres, R.O., Henry, A. (2018) Yield stability of selected rice breeding lines and donors across conditions of mild to moderately severe drought stress. *Field Crops Research* 220: 37–45.

Tripathi, R., Agrawal, S.B. (2012) Effects of ambient and elevated level of ozone on Brassica campestris L. with special reference to yield and oil quality parameters. *Ecotoxicology and Environmental Safety* 85: 1–12.

Tripathi, R., Agrawal, S.B. (2013) Interactive effect of supplemental ultraviolet B and elevated ozone on seed yield and oil quality of two cultivars of linseed (*Linum usitatissimum* L.) carried out in open top chambers. *Journal of the Science of Food and Agriculture* 93, no. 5: 1016–1025.

Tripathy, J.N., Zhang, J., Robin, S., Nguyen, T.T., Nguyen, H.T. (2000) QTLs for cell-membrane stability mapped in rice (Oryza sativa L.) under drought stress. *Theoretical and Applied Genetics* 100, no. 8: 1197–1202.

Tsay, C.C., Wang, L.W., Chen, Y.R. (1995) Plant in response to copper toxicity. *Taiwania* 40: 173–181.

Tsuji, H., Nakazono, M., Saisho, D., Tsutsumi, N., Hirai, A. (2000) Transcript levels of the nuclear-encoded respiratory genes in rice decrease by oxygen deprivation: Evidence for involvement of calcium in expression of the alternative oxidase 1a gene. *FEBS Letters* 471, no. 2–3: 201–204.

Turk, K.J., Hall, A.E., Asbell, C.W. (1980) Drought adaptation of cowpea. I. Influence of drought on seed yield 1. *Agronomy Journal* 72, no. 3: 413–420.

Ulm, R., Ichimura, K., Mizoguchi, T., Peck, S.C., Zhu, T., Wang, X., Shinozaki, K., et al. (2002) Distinct regulation of salinity and genotoxic stress responses by Arabidopsis MAP kinase phosphatase 1. *The EMBO Journal* 21, no. 23: 6483–6493.

UN DESA. (2017) United Nations Department of Economic and Social Affairs. www ... Accessed on 10th November, 2018.

Vardar, F., Ünal, M. (2007) Aluminum toxicity and resistance in higher plants. *Advances in Molecular Biology* 1: 1–12.

Vaultier, M.N., Cantrel, C., Vergnolle, C., Justin, A.M., Demandre, C., Benhassaine-Kesri, G., Çiçek, D., et al. (2006) Desaturase mutants reveal that membrane rigidification acts as a cold perception mechanism upstream of the diacylglycerol kinase pathway in Arabidopsis cells. *FEBS Letters* 580, no. 17: 4218–4223.

Volz-Thomas, A., Mihelic, D. (1990) Ozonproduktion in Reinluftgebieten. In *Einfluß von Schadstoff-Konzentrationen, Gesellschaft Österreichischer Chemiker (Hrsg.), Tagungsband zum Symposium "Bodennahes Ozon"* (Vol. 11). Salzburg, Austria.

Wahid, A., Maggs, R.S.R.A., Shamsi, S.R. A., Bell, J.N.B., Ashmore, M.R. (1995) Air pollution and its impacts on wheat yield in the Pakistan Punjab. *Environmental Pollution* 88, no. 2: 147–154.

Wahid, A. (2006a) Influence of atmospheric pollutants on agriculture in developing countries: a case study with three new wheat varieties in Pakistan. *Science of the Total Environment* 371, no. 1–3: 304–313.

Wahid, A. (2006b) Productivity losses in barley attributable to ambient atmospheric pollutants in Pakistan. *Atmospheric Environment* 40, no. 28: 5342–5354.

Wahid, A., Gelani, S., Ashraf, M., Foolad, M.R. (2007) Heat tolerance in plants: An overview. *Environmental and Experimental Botany* 61, no. 3: 199–223.

Wang, X., Komatsu, S. (2017) Proteomic analysis of calcium effects on soybean root tip under flooding and drought stresses. *Plant and Cell Physiology* 58, no. 8: 1405–1420.

Wang, W., Vinocur, B., Altman, A. (2003) Plant responses to drought, salinity and extreme temperatures: Towards genetic engineering for stress tolerance. *Planta* 218, no. 1: 1–14.

Wang, S.L., Lin, C.Y., Guo, Y.L., Lin, L.Y., Chou, W.L., Chang, L.W. (2004) Infant exposure to polychlorinated dibenzo-p-dioxins, dibenzofurans and biphenyls (PCDD/Fs, PCbs)—Correlation between prenatal and postnatal exposure. *Chemosphere* 54, no. 10: 1459–1473.

Wang, X., Li, W., Li, M., Welti, R. (2006) Profiling lipid changes in plant response to low temperatures. *Physiologia Plantarum* 126, no. 1: 90–96.

Wang, Q., Wu, J., Lei, T., He, B., Wu, Z., Liu, M., Mo, X., et al. (2014) Temporal-spatial characteristics of severe drought events and their impact on agriculture on a global scale. *Quaternary International* 349: 10–21.

Wani, P.A., Khan, M.S., Zaidi, A. (2007) Impact of heavy metal toxicity on plant growth, symbiosis, seed yield and nitrogen and metal uptake in chickpea. *Australian Journal of Experimental Agriculture* 47, no. 6: 712–720.

Wani, P.A., Khan, M.S., Zaidi, A. (2008) Effects of heavy metal toxicity on growth, symbiosis, seed yield and metal uptake in pea grown in metal amended soil. *Bulletin of Environmental Contamination and Toxicology* 81, no. 2: 152–158.

Wei, Q., Ji-wang, Z., Kong-jun, W., Peng, L., Shu-ting, D. (2010) Effects of drought stress on the grain yield and root physiological traits of maize varieties with different drought tolerance. *Yingyong Shengtai Xuebao* 21, no. 1: 48–52.

Wellburn, A.R., Majernik, O., Wellburn, F.A.M. (1972) Effects of SO2 and NO2 polluted air upon the ultrastructure of chloroplasts. *Environmental Pollution (1970)* 3, no. 1: 37–49.

Wery, J., Silim, S.N., Knights, E.J., Malhotra, R.S., Cousin, R. (1994) Screening techniques and sources of tolerance to extremes of moisture and air temperature in cool season food legumes. In *Expanding the production and use of cool season food legumes*, 439–456. Springer: Dordrecht, the Netherlands.

Wilhite, D.A. (2000) *Drought as a natural hazard: Concepts and definitions*. University of Nebraska: Lincoln, NE.

Wu, X., Tang, Y., Li, C., McHugh, A.D., Li, Z., Wu, C. (2018) Individual and combined effects of soil waterlogging and compaction on physiological characteristics of wheat in south western China. *Field Crops Research* 125: 163–172.

Xu, M., Ma, H., Zeng, L., Cheng, Y., Lu, G., Xu, J., Zhang, X., et al. (2015) The effect of waterlogging on yield and seed quality at the early flowering stage in *Brassica napus* L. *Field Crops Research* 180: 238–245.

Yadav, S.K. (2010) Heavy metals toxicity in plants: An overview on the role of glutathione and phytochelatins in heavy metal stress tolerance of plants. *South African Journal of Botany* 76, no. 2: 167–179.

Yan, B., Dai, Q., Liu, X., Huang, S., Wang, Z. (1996) Flooding-induced membrane damage, lipid oxidation and activated oxygen generation in corn leaves. *Plant and Soil* 179, no. 2: 261–268.

Yeo, A.R., Flowers, T.J. eds. (2008) *Plant solute transport*. Blackwell publications, New Jersey, U.S.A.

Yin, X., Sakata, K., Nanjo, Y., Komatsu, S. (2014) Analysis of initial changes in the proteins of soybean root tip under flooding stress using gel-free and gel-based proteomic techniques. *Journal of Proteomics* 106: 1–16.

Yordanova, R.Y., Christov, K.N., Popova, L.P. (2004) Antioxidative enzymes in barley plants subjected to soil flooding. *Environmental and Experimental Botany* 51, no. 2: 93–101.

Youn, J.T., Van, K.J., Lee, J.E., Kim, W.H., Yun, H.T., Kwon, Y.U., Ryu, Y.H., et al. (2008) Waterlogging effects on nitrogen accumulation and N2 fixation of supernodulating soybean mutants. *Journal of Crop Science and Biotechnology* 11: 111–118.

Yousuf, P.Y., Hakeem, K.U.R., Chandna, R., Ahmad, P. (2012) Role of glutathione reductase in plant abiotic stress. In *Abiotic stress responses in plants*, 149–158. Springer, New York, NY.

Yu, Z.L., Zhang, J.G., Wang, X.C., Chen, J. (2008) Excessive copper induces the production of reactive oxygen species, which is mediated by phospholipase D, nicotinamide adenine dinucleotide phosphate oxidase and antioxidant systems. *Journal of Integrative Plant Biology* 50, no. 2: 157–167.

Yuan, J.S., Himanen, S.J., Holopainen, J.K., Chen, F., Stewart Jr., C.N. (2009) Smelling global climate change: Mitigation of function for plant volatile organic compounds. *Trends in Ecology and Evolution* 24, no. 6: 323–331.

Yurekli, F.., Porgali, Z.B. (2006) The effects of excessive exposure to copper in bean plants. *Acta Biologica Cracoviensia Series Botanica* 48, no. 2: 7–13.

Zaidi, P.H., Rafique, S., Singh, N.N. (2003) Response of maize (Zea mays L.) genotypes to excess soil moisture stress: Morpho-physiological effects and basis ssof tolerance. *European Journal of Agronomy* 19, no. 3: 383–399.

Zargar, S.M., Nagar, P., Deshmukh, R., Nazir, M., Wani, A.A., Masoodi, K.Z., Agrawal, G.K., et al. (2017) Aquaporins as potential drought tolerance inducing proteins: Towards instigating stress tolerance. *Journal of Proteomics* 169: 233–238.

Zhang, W., Qin, C., Zhao, J., Wang, X. (2004) Phospholipase Dα1-derived phosphatidic acid interacts with ABI1 phosphatase 2C and regulates abscisic acid signaling. *Proceedings of the National Academy of Sciences of the United States of America* 101, no. 25: 9508–9513.

Zhang, A., Jiang, M., Zhang, J., Tan, M., Hu, X. (2006) Mitogen-activated protein kinase is involved in abscisic acid-induced antioxidant defense and acts downstream of reactive oxygen species production in leaves of maize plants. *Plant Physiology* 141, no. 2: 475–487.

Zhang, J., Yang, J., An, P., Ren, W., Pan, Z., Dong, Z., Han, G., *et al.* (2017a) Enhancing soil drought induced by climate change and agricultural practices: Observational and experimental evidence from the semiarid area of northern China. *Agricultural and Forest Meteorology* 243: 74–83.

Zhang, L., Zhang, H., Zhang, Q., Li, Y., Zhao, J. (2017b) On the potential application of land surface models for drought monitoring in China. *Theoretical and Applied Climatology* 128, no. 3–4: 649–665.

Zhang, M., Wang, L.F., Zhang, K., Liu, F.Z., Wan, Y.S. (2017c) Drought-induced responses of organic osmolytes and proline metabolism during pre-flowering stage in leaves of peanut (*Arachis hypogaea* L.). *Journal of Integrative Agriculture* 16, no. 10: 2197–2205.

Zhang, W., Feng, Z., Wang, X., Liu, X., Hu, E. (2017d) Quantification of ozone exposure-and stomatal uptake-yield response relationships for soybean in Northeast China. *Science of the Total Environment* 599–600: 710–720.

Zhao, D., Reddy, K.R., Kakani, V.G., Read, J.J., Sullivan, J.H. (2003) Growth and physiological responses of cotton (*Gossypium hirsutum* L.) to elevated carbon dioxide and ultraviolet-B radiation under controlled environmental conditions. *Plant, Cell and Environment* 26, no. 5: 771–782.

Zhao, G., Xu, H., Zhang, P., Su, X., Zhao, H. (2017) Effects of 2,4-epibrassinolide on photosynthesis and Rubisco activase gene expression in *Triticum aestivum* L. seedlings under a combination of drought and heat stress. *Plant Growth Regulation* 81, no. 3: 377–384.

Zhen, A., Bie, Z., Huang, Y., Liu, Z., Lei, B. (2011) Effects of salt-tolerant rootstock grafting on ultrastructure, photosynthetic capacity, and H2O2-scavenging system in chloroplasts of cucumber seedlings under NaCl stress. *Acta Physiologiae Plantarum* 33, no. 6: 2311–2319.

Zhu, Y., Zhao, Z., Duan, C., Zheng, Y., Wu, R. (2015) Contribution rate to yield per spike of different green organs of winter wheat under ozone stress. *Acta Ecologica Sinica* 35, no. 3: 10–16.

Zhu, D., Wei, H., Guo, B., Dai, Q., Wei, C., Gao, H., Hu, Y., et al. (2017) The effects of chilling stress after anthesis on the physicochemical properties of rice (*Oryza sativa* L) starch. *Food Chemistry* 237: 936–941.

Zouzoulas, D., Koutroubas, S.D., Vassiliou, G., Vardavakis, E. (2009) Effects of ozone fumigation on cotton (*Gossypium hirsutum* L.) morphology, anatomy, physiology, yield and qualitative characteristics of fibers. *Environmental and Experimental Botany* 67, no. 1: 293–303.

Zribi, L., Fatma, G., Fatma, R., Salwa, R., Hassan, N., Néjib, R.M. (2009) Application of chlorophyll fluorescence for the diagnosis of salt stress in tomato "*Solanum Lycopersicum* (variety Rio Grande)". *Scientia Horticulturae* 120, no. 3: 367–372.

Zuk-Golaszewska, K., Upadhyaya, M.K., Golaszewski, J. (2011) The effect of UV-B radiation on plant growth and development. *Plant, Soil and Environment* 49, no. 3: 135–140.

2 Impacts of Climate Change on Crop Production, with Special Reference to Southeast Asia

Jong Ahn Chun, Christianne M. Aikins, Daeha Kim,
Sanai Li, Wooseop Lee, and Eun-Jeong Lee

CONTENTS

2.1 INTRODUCTION

As crop production is heavily reliant on hydroclimatic variables, it is widely recognized that climate change presents a significant risk to agriculture globally. This challenge is particularly salient in Southeast Asia, where agriculture is a major source of livelihood, with approximately 114 million ha of land devoted to the production of five major crops (rice, maize, oil palm, natural rubber, and coconut), yet it is also one of the regions most vulnerable to the impacts of climate change (ADB, 2009). How plants will cope with these new environmental stresses and what this will mean for food security are central concerns from the local to regional scale. Rice (*Oryza sativa* L.), as the staple food of about 557 million people (Manzanilla et al., 2011), cultivated on over 50% of agricultural land (excepting Myanmar, at roughly 45%), and representing 20–48% of the GDP in countries on the Indochinese peninsula (Manzanilla et al., 2011; FAO, n.d.; WBD, 2005), is undoubtedly the most significant crop in the region. Because of its climate vulnerability – rice is still primarily cultivated in rainfed paddy fields – as well as economic and cultural significance, an investigation on rice provides critical insights to the complexities of crop production under

a changing climate. Climate change has been and will continue to be a critical factor affecting rice productivity on the Indochinese peninsula, with adverse impacts already increasing (Masutomi et al., 2009), necessitating urgent action. Crop modeling can help reveal the most advantageous adaptation strategies under climate change scenarios by assessing the impacts of climate change on crop production. (e.g., Masutomi et al., 2009; Challinor et al., 2004; Kim et al., 2013; Lehmann et al., 2013). These insights can be critical for agricultural management practices for both farmers and policymakers.

It is first essential to generate key climate variables using climate simulation models. Global Climate Models (GCM) reveal large-scale patterns of change, and outputs are relatively coarse, at approximately 150–300 km by 150–300 km. To meet agricultural purposes, these models can be downscaled through Regional Climate Downscaling (RCD) techniques to the local or regional level, generating important climate change scenarios. In this chapter, higher resolution precipitation and temperature datasets over the Indochinese peninsula are generated through the Coordinated Regional Downscaling Experiment (CORDEX)-East Asia. Understanding projected trends in temperature and precipitation change is critical to then investigating the impact on crop production using crop models.

Due to the myriad of factors concerning crop growth and productivity, a number of crop models exist on a range of spatial scales, each with relative advantages and disadvantages, that can be used to investigate climate change impacts. An understanding of the evolution of crop models can provide insight into their functional limitations in the simulation of environmental stresses on crops. For example, Masutomi et al. (2009) simulated changes in rice yield in East and South Asia under climate change using a global-scale crop model, M-GAEZ (Global Agro-Ecological Zones). However, as detailed soil and crop physiological process are not considered in the M-GAEZ model, inaccuracies can arise through this limitation. One way to help mitigate such limitations is to balance crop models on various scales (field-, regional-, and global-scales). This chapter applies two crop models using the climate variables generated from CORDEX: the General Large-Area Model for annual crops (GLAM), a regional-scale model, and the CERES-Rice model, a field-scale crop model. GLAM has been widely used to simulate crop yields, including China (Li et al., 2010) and India (Challinor et al., 2004; Challinor et al., 2005). While the GLAM model does not consider the effects of weeds, pests and diseases, and air pollution (Challinor et al., 2009), the CERES-Rice model can simulate nutrients and water processing in the soil at field-scale (Vaghefi et al., 2013), indicating trade-offs to both methods. National- and farmer-level adaptation strategies may thus be developed by combining the advantages from regional- and field-scale crop models.

Climate change scenarios in Southeast Asia using higher resolution of climate projections provided by a regional climate downscaling method can reveal important insights, not only to future conditions but also the potential effectiveness of adaptation techniques. These methods enable a better understanding of climate change on the Indochinese peninsula and its impacts on rice yields, leading to the enhancement of food security

2.2 CLIMATE CHANGE FOR SOUTHEAST ASIA

2.2.1 CLIMATE MODELS

Global climate simulations indicate large-scale patterns of change associated with natural and anthropogenic climate forcing. However, they may not adequately capture the effects of narrow mountain ranges, complex land/water interactions, or regional variations in land use.

Therefore, the outputs from GCMs may be relatively coarse for agricultural applications at the local and regional scales. High-resolution precipitation datasets are required for detailed agricultural and hydrological studies. The spatial resolution of the historical and climate change scenario datasets generated from the GCM modeling runs is approximately 150–300 km by 150–300 km.

The availability of spatially distributed data information, such as meteorological, hydrological, and topographical (soil type and land cover characteristics) gridded datasets, has contributed to the development of modeling for natural resources and environmental applications, including agricultural studies, to better reflect the spatial movement of hydrologic components. One of the major factors in distributed hydrologic modeling is spatial and temporal precipitation inputs, which can be obtained through radar precipitation data (Beven, 2002; Boyle et al., 2001; Haddeland et al., 2002). However, the coarse resolution (over 4 km by 4 km in watershed modeling) of this radar precipitation data often fails to capture the detailed rainfall-runoff processes from various rainfall distributions for environmental modeling (e.g., Seo and Smith, 1996; Carpenter et al., 1999a,b; Georgakakos, 2000; Carpenter et al., 2001; Krajewski and Smith, 2002). In this chapter, the RCD technique was used to generate higher resolution precipitation datasets.

The Coordinated Regional Downscaling Experiment (CORDEX) initiative, established by the World Climate Research Programme (WCRP) in 2009, generated regional climate change projections for all global land areas in step with the timeline of the Fifth Assessment Report (AR5). The CORDEX initiative aims to provide an interface for climate change projects and programs, to establish a framework for evaluating coordinated modeling experiments, and to project future climate. CORDEX-EA (East Asia), the East-Asian branch of the CORDEX initiative, aims at improving coordination of international efforts in RCD research and includes the Indochinese peninsula and Northeast Asian countries (http://cordex-ea.climate.go.kr/).

2.2.2 REGIONAL CLIMATE DOWNSCALING FOR THE INDOCHINESE PENINSULA

Five regional climate models developed by several institutes participated in CORDEX-EA: (1) the Hadley Centre Global Environment Model version 3 (HadGEM3-RA), an atmospheric regional climate model developed by the National Institute of Meteorological Research (NIMR), (2) the Regional

Climate Model version 4 (RegCM4) developed by the International Centre for Theoretical Physics (ICTP), (3) the Seoul National University Meso-scale Model version 5 (SNU-MM5), (4) the Seoul National University Weather Research and Forecasting model (SNU-WRF), and (5) the YonSei University-Regional Spectral Model (YSU-RSM). These models have different dimensions, both spatially and temporally. For example, the numbers of days in a year vary from the calendar year, 365 days or 360 days. Of the five CORDEX-EA models, only HadGEM3-RA, RegCM4, and SNU-WRF are used here. The dynamic frameworks and physical schemes of the models used in the project are summarized in Table 2.1.

The following variables and periods were used: daily mean, maximum and minimum Near-Surface Air Temperature, surface downwelling shortwave radiation (RSDS) and precipitation (PR) during the historical period (1980–2005), and the Representative Concentration Pathways (RCP) 4.5 and 8.5 scenarios (2006–2049). The horizontal resolution is 0.44° by 0.44°. The domain of each model in CORDEX-EA differed but included the Indochinese peninsula. The original dataset

was interpolated from 5–30°N, 90–110°E, and a horizontal resolution of 0.25° by 0.25°.

The annual mean spatial patterns for five climate variables during the historical period of HadGEM3-RA, RegCM4, and SNU-MM5 were compared over Southeast Asia. Similarities include a high temperature over the Indochinese peninsula across models, while the Daily Maximum Near-Surface Air Temperature shows a clear distinction between them. HadGEM3-RA and SNU-MM5 simulate much higher temperatures than RegCM4 over the Indochinese peninsula.

In areas where there are precipitable clouds, downwelling shortwave radiation to the surface is reduced, resulting in RSDS patterns that have a partly negative relationship to PR patterns. However, RSDS and PR patterns differ from each other. Particularly, RSDS over Cambodia is higher than over Vietnam and Thailand. Over the Andaman Sea and the Borneo Sea, all three models simulate more rainfall than over the peninsula. In general, the spatial characteristics of climate variables over Cambodia indicate that there is more shortwave radiation and higher temperature than over Thailand and Vietnam.

TABLE 2.1
Main Characteristics of the Three RCMs Used in This Chapter

Parameters	HadGEM3-RA	RegCM4	SNU-MM5
No. of grid points (lat × lon)	183 × 220	197 × 243	197 × 233
Vertical levels	38	σ-18	σ-24
Dynamic framework	Nonhydrostatic	Hydrostatic	Nonhydrostatic
PBL scheme	Nonlocal mixing scheme for unstable layers (Lock et al., 2000). Local Richardson number scheme for stable layers (Smith, 1990)	Holtslag (Holtslag et al., 1990)	YSU (Cha et al., 2008)
Convective scheme	Revised mass flux scheme from Gregory and Rowntree (1990) including triggering of deep and shallow cumulus convection based on the boundary layer scheme, parameterized entrainment/detrainment rates for shallow convection (Grant and Brown, 1999), and the treatment of momentum transports by deep and shallow convection based on an eddy viscosity model	MIT-Emanuel (Emanuel, 1991)	Kain-Fritch (Kain and Fritsch, 1990)
Land surface	MOSES-II (Essery et al., 2001); nine surface title types plus coastal tiling	NCAR CLM 3.5 (Oleson et al., 2008)	NCAR CLM3 (Bonan et al., 2002)
Radiation	General 2-stream radiation (Cusack et al., 1999; Edwards and Slingo, 1996)	NCAR CCM3 (Kiehl et al., 1996)	NCAR CCM2 (Briegleb, 1992)
Simulation period	Current: 1950–2005 Future: 2006–2100 (RCP4.5/8.5)	Current: 1979–2005 Future: 2006–2050 (RCP4.5/8.5)	Current: 1980–2005 Future: 2006–2049 (RCP4.5/8.5)

PRCP (RCP8.5-HIS)

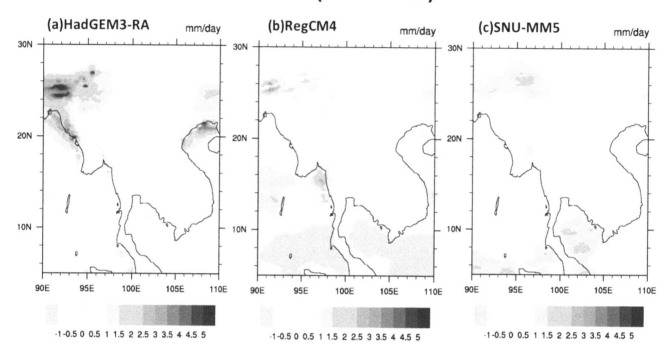

FIGURE 2.1 The difference between the RCP8.5 scenario and historical precipitation for the (a) HadGEM3-RA, (b) RegCM4, and (c) SNU-MM5 models over the Indochinese peninsula. The unit is mm per day.

Future changes of climate variables over the Indochinese peninsula were investigated between the RCP 8.5 scenario and the historical mean. A clear change in precipitation patterns over the Indochinese peninsula is not shown (Figure 2.1). HadGEM3-RA and RegCM4 show little decrease or increase and SNU-MM5 shows a slight increase in precipitation. Over the ocean, there is a robust decrease shown by RegCM4 and a slight increase shown by SNU-MM5. However, HadGEM3-RA showed great changes in temperature. While the Daily Maximum Near-Surface Air Temperature appeared to increase to above 3K near Cambodia in the HadGEM3-RA simulations, there was only a small increasing signal in RegCM4 and SNU-MM5 (Figure 2.2). Large changes in the temperature fields seem to be a result of RSDS. It should be noted that the increasing signal for the Daily Minimum Near-Surface Air Temperature over the Indochinese peninsula is greater than that of Daily Maximum Near-Surface Air Temperature in RegCM4 and SNU-MM5 (Figure 2.3). This means that the daily temperature range may become reduced.

Similarly, the projected changes in average total seasonal precipitation and temperature for the Indochinese peninsula from HadGEM3-RA, RegCM4, and SNU-MM5 were investigated by Li et al. (2017) at a 0.25° by 0.25° scale. Relative to the baseline 1990s (1991–2000), a small increase (0–15%) in projected precipitation from May to October for the RCP4.5 and RCP8.5 climate scenarios is projected for the 2020s, while a larger increase (0–30%) relative to the baseline is seen in the 2040s (Figure 2.4a–d). In the 2020s, a slightly greater increase (5–15%) in rainfall was predicted for RCP8.5 compared to the RCP4.5 scenario (0–5%), with the largest increases found in the north of Vietnam, Thailand, and northwest of Myanmar. In the 2040s, a large increase (10–15%) in rainfall is projected under RCP4.5 in Thailand and a portion of north Myanmar and Vietnam. In the RCP8.5 scenario, the greatest increase (15–30%) in precipitation is projected in the north of Myanmar, with a small decrease in precipitation (−5%) in a portion of the south of Vietnam, showing increased spatial variation compared to the 2020s. In the examination of the projected average temperature from May to October, Li et al. (2017) found a slight average increase (0.5–1.5°C) for RCP4.5 and RCP8.5 across the Indochinese peninsula compared to the baseline in the 2020s, with the 2040s showing a larger increases, 1.5°C and 2°C for RCP4.5 and RCP8.5, respectively (Figure 2.4e–h). In particular, by the 2020s, Thailand had the largest increase in seasonal temperatures in the RCP4.5 scenario. By the 2040s, the largest temperatures increases are seen in Thailand and Vietnam, while the least change is seen in the northwest of Myanmar for the RCP8.5 scenario.

Max_temp (RCP8.5-HIS)

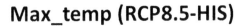

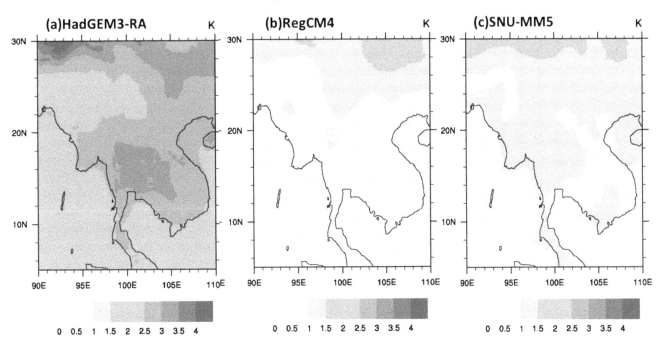

FIGURE 2.2 The difference between the RCP8.5 scenario and the historical maximum surface temperature for the (a) HadGEM3-RA, (b) RegCM4, and (c) SNU-MM5 models over the Indochinese peninsula. The unit is K.

Min_temp (RCP8.5-HIS)

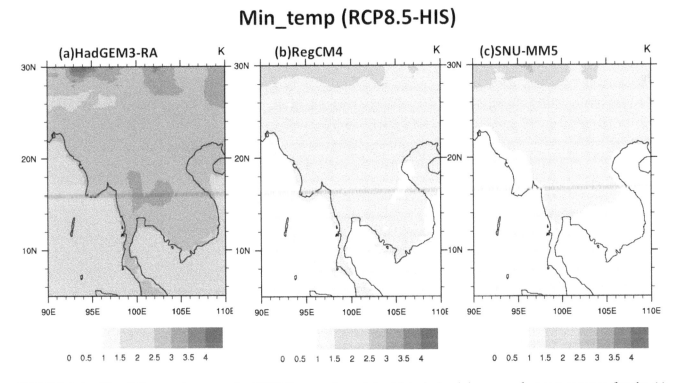

FIGURE 2.3 The difference between the RCP8.5 scenario and the historical minimum surface temperature for the (a) HadGEM3-RA, (b) RegCM4, and (c) SNU-MM5 models over the Indochinese peninsula. The unit is K.

The use of CORDEX-EA models, HadGEM3-RA, RegCM4, and SNU-WR, can thus bring insight to predicted changes of key hydroclimatic variables, essential for agricultural activities. These detailed climate change scenarios provide the necessary basis for an assessment of the impacts on rice yields in the region through the use of crop models.

2.3 USE OF CROP MODELS

2.3.1 A BRIEF HISTORY OF CROP MODELS

The earliest attempt to model agricultural production was the combination of physical and biological principles by de Wit of Wageningen University (Jones et al., 2016). Duncan et al. (1967) performed another pioneer study for modeling canopy photosynthesis that was later linked to crop-specific simulation models for corn, cotton, and peanut (Duncan, 1972). While most agricultural scientists were skeptical about the quantitative and deterministic approaches during the early period of crop models, the large import of wheat by the Soviet Union, which caused an unforeseen price rise and global wheat shortage (Pinter et al., 2003), boosted the development of crop models (Jones et al., 2016).

The United States government funded new research programs to create crop models that incorporate remote sensing information in global-scale yield predictions for internationally traded crops. These programs led to the development of CERES-Wheat (Ritchie and Otter, 1984) and CERES-Maize (Jones and Kiniry, 1986) that are now included in the DSSAT suite (Jones et al., 2003).

Crop models have evolved with programs to overcome challenges in agricultural productivity (e.g., the pests and diseases of plantation crops in Malaysia; Conway, 1987). Importantly, widespread availability of personal computers and the World Wide Web in the 1980s led to many innovations in computing skills, new communication tools, and improved data accessibility. With the emergence of the open source concept, sophisticated crop modeling systems, such as APSIM (http://www.apsim.info/) and DSSAT (https://dssat.net/), are now freely available online. Many reliable crop models have been developed with diverse model complexity since the 1980s (e.g., DDSAT, APSIM, EPIC, Oryza, CropSyst, AquaCrop, and SWAP among numerous examples) and used for various purposes from field crop management (e.g., He et al., 2012) to supporting decision/policy making (e.g., Kim and Kaluarachchi, 2016; García-Vila and Fereres, 2012).

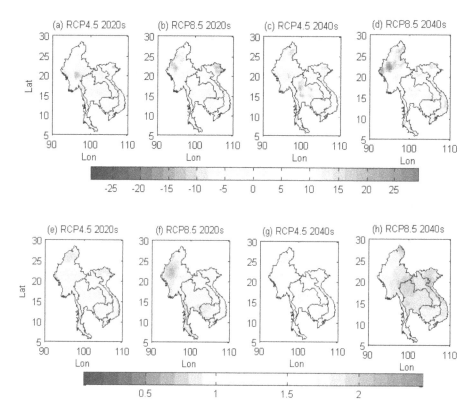

FIGURE 2.4 Changes in the average rainfall (%, a to d) and changes in average temperature (°C, e to h) from May to October from HadGEM3, YSU-RSM and RegCM4 for the 2020s and 2040s under the RCP4.5 and RCP8.5 scenarios, relative to the baseline 1990s (1991–2000) (Li et al., 2017). Reprinted with permission, *International Journal of Climatology*, John Wiley & Sons, Inc.

2.3.2 Types of Crop Models

The simplest method to predict crop productivity would be to develop statistical models using multi-year crop yield records and explanatory variables (e.g., weather conditions). Thompson (1969), for instance, performed a regression analysis between crop yields and weather conditions to find a general trend in regional crop yields. This approach is suitable for yield predictions at a regional or a national scale in which yield survey data are typically collected. It assumes that the regression model can predict regional crop yields in upcoming years under different weather conditions. However, the weakness of statistical approaches is that the models are only valid within the ranges of samples; yield estimates can be significantly biased if extrapolated beyond the sampled ranges (Jones et al., 2016). Indeed, statistical models are unable to represent unobserved environmental changes (e.g., climate change) or to consider various natural and human-made influences (e.g., fertilizer managements, pests and diseases, and changing soil conditions).

In contrast to statistical models, dynamic models have been widely used to simulate synthetic crop responses to environmental changes and management practices (Wallach et al., 2014). The dynamic crop models have functions to imitate actual crop behaviors in response to the external drivers. Prominent examples include the models in the DSSAT suite (Jones et al., 2003), APSIM (Keating et al., 1991), CropSyst (Stöckle et al., 2003), AquaCrop (Steduto et al., 2009), EPIC (Williams et al., 1989), and Oryza (Penning de Vries et al., 1989) among many others. Di Paola et al. (2016) provided a comprehensive review on more than 50 dynamic crop models in regards of their types, biogeochemical sub-components, spatial scales applied, and temporal scales, suggesting major advantages of dynamic crop models: (1) to support farm-level decision-making (e.g., selection of appropriate cultivars and practices under climatic and disease risks), (2) to better understand complex processes interacting within the atmosphere-plant-soil system in consideration of human-made influences, and (3) to guide high-level policymakers towards suitable planning for agro-environmental systems under climatic change.

Di Paola et al. (2016) classified the dynamic crop models into two categories of predictive and explanatory models. The predictive models usually aim to provide yield estimates under given environmental conditions. In some cases, the predicted yields are used for high-level agricultural planning and decision-making, and thus require high predictive performance with robust simulations. For practical uses, the predictive models often embed empirical functions in their structures that require a relatively small number of parameters for robust predictions (e.g., the basal crop coefficient and the normalized water productivity in AquaCrop). On the other hand, the explanatory models reflect specific interactions between plants and environments. These types of models are usually process-oriented to describe specific components in the atmosphere-plant-soil system with substantial details (e.g., crop responses to light and nutrient availability in CERES-Rice), thereby requiring many inputs and parameters. Some models are of intermediate complexity, falling between the predictive and the explanatory models, to balance model complexity and predictive performance (e.g., EPIC). However, mechanistic (process-oriented) models are not always more complex than predictive models based on empirical functions. Some predictive (empirical) models may simulate more system processes than mechanistic models with relatively simple structures (Di Paola et al., 2016).

For appropriate selections of crop models, Di Paola et al. (2016) suggested screening crop models with the model types (i.e., empirical, mechanistic, and hybrid) suitable for modeling purposes. Then, it is recommended to consider the biogeochemical processes included in crop models. For example, when defining crop yield response to water in arid regions where water availability is the dominant factor determining agricultural productivity, a simple empirical model (e.g., AquaCrop) is an attractive option for acceptable and robust predictive performance for policy development. However, this choice is likely unsuitable for regions where light availability limits crop yield. In this case, CropSyst with similar complexity to AquaCrop may be a better choice to consider both water and light availability in crop growth simulations. In the case that the water-nitrogen-plant interactions should be simulated in regions under deficient water and fertility conditions, modelers need to select more complex models including relevant biogeochemical components (e.g., EPIC). For appropriate model selections, specific information on dynamic crop models is comprehensively provided in Table S1 of Di Paola et al. (2016).

2.3.3 Spatial Scales of Crop Growth Modeling

The scope of the system analysis determines the spatial scales of crop models. Jones et al. (2016) linked spatial scales of crop growth modeling to relevant stakeholders and decisions (Jones et al., 2016). For example, one may prefer a field-scale model for the best management practices at a farm level (e.g., Geerts and Raes, 2009), while a regional-scale analysis would be efficient for a continental-scale assessment of climate change impacts (e.g., Chun et al., 2016). Since the scales of crop modeling are determined by questions being asked, and decisions and policies to be supported, stakeholder participation plays a meaningful role in setting modeling objectives and refining simulations (Thornton and Herrero, 2001).

Typical field-scale models consider the vertical heterogeneity of soil properties yet assume homogeneous climatic and soil conditions across the field (i.e., horizontal homogeneity). Jones et al. (2016) referred to this type of model as point-scale models that can be upscaled to national or even continental levels by aggregating model outputs. Hence, when required input data are available with sufficient computing power, field-scale models can be of higher flexibility in spatial scales than farm- and regional-level models (e.g., Elliott et al., 2014). However, Palosuo et al. (2011) raised concerns about field-scale crop growth modeling applied on a larger scale, highlighting that modelers often use improper parameterization and ignore conditions under which crop models were formulated. It is worth noting that a crop model may require inputs representing characteristics within the system boundaries rather than data for capturing climatological variability outside of the system (Ahuja and Ma, 2002); hence, upscaling a field-level model would result in significant uncertainty.

When using crop models at broader scales than field-level ones, modelers may define relatively large agricultural areas for simulating lumped crops' behaviors. The models produce lumped estimates of crop yields forced by average climatological and soil inputs within the specified areas. They are useful to evaluate policy and decision options at regional or broader levels (e.g., regional strategies for adaptation to climate change). Recently, the farm- or regional-level models often simulated biophysical processes under varying socioeconomic and environmental conditions (Jones et al., 2016); this thereby promoted understanding the interactions between agro-ecosystems and economic policies. As field-scale models, smaller-scale models can be adapted for larger-scale estimates by aggregating the simulations. However, due to their simplicity, crop models with large scales often have an inability to consider specific factors affecting crop productivity such as weeds, and pest and diseases, leading to a significant discrepancy between modeled and actual yields (e.g., GLAM).

2.3.4 APPLICATION OF CROP MODELS

Various crop models have been widely used to assess the impacts of climate change on crop yields. The GLAM-Rice model, a regional-scale crop model, was developed to simulate rice development and growth processes by incorporating soil and water dynamics under

flooding from ORYZA2000 (Bouman et al., 2001) into the GLAM-Wheat framework (Li, 2013). The GLAM-Rice model was used for the assessment of the impacts of climate change on rice yields and for efficacious adaptation strategies to climate change in Southeast Asia – especially five countries (Cambodia, Laos, Myanmar, Thailand, and Vietnam) in the Indochinese peninsula (Chun et al., 2016; Li et al., 2017). Figure 2.5 shows the rice yield changes (%) relative to a baseline period (1991–2000) for the RCP4.5 and RCP 8.5 scenarios. The predicted differences between rice yield changes under the RCP4.5 and RCP8.5 scenarios were minimal in the 2040s. However, the predicted differences were higher in the 2080s than in 2040s. Li et al. (2017) reported that the highest reduction of rice yield in the 2040s under the RCP 8.5 scenario was predicted over Cambodia (−12.2%) using the GLAM-Rice model, followed by Vietnam (-8.3%), Thailand (−6.6%), Myanmar (−6.1%), and Laos (−2.8%). These results imply that efficacious adaptation measures should be implemented to offset the negative impacts of climate change on rice yields in Southeast Asia. A selection of agricultural management practices were assessed for potential adaptation strategies including planting date adjustment and use of irrigation (Chun et al., 2016; Li et al., 2017). They reported that while a minor beneficial impact on rice yields from the use of irrigation was predicted in Myanmar and Laos a larger benefit (up to 7.9–19.8% of increases in rice yields) was predicted for Cambodia, Thailand, and Vietnam. They also showed that shifting the planting date later by 20 days later would slightly increase rice yields in Cambodia, Myanmar, and Vietnam (up to approximately 4.7%). However, they noted that up to 6.8% of the decrease in rice yield was predicted in Laos and Thailand with the shifting date.

Chun et al., 2016 proposed a multi-scale crop modeling approach by combining the advantages from regional- and field-scale crop models to develop efficacious adaptation strategies suited to farmers and governments, respectively, in Southeast Asia. The implementation of the multi-scale crop modeling approach is based on a multi-stage research approach (Figure 2.6). First, a country with the most reduced rice yield without adaptation strategies under climate change is identified, and adaptation strategies for the country are assessed using the GLAM-Rice model. Second, the most efficacious adaptation strategy among the assessed adaptations is identified based on predicted changes in rice yield relative to a baseline period. Third, the most benefited grid cell from the adaptation is identified through the further investigation of the spatial distribution of the effects of the adaptation. Finally, using the CERES-Rice model, a

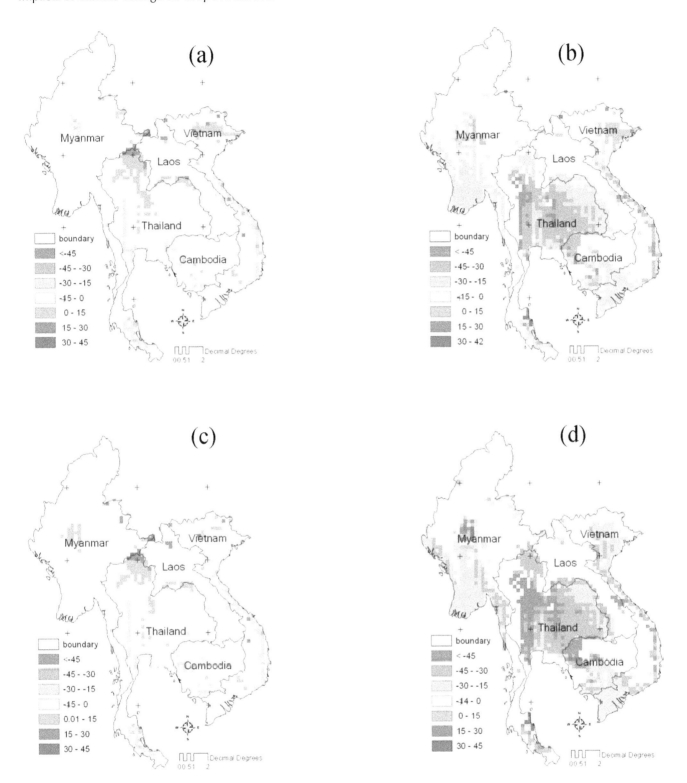

FIGURE 2.5   The projected rice yield change (%) relative to the baseline (1991–2000) for the RCP4.5 and RCP8.5 climate change scenarios: changes in rice yield under (a) and (b) RCP4.5, (c) and (d) RCP8.5; changes in rice yield in the (a) 2040s with CO_2 effect, (b) 2080s with CO_2 effect, (c) 2040s with CO_2 effect, and (d) 2080s with CO_2 effect (Chun et al., 2016). Reprinted by permission, *Agricultural Systems*, Elsevier Press.

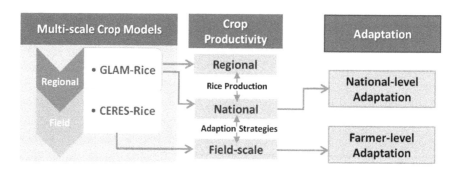

FIGURE 2.6 The proposed multi-scale crop modeling approach.

field-scale crop model, the best combination of adapta-tion measures can be determined by simulating various agricultural management practices including adaptation. Chun et al., 2016 found that the use of irrigation and Cambodia were identified as the most reduced rice yield and a country with the most reduced rice yield, respec-tively, among the five countries in Southeast Asia. They determined the best combination of adaptation strategies at the most benefited grid cell from irrigation (12.6°N and 103.8°E). For the Sen Pidao cultivar, simply dou-bling the recommended nitrogen fertilizer application

rate with no additional adaptation (i.e., 100 kg N ha–¹) would increase rice yield in the 2080s under RCP4.5 (Figure 2.7a). However, planting date adjustment would be additionally required to increase rice yield in the 2080s under RCP4.5 for the Phka Rumduol (Figure 2.7c). Unlike these results under RCP4.5, for both cultivars, 100 kg N ha–¹ and planting date adjustment might not be sufficient to offset the adverse impacts of climate change on rice yields at the location in the 2080s under RCP8.5 (Figure 2.7b and d). They concluded that the proposed multi-scale crop modeling approach could

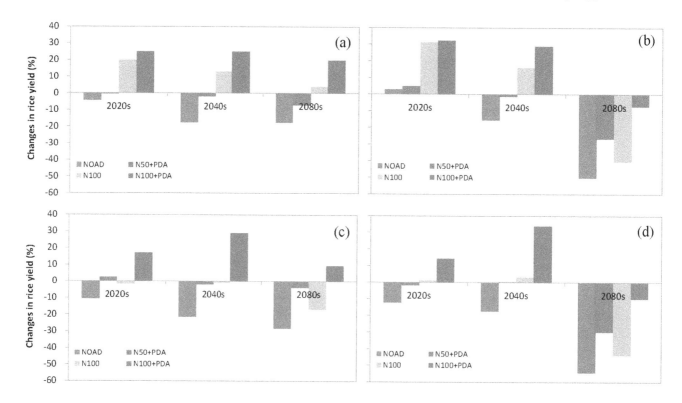

FIGURE 2.7 Changes in rice yields (%) (relative to the baseline period) using CERES-Rice with various combinations of adaptation strategies for the location (12.6 °N and 103.8 °E): changes in rice yield for (a) Sen Pidao under RCP4.5, (b) Sen Pidao under RCP8.5, (c) Phka Rumduol under RCP4.5, and (d) Phka Rumduol under RCP8.5: NOAD, 50 kg N ha–¹ and no water stress; N50+PDA, 50 kg N ha–¹, no water stress, and planting adjustment; N100, 100 kg N ha–¹ and no water stress; N100+PDA, 100 kg N ha–¹, no water stress, and planting adjustment (Chun et al., 2016). Reprinted by permission, *Agricultural Systems*, Elsevier Press.

be useful to improve food security in Southeast Asia through the development of national- and farmer-level adaptation strategies.

2.4   SUMMARY AND CONCLUSION

The projected changes in climate variables in Southeast Asia were investigated using the CORDEX-EXA. Unlike temperature, no clear change in precipitation patterns over the Indochinese peninsula is shown. An increase (–30%) in projected precipitation from May to October for the RCP4.5 and RCP8.5 climate scenarios relative to the baseline (1991–2000) is higher in the 2040s than the 2020s (Li et al., 2017). The largest increases in seasonal precipitation are projected in the north of Vietnam, Thailand, and northwest of Myanmar. While the least change in seasonal temperature is projected in the northwest of Myanmar for the RCP8.5 scenario, the largest increases in seasonal temperature are projected in Thailand and Vietnam by the 2040s.

Cambodia (–12.2%) was predicted as the country with the highest reduction of rice yield in the 2040s under the RCP 8.5 scenario by Li et al. (2017) using the GLAM-Rice model without any adaptation measures, followed by Vietnam (–8.3%), Thailand (–6.6%), Myanmar (–6.1%), and Laos (–2.8%). These results imply that efficacious adaptation measures should be implemented to offset the negative impacts of climate change on rice yields in Southeast Asia. Increases in rice yields differed between countries on the Indochinese peninsula (Chun et al., 2016; Li et al., 2017). While a minor beneficial impact on rice yields from the use of irrigation was predicted in Myanmar and Laos, a larger benefit (up to 7.9–19.8% of increases in rice yields) was predicted for Cambodia, Thailand, and Vietnam. Chun et al., 2016 concluded that a multi-scale crop modeling approach proposed by them could be useful to improve food security in Southeast Asia through the development of national- and farmer-level adaptation strategies.

It is concluded that this chapter can provide a better understanding of climate change and its impacts on rice yields in Southeast Asia, which can lead to the enhancement of food security by developing efficacious adaptation measures to climate change in the region.

REFERENCES

Ahuja, L.R., Ma, L., 2002. Parameterization of agricultural system models: current approaches and future needs. *In Agricultural System Models in Field Research and Technology Transfer.* Ahuja, L.R., Ma, L., Howell, T.A. (Eds.), 273–316. Lewis Publishers, London.

Asian Development Bank (2009) *The Economics of Climate Change in Southeast Asia: A Regional Review* Asian Development Bank: Philippines.

Beven, K. (2002) Towards an alternative blueprint for a physically based digitally simulated hydrologic response modelling system. *Hydrological Processes* 16(2): 189–206.

Bonan, G.B., Levis S., Kergoat, L., Oleson K.W. (2002) Landscapes as patches of plant functional types: An integrating concept for climate and ecosystem models. *Global Biogeochemical Cycles* 16(2): 1021.

Bouman, B.A.M., Kropff, M.J., Tuong, T.P., Woperis, M.C.S.H. Ten Berge, F.M. van Laar, H.H. (2001). *Modelling Lowland Rice.* 235. International Rice Research Institute, LosBanos, Philippines; and Wageningen University and Research Centre, Wageningen, the Netherlands.

Boyle, D.P., Gupta, H.V., Sorooshian, S., Koren, V., Zhang, Z., Smith, M. (2001) Toward improved streamflow forecast: Value of semidistributed modeling. *Water Resources Research* 37(11): 2749–2759.

Briegleb, B.P. (1992) Delta-Edington approximation for solar radiation in the NCAR community climate model. *Journal of Geophysical Research* 97: 7603–7612.

Carpenter, T.M., Georgakakos, K.P., Sperfslage, J.A. (1999a) *Distributed Hydrologic Modeling for Operational Use,* HRC Tech. Report No. 3. Hydrological Research Center, San Diego, CA.

Carpenter, T.M., Sperfslage, J.A., Georgakakos, K.P., Sweeney, T., Fread, D.L. (1999b) National threshold runoff estimation utilizing GIS in support of operational flash flood warning systems. *Journal of Hydrology* 224(1–2): 21–44.

Carpenter, T.M., Georgakakos, K.P., Sperfslage, J.A. (2001) On the parametric and NEXRAD-radar sensitivities of a distributed hydrologic model suitable for operational use. *Journal of Hydrology* 253(1–4): 169–193.

Cha, D.H., Lee, D.K., Hong, S.Y. (2008) Impact of boundary layer processes on seasonal simulation of the East Asian summer monsoon using a Regional Climate Model. *Meteorology and Atmospheric Physics* 100: 53–72.

Challinor, A.J., Wheeler, T.R., Craufurd, P.Q., Slingo, J.M., Grimes, D.I.F. (2004) Design and optimization of a large-area process-based model for annual crops. *Agricultural and Forest Meteorology* 124(1–2): 99–120.

Challinor, A.J., Slingo, J.M., Wheeler, T.R., Doblas-Reyes, F.J. (2005) Probabilistic simulations of crop yield over western India using the DEMETER seasonal hindcast ensembles. *Tellus A* 57: 498–512.

Challinor, A.J., Ewert, F., Arnold, S., Simelton, E.F., Fraser, E. (2009) Crops and climate change: Progress, trends, and challenges in simulating impacts and informing adaptation. *Journal of Experimental Botany* 60(10): 2775–2789.

Chun, J.A., Li, S., Wang, Q., Lee, W.-S., Lee, E.-J., Horstmann, N., Park, H., et al. (2016)Assessing rice productivity and adaptation strategies for Southeast Asia under climate change through multi-scale crop modeling. *Agricultural Systems* 143: 14–21.

Conway, G.R. (1987) The properties of agroecosystems. *Agricultural Systems* 24(2): 95–117.

Cusack, S., Edwards, J.M., Crowther, J.M. (1999) Investigating k-distribution methods for parameterizing gaseous absorption in the Hadley Centre climate model. *Journal of Geophysical Research* 104, 2051–2057.

Di Paola, A., Valentini, R., Santini, M. (2016) An overview of available crop growth and yield models for studies and assessments in agriculture. *Journal of the Science of Food and Agriculture* 96(3): 709–714.

Duncan, W.G. (1972) SIMCOT: A simulation of cotton growth and yield. In *Proceedings of a Workshop for Modeling Tree Growth*. Ed. C.M. Murphy, 115–118. Duke University, Durham, NC.

Duncan, W.G., Loomis, R.S., Williams, W.A., Hanau, R. (1967) A model for simulating photosynthesis in plant communities. *Hilgardia* 38(4): 181–205.

Edwards, J.M., Slingo, A. (1996) Studies with a flexible new radiation code. I: Choosing a configuration for a large-scale model. *Quarterly Journal of the Royal Meteorological Society* 122: 689–719.

Elliott, J., Kelly, D., Chryssanthacopoulos, J., Glotter, M., Jhunjhnuwala, K., Best, N., Wilde, M., Foster, I. (2014) The parallel system for integrating impact models and sectors (pSIMS). *Environmental Modelling & Software* 62: 509–516.

Emanuel, K.A. (1991) A scheme for representing cumulus convection in large-scale models. *Journal of Climate* 48: 2313–2335.

Essery, R., Best, M., Cox, P. (2001) MOSES 2.2 technical documentation. Hadley Centre Tech. Note 30, 30 [Available online at http://www.metoffice.gov.uk/research/hadley-centre/pubs/HCTN/index.html].

FAO, FAOSTAT Database Collections. Food and Agriculture Organization of the United Nations. n.d., Available at: http://faostat3.fao.org.

García-Vila, M., Fereres, E. (2012) Combining the simulation crop model AquaCrop with an economic model for the optimization of irrigation management at farm level. *European Journal of Agronomy* 36(1): 21–31.

Geerts, S., Raes, D. (2009) Deficit irrigation as an on-farm strategy to maximize crop water productivity in dry areas. *Agricultural Water Management* 96(9): 1275–1284.

Georgakakos, K.P. (2000) Covariance propagation and updating in the context of real-time radar data assimilation by quantitative precipitation forecast models. *Journal of Hydrology* 239(1–4): 115–129.

Grant, A. L.M., Brown, A.R. (1999) A similarity hypothesis for shallow-cumulus transports. *Quarterly Journal of the Royal Meteorological Society* 125, 1913–1936.

Gregory, D., Rowntree, P.R. (1990) A mass flux convection scheme with representation of cloud ensemble characteristics and stability-dependent closure. *Monthly Weather Review* 118: 1483–1506.

Haddeland, I., Matheussen, B.V., Lettenmaier, D.P. (2002) Influence of spatial resolution on simulated streamflow in a macroscale hydrologic model. *Water Resources Research* 38(7): 29.1–29.10.

He, J., Dukes, M.D., Hochmuth, G.J., Jones, J.W.G., Graham, W.D. (2012) Identifying irrigation and nitrogen best management practices for sweet corn production on sandy soils using CERES-maize model. *Agricultural Water Management* 109: 61–70.

Holtslag, A.A.M., De Bruijin, E.I.F., Pan, H.L. (1990) A high resolution air mass transformation model for short-range weather forecasting. *Monthly Weather Review* 118: 1561–1575.

Jones, C.A., Kiniry, J.R. (Eds.) (1986) *CERES-Maize: A Simulation Model of Maize Growth and Development*, 208. Texas A&M University Press, College Station, Texas.

Jones, J.W.G., Hoogenboom, C.H., Porter, K.J., Boote, W.D., Batchelor, L.A., Hunt, P.W.W., Singh, U., et al. (2003) The DSSAT cropping system model. *European Journal of Agronomy* 18: 235–265.

Jones, J.W., Antle, J.M., Basso, B., Kenneth, J.B., Conant, R.T., Foster, I., Godfray, H.C.J., et al. (2016) Brief history of agricultural systems modeling. *Agricultural Systems* 155: 240–254.

Kain, J.S., Fritsch, J.M. (1990) A one-dimensional entraining/detraining plume model and its application in convective parameterization. *Journal of the Atmospheric Sciences* 47: 2784–2802.

Keating, B.A., Godwin, D.C., Watiki, J.M. (1991) Optimization of nitrogen inputs under climatic risk. In *Climatic Risk in Crop Production – Models and Management for the Semi-Arid Tropics and Sub-Tropics*, 329–357. ed. Muchow, R.C. and Bellamy, J. A. CAB International, Wallingford, UK.

Kiehl, J.T., Hack, J.J., Bonan, G.B., Boville, B.A., Briegleb, B.P., Williamson, D.L., Rasch, P.J. (1996) Description of NCAR Community Climate Model(CCM3). NCAR Tech. Note NCAR/TN-420+STR, 152.

Kim, D., Kaluarachchi, J.J. (2016) A risk-based hydro-economic analysis for land and water management in water deficit and salinity affected farming regions. *Agricultural Water Management* 166: 111–122.

Kim, H.Y., Ko, J., Kang, S., Tenhunen, J. (2013) Impacts of climate change on paddy rice yield in a temperate climate. *Global Change Biology* 19(2): 548–562.

Krajewski, W.F., Smith, J.A. (2002) Radar hydrology: Rainfall estimation. *Advances in Water Resources* 25(8–12): 1387–1394.

Lehmann, N., Finger, R., Klein, T., Calanca, P.W., Walter, A. (2013) Adapting crop management practices to climate change: Modeling optimal solutions at the field scale. *Agricultural Systems* 117: 55–65.

Li, S.A. (2013) *Development of a Regional Rice Model for Assessing the Impact of Climate Change on Rice in South Korea. APCC Research Report*. APEC Climate Centre, Busan, Republic of Korea.

Li, S.A., Wheeler, T., Challinor, A., Lin, E.D., Xu, Y.L., Ju, H. (2010) Simulating the impacts of global warming on wheat in China using a large area crop model. *Acta Meteorologica Sinica* 24(1): 123–135.

Li, S., Wang, Q., Chun, J.A. (2017) Impact assessment of climate change on rice productivity in the Indochinese Peninsula using a regional-scale crop model. *International Journal of Climatology* 37: 1147–1160.

Lock, A.P., Brown, A.R., Bush, M.R., Martin, G.M., Smith, R.N.B. (2000) A new boundary layer mixing scheme. Part I: Scheme description and SCM tests. *Monthly Weather Review* 128: 3187–3199.

Manzanilla, D.O., Paris, T.R., Vergara, G.V., Ismail, A.M., Pandey, S., Labios, R.V., Tatlonghari, G.T., et al. (2011) Submergence risks and farmers' preferences: Implications for breeding Sub1 rice in Southeast Asia. *Agricultural Systems* 104(4): 335–347.

Masutomi, Y., Takahashi, K., Harasawa, H., Matsuoka, Y. (2009) Impact assessment of climate change on rice production in Asia in comprehensive consideration of process/parameter uncertainty in general circulation models. *Agriculture, Ecosystems & Environment* 131(3–4): 281–291.

Oleson, K.W., Niu, G.Y., Yang, Z.L., Lawrence, D.M., Thornton, P.E., Lawrence, P.J., Stöckli, R., Dickinson, R.E., Bonan G.B., Levis, S., Dai, A., Qian, T. (2008) Improvements to the community land model and their impact on the hydrological cycle. *Journal of Geophysical Research* 113: G01021.

Palosuo, T., Kersebaum, K.C., Angulo, C., Hlavinka, P., Moriondo, M., Olesen, J.E., Patil, R.H., et al. (2011) Simulation of winter wheat yield and its variability in different climates of Europe: A comparison of eight crop growth models. *European Journal of Agronomy* 35(3): 103–114.

Penning de Vries, F.W.T., Jansen, D.M., ten Berge, H.F.M., Bakema, A. eds. (1989) *Simulation of Ecophysiological Processes of Growth in Several Annual Crops.* PUDOC, Wageningen, the Netherlands.

Pinter Jr., P.J., Ritchie, J.C., Hatfield, J.L., Hart, G.F. (2003) The Agricultural Research Service's remote sensing program: An example of interagency collaboration. *Photogrammetric Engineering & Remote Sensing* 69(6): 615–618.

Ritchie, J.T., Otter, S. (1984) Description and performance of CERES-wheat: A user-oriented wheat yield model. In *Wheat Yield Project.* ed. A.R.S. 159–175. ARS-38. National Technical Information Service, Springfield, MO.

Seo, D.J., Smith, J.A. (1996) Characterization of the climatological variability of mean areal rainfall through fractional coverage. *Water Resources Research* 32(7): 2087–2095.

Smith, R.N.B. (1990) A scheme for predicting layer clouds and their water content in a general circulation model. *Quarterly Journal of the Royal Meteorological Society* 116: 435–460.

Steduto, P., Hsiao, T.C., Raes, D., Fereres, E. (2009) AquaCrop—The FAO crop model to simulate yield response to water: I. Concepts and underlying principles. *Agronomy Journal* 101(3): 426–437.

Stöckle, C.O., Donatelli, M., Nelson, R. (2003) CropSyst, a cropping systems simulation model. *European Journal of Agronomy* 18(3–4): 289–307.

Thompson, L.M. (1969) Weather and technology in the production of corn in the US Corn Belt. *Agronomy Journal* 61(3): 453–456.

Thornton, P.K., Herrero, M. (2001) Integrated crop-livestock simulation models for scenario analysis and impact assessment. *Agricultural Systems* 70(2–3): 581–602.

Vaghefi, N., Shamsudin, M.N., Radam, A.R., Rahim, K.A. (2013) Modeling the impact of climate change on rice production: An overview. *Journal of Applied Sciences* 13(24): 5649–5660.

Wallach, D., Makowski, D., Jones, J.W., Brun, F. (2014) *Working with Dynamic Crop Models: Methods, Tools and Examples for Agriculture and Environment* 2nd Edn. Academic Press, Waltham, MA.

WBD (2005) World Bank Data: Economy and Growth. Washington, DC. Available at: http://data.worldbank.org /indicator/NV.AGR.TOTL.ZS.

Williams, J.R., Jones, C.A., Kiniry, J.R., Spanel, D.A. (1989) The EPIC crop growth model. *Transactions of the American Society of Agricultural Engineers* 32(2): 497–511.

3 Plant Responses and Tolerance to Salt Stress

Babar Shahzad, Shah Fahad, Mohsin Tanveer, Shah Saud, and Imtiaz Ali Khan

CONTENTS

3.1 INTRODUCTION

Increasing levels of salts in the soil is one of the most overwhelming abiotic stresses limiting crop productivity and a serious threat to agricultural sustainability globally. The problem of salinity encompasses about 10% of the global land area and almost half of the irrigated lands, resulting in an overall loss of 12 billion US dollars with regard to agricultural production (Flowers et al., 2010). Reduction in available land will further put pressure on producing 70% more food for an additional 2.3 billion people by 2050 (Shabala, 2013). Soil salinity has numerous devastating effects on the morphological and physiological aspects of plants, including seed germination and seedling growth limiting overall crop productivity (Abbasi et al., 2012; Cuartero et al., 2006; Nabati et al., 2011). The mechanism of salt stress in plants governs through the low osmotic potential of soil imposing serious water scarcity to the plants. High salt contents cause an imbalance of nutrients due to specific ion effect, another detrimental effect of salt stress (Evelin et al., 2009). Other than low osmotic potential, salt stress generates toxic metabolites obstructing the photosynthetic rate, leading to cell and even whole plant death (Ashraf, 2004; Chartzoulakis and Psarras, 2005; Hasegawa et al., 2000; Sun et al., 2011). Soil salinity causes severe stress in plants, and perturbation of cellular organelles occurs due to the overproduction of reactive oxygen species (ROS). However, under stress conditions, plants produce certain enzymatic and non-enzymatic antioxidants to scavenge ROS, which otherwise causes disintegration of membranes and vital cellular structures (Anjum et al., 2015). Biosynthesis of antioxidants is pivotal for plant survival; however, in severe salt stress, these antioxidants are not produced enough to defend the plants against ROS.

Salt stress tolerance is the ability of plants to survive under high salt content. Salinity stress can cause specific ion toxicity or ROS production. Plants develop an internal system to tackle ROS. Plants internally operate internal mechanisms of antioxidant system scavenging excessive ROS. Plant growth regulators' application has numerous beneficial roles in augmenting salinity stress tolerance. The literature shows that plant growth regulators, including brassinosteroids (BRs), have great potential in alleviating abiotic stresses including salt stress, heavy metal stress and drought stress (Ali et al., 2007; Anuradha and Rao 2003; El-Mashad and Mohamed, 2012). In this chapter, we have tried a present a comprehensive overview regarding salt stress and its effects on different physiological and biochemical aspects of plants.

3.2 PLANT RESPONSES TO SALT STRESS

3.2.1 GERMINATION

Seed germination is one of the earliest critical stages for plant growth and establishment in a saline environment. Salinity affects several vital physiological processes during germination including imbibition, germination and root elongation. Higher concentration of salts inhibited imbibition and reduced germination percentage (Katembe et al., 1998; Kaymakanova, 2009). Authors suggested that salinity-induced reduction in germination and root elongation could be due to the combination of osmotic effect and specific ion toxicity (Katembe et al., 1998). Similarly, salinity caused a significant decline in the germination of cabbage (*Brassica oleracea* L.) under NaCl stress, and the time required for germination prolonged while increasing salt concentration (Jamil et al., 2007). Anuradha and Rao (2001) reported that salt stress reduced germination up to 130% compared with control. Meanwhile, wheat seeds grown under salt stress showed a significant reduction in germination (Lin et al., 2012). Surprisingly, authors noticed that non-germinated seeds were germinated when transferred to lower salt concentration or distilled water only, which shows that salt stress also delays the time for germination due to delayed catalysis of vital processes during germination.

3.2.2 SEEDLING GROWTH

Salinity-induced alterations in plants including seedling growth and establishment are crucial for plant survival. Salinity results in reduced plant biomass accumulation, leading to stunted growth (Takemura et al., 2000; Wang et al., 2003). However, the major effect of salt stress is the overall decline in leaf area expansion, thereby accelerating to the cessation of expansion due to higher salt concentrations (Wang and Nii, 2000), which ultimately reduces biomass accumulation (Chartzoulakis and Klapaki, 2000). Aziz and Khan (2001a) argued that *Rhizophora mucronata* showed optimum growth up to 50% of saline water; however, it started to decrease at higher salt contents. Jamil et al. (2007) showed negative linear regression between salinity concentration and seedling growth. Apart from negative effects of salt stress on plant growth, some authors showed that some plant species, such as *Salicornia rubra*, could grow better in term of fresh biomass accumulation at higher concentration, which could be attributed to the relative salt tolerance of different plant species.

Salinity reduces plant growth by limiting leaf area expansion, thereby reducing the light interception. Marcelis and Van Hooijdonk (1999) found that salinity reduced 80% of plant growth due to less leaf expansion and 20% due to stomatal conductance under high salt contents. In a similar manner, Kurban et al. (1999) reported that total fresh biomass of *Alhagi pseudoalhagi* was increased at a lower salt concentration (50 mM NaCl) but inclined to decrease with increasing salt concentration (100 and 200 mM NaCl). High salt contents caused a significant decrease in the root, shoot and leaf growth in cotton plants (Meloni et al., 2001). Hence, increasing salt concentrations inhibit plant growth and development, accompanied by significant reductions in plant canopy, seedling fresh and dry weights, plant height, number of leaves and branches per plant, root length and root surface area. Shahbaz et al. (2008), who found that salt stress decreased plant fresh and dry biomass and leaf area in wheat, presented similar findings. Anuradha and Rao (2001) reported an overall reduction of 107.15%, 61.43% and 70.31% in seedling length, seedling fresh weight and dry weight respectively.

3.2.3 WATER RELATIONS

Salt stress results in an increased water potential of the growth medium and induces osmotic stress (Tanveer and Shah, 2017). Excessive amounts of sodium and chloride ions increase the soil matrix osmotic potential and resultantly restrict the water entry into the plant cells. This reduction in water absorption leads to decreased water contents in plants cells and tissues and ultimately results in declined growth rate (Munns, 2002). Water potential and osmotic potential are directly linked with increasing salt concentration. Therefore, any increase in salt concentration in plant tissues will lead to increasing the water potential and osmotic potential, thus reducing cell turgidity (Hernandez et al., 1999; Khan et al., 1999; Khan, 2001; Meloni et al., 2001; Romero-Aranda et al., 2001).

Aziz and Khan (2001b) demonstrated that leaf osmotic and water potential in *R. mucronata* increased along with xylem tension in salinity media. However, Matsumura et al. (1998) reported a decline in leaf osmotic potential with increasing NaCl concentration in *Chrysanthemum*. This is consistent with Chaudhuri and Choudhuri (1997) who found that all the parameters regarding water relations decreased under a short-term salt stress in jute seedlings.

Other physiological effects of salt-stress-induced osmotic stress on water relations include stomatal closure, reduced photosynthetic activity, damage to photosynthetic apparatus and some others (reviewed in Munns, 1993, 2002; Xiong and Zhu, 2002). Water limitation due to salt stress imposes negative effects on photosynthetic apparatus and disrupts the biochemistry of thylakoid membranes with concomitant detriment effects on Calvin cycle enzymes (Farooq et al., 2015; Hussain and Reigosa, 2015). Stomatal closure due to salt stress further reduces the influx of CO_2 and limits photosynthesis (Flexas et al., 2006). Maintenance of plant water flux would be a key factor to prevent photosynthesis inhibition and plant growth. Recently, Kong et al. (2012) showed that the improved growth of cotton in non-uniform salinity conditions was due to increased plant water use that maintained leaf growth and photosynthesis. Therefore, for improved growth and development, maintenance of water relations is crucial; plants maintain their proper growth by increasing water use efficiency under salt stress conditions.

3.2.4 Leaf Anatomy

Salt stress causes modification and alteration in the ultrastructures of different tissues, organelles and even cell components (Parida et al., 2004). In leaves, salt stress increases epidermal thickness, mesophyll thickness, palisade cell length, palisade diameter and spongy cell diameter (Longstreth and Nobel, 1979; Sam et al., 2003). Other detrimental effects of salt stress on leaf anatomy includes (1) vacuolation development and partial swelling of endoplasmic reticulum, (2) decrease in mitochondrial cristae and swelling of mitochondria, (3) vesiculation and fragmentation of tonoplast and (4) degradation of cytoplasm by the mixture of cytoplasmic and vacuolar matrices in leaves of sweet potato (Mitsuya et al., 2000). Furthermore, salt stress also causes rounding of cells, smaller intercellular spaces and a reduction in chloroplast number (e.g., in tomato) (Bruns and Hecht-Buchholz, 1990). High concentration of salt stress in leaves reduces stomatal density and shape of stomata (Romero-Aranda et al., 2001). In another study,

it was noted that salt stress significantly influenced leaf cell organization and cell ultrastructure by altering the number and size of starch granules in chloroplasts, the number of electron-dense corpuscles in the cytoplasm, the structure of mitochondria and a number of plastoglobuli (Sam et al., 2003). Salt stress alters plant growth and development by altering numerous ultrastructures that can influence crop production.

3.2.5 Photosynthesis and Photosynthetic Pigments

Plants produce their food by a mechanism called 'photosynthesis', which is the main determinant of the dry matter accumulation and productivity of the crops. Salt stress limits photosynthesis either by stomatal or non-stomatal limitations (Qin et al., 2010; Tanveer and Shah, 2017). Stomatal limitation results from the limited stomatal opening or reduced CO_2 uptake and/or reduced stomatal conductance (Chaves et al., 2009). Non-stomatal restriction of net assimilation may originate from a reduced efficiency of the Calvin cycle with reduced chlorophyll contents (Lycoskoufis et al., 2005). Munns and Tester (2008) identified the reduction in stomatal aperture as the most dramatic and readily measurable whole-plant response to salinity and concluded that the osmotic effect of salts outside the roots induces stomatal responses. Salt stress affects stomatal conductance immediately due to perturbed water relations and shortly afterward due to the local synthesis of abscisic acid (Farooq et al., 2015). showed partial stomatal closure occurred with salinization, but reductions in photosynthesis were primarily non-stomatal in origin. This concludes that salt stress reduces photosynthesis by altering stomatal and non-stomatal conductance.

Salt stress also imposes negative effects on photosynthesis by reducing photosynthetic pigments or by inducing significant alterations in photobiochemistry (Munns, 2002). Reduction in the activity of Calvin cycle enzymes and disruption in thylakoid membranes are primary effects of salt stress in plants (Chaves et al., 2009; Wang and Nii, 2000). Furthermore, the imbalance between antioxidant defense system and reactive oxygen species production also reduces photosynthesis via oxidation of protein, membrane lipids and other cellular characteristics (Figure 3.1). Moreover, gas exchange analysis confirmed that reductions in net photosynthetic rate are connected with the limited availability of intercellular carbon dioxide due to reduced rates of transpiration and stomatal conductance in salt-treated maize plants (Omoto et al., 2012). Salt stress obstructed maximum quantum yield of photosystem II (PSII) photochemistry (Fv/Fm),

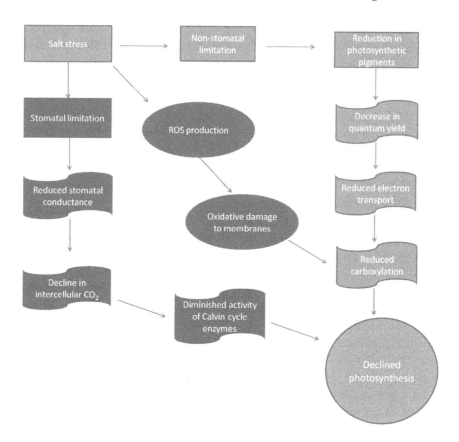

FIGURE 3.1 Mechanism of salt-stress-induced reduction in photosynthesis. Salt stress reduces photosynthesis either by stomatal limitation or by non-stomatal limitation. Salt stress also limits photosynthesis by producing reactive oxygen species that disrupt membranes and further decrease carboxylation reaction. Stomatal limitation results in reduced stomatal conductance and declined intercellular CO_2 accumulation. This causes decline in the activity of different Calvin cycle enzymes. Non-stomatal limitation includes the reduced production and breakdown of chlorophyll contents, which cause reduction in light harvesting with reduced maximum quantum yield. This results in reduced electron transport with concomitant reduction in carboxylation reaction. All these factors ultimately reduce photosynthesis under salt stress.

mainly due to damage at the receptor side of PSII, and decreased the density of active reaction centers and the structure performance (Shu et al., 2012).

Salt stress can reduce total chlorophyll contents by breaking down chlorophyll contents, and this breakdown is found to be linked with Na^+ toxicity (Li et al., 2010; Yang et al., 2011). Salt stress disrupts the thylakoid membrane integrity and stability. Microscopic analysis revealed that salt stress induced destruction of the chloroplast envelope and increased the number of plastoglobuli along with aberrations in thylakoid membranes (Shu et al., 2012). Salt-stress-induced reduction in chlorophyll contents could be due to impaired biosynthesis or accelerated pigment degradation. However, during the process of chlorophyll degradation, chlorophyll b may be converted into chlorophyll a, thus resulting in the increased content of chlorophyll a (Eckardt, 2009; Fang et al., 1998). Nonetheless, a series of studies documented that chlorophyll biosynthesis is more vulnerable to salt stress as compared with chlorophyll degradation.

In conclusion, salt stress reduces the photosynthetic efficiency of a plant by limiting and/or reducing the production of photosynthetic contents and efficiency of photosynthetic apparatus (Figure 3.1).

3.2.6 OSMOTIC STRESS

The initial effect of salt stress on plant growth is a limitation of water availability called osmotic stress (Munns, 2005; Rahnama et al., 2010). High concentration of salt in the root zone limits the water potential of the soil solution, which strictly reduces root water conductivity (Munns and Tester, 2008). As a result, cell membrane permeability drops and the influx of water to the plant is greatly reduced (Munns, 2002). In jute, relative water content, leaf water potential, water uptake, transpiration rate, water retention and water use efficiency reduced under salt stress (Chaudhuri and Choudhuri, 1997). Salt-stress-induced osmotic stress reduces turgor pressure and forces stomata to close following this decline in photosynthetic activity (Munns, 1993). Moreover, cell division and cell elongation were badly affected by the loss in turgor pressure (Shannon et al., 1998). Different studies revealed that cell growth is primarily correlated with turgor potential, and reduction in turgor pressure is one of the major causes of inhibition of plant growth under saline conditions, e.g., maize (Cramer et al., 1996), rice (Moons et al., 1995) and *Shepherdia argentea* (Qin et al., 2010). The adverse effect of salinity in the form of osmotic stress at the cellular level is well documented in a number of comprehensive reviews (Hasegawa et al., 2000; Munns, 2005; Munns and Tester, 2008). However, the extent of growth inhibition due to salt-induced osmotic stress depends on the type of plant tissue and the concentration of salts present in growing medium. In view of the aforementioned reports, it is clear that salinity causes osmotic stress to plants, but the extent of the effect of this stress varies from species to species. It is therefore necessary to understand the molecular mechanisms responsible for the salinity tolerance to find out whether their growth is limited by salt-induced osmotic stress or by the toxic effect of the salt within the plant.

3.2.7 SPECIFIC ION TOXICITY

Plants take up and accumulate certain toxic ions from soil that restrict plant growth. Salt stress results in the high accumulation of sodium (Na^+) and/or chloride (Cl^-) ion in soil, which then affects plant growth and causes specific ion toxicity. However, plant responses to specific toxic ions differ and depend on the type of species (Dogan et al., 2010). It is generally considered that excess amounts of Na^+ cause nutrient imbalance, thereby causing specific ion toxicity (Ashraf, 1994). Salt-sensitive species have no ability to control Na^+ transport. Na^+ appears to accumulate more rapidly to a toxic level than Cl^-; therefore, most studies have focused on Na^+ exclusion and the control of Na^+ transport within the plant (Munns and Tester, 2008). For example, salt stress increased the levels of Na^+ and Cl^- in all parts of guava, particularly in the leaves, thereby resulting in growth reduction. Similarly, high accumulation of Na^+ in the leaves of different cultivars of *Brassica napus* reduces photosynthetic capacity (Ulfat et al., 2007). Qasim and Ashraf (2006) showed that differential salt tolerance in canola cultivars was due to low accumulation of Na^+ in their leaves. In view of a huge number of published reports, Amtmann and Sanders (1998) were able to suggest that high Na^+ concentration in the cytoplasm interferes with normal ongoing metabolic processes. Consequently, plants try to avoid excessive

accumulation of Na^+ in the cytoplasm. Specific ion effect can be further assessed on salt-sensitive and salt-tolerant crop varieties. For example, leaf injuries and growth inhibition were observed in those cultivars that accumulate more Na^+ in their leaves, e.g., in radish, cabbage and canola (Jamil et al., 2007). In addition to Na^+ being a toxic ion, in some species, such as soybean, citrus and grapevine, Cl^- is considered to be a more toxic ion (Grattan and Grieve, 1998). Physiological basis of Cl^- toxicity on plant growth can be explained in view of the arguments of White and Broadley (2001) that Cl^- is taken up through roots and transported to shoots, where it causes damaging effects on photosynthesis and other metabolic processes. From these reports, it can be concluded that excessive amounts of cations or anions in growth medium can cause ion toxicity, which is genotype-specific. However, variation in specific ion toxicity at interspecific or intraspecific level could be due to some adaptations to toxic ions, which is species-specific.

3.2.8 OXIDATIVE STRESS

Salt stress results in the formation of different ROS and alters different plant processes. ROS are highly reactive and oxidize the substrate in the normal metabolism of the plant. In plant cells, ROS, mainly H_2O_2, O^{2-} and a hydroxyl ion OH^-, are generated in the cytosol, chloroplasts, mitochondria and the apoplastic space (Abbasi et al., 2014; Mittler, 2002). Under salt stress, the oxidation of protein, carbohydrates, lipids and nucleic acids is the most destructive feature of ROS (Mallik et al., 2011). ROS oxidize membrane permeability by inducing lipid peroxidation and protein oxidation (Tarchoune et al., 2010), which not only impairs membrane stability (Hajlaoui et al., 2009) but also disturbs membrane integrity (Panda and Khan, 2009). Besides these effects, OH^- causes severe damage to DNA via spoiling purine and pyrimidine bases (Halliwell and Gutteridge, 1999). Wiseman and Halliwell (1996) reported that oxides and peroxides also attack guanine. Furthermore, reduced amino acids synthesis, depolarization and instability of PM and damage to photosynthetic pigments are pronounced effects of hyper-salinization, which distresses the growth and development of the whole organism (Britt, 1999; Tuteja and Tuteja, 2001). Moreover, DNA damage due to salt stress results in numerous detrimental effects on plant growth and development, including the arrest of transcription, replication and signal transduction pathways with genomic instability (Cooke et al., 2003). Salt stress-induced oxidative stress results in the higher accumulation of malondialdehyde (MDA) concentration in cells by affecting the cell signaling and oxidation of integral structural components of cells. MDA has been considered as a clear indication of oxidative stress (Munns, 2002). Likewise, the peroxidation of lipids is also an indicator of the incidence of the oxidative effects of free radical reaction in plant tissues (Jaleel et al., 2007).

3.3 PHYSIOLOGICAL AND BIOCHEMICAL MECHANISMS OF SALT TOLERANCE

3.3.1 ION HOMEOSTASIS AND SALT TOLERANCE

Several plant species have developed an efficient method to maintain a low ion concentration in the cytoplasm. Therefore, maintaining ion homeostasis by ion uptake and compartmentalization is not only crucial for normal plant growth but is also a vital process for growth and development under salt stress conditions (Hasegawa, 2013; Niu et al., 1995; Serrano et al., 1999). Regardless of their nature, both glycophytes and halophytes cannot thrive under high salt concentrations in their cytoplasm. Nonetheless, excessive salt is transported to vacuoles or sequestered into the older leaves or tissue, which later are sacrificed, thereby protecting plants from the adversities of salinity (Reddy et al., 1992; Zhu, 2003). Plant cell membranes, along with their components, play a vital role in ion transport and maintaining ion concentration within the cytosol (Sairam and Tyagi, 2004). The transportation of ions is carried out through a number of proteins, channel proteins, antiporters and symporters. As NaCl is the major form of salt present in soil, the prime focus of ongoing research is mainly concerned with the study of the transport mechanism of Na^+ ions and their compartmentalization into the vacuoles (Blumwald et al., 2000). After entering into the cytoplasm, Na^+ ions are transported to the vacuoles via Na^+/H^+ antiporters. There are two type of antiporters present in the vacuolar membranes: vacuolar-type H^+-ATPase (formally known as V-ATPase) and vacuolar pyrophosphate (V-PPase) (Dietz et al., 2001; Otoch et al., 2001; Wang et al., 2001). Of these, V-ATPase is the most dominant proton pump present in the plant cells. Under normal conditions, this pump plays an important role in maintaining solute homeostasis, energizing secondary transport and facilitating in vesicle fusion. However, under stress conditions, the survival of plants completely relies on the activity of these V-ATPase channels (Dietz et al., 2001). Otoch et al. (2001) found that V-ATPase activity was increased under salt stress in *Vigna unguiculate*, but activity was inhibited under normal conditions. In contrast, *Suaeda salsa*, a halophyte, showed an upregulation in the activity of V-ATPase; however, V-PPase played only a minor role

in salt stress alleviation (Wang et al., 2001). Moreover, Chakraborty et al. (2016) found that increased expression and activity of plasma membrane *SOS1*-like Na^+/H^+ exchangers accelerated Na^+ extrusion from the root cells in *Brassica* species. This indicates that removal of Na^+ through roots is highly desirable under salt stress for conferring better salt stress tolerance.

Cytoplasmic potassium homeostasis plays an integral role in cell metabolism and normal functioning of plants. Several studies elucidated the dramatic decline in potassium concentration under salt stress along with the strong association of shoot K^+ concentration and plant for salt stress tolerance (Abbasi et al., 2014, 2015a). For normal cytoplasmic activity, the plant can maintain a high level of K^+ concentration within cells up to 100 mM, which is ideal for optimal enzymatic activity. The vacuolar concentration of K^+ ranges between 10 and 200 mM; therefore, the vacuole serves as a major reservoir of K^+ within the plant cell. This K^+ plays an important role in maintaining cell turgor. It is transported to the plant cell against the concentration gradient through membrane channels and K^+ transporters. K^+ transporters mediate high-affinity K^+ uptake mechanisms when extracellular concentration of K^+ is low, whereas K^+ channels facilitate low-affinity K^+ uptake when extracellular concentration of K^+ is high. Thus, the uptake and transport mechanism of K^+ is primarily determined by the K^+ concentration available in soil. Unlike K^+, a very low concentration of Na^+ (about 1 mM or less) is maintained in the cytosol. Na^+ concentration in soil is increased under salt stress conditions; therefore, Na^+ competes with K^+ for the same transporters due to their similar charge and transport mechanism, thereby reducing K^+ uptake (Munns and Tester, 2008; Sairam and Tyagi, 2004).

The majority of glycophytes confer salt stress tolerance through either a number of strategies operating simultaneously or in isolation, depending upon the severity of stress (Munns, 2002). Among all the other strategies, K^+ retention is an important feature for salinity tolerance in plants (Chakraborty et al., 2016). In an experiment, the intra-species differential sensitivity of salt was investigated using three *Brassica* species (*B. napus*, *B. juncea* and *B. oleracea*). Among the three species, *B. napus* conferred high salt stress tolerance through high K^+ retention in the roots, resulting from the stress-inducible activation of H^+-ATPase (Chakraborty et al., 2016). The ability of roots to retain more potassium has been verified as a key factor for salt stress tolerance in wheat, barley, maize, bean and lucerne (Abbasi et al., 2014; Chen et al., 2005, 2007; Cuin et al., 2012; Dawood et al., 2014; Smethurst et al., 2008). This has further verified with potassium fertilizer application, where significant improvement in plant growth was observed under salt stress (Abbasi et al., 2015b).

3.3.2 SYNTHESIS OF OSMOPROTECTANTS

Under stress conditions, plants synthesize specific organic compounds, such as sugars, free amino acids, proline and quaternary ammonium compounds, that are regarded as compatible osmolytes (Mansour, 2000; Yang et al., 2003). These osmolytes are synthesized and accumulated in different plant species depending upon the level of stress. There are some quaternary ammonium compounds such as beta-alanine betaine, which is only synthesized in a few members of *Plumbaginacease* family (Hanson et al., 1994), while amino acids like proline accumulation can be seen in a diverse set of plant species (Saxena et al., 2013). Compatible solute concentration within the cell is maintained either by irreversible synthesis of these compounds or by a combination of both synthesis and degradation. Several genes and biochemical pathways involved in these processes have been thoroughly studied (Jantaro et al., 2003; Slocum et al., 1984; Urano et al., 2003). As the accumulation of their concentration is proportional to the salt stress level, the majority of this osmolyte facilitate to protect the structure of the cells and to maintain osmotic balance via continuous water flux (Hasegawa et al., 2000).

3.3.2.1 Amino Acids

Free amino acids play an important role as a solute in an osmotic adjustment under salt stress conditions (Ashrafijou et al., 2010). Previously, studies indicated that osmotic adjustment does not give the physiological basis for salt stress tolerance (Munns, 1993). However, identification of certain solutes in plants under salt stress could prove valuable information for screening and identifying plants that are more tolerant to salt stress. Serval amino acids take part in the osmotic adjustment of the cell, including glycine, arginine, alanine, serine, valine, leucine and proline (Mansour, 2000). El-Shintinawy and El-Shourbagy (2001) reported that amino acids such as cysteine, arginine and methionine, constituting about 55% of total free amino acids, decreased under salinity stress, whereas proline concentration rose in response to salinity stress.

3.3.2.2 Proline

Generally, in higher plants, proline contents are higher and its contents increase under salt stress (Dogan et al., 2010; Nabati et al., 2011). It is well known that proline accumulation is a measure adopted for salinity stress alleviation (Ben Ahmed et al., 2010; Gadallah, 1999;

Matysik et al., 2002; Saxena et al., 2013). Increasing Na^+/Ca^{2+} of the growth solution increased proline contents markedly in four *Brassica* species under salinity stress (Ashraf and Naqvi, 1992). Kumar (1984) observed that higher concentration of proline was recorded in leaves by *B. juncea* under salt stress compared with salt-sensitive cultivars. Therefore, proline plays a significant role in decreasing lipid peroxidation in *B. juncea* (Alia et al., 1993). In another study, Ben Ahmed et al. (2010) discovered that olive plants (*Olea europaea*) supplemented with proline exhibited salt tolerance by ameliorating the activities of some antioxidants, photosynthesis and plant growth. It was further reported by Hoque et al. (2008), who found that proline application resulted in improved salt tolerance in *Nicotiana tabacum* by triggering the activates of key antioxidant enzymes. Meanwhile, pretreatment of rice seedlings with proline showed significant improvement in growth under salt stress (Deivanai et al., 2011).

3.3.2.3 Glycine Betaine

Glycine betaine is an organic amphoteric quaternary ammonium compound ubiquitously found in plants and having an imperative role in salt stress mitigation. It protects the plant cell through osmotic adjustment, stabilizes proteins and protects photosynthetic machinery against stress damages due to ROS production (Ashraf and Foolad, 2007; Chaum and Kirdmanee, 2010; Gadallah, 1999; Mäkelä et al., 2000). Positive effects of glycine betaine on the ultrastructure of rice seedlings were reported under salt stress (Rahman et al., 2002). Ultrastructure of seedlings under salt stress (150 mM NaCl) showed several damages including swelling of thylakoids, the disintegration of grana and intergranal lamellae and disruption of mitochondria. These structural disruptions were largely overcome with pretreatment of glycine betaine. Foliar application of glycine betaine stabilized photosynthetic pigments and improved photosynthesis under salt stress conditions (Ahmad et al., 2013; Chaum and Kirdmanee, 2010).

3.3.3 ANTIOXIDANT REGULATION AND SALT TOLERANCE

Salt stress is the combined consequence of water deficit and osmotic imbalances that poses a variety of effects the on metabolic activities in plants (Cheeseman, 1988; Greenway and Munns, 1980). Water deficit leads to the formation of ROS such as hydrogen peroxide (H_2O_2), superoxide (O^{2-}) and hydroxyl radical ($\cdot OH$) as well as singlet oxygen ($^1O^2$) (Elstner, 1987; Halliwell and Gutteridge, 1985). These cytotoxic activated oxygen species are highly toxic to cellular membranes and therefore disrupt normal metabolism, disintegrating proteins and nucleic acids through oxidative damage and lipid peroxidation (Fridovich, 1986; Wise and Naylor, 1987; Imlay and Linn, 1988; Anjum et al., 2015, 2016a,b,c; Shah et al., 2016; Shahzad et al., 2016, 2017, 2018). To tackle ROS manifestation, antioxidant metabolism, including enzymatic and non-enzymatic compounds, plays an integral role in detoxifying salinity-induced oxidative damage (Abbasi et al., 2014; Anjum et al., 2016c). Hence, salt stress tolerance is positively correlated with antioxidant enzyme activities such as catalase (CAT), superoxide dismutase (SOD), ascorbate peroxidase (APX), glutathione peroxidase (GPX) and glutathione reductase (GR) and non-enzymatic antioxidants such as tocopherols and carotenoids (Abbasi et al., 2014; Ali et al., 2011; Asada, 1999; Gupta et al., 2005). There are a couple of helicase proteins (DESD-box helicase and OsSUV3 dual helicase) that take part in salt stress tolerance, which was recently reported by Gill et al. (2013) and Tuteja et al. (2013). Salt-tolerant pea plants exhibited higher activities of mitochondrial Mn-SOD, chloroplastic CuZnSOD and APX than salt-sensitive plants under salt stress (Hernandez et al., 1995). Similarly, Kholova et al. (2010) reported that salt-tolerant maize genotypes showed enhanced activities of CAT, SOD, APX and GR compared with salt-sensitive maize genotypes under different salt stress levels.

Ascorbate: Ascorbate peroxidase (APX) is one of the major enzymatic antioxidants and the first line of defense against abiotic stresses. It has been reported that APX plays a vital role in abiotic stress tolerance including salt stress in scavenging ROS (Eltayeb et al., 2007; Lu et al., 2007). Hernandez et al. (2000) found higher activities of ascorbate under salt stress. In an experiment, pea plant exhibited an enhancement in both APX activity and S-nitrosylated APX under salt stress (150 mM NaCl) along with an increase of H_2O_2, NO and S-nitrosothiol (SNO) level, suggesting the induction of the APX activity (Begara-Morales et al., 2014). Proteomic studies also revealed that APX is one of the potential targets of post-translational modifications (PTMs) mediated by NO-derived molecules (Begara-Morales et al., 2014). Apart from endogenous synthesis under salt stress, exogenously applied ascorbates mitigate the harmful effects of salt stress in numerous plant species and attenuate plant recovery from stress (Agarwal and Shaheen, 2007; Munir and Aftab, 2011). Glutathione is another antioxidant functioning in stress conditions and assists ascorbate in mitigating the adversities of salt stress. It participates in the generation of ascorbate via the ascorbate–glutathione cycle (Foyer et al., 1997).

Exogenously applied glutathione and ascorbate effectively improved the plasma membrane permeability and cell viability under salt stress in *Allium cepa* (Aly-Salama and Al-Mutawa, 2009). Rawia et al. (2011) reported that, during salt stress, glutathione and ascorbate application effectively increased plant height, a number of branches, plant fresh and dry weight and contents of carbohydrates, phenols, mineral ions and the accumulation of xanthophyll pigment.

3.3.4 ROLE OF POLYAMINES IN SALT TOLERANCE

Polyamines (PAs) are ubiquitous in all organisms, including plants (Cohen, 1998; Tabor and Tabor, 1984; Tiburcio et al., 1990). The most common PAs are diamine putrescine (PUT), triamine spermidine (SPD) and tetraamine spermine (SPM), and they play a number of roles in plants under stress conditions, including osmotic adjustments, protection of cellular macromolecules, nitrogen storage, detoxification of cells, pH maintenance and scavenging of ROS. (Alcázar et al., 2011; Gill and Tuteja, 2010; Gupta et al., 2013; Hussain et al., 2011; Kovács et al., 2010; Kuznetsov and Shevyakova, 2007; Martin-Tanguy, 2001; Shu et al., 2012; Yang et al., 2007). Biosynthesis pathways of PAs have been thoroughly investigated in numerous plant species and reviewed in detail (Alcázar et al., 2005; Alet et al., 2012; Bortolotti et al., 2004; Illingworth et al., 2003; Knott et al., 2007; Kusano et al., 2007; Martin-Tanguy, 2001; Rambla et al., 2010).

Endogenous levels of polyamines increase when plants are exposed to salt stress. The intracellular level of PAs is regulated by polyamine catabolism. Polyamine oxidases (copper-binding diamine oxidase and FAD-binding polyamine oxidases) are responsible for polyamine catabolism and play a significant role in salt stress tolerance (Cona et al., 2006; Takahashi and Kakehi, 2010). It has been reported that exogenous application of PAs could alleviate salt stress, but this effect is concentration dependent as Duan et al. (2008) found that exogenously applied SPD improved polyamine metabolism and alleviated short-term salt stress by reducing salinity-induced membrane damage and photosynthetic inhibition in cucumber seedlings. Authors further suggested that mitigation of short-term salt stress by spermidine was attributed to enhanced activities of antioxidants and osmoticants (Duan et al., 2008). Similarly, Puyang et al. (2015) reported that exogenous application of SPD mitigated the oxidative damage by lowering MDA, H_2O_2 and O^{2-} contents under salt stress. Moreover, activities of antioxidants such as SOD, POD, CAT and APX were increased due to SPD application. Sorghum seedlings treated with 0.25 mM SPD showed improved growth and partial increase in POD and GR enzyme activities subjected to salt stress with a concomitant reduction in MDA contents (Chai et al., 2010). In another experiment, salt sensitive and tolerant *Chrysanthemum* cultivars were treated with different levels of SPD under salinity. Results indicated that SPD application reduced Na^+ accumulation, K^+ loss and MDA contents in response to salt stress. Moreover, antioxidant enzymes activities (SOD, POD, CAT and APX) were increased, resulting in improved photosynthesis efficiency, ROS scavenging and osmotic adjustment (Zhang et al., 2016).

3.3.5 Salt Stress Tolerance through the Exogenous Application of Brassinosteroids

Brassinosteroids (BRs) are polyhydroxy steroidal plant hormones that play pivotal roles in a wide range of developmental processes in plants (Clouse and Sasse, 1998). BRs are not only involved in numerous physiological and biochemical processes of plants but also induce tolerance against abiotic stresses (Ali et al., 2007, 2008a; Cao et al., 2005; Dhaubhadel et al., 1999; Hasan et al., 2008; Krishna, 2003; Nakashita et al., 2003; Rehman et al., 2018; Shahzad et al., 2018; Sharma et al., 2018; Steber and McCourt, 2001; Yu et al., 2004; Zhang et al., 2008). The potential application of BRs in agriculture to improve crop productivity under various stresses have been well documented (Khripach et al., 2000; Shahbaz et al., 2008; Xia et al., 2009).

BRs application improves growth of plants under salt stress. Foliar application of BRs resulted in restored chlorophyll contents and increased activity of nitrogen reductase enzyme (Anuradha and Rao, 2003). Nitrogen reductase activity is a key factor to supply nitrogen to the growing parts and development of plants. Salinity inhibits transport of nitrogen to shoots and xylem loading due to interference with nitrogen reductase (Anuradha and Rao, 2003). Exogenous application of 28-homobrassinolide (HBL) improved seedling growth, nucleic acid contents and seed yield of *Brassica juncea* under salt stress (Hayat et al., 2006, 2007a, b). Some authors proposed that BRs do not have any effect on ultrastructure under normal conditions but significantly decrease cellular damage induced by salt stress on nuclei and chloroplast (Kulaeva et al., 1991). Shahbaz et al. (2008) reported 24-epibrassinolide (EBL) application enhanced plant growth and photosynthesis efficiency of salt sensitive and tolerant wheat cultivars under salinity. Further, authors found that the ameliorative effect of BRs appeared more pronounced in salt-tolerant cultivars than in salt-sensitive cultivars. Shahid et al. (2015)

evaluated the drastic effects of salt stress on different growth parameters, water relations and ionic accumulation, and they found that EBL application significantly altered the water status and osmolyte accumulation in salt-stressed plants. Salt stress negatively affected all the growth-related traits, including fresh and dry weight, leaf area and intermodal distance as well as biochemical indices such as leaf water potential, osmotic potential, turgor potential, relative water contents and yield-related traits. However, EBL application recovered the salt-stressed plants and improved the growth and productivity of plants under salt stress.

BRs application resulted in improved plant stress tolerance, including salt stress (Ali et al., 2007). Seed germination and photosynthesis inhibition were overcome under BRs application in response to salt stress (Ali et al., 2008a, b; Hayat et al., 2007b; Kagale et al., 2007). BRs are involved in improving physiological pathways and photosynthetic apparatus protection under salt stress. Rubisco activity is an important photosynthetic factor responsible for photosynthetic efficiency; however, salt stress disturbs the activities of photosynthetic enzymes such as Rubisco, PEP-carboxylase and Sedoheptulose-1,7-bis-phosphate (Takahama and Oniki, 1992; Sasse, 1997; Nogués and Baker, 2000; Lefebvre et al., 2005; Yamori et al., 2006). BRs application has been reported in activating Rubisco (Yu et al., 2004). Shahbaz et al. (2008) reported that BRs application improved the quantum yield of PSII under salt stress in wheat and ameliorated the inhibitory effects of salt on photosynthetic apparatus, which could be the reason for growth improvement under BRs application (Anuradha and Rao, 2003). Salinity-stress-induced reduction is plant growth due to increased lipid peroxidation and decreased total phenols in cowpea plants subjected to NaCl stress. However, subsequent application of EBL reduced the toxic effects of salt stress and improved morphological attributes of plants. Moreover, EBL significantly lowered the MDA contents and recovered stressed plants (El-Mashad and Mohamed, 2012). Additionally, EBL application increased soluble protein contents in the leaves' modified antioxidant activities of enzymes, suggesting a protective role of EBL under salt stress.

3.4 CONCLUSION AND PERSPECTIVES

Salinity is a global issue that costs billion dollar losses each year due to its harmful effects on plant growth and development. Salinity affects the plants mainly by causing osmotic stress, ion specific toxicity and oxidative stress, as well as a nutrient imbalance. Tissue-specific toxicity leads to overproduction of ROS, resulting in oxidative damage. However, plants can regulate salinity through osmoregulation, ionic homeostasis and synthesis of antioxidants, genes and hormonal regulation as well as through the synthesis of stress-responsive proteins. Nonetheless, the relationship between these mechanisms should be identified at the molecular and genetic level to define research priorities for salt stress management. For future perspectives, strategies may include the targeting redox mechanisms, breeding for salt-tolerant cultivars, use of genetic transformations, application of molecular markers, nutrient management, as well as the application of PGRs. There is a need to conduct comprehensive studies of salt tolerance mechanisms to find out the physiological and biochemical basis under salt stress.

ACKNOWLEDGEMENT

This work was supported by The National Key Research and Development Program of China (2018YFD0300606).

REFERENCES

Abbasi, G.H., Akhtar, J., Haq, M.A., Ahmad, N. (2012) Screening of maize hybrids for salt tolerance at seedling stage under hydroponic condition. *Soil and Environment* 31(1): 83–90.

Abbasi, G.H., Akhtar, J., Anwar-ul-Haq, M., Ali, S., Chen, Z., Malik, W. (2014) Exogenous potassium differentially mitigates salt stress in tolerant and sensitive maize hybrids. *Pakistan Journal of Botany* 46(1): 135–146.

Abbasi, G. H., Akhtar, J., Ahmad, R., Jamil, M., Anwar-ul-Haq, M., Ali, S., Ijaz, M. (2015a) Potassium application mitigates salt stress differentially at different growth stages in tolerant and sensitive maize hybrids. *Plant Growth Regulation* 76(1): 111–125.

Abbasi, G.H., Akhtar, J., Anwar-ul-Haq, M., Malik, W., Ali, S., Chen, Z.H., Zhang, G. (2015b) Morpho-physiological and micrographic characterization of maize hybrids under NaCl and Cd stress. *Plant Growth Regulation* 75(1): 115–122.

Agarwal, S., Shaheen, R. (2007) Stimulation of antioxidant system and lipid peroxidation by abiotic stresses in leaves of *Momordica charantia*. *Brazilian Journal of Plant Physiology* 19(2): 149–161.

Ahmad, R., Lim, C.J., Kwon, S.Y. (2013) Glycine betaine: A versatile compound with great potential for gene pyramiding to improve crop plant performance against environmental stresses. *Plant Biotechnology Reports* 7(1): 49–57.

Alcázar, R., García-Martínez, J.L., Cuevas, J.C., Tiburcio, A.F., Altabella, T. (2005) Overexpression of ADC2 in Arabidopsis induces dwarfism and late-flowering through GA deficiency. *The Plant Journal* 43(3): 425–436.

Alcázar, R., Cuevas, J.C., Planas, J., Zarza, X., Bortolotti, C., Carrasco, P., Salinas, J., et al. (2011) Integration of polyamines in the cold acclimation response. *Plant Science* 180(1): 31–38.

Alet, A.I., Sánchez, D.H., Cuevas, J.C., Marina, M., Carrasco, P., Altabella, T., Tiburcio, A.F., et al. (2012) New insights into the role of spermine in *Arabidopsis thaliana* under long-term salt stress. *Plant Science* 182(1): 94–100.

Ali, B., Hayat, S., Ahmad, A. (2007) 28-homobrassinolide ameliorates the saline stress in chickpea (*Cicer arietinum* L.). *Environmental and Experimental Botany* 59(2): 217–223.

Ali, B., Hasan, S.A., Hayat, S., Hayat, Q., Yadav, S., Fariduddin, Q., Ahmad, A. (2008a) A role for brassinosteroids in the amelioration of aluminium stress through antioxidant system in mung bean (*Vigna radiata* L. Wilczek). *Environmental and Experimental Botany* 62(2): 153–159.

Ali, B., Hayat, S., Fariduddin, Q., Ahmad, A. (2008b) 24-Epibrassinolide protects against the stress generated by salinity and nickel in *Brassica juncea*. *Chemosphere* 72(9): 1387–1392.

Ali, S., Zeng, F., Cai, S., Qiu, B., Zhang, G. (2011) The interaction of salinity and chromium in the influence of barley growth and oxidative stress. *Plant, Soil and Environment* 57(4): 153–159.

Alia, Sardhi, P.P., Mohanty P. (1993) Proline in relation to free radical production in seedlings of *Brassica juncea* raised under sodium chloride stress. In *Plant Nutrition—From Genetic Engineering to Field Practice*, ed. Barrow, J., 731–734. Springer, Dordrecht, the Netherlands.

Aly-Salama, K.H., Al-Mutawa, M.M. (2009) Glutathione-triggered mitigation in salt-induced alterations in plasmalemma of onion epidermal cells. *International Journal of Agriculture and Biology* 11(5): 639–642.

Amtmann, A., Sanders, D. (1998) Mechanisms of Na+ uptake by plant cells. *Advances in Botanical Research* 29: 75–112.

Anjum, S.A., Tanveer, M., Hussain, S., Bao, M., Wang, L., Khan, I., Ullah, E., et al. (2015) Cadmium toxicity in Maize (*Zea mays* L.): Consequences on antioxidative systems, reactive oxygen species and cadmium accumulation. *Environmental Science and Pollution Research International* 22(21): 17022–17030.

Anjum, S.A., Tanveer, M., Ashraf, U., Hussain, S., Shahzad, B., Khan, I., Wang, L. (2016a) Effect of progressive drought stress on growth, leaf gas exchange, and antioxidant production in two maize cultivars. *Environmental Science and Pollution Research International* 23(17): 17132–17141.

Anjum, S.A., Tanveer, M., Hussain, S., ullah, E., Wang, L., Khan, I., Samad, R.A., et al. (2016b) Morpho-physiological growth and yield responses of two contrasting maize cultivars to cadmium exposure. *CLEAN—Soil, Air, Water* 44(1): 29–36.

Anjum, S.A., Tanveer, M., Hussain, S., Shahzad, B., Ashraf, U., Fahad, S., Hassan, W., et al. (2016c) Osmoregulation and antioxidant production in maize under combined cadmium and arsenic stress. *Environmental Science and Pollution Research International* 23(12): 11864–11875.

Anuradha, S., Rao, S.S.R. (2001) Effect of brassinosteroids on salinity stress induced inhibition of seed germination and seedling growth of rice (*Oryza sativa* L.). *Plant Growth Regulation* 33(2): 151–153.

Anuradha, S., Rao, S.S.R. (2003) Application of brassinosteroids to rice seeds (*Oryza sativa* L.) reduced the impact of salt stress on growth, prevented photosynthetic pigment loss and increased nitrate reductase activity. *Plant Growth Regulation* 40(1): 29–32.

Asada, K. (1999) The water-water cycle in chloroplasts: Scavenging of active oxygens and dissipation of excess photons. *Annual Review of Plant Physiology and Plant Molecular Biology* 50: 601–639.

Ashraf, M. (1994) Breeding for salinity tolerance in plants. *Critical Reviews in Plant Sciences* 13(1): 17–42.

Ashraf, M. (2004) Some important physiological selection criteria for salt tolerance in plants. *Flora—Morphology, Distribution, Functional Ecology of Plants* 199(5): 361–376.

Ashraf, M., Foolad, M.R. (2007) Roles of glycine betaine and proline in improving plant abiotic stress resistance. *Environmental and Experimental Botany* 59(2): 206–216.

Ashraf, M., Naqvi, M.I. (1992) Effect of varying Na/Ca ratios in saline sand culture on some physiological parameters of four *Brassica* species. *Acta Physiologiae Plantarum* 14(4): 197–205.

Ashrafijou, M., Sadat Noori, S.A., Izadi Darbandi, A., Saghafi, S. (2010) Effect of salinity and radiation on proline accumulation in seeds of canola (*Brassica napus* L.). *Plant, Soil and Environment* 56(7): 312–317.

Aziz, I., Khan, M.A. (2001a) Experimental assessment of salinity tolerance of *Ceriops tagal* seedlings and saplings from the Indus Delta, Pakistan. *Aquatic Botany* 70(3): 259–268.

Aziz, I., Khan, M.A. (2001b) Effect of seawater on the growth, ion content and water potential of *Rhizophora mucronata* Lam. *Journal of Plant Research* 114(3): 369–373.

Begara-Morales, J.C., Sánchez-Calvo, B., Chaki, M., Valderrama, R., Mata-Pérez, C., López-Jaramillo, J., Padilla, M.N., et al. (2014) Dual regulation of cytosolic ascorbate peroxidase (APX) by tyrosine nitration and S-nitrosylation. *Journal of Experimental Botany* 65(2): 527–538.

Ben Ahmed, C., Ben Rouina, B., Sensoy, S., Boukhriss, M., Ben Abdullah, F. (2010) Exogenous proline effects on photosynthetic performance and antioxidant defense system of young olive tree. *Journal of Agricultural and Food Chemistry* 58(7): 4216–4222.

Blumwald, E., Aharon, G.S., Apse, M.P. (2000) Sodium transport in plant cells. *Biochimica et Biophysica Acta* 1465(1–2): 140–151.

Bortolotti, C., Cordeiro, A., Alcázar, R., Borrell, A., Culiañez-Macià, F.A., Tiburcio, A.F., Altabella, T. (2004) Localization of arginine decarboxylase in tobacco plants. *Physiologia Plantarum* 120(1): 84–92.

Britt, A.B. (1999) Molecular Genetics of DNA repair in higher plants. *Trends in Plant Science* 4(1): 20–25.

Bruns, S., Hecht-Buchholz, C. (1990) Light and electron-microscope studies on the leaves of several potato cultivars after application of salt at various developmental stages. *Potato Research* 33(1): 33–41.

Cao, S., Xu, Q., Cao, Y., Qian, K., An, K., Zhu, Y., Binzeng, H., Zhao, H., Kuai, B. (2005) Loss-of-function mutations in *DET2* gene lead to an enhanced resistance to oxidative stress in *Arabidopsis*. *Physiologia Plantarum* 123(1): 57–66.

Chai, Y.Y., Jiang, C.D., Shi, L., Shi, T.S., Gu, W.B. (2010) Effects of exogenous spermine on sweet sorghum during germination under salinity. *Biologia Plantarum* 54(1): 145–148.

Chakraborty, K., Sairam, R.K., Bhaduri, D. (2016) Effects of different levels of soil salinity on yield attributes, accumulation of nitrogen, and micronutrients in *Brassica* spp. *Journal of Plant Nutrition* 39(7): 1026–1037.

Chartzoulakis, K., Klapaki, G. (2000) Response of two green house pepper hybrids to NaCl salinity during different growth stages. *Scientia Horticulturae* 86(3): 247–260.

Chartzoulakis, K., Psarras, G. (2005) Global change effect on crop photosynthesis and production in Mediterranean: The case of Crete, Greece. *Agriculture, Ecosystems & Environment* 106(2–3): 147–157.

Chaudhuri, K., Choudhuri, M.A. (1997) Effect of short-term NaCl stress on water relations and gas exchange of two jute species. *Biologia Plantarum* 40(3): 373–380.

Chaum, S., Kirdmanee, C. (2010) Effect of glycinebetaine on proline, water use, and photosynthetic efficiencies, and growth of rice seedlings under salt stress. *Turkish Journal of Agriculture and Forestry* 34(6): 517–527.

Chaves, M.M., Flexas, J., Pinheiro, C. (2009) Photosynthesis under drought and salt stress: Regulation mechanisms from whole plant to cell. *Annals of Botany* 103(4): 551–560.

Cheeseman, J.M. (1988) Mechanism of salinity tolerance in plants. *Plant Physiology* 87(3): 547–550.

Chen, Z., Newman, I., Zhou, M., Mendhum, N., Zhang, G., Shabala, S. (2005) Screening plants for salt tolerance by measuring K+ flux: A case study for barley. *Plant, Cell and Environment* 28(10): 1230–1246.

Chen, Z., Zhou, M., Newman, I.A., Mendham, N.J., Zhang, G., Shabala, S. (2007) Potassium and sodium relations in salinized barley tissues as a basis of differential salt tolerance. *Functional Plant Biology* 34(2): 150–162.

Clouse, S.D., Sasse, J.M. (1998) Brassinosteroids: Essential regulators of plant growth and development. *Annual Review of Plant Physiology and Plant Molecular Biology* 49(1): 427–451.

Cohen, S.S. (1998) *A Guide to the Polyamines*. Oxford University Press, New York, NY.

Cona, A., Rea, G., Angelini, R., Federico, R., Tavladoraki, P. (2006) Functions of amine oxidases in plant development and defense. *Trends in Plant Science* 11(2): 80–88.

Cooke, M.S., Evans, M.D., Dizdaroglu, M., Lunec, J. (2003) Oxidative DNA damage: Mechanisms, mutation, and disease. *FASEB Journal: Official Publication of the Federation of American Societies for Experimental Biology* 17(10): 1195–1214.

Cramer, G.R., Alberico, G.J., Schmidt, C. (1996) Salt tolerance is not associated with the sodium accumulation of two maize hybrids. *Functional Plant Biology* 21(5): 675–692.

Cuartero, J., Boların, M.C., Asins, M.J., Moreno, V. (2006) Increasing salt tolerance in the tomato. *Journal of Experimental Botany* 57(5): 1045–1058.

Cuin, T.A., Zhou, M., Parsons, D., Shabala, S. (2012) Genetic behavior of physiological traits conferring cytosolic K+/Na+ homeostasis in wheat. *Plant Biology* 14(3): 438–446.

Dawood, M.G., Abdelhamid, M.T., Schmidhalter, U. (2014) Potassium fertilizer enhances the salt-tolerance of common bean (*Phaseolus vulgaris* L.). *The Journal of Horticultural Science and Biotechnology* 89(2): 185–192.

Deivanai, S., Xavier, R., Vinod, V., Timalata, K., Lim, O.F. (2011) Role of exogenous proline in ameliorating salt stress at early stage in two rice cultivars. *Journal of Stress Physiology & Biochemistry* 7(4): 157–174.

Dhaubhadel, S., Chaudhary, S., Dobinson, K.F., Krishna, P. (1999) Treatment with 24-epibrassinolide, a brassinosteroid, increases the basic thermotolerance of *Brassica napus* and tomato seedlings. *Plant Molecular Biology* 40(2): 333–342.

Dietz, K.J., Tavakoli, N., Kluge, C., Mimura, T., Sharma, S.S., Harris, G.C., Chardonnens, A.N., Golldack, D. (2001) Significance of the V-type ATPase for the adaptation to stressful growth conditions and its regulation on the molecular and biochemical level. *Journal of Experimental Botany* 52(363): 1969–1980.

Dogan, M., Tipirdamaz, R., Demir, Y. (2010) Salt resistance of tomato species grown in sand culture. *Plant, Soil and Environment* 56(11): 499–507.

Duan, J., Li, J., Guo, S., Kang, Y. (2008) Exogenous spermidine affects polyamine metabolism in salinity-stressed *Cucumis sativus* roots and enhances short-term salinity tolerance. *Journal of Plant Physiology* 165(15): 1620–1635.

Eckardt, N.A. (2009) A new chlorophyll degradation pathway. *The Plant Cell* 21(3): 700.

El-Mashad, A.A.A., Mohamed, H.I. (2012) Brassinolide alleviates salt stress and increases antioxidant activity of cowpea plants (*Vigna sinensis*). *Protoplasma* 249(3): 625–635.

El-Shintinawy, F., El-Shourbagy, M.N. (2001) Alleviation of changes in protein metabolism in NaCl-stressed wheat seedlings by thiamine. *Biologia Plantarum* 44(4): 541–545.

Elstner, E.F. (1987) Metabolism of activated oxygen species. In *The Biochemistry of Plants, Biochemistry of Metabolism Vol. II*, ed. Davies, D.D., 252–315. Academic Press, San Diego, CA.

Eltayeb, A.E., Kawano, N., Badawi, G.H., Kaminaka, H., Sanekata, T., Shibahara, T., Inanaga, S., Tanaka, K. (2007) Overexpression of monodehydroascorbate reductase in transgenic tobacco confers enhanced tolerance to ozone, salt and polyethylene glycol stresses. *Planta* 225(5): 1255–1264.

Evelin, H., Kapoor, R., Giri, B. (2009) Arbuscular mycorrhizal fungi in alleviation of salt stress: A review. *Annals of Botany* 104(7): 1263–1280.

Fang, Z., Bouwkamp, J.C., Solomos, T. (1998) Chlorophyllase activities and chlorophyll degradation during leaf senescence in nonyellowing mutant and wild type of *Phaseolus vulgaris* L. *Journal of Experimental Botany* 49(320): 503–510.

Farooq, M., Hussain, M., Wakeel, A., Siddique, K.H.M. (2015) Salt stress in maize: Effects, resistance mechanisms, and management. A review. *Agronomy for Sustainable Development* 35(2): 461–481.

Flexas, J., Bota, J., Galmes, J., Medrano, H., Ribas-Carbó, M. (2006) Keeping a positive carbon balance under adverse conditions: Responses of photosynthesis and respiration to water stress. *Physiologia Plantarum* 127(3): 343–352.

Flowers, T.J., Galal, H.K., Bromham, L. (2010) Evolution of halophytes: Multiple origins of salt tolerance in land plants. *Functional Plant Biology* 37(7): 604–612.

Foyer, C.H., Lopez-Delgado, H., Dat, J.F., Scott, I.M. (1997) Hydrogen peroxide-and glutathione-associated mechanisms of acclimatory stress tolerance and signaling. *Physiologia Plantarum* 100(2): 241–254.

Fridovich, I. (1986) Biological effects of the superoxide radical. *Archives of Biochemistry and Biophysics* 247(1): 1–11.

Gadallah, M.A.A. (1999) Effects of proline and glycinebetaine on *Vicia faba* responses to salt stress. *Biologia Plantarum* 42(2): 249–257.

Gill, S.S., Tuteja, N. (2010) Polyamines and abiotic stress tolerance in plants. *Plant Signaling & Behavior* 5(1): 26–33.

Gill, S.S., Tajrishi, M., Madan, M., Tuteja, N. (2013) A DESD-box helicase functions in salinity stress tolerance by improving photosynthesis and antioxidant machinery in rice (*Oryza sativa* L. cv. PB1). *Plant Molecular Biology* 82(1–2): 1–22.

Grattan, S.R., Grieve, C.M. (1998) Salinity-mineral nutrient relations in horticultural crops. *Scientia Horticulturae* 78(1–4): 127–157.

Greenway, H., Munns, R. (1980) Mechanisms of salt tolerance in nonhalophytes. *Annual Review of Plant Physiology* 31(1): 149–190.

Gupta, K.J., Stoimenova, M., Kaiser, W.M. (2005) In higher plants, only root mitochondria, but not leaf mitochondria reduce nitrite to NO, in vitro and in situ. *Journal of Experimental Botany* 56(420): 2601–2609.

Gupta, K., Dey, A., Gupta, B. (2013) Plant polyamines in abiotic stress responses. *Acta Physiologiae Plantarum* 35(7): 2015–2036.

Hajlaoui, H., Denden, M., Elyeb, N. (2009) Changes in fatty acids composition, hydrogen peroxide generation and lipid peroxidation of salt-stressed corn (*Zea mays* L.) root. *Acta Physiologiae Plantarum* 31(4): 33–34.

Halliwell, B., Gutteridge, J.M.C. (1985) *Free Radicals in Biology and Medicine*. Clarendon Press, Oxford, UK.

Halliwell, B., Gutteridge, J.M.C. (1999) *Free Radicals in Biology and Medicine*, 3rd edn. Oxford University Press, Oxford, UK.

Hanson, A.D., Rathinasabapathi, B., Rivoal, J., Burnet, M., Dillon, M.O., Gage, D.A. (1994) Osmoprotective compounds in the *Plumbaginaceae*: A natural experiment in metabolic engineering of stress tolerance. *Proceedings of the National Academy of Sciences of the United States of America* 91(1): 306–310.

Hasan, S.A., Hayat, S., Ali, B., Ahmad, A. (2008) 28-homobrassinolide protects chickpea (*Cicer arietinum*) from cadmium toxicity by stimulating antioxidants. *Environmental Pollution* 151(1): 60–66.

Hasegawa, P.M. (2013) Sodium (Na+) homeostasis and salt tolerance of plants. *Environmental and Experimental Botany* 92: 19–31.

Hasegawa, P.M., Bressan, R.A., Zhu, J.K., Bohnert, H.J. (2000) Plant cellular and molecular responses to high salinity. *Annual Review of Plant Physiology and Plant Molecular Biology* 51(1): 463–499.

Hayat, S., Ali, B., Ahmad, A. (2006) Response of *Brassica juncea*, to 28-homobrassinolide, grown from the seeds exposed to salt stress. *Journal of Plant Biology* 33: 169–174.

Hayat, S., Ali, B., Ahmad, A. (2007b) Effect of 28-homobrassinolide on salinity-induced changes in *Brassica juncea*. *Turkish Journal of Biology* 31: 141–146.

Hayat, S., Ali, B., Hasan, S.A., Ahmad, A. (2007a) Brassinosteroid enhanced the level of antioxidants under cadmium stress in *Brassica juncea*. *Environmental and Experimental Botany* 60(1): 33–41.

Hernandez, J.A., Olmos, E., Corpas, F.J., Sevilla, F., Del Rio, L.A. (1995) Salt-induced oxidative stress in chloroplasts of pea plants. *Plant Science* 105(2): 151–167.

Hernandez, J.A., Campillo, A., Jimenez, A., Alarcon, J.J., Sevilla, F. (1999) Response of antioxidant systems and leaf water relations to NaCl stress in pea plants. *New Phytologist* 141(2): 241–251.

Hernandez, J.A., Jimenez, A., Mullineaux, P., Sevilla, F. (2000) Tolerance of pea plants (*Pisum sativum*) to long-term salt stress is associated with induction of antioxidant defences. *Plant, Cell & Environment* 23(8): 853–862.

Hoque, M.A., Banu, M.N.A., Nakamura, Y., Shimoishi, Y., Murata, Y. (2008) Proline and glycinebetaine enhance antioxidant defense and methylglyoxal detoxification systems and reduce NaCl-induced damage in cultured tobacco cells. *Journal of Plant Physiology* 165(8): 813–824.

Hussain, M.I., Reigosa, M.J. (2015) Characterization of xanthophyll pigments, photosynthetic performance, photon energy dissipation, reactive oxygen species generation and carbon isotope discrimination during artemisinin-induced stress in *Arabidopsis thaliana*. *PLOS ONE* 10(1): e0114826.

Hussain, S.S., Ali, M., Ahmad, M., Siddique, K.H.M. (2011) Polyamines: Natural and engineered abiotic and biotic stress tolerance in plants. *Biotechnology Advances* 29(3): 300–311.

Illingworth, C., Mayer, M.J., Elliott, K., Hanfrey, C., Walton, N.J., Michael, A.J. (2003) The diverse bacterial origins of the *Arabidopsis* polyamine biosynthetic pathway. *FEBS Letters* 549(1–3): 26–30.

Imlay, J.A., Linn, S. (1988) DNA damage and oxygen radical toxicity. *Science* 240(4857): 1302–1309.

Jaleel, C.A., Gopi, R., Manivannan, P., Panneerselvam, R. (2007) Antioxidative potentials as a protective mechanism in *Catharanthus roseus* (L.) G. Don. Plants under salinity stress. *Turkish Journal of Botany* 31: 245–251.

Jamil, M., Lee, K.B., Jung, K.Y., Lee, D.B., Han, M.S., Rha, E.S. (2007) Salt stress inhibits germination and early seedling growth in cabbage (*Brassica oleracea* capitata L.). *Pakistan Journal of Biological Sciences : PJBS* 10(6): 910–914.

Jantaro, S., Mäenpää, P., Mulo, P., Incharoensakdi, A. (2003) Content and biosynthesis of polyamines in salt and osmotically stressed cells of *Synechocystis* sp. PCC 6803. *FEMS Microbiology Letters* 228(1): 129–135.

Kagale, S., Divi, U.K., Krochko, J.E., Keller, W.A., Krishna, P. (2007) Brassinosteroid confers tolerance in *Arabidopsis thaliana* and *Brassica napus* to a range of abiotic stresses. *Planta* 225(2): 353–364.

Katembe, W.J., Ungar, I.A., Mitchell, J.P. (1998) Effect of Salinity on Germination and Seedling Growth of two *Atriplex* species (Chenopodiaceae). *Annals of Botany* 82(2): 167–175.

Kaymakanova, M. (2009) Effect of salinity on germination and seed physiology in bean (*Phaseolus vulgaris* L.). *Biotechnology & Biotechnological Equipment* 23(1): 326–329.

Khan, M.A., Ungar, I.A., Showalter, A.M. (1999) Effects of salinity on growth, ion content, and osmotic relations in *Halopyrum mocoronatum* (L.) Stapf. *Journal of Plant Nutrition* 22(1): 191–204.

Khripach, V., Zhabinskii, V., De Groot, A. (2000) Twenty years of brassinosteroids: Steroidal plant hormones warrant better crops for the XXI century. *Annals of Botany* 86(3): 441–447.

Knott, J.M., Romer, P., Sumper, M. (2007) Putative spermine synthases from *Thalassiosira pseudonana* and *Arabidopsis thaliana* synthesize thermospermine rather than spermine. *FEBS Letters* 581(16): 3081–3086.

Kong, X., Luo, Z., Dong, H., Eneji, A.E., Li, W. (2012) Effects of non-uniform root zone salinity on water use, Na+ recirculation, and Na+ and H+ flux in cotton. *Journal of Experimental Botany* 63(5): 2105–2116.

Kovács, Z., Simon-Sarkadi, L., Szűcs, A., Kocsy, G. (2010) Differential effects of cold, osmotic stress and abscisic acid on polyamine accumulation in wheat. *Amino Acids* 38(2): 623–631.

Krishna, P. (2003) Brassinosteroid-mediated stress responses. *Journal of Plant Growth Regulation* 22(4): 289–297.

Kulaeva, O.N., Burkhanova, E.A., Fedina, A.B., Khokhlova, V.A., Bokebayeva, G.A., Vorbrodt, H.M., Adam, G. (1991) Effect of brassinosteroids on protein synthesis and plant-cell ultrastructure under stress conditions. In *Brassinosteroids: Chemistry, Bioactivity and*

Applications ACS Symposium SER, ed. Cutler, H.G., Yokota, T., Adam, G., 141–155. American Chemical Society, Washington, DC.

Kumar, D. (1984) The value of certain plant parameters as an index for salt tolerance in Indian mustard (*Brassica juncea* L.). *Plant and Soil* 79(2): 261–272.

Kurban, H., Saneoka, H., Nehira, K., Adilla, R., Premachandra, G.S., Fujita, K. (1999) Effect of salinity on growth, photosynthesis and mineral composition in leguminous plant *Alhagi pseudoalhagi* (Bieb.). *Soil Science and Plant Nutrition* 45(4): 851–862.

Kusano, T., Yamaguchi, K., Berberich, T., Takahashi, Y. (2007) The polyamine spermine rescues *Arabidopsis* from salinity and drought stresses. *Plant Signaling and Behavior* 2(4): 251–252.

Kuznetsov, V.V., Shevyakova, N.I. (2007) Polyamines and stress tolerance of plants. *Plant Stress* 1(1): 50–71.

Lefebvre, S., Lawson, T., Zakhleniuk, O.V., Lloyd, J.C., Raines, C.A., Fryer, M. (2005) Increased sedoheptulose-1, 7-bisphosphatase activity in transgenic tobacco plants stimulates photosynthesis and growth from an early stage in development. *Plant Physiology* 138(1): 451–460.

Li, T., Zhang, Y., Liu, H., Wu, Y., Li, W., Zhang, H. (2010) Stable expression of *Arabidopsis* vacuolar Na+/H+ antiporter gene AtNHX1, and salt tolerance in transgenic soybean for over six generations. *Chinese Science Bulletin* 55(12): 1127–1134.

Lin, J., Li, X., Zhang, Z., Yu, X., Gao, Z., Wang, Y., Wang, J., Li, Z., Mu, C. (2012) Salinity–alkalinity tolerance in wheat: Seed germination, early seedling growth, ion relations and solute accumulation. *African Journal of Agricultural Research* 7(3): 467–474.

Longstreth, D.J., Nobel, P.S. (1979) Salinity effects on leaf anatomy: Consequences for photosynthesis. *Plant Physiology* 63(4): 700–703.

Lu, Z., Liu, D., Liu, S. (2007) Two rice cytosolic ascorbate peroxidases differentially improve salt tolerance in transgenic *Arabidopsis*. *Plant Cell Reports* 26(10): 1909–1917.

Lycoskoufis, I.H., Savvas, D., Mavrogianopoulos, G. (2005) Growth, gas exchange, and nutrient status in pepper (*Capsicum annuum* L.) grown in recirculating nutrient solution as affected by salinity imposed to half of the root system. *Scientia Horticulturae* 106(2): 147–161.

Mäkelä, P., Kärkkäinen, J., Somersalo, S. (2000) Effect of glycinebetaine on chloroplast ultrastructure, chlorophyll and protein content, and RuBPCO activities in tomato grown under drought or salinity. *Biologia Plantarum* 43(3): 471–475.

Mallik, S., Nayak, M., Sahu, B.B., Panigrahi, A.K., Shaw, B.P. (2011) Response of antioxidant enzymes to high NaCl concentration in different salt-tolerant plants. *Biologia Plantarum* 55(1): 191–195.

Mansour, M.M.F. (2000) Nitrogen containing compounds and adaptation of plants to salinity stress. *Biologia Plantarum* 43(4): 491–500.

Marcelis, L.F.M., Van Hooijdonk, J. (1999) Effect of salinity on growth, water use and nutrient use in radish (*Raphanus sativus* L.). *Plant and Soil* 215(1): 57–64.

Martin-Tanguy, J. (2001) Metabolism and function of polyamines in plants: Recent development (new approaches). *Plant Growth Regulation* 34(1): 135–148.

Matsumura, T., Kanechi, M., Inagaki, N., Maekawa, S. (1998) The effects of salt stress on ion uptake, accumulation of compatible solutes, and leaf osmotic potential in safflower, *Chrysanthemum paludosum* and sea aster. *Engei Gakkai Zasshi* 67(3): 426–431.

Matysik, J., Alia, Bhalu, B., Mohanty, P. (2002) Molecular mechanisms of quenching of reactive oxygen species by proline under stress in plants. *Current Science* 82(5): 525–532.

Meloni, D.A., Oliva, M.A., Ruiz, H.A., Martinez, C.A. (2001) Contribution of proline and inorganic solutes to osmotic adjustment in cotton under salt stress. *Journal of Plant Nutrition* 24(3): 599–612.

Mitsuya, S., Takeoka, Y., Miyake, H. (2000) Effects of sodium chloride on foliar ultrastructure of sweet potato (*Ipomoea batatas* Lam.) plantlets grown under light and dark conditions in vitro. *Journal of Plant Physiology* 157(6): 661–667.

Mittler, R. (2002) Oxidative stress, antioxidants and stress tolerance. *Trends in Plant Science* 7(9): 405–410.

Moons, A., Bauw, G., Prinsen, E., Van Montagu, M., Van Der Straeten, D. (1995) Molecular and physiological responses to abscisic acid and salts in roots of salt-sensitive and salt-tolerant *Indica* rice varieties. *Plant Physiology* 107(1): 177–186.

Munir, N., Aftab, F. (2011) Enhancement of salt tolerance in sugarcane by ascorbic acid pretreatment. *African Journal of Biotechnology* 10(80): 18362–18370.

Munns, R. (1993) Physiological processes limiting plant growth in saline soils: Some dogmas and hypotheses. *Plant, Cell & Environment* 16(1): 15–24.

Munns, R. (2002) Comparative physiology of salt and water stress. *Plant, Cell & Environment* 25(2): 239–250.

Munns, R. (2005) Genes and salt tolerance: Bringing them together. *The New Phytologist* 167(3): 645–663.

Munns, R., Tester, M. (2008) Mechanisms of salinity tolerance. *Annual Review of Plant Biology* 59: 651–681.

Nabati, J., Kafi, M., Nezami, A., Rezvani Moghaddam, P., Masomi, A., Zare Mehrjerdi, M. (2011) Effect of salinity on biomass production and activities of some key enzymatic antioxidants in kochia (*Kochia scoparia*). *Pakistan Journal of Botany* 43(1): 539–548

Nakashita, H., Yasuda, M., Nitta, T., Asami, T., Fujioka, S., Arai, Y., Sekimata, K., et al. (2003) Brassinosteroid functions in a broad range of disease resistance in tobacco and rice. *The Plant Journal : for Cell and Molecular Biology* 33(5): 887–898.

Niu, X., Bressan, R.A., Hasegawa, P.M., Pardo, J.M. (1995) Ion homeostasis in NaCl stress environments. *Plant Physiology* 109(3): 735–742.

Nogués, S., Baker, N.R. (2000) Effects of drought on photosynthesis in Mediterranean plants grown under enhanced UV-B radiation. *Journal of Experimental Botany* 51(348): 1309–1317.

Omoto, E., Taniguchi, M., Miyake, H. (2012) Adaptation responses in C4 photosynthesis of maize under salinity. *Journal of Plant Physiology* 169(5): 469–477.

Otoch, M.D.L.O., Sobreira, A.C.M., de Aragão, M.E.F., Orellano, E.G., Lima, M.D.G.S., de Melo, D.F. (2001) Salt modulation of vacuolar H+-ATPase and H+-Pyrophosphatase activities in *Vigna unguiculata*. *Journal of Plant Physiology* 158(5): 545–551.

Panda, S.K., Khan, M.H. (2009) Growth, oxidative damage and antioxidant responses in Greengram (*Vigna radiata* L.) under short-term salinity stress and its recovery. *Journal of Agronomy and Crop Science* 195(6): 442–454.

Parida, A.K., Das, A.B., Mittra, B. (2004) Effects of salt on growth, ion accumulation, photosynthesis and leaf anatomy of the mangrove, *Bruguiera parviflora*. *Trees—Structure and Function* 18(2): 167–174.

Puyang, X., An, M., Han, L., Zhang, X. (2015) Protective effect of spermidine on salt stress induced oxidative damage in two Kentucky bluegrass (*Poa pratensis* L.) cultivars. *Ecotoxicology and Environmental Safety* 117: 96–106.

Qasim, M., Ashraf, M. (2006) Time course of ion accumulation and its relationship with the salt tolerance of two genetically diverse lines of canola (*Brassica napus* L.). *Pakistan Journal of Botany* 38(3): 663.

Qin, J., Dong, W.Y., He, K.N., Yu, Y., Tan, G.D., Han, L., Dong, M., et al. (2010) NaCl salinity-induced changes in water status, ion contents and photosynthetic properties of *Shepherdia argentea* (Pursh) Nutt. Seedlings. *Plant, Soil & Environment* 56(7): 325–332.

Rahman, S., Miyake, H., Takeoka, Y. (2002) Effects of exogenous glycinebetaine on growth and ultrastructure of salt-stressed rice seedlings (*Oryza sativa* L.). *Plant Production Science* 5(1): 33–44.

Rahnama, A., James, R.A., Poustini, K., Munns, R. (2010) Stomatal conductance as a screen for osmotic stress tolerance in durum wheat growing in saline soil. *Functional Plant Biology* 37(3): 255–263.

Rambla, J.L., Vera-Sirera, F., Blázquez, M.A., Carbonell, J., Granell, A. (2010) Quantitation of biogenic tetraamines in *Arabidopsis thaliana*. *Analytical Biochemistry* 397(2): 208–211.

Rawia, A.E., Lobna, S.T., Soad, M.M.I. (2011) Alleviation of adverse effects of salinity on growth, and chemical constituents of marigold plants by using glutathione and ascorbate. *Journal of Applied Scientific Research* 7(5): 714–721.

Reddy, M.P., Sanish, S., Iyengar, E.R.R. (1992) Photosynthetic studies and compartmentation of ions in different tissues of *Salicornia brachiata* Roxb. under saline conditions. *Photosynthetica* 26(2): 173–179.

Rehman, S., Shahzad, B., Bajwa, A.A., Hussain, S., Rehman, A., Cheema, S.A., Abbas, T., et al. (2018) Utilizing the allelopathic potential of *Brassica* species for sustainable crop production: A review. *Journal of Plant Growth Regulation*: 1–14.

Romero-Aranda, R., Soria, T., Cuartero, J. (2001) Tomato plant-water uptake and plant-water relationships under saline growth conditions. *Plant Science: An International Journal of Experimental Plant Biology* 160(2): 265–272.

Sairam, R.K., Tyagi, A. (2004) Physiology and molecular biology of salinity stress tolerance in plants. *Current Science* 86(3): 407–421.

Sam, O., Ramírez, C., Coronado, M.J., Testillano, P.S., Risueño, M.C. (2003) Changes in tomato leaves induced by NaCl stress: Leaf organization and cell ultrastructure. *Biologia Plantarum* 47(3): 361–366.

Sasse, J.N. (1997) Recent progress in brassinosteroids research. *Acta Physiologia Plantarum* 100(3): 696–701.

Saxena, S.C., Kaur, H., Verma, P., Petla, B.P., Andugula, V.R., Majee, M. (2013) Osmoprotectants: Potential for crop improvement under adverse conditions. In *Plant Acclimation to Environmental Stress*, ed. Tuteja, N., Singh, S.G., 197–232. Springer, New York, NY.

Serrano, R., Mulet, J.M., Rios, G., Marquez, J.A., De Larrinoa, I.F., Leube, M.P., Mendizabal, I., et al. (1999) A glimpse of the mechanisms of ion homeostasis during salt stress. *Journal of Experimental Botany* 50(1): 1023–1036.

Shabala, S. (2013) Learning from halophytes: Physiological basis and strategies to improve abiotic stress tolerance in crops. *Annals of Botany* 112(7): 1209–1221.

Shah, A.N., Tanveer, M., Hussain, S., Yang, G. (2016) Beryllium in the environment: Whether fatal for plant growth? *Reviews in Environmental Science and Bio/Technology* 15(4): 549–561.

Shahbaz, M., Ashraf, M., Athar, H. (2008) Does exogenous application of 24-epibrassinolide ameliorate salt induced growth inhibition in wheat (*Triticum aestivum* L.)? *Plant Growth Regulation* 55(1): 51–64.

Shahid, M.A., Balal, R.M., Pervez, M.A., Abbas, T., Aqeel, M.A., Riaz, A., Mattson, N.S. (2015) Exogenous 24-epibrassinolide elevates the salt tolerance potential of pea (*Pisum sativum* L.) by improving osmotic adjustment capacity and leaf water relations. *Journal of Plant Nutrition* 38(7): 1050–1072.

Shahzad, B., Tanveer, M., Hassan, W., Shah, A.N., Anjum, S.A., Cheema, S.A., Ali, I. (2016) Lithium toxicity in plants: Reasons, mechanisms and remediation possibilities–A review. *Plant Physiology and Biochemistry: PPB* 107: 104–115.

Shahzad, B., Mughal, M.N., Tanveer, M., Gupta, D., Abbas, G. (2017) Is lithium biologically an important or toxic element to living organisms? An overview. *Environmental Science and Pollution Research International* 24(1): 103–115.

Shahzad, B., Tanveer, M., Che, Z., Rehman, A., Cheema, S.A., Sharma, A., Song, H., ur Rehman, S.U., Zhaorong, D. (2018) Role of 24-epibrassinolide (EBL) in mediating heavy metal and pesticide induced oxidative stress in plants: A review. *Ecotoxicology and Environmental Safety* 147: 935–944.

Shannon, M.C., Rhoades, J.D., Draper, J.H., Scardaci, S.C., Spyres, M.D. (1998) Assessment of salt tolerance in rice cultivars in response to salinity problems in California. *Crop Science* 38(2): 394–398.

Sharma, A., Kumar, V., Kumar, R., Shahzad, B., Thukral, A.K., Bhardwaj, R. (2018) Brassinosteroid-mediated pesticide detoxification in plants: A mini-review. *Cogent Food & Agriculture* 4(1): 1436212.

Shu, S., Guo, S.R., Sun, J., Yuan, L.Y. (2012) Effects of salt stress on the structure and function of the photosynthetic apparatus in *Cucumis sativus* and its protection by exogenous putrescine. *Physiologia Plantarum* 146(3): 285–296.

Slocum, R.D., Kaur-Sawhney, R., Galston, A.W. (1984) The physiology and biochemistry of polyamines in plants. *Archives of Biochemistry and Biophysics* 235(2): 283–303.

Smethurst, C.F., Rix, K., Garnett, T., Auricht, G., Bayart, A., Lane, P., Wilson, S.J., Shabala, S. (2008) Multiple traits associated with salt tolerance in lucerne: Revealing the underlying cellular mechanisms. *Functional Plant Biology* 35(7): 640–650.

Steber, C.M., McCourt, P. (2001) A role for brassinosteroids in germination in *Arabidopsis*. *Plant Physiology* 125(2): 763–769.

Sun, J.K., Li, T., Xia, J.B., Tian, J.Y., Lu, Z.H., Wang, R.T. (2011) Influence of salt stress on ecophysiological parameters of *Periploca sepium* Bunge. *Plant, Soil & Environment* 57(4): 139–144.

Tabor, C.W., Tabor, H. (1984) Polyamines. *Annual Review of Biochemistry* 53: 749–790

Takahama, U., Oniki, T. (1992) Regulation of peroxidase-dependent oxidation of phenolics in the apoplast of spinach leaves by ascorbate. *Plant and Cell Physiology* 33(4): 379–387.

Takahashi, T., Kakehi, J.I. (2010) Polyamines: Ubiquitous polycations with unique roles in growth and stress responses. *Annals of Botany* 105(1): 1–6.

Takemura, T., Hanagata, N., Sugihara, K., Baba, S., Karube, I., Dubinsky, Z. (2000) Physiological and biochemical responses to salt stress in the mangrove, *Bruguiera gymnorrhiza*. *Aquatic Botany* 68(1): 15–28.

Tanveer, M., Shah, A.N. (2017) An insight into salt stress tolerance mechanisms of *Chenopodium album*. *Environmental Science and Pollution Research International* 24(19): 16531–16535.

Tarchoune, I., Sgherri, C., Izzo, R., Lachaal, M., Ouerghi, Z., Navari-Izzo, F. (2010) Antioxidative responses of *Ocimum basilicum* to sodium chloride or sodium sulphate salinization. *Plant Physiology and Biochemistry: PPB* 48(9): 772–777.

Tiburcio, A.F., Kaur-Sawhney, R., Galston, A.W. (1990) Polyamine metabolism. In *The Biochemistry of Plants*, ed. Stumpf, P.K., Conn, E.E., 283–235. Academic Press, New York, NY.

Tuteja, N., Tuteja, R. (2001) Unraveling DNA repair in human: Molecular mechanisms and consequences of repair defect. *Critical Reviews in Biochemistry and Molecular Biology* 36(3): 261–290.

Tuteja, N., Sahoo, R.K., Garg, B., Tuteja, R. (2013) OsSUV3 dual helicase functions in salinity stress tolerance by maintaining photosynthesis and antioxidant machinery in rice (*Oryza sativa* L. cv. IR64). *The Plant Journal: For Cell and Molecular Biology* 76(1): 115–127.

Ulfat, M., Athar, H.R., Ashraf, M., Akram, N.A., Jamil, A. (2007) Appraisal of physiological and biochemical selection criteria for evaluation of salt tolerance in canola (*Brassica napus* L.). *Pakistan Journal of Botany* 39(5): 1593–1608.

Urano, K., Yoshiba, Y., Nanjo, T., Igarashi, Y., Seki, M., Sekiguchi, F., Yamaguchi-Shinozaki, K., Shinozaki, K. (2003) Characterization of *Arabidopsis* genes involved in biosynthesis of polyamines in abiotic stress responses and developmental stages. *Plant, Cell & Environment* 26(11): 1917–1926.

Wang, Y., Nii, N. (2000) Changes in chlorophyll, ribulose bisphosphate carboxylase-oxygenase, glycine betaine content, photosynthesis and transpiration in *Amaranthus tricolor* leaves during salt stress. *The Journal of Horticultural Science and Biotechnology* 75(6): 623–627.

Wang, B., Lüttge, U., Ratajczak, R. (2001) Effects of salt treatment and osmotic stress on V-ATPase and V-PPase in leaves of the halophyte *Suaeda salsa*. *Journal of Experimental Botany* 52(365): 2355–2365.

Wang, W., Vinocur, B., Altman, A. (2003) Plant responses to drought, salinity and extreme temperatures: Towards genetic engineering for stress tolerance. *Planta* 218(1): 1–14.

White, P.J., Broadley, M.R. (2001) Chloride in soils and its uptake and movement within the plant: A review. *Annals of Botany* 88(6): 967–988.

Wise, R.R., Naylor, A.W. (1987) Chilling-enhanced photooxidation: Evidence for the role of singlet oxygen and endogenous antioxidants. *Plant Physiology* 83(2): 278–282.

Wiseman, H., Halliwell, B. (1996) Damage to DNA by reactive oxygen and nitrogen species: Role in inflammatory disease and progression to cancer. *The Biochemical Journal* 313(1): 17–29.

Xia, X.J., Zhang, Y., Wu, J.X., Wang, J.T., Zhou, Y.H., Shi, K., Yu, Y.L., Yu, J.Q. (2009) Brassinosteroids promote metabolism of pesticides in cucumber. *Journal of Agricultural and Food Chemistry* 57(18): 8406–8413.

Xiong, L., Zhu, J.K. (2002) Molecular and genetic aspects of plant responses to osmotic stress. *Plant, Cell & Environment* 25(2): 131–139.

Yamori, W., Suzuki, K., Noguchi, K.O., Nakai, M., Terashima, I. (2006) Effects of RuBisCO kinetics and RuBisCO activation state on the temperature dependence of the photosynthetic rate in spinach leaves from contrasting growth temperatures. *Plant, Cell & Environment* 29(8): 1659–1670.

Yang, W.J., Rich, P.J., Axtell, J.D., Wood, K.V., Bonham, C.C., Ejeta, G., Mickelbart, M.V., Rhodes, D. (2003) Genotypic variation for glycinebetaine in sorghum. *Crop Science* 43(1): 162–169.

Yang, J., Zhang, J., Liu, K., Wang, Z., Liu, L. (2007) Involvement of polyamines in the drought resistance of rice. *Journal of Experimental Botany* 58(6): 1545–1555.

Yang, J.Y., Zheng, W., Tian, Y., Wu, Y., Zhou, D.W. (2011) Effects of various mixed salt-alkaline stresses on growth, photosynthesis, and photosynthetic pigment concentrations of *Medicago ruthenica* seedlings. *Photosynthetica* 49(2): 275–284.

Yu, J.Q., Huang, L.F., Hu, W.H., Zhou, Y.H., Mao, W.H., Ye, S.F., Nogues, S. (2004) A role of brassinosteroids in the regulation of photosynthesis in *Cucumis sativus*. *Journal of Experimental Botany* 55(399): 1135–1143.

Zhang, M., Zhai, Z., Tian, X., Duan, L., Li, Z. (2008) Brassinolide alleviated the adverse effect of water deficits on photosynthesis and the antioxidant of soybean (*Glycine max* L.). *Plant Growth Regulation* 56(3): 257–264.

Zhang, N., Shi, X., Guan, Z., Zhao, S., Zhang, F., Chen, S., Fang, W., Chen, F. (2016) Treatment with spermidine protects chrysanthemum seedlings against salinity stress damage. *Plant Physiology and Biochemistry : PPB* 105: 260–270.

Zhu, J.K. (2003) Regulation of ion homeostasis under salt stress. *Current Opinion in Plant Biology* 6(5): 441–445.

4 Plant Responses and Tolerance to Drought

Sumit Jangra, Aakash Mishra, Priti, Disha Kamboj,
Neelam R. Yadav, and Ram C. Yadav

CONTENTS

4.1 INTRODUCTION

While growing in their natural habitat, plants are exposed to several unfavorable environmental conditions which hamper proper growth and development and cause a great loss to yield (Seki et al., 2003; Kudo et al., 2017). Among them, drought is one of the greatest environmental constraints to agriculture worldwide (Boyer, 1982). The unavailability of water due to continuous lack of precipitation or lack of irrigation for a period of time long enough to exhaust soil moisture leads to agricultural drought (Manivannan et al., 2008; Mishra and Cherkauer, 2010; Jangra et al., 2017). According to breeders, drought can be defined as "a shortfall of water availability sufficient to cause a loss in yield" or "a period of no rainfall or irrigation that affects crop growth". India has also faced the problem of drought stress in recent years. In 2014–15, India had a 12% deficit in rainfall, followed by a 14% shortfall in 2015–16.

El Niño was the reason behind this deficiency in rainfall. Plants' growth and development are severely affected by drought stress, with a significant loss in growth rate and biomass accumulation. Drought mainly affects cell division and elongation, leaf development, stem elongation, root expansion, stomatal conductance, nutrient uptake, water uptake and productivity (Li et al., 2009; Farooq et al., 2009).

A series of physiological, biochemical and molecular changes are triggered by drought stress, which has a negative effect on plant growth and development. Physiological and biochemical responses include stomatal closure, reduction in cell growth and photosynthetic efficiency and respiration activation. Plants also respond to drought stress both at the cellular and molecular level by accumulating osmolytes and proteins involved in mitigating drought stress (Shinozaki and Yamaguchi-Shinozaki, 2006). A wide range of genes with diverse functions are upregulated or downregulated by drought stress (Shinozaki et al., 2003; Bartels and Sunkar, 2005; Yamaguchi-Shinozaki and Shinozaki, 2005). Production of a phytohormone, abscisic acid (ABA), is triggered by drought, which leads to closure of stomata and switches on several stress-related genes. However, evidence for both ABA-induced and ABA-independent, regulation of drought-inducible gene expression is available (Yamaguchi-Shinozaki and Shinozaki, 2005).

According to FAO reports, 815 million people were undernourished in 2016. In addition to this, the world population is increasing day by day at an alarming rate and is going to reach 9 billion by 2050, which will lead to food crisis worldwide. The present global climate change scenario has further increased the severity and frequency of drought stress (IPCC, 2014). Therefore, to maintain adequate food supply development to drought-tolerant and high-yielding crop varieties is the major challenge faced by researchers at present. Conventional breeding has long been employed to develop varieties with improved drought tolerance, but no significant progress has been made, and conventional breeding is rather time-consuming and labor-intensive (Ashraf, 2010). So, advanced biotechnological techniques like genetic engineering and marker-assisted breeding are required to meet the global food demands. In this review, we will be focusing on plant responses to drought stress and transgenics and molecular approaches for mitigating drought stress.

4.2 PLANT RESPONSES TO DROUGHT STRESS

A wide range of responses are triggered by drought stress, including alteration in gene expression, accumulation of osmotically active compounds or metabolites like abscisic acid and synthesis of specialized proteins like reactive oxygen species (ROS) scavenging protein and chaperones (Ramachandra Reddy et al., 2004).

4.2.1 Physiological and Biochemical Responses

At the cellular level, several physiological and biochemical alterations are associated with drought stress, including turgor loss, changes in protein–protein and lipid–protein interactions, alterations in solute concentration and alteration in membrane fluidity and composition (Chaves et al., 2003; DaMatta and Ramalho, 2006). During drought stress, plants maintain turgor by avoiding dehydration, tolerating dehydration or both (Bray, 2007). Developmental and morphological traits like root thickness, root penetration ability, root length and mass play

an important role in controlling these forms of stress resistance (Comas et al., 2013). Traits like root thickness are constitutive in nature and are present even in the absence of stress, but traits like osmotic adjustment and dehydration tolerance are adaptive in nature and arise only in response to stress (Serraj and Sinclair, 2002; Basu et al., 2016) Reduction in photosynthesis, organic acid and osmolyte accumulation and alteration in metabolism of sugars are characteristic physiological and biochemical responses to drought stress. Stomatal closure and decreased activity of photosynthetic enzymes are the major reasons behind the reduced photosynthetic activity (Anjum et al., 2011). Plants have evolved several mechanisms to adapt drought stress including synthesis of osmoprotectants, osmolytes or compatible solutes. These osmoprotectants or compatible solutes are readily regarded as the stress phase passes away (Tabaeizadeh, 1998; Khan et al., 2015). Amino acids, polyamines, quaternary ammonium compounds and tertiary sulfonium derivatives are the major classes of osmoprotectants used to engineer drought stress (Rontein et al., 2002; Nuccio et al., 1999; Chen and Murata, 2002; Anjum et al., 2017; Ebeed et al., 2017; Blum, 2017). Carbohydrate metabolism studies show that genes involved in the Calvin cycle are reduced by prolonged drought stress (Xue et al., 2008).

4.2.2 Molecular Responses

To maintain the water balance under drought stress, the plant undergoes many alterations at the molecular level. These molecular alterations lead to upregulation and downregulation of many transcription factors (TFs) and accumulation of several stress-related proteins (Kavar et al., 2008). The expression profiling of model plant (*Arabidopsis thaliana*) revealed that 277 genes were upregulated and 79 genes were downregulated under drought stress (Seki et al., 2002). In potato, a significant rise in CDSP 32 protein, which protects that plant from oxidative damage, was observed under drought stress (Broin et al., 2000). Several DREB genes are engaged in cellular signaling pathways in response to drought stress (Agarwal et al., 2006; Lata and Prasad, 2011; Zhang et al., 2009; Sun et al., 2014). These DREB transcription factors are involved in modulating the ABA-independent gene expression responses to drought stress. A subclass of DREB/CBF family, DREB2 is involved in articulating genes involved in drought stress (Seki et al., 2003). Signal transduction pathways to regulate growth are also induced in plants under drought stress. Hydrogen pump is activated on the membrane of root hairs of the plant as an early-warning response mechanism before the considerable decrease in plant relative water content (RWC), which triggers the synthesis of osmolytes such as proline and glycine betaine (GB) to maintain turgor pressure in plants (Gong et al., 2010). Several studies have been reported on molecular responses to drought stress which have increased our knowledge concerning the genes involved in drought tolerance and will help in developing engineered plants with superior yield and stress tolerance.

4.3 TRANSGENIC APPROACHES FOR DROUGHT TOLERANCE

The Green Revolution made us self-sufficient in food, but the increasing population and decreasing arable land are making us think to gain more from less. So, there is a need for high-yielding varieties which are able to perform well under environmental stress conditions. Conventional approaches are not sufficient to meet the global food demands. So, we are shifting towards genetically modified (GM) crops which express genes from different sources to impart stress tolerance. These genes include genes for various transcription factors, osmoprotectants, various hormones and ROS scavenging enzymes. The effect of expression or insertion of these genes in some crops is described below. Researchers from all over the globe are able to release a few drought-tolerant varieties using transgenic approaches; they are listed in Table 4.1.

4.3.1 TRANSCRIPTION FACTORS

Transcription factors, a type of regulatory protein, play a vital role in the cascade of signaling and harbor a major hub in the web of stresses. They perform their function by interacting with the *cis* elements present near promoter region of the genes (stress responsive) in the signaling pathway, thereby triggering the web of the signaling, where many genes work together to combat the stressed condition. Many transcription factors like WRKY (Phukan et al., 2016; Tripathi et al., 2014), MYB (Ambawat et al., 2013), bZIP (Liu et al., 2012), NAC (Gao et al., 2010), bHLH, DREB, MYC, Hsp (Aneja et al., 2015), CBF and eERF have been already studied and reported for enhancing the survival under stressed conditions. Raineri et al. (2015) had demonstrated that HaWRKY, a sunflower WRKY transcription factor, when introgressed in *Arabidopsis thaliana* by floral dip method of *in-planta* transformation, had increased biomass and yield as compared to non-transgenic plants. At the reproductive stage (25 days old), the plants were given conditions of drought stress, and it was found that transgenic plants were more tolerant and didn't have any effect on the yield as compared to non-transgenic plants. Rahman et al. (2016) studied that introgressed EcNAC67 from finger millet (*Eleusine coracana* L.) into a rice cultivar ASD16 through *Agrobacterium*-mediated transformation showed enhanced tolerance to drought and high-salt stress. Under drought conditions, the transgenic plants exhibited higher RWC and little reduction in grain

TABLE 4.1
Commercialized Varieties for Drought Tolerance

Crop Species	Developed By	Gene Introduced	Source of Gene	Product of Introduced Gene	Function
Zea mays L.	Monsanto company and BASF	*cspB*	*Bacillus subtilis*	Cold shock protein B	Sustains cellular function under drought stress
Triticum aestivum	Australia	–	–	–	Under field testing
Glycine max L.	Verdeca	*Hahb-4*	*Helianthus annuus*	Isolated nucleic acid molecule encoding the transcription factor Hahb-4	Binds to dehydration transcription regulating region of plant
Saccharum sp.	PT Perkebunan Nusantara XI (Persero)	*EcBetA*	*Escherichia coli*	Choline dehydrogenase	Catalyzes production of glycine betaine governing tolerance to water stress
Saccharum sp.	PT Perkebunan Nusantara XI (Persero)	*RmBetA*	*Rhizobium meliloti*	Choline dehydrogenase	Produces glycine betaine providing tolerance to water stress
Saccharum sp.	Sugarcane Research Institute, UP (UPCSR)	*DREB gene*	–	–	–

yield as compared to non-transgenic ASD16 rice lines. Some of the transcription factors involved in mitigating drought/abiotic stress are summarized in Table 4.2.

4.3.2 REACTIVE OXYGEN SPECIES

Reactive oxygen species are chemical species which contain oxygen and are chemically reactive in nature. These are present in plants naturally but are produced in excess amounts under stress conditions in chloroplasts, mitochondria and peroxisomes. ROS includes peroxides, hydroxyl radicals, superoxides and singlet oxygen. These serve as signaling ports in plants' mechanism to the stress by regulating plant metabolism and equilibrating the cellular energy. ROS like nitric oxide (NO), superoxides and hydrogen peroxide (H_2O_2), when produced in a large amount, lead to oxidative damages in the apoplast and the destruction of cellular membranes. Studies performed to date show that ROS plays dual regulatory function in drought and osmotic stress tolerance by perceiving the redox state of the cell and by performing retrograde signaling. ROS plays a vital role in developing an antioxidant pathway system in the plants under stress. Noctor et al. (2014) reported the function of ROS metabolism during drought stress. They concluded that singlet oxygen and hydrogen peroxide were the most active ROS produced in the chloroplast during stress condition (drought). Ren et al. (2016) demonstrated that soybean salt-tolerant gene GmST1 introduced in *Arabidopsis thaliana* gave tolerance to drought stress by reducing the production of ROS. The plasmid having construct 35S::GmST1 was first transformed into *Agrobacterium tumefaciens* GV3101 and then was introduced into *Arabidopsis* ecotype Col-0 by floral dip transformation method. Transgenic *Arabidopsis* plants with GmST1 gene survived 11 days of water-deprived conditions while non-transgenic plants died in the drought stress conditions. Some of the ROS-related genes are overexpressed in the transgenic scavenges' excessive ROS produced due to drought, thus providing tolerance to drought condition. Wu et al. (2015) reported that when mitogen-activated protein kinase (MAPK) cascades gene LcMKK isolated from *Lycium chinense* was introduced into tobacco plants through *Agrobacterium*-mediated transformation (pCAMBIA2300-LcMKK in *Agrobacterium tumefaciens* strain C58), it conferred tolerance to drought in transgenic tobacco plants by scavenging excessive ROS and regulating the expression of stress-responsive genes. It was observed that transgenic tobacco (*Nicotiana tabacum*) plants with overexpression of LcMKK produced higher transcript levels of the ROS-related genes (NtSOD, NtCAT and Nt POD), which scavenges the excessive

ROS produced due to dehydration conditions (drought). Similarly, Cai et al. (2014) found that when maize MAPK ZmMKK1 was overexpressed in transgenic *Arabidopsis thaliana*, it enhanced the ROS-scavenging enzymes and thus provided tolerance to drought stress by acting as a ROS-dependent protein kinase.

4.3.3 OSMOPROTECTANTS

Osmoprotectants are the compatible solutes that act as osmolytes so that an organism can survive in adverse stress conditions. In plants, these get accumulated to increase the survival rate during stress conditions. Singh et al. (2015) called the osmoprotectants universal molecules which help in regulating cellular osmotic adjustments and thus play an important role in drought tolerance. Osmoprotectants include quaternary active compounds like betaines, sugars, amino acids, etc. Sequera-Mutiozabal et al. (2017) had reported the role of polyamines (aliphatic organic compound) signaling during stress (drought and soil salinity) conditions. Tiburcio et al. (2014) had stated the function of polyamines in the life of a plant from its developmental stage to the stress conditions. Researchers have reported the role of many polyamines like putrescine (Put) (Espasandin et al., 2014), spermidine (Spd), and spermine (Spm) (Seyedsalehi et al., 2017) in plants for stress tolerance. Qin et al. (2017) developed transgenic soybeans by inserting BADH gene AcBADH from drought-tolerant *Atriplex canescens* into soybean cv. Jinong 17 by Agrobacterium-mediated transformation method. The transgenic plants were found to have enhanced tolerance to drought stress compared to non-transgenic plants. Also, proline content was increased by 12.5% to 16.6%, and peroxidase activity was increased to 7% from 1% in transgenic plants under drought stress as compared to non-transgenic plants. Duque et al. (2016) demonstrated that when oat Adc (Arginine decarboxylase) gene was introduced into *Medicago truncatula* through *Agrobacterium*-mediated transformation, the transgenic plants were found to have higher contents of polyamines viz. spermine, putrescine and norspermidine as compared to normal plants and were more tolerant to water deficit conditions. Adc gene is essential for the synthesis of polyamine. Transgenic plants had the ability to avoid tissue dehydration and also had increased photosynthetic rate under drought conditions. Ke et al. (2016) reported that when glycine betaine synthesizing gene *cod A* (choline deoxygenase) was expressed in transgenic poplar plants, transgenic plants gave tolerance to drought stress by accumulating higher content of GB as compared to normal poplar plants. Some of the plants engineered with osmoprotectants to mitigate drought stress are listed in Table 4.3.

TABLE 4.2

Various Transcription Factors Involved in Mitigating Abiotic Stress

Sl. No.	Transcription Factor	Function	Origin	Reference
		AP2/ERF Family		
1.	VrDREB2A	Transcriptional activator helps in increasing drought and salinity tolerance	*Vigna radiata* L.	Chen et al. (2016)
2.	ScDREB protein	Confers multiple abiotic stress tolerance to yeast	*Syntrichia caninervis*	Li et al. (2016)
3.	AtDREB1A	Confers drought tolerance	*Arabidopsis thaliana*	Wei et al. (2016)
4.	SpERF1	Acts as positive regulator by activating DRE/CRT elements in promoters of abiotic-stress-responsive genes	*Stipa purpurea*	Yang et al. (2016)
5.	TaAREB3	Participates in drought and freezing tolerance	*Triticum aestivum*	Wang et al. (2016a)
6.	FeDREB1	Enhances drought and freezing tolerance	*Fagopyrum esculentum*	Fang et al. (2015a)
7.	TaERF3	Promote drought and salinity tolerance	*Triticum aestivum*	Rong et al. (2014)
8.	AtDREB1D	Improve drought tolerance in soybean	*Arabidopsis thaliana*	Guttikonda et al. (2014)
		Basic Leucine-Zipper (bZIP) Protein		
1.	OsABF1	Improves drought tolerance	*Oryza sativa* L.	Zhang et al. (2017a)
2.	BnaABF2	Enhances salt and drought tolerance in transgenic *Arabidopsis*	*Brassica napus* L.	Zhao et al. (2016a)
3.	HD-Zip I	Confers drought tolerance when overexpressed	*Physcomitrella patens*	Romani et al. (2016)
4.	GhABF2	Confers salinity and drought tolerance in cotton	*Gossypium hirsutum* L.	Liang et al. (2016)
5.	AtTGA4	Confers drought response	*Arabidopsis thaliana*	Zhong et al. (2015)
6.	CaBZ1	Improves drought tolerance in tuber crops	*Capsicum annum*	Moon et al. (2015)
7.	TabZIP60	Improves tolerance to salt, drought and freezing stress	*Triticum aestivum*	Zhang et al. (2015)
8.	OsbZIP71	Plays vital role in ABA-mediated drought and salt tolerance	*Oryza sativa* L.	Liu et al. (2014a)
9.	OsbZIP52/RISBZ5	Negative regulator of drought and cold stress	*Oryza sativa* L.	Liu et al. (2012)
		Basic Helix-Loop-Helix (bHLH) Protein		
1.	VvbHLH1	Enhances drought and salt tolerance	*Vitis vinifera*	Wang et al. (2016b)
2.	EcbHLH57	Confers tolerance to drought, salt and oxidative stress	*Eleusine coracana* L.	Babitha et al. (2015)
3.	bHLH122	Positive regulator of salt (NaCl), drought and osmotic signaling	*Arabidopsis thaliana*	Liu et al. (2014b)
4.	PebHLH35	Positive regulator of drought stress response	*Populus euphratica*	Dong et al. (2014)
		MYB/MYC		
1.	FtMYB10	Novel negative regulator of drought and salt response in transgenic *Arabidopsis*	*Fagopyrum tataricum*	Gao et al. (2016)
2.	GbMYB5	Positive regulator in drought stress	*Gossypium barbadense*	Chen et al. (2014)
3.	OsMYB48-1	Enhances drought tolerance	*Oryza sativa* L.	Xiong et al. (2014)
4.	GmMYBJ1	Confers drought tolerance	*Glycine max* L.	Su et al. (2014)
5.	ScMYBAS1	Drought and salt tolerance	*Saccharum officinarum*	Prabu and Prasad (2012)
		NAC (NAM/ATAF1,2/CUC2)		
1.	NAC factor JUB1 (SlJIB1& AtJUB1)	Enhances drought tolerance	*Solanum lycopersicum* and *Arabidopsis thaliana*	Thirumalaikumar et al. (2017)
2.	OsNAC2	Regulates abiotic stress(drought and salinity) and ABA-mediated response	*Oryza sativa* L.	Shen et al. (2017)

(Continued)

TABLE 4.2 (CONTINUED)

Various Transcription Factors Involved in Mitigating Abiotic Stress

Sl. No.	Transcription Factor	Function	Origin	Reference
3.	MfNACsa	Essential regulator of plant tolerance to drought stress	*Medicago falcata*	Duan et al. (2017)
4.	AaNAC1	Increased tolerance to drought	*Artemisia annua*	Lv et al. (2016)
5.	MlNAC9	Enhanced tolerance to drought, cold and salinity stress	*Miscanthus lutarioriparus*	Zhao et al. (2016b)
6.	*Car*NAC4	Enhances drought and salt stress tolerance	*Cicer arietinum* L.	Yu et al. (2016a)
7.	*Car*NAC5	Enhances drought stress in transgenic *Arabidopsis*	*Cicer arietinum* L.	Yu et al. (2016b)
8.	EcNAC67	Confers drought and salinity tolerance in rice	*Eleusine coracana* L.	Rahman et al. (2016)
9.	ONAC022	Improves drought and salt tolerance	*Oryza sativa* L.	Hong et al. (2016)
10.	ZmNAC55	Confers drought resistance in transgenic *Arabidopsis*	*Zea mays* L.	Mao et al. (2016)
11.	SNAC3/ONAC003	Drought and heat tolerance	*Oryza sativa* L.	Fang et al. (2015b)
12.	NAC016	Promotes drought tolerance	*Arabidopsis thaliana*	Sakuraba et al. (2015)
13.	TaNAC29	Enhances drought tolerance	*Triticum aestivum*	Huang et al. (2015)
14.	HvSNAC1	Overexpression of transcription factor improves drought tolerance	*Hordeum vulgare* L.	Al Abdallat et al. (2014)
15.	TaNAC67	Overexpression leads to enhanced drought, freezing and salinity stress	*Triticum aestivum*	Mao et al. (2014)
16.	MuNAC4	Overexpression leads to drought tolerance in groundnut	*Macrotyloma uniflorum* Lam. Verdc.	Pandurangaiah et al. (2014)
WRKY				
1.	TaWRKY1	Transgenic *Arabidopsis* plants were more tolerant to drought stress	*Triticum aestivum* L.	He et al. (2016)
2.	HaWRKY67	Transgenic *Arabidopsis* plants gave increased yield and were more tolerant to drought stress	*Helianthus annus*	Raineri et al. (2015)
3.	GhWRKY27a	Negative regulator of drought tolerance in transgenic *Nicotiana benthamiana* plants	*Gossypium hirsutum*	Yan et al. (2015)
Others				
1.	CAMTA1	Tolerance to drought stress	*Arabidopsis thaliana*	Pandey et al. (2013)

4.3.4 PLANT HORMONES

Plant hormones also play an important role in regulatory processes during stress conditions. In a drought, there is a decrease in cytokinin, auxin and gibberellin contents while ethylene and ABA (abscisic acid) contents get increased. Gene expression in drought conditions is induced both by ABA-dependent and ABA-independent regulatory pathways (Zhang et al., 2006; Yoshida et al., 2014). For ABA biosynthesis, one of the key enzymes is 9-cis-epoxycarotenoid dioxygenase (NCED). In *Arabidopsis*, there are five genes which encode for NCED. Xu et al. (2016) found stress-responsive H2A histone variant gene from wheat (*Triticum aestivum*) TaH2A.7, when introduced into *Arabidopsis thaliana*, overexpresses, which leads to enhancement in the mRNA levels of genes involved in ABA pathway, thus promoting stomatal closure in drought conditions. Raineri et al. (2015) demonstrated that transgenic *Arabidopsis* plants having HaWRKY6, a transcription factor from sunflower, followed ABA-independent pathways to confer drought stress when treated with ABA. Transgenic plants had higher levels of ABA-related genes ABI3 and ABI5 compared to wild-type plants.

4.4 MOLECULAR APPROACHES FOR DROUGHT TOLERANCE

In spite of several advantages of GM crops, they are unable to solve the problem of abiotic stress tolerance mainly due to the multigenic nature of abiotic stress. Also, GM crops have to face the ethical problems due to which they are not publicly acceptable. So, to overcome

TABLE 4.3

Plant Engineered with Osmoprotectants to Mitigate Drought Stress

Sl. No.	Gene	Origin of Gene	Engineered Plant	Function	Reference
1.	BADH (Betaine aldehyde dehydrogenase)	*Atriplex canescens*	*Glycine max* cv. Jinong 17	Increased tolerance to drought stress	Qin et al. (2017)
2.	*cod A*, glycine betaine synthesizing gene	*Arthrobacter globiformis*	Poplar	Tolerance to drought stress	Ke et al. (2016)
3.	Arginine decarboxylase(ADC)	*Avena sativa*	*Medicago truncatula*	Transgenic plants were tolerant to drought	Duque et al. (2016)
4.	Betaine biosynthesizing enzymes (Mp*gsmt* and Mp*sdmt*)	*Methanohalophilus portucalensis*	*Arabidopsis thaliana*	Tolerant to drought and salinity stress	Lai et al. (2014)
5.	BADH (Betaine aldehyde dehydrogenase)	*Spinacia oleracea*	*Lycopersicon esculentum* cv. Moneymaker	Increases tolerance to heat-induced photoinhibition	Li et al. (2014)
6.	Choline oxidase (cod A)	*Arthrobacter globiformis*	*Solanum tubersoum* L. cv. Superior	Enhanced drought tolerance	Cheng et al. (2013)
7.	Galactinol synthase 2 gene (GolS2)	*Arabidopsis thaliana*	*Oryza sativa*	Improved drought tolerance as well as higher yield	Selvaraj et al. (2017)

these problems, molecular breeding is an alternative. With the advent of molecular markers in the late 1970s, it became feasible to select desirable traits more easily, which gave rise to a new era of plant breeding also referred to as "smart breeding". By the use of molecular markers in breeding, plant breeders can cut many cycles of breeding in the development of new variety. The molecular markers are the string of naturally occurring nucleic acid sequences which are located near the DNA sequence of the desired gene, which in turn are associated with specific phenotypes. Thus, marker-assisted breeding allows the selection of traits based on genotype using associated markers rather than the phenotype of the trait. When the correlation between a molecular marker and a trait is greater than the heritability of the trait, marker-assisted selection may be advantageous. Molecular markers are employed in various fields of plant science, e.g., genetic mapping, germplasm evaluation, map-based gene discovery, characterization of crop improvement, and thus are proved to be a powerful and reliable tool in crop improvement program. A first linkage map was constructed in tomato by using restriction fragment length polymorphisms (RFLPs) in 1986 (Bernatzky and Tanksley, 1986), and Paterson et al. (1988) were the first who used RFLP linkage map to resolve quantitative traits into discrete Mendelian factors. With the advent of the development of molecular

markers and statistical methodology for identifying quantitative trait loci (QTL), marker-assisted selection has become the choice of every breeder. By now, a stage has been reached where genomics research is focusing on generating functional markers that can help identify genes that underlie certain traits, thus facilitating their exploration in crop improvement programs. Molecular breeding approaches involved in combating drought stress in some crops are described below, and some lines/cultivars improved for drought tolerance using marker-assisted breeding are summarized in Table 4.4.

4.4.1 WHEAT

Wheat (*Triticum aestivum* L.) is the common cereal crop grown worldwide; it can be grown in cool, moist regions to tropical regions. Drought stress is the major abiotic stress affecting wheat productivity worldwide. Drought retards the performance of wheat at all growth stages, but during flowering and grain-filling phases (terminal drought), drought stress leads to considerable yield losses. For sustainable productivity of wheat in the water-stressed environment, there is a need to develop drought-tolerant wheat genotypes. A multidisciplinary partnership has been established to gain global expertise and resources by a program named Consultative Group on International Agriculture Research (CGIAR) on

TABLE 4.4

Improving Drought Tolerance in Lines/Cultivars through MAS

Crop	QTL Used	QTL Donor Line/ Cultivar	Recipient Line/ Cultivar	Traits Improved	Reference
Rice	QTL9 (on chromosome 9)	Azucena	Kalinga III	Root length and root thickness; drought tolerance	Steele et al. (2006)
	QTL9 (on chromosome 9)	Azucena	Kalinga III	Straw yield; drought tolerance	Steele et al. (2007)
	qDTY 12.1	IR84984-83-15-159B	Swarna sub1	Yield; drought tolerance	Sureshrao et al. (2016)
	qDTY 2.2, qDTY 3.1 and qDTY 12.1	IR 77298-14-1-2-10; IR 81896-B-B-195; IR 84984-83-15-18-B	MR219	Yield; drought tolerance	Shamsudin et al. (2016a)
	qDTY 2.2, qDTY 3.1 and qDTY 12.1	IR 77298-14-1-2-10; IR 81896-B-B-195; IR 84984-83-15-18-B	MRQ74	Grain yield; drought tolerance	Shamsudin et al. (2016b)
	qDTY 3.2 and qDTY 12.1	IR 77298-5-6-18; IR 74371-46-1-1	Sabitri	Yield; drought tolerance	Dixit et al. (2017)
	qVDT11	Samgang	Nagdong	Stable tiller; drought tolerance	Kim et al. (2017)
Pearl millet	Qgydt.icp-2.2 on LG2	863B	H-77/833-2	Panicle harvest index; grain dry mass; panicle dry mass; drought tolerance	Yadav et al. (2004)
	QTL on LG2	PRTL2/89-33	H-77/833-2	Yield; Drought tolerance	Serraj et al. (2005)
	QTL on LG2	PRTL2/89-33	H-77/833-2	Yield; drought tolerance	Bidinger et al. (2005)
	QTL on LG2	PRTL2/89-33	H-77/833-2	No effect on ROS scavenging	Kholová et al. (2011)
Maize	Five QTLs (on chromosome 1, 2, 3, 8 and 10)	Ac7643	CML 247	Yield; drought tolerance	Ribaut and Ragot (2006)
Cowpea	Drought tolerance QTLs	IT93K-503-1	IT97K-499-35	Drought tolerance	Benoit et al. (2016)
Chickpea	QTLs for root traits and drought tolerance	ICC 4958	JG 11	Drought tolerance	Varshney et al. (2013)
	Drought tolerance root traits	ICC 4958	ICCV 92944 and ICCV 00108	Yield; drought tolerance	Kosgei et al. (2017)
Barley	81 QTLs related to different traits were used	*Hordeum spontaneum*	*Hordeum vulgare*	Grain yield and grain filling	Baum et al. (2003), Talamè et al. (2004), and Tuberosa and Salvi (2006)
Sorghum	Stg 1, Stg 2, Stg 3 and Stg 4	B35	RT × 7000	Delayed leaf senescence and enhanced grain yield	Harris et al. (2007)
	Stg 1, Stg 2, Stg 3 and Stg 4	B35	R16	Drought tolerance	Kassahun et al. (2010)
	Stg 1, Stg 2, Stg 3 and Stg 4	B35	Tabat	Yield	Kamal et al. (2017)
	Stg 1, Stg 2, Stg 3 and Stg 4	B35	Sariaso09	Yield	Ouedraogo et al. (2017)

wheat (CRP WHEAT); a global Wheat Yield Consortium (WYC) has also been constituted to address the problem of productivity under abiotic stresses. Several physiological parameters have become available to allow efficient selection of drought-tolerant genotypes.

A number of reports for biparental and association mapping studies (~50) have been conducted for the identification and mapping of QTLs and their association with drought-related traits. In these studies, 800 QTLs were found, and these QTLs and associated markers are distributed on 21 chromosomes of wheat. Some literature searches have reported nine stable QTLs for agronomic traits (Quarrie et al., 2006; Kirigwi et al., 2007; Maccaferri et al., 2008; Pinto et al., 2010; Lopes et al., 2013; Shukla et al., 2015) and five stable QTLs for physiological traits (Salem et al., 2007; Bennett et al., 2012; Kumar et al., 2012) out of which two QTLs localized with meta-analysis. Further, these two QTLs and associated markers (*Xwmc420* and *Xgwm332*) make them suitable for use in marker-assisted selection (MAS).

It is well known that plants have developed complex stress response systems to deal with abiotic stresses, including drought stress, which involves the expression of several genes, including TFs, which enable plants to cope up with unfavorable conditions, and a marker associated with TFs can be utilized for marker-assisted selection for developing drought-tolerant lines. Markers associated with about 16 genes related to transcription factors have been listed by Budak et al. (2015), out of which ten markers were found associated with genes for DREB and WRKY transcription factors (Wei et al., 2009; Huseynova and Rustamova, 2010; Mondini et al., 2015; Edae et al., 2013). On the basis of the available sequence of DREB1 genes in common wheat and related species, five primer pairs were designed by Wei et al. (2009). polymerase chain reaction (PCR)-based *Dreb-B1* markers have been employed in various studies aimed at improving drought tolerance (Wei et al., 2009).

4.4.2 Rice

For the last decade, the International Rice Research Institute (IRRI) has concentrated on drought research for enhancing grain yield through direct selection under drought and led to the development of 17 drought-tolerant rice varieties with a yield advantage of 0.8–1.2 tons per hectare under drought. These varieties have been released in Africa, South Asia and Southeast Asia and are now being planted by farmers. For example, Sahod Ulan in the Philippines, Sahbhagi Dhan in India and the Sookha Dhan varieties in Nepal. Scientists have identified 14 QTLs for high yield in drought-susceptible

TABLE 4.5

Drought-Tolerant Varieties Released by IRRI through Marker-Assisted Breeding

Name	Year of Release	Country
Sukhadhan1	2011	Nepal
Sukhadhan4	2014	Nepal
DRR44	2014	India
Yaenelo4	2015	Myanmar
Yaenelo5	2016	Myanmar
Yaenelo6	2016	Myanmar
Yaenelo 7	2016	Myanmar

and high-yielding varieties like IR64, Swarna, Sambha and Mahsuri. Shamsudin et al. (2016a) performed marker-assisted breeding with three QTLs (qDTY2.2, q DTY3.1, qDTY12.1) with the aim of enhancing the grain yield of the MR219, a Malaysian rice cultivar under reproductive-stage drought stress. Selected introgressed lines were evaluated in drought stress and non-stress condition and gave yield advantage of 903 to 2500 kg ha^{-1} and 6900 kg/ha over MR219, respectively. Some drought-tolerant varieties of rice are developed by IRRI are mentioned in Table 4.5.

4.4.3 Pearl Millet

Pearl millet (*Pennisetum glaucum* (L) R. Br.) (2n = 2X=14) is the seventh most important global cereal and is a staple crop for millions in India and Africa, where dryland crop production is practiced. But terminal drought stress is a major threat associated with pearl millet, causing a great yield loss. Mapping of major QTLs in independent populations for terminal drought tolerance holds the hope for the development of improved varieties using molecular breeding. In 1991, the UK's overseas development agency, now the Department for International Development (DFID), decided to take the initiative for collaborative research and led the foundation of the DFID–JIC–ICRISAT project for molecular marker development for the improvement of pearl millet (Gale et al., 2005). Therefore, in the era of research of pearl millet, the construction of linkage map using RFLP by Liu et al. (1994) was the first milestone. A major QTL located on linkage group 2 (LG2) for terminal drought tolerance explains more than 30% of phenotypic variability for grain yield (Yadav et al., 2002; Bidinger et al., 2007). Sequentially, three major QTLs with high heritability for grain yield were identified in different post-flowering moisture environments (Bidinger et al., 2007). Yadav et al. (2002) identified a major drought-tolerant

QTL which has been currently found to be associated with reduced salt uptake (Sharma et al., 2014). Likewise, a QTL for low transpiration rates was co-mapped with QTL for the terminal drought tolerance and to support the fact that water conservation is important at the vegetative stage and for the post-flowering stage performance of pearl millet (Kholová et al., 2012). Expressed sequence tag–single sequence repeats (EST-SSRs) were identified by Senthilvel et al. (2008) from EST database and showed high polymorphism among the parental lines used for pearl millet linkage mapping (Rajaram et al., 2013). Supriya et al. (2011) constructed an improved linkage map by integration of diverse arrays technology (DArT) and SSR marker data in a mapping population of recombinant inbred lines (RILs) from cross H 77/833-2 × PRLT 2/89-33, which can be further helpful in the genetic analysis of important quantitative traits.

The progress in next-generation sequencing technologies brought about advancement in linkage mapping population from biparental to the second generation multiparent populations (Bohra, 2013), such as Nested Association Mapping (NAM) population (Yu et al., 2008) and Multiparent Advanced Generation Intercross (MAGIC) population (Kover et al., 2009), to exploit allelic diversity from different parents for fine mapping of QTLs. Use of these multiparent populations with advanced computer modeling systems provides opportunities in utilizing the advantages of both linkage and association mapping for crop improvement (Bevan and Uauy, 2013). A germplasm-based population "PMiGAP" (Pearl millet Inbred Germplasm Association Panel) has been developed in pearl millet to harness much more genetic variations of traits (Sehgal et al., 2015).

4.4.4 MAIZE

Maize (*Zea maize* L.) is among the world's most important food and staple crop for Sub-Saharan African (SSA) countries. About 20–25% worldwide maize production area is affected by drought (Heisey and Edmeades, 1997). Increase in the anthesis-silking interval (ASI) is the main effect of drought stress, which is the major cause of yield losses in maize (Sari-Gorla et al., 1999). So to identify the genomic region responsible for the expression of ASI, Ribaut et al. (1996) identified alleles at four QTLs contributed by drought-tolerant lines responsible for the reduction in ASI. Since the 1970s, the International Maize and Wheat Improvement Center (CIMMYT) has mainly been working on breeding strategies for improving maize performance under drought stress by direct selection of a large number of germplasms under stressed and well-watered conditions (Heisey and Edmeades, 1997).

Several QTLs related to drought stress have been identified by many researchers (Veldboom and Lee, 1996; Ribaut et al., 1997; Almeida et al., 2013; Semagn et al., 2013; Beyene et al., 2015). Marker-assisted recurrent selection (MARS) can harness minor effect genes or QTLs which exhibit the minor effect for the morphological traits. Genetic value of progenies can be predicted by incorporating all the marker information available into the model using genomic selection (Meuwissen et al., 2001). This strategy can count the small effect QTLs, which can be missed by the simple marker-assisted selection method (Guo et al., 2012).

Recently, Beyene et al. (2015) estimated the genetic gain in maize using genomic selection for grain yield in eight maize biparental populations under drought stress, and 13.4% and 18.9% gain in grain yield was found for hybrids derived from cycle 3 of genomic selection than cycle 0 and the best commercial check, respectively. Under the Drought-Tolerant Maize for Africa (DTMA) project, the CIMMYT Global Maize Program recently carried out large-scale genome-wide association study (GWAS) of tropical maize germplasm for grain yield and other secondary traits under drought conditions. An association panel comprising of 300 lines was evaluated over 11 locations in Kenya, Mexico, Zimbabwe and Thailand. The lines were genotyped using Illumina 55K chips as well as by GBS platform (Cornell). High-density single nucleotide polymorphism (SNP) positioning aided in achieving higher resolution of identified genomic regions. The identified drought-tolerant candidate genomic regions are currently validated independently through the diverse biparental population and other association panels of CIMMYT. Meta-analysis was conducted by Hao et al. (2010) on a subset of 239 QTLs and 160 QTLs identified under drought stress and well-irrigated conditions respectively and found 39 mQTLs and 36 mQTLs under drought and well-watered conditions. Likewise, a meta-analysis was conducted by Semagn et al. (2013) on 18 biparental maize populations under drought stress and well-irrigated conditions and reduced the number of QTLs to 68 from 183. These 68 mQTLs narrow down the confidence interval up to 12 fold and can be successfully employed for marker-assisted breeding for developing drought-tolerant varieties.

4.4.5 BARLEY

Barley (*Hordeum vulgare*) is a comparatively drought-tolerant crop with a simple genome structure. In 2012, the International Barley Genome Sequencing Consortium published the complete genome annotation of barley; this will enhance the possibility of finding QTLs associated with drought tolerance in mapping populations. In earlier studies of identification and mapping of drought-associated QTLs, mapping population used were derived from Tadmor cultivar of barley (drought and salt tolerant selected by ICARDA). Tadmor and Er/Apm derived mapping populations have been utilized by several researchers to identify QTLs for drought stress (Teulat et al., 1998, 2001a,b, 2002, Teulat et al., 2003; Diab et al., 2004). Some QTLs were found coupled with candidate genes, for example, QTL present on the long arm of chromosome 6H for Osmotic adjustment (OA) coincided with *Dhn*4 locus codes for dehydrin 4 and explained 17.7% of the phenotypic variation (Teulat et al., 1998; Teulat et al., 2003). A novel gene Hsdr4, localized on the long arm of 3H chromosome between markers Ebmac541 and Ebmag 705, related to drought tolerance (Suprunova et al., 2007), residing in a region harboring a QTL for osmotic potential (OP) and the RWC (Diab et al., 2004). von Korff et al. (2008) identified SSR marker pHva1 derived from a late embryogenesis abundant protein gene HVA1, and this marker can be further helpful in the selection of traits enhancing crop performance under drought stress. Zhang et al. (2017b), conducted a meta-analysis of 195 major QTLs and identified mQTLs that were further used to search for candidate genes for abiotic stresses in barley. Two QTLs, MQTL3H.4 and MQTL6H.2, that control drought tolerance in barley can be further employed in MAS for the development of drought-tolerant barley lines. Several studies of QTL identification controlling agronomic performance under water limiting conditions in the field and under control conditions (Baum et al., 2003; Teulat et al., 2003; von Korff et al., 2006; Worch et al., 2011; Tondelli et al., 2014; Wehner et al., 2015, 2016; Mikołajczak et al., 2016).

4.4.6 LEGUMES

Legume crops are the main source of nutrition and protein in Asia and Sub-Saharan Africa. But some legume's productivity is affected by abiotic stresses, mainly drought stress. Till 2005, genomic resources of these legumes were underdeveloped and aiding these resources employing molecular breeding for crop improvement was challenging. During 2005–2015, much innovation was led by International Crops Research Institute for the Semi-Arid Tropics (ICRISAT) and its research partner in developing genomic resources for these crops, like the development of molecular markers, trait mapping, molecular breeding.

Rehman et al. (2011) identified two QTLs (Q3-1 and Q1-1 on linkage group 3 (LG3) and linkage group 1, respectively) associated with many traits related to drought

from a cross between ILC588 (drought tolerant) and ILC 3279 (drought susceptible), which can be further exploited in drought studies. Varshney et al. (2013) introgressed a "QTL-hotspot" region containing QTLs for several drought tolerance related traits in leading chickpea variety JG11. Similarly, a "QTL-hotspot" genomic region has been introgressed in two more popular varieties of chickpea (Chefe and KAK 2) using MABC (Thudi et al., 2014). Similar attempts are under progress at the Indian Agriculture Research Institute, New Delhi (IARI), Indian Institute of Pulse Research, Kanpur (IIPR), Egerton University (Kenya) and the Ethiopian Institute of Agriculture Research (EIAR) to introgress a "QTL-hotspot" in other elite cultivars. A candidate gene for drought tolerance on Ca4 pseudomolecule in the QTL-hotspot region of chickpea was found by Kale et al. (2015), using QTL analysis and gene enrichment analysis. In both the approaches, 12 genes were found to be common, and four promising genes (Ca_04561, Ca_04562, Ca_04567 and Ca_04569) were found to be upregulated in roots under drought stress, which represents the importance of the "QTL-hotspot" region in chickpea. In a recent GWAS study (Hoyos-Villegas et al., 2017) of common bean, 27 significant marker-trait association were reported, where the panel comprised of 96 common bean genotypes under irrigated and rainfed conditions and the genetic diversity and ancestry of the panel was explored by using SNP data. Along with these marker-trait associations, one marker for leaf elongation rate on Chr 3 and one for a wilting score on Chr 11. Positional candidate genes, including PHVIL.011G102700 on Chr 11, associated with wilting were identified. The valuable QTLs found in this study can be utilized in marker-assisted breeding for improving drought tolerance in common bean. Likewise, in lentil (*Lens culinaris* Medik.), 18 QTLs were identified, controlling 14 roots and shoot traits associated with drought tolerance in 132 lentil recombinant inbred lines analyzed by 252 co-dominant and dominant markers (Idrissi et al., 2016)

4.5 HIGH THROUGHPUT "OMICS" TECHNOLOGIES FOR DROUGHT TOLERANCE

The advancement in the omics technologies has led to a better understanding of plant responses to drought stress. The term "omics" is a combination of high throughput techniques like genomics, transcriptomics, proteomics and metabolomics. The development of EST database and whole genome sequence information in rice and *Arabidopsis* have been supplemented with suitable information about gene discovery (Sreenivasulu et al., 2007). Several thousand genes induced under drought stress have been identified with the advancements in DNA microarray technology (Umezawa et al., 2006). Full-length cDNA microarray library was used to study expression profiling of several genes in *Arabidopsis* subjected to drought stress (Seki, 2001; Seki et al., 2002). The use of a microarray to study gene expression under abiotic stress for the first time in rice was reported by Kawasaki et al. (2001). A large number of ESTs from cDNA libraries were generated using functional genomics, and 589 genes involved in stress response were identified by Gorantla et al. (2005). cDNA library was compared using cDNA microarray to study gene expression between upland and lowland rice cultivars (Wang et al., 2007). Though the genome size of other cereals is much bigger and complex as compared to rice (Paterson, 2006), sequencing projects in some other cereals have been undertaken like in maize (http://www.maizegenome.org/), sorghum (https://jgi.doe.gov/) and wheat (https://www.wheatgenome.org/). Other than DNA, microarray techniques like serial analysis of gene expression (SAGE) and quantitative real-time PCR (qRT-PCR) have been used to assess several genes involved in drought response (Sreenivasulu et al., 2007).

The effect of drought on plants can be better understood by studying the protein composition (Barnabás et al., 2008). Therefore, the alterations in gene expression under drought stress in maize were studied using proteome analysis by Hu et al. (2010). Comparative proteomic studies were carried out by Cheng et al. (2016) to study the drought resistance mechanism in wheat. Comparative proteomic studies of drought-tolerant and drought-sensitive cotton were carried out by Zhang et al. (2016). They found that the proteins induced in drought tolerance are mainly associated with metabolism, transport, detoxification and cellular structure. Similar studies were carried out in barley by Wang et al. (2015), and it was found that proteins associated with drought tolerance are mainly involved in photosynthesis, metabolism, energy and amino-acid biosynthesis.

Comprehensive information about the metabolites in plants can be gained with the help of advanced omics techniques like metabolomics (Okazaki and Saito, 2012). Plants undergo various metabolic changes in response to drought stress, and complete profiling of these metabolic changes can lead to better understanding of plant stress tolerance mechanisms (Langridge et al., 2006). Metabolomics is an emerging area in the field of plant sciences, and it is assumed that when combined with genomics, proteomics and transcriptomics, it can be a valuable tool in the better understanding of complex biological processes, including drought tolerance (Langridge et al., 2006; Okazaki and Saito, 2012).

4.6 CONCLUSION

The complex attributes of drought tolerance have impeded the pace of development of drought-tolerant varieties. However, significant progress has been made in understanding the physiological, genetic and molecular aspects of drought tolerance. There are different avenues to combat the drought problem like genetic engineering and plant breeding, which are efficient ways to alter the genetic background of crops to enable them to perform well in drought conditions. Plants have evolved with complex response systems for drought stress, and these include drought-responsive genes including transcription factors, ROS scavenging enzymes and other functional proteins. Many genes involved in the regulatory process have been identified and genetically engineered for enhancing drought tolerance in crops by overexpression and suppression of transgenic technologies. In spite of gaining significant results from genetic engineering, this technology is not publicly acceptable worldwide. On the other hand, since the 1980s, with the advent of molecular markers, marker-assisted breeding is largely adapted by researchers to develop drought-tolerant lines/cultivars. The advancement in sequencing technology has provided high throughput genotyping platforms, which has facilitated the dissection of major and minor effect QTLs for diverse traits related to drought stress through MAS. These identified QTLs have been mapped and cloned in many crops. Multidisciplinary approaches for genetic engineering and molecular breeding will advance our knowledge to fathom the complex system underlying drought tolerance.

REFERENCES

Agarwal, P.K., Agarwal, P., Reddy, M.K., Sopory, S.K. (2006) Role of DREB transcription factors in abiotic and biotic stress tolerance in plants. *Plant Cell Reports* 25(12): 1263–1274.

Al Abdallat, A.M., Ayad, J.Y., Abu Elenein, J.M., Al Ajlouni, Z., Harwood, W.A. (2014) Overexpression of the transcription factor HvSNAC1 improves drought tolerance in barley (*Hordeum vulgare* L.). *Molecular Breeding* 33(2): 401–414.

Almeida, G.D., Makumbi, D., Magorokosho, C., Nair, S., Borém, A., Ribaut, J.-M., Bänziger, M., Prasanna, B.M., Crossa, J., Babu, R. (2013) QTL mapping in three tropical maize populations reveals a set of constitutive and adaptive genomic regions for drought tolerance. *Theoretical and Applied Genetics* 126(3): 583–600.

Ambawat, S., Sharma, P., Yadav, N.R., Yadav, R.C. (2013) MYB transcription factor genes as regulators for plant responses: An overview. *Physiology and Molecular Biology of Plants* 19(3): 307–321.

Aneja, B., Yadav, N.R., Kumar, N., Yadav, R.C. (2015) Hsp transcript induction is correlated with physiological changes under drought stress in Indian mustard. *Physiology and Molecular Biology of Plants* 21(3): 305–316.

Anjum, S.A., Xie, X.-Y., Wang, L.-C., Saleem, M.F., Man, C., Lei, W. (2011) Morphological, physiological and biochemical responses of plants to drought stress. *African Journal of Agricultural Research* 6(9): 2026–2032.

Anjum, S.A., Ashraf, U., Tanveer, M., Khan, I., Hussain, S., Shahzad, B., Zohaib, A., Abbas, F., Saleem, M.F., Ali, I., Wang, L.C. (2017) Drought induced changes in growth, osmolyte accumulation and antioxidant metabolism of three maize hybrids. *Frontiers in Plant Science* 8(69): 1–12.

Ashraf, M. (2010) Inducing drought tolerance in plants: Recent advances. *Biotechnology Advances* 28(1): 169–183.

Babitha, K.C., Vemanna, R.S., Nataraja, K.N., Udayakumar, M. (2015) Overexpression of EcbHLH57 transcription factor from *Eleusine coracana* L. in tobacco confers tolerance to salt, oxidative and drought stress. *PLoS ONE* 10(9): 1–21.

Barnabás, B., Jäger, K., Fehér, A. (2008) The effect of drought and heat stress on reproductive processes in cereals. *Plant, Cell and Environment* 31(1): 11–38.

Bartels, D., Sunkar, R. (2005) Drought and salt tolerance in plants. *Critical Reviews in Plant Sciences* 24(1): 23–58.

Basu, S., Ramegowda, V., Kumar, A., Pereira, A. (2016) Plant adaptation to drought stress. *F1000Research* 5: 1–10.

Baum, M., Grando, S., Backes, G., Jahoor, A., Sabbagh, A., Ceccarelli, S. (2003) QTLs for agronomic traits in the Mediterranean environment identified in recombinant inbred lines of the cross "Arta" x H. spontaneum 41–1. *Theoretical and Applied Genetics* 107(7):1215–1225.

Bennett, D., Izanloo, A., Edwards, J., Kuchel, H., Chalmers, K., Tester, M., Reynolds, M., Schnurbusch, T., Langridge, P. (2012) Identification of novel quantitative trait loci for days to ear emergence and flag leaf glaucousness in a bread wheat (*Triticum aestivum* L.) population adapted to southern Australian conditions. *Theoretical and Applied Genetics* 124(4): 697–711.

Benoit, J.B., Eric, D., Jean-Baptiste, T., Bao-Lam, H., Issa, D., Timothy, J.C., Kwadwo, O., Philip, R., Tinga, J.O. (2016) Application of marker-assisted backcrossing to improve cowpea (*Vigna unguiculata* L. Walp) for drought tolerance. *Journal of Plant Breeding and Crop Science* 8(12): 273–286.

Bernatzky, R., Tanksley, S.D. (1986) Toward a saturated linkage map in tomato based on isozymes and random cDNA sequences. *Genetics* 112(4): 887–898.

Bevan, M.W., Uauy, C. (2013) Genomics reveals new landscapes for crop improvement. *Genome Biology* 14(206): 1–11.

Beyene, Y., Semagn, K., Mugo, S., Tarekegne, A., Babu, R., Meisel, B., Sehabiague, P., Makumbi, D., Magorokosho, C., Oikeh, S., Gakunga, J., Vargas, M., Olsen, M., Prasanna, B.M., Banziger, M., Crossa, J. (2015) Genetic gains in grain yield through genomic selection in eight bi-parental maize populations under drought stress. *Crop Science* 55(1): 154–163.

Bidinger, F.R., Nepolean, T., Hash, C.T., Yadav, R.S., Howarth, C.J. (2007) Quantitative trait loci for grain yield in pearl millet under variable postflowering moisture conditions. *Crop Science* 47(3): 969–980.

Bidinger, F.R., Serraj, R., Rizvi, S.M.H., Howarth, C., Yadav, R.S., Hash, C.T. (2005) Field evaluation of drought tolerance QTL effects on phenotype and adaptation in pearl millet [*Pennisetum glaucum* (L.) R. Br.] topcross hybrids. *Field Crops Research* 94: 14–32.

Blum, A. (2017) Osmotic adjustment is a prime drought stress adaptive engine in support of plant production. *Plant, Cell, Environment* 40(1): 4–10.

Bohra, A. (2013) Emerging paradigms in genomics-based crop improvement. *The Scientific World Journal* 585467: 1–17.

Boyer, J.S. (1982) Plant productivity and environment. *Science* 218(4571): 443–448.

Bray, E.A. (2007) *Plant Response to Water-Deficit Stress. Encyclopedia of Life Sciences*, pp. 1–5. John Wiley, Sons, Ltd, Chichester, UK.

Broin, M., Cuiné, S., Peltier, G., Rey, P. (2000) Involvement of CDSP 32, a drought-induced thioredoxin, in the response to oxidative stress in potato plants. *FEBS Letters* 467(2–3): 245–248.

Budak, H., Hussain, B., Khan, Z., Ozturk, N.Z., Ullah, N. (2015) From genetics to functional genomics: Improvement in drought signaling and tolerance in wheat. *Frontiers in Plant Science* 6(1012): 1–13.

Cai, G., Wang, G., Wang, L., Liu, Y., Pan, J., Li, D. (2014) A maize mitogen-activated protein kinase kinase, ZmMKK1, positively regulated the salt and drought tolerance in transgenic Arabidopsis. *Journal of Plant Physiology* 171(12): 1003–1016.

Chaves, M.M., Maroco, J.P., Pereira, J.S. (2003) Understanding plant responses to drought—From genes to the whole plant. *Functional Plant Biology* 30: 239–264.

Chen, T., Li, W., Hu, X., Guo, J., Liu, A., Zhang, B. (2014) A cotton MYB transcription factor, GbMYB5, is positively involved in plant adaptive response to drought stress. *Plant and Cell Physiology* 56(5): 917–929.

Chen, H., Liu, L., Wang, L., Wang, S., Cheng, X. (2016) VrDREB2A, a DREB-binding transcription factor from Vigna radiata, increased drought and high-salt tolerance in transgenic *Arabidopsis thaliana*. *Journal of Plant Research* 129(2): 263–273.

Chen, T.H.H., Murata, N. (2002) Enhancement of tolerance of abiotic stress by metabolic engineering of betaines and other compatible solutes. *Current Opinion in Plant Biology* 5(3): 250–257.

Cheng, Y.-J., Deng, X.-P., Kwak, S.-S., Chen, W., Eneji, A.E. (2013) Enhanced tolerance of transgenic potato plants expressing choline oxidase in chloroplasts against water stress. *Botanical Studies* 54(30): 1–9.

Cheng, L., Wang, Y., He, Q., Li, H., Zhang, X., Zhang, F. (2016) Comparative proteomics illustrates the complexity of drought resistance mechanisms in two wheat

(*Triticum aestivum* L.) cultivars under dehydration and rehydration. *BMC Plant Biology* 16(188): 1–23.

Comas, L.H., Becker, S.R., Cruz, V.M.V., Byrne, P.F., Dierig, D.A. (2013) Root traits contributing to plant productivity under drought. *Frontiers in Plant Science* 4(442): 1–16.

DaMatta, F.M., Ramalho, J.D.C. (2006) Impacts of drought and temperature stress on coffee physiology and production: A review. *Brazilian Journal of Plant Physiology* 18(1): 55–81.

Diab, A.A., Teulat-Merah, B., This, D., Ozturk, N.Z., Benscher, D., Sorrells, M.E. (2004) Identification of drought-inducible genes and differentially expressed sequence tags in barley. *Theoretical and Applied Genetics* 109(7): 1417–1425.

Dixit, S., Yadaw, R.B., Mishra, K.K., Kumar, A. (2017) Marker-assisted breeding to develop the drought-tolerant version of Sabitri, a popular variety from Nepal. *Euphytica* 213(184): 1–16.

Dong, Y., Wang, C., Han, X., Tang, S., Liu, S., Xia, X., Yin, W. (2014) A novel bHLH transcription factor PebHLH35 from Populus euphratica confers drought tolerance through regulating stomatal development, photosynthesis and growth in *Arabidopsis*. *Biochemical and Biophysical Research Communications* 450(1): 453–458.

Duan, M., Zhang, R., Zhu, F., Zhang, Z., Gou, L., Wen, J., Dong, J., Wang, T. (2017) A Lipid-anchored NAC transcription factor is translocated into the nucleus and activates glyoxalase I expression during drought stress. *The Plant Cell* 29(7): 1748–1772.

Duque, A.S., López-Gómez, M., Kráčmarová, J., Gomes, C.N., Araújo, S.S., Lluch, C., Fevereiro, P. (2016) Genetic engineering of polyamine metabolism changes Medicago truncatula responses to water deficit. *Plant Cell, Tissue and Organ Culture* 127(3): 681–690.

Ebeed, H.T., Hassan, N.M., Aljarani, A.M. (2017) Exogenous applications of Polyamines modulate drought responses in wheat through osmolytes accumulation, increasing free polyamine levels and regulation of polyamine biosynthetic genes. *Plant Physiology and Biochemistry* 118: 438–448.

Edae, E.A., Byrne, P.F., Manmathan, H., Haley, S.D., Moragues, M., Lopes, M.S., Reynolds, M.P. (2013) Association mapping and nucleotide sequence variation in five drought tolerance candidate genes in spring wheat. *The Plant Genome* 6(2): 1–13.

Espasandin, F.D., Maiale, S.J., Calzadilla, P., Ruiz, O.A., Sansberro, P.A. (2014) Transcriptional regulation of 9-cis-epoxycarotenoid dioxygenase (NCED) gene by putrescine accumulation positively modulates ABA synthesis and drought tolerance in *Lotus tenuis* plants. *Plant Physiology and Biochemistry* 76: 29–35.

Fang, Y., Liao, K., Du, H., Xu, Y., Song, H., Li, X., Xiong, L. (2015a) A stress-responsive NAC transcription factor SNAC3 confers heat and drought tolerance through modulation of reactive oxygen species in rice. *Journal of Experimental Botany* 66(21): 6803–6817.

Fang, Z., Zhang, X., Gao, J., Wang, P., Xu, X., Liu, Z., Shen, S., Feng, B. (2015b) A Buckwheat (*Fagopyrum esculentum*) DRE-binding transcription factor gene, FeDREB1,

enhances freezing and drought tolerance of transgenic *Arabidopsis*. *Plant Molecular Biology Reporter* 33(5): 1510–1525.

Farooq, M., Basra, S.M.A., Wahid, A., Ahmad, N., Saleem, B.A. (2009) Improving the drought tolerance in rice (*Oryza sativa* L.) by exogenous application of salicylic acid. *Journal of Agronomy and Crop Science* 195(4): 237–246.

Gale, M.D., Devos, K.M., Zhu, J.H., Allouis, S., Couchman, M.S., Liu, H., Pittaway, T.S., Qi, X.Q., Kolesnikova-Allen, M., Hash, C.T. (2005) New molecular marker technologies for pearl millet improvement. *SAT eJournal* 1(1): 1–7.

Gao, F., Xiong, A., Peng, R., Jin, X., Xu, J., Zhu, B., Chen, J., Yao, Q. (2010) OsNAC52, a rice NAC transcription factor, potentially responds to ABA and confers drought tolerance in transgenic plants. *Plant Cell, Tissue and Organ Culture* 100(3): 255–262.

Gao, F., Yao, H., Zhao, H., Zhou, J., Luo, X., Huang, Y., Li, C., Chen, H., Wu, Q. (2016) Tartary buckwheat FtMYB10 encodes an R2R3-MYB transcription factor that acts as a novel negative regulator of salt and drought response in transgenic *Arabidopsis*. *Plant Physiology and Biochemistry* 109: 387–396.

Gong, D.S., Xiong, Y.C., Ma, B.L., Wang, T.M., Ge, J.P., Qin, X.L., Li, P.F., Kong, H.Y., Li, Z.Z., Li, F.M. (2010) Early activation of plasma membrane H+-ATPase and its relation to drought adaptation in two contrasting oat (*Avena sativa* L.) genotypes. *Environmental and Experimental Botany* 69(1): 1–8.

Gorantla, M., Babu, P.R., Lachagari, V.B.R., Feltus, F.A., Paterson, A.H., Reddy, A.R. (2005) Functional genomics of drought stress response in rice: Transcript mapping of annotated unigenes of an indica rice (*Oryza sativa* L. cv. Nagina 22). *Current Science* 89(3): 496–514.

Guo, Z., Tucker, D.M., Lu, J., Kishore, V., Gay, G. (2012) Evaluation of genome-wide selection efficiency in maize nested association mapping populations. *Theoretical and Applied Genetics* 124(2): 261–275.

Guttikonda, S.K., Valliyodan, B., Neelakandan, A.K., Tran, L.S.P., Kumar, R., Quach, T.N., Voothuluru, P., Gutierrez-Gonzalez, J.J., Aldrich, D.L., Pallardy, S.G., Sharp, R.E., Ho, T.H.D., Nguyen, H.T. (2014) Overexpression of AtDREB1D transcription factor improves drought tolerance in soybean. *Molecular Biology Reports* 41(12): 7995–8008.

Hao, Z., Li, X., Liu, X., Xie, C., Li, M., Zhang, D., Zhang, S. (2010) Meta-analysis of constitutive and adaptive QTL for drought tolerance in maize. *Euphytica* 174(2): 165–177.

Harris, K., Subudhi, P.K., Borrell, A., Jordan, D., Rosenow, D., Nguyen, H., Klein, P., Klein, R., Mullet, J. (2007) Sorghum stay-green QTL individually reduce post-flowering drought-induced leaf senescence. *Journal of Experimental Botany* 58(2): 327–338.

He, G.-H., Xu, J.-Y., Wang, Y.-X., Liu, J.-M., Li, P.-S., Chen, M., Ma, Y.-Z., Xu, Z.-S. (2016) Drought-responsive WRKY transcription factor genes TaWRKY1 and

TaWRKY33 from wheat confer drought and/or heat resistance in *Arabidopsis*. *BMC Plant Biology* 16(116): 1–16.

Heisey, P.W., Edmeades, G.O. (1997) World maize facts and trends 1997/98 maize production in drought-stressed environments: Technical options and research resource allocation. *Agricultural Economics* 74.

Hong, Y., Zhang, H., Huang, L., Li, D., Song, F. (2016) Overexpression of a stress-responsive NAC transcription factor gene ONAC022 improves drought and salt tolerance in rice. *Frontiers in Plant Science* 7(4): 1–19.

Hoyos-Villegas, V., Song, Q., Kelly, J.D. (2017) Genome-wide association analysis for drought tolerance and associated traits in common bean. *The Plant Genome* 10(1): 1–17.

Hu, X., Li, Y., Li, C., Yang, H., Wang, W., Lu, M. (2010) Characterization of small heat shock proteins associated with maize tolerance to combined drought and heat stress. *Journal of Plant Growth Regulation* 29(4): 455–464.

Huang, Q., Wang, Y., Li, B., Chang, J., Chen, M., Li, K., Yang, G., He, G. (2015) TaNAC29, a NAC transcription factor from wheat, enhances salt and drought tolerance in transgenic *Arabidopsis*. *BMC Plant Biology* 15(268): 1–15.

Huseynova, I., Rustamova, S. (2010) Screening for drought stress tolerance in wheat genotypes using molecular markers. *Proceedings of ANAS (Biological Sciences)* 65(5–6): 132–139.

Idrissi, O., Udupa, M.S., De Keyser, E., Van Damme, P., De Riek, J. (2016) Functional genetic diversity analysis and identification of associated simple sequence repeats and amplified fragment length polymorphism markers to drought tolerance in lentil (*Lens culinaris* ssp. culinaris medicus) landraces. *Plant Molecular Biology Reporter* 34(3): 659–680.

IPCC. (2014) *Climate Change 2014 Synthesis Report Summary Chapter for Policymakers. IPCC* 6: 31.

Jangra, S., Mishra, A., Kamboj, D., Yadav, N.R., Yadav, R.C. (2017) Engineering abiotic stress tolerance traits for mitigating climate change. *Plant Biotechnology: Recent Advancements and Developments*, pp. 59–73. Springer, Singapore.

Kale, S.M., Jaganathan, D., Ruperao, P., Chen, C., Punna, R., Kudapa, H., Thudi, M., Roorkiwal, M., Katta, M.A., Doddamani, D., Garg, V., Kishor, P.B.K., Gaur, P.M., Nguyen, H.T., Batley, J., Edwards, D., Sutton, T., Varshney, R.K. (2015) Prioritization of candidate genes in "QTL-hotspot" region for drought tolerance in chickpea (*Cicer arietinum* L.). *Scientific Reports* 5(15296): 1–14.

Kamal, N.M., Serag, Y., Gorafi, A., Mukhtar, A., Ghanim, A. (2017) Performance of sorghum stay-green introgression lines under post-flowering drought. *International Journal of Plant Research* 7(3): 65–74.

Kassahun, B., Bidinger, F.R., Hash, C.T., Kuruvinashetti, M.S. (2010) Stay-green expression in early generation sorghum [*Sorghum bicolor* (L.) Moench] QTL introgression lines. *Euphytica* 172(3): 351–362.

Kavar, T., Maras, M., Kidrič, M., Šuštar-Vozlič, J., Meglič, V. (2008) Identification of genes involved in the response of leaves of Phaseolus vulgaris to drought stress. *Molecular Breeding* 21(2): 159–172.

Kawasaki, S., Borchert, C., Deyholos, M., Wang, H., Brazille, S., Kawai, K., Galbraith, D., Bohnert, H.J. (2001) Gene expression profiles during the initial phase of salt stress in rice. *The Plant Cell* 13(4): 889–905.

Ke, Q., Wang, Z., Ji, C.Y., Jeong, J.C., Lee, H.-S., Li, H., Xu, B., Deng, X., Kwak, S.-S. (2016) Transgenic poplar expressing codA exhibits enhanced growth and abiotic stress tolerance. *Plant Physiology and Biochemistry* 100: 75–84.

Khan, M.S., Ahmad, D., Khan, M.A. (2015) Utilization of genes encoding osmoprotectants in transgenic plants for enhanced abiotic stress tolerance. *Electronic Journal of Biotechnology* 18: 257–266.

Kholová, J., Hash, C.T., Kočová, M., Vadez, V. (2011) Does a terminal drought tolerance QTL contribute to differences in ROS scavenging enzymes and photosynthetic pigments in pearl millet exposed to drought? *Environmental and Experimental Botany* 71(1): 99–106.

Kholová, J., Nepolean, T., Tom Hash, C., Supriya, A., Rajaram, V., Senthilvel, S., Kakkera, A., Yadav, R., Vadez, V. (2012) Water saving traits co-map with a major terminal drought tolerance quantitative trait locus in pearl millet [Pennisetum glaucum (L.) R. Br.]. *Molecular Breeding* 30(3): 1337–1353.

Kim, T.-H., Hur, Y.-J., Han, S.-I., Cho, J.-H., Kim, K.-M., Lee, J.-H., Song, Y.-C., Kwon, Y.-U., Shin, D. (2017) Drought-tolerant QTL qVDT11 leads to stable tiller formation under drought stress conditions in rice. *Plant Science* 256: 131–138.

Kirigwi, F.M., Van Ginkel, M., Brown-Guedira, G., Gill, B.S., Paulsen, G.M., Fritz, A.K. (2007) Markers associated with a QTL for grain yield in wheat under drought. *Molecular Breeding* 20(4): 401–413.

von Korff, M., Grando, S., Del Greco, A., This, D., Baum, M., Ceccarelli, S. (2008) Quantitative trait loci associated with adaptation to Mediterranean dryland conditions in barley. *Theoretical and Applied Genetics* 117(5): 653–669.

von Korff, M., Wang, H., Pillen, K. (2006) AB-QTL analysis in spring barley: II. Detection of favourable exotic alleles for agronomic traits introgressed from wild barley (*H. vulgare* ssp. spontaneum). *Theoretical and Applied Genetics* 112(7): 1221–1231.

Kosgei, A.J., Kimurto, P.K., Gaur, P.M., Yeboah, M.A., Offei, S.K., Danquah, E.Y., Muriuki, R.W., Thudi, M., Varshney, R.K. (2017) Introgression of drought tolerance traits into adapted Kenyan chickpea varieties using marker assisted backcrossing (MABC). *InterDrought-V*: 32.

Kover, P.X., Valdar, W., Trakalo, J., Scarcelli, N., Ehrenreich, I.M., Purugganan, M.D., Durrant, C., Mott, R. (2009) A multiparent advanced generation inter-cross to fine-map quantitative traits in *Arabidopsis thaliana* (ed R Mauricio). *PLoS Genetics* 5: e1000551.

Kudo, M., Kidokoro, S., Yoshida, T., Mizoi, J., Todaka, D., Fernie, A.R., Shinozaki, K., Yamaguchi-Shinozaki, K. (2017) Double overexpression of DREB and PIF transcription factors improves drought stress tolerance and cell elongation in transgenic plants. *Plant Biotechnology Journal* 15(4): 458–471.

Kumar, S., Sehgal, S.K., Kumar, U., Prasad, P.V.V., Joshi, A.K., Gill, B.S. (2012) Genomic characterization of drought tolerance-related traits in spring wheat. *Euphytica* 186(1): 265–276.

Lai, S.J., Lai, M.C., Lee, R.J., Chen, Y.H., Yen, H.E. (2014) Transgenic *Arabidopsis* expressing osmolyte glycine betaine synthesizing enzymes from halophilic methanogen promote tolerance to drought and salt stress. *Plant Molecular Biology* 85(4–5): 429–441.

Langridge, P., Paltridge, N., Fincher, G. (2006) Functional genomics of abiotic stress tolerance in cereals. *Briefings in Functional Genomics and Proteomics* 4(4): 343–354.

Lata, C., Prasad, M. (2011) Role of DREBs in regulation of abiotic stress responses in plants. *Journal of Experimental Botany* 62(14): 4731–4748.

Li, Y., Ye, W., Wang, M., Yan, X. (2009) Climate change and drought: A risk assessment of crop-yield impacts. *Climate Research* 39(1): 31–46.

Li, M., Li, Z., Li, S., Guo, S., Meng, Q., Li, G., Yang, X. (2014) Genetic engineering of glycine betaine biosynthesis reduces heat-enhanced photoinhibition by enhancing antioxidative defense and alleviating lipid peroxidation in tomato. *Plant Molecular Biology Reporter* 32(1): 42–51.

Li, H., Zhang, D., Li, X., Guan, K., Yang, H. (2016) Novel DREB A-5 subgroup transcription factors from desert moss (*Syntrichia caninervis*) confers multiple abiotic stress tolerance to yeast. *Journal of Plant Physiology* 194: 45–53.

Liang, C., Meng, Z., Meng, Z., Malik, W., Yan, R., Lwin, K.M., Lin, F., Wang, Y., Sun, G., Zhou, T., Zhu, T., Li, J., Jin, S., Guo, S., Zhang, R. (2016) GhABF2, a bZIP transcription factor, confers drought and salinity tolerance in cotton (*Gossypium hirsutum* L.). *Scientific Reports* 6: 35040.

Liu, C.J., Witcombe, J.R., Pittaway, T.S., Nash, M., Hash, C.T., Busso, C.S., Gale, M.D. (1994) An RFLP-based genetic map of pearl millet (*Pennisetum glaucum*). *Theoretical and Applied Genetics* 89(4): 481–487.

Liu, C., Wu, Y., Wang, X. (2012) BZIP transcription factor OsbZIP52/RISBZ5: A potential negative regulator of cold and drought stress response in rice. *Planta* 235(6): 1157–1169.

Liu, C., Mao, B., Ou, S., Wang, W., Liu, L., Wu, Y., Chu, C., Wang, X. (2014a) OsbZIP71, a bZIP transcription factor, confers salinity and drought tolerance in rice. *Plant Molecular Biology* 84(1–2): 19–36.

Liu, W., Tai, H., Li, S., Gao, W., Zhao, M., Xie, C., Li, W.X. (2014b) bHLH122 is important for drought and osmotic stress resistance in *Arabidopsis* and in the repression of ABA catabolism. *New Phytologist* 201(4): 1192–1204.

Lopes, M.S., Reynolds, M.P., McIntyre, C.L., Mathews, K.L., Jalal Kamali, M.R., Mossad, M., Feltaous, Y., Tahir, I.S.A., Chatrath, R., Ogbonnaya, F., Baum, M. (2013) QTL for yield and associated traits in the Seri/Babax population grown across several environments in Mexico, in the West Asia, North Africa, and South Asia regions. *Theoretical and Applied Genetics* 126(4): 971–984.

Lv, Z., Wang, S., Zhang, F., Chen, L., Hao, X., Pan, Q., Fu, X., Li, L., Sun, X., Tang, K. (2016) Overexpression of a novel NAC domain-containing transcription factor gene (AaNAC1) enhances the content of artemisinin and increases tolerance to drought and botrytis cinerea in artemisia annua. *Plant and Cell Physiology* 57(9): 1961–1971.

Maccaferri, M., Sanguineti, M.C., Corneti, S., Ortega, J.L.A., Salem, M.B., Bort, J., DeAmbrogio, E., del Moral, L.F.G., Demontis, A., El-Ahmed, A., Maalouf, F., Machlab, H., Martos, V., Moragues, M., Motawaj, J., Nachit, M., Nserallah, N., Ouabbou, H., Royo, C., Slama, A., Tuberosa, R. (2008) Quantitative trait loci for grain yield and adaptation of durum wheat (*Triticum durum* Desf.) across a wide range of water availability. *Genetics* 178(1): 489–511.

Manivannan, P., Jaleel, C.A., Somasundaram, R., Panneerselvam, R. (2008) Osmoregulation and antioxidant metabolism in drought-stressed *Helianthus annuus* under triadimefon drenching. *Comptes Rendus Biologies* 331(6): 418–425.

Mao, X., Chen, S., Li, A., Zhai, C., Jing, R. (2014) Novel NAC transcription factor TaNAC67 confers enhanced multi-abiotic stress tolerances in *Arabidopsis*. *PLoS ONE* 9(1): e84359.

Mao, H., Yu, L., Han, R., Li, Z., Liu, H. (2016) ZmNAC55, a maize stress-responsive NAC transcription factor, confers drought resistance in transgenic *Arabidopsis*. *Plant Physiology and Biochemistry* 105: 55–66.

Meuwissen, T.H., Hayes, B.J., Goddard, M.E. (2001) Prediction of total genetic value using genome-wide dense marker maps. *Genetics* 157(4): 1819–1829.

Mikołajczak, K., Ogrodowicz, P., Gudyś, K., Krystkowiak, K., Sawikowska, A., Frohmberg, W., Górny, A., Kędziora, A., Jankowiak, J., Józefczyk, D., Karg, G., Andrusiak, J., Krajewski, P., Szarejko, I., Surma, M., Adamski, T., Guzy-Wróbelska, J., Kuczyńska, A. (2016) Quantitative trait loci for yield and yield-related traits in spring barley populations derived from crosses between european and syrian cultivars. *PLoS ONE* 11: e0155938.

Mishra, V., Cherkauer, K.A. (2010) Retrospective droughts in the crop growing season: Implications to corn and soybean yield in the Midwestern United States. *Agricultural and Forest Meteorology* 150(7–8): 1030–1045.

Mondini, L., Nachit, M.M., Pagnotta, M.A. (2015) Allelic variants in durum wheat (*Triticum turgidum* L. var. durum) DREB genes conferring tolerance to abiotic stresses. *Molecular Genetics and Genomics* 290(2): 531–544.

Moon, S.J., Han, S.Y., Kim, D.Y., Yoon, I.S., Shin, D., Byun, M.O., Kwon, H.B., Kim, B.G. (2015) Ectopic expression of a hot pepper bZIP-like transcription factor in potato enhances drought tolerance without decreasing tuber yield. *Plant Molecular Biology* 89(4–5): 421–431.

Noctor, G., Mhamdi, A., Foyer, C.H. (2014) The roles of reactive oxygen metabolism in drought: Not so cut and dried. *Plant Physiology* 164(4): 1636–1648.

Nuccio, M.L., Rhodes, D., McNeil, S.D., Hanson, A.D. (1999) Metabolic engineering of plants for osmotic stress resistance. *Current Opinion in Plant Biology* 2(2): 128–134.

Okazaki, Y., Saito, K. (2012) Recent advances of metabolomics in plant biotechnology. *Plant Biotechnology Reports* 6(1): 1–15.

Ouedraogo, N., Sanou, J., Gracen, V., Tongoona, P. (2017) Incorporation of stay-green Quantitative Trait Loci (QTL) in elite sorghum (*Sorghum bicolor* L . Moench) variety through marker-assisted selection at early generation. *Journal of Applied Biosciences* 111: 10867–10876.

Pandey, N., Ranjan, A., Pant, P., Tripathi, R.K., Ateek, F., Pandey, H.P., Patre, U.V., Sawant, S.V. (2013) CAMTA 1 regulates drought responses in *Arabidopsis thaliana*. *BMC Genomics* 14(216): 1–23.

Pandurangaiah, M., Lokanadha Rao, G., Sudhakarbabu, O., Nareshkumar, A., Kiranmai, K., Lokesh, U., Thapa, G., Sudhakar, C. (2014) Overexpression of horsegram (*Macrotyloma uniflorum* Lam.Verdc.) NAC transcriptional factor (MuNAC4) in groundnut confers enhanced drought tolerance. *Molecular Biotechnology* 56(8): 758–769.

Paterson, A.H. (2006) Leafing through the genomes of our major crop plants: Strategies for capturing unique information. *Nature Reviews Genetics* 7: 174–184.

Paterson, A.H., Lander, E.S., Hewitt, J.D., Peterson, S., Lincoln, S.E., Tanksley, S.D. (1988) Resolution of quantitative traits into Mendelian factors by using a complete linkage map of restriction fragment length polymorphisms. *Nature* 335(6192): 721–726.

Phukan, U.J., Jeena, G.S., Shukla, R.K. (2016) WRKY transcription factors: Molecular regulation and stress responses in plants. *Frontiers in Plant Science* 7: 1–14.

Pinto, R.S., Reynolds, M.P., Mathews, K.L., McIntyre, C.L., Olivares-Villegas, J.-J., Chapman, S.C. (2010) Heat and drought adaptive QTL in a wheat population designed to minimize confounding agronomic effects. *Theoretical and Applied Genetics* 121(6): 1001–1021.

Prabu, G., Prasad, D.T. (2012) Functional characterization of sugarcane MYB transcription factor gene promoter (PScMYBAS1) in response to abiotic stresses and hormones. *Plant Cell Reports* 31(4): 661–669.

Qin, D., Zhao, C.-L., Liu, X.-Y., Wang, P.-W. (2017) Transgenic soybeans expressing betaine aldehyde dehydrogenase from *Atriplex canescens* show increased drought tolerance. *Plant Breeding* 136(5): 699–709.

Quarrie, S., Pekic Quarrie, S., Radosevic, R., Rancic, D., Kaminska, A., Barnes, J.D., Leverington, M., Ceoloni, C., Dodig, D. (2006) Dissecting a wheat QTL for yield present in a range of environments: From the QTL to candidate genes. *Journal of Experimental Botany* 57(11): 2627–2637.

Rahman, H., Ramanathan, V., Nallathambi, J., Duraialagaraja, S., Muthurajan, R. (2016) Over-expression of a NAC 67 transcription factor from finger millet (*Eleusine coracana* L.) confers tolerance against salinity and drought stress in rice. *BMC Biotechnology* 16(1): 8–20.

Raineri, J., Ribichich, K.F., Chan, R.L. (2015) The sunflower transcription factor HaWRKY76 confers drought and flood tolerance to *Arabidopsis thaliana* plants without yield penalty. *Plant Cell Reports* 34(12): 2065–2080.

Rajaram, V., Nepolean, T., Senthilvel, S., Varshney, R.K., Vadez, V., Srivastava, R.K., Shah, T.M., Supriya, A., Kumar, S., Ramana Kumari, B., Bhanuprakash, A., Narasu, M.L., Riera-Lizarazu, O., Hash, C.T. (2013) Pearl millet [*Pennisetum glaucum* (L.) R. Br.] consensus linkage map constructed using four RIL mapping populations and newly developed EST-SSRs. *BMC Genomics* 14(159): 1–15.

Ramachandra Reddy, A., Chaitanya, K.V., Vivekanandan, M. (2004) Drought-induced responses of photosynthesis and antioxidant metabolism in higher plants. *Journal of Plant Physiology* 161(11): 1189–1202.

Rehman, A.U., Malhotra, R.S., Bett, K., Tar'an, B., Bueckert, R., Warkentin, T.D. (2011) Mapping QTL associated with traits affecting grain yield in chickpea (*Cicer arietinum* L.) under terminal drought stress. *Crop Science* 51(2): 450–463.

Ren, S., Lyle, C., Jiang, G.-L., Penumala, A. (2016) Soybean salt tolerance 1 (GmST1) reduces ROS production, enhances ABA sensitivity, and abiotic stress tolerance in *Arabidopsis thaliana*. *Frontiers in Plant Science* 7(445): 1–14.

Ribaut, J.-M., Hoisington, D.A., Deutsch, J.A., Jiang, C., Gonzalez-de-Leon, D. (1996) Identification of quantitative trait loci under drought conditions in tropical maize. 1. Flowering parameters and the anthesis-silking interval. *Theoretical and Applied Genetics* 92(7): 905–914.

Ribaut, J.-M., Jiang, C., Gonzalez-de-Leon, D., Edmeades, G.O., Hoisington, D.A. (1997) Identification of quantitative trait loci under drought conditions in tropical maize. *Theoretical and Applied Genetics* 94(6–7): 887–896.

Ribaut, J.-M., Ragot, M. (2006) Marker-assisted selection to improve drought adaptation in maize: The backcross approach, perspectives, limitations, and alternatives. *Journal of Experimental Botany* 58(2): 351–360.

Romani, F., Ribone, P.A., Capella, M., Miguel, V.N., Chan, R.L. (2016) A matter of quantity: Common features in the drought response of transgenic plants overexpressing HD-Zip I transcription factors. *Plant Science* 251: 139–154.

Rong, W., Qi, L., Wang, A., Ye, X., Du, L., Liang, H., Xin, Z., Zhang, Z. (2014) The ERF transcription factor TaERF3 promotes tolerance to salt and drought stresses in wheat. *Plant Biotechnology Journal* 12(4): 468–479.

Rontein, D., Basset, G., Hanson, A.D. (2002) Metabolic engineering of osmoprotectant accumulation in plants. *Metabolic Engineering* 4(1): 49–56.

Sakuraba, Y., Kim, Y.-S., Han, S.-H., Lee, B.-D., Paek, N.-C. (2015) The *Arabidopsis* transcription factor NAC016 promotes drought stress responses by repressing AREB1 transcription through a trifurcate feed-forward regulatory loop involving NAP. *The Plant Cell* 27(6): 1771–1787.

Salem, K.F.M., Röder, M.S., Börner, A. (2007) Identification and mapping quantitative trait loci for stem reserve mobilisation in wheat (*Triticum aestivum* L.). *Cereal Research Communications* 35(3): 1367–1374.

Sari-Gorla, M., Krajewski, P., Di Fonzo, N., Villa, M., Frova, C. (1999) Genetic analysis of drought tolerance in maize by molecular markers. II. Plant height and flowering. *Theoretical and Applied Genetics* 99(1): 289–295.

Sehgal, D., Skot, L., Singh, R., Srivastava, R.K., Das, S.P., Taunk, J., Sharma, P.C., Pal, R., Raj, B., Hash, C.T., Yadav, R.S. (2015) Exploring potential of pearl millet germplasm association panel for association mapping of drought tolerance traits. *PLoS ONE* 10(5): e0122165.

Seki, M. (2001) Monitoring the expression pattern of 1300 *Arabidopsis* genes under drought and cold stresses by using a full-length cDNA microarray. *The Plant Cell Online* 13(1): 61–72.

Seki, M., Narusaka, M., Ishida, J., Nanjo, T., Fujita, M., Oono, Y., Kamiya, A., Nakajima, M., Enju, A., Sakurai, T., Satou, M., Akiyama, K., Taji, T., Yamaguchi-Shinozaki, K., Carninci, P., Kawai, J., Hayashizaki, Y., Shinozaki, K. (2002) Monitoring the expression profiles of 7000 *Arabidopsis* genes under drought, cold and high-salinity stresses using a full-length cDNA microarray. *The Plant Journal : For Cell and Molecular Biology* 31(3): 279–292.

Seki, M., Kamei, A., Yamaguchi-Shinozaki, K., Shinozaki, K. (2003) Molecular responses to drought, salinity and frost: Common and different paths for plant protection. *Current Opinion in Biotechnology* 14(2): 194–199.

Selvaraj, M.G., Ishizaki, T., Valencia, M., Ogawa, S., Dedicova, B., Ogata, T., Yoshiwara, K., Maruyama, K., Kusano, M., Saito, K., Takahashi, F., Shinozaki, K., Nakashima, K., Ishitani, M. (2017) Overexpression of an *Arabidopsis thaliana* galactinol synthase gene improves drought tolerance in transgenic rice and increased grain yield in the field. *Plant Biotechnology Journal* 15(11): 1465–1477.

Semagn, K., Beyene, Y., Warburton, M.L., Tarekegne, A., Mugo, S., Meisel, B., Sehabiague, P., Prasanna, B.M. (2013) Meta-analyses of QTL for grain yield and anthesis silking interval in 18 maize populations evaluated under water-stressed and well-watered environments. *BMC Genomics* 14(313): 1–16.

Senthilvel, S., Jayashree, B., Mahalakshmi, V., Kumar, P.S., Nakka, S., Nepolean, T., Hash, C. (2008) Development and mapping of simple sequence repeat markers for pearl millet from data mining of expressed sequence tags. *BMC Plant Biology* 8(119): 1–9.

Sequera-Mutiozabal, M., Antoniou, C., Tiburcio, A.F., Alcázar, R., Fotopoulos, V. (2017) Polyamines: Emerging hubs promoting drought and salt stress tolerance in plants. *Current Molecular Biology Reports* 3(1): 28–36.

Serraj, R., Sinclair, T.R. (2002) Osmolyte accumulation: Can it really help increase crop yield under drought conditions? *Plant, Cell and Environment* 25(2): 333–341.

Serraj, R., Hash, C.T., Rizvi, S.M.H., Sharma, A., Yadav, R.S., Bidinger, F.R. (2005) Recent advances in marker-assisted selection for drought tolerance in pearl millet. *Plant Production Science* 8(3): 334–337.

Seyedsalehi, M., Sharifi, P., Paladino, O., Hodaifa, G., Villegas, E.C., Osman, R.M. (2017) Variation in polyamine content among 12 pollinated wheat genotypes under drought stress condition. *Open Journal of Geology* 7: 1094–1109.

Shamsudin, N.A.A., Swamy, B.P.M., Ratnam, W., Sta. Cruz, M.T., Raman, A., Kumar, A. (2016a) Marker assisted pyramiding of drought yield QTLs into a popular Malaysian rice cultivar, MR219. *BMC Genetics* 17(30): 1–14.

Shamsudin, N.A.A., Swamy, B.P.M., Ratnam, W., Sta. Cruz, M.T., Sandhu, N., Raman, A.K., Kumar, A. (2016b) Pyramiding of drought yield QTLs into a high quality Malaysian rice cultivar MRQ74 improves yield under reproductive stage drought. *Rice* 9(21): 1–13.

Sharma, P.C., Singh, D., Sehgal, D., Singh, G., Hash, C.T., Yadav, R.S. (2014) Further evidence that a terminal drought tolerance QTL of pearl millet is associated with reduced salt uptake. *Environmental and Experimental Botany* 102(100): 48–57.

Shen, J., Lv, B., Luo, L., He, J., Mao, C., Xi, D., Ming, F. (2017) The NAC-type transcription factor OsNAC2 regulates ABA-dependent genes and abiotic stress tolerance in rice. *Scientific Reports* 7(40641): 1–14.

Shinozaki, K., Yamaguchi-Shinozaki, K., Seki, M. (2003) Regulatory network of gene expression in the drought and cold stress responses. *Current Opinion in Plant Biology* 6(5): 410–417.

Shinozaki, K., Yamaguchi-Shinozaki, K. (2006) Gene networks involved in drought stress response and tolerance. *Journal of Experimental Botany* 58(2): 221–227.

Shukla, S., Singh, K., Patil, R.V., Kadam, S., Bharti, S., Prasad, P., Singh, N.K., Khanna-Chopra, R. (2015) Genomic regions associated with grain yield under drought stress in wheat (*Triticum aestivum* L.). *Euphytica* 203(2): 449–467.

Singh, M., Kumar, J., Singh, S., Singh, V.P., Prasad, S.M. (2015) Roles of osmoprotectants in improving salinity and drought tolerance in plants: A review. *Reviews in Environmental Science and Biotechnology* 14: 407–426.

Sreenivasulu, N., Sopory, S.K., Kavi Kishor, P.B. (2007) Deciphering the regulatory mechanisms of abiotic stress tolerance in plants by genomic approaches. *Gene* 388(1–2): 1–13.

Steele, K.A., Price, A.H., Shashidhar, H.E., Witcombe, J.R. (2006) Marker-assisted selection to introgress rice QTLs controlling root traits into an Indian upland rice variety. *Theoretical and Applied Genetics* 112(2): 208–221.

Steele, K.A., Virk, D.S., Kumar, R., Prasad, S.C., Witcombe, J.R. (2007) Field evaluation of upland rice lines selected for QTLs controlling root traits. *Field Crops Research* 101: 180–186.

Su, L.T., Li, J.W., Liu, D.Q., Zhai, Y., Zhang, H.J., Li, X.W., Zhang, Q.L., Wang, Y., Wang, Q.Y. (2014) A novel MYB transcription factor, GmMYBJ1, from soybean confers drought and cold tolerance in *Arabidopsis thaliana*. *Gene* 538(1): 46–55.

Sun, J., Peng, X., Fan, W., Tang, M., Liu, J., Shen, S. (2014) Functional analysis of BpDREB2 gene involved in salt and drought response from a woody plant Broussonetia papyrifera. *Gene* 535(2): 140–149.

Supriya, A., Senthilvel, S., Nepolean, T., Eshwar, K., Rajaram, V., Shaw, R., Hash, C.T., Kilian, A., Yadav, R.C., Narasu, M.L. (2011) Development of a molecular linkage map of pearl millet integrating DArT and SSR markers. *Theoretical and Applied Genetics* 123(2): 239–250.

Suprunova, T., Krugman, T., Distelfeld, A., Fahima, T., Nevo, E., Korol, A. (2007) Identification of a novel gene (Hsdr4) involved in water-stress tolerance in wild barley. *Plant Molecular Biology* 64(1–2): 17–34.

Sureshrao, S.K., Rameshsing, C.N., Pradeeprao, K.T., Prasad, A., Verulkar, S.B. (2016) Enhancement of rice yield through introgression of drought tolerance QTL in Swarna Sub 1 (*Oryza sativa* L.) using MAS based approaches. *International Journal of Bio-Resource and Stress Management* 7(4): 791–797.

Tabaeizadeh, Z. (1998) Drought-induced responses in plant cells. *International Review of Cytology* 182: 193–247.

Talamè, V., Sanguineti, M.C., Chiapparino, E., Bahri, H., Salem, M.B., Forster, B.P., Ellis, R.P., Rhouma, S., Zoumarou, W., Waugh, R., Tuberosa, R. (2004) Identification of Hordeum spontaneum QTL alleles improving field performance of barley grown under rainfed conditions. *Annals of Applied Biology* 144(3): 309–319.

Teulat, B., This, D., Khairallah, M., Borries, C., Ragot, C., Sourdille, P., Leroy, P., Monneveux, P., Charrier, A. (1998) Several QTLs involved in osmotic-adjustment trait variation in barley (*Hordeum vulgare* L.). *Theoretical and Applied Genetics* 96(5): 688–698.

Teulat, B., Borries, C., This, D. (2001a) New QTLs identified for plant water status, water-soluble carbohydrate and osmotic adjustment in a barley population grown in a growth-chamber under two water regimes. *Theoretical and Applied Genetics* 103(1): 161–170.

Teulat, B., Merah, O., Souyris, I., This, D. (2001b) QTLs for agronomic traits from a Mediterranean barley progeny grown in several environments. *Theoretical and Applied Genetics* 103(5): 774–787.

Teulat, B., Merah, O., Sirault, X., Borries, C., Waugh, R., This, D. (2002) QTLs for grain carbon isotope discrimination in field-grown barley. *Theoretical and Applied Genetics* 106(1): 118–126.

Teulat, B., Zoumarou-Wallis, N., Rotter, B., Ben Salem, M., Bahri, H., This, D. (2003) QTL for relative water content in field-grown barley and their stability across Mediterranean environments. *Theoretical and Applied Genetics* 108(1): 181–188.

Thirumalaikumar, V.P., Devkar, V., Mehterov, N., Ali, S., Ozgur, R., Turkan, I., Mueller-Roeber, B., Balazadeh, S. (2017) NAC transcription factor JUNGBRUNNEN1 enhances drought tolerance in tomato. *Plant Biotechnology Journal* 16(2): 1–13.

Thudi, M., Gaur, P.M., Krishnamurthy, L., Mir, R.R., Kudapa, H., Fikre, A., Kimurto, P., Tripathi, S., Soren, K.R., Mulwa, R., Bharadwaj, C., Datta, S., Chaturvedi, S.K., Varshney, R.K. (2014) Genomics-assisted breeding for drought tolerance in chickpea. *Functional Plant Biology* 41(11): 1178.

Tiburcio, A.F., Altabella, T., Bitrián, M., Alcázar, R. (2014) The roles of polyamines during the lifespan of plants: From development to stress. *Planta* 240(1): 1–18.

Tondelli, A., Francia, E., Visioni, A., Comadran, J., Mastrangelo, A.M., Akar, T., Al-Yassin, A., Ceccarelli, S., Grando, S., Benbelkacem, A., van Eeuwijk, F.A., Thomas, W.T.B., Stanca, A.M., Romagosa, I., Pecchioni, N. (2014) QTLs for barley yield adaptation to Mediterranean environments in the "Nure" × "Tremois" biparental population. *Euphytica* 197(1): 73–86.

Tripathi, P., Rabara, R.C., Rushton, P.J. (2014) A systems biology perspective on the role of WRKY transcription factors in drought responses in plants. *Planta* 239(2): 255–266.

Tuberosa, R., Salvi, S. (2006) Genomics-based approaches to improve drought tolerance of crops. *Trends in Plant Science* 11(8): 405–412.

Umezawa, T., Fujita, M., Fujita, Y., Yamaguchi-Shinozaki, K., Shinozaki, K. (2006) Engineering drought tolerance in plants: Discovering and tailoring genes to unlock the future. *Current Opinion in Biotechnology* 17(2): 113–122.

Varshney, R.K., Gaur, P.M., Chamarthi, S.K., Krishnamurthy, L., Tripathi, S., Kashiwagi, J., Samineni, S., Singh, V.K., Thudi, M., Jaganathan, D. (2013) Fast-track introgression of "QTL-hotspot" for root traits and other drought tolerance traits in JG 11, an elite and leading variety of chickpea. *The Plant Genome* 6(3): 1–9.

Veldboom, L.R., Lee, M. (1996) Genetic mapping of quantitative trait loci in maize in stress and nonstress environments: I. Grain yield and yield components. *Crop Science* 36(5): 1310.

Wang, H., Zhang, H., Gao, F., Li, J., Li, Z. (2007) Comparison of gene expression between upland and lowland rice cultivars under water stress using cDNA microarray. *Theoretical and Applied Genetics* 115(8): 1109–1126.

Wang, N., Zhao, J., He, X., Sun, H., Zhang, G., Wu, F. (2015) Comparative proteomic analysis of drought tolerance in the two contrasting Tibetan wild genotypes and cultivated genotype. *BMC Genomics* 16(432): 1–19.

Wang, J., Li, Q., Mao, X., Li, A., Jing, R. (2016a) Wheat transcription factor TaAREB3 participates in drought and freezing tolerances in *Arabidopsis*. *International Journal of Biological Sciences* 12(2): 257–269.

Wang, F., Zhu, H., Chen, D., Li, Z., Peng, R., Yao, Q. (2016b) A grape bHLH transcription factor gene, VvbHLH1, increases the accumulation of flavonoids and enhances salt and drought tolerance in transgenic *Arabidopsis thaliana*. *Plant Cell, Tissue and Organ Culture* 125(2): 387–398.

Wehner, G.G., Balko, C.C., Enders, M.M., Humbeck, K.K., Ordon, F.F. (2015) Identification of genomic regions involved in tolerance to drought stress and drought stress induced leaf senescence in juvenile barley. *BMC Plant Biology* 15(125): 1–15.

Wehner, G., Balko, C., Humbeck, K., Zyprian, E., Ordon, F. (2016) Expression profiling of genes involved in drought stress and leaf senescence in juvenile barley. *BMC Plant Biology* 16(3): 1–12.

Wei, B., Jing, R., Wang, C., Chen, J., Mao, X., Chang, X., Jia, J. (2009) Dreb1 genes in wheat (*Triticum aestivum* L.): Development of functional markers and gene mapping based on SNPs. *Molecular Breeding* 23(1): 13–22.

Wei, T., Deng, K., Liu, D., Gao, Y., Liu, Y., Yang, M., Zhang, L., Zheng, X., Wang, C., Song, W., Chen, C., Zhang, Y. (2016) Ectopic expression of DREB transcription factor, AtDREB1A, confers tolerance to drought in transgenic salvia miltiorrhiza. *Plant and Cell Physiology* 57(8): 1593–1609.

Worch, S., Rajesh, K., Harshavardhan, V.T., Pietsch, C., Korzun, V., Kuntze, L., Börner, A., Wobus, U., Röder, M.S., Sreenivasulu, N. (2011) Haplotyping, linkage mapping and expression analysis of barley genes regulated by terminal drought stress influencing seed quality. *BMC Plant Biology* 11(1): 1–14.

Wu, D., Ji, J., Wang, G., Guan, W., Guan, C., Jin, C., Tian, X. (2015) LcMKK, a novel group A mitogen-activated protein kinase kinase gene in *Lycium chinense*, confers dehydration and drought tolerance in transgenic tobacco via scavenging ROS and modulating expression of stress-responsive genes. *Plant Growth Regulation* 76(3): 269–279.

Xiong, H., Li, J., Liu, P., Duan, J., Zhao, Y., Guo, X., Li, Y., Zhang, H., Ali, J., Li, Z. (2014) Overexpression of OsMYB48-1, a novel MYB-related transcription factor, enhances drought and salinity tolerance in rice. *PLoS ONE* 9(3): 1–13.

Xu, W., Li, Y., Cheng, Z., Xia, G., Wang, M. (2016) A wheat histone variant gene TaH2A.7 enhances drought tolerance and promotes stomatal closure in *Arabidopsis*. *Plant Cell Reports* 35(9): 1853–1862.

Xue, G.-P., McIntyre, C.L., Glassop, D., Shorter, R. (2008) Use of expression analysis to dissect alterations in carbohydrate metabolism in wheat leaves during drought stress. *Plant Molecular Biology* 67(3): 197–214.

Yadav, R.S., Hash, C.T., Bidinger, F.R., Cavan, G.P., Howarth, C.J. (2002) Quantitative trait loci associated with traits determining grain and stover yield in pearl millet under terminal drought-stress conditions. *Theoretical and Applied Genetics* 104(1): 67–83.

Yadav, R.S., Hash, C.T., Bidinger, F.R., Devos, K.M., Howarth, C.J. (2004) Genomic regions associated with grain yield and aspects of post-flowering drought tolerance in pearl millet across stress environments and tester background. *Euphytica* 136(3): 265–277.

Yamaguchi-Shinozaki, K., Shinozaki, K. (2005) Organization of cis-acting regulatory elements in osmotic- and cold-stress-responsive promoters. *Trends in Plant Science* 10(2): 88–94.

Yan, Y., Jia, H., Wang, F., Wang, C., Liu, S., Guo, X. (2015) Overexpression of GhWRKY27a reduces tolerance to drought stress and resistance to Rhizoctonia solani infection in transgenic Nicotiana benthamiana. *Frontiers in Physiology* 6: 1–16.

Yang, Y., Dong, C., Li, X., Du, J., Qian, M., Sun, X., Yang, Y. (2016) A novel Ap2/ERF transcription factor from Stipa purpurea leads to enhanced drought tolerance in *Arabidopsis thaliana*. *Plant Cell Reports* 35(11): 2227–2239.

Yoshida, T., Mogami, J., Yamaguchi-Shinozaki, K. (2014) ABA-dependent and ABA-independent signaling in response to osmotic stress in plants. *Current Opinion in Plant Biology* 21: 133–139.

Yu, J., Holland, J.B., McMullen, M.D., Buckler, E.S. (2008) Genetic design and statistical power of nested association mapping in maize. *Genetics* 178(1): 539–551.

Yu, X., Liu, Y., Wang, S., Tao, Y., Wang, Z., Mijiti, A., Wang, Z., Zhang, H., Ma, H. (2016a) A chickpea stress-responsive NAC transcription factor, CarNAC5, confers enhanced tolerance to drought stress in transgenic *Arabidopsis*. *Plant Growth Regulation* 79: 187–197.

Yu, X., Liu, Y., Wang, S., Tao, Y., Wang, Z., Shu, Y., Peng, H., Mijiti, A., Wang, Z., Zhang, H., Ma, H. (2016b) CarNAC4, a NAC-type chickpea transcription factor conferring enhanced drought and salt stress tolerances in *Arabidopsis*. *Plant Cell Reports* 35(3): 613–627.

Zhang, J., Jia, W., Yang, J., Ismail, A.M. (2006) Role of ABA in integrating plant responses to drought and salt stresses. *Field Crops Research* 97(1): 111–119.

Zhang, M., Liu, W., Bi, Y.-P. (2009) [Dehydration-responsive element-binding (DREB) transcription factor in plants and its role during abiotic stresses]. *Yi chuan = Hereditas* 31(3): 236–244.

Zhang, L., Zhang, L., Xia, C., Zhao, G., Liu, J., Jia, J., Kong, X. (2015) A novel wheat bZIP transcription factor, TabZIP60, confers multiple abiotic stress tolerances in transgenic *Arabidopsis*. *Physiologia Plantarum* 153(4): 538–554.

Zhang, H., Ni, Z., Chen, Q., Guo, Z., Gao, W., Su, X., Qu, Y. (2016) Proteomic responses of drought-tolerant and drought-sensitive cotton varieties to drought stress. *Molecular Genetics and Genomics* 291(3): 1293–1303.

Zhang, C., Li, C., Liu, J., Lv, Y., Yu, C., Li, H., Zhao, T., Liu, B., Zhang, J. (2017a) The OsABF1 transcription factor improves drought tolerance by activating the transcription of COR413-TM1 in rice. *Journal of Experimental Botany* 68(16): 4695–4707.

Zhang, X., Shabala, S., Koutoulis, A., Shabala, L., Zhou, M. (2017b) Meta-analysis of major QTL for abiotic stress tolerance in barley and implications for barley breeding. *Planta* 245(2): 283–295.

Zhao, B.-Y., Hu, Y.-F., Li, J., Yao, X., Liu, K. (2016a) BnaABF2, a bZIP transcription factor from rapeseed (*Brassica napus* L.), enhances drought and salt tolerance in transgenic *Arabidopsis*. *Botanical Studies* 57: 12.

Zhao, X., Yang, X., Pei, S., He, G., Wang, X., Tang, Q., Jia, C., Lu, Y., Hu, R., Zhou, G. (2016b) The *Miscanthus* NAC transcription factor MlNAC9 enhances abiotic stress tolerance in transgenic *Arabidopsis*. *Gene* 586(1): 158–169.

Zhong, L., Chen, D., Min, D., Li, W., Xu, Z., Zhou, Y., Li, A., Chen, M., Ma, Y. (2015) AtTGA4, a bZIP transcription factor, confers drought resistance by enhancing nitrate transport and assimilation in *Arabidopsis thaliana*. *Biochemical and Biophysical Research Communications* 457(3): 433–439.

5 Plants Signaling toward Drought Stress

Muhammad Jamil, Aamir Ali, Alvina Gul, Khalid Farooq Akbar, Abdul Aziz Napa, and A. Mujeeb-Kazi

CONTENTS

5.1 INTRODUCTION

Extreme temperature and water conditions, polluted soils along with high salt concentrations are the climatic conditions detrimental to plant growth and development, causing severe yield losses for food crops also. These adverse situations are declared as stress. Abiotic stress can deteriorate cellular architecture and retard physiological aspects. Turgor loss, deactivated proteins, increased reactive oxygen species (ROS) level and enhanced oxidative damages are reported consequences of drought stress. Looking into stress responses, there are two main phenomena: stress avoidance (variety of protective mechanisms preventing the negative impact of stress) and stress tolerance (aptitude of the plant to cope with stressful conditions). Both avoidance and tolerance mechanisms may be adaptive, acquired, inherited (reviewed by Krasensky and Jonak, 2012). Suzuki et al. (2014) reviewed abiotic and biotic stress recombinations in plants. Krasensky and Jonak (2012) reviewed stress-induced metabolic regulations and responses to drought.

Rhizobacteria is another tool as drought stress alleviators to ameliorate plant physiology in moisture-stressed land through a process called rhizobacterial-induced drought endurance and resilience (RIDER) that includes physio-biochemical changes (Kaushal and Wani, 2016). After drought stress signaling, transcriptional remodeling induces stress response using osmolytes, antioxidants, chaperons and some lipid metabolisms; this whole mechanism is presumably triggered by some beneficial plant growth promoting bacteria (Forni et al., 2017).

The objectives of this chapter are to make an undertaking about the physiological and biochemical changes in the plant when facing drought stress. This chapter encompasses cellular details at the molecular level, describing the alteration in cell functions under drought stress, the role of metabolites, plasma membrane and abscisic acid (ABA) and provides a brief highlight on retrograde signaling and transduction.

5.1.1 CELLULAR IMPACTS OF STRESS

Nearly all the levels of cellular organization are influenced by stress. Membrane permeability is readjusted, cell wall texture is altered, and amendments in the cell cycle are the responses at the cellular level. Metabolic variations include osmotic readjustment, accumulation of compatible compounds to stabilize proteins, removal of ROS and re-stabilization of redox potential (Janska et al., 2010). At the molecular level, gene regulation is revised, and this multigenic effect enables the plant to tolerate stress conditions (Hauser et al., 2011). Plants respond in a complicated manner when faced with sudden multiple stresses (Atkinson et al., 2015).

5.1.2 DROUGHT STRESS

Drought is a pronounced threat to such food crops as rice, wheat and maize. Stress-tolerant crops will be beneficial in stress-prone environments. Drought is a severe harm toward crop productivity and yield; it must be a major emphasis of research. About 27.6% crop yield reduction had to be faced in 2012 by the US. Livestock and farms are equally perished by drought (Tripathi et al., 2014). The drought has a massive impact on respiration rates and stomatal closure. The decrease in photosynthetic rates, deposition of appropriate solutes and synthesis of ascorbate, glutathione and alpha-tocopherol are the consequences of water stress (reviewed by Tripathi et al., 2014). Tolerance of medium dehydration up to such a level of moisture contents beneath which there is no bulk cytoplasmic water potential is supposed as drought tolerance (Hoekstra et al., 2001). Hirayama and Shinozaki (2010) reviewed the research on plant abiotic stress in the post-genome era.

5.2 ROLE OF METABOLITES IN DROUGHT SIGNALING

When the plant faces drought stress, a special set of metabolites known as osmoprotectants are deposited. Osmolytes, antioxidants and stress signals also assemble in response. Metabolites linked with sulfur metabolism have an integral role under drought stress. Hormones (Abscisic acid), reactive oxygen species (H_2O_2), carbohydrates (polyols, fructans, etc.), amines, peptides, vitamins, phosphonucleotides and pigments are the main metabolites involved in drought response (reviewed by Chan et al., 2013). In *Arabidopsis*, calcium elevation at the intracellular level is reported sensitive to abiotic and biotic stress by Johnson et al. (2014). Bajguz (2014) searched out the role of nitric oxide when the plants face abiotic stress. It has been emphasized that nitric oxide is also involved in the ABA-induced movement of guard cells, which activates antioxidant defense during environmental stress. During abiotic stress signaling, the role of polyamines gearing cation transport across the cell membrane in plants has been reviewed by Pottosin and Shabala (2014), and polyamine metabolic canalization under drought stress has been reported by Alcázar et al. (2011). De-Ollas et al. (2013) declared that jasmonate is required for ABA increase under drought stress in citrus plant roots. Seo et al. (2011) working on rice found some protein interacting with jasmonate signaling pathway leading to drought stress. Plant adaptation to moisture stress depends upon the osmotic adjustment along with aggregation of particular solutes harmonious to cellular machinery (Blum, 2017). When a plant faces dehydration, nitric oxide signaling and its crosstalk with other plant growth regulators are established (Asgher et al., 2017). Priming with promising chemicals (sodium nitroprusside, melatonin and polyamines) to confer multiple stress tolerance in plants can also be the effective methodology (Savvides et al., 2016).

5.3 REACTIVE OXYGEN SPECIES GENERATION DURING DROUGHT STRESS

Signaling through reactive oxygen species was highlighted by Suzuki et al. (2012). It was discussed that there must be a dynamics between respiratory and photosynthetic rates. Cell environment must not face ROS level elevated because, due to stress, it is usually scarce to avoid over synthesis of ROS. Hence, ROS are the stress indicators during water, heat or salt stress. ROS signaling in plants needs efficient coordination and equilibrium among signaling and metabolic pathways (Suzuki et al., 2012). ROS signaling has also been focused on by Miller et al. (2010). ROS performed a dual role in abiotic stress: one, as toxic agent by-product under stress; two, as prime signal transduction molecules. Jaspers and Kangasjärvi (2010) also reviewed the importance of ROS during abiotic stress as an indicator. Increased production of reactive oxygen species must be avoided, but there remains a dire need of ROS during extreme environmental conditions. These species maintain a dynamic balance between signaling and metabolic pathways during stress. Gilroy et al. (2014), focusing on calcium and ROS for rapid systemic signaling for abiotic stress, presented cell-to-cell communication in plants. Baxter et al. (2014) also claim ROS is a key player in plant stress signaling. The importance of ROS and the stomatal response of plants to acclimate drought stress has also been reviewed recently by Zandalinas et al. (2018).

5.4 ROLE OF THE PLASMA MEMBRANE IN DROUGHT SIGNALING

Huang et al. (2012) reviewed that plasma membrane is the cellular organelle that receives stress signals. Transmission of these signals starts signal transduction, and physiochemical properties of plasma membrane enable it to respond to abiotic stress.

5.5 REDOX SIGNALING IN CHLOROPLAST

It reveals that kinases and phosphatases are essential entities responsible for developing optimized photosynthetic activity and redox condition under stress (Pesaresi et al., 2010). On the other hand, the elevated reduction potential of photosynthesis (harmful to thylakoid membrane) is antagonized by mitochondrial respiration (Dinakar et al., 2010). Redox leveling in mitochondria is mainly performed by phosphorylation.

Suzuki et al. (2012) reviewed that ROS and redox retrograde (organelle to the nucleus) signaling during environmental stress is obvious because fluctuation in environmental conditions alters the manners of organelle biogenesis and their operational activities. Metabolism of ROS is the way of signaling between organelles and nucleus. Hence, redox disequilibrium is sensed by the organelles, and a signal is transmitted to the nucleus to post-transcriptionally regulate the expression of drought-sensitive genes (Galvez-Valdivieso and Mullineaux, 2010). Chloroplast, mitochondria, peroxisomes and cytosol are interrelated through metabolic pathways to maintain the higher order of cellular homeostasis and energy balance.

5.6 RETROGRADE SIGNALING IN DROUGHT

Pfannschmidt (2010) reported plastidial retrograde signaling between metabolically active plastid and nucleus. There is a regular communication between organelles (plastid, mitochondria) and nucleus because most of the enzymes catalyzing in these organelles translated from the nucleus. During the drought, decreased water (electron donor of photosystem II) availability reduces photosynthetic capacity. Stomatal closure occurs, causing reduced rate of CO_2 fixation. Electron emission to O_2 occurs greatly due to retarded Calvin cycle. Hence, oxidative destruction resulted along with the increased amount of ROS. Understanding these steps, it can be deduced that chloroplast–nucleus coordination is crucial. Leister (2012) also explained the retrograde signaling in plants.

Transcription factors (TFs) regulate metabolite concentration under drought stress because a number of genes translating for enzymes involved in cell metabolic pathways are variably expressed under stress.

5.7 SIGNAL TRANSDUCTION

Drought stress alters the biochemical features of the plants along with the operation of concerned genes and proteins. Molecular mechanisms such as signal transduction, activation of genes and synthesis of metabolic components are the main facets adapted by the plants during drought stress. Tripathi et al. (2014) reviewed that effector and regulatory proteins are responsible for signal transduction. Regulatory proteins consist of TFs, protein kinases and some signaling molecules. Regulatory proteins are involved in activation and deactivation of many downstream genes and thus can be utilized by modifying their configuration. Hence, these TFs are the key focus to enhance the efficiency of drought tolerance genes.

Drought stress regulatory network at the molecular level (Todaka et al., 2015) plays a role in the development of transgenic drought-tolerant plants. Moreover, tetrapyrrole-based drought signaling has been reported by Nagahatenna et al. (2015). Transgenic technology is helping crop breeders during the utilization of TFs to increase plant abiotic stress tolerance due to their role in mastering the regulation of many stress-responsive genes (Wang et al., 2017). So the TFs playing the vital role can be engineered to enhance the stress resilience.

Crosstalk among components of responsible signaling machinery is to be understood in detail, and genome-scale system biology can help in this regard (Verma et al., 2016). Energy sensing is the prime aptitude of the plants with the help of which stress signaling involves in the regulation of proteins (critical for ion and water transport) through reprogramming to attain the cellular stability under stress conditions (Zhu, 2016).

This necessitates the mapping and isolation of stress-related genes useful for drought stress tolerance in plants. It is a well-known fact that ABA has a significant role in stress tolerance in plants. Some TFs during abiotic stress (drought) also have regulatory functions depicting ABA-independent gene expression. These TFs trigger some target genes to be controlled during stress. Signal transduction manipulated by ABA (Weiner et al., 2010; Nakashima and Yamaguchi-Shinozaki, 2013) showed that some ABA-receptors ($SnRk_2$, PP_2Cs, and

RCAR/PYR/PYL) drive the signaling pathway. ABA-responsive element binding proteins/ABA-response element binding factors (AREBs/ABFs) are some additional receptors studied so far in land plants, reviewed by Miyakawa et al. (2013).

Xiong and Zhu (2001), working with *Arabidopsis*, used ABA-deficient mutants *los5/aba3* and *los6/aba1*. Stress-sensitive genes such as RD29A, RD22, COR15A, COR47 and P5CS found extremely reduced in *los5* mutant. ABA is produced and deposited during the response to drought stress and high-salinity stress rather as a result a result of cold shock. Yang et al. (2011) focused on *MbDREB1* gene and showed that this might enhance the plant tolerance to abiotic stresses.

The phytohormone abscisic acid regulates the expression of drought-responsive genes. ABA is also responsible for the synthesis of ROS and regulates the opening/closing of stomata. Many other drought-sensitive metabolites are integrated with sulfur metabolism, which indicates that it must actively be regulated (Chan et al., 2013). Cutler et al. (2010) stated that ABA is crucial for abiotic stress signaling. ABA mediates expression of different genes sensitive to stress. A reduced starch concentration and maltose deposition appear during stress at the cellular level. Accumulation of citrate, malate, succinate is ABA-independent during stress (Urano et al., 2009), while some amino acids (proline) are deposited and are dependent on ABA under drought stress.

5.8 ROLE OF ABSCISIC ACID (ABA) IN SIGNAL TRANSDUCTION

Yoshida et al. (2014) highlighted that abscisic acid is a vital hormone playing a role during drought stress signal transduction. Peleg and Blumwald (2011), reporting on hormone balance, described the role of cytokinins, brassinosteroids and auxins under abiotic stress. Although Colebrook et al. (2014) highlighted the role of gibberellin signaling also when plants faced stress and Ha et al. (2012) found the functions of cytokinins under abiotic stresses, research details have shown that ABA is of prime importance in stress regulation. Raghavendra et al. (2010) searched out perception of the endogenous messenger ABA under the effect of ABA binding receptors. Kuromori et al. (2014) analyzed the crosstalk complexity between ABA-dependent and ABA-independent pathways in drought stress signaling. The role of TFs in the plant during ABA-dependent and independent abiotic stress signaling was also highlighted by Agarwal and Jha (2010). Drought signaling pathway is basically of two types (Umezawa et al., 2006b) *viz.*

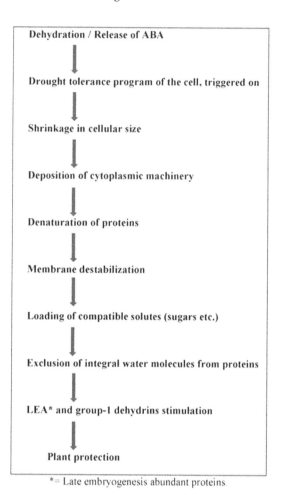

FIGURE 5.1 Cellular level mechanism of drought stress tolerance.

ABA-dependent pathway and ABA-independent pathway, also summarized in Figure 5.1.

5.8.1 ABA-DEPENDENT PATHWAY

ABA-dependent signaling is contributed by ABA-responsive elements (ABREs) which act as *cis*-acting elements under water stress. The base sequence in ABRE (CACGTG) is a binding site for both basic helix-loop-helix (bHLH) and NACs. AREBs and ABFs derived from Basic Leucine Zipper (bZIP) transcription factors seem to be phosphorylated prior to the activation (Kagaya et al., 2002). Stone (2014) highlighted the role of ubiquitin and 26S proteasome in abiotic stress signaling. Ubiquitin-dependent degradation also affects the production of ABA. If long-term drought stress is faced by the plant, it is reported by Sreenivasulu et al. (2012) that ABA-signaling is of vital importance under calcium-dependent and calcium-independent pathways. This phenomenon regulates transcriptional rearrangements under temporary stress responses.

ABRE are *cis*-acting elements along with a class of TFs; AREB/ABFs are involved in ABA-dependent gene expression. When drought is faced by the plant, increased level of AREB1/ABF2, AREB2/ABF4 and ABF3 indicates positive control of ABA-signaling and has been confirmed with overexpression studies (reviewed by Yoshida et al., 2014). It has been proved that AREBs/ABFs have a role of major and pivotal transcription factors jointly controlling ABRE-dependent gene expression as a result of desiccation (Yoshida et al., 2010).

AREB/ABF are the transcription factors equipped with conserved domains. These domains are often phosphorylated upon multiple sites by SNF1-related kinases 2 (SnRK2s) with an ABA-dependent pattern. Phosphorylation of these domains is responsible for the activation of AREB/ABF (Fujii et al., 2007). About nine SnRK2s are influenced by drought stress and firmly initiated by ABA; among them, three SnRK2s work in cooperation with AREB/ABFs in plant cell nuclei at the same place (Fujita et al., 2009).

SNF-related protein kinases 2 and proline deposition are ABA-dependently sensitive to water stress, and SnRK2.9 negatively regulate such accumulation (Fujii et al., 2011). SnRK3s integrate with come calcium-binding proteins (CBL). CBL interacting kinases (CIPKs) are also responsible for gearing up proline concentrations, increased under stress.

Umezawa et al. (2013) worked with *srk2d/e/i*, a triple mutant that was utilized in phosphoproteome studies. The purpose of the study was to mine out signaling networks mediated by subclass III SnRK2s. It was revealed that 32 phosphopeptides are there as candidates for SnRK2 substrates. There were also two phosphotpeptides related to AREB1/ABF2, AREB2/ABF4. SnRK2-substrate (SNS1) was detected as vital SnRK2 substrate. It was noted that ABA sensitivity was enhanced during post-germination stages by mutating *sns1*. The role of SNS1 is still not clear in ABA signaling. A further 84 phosphopeptides as possible SnRK2 substrates were detected in another set of experimentation. Eight proteins were categorized as a transcription factor and DNA-binding proteins (Wang et al., 2013).

Phosphoproteome studies revealed that AREB/ABFs are prime transcription factors. During ABA-signaling, these TFs are located on downstream of subclass III SnRK2s, and ABF1 is also a transcription factor that is operated downstream to the SnRK2s (Yoshida et al., 2014). It now becomes clear that SnRK2s control ABA-responsive gene expression under drought stress by integrating with further four AREB/ABF transcription factors.

Ito et al. (2012) detected that FBH3 is an SnRK2 substrate during ABA signaling. At the time of floral growth and development, *CONSTANS* (*CO*) expression is being triggered by an FBH3 homolog bHLH (TF). Floral activities are mediated by SnRK2 with the help of substrates like AREB/ABFs and FBH3; moreover, FBH3 and CO also play a role in the stomatal opening. But stomatal opening can also be inhibited by SnRK2-regulated ABA signaling at time of transcription and post-transcription. MYC and MYB proteins are an activator in an ABA-dependent system, and these proteins are produced along with deposition of endogenous ABA (Agarwal et al., 2006).

5.8.2 ABA-INDEPENDENT PATHWAY

On the other hand, a *cis*-element, dehydration-responsive elements/C-repeat (DRE/CRT) and DRE-/CRT-binding proteins 2 (DREB2) transcription factors are contributors to ABA-independent gene expression in drought stress.

Promoter analysis has exhibited that nine base pair sequence (TACCGACAT) called as DRE, a *cis*-elements controlling RD29A gene that is sensitive to drought. DRE sequence is also present in promoters of drought-sensitive genes following ABA-independent pathway.

C-repeat binding factor or dehydration-responsive element binding proteins (CBF/DREB1 proteins) are responsible during the transcriptional response to water stress (Thomashow, 2010).

DREB2s (genes) involved in drought response along with the protein transcribed under their control have a region if that region is removed it will compel DREB2A constitutively active and then it would be designated as DREB2Aca. Sakuma et al. (2006b) reported that this region in DREB2s has a negative regulatory domain (NRD). To remove the growth retardation affected by DREB2Aca when overexpressed for drought tolerance, DREB2ca was utilized by Engels et al. (2013). NRD sequence is needed for controlling proteins longevity coded by DREB2A. DREB1A mediates drought as well as freezing tolerance (Maruyama et al., 2009).

It has been noted that DREB1A and DREB2A have different ways to metabolize carbohydrates because in drought, genes for starch, alcohol and sucrose synthesis and break down works differently in both genes. Consequently, saccharides (mono, di and tri), sugars and alcohol are deposited in plant cells. DREB2 types of proteins play a role in heat and drought stress and can be taken from rice, sunflower, maize and wheat (Mizoi et al., 2013). Recently it has been reported that

ZmDREB2.7 has been found to take part in drought tolerance in maize (Liu et al., 2013).

In drought stress, DREB2A and DREB2B are involved in an ABA-independent manner through negative regulation during drought stress (Sakuma et al., 2006a). DREB2A triggers different proteins as a result of drought stress. Transcriptional regulatory mechanism of DREB2A has declared that Growth Regulatory Factor 7 (*GRF7*) acts as a repressor element against the genes (such as DREB2A) activated at the time of drought stress (Kim et al., 2012). If there is no drought, then DREB2A promoter sequence silences itself by integration with GRF7. So if *grf7* is mutated, it switches drought stress operative genes on through upregulation mechanism and acts as repressor regulator at the time of no stress. Another regulatory machinery for drought stress response is composed of DREB2A and DRIPs (DREB2A-interacting proteins). These are well-recognized TFs along with ubiquitin E3 ligase to operate drought stress-sensitive genes. Consisting of AtERF53 and two homologous C3HC4-type ring E3 ligase, ring domain ligase 2 (RGLG2) and RGLG1, another TFs combination has been known (Cheng et al., 2012) that negatively regulates plant drought stress. AtERF53 interrelates and is present with RGLG2 and RGLG1. AtERF53 and RGLGs can operate as positive as well as negative regulators during gene expression in drought stress. As subclass III SnRK2s seem to be triggered by autophosphorylation in ABA signaling but in drought stress, factual proteins to switch SnRK2 on have been yet unidentified.

Yoshida et al. (2014) reviewed that GRF7 is responsible for inhibiting both ABA-dependent and independent gene activation. Ma et al. (2014) identified two vital genes if mutated proved susceptible to abiotic stress. Maruyama et al. (2014) also claimed cytokinin signaling is reduced during drought stress in *Arabidopsis* and rice. Hence, it can be assumed that ABA and cytokinin have a function in water stress signaling. By creating medium moisture stress, Skirycz et al. (2010) highlighted the role of ethylene and gibberellin signaling in operating cell damage of young leaves of *Arabidopsis*. It creates a dire need to pinpoint the function performed by plant hormones other than ABA at the molecular level under drought stress.

5.8.3 ABA-Meditated Transcriptional Activities

Molecular analysis showed that ABA directs transcription of genes during drought stress. A *cis*-element for ABA-responsive gene expression is ABRE (Maruyama et al., 2012). In *Arabidopsis*, under drought stress, bZIP transcription factors (AREB/ABFs) controlling ABA-dependent gene expression act as major TFs and govern

signaling also at the vegetative growth stage. ABA-dependent phosphorylation regulates transcriptional activities of these stress-induced transcription factors, so ABA is mandatory for the proper functioning of AREB (Yoshida et al., 2010). Overexpression of AREB1in transgenic *Arabidopsis* increases drought tolerance and is efficiently controlled by ABA-dependent phosphorylation within the fixed domain (Barbosa et al., 2013). SnRK$_2$s phosphorylate AREB/ABFs (Umezawa et al., 2013) in ABA-dependent signaling network and group A PP$_2$Cs in land plants have a major contribution in the regulation of drought tolerance. PYL$_4$, a signaling factor, as well as a receptor, might be helpful if used to advance the drought tolerance (Pizzio et al., 2013).

Four transcription factors (AREB/ABF) are involved in gene expression during abscisic acid signaling due to osmotic stress (Yoshida et al., 2015). More recently it has been reported (Sack et al., 2018) that ABA accumulation in leaves under drought stress is related to a decline in cell volume rather turgor pressure. Abscisic acid can enhance respiration metabolism, amino acids and carbohydrates are involved in osmotic adjustment and salicylic acid is responsible for energy metabolism (Li et al., 2017). Abscisic acid-dependent bZIP is another transcription factor (Banerjee and Roychoudhury, 2017) that is activated by ABA-mediated signalosome. Different groups of kinases are involved to transduce the signals to phosphorylate bZIP TFs when a plant undergoes drought stress.

5.9 TRANSCRIPTION FACTORS (TFS) IN DROUGHT SIGNALING

Transcription factors are main elements in drought stress signaling, and few of these are major components of signaling network. MYB, bHLH, bZIP, ERF, NAC and WRKY are TFs notably important in cell signaling during drought stress. Hoang et al. (2014) explored the role of transcription factors in abiotic stress responses and highlighted their potential in crop improvement. Working with *Arabidopsis*, high-resolution profiling transcripts were utilized by Bechtold et al. (2013) to highlight drought stress signaling. Fujita et al. (2011) reviewed transcriptional regulation in response to osmotic stress. Golldack et al. (2011) also reported these TFs as stress-responsive agencies. Functional significance in the cellular transcriptional network has been pinpointed.

5.9.1 NAC Transcription Factors

NAC TFs is another class of proteins mediating gene expressions as a result of the drought. NAM, ATAF and

CUC TF proteins are separately present each in particular plant. Nakashima et al. (2012) reviewed over 100 NAC genes from *Arabidopsis* and rice. NAC TFs are responsible for the establishment of stress responses. Overexpression of NACs during drought stress with OsHox24 or OsNAC6 promoters from rice proved operative in inducing stress tolerance without harming plant growth (Takasaki et al., 2010). On the other hand, it has also been evaluated by Jeong et al. (2010) that root-specific promoter RCc3 can be utilized for the expression of SNACs to increase environmental stress tolerance. So it is said that SNACs have a part in regulating abiotic stress expressions. By using SNACs, drought tolerance can be enhanced with the help of effective promoters. Organ-specific promoters (which are of many types) suggested for roots and stomata can be a real device to operate the expression of drought-responsive factors influencing growth at actual position and time as well (Bang et al., 2013).

NAC domain-containing proteins are the regulators involved in hormonal regulation under stress. Glucose, fructose and sucrose (soluble sugars) along with proline seems to be elevated during stress conditions.

5.9.2 WRKY TRANSCRIPTION FACTORS

Rushton et al. (2010) reviewed that WRKY TFs are responsible for defense signaling and regulation of growth, developmental processes, senescence, biotic and abiotic stresses. These are equipped with DNA-binding domain with signature sequence (WRKYGQK) at the N-terminus along with 60 amino acids. Moreover, zinc finger motif is there at C-terminus. At the major groove of DNA, WRKY signature sequence binds with the appropriate DNA-binding site (TTGACC/T) also called W box (Yamasaki et al., 2013). It is now made well clear by Yamasaki et al. (2013) that a four-stranded beta sheet inserts the major groove of DNA, forming beta-wedge. Tao et al. (2011) reported the role of *OsWRKY 45* in the drought and cold tolerance in rice. Transcription factors such as $WRKY_{46}$, $WRKY_{54}$ and $WRKY_{70}$ are reportedly (Chen et al., 2017) involved in drought stress. Using RT-qPCR, the expression of *SlyWRKY*$_{75}$ has also been reported (López-Galiano et al., 2018) under drought stress.

5.9.3 INTERACTION AMONG TRANSCRIPTION FACTORS

Integration and coordination of TFs during signaling pathway has been reported by Lee et al. (2010), who described the crosstalk among elements of ABA-dependent and independent ways. Under drought stress, the DREB2A gene is influenced by SnRK2s, ABRE (promoter sequence) and AREB/ABF TFs. This is a

complicated interaction between AREB and DREB regulons at the level of gene and protein expression level as well (Kim et al., 2011).

Gene editing through CRISPR-Cas9 produced the variants of ARGOS8 for the improvement of maize grain yield under drought stress (Shi et al., 2017). Using this editing system (CRISPR/Cas9), wheat dehydration-responsive element binding protein 2 (*TaDREB2*) has recently been edited to regulate the signal transduction mechanism under drought stress (Kim et al., 2018).

NCED3 in *Arabidopsis* is a gene for the synthesis of ABA and is controlled by the interactions of SNAC, TF, ATAF1 (Jensen et al., 2013). It is further described by Nakashima et al. (2012) that SNAC TFs may gear up the expression of ABA-dependent genes of ABRE regulon while ABRE sequence has been found in SNAC genes. In *Arabidopsis* during drought stress as a response, AREB/ABF factors (ABF2/AREB1 and ABF4/AREB2) conjoins with ANAC096 (Xu et al., 2013). Hence, NAC and AREB/ABF regulons have composite interrelations. Another interaction by Cheng et al. (2013) has been described between DREB/CBFs and AP2/ERFs during the transcriptional level. ERF1 attaches to GCC box during pathogen attacks because it is an upstream TF in ethylene and jasmonate signaling. On the other hand, ERF1 integrate with DRE/CRT when drought stress is to be faced by the plant (reviewed by Nakashima et al., 2014).

Heptahelical proteins 1 (HHP 1) is a negative regulator in stresses. There are seven transmembrane domain structures in HHP1. HHP1 is also a vital constituent in the crosstalk between drought and cold stress signaling pathways. Gupta and Tuteja (2011) reported some chaperones and foldases in the endoplasmic reticulum as stress signal in plants. On the other hand, phyto-engineering of many drought-induced genes in a vast range of plants has been proposed by Thapa et al. (2011).

5.10 ROLE OF KINASES IN DROUGHT STRESS

Mitogen-activated protein kinases (MAPKs) are crucial enzymes for signal transduction. These enzymes interlink various receptors to a broad range of cellular responses in plants. After isolating a wide number of various components of MAPKs, it has now been declared that MAPKs play a vital role in responding abiotic stresses (Wu et al., 2011). Sinha et al. (2011) also elaborated the function of mitogen-activated protein kinase under abiotic stress through signal transduction. When the plants are exposed to diverse environmental stresses, mitogen-activated protein kinases (MAPKs) mediate the adaptations for survival by crosstalking with hormones and secondary messengers (Smékalová et al., 2014). The ABA is the core regulator for abiotic stress responses and initiates major alterations in gene expressions along with a major role of ABA-induced proteins kinases of the family SnRK to facilitate the responses of environmental stress. Nowadays, MAPKs have also been reported to mediate ABA-signaling (Danquah et al., 2014). It has been shown that when cereal signaling is under abiotic stress, then SnRK1 interacts with some negative regulators. This creates nutrition starvation signaling (Lin et al., 2014).

Calcium-dependent protein kinases (CDPKs) are also declared as a hub in plant stress signaling by Schulz et al. (2013). CDPKs are novel agencies being influenced by changes in intracellular calcium concentration. These exhibit overlapping and specific expression manners and play a unique role in the activation and inhibition of enzymes as well as transcription factors. In the complex immune and stress signaling network, recent advances have also been highlighted by Boudsocq and Sheen (2013). Calcium-dependent protein kinases (CDPKs) are known to have a role in abiotic stress responses. On the basis of gain-of-function and loss-of-function mutants, it has been shown that various CDPKs are involved in abiotic stress tolerance. CDPKs modify ABA signaling and minimize the assimilation of ROS (Asano et al., 2012) focuses on new approaches by in tackling drought stress. Receptor-like kinases (RLKs) can play a role to improve the stress tolerance of plant under drought. RLKs also perform in extreme environments if faced by the plants.

5.11 ROLE OF miRNAs AND siRNAs IN DROUGHT STRESS

The role of microRNAs (miRNAs) and small interfering RNAs (siRNAs) in the abiotic stress responses of plants has been shown by Khraiwesh et al. (2012). Micro RNAs regulate drought stress signaling, according to Ding et al. (2013). Recent molecular studies have been elaborated with miRNA-associated regulatory networks responsible for drought stress responses. Cabello et al. (2014) described the novel perspectives for the engineering of abiotic stress tolerance in plants. It has been focused that post-translational modifications along with overexpression of miRNAs mediate plant growth performance under stress. Related target genes are regulated by such major classes of small RNAs. These RNAs are a vital constituent in plant stress responses. Plants modulate their gene expression with the help of small RNAs.

Post-transcriptional gene regulation in plants has been described more recently by elucidating the role of microRNAs in plant stress responses. MicroRNAs are being expressed as a major plant growth regulator, such as miR398 influence the expression of the gene with known roles in stress tolerance (Sunkar et al., 2012). Recent studies identified non-coding RNAs known as circular RNAs (circRNAs) in leaves under drought stress (Wang et al., 2017). A class of microRNAs has been found differentially expressed in barley under drought stress and well-watered conditions (Ferdous et al., 2017). Moreover, the contribution of epigenetics has been uncovered in water-stressed environments, e.g. Mediterranean ecosystems have been emphasized by Balao et al. (2017).

5.12 CONCLUSION AND PROSPECTS

Drought stress signaling has been studied thoroughly at all the cellular levels. Signaling in stress also highlights the inherent capability of the plant to cope with the extreme environment. Such plant genotypes are better adapting as well as useful stock to widen the genetic pool. Tracing out stress signaling mechanisms can be helpful to genetically engineer such traits to other plant species. Morphological, physiological and biochemical alterations in plants remain a major focus. Although genes for abiotic stresses have been mapped and their promoter and transcriptional details have also been chalked out, phenomena of signaling are still to be explored. With the advancement in the field of proteomics, it will become more viable to assess the role of actual proteins and their function for stress resistance. As far as global food security is concerned regarding 2050, a lot of practical implications are still required. Under such a perspective, crops with improved abiotic stress tolerance should be within the reach of a common peasant rather than being exhibited in the labs only. It should be the goal of researchers, rather a social scientist, that high-yielding, stress tolerant crops should be the property of every farmer on earth. The implication of such a global vision can make the custodian of the earth (humans) prosperous.

REFERENCES

Agarwal, M., Hao, Y., Kapoor, A., Dong, C.H., Fujii, H., Zheng, X., Zhu, J.K. (2006) A R2R3 type MYB transcription factor is involved in the cold regulation of CBF genes and in acquired freezing tolerance. *Journal of Biological Chemistry* 281(49): 37636–37645.

Agarwal, P., Jha, B. (2010) Transcription factors in plants and ABA dependent and independent abiotic stress signaling. *Biologia Plantarum* 54(2): 201–212.

Alcázar, R., Bitrián, M., Bartels, D., Koncz, C., Altabella, T., Tiburcio, A.F. (2011) Polyamine metabolic canalization in response to drought stress in *Arabidopsis* and the resurrection plant *Craterostigma plantagineum*. *Plant Signaling and Behavior* 6(2): 243–250.

Asano, T., Hayashi, N., Kikuchi, S., Ohsugi, R. (2012) CDPK-mediated abiotic stress signaling. *Plant Signaling and Behavior* 7(7): 817–821.

Asgher, M., Per, T.S., Masood, A., Fatma, M., Freschi, L., Corpas, F.J., Khan, N.A. (2017) Nitric oxide signaling and its crosstalk with other plant growth regulators in plant responses to abiotic stress. *Environmental Science and Pollution Research* 24(3): 2273–2285.

Atkinson, N.J., Jain, R., Urwin, P.E. (2015) The response of plants to simultaneous biotic and abiotic stress. In *Combined Stresses in Plants*, ed. Ramamurthy Mahalingam, 181–201. Springer: Cham, Switzerland.

Bajguz, A. (2014) Nitric oxide: Role in plants under abiotic stress. In *Physiological Mechanisms and Adaptation Strategies in Plants Under Changing Environment*, eds. Peter, A., John, B., 137–159. Berlin, Germany: Springer.

Balao, F., Paun, O., Alonso, C. (2017) Uncovering the contribution of epigenetics to plant phenotypic variation in Mediterranean ecosystems. *Plant Biology* 20(1): 38–49.

Banerjee, A., Roychoudhury, A. (2017) Abscisic-acid-dependent basic leucine zipper (bZIP) transcription factors in plant abiotic stress. *Protoplasma* 254(1): 3–16.

Bang, S.W., Park, S-H., Jeong, J.S., Kim, Y.S., Jung, H., Ha, S-H., Kim, J-K. (2013) Characterization of the stress-inducible OsNCED3 promoter in different transgenic rice organs and over three homozygous generations. *Planta* 237(1): 211–224.

Barbosa, E.G.G., Leite, J.P., Marin, S.R.R., Marinho, J.P., Carvalho, F.C., Fuganti-Pagliarini, R., Farias, J.R.B., Neumaier, N., Marcelino-Guimarães, F.C., de-Oliveira, M.C.N. (2013) Overexpression of the ABA-dependent AREB1 transcription factor from *Arabidopsis* thaliana improves soybean tolerance to water deficit. *Plant Molecular Biology Reporter* 31(3): 719–730.

Baxter, A., Mittler, R., Suzuki, N. (2014) ROS as key players in plant stress signaling. *Journal of Experimental Botany* 65(5): 1229–1240.

Bechtold, U., Ott, S., Wild, D., Buchanan-Wollaston, V., Rand, D., Deynon, J., Smirnoff, N., Mullineaux, P. (2013) Using high-resolution profiling of transcripts to understand early drought stress signaling events. *BioTechnologia* 94(2): 149–151.

Blum, A. (2017) Osmotic adjustment is a prime drought stress adaptive engine in support of plant production. *Plant Cell and Environment* 40(1): 4–10.

Boudsocq, M., Sheen, J. (2013) CDPKs in immune and stress signaling. *Trends in Plant Science* 18(1): 30–40.

Cabello, J.V., Lodeyro, A.F., Zurbriggen, M.D. (2014) Novel perspectives for the engineering of abiotic stress tolerance in plants. *Current Opinion in Biotechnology* 26: 62–70.

Chan, K.X., Wirtz, M., Phua, S.Y., Estavillo, G.M., Pogson, B.J. (2013) Balancing metabolites in drought: The sulfur assimilation conundrum. *Trends in Plant Science* 18(1): 18–29.

Chen, J., Nolan, T.M., Ye, H., Zhang, M., Tong, H., Xin, P., Chu, J., Chu, C., Li, Z., Yin, Y. (2017) *Arabidopsis* WRKY46, WRKY54, and WRKY70 transcription factors are involved in brassinosteroid-regulated plant growth and drought responses. *The Plant Cell* 29(6): 1425–1439.

Cheng, M.C., Hsieh, E.J., Chen, J.H., Chen, H.Y., Lin, T.P. (2012) *Arabidopsis* RGLG2, functioning as a RING E3 ligase, interacts with AtERF53 and negatively regulates the plant drought stress response. *Plant Physiology* 158(1): 363–375.

Cheng, M.C., Liao, P.M., Kuo, W.W., Lin, T.P. (2013) The *Arabidopsis* Ethylene Response Factor1 regulates abiotic stress-responsive gene expression by binding to different cis-acting elements in response to different stress signals. *Plant Physiology* 162(3): 1566–1582.

Colebrook, E.H., Thomas, S.G., Phillips, A.L., Hedden, P. (2014) The role of gibberellin signaling in plant responses to abiotic stress. *The Journal of Experimental Biology* 217(1): 67–75.

Cutler, S.R., Rodriguez, P.L., Finkelstein, R.R., Abrams, S.R. (2010) Abscisic acid: Emergence of a core signaling network. *Annual Review of Plant Biology* 61: 651–679.

Danquah, A., de-Zelicourt, A., Colcombet, J., Hirt, H. (2014) The role of ABA and MAPK signaling pathways in plant abiotic stress responses. *Biotechnology Advances* 32(1): 40–52.

De-Ollas, C., Hernando, B., Arbona, V., Gómez-Cadenas, A. (2013) Jasmonic acid transient accumulation is needed for abscisic acid increase in citrus roots under drought stress conditions. *Physiologia Plantarum* 147(3): 296–306.

Dinakar, C., Abhaypratap, V., Yearla, S.R., Raghavendra, A.S., Padmasree, K. (2010) Importance of ROS and antioxidant system during the beneficial interactions of mitochondrial metabolism with photosynthetic carbon assimilation. *Planta* 231(2): 461–474.

Ding, Y., Tao, Y., Zhu, C. (2013) Emerging roles of microRNAs in the mediation of drought stress response in plants. *Journal of Experimental Botany* 64(11): 3077–3086.

Engels, C., Fuganti-Pagliarini, R., Marin, S.R.R., Marcelino-Guimarães, F.C., Oliveira, M.C.N., Kanamori, N., Mizoi, J., Nakashima, K., Yamaguchi-Shinozaki, K., Nepomuceno, A.L. (2013) Introduction of the rd29A: AtDREB2A CA gene into soybean (*Glycine max* L. Merril) and its molecular characterization in leaves and roots during dehydration. *Genetics and Molecular Biology* 36(4): 556–565.

Ferdous, J., Sanchez-Ferrero, J.C., Langridge, P., Milne, L., Chowdhury, J., Brien, C., Tricker, P.J. (2017) Differential expression of microRNAs and potential targets under drought stress in barley. *Plant Cell and Environment* 40(1): 11–24.

Forni, C., Duca, D., Glick, B.R. (2017) Mechanisms of plant response to salt and drought stress and their alteration by Rhizobacteria. *Plant and Soil* 410(1–2): 335–356.

Fujii, H., Verslues, P.E., Zhu, J-K. (2007) Identification of two protein kinases required for abscisic acid regulation of seed germination, root growth, and gene expression in *Arabidopsis*. *Plant Cell* 19: 485–494.

Fujii, H., Verslues, P.E., Zhu, J-K. (2011) *Arabidopsis* decuple mutant reveals the importance of SnRK2 kinases in osmotic stress responses in vivo. *Proceedings of the National Academy of Sciences* 108(4): 1717–1722.

Fujita, Y., Fujita, M., Shinozaki, K., Yamaguchi-Shinozaki, K. (2011) ABA-mediated transcriptional regulation in response to osmotic stress in plants. *Journal of Plant Research* 124(4): 509–525.

Fujita, Y., Nakashima, K., Yoshida, T., Katagiri, T., Kidokoro, S., Kanamori, N., Umezawa, T., Fujita, M., Maruyama, K., Ishiyama, K. (2009) Three SnRK2 protein kinases are the main positive regulators of abscisic acid signaling in response to water stress in *Arabidopsis*. *Plant and Cell Physiology* 50(12): 2123–2132.

Galvez-Valdivieso, G., Mullineaux, P.M. (2010) The role of reactive oxygen species in signaling from chloroplasts to the nucleus. *Physiologia Plantarum* 138(4): 430–439.

Gilroy, S., Suzuki, N., Miller, G., Choi, W-G., Toyota, M., Devireddy, A.R., Mittler, R. (2014) A tidal wave of signals: Calcium and ROS at the forefront of rapid systemic signaling. *Trends in Plant Science* 19(10): 623–630.

Golldack, D., Lüking, I., Yang, O. (2011) Plant tolerance to drought and salinity: Stress regulating transcription factors and their functional significance in the cellular transcriptional network. *Plant cell Reports* 30(8): 1383–1391.

Gupta, D., Tuteja, N. (2011) Chaperones and foldases in endoplasmic reticulum stress signaling in plants. *Plant Signaling Behavior* 6(2): 232–236.

Ha, S., Vankova, R., Yamaguchi-Shinozaki, K., Shinozaki, K., Tran, L-SP. (2012) Cytokinins: Metabolism and function in plant adaptation to environmental stresses. *Trends in Plant Science* 17(3): 172–179.

Hauser, M-T., Aufsatz, W., Jonak, C., Luschnig, C. (2011) Transgenerational epigenetic inheritance in plants. *(BBA)-Gene Regulatory Mechanisms* 1809(8): 459–468.

Hirayama, T., Shinozaki, K. (2010) Research on plant abiotic stress responses in the post-genome era: Past, present and future. *The Plant Journal* 61(6): 1041–1052.

Hoang, X.L.T., Thu, N.B.A., Thao, N.P., Tran, L-SP. (2014) Transcription factors in abiotic stress responses: Their potentials in crop improvement. In *Improvement of Crops in the Era of Climatic Changes*, eds. Ahmed, P., Wani, M.R., Azooz, M.M., Tran, P.L.-S., 337–366. New York, NY: Springer.

Hoekstra, F.A., Golovina, E.A., Buitink, J. (2001) Mechanisms of plant desiccation tolerance. *Trends in Plant Science* 6(9): 431–438.

Huang, G-T., Ma, S-L., Bai, L-P., Zhang, L., Ma, H., Jia, P., Liu, J., Zhong, M., Guo, Z.-F. (2012) Signal transduction during cold, salt, and drought stresses in plants. *Molecular Biology Reports* 39(2): 969–987.

Ito, S., Song, Y.H., Josephson-Day, A.R., Miller, R.J., Breton, G., Olmstead, R.G., Imaizumi, T. (2012) FLOWERING BHLH transcriptional activators control expression of the photoperiodic flowering regulator CONSTANS in *Arabidopsis*. *Proceedings of the National Academy of Sciences* 109(9): 3582–3587.

Janska, A., Maršík, P., Zelenková, S., Ovesná, J. (2010) Cold stress and acclimation–what is important for metabolic adjustment? *Plant Biology* 12(3): 395–405.

Jaspers, P., Kangasjärvi, J. (2010) Reactive oxygen species in abiotic stress signaling. *Physiologia Plantarum* 138(4): 405–413.

Jensen, M.K., Lindemose, S., Masi, Fd., Reimer, J.J., Nielsen, M., Perera, V., Workman, C.T., Turck, F., Grant, M.R., Mundy, J. (2013) ATAF1 transcription factor directly regulates abscisic acid biosynthetic gene *NCED3* in *Arabidopsis thaliana*. *FEBS Open Biology* 3(1): 321–327.

Jeong, J.S., Kim, Y.S., Baek, K.H., Jung, H., Ha, S-H., Choi, Y., Kim, M., Reuzeau, C., Kim, J-K. (2010) Root-specific expression of OsNAC10 improves drought tolerance and grain yield in rice under field drought conditions. *Plant Physiology* 153(1): 185–197.

Johnson, J.M., Reichelt, M., Vadassery, J., Gershenzon, J., Oelmüller, R. (2014) An *Arabidopsis* mutant impaired in intracellular calcium elevation is sensitive to biotic and abiotic stress. *BMC Plant Biology* 14: 162.

Kagaya, Y., Hobo, T., Murata, M., Ban, A., Hattori, T. (2002) Abscisic acid–induced transcription is mediated by phosphorylation of an abscisic acid response element binding factor, TRAB1. *The Plant Cell* 14(12): 3177–3189.

Kaushal, M., Wani, S.P. (2016) Plant-growth-promoting rhizobacteria: Drought stress alleviators to ameliorate crop production in drylands. *Annals of Microbiology* 66(1): 35–42.

Khraiwesh, B., Zhu, J-K., Zhu, J. (2012) Role of miRNAs and siRNAs in biotic and abiotic stress responses of plants. *(BBA)-Gene Regulatory Mechanisms* 1819(2): 137–148.

Kim, D., Alptekin, B., Budak, H. (2018) CRISPR/Cas9 genome editing in wheat. *Functional Integrative Genomics* 18(1): 31–41.

Kim, J-S., Mizoi, J., Kidokoro, S., Maruyama, K., Nakajima, J., Nakashima, K., Mitsuda, N., Takiguchi, Y., Ohme-Takagi, M., Kondou, Y. (2012) *Arabidopsis* GROWTH-REGULATING FACTOR7 functions as a transcriptional repressor of abscisic acid–and osmotic stress–responsive genes, including DREB2A. *The Plant Cell* 24(8): 3393–3405.

Kim, J-S., Mizoi, J., Yoshida, T., Fujita, Y., Nakajima, J., Ohori, T., Todaka, D., Nakashima, K., Hirayama, T., Shinozaki, K. (2011) An ABRE promoter sequence is involved in osmotic stress-responsive expression of the DREB2A gene, which encodes a transcription factor regulating drought-inducible genes in *Arabidopsis*. *Plant and Cell Physiology* 52(12): 2136–2146.

Krasensky, J., Jonak, C. (2012) Drought, salt, and temperature stress-induced metabolic rearrangements and regulatory networks. *Journal of Experimental Botany* 63(4): 1593–1608.

Kuromori, T., Mizoi, J., Umezawa, T., Yamaguchi-Shinozaki, K., Shinozaki, K. (2014) Drought Stress Signaling Network. In *Molecular Biology*, ed. Howell, S.H., 383–409. New York, NY: Springer.

Lee, S-J., Kang, J-Y., Park, H-J., Kim, M.D., Bae, M.S., Choi, H-I., Kim, S.Y. (2010) DREB2C interacts with ABF2, a bZIP protein regulating abscisic acid-responsive gene expression, and its overexpression affects abscisic acid sensitivity. *Plant Physiology* 153(2): 716–727.

Leister, D. (2012) Retrograde signaling in plants: From simple to complex scenarios. *Frontiers in Plant Science* 3(135): 1–9.

Li, Z., Yu, J., Peng, Y., Huang, B. (2017) Metabolic pathways regulated by abscisic acid, salicylic acid and γ-aminobutyric acid in association with improved drought tolerance in creeping bentgrass (*Agrostis stolonifera*). *Physiologia Plantarum* 159(1): 42–58.

Lin, C-R., Lee, K-W., Chen, C-Y., Hong, Y-F., Chen, J-L., Lu, C-A., Chen, K-T., Ho, T-H.D., Yu, S-M. (2014) SnRK1A-interacting negative regulators modulate the nutrient starvation signaling sensor SnRK1 in source-sink communication in cereal seedlings under abiotic stress. *The Plant Cell* 26(2): 808–827.

Liu, S., Wang, X., Wang, H., Xin, H., Yang, X., Yan, J., Li, J., Tran, L-S.P., Shinozaki, K., Yamaguchi-Shinozaki, K. (2013) Genome-wide analysis of ZmDREB genes and their association with natural variation in drought tolerance at seedling stage of *Zea mays* L. *PLoS Genetics* 9(9): e1003790.

López-Galiano, M.J., González-Hernández, A.I., Crespo-Salvador, O., Rausell, C., Real, M.D., Escamilla, M., Camañes, G., García-Agustín, P., González-Bosch, C., García-Robles, I. (2018) Epigenetic regulation of the expression of WRKY75 transcription factor in response to biotic and abiotic stresses in Solanaceae plants. *Plant Cell Reports* 37(1): 167–176.

Ma, C., Xin, M., Feldmann, K.A., Wang, X. (2014) Machine learning–based differential network analysis: A study of stress-responsive transcriptomes in *Arabidopsis*. *The Plant Cell* 26(2): 520–537.

Maruyama, K., Takeda, M., Kidokoro, S., Yamada, K., Sakuma, Y., Urano, K., Fujita, M., Yoshiwara, K., Matsukura, S., Morishita, Y. (2009) Metabolic pathways involved in cold acclimation identified by integrated analysis of metabolites and transcripts regulated by DREB1A and DREB2A. *Plant Physiology* 150(4): 1972–1980.

Maruyama, K., Todaka, D., Mizoi, J., Yoshida, T., Kidokoro, S., Matsukura, S., Takasaki, H., Sakurai, T., Yamamoto, Y.Y., Yoshiwara, K. (2012) Identification of cis-acting promoter elements in cold-and dehydration-induced transcriptional pathways in *Arabidopsis*, rice, and soybean. *DNA Research* 19(1): 37–49.

Maruyama, K., Urano, K., Yoshiwara, K., Morishita, Y., Sakurai, N., Suzuki, H., Kojima, M., Sakakibara, H., Shibata, D., Saito, K. (2014) Integrated analysis of the effects of cold and dehydration on rice metabolites, phytohormones, and gene transcripts. *Plant Physiology* 164(4): 1759–1771.

Miller, G., Suzuki, N., Ciftci-Yilmaz, S., Mittler, R. (2010) Reactive oxygen species homeostasis and signalling during drought and salinity stresses. *Plant Cell and Environment* 33(4): 453–467.

Miyakawa, T., Fujita, Y., Yamaguchi-Shinozaki, K., Tanokura, M. (2013) Structure and function of abscisic acid receptors. *Trends in Plant Science* 18(5): 259–266.

Mizoi, J., Ohori, T., Moriwaki, T., Kidokoro, S., Todaka, D., Maruyama, K., Kusakabe, K., Osakabe, Y., Shinozaki, K., Yamaguchi-Shinozaki, K. (2013) GmDREB2A; 2, a canonical dehydration-responsive element-binding protein2-type transcription factor in soybean, is post-translationally regulated and mediates dehydration-responsive element-dependent gene expression. *Plant Physiology* 161(1): 346–361.

Nagahatenna, D.S., Langridge, P., Whitford, R. (2015) Tetra-pyrrole-based drought stress signalling. *Plant Biotechnology Journal* 13(4): 447–459.

Nakashima, K., Takasaki, H., Mizoi, J., Shinozaki, K., Yamaguchi-Shinozaki, K. (2012) NAC transcription factors in plant abiotic stress responses. *(BBA)-Gene Regulatory Mechanisms* 1819(2): 97–103.

Nakashima, K., Yamaguchi-Shinozaki, K. (2013) ABA signaling in stress-response and seed development. *Plant Cell Reports* 32(7): 959–970.

Nakashima, K., Yamaguchi-Shinozaki, K., Shinozaki, K. (2014) The transcriptional regulatory network in the drought response and its crosstalk in abiotic stress responses including drought, cold, and heat. *Frontiers in Plant Science* 5: 1–7.

Peleg, Z., Blumwald, E. (2011) Hormone balance and abiotic stress tolerance in crop plants. *Current Opinion in Plant Biology* 14(3): 290–295.

Pesaresi, P., Hertle, A., Pribi, M., Schneider, A., Kleine, T., Leister, D. (2010) Optimizing photosynthesis under fluctuating light: The role of the *Arabidopsis* STN7 kinase. *Plant Signaling Behavior* 5(1): 21.

Pfannschmidt, T. (2010) Plastidial retrograde signaling – a true "plastid factor" or just metabolite signatures? *Trends in Plant Science* 15(8): 427–435.

Pizzio, G.A., Rodriguez, L., Antoni, R., Gonzalez-Guzman, M., Yunta, C., Merilo, E., Kollist, H., Albert, A., Rodriguez, P.L. (2013) The PYL4 A194T mutant uncovers a key role of PYR1-LIKE4/PROTEIN PHOSPHATASE 2CA interaction for abscisic acid signaling and plant drought resistance. *Plant Physiology* 163(1): 441–455.

Pottosin, I., Shabala, S. (2014) Polyamines control of cation transport across plant membranes: Implications for ion homeostasis and abiotic stress signaling. *Frontiers in Plant Science* 5(154): 1–16.

Raghavendra, A.S., Gonugunta, V.K., Christmann, A., Grill, E. (2010) ABA perception and signalling. *Trends in Plant Science* 15: 395–401.

Rushton, P.J., Somssich, I.E., Ringler, P., Shen, Q.J. (2010) WRKY transcription factors. *Trends in Plant Science* 15(7): 247–258.

Sack, L., John, G.P., Buckley, T.N. (2018) ABA Accumulation in dehydrating leaves is associated with decline in cell volume, not turgor pressure. *Plant Physiology* 176(1): 489–495.

Sakuma, Y., Maruyama, K., Osakabe, Y., Qin, F., Seki, M., Shinozaki, K., Yamaguchi-Shinozaki, K. (2006a)

Functional analysis of an *Arabidopsis* transcription factor, DREB2A, involved in drought-responsive gene expression. *The Plant Cell* 18(5): 1292–1309.

Sakuma, Y., Maruyama, K., Qin, F., Osakabe, Y., Shinozaki, K., Yamaguchi-Shinozaki, K. (2006b) Dual function of an *Arabidopsis* transcription factor DREB2A in water-stress-responsive and heat-stress-responsive gene expression. *Proceedings of the National Academy of Sciences USA* 103(49): 18822–18827.

Savvides, A., Ali, S., Tester, M., Fotopoulos, V. (2016) Chemical priming of plants against multiple abiotic stresses: Mission possible? *Trends in Plant Science* 21(4): 329–340.

Schulz, P., Herde, M., Romeis, T. (2013) Calcium-dependent protein kinases: Hubs in plant stress signaling and development. *Plant Physiology* 163: 523–530.

Seo, J.S., Joo, J., Kim, M.J., Kim, Y.K., Nahm, B.H., Song, S.I., Cheong, J.J., Lee, J.S., Kim, J.K., Choi, Y.D. (2011) OsbHLH148, a basic helix-loop-helix protein, interacts with OsJAZ proteins in a jasmonate signaling pathway leading to drought tolerance in rice. *The Plant Journal* 65(6): 907–921.

Shi, J., Gao, H., Wang, H., Lafitte, H.R., Archibald, R.L., Yang, M., Hakimi, S.M., Mo, H., Habben, J.E. (2017) ARGOS8 variants generated by CRISPR-Cas9 improve maize grain yield under field drought stress conditions. *Plant Biotechnology Journal* 15(2): 207–216.

Sinha, A.K., Jaggi, M., Raghuram, B., Tuteja, N. (2011) Mitogen-activated protein kinase signaling in plants under abiotic stress. *Plant Signaling and Behavior* 6(2): 196–203.

Skirycz, A., De-Bodt, S., Obata, T., De-Clercq, I., Claeys, H., De-Rycke, R., Andriankaja, M., Van-Aken, O., Van-Breusegem, F., Fernie, A.R. (2010) Developmental stage specificity and the role of mitochondrial metabolism in the response of *Arabidopsis* leaves to prolonged mild osmotic stress. *Plant Physiology* 152(1): 226–244.

Smékalová, V., Doskočilová, A., Komis, G., Šamaj, J. (2014) Crosstalk between secondary messengers, hormones and MAPK modules during abiotic stress signaling in plants. *Biotechnology Advances* 32(1): 2–11.

Sreenivasulu, N., Harshavardhan, V.T., Govind, G., Seiler, C., Kohli, A. (2012) Contrapuntal role of ABA: Does it mediate stress tolerance or plant growth retardation under long-term drought stress? *Genes* 506(2): 265–273.

Stone, S.L. (2014) The role of ubiquitin and the 26S proteasome in plant abiotic stress signaling. *Frontiers in Plant Science* 5: 135.

Sunkar, R., Li, Y-F., Jagadeeswaran, G. (2012) Functions of microRNAs in plant stress responses. *Trends in Plant Science* 17(4): 196–203.

Suzuki, N., Koussevitzky, S., Mittler, R., Miller, G. (2012) ROS and redox signaling in the response of plants to abiotic stress. *Plant and Cell Environment* 35(2): 259–270.

Suzuki, N., Rivero, R.M., Shulaev, V., Blumwald, E., Mittler, R. (2014) Abiotic and biotic stress combinations. *New Phytologist* 203(1): 32–43.

Takasaki, H., Maruyama, K., Kidokoro, S., Ito, Y., Fujita, Y., Shinozaki, K., Yamaguchi-Shinozaki, K., Nakashima, K. (2010) The abiotic stress-responsive NAC-type transcription factor OsNAC5 regulates stress-inducible genes and stress tolerance in rice. *Molecular Genetics and Genomics* 284(3): 173–183.

Tao, Z., Kou, Y., Liu, H., Li, X., Xiao, J., Wang, S. (2011) OsWRKY45 alleles play different roles in abscisic acid signaling and salt stress tolerance but similar roles in drought and cold tolerance in rice. *Journal of Experimental Botany* 62(14): 4863–4874.

Thapa, G., Dey, M., Sahoo, L., Panda, S. (2011) An insight into the drought stress induced alterations in plants. *Biologia Plantarum* 55(4): 603–613.

Thomashow, M.F. (2010) Molecular basis of plant cold acclimation: Insights gained from studying the CBF cold response pathway. *Plant Physiology* 154(2): 571–577.

Todaka, D., Shinozaki, K., Yamaguchi-Shinozaki, K. (2015) Recent advances in the dissection of drought-stress regulatory networks and strategies for development of drought-tolerant transgenic rice plants. *Frontiers in Plant Science* 6(84): 1–20.

Tripathi, P., Rabara, R.C., Rushton, P.J. (2014) A systems biology perspective on the role of WRKY transcription factors in drought responses in plants. *Planta* 239(2): 255–266.

Umezawa, T., Okamoto, M., Kushiro, T., Nambara, E., Oono, Y., Seki, M., Kobayashi, M., Koshiba, T., Kamiya, Y., Shinozaki, K. (2006) CYP707A3, a major ABA 8'-hydroxylase involved in dehydration and rehydration response in *Arabidopsis* thaliana. *The Plant Journal* 46(2): 171–182.

Umezawa, T., Sugiyama, N., Takahashi, F., Anderson, J.C., Ishihama, Y., Peck, S.C., Shinozaki, K. (2013) Genetics and phosphoproteomics reveal a protein phosphorylation network in the abscisic acid signaling pathway in *Arabidopsis* thaliana. *Science Signaling* 6(270): rs8.

Urano, K., Maruyama, K., Ogata, Y., Morishita, Y., Takeda, M., Sakurai, N., Suzuki, H., Saito, K., Shibata, D., Kobayashi, M. (2009) Characterization of the ABA-regulated global responses to dehydration in *Arabidopsis* by metabolomics. *The Plant Journal* 57(6): 1065–1078.

Verma, V., Ravindran, P., Kumar, P.P. (2016) Plant hormone-mediated regulation of stress responses. *BMC Plant Biology* 16: 86.

Wang, P., Xue, L., Batelli, G., Lee, S., Hou, Y-J., Oosten, M.J., Zhang, H., Tao, W.A., Zhu, J-K. (2013) Quantitative phosphoproteomics identifies SnRK2 protein kinase substrates and reveals the effectors of abscisic acid action. *Proceedings of the National Academy of Sciences USA* 110(27): 11205–11210.

Wang, Y., Yang, M., Wei, S., Qin, F., Zhao, H., Suo, B. (2017) Identification of circular RNAs and their targets in leaves of *Triticum aestivum* L. under dehydration stress. *Frontiers in Plant Science* 7(2024): 1–10.

Weiner, J.J., Peterson, F.C., Volkman, B.F., Cutler, S.R. (2010) Structural and functional insights into core ABA signaling. *Current Opinion in Plant Biology* 13(5): 495–502.

Wu, T., Kong, X-P., Zong, X-J., Li, D-P., Li, D-Q. (2011) Expression analysis of five maize MAP kinase genes in response to various abiotic stresses and signal molecules. *Molecular Biology Reports* 38(6): 3967–3975.

Xiong, L., Zhu, J.K. (2001) Abiotic stress signal transduction in plants: Molecular and genetic perspectives. *Physiologia Plantarum* 112(2): 152–166.

Xu, Z-Y., Kim, S.Y., Kim, D.H., Dong, T., Park, Y., Jin, J.B., Joo, S-H., Kim, S-K., Hong, J.C., Hwang, D. (2013) The *Arabidopsis* NAC transcription factor ANAC096 cooperates with bZIP-type transcription factors in dehydration and osmotic stress responses. *The Plant Cell* 25(11): 4708–4724.

Yamasaki, K., Kigawa, T., Seki, M., Shinozaki, K., Yokoyama, S. (2013) DNA-binding domains of plant-specific transcription factors: Structure, function, and evolution. *Trends in Plant Science* 18(5): 267–276.

Yang, W., Liu, X-D., Chi, X-J., Wu, C-A., Li, Y-Z., Song, L-L., Liu, X-M., Wang, Y-F., Wang, F-W., Zhang, C. (2011) Dwarf apple MbDREB1 enhances plant tolerance to low temperature, drought, and salt stress via both ABA-dependent and ABA-independent pathways. *Planta* 233(2): 219–229.

Yoshida, T., Fujita, Y., Maruyama, K., Mogami, J., Todaka, D., Shinozaki, K., Yamaguchi-Shinozaki, K. (2015) Four *Arabidopsis* AREB/ABF transcription factors function predominantly in gene expression downstream of SnRK2 kinases in abscisic acid signalling in response to osmotic stress. *Plant Cell and Environment* 38(1): 35–49.

Yoshida, T., Fujita, Y., Sayama, H., Kidokoro, S., Maruyama, K., Mizoi, J., Shinozaki, K., Yamaguchi-Shinozaki, K. (2010) AREB1, AREB2, and ABF3 are master transcription factors that cooperatively regulate ABRE-dependent ABA signaling involved in drought stress tolerance and require ABA for full activation. *The Plant Journal* 61(4): 672–685.

Yoshida, T., Mogami, J., Yamaguchi-Shinozaki, K. (2014) ABA-dependent and ABA-independent signaling in response to osmotic stress in plants. *Current Opinion in Plant Biology* 21: 133–139.

Zandalinas, S.I., Mittler, R., Balfagón, D., Arbona, V., Gómez-Cadenas, A. (2018) Plant adaptations to the combination of drought and high temperatures. *Physiologia Plantarum* 162(1): 2–12.

Zhu, J.K. (2016) Abiotic stress signaling and responses in plants. *Cell* 167(2): 313–324.

6 Variability in Physiological, Biochemical, and Molecular Mechanisms of Chickpea Varieties to Water Stress

Nataša Čerekovič, Nadia Fatnassi, Angelo Santino, and Palmiro Poltronieri

CONTENTS

6.1 INTRODUCTION

Chickpea (*Cicer arietinum* L.), the second most important grain crop in the world, is widely grown across the Mediterranean basin, East Africa, India, the Americas, and Australia. Most chickpea-producing areas are in arid and semi-arid zones, and approximately 90% of the world's chickpea is grown under rainfed conditions. Drought is one of the major constraints for crop productivity and is typical of the post-rainy season in semi-arid tropical regions. Drought is caused by variability in rainfall, by evaporation during crop season, and by soil characteristics.

Plants perceive, process, and translate different stimuli into adaptive responses. To support food availability and security, knowledge of plant stress response is vital for the development of breeding and biotechnological strategies to improve stress tolerance in crops and consequently crop yields.

Nowadays, one of the parameters used for enabling cropping systems under water shortage is the establishment of short duration chickpea cropping, enabling high yields even in arid regions. Under terminal drought, early crop duration and the yield potential have been shown to contribute to crop yield in dry climates.

There are great differences between Desi-type (small seeded, microcarpa) and Kabuli-type (large seeded, macrocarpa) chickpea varieties (Nayyar et al., 2006a,b). Desi-type chickpeas show a higher tolerance to water stress. Under terminal drought, the growth duration of chickpea is reduced, especially the reproductive phase, with effects more pronounced in Kabuli-type varieties.

Seed composition is slightly different between the two types, for example, Kabuli types BG-1053 and L-551, and Desi types GPF-2, PBG-1, and PDG-4 (Singh et al., 2008). Desi-type varieties are more tolerant to drought stress than Kabuli-type varieties. Currently, a selected germplasm for water stress tolerance is available for both types (Lakshmanan et al., 2010).

Several elite chickpea cultivars (ICC 4958, C 214, JG 74, Pusa 256, Phule G12, and Annigeri-1) from different agroclimatic zones have been established and used in breeding programs. For drought tolerance, several lead chickpea varieties, such as JG 11, Chefe, KAK2, Arerti, ICCV10, ICCV95423, ICCV97105, Ejere, and DCP92-3, have been established and used in field studies, with positive results also for root traits and agronomic performance (Varshney et al., 2012).

The genomes of Desi-type ICC 4958 (Parween et al., 2015) and other varieties were recently made available, as was the Kabuli CDC Frontier variety (Hiremath et al., 2011). A set of 100 chickpea varieties were sequenced: the genome sequences of 26 Kabuli varieties were compared with 58 Desi-type genomes and 5 wild chickpea genomes (Thudi et al., 2017). The resequencing of 29 varieties made in earlier studies (Edwards, 2016a, b) was also performed. Copy number variations and presence–absence variations have the potential to drive phenotypic variations for trait improvement. This study has provided an analysis of the genetic and genomic changes reflecting the history of breeding, while breeding signatures at selected loci have provided target sequences to perform crop improvement studies. This study showed great diversity in both the Desi and Kabuli varieties as a result of recent chickpea breeding efforts.

This review discusses the knowledge on plant signaling pathways in response to environmental challenges such as drought, high salinity, temperature fluctuations, nutrient deprivation, CO_2, and osmotic stress, and how the two varieties respond by activating and coordinating defense signals based on osmoprotectants, enzyme systems scavenging reactive oxygen intermediates, and activation of transcription factors (TFs) switching on the expression of drought response genes.

6.2 IN-FIELD CHICKPEA STUDIES

The growth duration of chickpea under terminal drought is reduced, especially the reproductive phase (Lakshmanan et al., 2013; Pushpavalli et al., 2014; Bantilan et al., 2014), with the effects more pronounced in Kabuli-type varieties. Several studies (Leport et al., 1999; Nayyar et al., 2006a, b) have shown a greater yield inhibition in Kabuli compared with Desi following water stress and have attributed this to a less effective remobilization of the assimilates toward developing seeds under stress conditions (Davies et al., 2000).

Seed filling and yield (Palta et al., 2005; Turner et al., 2005; Kashiwagi et al., 2006; Zaman-Allah et al., 2011b; Behboudian et al., 2001; Ulemale et al., 2013; Yaqoob et al., 2013; Awasthi et al., 2014) and yield-related traits (seed yield, 100 seed weight, days to 50% flowering, and days to maturity) were studied in field conditions for 320 elite varieties, under rainfed and irrigated conditions, applying genome selection and prediction models for yield-related traits (Roorkiwal et al., 2016). A repeat length variation in the *C. arietinum* myo-inositol

monophosphatase (*CaIMP*) gene was found to determine seed size (Dwivedi et al., 2017).

Physiological indices have been used to compare different varieties and their ability to cope with abiotic stress (Ulemale et al., 2013). The most studied parameters are the partitioning coefficient (Lakshmanan et al., 2013); pod production (Leport et al., 2006; Fang et al., 2010); pod production and estimation of abscisic acid (ABA) (Liu et al., 2005; Mantri et al., 2007); redox potential (Foyer, 2005); levels of proline (Kavikishore et al., 1995; Mafakheri et al., 2010; Mathur et al., 2009; Szabados and Savouré, 2009); glycine betaine (Chen and Murata, 2002; Ashraf and Foolad, 2007); and polyamines, polyols, and sugar osmolytes (Poltronieri et al., 2011).

Several characteristics were shown to be important in drought stress tolerance, such as early flowering and cropping, root depth, and the conservative pattern of water use (Zaman-Allah et al., 2011a, 2011b).

Stomatal closure and a reduction in transpiration are the principal mechanisms that plants adopt to respond to a water deficit; however, this leads to low oxygen availability, which translates into a reduction in nitrogen fixation in nodules (Soussi et al., 1999; Ltaief et al., 2007).

Among the physiological parameters (Gupta et al., 2010; Mir et al., 2012) included are the dry plant biomass (DPB), shoot dry weight (SDW), root dry weight (RDW), root to shoot ratio (RSR), carbon allocation (Hasibeder et al., 2015), relative water content (RWC); water use efficiency (WUE) (Siddique et al., 2001; Zaman-Allah et al., 2011a), estimated by the ratio of net CO_2 assimilation (A) and transpiration (E) (A/E) (Gonzalez et al., 1995); leaf gas exchange rates (Leport et al., 1998, 1999; Basu et al., 2007), leaf water potential (Karamanos and Papatheohari, 1999), stomatal conductance (Farquhar and Sharkey, 1982), reduction in transpiration in response to high vapor pressure deficit (VPD) (Ramamoorthy et al., 2015), leaf electrolyte leakage, photosynthesis rates, and levels of chlorophyll (Nayyar et al., 2006a). Recently, the drought response in Desi lines under field conditions was assessed measuring the canopy temperature depression (CTD) using thermal infrared thermography and the partitioning coefficient in elite varieties under water stress (Kashiwagi et al., 2008; Lakshmanan et al., 2013; Ramamoorthy and Lakshmanan, 2014; Ramamoorthy et al., 2015).

In a report by researchers at the International Crops Research Institute for the Semi-Arid Tropics (ICRISAT), India, genotype ICCV03408 with high CTD and high yield under terminal drought conditions has been identified as a drought-tolerant line, while lines ICCV03104, ICCV00202, ICCV01102, ICCV10112, and ICCV04106 with high CTD values have been identified as donors for drought tolerance (Bharadwaj et al., 2015). As regards

Kabuli-type varieties studied for drought tolerance, INRAT-73 (also known as Beja 1) is salt and drought tolerant, and Amdoun 1 is drought sensitive (Labidi et al., 2009). In their study, a change in RSR, loss of chlorophylls, and nodule mortality were considered an indicator of stress tolerance, with Beja and Kesseb the varieties with the best performances. However, RSR may be negatively affected in cases when the shoot biomass is reduced.

A few reports (Leport et al., 1999) have shown significantly higher levels of carotenoids and chlorophyll, with a reduction in the electron transport rate (ETR), photosynthesis efficiency, quantum yield, and photosystem-I activity in Desi varieties compared to Kabuli varieties grown under similar condition. The low values of the net photosynthesis rate are probably related to the low stomatal conductance, but they can also be a response to low carboxylation efficiency or low water potential. According to the response shown by the photosynthetic rate, Kabuli-type varieties are more sensitive to water deprivation than the Desi type, especially for drought-tolerant varieties.

In a greenhouse study on various varieties under drought conditions, the tolerant Desi-type ICC 6098 was able to maintain its photosynthesis rate for 1 week, dropping in respect to the control only after 10 days of water stress (Fatnassi et al., 2018). Several authors reported a decrease in these variables under field conditions. Therefore, it seems that Kabuli plants show a faster decrease in growth parameters during drought stress.

Kabuli varieties show a faster decrease in leaf water potential after exposure to drought stress (Nayyar et al., 2006a; Larrainzar et al., 2007). The water tolerance of Desi-type plants may be related to their superior ability to maintain their plant water status for longer periods, which results in less oxidative damage. The higher tolerance of Desi-type varieties to osmotic stress may be linked to the Desi plants' ability to maintain the functional integrity of the photosynthetic system and to develop an abundant nodule network, which in turn favors the rate of symbiotic nitrogen fixation (SNF).

Salt sensitivity has been studied in chickpea using various approaches: studying the expression changes in genes such as lipid transfer proteins (CarLTPs), late embryogenesis abundant 1 and 2 (CarLEA1, CarLEA2) (Romo et al., 2001), early responsive to dehydration (ERD1 to ERD16) (Alves et al., 2011), and LEA-related hydrophilins (Battaglia et al., 2008); plasma membrane (PM) H^+-ATPase (SOS1) and vacuolar (V) H^+-ATPase and H^+-PPase; osmotic stress-induced calcium-dependent protein kinases (CDPKs) and calcineurin B-like (CBL) proteins (for Ca^{2+}-dependent pathways for the regulation of ion homeostasis and plant salt tolerance through Na^+ extrusion) salt-regulated TFs such as WRKY33 and MYB51, salt tolerance zinc finger (STZ), and dehydration responsive element (DRE)-binding TFs (DREBs); physiological parameters, modifications such as loss of water content, biomass and shoot elongation reduction, in sensitive varieties, with tolerance ranging from 25 mM NaCl to highly resistant genotypes surviving up to 90 mM NaCl under hydroponic conditions (Flowers et al., 2010; Kotula et al., 2015; Ahmad et al., 2016; Sen et al., 2017; Kaashyap et al., 2017).

In an ICRISAT study identifying markers linked to salinity tolerance, for marker-assisted selection (MAS) and marker-assisted backcross (MABC) programs, during salt stress, tolerant chickpea varieties such as Pusa 362, Pusa 1103, and Pusa 72 outperformed other varieties, showing a reduction of 35% in shoot growth and 5% in root growth, while sensitive plants showed 73% in shoot growth and 51% in root growth compared to the unstressed conditions (Kumar et al., 2015).

6.3 STOMATA CLOSURE AND ABSCISIC ACID

In plants, a water deficit induces a signaling network, causing stomata closure through the movement of signaling compounds up to the guard cells to reduce the evaporation rates. The ABA concentration in guard cells increases when humidity is low or when salt levels are high. Low water levels, or high salt levels, slowly increase the concentration of ABA in the roots. Huang et al. (2008) showed that drought enhanced both ABA biosynthesis and catabolism, resulting in an increase in ABA and catabolites. ABA foliar application (15 µmol/L) under salinity conditions improved the shoot dry matter, photosynthesis rate, peroxidase and catalase (CAT) activity, and shoot K+ concentration, while decreasing the shoot Na^+ concentration in canola (Farhoudi and Saeedipour, 2011). Drought enhances both ABA biosynthesis and catabolism, resulting in an increase in ABA and the derived catabolites.

Sulfate, mobilized by the action of a sulfate transporter, acts as a long-distance signal moving through the sap to induce ABA biosynthesis in leaves. ABA moves through the xylem until it reaches the stomata to reduce the transpiration rate, an activity mediated by the release of nitric oxide (NO). An increase in endogenous NO levels via the application of an exogenous NO donor (sodium nitroprusside, SNP) has been shown to confer resistance to drought and salinity (Tanou et al., 2009).

Costimulation with ABA, ethylene (ET), and sulfate produces an additive increase in stomata closure, reinforcing the block of transpiration for an extended period.

Stomata are the organs that respond to abiotic and biotic stress, as they are the entryway of various pathogens into the plant. Therefore, virulence factors trigger stomatal reopening (coronatine, syringolin A, fusicoccin) and subvert the hormone regulation of stomata (Melotto et al., 2017).

6.4 HORMONES CROSSTALK AND DROUGHT TOLERANCE

6.4.1 JASMONATES

Jasmonates (JAs) promote stomatal closure. A crosstalk of ABA in methyl-jasmonate (MeJA)-induced stomatal closure in *Arabidopsis* has been found (Hossain et al., 2011). The exogenous application of the volatile MeJA induced stomatal closure, and this activity was found to be mediated by endogenous ABA, leading to stomatal closure in *Arabidopsis* guard cells (Hossain et al., 2011).

Several authors have shown that the genes in the JA biosynthesis pathway are upregulated at an early stage of drought stress in drought-tolerant chickpea varieties (Molina et al., 2008; De Domenico et al., 2012). However, the synthesis is dependent on the specific gene isoforms of *13-LOX* genes, and also continues at later stages of stress sensing (Fatnassi et al., 2018).

12-Oxo-phytodienoic acid (OPDA), an intermediate in the JA synthesis, is a drought-responsive regulator of stomatal closure that may act effectively with ABA (Wasternack, 2014; Wasternack and Strnad, 2016). In response to 100 mM NaCl treatment, the detection of JA, OPDA, 11-hydroxyjasmonate (11-OH-JA), 12-hydroxyjasmonate (12-OH-JA), and MeJA was shown in seedlings of tomato cv. Moneymaker (Andrade et al., 2005).

The jasmonic acid–derived active metabolite is the JA-isoleucine (JA-Ile) conjugate, which binds to a jasmonate-zim (JAZ) repressor, which is targeted for proteolysis, releasing the MYC TF from the inhibition complex.

MYC2 has been defined as an ABA and drought-responsive gene. Constitutive expression of MYC2 and gene disruption by knockout results in enhanced and reduced sensitivity to ABA, respectively. In addition, MYC2 activates the expression of the NAC domain TF RD22, suggesting a positive role in ABA signaling. Dehydration-responsive RD26 and RD22 are induced by JA, H_2O_2, drought, salinity, and ABA. The finding of JA involvement in plant responses to drought raises interest in the ABA–JA crosstalk (de Ollas et al., 2015; Llanes et al., 2016; Nir et al., 2017).

ABA seems to suppress the ERF1/PDF1.2 branch of the JA pathway. However, ABA strongly induces the YC2-VSP1 branch of the JA signaling pathway. It has been shown that the NAC domain–based ANAC019 and ANAC055 TFs also induce this pathway, and this pathway is dependent on MYC2. These TFs are also activated by JA, which suggests that ANAC019 and ANAC055 act downstream of both JA and ABA. A recent study has shown that ANAC019, ANAC055, and the homologous ANAC072 play synergistic and antagonistic roles in ABA signaling and osmotic stress (Nguyen et al., 2015).

Hormonal crosstalk is central to the adaptation of the plant response to the type of abiotic stress, and NO has been shown central to the activation/inhibition of various enzymes and TFs (Poltronieri et al., 2013; Hu et al., 2017).

6.4.2 ETHYLENE

ET production in an earlier period of water stress may be a signal, helping the plants to sense the stress condition and to make some adaptive physiological response in advance (Kazan, 2015). In response to salinity, ET appears to negatively affect salt tolerance. Comparing tolerant and sensitive varieties, plants that produce low ET are more tolerant to the harmful effects of environmental stresses as compared to those that produce higher levels (Leonard et al., 2005). The hormone activates ET-responsive elements (ERE) upstream of promoters of TFs, such as ET-responsive factor 6 (ERF6), a TF situated upstream of a stress-related network of TFs. ERF6 was shown to directly activate the expression of genes encoding the stress tolerance–related TFs, such as NAC TFs, basic Leucine Zipper (bZIP) TFs, MYB DOMAIN PROTEIN51 (MYB51), STZ finger protein (ZFP), and WRKY33. Thus, in addition to inhibiting *Arabidopsis* leaf growth, ERF6 also induces, independently of the GA/DELLA pathway, genes involved in stress tolerance (Dubois et al., 2013).

6.4.3 STRIGOLACTONES

Strigolactones (SLs) have been shown to be positive regulators of plant response to drought. Endogenous SLs have been shown to positively modulate both drought and salt stress responses through ABA-dependent and ABA-independent pathways. This complexity is reflected in the functions of ABA, cytokinins (CKs), and SLs on the regulation of stomatal closure and leaf senescence (Czarnecki et al., 2013). Several studies show the beneficial effects of an increase in CK levels associated with stomatal conductance, transpiration, and photosynthesis under water stress (Riefler et al., 2006).

Foo and Reid (2013) demonstrated the potential use of genetic engineering to improve the drought and salt tolerance of plants by manipulating the endogenous SLs levels and/or SLs signaling.

6.4.4 AUXINS: INDOLE-3-ACETIC ACID AND RELATED COMPOUNDS

The application of indole-3-butyric acid (IBA) increased the endogenous indole-3-acetic acid (IAA) level, which improved the drought tolerance of tall fescue (Zhang et al., 2009). Similarly, the activation of *YUCCA*7 and *YUCCA*6 (Du et al., 2013) genes, encoding a flavin monooxygenase belonging to the tryptophan-dependent auxin biosynthetic pathway, increased the endogenous Aux levels and enhanced drought tolerance in *Arabidopsis*.

6.4.5 BRASSINOSTEROIDS (BRs)

BR treatment has also been reported to increase the seedling growth of *Sorghum vulgare* under osmotic stress (Vardhini and Rao, 2003), and improve the drought tolerance of *Phaseolus vulgaris* (Upreti and Murti, 2004). Farooq and colleagues (2009) found that BR application in rice improved the leaf water economy and CO_2 assimilation, and enabled it to withstand drought. In addition, it has been shown that treatment with another BR, 24-epibrassinolide (24-EBR), increased the survival rate of *Arabidopsis thaliana* and *Brassica napus* seedlings subjected to drought stress (Kagale et al., 2007). Similarly, the exogenous application of BRs can also alleviate the adverse effects of salinity. Hayat and colleagues (2007) reported that the application of a 28-homobrassinolide (28-HomoBL) spray to the foliage or applied through the roots of *Brassica juncea* plants generated from seeds soaked in NaCl enhanced the growth and seed yield under normal conditions. However, this compound reduced the damage induced by salt stress on chloroplast. 24-EBR significantly enhanced the growth and photosynthetic capacity of salt-tolerant and salt-sensitive wheat plants (Shahbaz et al., 2008).

6.5 PLANT RESPONSE TO OXIDATIVE STATE AND TO REACTIVE OXYGEN SPECIES (ROS) PRODUCTS

In plants, the major ROS-scavenging pathways consist of (i) water–water cycle (Asada, 1999; Rizhsky et al., 2003; Weng et al., 2007), based on superoxide dismutase (SOD) and thylakoid-bound ascorbate peroxidase (t-APX) in chloroplasts; (ii) ascorbate–glutathione cycle; (iii) glutathione peroxidase (GPX) cycle based on glutathione reductase (GR), glutathione (GSH), oxidized glutathione (GSSG), and ascorbic acid (AsA), dehydroascorbate (DHA), and dehydroascorbate reductase (DHAR); and (iv) CAT. Among these, the ascorbate–glutathione cycle plays a crucial role in regulating ROS based on its wide distribution as well as the high affinity of ascorbate peroxidase (APX) for H_2O.

In tolerant varieties, several genes coding for ROS scavenging enzymes were found highly expressed at prolonged times of stress exposure, i.e. SOD, APX, GPX and CAT.

The ROS signal transduction of plants includes the MAPKKK cascade, based on ANP1 (also NPK1), the MAPKs, MPK3/6, p46MAPK, and calmodulin/CaM kinases, resulting in the regulation of the expression of various TFs belonging to the WRKY, DREBA, ERFs, MYB, and AP1 families.

Arginine is metabolized to polyamines through the actions of arginase (ARG), ornithine decarboxylase (ODC), and arginine decarboxylase (ADC), with ornithine and agmatine as intermediates, and is regulated by MeJA (Zhang et al., 2012). Ornithine could also be converted into γ-aminobutyric acid by delta 1-pyrroline-5-carboxylate synthase (P5CS) and glutamate decarboxylase (GAD), or into proline by pyrroline-5-carboxylate reductase (P5CR); γ-aminobutyric acid (GABA) is involved in osmotic regulation, the scavenge of ROS, and intracellular signal transduction (Kinnersley and Turano, 2000; Mittler, 2002; Knight and Knight, 2001; Zhang et al., 2012; Gustafsson, 2012); proline acts as a protein-compatible hydrotrope or as a hydroxyl radical scavenger able to regulate the NAD+/NADH ratio. Adenosine triphosphate (ATP) deficit could induce membrane lipid peroxidation and subsequently generate more free radicals, which would attack the cellular membrane.

6.6 ROOT SYSTEM, SYMBIOTIC NITROGEN FIXATION, AND PLANT–BACTERIA INTERACTIONS

Roots are the primary organ sensing the soil environment. Chickpea growth and development are dependent on the plant root system, due to its role in water and mineral uptake and in the nodulation and symbiotic relationship (Aouani et al., 2001; Mhadhbi et al., 2008; Nasr Esfahani et al., 2014a, b). A strong root system allows more soil water to be captured during the growth period (Kashiwagi et al., 2006). Plants spend energy on root production in search of water and/or reducing water loss. This parameter has been considered a criterion for adaptation to drought (Sassi et al., 2010), although some findings have shown no clear correlation between root trait and water extraction ability. However, if water is unavailable deeper in the soil profile, longer roots may reduce the SDW and harvest index by allowing the preferential partitioning of carbohydrates to roots at the expense of shoots (Pace et al., 1999). Moreover, the ability of plants to maintain root growth under osmotic stress is crucial for nodulation. Drought impairs the development of root hairs and the site of entry of rhizobia into the host, resulting in poor or no nodulation.

The existence of a large diversity of root biomass, root prolificacy, and rooting depth in chickpea mini-core germplasm accessions (Kashiwagi et al., 2005) prompted new efforts to improve germplasm through the selection of characteristics such as enhanced absorption.

Among the varieties most studied, ICC 4958 is drought tolerant, but salt sensitive. Varieties with a "deep root system" such as ICC 4958 grow well, the accumulation of seed mass after flowering is faster, and plants accumulate a large seed mass before the soil moisture recedes. Varieties such as ICC 4958 have been used in breeding programs for crossing with other drought-tolerant varieties to produce elite hybrids. In this way, many pedigree hybrids have been produced using ICC 4958 as the parent variety.

Plant growth promoting (PGP) fungi and bacteria are able to elicit induced systemic tolerance to drought in plant roots. PGP microorganisms can increase plant productivity and immunity through the induction of a higher antioxidative state, the production of leghemoglobins (binding NO and oxygen), the production of volatile compounds such as NO, CO, and H_2S, affecting stomatal closure, hormones such as IAA and 2,3 butanediol, and enzymes, degrading ET and inducing repression of ET synthesis genes.

In studies on nodules (Gonzalez et al., 1995; Ramos et al., 1999; Mhadhbi et al., 2008; Larrainzar et al., 2007; Sohrabi et al., 2012; Ramirez et al., 2013), several parameters have been shown to influence plant response to drought, such as fresh and dry nodule biomass, a decline in nitrogenase (N2ase) activity as a measure of the metabolic limitation of bacteroids, O_2 consumption, leghemoglobin content, energy charge and respiratory capacity, sucrose synthase activity, sucrose content, and ATP and organic acids content (Talbi et al., 2012). These authors showed the effective improvement of *Rhizobium etli* overexpressing *cbb3* oxidase under drought conditions in the common bean and the better ability of SNF. As shown in various reports (Naya et al., 2007; Labidi et al., 2009), SNF is sensitive to osmotic stress, while *Rh. etli* CFNX713 strain, overexpressing *cbb3* oxidase, showing higher respiratory capacity, was

able to improve the energy charge in nodules (Talbi et al., 2012).

Nayyar used rhizobium-inoculated Desi-type GPF2 and Kabuli-type L550 varieties, water stressed at the reproductive stage (plants with seven to nine pods), monitoring the water stress for 14 days. The greater stress tolerance of the Desi type may be related to its greater capacity to deal with oxidative stress, ascribed to its superior ability to maintain a better water status, which results in less oxidative damage. In addition, they conducted osmotic stress studies by subjecting both types of chickpea to similar levels of polyethylene glycol–induced water stress and to 10 µmol/L ABA, involved in oxidative injury, showing that Kabuli-type shoots were more inhibited than Desi-type plants (Nayyar et al., 2006a, b).

The decrease in N2ase activity in chickpea with progressive drought may be caused by a limitation in the O_2 availability and by the oxidative damage of nodule cells (Naya et al., 2007). Naya discussed the possibility that oxidative damage was the main cause for the decrease in N2ase fixation in indeterminate nodules before any effect on stochastic spineless (SS) expression was detected. A decrease in SS and N2ase activities was also observed in the common bean (Ramos et al., 1999), pea and soybean (Gonzalez et al., 1995; Maccarone et al., 1995). In addition, in chickpea nodules, N2ase activity decreased up to 50% after 10 days of drought stress. The active import of sucrose from the shoot and the limited sucrose consumption in the nodules, due to damage to respiratory activity, correlated with the increase in sucrose synthase mRNA and in sucrose content observed for soybean in drought stress conditions. However, the condition of reduced carbon did not limit the N2ase activity.

The limitation of oxygen in the nodules, an inhibition dependent on nitrogen availability, and the limitation of carbon flow to the bacteria are the main factors supposed to affect the N2ase. In alfalfa and *Medicago truncatula*, drought caused an accumulation of carbonated compounds in the nodules, which indicates that the regulation of bacteria-fixed nitrogen (BFN) in these species is produced independently of nodular carbon metabolism. BFN is a process that is highly sensitive to drought, to such an extent that it is rapidly inhibited in water stress conditions, thereby causing significant loss of leguminous crops.

Root nodules are highly specialized sink tissues in which at least one sucrose synthase isoform is strongly induced (Morell and Copeland, 1985): the reduction in sucrose synthase activity is considered one of the key factors responsible for the inhibition of SNF during drought (Gonzalez et al., 1995; Galvez et al., 2005). Under drought conditions, a decrease in sucrose synthesis activity in nodules was observed. This drop occurred simultaneously with a decrease in nitrogen fixation, with high correlation between both processes in adverse conditions. A drop in the concentration of phosphate, sugars, and organic acids was observed, indicating a decrease in carbon flow in the nodules, a drop that limited the supply of carbon to the bacteroid and the capacity of the bacteroid to fix nitrogen in various leguminous plants (Galvez, 2005; Ladrera Fernández, 2008). The stress occurred rapidly and intensely, while another pathway, dependent on ABA, involving control through leghemoglobin/oxygen, was activated. Treatment with exogenous ABA, carried out under conditions of water stress, played a beneficial role in the protein content of the plant. Likewise, recovery of the total activity of the metabolism of carbon and nitrogen enzymes in the nodules was observed. However, the application of ABA did not reverse the negative effect of the water stress and it was not possible to relate species of activated oxygen with the regulation of nitrogen fixation. In this context, the decrease in nitrogen fixation occurred in association with a limitation in the carbon flow in the nodules, caused by the inhibition of sucrose synthase activity under these conditions.

Oxidative stress can play a role in the inhibition of drought-induced N_2 fixation (Kurdali et al., 2013). Lower H_2O_2 levels and, hence, minor oxidative damage in the Desi type are possibly linked to a greater APX activity, reported to be differentially regulated in response to salt in legumes (Jebara et al., 2005; Mohammadi et al., 2011).

Different Mesorhizobium–chickpea associations were shown to influence symbiotic performances (Nasr Esfahani et al., 2016). *Mesorhizobium ciceri* CP-31 (McCP-31)-chickpea and *M. mediterranum* SWRI9 (MmSWRI9)-chickpea plantlets were studied under control and low Pi conditions. Under low Pi availability, MmSWRI9-chickpea showed a symbiotic efficiency lower than McCP-31-chickpea, with reduced growth parameters and downregulation of *nifD* and *nifK*. The differences were attributed to a decrease in the Pi level in MmSWRI9-induced nodules under low Pi stress, and the upregulation of several key Pi starvation-responsive genes, with an accumulation of asparagine in nodules: during Pi starvation, the amino acid levels in Pi-deficient leaves of MmSWRI9-inoculated plants exceeded the shoot nitrogen requirement, due to nitrogen feedback inhibition. On the other side, Pi levels increased in the nodules of Pi-stressed McCP-31-inoculated plants, thanks to various metabolic and biochemical strategies to maintain nodular Pi homeostasis under Pi deficiency. The adaptation was based on the activation of alternative pathways of carbon metabolism, the enhanced production and exudation of organic acids from roots into the

rhizosphere, and the ability to protect nodule metabolism against Pi deficiency–induced oxidative stress. The adaptation under Pi deficiency was due to extensive reprogramming of whole-plant metabolism (Nasr Esfahani et al., 2016).

Pseudomonas putida MTCC5279 was shown to improve drought stress response in cv. BG-362 (Desi) and cv. BG-1003 (Kabuli) chickpea cultivars under in vitro and greenhouse conditions. Polyethylene glycol–induced drought stress (osmotic stress) severely affected seed germination in both cultivars, which was considerably improved on *P. putida* inoculation (Tiwari et al., 2016).

Drought stress significantly affected various growth parameters, such as water status, membrane integrity, osmolyte accumulation, ROS scavenging ability, and stress-responsive gene expressions, which were positively modulated upon application of *P. putida* in both chickpea cultivars. Quantitative real-time polymerase chain reaction (qRT-PCR) analysis showed differential expression of genes for TFs, such as DRE/C-repeat (CRT) *cis*-acting element binding the TF (DREB1A) of the ET response factor (ERF) family, containing the conserved DNA-binding domain of ERF/Apetala2 (AP2) (Anbazhagan et al., 2014; Per et al., 2018); NAC (NAM, AFAT, and CUC) TF proteins such as NAC1; MYC2 TF (signaling downstream of JA); stress-responsive proteins, such as late embryogenesis abundant (LEA) and dehydrins (DHN) acting in protein–protein interaction networks (Saibi et al., 2016); ROS scavenging enzymes CAT, APX, glutathione-S-transferase (GST), genes involved in ET biosynthesis (ACO and ACS), in chickpea cultivars exposed to drought stress and recovery in the presence or absence of the strain of *P. putida*. The observations imply that the strain confers drought tolerance on chickpea by altering various physical, physiological, and biochemical parameters, as well as by modulating the differential expression of at least 11 stress-responsive genes. The analysis of chickpea growth promotion and stress alleviation showed a prolonged stress tolerance in one-month-old Desi and Kabuli plants under drought stress for 0, 1, 3, and 7 days and after water recovery.

6.7 IN PLANTA NETWORK OF SIGNALS MOVING FROM ROOTS TO LEAVES AND BACK

Understanding plant coordinated responses involves a fine description of the mechanisms occurring at the cellular and molecular level. These mechanisms involve numerous components that are organized into complex transduction pathways and networks, from signal perception to physiological responses. The major challenges of plant signaling are to understand the kinds of signals that cells receive, how these signals are recognized, and how cells respond spatially and temporally to these signals to program a specific response at the organism level. Furthermore, signal transduction cascades that involve a large array of molecular and cellular processes have been depicted in several studies showing crosstalk between hormones and signaling pathways (Per et al., 2018).

Previous research has focused on the components of the signaling cascades (receptors/sensors, Ca²⁺, MAP kinases, NO and ROS synthesis components, ion fluxes) and their role under different stress conditions (Fukami et al., 2017), changes in intracellular compartmentalization, phytohormone retrograde signaling, source-to-sink transport of sugars through the phloem (Lemoine et al., 2013), and other compounds (lipids, hormones, peptides, RNAs) moving through the xylem or phloem (Poltronieri et al., 2011). The information collected during these years has provided a link between the molecular cell signal transduction cascades and the plant response at the whole organism level.

6.8 TRANSCRIPTOME STUDIES

In the years before the chickpea genome sequence data, several approaches have been used in the identification of sequences of expressed genes. For instance, expression sequence tags (ESTs) have been retrieved from chickpeas under various stress conditions (Deokar et al., 2011). ESTs were classified and grouped from abiotic stressed chickpea plants (Jayashree et al., 2005; Varshney et al., 2009) using ICC 4958 (drought tolerant) and ICC 1882 (sensitive) varieties.

Other groups contributed by performing tissue-specific transcriptome analysis in chickpea with massively parallel pyrosequencing (Garg et al., 2011). With the availability of next-generation sequencing (NGS) platforms, many researchers used ICC 4958 for transcriptomic studies and tissue-specific gene expression analysis (Garg et al., 2011; Varshney et al., 2012).

The availability of genome sequences has made possible the realization of transcriptome studies through the sequencing of cDNA pools, as well as by RNA sequencing (RNA-seq) based on NGS technology (Varshney et al., 2014; Garg et al., 2016). RNA-seq has provided the means to analyze in great detail the whole transcriptome output of tissues and even specialized cell types, through laser capture microdissection or even at the single-cell level, through cell sorting using magnetic antibodies. Finally, "omics" techniques such as proteomics (Pandey

et al., 2008), genomics (Varshney et al., 2009), transcriptomics (Garg et al., 2011, 2015; Hiremath et al., 2011; Jain and Chattopadhyay, 2010; Agarwal et al., 2012), systems biology, and network modeling have recently been added to the tools in understanding the complexity of intra and intercellular signal transduction to unveil the signaling regulons during abiotic stress, and the changes in gene expression (upregulation or downregulation), recruitment of TFs (Nguyen et al., 2015; Ramalingam et al., 2015; Anbazhagan et al., 2014), regulation of chromatin organization and transcription machinery components such as Mediator25, interacting with DREB1A, ERF1, and MYC2; the epigenetic mechanisms for stress memory, and post-transcriptional control by small RNAs (Poltronieri et al., 2015; Tiwari et al., 2017).

Several researchers performed transcriptome analyses of chickpea genotypes belonging to drought-sensitive and drought-resistant varieties, to investigate the molecular basis of drought and salinity stress response/adaptation. Phenotypic analyses confirmed the contrasting responses of the chickpea genotypes to drought or salinity stress. RNA-seq of the roots of drought- and salinity-related genotypes was carried out under control and stress conditions at vegetative and/or reproductive stages. Comparative analysis of the transcriptomes revealed divergent gene expression in the chickpea genotypes at different developmental stages. Garg and colleagues found 4950 and 5545 genes exclusively regulated in drought- and salinity-tolerant genotypes, respectively. A significant fraction (higher than 45%) of the TF-encoding genes showed differential expression under stress. Several enzymes involved in metabolism, such as carbohydrate synthesis, photosynthesis, lipid metabolism, the generation of energy, protein modification, redox homeostasis, and cell wall component biogenesis, were affected by drought and salinity stresses. Various transcript isoforms showed expression specificity across the chickpea genotypes and developmental stages, such as the AP2-ERF family members. These findings may provide insights into the transcriptome dynamics and identify components of the regulatory network (regulons) associated with drought and salinity stress responses in chickpea.

Agarwal et al. (2012) performed a comparative analysis of Kabuli chickpea transcriptome (ICCV2) with Desi and wild chickpea, collecting root and shoot tissue samples from 15-day-old seedlings, determining the root, shoot, and floral bud transcriptomes.

Ramalingam et al. (2015) analyzed JA 11, an elite, drought-tolerant cultivar and an introgression line (JG 11+) developed by crossing ICC4958 and JG11 (ICC4958 × JG11), using greenhouse conditions. Slow drought stress was imposed on the four genotypes, and roots were collected when the transpiration ratio decreased tenfold.

ICC 4958 was used to study root transcripts under drought stress (Hiremath et al., 2011), compared to transcripts in the sensitive variety ICC 1882. Hiremath and colleagues used 22 different tissues of the ICC 4958 variety, representing different developmental stages as well as drought and salinity stressed roots, harvested up to several days after stress onset.

A few studies have analyzed the changes in gene expression during prolonged water stress: the longest period of drought treatment was up to 12 days in chickpea lines ICCV2 and PUSABGD72 (Jain and Chattopadhyay, 2010). The researchers studied the drought-tolerant PUSABGD72 and drought-sensitive ICCV2 varieties (Jain and Chattopadhyay, 2010). As for drought treatment, soil-grown 12-day-old plants were subjected to progressive drought by withholding water for 3, 6, and 12 days, respectively, a time point when the soil moisture content decreased from approximately 50% to approximately 15% at 12 days.

Previously, Mantri et al. (2007) performed transcripts profiling using abiotic stress-tolerant and stress-sensitive chickpea varieties, but used only the flower and leaf tissues due to the poor quality of the root RNA. Mantri showed that drought tolerant-2 (BG 362) had twice the number of repressed transcripts as tolerant-1 (BG 1103), drought susceptible-1 (Kaniva), and susceptible-2 (Genesis 508), so that differentially expressed (DE) transcripts would affect the response to drought stress (desiccation); in parallel, they determined the gene expression profiles of chickpea varieties in response to cold (Sonali, ILC 01276, Amethyst, DOOEN varieties) and high salinity (250 mM NaCl) (CPI 060546, CPI 60527, ICC 06474, ICC 08161 varieties). Leaf/shoots, roots, and flower/buds were collected and analyzed separately.

Comparative analyses of transcriptomes between *Cicer microphyllum* and cultivated Desi cv. ICC 4958 detected 12,772 transcripts, composed of 3,242 root- and 1,639 shoot-specific *C. microphyllum* genes, differentially expressed predominantly in ICC 4958 roots under drought stress. Among these genes, five *C. microphyllum* root-specific genes with non-synonymous and regulatory SNPs were found to determine drought-responsive yield traits (Srivastava et al., 2016).

Deep SuperSAGE (serial analysis of gene expression) studies, combined with NGS, have been used to identify and quantify the RNA transcriptome in Beja 1 nodules (Afonso-Grunz et al., 2014), in studies on chickpea roots and nodules during drought stress (Molina et al., 2008), and on roots and nodules during salt stress response (Molina et al., 2011; Kahl et al., 2011).

Using the Desi-type water stress–tolerant ILC 588 variety, subjected to 2 hours of water stress, Molina et al. (2008) studied gene expression changes showing the main deregulated transcripts in roots. They analyzed the response to water stress after 2 hours in the roots of the drought-tolerant variety ILC 588. JA biosynthesis and ROS scavenging components were the main gene activated during water stress response. The identified transcripts are considered to protect cells from water deficit not only through producing metabolic proteins but also by regulating genes involved in signal transduction. Considering the ROS, Molina and colleagues identified transcripts involved in the scavenging of oxygen radicals, showing that stress response in chickpea is linked to the production and management of ROS.

In a subsequent study by the same teams on chickpea roots and nodules in response to salt stress, using Beja 1 (salt tolerant), Amdoun 1 (salt sensitive), ICC 4958 (salt sensitive), and ICC 6098 (salt weakly tolerant) varieties under salt stress conditions, several isoforms of lipoxygenase (*LOX*) and other genes in the JA synthesis pathway were found upregulated or downregulated at 2, 8, 24, and 72 hours of salt stress (Molina et al., 2011; Kahl et al., 2011). Taqman probes were designed, and *LOX* isoforms and/or splicing variants were monitored by reverse transcription and qPCR (De Domenico et al., 2012; Fatnassi et al., 2018). These studies showed that the early events of response to water deficit and salt stress in chickpea are related to the production and management of ROS, JA accumulation, and hormone crosstalk, which are able to improve the plant response to water stress in the first week (De Domenico et al., 2012; Fatnassi et al., 2018).

6.9 CONCLUSION

The Desi-type and Kabuli-type varieties show differences in the timing of induction in water stress: tolerant Desi-type ICC 6098 was shown to withstand longer periods of water deprivation.

An unpublished study showed an increase in the expression of genes, sustaining plant physiology performances (reactive species buffering, water compartmentalization). Furthermore, a huge increase in the expression of genes linked to alarm signals, such as sucrose synthase, characterized the stress-sensitive varieties (Fatnassi, personal communication). The study confirmed that the Desi-tolerant variety responds by maintaining its physiological parameters for longer times under water stress challenge and induces the expression of genes associated with stress resistance.

In conclusion, these results may be useful for analyses of water stress tolerance in chickpea. These findings may be exploited to classify and to phenotype the characters of hybrids showing optimized expression of the required genes sustaining stress tolerance. The finding that certain genes are induced only after prolonged water stress may be used as a marker of the ability of cultivars to cope with moderate water stress, and to grow after rewatering. The selected genes may be tested in hybrids of elite varieties with optimized performances under stress conditions, to evaluate their ability to activate the expression of genes involved in ROS scavenging, signaling pathways, water potential maintenance, nodule biomass, and nitrogen fixation.

There is a need for selection programs for stress-resistant parent varieties on the basis of biomarker genes showing high or rapid inducible expression, to be used in breeding programs to select the pedigrees for crossing and to monitor the ability of hybrids to tolerate better abiotic stress challenges (Kumar et al., 2008). In an effort to produce plants that adapt to climate changes, new elite varieties and inbred lines of either Desi- or Kabuli-type chickpeas have been made available to growers (Hamwieh et al., 2013). It is thus probable that some varieties will confirm expectations and will constitute new resources for sustainable agriculture.

REFERENCES

Afonso-Grunz, F., Molina, C., Hoffmeier, K., Rycak, L., Kudapa, H., Varshney, R.K., Drevon, J.J., Winter, P., Kahl, G. (2014) Genome-based analysis of the transcriptome from mature chickpea root nodules. *Frontiers in Plant Science* 5:235.

Agarwal, G., Jhanwar, S., Priya, P., Singh, V.K., Saxena, M.S., Parida, S.K., Garg, R., Tyagi, A.K., Jain, M. (2012) Comparative analysis of Kabuli chickpea transcriptome with Desi and wild chickpea provides a rich resource for development of functional markers. *PLoS One* 7(12):e52443.

Ahmad, P., Abdel Latef, A.A., Hashem, A., Abd Allah, E.F., Gucel, S., Tran, L.S. (2016) Nitric oxide mitigates salt stress by regulating levels of osmolytes and antioxidant enzymes in chickpea. *Frontiers in Plant Science* 7:347.

Alves, M.S., Fontes, E.P.B., Fietto, L.G. (2011) Early responsive to dehydration 15, a new transcription factor that integrates stress signaling pathways. *Plant Signaling and Behaviour* 6(12):1993–1996.

Anbazhagan, K., Bhatnagar-Mathur, P., Valdez, V., Dumbala, S.R., Kishor, P.B., Sharma, K.K. (2014) DREB1A overexpression in transgenic chickpea alters key traits influencing plant water budget across water regimes. *Plant Cell Reports* 34:199–210.

Andrade, A., Vigliocco, A., Alemano, S., Miersch, O., Botella, M.A., Abdala, G. (2005) Endogenous Jasmonates and Octadecanoids in hypersensitive tomato mutants during germination and seedling development in response to abiotic stress. *Seed Science Research* 15:309–318.

Aouani, M.E., Mhamdi, R., Jebara, M., Amarger, N. (2001) Characterization of rhizobia nodulating chickpea in Tunisia. *Agronomie* 21:577–581.

Asada, K. (1999) The water-water cycle in chloroplasts: Scavenging of active oxygens and dissipation of excess photons. *Annual Reviews Plant Physiology and Plant Molecular Biology* 50:601–639.

Ashraf, M., Foolad, M.R. (2007) Roles of glycine betaine and proline in improving plant abiotic stress tolerance. *Environmental and Experimental Botany* 59:206–216.

Awasthi, R., Kaushal, N., Vadez, V., Turner, N.C., Berger, J., Siddique, K.H.M., Nayyar, H. (2014) Individual and combined effects of transient drought and heat stress on carbon assimilation and seed filling in chickpea. *Functional Plant Biology* 41:1148–1167.

Bantilan, C., Kumara Charyulu, D., Gaur, P., Moses Shyam, D., Davis, J. (2014) *Short-duration Chickpea Technology: Enabling Legumes Revolution in Andhra Pradesh*, India. Res. Report No. 23, ICRISAT, India.

Bartels, D., Sunkar, R. (2005) Drought and salt tolerance in plants. *Critical Reviews in Plant Science* 24:23–58.

Basu, P.S., Berger, J.D., Turner, N.C., Chaturvedi, S.K., Ali, M., Siddique, K.H.M. (2007) Osmotic adjustment of chickpea (*Cicer arietinum*) is not associated with changes in carbohydrate composition or leaf gas exchange under drought. *Annals of Applied Biology* 150:217–225.

Battaglia, M., Olvera-Carrillo, Y., Garciarrubio, A., Campos, F., Covarrubias, A.A. (2008) The enigmatic LEA proteins and other hydrophilins. *Plant Physiology* 148:6–24.

Behboudian, M.H., Ma, Q., Turner, N.C., Palta, J.A. (2001) Reactions of chickpea to water stress: Yield and seed composition. *Journal of Science of Food and Agriculture* 81:1288–1291.

Bharadwaj, C., Kumar, S., Singhal, T., Kumar, T., Singh, P., Tripathi, S., Pal, M., Hegde, V.S., Jain, P.K., Chauhan, S., Kumar Verma, A., Roorkiwal, M., Gaur, P.M., Varshney, R.K. (2015) *Genomic Approaches for Breeding Drought-Tolerant Chickpea*. Conference Paper, NGG-P11. 5th International Conference on Next Generation Genomics and Integrated Breeding for Crop Improvement, ICRISAT, Patancheru, India.

Chen, T.H., Murata, N. (2002) Enhancement of tolerance of abiotic stress by metabolic engineering of betaines and other compatible solutes. *Current Opinion in Plant Biology* 5:250–257.

Czarnecki, O., Yang, J., Weston, D., Tuskan, G., Chen, J. (2013) A dual role of strigolactones in phosphate acquisition and utilization in plants. *International Journal of Molecular Sciences* 14:7681–7701.

Davies, S.L., Turner, N.C., Palta, J.A., Siddique, K.H.M., Plummer, J.A. (2000) Remobilisation of carbon and nitrogen supports seed filling in chickpea subjected to water deficits. *Australian Journal of Agricultural Research* 51:855–866.

De Domenico, S., Bonsegna, S., Horres, R., Pastor, V., Taurino, M., Poltronieri, P., Imtiaz, M., Kahl, G., Flors, V., Winter, P., Santino, A. (2012) Transcriptomic analysis of oxylipin biosynthesis genes and chemical profiling reveal an early induction of jasmonates in chickpea roots under drought stress. *Plant Physiology and Biochemistry* 62:115–122.

Deokar, A.A., Kondawar, V., Jain, P.K., Karuppayil, S.M., Raju, N.L. (2011), Comparative analysis of expressed sequence tags (ESTs) between drought-tolerant and susceptible genotypes of chickpea under terminal drought stress. *BMC Plant Biology* 11:70.

de Ollas, C., Arbona, V., Gomez-Cadenas, A. (2015) Jasmonic acid interacts with abscisic acid to regulate plant responses to water stress conditions. *Plant Signaling and Behaviour* 10:e1078953.

Du, H., Liu, H., Xiong, L. (2013) Endogenous auxin and jasmonic acid levels are differentially modulated by abiotic stresses in rice. *Frontiers in Plant Science* 4:389–397.

Dubois, M., Van den Broeck, L., Claeys, H., Van Vlierberghe, K., Matsui, M., Inzé, D. (2015) The ethylene response factors ERF6 and ERF11 antagonistically regulate mannitol-induced growth inhibition in Arabidopsis. *Plant Physiology* 169:166–179.

Dwivedi, V., Parida, S.K., Chattopadhyay, D. (2017) A repeat length variation in myo-inositol monophosphatase gene contributes to seed size trait in chickpea. *Scientific Reports* 7:4764.

Edwards, D. (2016a) Improved desi reference genome. *CyVerse Data Commons*. Dataset. Doi: 10.7946/P2KW2Q. http://www.cicer.info/databases.php.

Edwards, D. (2016b) Improved kabuli reference genome. *CyVerse Data Commons*. Dataset. Doi: 10.7946/P2G596. http://www.cicer.info/databases.php.

Fang, X., Turner, N.C., Yan, G., Li, F., Siddique, K.H.M. (2010) Flower numbers, pod production, pollen viability, and pistil function are reduced and flower and pod abortion increased in chickpea (*Cicer arietinum* L.) under terminal drought. *Journal of Experimental Botany* 61:335–345.

Farhoudi, R., Saeedipour, S. (2011) Effect of exogenous abscisic acid on antioxidant activity and salt tolerance in rapeseed (*Brassica napus*) cultivars. *Research on Crops* 12:122–130.

Farooq, M., Wahid, A., Basra, S.M.A., Din, I.D. (2009) Improving water relations and gas exchange with Brassinosteroids in rice under drought stress. *Journal of Agronomy and Crop Science* 195:262–269.

Farooq, M., Gogoi, N., Barthakur, S., Baroowa, B., Bharadwaj, N., Alghamdi, S.S., Siddique, K.H.M. (2017) Drought stress in grain legumes during reproduction and grain filling. *Journal of Agronomy and Crop Science* 203:81–102.

Farquhar, G.D., Sharkey, T.D. (1982) Stomatal conductance and photosynthesis. *Annual Reviews in Plant Physiology* 33:317–345.

Fatnassi, N., Horres, R., Čerekovič, N., Santino, A., Poltronieri, P. (2018) Differences in adaptation to water stress in stress sensitive and resistant varieties of Kabuli and Desi type chickpea. In: *Metabolic Adaptations in Plants During Abiotic Stress*, eds. Ramakrishna, A., Gill, S.S, 403–412. Boca Raton, FL: CRC Press.

Flowers, T.J., Gaur, P.M., Gowda, C.L., Lakshmanan, K., Samineni, S., Siddique, K.H., Turner, N.C., Vadez, V., Varshney, R.K., Colmer, T.D. (2010) Salt sensitivity in chickpea. *Plant Cell and Environment* 33:490–509.

Foo, E., Reid, J.B. (2013) Strigolactones: New physiological roles for an ancient signal. *Journal of Plant Growth Regulation* 32:429–442.

Foyer, H.C. (2005) Redox homeostasis and antioxidant signaling: A metabolic interface between stress perception and physiological responses. *The Plant Cell* 17:1866–1875.

Fukami, J., Ollero, F.J., Megías, M., Hungria, M. (2017) Phytohormones and induction of plant-stress tolerance and defense genes by seed and foliar inoculation with *Azospirillum brasilense* cells and metabolites promote maize growth. *AMB Express* 7(1):153.

Galvez, L., Gonzalez, E.M., Arrese-Igor, C. (2005) Evidence for carbon flux shortage and strong carbon/nitrogen interactions in pea nodules at early stages of water stress. *Journal of Experimental Botany* 56:2551–2561.

Garg, R., Batthacharjee, A., Jain, M. (2015) Genomic scale transcriptomic insights into molecular aspects of abiotic stress response in chickpea. *Plant Molecular Biology Reports* 33:388–400.

Garg, R., Patel, R.K., Jhanwar, S., Priya, P., Bhattacharjee, A., Yadav, G., Bhatia, S., Chattopadhyay, D., Tyagi, A.K., Jain, M. (2011) Gene discovery and tissue-specific transcriptome analysis in chickpea with massively parallel pyrosequencing and web resource development. *Plant Physiology* 156:1661–1678.

Garg, R., Shankar, R., Thakkar, B., Kudapa, H., Lakshmanan, K., Mantri, N., Varshney, R.K., Bhatia, S., Jain, M. (2016) Transcriptome analyses reveal genotype- and developmental stage-specific molecular responses to drought and salinity stresses in chickpea. *Scientific Reports* 6:19228.

Gonzalez, E.M., Gordon, A.J., James, C.L., Arrese-Igor, C. (1995) The role of sucrose synthase in the response of soybean nodules to drought. *Journal of Experimental Botany* 46:1515–1523.

Gupta, V., Bhatia, S., Mohanty, N.A., Sethy, N., Tripathy, B.C. (2010) Comparative analysis of photosynthetic and biochemical characteristics of Desi and Kabuli gene pools of chickpea (*Cicer arietinum* L.). *International Journal of Genetic Engineering and Biotechnology* 1:65–76.

Gustafsson, H. (2012) Signal transduction during cold, salt, and drought stresses in plants. *Molecular Biology Reports* 39:969–987.

Hamwieh, A., Imtiaz, M., Malhotra, R.S. (2013) Multi-environment QTL analyses for drought-related traits in a recombinant inbred population of chickpea (*Cicer arietinum* L.). *Theoretical and Applied Genetics* 126:1025–1038.

Hasibeder, R., Fuchslueger, L., Richter, A., Bahn, M. (2015) Summer drought alters carbon allocation to roots and root respiration in mountain grassland. *New Phytologist* 205:1117–1127.

Hayat, S., Ali, B., Hassan, S.A., Ahmad, A. (2007) Brassinosteroids enhanced antioxidants under cadmium stress in *Brassica juncea*. *Environmental and Experimental Botany* 60:33–41.

Hiremath, P.J., Farmer, A., Cannon, S.B., Woodward, J., Kudapa, H., Tuteja, R., Kumar, A., Bhanuprakash, A., Mulaosmanovic, B., Gujaria, N., Lakshmanan, K., Gaur, P.M., Kavikishor, P.B., Shah, T., Srinivasan, R., Lohse, M., Xiao, Y., Town, C.D., Cook, D.R., May, G.D., Varshney, R.K. (2011) Large-scale transcriptome analysis in chickpea (*Cicer arietinum* L.), an orphan legume crop of the semi-arid tropics of Asia and Africa. *Plant Biotechnology Journal* 9:922–931.

Hossain, M.A., Munemasa, S., Uraji, M., Nakamura, Y., Mori, I.C., Murata, Y. (2011) Involvement of endogenous abscisic acid in methyl jasmonate-induced stomatal closure in Arabidopsis. *Plant Physiol* 156(1):430–438.

Hu, J., Yang, H., Mu, J., Lu, T., Peng, J., Deng, X., Kong, Z., Bao, S., Cao, X., Zuo, J. (2011) Nitric oxide regulates protein methylation during stress responses in plants. *Mol. Cell* 67:702–710.

Huang, D., Wu, W., Abrams, S.R., Cutler, A.J. (2008) The relationship of drought-related gene expression in *Arabidopsis thaliana* to hormonal and environmental factors. *Journal of Experimental Botany* 11:2991–3007.

Jain, D., Chattopadhyay, D. (2010) Analysis of gene expression in response to water deficit of chickpea (*Cicer arietinum* L.) varieties differing in drought tolerance. *BMC Plant Biology* 10:24.

Jayashree, B., Buhariwalla, H.K., Shinde, S., Crouch, J.H. (2005) A legume genomics resource: The chickpea root expressed sequence tag database. *Electronic Journal of Biotechnology* 8:128–133.

Jebara, S., Jebara, M., Limam, F., Aouani, M.E. (2005) Changes in ascorbate peroxidase, catalase, guaiacol peroxidase and superoxide dismutase activities in common bean (*Phaseolus vulgaris*) nodules under salt stress. *Journal of Plant Physiology* 162:929–936.

Kaashyap, M., Ford, R., Bohra, A., Kuvalekar, A., Mantri, N. (2017) Improving salt tolerance of chickpea using modern genomics tools and molecular breeding. *Current Genomics* 18:557–567.

Kagale, S., Divi, U.K., Krochko, J.E., Keller, W.A., Krishna, P. (2007) Brassinosteroid confers tolerance in *Arabidopsis thaliana* and *Brassica napus* to a range of abiotic stresses. *Planta* 225:353–364.

Kahl, G., Molina, C., Medina, C., Winter, P. (2011) Functional genomics-transcriptomics for legumes: Background, tools and insights. In: *Genetics, Genomics and Breeding of Cool Season Grain Legumes*, eds. Perez de la Vega, M., Torres, A.M., Cubero, J.I., Kole, C., 237–284. Boca Raton, FL: CRC Press.

Karamanos, A.J., Papatheohari, A.Y. (1999) Assessment of drought resistance of crop genotypes by means of the water potential index. *Crop Science* 39:1792–1797.

Kashiwagi, J., Lakshmanan, K., Crouch, J.H., Serraj, R. (2006) Variability of root characteristics and their contribution to seed yield in chickpea (*Cicer arietinum* L.) under terminal drought stress. *Field Crop Research* 95:171–181.

Kashiwagi, J., Lakshmanan, K., Upadhyaya, H.D., Gaur, PM. (2008) Rapid screening technique for canopy temperature status and its relevance to drought tolerance

improvement in chickpea. *Journal of SemiArid Tropical Agricultural Research* 6:1–4.

Kashiwagi, J., Lakshmanan, K., Upadhyaya, H.D., Krishna, H., Chandra, S., Vadez, V., Serraj, R. (2005) Genetic variability of drought-avoidance root traits in the minicore germplasm collection of chickpea (*Cicer arietinum* L.). *Euphytica* 146:213–222.

Kavikishore, P.B., Hong, Z., Miao, G.H., Hu, C.A.A., Verma, D.P.S. (1995) Overexpression of D1-pyrroline-5-carboxylate synthetase increases proline production and confers osmotolerance in transgenic plants. *Plant Physiology* 108:1387–1394.

Kazan, K. (2015) Diverse roles of jasmonates and ethylene in abiotic stress tolerance. *Trends in Plant Science* 20(4):219–229.

Kinnersley, A.M., Turano, F.J. (2000) Gamma aminobutyric acid (GABA) and plant responses to stress. *Critical Reviews in Plant Science* 19(6):479–509.

Knight, H., Knight, M.R. (2001) Abiotic stress signalling pathways: Specificity and cross-talk. *Trends in Plant Science* 6(6):262–267.

Kotula, L., Khan, H.A., Quealy, J., Turner, N.C., Vadez, V., Siddique, K.H., Clode, P.L., Colmer, T.D. (2015) Salt sensitivity in chickpea (*Cicer arietinum* L.): Ions in reproductive tissues and yield components in contrasting genotypes. *Plant Cell and Environment* 38(8):1565–77.

Kumar, A., Bernier, J., Verulkar, S., Lafitte, H.R., Atlin, G.N. (2008) Breeding for drought tolerance: Direct selection for yield, response to selection and use of drought-tolerant donors in upland and lowland-adapted populations. *Field Crops Research* 107:221–231.

Kumar, T., Bharadwaj, C., Rizvi, A.H., Sarker, A., Tripathi, S., Alam, A., Chauhan, S.K. (2015) Chickpea landraces: A valuable and divergent source for drought tolerance. *International Journal of Tropical Agriculture* 33(3):1–6.

Kurdali, F., Al-Chammaa, M., Mouasess, A. (2013) Growth and nitrogen fixation in silicon and/or potassium fed chickpeas grown under drought and well-watered conditions. *Journal of Stress Physiology and Biochemistry* 9:385–406.

Labidi, N., Mahmoudi, H., Dorsaf, M., Slama, I., Abdelly, C. (2009) Assessment of intervarietal differences in drought tolerance in chickpea using both nodule and plant traits as indicators. *Journal of Plant Breeding and Crop Science* 1(4):080–086.

Ladrera Fernández, R. (2008) Models of regulation of nitrogen fixation in response to drought: Soya and Medicago. PhD thesis, Department of Environmental Sciences, Public University of Navarre, Spain.

Lakshmanan, K., Kashiwagi, J., Gaur, P.M., Upadhyaya, H.D., Vadez, V. (2010) Sources of tolerance to terminal drought in the chickpea (*Cicer arietinum* L.) minicore germplasm. *Field Crops Research* 119:322–330.

Lakshmanan, K., Kashiwagi, J., Upadhyaya, H.D., Gowda, C.L.L., Gaur, P.M., Singh, S., Ramamoorthy, P., Varshney, R.K. (2013) Partitioning coefficient: A trait that contributes to drought tolerance in chickpea. *Field Crops Research* 149:354–365.

Larrainzar, E., Wienkoop, S., Scherling, C., Kempa, S., Ladrera, R., Arres-Igor, C., Gonzalez, E.M. (2007) *Medicago truncatula* root nodule proteome analysis reveals differential plant and bacteroid responses to drought stress. *Plant Physiology* 144:1495–1507.

Lemoine, R., La Camera, S., Atanassova, R., Dédaldéchamp, F., Allario, T., Pourtau, N., Bonnemain, J.-L., Laloi, M., Coutos-Thévenot, P., Maurousset, L., Faucher, M., Girousse, C., Lemonnier, P., Parrilla, J., Durand, M. (2013) Source-to-sink transport of sugar and regulation by environmental factors. *Front Plant Science* 4:272.

Leonard, R.T., Nell, T.A., Hoyer, L. (2005) Response of potted rose varieties to short-term Ethylene exposure. In: *8th International Symposium on Postharvest Physiology of Ornamental Plants*, eds. van Meeteren, U., Marissen, N., van Doorn, W.G., 373–380. Leuven, Belgium: International Society for Horticultural Science.

Leport, L., Turner, N.C., Davies, S.L., Siddique, K.H.M. (2006) Variation in pod production and abortion among chickpea cultivars under terminal drought. *European Journal of Agronomy* 24:236–246.

Leport, L., Turner, N.C., French, R.J., Barr, M.D., Duda, R., Davies, S.L., Tennant, D., Siddique, K.H.M. (1999) Physiological responses of chickpea genotypes to terminal drought in a Mediterranean-type environment. *European Journal of Agronomy* 11:279–291.

Leport, L., Turner, N.C., French, R.J., Tennant, D., Thomson, B.D., Siddique, K.H.M. (1998) Water relations, gas-exchange, and growth of cool-season grain legumes in a Mediterranean-type environment. *European Journal of Agronomy* 9:295–303.

Liu, F., Jensen, C.R., Andersen, M.N. (2005) A review of drought adaptation in crop plants: Changes in vegetative and reproductive physiology induced by ABA-based chemical signals. *Australian Journal of Agricultural Research* 56:1245–1252.

Llanes, A., Andrade, A., Alemano, S., Luna, V. (2016) Alterations of endogenous hormonal levels in plants under drought and salinity. *American Journal of Plant Science* 7:1357–1371.

Ltaief, B., Sifi, B., Zaman-Allah, M., Drevon, J.J., Lachaâl, M. (2007) Effect of salinity on root-nodule conductance to the oxygen diffusion in the *Cicer arietinum–Mesorhizobium ciceri* symbiosis. *Journal of Plant Physiology* 164:1028–1036.

Maccarrone, M., Veldink, G.A., Agro, A.F., Vliegenthart, J.F. (1995) Modulation of soybean lipoxygenase expression and membrane oxidation by water deficit. *FEBS Letters* 371:223–226.

Mafakheri, A., Siosemardeh, A., Bahramnejad, B., Struik, P.C., Sohrabi, Y. (2010) Effect of drought stress on yield, proline and chlorophyll contents in three chickpea cultivars. *Australian Journal of Crop Science* 4:580–585.

Mantri, N.L., Ford, R., Coram, T.E., Pang, E.C.K. (2007) Transcriptional profiling of chickpea genes differentially regulated in response to high-salinity, cold and drought. *BMC Genomics* 8:303.

Mathur, P.B., Vadez, V., Jyotsna Devi, M., Lavanya, M., Vani, G., Sharma, K.K. (2009) Genetic engineering of chickpea (*Cicer arietinum* L.) with the P5CSF129A gene for osmoregulation with implications on drought tolerance. *Molecular Breeding* 23:591–606.

Melotto, M., Zhang, L., Oblessuc, P.R., He, S.Y. (2017) Stomatal defense a decade later. *Plant Physiology* 174:561–571.

Mhadhbi, H., Jebara, M., Zitoun, A., Limam, F., Aouani, M.E. (2008) Symbiotic effectiveness and response to mannitol-mediated osmotic stress of various chickpea–rhizobia associations. *World Journal of Microbiology and Biotechnology* 24:1027–1035.

Mir, R.R., Zaman-Allah, M., Sreenivasulu, N., Trethowan, R., Varshney, R.K. (2012) Integrated genomics, physiology and breeding approaches for improving drought tolerance in crops. *Theoretical and Applied Genetics* 125:625–645.

Mittler, R. (2002) Oxidative stress, antioxidants and stress tolerance. *Trends Plant Science* 7(9):405–410.

Mohammadi, A., Habibi, D., Rohami, M., Mafakheri, S. (2011) Effect of drought stress on antioxidant enzymes activity of some chickpea cultivars. *Am-Eurasian Journal of Agriculture and Environmental Science* 11:782–785.

Molina, C., Rotter, B., Horres, R., Udupa, S.M, Besser, B., Bellarmino, L., Baum, M., Matsumura, H., Terauchi, R., Kahl, G., Winter, P. (2008) SuperSAGE: The drought stress-responsive transcriptome of chickpea roots. *BMC Genomics* 9:553.

Molina, C., Zaman-Allah, M., Khan, F., Fatnassi, N., Horres, R., Rotter, B., Steinhauer, D., Amenc, L., Drevon, J.-J., Winter, P., Kahl, G. (2011) The salt-responsive transcriptome of chickpea roots and nodules via deepSuperSAGE. *BMC Plant Biology* 11:31.

Morell, M. and Copeland, L. (1985) Sucrose Synthase of Soybean Nodules. *Plant Physiol.* 78(1):149–154.

Nasr Esfahani, M., Kusano, M., Nguyen, K.H., Watanabe, Y., Ha, C.V., Saito, K., Sulieman, S., Herrera-Estrella, L., Tran, L.-S. (2016) Adaptation of the symbiotic Mesorhizobium–chickpea relationship to phosphate deficiency relies on reprogramming of whole-plant metabolism. *Proceedings of the National Academy of Science USA* 113(32):E4610–E4619.

Nasr Esfahani, M., Sulieman, S., Schulze, J., Yamaguchi-Shinozaki, K., Shinozaki, K., Tran, L.S. (2014a) Mechanisms of physiological adjustment of N$_2$ fixation in chickpea (*Cicer arientum* L.) during early stages of water deficit: Single or multi-factor controls? *The Plant Journal* 79:964–980.

Nasr Esfahani, M., Sulieman, S., Schulze, J., Yamaguchi-Shinozaki, K., Shinozaki, K., Tran, L.S. (2014b) Approaches for enhancement of N$_2$ fixation efficiency of chickpea (*Cicer arietinum* L.) under limiting nitrogen conditions. *Plant Biotechnology Journal* 12(3):387–397.

Naya, L., Ladrera, R., Ramos, J., Gonzalez, E.M., Arrese-Igor, C., Minchin, F.R., Becana, M. (2007) The responses of carbon metabolism and antioxidant defenses of alfalfa nodules to drought stress and to subsequent recovery of plants. *Plant Physiology* 144:1104–1114.

Nayyar, H., Kaur, S., Singh, S., Upadhyaya, H.D. (2006b) Differential sensitivity of Desi (small-seeded) and Kabuli (large-seeded) chickpea genotypes to water stress during seed filling: Effects on accumulation of seed reserves and yield. *Journal of Science of Food and Agriculture* 86(13):2076–2082.

Nayyar, H., Singh, S., Kaur, S., Kumar, S., Upadhyaya, H.D. (2006a) Differential sensitivity of *Macrocarpa* and *Microcarpa* types of chickpea (*Cicer arietinum* L.) to water stress: Association of contrasting stress response with oxidative injury. *Journal of Integrative Plant Biology* 48:1318–1329.

Nguyen, K.H., Ha, C.V., Watanabe, Y., Tran, U.T., Nasr Esfahani, M., Van Nguyen, D., Tran, L.S. (2015) Correlation between differential drought tolerability of two contrasting drought-responsive chickpea cultivars and differential expression of a subset of *CaNAC* genes under normal and dehydration conditions. *Frontiers in Plant Science* 6:449.

Nir, I., Shohat, H., Panizel, I., Olszewski, N., Aharoni, A., Weiss, D. (2017) The tomato DELLA protein PROCERA acts in guard cells to promote stomatal closure. *The Plant Cell* 29:3186–3197.

Pace, P.F., Crale, H.T., El-Halawany, S.H.M., Cothren, J.T., Senseman, S.A. (1999) Drought induced changes in shoot and root growth of young cotton plants. *Journal of Cotton Science* 3:183–187.

Palta, J.A., Nandwal, A.S., Kumari, S., Turner, N.C. (2005) Foliar nitrogen applications increase the seed yield and protein content in chickpea (*Cicer arietinum* L.) subject to terminal drought. *Australian Journal of Agricultural Research* 56:105–112.

Pandey, A., Chakraborty, S., Datta, A., Chakraborty, N. (2008) Proteomics approach to identify dehydration responsive nuclear proteins from chickpea (*Cicer arietinum* L). *Molecular and Cellular Proteomics* 7:88–107.

Parween, S., Nawaz, K., Roy, R., Pole, A.K., Venkata Suresh, B., Misra, G., Jain, M., Yadav, G., Parida, S.K., Tyagi, A.K., Bhatia, S., Chattopadhyay, D. (2015) An advanced draft genome assembly of a desi type chickpea (*Cicer arietinum* L.). *Scientific Reports* 5:12806.

Per, T.S., Khan, M.I.R., Anjum, N.A., Masood, A., Hussain, S.J., Khan, N.A. (2018) Jasmonates in plants under abiotic stresses: Crosstalk with other phytohormones matters. *Environmental and Experimental Botany* 145:104–120.

Poltronieri, P., Bonsegna, S., De Domenico, S., Santino, A. (2011) Molecular mechanisms of abiotic stress response in plants. *Field and Vegetable Crops Research/Ratarstvo Povrtartsvo* 48:15–24.

Poltronieri, P., Taurino, M., Bonsegna, S., De Domenico, S., Santino, A. (2015) Monitoring the activation of jasmonate biosynthesis genes for selection of chickpea hybrids tolerant to drought stress. In: *Abiotic Stresses in Crop Plants*, eds. Chakraborty, U., Chakraborty, B., 54–70. Wallingford: CABI.

Poltronieri, P., Taurino, M., De Domenico, S., Bonsegna, S., Santino, A. (2013) Activation of the jasmonate biosynthesis pathway in roots in drought stress. In: *Climate Change and Abiotic Stress Tolerance*, eds. Tuteja, N., Gill, S.S., 325–342. Weinheim, Germany: Wiley-VCH.

Pushpavalli, R., Zaman-Allah, M., Turner, N.C., Baddam, R., Rao, M.V., Vadez, V. (2014) Higher flower and seed number leads to higher yield under water stress conditions imposed during reproduction in chickpea. *Functional Plant Biology* 42:162–174.

Ramamoorthy, P., Lakshmanan, K. (2014) Timing of sampling for the canopy temperature depression can be critical for the best differentiation of drought tolerance in chickpea. *Journal of SemiArid Tropical Agricultural Research* 12:1–8.

Ramamoorthy, P., Lakshmanan, K., Upadhyaya, H.D., Vadez, V., Varshney, R.K. (2016) Shoot traits and their relevance in terminal drought tolerance of chickpea (*Cicer arietinum* L.). *Field Crops Research* 197:10–27.

Ramamoorthy, P., Thudi, M., Lakshmanan, K., Upadhyaya, H.D., Kashiwagi, J., Gowda, C.L.L., Varshney, R.K. (2015) Association of mid-reproductive stage canopy temperature depression with the molecular markers and grain yields of chickpea (*Cicer arietinum* L.) germplasm under terminal drought. *Field Crops Research* 174:1–11.

Ramamoorthy, P., Upadhyaya, H.D., Gaur, P.M., Gowda, C.L.L., Lakshmanan, K. (2014) Kabuli and desi chickpeas differ in their requirement for reproductive duration. *Field Crops Research* 163:4–31.

Ramalingam, A., Kudapa, H., Pazhamala, L.T., Garg, V., Varshney, R.K. (2015) Gene expression and yeast two-hybrid studies of 1R-MYB transcription factor mediating drought stress response in chickpea (*Cicer arietinum* L.). *Frontiers in Plant Science* 6:1117.

Ramirez, M., Guillen, G., Fuentes, S.I., Iniguez, L.P., Aparicio-Fabre, R., Zamorano-Sanchez, D., Encarnacion-Guevara, S., Panzeri, D., Castiglioni, B., Cremonesi, P., Strozzi, F., Stella, A., Girard, L., Sparvoli, F., Hernandez, G. (2013) Transcript profiling of common bean nodules subjected to oxidative stress. *Physiologia Plantarum* 149:389–407.

Ramos, M.L.G., Gordon, A.J., Minchen, F.R., Sprent, J.I., Parsons, R. (1999) Effect of water stress on nodule physiology and biochemistry of a drought tolerant cultivar of common bean (*Phaseolus vulgaris* L.). *Annals of Botany* 83:57–63.

Riefler, M., Novak, O., Strnad, M., Schmulling, T. (2006) Arabidopsis cytokinin receptor mutants reveal functions in shoot growth, leaf senescence, seed size, germination, root development, and cytokinin metabolism. *The Plant Cell* 18:40–54.

Rizhsky, L., Liang, H., Mittler, R. (2003) The water-water cycle is essential for chloroplast protection in the absence of stress. *Journal of Biological Chemistry* 278:38921–38925.

Romo, S., Labrador, E., Dopico, B. (2001) Water stress-regulated gene expression in *Cicer arietinum* seedlings and plants. *Plant Physiology and Biochemistry* 39:1017–1026.

Roorkiwal, M., Rathore, A., Dasm, R.R., Singh, M.K., Jain, A., Srinivasan, S., Gaur, P.M., Chellapilla, B., Tripathi, S., Li, Y., Hickey, J.M., Lorenz, A., Sutton, T., Crossa, J., Jannink, J.L., Varshney, R.K. (2016) Genome-enabled prediction models for yield related traits in chickpea. *Frontiers in Plant Science* 7:1666.

Saibi, W., Zouari, N., Masmoudi, K., Brini, F. (2016) Role of the durum wheat dehydrin in the function of proteases conferring salinity tolerance in *Arabidopsis thaliana* transgenic lines. *International Journal of Biological Macromolecules* 85:311–316.

Sassi, S., Aydi, S., Gonzalez, E.M., Arrese-Igor, C., Abdelly, C. (2010) Understanding osmotic stress tolerance in leaves and nodules of two *Phaseolus vulgaris* cultivars with contrasting drought tolerance. *Symbiosis* 52:1–10.

Sen, S., Chakraborty, J., Ghosh, P., Basu, D., Das, S. (2017) Chickpea WRKY70 regulates the expression of a Homeodomain-Leucine Zipper (HD-Zip) I Transcription Factor CaHDZ12, which confers abiotic stress tolerance in transgenic tobacco and chickpea. *Plant and Cell Physiology* 58:1934–1952.

Shahbaz, M., Ashraf, M., Athar, H.R. (2008) Does exogenous application of 24-Epibrassinolide ameliorate salt induced growth inhibition in wheat (*Triticum aestivum* L.)? *Plant Growth Regulation* 55:51–64.

Singh, G.D., Wani, A.A., Kaur, D., Sogi, D.S. (2008) Characterisation and functional properties of proteins of some Indian chickpea (*Cicer arietinum*) cultivars. *Journal of Science of Food and Agriculture* 88:778–786.

Siddique, K.H.M., Regan, K.L., Tennant, G., Thomson, B.D. (2001) Water use and water use efficiency of cool season grain legumes in low rainfall Mediterranean-type environments. *European Journal of Agronomy* 15:267–280.

Sohrabi, Y., Heidari, G., Weisany, W., Ghasemi Golezani, K., Mohammadi, K. (2012) Some physiological responses of chickpea cultivars to arbuscular mycorrhiza under drought stress. *Russian Journal of Plant Physiology* 59:708–716.

Soussi, M., Lluch, C., Ocana, A., Norara, A.L. (1999) Comparative study of nitrogen fixation and carbon metabolism in two chickpea (*Cicer arietinum L.*) cultivars under salt stress. *Journal of Experimental Botany* 50:1701–1708.

Srivastava, R., Bajaj, D., Malik, A., Singh, M., Parida, S.K. (2016) Transcriptome landscape of perennial wild *Cicer microphyllum* uncovers functionally relevant molecular tags regulating agronomic traits in chickpea. *Scientific Reports* 6:33616.

Szabados, L., Savouré, A. (2009) Proline: A multifunctional amino acid. *Trends in Plant Science* 15:89–97.

Talbi, C., Sanchez, C., Hidalgo-Garcia, A., Gonzalez, E.M., Arrese-Igor, C., Girard, L., Bedmar, E.J., Delgado, M.J. (2012) Enhanced expression of *Rhizobium etli* cbb3 oxidase improves drought tolerance of common bean symbiotic nitrogen fixation. *Journal of Experimental Botany* 63:5035–5043.

Tanou, G., Job, C., Rajjou, L., Arc, E., Belghzi, M., Diamantidis, G., Molassiotis, A., Job, D. (2009) Proteomics reveal the overlapping roles of hydrogen peroxide and nitric oxide in the acclimation of citrus plants to salinity. *The Plant Journal* 60:795–804.

Thudi, M., Chitikineni, A., Liu, X., He, W., Roorkiwal, M., Yang, W., Jian, J., Doddamani, D., Gaur, P.M., Rathore, A., Samineni, S., Saxena, R.K., Xu, D., Singh, N.P., Chaturvedi, S.K., Zhang, G., Wang, J., Datta, S.K., Xu, X., Varshney, R.K. (2017) Recent breeding programs enhanced genetic diversity in both desi and kabuli varieties of chickpea (*Cicer arietinum* L.). *Scientific Reports* 6:38636.

Tiwari, S., Lata, C., Chauhan, P.S., Nautiyal, C.S. (2016) *Pseudomonas putida* attunes morphophysiological, biochemical and molecular responses in *Cicer arietinum* L. during drought stress and recovery. *Plant Physiology and Biochemistry* 99:108–117.

Tiwari, S., Lata, C., Chauhan, P.S., Prasad, V., Prasad, M. (2017) A functional genomic perspective on drought signalling and its crosstalk with phytohormone-mediated signalling pathways in plants. *Current Genomics* 18(6):471–482.

Turner, N.C., Davies, S.L., Plummer, J.A., Siddique, K.H.M. (2005) Seed filling in grain legumes (pulses) under water deficits with emphasis on chickpea (*Cicer arietinum* L.). *Advances in Agronomy* 87:211–250.

Ulemale, C.S., Mate, S.N., Deshmukh, D.V. (2013) Physiological indices for drought tolerance in chickpea (*Cicer arietinum* L.). *World Journal of Agricultural Sciences* 9:123–131.

Upreti, K.K., Murti, G.S. (2004) Effects of Brassinosteroids on growth, nodulation, phytohormone content and nitrogenase activity in French Bean under water stress. *Biologia Plantarum* 48:407–411.

Vardhini, B.V., Rao, S.S.R. (2003) Amelioration of osmotic stress by Brassinosteroids on seed germination and seedling growth of three varieties of sorghum. *Plant Growth Regulation* 41:25–31.

Varshney, R.K., Hiremath, P.J., Lekha, P., Kashiwagi, J., Balaji, J., Deokar, A.A., Vadez, V., Xiao, Y., Srinivasan, R., Gaur, P.M., Siddique, K.H.M., Town, C.D., Hoisington, D.A. (2009) A comprehensive resource of drought- and salinity-responsive ESTs for gene discovery and marker development in chickpea (*Cicer arietinum* L.). *BMC Genomics* 10:523.

Varshney, R.K., Kudapa, H., Roorkiwal, M., Thudi, M., Pandey, M.K., Saxena, R.K., Chamarti, S., Sabbavarapu, M.M., Mallikarjuna, N., Upadhyaya, H.D., Gaur, P.M., Lakshmanan, K. (2012) Advances in genetics and molecular breeding of three legume crops of semi-arid tropics using next-generation sequencing and high-throughput genotyping technologies. *Journal of Biosciences* 37(5):811–820.

Varshney, R.K., Thudi, M., Nayak, S.N., Gaur, P.M., Kashiwagi, J., Lakshmanan, K., Jaganathan, D., Koppolu, J., Bohra, A., Tripathi, S., Rathore, A., Jukanti, A.K., Jayalakshmi, V., Vemula, A., Singh, S.J., Yasin, M., Sheshshayee,

M.S., Viswanatha, K.P. (2014) Genetic dissection of drought tolerance in chickpea (*Cicer arietinum* L.). *Theoretical and Applied Genetics* 127(2):445–462.

Wasternack, C. (2014) Action of jasmonates in plant stress responses and development-applied aspects. *Biotechnology Advances* 32:31–39.

Wasternack, C., Strnad, M. (2016) Jasmonate signaling in plant stress responses and development: Active and inactive compounds. *New Biotechnology* 33:604–613.

Weng, X.Y., Zheng, C.J., Xu, H.X., Sun, J.Y. (2007) Characteristics of photosynthesis and functions of the water-water cycle in rice (*Oryza sativa*) leaves in response to potassium deficiency. *Physiologia Plantarum* 131(4):614–621.

Yaqoob, M., Hollington, P.A., Mahar, A.B., Gurmani, Z.A. (2013) Yield performance and responses studies of chickpea (*Cicer arietinum* L.) genotypes under drought stress. *Emirates Journal of Food and Agriculture* 25:117–123.

Zaman-Allah, M., Jenkinson, D.M., Vadez, V. (2011a) A conservative pattern of water use, rather than deep or profuse rooting, is critical for the terminal drought tolerance of chickpea. *Journal of Experimental Botany* 62:4239–4252.

Zaman-Allah, M., Jenkinson, D.M., Vadez, V. (2011b) Chickpea genotypes contrasting for seed yield under terminal drought stress in the field differ for traits related to the control of water use. *Functional Plant Biology* 38:270–281.

Zhang, H., Tan, G., Wang, Z., Yang, J., Zhang, J. (2009) Ethylene and ACC levels in developing grains are related to the poor appearance and milling quality of rice. *Plant Growth Regulation* 58:85–96.

Zhang, X.H., Sheng, J.P., Li, F.J., Meng, D., Shen, L. (2012) Methyl jasmonate alters arginine catabolism and improves postharvest chilling tolerance in cherry tomato fruit. *Postharvest Biology and Technology* 64:160–167.

7 Plant Responses and Mechanisms of Tolerance to Cold Stress

Aruna V. Varanasi, Nicholas E. Korres, and Vijay K. Varanasi

CONTENTS

7.1 INTRODUCTION

Plants endure a variety of environmental stresses, biotic as well as abiotic, during their growth cycle. Abiotic factors affecting plant growth and performance include cold, heat, salinity, and drought. Among these factors, low temperature stress is considered one of the major threats limiting crop production an3d productivity (Yadav, 2010). Low temperature conditions, which are common in temperate climates, limit crop production on most of the world's 1.4 billion hectares of agricultural land. Plants growing in temperate climates have evolved several adaptive traits to overcome extremely low temperatures and fluctuations throughout the growing season. For this reason, temperate plants exhibit increased cold tolerance compared to plants growing in tropical and subtropical climates (Yamaguchi-Shinozaki and Shinozaki, 2006). One of the most important traits acquired by temperate plants is their ability to tolerate low temperatures (0°C–15°C) without injury and gradually acclimatize to cold temperatures (Somerville, 1995; Thomashow, 1999). Temperate plants have undergone this gradual acclimatization through various cellular modifications, including anatomical, physiological, and metabolic changes when exposed to low, non-freezing temperatures (Levitt, 1980), thereby contributing to their improved fitness and survival under extreme conditions. On the other hand, plants of tropical and subtropical regions are typically sensitive to low temperatures (0°C–10°C), which renders them incapable of cold acclimation.

Plants exposed to cold stress show various symptomology, including effects on plant morphology, physiology, and reproduction (Jiang et al., 2002; Chinnusamy et al., 2007; Yadav, 2010). Cold temperatures may adversely affect crop cultivation by reducing both crop quality and productivity (Thomashow, 1998). Worldwide annual losses in crop production due to cold stress damage amount to approximately $2 billion (Sanghera et al., 2011). In order to mitigate the effects of cold stress on crop productivity and find better options to combat abiotic stresses such as low temperatures, an understanding of cold stress response and the underlying mechanisms in plants is needed. This chapter is an attempt to review some of the major findings in cold stress effects on plant morphology and physiology, mechanisms and regulatory networks underlying cold stress response, signal trans-duction pathways, and approaches to improve crop toler-ance to freezing temperatures.

7.2 ACCLIMATION AND ADAPTATION TO COLD STRESS

Plants develop tolerance to low temperature stress in three distinct phases: cold acclimation, hardening, and finally plant recovery after a period of cold stress (Li et al., 2008). Cold acclimation can also be called the pre-hardening phase, which occurs during periods of low, but non-freezing temperatures. It is especially observed in plants growing in temperate climates that exhibit a variable degree of chilling tolerance. During cold accli-mation, plants initiate multiple mechanisms to mini-mize potential damage during cold stress (Thomashow, 1999). In the second stage (hardening), which occurs during periods of sub-zero temperatures, plants become hardy by increasing their freezing tolerance through a cascade of biochemical and physiological changes that, in turn, result from the reprogramming of gene expres-sion (Heidarvand and Amiri, 2010). Through this pro-cess, plants are able to minimize irreversible freeze damage and enhance overall plant fitness to survive and adapt to low temperatures (Levitt, 1980). During the final recovery phase, a large number of metabolites are redistributed and partitioned between different cel-lular components, which not only protects the cellular structures from freeze damage but also re-establishes the metabolic pathways to a functionally active state, thereby enabling the plants to adapt to the new thermal conditions (Hoermiller et al., 2017). Some plants, espe-cially trees, require multiple environmental cues, such as low temperatures and short photoperiods, to fully develop cold tolerance (Juntilla and Robberecht, 1999; Kacperska, 1999). In these cases, tolerance may be reverted when exposed to above-zero temperatures and longer photoperiods.

Metabolic alterations during plant response to cold stress are mainly observed in the plastid and cytosol compartments through the regulation of the movement of cryoprotective proteins and metabolites such as sol-uble sugars, amino acids, and excess redox equivalents, eventually signaling the nucleus to initiate gene expres-sion for cold acclimation (Huner et al., 1998). During the process of metabolite reallocation, growth repres-sion may occur temporarily in some plants (Scott et al., 2004). Changes in the metabolic profile during cold stress are also reflected in the cell structure and compo-sition, leaf architecture, and other plant morphological

characteristics, thus affecting plants very broadly (Pino et al., 2008). Such overall alterations in plant biology are necessary to support the recovery of cellular metabolism as well as for the functioning of essential plant processes such as photosynthesis and reproduction (Lundmark et al., 2006).

7.3 PLANT RESPONSES TO LOW TEMPERATURES

Plant adjustments to cold environments are related to both the extremes or to the gradual influences of low tem-peratures (Korner, 2016), the former expressed through the development of resistance to freezing temperatures, an attribute related to plant phenology and acclima-tion. According to Hasanuzzaman et al. (2013), plants exhibit various tolerance levels to chilling (0°C–15°C) and freezing (<0°C) temperatures, both comprised of what is defined as cold stress. Obviously, resistant plants to chilling differ from resistant plants to freezing and exhibit different ecological attributes in terms of habitat type, morphology, and life expectancy (Larcher et al., 2010; Korner, 2016).

7.3.1 MORPHOLOGY

Species adapted to cold environments have evolved a number of physiological and morphological character-istics for their survival against extended cold periods (Guy, 1999). These species usually exhibit a short stature with a low leaf area index and a high root:shoot ratio. Due to their growth habit, they obtain full advantage of the heat emitted from the ground during the day against night chilling as the air temperature is maintained most effectively near the soil surface (Nilsen and Orcutt, 1996). Plant organs differ in their level of tolerance, with the roots being more sensitive than the crown (McKersie and Leshem, 1994), a comprehensible response, as a meristematic activity for the production of new roots and shoots at the end of the cold period is accommodated in the crown zone.

7.3.2 REPRODUCTION

Temperate plants, in an effort to reduce the negative effects of winter temperatures, synchronize their cold stress vulnerable reproductive stages with the favor-able environmental conditions of the spring and sum-mer (Nishiyama, 1995; King and Heide, 2009). Spring annuals germinate, reproduce, and senesce during more favorable seasons. The seeds of spring annuals, for example, remain dormant during the winter, only to germinate in response to inductive temperatures in the spring (Hemming and Trevaskis, 2011). On the contrary, winter annuals set seed and germinate in autumn, overwinter vegetative growth, and flower in the spring. The ability of the plants to respond to extended periods of cold and attain a rapid flowering capability afterward is known as vernalization (Chouard, 1960).

In the case of herbaceous perennials, germination, reproduction, and senescence occur during the warm seasons. Herbaceous perennials are also capable, at the end of winter, of secondary vegetative growth from dormant underground meristems (e.g., rhizomes), which serve as an alternate survival strategy to cold stress (Preston and Sandve, 2013). By contrast, woody perennials often delay flowering for several years until a critical biomass is achieved (Rohde and Bhalerao, 2007). Cold stress can delay flowering and bud abscission, and reduce the kernel filling rate or produce small, unfilled, or aborted seeds (Jiang et al., 2002; Thakur et al., 2010).

7.3.3 YIELD

Cold stress-induced yield reduction is a common phenomenon observed in many crops, e.g., rice, wheat, and cucumbers (Nahar et al., 2009; Kalbarczyk, 2009; Riaz-ud-din Subhani et al., 2010), by directly damaging crops at critical developmental stages, such as temperature thresholds during flowering, pollen and ovule formation, and seed filling, or during seed setting (Antle et al., 2004; Porter and Semenov, 2005; Thakur et al., 2010).

7.3.4 INJURIES DUE TO COLD STRESS

The common injury symptoms of various species due to low temperatures include reduced seedling growth and discoloration, leaf yellowing and whitening, suppressed tillering capacity, chlorotic and/or necrotic surface lesions on the leaf surface, water-soaked appearance of tissues, reduced leaf expansion, wilting, abnormal tillers, and bushy plant stature (Angadi et al., 2000; Solanke and Sharma, 2008; Nahar et al., 2009, 2012). Nevertheless, according to Levitt (1972), the degree of plant injury due to frost depends on a number of factors including re-cooling and warming rates, relative humidity of the air, cold-hardening of plant tissue, and the minimum temperature at which plant tissue is exposed. Teutonico and Osborn (1995) reported the importance of cold stress frequency (i.e., diurnally or seasonally) on the strategies adapted by plants in a given environment despite evidence for genetic response overlaps to cold (Dhillion et al., 2010; Greenup et al., 2011).

7.4 MECHANISMS UNDERLYING COLD TOLERANCE

Cold stress response in plants is a highly complex process that is mediated through a series of physiological and biochemical modifications that arise from alterations of several gene expression patterns (Hannah et al., 2005; Chinnusamy et al., 2010). The regulatory networks underlying the mechanisms in cold acclimation involve the interaction of several gene-signaling pathways. The mechanisms in cold acclimation and freezing tolerance have been extensively studied in the model plant *Arabidopsis* and other winter cereals (Sanghera et al., 2011). In *Arabidopsis*, an estimated 3%–20% of the transcriptional changes of genes occur in response to cold stress (Chinnusamy et al., 2007; Matsui et al., 2008). Much attention has been devoted to the different components involved in plant responses to cold stress, which include several signal transduction networks and a multitude of enzymes, other functional proteins, and hormones.

7.4.1 COLD STRESS REGULATORY NETWORKS AND SIGNAL TRANSDUCTION

Plant responses to environmental cues are mediated through a series of cellular reactions referred to as signal transduction. Molecular responses to cold stress are triggered first by perceiving the low temperature signal and subsequently relaying the information to various subcellular components through signal transduction. During this process, several signaling pathways are triggered to reprogram gene expression (Guy et al., 1985) and initiate the production of cold-responsive (COR) proteins (Park et al., 2015). Several COR genes have been identified and characterized in plants and COR gene expression was found to be critical for both cold acclimation and chilling tolerance (Thomashow, 1998; Yamaguchi-Shinozaki and Shinozaki, 2006).

The plant cytoskeleton possibly acts as the first sensor of low temperature (Thion et al., 1996; Dodd et al., 2006), while the plasma membrane has been considered as a major site for the perception of temperature change and the transduction of this stimulus (Sangwan et al., 2002a; Uemura et al., 2006; Vaultier et al., 2006; Wang et al., 2006). Microtubules and microfilaments are suggested to play a crucial role in signal transduction by acting as downstream targets for transducing signals to various cellular processes. This hypothesis is further supported by their role in modulating the Ca^{2+} channels following membrane rigidification during the signal transduction process (Knight et al., 1991; Thion et al., 1996;

Örvar et al., 2000). Phospholipase D, one of the plasma membrane proteins, not only connects the cytoskeleton and plasma membrane structurally, but also acts as a signaling link between them (Gardiner et al., 2003; Drøbak et al., 2004; Hong et al., 2008). This interaction of cytoskeleton and plasma membrane is regulated by phospholipase D through Ca^{2+} channels in a stress-dependent manner (Dhonukshe et al., 2003).

Ca^{2+} is the second messenger in plants and influences the signaling network in cells through a change in the cytosolic Ca^{2+} concentrations (Dodd et al., 2006; Klimecka and Muszyńska, 2007). Each environmental stress (e.g., low temperature) can stimulate a specific Ca^{2+} signature, which is further recognized by different calcium-sensing elements (Sanders et al., 2002; Klimecka and Muszyńska, 2007). The calcium signatures are then transduced to affect various downstream processes such as gene expression, protein phosphorylation, and cell wall remodeling (Sanders et al., 2002; McAinsh and Pittman, 2009). The plasma membrane of plant cells is composed of three such Ca^{2+} permeable channels that function in the detection of chemical and physical stimuli (e.g., heat and cold), reactive oxygen species (ROS), and other mechanical forces (Chinnusamy et al., 2004; Carpaneto et al., 2007; McAinsh and Pittman, 2009). They are the mechanosensitive Ca^{2+} channel, the depolarization-activated Ca^{2+} channel, and the hyperpolarization-activated Ca^{2+} channel. Other Ca^{2+} permeable channels, such as the inositol 1,4,5-trisphosphate (InsP3)- and cyclic ADP-ribose (cADPR)-gated channels, have been identified in the vacuolar membrane (Sanders et al., 2002; McAinsh and Pittman, 2009). The different Ca^{2+} signatures transduced in the cell are translated to specific cellular responses depending on the stimulus perceived. This is achieved through Ca^{2+} decoders, such as calmodulin (CaM), Ca^{2+}-dependent protein kinases (CDPKs), Ca^{2+}- and Ca^{2+}/CaM-dependent protein kinases (CCaMKs), CaM-binding transcription activator (CAMTA), calcineurin B-like proteins (CBLs), and CBL-interacting protein kinases (CIPKs) (Knight et al., 1991; Knight and Knight, 2012; Barrero-Gil and Salinas, 2013). CaM, one of the most conserved Ca^{2+}-binding proteins in eukaryotes (Kim et al., 2009), is a negative regulator of genes associated with cold tolerance (Townley and Knight, 2002). On the other hand, CDPKs act as positive regulators (Saijo et al., 2000) and are specifically expressed in plants in response to cold stress. CDPKs catalyze the Ca^{2+}-dependent phosphorylation and dephosphorylation processes involved in activating the cold response machinery (Guy, 1999; Chinnusamy et al., 2004; Klimecka and Muszyńska, 2007). CBLs transmit the Ca^{2+} signals by interacting with CIPKs.

Mitogen-activated protein kinases (MAPKs), belonging to the serine/threonine protein kinase family, were also reported to enhance their activity in response to cold stress in plants such as alfalfa (Jonak et al., 1996; Sangwan et al., 2002b) and *Arabidopsis* (Ichimura et al., 2000). MAPK cascades are known to induce different transcription factors and regulate gene expression patterns specific to the stimulus perceived (Zhu et al., 2007; Knight and Knight, 2012; Barrero-Gil and Salinas, 2013). Thus, a variety of signaling pathways are triggered during the cold acclimation process, all of which activate the transcription factors and the subsequent production of COR proteins. The COR proteins in plant cells comprise both regulatory proteins that control signal transduction and functional proteins such as dehydrins, antifreeze proteins (AFPs), molecular chaperones, lipid-transfer proteins, and detoxification enzymes, which modify the cellular structures and compositions (Wang et al., 2003; Grennan, 2006; Yamaguchi-Shinozaki and Shinozaki, 2006).

Among the different cold signaling pathways, the C-repeat binding factor/dehydration-responsive element-binding factor (CBF/DREB1)-dependent cold signaling pathway has been extensively studied in the regulation of COR gene expression (Chinnusamy et al., 2007). The CBF/DREB1 cold response pathway is best characterized in *Arabidopsis*, and a number of homologous components of this pathway have been identified and functionally tested in many other plants, including canola, tomato, and poplar (Yamaguchi-Shinozaki and Shinozaki, 2006; Thomashow, 2010). CBFs belong to the APETALA2 domain transcription factors (Stockinger et al., 1997), whereas DREB1 belongs to the DREB subfamily (Mizoi et al., 2012). They are associated with resistance against several abiotic stress signals such as freezing, drought, and high salt exposure. DREB1s are encoded by genes belonging to the A1 subgroup and are induced by cold stress, whereas DREB2s belong to the A2 subgroup of genes and are mainly induced in response to osmotic stress (Shinozaki and Yamaguchi-Shinozaki, 2007; Nakashima et al., 2009; Chen et al., 2009). CBF/DREB1s bind to the cold- and dehydration-responsive DNA regulatory elements (DREs), also known as C-repeats (CRTs) to regulate the expression of COR genes. CRT/DREs are *cis*-elements with a conserved A/GCCGAC core sequence in the promoter regions of COR genes (Maruyama et al., 2004; Thomashow, 2010). In *Arabidopsis*, the concerted action of three CBF/DREB1s is required to induce the CBF/DREB1 regulatory pathway and the COR gene expression during the cold acclimation process (Gilmour et al., 2004; Chinnusamy et al., 2007; Novillo et al., 2007).

The three different CBFs, CBF1, 2, and 3, are arranged tandemly on the chromosome and are regulated by feedback loops where CBF2 represses CBF1 and CBF3 expression to alter their mRNA levels (Novillo et al., 2004; Park et al., 2015).

The expression of CBF/DREB1 genes is also regulated by a transcription factor gene termed an inducer of CBF expression 1 (ICE1). ICE1 is localized in the nucleus and is expressed constitutively, but induces the expression of CBF transcripts only under cold stress (Chinnusamy et al., 2003; Fursova et al., 2009; Doherty et al., 2009). ICE1 encodes a MYC-like basic helix-loop-helix (bHLH) protein that increases the expression of CBF3/DREB1A by specifically binding to the MYC cis-elements in the CBF3/DREB1A promoter, thereby activating the expression of downstream genes required to enhance freezing tolerance (Chinnusamy et al., 2003; Chen et al., 2009). In addition, several post-transcriptional and post-translational modifications, such as ubiquitylation and sumoylation, are regulated by ICE1 in response to cold signaling. Sumoylation of ICE1 is catalyzed by the SUMO E3 ligase, whereas ubiquitination is catalyzed by the E3 ubiquitin ligase (Dong et al., 2006). ICE1 protein abundance is controlled by phosphorylation (Miura et al., 2007; Ding et al., 2015).

CBF/DREB1 expression is also controlled by another direct regulator, MYB15 (an R2R3-MYB family protein), identified in *Arabidopsis*. MYB15 is an upstream transcription factor that binds to the MYB recognition elements (MYBRS) in the CBF/DREB1 promoter regions to repress expression and negatively regulate freezing tolerance (Agarwal et al., 2006). Unlike ICE1, MYB15 is expressed even in the absence of cold stress (Chinnusamy et al., 2007). However, ICE1 was found to negatively regulate MYB15 by directly binding to the MYB15 promoter or indirectly through its downstream genes (Badawi et al., 2008).

In addition to the CBF pathway, COR genes are also regulated by another small-group gene known as C_2H_2 zinc finger protein, ZAT12, which responds to different abiotic stresses, including cold stress (Vogel et al., 2005). The ZAT12 gene might be activated by ROS such as hydrogen peroxide (H_2O_2), which accumulates in the cell in response to several stress conditions. Therefore, ZAT12 is implicated in a much broader response to almost all abiotic stress conditions (Dat et al., 2000; Apel and Hirt, 2004; Mittler et al., 2004; Davletova et al., 2005). CBF and ZAT12 cold response pathways possibly interact due to the overlap in the genes comprising the ZAT12 and CBF regulons. Thus, the ZAT12 and CBF transcripts may coordinately regulate the expression of certain COR genes. Furthermore, constitutive expression of ZAT12 was found to weaken the induction of CBF genes in response to cold stress, which indicates that in addition to MYB15, ZAT12 also negatively regulates CBF expression (Vogel et al., 2005; Chinnusamy et al., 2007).

7.4.2 Epigenetic Regulation in Cold Stress Response

In addition to the molecular pathways previously discussed, epigenetic regulation, mediated by histone modification, DNA methylation, and RNA-mediated regulation, is also correlated with resistance to several abiotic stresses including cold stress (Kim et al., 2008; Luo et al., 2012). Some epigenetic regulators such as histone proteins and histone modifier proteins are known to upregulate under cold stress, which further induce epigenetic changes in the transcription of COR genes. Histone modifiers are enzymes such as histone acetyltransferases (HATs), histone deacetylases (HDACs), histone methyltransferases (HMTs), and histone demethylases (HDMs), which catalyze the post-translational modification of basic residues on histone tails through processes such as methylation, acetylation, phosphorylation, sumoylation, and ubiquitination (Kim et al., 2015). Depending on the type of modification, they differentially affect chromatin regulation by altering the activity of core histones, which are structural proteins required for chromosome condensation in the nucleus. Histone modifiers are thought to play an important role in COR gene activation and repression during cold acclimation. In *Arabidopsis*, the expression of a histone modifier gene, histone deacetylase 6 (HDA6), was induced by low temperature treatments, and plants with a mutated HDA6 gene exhibited hypersensitivity to cold stress (To et al., 2011). The expression of HDACs was also upregulated in maize during cold acclimation, followed by genome-wide deacetylation at core histones, H3 and H4 (Hu et al., 2012). Likewise, in rice, cold stress was reported to induce histone acetylation at the DREB1 gene, which subsequently increased the COR gene expression (Roy et al., 2014). A WD40-repeat protein, termed HOS15, was found to be induced by cold stress. HOS15 was shown to activate a nucleus-localized repressor protein that functions in the deacetylation at histone H4 (Zhu et al., 2008).

RNA-mediated regulation of the COR gene expression occurs through post-transcriptional mechanisms such as alternative splicing, pre-mRNA processing, RNA silencing, and RNA export from the nucleus during cold acclimation (Han et al., 2011). At low temperatures, plants regulate the export of mRNA from the nucleus, selectively translate COR genes, and tend to increase the stability of these selected transcripts (Ambrosone et al., 2012).

RNA-binding proteins function as RNA chaperones and stabilize the native conformation of misfolded RNA molecules. Proteins such as glycine-rich protein, GRP7, and RNA helicase LOS4 play an important role in the nuclear mRNA export under cold stress conditions (Gong et al., 2005; Kim et al., 2008). RNA silencing is processed through small non-coding RNAs known as micro-RNAs (miRNAs) and small interfering RNAs (siRNAs), which act as repressors of gene expression (Ghildiyal and Zamore, 2009). These small RNA molecules are suggested to regulate abiotic stress responses in plants. Micro RNAs target mRNAs through imperfect sequence complementation and mediate post-transcriptional gene silencing by the cleavage of specific mRNAs, thereby repressing protein translation (Sunkar et al., 2012). Cold-induced miRNAs have been identified in several plant species, including *Arabidopsis* (Zhou et al., 2008), poplar (Chen et al., 2012), and rice (Zhang et al., 2009). Alternative splicing of pre-mRNA is another RNA-mediated regulation induced as a stress response in plants. In *Arabidopsis*, approximately 42% of genes are regulated by alternative splicing (Filichkin et al., 2010), while in rice about 21% of the expressed genes undergo alternative splicing to produce different proteins (Wang and Brendel, 2006). Several genes encoding protein kinases and other transcription factors undergo alternative splicing during abiotic stress responses (Mastrangelo et al., 2012). Alternative splicing of serine/arginine-rich proteins that function in the regulation of mRNA splicing under cold and heat stresses was reported in *Arabidopsis* (Palusa et al., 2007).

7.4.3 Metabolic Signals

Although metabolism and the redistribution of metabolites is the major target of cold signaling, certain metabolites can themselves act as signal molecules to regulate cold stress response and initiate the COR gene expression (Svensson et al., 2006; Zhu et al., 2007). Soluble sugars are one such cell metabolite that regulate plant acclimation to cold stress (Rekarte-Cowie et al., 2008). Plants use the sugar status in the cells as a signal to bring necessary changes in growth and development that facilitate adaptation to different abiotic stresses including cold stress (Uemura et al., 2006). Sugar metabolism is affected during cold acclimation, and certain sugars, including sucrose and raffinose, accumulate in plants. This change in sugar concentrations triggers synthesis, metabolism, and the transport of sugars, resulting in the upregulation of genes encoding enzymes such as β-amylases and sucrose synthase in response to low temperatures (Kaplan and Guy, 2004; Rekarte-Cowie et al., 2008).

In addition to sugar metabolism, chloroplast may also play a role in sensing temperature changes. In plants exposed to low temperatures, an energy imbalance may be created due to the difference in the light energy harvested and the energy dissipated through metabolic activity, which, in turn, induces high photosystem II excitation pressure, resulting in the generation of ROS. The redox status of photosynthesis could thus be an indicator of ambient temperature and acts as an important signaling mechanism, especially during low temperature stress. In barley, mutants with impaired chloroplasts were found to be completely susceptible to cold and frost damage (Zhu et al., 2007).

Another group of metabolites that possibly act as ubiquitous metabolic signals to different abiotic stresses such as cold, drought, and salinity is ROS. ROS are generated when molecular oxygen is partially reduced under environmental stress conditions. Studies suggest that ROS may have a dual role in plant metabolism as both toxic by-products and signal molecules (Mittler et al., 2004; Hung et al., 2005; Ouellet and Charron, 2013). ROS modulates many cellular processes through a network of signaling pathways during cold responses (Renaut et al., 2008). ROS is managed by a highly dynamic network of genes that encode both ROS-producing and ROS-scavenging proteins (Mittler et al., 2004). H_2O_2 is an important member of ROS that can regulate gene expression by upregulation or repression (Desikan et al., 2001; Hung et al., 2005). ROS signals affect several cellular processes such as the activation of redox-responsive proteins, including transcription factors and protein kinases (Chinnusamy et al., 2007), Ca^{2+} and Ca^{2+}-binding proteins (e.g., CaM), G-proteins, phospholipid signaling (Mittler et al., 2004), and polyunsaturated fatty acid peroxidation of membrane lipids (Maali-Amiri et al., 2007).

7.4.4 Enzymes Induced During Cold Stress

A diverse array of enzymes involved in respiration and the metabolism of carbohydrates, lipids, proteins, antioxidants, and molecular chaperones are known to function in the cold response machinery (Heidarvand and Amiri, 2010). Furthermore, other enzymes involved in the metabolism of lignins (caffeic acid 3-*O*-methyltransferase), osmolytes, starch, and the biosynthesis of sterols and oligosaccharides of the raffinose family (myo-inositol-1-phosphate synthase and galactinol synthase) are all important components in the global response to cold stress (Fowler and Thomashow, 2002; Renaut et al., 2006). Scavenging enzymes play an important role in cold tolerance to activate protective mechanisms that combat oxidative stress, which is typically accompanied by cold stress (Heidarvand and Amiri, 2010).

Galactinol synthase is a key enzyme in the synthesis of raffinose oligosaccharides, which catalyzes the first committed step in raffinose synthesis. It is found to play a vital role in plant response to low temperatures through saccharide metabolism, especially the raffinose oligosaccharide pathway, which results in the accumulation of monosaccharides and disaccharides such as glucose, fructose, sucrose, galactinol, melibiose, and raffinose (Hannah et al., 2006; Usadel et al., 2008). Among these sugars, galactinol and raffinose possibly act as scavengers of ROS (Nishizawa et al., 2008), whereas sucrose and other simple sugars likely play a role in the stabilization of cell membranes by interacting with phospholipids and proteins in the plasma membrane, and supporting the structure and function of cell membranes (Uemura and Steponkus, 2003; Yano et al., 2005). In addition, other sugar metabolism enzymes such as sucrose phosphate synthase and invertase are also involved in the cold response. Changes in sugar concentrations catalyzed by yeast invertase in potato plants resulted in a higher tolerance to low temperatures (Deryabin et al., 2005). In *Arabidopsis*, while the transcript levels of sucrose phosphate synthase are induced by low temperatures, several genes from the invertase family are suppressed (Usadel et al., 2008). On the contrary, in wheat and tomato, invertase activity was found to be upregulated as the temperature decreased, although the response in chilling-tolerant accessions was weaker (Artuso et al., 2000; Vargas et al., 2007).

Enzymes involved in the metabolism of amino acid and polyamine compounds are also responsive to low temperature stress (Kaplan et al., 2004; Usadel et al., 2008; Davey et al., 2009). The transcript levels of these enzymes in *Arabidopsis* were found to be regulated by cold stress, with some enzymes such as those associated with proline, cysteine, and polyamine biosynthesis, and the glutamate and ornithine pathways, being induced, while others responsible for branched-chain amino acid degradation were repressed (Kaplan et al., 2007; Usadel et al., 2008).

The effect of low temperature stress on lipid metabolism is negatively correlated as the transcript levels of lipid metabolism genes are generally repressed (Hannah et al., 2006). However, some lipid catabolism enzymes such as phospholipase A and D may be activated by low temperature, followed by an increase in the amount of free fatty acids (Wang et al., 2006; Usadel et al., 2008). Phospholipase D functions in the degradation of membrane lipids and its suppression may be necessary to enhance freezing tolerance (Rajashekar, 2000). Galactolipases, belonging to the hydrolase family, are also induced during cold acclimation but at considerably lower levels compared to the phospholipases (Kaniuga, 2008). At low temperatures, acyl-lipid desaturases, an important class of desaturases, introduce double bonds into fatty acids esterified to glycerolipids to enhance the molecular activities in the lipids of both plastid and endoplasmic reticulum membranes (Jin et al., 2001; Maali et al., 2007). This action regulates the level of unsaturation of membrane lipids and remodels the cell membrane structure and function to facilitate adaptation to temperature changes (Somerville and Browse, 1991; Wada et al., 1994; Murata and Wada, 1995).

Another group of enzymes showing increased activity during cold stress are the antioxidative enzymes that include superoxide dismutase, glutathione peroxidase, glutathione reductase, ascorbate peroxidase, and catalase. Furthermore, tripeptide thiol, glutathione, ascorbic acid (vitamin C), and α-tocopherol (vitamin E) are other non-enzymatic antioxidants notably induced during cold stress (Chen and Li, 2002). Vitamin E was found to be a vital component of the protective mechanism against oxidative stress and chilling tolerance in *Arabidopsis* (Maeda et al., 2006; Zhu et al., 2007).

7.4.5 ROLE OF FUNCTIONAL PROTEINS

Proteins involved in cold response machinery are associated with metabolites that show differential accumulation during cold stress. These proteins play a vital role in several processes, such as photosynthesis, photorespiration, primary and secondary metabolism, signal transduction, redox homeostasis, and nucleotide processing (Yan et al., 2006). Apart from the enzymes discussed previously, some of the other functional proteins induced during cold acclimation and freezing tolerance include dehydrins or late embryogenesis abundant (LEA) proteins, heat shock proteins (HSPs), antifreeze proteins (AFPs)/pathogen-related proteins (PRs), and cold-shock domain proteins (CSDPs).

Dehydrins, also known as Group 2 LEA proteins, are a group of heat-stable and glycine-rich proteins that significantly accumulate in response to abiotic stresses causing cell dehydration such as low temperature, drought, and salinity (Renaut et al., 2004; Rorat, 2006; Kosová et al., 2007). Dehydrins have highly conserved domains, known as the K, Y, and S segments, whose number and order group the different dehydrins into subclasses. In wheat, three COR genes encoding Wcs120 (K6), Wcor410 (SK3), and Wcor14 are identified as dehydrins (Ganeshan et al., 2008). Barley has 13 dehydrin genes, *dhn1* to *dhn13*, of which *dhn5* and *dhn8* genes are induced at the transcription level. The intermediate and winter cultivars of barley tend to accumulate higher levels of *dhn5* than the spring cultivars during freezing tolerance (Kosová et al., 2008). Dehydrins function in membrane stabilization and act as protectants of cytoplasmic proteins from denaturation. While some of them are important for preventing protein aggregation (Nakayama et al., 2008), others such as ERD10 (early response to dehydration) and ERD14 act as chaperones and interact with phospholipid vesicles (Kovacs et al., 2008). Some of the dehydrin genes and their specific interactions are thought to provide a reliable indication of cold and drought stress (Tommasini et al., 2008). They are shown to exhibit *in vitro* cryoprotective activity and *in vivo* antifreeze activity in several plant species. Dehydrins also possibly play a role in osmoregulation and radical scavenging.

HSPs are more commonly induced during high temperature stress, but some studies indicate that some HSPs (Hsps104 and 90 m of the Hsp70 family, other small HSPs, and chaperonins 60 and 20) are also induced in response to low temperatures (Guy, 1999; Ukaji et al., 1999). In the *Arabidopsis* Hsp70 family, genes encoding four cytoplasmic and two mitochondrial proteins were found to upregulate in response to cold temperature. Similar to dehydrins, HSPs also play a pivotal role in cryoprotection and their increased activity is essential to prevent protein aggregation, facilitate refolding of the denatured proteins (Yan et al., 2006; Timperio et al., 2008), and aid in the translocation of proteins into organelles during chilling stress (Nover and Scharf, 1997).

Another class of proteins that possibly show increased activity following low temperature exposure is the CSDPs, which belong to a superfamily of proteins that contain the cold-shock domain. This domain is generally found in proteins that bind to nucleic acids. CSDPs function as RNA chaperones (Guy, 1999; Somerville, 1999) and are conserved in prokaryotes and eukaryotes. The expression patterns of CSDP genes in *Arabidopsis* are strongly correlated with plant development under stress conditions. While some of them were specifically correlated to abiotic stresses, high levels of CSDP activity were also detected in meristematic tissues under stress-free conditions (Nakaminami et al., 2009; Park et al., 2009). Although the precise function of plant CSDPs remains to be elucidated, they probably possess multiple functions that may be important under both stress and non-stress conditions.

AFPs are another important group of proteins from the COR gene family that inhibit the activity of ice nucleators. These proteins are highly similar to plant PR and accumulate in abundance in response to abiotic stresses such as cold and drought (Moffatt et al., 2006).

AFPs assemble as oligomers and inhibit the expansion of ice crystals. They bind to a newly formed ice crystal through a non-colligative mechanism and prevent the coalition of water molecules, thereby influencing the shape and growth of ice crystals. This is accompanied by the downregulation of aquaporins (water channels), which is important for cold acclimation as extracellular freezing can cause cell dehydration (Peng et al., 2008). AFPs inhibit not only the formation of new large ice crystals but also recrystallization during thawing, which can cause greater physical damage to the frozen plant tissues (Griffith et al., 1997). The antifreeze activity of these proteins is regulated by Ca^{2+} released from pectin or other specific proteins. Their activity is generally detected in the apoplastic space of monocotyledons and dicotyledons that exhibit freezing tolerance after cold acclimation. Some of the PR proteins synthesized in overwintering monocots under cold stress were found to exhibit antifreeze activities (Atici and Nalbantoglu, 2003; Seo et al., 2008). Among these, β-1,3-glucanase (PR-1 and -2), chitinases (PR-11), and thaumatin-like proteins (PR-5) are found in *Arabidopsis* (Seo et al., 2008). In some other species, Bet v-1 homologues (PR-10) and lipid transfer proteins (PR-14) were also found to be cold inducible (Liu et al., 2003; Pak et al., 2009; Takenaka et al., 2009; Zhang et al., 2010; Lee et al., 2012). These PR proteins are thought to have a dual function as they protect plants against ice crystallization and pathogen attack during winter. It was reported that many freeze-tolerant grasses exhibited increased resistance to snow molds, powdery mildews, leaf spots, and rusts during cold acclimation (Hon et al., 1995; Guy, 1999).

7.4.6 HORMONAL REGULATION OF COLD STRESS

While several nuclear pathways are shown to directly control the COR gene expression, phytohormones have been suggested to act as signal molecules that govern the upstream regulatory modes of these nuclear pathways during cold stress. Phytohormones are low molecular weight chemicals produced in plants in very low concentrations; however, they play a crucial role in transducing various signals from the site of synthesis to the site of action. Hormone homeostasis, maintained in the cells by modulating biosynthesis, catabolism, transport, and the interaction of various phytohormones, is required to initiate signal transduction events that affect downstream cellular processes (Santner et al., 2009). Phytohormones are divided into different classes based on their specific functions. Furthermore, different phytohormones may have overlapping functions with synergistic or antagonistic effects. Thus, a complex network of hormonal interaction is created that facilitates the signaling of external information not only to endogenous developmental programs but also to activate biochemical and regulatory pathways that induce resistance to various stress conditions, including low temperature stress. Among the different classes of phytohormones, abscisic acid (ABA), gibberellic acid (GA), salicylic acid (SA), and ethylene have been implicated in the regulation of plant response to cold stress.

ABA is an isoprenoid hormone that is important for many endogenous processes such as seed dormancy and abscission as well as abiotic stress signaling (Nakashima et al., 2014). ABA is probably the most studied phytohormone for its role in cold stress responses. ABA levels have been shown to increase in response to low temperatures in many plant species (Lang and Palva, 1992; Lissarre et al., 2010), including *Arabidopsis* and rice (Cuevas et al., 2008; Baron et al., 2012; Maruyama et al., 2014). The COR gene expression is also regulated by ABA-dependent pathways through CBF transcription (Chinnusamy et al., 2004; Knight et al., 2004). Low temperature exposure increases the ABA content in many plant species, including herbaceous and woody plants. This increase in ABA is associated with the induction of the cold acclimation process and freezing tolerance. The exogenous application of ABA was found to induce CBF genes and promote chilling and freezing tolerance at the normal growth temperature (Guy, 1999; Knight et al., 2004). During cold acclimation, ABA activates a serine/threonine protein kinase (Mustilli et al., 2002), which phosphorylates ICE1 to promote its transcriptional activity and induce COR gene expression. The promoters of some COR genes have ABA response elements (ABREs), which are *cis*-acting regulators that bind to ABRE-binding proteins (AREBs or ABFs) to activate ABA-dependent COR gene expression. Through a common ABRE, ABA induces different ABF genes (ABF1–4) depending on the type of abiotic stress. While ABF1 is induced by cold, ABF2 and ABF3 are activated by high salt concentration, and ABF4 is induced by several abiotic stresses such as cold, high salt concentration, and drought (Choi et al., 2000).

Among the growth-promoting hormones, GAs play a key role in the adaptation of plants to changes in light and temperature (Franklin, 2009). GAs promote cell elongation under warm temperatures (Penfield, 2008) and are produced as bioactive GAs by the activity of oxygenases such as GA 20-oxidase, GA 3-oxidase, and KAO-oxidase (Eremina et al., 2016). In *Arabidopsis* plants exposed to low temperatures, a GA-catabolizing enzyme, GA 2-oxidase, is induced and increases the hydroxylation and inactivation of bioactive GAs, while GA 20-oxidase, which is required for GA biosynthesis, is repressed. GA-deficient mutants of *Arabidopsis* (Achard et al., 2008; Richter et al., 2013) and rice (Richter et al., 2010) showed altered chilling and freezing tolerance, indicating the importance of GA signaling in cold stress response. Likewise, a wheat variety (Rht3) with defective GA signaling was found to have a reduced GA content, leading to a dwarf plant phenotype under cold stress (Tonkinson et al., 1997). Cold stress affects both the metabolism and the signaling of GAs possibly through CBF expression (Achard et al., 2008; Richter et al., 2013). CBF transcripts when overexpressed in *Arabidopsis* (Hsieh et al., 2002) and tobacco (Zhou et al., 2014) reduced the bioactive GA levels, which caused growth suppression and late flowering, although these effects were reversed by exogenous GA application (Shan et al., 2007; Achard et al., 2008). CBF expression may also be mediated by GAs through another transcription factor known as PIF4 (phytochrome-interacting factor), which binds directly to the CBF promoters and represses their activity under long-day conditions (Lee and Thomashow, 2012). The activity of PIF4 is controlled by phytochromes, DELLAs (nuclear growth-repressing proteins), and brassinosteroid phytohormones (Achard et al., 2008; Feng et al., 2008; de Lucas et al., 2008; Bernardo-García et al., 2014).

SA is another phytohormone that may affect plant growth under low temperature conditions. SA is known for triggering a hypersensitive response against pathogen attack as part of the plant defense mechanism. SA also possibly affects the cell cycle in plants, thus contributing to SA-induced growth retardation during cold stress (Wolters and Jürgens, 2009). SA levels increased during cold acclimation in several chilling-sensitive and freezing-tolerant plant species (Scott et al., 2004; Huang et al., 2010; Kosová et al., 2012; Kim et al., 2013; Dong et al., 2014). While endogenous free SA and glucosyl SA were reported to increase following cold exposure in *Arabidopsis*, wheat, and grape berries (Scott et al., 2004; Kim et al., 2013; Kosová et al., 2012; Wan et al., 2009), exogenous SA applications were found to enhance cold tolerance in rice, wheat, and potato (Kang and Saltveit, 2002; Mora-Herrera et al., 2005; Tasgin et al., 2006; Hara et al., 2012).

Another phytohormone, ethylene is a gaseous hormone known to play a major role in abiotic stress resistance, especially in the regulation of cold stress response in plants (Yoo et al., 2009). However, the precise function of ethylene, whether as a positive or a negative regulator of cold response, remains to be elucidated. While an increase in ethylene concentrations in response to cold has been reported in several plant species including tomato, rye, bean, alfalfa, and wheat (Guye et al., 1987; Ciardi et al., 1997; Yu et al., 2001; Kosová et al., 2012; Guo et al., 2014), some plants (e.g., barrel clover) showed a reduction in ethylene levels during cold stress (Zhao et al., 2014).

7.5 PROSPECTS OF COLD TOLERANCE IN AGRICULTURE

Low temperature is one of the most important abiotic stresses that not only limits the geographical distribution of agricultural crops but also affects growth and crop productivity (Xin and Browse, 2001). Significant yield losses can occur from sudden frosts and unusual freezing temperatures in a particular growing season. The study of cold stress responses in plants is therefore required to enhance the tolerance of crops to cold and freezing temperatures. Conventional breeding methods offered limited success in improving freezing tolerance in crop plants, especially through interspecific and inter-generic hybridization, as limited information on the mechanisms underlying cold acclimation and freezing tolerance was available. Therefore, the first step toward establishing a good crop improvement strategy to cold stress is to understand the complex biological networks involved in the initiation of cold acclimation and the induction of freezing tolerance. This was achieved to a certain extent through the application of molecular genetic tools and biotechnology that improve the efficiency of traditional breeding methods through more accurate and faster incorporation of stress-tolerant genes in crops. Furthermore, the application of genomic approaches and gene knockout strategies helped elucidate some of the mechanisms underlying the complex molecular and cellular machinery involved in abiotic stress tolerance (Kumar and Bhatt, 2006). Molecular studies involving genetic engineering and transformation experiments in which beneficial genes are introduced and/or overexpressed have enabled plants to be more tolerant not only to low temperatures but also to other abiotic stress factors (Wani et al., 2008; Gosal et al., 2009; Wani and Gosal, 2011; Wang et al., 2016). For example, chilling tolerance was achieved in several transgenic crops by overexpressing the glycerol-3-phosphate acyl transferase, which altered the unsaturation of fatty acids (Ariizumi et al., 2002; Sakamoto et al., 2003; Sui et al., 2007). Likewise, the overexpression of the gene encoding chloroplast $w3$ fatty acid desaturase conferred chilling tolerance in transgenic tobacco (Kodama et al., 1994). Furthermore, transgenic rice plants expressing bacterial cold shock proteins, CspA and CspB, exhibited improved tolerance to a number of abiotic stresses, including cold, heat, and water deficits (Castiglioni et al., 2008). Other numerous studies on cold response mechanisms in *Arabidopsis* helped to identify related genes that can be used as potential targets to improve cold stress tolerance in crops (Kodama et al., 1994; Jaglo-Ottosen et al., 1998, 2001; Ariizumi et al., 2002; Sakamoto et al., 2003; Sui et al., 2007). Constitutive overexpression of *Arabidopsis* cold stress response genes such as the genes associated with the CBF/DREB1 pathway were successfully used to engineer cold stress tolerance in many crops (Hsieh et al., 2002; Zhang et al., 2004; Yamaguchi-Shinozaki and Shinozaki, 2006). For example, the CBF1 transcript introduced into tomato under the control of a cauliflower mosaic virus 35S promoter improved tolerance not only to low temperatures but also to drought and salt stress, although these plants exhibited a dwarf phenotype and a reduced fruit set (Hsieh et al., 2002). Although the CaMV35S promoter facilitated constitutive overexpression of the CBF genes, undesirable agronomic traits have also been reported in plants. In *Arabidopsis*, CBF overexpression resulted in stunted growth, flowering delays, and reduced yield (Gilmour et al., 2000). To overcome this difficulty, the regulation of transgene expression was achieved by using another promoter known as responsive to dehydration 29A (RD29A). RD29A is a stress-inducible promoter derived from *Arabidopsis* and is reported to minimize the adverse effects of transgenes on plant growth (Kasuga et al., 2004). The *Arabidopsis* DREB1A gene overexpressed *via* the RD29A promoter improved cold stress tolerance in tobacco (Kasuga et al., 2004) and bread wheat (Pellegrineschi et al., 2004) without any major detrimental effects on plant growth and development. Transcription factors belonging to other families, including basic Leucine Zippers (bZIPs), WRKYs, MYBs, and nascent polypeptide–associated complexes (NACs), have also been successfully used for breeding multiple stress-tolerant crops using transgenic technology (Su et al., 2010; Gao et al., 2011; Mao et al., 2014; Meng et al., 2014; Wang et al., 2014; Qin et al., 2015; Yang et al., 2015; Zhang et al., 2015). In addition to the transcription factors, transgenic attempts with several other important genes such as TPP1 (trehalose-6-phosphate phosphatase) (Ge et al., 2008; Jang et al., 2003), chloroplast GPAT (glycerol-3-phosphate acyltransferase) (Murata et al., 1992), β-galactosidase (Pennycooke et al., 2003), LEA proteins (Hara et al., 2003; Houde et al., 2004), and zinc finger protein (Kim et al., 2007) achieved success in enhancing cold tolerance in several plant species. Thus, the identification and functional testing of stress-tolerant genes is of great importance for elucidating the regulatory and structural networks of stress responses as well as for sustaining agricultural crop productivity through breeding stress-tolerant crops.

However, studies examining plant responses to any one particular abiotic stress under laboratory conditions for short periods may not be sufficient to completely unravel the information required for sustainable crop improvement. Cold stress response in plants is a complex multigene trait involving several molecular and metabolic pathways in different cell compartments (Thomashow, 2001; Hannah et al., 2005). Furthermore, they interact with other abiotic stress responses, thereby causing unexpected physiological changes in plant cells under field conditions (Larkindale et al., 2005; Mittler, 2006). Therefore, it is essential to fully understand the stress-response system of plants in a comprehensive manner using a combination of genetic, molecular, and bioinformatic tools. These tools are also useful in determining the genetic potential of plants and identifying specific crop varieties that exhibit natural tolerance to cold stress and other abiotic stresses.

7.6 CONCLUSIONS

As one of the important abiotic stress factors, cold stress can have a major impact on limiting agricultural crop production and productivity. Cold stress affects the physiological, biochemical, and molecular functions in plants, eventually leading to impairment of their growth and development aspects. Recent advances in genomic and molecular technologies have enabled us to dissect the cold acclimation pathway and identify several genes associated with cold responses in plants. A complex network of signal transduction pathways is involved in sensing and responding to cold stress. Some of the major components of the cold signaling mechanism include protein and lipid signaling cascades, ROS, phytohormones, molecular chaperones, osmolytes, and several cold-regulated genes such as CBFs, ICE1, DREB1, etc. More recent studies have indicated the involvement of epigenetic regulatory processes such as histone modifications and DNA methylation in plants. In addition, post-transcriptional regulatory mechanisms involving mRNA splicing, export, and degradation also play important roles in cold stress response and signal transduction. Furthermore, some post-translational mechanisms such as ubiquitination, proteolysis, and sumoylation are implicated in the cold acclimation and tolerance process. While the scientific information acquired through all these studies has been useful to engineer cold resistance in some crops, additional research is required to identify potential genes that play a central role in imparting tolerance to several abiotic/biotic stresses in order to sustain agricultural productivity under changing weather patterns.

REFERENCES

Achard, P., Gong, F., Cheminant, S., Alioua, M., Hedden, P., Genschik, P. (2008) The cold-inducible CBF1 factor-dependent signaling pathway modulates the accumulation of the growth-repressing DELLA proteins via its effect on gibberellin metabolism. *Plant Cell* 20: 2117–2129.

Agarwal, M., Hao, Y., Kapoor, A., Dong, C.H., Fujii, H., Zheng, X., Zhu, J.K. (2006) A R2R3 type MYB transcription factor is involved in the cold regulation of CBF genes and in acquired freezing tolerance. *Journal of Biological Chemistry* 281: 37636–37645.

Ambrosone, A., Costa., A., Leone, A., Grillo, S. (2012) Beyond transcription: RNA-binding proteins as emerging regulators of plant response to environmental constraints. *Plant Sciences* 182: 12–18.

Angadi, S.V., Cutforth, H.W., McConkey, B.G. (2000) Seeding management to reduce temperature stress in Brassica species. *Saskatchewan Soils and Crops Proceedings.* http://www.usask.ca/soilsncrops/conference-proceedings/previous_years/Files/2000/2000docs/435.pdf, Accessed 3/8/2018.

Antle, J.M., Capalbo, S.M., Elliott, E.T., Paustian, K.H. (2004) Adaptation, spatial heterogeneity, and the vulnerability of agricultural systems to climate change and CO 2 fertilization: An integrated assessment approach. *Climate Change* 64: 289–315.

Apel, K., Hirt, H. (2004) Reactive oxygen species: Metabolism, oxidative stress, and signal transduction. *Annual Review of Plant Biology* 55: 373–399.

Ariizumi, T., Kishitani, S., Inatsugi, R., Nishida, I., Murata, N., Toriyama, K. (2002) An increase in unsaturation of fatty acids in phosphatidylglycerol from leaves improves the rates of photosynthesis and growth at low temperatures in transgenic rice seedlings. *Plant Cell Physiology* 43: 751–758.

Artuso, A., Guidi, L., Soldatini, G.F., Pardossi, A., Tognoni, F. (2000) The influence of chilling on photosynthesis and activities of some enzymes of sucrose metabolism in *Lycopersicon esculentum* Mill. *Acta Physiologiae Plantarum* 22: 95–101.

Atici, O., Nalbantoglu, B. (2003) Antifreeze proteins in higher plants. *Phytochemistry* 64: 1187–1196.

Badawi, M., Reddy, Y.V., Agharbaoui, Z., Tominaga, Y., Danyluk, J., Sarhan, F., Houde, M. (2008) Structure and functional analysis of wheat ICE (inducer of CBF expression) genes. *Plant Cell Physiology* 49: 1237–1249.

Baron, K.N., Schroeder, D.F., Stasolla, C. (2012) Transcriptional response of abscisic acid (ABA) metabolism and transport to cold and heat stress *Arabidopsis thaliana*. *Plant Sciences* 188–189: 48–59.

Barrero-Gil, J., Salinas, J. (2013) Post-translational regulation of cold acclimation response. *Plant Sciences* 205–206: 48–54.

Bernardo-García, S., de Lucas, M., Martínez, C., Espinosa-Ruiz, A., Davière, J.M., Prat, S. (2014) BR-dependent phosphorylation modulates PIF4 transcriptional activity and shapes diurnal hypocotyl growth. *Genes Development* 28: 1681–1694.

Carpaneto, A., Ivashikina, N., Levchenko, V., Krol, E., Jeworutzki, E., Zhu, J.K., Hedrich, R. (2007) Cold transiently activates calcium-permeable channels in *Arabidopsis* mesophyll cells. *Plant Physiology* 143: 487–494.

Castiglioni, P., Warner, D., Bensen, R.J., Anstrom, D.C., Harrison, J., Stoecker, M., Abad, M., Kumar, G., Salvador, S., D'Ordine, R., Navarro, S., Back, S., Fernandes, M., Targolli, J., Dasgupta, S., Bonin, C., Luethy, M.H., Heard, J.E. (2008) Bacterial RNA chaperones confer abiotic stress tolerance in plants and improved grain yield in maize under water-limited conditions. *Plant Physiology* 147: 446–455.

Chen, W.P., Li, P.H. (2002) Attenuation of reactive oxygen production during chilling in ABA-treated maize cultured cells. In *Plant Cold Hardiness*, ed. Li, C. and Palva, E.T., 223–233. Dordrecht, The Netherlands: Kluwer Academic.

Chen, M., Xu, Z., Xia, L., Li, L., Cheng, X., Dong, J., Wang, Q., Ma, Y. (2009) Cold-induced modulation and functional analyses of the DRE-binding transcription factor gene, GmDREB3, in soybean (*Glycine max* L.). *Journal of Experimental Botany* 60: 121–135.

Chen, L., Zhang, Y., Ren, Y., Xu, J., Zhang, Z., Wang, Y. (2012) Genome-wide identification of cold-responsive and new microRNAs in *Populus tomentosa* by high-throughput sequencing. *Biochemical and Biophysical Research Communications* 417: 892–896.

Chinnusamy, V., Ohta, M., Kanrar, S., Lee, B.H., Hong, X., Agarwal, M., Zhu, J.K. (2003) ICE1: A regulator of cold-induced transcriptome and freezing tolerance in *Arabidopsis*. *Genes Development* 17: 1043–1054.

Chinnusamy, V., Schumaker, K., Zhu, J.K. (2004) Molecular genetic perspectives on cross-talk and specificity in abiotic stress signaling in plants. *Journal of Experimental Botany* 55: 225–236.

Chinnusamy, V., Zhu, J.K., Sunkar, R. (2010) Gene regulation during cold stress acclimation in plants. *Methods in Molecular Biology* 639: 39–55.

Chinnusamy, V., Zhu, J., Zhu, J.K. (2007) Cold stress regulation of gene expression in plants. *Trends in Plant Sciences* 12: 444–451.

Choi, H.I., Hong, J.H., Ha, J.O., Kang, J.Y., Kim, S.Y. (2000) ABFs, a family of ABA-responsive element binding factors. *Journal of Biological Chemistry* 275: 1723–1730.

Chouard, P. (1960) Vernalization and its relations to dormancy. *Annual Review of Plant Physiology* 11: 191–238.

Ciardi, J.A., Deikman, J., Orzolek, M.D. (1997) Increased ethylene synthesis enhances chilling tolerance in tomato. *Plant Physiology* 101: 333–340.

Cuevas, J.C., López-Cobollo, R., Alcázar, R., Zarza, X., Koncz, C., Altabella, T., Salinas, J., Tiburcio, A.F., Ferrando, A. (2008) Putrescine is involved in *Arabidopsis* freezing tolerance and cold acclimation by regulating abscisic acid levels in response to low temperature. *Plant Physiology* 148: 1094–1105.

Dat, J., Vandenabeele, S., Vranova, E., Van Montagu, M., Inzé, D., Van Breusegem, F. (2000) Dual action of the active oxygen species during plant stress responses. *Cell and Molecular Life Sciences* 57: 779–795.

Davey, M.P., Woodward, F.I., Quick, W.P. (2009) Intraspecific variation in cold-temperature metabolic phenotypes of *Arabidopsis lyrata* ssp *petraea*. *Metabolomics* 5: 138–149.

Davletova, S., Schlauch, K., Coutu, J., Mittler, R. (2005) The zinc-finger protein Zat12 plays a central role in reactive oxygen and abiotic stress signaling in *Arabidopsis*. *Plant Physiology* 139: 847–856.

de Lucas, M., Davière, J.M., Rodríguez-Falcón, M., Pontin, M., Iglesias-Pedraz, J.M., Lorrain, S., Fankhauser, C., Blázquez, M.A., Titarenko, E., Prat, S. (2008) A molecular framework for light and gibberellin control of cell elongation. *Nature* 451: 480–484.

Deryabin, A.N., Dubinina, I.M., Burakhanova, E.A., Astakhova, N.V., Sabel'nikova, E.P., Trunova, T.I. (2005) Influence of yeast-derived invertase gene expression in potato plants on membrane lipid peroxidation at low temperature. *Journal of Thermal Biology* 30: 73–77.

Desikan, R., Mackerness, S.A.H., Hancock, J.T., Neill, S.J. (2001) Regulation of the *Arabidopsis* transcriptome by oxidative stress. *Plant Physiology* 127: 159–172.

Dhillion, T., Pearce, S.P., Stockinger, E.J., Distelfeld, A., Li, C., Knox, A.K., Vashegyi, I., Vágújfalvi, A., Galiba, G., Dubcovsky, J. (2010) Regulation of freezing tolerance and flowering in temperate cereals: The VRN-1 connection. *Plant Physiology* 153: 1846–1858.

Dhonukshe, P., Laxalt, A.M., Goedhart, J., Gadella, T.W.J., Munnik, T. (2003) Phospholipase D activation correlates with microtubule reorganization in living plant cells. *Plant Cell* 15: 2666–2679.

Ding, Y., Li, H., Zhang, X., Xie, Q., Gong, Z., Yang, S. (2015) OST1 kinase modulates freezing tolerance by enhancing ICE1 stability in *Arabidopsis*. *Developmental Cell* 32: 278–289.

Dodd, A.N., Jakobsen, M.K., Baker, A.J., Telzerow, A., Hou, S.W., Laplaze, L., Barrot, L., Poethig, R.S., Haseloff, J., Webb, A.A.R. (2006) Time of day modulates low-temperature Ca^{2+} signals in *Arabidopsis*. *Plant Journal* 48: 962–973.

Doherty, C.J., Van Buskirk, H.A., Myers, S.J., Thomashow, M.F. (2009) Roles for *Arabidopsis* CAMTA transcription factors in cold-regulated gene expression and freezing tolerance. *Plant Cell* 21: 972–984.

Dong, C.H., Agarwal, M., Zhang, Y., Xie, Q., Zhu, J.K. (2006) The negative regulator of plant cold responses, HOS1, is a RING E3 ligase that mediates the ubiquitination and degradation of ICE1. *Proceedings of National Academic Sciences of USA* 103: 8281–8286.

Dong, C.J., Li, L., Shang, Q.M., Liu, X.Y., Zhang, Z.G. (2014) Endogenous salicylic acid accumulation is required for chilling tolerance in cucumber (*Cucumis sativus* L.) seedlings. *Planta* 240: 687–700.

Drøbak, B.K., Franklin-Tong, V.E., Staiger, C.J. (2004) The role of the actin cytoskeleton in plant cell signaling. *New Phytology* 163: 13–30.

Eremina, M., Rozhon, W., Poppenberger, B. (2016) Hormonal control of cold stress responses in plants. *Cellular and Molecular Life Sciences* 73: 797–810.

Feng, S., Martinez, C., Gusmaroli, G., Wang, Y., Zhou, J., Wang, F., Chen, L., Yu, L., Iglesias-Pedraz, J.M., Kircher, S., Schäfer, E., Fu, X., Fan, L.M., Deng, X.W. (2008) Coordinated regulation of *Arabidopsis thaliana* development by light and gibberellins. *Nature* 451: 475–479.

Filichkin, S.A., Priest, H.D., Givan, S.A., Shen, R., Bryant, D.W., Fox, S.E., Wong, W.K., Mockler, T.C. (2010) Genome-wide mapping of alternative splicing in *Arabidopsis thaliana. Genome Research* 20: 45–58.

Fowler, S., Thomashow, M.F. (2002) *Arabidopsis* transcriptome profiling indicates that multiple regulatory pathways are activated during cold acclimation in addition to the CBF cold response pathway. *The Plant Cell* 14: 1675–1690.

Franklin, K.A. (2009) Light and temperature signal crosstalk in plant development. *Current Opinion in Plant Biology* 12: 63–68.

Fursova, O.V., Pogorelko, G.V., Tarasov, V.A. (2009) Identification of ICE2, a gene involved in cold acclimation which determines freezing tolerance in *Arabidopsis thaliana. Gene* 429: 98–103.

Ganeshan, S., Vitamvas, P., Fowler, D.B., Chibbar, R.N. (2008) Quantitative expression analysis of selected COR genes reveals their differential expression in leaf and crown tissues of wheat (*Triticum aestivum* L.) during an extended low temperature acclimation regimen. *Journal of Experimental Botany* 59: 2393–2402.

Gao, S.Q., Chen, M., Xu, Z.S., Zhao, C.P., Li, L., Xu, H.J., Tang, Y.M., Zhao, X., Ma, Y.Z. (2011) The soybean GmbZIP1 transcription factor enhances multiple abiotic stress tolerances in transgenic plants. *Plant Molecular Biology* 75: 537–553.

Gardiner, J., Collings, D.A., Harper, J.D.I., Marc, J. (2003) The effects of the phospholipase D-antagonist 1-butanol on seedling development and microtubule organisation in *Arabidopsis. Plant and Cell Physiology* 44: 687–696.

Ge, L.F., Chao, D.Y., Shi, M., Zhu, M.Z., Gao, J.P., Lin, H.X. (2008) Overexpression of the trehalose-6-phosphate phosphatase gene *OsTPP1* confers stress tolerance in rice and results in the activation of stress responsive genes. *Planta* 228: 191–201.

Ghildiyal, M., Zamore, P.D. (2009) Small silencing RNAs: An expanding universe. *Nature Reviews Genetics* 10: 94–108.

Gilmour, S.J., Fowler, S.G., Thomashow, M.F. (2004) *Arabidopsis* transcriptional activators CBF1, CBF2, and CBF3 have matching functional activities. *Plant Molecular Biology* 54: 767–781.

Gilmour, S.J., Sebolt, A.M., Salazar, M.P., Everard, J.D., Thomashow, M.F. (2000) Overexpression of the *Arabidopsis* CBF3 transcriptional activator mimics multiple biochemical changes associated with cold acclimation. *Plant Physiology* 124: 1854–1865.

Gong, Z., Dong, C.H., Lee, H., Zhu, J., Xiong, L., Gong, D., Stevenson, B., Zhu, J.K. (2005) A DEAD box RNA helicase is essential for mRNA export and important for development and stress responses in *Arabidopsis. Plant Cell* 17: 256–267.

Gosal, S.S., Wani, S.H., Kang, M.S. (2009) Biotechnology and drought tolerance. *Journal of Crop Improvement* 23: 19–54.

Greenup, A.G., Sasani, S., Oliver, S.N., Walford, S.A., Millar, A.A., Trevaskis, B. (2011) Transcriptome analysis of the vernalization response in barley (*Hordeum vulgare*) seedlings. *PLoS ONE* 6: e17900.

Grennan, A.K. (2006) Abiotic stress in rice. An "omic" approach. *Plant Physiology* 140: 1139–1141.

Griffith, M., Antikainen, M., Hon, W.C., Pihakaski-Maunsbach, K., Yu, X.M., Chun, J.U., Yang, D.S.C. (1997) Antifreeze proteins in winter rye. *Physiologia Plantarum* 100: 327–332.

Guo, Z., Tan, J., Zhuo, C., Wang, C., Xiang, B., Wang, Z. (2014) Abscisic acid, H_2O_2 and nitric oxide interactions mediated cold-induced *S*-adenosylmethionine synthetase in *Medicago sativa* subsp. *falcata* that confers cold tolerance through up-regulating polyamine oxidation. *Plant Biotechnology Journal* 12: 601–612.

Guy, C. (1999) The influence of temperature extremes on gene expression, genomic structure, and the evolution of induced tolerance in plants. In *Plant Responses to Environmental Stresses. From Phytohormones to Genome Reorganization*, ed. Lerner, H.R., 497–548. New York/Basel: Marcel Dekker, Inc.

Guy, C.L., Niemi, K.J., Brambl, R. (1985) Altered gene-expression during cold acclimation of spinach. *Proceedings of the National Academy of Sciences of the United States of America* 82: 3673–3677.

Guye, M.G., Vigh, L., Wilson, L.M. (1987) Chilling-induced ethylene production in relation to chill-sensitivity in *Phaseolus* spp. *Journal of Experimental Botany* 38: 680–690.

Han, J., Xiong, J., Wang, D., Fu, X.D. (2011) Pre-mRNA splicing: Where and when in the nucleus. *Trends in Cell Biology* 21: 336–343.

Hannah, M.A., Heyer, A.G., Hincha, D.K. (2005) A global survey of gene regulation during cold acclimation in *Arabidopsis thaliana. PLoS Genetics* 1: e26. Doi: 10.1371/journal.pgen.0010026.

Hannah, M.A., Wiese, D., Freund, S., Fiehn, O., Heyer, A.G., Hincha, D.K. (2006) Natural genetic variation of freezing tolerance in *Arabidopsis. Plant Physiology* 142: 98–112.

Hara, M., Furukawa, J., Sato, A., Mizoguchi, T., Miura, K. (2012) Abiotic stress responses in plant. In *Abiotic Stress Responses in Plants*, ed. Parvaiz, A. and Prasad, M.N.V., 235–251. New York: Springer.

Hara, M., Terashima, S., Fukaya, T., Kubol, T. (2003) Enhancement of cold tolerance and inhibition of lipid peroxidation by citrus dehydrin in transgenic tobacco. *Planta* 217: 290–298.

Hasanuzzaman, M., Nahar, K., Fujita, M. (2013) Extreme temperature response, oxidative stress and antioxidant defense in plants. In *Plant Responses and Applications in Agriculture*, ed. Vahdati, K. and Leslie, C., 169–205. Rijeka, Croatia: InTech.

Heidarvand, L., Amiri, R.M. (2010) What happens in plant molecular responses to cold stress? *Acta Physiologiae Plantarum* 32: 419–431.

Hemming, M.N., Trevaskis, B. (2011) Make hay when the sun shines: The role of MADS-box genes in temperature-dependent seasonal flowering responses. *Plant Science* 180: 447–453.

Hoermiller, I.I., Naegele, T., Augustin, H., Stutz, S., Weckwerth, W., Heyer, A.G. (2017) Subcellular reprogramming of metabolism during cold acclimation in *Arabidopsis thaliana*. *Plant, Cell and Environment* 40: 602–610.

Hon, W.C., Griffith, M., Mlynarz, A., Kwok, Y.C., Yang, D.S.C. (1995) Antifreeze proteins in winter rye are similar to pathogenesis-related proteins. *Plant Physiology* 109: 879–889.

Hong, Y., Zheng, S., Wang, X. (2008) Dual functions of phospholipase Dα1 in plant response to drought. *Molecular Plant* 1: 262–269.

Houde, M., Dallaire, S., N'dong, D., Sarhan, F. (2004) Overexpression of the acidic dehydrin WCOR410 improves freezing tolerance in transgenic strawberry leaves. *Plant Biotechnology Journal* 2: 381–387.

Hsieh, T.H., Lee, J.T., Yang, P.T., Chiu, L.H., Charng, Y.Y., Wang, Y.C., Chan, M.T. (2002) Heterology expression of the *Arabidopsis* C-repeat/dehydration response element binding factor 1 gene confers elevated tolerance to chilling and oxidative stresses in transgenic tomato. *Plant Physiology* 129: 1086–1094.

Hu, Y., Zhang, L., He, S., Huang, M., Tan, J., Zhao, L., Yan, S., Li, H., Zhou, K., Liang, Y., Li, L. (2012) Cold stress selectively unsilences tandem repeats in heterochromatin associated with accumulation of H3K9ac. *Plant Cell and Environment* 35: 2130–2142.

Huang, X., Li, J., Bao, F., Zhang, X., Yang, S. (2010) A gain-of-function mutation in the *Arabidopsis* disease resistance gene *RPP4* confers sensitivity to low temperature. *Plant Physiology* 154: 796–809.

Huner, N.P.A., Oquist, G., Sarhan, F. (1998) Energy balance and acclimation to light and cold. *Trends in Plant Science* 3: 224–230.

Hung, S.H., Yu, C.W., Lin, C.H. (2005) Hydrogen peroxide functions as a stress signal in plants. *Botanical Bulletin of Academia Sinica* 46: 1–10.

Ichimura, K., Mizoguchi, T., Yoshida, R., Yuasa, T., Shinozaki, K. (2000) Various abiotic stresses rapidly activate *Arabidopsis* MAP kinases ATMK4 and ATMK6. *Plant Journal* 24: 655–665.

Jaglo-Ottosen, K.R., Gilmour, S.J., Zarka, D.G., Schabenberger, O., Thomashow, M.F. (1998) *Arabidopsis* CBF1 overexpression induces COR genes and enhances freezing tolerance. *Science* 280: 104–106.

Jaglo-Ottosen, K.R., Kleff, S., Amundsen, K.L., Zhang, X., Haake, V., Zhang, J.Z., Detis, T., Thomashow, M.F. (2001) Components of the *Arabidopsis* C-repeat/dehydration-responsive element binding factor cold-response pathway are conserved in *Brassica napus* and other plant species. *Plant Physiology* 127: 910–917.

Jang, I.C., Oh, S.J., Seo, J.S., Choi, W.B., Song, S.I., Kim, C.H., Kim, Y.S., Seo, H.S., Choi, Y.D., Nahm, B.H., Kim, J.K. (2003) Expression of a bifunctional fusion of the *Escherichia coli* genes for trehalose-6-phosphate synthase and trehalose-6-phosphate phosphatase in transgenic rice plants increases trehalose accumulation and abiotic stress tolerance without stunting growth. *Plant Physiology* 131: 516–524.

Jiang, Q.W, Kiyoharu, O., Ryozo, I. (2002) Two novel mitogen-activated protein signaling components, *OsMEK1* and *OsMAP1*, are involved in a moderate low-temperature signaling pathway in rice. *Plant Physiology* 129: 1880–1891.

Jin, U.H., Lee, J.W., Chung, Y.S., Lee, J.H., Yi, Y.B., Kim, Y.K., Hyung, N.I., Pyee, J.H., Chung, C.H. (2001) Characterization and temporal expression of a ω-6 fatty acid desaturase cDNA from sesame (*Sesamum indicum* L.) seeds. *Plant Science* 161: 935–941.

Jonak, C., Kiegerl, S., Ligterink, W., Barker, P.J., Huskisson, N.S., Hirt, H. (1996) Stress signaling in plants: A mitogen-activated protein kinase pathway is activated by cold and drought. *Proceedings of National Academy of Sciences USA* 93: 11274–11279.

Juntilla, O., Robberecht, R. (1999) Ecological aspects of cold-adapted plants with special emphasis on environmental control of cold hardening and dehardening. In *Cold-Adapted Organisms: Ecology, Physiology, Enzymology and Molecular Biology*, ed. Margesin, R. and Schinner, F., 57–77. Berlin: Springer-Verlag.

Kacperska, A. (1999) Plant response to low temperature: Signaling pathways involved in plant acclimation. In *Cold-Adapted Organisms: Ecology, Physiology, Enzymology and Molecular Biology*, ed. Margesin, R. and Schinner, F., 79–103. Berlin: Springer-Verlag.

Kalbarczyk, R. (2009) Potential reduction in cucumber yield (*Cucumis sativus* l.) in Poland caused by unfavourable thermal conditions of soil. *Acta Scientiarum Polonorum Hortorum Cultus* 8: 45–58.

Kang, H.M., Saltveit, M.E. (2002) Chilling tolerance of maize, cucumber and rice seedling leaves and roots are differentially affected by salicylic acid. *Physiologia Plantarum* 115: 571–576.

Kaniuga, Z. (2008) Chilling response of plants: Importance of galactolipase, free fatty acids and free radicals. *Plant Biology* 10: 171–184.

Kaplan, F., Guy, C.L. (2004) β-Amylase induction and the protective role of maltose during temperature shock. *Plant Physiology* 135: 1674–1684.

Kaplan, F., Kopka, J., Haskell, D.W., Zhao, W., Schiller, K.C., Gatzke, N., Sung, D.Y., Guy, C.L. (2004) Exploring the temperature-stress metabolome of *Arabidopsis*. *Plant Physiology* 136: 4159–4168.

Kaplan, F., Kopka, J., Sung, D.Y., Zhao, W., Popp, M., Porat, R., Guy, C.L. (2007) Transcript and metabolite profiling during cold acclimation of *Arabidopsis* reveals an intricate relationship of cold-regulated gene expression with modifications in metabolite content. *The Plant Journal* 50: 967–981.

Kasuga, M., Miura, S., Shinozaki, K., Yamaguchi-Shinozaki, K. (2004) A combination of the *Arabidopsis* DREB1A gene and stress-inducible *rd29A* promoter improved drought and low-temperature stress tolerance in tobacco by gene transfer. *Plant Cell Physiology* 45: 346–350.

Kim, M.C., Chung, W.S., Yun, D.J., Cho, M.J. (2009) Calcium and calmodulin-mediated regulation of gene expression in plants. *Molecular Plant* 2: 13–21.

Kim, Y., Park, S., Gilmour, S.J., Thomashow, M.F. (2013) Roles of CAMTA transcription factors and salicylic acid in configuring the low-temperature transcriptome and freezing tolerance of *Arabidopsis*. *Plant Journal* 75: 364–376.

Kim, J.S., Park, S.J., Kwak, K.J., Kim, Y.O., Kim, J.Y., Song, J., Jang, B., Jung, C.H., Kang, H. (2007) Cold shock domain proteins and glycine-rich RNA-binding proteins from *Arabidopsis* thaliana can promote the cold adaptation process in *Escherichia coli*. *Nucleic Acids Research* 35: 506–516.

Kim, J.M., Sasaki, T., Ueda, M., Sako, K., Seki, M. (2015) Chromatin changes in response to drought, salinity, heat, and cold stresses in plants. *Frontiers in Plant Science* 6: 114.

Kim, J.M., To, T.K., Ishida, J., Morosawa, T., Kawashima, M., Matsui, A., Toyoda, T., Kimura, H., Shinozaki, K., Seki, M. (2008) Alterations of lysine modifications on the histone H3 N-tail under drought stress conditions in *Arabidopsis thaliana*. *Plant and Cell Physiology* 49: 1580–1588.

King, R.W., Heide, O.M. (2009) Seasonal flowering and evolution: The heritage from Charles Darwin. *Functional Plant Biology* 36: 1027–1036.

Klimecka, M., Muszyńska, G. (2007) Structure and functions of plant calcium-dependent protein kinases. *Acta Biochimica Polonica* 54: 219–233.

Knight, M.R., Campbell, A.K., Smith, S.M., Trewavas, A.J. (1991) Transgenic plant aequorin reports the effects of touch and cold-shock and elicitors on cytoplasmic calcium. *Nature* 352: 524–526.

Knight, M.R., Knight, H. (2012) Low-temperature perception leading to gene expression and cold tolerance in higher plants. *New Phytologist* 195: 737–751.

Knight, H., Zarka, D.G., Okamoto, H., Thomashow, M.F., Knight, M.R. (2004) Abscisic acid induces *CBF* gene transcription and subsequent induction of cold-regulated genes via the CRT promoter element. *Plant Physiology* 135: 1710–1717.

Kodama, H., Hamada, T., Horiguchi, G., Nishimura, M., Iba, K. (1994) Genetic enhancement of cold tolerance by expression of a gene for chloroplast *w*-3 fatty acid desaturase in transgenic tobacco. *Plant Physiology* 105: 601–605.

Korner, C. (2016) Plant adaptation to cold climates. *F1000Research* 5: 2769. Doi: 10.12688/f1000research.9107.1.

Kosová, K., Holkova, L., Prasil, I.T., Prasilova, P., Bradacova, M., Vitamvas, P., Capkova, V. (2008) Expression of dehydrin 5 during the development of frost tolerance in barley (*Hordeum vulgare*). *Journal of Plant Physiology* 165: 1142–1151.

Kosová, K., Prášil, I.T., Vítámvás, P., Dobrev, P., Motyka, V., Floková, K., Novák, O., Turečková, V., Rolčik, J., Pešek, B., Trávničková, A., Gaudinová, A., Galiba, G., Janda, T., Vlasáková, E., Prášilová, P., Vanková, R. (2012) Complex phytohormone responses during the cold acclimation of two wheat cultivars differing in cold tolerance, winter Samanta and spring Sandra. *Journal of Plant Physiology* 169: 567–576.

Kosová, K., Vitamvas, P., Prasil, I.T. (2007) The role of dehydrins in plant response to cold. *Biologia Plantarum* 51: 601–617.

Kovacs, D., Kalmar, E., Torok, Z., Tompa, P. (2008) Chaperone activity of ERD10 and ERD14, two disordered stress-related plant proteins. *Plant Physiology* 147: 381–390.

Kumar, N., Bhatt, R.P. (2006) Transgenics: An emerging approach for cold tolerance to enhance vegetables production in high altitude areas. *Indian Journal of Crop Sciences* 1: 8–12.

Lang, V., Palva, E.T. (1992) The expression of a rab-related gene, rab18, is induced by abscisic acid during the cold acclimation process of *Arabidopsis thaliana* (L.) Heynh. *Plant Molecular Biology* 20: 951–962.

Larcher, W., Kainmüller, C., Wagner, J. (2010) Survival types of high mountain plants under extreme temperatures. *Flora* 205: 3–18.

Larkindale, J., Hall, J.D., Knight, M.R., Vierling, E. (2005) Heat stress phenotypes of *Arabidopsis* mutants implicate multiple signaling pathways in the acquisition of thermotolerance. *Plant Physiology* 138: 882–897.

Lee, O.R., Pulla, R.K., Kim, Y.J., Balusamy, S.R., Yang, D.C. (2012) Expression and stress tolerance of PR10 genes from *Panax ginseng* C.A. Meyer. *Molecular and Biological Reports* 39: 2365–2374.

Lee, C.M., Thomashow, M.F. (2012) Photoperiodic regulation of the C-repeat binding factor (CBF) cold acclimation pathway and freezing tolerance in *Arabidopsis thaliana*. *Proceedings of National Academic Sciences of USA* 109: 15054–15059.

Levitt, J. (1972) *Responses of Plants to Environmental Stress*. New York: Academic Press.

Levitt, J. (1980) *Responses of Plants to Environmental Stress, Chilling, Freezing and High Temperature Stresses*. New York: Academic Press.

Li, W., Wang, R., Li, M., Li, L., Wang, C., Welti, R., Wang, X. (2008) Differential degradation of extraplastidic and plastidic lipids during freezing and post-freezing recovery in *Arabidopsis thaliana*. *The Journal of Biological Chemistry* 283: 461–468.

Lissarre, M., Ohta, M., Sato, A., Miura, K. (2010) Cold-responsive gene regulation during cold acclimation in plants. *Plant Signaling and Behavior* 5: 948–952.

Liu, J.J., Ekramoddoullah, A.K.M., Yu, X. (2003) Differential expression of multiple PR10 proteins in western white pine following wounding, fungal infection and cold-hardening. *Physiologia Plantarum* 119: 544–553.

Lundmark, M., Cavaco, A.M., Trevanion, S., Hurry, V. (2006) Carbon partitioning and export in transgenic *Arabidopsis thaliana* with altered capacity for sucrose synthesis grown at low temperature: A role for metabolite transporters. *Plant, Cell and Environment* 29: 1703–1714.

Luo, M., Liu, X., Singh, P., Cui, Y., Zimmerli, L., Wu, K. (2012) Chromatin modifications and remodeling in plant abiotic stress responses. *Biochimica et Biophysica Acta* 1819: 129–136.

Maali, R., Schimschilaschvili, H.R., Pchelkin, V.P., Tsydendambaev, V.D., Nosov, A.M., Los, D.A., Goldenkova-Pavlova, I.V. (2007) Comparative expression in *Escherichia coli* of the native and hybrid genes for acyl-lipid Δ^9-desaturase. *Russian Journal of Genetics* 43: 121–126.

Maali-Amiri, R, Goldenkova-Pavlova, I.V., Yur'eva, N.O., Pchelkin, V.P., Tsydendambaev, V.D., Vereshchagin, A.G., Deryabin, A.N., Trunova, T.I., Los, D.A., Nosov, A.M. (2007) Lipid fatty acid composition of potato plants transformed with the $\Delta12$-desaturase gene from cyanobacterium. *Russian Journal of Plant Physiology* 54: 600–606.

Maeda, H., Song, W., Sage, T.L., DellaPenna, D. (2006) Tocopherols play a crucial role in low-temperature adaptation and phloem loading in *Arabidopsis*. *Plant Cell* 18: 2710–2732.

Mao, X., Chen, S., Li, A., Zhai, C., Jing, R. (2014) Novel NAC transcription factor TaNAC67 confers enhanced multi-abiotic stress tolerances in *Arabidopsis*. *PLoS ONE* 9: e84359. Doi: 10.1371/journal.pone.0084359.

Maruyama, K., Sakuma, Y., Kasuga, M., Ito, Y., Seki, M., Goda, H., Shimada, Y., Yoshida, S., Shinozaki, K., Yamaguchi-Shinozaki, K. (2004) Identification of cold-inducible downstream genes of the *Arabidopsis* DREB1A/CBF3 transcriptional factor using two microarray systems. *The Plant Journal* 38: 982–993.

Maruyama, K., Urano, K., Yoshiwara, K., Morishita, Y., Sakurai, N., Suzuki, H., Kojima, M., Sakakibara, H., Shibata, D., Saito, K., Shinozaki, K., Yamaguchi-Shinozaki, K. (2014) Integrated analysis of the effects of cold and dehydration on rice metabolites, phytohormones, and gene transcripts. *Plant Physiology* 164: 1759–1771.

Mastrangelo, A.M., Marone, D., Laido, G., de Leonardis, A.M., de Vita, P. (2012) Alternative splicing: Enhancing ability to cope with stress via transcriptome plasticity. *Plant Science* 185–186: 40–49.

Matsui, A., Ishida, J., Morosawa, T., Mochuzuki, Y., Kaminuma, E., Endo, T.A., Okamoto, M., Nambara, E., Nakajima, M., Kawashima, M., Satou, M., Kim, J.M., Kobayashi, N., Toyoda, T., Shinozaki, K., Seki, M. (2008) *Arabidopsis* transcriptome analysis under drought, cold, high-salinity and ABA treatment conditions using a tiling array. *Plant and Cell Physiology* 49: 1135–1149.

McAinsh, M.R., Pittman, J.K. (2009) Shaping the calcium signature. *New Phytologist* 181: 275–294.

McKersie, B.D., Leshem, Y.Y. (1994) *Stress and Stress Coping in Cultivated Plants*. Dordrecht, The Netherlands: Kluwer Academic.

Meng, X., Yin, B., Feng, H.L., Zhang, S., Liang, X.Q., Meng, Q.W. (2014) Overexpression of R2R3-MYB gene leads to accumulation of anthocyanin and enhanced resistance to chilling and oxidative stress. *Biologia Plantarum* 58: 121–130.

Mittler, R. (2006) Abiotic stress, the field environment and stress combination. *Trends in Plant Science* 11: 15–19.

Mittler, R., Vanderauwera, S., Gollery, M., Van Breusegem, F. (2004) Reactive oxygen gene network of plants. *Trends in Plant Science* 9: 490–498.

Miura, K., Jin, J.B., Lee, J., Yoo, C.Y., Stirm, V., Miura, T., Ashworth, E.N., Bressan, R.A., Yun, D.J., Hasegawa, P.M. (2007) SIZ1-mediated sumoylation of ICE1 controls *CBF3/DREB1A* expression and freezing tolerance in *Arabidopsis*. *Plant Cell* 19: 1403–1414.

Mizoi, J., Shinozaki, K., Yamaguchi-Shinozaki, K. (2012) AP2/ERF family transcription factors in plant abiotic stress responses. *Biochimica et Biophysica Acta* 1819: 86–96.

Moffatt, B., Ewart, V., Eastman, A. (2006) Cold comfort: Plant antifreeze proteins. *Physiologia Plantarum* 126: 5–16.

Mora-Herrera, M.E., López-Delgado, H., Castillo-Morales, A., Foyer, C.H. (2005) Salicylic acid and H_2O_2 function by independent pathways in the induction of freezing tolerance in potato. *Physiologia Plantarum* 125: 430–440.

Mustilli, A.C., Merlot, S., Vavasseur, A., Fenzi, F., Giraudat, J. (2002) *Arabidopsis* OST1 protein kinase mediates the regulation of stomatal aperture by abscisic acid and acts upstream of reactive oxygen species production. *Plant Cell* 14: 3089–3099.

Murata, N., Ishizaki-Nishizawa, O., Higashi, S., Hayashi, H., Tasaka, Y., Nishida, I. (1992) Genetically engineered alteration in the chilling sensitivity of plants. *Nature* 356: 710–713.

Murata, N., Wada, H. (1995) Acyl-lipid desaturases and their importance in the tolerance and acclimatization to cold of Cyanobacteria. *Biochemical Journal* 308: 1–8.

Nahar, K., Biswas, J.K., Shamsuzzaman, A.M.M. (2012) *Cold Stress Tolerance in Rice Plant: Screening of Genotypes based on Morphophysiological Traits*. Berlin: Lambert Academic Publishing.

Nahar, K., Biswas, J.K., Shamsuzzaman, A.M.M., Hasanuzzaman, M., Barman, H.N. (2009) Screening of indica rice (*Oryza sativa* L.) genotypes against low temperature stress. *Botany Research International* 2: 295–303.

Nakaminami, K., Hill, K., Perry, S.E., Sentoku, N., Long, J.A., Karlson, D.T. (2009) *Arabidopsis* cold shock domain proteins: Relationships to floral and silique development. *Journal of Experimental Botany* 60: 1047–1062.

Nakashima, K., Ito, Y., Yamaguchi-Shinozaki, K. (2009) Transcriptional regulatory networks in response to abiotic stresses in *Arabidopsis* and grasses. *Plant Physiology* 149: 88–95.

Nakashima, K., Yamaguchi-Shinozaki, K., Shinozaki, K. (2014) The transcriptional regulatory network in the drought response and its crosstalk in abiotic stress responses including drought, cold, and heat. *Frontiers in Plant Science* 5: 170.

Nakayama, K., Okawa, K., Kakizaki, T., Inaba, T. (2008) Evaluation of the protective activities of a late embryogenesis abundant (LEA) related protein, Cor15am, during various stresses in vitro. *Bioscience, Biotechnology and Biochemistry* 72: 1642–1645.

Nilsen, E.T., Orcutt, D.M. (1996) *The Physiology of Plants Under Stress: Abiotic Factors.* New York: John Wiley and Sons, Inc.

Nishiyama, I. (1995) Damage due to extreme temperatures. In *Science of the Rice Plant*, ed. Matsuo, T., Kumazawa, K., Ishii, R., Ishihara, H., Hirata, H., 769–812. Tokyo, Japan: Food and Agriculture Policy Research Center.

Nishizawa, A., Yabuta, Y., Shigeoka, S. (2008) Galactinol and raffinose constitute a novel function to protect plants from oxidative damage. *Plant Physiology* 147: 1251–1263.

Nover, L., Scharf, K.D. (1997) Heat stress proteins and transcription factors. *Cellular and Molecular Life Sciences* 53: 80–103.

Novillo, F., Alonso, J.M., Ecker, J.R., Salinas, J. (2004) CBF2/DREB1C is a negative regulator of *CBF1/DREB1B* and *CBF3/DREB1A* expression and plays a central role in stress tolerance in *Arabidopsis. Proceedings of National Academy of Sciences USA* 101: 3985–3990.

Novillo, F., Medina, J., Salinas, J. (2007) *Arabidopsis* CBF1 and CBF3 have a different function than CBF2 in cold acclimation and define different gene classes in the CBF regulon. *Proceedings of National Academy of Science USA* 104: 21002–21007.

Örvar, B.L., Sangwan, V., Omann, F., Dhindsa, R.S. (2000) Early steps in cold sensing by plant cells: The role of actin cytoskeleton and membrane fluidity. *Plant Journal* 23: 785–794.

Ouellet, F., Charron, J.B. (2013) *Cold Acclimation and Freezing Tolerance in Plants.* Chichester: John Wiley and Sons Ltd.

Pak, J.H., Chung, E.S., Shin, S.H., Jeon, E.H., Kim, M.J., Lee, H.Y., Jeung, J.U., Hyung, N.I., Lee, J.H., Chung, Y.S. (2009) Enhanced fungal resistance in *Arabidopsis* expressing wild rice PR-3 (OgChitIVa) encoding chitinase class IV. *Plant Biotechnology Reports* 3: 147–155.

Palusa, S.G., Ali, G.S., Reddy, A.S. (2007) Alternative splicing of pre-mRNAs of *Arabidopsis* serine/arginine-rich proteins: Regulation by hormones and stresses. *Plant Journal* 49: 1091–1107.

Park, S.J., Kwak, K.J., Oh, T.R., Kim, Y.O., Kang, H. (2009) Cold shock domain proteins affect seed germination and growth of *Arabidopsis thaliana* under abiotic stress conditions. *Plant Cell Physiology* 50: 869–878.

Park, S., Lee, C.M., Doherty, C.J., Gilmour, S.J., Kim, Y., Thomashow, M.F. (2015) Regulation of the *Arabidopsis* CBF regulon by a complex low-temperature regulatory network. *Plant Journal* 82: 193–207.

Pellegrineschi, A., Reynolds, M., Pacheco, M., Brito, R.M., Almeraya, R., Yamaguch-Shinozaki, K., Hoisington, D. (2004) Stress-induced expression in wheat of the *Arabidopsis thaliana* DREB1A gene delays water stress symptoms under greenhouse conditions. *Genome* 47: 493–500.

Penfield, S. (2008) Temperature perception and signal transduction in plants. *New Phytology* 179: 615–628.

Peng, Y., Arora, R., Li, G., Wang, X., Fessehae, A. (2008) *Rhododendron catawbiense* plasma membrane intrinsic proteins (RcPIPs) are aquaporins and their overexpression compromises constitutive freezing tolerance and cold acclimation ability of transgenic *Arabidopsis* plants. *Plant, Cell and Environment* 31: 1275–1289.

Pennycooke, J.C., Jones, M.L., Stushnoff, C. (2003) Down-regulating α-Galactosidase enhances freezing tolerance in transgenic Petunia. *Plant Physiology* 133: 901–909.

Pino, M.T., Skinner, J.S., Jeknic, Z., Hayes, P.M., Soeldner, A.H., Thomashow, M.F., Chen, T.H.H. (2008) Ectopic AtCBF1 over-expression enhances freezing tolerance and induces cold acclimation-associated physiological modifications in potato. *Plant Cell and Environment* 31: 393–406.

Porter, J.R., Semenov, M.A. (2005) Crop responses to climatic variation. *Philosophical Transactions Royal Society London B* 360: 2021–2035.

Preston, J.C., Sandve, S.R. (2013) Adaptation to seasonality and the winter freeze. *Frontiers in Science* 4: 167.

Qin, Y., Tian, Y., Liu, X. (2015) A wheat salinity-induced WRKY transcription factor TaWRKY93 confers multiple abiotic stress tolerance in *Arabidopsis thaliana. Biochemical and Biophysical Research Communications* 464: 428–433.

Rajashekar, C.B. (2000) Cold response and freezing tolerance in plants. In *Plant–Environment Interactions*, ed. Wilkinson, R.E., 321–341. New York: Marcel Dekker, Inc.

Rekarte-Cowie, I., Ebshish, O.S., Mohamed, K.S., Pearce, R.S. (2008) Sucrose helps regulate cold acclimation of *Arabidopsis thaliana. Journal of Experimental Botany* 59: 4205–4217.

Renaut, J., Hausman, J.F., Bassett, C., Artlip, T., Cauchie, H.M., Witters, E., Wisniewski, M. (2008) Quantitative proteomic analysis of short photoperiod and low-temperature responses in bark tissues of peach (*Prunus persica* L. Batsch). *Tree Genetics and Genomes* 4: 589–600.

Renaut, J., Hausman, J.F., Wisniewski, M.E. (2006) Proteomics and low temperature studies: Bridging the gap between gene expression and metabolism. *Physiologia Plantarum* 126: 97–109.

Renaut, J., Lutts, S., Hoffmann, L., Hausman, J.F. (2004) Responses of poplar to chilling temperatures: Proteomic and physiological aspects. *Plant Biology* 6: 81–90.

Riaz-ud-din Subhani, G.M., Ahmad, N., Hussain, M., Rehman, A.U. (2010) Effect of temperature on development and grain formation in spring wheat. *Pakistan Journal of Botany* 42: 899–906.

Richter, R., Bastakis, E., Schwechheimer, C. (2013) Cross-repressive interactions between SOC1 and the GATAs GNC and GNL/CGA1 in the control of greening, cold tolerance, and flowering time in *Arabidopsis. Plant Physiology* 162: 1992–2004.

Richter, R., Behringer, C., Müller, I.K., Schwechheimer, C. (2010) The GATA type transcription factors GNC and GNL/CGA1 repress gibberellin signaling downstream from DELLA proteins and phytochrome interacting factors. *Genes and Development* 24: 2093–2104.

Rohde, A., Bhalerao, R.P. (2007) Plant dormancy in the perennial context. *Trends in Plant Science* 12: 217–223.

Rorat, T. (2006) Plant dehydrins: Tissue location, structure and function. *Cell Molecular and Biology Letters* 11: 536–556.

Roy, D., Paul, A., Roy, A., Ghosh, R., Ganguly, P., Chaudhuri, S. (2014) Differential acetylation of histone H3 at the regulatory region of OsDREB1b facilitates chromatin remodeling and transcription activation during cold stress. *PLoS ONE* 9: e100343. Doi: 10.1371/journal.pone.0100343.

Saijo, Y., Hata, S., Kyozuka, J., Shimamoto, K., Izui, K. (2000) Over-expression of a single Ca^{2+}-dependent protein kinase confers both cold and salt/drought tolerance on rice plants. *Plant Journal* 23: 319–327.

Sakamoto, A., Sulpice, R., Hou, C.X., Kinoshita, M., Higashi, S.I., Kanaseki, T., Nonaka, H., Moon, B.Y., Murata, N. (2003) Genetic modification of the fatty acid unsaturation of phosphatidylglycerol in chloroplasts alters the sensitivity of tobacco plants to cold stress. *Plant Cell and Environment* 27: 99–105.

Sanders, D., Pelloux, J., Brownlee, C., Harper, J.F. (2002) Calcium at the crossroads of signaling. *Plant Cell* 14: S401–S417.

Sanghera, G.S., Wani, S.H., Hussain, W., Singh, N.B. (2011) Engineering cold stress tolerance in crop plants. *Current Genomics* 12: 30.

Sangwan, V., Örvar, B.L., Beyerly, J., Hirt, H., Dhindsa, R.S. (2002a) Opposite changes in membrane fluidity mimic cold and heat stress activation of distinct plant MAP kinase pathways. *Plant Journal* 31: 629–638.

Sangwan, V., Örvar, B.L., Dhindsa, R.S. (2002b) Early events during low temperature signaling. In *Plant Cold Hardiness*, ed. Li, C., Palva, E.T., 43–53. Dordrecht, The Netherlands : Kluwer Academic Publishers.

Santner, A., Calderon-Villalobos, L.I., Estelle, M. (2009) Plant hormones are versatile chemical regulators of plant growth. *Nature Chemical Biology* 5: 301–307.

Scott, I.M., Clarke, S.M., Wood, J.E., Mur, L.A. (2004) Salicylate accumulation inhibits growth at chilling temperature in *Arabidopsis*. *Plant Physiology* 135: 1040–1049.

Seo, P.J., Lee, A.K., Xiang, F., Park, C.M. (2008) Molecular and functional profiling of *Arabidopsis* pathogenesis-related genes: Insights into their roles in salt response of seed germination. *Plant Cell Physiology* 49: 334–344.

Shan, D.P., Huang, J.G., Yang, Y.T., Guo, Y.H., Wu, C.A., Yang, G.D., Gao, Z., Zheng, C.C. (2007) Cotton GhDREB1 increases plant tolerance to low temperature and is negatively regulated by gibberellic acid. *New Phytology* 176: 70–81.

Shinozaki, K., Yamaguchi-Shinozaki, K. (2007) Gene networks involved in drought stress response and tolerance. *Journal of Experimental Botany* 58: 221–227.

Solanke, A.U., Sharma, A.K. (2008) Signal transduction during cold stress in plants. *Physiology and Molecular Biology of Plants* 14: 69–79.

Somerville, C. (1995) Direct tests of the role of membrane lipid composition in low temperature-induce photoinhibition and chilling sensitivity in plants and cyanobacteria. *Proceedings of the National Academy of Science USA* 92: 6215–6218.

Somerville, J. (1999) Activities of cold-shock domain proteins in translation control. *Bio Essays* 21: 319–325.

Somerville, C., Browse, J. (1991) Plant lipids: Metabolism, mutants, and membranes. *Science* 252: 80–87.

Stockinger, E.J., Gilmour, S.J., Thomashow, M.F. (1997) *Arabidopsis thaliana CBF1* encodes an AP2 domain-containing transcriptional activator that binds to the C-repeat/DRE, a *cis*-acting DNA regulatory element that stimulates transcription in response to low temperature and water deficit. *Proceedings of the National Academy of Science USA* 94: 1035–1040.

Su, C.F., Wang, Y.C., Hsieh, T.H., Lu, C.A., Tseng, T.H., Yu, S.M. (2010) A novel MYBS3-dependent pathway confers cold tolerance in rice. *Plant Physiology* 153: 145–158.

Sui, N., Li, M., Zhao, S.J., Li, F., Liang, H., Meng, Q.W. (2007) Overexpression of glycerol-3-phosphate acyl transferase gene improves chilling tolerance in tomato. *Planta* 226: 1097–1108.

Sunkar, R., Li, Y.F., Jagadeeswaran, G. (2012) Functions of microRNAs in plant stress responses. *Trends in Plant Sciences* 17: 196–203.

Svensson, J.T., Crosatti, C., Campoli, C., Bassi, R., Stanca, A.M., Close, T.J., Cattivelli, L. (2006) Transcriptome analysis of cold acclimation in barley *albina* and *xantha* mutants. *Plant Physiology* 141: 257–270.

Takenaka, Y., Nakano, S., Tamoi, M., Sakuda, S., Fukamizo, T. (2009) Chitinase gene expression in response to environmental stresses in *Arabidopsis thaliana*: Chitinase inhibitor allosamidin enhances stress tolerance. *Bioscience Biotechnology Biochemistry* 73: 1066–1071.

Tasgin, E., Atici, O., Nalbantoglu, B., Popova, L.P. (2006) Effects of salicylic acid and cold treatments on protein levels and on the activities of antioxidant enzymes in the apoplast of winter wheat leaves. *Phytochemistry* 67: 710–715.

Teutonico, R.A., Osborn, T.C. (1995) Mapping loci controlling vernalization requirement in *Brassica rapa*. *Theoretical Applied Genetics* 91: 1279–1283.

Thakur, P., Kumara, S., Malika, J.A., Bergerb, J.D., Nayyar, H. (2010) Cold stress effects on reproductive development in grain crops: An overview. *Environmental and Experimental Botany* 67: 429–443.

Thion, L., Mazars, C., Thuleau, P., Graziana, A., Rossignol, M., Moreau, M., Ranjeva, R. (1996) Activation of plasma membrane voltage-dependent calcium-permeable channels by disruption of microtubules in carrot cells. *FEBS Letters* 393: 13–18.

Thomashow, M.F. (1998) Role of cold-responsive genes in plant freezing tolerance. *Plant Physiology* 118: 1–8.

Thomashow, M.F. (1999) Plant cold acclimation: Freezing tolerance genes and regulatory mechanisms. *Annual Review of Plant Physiology* 50: 571–599.

Thomashow, M.F. (2001) So what's new in the field of plant cold acclimation? Lots! *Plant Physiology* 125: 89–93.

Thomashow, M.F. (2010) Molecular basis of plant cold acclimation: Insights gained from studying the CBF cold response pathway. *Plant Physiology* 154: 571–577.

Timperio, A.M., Egidi, M.G., Zolla, L. (2008) Proteomics applied on plant abiotic stresses: Role of heat shock proteins (HSP). *Journal of Proteomics* 71: 391–411.

To, K.T., Nakaminami, K., Kim, J.M., Morosawa, T., Ishida, J., Tanaka, M., Yokoyama, S., Shinozaki, K., Seki, M. (2011) *Arabidopsis HDA6* is required for freezing tolerance. *Biochemical Biophysics Research Communications* 406: 414–419.

Tommasini, L., Svensson, J.T., Rodriguez, E.M., Wahid, A., Malatrasi, M., Kato, K., Wanamaker, S., Resnik, J., Close, T.J. (2008) Dehydrin gene expression provides an indicator of low temperature and drought stress: transcriptome-based analysis of barley (*Hordeum vulgare* L.). *Functional and Integrative Genomics* 8: 387–405.

Tonkinson, C.L., Lyndon, R.F., Arnold, G.M., Lenton, J.R. (1997) The effects of temperature and the Rht3 dwarfing gene on growth, cell extension, and gibberellin content and responsiveness in the wheat leaf. *Journal of Experimental Botany* 48: 963–970.

Townley, H.E., Knight, M.R. (2002) Calmodulin as a potential negative regulator of *Arabidopsis* COR gene expression. *Plant Physiology* 128: 1169–1172.

Uemura, M., Steponkus, P.L. (2003) Modification of the intracellular sugar content alters the incidence of freeze-induced membrane lesions of protoplasts isolated from *Arabidopsis thaliana* leaves. *Plant Cell and Environment* 26: 1083–1096.

Uemura, M., Tominaga, Y., Nakagawara, C., Shigematsu, S., Minami, A., Kawamura, Y. (2006) Responses of the plasma membrane to low temperatures. *Physiologia Plantarum* 126: 81–89.

Ukaji, N., Kuwabara, C., Takezawa, D., Arakawa, K., Yoshida, S., Fujikawa, S. (1999) Accumulation of small heat-shock protein homologs in the endoplasmic reticulum of cortical parenchyma cells in mulberry in association with seasonal cold acclimation. *Plant Physiology* 120: 481–489.

Usadel, B., Blasing, O.E., Gibon, Y., Retzlaff, K., Hoehne, M., Gunther, M., Stitt, M. (2008) Multilevel genomic analysis of the response of transcripts, enzyme activities and metabolites in *Arabidopsis* rosettes to a progressive decrease of temperature in the non-freezing range. *Plant, Cell and Environment* 31: 518–547.

Vargas, W.A., Pontis, H.G., Salerno, G.L. (2007) Differential expression of alkaline and neutral invertases in response to environmental stresses: Characterization of an alkaline isoform as a stress-response enzyme in wheat leaves. *Planta* 226: 1535–1545.

Vaultier, M.N., Cantrel, C., Vergnolle, C., Justin, A.M., Demandre, C., Benhassaine-Kesri, G., Cicek, D., Zachowski, A., Ruelland, E. (2006) Desaturase mutants reveal that membrane rigidification acts as a cold perception mechanism upstream of the diacylglycerol kinase pathway in *Arabidopsis* cells. *FEBS Letters* 580: 4218–4223.

Vogel, J.T., Zarka, D.G., Van Buskirk, H.A., Fowler, S.G., Thomashow, M.F. (2005) Roles of the CBF2 and ZAT12 transcription factors in configuring the low temperature transcriptome of *Arabidopsis*. *Plant Journal* 41: 195–211.

Wada, H., Gombos, Z., Murata, N. (1994) Contribution of membrane lipids to the ability of the photosynthetic machinery to tolerate temperature stress. *Proceedings of the National Academy of Science USA* 91: 4273–4277.

Wan, S.B., Tian, L., Tian, R.R., Pan, Q.H., Zhan, J.C., Wen, P.F., Chen, J.Y., Zhang, P., Wang, W., Huang, W.D. (2009) Involvement of phospholipase D in the low temperature acclimation-induced thermo tolerance in grape berry. *Plant Physiology and Biochemistry* 47: 504–510.

Wang, B.B., Brendel, V. (2006) Genome wide comparative analysis of alternative splicing in plants. *Proceedings of the National Academy of Science USA* 103: 7175–7180.

Wang, R.K., Cao, Z.H., Hao, Y.J. (2014) Overexpression of a R2R3 MYB gene MdSIMYB1 increases tolerance to multiple stresses in transgenic tobacco and apples. *Physiologia Plantarum* 150: 76–87.

Wang, X., Li, W., Li, M., Welti, R. (2006) Profiling lipid changes in plant response to low temperatures. *Physiologia Plantarum* 126: 90–96.

Wang, W., Vinocur, B., Altman, A. (2003) Plant responses to drought, salinity and extreme temperatures: Towards genetic engineering for stress tolerance. *Planta* 218: 1–14.

Wang, H., Wang, H., Shao, H., Tang, X. (2016) Recent advances in utilizing transcription factors to improve plant abiotic stress tolerance by transgenic technology. Frontiers in Plant Science 7: 67.

Wani, S.H., Gosal, S.S. (2011) Introduction of OsglyII gene into Indica rice through particle bombardment for increased salinity tolerance. *Biologia Plantarum* 55: 536–540.

Wani, S.H., Sandhu, J.S., Gosal, S.S. (2008) Genetic engineering of crop plants for abiotic stress tolerance. In *Advanced Topics in Plant Biotechnology and Plant Biology*, ed. Malik, C.P., Kaur, B., Wadhwani, C., 149–183. New Delhi: MD Publications.

Wolters, H., Jürgens, G. (2009) Survival of the flexible: Hormonal growth control and adaptation in plant development. *Nature Review Genetics* 10: 305–317.

Xin, Z., Browse, J. (2001) Cold comfort farm: The acclimation of plants to freezing temperatures. *Plant Cell and Environment* 23: 893–902.

Yadav, S.K. (2010) Cold stress tolerance mechanisms in plants: A review. *Agronomy for Sustainable Development* 30: 515–527.

Yamaguchi-Shinozaki, K., Shinozaki, K. (2006) Transcriptional regulatory networks in cellular responses and tolerance to dehydration and cold stresses. *Annual Reviews of Plant Biology* 57: 781–803.

Yan, S.P., Zhang, Q.Y., Tang, Z.C., Su, W.A., Sun, W.N. (2006) Comparative proteomic analysis provides new insights into chilling stress responses in rice. *Molecular and Cell Proteomics* 5: 484–496.

Yang, X., Wang, X., Lu, J., Yi, Z., Fu, C., Ran, J., Hu, R., Zhou, G. (2015) Overexpression of a *Miscanthus lutarioriparius* NAC gene MlNAC5 confers enhanced drought and cold tolerance in *Arabidopsis*. *Plant Cell Reports* 34: 943–958.

Yano, R., Nakamura, M., Yoneyama, T., Nishida, I. (2005) Starch-related α-glucan/water dikinase is involved in the cold-induced development of freezing tolerance in *Arabidopsis*. *Plant Physiology* 138: 837–846.

Yoo, S.D., Cho, Y., Sheen, J. (2009) Emerging connections in the ethylene signaling network. *Trends in Plant Sciences* 14: 270–279.

Yu, X.M., Griffith, M., Wiseman, S.B. (2001) Ethylene induces antifreeze activity in winter rye leaves. *Plant Physiology* 126: 1232–1240.

Zhang, J.Z., Creelman, R.A., Zhu, J.K. (2004) From laboratory to field. Using information from *Arabidopsis* to engineer salt, cold, and drought tolerance in crops. *Plant Physiology* 135: 615–621.

Zhang, J., Xu, Y., Huan, Q., Chong, K. (2009) Deep sequencing of *Brachypodium* small RNAs at the global genome level identifies microRNAs involved in cold stress response. *BMC Genomics* 10: 449.

Zhang, R., Wang, Y., Liu, G., Li, H. (2010) Cloning and characterization of a pathogenesis-related gene (ThPR10) from *Tamarix hispida*. *Acta Biologica Cracoviensia Series Botanica* 52: 17–25.

Zhang, L., Zhang, L., Xia, C., Zhao, G., Liu, J., Jia, J., Kong, X. (2015) A novel wheat bZIP transcription factor, TabZIP60, confers multiple abiotic stress tolerances in transgenic *Arabidopsis*. *Physiologia Plantarum* 153: 538–554.

Zhao, M., Liu, W., Xia, X., Wang, T., Zhang, W.H. (2014) Cold acclimation-induced freezing tolerance of *Medicago truncatula* seedlings is negatively regulated by ethylene. *Physiologia Plantarum* 152: 115–129.

Zhou, M., Xu, M., Wu, L., Shen, C., Ma, H., Lin, J. (2014) *CbCBF* from *Capsella bursa-pastoris* enhances cold tolerance and restrains growth in *Nicotiana tabacum* by antagonizing with gibberellin and affecting cell cycle signaling. *Plant Molecular and Biology* 85: 259–275.

Zhou, X., Wang, G., Sutoh, K., Zhu, J.K., Zhang, W. (2008) Identification of cold-inducible microRNAs in plants by transcriptome analysis. *Biochemistry Biophysics Acta* 1779: 780–788.

Zhu, J., Dong, C.H., Zhu, J.K. (2007) Interplay between cold-responsive gene regulation, metabolism and RNA processing during plant cold acclimation. *Current Opinion in Plant Biology* 10: 290–295.

Zhu, J., Jeong, J.C., Zhu, Y., Sokolchik, I., Miyazaki, S., Zhu, J.K., Hasegawa, P.M., Bohnert, H.J., Shi, H., Yun, D.J., Bressan, R.A. (2008) Involvement of *Arabidopsis* HOS15 in histone deacetylation and cold tolerance. *Proceedings of the National Academy of Science USA* 105: 4945–4950.

8 Unraveling the Molecular and Biochemical Mechanisms of Cold Stress Tolerance in Rice

Joseph Msanne, Lymperopoulos Panagiotis, Roel C. Rabara, and Supratim Basu

CONTENTS

8.1   INTRODUCTION

Climate variability is the paramount factor limiting the primary productivity of terrestrial plants and has major impacts on the growth and productivity of crops worldwide. The importance of crop resistance to the resulting abiotic stress, including drought and temperature extremes, is likely to increase further as the range of environments in which crops are cultivated expands, mostly due to a continuous increase in the human population. Crops are now cultivated on marginal lands and are increasingly farmed at higher altitudes and latitudes that encounter more extreme seasonal temperature

variations (Dong et al., 2004; Funatsuki and Ohnishi, 2009; Zinn et al., 2010). Several plant species of tropical and subtropical origin experience cold stress when the ambient temperature drops below the optimal growth temperature, even at moderate, non-freezing temperatures (Graham and Patterson, 1982; Hahn and Walbot, 1989). Under stress conditions, morphological, physiological, biochemical, and molecular alterations occur in plants and often result in growth retardation and yield reduction. Due to climate change, more frequent and unpredictable extreme temperature events are expected worldwide (IPCC, 2007). The regular occurrence of cold waves, which vary in severity and duration in various geographical locations, is negatively affecting agricultural production (Marengo and Camargo, 2008; Zinn et al., 2010). Low temperatures are particularly damaging to the rice (*Oryza sativa*) plant, owing to its origin in tropical regions (Xie et al., 2012). Rice is the staple cereal food source for more than half of the world's population (da Cruz et al., 2013). Depending on the cultivar, growth stage, and duration (Schwender et al., 2004; Li et al., 1981), cold stress can severely limit rice yields, and negatively affect grain quality (Jena et al., 2012). It is well known that intraspecific variability for cold susceptibility exists among rice varieties. These significantly differ in their ability to tolerate low temperatures, with *japonica* subspecies more tolerant to cold stress than *indica* (Hahn and Walbot, 1989). *Japonica* cultivars are better adapted to temperate and high altitude regions, frequently experiencing lower temperatures (Mackill and Lei, 1997).

Cold stress can negatively impact rice growth at any developmental stage between germination, maturity, and grain fill (Ye et al., 2009; da Cruz et al., 2013). The optimal growth temperatures for rice cultivation range between 25°C and 35°C (De Los Reyes et al., 2003). In some cultivars, exposure to temperatures lower than 20°C can have damaging effects on seed germination. It was also reported that in some cold-sensitive cultivars, temperatures lower than 18°C caused about 50% sterility, which reached 100% at 10°C, leading to serious yield losses (Ndour et al., 2016). Some of the phenotypic indications of cold stress include reduced growth and leaf expansion (Ali et al., 2006), reduced tillering (Shimono et al., 2002), chlorosis, and increased plant mortality (Farrell et al., 2006). In addition, rice plants experiencing low temperatures, from 0°C to 18°C, exhibit physiological and metabolic alterations. The resulting major damages include loss of membrane fluidity leading to ion leakage and the disruption of electron transport across thylakoids, which negatively affect photosynthesis and CO_2 uptake (Graham and Patterson, 1982; Hahn and Walbot, 1989; Zinn et al., 2010). The accumulation of reactive oxygen species (ROS) and malondialdehyde (MDA) under cold stress leads to cellular oxidative damage (Xie et al., 2009; Nakashima et al., 2007). This results in changes in the structure and function of enzymes, while destabilizing cell membranes and other cellular components (Kubien et al., 2003). ROS can also play a signaling role by triggering transcriptional changes and the reprogramming of gene expression in response to cold stress conditions (Cook et al., 2004; Zinn et al., 2010). Some rice cultivars can cope by adopting strategies that allow them to enhance their tolerance to cold stress (Kandpal and Rao, 1985). For example, cold-treated rice can accumulate various metabolites and osmolytes including sugars, amino acids, polyamines, and lipids (Nayyar et al., 2005; Farooq et al., 2009), which can regulate water content and cell dehydration (Kandpal and Rao, 1985), and improve plant survival under abiotic stress.

Rice cultivars are grown in vast regions across the world, which makes tolerance to abiotic stress a necessity. Cold-tolerant rice cultivars need to be bred and/or engineered for a particular region. The use of molecular markers and linkage maps has enabled the identification of major quantitative trait loci (QTLs) on the rice genome that may be associated with cold tolerance (Andaya and Mackill, 2003b; Zhang et al., 2005; Jiang et al., 2008; da Cruz et al., 2013). This can facilitate the breeding of new cold-tolerant cultivars. Although conventional plant-breeding programs may improve yields for rice grown in stressful environments, there is a growing belief that further gains can only be achieved through targeted manipulation of genes involved in stress resistance. In rice, various genes have been found to be regulated in response to cold stress, including several transcription factors (TFs) that play a major role in improving cold tolerance when overexpressed in rice plants (Beer and Tavazoie, 2004; Morsy et al., 2005; Benedict et al., 2006). Other genes used for engineering cold stress tolerance in rice include enzymes involved in osmolytes biosynthesis (McNeil et al., 1999), components of signal transduction pathways (Xie et al., 2012), chaperones, and membrane transporters (Mittal et al., 2009). Cold tolerance is a complex trait. This chapter seeks to update our knowledge and understanding of cold tolerance in rice plants at different developmental stages. We will discuss the response mechanisms of rice under cold stress, with emphasis on the role of various metabolites and enzymes in cold tolerance. We will also review the QTLs and various genes identified from rice cultivars, which are used to facilitate the production of cold-tolerant rice.

8.2 MORPHOLOGICAL AND PHYSIOLOGICAL RESPONSES OF RICE PLANTS TO COLD STRESS

Throughout their life cycle, land plants are exposed to various environmental stress conditions that negatively affect their growth and development. Low temperature is considered a major abiotic factor that can limit crop geographical distribution and productivity (Zhu et al., 2007). In plants, low temperature stress can result in chilling injury (0°C–12°C) or freezing stress, when the temperature drops below 0°C (Zhu et al., 2007). Understanding the response mechanisms of plants to environmental stress conditions at the morphological, physiological, and biochemical levels is essential to overcome the adverse effects of low temperatures. Rice is grown in areas ranging from tropical and subtropical to temperate (Sharifi, 2010; Kim and Tai, 2011). Despite its wide adaptability to cold and the availability of cold-tolerant cultivars, low temperatures still have major damaging effects on rice in several regions of the world (Datta and Datta, 2006), affecting grain yield and quality (Ndour et al., 2016). Studies using a large number of cultivars belonging to both rice subspecies have shown clear varietal differences in the extent of cold tolerance, with *japonica* cultivars showing better adaptation to low temperature stress than *indica* when challenged at different phenological stages including germination, vegetative stage, flowering, and seed set (Li et al., 1981; Yoshida, 1981; Mackill and Lei, 1997; Sharifi, 2010; Kim and Tai, 2011; da Cruz et al., 2013; Zhang et al., 2014). Although the critical temperature varies among different rice cultivars, cold stress is most damaging when it occurs at the pollination stage, resulting in partial or complete pollen sterility, which can cause severe yield loss especially in sensitive cultivars (Hatfield and Prueger, 2015). Cold stress-induced male sterility in rice may result from microsporogenesis inhibition at the booting stage, leading to pollen grain degeneration (Sharifi, 2010; da Cruz et al., 2013; Ndour et al., 2016). Besides reducing rice yields, low temperatures occurring at the reproductive stage induce poor grain filling, lowering grain quality and affecting milling properties (Ndour et al., 2016). It was also found that exposure to low temperatures during grain filling led to an increase in the ratio of amylose to amylopectin (Wilson et al., 2004). Cold tolerance in rice has often been evaluated by screening and identifying cultivars and genotypes with higher survival rates at the seedling stage (Morsy et al., 2007), reduced chlorosis and leaf withering as symptoms of chilling injuries during the vegetative growth stage (Nagamine, 1991), and reduced pollen sterility during the reproductive and flowering stages (Satake, 1969). Such cold-tolerant genotypes may offer great agronomic potential, and can serve as genetic donors in breeding programs.

8.2.1 Effects of Low Temperature Stress on Photosynthesis, and Chlorophyll Content and Fluorescence

Low temperature stress can negatively affect plant growth and development by impairing photosynthesis and CO_2 assimilation, thereby reducing yields by decreasing the level of carbohydrates needed for grain filling (Smillie et al., 1988). Exposure to low temperatures over extended periods can ultimately result in loss of membrane integrity, which impairs photosynthesis and general metabolic processes. In rice, cold stress has detrimental effects on chlorophyll content and fluorescence (Kanneganti and Gupta, 2008; Kim et al., 2009). During the vegetative growth, cold exposure can inhibit chloroplast formation and significantly decrease chlorophyll content in the leaves of cold-sensitive rice cultivars (Aghaee et al., 2011; da Cruz et al., 2013). Gas exchange measurements show that under these conditions the photosynthetic rate and stomatal conductance are considerably reduced (Wang and Guo, 2005; Saad et al., 2010). Cold exposure progressively decreases the abundance of the mRNA transcripts for both Rubisco subunits, which decreases the protein synthesis in the leaves of cold-sensitive rice (Hahn and Walbot, 1989). Under high light conditions, cold stress selectively inhibits Photosystem II (PSII), while PSI is more stable (Paredes and Quiles, 2015). In addition, non-photochemical quenching (NPQ) and photosynthetic electron transport decrease in rice seedlings when shoots, but not roots, are subjected to cold stress (Suzuki et al., 2011). Changes in chlorophyll fluorescence also indicate whether cold stress has compromised the plant photosynthetic properties (McFarlane et al., 1980). This parameter is assessed by measuring the ratio of variable fluorescence to maximum fluorescence (Fv/Fm), also known as the maximum quantum efficiency of PSII (McFarlane et al., 1980). During cold stress, Fv/Fm values significantly decrease in cold-sensitive compared to cold-tolerant rice cultivars (Zahedi and Alahrnadi, 2007; Bonnecarrère et al., 2011). Chlorophyll content and fluorescence are commonly used to quantify the degree of tolerance or sensitivity of rice cultivars and transgenic lines to low temperature stress during the vegetative stage (Tian et al., 2011), and to assess rice plant recovery after cold exposure (Kuk et al., 2003).

8.2.2 Cold Acclimation in Rice through Osmotic Adjustment and Antioxidant Response Mechanisms

Mass spectrometry (MS)-based analyses of metabolomic changes have identified and characterized several metabolites that accumulate in plants under stress conditions (Urano et al., 2010; Obata and Fernie, 2012; Maruyama et al., 2014). Some of the cold response mechanisms include the accumulation of cryoprotectants, soluble sugars, and proline, an increase in antioxidant activities, and changes in the lipid composition of the cell membranes (Xin and Browse, 2000). These modifications have a primary function in enhancing the stability of cellular membranes, as well as adjusting the osmotic potential and protecting the plant from further damage due to cold stress (Mahajan and Tuteja, 2005). In *Arabidopsis thaliana*, the levels of several soluble sugars and amino acids increased under low temperature stress, significantly enhancing the plant cold stress tolerance (Cook et al., 2004; Hannah et al., 2006).

In rice, low temperature stress initiates physiological changes in cold-treated plants compared to untreated plants. Some modifications include an increased accumulation of metabolites such as glucose 6-phosphate, glucose, fructose, sucrose, trehalose, and raffinose (Maruyama et al., 2014). These soluble metabolites serve as osmoprotectants against freezing-dehydration damage (Nagao et al., 2005; Yuanyuan et al., 2010; Zhang et al., 2014). The abundance of several metabolites in contrasting rice cultivars has been linked to the variations in their capacity to tolerate low temperatures (da Cruz et al., 2013). In cold-sensitive rice cultivars, the increase in the accumulation of hexoses and sucrose in anthers is linked to starch degradation under stress conditions, which increases pollen sterility (da Cruz et al., 2013). Starch degradation enzymes are upregulated in rice plants upon exposure to cold or drought stress (Maruyama et al., 2014). In cold-tolerant rice cultivars, sugar accumulation is not observed in anthers upon exposure to cold stress (da Cruz et al., 2013). The accumulation of several amino acids including valine, leucine, isoleucine, and proline, as well as quaternary ammonium compounds such as glycine-betaine is also enhanced by cold stress (Maruyama et al., 2014). The increased proline content observed in rice varieties under low temperature correlates with cold stress tolerance (Kim and Tai, 2011). Proline concentration is essential for the osmotic adjustment of the cells under stress, which can lead to dehydration by osmotic pressure (da Cruz et al., 2013). Proline is also known to protect proteins and enzymes from denaturation (Shah and Dubey, 1997). To enhance cold tolerance, the overexpression of various genes encoding for enzymes involved in osmolytes biosynthesis has been tested in rice (McNeil et al., 1999; Ito et al., 2006; da Cruz et al., 2013). The overexpression of *OsPRP3* increased cold tolerance by accumulating free proline in transgenic rice (Gothandam et al., 2010), while the overexpression of *OsTPP1*, encoding for trehalose-6P phosphatase in rice, enhanced cold tolerance by increasing trehalose accumulation and the activation of stress-responsive genes (Garg et al., 2002).

During cold stress, ROS also accumulate in rice plants (Saruyama and Tanida, 1995; Song et al., 2011; da Cruz et al., 2013). Some of the resulting damage includes degradation of the membrane lipids forming toxic MDA and increased ion leakage, subsequently affecting photosynthesis and causing cell injury (Pamplona, 2011; da Cruz et al., 2013; Zhang et al., 2014). ROS molecules including superoxide ($\cdot O_2^-$), hydroxyl radical (HO^-), and hydrogen peroxide (H_2O_2) are normally produced at low levels as metabolic by-products mainly in the chloroplast and mitochondria. However, under stress conditions, ROS significantly accumulate in the cell leading to severe damage from oxidative stress (Apel and Hirt, 2004; Skopelitis et al., 2006). The evaluation of membrane damage under cold stress has been used to assess the degree of cold tolerance or sensitivity among rice cultivars at different developmental stages (Lee et al., 2004; Morsy et al., 2007; Song et al., 2011; Zhang et al., 2011; Yang et al., 2012a, da Cruz et al., 2013). It was found that under low temperature stress, *japonica* cultivars exhibited less membrane damage and ion leakage compared to *indica* cultivars, which showed higher MDA levels (Kim and Tai, 2011). Differences in the content of saturated and unsaturated fatty acids (FAs) in cellular membranes can be linked to their stability upon exposure to low temperatures. Tolerant cultivars exhibit higher levels of unsaturated FAs (e.g., linolenic acid) and lower saturated FAs compared to sensitive cultivars (da Cruz et al., 2013). The accumulation of ROS and MDA can mediate cold sensing in rice by regulating the cold-responsive signaling network via the *OsMKK6–OsMPK3* pathway (Xie et al., 2009). This leads to the expression of various genes including those encoding for antioxidant enzymes and ROS scavengers (Theocharis et al., 2012). Rice exposed to low temperatures activates the antioxidant systems crucial for plant defense against oxidative stress (Noctor and Foyer, 1998; Mittler, 2002). Such antioxidants occurring at high concentrations in different cell compartments include enzymes such as superoxide dismutase (SOD), catalase (CAT), peroxidase (POD), and ascorbate peroxidase (APX), as well as nonenzymatic compounds such as ascorbic acid (AsA) and glutathione (GSH) (Xie et al., 2009; Kim and Tai, 2011). Engineering increased expression of components from the antioxidant system has been used to improve cold tolerance. Transgenic rice plants overexpressing various antioxidant enzymes, including *OsAPXa*, *OsPOX1*, and *Sodc1*, and the kinase *OsTrx23* (Lee et al., 2009; Kim et al., 2011; Sato et al., 2011a; Xie et al., 2012), showed improved tolerance to cold and oxidative stress, with increased ROS scavenging activity and improved spikelet fertility (Sato et al., 2011a).

8.3 TRANSCRIPTION FACTORS: MAJOR REGULATORS OF COLD TOLERANCE IN RICE

Under stress conditions, plants alter their morphological, physiological, biochemical, and molecular makeup to deter growth retardation and yield reduction. Among the molecular adaptation mechanisms of plants is the expression of TFs, the major regulator genes that respond to abiotic and biotic stresses upsetting plant productivity and survival. As master regulators of many cellular processes, TFs are DNA-binding proteins that interact with other transcriptional regulators, including chromatin remodeling/modifying proteins, to recruit or block transcriptions (Century et al., 2008). Various TF families that respond to biotic and abiotic stresses, as well as the diverse functions of TFs in plant development and those specific to drought tolerance, have been appraised (Rabara et al., 2014), including TFs that regulate defense responses to pests and diseases. Subsequent discussions, however, will focus on TF families that play a principal role in cold tolerance defense and improvement when overexpressed in rice plants. Of the 2604 genes upregulated in response to chilling in *japonica* rice, about 6% (148) have been estimated to be TFs (Zhang et al., 2012a). Additionally, these putative TFs, belonging to APETALA 2/ethylene-responsive element-binding factor (AP2/ERF), basic helix-loop-helix (bHLH), basic region leucine zipper (bZIP), MYB, NAC, and WRKY families, were characterized by waves of induction at different time periods. An earlier study showed that 196 (8.2%) of the 2384 annotated genes in the rice genome were differentially regulated by cold with TFs identified as 26 AP2/EREBP, 21 bHLH, 19 MYB, 12 NAC, and 17 WRKY genes. About 14 bZIP TFs in rice have been identified or functionally characterized (Nijhawan et al., 2008).

8.3.1 AP2/ERF FAMILY OF TRANSCRIPTION FACTORS

The AP2/ERF family is one of the largest families of TFs involved in plant response to abiotic stress (Rushton et al., 2012; Rabara et al., 2014). The AP2/ERF family is composed of several subfamilies including AP2, RAV, ERF, DREB (dehydration-responsive element-binding) protein, and others (AL079349) (Mizoi et al., 2012). The ERF family is further itemized into 14 groups (I–XIV) based on gene structure, phylogeny, and conserved motifs. About 70 amino acids, known as the AP2/ERF domain, are involved in DNA binding (Nakano et al., 2006). About 163 loci in rice encode family members of the AP2/ERF-type DNA-binding domains (Sharoni

et al., 2011). The same study reported the identification of 52 and 53 non-redundant AP2/ERF genes that were upregulated in response to abiotic and biotic stresses, respectively, based on 44 K and 22 K microarray analysis results. Through reverse transcription polymerase chain reaction (RT-PCR) analysis, three rice genes (*Os04g32620*, *Os09g13940*, and *Os04g34970*) were upregulated and one (*Os07g22730*) was downregulated while rice seedling samples were under cold stress treatments for 24 h, 48 h, and 72 h. Moreover, a study on the transcriptional regulatory network responding to chilling stress revealed that about 18% (31) of AP2/ERF genes in rice were induced by chilling and most ERF-type belonged to nine phylogenetic groups (Yun et al., 2010). A recent review has identified studies on the AP2/ERF genes, summarized information about their physiological mechanisms under stress conditions in rice, and concluded that available information was still limited, suggesting that further physiological studies be conducted in rice to identify additional features of this crucial gene family.

8.3.2 bHLH FAMILY OF TRANSCRIPTION FACTORS

bHLH is one of the largest TF families found in plants as well as in animals and fungi, and represents key regulatory components in transcriptional networks controlling several biological processes (Carretero-Paulet et al., 2010). bHLH proteins play crucial roles in cell proliferation, determination, and differentiation in animals, plants, and yeast, like transcription activation activity in yeast and plants. In a genome-wide analysis of the bHLH TF family in rice and *Arabidopsis*, 167 bHLH genes were identified in the rice genome (Li et al., 2006). Their phylogenetic analysis indicates the formation of well-supported clades, which are defined as subfamilies. Although the study did not reveal any bHLH gene particular to cold tolerance in rice, bioinformatics analysis suggests that rice bHLH proteins can potentially participate in a variety of combinatorial interactions, endowing them with the capacity to regulate a multitude of transcriptional programs. Specific to cold stress, one bHLH-type gene, *OsbHLH1*, isolated from rice was found to have a putative nuclear-localization signal and a putative DNA-binding domain bHLH-ZIP (Wang et al., 2003). Additionally, *OsbHLH1* was reported to have dimerization ability and TF function in a cold signal transduction pathway. In another study, the bHLH protein gene *OrbHLH001* isolated from Dongxiang wild rice was characterized and expressed in transgenic *Arabidopsis*. *OrbHLH001* expression analyses showed enhance tolerance to freezing as well as salt tolerance of transgenic *Arabidopsis*. A native of China's Jiangxi Province, Dongxiang wild rice (*Oryza rufipogon*) is known for its high tolerance to cold stress and to winter temperatures as low as –12.8°C (Li et al., 2010), making it a valuable germplasm for cold tolerance in rice (Table 8.1).

8.3.3 bZIP FAMILY OF TRANSCRIPTION FACTORS

The bZIP family is exclusively present in eukaryotes and reportedly regulates diverse plant functions like stress responses, growth, and development (Deppmann et al., 2004). The bZIP TF family is characterized by a 60–80 amino acid long conserved domain composed of two motifs: (1) the basic region that functions as the DNA-binding domain and (2) a leucine zipper that is responsible for the TF dimerization (Corrêa et al., 2008). In rice, 13 groups of TFs are classified as bZip, among these are groups A, C, and S reported to participate in abiotic stress signaling. Group A members are involved in abscisic acid (ABA) and stress signaling and have been studied extensively (Corrêa et al., 2008). Group C has three members that respond to abiotic stress, namely, *OsbZIP20*, *OsbZIP33*, and *OsbZIP88* (Nijhawan et al., 2008). Although the largest and most poorly characterized, Group S has *lip19/OsbZIP38* and *OBF1/OsbZIP37*, both well-known participants in the cold signaling pathways in rice. The results of a genomic survey and gene expression analysis of the bZIP TF family in rice revealed that two *OsbZIP* genes, *OsbZIP23* and *OsbZIP45*, were upregulated under dehydration, salinity, and cold stress conditions (Nijhawan et al., 2008). Findings revealed that *OsbZIP45* is a closely related ortholog of maize GBF1, known to be induced by hypoxia, while *OsbZIP23* was found to be downregulated in panicle and seed development stages. Moreover, two each of up- and downregulated *OsbZIP* genes showed differential expression under cold stress. For example, *OsbZIP38* (upregulated) resembled the low temperature induced *LIP19* gene identified previously in rice (Nijhawan et al., 2008). Recently, a study analyzing the biological function of *OsbZIP52 in vivo*, isolated from panicles of rice (Zhonghua 11) and introduced into rice plants, found that rice seedlings overexpressing the gene showed sensitivity to cold and drought stresses as well as downregulation of some stress-related genes in response to abiotic stresses; thus, this gene could also function as a negative regulator of cold and drought stresses (Liu et al., 2012). Aside from cold tolerance, several members of bZIP TFs in rice, such as *OsbZIP16*, *OsbZIP23*, and *OsbZIP46*, have shown

the capability of improving the drought response in transgenic rice (Table 8.1) (Rabara et al., 2014).

8.3.4 MYB FAMILY OF TRANSCRIPTION FACTORS

Similar to bHLH TFs, the MYB family of TFs occur plentifully in plants, animals, and fungi; however, they were first recognized as oncogenes in animals, functioning primarily in cell-cycle control (Yang et al., 2012b). About 51–53 amino acids comprise MYB proteins with one, two, or three imperfect repeats in their DNA-binding domain, and are classified further into three subfamilies, types MYBR2R3 and MYBR1R2R3, and those MYB related based on the number of repeats (Yanhui et al., 2006). The genome-wide analysis identified about 155 MYB genes in rice (Katiyar et al., 2012). R2R3-type MYB proteins were reported to be involved in the environmental stress responses of plants, especially in *Arabidopsis*. Several studies on transgenic plants showed that overexpressing MYB genes can greatly enhance tolerance to cold stress as well as salinity and drought (Dubouzet et al., 2003). In rice, MYB TFs have approximately 183 members but only a few MYB genes have been characterized for regulatory functions in stress tolerance or stress response. Three MYB proteins, *OsMYB4*, *OsMYBS3*, and *OsMYB3R-2*, were reported to be involved in cold stress response in rice (Deng et al., 2017; Ma et al., 2009; Soltész et al., 2012). Overexpression of *OsMYB4* significantly confers tolerance to chilling and freezing stress in transgenic *Arabidopsis* (Vannini et al., 2004), enhances tolerance to frost, and improves germination under unfavorable conditions in transgenic barley plants (Soltész et al., 2012) (Table 8.1). *OsMYBS3* confers tolerance of rice plants to cold stress (Su et al., 2010) and *OsMYB3R-2* participates in the cold signaling pathway by targeting the cell cycle and a putative DREB/C-repeat binding factor (CBF) (Ma et al., 2009). When exposed to 4°C for a week in normal field conditions, transgenic rice overexpressing *OsMYBS3* showed no yield penalty (Su et al., 2010). Its transcription profile revealed many genes in the *OsMYBS3*-mediated cold signaling pathway and that its slow response to cold stress implied distinct pathways for short- and long-term cold stress adaptation in rice. *OsMYB2*, an R2R3-MYB TF isolated in rice, was tested and characterized for its role and response to salt, cold, and dehydration stress through generated transgenic plants with overexpressing and RNA interference (RNAi) *OsMYB2*. Study results showed that *OsMYB2* TF is a positive regulator to mediate the tolerance of rice seedlings to salt, cold, and dehydration stress and that its overexpression did not affect the rice seedlings phenotypes (Yang et al., 2012b).

TABLE 8.1

List of Genes Identified and Functionally Characterized in Response to Cold Stress in Rice

Gene Name	Locus ID	Putative Function	Transgenic Analysis	Reference
Os14-3-3f	Os03g50290	May be involved in different physiological processes	N/A	Yashvardhini et al. (2017)
Os14-3-3g	Os01g11110		N/A	
OsIMP	Os03g39000	ROS scavenging due to increased antioxidant enzymes	Tobacco	Zhang et al. (2017a)
OsbZIP46	Os06g10880	Modulation of the ABA signaling pathway	Co-overexpressed in O. sativa	Chang et al. (2017)
CA1 SAPK6	Os02g34600			
OsICE1	Os11g32100	Interacts with OsMYBS3 to increase tolerance	A. thaliana	Deng et al. (2017)
OsICE2	Os01g70310	Interacts with OsMYBS3 to increase tolerance	A. thaliana	Deng et al. (2017)
OsPUB2	Os05g39930	Improves chlorophyll content and electrolyte leakage	O. sativa	Byun et al. (2017)
OsPUB3	Os01g60860			
OsCPK17	Os07g06740	Affects the activity of membrane channels and sugar metabolism	O. sativa	Almadanim et al. (2017)
OsPIL16	Os05g04740	OsPIL16 regulates OsDREB1 in the absence of phyB to increase the integrity of membrane and reduce MDA concentration	O. sativa	He et al. (2016)
OsMYB30	Os02g41510	Negatively regulates β-amylase genes by interacting with OsJAZ9	O. sativa	Lv et al. (2017)
ONAC095	Os06g51070	Increased cold sensitivity due to the accumulation of ROS and the downregulation of cold-inducible genes	O. sativa	Huang et al. (2016)
ROC1	Os08g08820	Interacts with CBF1/CBF3 to improve cold tolerance	O. sativa	Dou et al. (2016)
OsSRFP1	Os03g22680	Increases cold sensitivity by reducing proline and antioxidant enzymes	O. sativa	Fang et al. (2016)
ZFP185	Os02g10200	Reduces cold tolerance by affecting ABA and GA biosynthesis	O. sativa	Zhang et al. (2016a)
OsNAP	Os03g21060	Reduced water loss contributing to cold tolerance	O. sativa	Chen et al. (2014)
OsSPX1	Os06g40120	Enhanced protection against ROS enabling cold tolerance	A. thaliana, tobacco	Wang et al. (2013)
OsSRO1c	Os03g12820	Tolerance induced by different metabolic pathways	O. sativa	You et al. (2014)
OsBURP16	Os10g26940	Increased transpiration, degradation of pectin damaging membrane integrity leading to cold sensitivity	O. sativa	Liu et al. (2014)
OsWRKY76	Os09g25060	Improved cold tolerance due to the induction of cold tolerance genes	O. sativa	Yokotani et al. (2013)
PSY2	Os12g43130	Damages membrane integrity, increased IAA content leading to cold sensitivity	O. sativa	Du et al. (2013)
OsGSTL2	Os03g17470	ROS detoxification	A. thaliana	Kumar et al. (2013)
OsGH3-2	Os01g55940	Tolerance due to a reduction in free IAA, membrane permeability and alleviation of oxidative damage	O. sativa	Du et al. (2012)
ZFP182	Os03g60560	Increased accumulation of proline	O. sativa	Huang et al. (2012)
OsMYB2	Os03g25550	ROS detoxification and the induction of stress tolerance genes	O. sativa	Yang et al. (2012b)
Osmyb4	Os04g43680	Improves germination vigor	H. vulgare	Soltész et al. (2012)
OsTPS1	Os01g23530	Cold tolerance due to the increased accumulation of trehalose and proline	O. sativa	Li et al. (2011)
OsSFR6	Os10g35560	Increased expression of COR genes	A. thaliana	Wathugala et al. (2011)
OsRAN2	Os05g49890	Facilitates cold tolerance through the maintenance of cell division	O. sativa	Chen et al. (2011)
OsDREB1D	Os06g06970	Tolerance through CBF pathway	A. thaliana	Zhang et al. (2009)
OsMYB3R-2	Os01g62410	Tolerance through the regulation of cell cycle	O. sativa	Ma et al. (2009)
OsDREB1F	Os01g73770	ABA-dependent pathway	O. sativa, A. thaliana	Wang et al. (2008)

(Continued)

TABLE 8.1 (CONTINUED)
List of Genes Identified and Functionally Characterized in Response to Cold Stress in Rice

Gene Name	Locus ID	Putative Function	Transgenic Analysis	Reference
OsTPP1	Os02g44230	Tolerance through the activation of stress responsive genes	*O. sativa*	Ge et al. (2008)
OsbHLH001	Os01g70310	Tolerance through CBF/DREB independent pathway	*A. thaliana*	Li et al. (2010)

Additionally, cold treatment induced the expression of *OsMYB2* where its expression peaked at 5 h after exposure and declined thereafter. Similarly, overexpression of *OsMYB3R-2*, an R1R2R3 MYB gene, showed increased tolerance to freezing, drought, and salt stress in transgenic *Arabidopsis* (Dai et al., 2007). Under cold stress, *OsMYB30* negatively regulated beta-amylase (BMY) genes at the transcription level and totally decreased BMY activity and maltose content, where maltose exhibited a protective role in cell membranes under cold stress conditions in rice (Lv et al., 2017). Additionally, *OsJAZ9* was identified as an *OsMYB30*-interacting protein that *OsMYB30* upregulated at the transcriptional level and conferred the negative regulation dependent on *OsMYB30*. In short, *OsMYB30* was found to negatively regulate cold tolerance by suppressing the BMY genes via interaction with *OsJAZ9* (Lv et al., 2017).

8.3.5 NAC FAMILY OF TRANSCRIPTION FACTORS

One of the largest families among plant-specific TFs, the NAC family proteins are plant-specific TFs (Nakashima et al., 2012; Shen et al., 2017). The name NAC is derived from three genes: (1) no apical meristem (**NAM**) gene from petunia hybrid, (2) **ATAF1/ATAF2**, and (3) cup-shaped Cotyledon2 (**CUC2**) genes from *Arabidopsis* located in the N-terminal region (Shen et al., 2017; Hu et al., 2008). Specific to higher plants, NAC TF family members can bind to promoter DNA as a dimer and induce gene expression. The NAC DNA-binding domain comprises five subdomains labeled A–E and typically has about 150 amino acids in length; it can act as either a transcriptional activator or repressor because it possesses a highly variable C-terminal transcriptional regulatory region (Tran et al., 2010). In general, the NAC family of TFs is mainly involved in plant growth and development as well as abiotic and biotic stress responses. NAC family members play crucial roles in cell division and extension, flower development and flowering, senescence, and seed germination. Their study examined the expression of *OsNAC2* responding to low temperature as well as dehydration, ABA, and NaCl, and found that *OsNAC2* expression decreased slightly during 12-h cold treatment

based on a qRT-PCR analysis. No other result or discussion on cold or temperature-related stress was presented. Continual research about the NAC TF genes in rice shows abiotic stressors such as cold or low temperature can induce their expressions. In a study involving NAC TF, named *SNAC2*, the transgenic rice overexpressing *SNAC2* showed significant tolerance to cold, including salinity and dehydration stress (Hu et al., 2008). *SNAC2* was isolated from IRAT109, an upland rice. Using the Affymetrix DNA chip for a genomic expression profile of the *SNAC2* overexpressed in rice revealed 36 upregulated and 9 downregulated genes in rice seedlings under both normal and cold stress conditions. Of the 45 genes, 26 genes contained both NAC recognition sequences (NACRS) and core DNA-binding sequences in their promoter regions, hinting that some of the genes might be *SNAC2* transcriptional targets. Overall, the study results showed that overexpressing *SNAC2* in rice had no detrimental effects on plant growth and development but significantly boosted both cold and salt tolerance. Meanwhile, *SNAC1* gene when overexpressed in rice considerably improved drought and salt resistance (Hu et al., 2006). At the overexpressed state of *SNAC1* gene in transgenic rice, stomata closure and drought resistance increased with no side effects on the photosynthetic rate and yield of transgenic plants. Two more NAC genes, *OsNAC5* and *OsNAC6*, were found responsive to cold stress, but their promoters do not contain putative cold-responsive elements (Takasaki et al., 2010). However, when both genes were overexpressed in transgenic rice, plant growth was delayed with *OsNAC6* but was normal with *OsNAC5* and appeared similar to the control. Thus, the authors concluded that *OsNAC5* may be useful for improving cold stress tolerance as it does not affect plant growth. Likewise, the *OsNAC5* expression is induced by other stresses such as high salinity, drought, ABA, and methyl jasmonic acid (Takasaki et al., 2010).

8.3.6 WRKY FAMILY OF TRANSCRIPTION FACTORS

WRKY is a large family of TFs (Rushton et al., 2010) first isolated in plants (Wu et al., 2005). Its name comes from the WRKY domain composed of about 60

TABLE 8.2

List of QTLs Identified from Different Growth Stages in Response to Cold Stress in Rice

QTL	Bordering Markers	Chromosome	Parents (Recipient)	Parents (Donor)	Traits Associated With	Reference
qCTG6	MGR3332-Wx	6	Hokuriku-142 or Yume-Toiro (HOK) *indica* subspecies	HGKN (*japonica* var.)	Percentage rate of germination	Ranawake et al. (2014)
qCTG7-1	RM6728–RM125	7				
qCTG7-2	RM125–RM1973	7				
qCTG8	RM7049–RM284	8				
qCTG11	RM1761–RM167	11				
qCTS5(1)	RM163–RM4501	5			Seedling recovery after cold stress	
qCTS6(1)	Wx-RM225	6				
qCTS11(1)-1	RM167–RM202	11				
qCTS11(1)-2	RM224–RM206	11				
qCTS2(2)	RM109–RM4355	2				
qCTS7(2)	RM6767–RM2752	7				
qCTS8(2)	RM1235–RM72	8				
qCTS11(2)-1	RM229–RM21	11				
qCTS11(2)-2	RM21–RM206	11				
qCTS4-	RM255–RM348	4	IR50 (*indica* subspecies)	M202 (*japonica* var.)	Cold-induced wilting tolerance (CIWT)	Andaya and Mackill (2003)
qCTS6-2	RM3–RM325a	6				
qCTS8-2a	RM223–RM284	8				
qCTS11-1	RM20–RM4	11				
qCTS12a	RM101–RM292	12				
qCTS1	RM297–RM319	1			Cold tolerance (CT)	Ranawake et al. (2014); Pan et al. (2015); Zhang et al. (2014)
qCTS3	RM200–RM85	3				
qCTS4-1	RM335–RM261	4				
qCTS6-1	RM253–RM50	6				
qCTS8-1	RM284–RM230	8				
qCTS10	RM239–RM284	10				
qCTS 8-2b	RM223–RM284	8			Cold-induced necrosis tolerance (CINT)	Andaya and Mackill (2003a)
qCTS11-2	RM254–RM330a	11				
qCTS12b	RM101–RM292	12				
qCTS4-3	RM241–RM317	4			Cold-induced yellowing tolerance (CIYT)	Xiao et al. (2015)
qLOP2	RM221–RS8	2	*Indica* rice	Dongxiang	Root conductivity	Xiao et al. (2014)
qPSR2-1	RM221–RS8	2				
qRC10-1	RM171–RM1108	10	Nanjing 11 (NJ) (*indica* var.)	Dongxiang wild rice (DX)	Chilling stress tolerance (CST) at the seedling stage	Fujino et al. (2004); Xiao et al. (2014)

(*Continued*)

TABLE 8.2 (CONTINUED)

List of QTLs Identified from Different Growth Stages in Response to Cold Stress in Rice

QTL	Bordering Markers	Chromosome	Parents (Recipient)	Parents (Donor)	Traits Associated With	Reference
qRC10-2	RM25570–RM304	10				
qLTG3–1	GBR3001–GBR3002	3	Hayamasari	Italica Livorno	Low temperature germinability	Biswas et al. (2017)
qCTSL-6-1	RM19996–RM3	6	BRRI dhan28	Hbj.BVI	Leaf discoloration (LD)	Suh et al. (2010)
qCTSL-8-1	RM7027–RM339	6				
qCTSL-12-1	RM247–RM2529	12			% Survivability	Andaya and Mackill (2003b)
qCTSS-8-1	RM7027–RM339	8				
qCTSS-11-1	RM26324–RM7283	11				
qCTSS-12-1	RM247–RM2529	12				
qPSST-3	RM569	3	Geumobyeo	IR66160-121-4-4-2	Seed set	Suh et al. (2010)
qPSST-7	RM1377	7				
qPSST-9	RM24545	9				
qCTB2a	RM324–RM301	2	IR50	M202	% Spikelet fertility	Andaya and Mackill (2003)
qCTB9	RM257–RM242	9				
qCTB1	RM151–RM259	1				
qCTB2b	RM324–RM301	2				
qCTB3	RM156–RM214	3				
qCTB5	RM26–RM334	5			% Undeveloped spikelet (USP) (15.9)	Andaya and Mackill (2003)
qCTB6	RM50–RM173	6				
qCTB7	RM129–RM81	7				
qDTH3	RM85–RM168	3	Hapcheonaengmi	Milyang23	Days to heading	Oh et al. (2004)
qDTH6	RM527–RM539	6				
qDTH7	RM234–RM351	7				
qDDH3	RM85–RM168	3			Difference in days to heading	
qDDH8	RM152–RM506	8				
qCL1	RM128–PBC121	1			Culm length	
qCL9	RM285–RM434	9				
qFER11	RM3701–RM522	11			Spikelet fertility	
qPE1	RM9–PBC121	1			Panicle exsertion	
qPE11	RM558–RM3747	11				
qDC5	RM7568–RM430	5			Discoloration	
qDC11	RM552–RM3137	11				

amino acid residues with high affinity for the W-box cis-regulatory element (TTGACC/T) (Rushton et al., 2010). The N-terminus of the WRKY domain shows a few reported variants such as WIKY, WRRY, WKRY, WSKY, and WKKY upon replacing WRKY amino acid sequences (Rabara et al., 2014). To date, findings have shown that WRKY proteins are activators as well as repressors in important plant processes and that family members play roles in both repression and depression (Rushton et al., 2010).

Several studies demonstrated that WRKY TF family members play major roles in biotic and abiotic stress responses. For example, *OsWRKY11*, *OsWRKY30*, and *OsWRKY45* may have roles in drought tolerance improvement through genetic engineering (Rabara et al., 2014). About 15 WRKY TF genes in rice were reportedly repressed by cold with spatial expression located in various plant parts and growth stages. Recent studies of two genes on the list, *OsWRKY74* and *OsWRKY71*, further elucidated their roles in cold stress (Dai et al., 2016). *OsWRKY74* showed that the gene regulated cold stress and multiple nutrient starvation responses including possible crosstalk between P and Fe, and P and cold stress (Dai et al., 2016). In a functional study, two *OsWRKY71* transgenic rice lines under cold treatment of 4°C recovered better than the control rice lines in terms of survival rate, photosynthetic ability, and fresh and dry weights. RT-PCR analyses confirmed the increased expression of *OsTGR* and *WS176* in *OsWRKY71* transgenic lines under colder stress (Kim et al., 2016). A study of the *OsWRKY82* TF gene reveals that *OsWRKY82* is a multiple stress-inducible gene including cold stress and may be involved in the regulation of a defense response to pathogens and tolerance against abiotic stresses by the jasmonic acid/ethylene-dependent signaling pathway (Li et al., 2013).

8.4 FUTURE OF TF-BASED COLD STRESS IMPROVEMENT

Cold stress is one of many environmental factors that can stunt or delay plant growth and development as well as reduce plant productivity, particularly yield potential. Plants, particularly rice, have innate capacities to withstand or cope with cold stress. For example, *OrbHLH001* and *SNAC2* genes isolated from two rice germplasms, Dongxiang wild rice and upland rice IRAT109, respectively, have been used to confer cold stress tolerance in transgenic rice lines and, to date, have been further explored genetically for hybridization and tolerance to various abiotic stresses (Zhang et al., 2016b). Moreover, several studies attest that various TF families regulate cold stress as well as other abiotic stresses occurring either singly or multiple times. Hence, the transcriptional control of genes related to cold regulation in plants is a critical part of plant responses to cold stress. In the past decades, owing to advances in molecular and biotechnology tools, much progress has been made in characterizing the TF families (AP2/ERF, bHLH, bZIP, MYB, NAC, and WRKY) that regulate cold stress and other abiotic stresses in rice. For example, *OsMYB4* has been shown to improve tolerance or resistance to various stresses when expressed ectopically in both monocots and dicots such as tobacco, potato, tomato, and apple among others (Soltész et al., 2012). This illustrates the importance and versatility of TFs as major regulators of stress tolerance, including cold stress. Advances in

transcriptome research would further pinpoint strategies toward improving crop tolerance to various environmental stresses. The identification of genes that regulate cold and other stresses is essential for more efficient methods to confer cold tolerance or resistance in new high-yielding rice varieties.

8.5 MEMBRANE FLUIDITY INITIATES CELLULAR COLD RESPONSIVE

The functional characterization of stress-activated MAPK (*SAMK*) revealed that an MAPK signaling cascade is triggered by increased membrane rigidity and altered ion conductance within cells and tissues in response to cold stress (Sangwan et al., 2002). Also, the influx of Ca^{+2} into the cytoplasm, an early event in cold stress, may be mediated by Ca^{+2} channels that are activated by membrane rigidification, ligands, or mechanical stimuli (Chinnusamy et al., 2007). The induction of the Ca^{2+} signaling cascade, in turn, activates the ABA-independent pathway of cold tolerance that is mediated by the dehydration-responsive element-binding proteins-C-repeat/dehydration-responsive elements (DREB-CRT/DRE) (Zhang et al., 2013). *ROC1*, an indeterminate domain (IDD) protein, which is a CBF1 regulator, was identified by yeast one-hybrid assay. After transcript analysis, it was found that *ROC1* was induced by exogenously treated auxin, but not altered by the cold. However, *ROC1* mutants exhibited chilling-sensitive symptoms due to the inhibition of *CBF1* and *CBF3*, suggesting *ROC1* as a positive regulator of cold stress responses (Dou et al., 2016).

8.6 ROS AND MDA MEDIATE COLD DAMAGE AND COLD SENSING IN RICE

ROS are chemically reactive molecules that contain oxygen, including superoxide (O_2^-), hydrogen peroxide (H_2O_2), or the hydroxyl radical (HO^-), which are produced at low levels as normal by-products of plant cellular metabolism, mainly in organelles such as chloroplasts, mitochondria, and peroxisomes. However, both abiotic and biotic stresses can lead to excessive production of ROS that can then react rapidly with proteins, DNA, and lipids to cause cellular oxidative damage (Skopelitis et al., 2006; Mittal et al., 2012). ROS degrade polyunsaturated lipids to form MDA, a reactive aldehyde that initiates toxic stress in cells and subsequently causes cellular dysfunction and tissue damage (Pamplona et al., 2011). AsA and GSH, the primary antioxidant compounds found predominantly in the chloroplast or other compartments in the cell, are known to provide protection against ROS-induced damage. Overexpression of *OsAPXa* improved tolerance to cold stress by enhancing the ROS scavenging activity, thereby leading to a reduction in the hydrogen peroxide level and lipid peroxidation–induced membrane damage (Sato et al., 2011b). In another study, it has been observed that the silencing of *OsSPX1* leads to increased ROS-induced damage owing to the increased accumulation of hydrogen peroxide. This observation was further validated by the downregulation of genes involved in ROS scavenging pathways such as GSH S-transferase or *OsPHO2* involved in phosphate signaling (Wang et al., 2013). In addition, it has also been observed that overexpression of *OsIMP* (1-*myo*-inositol monophosphatase), *ZFP182* (TFIIIA-type zinc finger protein), and *OsTPS1* (trehalose-6-phosphate synthase) conferred cold tolerance through the increased accumulation of proline or trehalose (Table 8.2) (Zhang et al., 2012b, 2017a; Li et al., 2011).

8.7 *VIRESCENT* MUTANTS AT LOW TEMPERATURES DURING EARLY LEAF DEVELOPMENT

Numerous genes responsible for *virescent* mutations have been identified in rice, and have been shown to be involved in transcription, translation, and nucleotide metabolism. Among them, *NUS1*, *RNRS1*, and *RNRL1* have been identified through the genetic and functional analysis of *virescent* mutants in rice (*v1*, *v2*, *v3*, and *st1*) and are hypothesized to be involved in chloroplast biogenesis under low temperatures. *Virescent-1* (*V1*) was shown to encode a novel chloroplast RNA-binding protein, named *NUS1* (Kusumi et al., 2011). This research group has shown that *NUS1* expression specifically occurs in the developing leaves at the P4 stage, and is enhanced by low temperature treatment. Also, they found that in the *v1* seedlings grown at low temperatures, not only the processing and accumulation of chloroplast rRNA but also transcription as well as translation was severely suppressed. Recently, several other genes have been isolated by screening low temperature conditional, chloroplast-deficient mutants of rice, such as *OsV4* (*virescent 4*), *wlp1* (white leaf and panicles 1), and *tcd9* (thermosensitive chloroplast development 9) (Gong et al., 2014; Jiang et al., 2014; Song et al., 2014). A possibility exists that these factors are involved in a related mechanism to chloroplast protein expression and assembly for low temperature tolerance and are good candidates for further analysis.

8.8 PENTATRICOPEPTIDE REPEAT (PPR) PROTEINS

8.8.1 TCD10

PPR proteins are one of the largest protein families in plants. *OsPPR1*, carrying 11 PPR motifs, is the first identified PPR protein responsible for rice development (Gothandam et al., 2005). Wu et al. (2016) have characterized a thermosensitive chlorophyll-deficient mutant called *tcd10*, which displays the albino phenotype below 20°C and *TCD10* encodes a novel PPR protein, containing 27 PPR motifs, required for chloroplast development and photosynthesis in rice under cold stress. They determined the expression pattern of the *TCD10* gene in various tissues (root, stem, young leaf, and panicle) and found that in the flag leaves *TCD10* was expressed higher than other tissues. The expression levels were increased together with leaf age, indicating the importance of *TCD10* for leaf chloroplast development under cold stress (Wang et al., 2016; Wu et al., 2016). They have also speculated that no survival could be achieved for *tcd10* mutants after the five-leaf stage at 20°C because of the loss of photosynthesis due to the block of electron transport in photosystems under cold stress.

8.8.2 TCD5

Recently, Wang et al. (2016) characterized the temperature-sensitive chlorophyll-deficient rice mutant *tcd5*, which develops albino leaves at a low temperature of 20°C. They found that *TCD5* encodes a conserved plastid-targeted monooxygenase family protein that has not been previously reported to be associated with a temperature-sensitive albino phenotype in plants. *TCD5* is abundantly expressed in young leaves and immature spikes, and low temperatures increase this expression. In conclusion, physiological and molecular analyses suggested that the *TCD5* gene is essential for chloroplast differentiation during early development under cold stress (Wang et al., 2016).

8.9 UBIQUITINATION PATHWAY AS A TARGET TO DEVELOP COLD TOLERANCE IN RICE

Ubiquitin, found in both cytosol and the nucleus of eukaryotic cells, can be covalently bound to other proteins to regulate the stability, function, or location of the modified protein. Ubiquitin is recognized by specific receptors that contain one or more ubiquitin-binding domains (Dikic et al., 2009). Usually, these domains bind to ubiquitin with low affinity, which makes this bond highly dynamic. Therefore, the ubiquitin coupling and uncoupling system mediates several cell processes involved in the growth and development of plants, such as embryogenesis, photomorphogenesis, and hormone regulation. Stress caused by extreme temperatures disturbs the cell homeostasis, resulting in a delay in plant development, as it affects seed germination, photosynthesis, respiration, and plasma membrane stability. E3 ubiquitin ligases play a crucial role in the specific recognition of appropriate target proteins and the attachment of a poly-ubiquitin chain (Chen and Hellmann, 2013). They are divided into two groups based on their structures. The U-box E3 ubiquitin ligases contain a modified Really Interesting New Gene (RING) domain and widely exist in eukaryotic organisms. An increased number of U-box proteins (PUBs) in higher plants might indicate their important role in diverse cellular processes. Research carried out by Byun et al. (2017) has shown that two homologous U-box type E3 ubiquitin ligases, *OsPUB2* and *OsPUB3*, function in coordination to improve cold stress tolerance. Byun et al. (2017) proved this by overexpression analysis where it was observed that transgenic rice plants showed improved cold tolerance assessed through chlorophyll content or ion leakage. Another research has shown that silencing *OsSRFP1* (stress-related RING finger protein 1) leads to improved cold tolerance as opposed to the overexpressing lines, suggesting a negative role in cold tolerance. From the research of Fang et al. (2015), it was concluded that the cold tolerance of RNAi lines was achieved primarily through improved antioxidant defense machinery (Fang et al., 2016).

8.10 QUANTITATIVE TRAIT LOCI MAPPING OF COLD TOLERANCE AT GERMINATION STAGE

Cold stress tolerance in plants is a relatively complex trait associated with a great number of physiological and biochemical changes (Thomashow, 1999). Perception and cold tolerance in plants is governed by different biological mechanisms such as cold sensing and transcriptional and post-transcriptional processing (Sanghera et al., 2011). Recently, many studies have been conducted using several mapping populations or bioassays that have focused on the different developmental stages of rice including germination (CTG) as well as seedling stages (CTS). Several QTLs have been identified for CTS and CTG in rice, namely, *qCTS12*, *qCTS4*, *qLTG3-1*, and *qCtss11*, which have been mapped to the rice genome, but their functions are still unknown (Andaya and Tai,

2006; Fujino et al., 2008; Koseki et al., 2010; Ranawake et al., 2014). Fine mapping performed using recombinant inbred lines (RILs) resulting from a cross between *japonica* M202 and *indica* IR24 identified QTLs on chromosomes 11 and 12 for CTS and on chromosomes 2 and 3 for the booting stage (Andaya and Mackill, 2003a, 2003c). Specifically, temperatures ranging from 10°C to 25°C have been used for CTG while 6°–10°C have been used for CTS. Several QTLs have been suggested for CTG and CTS using the same mapping populations; however, one thing that needs to be considered here is that the severity of cold stress is a determining factor in identifying QTLs (Han et al., 2006). $F_{2:3}$ populations derived from a cross between Milyang 23× Jileng 1 were used in the detection of QTLs for the low temperature vigor of germination (LVG). This study identified QTL *qLVG2* between RM29 and RM262 on chromosome 2, while *qLVG7-2* and *qCIVG7-2* were located between RM336 and RM118 on chromosome 7 (Han et al., 2006). *qLTG3-1* identified on chromosome 3 from backcross inbred lines (BIL) derived from a cross between Italica Livorno and Hayamasari was predicted to be responsible for the weakening of the tissues during germination under cold stress (Fujino et al., 2008). From a natural population of cultivated (*O. sativa*) and 23 wild (*O. rufipogon*) rice strains, *qCTP11* and *qCTP12* QTLs responsible for cold tolerance at the plumule stage were identified (Zhang et al., 2014). QTL *qCTB-5-1*, *qCTB-5-2*, *qCTB-5-3*, and *qCTB-7* on chromosome 5 and 7, respectively, were derived from chromosome-segment substitution lines (CSSL) from a cross between *indica* rice accession 9311 and *japonica* rice cultivar Nipponbare for tolerance at the bud burst stage (CTB). Research conducted by Xiao et al. (2015) identified two QTLs, *qLOP2 and qPSR2-1*, for chilling tolerance at the seedling stage using a mapping population derived by crossing chilling-tolerant Dongxiang wild rice (*O. rufipogon*) and a chilling-sensitive *indica* rice as the recipient parent. They went on to identify the gene *Os02g067730* for the loci and reported it to be inducible by cold (Xiao et al., 2015).

8.11 SEEDLING STAGE QTLs FOR COLD TOLERANCE

RILs derived from a cross between IR50 and M202 enabled the identification of QTL *qCTS12a* on chromosome 12 conferring cold tolerance at the seedling stage (Andaya and Mackill, 2003a). With the help of microsatellite markers and open reading frame analysis, two putative candidate genes, *OsGSTZ1* and *OsGSTZ2*, were mapped. *OsGSTZ1* overexpression in rice increased the cold tolerance as evidenced by the improved germination and growth of seedlings at low temperature (Zhang et al., 2014). Furthermore, using the same mapping population, *qCTS4*, which is associated with the stunting and yellowing of rice seedlings, was mapped onto chromosome 4. *qCTS-2*, a major QTL for cold tolerance located on chromosome 2, was identified from double haploid lines derived from a cross between a cold-tolerant *japonica* variety (AAV002863) and a cold-sensitive *indica* cultivar (Zhenshan97B) (Collins et al., 2008). Another QTL, *qCtss11*, for cold tolerance at the seedling stage was mapped onto chromosome 11 and contains six annotated genes of which *Os11g0615600* was expressed from the GLA4 allele that came from the sensitive parent Guang-lu-ai 4 (GLA4), while *Os11g0615600* from the GLA4 haplotype had a premature stop codon. These observations clearly suggest that these genes might have a role in regulating cold tolerance at the seedling stage (Koseki et al., 2010). *Os11g0615900*, another likely candidate gene identified from the QTL *qCtss11*, contains the NB-ARC (nucleotide-binding adaptor shared by *APAF-1*, R protein, and *CED-4*) domain and could be a likely candidate that functions as a bridge for hormonal crosstalk between biotic and abiotic stress. Furthermore, another important QTL, *COLD* (chilling-tolerance divergence) was identified on the long arm of chromosome 4 from the RILs obtained from chilling-tolerant Nipponbare (*japonica*) and chilling-sensitive 93-11 (*indica*) cultivars. In addition, Ma et al. went on to map the gene COLD1, which improved cold tolerance by enhancing the G-protein GTPase activity when overexpressed in rice. Moreover, Ma et al. also identified a single nucleotide polymorphism (SNP; T/C versus A) in the chilling-sensitive cultivars in comparison to the tolerant *japonica* or wild rice. The presence of the SNP that results in Met[187]/Thr[187] in sensitive *indica* rice is responsible for the sensitivity (Ma et al., 2015). In their research, Schläppi et al. (2017) identified two novel low temperature seedling survivability (LTSS) QTLs, *qLTSS3-4* and *qLTSS4-1*, of which *japonica* subspecies contributed the alleles for *qLTSS4-1* and exhibited positive phenotypic effects. The putative candidate genes identified from the QTL analysis could be a likely candidate for functional analysis in cold-sensitive and cold-tolerant rice populations.

8.12 QTLs FOR COLD TOLERANCE AT REPRODUCTIVE STAGE

The yield of rice is severely affected by cold stress at the reproductive stage, more specifically at the booting stage. Using a set of near-isogenic lines (NIL) obtained from backcrossing Kirara397/Norin-PL8/Kirara397, two QTLs, *Ctb1* and *Ctb2*, were identified on chromosome

4 (Miura et al., 2011). The F-box protein encoded by *Ctb1* was identified and was hypothesized to be a part of the E3 ubiquitin ligase complex (Ma et al., 2015). RILs derived from a cross between cold-tolerant *japonica* rice, M-202, and a tropical *indica* variety, IR50, enabled the identification of *qCTB2a* and *qCTB3* responsible for cold tolerance at the booting stage (Andaya and Mackill, 2003c). In addition, several QTLs, such as *cl* (culm length), *Dth* (days to heading), *pe* (panicle neck exsertion), *fer* (spikelet fertility), and *dc* (discoloration), were also identified (Zhang et al., 2014). Besides the aforementioned, eight QTLs were mapped onto chromosomes 1, 4, 5, 10, and 11, respectively, based on their variation in spikelet sterility (Table 8.2) and all the alleles for cold tolerance came from cold-tolerant *japonica* landrace, Kunmingxiaobaigu (KMXBG) (Yang et al., 2013). *qPSST-3*, *qPSST-7*, and *qPSST-9* were identified on chromosome 3, 7, and 9, respectively, by using simple sequence repeat (SSR) markers and composite interval mapping and their effect was assessed by growth under cold stress and spikelet fertility (Suh et al., 2010). Cold tolerance related to qLTB3 on chromosome 3 identified from F_2 and BC_1F_2 populations of Ukei 840 and Hitomebore was measured by the fertility of the seeds under cold stress. In an effort to understand the cold tolerance at the reproductive stage, NIL were developed by crossing KMXBG (cold tolerant) and Towada (cold sensitive) from which NIL1913 was selected, which exhibited more cold tolerance and included the QTL, *qCTB4-1*. Li et al. (2017) mapped the cold tolerance QTL *qCTB10-2* for the booting stage to a 132.5 kb region that contained 17 candidate genes, of which 4 genes were inducible by cold stress identified using NIL ZL31-2. They further suggested that of the five genes, LOC_Os10g11730, LOC_Os10g11770, and LOC_Os10g11810 were highly cold inducible in ZL31-2 in comparison to cold-sensitive Towada, while LOC_Os10g11820 showed constitutive expression and LOC_Os10g11750 showed no significant changes. These results can be the basis for identifying genes in the QTL and improving the cold tolerance at the booting stage by marker-assisted selection. In a further study, Zhang et al. developed a backcross population and used it to map and clone the target gene *CTB4a* that encodes leucine-rich repeat receptor-like protein kinase (LRR-RLK). The overexpression of *CTB4a* in rice not only improved cold tolerance at the vegetative stage but it also improved seed setting and enhanced yield under cold stress at the reproductive stage. Further, it was also observed that the gene expression increased with a simultaneous increase in adenosine triphosphate (ATP) synthase activity and ATP concentration, thereby suggesting that it interacts with the beta subunit of ATP

synthase (Zhang et al., 2017b). Shakiba et al. (2017) screened Rice Diversity Panel 1 (RDP1) for cold tolerance at the seedling stage and identified 42 QTLs of which the majority were contributed by the *japonica* subspecies. At the reproductive stage, they identified 29 QTLs that co-localized with 15 previously identified QTLs and they correlated with grain yield components such as seed weight per panicle or per plant; a gene ontology search identified that they were associated with processes such as response to stress, hormonal crosstalk between stresses, and lipid metabolism (Shakiba et al., 2017). Thus, these results clearly suggest that the *japonica* germplasm can be a good resource for improving the cold tolerance of *indica* rice, which feeds a major part of the Asian population. However, the identified QTLs on chromosome 2, 7, 8, and 11 and the identified putative candidate genes need to be studied in detail to understand their contribution to cold tolerance. To conclude, it can be said that the cold tolerance mechanism in rice is a complex phenomenon as evidenced by the identification of cold tolerance QTLs at different developmental and growth stages (Shirasawa et al., 2012).

8.13   FUTURE PERSPECTIVE

In the last decade, advances in molecular genomics and systematics and their applications in research have helped to improve knowledge on the physiological and genetic bases of cold tolerance in rice. Cold stress induces various changes in the physiological properties of plants. Central to this response are plants cold-sensing system and cold-responsive signaling network. To date, no standard evaluation protocol for cold tolerance in rice cultivars has been established despite the prior use of many criteria. The most promising criteria for cold tolerance screening of rice cultivars that should be considered are the seedling survival rate and the reproductive stage, particularly spikelet fertility. Seedlings have more actual experience of cold stress in the field while cold tolerance at the reproductive stage would help ensure achieving the yield potential of plants. QTL or gene expression analysis has been used to search for potential genes responsible for cold tolerance in rice plants. Various QTLs were identified as being induced by cold stress at various stages of the rice plant development. Future studies on QTLs will further ascertain their characteristics in conferring cold tolerance, which combined with genetic engineering tools could lead to the successful development of cold-tolerant rice cultivars. Additionally, transgenic technology has shown promise as a tool for cold tolerance improvement in rice plants through overexpression or silencing of target genes. TFs have been identified as

conferring cold tolerance and most of these genes likewise enhance tolerance to other abiotic stresses. Studies on improving cold tolerance in rice through genetic engineering is one of the available approaches that researchers could tap into to address the pressing issue on cold stress. The challenge would be to identify the genes that regulate this quantitative trait. With the availability of next-generation sequencing (NGS) technology, high-throughput transcriptome profiling can be utilized to generate a large transcriptome dataset that will provide a comprehensive overview of plant responses to cold stress at the molecular level. Through this approach, putative target genes can be identified and genetically engineered for improving rice response to cold stress.

REFERENCES

Aghaee, A., Moradi, F., Zare-Maivan, H., Zarinkamar, F., Pour Irandoost, H., Sharifi, P. (2011) Physiological responses of two rice (*Oryza sativa* L.) genotypes to chilling stress at seedling stage. *African Journal of Biotechnology* 10: 7617–7621.

Ali, M. G., Naylor, R. E. L., Matthews, S. (2006) Distinguishing the effects of genotype and seed physiological age on low temperature tolerance of rice (*Oryza sativa* L.). *Experimental Agriculture* 42(3): 337–349.

Almadanim, M. C., Alexandre, B. M., Rosa, M. T. G., Sapeta, H., Leitão, A. E., Ramalho, J. C., Lam, T. T., Negrão, S., Abreu, I. A., Oliveira, M. M. (2017) Rice calcium-dependent protein kinase OsCPK17 targets plasma membrane intrinsic protein and sucrose-phosphate synthase and is required for a proper cold stress response. *Plant Cell and Environment* 40(7): 1197–1213.

Andaya, V. C., Mackill, D. J. (2003a) Mapping of QTLs associated with cold tolerance during the vegetative stage in rice. *Journal of Experimental Botany* 54: 2579–2585.

Andaya, V. C., Mackill, D. J. (2003b) Mapping of QTLs associated with cold tolerance during the vegetative stage in rice. *Journal of Experimental Botany* 54(392): 2579–2585.

Andaya, V. C., Mackill, D. J. (2003c) QTLs conferring cold tolerance at the booting stage of rice using recombinant inbred lines from a *japonica* x *indica* cross. *Theoretical and Applied Genetics* 106(6): 1084–1090.

Andaya, V. C., Tai, T. H. (2006) Fine mapping of the qCTS12 locus, a major QTL for seedling cold tolerance in rice. *Theoretical and Applied Genetics* 113(3): 467–475.

Apel, K., Hirt, H. (2004) Reactive oxygen species: Metabolism, oxidative stress, and signal transduction. *Annual Review of Plant Biology* 55: 373–399.

Beer, M. A., Tavazoie, S. (2004) Predicting gene expression from sequence. *Cell* 117: 185–198.

Benedict, C., Geisler, M., Trygg, J., Huner, N., Hurry, V. (2006) Consensus by democracy. Using metaanalyses of microarray and genomic data to model the cold acclimation signaling pathway in *Arabidopsis*. *Plant Physiology* 141: 1219–1232.

Bonnecarrère, V., Borsani, O., Díaz, P., Capdevielle, F., Blanco, P., Monza, J. (2011) Response to photooxidative stress induced by cold in *japonica* rice is genotype dependent. *Plant Science* 180(5): 726–732.

Byun, M. Y., Cui, L. H., Oh, T. K., Jung, Y. J., Lee, A., Park, K. Y., Kang, B. G., Kim, W. T. (2017) Homologous U-box E3 ubiquitin ligases OsPUB2 and OsPUB3 are involved in the positive regulation of low temperature stress response in rice (*Oryza sativa* L.). *Frontiers in Plant Science* 8: 16.

Carretero-Paulet, L., Galstyan, A., Roig-Villanova, I., Martínez-García, J. F., Bilbao-Castro, J. R., Robertson, D. L. (2010) Genome-wide classification and evolutionary analysis of the bHLH family of transcription factors in *Arabidopsis*, poplar, rice, moss, and algae. *Plant Physiology* 153(3): 1398–1412.

Century, K., Reuber, T. L., Ratcliffe, O. J. (2008) Regulating the regulators: The future prospects for transcription-factor-based agricultural biotechnology products. *Plant Physiology* 147(1): 20–29.

Chang, Y., Nguyen, B. H., Xie, Y., Xiao, B., Tang, N., Zhu, W., Mou, T., Xiong, L. (2017) Co-overexpression of the constitutively active form of OsbZIP46 and ABA-activated protein kinase SAPK6 improves drought and temperature stress resistance in rice. *Frontiers in Plant Science* 8: 1102.

Chen, L., Hellmann, H. (2013) Plant E3 ligases: Flexible enzymes in a sessile world. *Molecular Plant* 6(5): 1388–1404.

Chen, N., Xu, Y., Wang, X., DU, C., DU, J., Yuan, M., Xu, Z., Chong, K. (2011) OsRAN2, essential for mitosis, enhances cold tolerance in rice by promoting export of intranuclear tubulin and maintaining cell division under cold stress. *Plant Cell and Environment* 34(1): 52–64.

Chen, X., Wang, Y., Lv, B., Li, J., Luo, L., Lu, S., Zhang, X., Ma, H., Ming, F. (2014) The NAC family transcription factor OsNAP confers abiotic stress response through the ABA pathway. *Plant and Cell Physiology* 55(3): 604–619.

Chinnusamy, V., Zhu, J., Zhu, J. K. (2007) Cold stress regulation of gene expression in plants. *Trends in Plant Science* 12(10): 444–451.

Collins, N. C., Tardieu, F., Tuberosa, R. (2008) Quantitative trait loci and crop performance under abiotic stress: Where do we stand? *Plant Physiology* 147(2): 469–486.

Cook, D., Fowler, S., Fiehn, O., Thomashow, M. F. (2004) A prominent role for the CBF cold response pathway in configuring the low temperature metabolome of *Arabidopsis*. *Proceedings of the National Academy of Sciences USA* 101: 15243–15248.

Corrêa, L. G., Riaño-Pachón, D. M., Schrago, C. G., dos Santos, R. V., Mueller-Roeber, B., Vincentz, M. (2008) The role of bZIP transcription factors in green plant evolution: Adaptive features emerging from four founder genes. *PLoS One* 3(8): e2944.

da Cruz, R. P., Sperotto, R. A., Cargnelutti, D., Adamski, J. M., de FreitasTerra, T., Fett, J. P. (2013) Avoiding damage and achieving cold tolerance in rice plants. *Food and Energy Security* 2(2): 96–119.

Dai, X., Wang, Y., Zhang, W. H. (2016) OsWRKY74, a WRKY transcription factor, modulates tolerance to phosphate starvation in rice. *Journal of Experimental Botany* 67(3): 947–960.

Dai, X., Xu, Y., Ma, Q., Xu, W., Wang, T., Xue, Y., Chong, K. (2007) Overexpression of an R1R2R3 MYB gene, OsMYB3R-2, increases tolerance to freezing, drought, and salt stress in transgenic *Arabidopsis*. *Plant Physiology* 143(4): 1739–1751.

Datta, K., Datta, S. K. (2006) *Indica* rice (*Oryza sativa*, BR29 and IR64). *Methods in Molecular Biology* 343: 201–212.

De Los Reyes, B. G., Myers, S. J., McGrath, J. M. (2003) Differential induction of glyoxylate cycle enzymes by stress as a marker for seedling vigor in sugar beet (*Beta vulgaris*). *Molecular Genetics and Genomics* 269(5): 692–698.

Deng, C., Ye, H., Fan, M., Pu, T., Yan, J. (2017) The rice transcription factors OsICE confer enhanced cold tolerance in transgenic *Arabidopsis*. *Plant Signal Behaviour* 12(5): e1316442.

Deppmann, C. D., Acharya, A., Rishi, V., Wobbes, B., Smeekens, S., Taparowsky, E. J., Vinson, C. (2004) Dimerization specificity of all 67 B-ZIP motifs in *Arabidopsis thaliana*: A comparison to Homo sapiens B-ZIP motifs. *Nucleic Acids Research* 32(11): 3435–3445.

Dikic, I., Wakatsuki, S., Walters, K. J. (2009) Ubiquitin-binding domains: From structures to functions. *Nature Reviews Molecular Cell Biology* 10(10): 659–671.

Dong, Y. S., Zhao, L. M., Liu, B., Wang, Z. W., Jin, Z. Q., Sun, H. (2004) The genetic diversity of cultivated soybean grown in China. *Theoretical and Applied Genetics* 108: 931–936.

Dou, M., Cheng, S., Zhao, B., Xuan, Y., Shao, M. (2016) The indeterminate domain protein ROC1 regulates chilling tolerance via activation of DREB1B/CBF1 in rice. *International Journal of Molecular Sciences* 17(3): 233.

Du, H., Wu, N., Chang, Y., Li, X., Xiao, J., Xiong, L. (2013) Carotenoid deficiency impairs ABA and IAA biosynthesis and differentially affects drought and cold tolerance in rice. *Plant Molecular Biology* 83(4–5): 475–488.

Du, H., Wu, N., Fu, J., Wang, S., Li, X., Xiao, J., Xiong, L. (2012) A GH3 family member, OsGH3-2, modulates auxin and abscisic acid levels and differentially affects drought and cold tolerance in rice. *Journal of Experimental Botany* 63(18): 6467–6480.

Dubouzet, J. G., Sakuma, Y., Ito, Y., Kasuga, M., Dubouzet, E. G., Miura, S., Seki, M., Shinozaki, K., Yamaguchi-Shinozaki, K. (2003) OsDREB genes in rice, *Oryza sativa* L., encode transcription activators that function in drought-, high-salt- and cold-responsive gene expression. *Plant Journal* 33(4): 751–763.

Fang, H., Meng, Q., Zhang, H., Huang, J. (2016) Knockdown of a RING finger gene confers cold tolerance. *Bioengineered* 7(1): 39–45.

Farooq, M., Wahid, A., Kobayashi, N., Fujita, D., Basra, S. M. A. (2009) Plant drought stress: Effects, mechanisms and management. *Agronomy for Sustainable Development* 29: 185–212.

Farrell, T. C., Fox, K. M., Williams, R. L., Fukai, S., Lewin, L. G. (2006) Minimising cold damage during reproductive development among temperate rice genotypes. II. Genotypic variation and flowering traits related to cold tolerance screening. *Australian Journal of Agricultural Research* 57(1): 89–100.

Fujino, K., Sekiguchi, H., Matsuda, Y., Sugimoto, K., Ono, K., Yano, M. (2008) Molecular identification of a major quantitative trait locus, qLTG3-1, controlling low-temperature germinability in rice. *Proceedings of the National Academy of Sciences USA* 105(34): 12623–12628.

Funatsuki, H., Ohnishi, S. (2009) Recent advances in physiological and genetic studies on chilling tolerance in soybean. *Jarq-Japan Agricultural Research Quarterly* 43: 95–101.

Garg, A. K., Kim, J. K., Owens, T. G., Ranwala, A. P., Choi, Y. D., Kochian, L. V., Wu, R. J. (2002) Trehalose accumulation in rice plants confers high tolerance levels to different abiotic stresses. *Proceedings of the National Academy of Sciences USA* 99: 15898–15903.

Ge, L. F., Chao, D. Y., Shi, M., Zhu, M. Z., Gao, J. P., Lin, H. X. (2008) Overexpression of the trehalose-6-phosphate phosphatase gene OsTPP1 confers stress tolerance in rice and results in the activation of stress responsive genes. *Planta* 228(1): 191–201.

Gong, X., Su, Q., Lin, D., Jiang, Q., Xu, J., Zhang, J., Teng, S., Dong, Y. (2014) The rice OsV4 encoding a novel pentatricopeptide repeat protein is required for chloroplast development during the early leaf stage under cold stress. *Journal of Integrative Plant Biology* 56(4): 400–410.

Gothandam, K. M., Kim, E. S., Cho, H., Chung, Y. Y. (2005) OsPPR1, a pentatricopeptide repeat protein of rice is essential for the chloroplast biogenesis. *Plant Molecular Biology* 58(3): 421–433.

Gothandam, K. M., Nalini, E., Karthikeyan, S., Shin, J. S. (2010) OsPRP3, a flower specific proline-rich protein of rice, determines extracellular matrix structure of floral organs and its overexpression confers cold-tolerance. *Plant Molecular Biology* 72: 125–135.

Graham, D., Patterson, B. D. (1982) Responses of plants to low, nonfreezing temperatures: Proteins, metabolism, and acclimation. *Annual Review of Plant Physiology* 33: 347–372.

Hahn, M., Walbot, V. (1989) Effects of cold-treatment on protein synthesis and mRNA levels in rice leaves. *Plant Physiology* 91: 930–938.

Han, L. Z., Zhang, Y. Y., Qiao, Y. L., Cao, G. L., Zhang, S. Y., Kim, J. H., Koh, H. J. (2006) Genetic and QTL analysis for low-temperature vigor of germination in rice. *Yi Chuan Xue Bao* 33(11): 998–1006.

Hannah, M. A., Wiese, D., Freund, S., Fiehn, O., Heyer, A. G., Hincha, D. K. (2006) Natural genetic variation of freezing tolerance in *Arabidopsis*. *Plant Physiology* 142: 98–112.

Hatfield, J. L., Prueger, J. H. (2015) Temperature extremes: Effect on plant growth and development. *Weather and Climate Extremes* 10: 4–10.

He, Y., Li, Y., Cui, L., Xie, L., Zheng, C., Zhou, G., Zhou, J., Xie, X. (2016) Phytochrome B negatively affects cold tolerance by regulating OsDREB1 gene expression through phytochrome interacting factor-like protein OsPIL16 in rice. *Frontiers in Plant Sciences* 7: 1963.

Hu, H., Dai, M., Yao, J., Xiao, B., Li, X., Zhang, Q., Xiong, L. (2006) Overexpressing a NAM, ATAF, and CUC (NAC) transcription factor enhances drought resistance and salt tolerance in rice. *Proceedings of the National Academy of Sciences USA* 103(35): 12987–12992.

Hu, H., You, J., Fang, Y., Zhu, X., Qi, Z., Xiong, L. (2008) Characterization of transcription factor gene SNAC2 conferring cold and salt tolerance in rice. *Plant Molecular Biology* 67(1–2): 169–181.

Huang, J., Sun, S., Xu, D., Lan, H., Sun, H., Wang, Z., Bao, Y., Wang, J., Tang, H., Zhang, H. (2012) A TFIIIA-type zinc finger protein confers multiple abiotic stress tolerances in transgenic rice (*Oryza sativa* L.). *Plant Molecular Biology* 80(3): 337–350.

Huang, L., Hong, Y., Zhang, H., Li, D., Song, F. (2016) Rice NAC transcription factor ONAC095 plays opposite roles in drought and cold stress tolerance. *BMC Plant Biology* 16(1): 203.

IPCC (2007) Climate change 2007: Impacts, adaptation and vulnerability. In *Contribution of Working Group II to Fourth Assessment Report of the Intergovernmental Panel on Climate Change*, Parry M. L., Canziani O. F., Palutikof J. P., van der Linden, P. J., and Hanson, C. E. (eds). Cambridge, UK: Cambridge University Press.

Ito, Y., Katsura, K., Maruyama, K., Taji, T., Kobayashi, M., Seki, M., Shinozaki, K., Yamaguchi-Shinozaki, K. (2006) Functional analysis of rice DREB1/CBF-type transcription factors involved in cold-responsive gene expression in transgenic rice. *Plant and Cell Physiology* 47: 141–153.

Jena, K. K., Kim, S. M., Suh, J. P., Yang, C. I., Kim, Y. J. (2012) Identification of cold-tolerant breeding lines by quantitative trait loci associated with cold tolerance in rice. *Crop Science* 51(2): 517–523.

Jiang, L., Xun, M., Wang, J., Wan, J. (2008) QTL analysis of cold tolerance at seedling stage in rice (*Oryza sativa* L.) using recombination inbred lines. *Journal of Cereal Science* 48: 173–179.

Jiang, Q., Mei, J., Gong, X. D., Xu, J. L., Zhang, J. H., Teng, S., Lin, D. Z., Dong, Y. J. (2014) Importance of the rice TCD9 encoding α subunit of chaperonin protein 60 (Cpn60α) for the chloroplast development during the early leaf stage. *Plant Science* 215–216: 172–179.

Kandpal, R. P., Rao, N. A. (1985) Alterations in the biosynthesis of proteins and nucleic acids in finger millet (*Eleucine coracana*) seedlings during water stress and the effect of proline on protein biosynthesis. *Plant Science* 40(2): 73–79.

Kanneganti, V., Gupta, A. K. (2008) Overexpression of OsiSAP8, a member of stress associated protein (SAP) gene family of rice confers tolerance to salt, drought and cold stress in transgenic tobacco and rice. *Plant Molecular Biology* 66(5): 445–462.

Katiyar, A., Smita, S., Lenka, S. K., Rajwanshi, R., Chinnusamy, V., Bansal, K. C. (2012) Genome-wide classification and expression analysis of MYB transcription factor families in rice and *Arabidopsis*. *BMC Genomics* 13: 544.

Kim, S.-H., Choi, H.-S., Cho, Y.-C., Kim, S.-R. (2011) Cold-responsive regulation of a flower-preferential class III peroxidase gene, OsPOX1, in rice (*Oryza sativa* L.). *Journal of Plant Biology* 55(2): 123–131.

Kim, S.-I., Tai, T. H. (2011) Evaluation of seedling cold tolerance in rice cultivars: A comparison of visual ratings and quantitative indicators of physiological changes. *Euphytica* 178(3): 437–447.

Kim, S. J., Lee, S. C., Hong, S. K., An, K., An, G., Kim, S. R. (2009) Ectopic expression of a cold-responsive OsAsr1 cDNA gives enhanced cold tolerance in transgenic rice plants. *Molecular Cells* 27(4): 449–458.

Koseki, M., Kitazawa, N., Yonebayashi, S., Maehara, Y., Wang, Z. X., Minobe, Y. (2010) Identification and fine mapping of a major quantitative trait locus originating from wild rice, controlling cold tolerance at the seedling stage. *Molecular Genetics and Genomics* 284(1): 45–54.

Kubien, D. S., von Caemmerer, S., Furbank, R. T., Sage, R. F. (2003) C4 photosynthesis at low temperature. A study using transgenic plants with reduced amounts of Rubisco. *Plant Physiology* 132: 1577–1585.

Kuk, Y. I., Shin, J. S., Burgos, N. R., Hwang, T. E., Han, O., Cho, B. H., Jung, S., Guh, J. O. (2003) Antioxidative enzymes offer protection from chilling damage in rice plants. *Crop Science* 43: 2109–2117.

Kumar, S., Asif, M. H., Chakrabarty, D., Tripathi, R. D., Dubey, R. S., Trivedi, P. K. (2013) Expression of a rice Lambda class of glutathione S-transferase, OsGSTL2, in *Arabidopsis* provides tolerance to heavy metal and other abiotic stresses. *Journal of Hazardous Materials* 248–249: 228–237.

Lee, S. C., Huh, K. W., An, K., An, G., Kim, S. R. (2004) Ectopic expression of a cold-inducible transcription factor, CBF1/DREB1b, in transgenic rice (*Oryza sativa* L.). *Molecular Cells* 18: 107–114.

Lee, S. C., Kwon, S. Y., Kim, S. R. (2009) Ectopic expression of a cold-responsive CuZn superoxide dismutase gene, SodCc1, in transgenic rice (*Oryza sativa* L.). *Journal of Plant Biology* 52(2): 154–160.

Li, F., Guo, S., Zhao, Y., Chen, D., Chong, K., Xu, Y. (2010) Overexpression of a homopeptide repeat-containing bHLH protein gene (OrbHLH001) from Dongxiang wild rice confers freezing and salt tolerance in transgenic *Arabidopsis*. *Plant Cell Reports* 29(9): 977–986.

Li, H. W., Zang, B. S., Deng, X. W., Wang, X. P. (2011) Overexpression of the trehalose-6-phosphate synthase gene OsTPS1 enhances abiotic stress tolerance in rice. *Planta* 234(5): 1007–1018.

Li, J., Besseau, S., Törönen, P., Sipari, N., Kollist, H., Holm, L., Palva, E. T. (2013) Defense-related transcription factors WRKY70 and WRKY54 modulate osmotic stress tolerance by regulating stomatal aperture in *Arabidopsis*. *New Phytologist* 200(2): 457–472.

Li, T. G., Visperas, R. M., Vergara, B. S. (1981) Correlation of cold tolerance at different growth stages in rice. *Acta Botanica Sinica* 23: 203–207.

Li, X., Duan, X., Jiang, H., Sun, Y., Tang, Y., Yuan, Z., Guo, J., Liang, W., Chen, L., Yin, J., Ma, H., Wang, J., Zhang, D. (2006) Genome-wide analysis of basic/helix-loop-helix transcription factor family in rice and *Arabidopsis*. *Plant Physiology* 141(4): 1167–1184.

Liu, C., Wu, Y., Wang, X. (2012) BZIP transcription factor OsbZIP52/RISBZ5: A potential negative regulator of cold and drought stress response in rice. *Planta* 235(6): 1157–1169.

Liu, H., Ma, Y., Chen, N., Guo, S., Guo, X., Chong, K., Xu, Y. (2014) Overexpression of stress-inducible OsBURP16, the β subunit of polygalacturonase 1, decreases pectin content and cell adhesion and increases abiotic stress sensitivity in rice. *Plant Cell and Environment* 37(5): 1144–1158.

Lv, Y., Yang, M., Hu, D., Yang, Z., Ma, S., Li, X., Xiong, L. (2017) The OsMYB30 transcription factor suppresses cold tolerance by interacting with a JAZ protein and suppressing β-amylase expression. *Plant Physiology* 173(2): 1475–1491.

Ma, Q., Dai, X., Xu, Y., Guo, J., Liu, Y., Chen, N., Xiao, J., Zhang, D., Xu, Z., Zhang, X., Chong, K. (2009) Enhanced tolerance to chilling stress in OsMYB3R-2 transgenic rice is mediated by alteration in cell cycle and ectopic expression of stress genes. *Plant Physiology* 150(1): 244–256.

Ma, Y., Dai, X., Xu, Y., Luo, W., Zheng, X., Zeng, D., Pan, Y., Lin, X., Liu, H., Zhang, D., Xiao, J., Guo, X., Xu, S., Niu, Y., Jin, J., Zhang, H., Xu, X., Li, L., Wang, W., Qian, Q., Ge, S., Chong, K. (2015) COLD1 confers chilling tolerance in rice. *Cell* 160(6): 1209–1221.

Mackill, D. J., Lei, X. (1997) Genetic variation for traits related to temperate adaptation of rice cultivars. *Crop Science* 37: 1340–1346.

Mahajan, S., Tuteja, N. (2005) Cold, salinity and drought stresses: An overview. *Archives of Biochemistry and Biophysics* 444: 139–158.

Marengo, J. A., Camargo, C. (2008) Surface air temperature trends in Southern Brazil for 1960–2002. *International Journal of Climatology* 28: 893–904.

Maruyama, K., Urano, K., Yoshiwara, K., Morishita, Y., Sakurai, N., Suzuki, H., Kojima, M., Sakakibara, H., Shibata, D., Saito, K., Shinozaki, K., Yamaguchi-Shinozaki, K. (2014) Integrated analysis of the effects of cold and dehydration on rice metabolites, phytohormones, and gene transcripts. *Plant Physiology* 164: 1759–1771.

McFarlane, J., Watson, R., Theisen, A., Jackson, R. D., Ehrler, W., Pinter, P. J., Idso, S. B., Reginato, R. (1980) Plant stress detection by remote measurement of fluorescence. *Applied Optics* 19(19): 3287–3289.

McNeil, S. D., Nuccio, M. L., Hanson, A. D. (1999) Betaines and related osmoprotectants. Targets for metabolic engineering of stress resistance. *Plant Physiology* 120: 945–949.

Mittal, D., Chakrabarti, S., Sarkar, A., Singh, A., Grover, A. (2009) Heat shock factor gene family in rice: Genomic organization and transcript expression profiling in response to high temperature, low temperature and oxidative stresses. *Plant Physiology and Biochemistry* 47: 785–795.

Mittler, R. (2002) Oxidative stress, antioxidants and stress tolerance. *Trends in Plant Science* 7(9): 405–410.

Miura, K., Ashikari, M., Matsuoka, M. (2011) The role of QTLs in the breeding of high-yielding rice. *Trends in Plant Sciences* 16(6): 319–326.

Mizoi, J., Shinozaki, K., Yamaguchi-Shinozaki, K. (2012) AP2/ERF family transcription factors in plant abiotic stress responses. *Biochimica et Biophysica Acta* 1819(2): 86–96.

Morsy, M. R., Almutairi, A. M., Gibbons, J., Yun, S. J., de Los Reyes, B. G. (2005) The OsLti6 genes encoding low-molecular-weight membrane proteins are differentially expressed in rice cultivars with contrasting sensitivity to low temperature. *Gene* 344: 171–180.

Morsy, M. R., Jouve, L., Hausman, J. F., Hoffmann, L., Stewart, J. D. (2007) Alteration of oxidative and carbohydrate metabolism under abiotic stress in two rice (*Oryza sativa* L.) genotypes contrasting chilling tolerance. *Journal of Plant Physiology* 164: 157–167.

Nagamine, T. (1991) Genetic control of tolerance to chilling injury at seedling in rice, *Oryza sativa* L. *Japanese Journal of Breeding* 41: 35–40.

Nagao, M., Minami, A., Arakawa, K., Fujikawa, S., Takezawa, D. (2005) Rapid degradation of starch in chloroplasts and concomitant accumulation of soluble sugars associated with ABA-induced freezing tolerance in the moss *Physcomitrella patens*. *Journal of Plant Physiology* 162(2): 169–180.

Nakano, T., Suzuki, K., Fujimura, T., Shinshi, H. (2006) Genome-wide analysis of the ERF gene family in *Arabidopsis* and rice. *Plant Physiology* 140(2): 411–432.

Nakashima, K., Takasaki, H., Mizoi, J., Shinozaki, K., Yamaguchi-Shinozaki, K. (2012) NAC transcription factors in plant abiotic stress responses. *Biochimica et Biophysica Acta* 1819(2): 97–103.

Nakashima, K., Tran, L. S., Van Nguyen, D., Fujita, M., Maruyama, K., Todaka, D., Ito, Y., Hayashi, N., Shinozaki, K., Yamaguchi-Shinozaki, K. (2007) Functional analysis of a NAC-type transcription factor OsNAC6 involved in abiotic and biotic stress-responsive gene expression in rice. *Plant Journal* 51(4): 617–630.

Nayyar, H., Chander, K., Kumar, S., Bains, T. (2005) Glycine betaine mitigates cold stress damage in chickpea. *Agronomy for Sustainable Development* 25: 381–388.

Ndour, D., Diouf, D., Bimpong, I. K., Sow, A., Manneh, B. (2016) Agro-morphological evaluation of rice (*Oryza sativa* L.) for seasonal adaptation in the Sahelian environment. *Agronomy* 6: 8.

Nijhawan, A., Jain, M., Tyagi, A. K., Khurana, J. P. (2008) Genomic survey and gene expression analysis of the basic leucine zipper transcription factor family in rice. *Plant Physiology* 146(2): 333–350.

Noctor, G., Foyer, C. H. (1998) Ascorbate and glutathione: Keeping active oxygen under control. *Annual Review of Plant Biology* 49(1): 249–279.

Obata, T., Fernie, A. R. (2012) The use of metabolomics to dissect plant responses to abiotic stresses. *Cellular and Molecular Life Sciences* 69: 3225–3243.

Pamplona, R. (2011) Advanced lipoxidation end-products. *Chemico-Biological Interactions* 192(1): 14–20.

Paredes, M., Quiles, M. J. (2015) The effects of cold stress on photosynthesis in Hibiscus plants. *PLoS ONE* 10(9): e0137472.

Rabara, R. C., Tripathi, P., Rushton, P. J. (2014) The potential of transcription factor-based genetic engineering in improving crop tolerance to drought. *OMICS* 18(10): 601–614.

Ranawake, A. L., Manangkil, O. E., Yoshida, S., Ishii, T., Mori, N., Nakamura, C. (2014) Mapping QTLs for cold tolerance at germination and the early seedling stage in rice (*Oryza sativa* L.). *Biotechnology and Biotechnological Equipment* 28(6): 989–998.

Rushton, D. L., Tripathi, P., Rabara, R. C., Lin, J., Ringler, P., Boken, A. K., Langum, T. J., Smidt, L., Boomsma, D. D., Emme, N. J., Chen, X., Finer, J. J., Shen, Q. J., Rushton, P. J. (2012) WRKY transcription factors: Key components in abscisic acid signalling. *Plant Biotechnology Journal* 10(1): 2–11.

Rushton, P. J., Somssich, I. E., Ringler, P., Shen, Q. J. (2010) WRKY transcription factors. *Trends in Plant Science* 15(5): 247–258.

Saad, B. R., Zouari, N., Ramdhan, B. W., Azaza, J., Meynard, D., Guiderdoni, E., Hassairi, A. (2010) Improved drought and salt stress tolerance in transgenic tobacco overexpressing a novel A20/AN1 zinc-finger AlSAP gene isolated from the halophyte grass *Aeluropus littoralis*. *Plant Molecular Biology* 72: 171–190.

Sanghera, G. S., Wani, S. H., Hussain, W., Singh, N. B. (2011) Engineering cold stress tolerance in crop plants. *Current Genomics* 12(1): 30–43.

Sangwan, V., Orvar, B. L., Beyerly, J., Hirt, H., Dhindsa, R. S. (2002) Opposite changes in membrane fluidity mimic cold and heat stress activation of distinct plant MAP kinase pathways. *Plant Journal* 31(5): 629–638.

Saruyama, H., Tanida, M. (1995) Effect of chilling on activated oxygen-scavenging enzymes in low temperature-sensitive and -tolerant cultivars of rice (*Oryza sativa* L.). *Plant Science* 109: 105–113.

Satake, T. (1969) Research on cold injury of paddy rice plants in Japan. *Japanese Agricultural Research Quarterly* 4: 5–10.

Sato, Y., Masuta, Y., Saito, K., Murayama, S., Ozawa, K. (2011a) Enhanced chilling tolerance at the booting stage in rice by transgenic overexpression of the ascorbate peroxidase gene, OsAPXa. *Plant Cell Reports* 30(3): 399–406.

Schläppi, M. R., Jackson, A. K., Eizenga, G. C., Wang, A., Chu, C., Shi, Y., Shimoyama, N., Boykin, D. L. (2017) Assessment of five chilling tolerance traits and GWAS mapping in rice using the USDA mini-core collection. *Frontiers in Plant Sciences* 8: 957.

Schwender, J., Ohlrogge, J., Shachar-Hill, Y. (2004) Understanding flux in plant metabolic networks. *Current Opinion in Plant Biology* 7: 309–317.

Shah, K., Dubey, R. (1997) Effect of cadmium on proline accumulation and ribonuclease activity in rice seedlings: Role of proline as a possible enzyme protectant. *Biologia Plantarum* 40(1): 121–130.

Shakiba, E., Edwards, J. D., Jodari, F., Duke, S. E., Baldo, A. M., Korniliev, P., McCouch, S. R., Eizenga, G. C. (2017) Genetic architecture of cold tolerance in rice (*Oryza sativa*) determined through high resolution genome-wide analysis. *PLoS One* 12(3): e0172133.

Sharifi, P. (2010) Evaluation on sixty-eight rice germplasms in cold tolerance at germination stage. *Rice Science* 17(1): 77–81.

Sharoni, A. M., Nuruzzaman, M., Satoh, K., Shimizu, T., Kondoh, H., Sasaya, T., Choi, I. R., Omura, T., Kikuchi, S. (2011) Gene structures, classification and expression models of the AP2/EREBP transcription factor family in rice. *Plant Cell Physiology* 52(2): 344–360.

Shen, J., Lv, B., Luo, L., He, J., Mao, C., Xi, D., Ming, F. (2017) The NAC-type transcription factor OsNAC2 regulates ABA-dependent genes and abiotic stress tolerance in rice. *Scientific Reports* 7: 40641.

Shimono, H., Hasegawa, T., Iwama, K. (2002) Response of growth and grain yield in paddy rice to cool water at different growth stages. *Field Crops Research* 73(2–3): 67–79.

Shirasawa, S., Endo, T., Nakagomi, K., Yamaguchi, M., Nishio, T. (2012) Delimitation of a QTL region controlling cold tolerance at booting stage of a cultivar, Lijiangxintuanheigu, in rice, *Oryza sativa* L. *Theoretical and Applied Genetics* 124(5): 937–946.

Skopelitis, D. S., Paranychianakis, N. V., Paschalidis, K. A., Pliakonis, E. D., Delis, I. D., Yakoumakis, D. I., Kouvarakis, A., Papadakis, A. K., Stephanou, E. G., Roubelakis-Angelakis, K. A. (2006) Abiotic stress generates ROS that signal expression of anionic glutamate dehydrogenases to form glutamate for proline synthesis in tobacco and grapevine. *Plant Cell* 18(10): 2767–2781.

Smillie, R. M., Hetherington, S. E., He, J., Nott, R. (1988) Photoinhibition at chilling temperatures. *Australian Journal of Plant Physiology* 15: 207–222.

Soltész, A., Vágújfalvi, A., Rizza, F., Kerepesi, I., Galiba, G., Cattivelli, L., Coraggio, I., Crosatti, C. (2012) The rice Osmyb4 gene enhances tolerance to frost and improves germination under unfavourable conditions in transgenic barley plants. *Journal of Applied Genetics* 53(2): 133–143.

Song, J., Wei, X., Shao, G., Sheng, Z., Chen, D., Liu, C., Jiao, G., Xie, L., Tang, S., Hu, P. (2014) The rice nuclear gene WLP1 encoding a chloroplast ribosome L13 protein is needed for chloroplast development in rice grown under low temperature conditions. *Plant Molecular Biology* 84: 301–314.

Song, S. Y., Chen, Y., Chen, J., Dai, X. Y., Zhang, W. H. (2011) Physiological mechanisms underlying OsNAC5-dependent tolerance of rice plants to abiotic stress. *Planta* 234: 331–345.

Su, C. F., Wang, Y. C., Hsieh, T. H., Lu, C. A., Tseng, T. H., Yu, S. M. (2010) A novel MYBS3-dependent pathway confers cold tolerance in rice. *Plant Physiology* 153(1): 145–158.

Suh, J. P., Jeung, J. U., Lee, J. I., Choi, Y. H., Yea, J. D., Virk, P. S., Mackill, D. J., Jena, K. K. (2010) Identification and analysis of QTLs controlling cold tolerance at the reproductive stage and validation of effective QTLs in cold-tolerant genotypes of rice (*Oryza sativa* L.). *Theoretical and Applied Genetics* 120(5): 985–995.

Suzuki, K., Ohmori, Y., Ratel, E. (2011) High root temperature blocks both linear and cyclic electron transport in the dark during chilling of the leaves of rice seedlings. *Plant and Cell Physiology* 52: 1697–1707.

Takasaki, H., Maruyama, K., Kidokoro, S., Ito, Y., Fujita, Y., Shinozaki, K., Yamaguchi-Shinozaki, K., Nakashima, K. (2010) The abiotic stress-responsive NAC-type transcription factor OsNAC5 regulates stress-inducible genes and stress tolerance in rice. *Molecular Genetics and Genomics* 284(3): 173–183.

Theocharis, A., Clement, C., Barka, E. A. (2012) Physiological and molecular changes in plants at low temperatures. *Planta* 235: 1091–1105.

Thomashow, M. F. (1999) Plant cold acclimation: Freezing tolerance genes and regulatory mechanisms. *Annual Review of Plant Physiology and Plant Molecular Biology* 50: 571–599.

Tian, Y., Zhang, H., Pan, X., Chen, X., Zhang, Z., Lu, X., Huang, R. (2011) Overexpression of ethylene response factor TERF2 confers cold tolerance in rice seedlings. *Transgenic Research* 20: 857–866.

Tran, L. S., Nishiyama, R., Yamaguchi-Shinozaki, K., Shinozaki, K. (2010) Potential utilization of NAC transcription factors to enhance abiotic stress tolerance in plants by biotechnological approach. *GM Crops* 1(1): 32–39.

Urano, K., Kurihara, Y., Seki, M., Shinozaki, K. (2010) Omics analyses of regulatory networks in plant abiotic stress responses. *Current Opinion in Plant Biology* 13: 132–138.

Vannini, C., Locatelli, F., Bracale, M., Magnani, E., Marsoni, M., Osnato, M., Mattana, M., Baldoni, E., Coraggio, I. (2004) Overexpression of the rice Osmyb4 gene increases chilling and freezing tolerance of *Arabidopsis thaliana* plants. *Plant Journal* 37(1): 115–127.

Wang, C., Wei, Q., Zhang, K., Wang, L., Liu, F., Zhao, L., Tan, Y., Di, C., Yan, H., Yu, J., Sun, C., Chen, W. J., Xu, W., Su, Z. (2013) Down-regulation of OsSPX1 causes high sensitivity to cold and oxidative stresses in rice seedlings. *PLoS One* 8(12): e81849.

Wang, G., Guo, Z. (2005) Effects of chilling stress on photosynthetic rate and chlorophyll fluorescence parameter in seedlings of two rice cultivars differing cold tolerance. *Rice Science* 12: 187–191.

Wang, Q., Guan, Y., Wu, Y., Chen, H., Chen, F., Chu, C. (2008) Overexpression of a rice OsDREB1F gene increases salt, drought, and low temperature tolerance in both *Arabidopsis* and rice. *Plant Molecular Biology* 67(6): 589–602.

Wang, Y., Zhang, J., Shi, X., Peng, Y., Li, P., Lin, D., Dong, Y., Teng, S. (2016) Temperature-sensitive albino gene TCD5, encoding a monooxygenase, affects chloroplast development at low temperatures. *Journal of Experimental Botany* 67(17): 5187–5202.

Wang, Y. J., Zhang, Z. G., He, X. J., Zhou, H. L., Wen, Y. X., Dai, J. X., Zhang, J. S., Chen, S. Y. (2003) A rice transcription factor OsbHLH1 is involved in cold stress response. *Theoretical and Applied Genetics* 107(8): 1402–1409.

Wathugala, D. L., Richards, S. A., Knight, H., Knight, M. R. (2011) OsSFR6 is a functional rice orthologue of SENSITIVE TO FREEZING-6 and can act as a regulator of COR gene expression, osmotic stress and freezing tolerance in *Arabidopsis. New Phytologist* 191(4): 984–995.

Wilson, L. M., Whitt, S. R., Ibanez, A. M., Rocheford, T. R., Goodman, M. M., Buckle, E. S. (2004) Dissection of maize kernel composition and starch production by candidate gene association. *Plant Cell* 16: 2719–2733.

Wu, K. L., Guo, Z. J., Wang, H. H., Li, J. (2005) The WRKY family of transcription factors in rice and *Arabidopsis* and their origins. *DNA Research* 12(1): 9–26.

Wu, L., Wu, J., Liu, Y., Gong, X., Xu, J., Lin, D., Dong, Y. (2016) The rice pentatricopeptide repeat gene TCD10 is needed for chloroplast development under cold stress. *Rice (N Y)* 9(1): 67.

Xiao, N., Huang, W. N., Li, A. H., Gao, Y., Li, Y. H., Pan, C. H., Ji, H., Zhang, X. X., Dai, Y., Dai, Z. Y., Chen, J. M. (2015) Fine mapping of the qLOP2 and qPSR2-1 loci associated with chilling stress tolerance of wild rice seedlings. *Theoretical and Applied Genetics* 128(1): 173–185.

Xie, G., Kato, H., Imai, R. (2012) Biochemical identification of the OsMKK6-OsMPK3 signalling pathway for chilling stress tolerance in rice. *Biochemical Journal* 443(1): 95–102.

Xie, G., Kato, H., Sasaki, K., Imai, R. (2009) A cold-induced thioredoxin h of rice, OsTrx23, negatively regulates kinase activities of OsMPK3 and OsMPK6 in vitro. *FEBS Letters* 583(17): 2734–2738.

Xin, Z., Browse, J. (2000) Cold comfort farm: The acclimation of plants to freezing temperatures. *Plant, Cell and Environment* 23: 893–902.

Yang, A., Dai, X., Zhang, W. H. (2012a) A R2R3-type MYB gene, OsMYB2, is involved in salt, cold, and dehydration tolerance in rice. *Journal of Experimental Botany* 63: 2541–2556.

Yang, A., Dai, X., Zhang, W. H. (2012b) A R2R3-type MYB gene, OsMYB2, is involved in salt, cold, and dehydration tolerance in rice. *Journal of Experimental Botany* 63(7): 2541–2556.

Yang, Z., Huang, D., Tang, W., Zheng, Y., Liang, K., Cutler, A. J., Wu, W. (2013) Mapping of quantitative trait loci underlying cold tolerance in rice seedlings via high-throughput sequencing of pooled extremes. *PLoS One* 8(7): e68433.

Yanhui, C., Xiaoyuan, Y., Kun, H., Meihua, L., Jigang, L., Zhaofeng, G., Zhiqiang, L., Yunfei, Z., Xiaoxiao, W., Xiaoming, Q., Yunping, S., Li, Z., Xiaohui, D., Jingchu, L., Xing-Wang, D., Zhangliang, C., Hongya, G., Li-Jia, Q. (2006) The MYB transcription factor superfamily of *Arabidopsis*: Expression analysis and phylogenetic comparison with the rice MYB family. *Plant Molecular Biology* 60(1): 107–124.

Yashvardhini, N., Bhattacharya, S., Chaudhuri, S., Sengupta, D. N. (2017) Molecular characterization of the 14-3-3 gene family in rice and its expression studies under abiotic stress. *Planta* 247: 229–253.

Ye, H., Du, H., Tang, N., Li, X., Xiong, L. (2009) Identification and expression profiling analysis of TIFY family genes involved in stress and phytohormone responses in rice. *Plant Molecular Biology* 71: 291–305.

Yokotani, N., Sato, Y., Tanabe, S., Chujo, T., Shimizu, T., Okada, K., Yamane, H., Shimono, M., Sugano, S., Takatsuji, H., Kaku, H., Minami, E., Nishizawa, Y. (2013) WRKY76 is a rice transcriptional repressor playing opposite roles in blast disease resistance and cold stress tolerance. *Journal of Experimental Botany* 64(16): 5085–5097.

Yoshida, S. (1981) Physiological analysis of rice yield. In *Fundamentals of Rice Crop Science*, Yoshida, S. (ed.), 231–251. Los Banos, Philippines: International Rice Research Institute.

You, J., Zong, W., Du, H., Hu, H., Xiong, L. (2014) A special member of the rice SRO family, OsSRO1c, mediates responses to multiple abiotic stresses through interaction with various transcription factors. *Plant Molecular Biology* 84(6): 693–705.

Yuanyuan, M., Yali, Z., Jiang, L., Hongbo, S. (2010) Roles of plant soluble sugars and their responses to plant cold stress. *African Journal of Biotechnology* 8(10): 2004–2010.

Yun, K. Y., Park, M. R., Mohanty, B., Herath, V., Xu, F., Mauleon, R., Wijaya, E., Bajic, V. B., Bruskiewich, R., de Los Reyes, B. G. (2010) Transcriptional regulatory network triggered by oxidative signals configures the early response mechanisms of *japonica* rice to chilling stress. *BMC Plant Biology* 10: 16.

Zahedi, H., Alahrnadi, S. M. J. (2007) Effects of drought stress on chlorophyll fluorescence parameters, chlorophyll content and grain yield of wheat cultivars. *The Journal of Biological Sciences* 7(6): 841–847.

Zhang, F., Huang, L., Wang, W., Zhao, X., Zhu, L., Fu, B., Li, Z. (2012a) Genome-wide gene expression profiling of introgressed *indica* rice alleles associated with seedling cold tolerance improvement in a *japonica* rice background. *BMC Genomics* 13: 461.

Zhang, H., Ni, L., Liu, Y., Wang, Y., Zhang, A., Tan, M., Jiang, M. (2012b) The C_2H_2-type zinc finger protein ZFP182 is involved in abscisic acid-induced antioxidant defense in rice. *Journal of Integrative Plant Biology* 54(7): 500–510.

Zhang, J., Li, J., Wang, X., Chen, J. (2011) OVP1, a vacuolar H^+-translocating inorganic pyrophosphatase (V-PPase), overexpression improved rice cold tolerance. *Plant Physiology and Biochemistry* 49: 33–38.

Zhang, Q., Chen, Q., Wang, S., Hong, Y., Wang, Z. (2014) Rice and cold stress: Methods for its evaluation and summary of cold tolerance-related quantitative trait loci. *Rice* 7: 24.

Zhang, R. X., Qin, L. J., Zhao, D. G. (2017a) Overexpression of the OsIMP gene increases the accumulation of inositol and confers enhanced cold tolerance in tobacco through modulation of the antioxidant enzymes activities. *Genes (Basel)* 8(7): 179.

Zhang, Y., Chen, C., Jin, X. F., Xiong, A. S., Peng, R. H., Hong, Y. H., Yao, Q. H., Chen, J. M. (2009) Expression of a rice DREB1 gene, OsDREB1D, enhances cold and high-salt tolerance in transgenic *Arabidopsis*. *BMB Reports* 42(8): 486–492.

Zhang, Y., Lan, H., Shao, Q., Wang, R., Chen, H., Tang, H., Zhang, H., Huang, J. (2016a) An A20/AN1-type zinc finger protein modulates gibberellins and abscisic acid contents and increases sensitivity to abiotic stress in rice (*Oryza sativa*). *Journal of Experimental Botany* 67(1): 315–326.

Zhang, Z., Li, J., Pan, Y., Zhou, L., Shi, H., Zeng, Y., Guo, H., Yang, S., Zheng, W., Yu, J., Sun, X., Li, G., Ding, Y., Ma, L., Shen, S., Dai, L., Zhang, H., Guo, Y., Li, Z. (2017b) Natural variation in CTB4a enhances rice adaptation to cold habitats. *Nature Communications* 8: 14788.

Zhang, Z. F., Li, Y. Y., Xiao, B. Z. (2016b) Comparative transcriptome analysis highlights the crucial roles of photosynthetic system in drought stress adaptation in upland rice. *Scientific Reports* 6: 19349.

Zhang, Z. H., Su, L., Li, W., Chen, W., Zhu, Y. G. (2005) A major QTL conferring cold tolerance at the early seedling stage using recombinant inbred lines of rice (*Oryza sativa* L.). *Plant Science* 168: 527–534.

Zhu, J., Dong, C. H., Zhu, J. K. (2007) Interplay between cold-responsive gene regulation, metabolism and RNA processing during plant cold acclimation. *Current Opinion in Plant Biology* 10(3): 290–295.

Zinn, K. E., Tunc-Ozdemir, M., Harper, J. F. (2010) Temperature stress and plant sexual reproduction: Uncovering the weakest links. *Journal of Experimental Botany* 61(7): 1959–1968.

9 Heavy Metal Toxicity in Plants and Its Mitigation

Roomina Mazhar and Noshin Ilyas

CONTENTS

9.1 INTRODUCTION

The term heavy metal (HM) is generally used for those metals and metalloids that have 4 g/cm^3 or more atomic density; however, this group is not well defined and includes elements with metallic properties, such as lanthanides, actinides, and metalloids, and transition elements (Hawkes, 1997). Overall, metallic material or substances having high density and toxicity, even in low concentrations, are called heavy metal. On the basis of metal coordination chemistry and in relation to biological systems, metals are divided into three groups. These three groups are Class A with an affinity for O-containing ligands, Class B with an affinity for N- and S-containing ligands, and the third is borderline, meaning an intermediate between Class A and Class B. Metal ion interaction is reflected by this classification.

According to another classification based on the requirements of plants, metals are divided into two groups: essential and non-essential metals. The first group of metals act as micronutrients but only when present in optimum concentrations (Yusuf et al., 2012). At the optimum level, they play a role as a cofactor for various enzymes in which they are vital for different cellular reactions. If their concentration is lower than the optimum level, plants will show different deficiency symptoms, which ultimately results in growth retardation. While at higher concentrations, metals produce toxicity that negatively affects plant growth. Reducible metals are more toxic because they are soluble in an aqueous environment (Boughriet et al., 2007).

Furthermore, HMs play an essential role in the growth and development of plants (Higgins and Burns, 1975).

HMs have several important roles in various biochemical reactions up to certain limits; however, beyond these limits, the formation of different toxic substances leads to cytotoxicity (Appenroth, 2010). Not all HMs are biologically important; for example, gallium, tin, cerium, zirconium, and thorium have very little or no biological importance. Manganese, iron, and molybdenum are essential trace elements and are less toxic in biological systems, while chromium, cobalt, copper, tungsten, zinc, and vanadium are of great importance as trace elements and are also toxic. The metals with limited biological importance and high toxicity are cadmium, lead, antimony, silver, arsenic, and uranium (Appenroth, 2010).

9.2 SOURCES OF HEAVY METAL BUILDUP IN ENVIRONMENT/ AGRICULTURAL SOILS

Industrial effluents are the main source of HMs and their accumulation in soil (Glick, 2007b). Human activities such as mining, metallurgical processes, disposal of metals, and transportation problems play a major role in HM pollution (Zhang et al., 2011). Both natural and anthropogenic sources, including agricultural pollutants, domestic effluent, atmospheric sources, combustion of fossil fuels, disposal of coal ash residue, disposal of commercial products on land, electroplating, the production of different chemicals such as fertilizers, pesticide applications, and industrial sources continuously add HMs into the environment. A natural source of HM includes weathering of rocks, especially those that are close to HM mines (Cabala and Teper, 2007). Large areas of different countries, including China, Japan, and Indonesia, are being contaminated by metals such as Zn, Cd, and Cu, which arise from agriculture, smelting, and mining sources.

Agricultural practices, such as the use of fertilizers and pesticides, sewage sludge, industrial activities, and anthropogenic sources, are a continuous source of toxic HMs such as zinc, nickel, mercury, and cadmium in soil. The addition of these metals to soil leads to their transfer to plants, ultimately causing health hazards and food security threats. As these toxic metals persist in the soil, the use of food from these contaminated soils and water may result in their accumulation at the organism level. The accumulation of HMs in soil also affects the physical and chemical properties of soil. The results from extreme soil contamination are referred to as harmful wastes (Berti and Jacob, 1996).

The reclamation of HM-contaminated soils is very important because these soils are not suitable for human or agricultural purposes. Recently, such contaminated soils have received attention for their remediation in developing countries (Yanez et al., 2002). Bioremediation including phytoremediation, associated with microbes and other agronomic applications, proved beneficial in the treatment of soil contamination. In this review, we will discuss some remediation techniques, particularly bioremediation involving phytoremediation in conjunction with microbes. This review will emphasize the different strategies and the role of this technique in the remediation of HM-contaminated soil.

9.3 HEAVY METAL CYTOTOXICITY AND RESPONSES IN PLANTS

HMs have caused great ecological concerns due to their toxicity, which is why they are considered serious soil pollutants, as they induce severe effects on plant health. The toxic effects of metals depend on the metal's availability, its quantity in soil, its exposure to plants, and also the genotype of a plant. HMs are taken up by the plant roots and affect various physiological, morphological, biochemical, and molecular processes of plants (Malec et al., 2008, 2009; Maleva et al., 2009; Wan et al., 2011; Gangwar et al., 2011; Kumar et al., 2012). When the level of a specific metal in plant tissue reaches a toxic threshold, reactive oxygen species (ROS) production causes cellular injury to the plant. The redox properties of metals aids in their escape from cellular control mechanisms such as active transport, cellular homeostasis, and compartmentalization (Leonard et al., 2004). Such metal invasion leads to cellular toxicity (Flora et al., 2008). Various HMs such as iron, copper, cadmium,

mercury, nickel, lead, and arsenic are involved in the generation of reactive radicals, which are major sources of interruption in cellular activities and also deplete enzyme activities. Due to the toxic effects of HMs on plant physiology and normal functioning, HMs should be kept in a narrow range for the proper functioning of plant systems. For this purpose, plants have several mechanisms to avoid metallic stress up to a certain limit; however, when toxicity exceeds the maximum limit, plants are unable to cope up with the injurious effects (Appenroth, 2010). Non-essential HMs such as Cd, Pb, Tl, and Hg have a biological role in plant functioning (Sari and Tuzen 2009). Cd is a non-essential HM with a role in the degeneration of mitochondria, as an enzyme inhibitor, cellular inhibition, and chromosomal aberration. Such metals cause a very rapid response in the plant even in small quantities (Appenroth, 2010). Zn, an essential element, plays a vital role in various cellular processes such as transcription, translation, cellular signaling, and enzyme activity (Hong-Bo et al., 2010). Plants exposed to zinc metal for a long time undergo disintegration of the cellular membranes, ion leakage, nucleic acid disintegration, and eventually cell death.

Plants adopt various avoidance or tolerance mechanisms to cope with metallic stress, for example, ionic compartmentalization, control of metal influx and efflux, chelate metal ion, a reduction in the bioavailability of metal, and metal ion detoxification by metal-induced ROS. Furthermore, plants can use one or more ways to avoid stress conditions (Gupta et al., 2010), depending on their genetic makeup and the prevailing environmental conditions. HM-sensitive plants develop phytotoxicity symptoms on exposure to metals. Elevated levels of HMs result in excessive uptake by plants, which results in growth reduction and disturbed metabolism. The direct effect of the accumulation of metal ions is the impairment of metabolic activities. Jonak et al. (2004) reported that metal toxicity might affect plant signaling pathways that initiate metal toxicity responses.

During stress conditions, there is an increase in the production of ROS, which causes damage to many cellular membranes and photosynthetic pigments (Mittler et al., 2004; Flora et al., 2008). Enzymes sensitive to metals are mostly affected by As, Cd, and Hg, and distortion of these enzymes leads to retarded growth and death of the organism. Enzymatic antioxidants include peroxidases (POD), catalase (CAT), and superoxide dismutases (SOD), while non-enzymatic antioxidants include glutathione, carotenoids, and ascorbate. The activities of free radicals increase as a result of stress conditions. In particular, HMs induce the activities of CAT, POD, and SOD in many plant species (Lee et al., 2007; Li et al., 2007).

HMs induce ROS production and the production of proteins that stimulate rapid signal transduction in a mitogen-activated protein kinase (MAPK) cascade. An MAPK cascade transduces a signal that is sent to transcriptional factors, which bind with a specific element that activates metal-responsive gene expression, macromolecules damaged by oxidation, impaired DNA repair, and poor protein folding (Jonak et al., 2002).

Furthermore, the cytotoxicity of HMs results in the production of reactive nitrogen species (NO) along with ROS. It also results in a change in cofactors, redox imbalance, ionic transport imbalance, DNA damage, and transcriptional factors (Aravind et al., 2009; Cuypers et al., 2011; Gangwar et al., 2011). Enzymes sensitive to metals are mostly affected by As, Cd, and Hg. In metal-sensitive plants, the MAPK cascade does not provide sufficient tolerance, while metal-tolerant plants successfully activate different processes such as a hormonal signaling pathway, a decrease in photosynthesis, senescence, and programmed cell death induction. Photosynthesis decreases by restraining photosystem II (PSII) activity, inhibiting the PSII photoreaction, decreasing photophosphorylation, reducing the activity of chloroplast enzymes RuBPC and phosphoribulokinase, decreasing photosynthetic pigments (such as total chlorophyll content and chlorophyll a/b ratio), decreasing net leaf photosynthesis, and reducing chloroplast metabolism (Huang et al., 2012).

9.4 BIOREMEDIATION/ PHYTOREMEDIATION AND COMBINED APPLICATION FOR HEAVY METALS

Bioremediation is a technology that relies on living organisms to deteriorate or degenerate pollutants into their lowest hazardous form. Regardless of the toxicity of HMs, certain plants are able to grow under such conditions and either accumulate or exclude them. Such plants have various adaptive strategies to resist the toxic effects of HMs (Sharma et al., 2008). The techniques that rely on plants for bioremediation are known as phytoremediation (Macek et al., 2008). Plants that can endure and store large amounts of metals are called hyperaccumulators. Perfect hyperaccumulators for remediation need to be fast growing and have a large concentration of mass. Nevertheless, several hyperaccumulators have sluggish development and repressed growth due to the existence of large amounts of toxic metals. The decontamination of metals by plants, which is a prerequisite for elimination from surroundings, is problematic as well. The basic restrictions on many

methods of metal removal by plants are the presence of metal(s) and the capability of many plants to store metals in their shoots (Raskin and Ensley, 2000; Sari and Tuzen 2009). Researchers have attempted to enhance the existence of metals by the application of many binding factors, an approach that is frequently useful at low range in the laboratory but it is not very efficient in the field. Numerous plants have been studied for their ability to absorb large amount of metals and, after absorption, to transfer the metals from the roots to the shoots and leaves; however, most of them are known as hyperaccumulating plants that cannot synthesize enough biomass to make such a method effectual in the field (Raskin and Ensley, 2000; McGrath et al., 2001). Furthermore, a huge amount of metal pollutants are inaccessible to the plant by root intake because HMs in soils usually have inorganic and organic soil components, or instead, are available as unsolvable substances. Thus, the accessibility of metals in soil by plants is a serious concern for the success of phytoremediation (Kukier et al., 2004; Mulligan et al., 2001).

Plant growth-promoting rhizobacteria (PGPR) have a strong potential to improve plant growth in various abiotic stresses such as salt, drought, and HM stress, previously documented by various studies (Mazhar et al., 2016; Saeed et al., 2016; Kanwal et al., 2017). Furthermore, the use of soil microbes (such as plant growth-promoting bacteria) also aids in the biodegradation of metals and could considerably stimulate plant growth in the presence of metals. The remediation of HMs with PGPR involves several mechanisms: rhizofiltration, phytostabilization, phytovolatilization, and phytoextraction (Glick et al., 1999). Root microbes could benefit the plant by nutrient uptake, enhancing the plant growth and the quality of soil (Barea et al., 2002). Certain bacteria could improve the growth of plants, metal and P intake when inoculated with hyperaccumulating plant, e.g., *Sedum alfredii* (Li et al., 2007). Much study of the microbe–plant association has been carried out during the last 50 years. The rhizospheric microbe–plant interaction is a necessary metabolism. In addition, the composition of rhizospheric bacteria largely depends on the type of plant, its concentration, and the components of root secretions and various root regions (Marschner et al., 2004; Yang et al., 2009). PGPR, which have been used in research, are usually applied in the reclamation of poisonous metal(s) and after they are frequently chosen for the existence of 1-aminocyclopropane-1-carboxylate (ACC) deaminase and their potential to produce siderophores and indole acetic acid (IAA) (Elobeid and Polle, 2012). Such investigations have been carried out to examine the probability that metal-resistant bacteria are capable of either increasing or stimulating plant growth in noxious elements.

9.5 MECHANISMS OF PGPR-INDUCED REMEDIATION OF HEAVY METALS

The involvement of PGPR in phytoremediation techniques is being studied extensively as PGPR can improve plant growth in polluted areas (Burd et al., 2000; Lucy et al., 2004) and can also help in land reclamation (Mayak et al., 2004). Microbes in metal-contaminated soils have been analyzed with special reference to their structure and population in the rhizosphere (Dell'Amico et al., 2005). In HM-contaminated soil, plant growth-promoting bacteria can help to improve the growth of a plant by direct as well as indirect mechanisms (Glick et al., 1999). Certain reports have recognized the particular procedure(s) for stimulating plant growth in the presence of metal, e.g., the production of IAA, ACC deaminase, and siderophores. IAA-producing bacteria can stimulate the growth of plants (Patten and Glick, 2002); ACC deaminase inhibits ethylene production (Glick et al., 2007); and siderophores help plants to obtain enough iron under HM stress (Burd et al., 2000). Furthermore, in addition to the above-mentioned methods for plant growth promotion, many additional bacterial features may help to improve metal bioremediation. Scientists have genetically modified bacteria to utilize the bacterial genetic potential to synthesize several metal-chelating proteins (Wu et al., 2006a; Ike et al., 2007). Moreover, numerous studies have proven that bacteria that improve phytoremediation have a dynamic ability to solubilize phosphate in contaminated soil. For instance, microbial isolates that help in bioremediation have the ability to synthesize biosurfactants, thereby facilitating the availability of metals (Sheng et al., 2008a).

As free-living and synergetic, PGPR can also directly improve plant growth by phosphorus solubilization for plant intake; nitrogen fixation; the absorption of micro elements such as iron by siderophores; the release of phytohormones as cytokinins, auxins, and gibberellins; and the decrease in ethylene production. PGPR can also improve plant growth by several indirect mechanisms such as antibiotic synthesis, exhaustion of iron by rhizospheres, induced systemic resistance, synthesis of cell wall–degrading enzymes, and the struggle for association with roots by quorum sensing (Glick et al., 2007; Daniels et al., 2004). Nitrogen-fixing bacteria chemically stimulate rhizobium nodule genes, Nod factors, and acyl homoserine lactone (AHL) production for phytoremediation. Dell'Amico et al. (2005) stated that

in the presence of PGPR in relation to rhizoplane and rhizosphere, several populations of microbes are capable of tolerating large HM contaminants. Soil microbes are famous for enabling the mobility of metals in plants as well as oxidoreduction modification or with the synthesis of iron chelators and siderophores for confirming the presence of iron, and assembling the phosphates (Burd et al., 2000; Guan et al., 2001). Likewise, ethylenediaminetetraacetic acid (EDTA) and EGTA are also better chelators to stimulate the accessibility of metals in plants, while such chelators might cause negative impacts such as metal leakage and fewer microbes action (Römkens et al., 2002).

PGPR has the strong potential to sorb the metals from the soil. Kumar and Nagendran (2009) documented that such a procedure is beneficial for the effective reduction of HMs by land and, additionally, that they might be disposed of securely. A well-recognized study showed the use of *Thiobacillus*, which is capable of performing microbial percolating of elements such as silver, zinc, copper, and uranium with oxidation of constituents. Mishra et al. (2008) reported another use of this procedure: metals suspension from lithium elements of secondary batteries utilizing *Acidithiobacillus ferrooxidans*. Ginn and Fein (2008) analyzed the influence of bacterial species variation on metal uptake. This research determined Pb, Cd, and Ni absorption by bacteria *Flavobacterium aquatile*, *Thermus thermophilus*, *Acidophilus mangustum*, *Deinococcus radiodurans*, and *Flavobacterium hibernum* and measured the thermodynamic constancy by essential apparent multiplexes. While certain microbes are capable of reducing the solubilization of metals and their flexibility, reducing their toxicity to plants (Lasat, 2002) could be linked to biosurfactant synthesis. Juwarkar et al. (2007) used biosurfactant synthesis by *Pseudomonas aeruginosa* in the decontamination of lead- and cadmium-polluted land. Benisse et al. (2004) has documented that certain rhizospheric microbial species reduced zinc toxicity. In short, the bacterial inoculates and isolates from HM-contaminated sites have the strong potential to degrade metals and can help to improve plant growth due to various PGPR characteristics.

9.6 CONCLUSION

It is necessary to note that in natural surroundings about 90% of land plants are associated with PGPR, making them ideal for bioremediation in combination with plants. Thus, plant-linked microorganisms have the ability to stimulate the growth of a plant by various PGPR mechanisms such as siderophores production, ACC deaminase synthesis, phytohormones production, phosphate solubilization, and sequestration or biosorption of metals in ionic form, which enable the bacterial cells and plants to survive in contaminated soil. Furthermore, PGPR plant association can aid metal mobility/immobility in metal-polluted land.

REFERENCES

Appenroth, K.J. (2010) What are "heavy metals" in plant sciences? *Acta Physiologiae Plantarum* 32(4):615–619.

Aravind, P., Prasad, M.N.V., Malec, P., Waloszek, A., Strzalka, K. (2009) Zinc protects *Ceratophyllum demersum* L. (free floating hydrophyte) against reactive oxygen species induced by cadmium. *Journal of Trace Elements and Medicinal Biology* 23(1):50–60.

Barea, J.M., Azcón, R., Azcón-Aguilar, C. (2002) Mycorrhizosphere interactions to improve plant fitness and soil quality. *Antonie Van Leeuwenhoek* 81(1–4):343–351.

Benisse, R., Labat, M., Elasli, A., Brhada, F., Chandad, F., Liegbott, P.P., Hibti, M., Qatibi, A. (2004) Rhizosphere bacterial populations of metallophyte plants in heavy metal-contaminated soils from mining areas in semi-arid climate. *World Journal of Microbiology & Biotechnology* 20(7):759–766.

Berti, W.R., Jacob, L.W. (1996) Chemistry and phytotoxicity of soil trace elements from repeated sewage sludge applications. *Journal of Environment Quality* 25(5):1025–1032.

Boughriet, A., Proix, N., Billon, G., Recourt, P., Ouddane, B. (2007) Environmental impacts of heavy metal discharges from a smelter in Deule-canal sediments (Northern France): Concentration levels and chemical fractionation. *Water, Air, and Soil Pollution* 180(1–4):83–95.

Burd, G.I., Dixon, D.G., Glick, B.R. (2000) Plant growth-promoting bacteria that decrease heavy metal toxicity in plants. *Canadian Journal of Microbiology* 46(3):237–245.

Cabala, J., Teper, L. (2007) Metalliferous constituents of rhizosphere soils contaminated by Zn–Pb mining in Southern Poland. *Water, Air, and Soil Pollution* 178(1–4):351–362.

Cuypers, A., Smeets, K., Ruytinx, J., Opdenakker, K., Keunen, E., Remans, T., Horemans, N., Vanhoudt, N., Van Sanden, S., Van Belleghem, F., Guisez, Y., Colpaert, J., Vangronsveld, J. (2011) The cellular redox state as a modulator in cadmium and copper responses in *Arabidopsis thaliana* seedlings. *Journal of Plant Physiology* 168(4):309–316.

Daniels, R., Vanderleyden, J., Michiels, J. (2004) Quorum sensing and swarming migration in bacteria. *FEMS Microbiology Reviews* 28(3):261–289.

Dell'Amico, E., Cavalca, L., Andreoni, V. (2005) Analysis of rhizobacterial communities in perennial Graminaceae from polluted water meadow soil, and screening of metal-resistant, potentially plant growth-promoting bacteria. *FEMS Microbiology Ecology* 52(2):153–162.

Elobeid, M., Polle, A. (2012) Interference of heavy metal toxicity with auxin physiology. *Metal. Toxicity in Plants: Perception, Signaling and Remediation* 199:249–259.

Flora, S.J.S., Mittal, M., Mehta, A. (2008) Heavy metal induced oxidative stress and its possible reversal by chelation therapy. *Indian Journal of Medicinal Research* 128:501–523.

Gangwar, S., Singh, V.P., Srivastava, P.K., Maurya, J.N. (2011) Modification of chromium (VI) Phytotoxicity by exogenous gibberellic acid application in *Pisum sativum* (L.) seedlings. *Acta Physiologiae Plantarum* 33(4):1385–1397.

Ginn, B.R., Fein, J.B. (2008) The effect of species diversity on metal adsorption onto bacteria. *Geochimica et Cosmochimica Acta* 72(16):3939–3948.

Glick, B.R., Cheng, Z., Czarny, J., Duan, J. (2007) Promotion of plant growth by ACC deaminase containing soil bacteria. *European Journal of Plant Pathology* 119(3):329–339.

Glick, B.R., Patten, C.L., Holguin, G., Penrose, D.M. (1999) *Biochemical and Genetic Mechanisms Used by Plant Growth-Promoting Bacteria*. London: Imperial College Press.

Guan, L.L., Kanoh, K., Kamino, K. (2001) Effect of exogenous siderophores on iron uptake activity of marine bacteria under iron limited conditions. *Applied and Environmental Microbiology* 67(4):1710–1717.

Gupta, S.C., Sharma, A., Mishra, M., Mishra, R.K., Chowdhuri, D.K. (2010) Heat shock proteins in toxicology: How close and how far? *Life Sciences* 86(11–12):377–384.

Hawkes, S.J. (1997) Heavy metals. *Journal of Chemical Education* 74(11):1369–1374.

Higgins, I.J., Burns, R.G. (1975) *The Chemistry and Microbiology of Pollution*. London: Academic Press.

Hong-Bo, S., Li-Ye, C., Cheng-Jiang, R., Hua, L., Dong-Gang, G., Wei-Xiang, L. (2010) Understanding molecular mechanisms for improving phytoremediation of heavy metal-contaminated soils. *Critical Reviews in Biotechnology* 30(1):23–30.

Huang, H., Gupta, D.K., Tian, S., Yang, X.E., Li, T. (2012) Lead tolerance and physiological adaptation mechanism in roots of accumulating and non-accumulating ecotypes of Sedum alfredii. *Environmental Science and Pollution Research International* 19(5): 1640–1651.

Ike, A., Sriprang, R., Ono, H., Murooka, Y., Yamashita, M. (2007) Bioremediation of cadmium contaminated soil using symbiosis between leguminous plant and recombinant rhizobia with the MTL4 and the PCS genes. *Chemosphere* 66(9):1670–1676.

Jonak, C., Nakagami, H., Hirt, H. (2004) Heavy metal stress. Activation of distinct mitogen-activated protein kinase pathways by copper and cadmium. *Plant Physiology* 136(2):3276–3283.

Jonak, C., Ökrész, L., Bögre, L., Hirt, H. (2002) Complexity, cross talk and integration of plant MAP kinase signalling. *Current Opinion in Plant Biology* 5(5): 415–424.

Juwarkar, A.A., Nair, A., Dubey, K.V., Singh, S.K., Devotta, S. (2007) Biosurfactant technology for remediation of cadmium and lead contaminated soils. *Chemosphere* 68(10):1996–2002.

Kanwal, S., Noshin, I., Sumera, S., Maimona, S., Robina, G., Maryum, Z., Nazima, B., Roomina, M. (2017) Application of biochar in mitigation of negative effects of salinity stress in wheat (*Triticum aestivum* L.). *Journal of Plant Nutrition* 41(4): 526–538.

Kukier, U., Peters, C.A., Chaney, R.L., Angle, J.S., Roseberg, R.J. (2004) The effect of pH on metal accumulation in two alyssum species. *Journal of Environmental Quality* 33(6):2090–2102.

Kumar, A., Prasad, M.N.V., Sytar, O. (2012) Lead toxicity, defense strategies and associated indicative biomarkers in *Talinum triangulare* grown hydroponically. *Chemosphere* 89(9):1056–1065.

Kumar, R.N., Nagendran, R. (2009) Fractionation behavior of heavy metals in soil during bioleaching with *Acidithiobacillus* thiooxidans. *Journal of Hazardous Materials* 169(1–3):1119–1126.

Lasat, M.M. (2002) Phytoextraction of toxic metals: A review of biological mechanisms. *Journal of Environmental Quality* 31(1):109–120.

Lee, K.P., Kim, C., Landgraf, F., Apel, K. (2007) EXECUTER1-and EXECUTER2-dependent transfer of stress-related signals from the plastid to the nucleus of *Arabidopsis thaliana*. *Proceedings of the National Academy of Sciences USA* 104(24):10270–10275.

Leonard, S.S., Harris, G.K., Shi, X.L. (2004) Metal-induced oxidative stress and signal transduction. *Free Radical Biology and Medicine* 37(12):1921–1942.

Li, W.C., Ye, Z.H., Wong, M.H. (2007) Effects of bacteria on enhanced metal uptake of the Cd/Zn hyperaccumulating plant, *Sedum alfredii*. *Journal of Experimental Botany* 58(15–16):4173–4182.

Lucy, M., Reed, E., Glick, B.R. (2004) Application of free living plant growth promoting rhizobacteria. *Antonie Van Leeuwenhoek* 86(1):1–25.

Macek, T., Kotrba, P., Svatos, A., Novakova, M., Demnerova, K., Mackova, M. (2008) Novel roles for genetically modified plants in environmental protection. *Trends in Biotechnology* 26(3):146–152.

Malec, P., Maleva, M.G., Prasad, M.N.V., Strzałka, K. (2009) Identification and characterization of Cd-induced peptides in *Egeria densa* (water weed): Putative role in Cd detoxification. *Aquatic Toxicology* 95(3):213–221.

Malec, P., Waloszek, K.A., Prasad, M.N.V., Strzałka, K. (2008) Zinc reversal of Cd-induced energy transfer changes in photosystem II of *Ceratophyllum demersum* L. as observed by whole-leaf 77K fluorescence. *Plant Stress* 2:121–126.

Maleva, M.G., Nekrasova, G.F., Malec, P., Prasad, M.N.V., Strzałka, K. (2009) Ecophysiological tolerance of *Elodea canadensis* to nickel exposure. *Chemosphere* 77(3):392–398.

Marschner, P., Crowley, D., Yang, C.H. (2004) Development of specific rhizosphere bacterial communities in relation to plant species, nutrition and soil type. *Plant and Soil* 261(1/2):199–208.

Mayak, S., Tirosh, S., Glick, B.R. (2004) Plant growth promoting bacteria that confer resistance to water stress in tomatoes and peppers. *Plant Physiology* 166(2):525–530.

Mazhar, R., Ilyas, N., Saeed, M., Bibi, F., Batool, N. (2016) Biocontrol and salinity tolerance potential of *Azospirillum lipoferum* and its inoculation effect in wheat crop. *International Journal of Agriculture and Biology* 18(3):494–500.

McGrath, S.P., Zhao, F.J., Lombi, E. (2001) Plant and rhizosphere processes involved in phytoremediation of metal-contaminated soils. *Plant and Soil* 232(1–2):207–214.

Mishra, D., Kim, D.J., Ralph, D.E., Ahn, J.G., Rhee, Y.A. (2008) Bioleaching of metals from spent lithium ion secondary batteries using *Acidithiobacillus ferrooxidans*. *Waste Management* 28:33.

Mulligan, C.N., Young, R.N., Gibbs, B.F. (2001) Remediation technologies for metal-contaminated soils and groundwater: An evaluation. *Engineering Geology* 60(1–4):193–207.

Patten, C.L., Glick, B.R. (2002) The role of bacterial indole-acetic acid in the development of the host plant root system. *Applied and Environmental Microbiology* 68(8):3795–3801.

Raskin, I., Ensley, B.D. (2000) *Phytoremediation of Toxic Metals: Using Plants to Clean Up the Environment.* New York: Wiley-Interscience.

Römkens, P., Bouwman, L., Japenga, J., Draaisma, C. (2002) Potentials and drawbacks of chelate-enhanced phytoremediation of soils. *Environmental Pollution* 116(1):109–121.

Saeed, M., Ilyas, N., Mazhar, R., Bibi, F., Batool, N. (2016) Drought mitigation potential of *Azospirillum* inoculation in canola (*Brassica napus*). *Journal of Applied Botany and Food Quality* 89:270–278.

Sari, A., Tuzen, M. (2009) Kinetic and equilibrium studies of biosorption of Pb(II) and Cd(II) from aqueous solution by macrofungus (*Amanita rubescens*) biomass. *Journal of Hazardous Materials* 164(2–3):1004–1011.

Sharma, S.K., Goloubinoff, P., Christen, P. (2008) Heavy metal ions are potent inhibitors of protein folding. *Biochemical and Biophysical Research Communications* 372(2):341–345.

Sheng, X.F., Xia, J.J., Jiang, C.Y., He, L.Y., Qian, M. (2008a) Characterization of heavy metal-resistant endophytic bacteria from rape (*Brassica napus*) roots and their potential in promoting the growth and lead accumulation of rape. *Environmental Pollution* 156(3):1164–1170.

Wan, G., Najeeb, U., Jilani, G., Naeem, M.S., Zhou, W. (2011) Calcium invigorates the cadmium-stressed *Brassica napus* L. plants by strengthening their photosynthetic system. *Environmental Science and Pollution Research International* 18(9):1478–1486.

Wu, C.H., Wood, T.K., Mulchandani, A., Chen, W. (2006a) Engineering plant–microbe symbiosis for rhizoremediation of heavy metals. *Applied and Environmental Microbiology* 72(2):1129–1134.

Yanez, L., Ortiz, D., Calderon, J., Batres, L., Carrizales, L., Mejia, J., Martínez, L., García-Nieto, E., Díaz-Barriga, F. (2002) Overview of human health and chemical mixtures: Problems facing developing countries. *Environmental Health Perspectives* 110:901–909.

Yang, J., Kloepper, J.W., Ryu, C.M. (2009) Rhizosphere bacteria help plants tolerate abiotic stress. *Trends in Plant Science* 14(1):1–4.

Yusuf, M., Fariduddin, Q., Varshney, P., Ahmad, A. (2012) Salicylic acid minimizes nickel and/or salinity-induced toxicity in Indian mustard (*Brassica juncea*) through an improved antioxidant system. *Environmental Science and Pollution Research International* 19(1): 8–18.

Zhang, Y.F., He, L.Y., Chen, Z.J., Zhang, W.H., Wang, Q.Y., Qian, M., Sheng, X.F. (2011) Characterization of lead-resistant and ACC deaminase-producing endophytic bacteria and their potential in promoting lead accumulation of rape. *Journal of Hazardous Materials* 186(2–3):1720–1725.

10 Nutrient Deficiency and Toxicity Stress in Crop Plants
Lessons from Boron

Himanshu Bariya, Durgesh Nandini, and Ashish Patel

CONTENTS

10.1 INTRODUCTION

The field of plant abiotic stress contains all studies on abiotic factors or stress conditions in the environment that can inflict stress on a variety of species (Sulmon et al., 2015). This stress includes acute levels of light (high and low), radiation (UV-B and UV-A), temperature (high and low [chilling, freezing]), and water (drought, flooding, and submergence), chemical factors (heavy metals and pH), salinity due to excessive Na^+, nutrients stress (deficient or excessive essential nutrients), gaseous pollutants (ozone, sulfur dioxide), mechanical factors, and other less-frequently occurring stressors. Since combinations of these stresses, such as heat and drought, frequently occur under field conditions and can cause unique effects that cannot be predicted from individual stress (Suzuki et al., 2014), a multiplicity of physiological interactions can be expected, needing individual novel solutions.

One of the most important stresses is nutrient stress. Essential mineral nutrients play key roles in many aspects of plant metabolism, growth, and development. There is, therefore, a very wide spectrum of responses to nutrient stresses. Typically, the nature of the responses to a particular nutrient stress depends upon which plant processes are most sensitive to that stress and the severity of the stress. These may vary between species. They may also vary between genotypes within the same species and so reflect genotypic differences in nutrient efficiency related to differences in nutrient acquisition and/or utilization. Such differences provide plants with avenues that enable them to adapt to moderate nutrient stresses. Plant breeders are successfully exploiting these adaptive traits to produce

179

more nutrient-efficient genotypes. As might be expected from complex traits, they are commonly multigenic.

Although most nutrient stresses decrease the plant growth rate, the effects on metabolic processes differ between nutrients. For some nutrients, changes in the rates of specific assimilatory and biochemical reactions are the predominant responses to a deficiency of that nutrient (Marschner, 1995a). However, plants have adapted to some nutrient stresses by increasing their capacity to acquire more nutrient from the soil. These adaptations have arisen during evolution and apply to nutrients that have been poorly available to plants for a long time. They may give rise to major morphological and biochemical adaptations. For nutrients such as phosphate, zinc, and nitrogen, many plant species have also acquired or retained the capacity to form symbiotic associations with other organisms in order to gain that essential nutrient in stress situations.

Boron is a necessary micronutrient for the regular development of higher plants (Blevins and Lukaszewski, 1998; Bariya et al., 2014). Its main role is to make borate-diol ester bonds link two rhamnogalacturonan II (RG-II) chains of pectic polysaccharide (Pérez-Castro et al., 2012; Funakawa and Miwa, 2015). Recent research has shown that boron also cross-links glycosyl inositol phosphoryl ceramides (GIPCs) of the plasma membrane (PM) with arabinogalactan proteins (AGPs) of the cell wall, in this way attaching the membrane to the cell wall (Voxeur and Fry, 2014). Thus, boron is recognized as influencing the involuntary properties of the cell wall. There is a very narrow range of optimum boron concentrations for plant development. Anomalous levels of boron can be toxic or can trigger deficiency symptoms (Pérez-Castro et al., 2012). The optimum boron concentration for one species can be either toxic or deficient for other species (Blevins and Lukaszewski, 1998).

10.2 BORON TOXICITY IN CROP PLANT

Phenotypic symptoms of B toxicity comprise retardation of root and shoot growth, necrosis of shoots, and inadequate production of chlorophyll (Nable et al., 1990). The primary cause of these developmental changes may be linked to the interruption of a cascade (slowing down or stopping) of physiological processes, including hang-up of photosynthesis, lower stomatal conductance, augmented peroxidation of lipids, changes in enzymes within antioxidation pathways, induced membrane leakiness (Karabal et al., 2003), and reduced proton flow from roots (Roldán et al., 1992). Increased synthesis and the deposition of suberin and lignin have also been reported (Ghanati et al., 2005). B toxicity symptoms are generally interrelated with the accumulation of high levels of B in

shoots, which is associated with the soil boron content and the period of exposure. Early remarks by Oertli and Kohl (1961) on 29 plant species, which included various plant species, recognized that, in general, chlorosis of the leaves occurred at approximately 1000 mg kg^{-1} DW and necrosis between 1500 and 2000 mg kg^{-1} DW. The pattern of necrosis was correlated with venation, such that symptoms initially developed at the ends of the veins. From this, it was concluded that excess boron remained in the xylem and therefore accumulated where the xylem vessels terminated. This means that for grasses, which have parallel venation, toxicity is first observed at the leaf tip, whereas for dicots, which generally have reticulate venation, toxicity is observed around the leaf margins.

Moreover, Cervilla et al. (2012) examined different abiotic stress indicators to select the parameters most indicative of B toxicity in two tomato genotypes, characterized by different sensitivity to B excess. They suggested that O^{2-} and anthocyanin levels in leaves, GPX activity, chlorophyll b, and proline content are the best indicators of B stress in tomato. In B-mobile species (e.g., Prunus, Malus, Pyrus), B accumulation has been observed in developing sinks rather than at the end of the transpiration stream. In these plants, the symptoms of B toxicity are expressed as fruit disorders (gummy nuts, internal necrosis), bark necrosis, death of the cambial tissues, and stem dieback (Brown and Hu, 1996). In particular, in stone-fruit trees, B toxicity caused a reduction in flower bud formation, poor fruit set, and malformed fruit that is particularly poor in flavor (Suarez, 2012). In contrast, in rice, a B-mobile species, B toxicity caused similar foliar symptoms as barley (Bellaloui et al., 2003).

A direct relationship between the B content in leaves and the severity of toxicity symptoms has been demonstrated. The leaf B concentrations of sensitive and tolerant species have been reported to greatly vary up to tenfold (Furlani et al., 2003). For this reason, the diagnosis of B toxicity has been extensively evaluated using tissue B-content analysis in leaves rather than in shoots (Reid, 2013).

Moreover, the critical toxicity values of tissue B concentrations have been established in many plant species, because B concentrations also vary greatly in relation to different plant tissues and/or plant developmental stages.

10.3 EFFECTS OF EXCESS BORON STRESS ON THE PLANT

10.3.1 PHOTOSYNTHESIS

High B concentration induced damage in photosynthesis, although the mechanisms of toxicity remain unclear.

Under high B stress, the edge of the leaf died (Fang, 2001), and the photosynthetic area, chlorophyll content, and consequently the photosynthetic rate were reduced (Han et al., 2009; Chen et al., 2013). A complementary outcome was observed only in barley leaves in which photosynthesis was not predominantly sensitive to B excess because it was untouched by 50 mM and hardly inhibited (23%) at a 100 mM B concentration (Reid et al., 2004). Landi et al. (2013) found that B excess was responsible for decreasing the chlorophyll a/b ratio simultaneously with downregulation of the photosystem II photochemical efficiency in cucurbits (*Cucumis sativus* L. and *Cucurbita pepo* L.). In many species, B excess significantly reduced the Fv/Fm ratio (maximum quantum output of chlorophyll fluorescence), which indicated that leaves were photo inhibited, a condition that can lead to reactive oxygen species (ROS) generation (Velez-Ramirez et al., 2011). This event could also explain the drop off in chlorophyll content (Han et al., 2009; Chen et al., 2012). Furthermore, the inhibition in electron transport rate was also associated with the decrease in the activity of some enzymes involved in CO_2 assimilation (carboxylase/oxygenase, ribulose-1, 5-bisphosphate, and fructose-1,6-bisphosphate phosphatase), determining a reduction in nicotinamide adenine dinucleotide phosphate (NADPH) and adenosine triphosphate (ATP) utilization (Han et al., 2009). Recently, Chen et al. (2013), through a proteomic approach, investigated the protein profiles in *Arabidopsis* leaves in response to B excess. Interestingly, proteins involved in both the light and CO_2 fixation reactions of the photosynthesis process were affected by excess B, before the manifestation of visible symptoms in leaves and the decrease in chlorophyll content, total cell protein, and growth.

10.3.2 Nitrogen Metabolism

Plants require large amounts of nitrogen (N) for the biosynthesis of amino acids, nucleic acids, proteins, and secondary metabolites. In agricultural soils, nitrate is the dominant form of nitrogen, essential for plant growth and productivity (Crawford, 1995). Once nitrate is acquired from roots, it is reduced to nitrite (NO_2^-) by nitrate reductase (NR), the key step in the nitrate assimilation process, which in turn is transformed into ammonium (NH_4^+) by nitrite reductase and then integrated into carbon skeletons for amino acid biosynthesis. Boron is responsible for the alteration of N metabolism in many crops. Both NR and glutamate dehydrogenase (GDH) activities are affected in the leaf and root tissues of barley and wheat under B toxicity (Mahboobi et al., 2002). They found a decrease in NR activity (16%) in

the leaf and root tissues of both tolerant and sensitive species, jointly with an increase (30% in leaf and 81% in root tissues) in GDH activity. The effect of nitrogen metabolism on B toxicity in a tomato crop has also been reported (Cervilla et al., 2009). The researchers found that enzyme glutamine synthase, glutamate synthetase, and GDH were induced in tomato leaves under B toxicity, while NR and nitrite reductase activities were significantly decreased. They concluded that B toxicity was responsible for the inhibition of nitrate reduction and increasing ammonium assimilation in a tomato plant. Recently, it has been suggested that B excess can also affect nitrate uptake by roots, the first important step in nitrogen metabolism. In a susceptible tomato plant, B excess reduced the net nitrate uptake affecting the PM H^+-ATPase activity (Princi et al., 2013). Finally, the possibility of alleviating B stress through improving N fertilization has been evaluated. Tepe and Aydemir (2011) observed that the NH_4^+ supply in lentil and barley had a smaller amount of oxidative damage and yield reduction under B stress in comparison with plants supplied with NO_3^- and urea.

10.3.3 Antioxidant Pathways

Abiotic stress normally induces oxidative stress, which causes ROS accumulation, such as hydroxyl radicals (OH^-), superoxide radicals (O_2^-), and hydrogen peroxide (H_2O_2), responsible for proteins, nucleic acids, and lipids damage, which ultimately results in cell death. B excess induced ROS accumulation in barley (Karabal et al., 2003) and oxidative damage by lipid peroxidation and hydrogen peroxide accumulation in grapevine (*Vitis vinifera*) and *Artemisia annua* (Gunes et al., 2006). Antioxidant compounds such as ascorbate and glutathione, and enzymes such as ascorbate peroxidase, catalase, and superoxide dismutase are considered an important resistance mechanism against free radicals (Sharma et al., 2012; Hossain et al., 2015). Thus, they have been studied in different crops under B excess (Karabal et al., 2003; Cervilla et al., 2007; Aftab et al., 2011). This condition repressed the synthesis of tocopherol in orange (*Citrus × sinensis* L. Osbeck), where it also induced ascorbate, glucose, and fructose levels (Keles et al., 2004). Moreover, in two citrus species, *C. sinensis* and *C. grandis*, Sang et al. (2015) confirmed that B toxicity differentially increased the occurrence of numerous proteins involved in various biological functions of the cell. In apple rootstock, the glutathione and ascorbate content increased with elevating levels of B concentrations in the culture medium. The researchers also found that a decrease in the proline content of leaves, important for ROS detoxification, could contribute to more lipid peroxidation under B excess (Molassiotis et al., 2006). Finally, B excess–tolerant chickpea (*Cicer arietinum* L.) and basil (*Ocimum basilicum* L.) genotypes seem to survive with B stress–enhancing antioxidant machinery, but the signaling and coordination of responses remain unclear (Landi et al., 2013).

10.3.4 CARBOHYDRATE METABOLISM

Alterations in sucrose levels are very common in plant responses to various environmental stresses (Rosa et al., 2009), including B stress. Several studies showed that B had a variable effect on plant glycosides biosynthesis including sucrose. For example, a decline in glucose in both leaf and root sugar beet sap under B toxicity was observed (Bonilla et al., 1980). Furthermore, B inhibited the formation of starch from sugar. An increase in reducing sugars (RS) has also been found in the root tip under B excess in soil (Marschner, 1995b; McDonald et al., 2003). Recently, an increase in the invertase activity in the root tip collectively with a simultaneous increase in RS content, glucose, and fructose, was reported intolerant to barley under excess B. This change in carbohydrate metabolism would help root development, maintaining plant growth under B toxicity (Choi et al., 2007). Recently, genome regions (quantitative trait locus [QTL]) linked with RS content have been mapped using a segregant population resulting from a cross between B-tolerant and B-susceptible barley cultivars, grown in a high B environment (Huynh et al., 2009). The relationship between B tolerance and an increased level of RS in the root tip under B excess could prove the role of RS in B tolerance.

10.3.5 PLANT ROOT SYSTEM

A most important phenotypic effect of B toxicity is root growth retardation often associated with a reduced plant dry weight (Turan et al., 2009) and an increased B level in root tissues. A decrease in root growth has been reported in different crops such as tomato (Cervilla et al., 2009), wheat (Turan et al., 2009), and grapevine (Gunes et al., 2006). In particular, B toxicity is responsible for an abnormal cell division in the root meristem of the broad bean (Liu et al., 2000) and the synthesis of hypodermis together with a progressive suberin deposition in the cortical cell wall of soybean root (Ghanati et al., 2005). However, lignification was not considered an essential factor for B-induced root growth inhibition in tomato (Cervilla et al., 2009). Further, Reid et al. (2004) observed a restricted inhibitory response to

high B concentration in wheat root tips but not in older root zones. Boron excess promoted cytotoxic effects on root tip cells during mitosis similar to that of colchicine, forming bridges, fragments, and stickiness in chromosomes and micronuclei growth (Liu et al., 2000; Konuk et al., 2007). Using a mathematical model, it was assumed that the highest B concentration is near the tip and lowest in the farthest region of the meristem zone. Furthermore, the model also assumed that B taken up in the root tip was not competently translocated to shoots, suggesting that B absorbed by the root tip was perhaps used for local root growth, while the more mature root regions were responsible for transporting B toward the shoot (Shimotohno et al., 2015).

Recently, Aquea et al. (2012) reported that excess B in *Arabidopsis* caused root growth inhibition. They also observed that B toxicity promoted the expression of genes involved in abscisic acid signaling, abscisic acid response, and cell wall modifications. Furthermore, B repressed the expression of genes encoding water transporters. The authors concluded that B toxicity triggered a water-stress response associated with root growth inhibition. Considering the role of the root system in the B-excess response, genotypic variation in root elongation has been well accepted as a pointer of B tolerance (Hayes and Reid, 2004; Choi et al., 2006). Indeed, Choi et al. (2007) observed that B tolerance in barley is associated with root morphological changes, leading to an increase in branching and finer root development, which allows better soil exploitation as a result of the osmotic adjustment. More recently, Princi et al. (2013) reported that short-term B-excess treatment had an evident effect on various root morphological traits. In particular, under B excess, tolerant tomato showed a longer and thinner root system compared to a susceptible one.

10.4 LOW BORON STRESS PLANT RESPONSES

B deficiency is a worldwide problem in both agriculture and forestry, particularly in sandy and alkaline soils (Bell and Dell, 2008). Areas where low B soils are found include South and Southeast Asia, Eastern Australia and New Zealand, Africa, North and South America, and Northern Europe (reviewed by Lehto et al., 2010). Low B causes significant losses of yield or quality by influencing vegetative or reproductive growth in forest trees (Lehto et al., 2010) and fruit trees (Kumar, 2011; Ganie et al., 2013; Liu et al., 2013a). However, compared to model plants, the understanding of B-deficiency responses and tolerance mechanisms is limited in woody trees. Recently, advances have been made in the regulation of B transport both in herbaceous and other plants. For instance, the B transport system and its transporters are better understood (Miwa et al., 2013; Tanaka et al., 2013; Chatterjee et al., 2014) and studies on the signal transduction of B malnourishment responses have been initiated (Quiles-Pando et al., 2013; González-Fontes et al., 2014). These new conclusions provide significant insights into understanding B deficiency in plants.

10.4.1 PLANT GROWTH AND VISUAL SYMPTOMS

Boron-deficient plants exhibit various noticeable symptoms in their vegetative and reproductive organs. The scarcity of boron firstly reduces the elongation of growing points due to limited cell wall deposition and then induces necrosis of these tissues due to cell death. This negative effect directly results in root growth, particularly the lateral roots (Mei et al., 2011; Zhou et al., 2014). Boron deficiency also reduces growth in the aerial parts, such as the plant height and leaf area (Möttönen et al., 2001; Wojcik et al., 2008). If B deficiency lasts for many years, it results in the underdeveloped appearance of trees. In some forest trees, long-term B deficiency can reduce the quality and utility of the wood. Loblolly pine (*Pinus taeda*), for example, can grow normally for the first 3 years and then experience dieback under low B conditions (Vail et al., 1961). Similar symptoms of dieback are observed in many tree species. When grape (*V. vinifera*) was cultured under low B conditions, diffuse yellowing of the young leaves, brownish areas of the apical tendrils, and cupping of the third and fourth leaves from the shoot tips were observed in the early stage. Over time, the leaves became more cupped and chlorotic, and the tendrils developed transverse cracks and necrosis (Scott and Schrader, 1947). Mulberry (*Morus alba*), whose leaves are used to feed silkworms, changed to cup-shaped leaves with bent and cracked veins under B limitation (Tewari et al., 2010). In addition, compared to vegetative growth, reproductive growth, especially flowering, fruit set, and yield, is more sensitive to low B (Dell and Huang, 1997). In grape, flower and fruit cluster necrosis and small "shot berries" that are round to pumpkin shaped often appear due to B starvation (Christensen et al., 2006). The papaya (*Carica papaya*) fruit is often affected by B deficiency with latex secretion and deformity (Wang and Ko, 1975). Usually, milky latex secretion appears on the fruit surface at the early stage, after which the milky latex becomes brown. Finally, the fruit surface becomes rough and deformed (Wang and Ko, 1975).

Taken together, B deficiency symptoms in trees can be divided into two main groups. One is the inhibition, even necrosis, of growing points, such as the root tip, bud, flower, and young leaf. Light microscopy observation has shown that cell death occurs in the B-deficient Norway spruce (*Picea abies*) needle buds (Sutinen et al., 2006, 2007) probably due to the B function in the cell wall. The other symptom is the deformity of some organs, such as the shoot, leaf, and fruit. Relevant anatomical studies have demonstrated that B deficiency could severely damage the vascular tissues and induce hypertrophy at the tissue/cellular level. A disorganized vascular tissue was induced by B deficiency in coffee (*Coffea arabica*), and the discontinuity of both the xylem and phloem vessels was observed in the B-deficient stem tip and young leaf (Rosolem and Leite, 2007). Boron deficiency also reduced the length of the xylem vessel in both the leaf and fruit vascular bundles and reduced the diameter of the xylem vessel in only the leaf vascular bundle in pumelo (*C. grandis*; Liu et al., 2013a). A consistent observation that has been reported is that B deficiency can increase vascular cross-sectional areas in Norway spruce needle (Sutinen et al., 2006, 2007), pumelo leaf and fruit vascular tissues (Liu et al., 2013a), and sweet orange (*C. sinensis*) leaf veins (Yang et al., 2013). These results suggest that B deficiency can increase the size but weaken the function of the vascular tissue in trees.

10.4.2 CELL WALL AND MEMBRANE

Boron plays a crucial role in the cell wall structure (Brown et al., 2002; Goldbach and Wimmer, 2007). In B-efficient plants, the structures of the cell wall are greatly altered at both the macroscopic and microscopic levels (Shorrocks, 1997). At the subcellular level, B starvation usually results in abnormally formed walls that are often thick and brittle as a consequence of altered mechanical properties and abnormal expansion (Brown et al., 2002). It has been suggested that B may be necessary for cell-to-wall adhesion and the organization of the architectural integrity of the cell (Lord and Mollet, 2002; Bassil et al., 2004). This is further supported by an altered cell wall porosity and tensile strength under B deficiency (Fleischer et al., 1999; Ryden et al., 2003). In citrus and tea trees, B deprivation not only resulted in a heavily thickened and folded cell wall in the roots (Zhou et al., 2015) but also increased the portion of the cell wall relative to the whole cell (Hajiboland et al., 2013a; Liu et al., 2013b) and induced changes in both the amount and assembly of its component polymers in leaves (Liu et al., 2014). In forest trees, B deficiency impaired the primary cell wall and interrupted the structural development of organs and whole plants, resulting in adverse impacts on tree formation, wood quality, and cold tolerance (Lehto et al., 2010).

In general, the responses of membranes to B deficiency are faster than those of the cell wall (Goldbach et al., 2001; Brown et al., 2002). For example, within minutes of B deprivation, inhibition of PM-bound oxidoreductase activity was frequently observed (Barr et al., 1993). In addition, B-efficient plants often exhibit lower ion uptake rates in roots but higher efflux of potassium and organic compounds in leaves than B-sufficient plants (Cakmak et al., 1995; Goldbach and Wimmer, 2007). The membrane-bound ATPase activity was reduced by B deficiency, but within 1 h after resupplying B, the activity was restored to the same level as that in the B-sufficient bean (*Phaseolus vulgaris*) and maize (*Zea mays*) roots (Pollard et al., 1977). These results suggest that B deficiency might disturb the membrane transport process, the activity of membrane-located proteins, and the integrity and functioning of the PM (Brown et al., 2002; Goldbach and Wimmer, 2007; Camacho-Cristóbal et al., 2008).

There is increasing evidence that B may play structural roles in the cell wall–cell membrane interface (O'Neill et al., 2001; Brown et al., 2002; Goldbach and Wimmer, 2007; Voxeur and Fry, 2014). Bassil et al. (2004) proposed that B may function in transvacuolar cytoplasmic strands and cell-to-wall adhesion. Recently, these predictions were partly confirmed by Voxeur and Fry (2014), who found that B can bind both the RG-II of the cell wall and the GIPCs of the cell membrane, thus forming a GIPC–B–RG-II complex. As a result, B serves as a bridge to connect the cell wall and the PM, which opens a possible avenue to probe the relationship between the cell wall and the membrane via the B-bridge. The GIPC–B–RG-II complex may also explain, at least partly, why both the cell wall and membrane are influenced by B deficiency.

10.4.3 PLANT METABOLISM

Early detectable changes in B-deficient plants are considered to be reflected by damage to the cell membrane or disturbances in the hormonal metabolism (Blevins and Lukaszewski, 1998; Martín-Rejano et al., 2011; Camacho-Cristóbal et al., 2015), but the primary reaction remains unclear. However, an accumulation of phenols has repeatedly been observed in B-deficient plants (Cakmak and Römheld, 1997; Marschner, 2012). It is believed that the accumulation of phenolic compounds is an indirect long-term effect of B deficiency. Initially, boron starvation damages the integrity of the cell wall and membrane, disrupting the phenol metabolism-related enzyme systems, such as phenylalanine-ammonium-lyase (PAL); Cakmak et al., 1995; Brown et al., 2002), resulting in the accumulation of phenols and related alterations of lignin synthesis from phenol-alcohols

(Yang et al., 2013; Zhou et al., 2015). In tea and olive trees, a significant accumulation of phenolic compounds has been detected in B-deficient leaves (Liakopoulos et al., 2005; Hajiboland et al., 2013a). An excessive accumulation of phenols probably leads to tissue necrosis. The poorly lignified branches of woody trees due to B deficiency may be unable to support the weight of leaves (Dell and Huang, 1997). Additionally, changes in phenol and lignin may also affect plant defense systems against herbivores and pathogens (Lehto et al., 2010).

There is no evidence that B plays a direct role in photosynthesis (Dell and Huang, 1997). However, B deficiency limits root growth and results in a weak vascular tissue, which may restrain water uptake and transport within the plant and further alter leaf function (e.g., the reduction in stomata numbers and abnormal shapes, reviewed by Wimmer and Eichert, 2013). Moreover, a wealth of information is available to suggest that B deficiency may indirectly affect photosynthesis by decreasing the photosynthetic area and altering the leaf constituents (Dell and Huang, 1997; Brown et al., 2002). As in herbaceous species (Dell and Huang, 1997), B deficiency of leaves in trees reduces the content of chlorophyll, CO_2 assimilation, and stomatal conductance, as well as the activities of photosynthetic enzymes and catalase, but enhances the production of ROS and intercellular CO_2 concentration, resulting in decreased photosynthetic capability (Han et al., 2008; Wojcik et al., 2008; Tewari et al., 2010). Moreover, the accumulation of soluble sugars in B-deficient leaves of trees may also produce feedback inhibition to net photosynthesis (Han et al., 2008; Ruuhola et al., 2011; Hajiboland et al., 2013b; Lu et al., 2014).

10.5 MECHANISMS FOR TOLERANCE TO B DEFICIENCY

Boron efficiency as used here refers to the extent of variation in response to low B among genotypes within one species and plant species (Rerkasem and Jamjod, 1997). Boron-efficient genotypes are those that are able to grow well in soils in which other genotypes are adversely affected by B deficiency, and the opposite is the case for B-inefficient genotypes. As in herbaceous plants (Rerkasem and Jamjod, 2004; Zhang et al., 2014), differential B efficiencies have also been observed in a variety of trees (Rerkasem and Jamjod, 1997; Xiao et al., 2007). It is widely accepted that the wide range of B efficiency among genotypes is associated with B uptake rate (B uptake efficiency), B translocation (B transport efficiency), and B utilization within the plants (B use efficiency; Marschner, 2012).

10.5.1 B Uptake

At adequate to high B supply, B uptake occurs via passive diffusion across the lipid bilayer, whereas at low B supply, B in the external medium is initially taken up into the root symplasm through a passive facilitated transport process (Dannel et al., 2002; Miwa and Fujiwara, 2010). There are genotype-related differences in B uptake efficiency among trees. For example, under B-deficiency conditions, the sweet orange scion grafted onto *Carrizo citrange* (*C. sinensis* × *Poncirus trifoliata*) had a higher newly acquired B concentration in leaves than those grafted onto trifoliate orange, suggesting that the rootstock *C. citrange* has a greater B uptake efficiency than trifoliate orange (Liu et al., 2012). The boron uptake efficiency of trees has been suggested to be associated with root morphology (Mei et al., 2011) and mycorrhizas. Under low B conditions, B-efficient tree cultivars usually show less depression of root length and number (Mei et al., 2011), thereby exhibiting a higher B absorption rate (Wojcik et al., 2003; Han et al., 2012) compared to B-inefficient cultivars.

Mycorrhizas, which often exist in symbiosis with trees, may also play an important role in B uptake efficiency. In silver birch (*Betula pendula*), the B uptake rate was higher in Laccaria inoculated compared with Laccaria non-inoculated seedlings (Ruuhola and Lehto, 2014). At the molecular level, the AtNIP5;1 gene is a boric acid channel that is involved in the initial uptake process in root cells (Takano et al., 2006). The overexpressed lines for this gene have greater root elongation under B-limited conditions (Kato et al., 2009). These results suggest that a greater accumulation of transcripts of the boric acid channel gene could increase tolerance to B starvation by enhancing the initial uptake process. This suggestion is further supported by the fact that the CiNIP5 transcript level in the roots of B-efficient *C. citrange* increased continuously up to 7.7 times at 48 h after B-deficiency treatment compared to the initial level, where as that of B-inefficient fragrant citrus (*C. junos*) increased only up to 4.4 times at 24 h and then decreased (An et al., 2012). A similar observation was also recently described in that the level of NIP5;1 mRNA in roots after 12 h of B deficiency was upregulated 5.2 times for B-efficient *C. citrange*, but only 3.8 times for B-inefficient trifoliate orange (Zhou et al., 2015).

10.5.2 B Translocation and Retranslocation

At adequate to toxic B supply, a substantial retention of B in the root symplasm occurs at xylem loading; at low B supply, the B retained in the symplasm can be loaded into the xylem by an active transport system (Dannel et al., 2002). Once loaded into the xylem, B can be translocated from the root to the shoot by transpiration under high B conditions, but by active transport processes at low B supply (Raven, 1980; Shelp et al., 1995; Eichert and Goldbach, 2010; Miwa and Fujiwara, 2010), both of which are influenced by water use efficiency (Wimmer and Eichert, 2013). In general, B translocation efficiency is evaluated by the ratio between B concentration in the root cell sap and xylem exudates using table isotope 10 B tracer. Under B-deficiency conditions, B-efficient genotypes usually have relatively higher B concentrations in xylem exudates than B-inefficient genotypes, probably due to the greater ability to translocate B from the root to the shoot.

Furthermore, "Rutgers" was more efficient than "T3238" in translocating 10 B from the root to the shoot (Bellaloui and Brown, 1998). For grafted trees, in addition to functioning in B uptake by roots, the rootstock may also play a role in B translocation from the root to the scion (Papadakis et al., 2003; Wojcik et al., 2003; Boaretto et al., 2008; Sheng et al., 2009b; Wang et al., 2014). Under B-deficiency conditions, the ratio of B concentration in the scion stem to the rootstock stem increased as the B efficiency of citrus combinations increased (Wang et al., 2014). This implies, at least in part, that B-efficient grafted combinations possess a greater ability to translocate B from the rootstock (root) to the scion (shoot).

10.6 EXCESS BORON TOLERANCE

10.6.1 Previous Deliberation

For many years, soil amendment such as leaching B with water and the application of organic compounds to inactivate or immobilize B in soil has been considered the main approach to solving the B toxicity issue. Nowadays, it does not appear feasible on a large scale in B-toxic areas due to economic and practical considerations. On the contrary, the most realistic and effective method to increase crop yields in B-rich soils could be the selection of B excess–tolerant genotypes (Nable et al., 1997). Since the 1980s, genetic variation for B excess tolerance has been assessed in many crop species (Schnurbusch et al., 2010b; Bogacki et al., 2013). Therefore, tolerant varieties can be easily bred, offering a most hopeful approach to minimizing decreases in crop yield in areas with high B concentrations in the soil.

10.6.2 Tolerance Mechanisms Rewind

Physiological mechanisms related to B excess tolerance are not yet well understood. The mechanisms of tolerance in vascular plants include B uptake from soil and its mobility within the plant, B accumulation at the end of the transpiration stream, tissue B contents, and concentration gradient within a leaf (Reid et al., 2004). The boron tolerance model assumes (1) the existence of binding B compounds once B concentration reaches toxic concentrations within the cell; (2) B compartmentation; and (3) an active B efflux by transporters (Hayes and Reid, 2004). Moreover, B accumulation at lower concentrations in tolerant cultivars compared with sensitive ones underlined the predominant role of efflux-type borate transporter(s) in roots rather than internal tolerance mechanisms (B-binding complexes or B compartmentation in vacuoles) (Reid, 2007). Taken together, the basis of tolerance to B excess postulates a more limited tissue B concentration involving B uptake reduction or an active efflux of the micronutrient, at least partly, from the roots (Reid, 2014).

As previously discussed, AtBOR1 and AtNIP5;1 are required for efficient B uptake when its availability in the soil is limited (Takano et al., 2002, 2006). However, in excess, B uptake is mainly regulated through the transcriptional regulation of AtNIP5;1 (Takano et al., 2006) or by the endocytosis and degradation of AtBOR1 (Takano et al., 2005). Besides, *AtBOR1* overexpression does not result in better plant growth under toxic B concentrations (Miwa et al., 2007). Further, the degradation of AtNIP5;1 messenger RNA under B excess is controlled by the 50 untranslated region of AtNIP5;1, suggesting that both AtBOR1 and AtNIP5;1 do not appear to be involved in B tolerance (Tanaka et al., 2011). On the contrary, Mishra et al. (2015) concluded that BOR1 and NIP5;1 activity could be used as markers to identify plant genotypes with increased tolerance to B stress in barley. Recently, many other B transporters, as well as aquaporins, have been identified in plants, and the involvement of some in B tolerance mechanisms has been proposed (Miwa and Fujiwara, 2010). Miwa et al. (2007) found that AtBOR4, one of the six BOR1 paralogs in the *Arabidopsis* genome, showed B efflux activity in yeast cells. Using a green fluorescent protein construct, BOR4 protein was detected on the outer (soil-facing) membranes of root epidermal cells in *Arabidopsis*. This localization is important for B directional export to the soil, avoiding high B concentrations in growing cells and xylem. AtBOR4 overexpression significantly improved plant growth under B excess, suggesting that it avoids the post-translational degradation mechanism (reported for AtBOR1), being on the contrary highly inducible in this condition (Miwa et al., 2007, 2014). Further, transgenic rice plants expressing AtBOR4 showed a high tolerance to B toxicity (Kajikawa et al., 2011). The growth enhancement was attributed to effective B export from

the roots, as a consequence of the optimal B concentration maintained within the plant. Thus, the difference in BOR1 and BOR4 genes regulation suggests the presence of complex mechanisms for the perception and control of B homeostasis.

Recently, aquaporin isoforms, involved in water and ion transport, appeared to improve tolerance toward many abiotic stresses (Pang et al., 2010). Indeed, the overexpression of AtTIP5;1, a tonoplast aquaporin, resulted in increased tolerance to moderately high B levels in the growing medium involved in borate compartmentation in the vacuole (Pang et al., 2010). Further, two aquaporin rice genes, OsPIP2;4 and OsPIP2;7, have been found to be involved in B permeability and tolerance (Kumar et al., 2014). Both genes, responsible for exporting B from roots under B excess, were downregulated in shoots and strongly upregulated in roots, whose higher expression avoided B toxicity. Furthermore, efflux B assay in roots indicated that, after 1 h of exposure, 10 B was excluded from the roots in *Arabidopsis* transgenic plants overexpressing OsPIP2;4 or OsPIP2;7 (Kumar et al., 2014). Recently, a gene encoding a nascent polypeptide–associated complex (NAC)-like transcription factor (TF) with a single nucleotide polymorphism between the sensitive and tolerant rice cultivars has been identified using recombinant inbred lines (Ochiai et al., 2011). The change in a single nucleotide appeared to provide tolerance to B toxicity in rice by disrupting the gene, which was named BET1 (Boron Excess Tolerant 1). This mechanism could be independent from B efflux because differences in root and/or shoot B concentrations between sensitive and tolerant rice cultivars were not observed (Ochiai et al., 2011).

To identify novel mechanisms involved in B tolerance, two *Arabidopsis* mutants, defecting in genes related to B absorption, have also been studied (Sakamoto et al., 2011). Thus, heb1-1 and heb2-1 (hypersensitive to excess B) mutants, showing growth defects only under B excess, could not encode for two subunits of the chromosomal protein complex known as "condensin II." Although both heb mutants contained less B than wild type, their sensitivity to B excess was much greater. These findings confirmed the existence of tolerance mechanisms different from the B efflux. Condensin II seemed to act in DNA double-strand break improvements and to maintain the replication process, both functions considered to be required for plant B excess tolerance (Sakamoto et al., 2011). Throughout RNA-seq analysis, Tombuloglu et al. (2015) profiled differential transcripts in response to B excess in tolerant barley. They revealed that genes related to the cell wall, PM and cytoskeleton construction, Ca2þ/calmodulin system, phospholipase activity, and signal transduction played a crucial role under B excess (Tombuloglu et al., 2015).

10.6.3 GENES AND QTL CONCERNED WITH B TOLERANCE

One of the first examples of QTL analysis showed that the B excess tolerance of *japonica* rice cultivars was greater than *indica* because of a major QTL that accounted for the phenotypic variation (Ochiai et al., 2008). This difference was evident even though the B content in the root and shoot of both tolerant and susceptible rice genotypes did not significantly vary, highlighting the potential role of molecular tools for selecting novel B-tolerant genotypes (Ochiai et al., 2008). In any case, QTL detection has also been useful in isolating genes involved in this genetic complex trait. The identification of QTL regions and cloning genes conferring B toxicity tolerance is potentially the major challenge for the development of varieties able to grow in high soil B levels. In barley, four QTLs associated with B toxicity tolerance were detected on chromosome 2, 3, 4, and 6H. Thus, HvBot1, an AtBOR1-like gene, was detected in the QTL of chromosome 4H and then cloned (Sutton et al., 2007). It was the first B toxicity tolerance gene identified in plants, playing a role in limiting the net B uptake into the root and in the disposal of B from leaves. The B tolerance mechanism in the tolerant cultivar Sahara, related to an increase in the copy number of HvBot1 gene and the abundance of messenger RNA transcript, has been demonstrated (Sutton et al., 2007). Another QTL on barley chromosome 3H was identified to control relative root length at toxic B concentrations having a lesser effect than 4H chromosome QTL, but they cooperate additively to confer tolerance. Moreover, a gene encoding an NIP-like aquaporin dHvNIP 2; 1d has been identified in barley and mapped to B tolerance QTL on 6H (Schnurbusch et al., 2010a). Finally, Hassan et al. (2010) found the chromosome 2H QTL region encoding S-adenosyl methionine decarboxylase precursor (SAMDC), involved in antioxidative response. Yeast overexpressing was able to provide tolerance to high B. In bread wheat, at least three unlinked Bo1, Bo2, and Bo3 genes controlled tolerance to B toxicity mapped on chromosomes 4 and 7 (Paull et al., 1991). They additively controlled yield and tissue B concentrations under excess B condition (Jefferies et al., 2000) and one of these genes mapped in the 7B chromosome was considered to play the main role in crop yield under B toxicity (Nable et al., 1997).

Recently, Pallotta et al. (2014) described the identification of near-identical, root-specific B transporter genes underlying two major effects (QTL) for B tolerance in wheat, Bo1, and Bo4. They showed that tolerance to high B concentration was associated with multiple genomic changes including dispersed gene duplication, tetraploid introgression, and variation in gene structure and transcript level. A distinct pattern of gene variant distribution correlated to B levels in soils from different geographical regions was also observed. These findings could support wheat breeders with molecular tools to select for variants of the tolerance gene required for specific environments. Thus, the characterization of B tolerance in wheat well highlights the power of the new genomic technologies to define key adaptive processes underpinning crop improvement (Pallotta et al., 2014).

10.7 CONCLUSION AND FUTURE PROSPECTS

In summary, at a physiological level, B efficiency in trees is mainly attributed to four mechanisms: (i) the ability to absorb B from the soil/medium, which depends on root morphology and mycorrhiza; (ii) B translocation from root to shoot as indicated by the composition of root cell sap and xylem exudates and likely influenced by xylem vessel characteristics and water use efficiency; (iii) B retranslocation through xylem-to-phloem transfer and the formation of complexes with hydroxyl groups in phloem; and (iv) the B requirement in cell wall construction and cell membrane composition. At a molecular level, the tolerance to B deficiency can be improved by the higher or stronger expression of NIPs, BORs, and S6PDH to facilitate B uptake, B translocation, and retranslocation, as well as B utilization processes. These molecular results indicate that genetically modified trees with B-deficiency-tolerance-related genes may be useful in forestry or other tree industries in the future. Furthermore, the bor1 bor2 double mutants exhibited more severe growth defects under B-limited conditions than bor1 or bor2 single mutants in *Arabidopsis* (Miwa et al., 2013), indicating that B-deficiency-tolerance-related genes are probably dosage dependent. That is, the differences in B efficiency probably originate from a combined effect of the four mechanisms previously mentioned. However, previous studies were mainly limited to one single mechanism of B efficiency. Consequently, a more systematic study is needed on B-deficiency tolerance mechanisms in trees, including uptake, translocation, retranslocation, and utilization, spanning investigations from the physiological to the molecular level.

REFERENCES

Aftab, T., Khan, M.M.A., Idrees, M., Naeem, M., Moinuddin, N.H., Hashmi, N. (2011) Methyl jasmonate counteracts boron toxicity by preventing oxidative stress and regulating antioxidant enzyme activities and artemisinin biosynthesis in *Artemisia annua* L. *Protoplasma* 248(3): 601–612.

An, J.C., Liu, Y.Z., Yang, C.Q., Zhou, G.F., Wei, Q.J., Peng, S.A. (2012) Isolation and expression analysis of CiNIP5, a citrus boron transport gene involved in tolerance to boron deficiency. *Scientia Horticulturae* 142: 149–154.

Aquea, F., Federici, F., Moscoso, C., Vega, A., Jullian, P., Haseloff, J., Arce-Johnson, P. (2012) A molecular framework for the inhibition of *Arabidopsis* root growth in response to boron toxicity. *Plant, Cell and Environment* 35(4): 719–734.

Bariya, H.S., Snehal, B., Patel, A.D. (2014) Boron: A promising nutrient for increasing growth and yield of plant. In *Nutrient Use Efficiency in Plants: Concepts and Approaches*, ed. Hawkesford M. J., 153–170. Springer, New York.

Barr, R., Böttger, M., Crane, F.L. (1993) The effect of boron on plasma membrane electron transport and associated proton secretion by cultured carrot cells. *Biochemistry and Molecular Biology International* 31(1): 31–39.

Bassil, E., Hu, H., Brown, P.H. (2004) Use of phenyl boronic acids to investigate boron function in plants. Possible role of boron in transvacuolar cytoplasmic strands and cell-to-wall adhesion. *Plant Physiology* 136(2): 3383–3395.

Bell, R.W., Dell, B. (2008) *Micronutrients for Sustainable Food, Feed, Fibre and Bioenergy Production*. International Fertilizer Industry Association (IFA), Paris.

Bellaloui, N., Brown, P. H. (1998) Cultivar differences in boron uptake and distribution in celery (*Apium graveolens*), tomato (*Lycopersicon esculentum*) and wheat (*Triticum aestivum*). *Plant and Soil* 198: 153–158.

Bellaloui, N., Yadavc, R.C., Chern, M.S., Hu, H., Gillenb, A.M., Greve, C., Dandekar, A.M., Ronald, P.C., Brown, P.H. (2003) Transgenically enhanced sorbitol synthesis facilitates phloem-boron mobility in rice. *Physiologia Plantarum* 117(1): 79–84.

Blevins, D.G., Lukaszewski, K.M. (1998) Boron in plant structure and function. *Annual Review of Plant Physiology and Plant Molecular Biology* 49: 481–500.

Boaretto, R.M., Quaggio, J.A., MourãoFilho, F.A., Giné, M.F., Boaretto, A.E. (2008) Absorption and mobility of boron in young citrus plants. *Communications in Soil Science and Plant Analysis* 39(17–18): 2501–2514.

Bogacki, P., Peck, D.M., Nair, R.M., Klaus, J.H., Oldach, K.H. (2013) Genetic analysis of tolerance to boron toxicity in the legume *Medicago truncatula*. *BMC Plant Biology* 13: 54–64.

Bonilla, I., Cadahia, C., Carpena, O., Hernando, V. (1980) Effects of boron on nitrogen metabolism and sugar levels of sugar beet. *Plant and Soil* 57(1): 3–9.

Brown, P.H., Bellaloui, N., Wimmer, M.A., Bassil, E.S., Ruiz, J., Hu, H., Pfeffer, H., Dannel, F., Römheld, V. (2002) Boron in plant biology. *Plant Biology* 4(2): 205–223.

Brown, P.H., Hu, H. (1996) Phloem mobility of boron is species dependent: Evidence for phloem mobility in sorbitol-rich species. *Annals of Botany* 77(5): 497–506.

Cakmak, I., Kurz, H., Marschner, H. (1995) Short-term effects of boron, germanium and high light intensity on membrane permeability in boron deficient leaves of sunflower. *Physiologia Plantaram* 95: 11–18.

Cakmak, I., Römheld, V. (1997) Boron deficiency-induced impairments of cellular functions in plants. *Plant and Soil* 193(2): 71–83.

Camacho-Cristóbal, J.J., Martín-Rejano, E.M., Herrera-Rodríguez, M.B., Navarro-Gochicoa, M.T., Rexach, J., González-Fontes, A. (2015) Boron deficiency inhibits root cell elongation via an ethylene/auxin/ROS- dependent pathway in *Arabidopsis* seedlings. *Journal of Experimental Botany* 66(13): 3831–3840.

Camacho-Cristóbal, J.J., Rexach, J., Gonzalez-Fontes, A. (2008) Boron in plants: Deficiency and toxicity. *Journal of Integrative Plant Biology* 50(10): 1247–1255.

Cervilla, L.M., Blasco, B., Ríos, J.J., Romero, L., Ruiz, J.M. (2007) Oxidative stress and antioxidants in tomato (*Solanum lycopericum*) plants subjected to boron toxicity. *Annals of Botany* 100(4): 747–756.

Cervilla, L.M., Blasco, B., Ríos, J.J., Rosales, M.A., Rubio-Wilhelmi, M.M., Sánchez-Rodríguez, E., Romero, L., Ruiz, J.M. (2009) Response of nitrogen metabolism to boron toxicity in tomato plants. *Plant Biology* 11(5): 671–677.

Cervilla, L.M., Blasco, B., Ríos, J.J., Rosales, M.A., Sánchez-Rodríguez, E., Rubio-Wilhelmi, M.M., Romero, L., Ruiz, J.M. (2012) Parameters symptomatic for boron toxicity in leaves of tomato plants. *Journal of Botany* 2012: 1–17.

Chatterjee, M., Tabi, Z., Galli, M., Malcomber, S., Buck, A., Muszynski, M., Gallavotti, A. (2014) The boron efflux transporter ROTTENEAR is required for maize inflorescence development and fertility. *The Plant Cell* 26(7): 2962–2977.

Chen, L.S., Han, S., Qi, Y.P., Yang, L.T. (2012) Boron stresses and tolerance in citrus. *African Journal of Biotechnology* 11(22): 5961–5969.

Chen, M., Mishra, S., Heckathorn, S.A., Frantz, J.M., Krause, C. (2014) Proteomic analysis of *Arabidopsis thaliana* leaves in response to acute boron deficiency and toxicity reveals effects on photosynthesis, carbohydrate metabolism, and protein synthesis. *Journal of Plant Physiology* 171(3–4): 235–242.

Choi, E.Y., Kolesik, P., Mcneill, A., Collins, H., Zhang, Q., Huynh, B.L., Graham, R., Stangoulis, J. (2007) The mechanism of boron tolerance for maintenance of root growth in barley (*Hordeum vulgare* L.). *Plant, Cell and Environment* 30(8): 984–993.

Choi, E.Y., McNeill, A.M., Coventry, D., Stangoulis, J.C.R. (2006) Whole plant response of crop and weed species to high subsoil boron. *Australian Journal of Agricultural Research* 57(7): 761–770.

Christensen, L.P., Beede, R.H., Peacock, W.L. (2006) Fall foliar sprays prevent boron-deficiency symptoms in grapes. *California Agriculture* 60(2): 100–103.

Crawford, N.M. (1995) Nitrate: Nutrient and signal for plant growth. *The Plant Cell* 7(7): 859–868.

Dannel, F., Pfeffer, H., Römheld, V. (2002) Update on boron in higher plants: Uptake, primary translocation and compartmentation. *Plant Biology* 4(2): 193–204.

Dell, B., Huang, L. (1997) Physiological response of plants to low boron. *Plant and Soil* 193: 103–120.

Eichert, T., Goldbach, H. E. (2010) Transpiration rate affects the mobility of foliar-applied boron in *Ricinus communis* L. cv. Impala. *Plant and Soil* 328: 165–174.

Fang, Y.H. (2001) Study on effect of high boron stress on photosynthesis of oilseed rape. *Plant Nutrition and Fertilizer Science* 7(1): 109–112.

Fleischer, A., O'Neill, M.A., Ehwald, R. (1999) The pore size of non-graminaceous plant cell walls is rapidly decreased by borate ester cross-linking of the pectic polysaccharide rhamnogalacturonan II. *Plant Physiology* 121(3): 829–838.

Funakawa, H., Miwa., K. (2015) Synthesis of borate cross-linked rhamnogalacturonan II. *Frontiers in Plant Science* 6: 223.

Furlani, A.M.C., Carvalho, C.P., De Freitas, J.G., Verdial, M.F. (2003) Wheat cultivar tolerance to boron deficiency and toxicity in nutrient solution. *Science and Agriculture* 60: 359–370.

Ganie, M. A., Akhter, F., Bhat, M., Malik, A., Junaid, J. M., Shah, M. A., et al. (2013) Boron-a critical nutrient element for plant growth and productivity with reference to temperate fruits. *Current Science* 104: 76–85.

Ghanati, F., Morita, A., Yokota, H. (2005) Deposition of suberin in roots of soybean induced by excess boron. *Plant Science* 168(2): 397–405.

Goldbach, H.E., Wimmer, M.A. (2007) Boron in plants and animals: Is there a role beyond cell wall structure? *Journal of Plant Nutrition and Soil Science* 170(1): 39–48.

Goldbach, H.E., Yu, Q., Wingender, R., Schulz, M., Wimmer, M.A., Findeklee, P., Baluška, F. (2001) Rapid response reactions of roots to boron deprivation. *Journal of Plant Nutrition and Soil Science* 164(2): 173–181.

González-Fontes, A., Navarro-Gochicoa, M.T., Camacho-Cristóbal, J.J., Herrera-Rodríguez, M.B., Quiles-Pando, C., Rexach, J. (2014) Is Ca^{2+} involved in the signal transduction pathway of boron deficiency? New hypotheses for sensing boron deprivation. *Plant Science* 217–218: 135–139.

Gunes, A., Soylemezoglu, G., Inal, A., Bagci, E.G., Coban, S., Sahin, O. (2006) Antioxidant and stomatal responses of grapevine (*Vitis vinifera* L) to boron toxicity. *Scientia Horticulturae* 110(3): 279–284.

Han, J., Zhou, G. F., Li, Q. H., Liu, Y. Z., Peng, S. A. (2012) Effects of magnesium, iron, boron deficiency on the growth and nutrition absorption of four major citrus rootstocks. *Acta Horticulturae Sinica* 39: 2105–2112.

Hajiboland, R., Bahrami-Rad, S., Bastani, S. (2013a) Phenolics metabolism in boron-deficient tea [Camellia sinensis (L.) O. Kuntze] plants. *Acta Biologica Hungarica* 64: 196–206.

Hajiboland, R., Bahrami-Rad, S., Bastani, S., Tolrà, R., Poschenrieder, C. (2013b) Boron re-translocation in tea (*Camellia sinensis* (L.) O. Kuntze) plants. *Acta Physiologiae Plantarum* 35(8): 2373–2381.

Han, S., Chen, L.S., Jiang, H.X., Smith, B.R., Yang, L.T., Xie, C.Y. (2008) Boron deficiency decreases growth and photosynthesis, and increases starch and hexoses in leaves of citrus seedlings. *Journal of Plant Physiology* 165(13): 1331–1341.

Han, S., Tang, N., Jiang, H.X., Yang, L.T., Li, Y., Chen, L.S. (2009) CO_2 assimilation, photosystem II photochemistry, carbohydrate metabolism and antioxidant system of citrus leaves in response to boron stress. *Plant Science* 176(1): 143–153.

Hassan, M., Oldach, K., Baumann, U., Langridge, P., Sutton, T. (2010) Genes mapping to boron tolerance QTL in barley identified by suppression subtractive hybridization. *Plant, Cell and Environment* 33(2): 188–198.

Hayes, J.E., Reid, R.J. (2004) Boron tolerance in barley is mediated by efflux of boron from the roots. *Plant Physiology* 136(2): 3376–3382.

Hossain, M.F., Shenggang, P., Meiyang, D., Zhaowen, M., Karbo, M.B., Bano, A., Xiangru, T.A.N.G. (2015) Photosynthesis and antioxidant response to winter rapeseed (*Brassica napus* L.) as affected by boron. *Pakistan Journal of Botany* 47(2): 675–684.

Huynh, B.L., Pallotta, M., Choi, E.Y., Garnett, T., Graham, R., Stangoulis, J. (2009) Quantitative trait loci for reducing sugar concentration in the barley root tip under boron toxicity. In: *The Proceedings of the International Plant Nutrition Colloquium XVI*, Department of Plant Sciences, UC Davis, Davis, CA.

Jefferies, S.P., Pallotta, M.A., Paull, J.G., Karakousis, A., Kretschmer, J.M., Manning, S., Islam, A.K.M.R., Langridge, P., Chalmers, K.J. (2000) Mapping and validation of chromosome regions conferring boron toxicity tolerance in wheat (*Triticum aestivum* L.). *TAG Theoretical and Applied Genetics* 101(5–6): 767–777.

Kajikawa, M., Fujibe, T., Uraguchi, S., Miwa, K., Fujiwara, T. (2011) Expression of the *Arabidopsis* borate efflux transporter gene, AtBOR4, in rice affects the xylem loading of boron and tolerance to excess boron. *Bioscience, Biotechnology, and Biochemistry* 75(12): 2421–2423.

Karabal, E., Yucel, M., Oktem, H.A. (2003) Antioxidant responses of tolerant and sensitive barley cultivars to boron toxicity. *Plant Science* 164(6): 925–933.

Kato, Y., Miwa, K., Takano, J., Wada, M., Fujiwara, T. (2009) Highly boron deficiency-tolerant plants generated by enhanced expression of NIP5;1, a boric acid channel. *Plant and Cell Physiology* 50(1): 58–66.

Keles, Y., Öncel, I., Yenice, N. (2004) Relationship between boron content and antioxidant compounds in citrus leaves taken from fields with different water source. *Plant and Soil* 265(1–2): 345–353.

Konuk, M., Liman, R., Ciğerci, H. (2007) Determination of genotoxic effect of boron on *Allium cepa* root meristematic cells. *Pakistan Journal of Botany* 39(1): 73–79.

Kumar, K., Mosa, K.A., Chhikara, S., Musante, C., White, J.C., Dhankher, O.P. (2014) Two rice plasma membrane intrinsic proteins, OsPIP2;4 and OsPIP2;7, are involved in transport and providing tolerance to boron toxicity. *Planta* 239(1): 187–198.

Kumar, R. (2011) Boron deficiency disorders in mango (*Mangifera indica*): Field screening, nutrient composition and amelioration by boron application. *Indian Journal of Agriculture Science* 51: 751–754.

Landi, M., Pardossi, A., Remorini, D., Guidi, L. (2013) Antioxidant and photosynthetic response of a purple-leaved and a green-leaved cultivar of sweet basil (*Ocimum basilicum*) to boron excess. *Environmental and Experimental Botany* 85: 64–75.

Lehto, T., Ruuhola, T., Dell, B. (2010) Boron in forest trees and forest ecosystems. *Forest Ecology and Management* 260(12): 2053–2069.

Liakopoulos, G., Stavrianakou, S., Filippou, M., Fasseas, C., Tsadilas, C., Drossopoulos, I., Karabourniotis, G. (2005) Boron remobilization at low boron supply in olive (*Olea europaea*) in relation to leaf and phloem mannitol concentrations. *Tree Physiology* 25(2): 157–165.

Liu, D., Jiang, W., Zhang, L., Li, L. (2000) Effects of boron ions on root growth and cell division of broadbean (*Vicia faba* L.). *Israel Journal of Plant Sciences* 48(1): 47–51.

Liu, Y.Z., Li, S., Yang, C.Q., Peng, S.A. (2013a) Effects of boron-deficiency on anatomical structures in the leaf main vein and fruit mesocarp of pummelo [*Citrus grandis* (L.) Osbeck]. *Korean Journal of Horticultural Science* 88: 693–700.

Liu, G. D., Wang, R. D., Wu, L. S., Peng, S. A., Wang, Y. H., and Jiang, C. C. (2012). Boron distribution and mobility in navel orange grafted on citrange and trifoliate orange. *Plant Soil* 360: 123–133.

Liu, G.D., Dong, X.C., Liu, L.C., Wu, L.S., Peng, S.A., Jiang, C.C. (2014) Boron deficiency is correlated with changes in cell wall structure that lead to growth defects in the leaves of navel orange plants. *Scientia Horticulturae* 176: 54–62.

Liu, G.D., Wang, R.D., Liu, L.C., Wu, L.S., Jiang, C.C. (2013b) Cellular boron allocation and pectin composition in two citrus root stock seedlings differing in boron-deficiency response. *Plant and Soil* 370(1–2): 555–565.

Lord, E.M., Mollet, J.C. (2002) Plant cell adhesion: A bioassay facilitates discovery of the first pectin biosynthetic gene. *Proceedings of the National Academy of Sciences USA* 99(25): 15843–15845.

Mahboobi, H., Yucel, M., Oktem, H.A. (2002) Nitrate reductase and glutamate dehydrogenase activities of resistant and sensitive cultivars of wheat and barley under boron toxicity. *Journal of Plant Nutrition* 25(8): 1829–1837.

Marschner, H. (1995a) *Mineral Nutrition of Higher Plants*. 2nd edn. Academic Press, San Diego, CA.

Marschner, H. (1995b) Functions of mineral nutrients: Micronutrients. *In Mineral Nutrition of Higher Plants*, ed. Marschner, H., 313–404. Academic Press, London.

Marschner, P. (2012) *Marschner's Mineral Nutrition of Higher Plants*. Academic Press, London.

Martín-Rejano, E.M., Camacho-Cristobal, J.J., Herrera-Rodriguez, M.B., Rexach, J., Navarro-Gochicoa, M.T., González-Fontes, A. (2011) Auxin and ethylene are involved in the responses of root system architecture to low boron supply in *Arabidopsis* seedlings. *Physiologia Plantarum* 142(2): 170–178.

McDonald, G.K., Stangoulis, J.C.R., Genc, Y., Lewis, J., Robin, D.G. (2003) Boron toxicity, micronutrient deficiency and salt: Overcoming the trifecta of nutritional problems in alkaline soils. In *The Meeting on "Genetic Solutions for Hostile Crops."* CSIRO Plant Industry, Canberra, Australia.

Mei, L., Sheng, O., Peng, S.A., Zhou, G.F., Wei, Q.J., Li, Q.H. (2011) Growth, root morphology and boron uptake by citrus root stock seedlings differing in boron-deficiency responses. *Scientia Horticulturae* 129(3): 426–432.

Mishra, S., Heckathorn, S., Krause, C. (2015) The levels of Boron-Uptake proteins in roots are correlated with tolerance to Boron stress in barley. *Crop Science* 55(4): 1741–1748.

Miwa, K., Takano, J., Omori, H., Seki, M., Shinozaki, K., Fujiwara, T., (2007) Plants tolerant of high boron levels. *Science* 318: 1417.

Miwa, K., Fujiwara, T. (2010) Boron transport in plants: coordinated regulation of transporters. *Annual review in Botany* 105: 1103–1108

Miwa, K., Wakuta, S., Takada, S., Ide, K., Takano, J., Naito, S., Omori, H., Matsunaga, T., Fujiwara, T. (2013) Roles of BOR2, a boron exporter, in crosslinking of rhamnogalacturonan II and root elongation under boron limitation in *Arabidopsis*. *Plant Physiology* 163(4): 1699–1709.

Miwa, K., Aibara, I., Fujiwara, T., (2014) *Arabidopsis thaliana* BOR4 is upregulated under high boron conditions and confers tolerance to high boron. *Soil Science and Plant Nutrition* 60: 349–355.

Molassiotis, A., Sotiropoulos, T., Tanou, G., Diamantidis, G., Therios, I. (2006) Boron-induced oxidative damage and antioxidant and nucleolytic responses in shoot tips culture of the apple rootstock EM9 (*Malus domestica* Borkh). *Environmental and Experimental Botany* 56(1): 54–62.

Möttönen, M., Aphalo, P.J., Lehto, T. (2001) Role of boron in drought resistance in *Norway spruce* (*Picea abies*) seedlings. *Tree Physiology* 21(10): 673–681.

Nable, R.O., Banuelos, G.S., Paull, J.G. (1997) Boron toxicity. *Plant and Soil* 193(2): 181–198.

Nable, R.O., Cartwright, B., Lance, R.C.M. (1990) Genotypic differences in boron accumulation in barley: Relative susceptibilities to boron deficiency and toxicity. In *Genetic Aspects of Plant Mineral Nutrition*, ed. El Bassam, N., 243–251. Kluwer Academic Publishers, Dordrecht, The Netherlands.

Ochiai, K., Shimizu, A., Okumoto, Y., Fujiwara, T., Matoh, T. (2011) Suppression of a NAC-like transcription factor gene improves boron-toxicity tolerance in rice. *Plant Physiology* 156(3): 1457–1463.

Ochiai, K., Uemura, S., Shimizu, A., Okumoto, Y., Matoh, T. (2008) Boron toxicity in rice (*Oryza sativa* L.) I. Quantitative trait locus (QTL) analysis of tolerance to boron toxicity. *TAG. Theoretical and Applied Genetics. Theoretische und Angewandte Genetik* 117(1): 125–133.

Oertli, J. J., Kohl, H. C. (1961) Some consideration about the tolerance of various plant soecies to excessive supplies of Boron. *Soil science* 92: 243-247.

Pallotta, M., Schnurbusch, T., Hayes, J., Hay, A., Baumann, U., Paul, J., Langridge, P., Sutton, T. (2014) Molecular basis of adaptation to high soil boron in wheat landraces and elite cultivars. *Nature* 514(7520): 88–91.

Pang, Y., Li, L., Ren, F., Lu, P., Wei, P., Cai, J., Xin, L., et al. (2010) Overexpression of the tonoplast aquaporin AtTIP5;1 conferred tolerance to boron toxicity in *Arabidopsis*. *Journal of Genetics and Genomics* 37: 389–397.

Papadakis, I.E., Dimassi, K.N., Therios, I.N. (2003) Response of two citrus genotypes to six boron concentrations: Concentration and distribution of nutrients, total absorption, and nutrient use efficiency. *Australian Journal of Agricultural Research* 54(6): 571–580.

Paull, J.G., Rathjen, A.J., Cartwright, B. (1991) Major gene control of tolerance of bread wheat (*Triticum aestivum* L.) to high concentrations of soil boron. *Euphytica* 55(3): 217–228.

Pérez-Castro, R., Kasai, K., Gainza-Cortés, F., Ruiz-Lara, S., Casaretto, J.A., Peña-Cortés, H., Tapia, J., Fujiwara, T., González, E. (2012) VvBOR1, the grapevine ortholog of AtBOR1, encodes an efflux boron transporter that is differentially expressed throughout reproductive development of *Vitis vinifera* L. *Plant & Cell Physiology* 53(2): 485–494.

Pollard, A.S., Parr, A.J., Loughman, B.C. (1977) Boron in relation to membrane function in higher plants. *Journal of Experimental Botany* 28(4): 831–841.

Princi, M.P., Lupini, A., Araniti, F., Sunseri, F., Abenavoli, M.R. (2013) Short-term effects of boron excess on root morphological and functional traits in tomato. In *XVII International Plant Nutrition Colloquium, Boron Satellite Meeting—Proceedings Book*, 1150–1151. August 17–18, 2013, Istanbul, Turkey.

Quiles-Pando, C., Rexach, J., Navarro-Gochicoa, M.T., Camacho-Cristóbal, J.J., Herrera-Rodríguez, M.B., González-Fontes, A. (2013) Boron deficiency increases the levels of cytosolic Ca^{2+} and expression of Ca^{2+}-related genes in *Arabidopsis thaliana* roots. *Plant Physiology and Biochemistry* 65: 55–60.

Raven, J. (1980) Short- and long-distance transport of boric acid in plants. *New Phytologist* 84: 231–249

Reid, R. (2007) Identification of boron transporter genes likely to be responsible for tolerance to boron toxicity in wheat and barley. *Plant and Cell Physiology* 48(12): 1673–1678.

Reid, R. (2013) Boron toxicity and tolerance in crop plants. *In Crop Improvement under Adverse Conditions*, ed. Tuteja, N., Gill, S., 333–346. Springer, New York.

Reid, R. (2014) Understanding the boron transport network in plants. *Plant and Soil* 385(1–2): 1–13.

Reid, R.J., Hayes, J.E., Post, A., Stangoulis, J.C.R., Graham, R.D. (2004) A critical analysis of the causes of boron toxicity in plants. *Plant, Cell and Environment* 27(11): 1405–1414.

Rerkasem, B., Jamjod, S. (1997) Genotypic variation in plant response to low boron and implications for plant breeding. *Plant and Soil* 193(2): 169–180.

Rerkasem, B., Jamjod, S. (2004) Boron deficiency in wheat: A review. *Field Crops Research* 89(2–3): 173–186.

Roldán, M., Belver, A., Rodríguez-Rosales, P., Ferrol, N., Donaire, J.P. (1992) In vivo and in vitro effects of boron on the plasma membrane proton pump of sunflower roots. *Physiologia Plantarum* 84: 49–54.

Rosa, M., Prado, C., Podazza, G., Interdonato, R., González, J.A., Hilal, M., Prado, F.E. (2009) Soluble sugars metabolism, sensing and abiotic stress. A complex network in the life of plants. *Plant Signaling and Behavior* 4(5): 388–393.

Rosolem, C.A., Leite, V.M. (2007) Coffee leaf and stem anatomy under boron deficiency. *Revista Brasileira de Ciência Do Solo* 31(3): 477–483.

Ruuhola, T., Keinänen, M., Keski-Saari, S., Lehto, T. (2011) Boron nutrition affects the carbon metabolism of silver birch seedlings. *Tree Physiology* 31(11): 1251–1261.

Ruuhola, T., Lehto, T. (2014) Do ectomycorrhizas affect boron uptake in *Betula pendula*? *Canadian Journal of Forest Research* 44(9): 1013–1019.

Ryden, P., Sugimoto-Shirasu, K., Smith, A.C., Findlay, K., Reiter, W.D., McCann, M.C. (2003) Tensile properties of *Arabidopsis* cell walls depend on both a xyloglucan cross-linked microfibrillar network and rhamnogalacturonan-II borate complexes. *Plant Physiology* 132(2): 1033–1040.

Sakamoto, T., Inui, Y.T., Uraguchi, S., Yoshizumi, T., Matsunaga, S., Mastui, M., Umeda, M., Fukui, K., Fujiwara, T. (2011) Condensin II alleviates DNA damage and is essential for tolerance of boron overload stress in *Arabidopsis*. *The Plant Cell* 23(9): 3533–3546.

Sang, W., Huang, Z.R., Qi, Y.P., Yang, L.T., Guo, P., Chen, L.S. (2015) An investigation of B-toxicity in leaves of two citrus species differing in boron-tolerance using comparative proteomics. *Journal of Proteomics* 123: 128–146.

Schnurbusch, T., Hayes, J., Hrmova, M., Baumann, U., Ramesh, S.A., Tyerman, S.D., Langridge, P., Sutton, T. (2010a) Boron toxicity tolerance in barley through reduced expression of the multifunctional aquaporin HvNIP2;1. *Plant Physiology* 153(4): 1706–1715.

Schnurbusch, T., Hayes, J., Sutton, T. (2010b) Boron toxicity tolerance in wheat and barley: Australian perspectives. *Breeding Science* 60(4): 297–304.

Scott, L.E., Schrader, A.L. (1947) Effect of alternating conditions of boron nutrition upon growth and boron content of grape vines in sand culture. *Plant Physiology* 22(4): 526–537.

Sharma, P., Jha, A.B., Dubey, R.S., Pessarakli, M. (2012) Reactive oxygen species, oxidative damage, and anti-oxidative defense mechanism in plants under stressful conditions. *Journal of Botany* 2012: 1–26.

Shelp, B. J., Marentes, E., Kitheka, A. M., Vivekanandan, P. (1995) Boron mobility in plants. *Physiologia Plantarum* 94: 356–361.

Sheng, O., Song, S. W., Peng, S. A., Deng, X. X. (2009b) The effects of low boron on growth, gas exchange, boron concentration and distribution of 'Newhall' navel orange (*Citrus sinensis* Osb.) plants grafted on two rootstocks. *Scientia Horticulturae* 121: 278–283.

Shimotohno, A., Sotta, N., Sato, T., De Ruvo, M., Marée, A.F.M., Grieneisen, V.A., Fujiwara, T. (2015) Mathematical modelling and experimental validation of spatial distribution of boron in the root of *Arabidopsis thaliana* identify high boron accumulation in the tip and predict a distinct root tip uptake function. *Plant and Cell Physiology* 56(4): 620–630.

Shorrocks, V.M. (1997) The occurrence and correction of boron deficiency. *Plant and Soil* 193(2): 121–148.

Suarez, D.L. (2012) Irrigation water quality assessments. In *ASCE Manual and Reports on Engineering Practice No. 71 Agricultural Salinity Assessment and Management*, ed. Wallender, W.W. Tanji, K.K., 343–370. ASCE, Reston, VA.

Sulmon, C., van Baaren, J., Cabello-Hurtado, F., Gouesbet, G., Hennion, F., Mony, C., Renault, D., et al. (2015) Abiotic stressors and stress responses: What commonalities appear between species across biological organization levels? *Environmental Pollution* 202: 66–77.

Sutinen, S., Aphalo, P.J., Lehto, T. (2007) Does timing of boron application affect needle and bud structure in Scots pine and Norway spruce seedlings? *Trees* 21(6): 661–670.

Sutinen, S., Vuorinen, M., Rikala, R. (2006) Developmental disorders in buds and needles of mature Norway spruce, *Picea abies* (L.) Karst., in relation to needle boron concentrations. *Trees* 20(5): 559–570.

Sutton, T., Baumann, U., Hayes, J., Collins, N.C., Shi, B.J., Schnurbusch, T., Hay, A., et al. (2007) Boron-toxicity tolerance in barley arising from efflux transporter amplification. *Science* 318(5855): 1446–1449.

Suzuki, N., Rivero, R.M., Shulaev, V., Blumwald, E., Mittler, R. (2014) Abiotic and biotic stress combinations. *The New Phytologist* 203(1): 32–43.

Tanaka, M., Takano, J., Chiba, Y., Lombardo, F., Ogasawara, Y., Onouchi, H., Naito, S., Fujiwara, T., (2011) Boron-dependent degradation of NIP5;1 mRNA for acclimation to excess boron conditions in *Arabidopsis*. *Plant Cell* 23: 3547–3559.

Takano, J., Miwa, K., Yuan, L., von Wirén, N., Fujiwara, T. (2005) Endocytosis and degradation of BOR1, a boron transporter of *Arabidopsis thaliana*, regulated by boron availability. *Proceedings of the National Academy of Sciences USA* 102(34): 12276–12281.

Takano, J., Noguchi, K., Yasumori, M., Kobayashi, M., Gajdos, Z., Miwa, K., Hayashi, H., Yoneyama, T., Fujiwara, T. (2002) *Arabidopsis* boron transporter for xylem loading. *Nature* 420(6913): 337–340.

Takano, J., Wada, M., Ludewig, U., Schaaf, G., vonWirén, N., Fujiwara, T. (2006) The *Arabidopsis* major intrinsic protein NIP5;1 is essential for efficient boron uptake and plant development under boron limitation. *The Plant Cell* 18(6): 1498–1509.

Tanaka, N., Uraguchi, S., Saito, A., Kajikawa, M., Kasai, K., Sato, Y., Nagamura, Y., Fujiwara, T. (2013) Roles of pollen-specific boron efflux transporter, OsBOR4, in the rice fertilization process. *Plant and Cell Physiology* 54(12): 2011–2019.

Tepe, M., Aydemir, T. (2011) Antioxidant responses of lentil and barley plants to boron toxicity under different nitrogen sources. *African Journal of Biotechnology* 53: 10882–10891.

Tewari, R.K., Kumar, P., Sharma, P.N. (2010) Morphology and oxidative physiology of boron-deficient mulberry plants. *Tree Physiology* 30(1): 68–77.

Tombuloglu, G., Tombuloglu, H., Sakcali, M.S., Unver, T. (2015) High-throughput transcriptome analysis of barley (*Hordeum vulgare*) exposed to excessive boron. *Gene* 557(1): 71–81.

Turan, M., Taban, N., Taban, S. (2009) Effect of calcium on the alleviation of boron toxicity and localization of boron and calcium in cell wall of wheat. *Notulae Botanicae Horti Agrobotanici Cluj-Napoca* 37(2): 99–103.

Vail, J.W., Parry, M.S., Calton, W.E. (1961) Boron-deficiency dieback in pines. *Plant and Soil* 14(4): 393–398.

Velez-Ramirez, A.I., van Ieperen, W., Vreugdenhi, D., Millenaar, F.F. (2011) Plants under continuous light. *Trends in Plant Science* 16(6): 310–318.

Voxeur, A., Fry, S.C. (2014) Glycosyl inositol phosphoryl ceramides (GIPCs) from Rosa cell cultures are boron-bridged in the plasma membrane and form complexes with rhamnogalacturonan-II. *Plant Journal* 79: 139–149.

Wang, D.N., Ko, W. (1975) Relationship between deformed fruit disease of papaya and boron deficiency. *Phytopathology* 65(4): 445–447.

Wang, N. N., Yan, T. S., Fu, L. N., Zhou, G. F., Liu, Y. Z., Peng, S. A. (2014) Differences in boron distribution and forms in four citrus scion-rootstock combinations with contrasting boron efficiency under boron deficient conditions. *Trees* 28: 1589–1598.

Wimmer, M.A., Eichert, T. (2013) Review: Mechanisms for boron deficiency-mediated changes in plant water relations. *Plant Science* 203–204: 25–32.

Wojcik, P., Wojcik, M., Klamkowski, K. (2008) Response of apple trees to boron fertilization under conditions of low soil boron availability. *Scientia Horticulturae* 116(1): 58–64.

Wojcik, P., Wojcik, M., Treder, W. (2003) Boron absorption and translocation in apple rootstocks under conditions of low medium boron. *Journal of Plant Nutrition* 26: 961–968.

Xiao, J.X., Yan, X., Peng, S.A., Fang, Y.W. (2007) Seasonal changes of mineral nutrients in fruit and leaves of "Newhall" and "Skagg's Bonanza" navel oranges. *Journal of Plant Nutrition* 30(5): 671–690.

Yang, C.Q., Liu, Y.Z., An, J.C., Li, S., Jin, L.F., Zhou, G.F., Wei, Q.J., et al. (2013) Digital gene expression analysis of corky split vein caused by boron deficiency in "Newhall" navel orange (*Citrus sinensis* Osbeck) for selecting differentially expressed genes related to vascular hypertrophy. *PLOS ONE* 8(6): e65737.

Zhang, D.D., Zhao, H., Shi, I.., Xu, F.S. (2014) Physiological and genetic responses to boron deficiency in *Brassica napus*: A review. *Soil Science and Plant Nutrition* 60(3): 304–313.

Zhou, G.F., Liu, Y.Z., Sheng, O., Wei, Q.J., Yang, C.Q., Peng, S.A. (2015) Transcription profiles of boron-deficiency responsive genes in citrus root stock root by suppression subtractive hybridization and cDNA microarray. *Frontiers of Plant Science* 5: 795–815.

Zhou, G.F., Peng, S.A., Liu, Y.Z., Wei, Q.J., Han, J., Islam, M.Z. (2014) The physiological and nutritional responses of seven different citrus root stock seedlings to boron deficiency. *Trees* 28(1): 295–307.

11 Plant Responses to Ozone Stress
Actions and Adaptations

Santisree Parankusam, Srivani S. Adimulam, Pooja
Bhatnagar-Mathur, and Kiran K. Sharma

CONTENTS

11.1 INTRODUCTION

11.1.1 WHAT IS OZONE AND ITS LOCATION?

O_3 is a blue-colored gaseous molecule naturally present in the Earth's stratosphere region at about 20–30 km above the Earth's surface. O_3 was naturally formed in the stratosphere about 200 billion years ago by photolysis of molecular oxygen by UV radiations from the Sun and chemical recombination with oxygen molecules. The stratosphere contains about 90% of the atmospheric O_3 that absorbs UV –B radiations of 280–300 nm range and protects biota from the harmful effects of the UV radiation such as DNA damage. O_3 spreads in the tropospheric region, which is present 1.5–2 km lower than the stratosphere (Ainsworth, 2017). The presence of O_3 in the troposphere was recognized early in the 1940s, and it was believed to accumulate in the troposphere due to the exchange from the stratosphere region (Dütsch, 1971). Although O_3 is predominantly formed in the tropical stratosphere, the high O_3 levels are observed at high latitudes in both hemispheres. Eventually, tropospheric O_3 referred to causes serious damage to plant and animal life (Kalabokas and Repapis, 2004). Although the half-life of the tropospheric O_3 is 22–23 days (Young et al., 2013),

the global temperature change since preindustrial times is said to be due to the increase in the ambient O_3 concentration. Surface O_3 concentrations have more than doubled since the Industrial Revolution (Monks et al., 2015). According to IPCC assessment, current surface O_3 concentrations during the summer months over the Northern Hemisphere are approximately 30–40 ppb, and the predicted values will be 45–50 ppb in 2030, 60 ppb in 2060 and 70 ppb in 2100. An annual increment of 0.5% to 2.5% was observed in tropospheric O_3, especially in the Northern Hemisphere (Hertstein et al., 1995). If the increasing trend of O_3 is not controlled, then O_3 pollution will be the major concern in the coming decades. Despite the general negative effects of O_3, O_3 sensitivity has been found to vary considerably between species, some species being more sensitive and others being barely affected at the same external O_3 concentration (Scebba et al., 2006; Li et al., 2016).

The surface levels of O_3 are steadily increasing due to urbanization, rapid consumption of fossil fuels and growing transportation networks. O_3 pollution is emerging as a serious problem both for human health and vegetation. Being an important part of greenhouse gases spread over large parts of the globe, O_3 contributes to air pollution on the one hand and impacts crop productivity through direct oxidative damage on the other hand. High levels of O_3 has been shown to be detrimental to the growth and yield of economically important crops, depending on species and cultivars (Black et al., 2000). Although there is limited information on the effect of O_3 on crop yields, a study in Europe recorded a loss of crop amounting to 4.3 billion pounds (Holland et al., 2006). Depending on the frequency of stress episodes, O_3 causes short-term responses in plants, such as the development of visible fine bronze or pale yellow specks on the upper surface of leaves or reductions in photosynthesis. Frequent O_3 stress episodes result in longer-term responses such as reductions in growth and yield, and early dieback can also occur. O_3 sensitivity is often linked to low leaf mass per area and low leaf area-based antioxidant levels or stomatal conductance.

The damaging effects of O_3 on plant health were reported as early as 1905 when there was severe vegetable crop damage in Los Angeles. After that, studies have shown that O_3 concentrations in Southern California are sufficient to reduce photosynthetic capacity, cause premature senescence and predispose trees to insect and pathogen attacks (Duriscoe and Stolte, 1989). On exposure to 140 ppb O_3 for 6 days, plants display external damage symptoms as well as physiological changes such as reduced leaf water content, increased reactive oxygen species (ROS) and antioxidants and reduced nitric oxide (NO) levels (Iyer et al., 2013). The increase in the surface O_3 concentration negatively impacts the plant yield and thereby impacts the global food production; it also casts its effect on the natural vegetation, thereby reducing the carbon dioxide (CO_2) uptake by the plants, in turn reducing the biodiversity, which, in turn, plays a key role in global warming (Sitech et al., 2007). Moreover, plant sensitivity to O_3 is also influenced by varied environmental factors like temperature, humidity or photosynthetically active radiations (Dunning and Heck, 1977). For instance, drought, low humidity and low temperature decrease plant sensitivity to O_3 (Otto and Daines, 1969; McLaughlin and Taylor, 1981; Tingey and Hogsett, 1985). Although plants usually survive polluted environments through the development of tolerance mechanisms, still O_3 tolerance pathway is yet to decipher fully.

Although generating projections for O_3 is complex, the O_3 problem is predicted to escalate given climate change and increasing O_3 precursor emissions in many areas (Young et al., 2013; IPCC, 2013). A modeling study in Europe indicated the reduction in mean annual gross primary production by 22% and leaf index area by 15–20% (Anav et al., 2011). There was also a 30% decrease in the gross primary productivity of European forest sites from 2000 to 2010 and 14–23% during 1901–2100 (Proietti et al., 2016).

For angiosperms, the total reduction in the biomass was 23% at an O_3 concentration of 74 parts per billion (ppb) and for gymnosperms, it was 7% at 92 ppb (Wittig et al., 2009). The estimated loss in 23 common crop species accounts for approximately 4.5 billion euro due to surface O_3 exposure by 2020 (Holland et al., 2006). There will be a global loss in the yield of 4–15% *Triticum aestivum*, 3–4% *Oryza sativa*, 2–5% *Zea mays* and 5–15% *Glycine max* due to O_3 every year (McGrath et al., 2015; Tai and Martin, 2017).

11.1.2 FORMATION, TRANSPORT, DEPOSITION OF O_3

O_3 is now a global problem that exerts greater impact worldwide. O_3 concentration has been steadily increasing since World War II (Oltmans et al., 2006). From the last two decades, O_3 levels have been increasing at an approximate rate of 1 to 2% per year (Hough and Derwent, 1990). The formation of O_3 occurs through a complex series of photochemical reactions of its precursors emitted from a variety of anthropogenic activities (Crutzen and Zimmermann, 1991). In the sunlight, the primary pollutants like nitrogen oxides (NO_x) and hydrocarbons form secondary pollutants such as peroxyacetyl nitrate (PAN), O_3, other aldehydes and ketones in the troposphere (Becker et al., 1998). For instance, photodissociation of NO_x by short radiations (between 280 and 430 nm) produce free oxygen atoms along with nitrogen monoxide. This excited free oxygen atom reacts with O_2, leading to the formation of O_3. However, in a non-polluted area, where the NO_2/NO ratio is low, O_3 gets disrupted by the reaction with NO. At times, the additional radicles formed by the photolysis of O_3 react with CO, resulting in additional O_3 production (The Royal Society, 2008). Similarly, volatile organic compounds also promote O_3 accumulation due to their ability to oxidize NO and thus increase the NO_2/NO ratio. Once O_3 is formed, its lifetime in the free troposphere may vary from weeks to months, allowing a long-range transport.

Stratospheric O_3 is in photochemical equilibrium, where it adapts instantaneously to variations in chemical production or loss. Below this altitude, air mass transport processes mainly determine the O_3 amount at a particular location. Often, O_3 can also be formed at a region far away from its site of origin due to the downwind of polluted air masses to a region that is suitable for O_3 formation (Lin et al., 2012). After the formation, the surface O_3 can generally travel downwind with a range of 240–800 km. O_3 molecules have been found to be transported long distances as far as thousands of kilometers away from their source regions (Figure 11.1). Due to this reason, sometimes O_3 levels have been reported

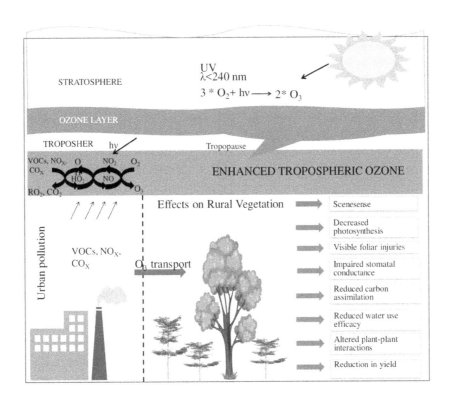

FIGURE 11.1   An illustration depicting ozone formation, transport and impacts on plants.

higher in many remote areas, even though away from industrial zones and anthropogenic pollutions (Prather et al., 2003). The O_3 accumulation also depends on seasonal as well as diurnal variation. Usually, O_3 levels are higher during summer and autumn seasons, while humid and winter months exhibit minimum values (Jain et al., 2005). This is because of the buildup of photochemical smog due to low wind and high levels of sunlight during summers and vice versa during winters. Similarly, the recorded O_3 levels are higher during daytime compared to night levels of O_3. Studies also show that plants in northern regions can be more O_3-sensitive than plants at lower latitudes, despite lower ambient O_3 levels (De Temmerman et al., 2002a). It has been suggested that the conditions of summers at higher latitudes, characterized by moderate temperatures, moist climate and long days, are favorable for stomatal opening, thereby increasing stomatal O_3 uptake and damage (Pleijel et al., 2000). It has also been proposed that the shorter summer nights in the northern part of Europe provide a too short period of overnight recovery of O_3 damage between naturally occurring daytime O_3 exposures (De Temmerman et al., 2002b). Moreover, the responses to O_3 also depend on the inherent genetic makeup and growth stage of the plants. Studies suggested that when plants are exposed to high O_3 concentration during a reproductive phase like the period between flowering and pod setting, it causes greater impact on the yield than at the vegetative phase

(Lee et al., 1988). When the *G. max* plants were exposed to 72 ppb of O_3 continuously for 5 days, the symptoms of injury varied with leaf position, supporting that O_3 effects vary with the age and position of the leaf (Burton et al., 2016).

The continuous production of O_3 by solar radiation is restricted by O_3 chemical breakdown, through cycles that involve hydrogen, nitrogen, chlorine and bromine compounds. The accumulation of O_3 in the surface is also restricted due to its depositions. Forests are the major sinks for O_3 dry deposition. The major portion of O_3 deposition occurs through stomatal absorption while cuticular deposition also contributes to some extent in plants (Ainsworth, 2017).

11.1.3   Ozone Sensing and Signaling

O_3 uptake is generally associated with stomatal conductance, which is genus-specific, where the difference appears in the absorption capacity of the plant by other external surfaces like stems and cuticles and also the absorption of the O_3 into the apoplast of the leaves after entering the stomata (Low et al., 2006). O_3 can also directly diffuse into the cytosol through the cell membrane and generate ROS; in turn, the generated ROS affects the stomatal conductance (Ainsworth, 2017). Stomatal O_3 flux depends on the opening and closing of the stomata and varies with the age of the leaves, environmental conditions like temperature, light intensity, humidity and relative soil water content. Stomatal conductance is high in humid conditions compared to dry and hot conditions.

The O_3 sensitivity in plants is indistinct due to the lake of clear visible symptoms in all the scenarios. While the O_3 sensitivity in foliage crops appear as visible foliar injury symptoms during early growth stages, other crops (such as leafy salad crops and *Medicago spp.*) do not necessarily give rise to an equivalent negative impact on yield in fully developed crops. For instance, few cultivars in staple crops such as *T. aestivum* and *O. sativa* when O_3-stressed exhibited least visible injury symptoms to foliage but had greater grain yield loss. The observed variation in the appearance of visible injury symptoms may be attributed to genotypic variation in a stomatal moment in response to O_3 (Picchi et al., 2010). Genotypic variability in O_3 sensitivity can arise in several ways. Generally, O_3-insensitive genotypes close their stomata even after exposure to a very little dose of O_3 and thus prevent further foliar injury compared to O_3-sensitive genotypes. However, prolonged stomatal closure reduces the photosynthetic assimilates. Hence, there is scope for the developing genotypes with greater O_3 tolerance concerning yield that mitigate food security.

For any plant to accustom to O_3 stress, the stress signal has to be perceived first at the cellular level and relayed to the nucleus to convert this signal into defense through a series of cell reprogramming events. Three main O_3-sensing localizations have been suggested in the literature, including epicuticular waxes, the cell wall and the plasma membrane. Later on, different early signaling components related to these O_3-sensing sites such as lipids, various membrane proteins like G proteins, nicotinamide adenine dinucleotide phosphate (NADPH) oxidases, ion channels and mitogen-activated protein kinases (MAPK) were also known to moderate O_3 signaling in plants. Epicuticular waxes protect the underlying cells from the influence of climate adversaries. Although O_3 has no significant effect on wax composition, O_3 exposure has a negative influence on wax quality, especially in O_3-sensitive clones, as evidenced by the increased rust colonization on leaf surfaces of *populous termuloides* under O_3 stress (Maňkovska et al., 2005). Comparative metabolite profiling of two *Betula pendula* genotypes varied in their O_3 sensitivity and exposed to O_3 for nearly 7 years resulted in the accumulation of compounds related to leaf cuticular wax formation ,such as triterpenoids, 1-hexacosanol, squalene, dammaran-3-ol, 1-dotriacontanol, etc., possibly linked with the O_3 tolerance (Kontunen-Soppela et al., 2007). Moreover, a combination of high CO_2 and O_3 substantially modified the fatty acid and primary alcohol composition of wax when compared to O_3 exposure alone (Casteel et al., 2008).

Cell walls are other direct contacts of O_3 in plants. Long-term exposure to O_3 increases the lignification of the cell wall (Cabané et al., 2012). In most cases, like in *B. pendula*, the perceived O_3 is immediately (within 2 hours) degraded into ROS in the apoplastic space due to cell wall peroxidase (POD) activity (Pellinen et al., 1999; Ranieri et al., 2003). Plasma membrane-localized kinase proteins and other apoplastic receptors present at the interface between the cell wall and cell seem to be initial candidates of stress signal perception. Being the first stress signal, apoplastic ROS, in turn, induce transcription of genes coding for cysteine-rich receptor-like kinases, with a transmembrane domain (Wrzaczek et al., 2010). This is confirmed by the doubled mRNA abundance immediately after 1 hour of 250 ppb O_3 exposure (Wrzaczek et al., 2010).

Being powerful oxidants, O_3 and O_3-induced ROS cause strong lipid peroxidation. This lipid oxidation not only damages the membrane components but also releases a cascade of biologically active ozonation products as stress signals. O_3 exposure leads to higher membrane permeability due to rapid loss of K^+ in *Chlorella* plasma membrane (Heath and Frederick, 1979). O_3-induced membrane oxidation rapidly alters the composition and conformation of the phospholipids that increase the permeability, much before any visible injury (Farmer and Mueller, 2013). For instance, 150 ppb O_3 increased the phospholipid content within 2–4 hours after exposure in *Pharbitis nil*, followed by an increment in structural phospholipids such as phosphatidylcholine, phosphatidylglycerol, phosphatidylinositol and phosphatidylethanolamine and decreased gradually after that (Nouchi and Toyama, 1988). In O_3-exposed *Lens culinaris* seedlings, an increased lipoxygenase activity and lipoxygenase gene correlated with lipid peroxidation (Maccarrone et al., 1997). Similarly, 30 min after exposure to a flux of O_3 enhanced the gene transcription and specific activity of lipoxygenases 1 and 2 in *G. max* seedlings (Maccarrone et al., 1992). O_3 also stimulated the monogalactosyldiacylglycerols (MGDGs) hydrolysis, and free fatty acid production was also enhanced and even in the chloroplast of *Arabidopsis thaliana* leaves. Moreover, the requirement of plastid trienoic fatty acids in O_3-induced superoxide accumulation has also been studied in detail through genetic analysis in O_3-exposed leaves of the *A. thaliana* (Sharma et al., 2006).

The MAPKs play diverse roles in intra and extracellular signaling in response to various stresses in plants. Members of MAPK are known to get activated under O_3 exposure in a wide number of plants, including

Nicotiana tabacum (Samuel et al., 2000; Gomi et al., 2005), *A. thaliana* (Ahlfors et al., 2004) and suspension cultures of *Populus tricocarpa* X *Populus deltoides* (Hamel et al., 2005). O_3 also induce the transcription of MAPK family genes in *O. sativa* (Agrawal et al., 2002). Besides MAP kinases, other cell wall associated kinases, lipid rafts, ion channels and calcium have been suspected to be involved in O_3 signaling. External O_3 stimulates apoplastic ascorbate (AsA), diffusing across the cytosol as AsA, or dehydro AsA, at the expense of glutathione (GSH), acts as the signal for the activation of the O_3 stress responses (Venkatesh and Park, 2014). An increase in the biphasic intracellular Ca^{+2} concentration leading to the expression of GSH *S*-transferase was observed when the atmospheric O_3 concentration crossed 70 ppb, thereby acting as ROS sensors affecting the stomatal ion conductance (Clayton et al., 1999). However, detailed research is needed to better decipher the O_3 specificity and complete plant signaling networks.

11.1.3.1 Deploying Ozone in Plant Studies

To understand the plant responses to O_3 stress, many studies have been conducted simulating experiments using external O_3 source (Table 11.1). In most cases, the impact of O_3 depends on dose rather than actual concentration. For external supplementation, O_3 dose is usually calculated by multiplying mean O_3 concentration by the duration of exposure, which can be expressed as ppb or L L^{-1} h^{-1} (Sandermann, 1996). Based on the O_3 concentration in the plant vicinity, O_3 stress can be differentiated into two distinct types, namely, chronic and acute stress. In natural conditions, O_3 peaks occur up to around 120 ppb in some polluted area while the ambient air carries approximately 40–50 nL L^{-1}. Hence, chronic exposure refers to long-term exposures to O_3 concentrations below 120 ppb (The Royal Society, 2008). Acute O_3 stress corresponds to high O_3 concentrations with values above 200 ppb and up to 1000 ppb within a short time frame, which can occur several times during the growth season around the world. However, the acute dose is unlikely to present in natural conditions, but these high O_3 doses are used as an external supplement in growth chambers in order to trigger and explore O_3 signaling components. Both chronic and acute O_3 exposures were found to induce several plant responses at the biochemical and molecular levels in model legume *Medicago truncatula* (Puckette et al., 2007). For example, UNECE introduced the critical O_3 concept which is of 40 ppb (AOT40). A specific European program was set up by ICP Forests monitoring network for the validation of O_3 injury in 2000. They reported that exceeding AOT40 by even 5 parts per

million (ppm) in one growing season would cause growth reduction in forest trees. However, it is very difficult to assess plant responses to O_3 in an open environment due to the uncontrolled atmospheric O_3 because it is difficult to determine whether the observed effect is due to O_3 or to any other growth condition. Hence, many studies have opted to grow plants inside controlled enclosures in which O_3 is continually released in known concentrations that mimic the effect of the ambient O_3 variations in the natural environments (Sandermann, 1996). The enclosure system in which all other growth parameters were kept was controlled to allow O_3 enrichment. These simulating studies give an opportunity not just to evaluate the effect of several concentrations of O_3 but also the impact of simultaneous addition of other gases or pollutants as in studies conducted on *B. pendula*, *B. pubescences* and *T. aestivum* (Pääkkönen et al., 1997; Pleijel et al., 1991; Kontunen-Soppela et al., 2010). Free-air CO_2 enrichment (FACE) technology has also been adapted for O_3, enabling crops to be grown under standard field conditions with elevated O_3 levels. In another approach, called the zonal air pollution system, a series of pipes facilitated to continuously release O_3 at various rates into the plant canopy present in different plots (Runeckles et al., 1990). Although this method eliminates the issues with artificial conditions inside chambers, it is often very expensive. Moreover, due to the continuous mixing with ambient air, only low maximum enrichment can be achieved by this method. Hence, most of the knowledge on O_3 stress was obtained from experiments in controlled conditions or free-air fumigation experiments, open-top chambers and modeling studies with slightly higher O_3 exposure (up to 120 ppb) applied at regular intervals for a shorter time. Nonetheless, there are a number of ambiguities that exist in these global predictions of surface O_3 levels and crop loss models to O_3 depending on the region or time of year (Van Dingenen et al., 2009; Avnery et al, 2011). In spite of uncertainties, the simulating and modeling studies provide strong evidence for the heavy baring damaging effects of O_3 on global agricultural production in a bigger picture (Lobell and Field, 2007).

11.1.3.2 Impacts of Ozone Stress on Plants

Tropospheric O_3 is a greenhouse gas and phytotoxic pollutant. The effect of O_3 on plants depends upon the concentration of O_3 and the duration of the exposure. The rising O_3 level in the atmosphere affects the plant growth rate and induces visible foliar injury. However, the severity of O_3 stress is dependent on the combination of O_3 uptake into the leaf and the induced plant defense mechanisms in plant tissues. Generally, chronic exposures

TABLE 11.1

Summary of Up-to-Date Research Focusing on the Ozone Responses in Various Plant Species

Plant Species	Stress Induction	Response	References
Medicago truncatula	40 nmol mol^{-1} 300 ppb	Visible injuries, decrease in the net photosynthetic rate and stomatal conductance; jasmonic acid signaling and cell death	Iyer et al. (2013), Rao et al. (2000)
Sansevieria trifasciata	500 ppb	Effective in mitigating indoor ozone levels	Papinchak et al. (2009)
Chlorophytum comosum	500 ppb	Effective in mitigating indoor ozone levels	Papinchiak et al. (2009)
Epipremnum aureum	500 ppb	Effective in mitigating indoor ozone levels	Papinchiak et al. (2009)
Glycine max	72, 150 ppb	Pod formation and flower production	Leisner et al. (2014) and Bruton et al. (2016)
Betula pendula	130 nmol. mol^{-1} Ambient ozone	Drought and ozone stress combination responses; ozone on cuticular wax development	Pääkkönen et al. (1998) and Kontunen-Soppela et. al. (2007)
Quercus pubescens	100–300 ppb	Isoprene production	Pinelli and Tricoli (2008)
Populus x canescens	800 ppb	Negative impact on S-nitrosylation and increase in the phenylalanine ammonia-lyase activity	Vanzo et al. (2014)
Trifolium subterraneum	70 ppb	Visible foliar injury, chlorotic and necrotic spots, importance of photoperiod	Futsæther et al. (2015)
Glycine max	200 ppb, 5 ml.min^{-1}	Gene expression, Lypoxigenase activity, early senescence	Moon et al. (2013), Maccarrone et al. (1992)
Fagus crenata	60 nmol mol^{-1}	Stomatal conductance	Hoshika et al. (2013)
Arabidopsis thaliana	300 ppb	Oxidative stress	Rao et al. (2000)
Pinus ponderosa	Ambient ozone	Root growth rate and carbohydrate assimilation	Andersen et al. (1997)
Nicotiana plumbaginifolia	59 nl L^{-1}	Antioxidant defense	Van Champ et al. (1993)
Triticum aestivum	120 ppb, 80 ppb 6–4 & 90 μg m^{-3}	Exogenous application of spermidine and ethyldiurea, antioxidant defense Combination of O_3, SO_2 and NO_2 on yield	Van Champ et al. (1993) Ashrafuzzaman et al. (2017)
Nicotiana tabacum, Ipomoea nil, Psidium guajava	Ambient ozone	Epicuticular wax degradation and stomatal movement	Alves et al. (2016)
Pisum sativum, Glycine max, Phaseolus vulgare	151.2 nL L^{-1} 150 nL L^{-1}	Vegetative growth, foliar injury and physiological responses under different stress durations	Yendrek et al. (2015) and Mehelhorn et al. (1991)
Petroselinum crispum	200 nl L^{-1}	Lesions and necrosis, accumulation of flavone glycosides	Eckey-Kaltenbach et al. (1994)
Melissa officinalis	200 ppb	PSII efficiency, photosynthesis	Pellegrini et al. (2011)
Lycopersicon esculentum	0.1 to 0.4 ppm	Dry weight, plant biomass, resource allocation to root, shoot and leaves	Sudhaar et al. (2008)
Solanum lycopersicum	500 μg m^{-3}	Age-dependent ozone sensitivity, photo inhibition of PSII activity	Thwe et al. (2014)
Populus termoides	70 ppb	Stomatal conductance	Kovska et al. (2005)
Pharbitis nil	150 ppb	Phospholipid content	Nouchi and Toyama (1988)
Lens culinaris	5 ml.min^{-1}	Gene expression	Maccarrone et al. (1997)
Nicotiana tabacum	500 ppb	MAPK activity, jasmonic acid signaling and stomatal movement	Samuel et al. (2000) and Gomi et al. (2005)

do not lead to visible damage but result in a decline of photosynthesis, growth inhibition and premature senescence (Krupa, 2003). In contrast, acute exposure leads to induction of cell death accompanied by visible lesions in sensitive plants (Kangasjärvi et al., 1994, 2005; Rao et al., 2000a,b).

11.1.3.2.1 Phenotypic Responses of Plants to Ozone Stress

Depending on the severity of the stress, O_3 causes visible injury symptoms to foliage. Some of the well-known O_3 stress symptoms include enhanced leaf senescence, visible foliar injuries, reduced carbon assimilation, impaired stomatal conductance and reduced water-use efficiency (Joo et al., 2005). These phenotypic responses help to differentiate plant genotypes into O_3-tolerant and O_3-sensitive ones (Overmyer et al., 2000; Vahisalu et al., 2008). For instance, phenotypic symptoms in response to O_3 develop faster in sensitive plants like herbaceous plants or maybe slow down by many months in conifers; this might be due to the difference in the stomatal conductance (Davison and Barnes, 1998). O_3 stress results in two-stage stress symptoms in plants where the first phase includes lowered metabolism that includes photosynthesis and may or may not be followed by a second stage comprising of visible symptoms such as chlorotic lesion development (Heath, 1994). O_3 induces early senescence and abscission of leaves and thus can moderate biomass growth via carbon allocation to edible plant parts. This directly impacts yields, especially in leafy biomass crops and ornamental plants such as *Lactuca sativa*, *Spinacea oleracea*, etc. O_3-induced visible damage to the leaves decreases the economic value of the crops (Ashmore, 2005). These foliar injuries and reduced leaf area also affect yield in non-foliage crops, due to the reduction in the amount of green leaf available for carbon fixation. The reduced carbon fixation indirectly affects the biomass and grain filling in *T. aestivum* (Mulholland et al., 1998; McKee and Long, 2001). O_3 also affect the grain quality by reducing the starch content and increasing protein and nutrient content in crops like *T. aestivum* and *O. sativa* (Broberg et al., 2015; Frei, 2015). Additionally, O_3 exposure significantly decreased the grain size in these crops, drastically affecting the consumer acceptance (Ainsworth, 2017).

11.1.3.2.2 Ozone and Stomatal Conductance

The entry of O_3 into the leaf through the stomata triggers a cellular response and ultimately results in damage to crop productivity (Ainsworth, 2017). When plants sense the O_3 stress in their vicinity, reduction of stomatal conductance may be offered as a physical resistance

mechanism to O_3 (Mansfield and Freer-Smith, 1984; Reiling and Davison, 1995). Night peaks in O_3 hinder the guard cell openings and lead to a reduction of biomass in plants (McCurdy, 1994). With the increase in the O_3 concentration, the stomatal conductance and stomatal pore area were reduced in *Solanum lycopersicon* (Thwe et al., 2014). According to Levitt (1972), plants avoid O_3 stress by decreasing stomatal conductance during O_3 exposure (Tausz et al., 2007). This is because stomata are the principal interface for entry of O_3 into plants. Often, O_3-induced inhibition of carbon assimilation by chloroplasts increases in internal CO_2 concentration, also leading to stomatal closure (Reich, 1987; Weber et al., 1993; Heath and Taylor, 1997). Ainsworth et al. (2012) also suggested the reduction in stomatal conductance in plants after exposure to chronic O_3 stress is possibly due to a direct effect of O_3 on photosynthesis and to a resultant increase in internal CO_2 concentration. As a matter of fact, the rapidity of stomatal closure to environmental stimuli has also been shown to be impaired in O_3-exposed plants (McAinsh et al., 2002; Reich et al., 1984). For example, O_3 fumigation for 36 days at a concentration of 130 nmol mol≤ to *B. pendula* caused decreased stomatal conductance, net photosynthesis rate and Ribulose-1,5-bisphosphate carboxylase/oxygenase (RuBisCo) content while increasing the stomatal number and structural chloroplast injuries (Pääkkönen et al., 1998). Similarly, in *Fagus crenata*, the efficiency of O_3 stress avoidance is linked to a decrease in stomatal conductance and an O_3-induced decrease in photosynthesis (Hoshika et al., 2013).

The differences in the plant's sensitivity to O_3 varies partly with the difference in the stomatal conductance of the plant (Reich, 1987). It is proven that drought-induced stomatal closure would protect plants from O_3 stress based on the O_3 uptake models (Panek and Goldstein, 2001; Grünhage and Jäger, 2003). These studies also contradicted that O_3 caused stomatal 'sluggishness', which led to incomplete closure of stomata and hence exacerbated the effects of drought (Grulke et al., 2003; Karnosky et al., 2005). In addition, a plasma membrane slow anion channel, *slow anion channel1* (*SLAC1*), specific to guard cells has also been shown to be involved in O_3-induced stomatal closure (Vahisalu et al., 2008). Activation of SLAC1 leads to anion efflux and depolarization of the cell membranes, which subsequently activates potassium-efflux channels. Recently, repression of GOLDEN 2-LIKE1 and 2 (*GLK1, GLK2*) transcriptional factors resulted in reduced stomatal aperture and water loss under O_3 stress, resulting in greater O_3 tolerance (Nagatoshi et al., 2016). The knockout mutations in open stomata 1 (OST1), which phosphorylates and activates

SLAC1, leads to the loss of stomatal closure in response to O_3 exposure (Vahisalu et al., 2010). Besides, O_3 also modulates K^+ channels that lead to altered Ca^{2+} homeostasis of guard cells which can directly induce stomatal closure (Torsethaugen et al., 1999; Vahisalu et al., 2010; McAinsh et al., 2002). Clayton et al. (1999) used intracellular Ca^{+2} indicators to demonstrate the O_3-induced elevation of intracellular Ca^{+2} levels in *A. thaliana*. This is due to the oxidative activation of Ca^{+2} channels when the O_3 concentration in the ambiance crossed 70 ppb. These accumulated Ca^{+2} ions function as ROS sensors, thereby affecting the stomatal ion conductance.

11.1.3.2.3 Carbon Assimilation and Leaf Senescence

A high level of surface O_3 is posing a great threat to agricultural production. Several reports across the world have suggested the serious economic consequences due to O_3-induced reductions in crop yield (Tonneijck, 1989; Heck et al., 1982). Reich (1983) characterized the growth defects in O_3-exposed greenhouse-grown trees and suggested a direct correlation between the O_3 exposure and reductions in net CO_2 assimilation rate. This was further confirmed across several tree species grown in fumigation chamber, open-top chamber and field fumigation systems. Furthermore, O_3-induced significant decrease in both photosynthesis and stomatal conductance of trees negatively affects both carbon sequestration and transpiration (Wittig et al., 2007). In another study, high O_3 caused a reduction in net photosynthetic rate in 29 deciduous and evergreen woody species. Evergreen species were found to be more tolerant of O_3 than deciduous species (Li et al., 2016). When exposed to O_3 stress, three legume crops *Pisum sativum, G. max, Phaseolus vulgare* exhibited a decline in the photosynthetic parameters and leaf longevity (Yendrek et al., 2015). Similarly, in *Quercus pubescent,* acute O_3 concentrations of 300 ppb caused long-lasting inhibition of photosynthesis and isoprene emission (Pinelli and Tricoli, 2008). Strong negative effects on photosynthetic ability were observed due to the decrease in stomatal conductance and increased intercellular CO_2 concentration. Further, the defects in growth and photosynthesis in O_3-sensitive birch clones have been found to be akin to O_3-induced ultrastructural injuries, especially in chloroplasts (Pääkkönen et al., 1996). O_3 exposure also reduces chloroplast size and cell starch content, as observed in field-grown *P. termuloides* and *B. papyrifera* (Oksanen et al., 2004). The reduced chlorophyll (chl) and carotenoid levels have been shown to impact photochemistry and a reduced electron transport rate in *Populus spp.* (Bagard et al., 2008). Similarly, *Melissa officinalis* plants fumigated with O_3 for 24 and 48 h showed a significant reduction in the chl content,

suggesting the effect on the chl-binding proteins of the light-harvesting complex, thereby reducing the photosystem II (PSII) efficiency and electron transport rate through PSII. Moreover, these plants showed a significant decrease in CO_2 fixation ability, possibly correlated with a strong reduction in grain protein yield with a concomitant increase in confidence intervals. Interestingly, in most of the crops, a greater reduction in photosynthesis occurs due to high O_3 when compared with vegetative phases of development due to the longer time exposure at the top of the canopy (Feng et al., 2008). However, the reduced pigment content may protect the PSII from photoinhibition through a reduction of the number of light-harvesting antennae. The accumulating studies suggest the principal mechanism of decrease in photosynthetic rate either due to direct effects of O_3 exposure on light or dark or both reactions of photosynthesis (Power and Ashmore, 2002) or through an indirect stomatal closure effect (Noormets et al., 2001).

In addition to its impact on photochemistry activity, O_3 has also been shown to affect the Calvin–Benson cycle (Saxe, 2002). A clear decline in RuBisCo carboxylation efficiency, CO_2 assimilation, gs and chl content has been observed in *Populus spp.* plants when exposed to O_3 stress (Renaut et al., 2009; Wittig et al., 2009). A good number of studies have shown the correlation between O_3 exposure and a decrease in RuBisCo activity and content (Dizegremel et al., 1994; Lütz et al., 2000; Gaucher et al., 2003). Similarly, both RuBisCo and RuBisCo activase levels were reduced in *Pinus halepensis* exposed to O_3 (Pelloux et al., 2001). Proteomic analysis of *Populus spp.* leaves exposed to chronic O_3 also confirmed the downregulation of a large number of proteins involved in the Calvin–Benson cycle and electron transport (Bohler et al., 2007). Since O_3 reduces carbon gain by lowering RuBisCo activity and inducing early senescence, it, in turn, increases the carbon cost for tissue repair and antioxidant synthesis. Additionally, an increase in source strength and a decrease in carbon export to sink tissues has also been suggested due to chronic O_3 exposure (Friend and Tomlinson, 1992; Grantz and Farrar, 1999; Grantz and Yang, 2000).

Although multiple studies attributed the reduction in gross CO_2 assimilation rate to RuBisCo, a few studies also suggested enhanced CO_2 losses through an increased respiration in trees like *Populus spp., B. pubescence*, etc. (Bagard et al., 2008; Matyssek et al., 1997; Noormets et al., 2001; Kitao et al., 2009). In support of this notion, the enhanced CO_2 efflux from respiration is also supported by enhanced glycolysis, pentose-phosphate pathway and TCA cycle activity (Dizengremel et al., 2012). Furthermore, the phosphoenolpyruvate carboxylase is strongly upregulated in the leaves of a wide range of species after O_3 exposure (Dizengremel et al., 2012). Phosphoenolpyruvate carboxylase induced pathways also supply additional carbon and NADPH to detoxification processes thus enhance O_3 tolerance (Dizengremel et al., 2009). Moreover, O_3-induced changes in carbon assimilation and respiration further reduce plant growth rates, above- and belowground biomass (Ainsworth, 2017).

11.1.3.2.4   Cell Wall Modification

The cell wall has been regarded as one of the early targets of O_3 prior to reaching the membrane. Of the complex constituents of the cell wall, the aromatic compounds, lignin and proteins are most vulnerable to O_3 damage. A range of plant species has been shown to have strong rearrangements in the cell wall organization and in its component biosynthesis in response to O_3 exposure (Le Gall et al., 2015). O_3 increased both enzyme activities and related gene transcript levels involved in lignin biosynthesis in the leaves of a number of tree species such as *Pinus sylvestris, Populus, B. pendula* and many other species (Zinser et al., 1998; Pääkkönen et al., 1998; Di Baccio et al., 2008). Similarly, lignin content was also increased in stems of O_3-fumigated *Populus* and *Betula* (Kaakinen et al., 2004) in a 3-year open-air fumigation while the increase was not there after 5 years exposure (Kostiainen et al., 2008). The observed differences are partly due to the uncontrolled climatic variations in open-air experiments. In several species such as *Fraxinus ornus*, the O_3 injury was evident as punctures within the cell wall, which is followed by the thickening of palisade mesophyll cell walls (Paoletti et al., 2010). Thickening of the cell wall helps to increase mechanical resistance and aid in cellular detoxification processes of O_3. Moreover, leaf anatomy analysis of deciduous trees showing visible stress symptoms confirmed the thickened cell walls with pectinaceous projections. In addition, both acute and chronic O_3 exposure has been shown to promote phenylpropanoid metabolism in the leaves of many tree species. In another study, 7-day exposure to O_3 increased cell wall sugar content, possibly due to increased pectin content in the *Fragaria ananassa* plant (Le Gall et al., 2015). Fourier transform infrared spectroscopy (FTIR) analysis of the ragweed (*Ambrosia artemisiifolia* L.) pollen exhibited a reduced phenolic content and increased pectin components (rhamnogalacturonic acid, arabinan) in response to O_3 (Kanter et al., 2013). Moreover, a remarkable decrease in acetyl ester groups, suggesting the de-esterification of pectin, was also observed. This observation was consistent with the transcriptome analysis that confirmed the strong upregulation of pectin methylesterase and pectate lyase encoding

genes when exposed to O_3. O_3 treatment delayed fruit ripening in *Actinidia deliciosa* fruit due to reduced cell wall disassembly of the fruits at room temperature, which was possibly attributed to its negative impact on ethylene-mediated cell wall oxidation (Minas et al., 2014). O_3 treatment reduced the activities of enzymes involved in pectin and hemicellulose degradation such as peptidoglycan and endo-1,4-β-glucanase compared with the control fruits. Similarly, the expression of the genes, including cellulose, β-galactosidase, cell wall invertase inhibitor or glycine-rich protein, were upregulated while the expression of XET and COBRA genes was downregulated in an O_3-sensitive rice cultivar when compared to insensitive cultivar (Cho et al., 2013). O_3 was observed to repress cambial growth and xylem differentiation, which was correlated with the reduction in cellulose and lignin biosynthesis in *Populus* wood (Richet et al., 2011). This, in turn, resulted in the reduction of phenylpropanoid pathway and modifying the cellulose to lignin ratio by increasing cell wall lignin content (Richet et al., 2011). A clear enrichment in condensed lignin due to the accumulation of H-units was observed near the foliar necrotic area in these plants (Cabané et al., 2004). Although lignin and pectin are the most studied cell wall components under O_3 stress, another component like cell wall POD was also induced by O_3 in *Chrysanthemum morifolium* leaves, resulting in cell wall stiffening, thereby reducing O_3 penetration into the cell (Ranieri et al., 2000). The quantity of the wall-bound benzaldehyde isomer was also shown to reduce within 2–3 h of O_3 exposure in *S. lycopersicon* leaves (Wiese and Pell, 2003). With all this, it is clear that O_3 can induce long-term modifications of cell wall structure and composition and is thus critical to decide the wood quality and quantity in tree species.

Like all other stresses, O_3 treatment also damages the plasma membrane. The leaves of *Medicago officinalis* showed typical symptoms represented by roundish dark-blackish necrosis within 48 h of O_3 treatment. Prior to the presence of visible injury, there was an increase in membrane damage, which lasted during the recovery period. Similarly, in many other species, even short exposure to O_3 can also induce a deleterious effect on function (Guidi et al., 2001), integrity (Calatayud et al., 2002; Francini et al., 2007), conformation (Ranieri et al., 2001) and transport capacity of membranes (Płażek et al., 2000; Ranieri et al., 2001).

11.1.3.2.5   Ozone as an Oxidant

As a strong oxidant, O_3 causes several types of visible injury, including chlorosis and necrosis. Sometimes, even though the O_3 stress had no visible symptoms of injury, it had drastic damage at the cellular level

(Ainsworth, 2017). O_3 interacts with the adjoining cell membranes to form ROS, resulting in oxidative bursts (Fiscus et al., 2005) within adjoining cells. Acute O_3 exposure (350 ppb) of *A. thaliana* leaves for 6 h induced a biphasic oxidative burst with a rapid first peak after 1–1.5 h followed by a second peak which occurred after 12–24 h (Joo et al., 2005). The reactive oxygen species as superoxide, hydrogen peroxide (H_2O_2) and NO, generated due to the O_3 degradation in the apoplast, has been attributed to most of the observed O_3-induced damage in plants (Ahlfors et al., 2009). The unscavenged ROS result in foliar injury and programmed cell death, as evidenced by visible symptoms such as interveinal necrosis, early senescence in 27 crop species growing in fields across Europe (Mills et al., 2011). Similarly, O_3 promoted the winter *T. aestivum* leaves to senesce faster through the promotion of membrane lipid peroxidation and inhibiting the function of the antioxidation system (Zheng et al., 2005). In several studies, O_3 induced oxidative burst that leads to a hypersensitive response in a fashion akin to pathogen defense pathways (Kangasjärvi et al., 2005). The involvement of plasma-membrane-bound NADPH oxidase (RBOH) in the O_3-induced oxidative burst has been suggested by many studies (Bettinin et al., 2008). Seven days of chronic O_3 exposure significantly increased RBOH activity in the pollen of *A. artemisiifolia* (Pasqualini et al., 2011). Furthermore, the reduced ROS accumulation and leaf damage in *radical-induced cell death 1* (*rcd1*) mutant of *A. thaliana* treated with NADPH oxidase inhibitor diphenylene iodonium confirm that O_3 initiates active cellular ROS production through NADPH oxidase (Overmyer et al., 2000). Moreover, chloroplastic ROS signaling also participates in the activation of NADPH oxidases in the O_3-induced oxidative stress response (Joo et al., 2005).

Accute O_3 exposure often induces cell death due to the uncontrolled generation of reactive free radicals. In addition, O_3-induced oxidative burst in *N. tabacum* leaves, accompanied by transient increase of transcript levels of pathogenesis-related-1a, increased protease activity and chromatin condensation, suggesting the programmed cell death initiated by O_3-induced oxidative stress (Pasqualini et al., 2003). O_3 injury in plants is also dependent on photoperiod and antioxidant content of the tissue (Futsæther et al., 2015). For example, an NADPH oxidase-dependent burst led to cell death in transgenic *N. tabacum* plants having reduced catalase (CAT) activity (Dat et al., 2003). Similarly, several reports supported the detoxification role of apoplastic AsA that even account for scavenging up to 50% of O_3 entering the leaves. The role of AsA in O_3 detoxification

was further confirmed by isolation of ascorbic acid biosynthetic mutants based on their enhanced O_3 sensitivity (Conklin et al., 1999). Accordingly, increasing the AsA content in *Raphanus sativus* by externally supplied biosynthetic precursor l-galactono-1,4-lactone increased the O_3 tolerance (Maddison et al., 2002). Although O_3 exposure did not bring in any visible injuries in winter wheat, clear shifts in the antioxidant levels have been observed in the leaves. While chronic ozone fumigation did not alter GSH reductase activity, POD levels increased and superoxide dismutase (SOD) and CAT decreased. However, the activity of all these antioxidant enzymes was increased with the increment in O_3 concentration (Liu et al., 2015). The accumulating evidence also suggests the involvement of membrane-bound heterotrimeric G proteins in the O_3-induced oxidative burst (Joo et al., 2005). But exposure to chronic O_3 stress did not show any phenotype in *A. thaliana* null mutants of G proteins (Booker et al., 2012). Extracellular ROS generated through the oxidation at the cell surface rather than the ROS produced by O_3 dissolution in the apoplast is required to activate the heterotrimeric G protein directly or indirectly. The O_3-induced oxidative burst also alters the activity cysteine-rich receptor-like kinases of plasma membrane and many apoplastic proteins through redox modifications (Ainsworth, 2017). Further, O_3-triggered ROS interact directly with plasma-membrane-bound receptors and trigger downstream events in the cytosol.

11.1.3.2.6 Ozone and Ion Channels

O_3-induced oxidative stress often leads to general redox-dependent alterations in ion conductance in many plant species. The increased intracellular Ca^{2+} concentration has been recorded using an intracellular Ca^{2+} indicator with the increase in the atmospheric ozone above 70 ppb in intact *A. thaliana* plants followed by an increase in expression of glutathione *S*-transferase (GST) gene (Clayton et al., 1999). This O_3 induced a temporal biphasic Ca^{+2} surge directly proportional to O_3 concentrations. The increased Ca^{+2} can lead to the changes in protein phosphorylation pattern through the activation of MAPK cascades (Baier et al., 2005). O_3 elicits the activity of RBOH, which subsequently activates ion channels in guard cells, leading to the changes in stomatal movement. On the other hand, O_3 has been shown to inhibit K^+ channels in guard cells of *Vicia faba* that lead to reduced K^+ uptake stomatal opening (Torsethaugen et al., 1999). In another study, O_3-induced oxidative stress retarded the activation of K^+ outward-rectifying channels and ultimately led to programmed cell death (Tran et al., 2013). Moreover, the expression of genes encoding plasma membrane and vacuolar channels such as proton

ATPase (AHA11) and (calcium channels) CAX1 and CAX3 were downregulated by O_3 (Dumont et al., 2014). Although the occurrence of ion channel changes due to O_3 exposure has been established, the actual biological significance needs deeper study.

11.1.3.2.7 Ozone and Secondary Metabolites

O_3 fumigation doubled the content of flavones, glycosides and furanocoumarins along with necrotic lesions in *Petroselinum crispum* (Eckey-Kaltenbach et al., 1994). Similarly, *Prunus avium* and *Viburnum lantana* accumulated anthocyanins under 92% of ambient O_3 (Gravano et al., 2004). O_3 stress results in the increase in the expression of the genes involved in flavanoid and phenylpropanoid pathways, salicylic acid (SA)-signaling, ethylene (ET) and jasmonic acid (JA) biosynthesis as well as encoding pathogenesis-related (PR) proteins and antioxidant enzymes (Tamaoki, 2008).

11.1.3.2.8 Root Growth and Resource Allocation

In many species, O_3 reduced root growth more than shoot growth (Hogsett et al., 1985). Elevated O_3 affects the carbohydrate allocation to the belowground parts and, hence, can change biomass accumulation patterns (McCool and Menge, 1983; Gorissen and van Veen, 1988; Spence et al., 1990; Edwards et al., 1992). Several works have shown a decrease in allocation to roots and root to shoot biomass ratio in response to O_3 (Wittig et al., 2009). Similarly, reduced carbon transport to roots decreases nutrient and water uptake as well as affects anchorage. Moreover, O_3-induced early senescence would primarily affect root growth inclined that mature leaves preferentially allocate carbon resources to stems and roots (Matyssek et al., 2010). O_3 also reduces belowground growth due to its ability to alter root physiology (McLaughlin and McConathy, 1983; Andersen et al., 1991). For instance, O_3-exposed *Pinus ponderosa* seedlings had significantly less root starch content and, hence, developed coarse roots and reduced root formation (Andersen et al., 1997). Furthermore, *P. ponderosa* seedlings exposed to the highest O_3 level for one season had 34% less lateral root biomass and 65% less new root biomass in the following spring due to the significant decrease in root storage carbohydrate pools. These kind of carry-over effects of O_3 stress from one season to another are significant to consider in long-lived perennial species that are subjected to O_3 every year (Andersen et al., 1997). Elevated-O_3-elicited decrease in carbon allocation may trigger indirect effects on mycorrhizas, extramatrical mycelium and soil processes (Matyssek et al., 2010). Such changes in the root morphology may have drastic impacts on rhizosphere organ functions,

nutrient cycling and plant survival to other environmental constraints such as drought (Agathokleous et al., 2016). In beech and spruce, chronic O_3-exposure reduced nitrogen acquisition and allocation that was predicted to accumulate in the long term with unfavorable consequences for nitrogen storage and tree growth (Weigt et al., 2012). Apart from belowground impacts, detrimental effects of O_3 are often seen on vegetation as reductions in plant growth, biomass, leaf area and seed yield, as well as early senescence and visible damage to leaves (Ashmore, 2005).

11.1.3.2.9 Ozone and Plant Reproduction

Reproductive development is the critical factor to decide plant productivity and propagation of species. Elevated O_3 casts a negative impact on the plant reproductive system mainly due to the altered carbon allocation from tissues and also through direct effects on the reproductive tissues and reproductive processes of the plant (Black et al., 2000; Ashmore, 2005). For example, the study of *Petunia x hybrida* suggested that O_3 exposure might alter the topography of the stigmatal surface (Black et al., 2000). Several studies demonstrated the impact of O_3 on floral development, seed quality and seedling vigor with substantial genotypic variation (Black et al., 2000). O_3 has been reported to delay flowering in various species such as *G. max* and *Gossypium hirsutum*. Elevated O_3 also increased ET emission in *G. max* plants, thereby indirectly increasing abscission of flowers and pods (Betzelberger et al., 2012). Six hours of O_3 exposure to the inflorescences of *Brassica napus* was reported to be sufficient to increase flower bud abortion. Moreover, the number of inflorescences that were produced and that reached maturity was reduced in *Buddleia* and dogbane due to O_3 treatment (Black et al., 2000). *G. max*, when exposed to 150 ppb O_3 in O_3 chambers, showed significant effects on pod formation without any significant impact on flower production. (Leisner et al., 2014). However, its negative consequences were observed by a decrease in the seed number and seed size as well as a decrease in the fruit number and fruit size (Leisner et al., 2012).

A clear reduction in pollen tube elongation was reported in *N. tabacum* and *Z. mays* (Mumford et al., 1972). Pollen from *S. lycopersicum* plants grown in the O_3-enriched air was significantly slower to germinate and also had reduced pollen tube lengths compared with respective control plant pollen. O_3 elicited harm to the pollen viability of *S. lycopersicon* and led to reduced fruit weight, size and quality (Gillespie et al., 2015). In another study in *Z. mays*, biochemical analysis of pollen has shown an enhanced free amino-acid level while reducing and neutral sugar contents were decreased following O_3 exposure for 5 h (Mumford et al., 1972). Similarly, the decreased rate of photosynthesis due to O_3 exposure in *Brassica campestris* had an impact on carbohydrate accumulation within the pollen (Stewart, 1998). As the carbohydrates serve as an energy source in germinating pollen, the reduction in carbohydrate content might have a definite impact on the pollen germination. Further, O_3 has also been shown to influence allergen release from pollen grains. O_3 increases the content of group 5 allergenic proteins of *Lolium perenne* (Masuch et al., 1997). In another investigation, O_3 fumigation decreased pollen viability due to the direct damage to the pollen membrane caused by enhanced ROS and NO contents and significant enhancement of the ROS-generating enzyme NADPH oxidase in ragweed pollen (*A. artemisiifolia*) that enhanced pollen allergenicity as well (Pasqualini et al., 2011). O_3 reduces grain size, grain weight, grain nutritional quality and grain number in field-grown crops such as *T. aestivum, O. sativa, Z. mays* and *G. max* (Biswas and Jiang, 2011) O_3 also has detrimental effects on crop quality, for example, protein or oil content of pods or grains (Fuhrer, 2009). Tuber crops such as *Solanum tuberosum* show a reduction in tuber size rather than tuber number, while in pod crops, pods are smaller, and lesser in number. It was showed that a single 6 h exposure of the terminal inflorescence of *B. napus* reduced the following root and shoot growth of seedlings (Bosac et al., 1998). This effect has been attributed to O_3-induced reductions in seed weight and storage reserve available for seedling growth. However, the exact sites of action and the mechanisms responsible for the observed effects of O_3 on reproductive growth remain unclear.

11.2 IMPACTS OF OZONE ON ECOSYSTEMS

Under natural conditions, plants constitute a significant part of the ecosystem. Practically any change in a plant can have a great impact on associated ecosystems. Although many plant species coexist in natural systems facing the similar ambient O_3, the tolerance to O_3 may differ greatly. Indeed, it was found that O_3 pollution was the third strongest driver after inorganic nitrogen deposition and annual evapotranspiration of plant community composition change in calcifuge grasslands in the UK (Payne et al., 2011). Ambient O_3 stress can bring shifts in species composition in diverse plant communities. Results from open-top chamber experiments revealed that elevated O_3 could cause a shift in the biomass towards more O_3 tolerant species in a co-existing plant species (Fuehrer and Achermann, 1994). However, in

such studies, it is essential to eliminate the external environmental factors influencing the O_3 stress responses by plants. Another O_3 stress experiment was carried out by growing *Trisetum flavescens* in series mixtures with *Centaurea jacea* or *Trifolium pretense* under full and reduced irrigation conditions disclose genotype specific competitive ratios for irrigation (Nussbaum and Fuhrer, 2000). Depending on the severity of stress, O_3 has been shown to both induce and reduce volatile emissions from different plant species (Blande et al., 2014). For example, O_3 exposure reduced the emission of two monoterpenes, sabinene and δ-3-carene, in *B. napus* plants while emission of stress volatiles, including green leaf volatiles, methyl salicylate and sesquiterpene has been strongly induced in *N. tabacum* (Blande et al., 2014). This effect of O_3 on volatile emission has been shown to reduce the distance of communication between plants, as observed in the case of volatile-mediated signaling between *Phaseolus lunatus* plants in chamber experiments. Similarly, O_3 has been observed to decrease the associational susceptibility of *Brassica oleracea var italica* plants mediated by the passive adsorption of sesqiterpines among adjacent plants (Li and Blande, 2015).

The O_3 effects on host nutritional quality derived from changes in metabolism and plant morphology due to increased toughness can cause greater impact on host–pathogen relations. O_3-exposed *P. ponderosa* and eastern white pines (*Pinus strobus*) became more susceptible to invasion by pine beetles (*Dendroctonus brevicomis*) due to weakened stems (Skelly et al., 1983). The altered volatile composition due to the increase in O_3 has an indirect effect on plant and herbivore interactions. It was proved that 80 ppb O_3 is sufficient to reduce the plant signaling to herbivores by altering the volatile organic compounds (Blande et al., 2010). When *Brassica nigra* was treated with 120 ppb of O_3, the subsequent degradation of floral volatiles and floral scent thus reduced the interaction with generalist pollinator *Bombus terrestris* (Ferré-Armengol et al., 2015). O_3 tends to change the host-finding ability of the cucumber beetle (*Acalymma vittatum*) and is also known to degrade the aggregation pheromone of *Drosophila melanogaster*.

Apart from insects, many pathogens that are associated with plants are known to be affected by prevalent O_3 toxicity. O_3 significantly reduced the growth of uredia of *Phytophthora coronata* and sporulation of the *T. aestivum* stem rust fungus *P. coronata*. Similarly, ozone-injured potato foliage has severe attacks of *Botrytis cinerea* that resulted in early senescence and loss of yields. O_3 delayed the growth of *Fusarium oxysporum f. sp. lycopersici* and delayed the onset of vascular wilt disease by several weeks in young *S. lycopersicon* plants.

Similarly, *P. ponderosa* in California with O_3 injury were more susceptible to root rot caused by *Fomes annosus* (James et al., 1980b). Dry bean plants with infections caused by *Xanthomonas phaseoli* were less susceptible to O_3 injury in the field; however, O_3 injury had no significant effect on infection. Bacterial lesions of *G. max* leaves by *Pseudomonas glycinea* were reduced by O_3 (Laurence and Wood, 1978) due to the overlap in the defense pathways activated by pathogens and O_3 stress. O_3-exposed *A. thaliana* plants exhibited less visible symptoms like chlorotic lesions in response to the virulent strain of *Pseudomonas syringae*, turnip crinkle and tobacco mosaic viruses (Malamy et al., 1990; Yalpani et al., 1994). The accumulating information underlines the need for a holistic view of the O_3 impact on ecological systems.

11.3 MOLECULAR RESPONSES TO OZONE STRESS

Despite its remarkable impact on the plant metabolism, O_3 also has few molecular signatures. Characterizing these underlying molecular changes that determine the metabolic responses to elevated O_3 has been an active area of research for decades. Although both chronic as well as acute O3 exposures impact gene regulation, acute O_3 exposures provoke more important changes than chronic episodes (Ainsworth et al., 2012). Genes involved in senescence and protein turnover are upregulated in *P. tremuloides* leaves exposed to O_3 (Gupta et al., 2005). Gene expression in *G. max* is altered in reproductive tissues as a response to elevated O_3. There was significant variation in the genes expressed in flowers and pod tissues of *G. max* when O_3 stress was imposed (Leisner et al., 2014). Approximately 4,595 genes were shown to be responsive to O_3, some of which were matrix metalloproteinase genes, differentially expressed in flower, and 1,375, along with some *xyloglucan endotransglucosylase* genes, expressed in pods while 277 of these genes were expressed in both flowers and pods. Most of these transcripts involved in signaling, development, transport, stress, protein and RNA expression (Leisner et al., 2014). Similarly, 408 genes were found to be O_3 responsive in 200 ppb O_3-fumigated *G. max* leaves in an airtight greenhouse, of which 153 genes were upregulated and 223 genes were downregulated by O_3 in these leaves (Moon et al., 2013). It is interesting to note that O_3 has species-specific molecular signature in legumes. A transcriptome-wide comparison of differentially expressed genes revealed that three legume species including *G. max*, *P. sativum* and *Phaseolus radiatus* had a distinct response to elevated O_3. Several stress-related genes,

transcription factors, POD and receptor-like kinases were turned on by O_3 exposure in all three legumes with different extents of expression (Yendrek et al., 2015). Increase in the abundance of some key enzymes in the phenylpropanoid pathway, such as phenylalanine ammonia lyase, chalcone synthase, isoflavone reductase and dihydroflavonol 4-reductase, suggest the O_3-mediated stress response (Yendrek et al., 2015).

Molecular events related to ROS scavenging in plants has been one of the emerging areas in O_3 research (Rao et al., 2000a,b; Wohlgemuth et al., 2002). The enhanced antioxidant levels are vital for the plants to cope with O_3 toxicity. Hence, plants initiate the transcription of enzymes comprising the AsA-GSH cycle (Noctor and Foyer, 1998). Global gene expression and proteomics studies have confirmed the abundance of antioxidant transcripts and proteins in response to elevated O_3 (Agrawal et al., 2002; Baier et al., 2005). As another evidence, the transcripts encoding GSH peroxidase and SOD genes related to respiration were found increased in *G. max* in response to O_3 stress. (Yendrek et al., 2015). The genes associated with sucrose biosynthesis, cell wall metabolism, glycolysis, pentose phosphate pathway, lipid metabolism, amino acid and protein metabolism, photorespiration, the shikimate pathway and its derivates, as well as genes coding for enzymes of the tricarboxylic acid cycle (TCA) cycle, were affected significantly in response to O_3.

The induction of hormone biosynthesis genes is another important plant response to O_3 stress (Kangasjärvi et al., 2009). In *A. thaliana*, O_3 induced the transcription of ET biosynthesis genes including *AtACS2*, *AtACS6* and *AtACS10*. However, the reduced accumulation of *ACS2* transcript in the O_3-treated *ABI1* knockout plants needs thorough validation (Ludwikow and Sadowski, 2008). Similarly, O_3 modulates JA biosynthesis via the modulation of gene expression of lipoxygenases (LOX1-3), allene oxide synthase (AOS) and 12-oxo-phytodienoate reductases (OPR) (Tosti et al., 2006). Other genes that emerged from available O_3 studies are *S*-adenosyl-L methionine, jasmonic acid carboxyl methyltransferase (JMT), OPC-8:0 CoA ligase1 (OPCL1), 3-ketoacyl-CoA thiolase (PED1), stearoyl-ACP desaturase and allene oxide cyclases (Ludwikow and Sadowski, 2008). Besides promoting various hormone biosynthesis genes, O_3 also upregulate other genes involved in plant defense mechanism including pathogen-responsive genes PR1, PR2 and Eli 16 in *P. crispum* while β-1,3-glucanase, chitinase, and PR protein 1b are induced in *N. tabacum*, increase in the stilbene phytoalexins and stilbene synthase observed in *P. ponderosa* after O_3 treatment (Ernst et al., 1992). These data clearly indicate the cross-induction of defensive pathways by O_3 (Eckey-Kaltenbach et al., 1994).

11.4 COMBINATION OF OZONE STRESS WITH OTHER POLLUTANTS

Along with O_3, plants are often exposed to a number of additional pollutants such as sulfur dioxide (SO_2), NO, nitrogen dioxide (NO_2) and persistent organic pollutants. The response of plants to a combination of air pollutants has been reported to be different from the response of plants to individual pollutants. However, the air composition pattern varies both spatially and temporally. The photochemical smog that causes severe oxidative burst is due to the combination of the most common pollutants, O_3 and NO_2 (Ainsworth, 2017). An open-top chamber study was conducted using near-ambient concentrations of O_3, SO_2 and NO_2 in *B. napus*. This study revealed that a combination of O_3 with either SO_2 or NO_2 resulted in antagonistic effects, even though positive effects on yield were observed under individual treatments (Adaros et al., 1991). Any combination of these three gaseous pollutants, SO_2, NO_2 or O_3, was found to be adverse in *Hordeum vulgare* and *T. aestivum* (Adaros et al., 1991). The interactive effects of CO_2 and tropospheric O_3 on the seed yield, quality, antioxidant metabolism and gene expression patterns were reported in several plant species, including *P. tremuloides*, *B. pendula* and *G. max* plants (Gupta et al., 2005; Burkey et al., 2007; Kontunen-Soppela et al., 2010; Gillespie et al., 2012).

As mentioned earlier, elevated O_3 also alters plant responses to other stress conditions. The interactive effects of O_3 and drought have been well studied in plants and reported to be contradictory (Dobson et al., 1990). For example, soil water deficit enhanced the O_3-induced yield loss in *G. max* (Heggestad et al., 1985). In contrast, some studies have indicated that plants subjected to drought stress are protected from O_3 damage due to the reduced stomatal conductance (Tingey and Hogsett, 1985; Beyers et al., 1992; Reiner et al., 1996). Water stress reduced the O_3-induced loss of needle biomass in *P. ponderosa* (Beyers et al., 1992) and O_3-related growth reductions and visible foliar symptoms in *Picea abies* (Karlsson et al., 1995). Combination of O_3 and drought resulted in higher rates of photosynthesis and reduced chl loss compared to well-watered controls (Le Thiec et al., 1994). Thus, single stress and combined stresses synergize or antagonize signaling pathways that ultimately trigger distinct physiological responses (Iyer et al., 2013).

11.5 PLANT ADAPTATION MECHANISM TO OZONE STRESS

Plants adjust their metabolism to ensure adaptation to the prevailing O_3 stress in the ambiance. The first and foremost defense action adapted by plants is to restrict the entry of O_3 into the cell. The first line of defense against O_3 stress is the stomatal closure. The signaling events associated with the stomatal closure get induced immediately to restrict O_3 entry. Events such as induced expression of SLAC1 and oxidative activation of the calcium ion channels are known to be involved in O_3-induced stomatal closure. Stomata not only reduce O_3 influx but also simultaneously control the balance between carbon gain and water loss (Medlyn et al., 2011). On the other hand, enhanced cell wall thickenings and modification of the cellulose to lignin ratio allow plant growth at reduced carbon cost and enhance the toughness of plants to minimize O_3 stress injury (Richet et al., 2011).

Although shutting stomata down can temporarily reduce O_3 uptake and injury, plants exhibit a variety of other defense mechanisms to detoxify O_3 or repair injured tissue. Once O_3 enters the leaf through the stomata, the second line of defense is the O_3 detoxification in the apoplast. After entering into the plant cell through the stomata, O_3 breaks down into different ROS, thereby increasing the oxidative load to the apoplastic fluid (Kangasjärvi et al., 2005). The accumulating ROS elicits a series of signaling events leading to the adjustment of plant metabolism, defense and cell signaling. ROS-induced activation of NADPH in the plasma membrane results in ROS accumulation in the adjacent cells. Hence, the ability to limit ROS accumulation is an important aspect in O_3 tolerance. For this reason, the antioxidant enzyme activities in the cell would increase to detoxify free radicals to reduce damage to membranes and vital structures. Meanwhile, the content of permeable mass would be regulated by the cell to maintain osmotic potential balance while changing the metabolic pathway to reduce metabolic rate and also to ensure enough energy reserves. A large panel of metabolites, antioxidant enzymes, have been known to be involved in the defense mechanisms to decrease the ROS level or to increase antioxidant by regenerating the reduced forms. In addition to metabolites, apoplastic AsA extracellular enzymes like SOD and POD play a significant role in the mitigation of ROS generation in a number of plants (Ainsworth, 2017).

Furthermore, the accumulation of some secondary metabolites, including phenolic compounds, is another essential step to cope with O_3 stress (Fares et al., 2009). The involvement of condensed tannins, isoprene, polyamines, the induction of the shikimate and phenylpropanoid pathway has been well documented under O_3 exposure. Additionally, elevated O_3 also induces changes in the expression level of several genes connected with signal transduction. O_3 treatment induces the expression of ET biosynthesis-related genes, genes coding for MAPK and other genes involved in systemic acquired resistance. MAP cascade, in turn, evokes a number of signaling events either independently or dependent on complex interactions with stress hormones including SA, abscisic acid (ABA), ET and JA signaling, thereby leading to resistance or cell by acute O_3 exposure (Gupta et al., 2005; Kangasjärvi et al., 2005; Figure 11.2).

Free radical burst induced by O_3, in turn, affects phytohormone signaling and their interactions. (Rao and Davis, 2001). ABA has a role in the stomatal conductance, thereby influencing the O_3 influx into the leaves. It has been shown that any increase in ABA content in leaves due to foliar applications or by water stress induces stomatal closure, resulting in decreased O_3 uptake and injury to plants (Jeong et al., 1981). Interestingly, O_3 has been shown to suppress ABA-induced stomatal closure via an ET-dependent mechanism (Wilkinson and Davies, 2010). When *P. sativum* plants were exposed to O_3 stress for 21 days, they remained unaffected while plants exposed to O_3 for 2 days showed visible injury, partially due to the difference in ET release (Mehlhorn and Wellburn, 1987). The pretreatment with ET biosynthesis inhibitor aminoethoxyvinylglycine 1 day before the O_3 exposure abolished the visible injuries. However, O_3-induced increase in ET biosynthesis is dependent on a functional SA signaling pathway (Rao et al., 2002; Ahlfors et al., 2004). An antagonistic relationship between JA- and SA-signaling pathways has been reported during O_3-induced cell death in JA-signaling mutants *jar1* and *fad3/7/8*. Moreover, O_3-induced alterations in lipid composition, especially linoleic acid

composition in the plasma membrane, triggers the JA biosynthesis (Creelman and Mullet, 1997). Exogenous application of JA prior to O_3 exposure reduces O_3-induced cell death in *N. tabacum* and *Populus spp.*, further supporting the role of JA in mediating O_3-induced responses (Örvar et al., 1997). Moreover, exogenous application of Me-JA decreases the O_3-induced concentrations of H_2O_2 in *N. tabacum*, and free SA thereby abolished cell death in a dose-dependent manner (Schraudner et al., 1998). Similarly, NO is another important molecule known to modify O_3-induced cell death in *A. thaliana* (Ahlfors et al., 2009). Moreover, treatment with NO donor decreases the visible lesions in O_3 stressed *A. thaliana* plants (Ahlfors et al., 2009). While NO, together with ethylene, regulates the expression of O_3 responsive genes like AOX in *N. tabacum* plants, NO alone attenuates O_3-induced SA signaling/biosynthesis (Ahlfors et al., 2009). Even though growing evidence indicates the involvement of plant hormones ET, JA and SA in O_3-induced stress responses in plants, the actual mode of action that influences O_3-induced plant responses needs further study.

11.6 BREEDING AND BIOTECHNOLOGICAL APPROACHES FOR OZONE STRESS TOLERANCE

Crop species differ widely in their susceptibility to O_3 (Brosche et al., 2010). Recently, genetic loci associated

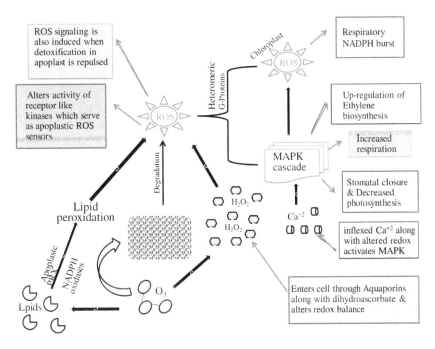

FIGURE 11.2 Schematic representation of ozone uptake and tolerance mechanisms in plants. Ozone enters into the cells through stomata or directly diffuses into the apoplastic space forming reactive oxygen species (ROS) serving as sensors to activate mitogen activated protein kinase (MAPK) cascades, influx of Ca^{+2} and various adaptive responses.

with O_3 resistance have been identified in major food crops including wheat, rice (Ainsworth, 2017; Frei et al., 2010) and additional crops including *G. max* (Betzelberger et al., 2010) and *P. vulgares* (Flowers et al., 2007). These studies are highly valuable to broaden the research focus from simple visible O_3 damages to the respective physiological traits and cellular markers. Several QTLs and subsequent analysis of the genes underpinning O_3 tolerance has been identified in *O. sativa*. However, additional research is required to confirm the importance of these QTL for O_3 tolerance in increasing crop yields in elite germplasm of other crops.

Apart from breeding success, some mutagenic and transgenic efforts were made to understand O_3 stress in plants (Table 11.2). Although most transgenic studies so far mainly focused on the over/underexpression of antioxidant enzyme isoforms, thereby altering detoxification capacity of plants to O_3 stress, a few attempts were also made to alter the signal transduction pathway. Transgenic approaches have been used to alter the abundance and/or redox state of the AsA and/or GSH pools in order to better understand the mechanisms of cellular O_3 response. Reduction in ascorbate levels in *O. sativa* due to a mutation in GDP-D-mannose 3-5-epimerase gene resulted in enhanced leaf damage compared to the wild-type under O_3 stress (Ferie et al., 2012). While T-DNA knock out mutation in AsA mannose pathway regulator 1 increased

O_3 tolerance in *A. thaliana* (Zhang et al., 2009), other mutations that significantly lowered the apoplastic AsA levels such as cytoplasmic dehydroascorbate reductase showed high sensitivity to O_3, further ascertaining the O_3 degradation in the apoplastic fluid. Consistent with this, *N. tabacum* plants overexpressing monodehydroascorbate reductase and dehydroascorbate reductase showed enhanced O_3 tolerance while ascorbate peroxidase overexpression increased O_3 susceptibility (Eltayed et al., 2007; Sanmartin et al., 2003). Further, overexpression of chloroplastic Mn-SOD gene in *N. tabacum* increased endogenous GSH levels and minimized the O_3 injury (Van Camp et al., 1994). However, no protection was observed by overexpression of a chloroplastic copper (Cu) or Zinc (Zn) SOD (Pitcher et al., 1991) due to the difference in the O_3 exposures. In another study, *A. thaliana* mutant vitamin C-1-deficient of AsA has shown to enhance the sensitivity to O_3-stress-induced oxidative stress (Conklin et al., 1996; Conlin and Barth, 2004). These studies have emphasized the complexity of O_3-induced oxidative signaling. Overexpression of *Pityriasis alba* isoprene synthase in *N. tabacum* drastically reduced visible O_3 damage due to the reduced ROS accumulation and membrane damage (Ainsworth, 2017).

In another approach, transgenic *A. thaliana* expressing salicylic hydroxylase show reduced accumulation of SA, and *nprl* mutant plants defective in SA-mediated

TABLE 11.2
Transgenic and Mutational Approaches to Mitigate Ozone Stress Tolerance in Plants

Plant Species	Gene Name	Gene Source	Response	Reference
Nicotiana tabacum	*Monohydroascorbate reductase*	*Arabidopsis thaliana*	Enhanced tolerance to ozone	Eltayeb et al. (2007)
Nicotiana tabacum	*Dehydroascorbate reductase*	*Arabidopsis thaliana*	Enhanced tolerance to ozone	Eltayeb et al. (2007)
Nicotiana tabacum	*Dehydrorascorbate reductase*	*Triticum aestivum*	Enhanced tolerance to ozone	Chen and Gallie (2005)
Nicotiana tabacum	*Ascorbate oxidase*	*Cucumis sativus*	Enhanced susceptibility to ozone	Sanmartin et al. (2003)
Nicotiana tabacum	*AsA mannose pathway regulator 1*	*Arabidopsis thaliana*	Enhanced tolerance to ozone	Zhang et al. (2009)
Nicotiana tabacum	*Isoprene synthase*	*Pityrisis alba*	Reduced visible symptoms and damage	Vickers et al. (2009)
Arabidopsis thaliana	*Vitamin C1*	*Vtc 1*	Low ascorbate content and hypersensitivity to ozone	Conklin and Barth (2004)
Oryza sativa	*GDP-D-mannose 3',-5'-epimerase gene*	*ND6172*	Low ascorbate content and enhanced visible leaf damage	Feri et al. (2012)
Arabidopsis thaliana	*Jasmonate resistant 1* *Omega-3 fatty acid desaturase*	*jar1* *fad3/7/8*	Jasmonic acid signaling, cell death	Bell et al. (1995)
Arabidopsis thaliana	*Regulatory protein/non expresser of PR genes1*	*npr 1*	Salicylic acid regulation and systemic acquired resistance	Bell et al. (1995)

systemic acquired resistance showed increased sensitivity to O_3 exposure. This study indicates that the O_3-induced resistance shares are signaling events with SA-mediated activation of systemic acquired resistance (Sharma et al., 1995). Blocking ET biosynthesis by reducing the expression of 1-aminocyclopropane-1-carboxylate synthase has also improved O_3 tolerance in *N. tabacum* and *Populus spp.* (Nakajima et al., 2002; Mohri et al., 2011). In the case of transgenic *S. tuberosum*, where the modified ET biosynthesis displays a correlation between visual leaf damage in greenhouse trials, it showed stunted growth in the field (Sinn et al., 2004). Shifting from controlled environments to FACE experiments will eventually eliminate discrepancies in genetic and phenotyping screening.

11.7 SUMMARY AND FUTURE PROSPECTS

Surface O_3 is one of the critical damaging phytotoxic air pollutants posing a serious threat to natural ecosystems. Crop plants experience O_3 stress when the concentration crosses 100 ppb in the ambiance, which subsequently leads to hypersensitive response and cell death. However, the response of plant species to O_3 stress is dependent on the growth stage of plants during O_3 exposure. In some species, such as bean, the reproductive stage was found to be relatively more sensitive than the vegetative stage, while in some other species, including *T. aestivum*, O_3 exposure during the vegetative stage was found to be more detrimental to yield as compared to the reproductive stage (Soja, 1996). This indicates that plants may be more sensitive to O_3 at some growth stage and tolerant at other stages, again subject to climatic variations. Hence, systematic studies on the effect of surface O_3 over the entire crop cycle covering different growth stages are highly critical in this context. Till now, most of the O_3 stress studies have been restricted only to growth and yield performance of plants. However, the co-occurrence of different abiotic stresses along with O_3 stress was shown to result in a high degree of complexity in plant responses. O_3 stress leads to differential susceptibility to other stresses such as drought, possibly due to its effects on stomatal movement in plants. However, the interactive impact of O_3 and other stress conditions such as drought, heat, salt and heavy metal stress on vegetation has not ascertained in detail.

Moreover, the gap between field versus controlled conditions has to be reduced further. Although simulation studies are informative, long-term field studies for assessing the impact of elevated surface O_3 on crops are imperative. The complete field-based understanding of the impacts of O_3 on plant species diversity and species composition requires further study. Moreover, each plant type may possess their defense mechanism, and, hence, direct extrapolation of results from model plants to others is not always possible. In addition, all the flux-based methodologies are based on stomatal O_3 flux to biomass loss through a series of models at the tissue, whole plant and community levels. These models provide estimates of O_3 flux exceeding the plant detoxification capacity under the given environmental conditions (Buker et al., 2015). Hence, it is crucial to develop a robust O_3 metric suitable for diverse genotypes and climatic conditions.

Most of the O_3 studies thus far conclude that O_3 enters plants through stomata and reacts with water in the apoplast to form reactive oxygen species, which alter physiological and biochemical processes in plants. However, it is essential to assess the impact of O_3 on other physiological functions of plants at the community level. Moreover, the actual source of reducing power to supply some of the free radical scavenging systems remains to be elucidated under O_3 stress. Besides, O_3 research should be focused on the gene regulation and related mechanism of the signal transmission network system of plants under O_3 stress. In addition, plants' breeding and transgenic methods such as a molecular marker, gene mapping and genomics approach to adapt to the elevated O_3 should be given priority. Moreover, differential expression of transcription factors associated with redox, defense signaling and chromatin modifications may be important in tweaking the metabolome and proteome networks during plant adaptation to O_3. Although substantial progress has been made, much work remains to be done to completely characterize the association between gene-expression patterns and the O_3-induced phenotype to have a closer systems-level picture in O_3 stress. Further investigations should shed light on how to assemble various known components involved in acute as well as chronic O_3 signaling in order to have complete stress-specific signaling network.

ACKNOWLEDGMENTS

We thank the Department of Science and Technology (DST), Government of India for providing the financial support [IFA12-LSPA-08]. This work was undertaken as part of the CGIAR Research Program on Grain Legumes.

REFERENCES

Adaros, G., Weigel, H.J., Jäger, H.J. (1991) Concurrent exposure to SO_2 and/or NO_2 alters growth and yield responses of wheat and barley to low concentrations of O3. *New Phytologist* 118(4): 581–591.

Agathokleous, E., Saitanis, C.J., Wang, X., Watanabe, M., Koike, T. (2016) A review study on past 40 years of research on effects of tropospheric O3 on belowground structure, functioning, and processes of trees: A linkage with potential ecological implications. *Water, Air, and Soil Pollution* 227(1): 33.

Agrawal, G.K., Rakwal, R., Yonekura, M., Kubo, A., Saji, H. (2002) Proteome analysis of differentially displayed proteins as a tool for investigating ozone stress in rice (*Oryza sativa* L.) seedlings. *Proteomics* 2(8): 947–959.

Ahlfors, R., Brosché, M., Kangasjär, J. (2009) Ozone and nitric oxide interaction in *Arabidopsis thaliana*: A role for ethylene? *Plant Signaling and Behavior* 4(9): 878–879.

Ahlfors, R., Macioszek, V., Rudd, J., Brosché, M., Schlichting, R., Scheel, D., et al. (2004) Stress hormone-independent activation and nuclear translocation of mitogen-activated protein kinases in *Arabidopsis thaliana* during ozone exposure. *The Plant Journal* 40(4): 512–522.

Ainsworth, E.A. (2017) Understanding and improving global crop response to ozone pollution. *The Plant Journal* 90(5): 886–897.

Ainsworth, E.A., Yendrek, C.R., Sitch, S., Collins, W.J., Emberson, L.D. (2012) The effects of tropospheric ozone on net primary productivity and implications for climate change. *Annual Review of Plant Biology* 63: 637–661.

Alves, E.S., Moura, B.B., Pedroso, A.N.V., Tresmondi, F., Machado, S.R. (2016) Cellular markers indicative of ozone stress on bioindicator plants growing in a tropical environment. *Ecological Indicators* 67: 417–424.

Anav, A., Menut, L., Khvorostyanov, D., Viovy, N. (2011) Impact of tropospheric ozone on the Euro-Mediterranean vegetation. *Global Change Biology* 17(17): 2342–2359.

Andersen, C.P., Hogsett, W.E., Wessling, R., Plocher, M. (1991) Ozone decreases spring root growth and root carbohydrate content in ponderosa pine the year following exposure. *Canadian Journal of Forest Research* 21(8): 1288–1291.

Andersen, C.P., Rygiewicz, P.T. (1995) Allocation of carbon in mycorrhizal Pinus ponderosa seedlings exposed to ozone. *New Phytologist* 131(4): 471–480.

Andersen, C.P., Wilson, R., Plocher, M., Hogsett, W.E. (1997) Carry-over effects of ozone on root growth and carbohydrate concentrations of ponderosa pine seedlings. *Tree Physiology* 17(12): 805–811.

Ashmore, M.R. (2005) Assessing the future global impacts of ozone on vegetation. *Plant, Cell and Environment* 28(8): 949–964.

Ashrafuzzaman, M., Lubna, F.A., Holtkamp, F., Manning, W.J., Kraska, T., Frei, M. (2017) Diagnosing ozone stress and differential tolerance in rice (*Oryza sativa* L.) with ethylenediurea (EDU). *Environmental Pollution* 230: 339–350.

Avnery, S., Mauzerall, D. L., Liu, J., Horowitz, L.W. (2011) Global crop yield reductions due to surface ozone exposure: Year 2000 crop production losses and economic damage. *Atmospheric Environment* 30: 2284–96.

Bagard, M., Le Thiec, D., Delacote, E., Hasenfratz-Sauder, M.P., Banvoy, J., Gérard, J., Dizengremel, P., Jolivet, Y. (2008) Ozone-induced changes in photosynthesis and photorespiration of hybrid poplar in relation to the developmental stage of the leaves. *Physiologia Plantarum* 134(4): 559–574.

Baier, M., Dietz, K.J. (2005) Chloroplasts as source and target of cellular redox regulation: A discussion on chloroplast redox signals in the context of plant physiology. *Journal of Experimental Botany* 56(416): 1449–1462.

Baier, M., Kandlbinder, A., Golldack, D., Dietz, K.J. (2005) Oxidative stress and ozone: Perception, signaling and response. *Plant, Cell and Environment* 28(8): 1012–102

Bassin, S., Volk, M., Fuhrer, J. (2007) Factors affecting the ozone sensitivity of temperate European grasslands: An overview. *Environmental Pollution* 146(3): 678–691.

Becker, G., Müller, R., McKenna, D.S., Rex, M., Carslaw, K.S. (1998) Ozone loss rates in the Arctic stratosphere in the winter 1991/92: Model calculations compared with Match results. *Geophysical Research Letters* 25(23): 4325–4328.

Bell, E., Mullet, J.E. (1993) Characterization of an *Arabidopsis* lipoxygenase gene responsive to methyl jasmonate and wounding. *Plant Physiology* 103(4): 1133–1137.

Betzelberger, A.M., Gillespie, K.M., Mcgrath, J.M., Koester, R.P., Nelson, R.L., Ainsworth, E.A. (2010) Effects of chronic elevated ozone concentration on antioxidant capacity, photosynthesis and seed yield of 10 soybean cultivars. *Plant, Cell and Environment* 33(9): 1569–1581.

Beyers, J.L., Riechers, G.H., Temple, P.J. (1992) Effects of long-term ozone exposure and drought on the photosynthetic capacity of ponderosa pine (*Pinus ponderosa* Laws.). *New Phytologist* 122(1): 81–90.

Biswas, D.K., Jiang, G.M. (2011) Differential drought-induced modulation of ozone tolerance in winter wheat species. *Journal of Experimental Botany* 62(12): 4153–4162.

Biswas, D.K., Xu, H., Li, Y.G., Liu, M.Z., Chen, Y.H., Sun, J.Z., Jiang, G.M. (2008) Assessing the genetic relatedness of higher ozone sensitivity of modern wheat to its wild and cultivated progenitors/relatives. *Journal of Experimental Botany* 59(4): 951–963.

Black, V.J., Black, C.R., Roberts, J.A., Stewart, C.A. (2000) Tansley Review No. 115 Impact of ozone on the reproductive development of plants. *The New Phytologist* 147(3): 421–447.

Blande, J.D., Holopainen, J.K., Li, T. (2010) Air pollution impedes plant-to-plant communication by volatiles. *Ecology Letters* 13(9): 1172–1181.

Bohler, S., Bagard, M., Oufir, M., Planchon, S., Hoffmann, L., Jolivet, Y., Hausman, J.F., Dizengremel, P., Renaut, J. (2007) A DIGE analysis of developing poplar leaves subjected to ozone reveals major changes in carbon metabolism. *Proteomics* 7(10): 1584–1599.

Booker, F.L., Burkey, K.O., Jones, A.M. (2012) Re-evaluating the role of ascorbic acid and phenolic glycosides in ozone scavenging in the leaf apoplast of *Arabidopsis thaliana* L. *Plant, Cell and Environment* 35(8): 1456–1466.

Bosac, C., Black, V.J., Roberts, J.A., Black, C.R. (1998) Impact of ozone on seed yield and quality and seedling vigour in oilseed rape (*Brassica napus* L.). *Journal of Plant Physiology* 153(1–2): 127–134.

Broberg, M.C., Feng, Z., Xin, Y., Pleijel, H. (2015) Ozone effects on wheat grain quality–A summary. *Environmental Pollution* 197: 203–213.

Brosche, M., Merilo, E.B.E., Mayer, F., Pechter, P., Puzorjova, I., Brader, G., Kangasjärvi, J., Kollist, H. (2010) Natural variation in ozone sensitivity among *Arabidopsis* thaliana accessions and its relation to stomatal conductance. *Plant, Cell and Environment* 33(6): 914–925.

Burkey, K.O., Booker, F.L., Pursley, W.A., Heagle, A.S. (2007) Elevated carbon dioxide and ozone effects on peanut: II. Seed yield and quality. *Crop Science* 47(4): 1488–1497.

Burton, A.L., Burkey, K.O., Carter, T.E., Orf, J., Cregan, P.B. (2016) Phenotypic variation and identification of quantitative trait loci for ozone tolerance in a Fiskeby III× Mandarin (Ottawa) soybean population. *Theoretical and Applied Genetics* 129(6): 1113–1125.

Cabané, M., Afif, D., Hawkins, S. (2012) Lignins and abiotic stresses. *Advances in Botanical Research* 61: 219–262.

Cabané, M., Pireaux, J.C., Léger, E., Weber, E., Dizengremel, P., Pollet, B., Lapierre, C. (2004) Condensed lignins are synthesized in poplar leaves exposed to ozone. *Plant Physiology* 134(2): 586–594.

Calatayud, A., Ramirez, J.W., Iglesias, D.J., Barreno, E. (2002) Effects of ozone on photosynthetic CO_2 exchange, chlorophyll a fluorescence and antioxidant systems in lettuce leaves. *Physiologia Plantarum* 116(3): 308–316.

Casteel, C.L., O'Neill, B.F., Zavala, J.A., Bilgin, D.D., Berenbaum, M.R. DeLucia, E.H. (2008) Transcriptional profiling reveals elevated CO2 and elevated O3 alter resistance of soybean (*Glycine max*) to Japanese beetles (*Popillia japonica*). *Plant, Cell and Environment* 31(4): 419–434.

Cho, K., Shibato, J., Kubo, A., Kohno, Y., Satoh, K., Kikuchi, S., Sarkar, A., Agrawal, G.K., Rakwal, R. (2013) Comparative analysis of seed transcriptomes of ambient ozone-fumigated 2 different rice cultivars. *Plant Signaling and Behavior* 8(11): e26300.

Clayton, H., Knight, M.R., Knight, H., McAinsh, M.R., Hetherington, A.M. (1999) Dissection of the ozone-induced calcium signature. *The Plant Journal* 17(5): 575–579.

Conklin, P.L., Barth, C. (2004) Ascorbic acid, a familiar small molecule intertwined in the response of plants to ozone, pathogens, and the onset of senescence. *Plant, Cell and Environment* 27(8): 959–970.

Conklin, P.L., Norris, S.R., Wheeler, G.L., Williams, E.H., Smirnoff, N., Last, R.L. (1999) Genetic evidence for the role of GDP-mannose in plant ascorbic acid (vitamin C) biosynthesis. *Proceedings of the National Academy of Sciences* 96(7): 4198–4203.

Conklin, P.L., Williams, E.H., and Last, R.L. (1996) Environmental stress sensitivity of an ascorbic acid-deficient *Arabidopsis* mutant. *Proceedings of the National Academy of Sciences* 93(18): 9970–9974.

Creelman, R.A., Mullet, J.E. (1997) Biosynthesis and action of jasmonates in plants. *Annual Review of Plant Biology* 48(1): 355–381.

Crutzen, P.J., Zimmermann, P.H. (1991) The changing photochemistry of the troposphere. *Tellus B* 43(4): 136–151.

Dat, J.F., Pellinen, R., Van De Cotte, B., Langebartels, C., Kangasjärvi, J., Inzé, D., Van Breusegem, F. (2003) Changes in hydrogen peroxide homeostasis trigger an active cell death process in tobacco. *The Plant Journal* 33(4): 621–632.

Davison, A.W., Barnes, J.D. (1998) Effects of ozone on wild plants. *The New Phytologist* 139(1): 135–151.

De Temmerman, L., Karlsson, G.P., Donnelly, A., Ojanperä, K., Jäger, H.J., Finnan, J., Ball, G. (2002a) Factors influencing visible ozone injury on potato including the interaction with carbon dioxide. *European Journal of Agronomy* 17: 291–302.

De Temmerman, L., Vandermeiren, K., D'Haese, D., Bortier, K., Asard, H., Ceulemans, R. (2002b) Ozone effects on trees, where uptake and detoxification meet. *Dendrobiology* 47: 9–19.

Di Baccio, D., Castagna, A., Paoletti, E., Sebastiani, L., Ranieri, A. (2008) Could the differences in O3 sensitivity between two poplar clones be related to a difference in antioxidant defense and secondary metabolic response to O3 influx? *Tree Physiology* 28(12): 1761–1772.

Dizengremel, P., Le Thiec, D., Hasenfratz-Sauder, M.P., Vaultier, M.N., Bagard, M., Jolivet, Y. (2009) Metabolic-dependent changes in plant cell redox power after ozone exposure. *Plant Biology* 11(1): 35–42.

Dizengremel, P., Vaultier, M.N., Le Thiec, D., Cabané, M., Bagard, M., Gérant, D., et al. (2012) Phosphoenolpyruvate is at the crossroads of leaf metabolic responses to ozone stress. *New Phytologist* 195(3): 512–517.

Dumont, J., Cohen, D., Gérard, J., Jolivet, Y., Dizengremel, P., Le Thiec, D. (2014) Distinct responses to ozone of abaxial and adaxial stomata in three Euramerican poplar genotypes. *Plant, Cell and Environment* 37(9): 2064–2076.

Dunning, J.A., Heck, W.W. (1977) Response of bean and tobacco to ozone: Effect of light intensity, temperature and relative humidity. *Journal of the Air Pollution Control Association* 27(9): 882–886.

Duriscoe, D.M., Stolte, K.W. (1989) Photochemical oxidant injury to ponderosa pine (*Pinus ponderosa* Laws.) and Jeffrey pine (*Pinus jeffreyi* Grev. and Balf.) in the national parks of the Sierra Nevada of California. In *Effects of Air Pollution on Western Forests*, ed. Olson, R.K. and Lefohn, A.S., 261–278. AWMA, Anaheim, CA.

Dütsch, H.U. (1971) Photochemistry of atmospheric ozone. *Advances in Geophysics* 15: 219–322.

Eckey-Kaltenbach, H., Ernst, D., Heller, W., Sandermann Jr, H. (1994) Biochemical plant responses to ozone (IV. Cross-induction of defensive pathways in parsley (Petroselinum crispum L.) plants). *Plant Physiology* 104(1): 67–74.

Edwards, G.S., Friend, A.L., O'Neill, E.G., Tomlinson, P.T. (1992) Seasonal patterns of biomass accumulation and carbon allocation in Pinus taeda seedlings exposed to ozone, acidic precipitation, and reduced soil Mg. *Canadian Journal of Forest Research* 22(5): 640–646.

Eltayeb, A.E., Kawano, N., Badawi, G.H., Kaminaka, H., Sanekata, T., Shibahara, T., et al. (2007) Overexpression of monodehydroascorbate reductase in transgenic tobacco confers enhanced tolerance to ozone, salt and polyethylene glycol stresses. *Planta* 225(5): 1255–1264.

Ernst, D., Schraudner, M., Langebartels, C., Sandermann, H. (1992) Ozone-induced changes of mRNA levels of β-1, 3-glucanase, chitinase and 'pathogenesis-related' protein 1b in tobacco plants. *Plant Molecular Biology* 20(4): 673–682.

Fares, S., Goldstein, A., Loreto, F. (2009) Determinants of ozone fluxes and metrics for ozone risk assessment in plants. *Journal of Experimental Botany* 61(3): 629–633.

Farmer, E.E., Mueller, M.J. (2013) ROS-mediated lipid peroxidation and RES-activated signaling. *Annual Review of Plant Biology* 64: 429–450.

Farré-Armengol, G., Peñuelas, J., Li, T., Yli-Pirilä, P., Filella, I., Llusia, J., Blande, J.D. (2016) Ozone degrades floral scent and reduces pollinator attraction to flowers. *New Phytologist* 209(1): 152–160.

Feng, Z., Kobayashi, K., Ainsworth, E.A. (2008) Impact of elevated ozone concentration on growth, physiology, and yield of wheat (*Triticum aestivum* L.): A meta-analysis. *Global Change Biology* 14(11): 2696–2708.

Fiscus, E.L., Booker, F.L., Burkey, K.O. (2005) Crop responses to ozone: Uptake, modes of action, carbon assimilation and partitioning. *Plant, Cell and Environment* 28: 997–1011.

Flowers, M.D., Fiscus, E.L., Burkey, K.O., Booker, F.L., Dubois, J.J.B. (2007) Photosynthesis, chlorophyll fluorescence, and yield of snap bean (*Phaseolus vulgaris* L.) genotypes differing in sensitivity to ozone. *Environmental and Experimental Botany* 61: 190–198.

Foyer, C.H., Noctor, G. (2005) Oxidant and antioxidant signaling in plants: A re-evaluation of the concept of oxidative stress in a physiological context. *Plant, Cell and Environment* 28(8): 1056–1071.

Francini, A., Nali, C., Picchi, V., Lorenzini, G. (2007) Metabolic changes in white clover clones exposed to ozone. *Environmental and Experimental Botany* 60: 11–19.

Frei, M. (2015) Breeding of ozone resistant rice: Relevance, approaches and challenges. *Environmental Pollution* 197: 144–155.

Frei, M., Tanaka, J.P., Chen, C.P., Wissuwa, M. (2010) Mechanisms of ozone tolerance in rice: Characterization of two QTLs affecting leaf bronzing by gene expression profiling and biochemical analyses. *Journal of Experimental Botany* 61(5): 1405–1417.

Friend, A.L., Tomlinson, P.T. (1992) Mild ozone exposure alters 14C dynamics in foliage of *Pinus taeda* L. *Tree Physiology* 11(3): 215–227.

Fuehrer, J., Achermann, B. (1994) Critical levels for ozone, a UN-ECE workshop report.

Fuhrer, J. (2009) Ozone risk for crops and pastures in present and future climates. *Naturwissenschaften* 96(2): 173–194.

Futsæther, C.M., Vollsnes, A.V., Kruse, O.M.O., Indahl, U.G., Kvaal, K., Eriksen, A.E.B. (2014) Daylength influences the response of three clover species (*Trifolium* spp.) to short-term ozone stress.

Gaucher, C., Costanzo, N., Afif, D., Mauffette, Y., Chevrier, N., Dizengremel, P. (2003) The impact of elevated ozone and carbon dioxide on young *Acer saccharum* seedlings. *Physiologia Plantarum* 117(3): 392–402.

Gillespie, C., Stabler, D., Tallentire, E., Goumenaki, E., Barnes, J. (2015) Exposure to environmentally-relevant levels of ozone negatively influence pollen and fruit development. *Environmental Pollution* 206: 494–501.

Gillespie, K.M., Xu, F., Richter, K.T., McGrath, J.M., Markelz, R.C., Ort, D.R., et al. (2012) Greater antioxidant and respiratory metabolism in field-grown soybean exposed to elevated O3 under both ambient and elevated CO2. *Plant, Cell and Environment* 35(1): 169–184.

Gomi, K., Ogawa, D., Katou, S., Kamada, H., Nakajima, N., Saji, H., et al. (2005) A mitogen-activated protein kinase NtMPK4 activated by SIPKK is required for jasmonic acid signaling and involved in ozone tolerance via stomatal movement in tobacco. *Plant and Cell Physiology* 46(12): 1902–1914.

Gorissen, A., van Veen, J.A. (1988) Temporary disturbance of translocation of assimilates in Douglas firs caused by low levels of ozone and sulfur dioxide. *Plant Physiology* 88(3): 559–563.

Grantz, D.A., Farrar, J.F. (1999) Acute exposure to ozone inhibits rapid carbon translocation from source leaves of Pima cotton. *Journal of Experimental Botany* 50(336): 1253–1262.

Grantz, D.A., Yang, S. (2000) Ozone impacts on allometry and root hydraulic conductance are not mediated by source limitation nor developmental age. *Journal of Experimental Botany* 51(346): 919–927.

Gravano, E., Bussotti, F., Strasser, R.J., Schaub, M., Novak, K., Skelly, J., et al. (2004) Ozone symptoms in leaves of woody plants in open-top chambers: Ultrastructural and physiological characteristics. *Physiologia Plantarum* 121(4): 620–633.

Grulke, N.E., Johnson, R., Esperanza, A., Jones, D., Nguyen, T., Posch, S., et al. (2003) Canopy transpiration of Jeffrey pine in mesic and xeric microsites: O3 uptake and injury response. *Trees* 17(4): 292–298.

Grünhage, L., Jäger, H.J. (2003) From critical levels to critical loads for ozone: A discussion of a new experimental and modelling approach for establishing flux–response relationships for agricultural crops and native plant species. *Environmental Pollution* 125(1): 99–110.

Guidi, L., Nali, C., Lorenzini, G., Filippi, F., Soldatini, G.F. (2001) Effect of chronic ozone fumigation on the photosynthetic process of poplar clones showing different sensitivity. *Environmental Pollution* 113(3): 245–254.

Gupta, P., Duplessis, S., White, H., Karnosky, D.F., Martin, F., Podila, G.K. (2005) Gene expression patterns of trembling aspen trees following long-term exposure to interacting elevated CO2 and tropospheric O3. *New Phytologist* 167(1): 129–142.

Hamel, L.P., Miles, G.P., Samuel, M.A., Ellis, B.E., Séguin, A., Beaudoin, N. (2005) Activation of stress-responsive mitogen-activated protein kinase pathways in hybrid poplar (*Populus trichocarpa* × *Populus deltoides*). *Tree Physiology* 25(3): 277–288.

Heath, R.L. (1994) Alterations of plant metabolism by ozone exposure. *Plant Responses to the Gaseous Environment* 121–145.

Heath, R.L., Frederick, P.E. (1979) Ozone alteration of membrane permeability in Chlorella. *Plant physiology* 64(3): 455–459.

Heath, R.L., Taylor, G.E. (1997) Physiological processes and plant responses to ozone exposure. In: *Forest Decline and Ozone. Ecological Studies*, ed. Sandermann, H., Wellburn, A.R., and Heath, R.L., Vol. 127, 317–368. Springer-Verlag, Berlin, Germany.

Heck, W.W. (1968) Factors influencing expression of oxidant damage to plants. *Annual Review of Phytopathology* 6(1): 165–188.

Heck, W.W., Taylor, O.C., Adams, R., Bingham, G., Miller, J., Preston, E., et al. (1982) Assessment of crop loss from ozone. *Journal of the Air Pollution Control Association* 32(4): 353–361.

Hertstein, U., Grünhage, L., Jäger, H.J. (1995) Assessment of past, present, and future impacts of ozone and carbon dioxide on crop yields. *Atmospheric Environment* 29(16): 2031–2039.

Hogsett, W.E., Plocher, M., Wildman, V., Tingey, D.T., Bennett, J.P. (1985) Growth response of two varieties of slash pine seedlings to chronic ozone exposures. *Canadian Journal of Botany* 63(12): 2369–2376.

Holland, M., Kinghorn, S., Emberson, L., Cinderby, S., Ashmore, M., Mills, G., Harmens, H. (2006) Development of a framework for probabilistic assessment of the economic losses caused by ozone damage to crops in Europe. *UNECE International Cooperative Programme on Vegetation, Project Report Number C02309, NERC/Centre for Ecology and Hydrology*: 50.

Holland, N.T., Duramad, P., Rothman, N., Figgs, L.W., Blair, A., Hubbard, A., et al. (2002) Micronucleus frequency and proliferation in human lymphocytes after exposure to herbicide 2, 4-dichlorophenoxyacetic acid in vitro and in vivo. *Mutation Research/Genetic Toxicology and Environmental Mutagenesis* 521(1): 165–178.

Hoshika, Y., Watanabe, M., Inada, N., Koike, T. (2013) Model-based analysis of avoidance of ozone stress by stomatal closure in Siebold's beech (*Fagus crenata*). *Annuls of Botany* 112(6): 1149–1158.

Hough, A.M., Derwent, R.G. (1990) Changes in the global concentration of tropospheric ozone due to human activities. *Nature* 344(6267): 645–648.

Iyer, N.J., Tang, Y., Mahalingam, R. (2013) Physiological, biochemical and molecular responses to a combination of drought and ozone in *Medicago truncatula*. *Plant, Cell and Environment* 36(3): 706–720.

Jain, S.L., Arya, B.C., Kumar, A., Ghude, S.D., Kulkarni, P.S. (2005) Observational study of surface ozone at New Delhi, India. *International Journal of Remote Sensing* 26(16): 3515–3524.

James, R.L., Cobb Jr, F.W., Miller, P.R., Parmeter Jr, J.R. (1980) Effects of oxidant air pollution on susceptibility of pine roots to *Fomes annosus*. *Phytopathology* 70(6): 560–563.

Jeong, Y.H., Nakamura, H., Ota, Y. (1981) Physiological Studies on Photochemical Oxidant Injury in Rice Plants: II. Effect of abscisic acid (ABA) on ozone injury and ethylene prosuction in rice plants. *Japanese Journal of Crop Science* 50(4): 560–565.

Joo, J.H., Wang, S., Chen, J.G., Jones, A.M., Fedoroff, N.V. (2005) Different signaling and cell death roles of heterotrimeric G protein α and β subunits in the *Arabidopsis* oxidative stress response to ozone. *The Plant Cell* 17(3): 957–970.

Kaakinen, S., Kostiainen, K., Ek, F., Saranpää, P., Kubiske, M.E., Sober, J., Karnosky, D.F., Vapaavuori, E. (2004) Stem wood properties of *Populus tremuloides, Betula papyrifera* and *Acer saccharum* saplings after 3 years of treatments to elevated carbon dioxide and ozone. *Global Change Biology* 10(9): 1513–1525.

Kalabokas, P.D., Repapis, C.C. (2004) A climatological study of rural surface ozone in central Greece. *Atmospheric Chemistry and Physics* 4(4): 1139–1147.

Kangasjärvi, J., Jaspers, P., Kollist, H. (2005) Signaling and cell death in ozone-exposed plants. *Plant, Cell and Environment* 28(8): 1021–1036.

Kangasjärvi, J., Talvinen, J., Utriainen, M., Karjalainen, R. (1994) Plant defence systems induced by ozone. *Plant, Cell and Environment* 17(7): 783–794.

Kanter, U., Heller, W., Durner, J., Winkler, J.B., Engel, M., Behrendt, H., Holzinger, A., Braun, P., Hauser, M., Ferreira, F., Mayer, K. (2013) Molecular and immunological characterization of ragweed (*Ambrosia artemisiifolia* L.) pollen after exposure of the plants to elevated ozone over a whole growing season. *PLoS One* 8(4): e61518.

Karlsson, P.E., Medin, E.L., Wickström, H., Selldén, G., Wallin, G., Ottosson, S., et al. (1995) Ozone and drought stress—Interactive effects on the growth and physiology of Norway spruce (*Picea abies* (L.) Karst.). *Water, Air, and Soil Pollution* 85(3): 1325–1330.

Karnosky, D.F., Pregitzer, K.S., Zak, D.R., Kubiske, M.E., Hendrey, G.R., Weinstein, D., et al. (2005) Scaling ozone responses of forest trees to the ecosystem level in a changing climate. *Plant, Cell and Environment* 28(8): 965–981.

Kitao, M., Löw, M., Heerdt, C., Grams, T.E., Häberle, K.H., Matyssek, R. (2009) Effects of chronic elevated ozone exposure on gas exchange responses of adult beech trees (*Fagus sylvatica*) as related to the within-canopy light gradient. *Environmental Pollution* 157(2): 537–544.

Kontunen-Soppela, S.A.R.I., Ossipov, V., Ossipova, S., Oksanen, E. (2007) Shift in birch leaf metabolome and carbon allocation during long-term open-field ozone exposure. *Global Change Biology* 13(5): 1053–1067.

Kontunen-Soppela, S.A.R.I., Riikonen, J., Ruhanen, H., Brosche, M., Somervuo, P., Peltonen, P., et al. (2010) Differential gene expression in senescing leaves of two silver birch genotypes in response to elevated CO2 and tropospheric ozone. *Plant, Cell and Environment* 33(6): 1016–1028.

Kostiainen, K., Kaakinen, S., Warsta, E., Kubiske, M.E., Nelson, N.D., Sober, J., et al. (2008) Wood properties of trembling aspen and paper birch after 5 years of exposure to elevated concentrations of CO2 and O3. *Tree Physiology* 28(5): 805–813.

Krupa, S.V. (2003) Joint effects of elevated levels of ultraviolet-B radiation, carbon dioxide and ozone on plants. *Photochemistry and Photobiology* 78(6): 535–542.

Laurence, J.A., Wood, F.A. (1978) Effects of ozone on infection of soybean by Pseudomonas glycinea. *Phytopathology* 68: 441–445.

Le Gall, H., Philippe, F., Domon, J.M., Gillet, F., Pelloux, J., Rayon, C. (2015) Cell wall metabolism in response to abiotic stress. *Plants* 4(1): 112–166.

Lee, E.H., Tingey, D.T., Hogsett, W.E. (1988) Evaluation of ozone exposure indices in exposure-response modeling. *Environmental Pollution* 53(1–4): 43–62.

Leisner, C.P., Ainsworth, E.A., Ming, R. (2014) Distinct transcriptional profiles of ozone stress in soybean (Glycine max) flowers and pods. *BMC Plant Biology* 14(1): 335.

Levitt, J. (1972) Response of plants to environmental stresses. *Water, Radiation, Salt and Other Stresses* 2.

Li, P., Calatayud, V., Gao, F., Uddling, J., Feng, Z. (2016) Differences in ozone sensitivity among woody species are related to leaf morphology and antioxidant levels. *Tree Physiology* 36(9): 1105–1116.

Li, T., Blande, J.D. (2015) Associational susceptibility in broccoli: Mediated by plant volatiles, impeded by ozone. *Global Change Biology* 21(5): 1993–2004.

Lin, M., Fiore, A.M., Cooper, O.R., Horowitz, L.W., Langford, A.O., Levy, H., et al. (2012) Springtime high surface ozone events over the western United States: Quantifying the role of stratospheric intrusions. *Journal of Geophysical Research: Atmospheres* 117(D21).

Liu, X., Sui, L., Huang, Y., Geng, C., Yin, B. (2015) Physiological and visible injury responses in different growth stages of winter wheat to ozone stress and the protection of spermidine. *Atmospheric Pollution Research* 6(4): 596–604.

Lobell, D.B., Field, C.B. (2007) Global scale climate–crop yield relationships and the impacts of recent warming. *Environmental Research Letters* 2(1): 014002.

Ludwikow, A., Sadowski, J. (2008) Gene networks in plant ozone stress response and tolerance. *Journal of Integrative Plant Biology* 50(10): 1256–1267.

Lütz, C., Anegg, S., Gerant, D., Alaoui-Sossé, B., Gérard, J., Dizengremel, P. (2000) Beech trees exposed to high CO2 and to simulated summer ozone levels: Effects on photosynthesis, chloroplast components and leaf enzyme activity. *Physiologia Plantarum* 109(3): 252–259.

Maňkovska, B., Percy, K.E., Karnosky, D.F. (2005) Impacts of greenhouse gases on epicuticular waxes of Populus tremuloides Michx.: Results from an open-air exposure and a natural O3 gradient. *Environmental Pollution-Kidlington* 137(3): 580–586.

Maccarrone, M., Veldink, G.A., Vliegenthart, J.F. (1992) Thermal injury and ozone stress affect soybean lipoxygenases expression. *FEBS Letters* 309(3): 225–230.

Maccarrone, M., Veldink, G.A., Vliegenthart, J.F., Finazzi Agrò, A. (1997) Ozone stress modulates amine oxidase and lipoxygenase expression in lentil (*Lens culinaris*) seedlings. *FEBS Letters* 408(2): 241–244.

Maddison, J., Lyons, T., Plöchl, M., Barnes, J. (2002) Hydroponically cultivated radish fed L-galactono-1, 4-lactone exhibit increased tolerance to ozone. *Planta* 214(3): 383–391.

Malamy, J., Carr, J.P., Klessig, D.F., Raskin, I. (1990) Salicylic acid: A likely endogenous signal in the resistance response of tobacco to viral infection. *Science* 250(4983): 1002.

Mansfield, T.A., Freer-Smith, P.H. (1984) The role of stomata in resistance mechanisms. In *Gaseous Air Pollutants and Plant Metabolism*, ed. Koziol, M.J. and Whatley, F.R., 131–145. Butterworths, London, UK.

Martin, M.J., Farage, P.K., Humphries, S.W., Long, S.P. (2000) Can the stomatal changes caused by acute ozone exposure be predicted by changes occurring in the mesophyll? A simplification for models of vegetation response to the global increase in tropospheric elevated ozone episodes. *Functional Plant Biology* 27(3): 211–219.

Masuch, G., Franz, J.T., Schoene, K., Müsken, H., Bergmann, K.C. (1997) Ozone increases group 5 allergen content of *Lolium perenne*. *Allergy* 52(8): 874–875.

Matyssek, R., Maurer, S., Günthardt-Goerg, M., Landolt, W., Saurer, M., Polle, A. (1997) Nutrition determines the "strategy" of Betula pendula for coping with ozone stress. *Phyton* 37: 157–168.

McAinsh, M.R., Evans, N.H., Montgomery, L.T., North, K.A. (2002) Calcium signaling in stomatal responses to pollutants. *New Phytologist* 153(3): 441–447.

McCool, P.M., Menge, J.A. (1983) Influence of ozone on carbon partitioning in tomato: Potential role of carbon flow in regulation of the mycorrhizal symbiosis under conditions of stress. *New Phytologist* 94(2): 241–247.

McCurdy, T.R. (1994) Human exposure to ambient ozone. *Tropospheric Ozone: Human Health and Agricultural Impacts* 85–127.

McGrath, J.M., Betzelberger, A.M., Wang, S., Shook, E., Zhu, X.G., Long, et al. (2015) An analysis of ozone damage to historical maize and soybean yields in the United States. *Proceedings of the National Academy of Sciences* 112(46): 14390–14395.

McKee, I.F., Long, S.P. (2001) Plant growth regulators control ozone damage to wheat yield. *New Phytologist* 152(1): 41–51.

McLaughlin, S.B., Taylor, G.E. (1981) Relative humidity: Important modifier of pollutant uptake by plants. *Science* 211(4478): 167–169.

Medlyn, B.E., Duursma, R.A., Eamus, D., Ellsworth, D.S., Prentice, I.C., Barton, C.V., et al. (2011) Reconciling the optimal and empirical approaches to modelling stomatal conductance. *Global Change Biology* 17(6): 2134–2144.

Mehlhorn, H., Wellburn, A.R. (1987) Stress ethylene formation determines plant sensitivity to ozone. *Nature* 327(6121): 417–418.

Menser, H.A., Heggestad, H.E., Street, O.E., Jeffrey, R.N. (1963) Response of plants to air pollutants. I. Effects of ozone on tobacco plants preconditioned by light and temperature. *Plant Physiology* 38(5): 605.

Mills, G., Hayes, F., Simpson, D., Emberson, L., Norris, D., Harmens, H., et al. (2011) Evidence of widespread effects of ozone on crops and (semi-) natural vegetation in Europe (1990–2006) in relation to AOT40-and flux-based risk maps. *Global Change Biology* 17(1): 592–613.

Minas, I.S., Vicente, A.R., Dhanapal, A.P., Manganaris, G.A., Goulas, V., et al. (2014) Ozone-induced kiwifruit ripening delay is mediated by ethylene biosynthesis inhibition and cell wall dismantling regulation. *Plant Science* 229: 76–85.

Mohri, T., Kogawara, S., Igasaki, T., Yasutani, I., Aono, M., Nakajima, N., et al. (2011) Improvement in the ozone tolerance of poplar plants with an antisense DNA for 1-aminocyclopropane-1-carboxylate synthase. *Plant Biotechnology* 28(4): 417–421.

Monks, P.S., Archibald, A.T., Colette, A., Cooper, O., Coyle, M., Derwent, R., et al. (2015) Tropospheric ozone and its precursors from the urban to the global scale from air quality to short-lived climate forcer. *Atmospheric Chemistry and Physics* 15(15): 8889–8973.

Moon, J.C., Lim, S.D., Yim, W.C., Song, K., Lee, B.M. (2013) Characterization of expressed genes under ozone stress in soybean. *Plant Breeding and Biotechnology* 1(3): 270–276.

Mumford, R.A., Lipke, H., Laufer, D.A., Feder, W.A. (1972) Ozone-induced changes in corn pollen. *Environmental Science and Technology* 6(5): 427–430.

Musselman, R.C., Lefohn, A.S., Massman, W.J., Heath, R.L. (2006) A critical review and analysis of the use of exposure-and flux-based ozone indices for predicting vegetation effects. *Atmospheric Environment* 40(10): 1869–1888.

Nagatoshi, Y., Mitsuda, N., Hayashi, M., Inoue, S.I., Okuma, E., Kubo, A., et al. (2016) GOLDEN 2-LIKE transcription factors for chloroplast development affect ozone tolerance through the regulation of stomatal movement. *Proceedings of the National Academy of Sciences* 113(15): 4218–4223.

Nakajima, N., Itoh, T., Takikawa, S., Asai, N., Tamaoki, M., Aono, et al. (2002) Improvement in ozone tolerance of tobacco plants with an antisense DNA for 1-aminocyclopropane-1-carboxylate synthase. *Plant, Cell and Environment* 25(6): 727–735.

Noctor, G., Foyer, C.H. (1998) Ascorbate and glutathione: Keeping active oxygen under control. *Annual Review of Plant Biology* 49(1): 249–279.

Noormets, A., Sober, A., Pell, E.J., Dickson, R.E., Podila, G.K., Sober, J., et al. (2001) Stomatal and non-stomatal limitation to photosynthesis in two trembling aspen (*Populus tremuloides* Michx.) clones exposed to elevated CO2 and/or O3. *Plant, Cell and Environment* 24(3): 327–336.

Nouchi, I., Toyama, S. (1988) Effects of ozone and peroxyacetyl nitrate on polar lipids and fatty acids in leaves of morning glory and kidney bean. *Plant Physiology* 87(3): 638–646.

Nussbaum, S., Fuhrer, J. (2000) Difference in ozone uptake in grassland species between open-top chambers and ambient air. *Environmental Pollution* 109(3): 463–471.

Oksanen, E., Häikiö, E., Sober, J., Karnosky, D.F. (2004) Ozone-induced H2O2 accumulation in field-grown aspen and birch is linked to foliar ultrastructure and peroxisomal activity. *New Phytologist* 161(3): 791–799.

Oltmans, S.J., Lefohn, A.S., Harris, J.M., Galbally, I., Scheel, H.E., Bodeker, et al. (2006) Long-term changes in tropospheric ozone. *Atmospheric Environment* 40(17): 3156–3173.

Örvar, B.L., McPherson, J., Ellis, B.E. (1997) Pre-activating wounding response in tobacco prior to high-level ozone exposure prevents necrotic injury. *The Plant Journal* 11(2): 203–212.

Otto, H.W., Daines, R.H. (1969) Plant injury by air pollutants: Influence of humidity on stomatal apertures and plant response to ozone. *Science* 163(3872): 1209–1210.

Overmyer, K., Brosché, M., Kangasjärvi, J. (2003) Reactive oxygen species and hormonal control of cell death. *Trends in Plant Science* 8(7): 335–342.

Overmyer, K., Tuominen, H., Kettunen, R., Betz, C., Langebartels, C., Sandermann, H., et al. (2000) Ozone-sensitive *Arabidopsis* rcd1 mutant reveals opposite roles for ethylene and jasmonate signaling pathways in regulating superoxide-dependent cell death. *The Plant Cell* 12(10): 1849–1862.

Pääkkönen, E., Holopainen, T., Kärenlampi, L. (1997) Variation in ozone sensitivity among clones of *Betula pendula* and *Betula pubescens*. *Environmental Pollution* 95(1): 37–44.

Pääkkönen, E., Seppänen, S., Holopainen, T., Kokko, H., Kärenlampi, S., Kärenlampi, L., et al. (1998) Induction of genes for the stress proteins PR-10 and PAL in relation to growth, visible injuries and stomatal conductance in birch (*Betula pendula*) clones exposed to ozone and/or drought. *The New Phytologist* 138(2): 295–305.

Pääkkönen, E., Vahala, J., Holopainen, T., Karjalainen, R., Kärenlampi, L. (1996) Growth responses and related biochemical and ultrastructural changes of the photosynthetic apparatus in birch (*Betula pendula*) saplings exposed to low concentrations of ozone. *Tree Physiology* 16(7): 597–605.

Panek, J.A., Goldstein, A.H. (2001) Response of stomatal conductance to drought in ponderosa pine: Implications for carbon and ozone uptake. *Tree Physiology* 21(5): 337–344.

Paoletti, E., Contran, N., Bernasconi, P., Günthardt-Goerg, M.S., Vollenweider, P. (2010) Erratum to "Structural and physiological responses to ozone in Manna ash (*Fraxinus ornus* L.) leaves of seedlings and mature trees under controlled and ambient conditions". *Science of the Total Environment* 408(8): 2014–2024.

Papinchak, H.L., Holcomb, E.J., Best, T.O., Decoteau, D.R. (2009) Effectiveness of houseplants in reducing the indoor air pollutant ozone. *HortTechnology* 19(2): 286–290.

Pasqualini, S., Meier, S., Gehring, C., Madeo, L., Fornaciari, M., Romano, B., Ederli, L. (2009) Ozone and nitric oxide induce cGMP-dependent and-independent transcription of defence genes in tobacco. *New Phytologist* 181(4): 860–870.

Pasqualini, S., Piccioni, C., Reale, L., Ederli, L., Della Torre, G., Ferranti, F. (2003) Ozone-induced cell death in tobacco cultivar Bel W3 plants. The role of programmed cell death in lesion formation. *Plant Physiology* 133(3): 1122–1134.

Pasqualini, S., Tedeschini, E., Frenguelli, G., Wopfner, N., Ferreira, F., D'Amato, G., et al. (2011) Ozone affects pollen viability and NAD (P) H oxidase release from *Ambrosia artemisiifolia* pollen. *Environmental Pollution* 159(10): 2823–2830.

Payne, R.J., Stevens, C.J., Dise, N.B., Gowing, D.J., Pilkington, M.G., Phoenix, G.K., et al. (2011) Impacts of atmospheric pollution on the plant communities of British acid grasslands. *Environmental Pollution* 159(10): 2602–2608.

Pellegrini, E., Carucci, M.G., Campanella, A., Lorenzini, G. Nali, C. (2011) Ozone stress in Melissa officinalis plants assessed by photosynthetic function. *Environmental and Experimental Botany* 73: 94–101.

Pellinen, R., Palva, T., KangasjaÈrvi, J. (1999) Subcellular localization of ozone-induced hydrogen peroxide production in birch (Betula pendula) leaf cells. *The Plant Journal* 20(3): 349–356.

Pelloux, J., Jolivet, Y., Fontaine, V., Banvoy, J., Dizengremel, P. (2001) Changes in Rubisco and Rubisco activase gene expression and polypeptide content in *Pinus halepensis* M. subjected to ozone and drought. *Plant, Cell and Environment* 24(1): 123–131.

Picchi, V., Iriti, M., Quaroni, S., Saracchi, M., Viola, P., Faoro, F. (2010) Climate variations and phenological stages modulate ozone damages in field-grown wheat. A three-year study with eight modern cultivars in Po Valley (Northern Italy). *Agriculture, Ecosystems and Environment* 135(4): 310–317.

Pinelli, P., Tricoli, D. (2004) A new approach to ozone plant fumigation: The Web-O3-Fumigation. Isoprene response to a gradient of ozone stress in leaves of Quercus pubescens. *Forest@* 1(2): 100–108.

Pinelli, P., Tricoli, D. (2008) A new approach to ozone plant fumigation: The Web-O3-Fumigation. Isoprene response to a gradient of ozone stress in leaves of *Quercus pubescens. iForest-Biogeosciences and Forestry* 1(1): 22–26.

Płażek, A., Hura, K., Rapacz, M., Żur, I. (2001) The influence of ozone fumigation on metabolic efficiency and plant resistance to fungal pathogens. *Journal of Applied Botany and Food Quality* 75(1): 8–13.

Pleijel, H., Danielsson, H., Karlsson, G.P., Gelang, J., Karlsson, P.E., Selldén, G. (2000) An ozone flux–response relationship for wheat. *Environmental Pollution* 109(3): 453–462.

Pleijel, H., Skärby, L., Wallin, G., Sellden, G. (1991) Yield and grain quality of spring wheat (*Triticum aestivum* L., cv. Drabant) exposed to different concentrations of ozone in open-top chambers. *Environmental Pollution* 69(2–3): 151–168.

Power, S.A., Ashmore, M.R. (2002) Responses of fen and fen-meadow communities to ozone. *New Phytologist* 156(3): 399–408.

Prather, M., Gauss, M., Berntsen, T., Isaksen, I., Sundet, J., Bey, I., Brasseur, G., Dentener, F., Derwent, R., Stevenson, D., Grenfell, L. (2003) Fresh air in the 21st century? *Geophysical Research Letters* 30(2).

Proietti, C., Anav, A., De Marco, A., Sicard, P., Vitale, M. (2016) A multi-sites analysis on the ozone effects on Gross Primary Production of European forests. *Science of the Total Environment* 556: 1–11.

Puckette, M.C., Weng, H., Mahalingam, R. (2007) Physiological and biochemical responses to acute ozone-induced oxidative stress in Medicago truncatula. *Plant Physiology and Biochemistry* 45(1): 70–79.

Ranieri, A., Castagna, A., Pacini, J., Baldan, B., Mensuali Sodi, A., Soldatini, G.F. (2003) Early production and scavenging of hydrogen peroxide in the apoplast of sunflower plants exposed to ozone. *Journal of Experimental Botany* 54(392): 2529–2540.

Ranieri, A., Castagna, A., Soldatini, G.F. (2000) Differential stimulation of ascorbate peroxidase isoforms by ozone exposure in sunflower plants. *Journal of Plant Physiology* 156(2): 266–271.

Rao, M.V., Davis, K.R. (2001) The physiology of ozone induced cell death. *Planta* 213(5): 682–690.

Rao, M.V., Lee, H.I., Creelman, R.A., Mullet, J.E., Davis, K.R. (2000a) Jasmonic acid signaling modulates ozone-induced hypersensitive cell death. *The Plant Cell* 12(9): 1633–1646.

Rao, M.V., Lee, H.I., Davis, K.R. (2002) Ozone-induced ethylene production is dependent on salicylic acid, and both salicylic acid and ethylene act in concert to regulate ozone-induced cell death. *The Plant Journal* 32(4): 447–456.

Rao, S.T., Hogrefe, C., Sistla, G., Wu, S.Y., Hao, W., Zalewsky, et al. (2000b) An integrated modeling and observational approach for designing ozone control strategies for the Eastern US. In *Air Pollution Modeling and Its Application XIII*, ed., Gryning, S.E. and Batchvarova, E. 3–18. Springer, Boston, MA.

Reich, P.B. (1983) Effects of low concentrations of O3 on net photosynthesis, dark respiration, and chlorophyll contents in aging hybrid poplar leaves. *Plant Physiology* 73(2): 291–296.

Reiling, K., Davison, A.W. (1995) Effects of ozone on stomatal conductance and photosynthesis in populations of *Plantago major* L. *New Phytologist* 129(4): 587–594.

Reiner, S., Wiltshire, J.J.J., Wright, C.J., Colls, J.J. (1996) The impact of ozone and drought on the water relations of ash trees (*Fraxinus excelsior* L.). *Journal of Plant Physiology* 148(1–2): 166–171.

Renaut, J., Bohler, S., Hausman, J.F., Hoffmann, L., Sergeant, K., Ahsan, N., et al. (2009) The impact of atmospheric composition on plants: A case study of ozone and poplar. *Mass Spectrometry Reviews* 28(3): 495–516.

Richet, N., Afif, D., Huber, F., Pollet, B., Banvoy, J., El Zein, et al. (2011). Cellulose and lignin biosynthesis is altered by ozone in wood of hybrid poplar (*Populus tremula× alba). Journal of Experimental Botany* 62(10): 3575–3586.

Runeckles, V.C., Wright, E.F., White, D. (1990) A chamber-less field exposure system for determining the effects of gaseous air pollutants on crop growth and yield. *Environmental Pollution* 63(1): 61–77.

Samuel, M.A., Ellis, B.E. (2002) Double jeopardy both over-expression and suppression of a redox-activated plant mitogen-activated protein kinase render tobacco plants ozone sensitive. *The Plant Cell* 14(9): 2059–2069.

Sandermann Jr, H. (1996) Ozone and plant health. *Annual Review of Phytopathology* 34(1): 347–366.

Sanmartin, M., Drogoudi, P.D., Lyons, T., Pateraki, I., Barnes, J., Kanellis, A.K. (2003) Over-expression of ascorbate oxidase in the apoplast of transgenic tobacco results in altered ascorbate and glutathione redox states and increased sensitivity to ozone. *Planta* 216(6): 918–928.

Saxe, H. (2002) Physiological responses of trees to ozone-interactions and mechanisms. *Current Topics in Plant Biology* 3: 27–55.

Scebba, F., Giuntini, D., Castagna, A., Soldatini, G., Ranieri, A. (2006) Analysing the impact of ozone on biochemical and physiological variables in plant species belonging to natural ecosystems. *Environmental and Experimental Botany* 57(1): 89–97.

Sharma, Y.K., León, J., Raskin, I., Davis, K.R. (1996) Ozone-induced responses in *Arabidopsis thaliana*: The role of salicylic acid in the accumulation of defense-related transcripts and induced resistance. *Proceedings of the National Academy of Sciences* 93(10): 5099–5104.

Sinn, J.P., Schlagnhaufer, C.D., Arteca, R.N., Pell, E.J. (2004) Ozone-induced ethylene and foliar injury responses are altered in 1-aminocyclopropane-1-carboxylate synthase antisense potato plants. *New phytologist* 164(2): 267–277.

Skelly, J.M., Yang, Y.S., Chevone, B.I., Long, S.J., Nellessen, J.E., Winner, W.E. (1983) Ozone concentrations and their influence on forest species in the Blue Ridge Mountains of Virginia. *Air Pollution and the Productivity of the Forest* 143–160.

Soja, G. (1996) Growth stage as a modifier of ozone response in winter wheat. In *Exceedance of Critical Loads and Levels. Conference Paper*, ed. Knoacher, M., Schneider, J. and Soja, G., Vol. 15, 155–163. Federal Environment Agency, Vienna, Austria.

Spence, R.D., Rykiel, E.J., Sharpe, P.J. (1990) Ozone alters carbon allocation in loblolly pine: Assessment with carbon-11 labeling. *Environmental Pollution* 64(2): 93–106.

Stewart, C.A. (1998) Impact of ozone on the reproductive biology of *Brassica campestris* L. and *Plantago major* L. Doctoral dissertation. Loughborough University, UK.

Tai, A.P., Martin, M.V. (2017). Impacts of ozone air pollution and temperature extremes on crop yields: Spatial variability, adaptation and implications for future food security. *Atmospheric Environment* 169: 11–21.

Tamaoki, M. (2008) The role of phytohormone signaling in ozone-induced cell death in plants. *Plant Signaling and Behavior* 3(3): 166–174.

Tausz, M., Grulke, N.E., Wieser, G. (2007) Defense and avoidance of ozone under global change. *Environmental Pollution* 147(3): 525–531.

The Royal Society (2008) Ground-level ozone in the 21st century: Future trends, impacts and policy implications. *Science Policy REPORT 15/08.* Retrieved from: http://royalsociety.org/policy/publications/2008/ground-level-ozone/ (January 2014, date last accessed). Google Scholar.

Thiec, D., Dixon, M., Garrec, J.P. (1994) The effects of slightly elevated ozone concentrations and mild drought stress on the physiology and growth of Norway spruce, *Picea abies* (L.) Karst. and beech, *Fagus sylvatica* L., in open-top chambers. *New Phytologist* 128(4): 671–678.

Thwe, A.A., Vercambre, G., Gautier, H., Gay, F., Phattaralerphong, J., Kasemsap, P. (2014) Response of photosynthesis and chlorophyll fluorescence to acute ozone stress in tomato (*Solanum lycopersicum* Mill.). *Photosynthetica* 52(1): 105–116.

Tingey, D.T., Hogsett, W.E. (1985) Water stress reduces ozone injury via a stomatal mechanism. *Plant Physiology* 77(4): 944–947.

Tonneijck, A.E.G. (1989) Evaluation of ozone effects on vegetation in the Netherlands. *Studies in Environmental Science* 35: 251–260.

Torsethaugen, G., Pell, E.J., Assmann, S.M. (1999) Ozone inhibits guard cell K+ channels implicated in stomatal opening. *Proceedings of the National Academy of Sciences* 96(23): 13577–13582.

Vahisalu, T., Kollist, H., Wang, Y.F., Nishimura, N., Chan, W.Y., Valerio, et al. (2008) SLAC1 is required for plant guard cell S-type anion channel function in stomatal signaling. *Nature* 452(7186): 487–491.

Vahisalu, T., Puzõrjova, I., Brosché, M., Valk, E., Lepiku, M., Moldau, et al. (2010) Ozone-triggered rapid stomatal response involves the production of reactive oxygen species, and is controlled by SLAC1 and OST1. *The Plant Journal* 62(3): 442–453.

Van Camp, W., Willekens, H., Bowler, C., Van Montagu, M., Inzé, D., Reupold-Popp, et al. (1994) Elevated levels of superoxide dismutase protect transgenic plants against ozone damage. *Nature Biotechnology* 12(2): 165–168.

Vanzo, E., Ghirardo, A., Merl-Pham, J., Lindermayr, C., Heller, W., Hauck, et al. (2014) S-nitroso-proteome in poplar leaves in response to acute ozone stress. *PloS One* 9(9): e106886.

Venkatesh, J., Park, S.W. (2014) Role of L-ascorbate in alleviating abiotic stresses in crop plants. *Botanical Studies* 55(1): 38.

Vickers, C.E., Possell, M., Cojocariu, C.I., Velikova, V.B., Laothawornkitkul, J., Ryan, A., Mullineaux, et al. (2009) Isoprene synthesis protects transgenic tobacco plants from oxidative stress. *Plant, Cell and Environment* 32(5): 520–531.

Weber, J.A., Clark, C.S., Hogsett, W.E. (1993) Analysis of the relationships among O3 uptake, conductance, and photosynthesis in needles of Pinus ponderosa. *Tree Physiology* 13(2): 157–172.

Weigt, R.B., Häberle, K.H., Millard, P., Metzger, U., Ritter, W., Blaschke, H., et al. (2012) Ground-level ozone differentially affects nitrogen acquisition and allocation in mature European beech (*Fagus sylvatica*) and Norway spruce (*Picea abies*) trees. *Tree Physiology* 32(10): 1259–1273.

Wiese, C.B., Pell, E.J. (2003) Oxidative modification of the cell wall in tomato plants exposed to ozone. *Plant Physiology and Biochemistry* 41(4): 375–382.

Wilkinson, S., Davies, W.J. (2010) Drought, ozone, ABA and ethylene: New insights from cell to plant to community. *Plant, Cell and Environment* 33(4): 510–525.

Wittig, V.E., Ainsworth, E.A., Long, S.P. (2007) To what extent do current and projected increases in surface ozone affect photosynthesis and stomatal conductance of trees? A meta-analytic review of the last 3 decades of experiments. *Plant, Cell and Environment* 30(9): 1150–1162.

Wittig, V.E., Ainsworth, E.A., Naidu, S.L., Karnosky, D.F., Long, S.P. (2009) Quantifying the impact of current and future tropospheric ozone on tree biomass, growth, physiology and biochemistry: A quantitative meta-analysis. *Global Change Biology* 15(2): 396–424.

Wohlgemuth, H., Mittelstrass, K., Kschieschan, S., Bender, J., Weigel, H.J., Overmyer, K., et al. (2002) Activation of an oxidative burst is a general feature of sensitive plants exposed to the air pollutant ozone. *Plant, Cell and Environment* 25(6): 717–726.

Wrzaczek, M., Brosché, M., Salojärvi, J., Kangasjärvi, S., Idänheimo, N., Mersmann, S., et al. (2010) Transcriptional regulation of the CRK/DUF26 group of receptor-like protein kinases by ozone and plant hormones in *Arabidopsis*. *BMC Plant Biology* 10(1): 95.

Yalpani, N., Enyedi, A.J., León, J., Raskin, I. (1994) Ultraviolet light and ozone stimulate accumulation of salicylic acid, pathogenesis-related proteins and virus resistance in tobacco. *Planta* 193(3), 372–376.

Yendrek, C.R., Koester, R.P., Ainsworth, E.A. (2015) A comparative analysis of transcriptomic, biochemical, and physiological responses to elevated ozone identifies species-specific mechanisms of resilience in legume crops. *Journal of Experimental Botany* 66(22):7101–7112.

Young, P.J., Archibald, A.T., Bowman, K.W., Lamarque, J.F., Naik, V., Stevenson, D.S., et al. (2013) Pre-industrial to end 21st century projections of tropospheric ozone from the Atmospheric Chemistry and Climate Model Intercomparison Project (ACCMIP). *Atmospheric Chemistry and Physics* 13(4): 2063–2090.

Zhang, W., Lorence, A., Gruszewski, H.A., Chevone, B.I., Nessler, C.L. (2009) AMR1, an *Arabidopsis* gene that coordinately and negatively regulates the mannose/l-galactose ascorbic acid biosynthetic pathway. *Plant Physiology* 150(2): 942–950.

Zheng, Q., Wang, X., Feng, Z., Song, W., Feng, Z. (2005) Ozone effects on chlorophyll content and lipid peroxidation in the in situ leaves of winter wheat. *Acta Botanica Boreali-occidentalia Sinica* 25(11): 2240–2244.

Zinser, C., Ernst, D., Sandermann Jr, H. (1998) Induction of stilbene synthase and cinnamyl alcohol dehydrogenase mRNAs in Scots pine (*Pinus sylvestris* L.) seedlings. *Planta* 204(2): 169–176.

12 Hydrocarbon Contamination in Soil and Its Amelioration

Maimona Saeed and Noshin Ilyas

CONTENTS

12.1 INTRODUCTION

Crude oil (petroleum) is a heterogeneous mixture consisting primarily of hydrocarbons and few substances of organic nitrogen and sulfur (Speight, 2010). Studies through Fourier transform ion cyclotron resonance mass spectrometry have shown that almost 20,000 different heteroatom substances of oxygen, nitrogen, and sulfur (CHNOS) are present in petroleum crude oil (Marshall and Rodgers, 2004). These chemical investigations into the components of crude oil and its products ought to have amplified a growing petroleum price; the reduction of ideal less "sweet crude" has resulted in an international need of denser and more high-sulfur sour crude, along with latest perfections in systematic approaches (Marshall and Rodgers, 2008).

Crude oil is made with catagenesis of bitumen and kerogen, an irretrievable method which happens at amassed temperature, pressure, and interment complexity in hydrocarbon source rocks and sandy basins (Speight, 2010). Crude oil can be classified into four parts –saturates, aromatics, resins, and asphaltenes (SARA) – built on their polarity and solubility characteristics (Harayama et al., 1999; Fan and Buckley, 2002). Saturates are the hydrocarbons that have one bond and can be categorized into aliphatic alkanes and cycloalkanes with common formulae C_nH_{2n+2} and C_nH_{2n}, individually; the aromatics contain one or more aromatic rings with benzene as the simplest form, while resins and asphaltenes contain non-hydrocarbon polar substances (composed of nitrogen, sulfur, and/or oxygen) and can simply form compound with heavy metals (Harayama et al., 1999; Grin'ko et al., 2012). Molecular investigations on the constituents of crude oil are orthodoxly approved by utilizing normal chromatographic separation techniques followed by high-resolution mass spectrometry analysis (Gaspar et al., 2012).

Even though the toxicity of all discrete components is well-known, the injurious effect of composite mixtures like crude oil and advanced products is tremendously difficult to evaluate because scientists study little about the synergistic, additive, or antagonistic effects of the several mixtures. Furthermore, the chemical structure of petroleum and crude oil goods differed considerably and affected a variety of life in the ecosystem. These variations in the harmful effects of petroleum products are the result of their different composition, and also the variant amount of chemical ingredients. On the basis of structure, the crude oil hydrocarbons are categorized as alkanes, cycloalkanes, and aromatics. Alkenes, which are the type of unsaturated hydrocarbons, are infrequent in crude oil; however, they are a result of the breakdown of various products of petroleum. The number of carbons in alkanes (homology), various bonding arrangement of carbon chains (isoalkanes), condensation of the ring, and grouping within class, e.g., phenylalkanes, contributes to the increasing hydrocarbons. Furthermore, lower numbers of oxygen- (phenols, naphthenic acids), nitrogen- (pyridine, pyrrole, indole), and sulfur- (alkyl-thiol, thiophene) containing substances, mutually chosen as resins‖ and partially oxygenated, highly condensed asphaltic part are also present in crude but not in purified form of petroleum (Sverdrup et al., 2003).

12.2 SOURCES OF HYDROCARBON AND ITS TOXICITY

Polyaromatic hydrocarbons (PAHs) are all over the place in both land and aquatic ecosystems (IARC, 2010). They are present obviously in petroleum and coal but are also formed by human actions (WHO, 2000). The increased concentration of PAHs in soils is frequently a result of the utilization of fossil fuel preservation and discarding, waste management, environmental deposis and the fertilizer and composts' utilization (Cerniglia, 1992; ATSDR, 1995; Macleod and Semple, 2002).

Soil pollution with the discharge of oil is the worldwide apprehension currently. Soil polluted with petroleum has an injurious effect on human life, results in water table contamination which resists its utilization, leads to a reduction in the economy, environmental issues, and decreases the productivity of soil (Wang et al., 2008). Each year almost 35 million barrels of oil is shipped across the seas around the world, leading to contamination by oil discharge that imposes danger to marine organisms as a whole (Macaulay, 2015). The increasing hazardous effects, through contact by the contaminated soil, pollutants fumes, and from minor pollution of water, are transported above and under soil. The poisonous effects of crude oil hydrocarbons on microbes, plants, animals, and humans are well documented. The harmful

effects of hydrocarbons on land crops and their utilization as weed killers have been attributed to the potential of oil to solubilize the lipid part of plasma membrane, thus facilitating components of oil to flow outside. Such instances of pollution in both soil and water have become quite common nowadays. These pose severe instant and long-lasting impact since many hydrocarbon components are toxic in nature. These pollutants stay in soil and water for a very long time, often decades. The more obvious causes of pollution are discharge by production and purification areas, oil-tanker leakage, and mishaps during oil transference. Crude oil is being delivered to larger pathways by soil pipes or tankers, both of which are susceptible to oil leakage and accidents (EPA). The major oil contamination is caused by the fact that most oil-production countries are not oil users. It shows that huge transport of petroleum has to be made from region of manufacturing to utilization. Oil spills due to seepages from point of contact between pipelines or accidents of tankers are common yet disturbing the environment. A collection of such hazardous contaminants in plant and animal tissues may result in mutation and even death in certain cases. Not only petroleum but also diesel contain mixtures of highly intense components that are toxic, which have highly harmful and fatal effect on humans and animals (Jain and Bajpai, 2012).

To remediate the contaminants from the polluted area of land and ocean, various physicochemical techniques have been established for treating the area. These techniques are not only expensive but also need manual labor, and there is a possible threat of provoking the circumstances and dispersion of the pollution (Salleh et al., 2003). Other possible methods are evaporation, burying, dispersion, and washing (Das and Chandran, 2011), although there are some limitations to these methods as they frequently result in partial decay of contaminants. So, a less complex and cost-effective technique for treating hydrocarbon is required by today's world.

12.3 BIOREMEDIATION

Bioremediation is the method of utilizing living organisms, microbes, to reduce contaminants from the ecosystem and convert them into less hazardous forms. Bioremediation is centered on the potential of microbes to reduce the hydrocarbons into substances that can be used by other microbes as food or can be securely reimbursed to the surrounding. Reduced pollutants are transformed into water, carbon dioxide, and other simple substances. Not only microorganisms but also plants are used in the biodegradation of hydrocarbons. A productive bioremediation requires enzymatic attack by microorganisms to convert contaminants into less toxic substances. Environmental parameters should be optimum to help the microorganisms to grow and degrade the pollutants at a faster rate (de la Cueva et al., 2016). The utilization of living organisms to reducing contaminants was primarily formed to reduce pollution of crude oil (Juwarkar et al., 2010).

Currently, there are more than 70 known genera of oil-degrading microorganisms, each capable of breaking down a specific group of molecules. They include bacteria such as *Achromobacter, Acinetobacter, Bacillus* and fungus-like *Allescheria, Aspergillus, Candida* and many others which are widely distributed in the soil (Joo et al., 2008; Rufino et al., 2013; Zanaroli et al., 2010). These are either native to the polluted soils mostly due to situations of severe contamination and may need only biostimulation to stimulate active growth of degraders as in the case of the Exxon Valdez oil clean up (Lindstrom et al., 1991). Others may need additional BA to enrich the overall capabilities of the degrading community of microorganisms, as has been reported successful in several studies (Zhang et al., 2010; Joo et al., 2008; Xu and Lu, 2010).

There are certain restrictions to this technology also, for example, chlorinated hydrocarbons or other high aromatic hydrocarbons are almost persistent to microbial degradation or are degraded at a really slow pace. But bioremediation techniques are somehow economical and can be widely implemented. Most of the techniques in bioremediation are aerobic, but anaerobic processes are also being developed to help degrade pollutants in oxygen deficit areas (Franchi et al., 2016).

12.3.1 Principle of Bioremediation

Crude oil is a complex mixture of many chemical substances. As the chemical constituent of all types of oil is different, there are many methods to reduce them with the help of microorganisms and plants. Bioremediation can happen in nature or also stimulated by the use of microorganisms and fertilizers.

The microorganisms in the soil first identify the oil components by producing biosurfactants and bioemulsifiers, then bind with hydrocarbon and utilize them as food (National Research Council). The less solubilization and adsorption of macro hydrocarbons resist their use by microorganisms. The use of biosurfactants increases the solubilization and discharge of such pollutants, enhancing the speed of oil degradation biodegradations (http://www.pollutionissues.com/Na-Ph/Petroleum.html).

The components of oil vary, specifically in volatility, volubility, and vulnerability to biodegradation. Some substances are easily broken down, some have certain limitation, and some cannot be broken down (Mukred et al., 2008). The biodegradation of various petroleum substances occurs instantaneously but at a different speed because various types of microorganisms preferably boust various substances. This results in the sequential loss of each constituent of petroleum over time (ISU).

Microbes form enzymes on the availability of carbon compounds which are accountable for breaking the hydrocarbon molecules. Mostly, various enzymes and metabolic pathways are used to break hydrocarbons present in petroleum. However, unavailability of the proper enzyme either limits breakdown or will behave as a resistance to the accomplishment of hydrocarbon degradation (Brubaker, n.d.).

12.3.2 Role of Oxygen and Other Factors in Bioremediation

Oxygen (O_2) is the most widely used electron receptor in aerobic bioremediation to help the degradation of pollutants (Dmytrenko, 2007). Even though oxygen is not the speed-limiting factor, it's one of many compulsory elements of hydrocarbon degradation. Organic and inorganic pollutants are broken down faster under aerobic condition because it enhances the performance of the majority of the microorganisms (Khorasanizadeh, 2014; Levi et al., 2014; Sanscartier et al., 2011). Although Merlin Christy et al. (2014) argued that the anaerobic process is more cost-effective, it is, however, only realistic in bioreactors and not as effective in field application, which is the contemporary challenge in bioremediation. A recent study by Walworth et al. (2013) attempted to develop optimal O_2 level for bioremediation. This was done by investigating the efficacy of different O_2 levels' terminal electron acceptor depletion and hydrocarbon degradation in an oil-contaminated soil. The result shows that degradation declined with increased consumption of O_2 during the entire processes. Moreover, O_2 levels above or below 10.4% tend to impede degradation of hydrocarbons. A recent study by Sihag and Pathak (2014) reported a similar percentage but further reported a range 10% to 40% as the optimum hydrocarbon oxygen requirement. Though the microbes occur in polluted soil, they are not present in the proper numbers needed for degradation. Their growth and activity must be monitored. Carbon is the major element needed by living organisms, while microbes also require major elements such as nitrogen and phosphorous to ensure useful degradation of the oil. The maximum nutrient stability needed for hydrocarbon degradation is Carbon: Nitrogen: Phosphorus equals 100:10:4. A research article by Mori et al. (2013) assesses whether a cost-effective procedure can be developed using biostimulation in soils with high volumes of macropores. The results indicated higher bioremediation in the unsaturated condition credited to non-clogging due to macropores and presence of air. Micropores might have reduced the toxic effects of contaminant concentration on the degrading microorganisms (Gogoi et al., 2003).

Other factors affecting bioremediation include bioavailability, pH, porosity, permeability, contaminant concentration and toxicity, temperature, and mineral nutrients (Khorasanizadeh, 2014; Sihag and Pathak, 2014).

12.3.3 BIODEGRADATION BY USING PGPR

Plant growth-promoting rhizobacteria (PGPR) are a group of rhizospheric bacteria that efficiently colonize the rhizosphere of plants and stimulates plant growth either directly and indirectly. The use of plant growth-promoting rhizobacteria (PGPR) accompanying plants for bioremediation has appeared as an auspicious field, though some field studies have been accomplished (Zhuang et al., 2007).

PGPR can benefit plant development by utilizing a broad spectrum of processes, such as the production of plant growth-stimulating substances, enhancing the intake of nutrients, and behaving as a biocontrol agent by decreasing plant pathogens in the surrounding of roots. Although, plant growth-promoting microorganisms could participate in the degradation method by many pathways (Amezcua-Allieri et al., 2010) because these microorganisms can both degrade and/or mineralize organic xenobiotic compounds, permitting them to act directly as pollutant degraders (Cherian and Oliveira, 2005) or along plants (Zhuang et al., 2007; Liu et al., 2013).

In fact, the mutualistic interaction of rhizospheric microbes and plant results in enhancing the supply of hydrophobic substances, stimulating their removal and/or degradation (Macková et al., 2007). However, during microbe-assisted phytoremediation, the plant growth-promoting microbial strains could stimulate the resistance of plants to pollutants, both stimulating plant germination and root biomass concentration with the help of phytohormone production (Huang et al., 2004) and breaking the xenobiotic substances before they harmfully affect the crops (Liu et al., 2007).

Indeed, bacteria have developed many methods to reduce the low solubility of PAHs, like bioemulsifier manufacture, membrane carriers, and the potential to grow as biofilms on hydrophobic surfaces (Grosser et al., 2000).

In fact, biofilms are an association of surface-attached microbes which are closed in an extracellular polymeric matrix composed of mainly polysaccharide material (Sutherland, 2001). They are supposed to be used as dominant microbes in an ecosystem (Davey and O'Toole, 2000; Donlan, 2002) and permit the bacteria to appreciate many benefits, particularly enhance susceptibility to antimicrobial substances (Costerton et al., 1999; Chen and Wen, 2011).

However, the utilization of isolates from other environment is not very useful due to the dependence of microbes' activity on certain factors, including the situation of the area, native microbes and amount, type, availability to microbes, and piousness of contaminants in soil (Megharaj et al., 2011). Furthermore, the addition of microbes, the period of availability, and method rely on the change in microbes' community concerning the composition and catabolic activity (Andreoni et al., 2004).

12.3.4 PHYTOREMEDIATION: BIOREMEDIATION BY PLANTS

Phytoremediation is an emerging trend of using plants for the reduction and removal of contaminants from the terrestrial and aquatic environment. It is a cost-effective remedial method utilizing sunlight by plants. It is quite effective for an area that is not very deep having fewer contaminants and where plant usage is the only remedial approach. Phytoremediation is gaining imputes owing to its benefits and resistance where microbes are no longer available. The utilization of such a method is normally an in situ method, leading to less damage to nearby areas and reducing the risk of contaminants. Exclusion of transport is another benefit of this method.

Microbial digestion of complex hydrocarbons is limited and has met a low amount of success, but the process of vegetation-based biodegradation shows a high potential for accumulating, immobilizing, and transforming persistent hydrocarbon pollutants. Plants act as filters and metabolize complex compounds for its growth.

12.4 TYPES OF PHYTOREMEDIATION TECHNIQUES BASED ON THE FATE OF CONTAMINANT

12.4.1 PHYTOEXTRACTION OR PHYTOACCUMULATION

This process is used by plants for the accumulation of contaminants in the roots and shoots or leaves. Usually, metals from the contaminated sites are accumulated that can either be taken up by the plants or transported for disposal.

12.4.2 Phytotransformation or Phytodegradation

This process is characterized by the uptake of the organic pollutants from soil or water or sediments, and then they are transformed into a more stable and less toxic state. It also transforms them into a less mobile state so that they do not spread to other locations.

12.4.3 Phytostabilization

This is a technique where the plants curtail the mobility and migration of pollutants; those that are leachable are absorbed into the plant cavity and bound so that they can form a less toxic stable mass.

12.4.4 Rhizodegradation or Phytodegradation

This is a process of degradation or breakdown of contaminants through the activity in the rhizosphere. The proteins and enzymes secreted by plants and microbes such as bacteria, yeast, and fungi help to degrade the contaminants. It also represents the symbiotic relationship between plants and microorganisms, where the plant provides shelter and nutrients and microorganisms help break complex compounds in the soil.

12.4.5 Rhizofiltration

This is a water-based technique of remediation where plants take up contaminants by its roots; it is very useful for contamination of wetlands and estuaries.

12.5 BIOREMEDIATION STRATEGIES

12.5.1 Bioaugmentation

This is a technique which uses native microbes or genetically modified microbes in the polluted area and remediation starts. These microbes help in the degradation of contaminants, and the growth of already present microbes can be stimulated. This technique is quite useful where indigenous microbes are unable to degrade individually.

12.5.2 Biostimulation

This is where the indigenous microorganisms are stimulated to grow with the addition of growth factors like nutrients. Occasionally, the proper removal of contaminants does not occur by native microbes. Effective remediation is not possible for indigenous microorganisms in normal circumstances; thus, they have to be stimulated by optimizing the surrounding environment of the contaminated site. So, by adding nutrient, oxygen, electron acceptor, the existing population is stimulated. The stimulants are added to the subsurface through injection wells.

12.5.3 Land Farming

This is a process where oil-contaminated soil is taken, and then it is spread over a bed, where it is stimulated by fertilizers to maximize the activity of microorganisms and enhance their chances of hydrocarbon degradation. The soil is rotated at a regular interval. The site specifications are a 3-feet distance between the surface and the groundwater table, the slope of the land not exceeding 8%. This is an ex situ method of bioremediation

12.5.4 Composting

This is an ex situ method of bioremediation. In this method, the contaminated soil is added to organic substances like agricultural waste and manure. These, in return, stimulate the microorganisms to degrade the hydrocarbon pollutants.

12.5.5 Anaerobic Degradation

The process of adding urea and ammonia, like fertilizers, is used in the bioremediation of oil spills; in such a method, oxygen demand can go high due to the oxidation of ammonia. There may be sites where the addition of oxygen is not feasible or not possible; thus, in these sites, those microorganisms are used which can grow in anaerobic environments and are used for bioremediation of oil-spill-contaminated sites.

12.6 CONCLUSION

Crude oil contamination is one of the major global concerns today. There are different physical and chemical methods to treat crude oil-contaminated soil, but these methods have several drawbacks. One eco-friendly and emerging method to reduce crude oil contamination is bioremediation. Bioremediation is based on the use of microbes and plants for treating contaminated soil. Plant growth-promoting rhizobacteria use different strategies to reduce hydrocarbon contamination, such as biosurfactant production, and act as bioemulsifier and make soil favorable for plant growth. Therefore, the use of plant growth-promoting rhizobacteria to treat crude oil-contaminated soil is one promising strategy to enhance crop production.

REFERENCES

Amezcua-Allieri, M.A., Rodríguez-Dorantes, A., Meléndez-Estrada, J. (2010) The use of biostimulation and bioaugmentation to remove phenanthrene from soil. *International Journal of Oil, Gas and Coal Technology* 3:39–59.

Andreoni, V., Cavalca, L., Rao, M.A. (2004) Bacterial communities and enzyme activities of PAHs polluted soils. *Chemosphere* 57:401–412.

ATSDR. (1995) *Toxicological Profile for Polycyclic Aromatic Hydrocarbons.* Atlanta, GA: Agency for Toxic Substances and Disease Registry.

Brubaker, T. (n.d.) *Bioremediation of Petroleum-Contaminated Soils and Wetlands by Indigenous Microorganisms and Bioaugmentation.* Retrieved from http://home.eng.iastate.edu/~tge/ce421-521/brubaker.pdf (accessed on 27th March 2011).

Cerniglia, C.E. (1992) Biodegradation of polycyclic aromatic hydrocarbons. *Biodegradation* 3:351–368.

Chen, L., Wen, Y. (2011) The role of bacterial biofilm in persistent infections and control strategies. *International Journal of Oral Science* 3:66–73.

Cherian, S., Oliveira, M.M. (2005) Transgenic plants in phytoremediation: Recent advances and new possibilities. *Environmental Science and Technology* 39:9377–9390.

Costerton, J.W., Stewart, P.S., Greenberg, E.P. (1999) Bacterial biofilms: A common cause of persistent infections. *Science* 284:1318–1322.

Das, N., Chandran, P. (2011) Microbial degradation of petroleum hydrocarbon contaminants: An overview, SAGE-Hindawi access to research. *Biotechnology Research International* 1–13.

Davey, M.E., O'Toole, G.A. (2000) Microbial biofilms: From ecology to molecular genetics. *Microbiology and Molecular Biology Reviews* 64:847–867.

de la Cueva, S.C., Rodríguez, C.H., Cruz, N.O.S., Contreras, J.A.R., Miranda, J.L. (2016) Changes in bacterial populations during bioremediation of soil contaminated with petroleum hydrocarbons. *Water, Air, and Soil Pollution* 227:1–12.

Dmytrenko, G.M. (2007) Regularities in the oxidising metabolism of bacteria. In *Bioremediation of Soil Contaminated with Aromatic Compounds*, Heipieper, H.J. (ed.). Amsterdam, the Netherlands: IOS Press, 51–57.

Donlan, R.M. (2002) Biofilms: Microbial life on surfaces. *Emerging Infectious Diseases* 8:881–890.

Fan, T., Buckley, J.S. (2002) Rapid and accurate SARA analysis of medium gravity crude oils. *Energy and Fuels* 16:1571–1575.

Franchi, E., Agazzi, G., Rolli, E., Borin, S., Marasco, R., Chiaberge, S., Barbafieri, M. (2016) Exploiting hydrocarbon-degrader indigenous bacteria for bioremediation and phytoremediation of a multi-contaminated soil. *Chemical Engineering and Technology* 39:1676–1684.

Gaspar, A., Zellermann, E., Lababidi, S., Reece, J., Schrader, W. (2012) Characterization of saturates, aromatics, resins, and asphaltenes heavy crude oil fractions by atmospheric pressure laser ionization Fourier transform ion cyclotron resonance mass spectrometry. *Energy and Fuels* 26:3481–3487.

Gogoi, B.K., Dutta, N.N., Goswami, P., Krishna Mohan, T.R. (2003) A case study of bioremediation of petroleum-hydrocarbon contaminated soil at a crude oil spill site. *Advances in Environmental Research* 7:767–782.

Grin'ko, A.A., Min, R.S., Sagachenko, T.A., Golovko, A.K. (2012) Aromatic sulfur-containing structural units of resins and asphaltenes in heavy hydrocarbon feedstock. *Petroleum Chemistry* 52:221–227.

Grosser, R.J., Friedrich, M., Ward, D.M., Inskeep, W.P. (2000) Effect of model sorptive phases on phenanthrene biodegradation: Different enrichment conditions influence bioavailability and selection of phenanthrene-degradingisolates. *Applied and Environmental Microbiology* 66:2695–2702.

Harayama, S., Kishira, H., Kasai, Y., Shutsubo, K. (1999) Petroleum biodegradation in marine environments. *Journal of Molecular Microbiology and Biotechnology* 1:63–70.

Huang, X., El-Alawi, Y., Penrose, D.M., Glick, B.R., Greenberg, B.M. (2004) A multi-process phytoremediation system for removal of polycyclic aromatic hydrocarbons from contaminated soils. *Environmental Pollution* 130:465–476.

IARC. (2010) *Monographs on the Evaluation of the Carcinogenic Risk of Chemicals to Humans, 92, Some Non-heterocyclic Polycyclic Aromatic Hydrocarbons and Some Related Exposures.* Lyon, France: International Agency for Research on Cancer Press.

ISU Engineering Information Technology.

Jain, P.K., Bajpai, V. (2012) Biotechnology of bioremediation – A review. *International Journal of Environmental Sciences* 3:535–549.

Joo, H., Ndegwa, P.M., Shoda, M., Phae, C. (2008) Bioremediation of oil-contaminated soil using *Candida catenulata* and food waste. *Environmental Pollution* 156:891–896.

Juwarkar, A.A., Singh, S.K., Mudhoo, A. (2010) A comprehensive overview of elements in bioremediation. *Reviews in Environmental Science and BioTechnology* 9: 215–288.

Khorasanizadeh, Z. (2014) *The Effect of Biotic and Abiotic Factors on Degradation of Polycyclic Aromatic Hydrocarbons (PAHs) by Bacteria in the Soil.* Ph.D. Thesis, University of Hertfordshire, UK.

Levi, S., Hybel, A., Bjerg, P.L., Albrechtsen, H. (2014) Stimulation of aerobic degradation of bentazone, mecoprop and dichlorprop by oxygen addition to aquifer sediment. *Science of the Total Environment* 473–474:667–675.

Lindstrom, J.E., Prince, R.C., Clark, J.C., Grossman, M.J., Yeager, T.R., Braddock, J.F. Brown, E.J. (1991) Microbial populations and hydrocarbon biodegradation potentials in fertilized shoreline sediments affected by the T/V Exxon Valdez oilspill. *Applied and Environmental Microbiology* 57:2514–2522.

Liu, L., Jiang, C., Liu, X., Wu, J., Han, J., Liu, S. (2007) Plant-microbe association for rhizoremediation of chloronitroaromatic pollutants with *Comamonas* sp. strain CNB-1. *Environmental Microbiology* 9:465–473.

Liu, W., Sun, J., Ding, L., Luo, Y., Chen, M., Tang, C. (2013) Rhizobacteria (*Pseudomonas* sp. SB) assist phytoremediation of oily-sludge-contaminated soil by tall fescue (*Festuca arundinacea* L.). *Plant and Soil* 371:533–542.

Macaulay, B.M. (2015) Understanding the behaviour of oil-degrading micro-organisms to enhance the microbial remediation of spilled petroleum. *Applied Ecology and Environmental Research* 13:247–262.

Macková, M., Vrchotová, B., Francová, K. (2007) Biotransformation of PCBs by plants and bacteria e consequences of plant-microbe interactions. European *Journal of Soil Biology* 43:233–241.

Macleod, C.J.A., Semple, K.T. (2002) The adaptation of two similar soils to pyrene catabolism. *Environmental Pollution* 119:357–364.

Marshall, A.G., Rodgers, R.P. (2004) Petroleomics: The next grand challenge for chemical analysis. *Accounts of Chemical Research* 3:53–59.

Marshall, A.G., Rodgers, R.P. (2008) Petroleomics: Chemistry of the underworld. *Proceedings of the National Academy of Sciences of the United States of America* 105:18090–18095.

Megharaj, M., Ramakrishnan, B., Venkateswarlu, K., Sethunathan, N., Naidu, R. (2011) Bioremediation approaches for organic pollutants: A critical perspective. *Environment International* 37:1362–1375.

Merlin Christy, P., Gopinath, L.R., Divya, D. (2014) A review on anaerobic decomposition and enhancement of biogas production through enzymes and microorganisms. *Renewable and Sustainable Energy Reviews* 34:167–173.

Mori, Y., Suetsugu, A., Matsumoto, Y., Fujihara, A., Suyama, K. (2013) Enhancing bioremediation of oil-contaminated soils by controlling nutrient dispersion using dual characteristics of soil pore structure. *Ecological Engineering* 51:237–243.

Mukred, M.A., Hamid, A.A., Hamzah, A., Yusoff, W.M.W. (2008) Development of three bacteria consortium for the bioremediation of crude petroleum-oil in contaminated water. *Online Journal of Biological Sciences* 8:1608–4217.

National Research Council, op. cit., footnote 1:17.

Pollution Issues. (n.d.) Petroleum. Retrieved from http://www.pollutionissues.com/Na-Ph/Petroleum.html, (accessed on 27th March 2011).

Rufino, R.D., Luna, J.M., Marinho, P.H.C., Farias, C.B.B., Ferreira, S.R.M., Sarubbo, L.A. (2013) Removal of petroleum derivative adsorbed to soil by biosurfactant Rufisan produced by Candida lipolytica. *Journal of Petroleum Science and Engineering* 109:117–122.

Salleh, A.B., Ghazali, F.M., Rahman, R. and Basri, M. (2003) Bioremediation of petroleum hydrocarbon pollution. *Indian Journal of Biotechonolgy* 2:411–425.

Sanscartier, D., Reimer, K., Zeeb, B., Koch, I. (2011) The effect of temperature and aeration rate on bioremediation of diesel-contaminated soil in solid-phase Bench-scale bioreactors. *Soil and Sediment Contamination* 20:353–369.

Sihag, S. and Pathak, H. (2014) Factors affecting the rate of biodegradation of polyaromatic hydrocarbons. *International Journal of Pure Applied Bioscience* 2:185–202.

Speight, J.G. (2010) *The Chemistry and Technology of Petroleum*. 4th edn. Taylor & Francis.

Sutherland, I.W. (2001) Biofilm exopolysaccharides: A strong and sticky framework. *Microbiology* 147:3–9.

Sverdrup, L.E., Krogh, P.H., Nielsen, T., Kjær, C., Stenersen, J. (2003) Toxicity of eight polycyclic aromatic compounds to red clover (*Trifolium pratense*), ryegrass (*Loliumperenne*), and mustard (*Sinapsisalba*). *Chemosphere* 53:993–1003.

Walworth, J., Harvey, P., Snape, I. (2013) Low temperature soil petroleum hydrocarbon degradation at various oxygen levels. *Cold Regions Science and Technology* 96:117–121.

Wang, J., Zhang, Z.Z., Su, M.Y., He, W., He, F., Song, G.H. (2008) Phytoremediation of petroleum polluted soil. *Petroleum Science* 5:167.

WHO. (2000) *Air Quality Guidelines*. 2nd edn. Copenhagen, Denmark: WHO Regional Office for Europe.

Xu, Y., Lu, M. (2010) Bioremediation of crude oil-contaminated soil: Comparison of different biostimulation and bioaugmentation treatments. *Journal of Hazardous Materials* 183:395–401.

Zanaroli, G., Toro, Di., Todaro, S., Varese, D.G.C., Bertolotto, A., Fava, F. (2010) Characterization of two diesel fuel degrading microbial consortia enriched from a non-acclimated, complex source of microorganisms. *Microbial Cell Factories* 9:233–239.

Zhang, Z., Gai, L., Hou, Z., Yang, C., Ma, C., Wang, Z., Sun, B., He, X., Tang, H., Xu, P. (2010) Characterization and biotechnological potential of petroleum-degrading bacteria isolated from oil-contaminated soils. *Bioresource Technology* 101:8452–8456.

Zhuang, X., Chen, J., Shim, H., Bai, Z. (2007) New advances in plant growth-promoting rhizobacteria for bioremediation. *Environment International* 33:406–413.

13 Abiotic Stress-Mediated Oxidative Damage in Plants
An Overview

Ruchi Rai, Shilpi Singh, Shweta Rai, Alka Shankar, Antra Chatterjee, and L.C. Rai

CONTENTS

13.1 INTRODUCTION

Abiotic stresses such as drought, temperature, radiation, salinity, heavy metals/metalloid and nutrient deprivation cause physiological and metabolic disorders and adversely affect plant phenological and developmental processes. It is an unavoidable limiting factor for agriculture, causing more than 50% of yield reduction (Acquaah, 2007; Hasanuzzaman et al., 2012) and is becoming a major problem in the modern era. One of the key consequences of abiotic stress is oxidative stress (Hasanuzzaman et al., 2013). Impaired stomatal conductance, disruption of the photosynthetic apparatus or of pigments, malfunctioning of the Calvin cycle and photosystem, inactivation of the enzymes of photosynthesis including RuBisCO, reductions in carboxylation reaction efficiency, electron transport chain (ETC) efficiency, regeneration of $NADP^+$ and increased photorespiration are some of the major reasons for the overproduction of reactive oxygen species (ROS) under abiotic stress (Hasanuzzaman et al., 2012; Tsukahara et al., 2015). Methylglyoxal (MG) is a highly reactive, dicarbonyl ketoaldehyde cytotoxin generated as a by-product of several metabolic pathways such as glycolysis, and it can be produced from photosynthesis intermediates glyceraldehyde-3-phosphate and dihydroxyacetone phosphate. MG increases oxidative stress by catalyzing the photoreduction of O_2 to O^{2-} at photosystem I (Saito et al., 2011). Both ROS and MG are highly reactive and capable of a complete disruption of cellular functions, including the lipid peroxidation, the oxidation of protein, the oxidation of fatty acids and the disruption of biomembrane structures and functions (Chaplen, 1998; Gill and Tuteja, 2010). Excessively produced ROS in plants has been scavenged by antioxidant defense systems composed of a range of non-enzymatic (ascorbic acid, glutathione (GSH), phenolic compounds, alkaloids, non-protein amino acids and α-tocopherols) and enzymatic components (superoxide dismutase (SOD), catalase (CAT), ascorbate peroxidase (APX), glutathione reductase (GR), monodehydroascorbate reductase (MDHAR), dehydroascorbate reductase (DHAR), glutathione peroxidase (GPx) and glutathione *S*-transferase (GST)) (Gill and Tuteja, 2010). In contrast, MG is detoxified into D-lactate via the glyoxalase system composed of glyoxalase I (gly I) and glyoxalase II (gly II) (Yadav et al., 2005). This process requires reduced GSH to act as a cofactor which recycled back at the end of the reaction. Plants develop strong defense mechanisms against different abiotic stresses which include an increase in ROS and cytosolic Ca^{2+}, activation of kinase cascades, increased concentrations of hormones such as salicylic acid (SA), jasmonic acid (JA), abscisic acid (ABA) and ethylene (Hirayama and Shinozaki, 2010). Reactive oxygen species, including hydrogen peroxide (H_2O_2), superoxide radical ($O^{2\bullet-}$), hydroxyl radical (OH•) and singlet oxygen ($^1O_2$) etc., resulting from excitation or incomplete reduction of molecular oxygen, are harmful by-products of basic cellular metabolism in aerobic organisms (Apel and Hirt, 2004; Miller et al., 2010). Under stress conditions, cell homeostasis is disrupted, leading to an imbalance in redox status, which further results in the increase of ROS concentration, activating the antioxidant defense mechanisms of plants for capturing excess ROS (Mittler et al., 2004). These include the involvement of the ascorbate–glutathione cycle enzymes, which are major antioxidative defense molecules, and of other antioxidant enzymes. This chapter mainly focuses on the involvement of ROS and MG in abiotic stress-mediated oxidative damage in higher plants and the coordinated role of antioxidative enzymes, glyoxalase systems and signaling pathways employed by plants to cope with it. *Omics* tool is also illustrated as an emerging approach to induce abiotic stress tolerance in plants via targeting specific genes, metabolites and proteins.

13.2 REACTIVE OXYGEN SPECIES AND METHYLGLYOXAL GENERATION: FIRST RESPONSE TOWARDS ABIOTIC STRESS

Plants continuously produce ROS as by-product of various metabolic pathways in different cellular compartments like chloroplast, mitochondria, peroxisome and apoplast (Dietz et al., 2016; Gilroy et al., 2016; Huang et al., 2016; Kerchev et al., 2016; Rodrıguez-Serrano et al., 2016; Takagi et al., 2016). In addition, ROS are also overproduced in non-photosynthetic tissues such as roots, hypocotyls or coleoptiles under stress. The equilibrium between production and scavenging of ROS via antioxidant defense systems may be disturbed by a number of adverse climatic factors. The rapid rise in the intracellular levels of ROS generates imbalance inside the cell, leading to oxidative stress. Higher plants have thus acquired diverse pathways to protect themselves from ROS toxicity, as well as to use ROS as signaling molecules (Foyer and Noctor, 2013; Vaahtera et al., 2014; Considine et al., 2015; Dietz, 2015; Mignolet-Spruyt et al., 2016). This incessant process of ROS production and ROS scavenging occurs at all cellular compartments of the cells and is controlled by the ROS gene network (Mittler et al., 2004). If kept unchecked, ROS concentrations will increase in cells and cause oxidative damage to membranes (lipid peroxidation), proteins, RNA and DNA molecules, and can even lead to the oxidative destruction of the cell in a process termed oxidative stress (Mittler, 2002). Plants also generate ROS by activating various oxidases and peroxidases that produce ROS in response to certain environmental changes (Bolwell et al., 1998, 2002; Doke, 1985; Schopfer et al., 2001). Under most stresses, inadequate energy indulgence in photosynthesis and excitation or incomplete reduction of molecular oxygen yields ROS including hydrogen peroxide (H_2O_2), superoxide radical ($O_2^{\bullet-}$), hydroxyl radical (OH•) and singlet oxygen ($^1O_2$), which are potentially hazardous and create oxidative stress. During this process, ground state oxygen may be converted to more reactive ROS forms either by energy transfer or by electron transfer reactions. The former leads to the formation of singlet oxygen, whereas the latter results in the sequential reduction to superoxide, hydrogen peroxide and hydroxyl radical (Figure 13.1; Scandalios, 2005).

Reduction of photosynthetic pigment content, inhibition of photochemistry efficiency or the uneven function of photosystem II (PS II), inhibition of biochemical metabolism or enzymatic activities are a few common consequences associated with excess ROS generation under various abiotic stresses (Nahar et al., 2015; Rajwanshi et al., 2016). However, the mode of ROS

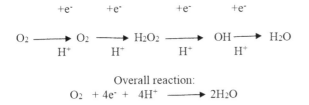

$$O_2 \xrightarrow[H^+]{+e^-} O_2 \xrightarrow[H^+]{+e^-} H_2O_2 \xrightarrow[H^+]{+e^-} OH \xrightarrow[H^+]{+e^-} H_2O$$

Overall reaction:

$$O_2 + 4e^- + 4H^+ \longrightarrow 2H_2O$$

FIGURE 13.1 Pathways in the univalent reduction of oxygen to water leading to generation of various intermediate reactive oxygen species (ROS).

production varies with stress types. Under salinity and drought stress, plants avoid excess water loss and reduce the stomatal conductance. Hence, internal CO_2 concentration decreases, causing less reduction of CO_2 by the Calvin cycle, which further hinders carbon fixation. Due to excessive excited energy and impaired electron transport procedures in the chloroplasts and mitochondria of plant cells, production of different ROS is increased (Gill and Tuteja, 2010). This excess generation of ROS during salt or drought stress results from the reduction of the activity in PS II, resulting in alterations in quantum yield. Under high-temperature stress, RuBisCO is able to generate H_2O_2 via oxygenase reactions (Kim and Portis, 2004). On the other hand, the solubility of a gas in plant cells increases under cold stress. As a result, O_2 concentration increases in cells and raises the threat of oxidative damage at low stress, which leads to the enhancement of ROS. Under waterlogging conditions, the photosynthetic ETC becomes over-reduced, leading to the generation of several ROS. Heavy metal (HM) stress affects the H_2O oxidizing system of PS II as they replace Ca^{2+} and Mn^{2+} ions in the PS II reaction center, thus hindering the reaction of PS II, which results in the uncoupling of the electron transport in the chloroplast. The redox-active HMs such as Fe, Cu, Cr, V are involved in the formation of OH^- from H_2O_2 via Haber–Weiss and Fenton reactions and initiate non-specific lipid peroxidation (Sharma and Dietz, 2008). Reactive oxygen species overproduction and the occurrence of oxidative stress in plants are the indirect consequence of cadmium (Cd) toxicity. However, the mechanism of Cd toxicity includes the interaction of Cd with the antioxidant system, causing induction of nicotinamide adenine dinucleotide phosphate (NADPH) oxidase, metabolism of essential plant nutrients (Srivastava et al., 2004; Qadir et al., 2004; Dong et al., 2006) and disruption of the ETC, leading to overproduction of ROS.

Similar to ROS, plants also produce a high amount of MG, a cytotoxic compound and by-product of glycolysis as well as the breakdown product of threonine and acetone (Thornalley et al., 2003) distributed across cell organelles like chloroplast, mitochondrion and cytosol (Kaur et al., 2015a,b). The formation of MG from triose phosphates occurs through the elimination of the phosphoryl group from 1,2-enediolate of these trioses, metabolism of aminoacetone and acetone (Kaur et al., 2014; Kalapos, 1999). The MG modifies protein, nucleic acid and basic phospholipid, producing advanced glycation end-products (Thornalley et al., 2003). This can damage cells or cell components and can even destroy DNA or cause mutation. Methylglyoxal is highly toxic, reacting with both DNA and protein and is detoxified by a two-step enzyme-catalyzed process, which shows some similarities to the formaldehyde dehydrogenase system. The two enzyme glyoxalase systems have been identified in plants as well as other living organisms to protect DNA and protein by converting MG into D-lactate. However, under abiotic stress, MG concentration increases at a rate that is usually higher than the rate of detoxification by the glyoxalase system. A higher accumulation of MG results in the inhibition of germination and cell proliferation and causes the glycation of proteins, disruption of the antioxidant defense system and other metabolic dysfunctions (Kaur et al., 2014; Hoque et al., 2016). Furthermore, MG inhibits the synthesis of protein and nucleic acid, causes a reduction in the growth of root and shoot and inhibition of photosynthesis. MG hampers photosynthesis by inactivating the CO_2 photoreduction by 17% (Mano et al., 2009). The major route for MG detoxification is through the glyoxalase system, universally present in mammals, yeasts, bacteria and plants (Hoque et al., 2016; Hossain et al., 2009). The glyoxalase enzymes viz. gly I and gly II act coordinately to detoxify MG by converting it into a nontoxic product using GSH as a cofactor. In conjunction with glyoxalase systems, MG can be detoxified via some minor routes via enzymes such as aldose/aldehyde reductase (ALR) or aldo-keto reductase (AKR). These enzymes are involved in oxido-reductions and detoxify MG by reducing it to α-oxoaldehyde. Turóczy et al. (2011) reported that overproduction of a rice aldo-keto reductase, OsAKR1, increases oxidative and heat stress tolerance by malondialdehyde and methylglyoxal detoxification. These minor routes of MG detoxification suggest the role of ALRs and AKRs in the reduction of MG and other MG-like aldehyde generated during stress (Hossain et al., 2011). The mechanism of abiotic stress-induced oxidative damage through the generation of ROS and MG and their consequent effects are summarized in Figure 13.2.

ROS lead to the unspecific oxidation of proteins and membrane lipids or may cause DNA injury. The tissue damaged by oxidative stress generally contain an elevated concentration of carbonylated proteins and

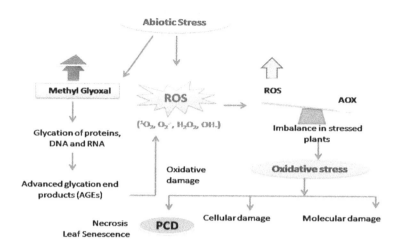

FIGURE 13.2 Abiotic stress creates an imbalance between reactive oxygen species (ROS) and antioxidants (AOX) in plants, causing oxidative stress, which subsequently leads to programmed cell death (PCD).

malondialdehyde and show an increased production of ethylene (Dean et al., 1993; Ames et al., 1993). ROS react with a large variety of biomolecules and may thus cause irreversible damage that can lead to tissue necrosis and may ultimately kill the plants. On the other hand, ROS influence the expression of a number of genes and signal transduction pathways. Recent studies have shown that under stress conditions chloroplast, peroxisomes and mitochondria can extend membrane structures (stromules, peroxides and matrixules, respectively) that will contact the nuclear envelope and could directly alter the ROS status of the nuclei (Noctor and Foyer, 2016). However, if labile Fe^{2+} exists in cells, ROS such as H_2O_2 can react with it to generate the highly toxic hydroxyl radical that would lead to oxidative stress and cell damage. The production of $_1O^2$ in chloroplasts can also cause reprogramming of nuclear gene expression, leading to chlorosis and programmed cell death (PCD), as well as induce a wide range of responses related to abiotic stresses through the function of EXECUTER1 (EX1) and EX2, two nuclear-encoded chloroplast proteins associated with thylakoid membranes (Wagner et al., 2004; Lee et al., 2007a; Kleine and Leister, 2016). Abiotic stress-induced PCD has the potential to significantly influence crop yield (Mittler and Blumwald, 2010). The severity of oxidative damage and senescence depends on the developmental stage of the plants. The mechanisms leading to and controlling PCD in response to adverse conditions in plants involve ROS as signaling mediators (Gechev and Hille, 2005). Programmed cell death is a genetically controlled process in which cells are selectively eliminated in a highly coordinated manner through the involvement of specific proteases and nucleases. Thus, only cells that are destined to die are

destroyed while neighboring cells remain unaffected. Although PCD is vital for normal growth and development of plants, it can be a consequence of severe abiotic stresses. Plants respond to various abiotic stresses in a different manner that depends on the nature of the stress factors. Although the mechanism of induction of ROS production (Jaspers and Kangasjärvi, 2010) is common for different stress factors, the utilized signal transduction networks, recruited transcription factors (TFs) and modulated genes are specific for each kind of stress.

13.2.1 Drought-Induced PCD

Drought is among the abiotic factors that most strongly affect plant productivity and cause serious losses to crop yield (Godfray et al., 2010; Tester and Langridge, 2010). It usually occurs in combination with heat, and their simultaneous action has a synergistic detrimental effect on plant productivity (Rampino et al., 2012). A prominent example of drought-induced increase in ROS concentrations and the subsequent PCD is microspore abortion resulting in male sterility (De Storme and Geelen, 2014). ROS-triggered PCD in response to drought can also be exemplified by triggered PCD in selected leaves during leaf senescence which allows nutrient remobilization. Strategies to restrict ROS-induced programmed cell death include several mechanisms:

- Downregulation of synthesis of chlorophyll and other components of the photosynthetic machinery (Farrant et al., 2007)
- Accumulation of compatible solutes like sucrose and raffinose which enhance the yield of reducing power (NADPH) required for the synthesis

of many ROS scavenging, non-enzymatic anti-oxidants (Bolouri-Moghaddam et al., 2010)

- Protection from oxidative stress by overexpressing galactinol synthase where elevated levels of galactinol and raffinose can scavenge HO• (Nishizawa et al., 2008) directly and protect plants from oxidative stress.

13.2.2 SALT-INDUCED PCD

During osmotic stress, NADPH oxidases and apoplastic diamine and PA oxidases are activated in mitochondrial respiration, which elevates ROS concentration (Abogadallah, 2010). Salt-induced PCD is strongly affected by ion disequilibrium resulting from the high concentration of Na^+ into the cytosol accompanied by a deficiency of K^+ (Shabala, 2009; Kim et al., 2014). The efflux of K^+ is influenced by hydroxyl radicals and in turn regulates different enzymes involved in PCD (Demidchik et al., 2010). Another mode of modulating the stress and PCD responses is through cysteine protease inhibitors (cystatins). The ROS minimization under salt stress is usually achieved by the antioxidant systems and upregulation of different antioxidants, e.g., synthesis of flavonoid compounds and increase in the carotenoid/chlorophyll ratio (Reginato et al., 2014). Another mechanism to keep ROS levels under control is the accumulation of compatible solutes like proline to keep proper osmotic potential in stressed tissues and also by alleviating oxidative stress (Liu et al., 2011; Duan et al., 2012).

13.2.3 HIGH-TEMPERATURE-INDUCED PCD

PCD induced by heat shock is triggered only when the temperature rises above a specific threshold (Marsoni et al., 2010). The major sites of ROS production during heat shock are photosynthetic apparatus and peroxisomes where ROS is generated during damage of PSII complex and as a photorespiratory product. Mitochondria also play an important role and release functionally active cytochrome c in a ROS-dependent manner. It leads to the activation of caspase-like proteases in the cytosol that execute the PCD program (Vacca et al., 2006). Rigorously heat-shock-treated cells cause downregulation of antioxidant proteins, leading to the perturbation of the redox homeostasis, and play a role in the PCD events. It has been proposed that heat shock transcription factors (HSFs) and heat shock proteins (HSPs) in coordination and other stress-related genes have also an important role in the ROS-mediated response to heat stress. Both the *HSF* and *HSP* genes are highly induced

by ROS (Gechev et al., 2005; Piterkova et al., 2013). The PCD cascade is triggered at elevated temperatures and starts with the accumulation of ROS and an increase of the cytosolic Ca^{2+} concentration. Calcium acts as a ubiquitous second messenger calmodulin (CaM3) which leads to the activation of MPK6. Finally, MPK6 upregulates γVPE expression and γVPE precursors are targeted to the vacuole, where they are activated to process a number of vacuolar hydrolases and proteases associated with PCD (Li et al., 2012). Low temperature (LT) can also induce PCD in plants on its own (Koukalova et al., 1997; Lyubushkina et al., 2014). LT is particularly damaging when combined with elevated light intensities. During such conditions, the photosynthetic electron transport chains are over-reduced, causing an increased generation of ROS leading to photoinhibition or/and cell death (Murata et al., 2007).

13.2.4 HIGH LIGHT AND HEAVY-METAL-INDUCED PCD

High light intensity induces photoinhibition by inactivating PSII complex which subsequently provokes massive accumulation of ROS. Excessively light-stressed PCD is either H_2O_2 or $^1O_2$-dependent. The propagation of excess excitation energy induced PCD in chloroplasts is dependent on the activities of several regulatory genes like *LSD1*, *EDS1*, *phytoalexin deficient 4* (*PAD4*) and *ethylene insensitive2* (*EIN2*). These protein products functioned upstream of ROS production (Muhlenbock et al., 2008; Karpinski et al., 2013) and initiated PCD by changing the redox state of the chloroplasts.

Heavy metals like Fe and Cu may participate in the Haber–Weiss reaction which leads to the production of the hyper-active hydroxyl radical ($HO^{•-}$; Kehrer, 2000) while Cd appears to trigger PCD by a pathway that involves the endoplasmic reticulum (ER), in which unfolded proteins accumulate (Xu et al., 2013). Ethylene also participates in the Cd stress since it stimulates the onset of PCD (Yakimova et al., 2006; Liu et al., 2008). In a study with pea plants, it was shown that one of the primary effects of Cd^{2+} treatment is the depletion of cytosolic Ca^{2+}, which negatively regulates components of the antioxidant system like SOD, causing an increase in H_2O_2 and $O^{•-2}$ concentrations. Reduction of Ca^{2+} further leads to the decrease in NO which can, in turn, promote the accumulation of $O^{•-2}$, further enhancing the oxidative stress. Aluminum stressed plants leads to mitochondrial dysfunction, i.e., activation of the caspase-like 3 protease or decrease in the mitochondrial enzyme alternative oxidase (AOX) through ROS burst which further induces

PCD (Liu et al., 2014; Huang et al., 2014). In a study with *Solanum nigrum* roots, the rise of NO was proven to be necessary for Zn-induced PCD as the inhibition of NO prevented ROS accumulation and subsequent cell death (Xu et al., 2010).

13.2.5 WATERLOGGING-INDUCED PCD

Flooding significantly reduced the availability of molecular oxygen which causes hypoxia or anoxia. To alleviate the symptoms related to hypo/anoxia, plants resort to a defense strategy which involves the formation of specific air channels in the submerged tissues called aerenchyma. Flooding-induced PCD is accompanied by mitochondrial dysfunction and the accumulation of metal ions, which further increase ROS concentration (Shabala et al., 2014). On the other hand, it is long known that ethylene synthesis is also enhanced in a low-oxygen environment, and, subsequently, ethylene triggers PCD (He et al., 1996a,b). According to Rajhi et al. (2011), a respiratory burst oxidase homolog (*Rboh*), which are major ROS producer genes, is known to upregulated during waterlogging in maize, subsequently leading to PCD. As in other cases of abiotic stress-induced PCD, NO may also play a role during exposure to low oxygen levels. NO is mainly produced in mitochondria, mostly by the reduction of nitrite by the electron transport chain (Gupta and Igamberdiev, 2011). Although NO is able to actually induce *AOX* expression, probably to reduce the ROS and NO burden (Fu et al., 2010), accumulation of ROS and NO from the non-mitochondrial system may occur and induce PCD (Gupta et al., 2012).

13.3 ANTIOXIDATIVE DEFENSE STRATEGIES

Plants employ various defense mechanisms to quench ROS, which includes both non-enzymatic and enzymatic antioxidant defenses. Non-enzymatic defenses include compounds of intrinsic antioxidant properties, such as vitamins C and E, glutathione and ß-carotene. The enzymic antioxidants include superoxide dismutase (SOD), catalase (CAT), enzymes of the ascorbate–glutathione (AsA-GSH) cycle such as ascorbate peroxidase (APX), monodehydroascorbate reductase (MDHAR), dehydroascorbate reductase (DHAR) and glutathione reductase (GR) (Noctor and Foyer, 1998). These antioxidative enzymes and their related isozymes are localized in different cell compartments and work in coordination upon exposure to stress (Table 13.1, Ahmad, 2014).

13.3.1 SUPEROXIDE DISMUTASE

Superoxide dismutase (SOD, EC 1.15.1.1) is ubiquitous and plays a central role in defense against oxidative stress in all aerobic organisms. The enzyme SOD belongs to the group of metalloenzymes and catalyzes the dismutation of superoxide molecules into hydrogen peroxide and oxygen (Alscher et al., 2002). It is present in most of the subcellular compartments that generate activated oxygen and has a metal cofactor, depending on which it can be classified in three different groups: (1) Fe SODs consisting of two species, one homodimer and one tetramer, found within both prokaryotes and eukaryotes. They are most abundantly localized inside plant chloroplasts, where they are indigenous. (2) Mn SODs consist of a homodimer and homotetramer species, each containing

TABLE 13.1

Important Enzymatic Antioxidative Systems, Their Function and Localization

Antioxidant Enzymes	Function	Subcellular Localization
SOD	$O_2^- + O_2^- + 2H+ \rightarrow H_2O_2 + O_2$	Cyt, Chl, Per and Mit
CAT	$2H_2O_2 \rightarrow O_2 + 2H_2O$	Per, Gly and Mit
APX	$H_2O_2 + ASC \rightarrow 2H_2O + DHA$	Cyt, Per, Chl and Mit
GR	$GSSG + NADPH \rightarrow 2GSH + NADP^+$	Cyt, Chl and Mit
MDHAR	$2MDHA + NADH \rightarrow 2ASC + NAD^+$	Chl, Mit and Cyt
DHAR	$DHA + 2GSH \rightarrow ASC + GSSG$	Chl, Mit and Cyt

APX, ascorbate peroxidase; ASC, ascorbic acid; CAT, catalase; Cyt, cytosol; Chl, chloroplast; DHA, dehydroascorbate; DHAR, dehydroascorbate reductase; Gly, glyoxisomes; GR, glutathione reductase; GSH, glutathione; GSSG, oxidized glutathione; H_2O, water; H_2O_2, hydrogen peroxide; MDHA, monodehydroascorbate; MDHAR, monodehydroascorbate reductase; Mit, mitochondria; O_2, oxygen; O_2^{-2}, superoxide radical; Per, peroxisomes; SOD, superoxide dismutase.

a single Mn (III) atom per subunit, found predominantly in mitochondrion and peroxisomes. (3) Cu/Zn-SODs are cyanide sensitive and have electrical properties very different from those of the other two classes, which are cyanide insensitive. These are concentrated in the chloroplast, cytosol and, in some cases, the extracellular space. The compartmentalization of different forms of SOD throughout the plant makes them counteract stress very effectively. All forms of SOD are nuclear encoded and targeted to their respective subcellular compartments by an amino-terminal targeting sequence (Bowler et al., 1992; Racchi et al., 2001). SOD activity has been reported to increase in plants exposed to various environmental stresses, including drought and metal toxicity (Sharma and Dubey, 2005; Borsetti et al., 2005). It was suggested that SOD can be used as an indirect selection criterion for screening drought-resistant plant materials (Zaefyzadeh et al., 2009). Overproduction of SOD has been reported to result in enhanced oxidative stress tolerance in plants (Gupta et al., 1993). SOD mRNA abundance increases whenever there is a chloroplast-localized oxidative stress, thus providing an insight into the way that each treatment affects the different subcellular compartments. Bowler and coworkers in 1991 developed transgenic tobacco plants from *Nicotiana plumbaginifolia* in which leaves of the transgenic plants showed reduced levels of membrane damage following exposure to methyl viologen (MV) and light. In addition, these plants were found to have increased protection from ozone damage (Bowler et al., 1991). Expression of the same chloroplastic Mn-SOD gene in alfalfa was found to provide increased tolerance to acifluorfen and freezing (McKersie et al., 1993, 1996).

13.3.2 ASCORBATE PEROXIDASE

Ascorbate peroxidase (APX, EC 1.1.11.1) is a central component of the AsA-GSH cycle and plays an essential role in the control of intracellular ROS levels. APX uses two molecules of AsA to reduce H_2O_2 to water with a concomitant generation of two molecules of MDHA. APX is a member of Class I superfamily of heme peroxidases (Welinder, 1992) and is regulated by redox signals and H_2O_2 (Patterson and Poulos, 1995). Based on amino acid sequences, five chemically and enzymatically distinct isoenzymes of APX have been found at different subcellular localization in higher plants. These are cytosolic, stromal, thylakoidal, mitochondrial and peroxisomal isoforms. APX found in organelles scavenges H_2O_2 produced within the organelles, whereas cytosolic APX eliminates H_2O_2 produced in the cytosol, apoplast or that diffused from organelles (Madhusudhan et al., 2003;

Sharma and Dubey, 2004). The chloroplastic and cytosolic APX isoforms are specific for AsA as electron donor, and the cytosolic isoenzymes are less sensitive to depletion of AsA than the chloroplastic isoenzymes, including stromal- and thylakoid-bound enzymes. APX is regarded as one of the most widely distributed antioxidant enzymes in plant cells, and isoforms of APX have a much higher affinity for H_2O_2 than CAT, making APXs efficient scavengers of H_2O_2 under stressful conditions (Ishikawa et al., 1998; Wang et al., 1999). APX differs from other peroxidases in its dependency on ASC as the source of reducing power and becomes unstable in its absence (Shigeoka et al., 2002). There are multigenic families of APX in higher plants, for example, *Arabidopsis* has nine APX genes (AtAPX1–AtAPX6, sAPX, tAPX, lAPX), and *Oryza sativa* has eight isozymes (OsAPX1–OsAPX8), having two APX each in cytosol, peroxisome, chloroplast and mitochondria (Chew et al., 2003). APXs have two important histidine residues, His[42] and His[163] and a K+ binding site which are required for APX activity (Chen and Asada, 1989). APX activity has been found to increase in the presence of other antioxidant enzymes like superoxide dismutase and glutathione reductase, indicating a crosstalk among various antioxidant enzymes. APX is unable to scavenge lipid hydroperoxides and is inhibited by cyanide, azide and thiol-modifiers. The whole family of APX shows inductive responses to ABA treatment, with the cytosolic one being the most induced one (Zhang et al., 2014). The existence of multiple molecular forms of APX within the cells and organelles signifies the important role played by them in developmental processes and stress tolerance (Ishikawa et al., 1998; Shigeoka et al., 2002). Many workers have reported enhanced activity of APX in response to abiotic stresses such as drought, salinity, chilling, metal toxicity and UV irradiation. Overexpression of a cytosolic APX gene derived from pea (*Pisum sativum* L.) in transgenic tomato plants (*Lycopersicon esculentum* L.) ameliorated oxidative injury induced by chilling and salt stress (Wang et al., 2005). Similarly, overexpression of the tApx gene in either tobacco or in *Arabidopsis* increased tolerance to oxidative stress (Yabuta et al., 2002).

13.3.3 CATALASE

Among antioxidant enzymes, catalase (CAT, 1.11.1.6) was the first enzyme to be discovered and characterized in tobacco leaf extracts by Loew (1901). It is a ubiquitous tetrameric heme-containing enzyme that catalyzes the dismutation of two molecules of H_2O_2 into water and oxygen. It has high specificity for H_2O_2 but weak activity against organic peroxides. Plants contain several types of H_2O_2-degrading enzymes; however, CATs are unique as they do not require cellular reducing equivalent. CATs have a very fast turnover rate but a much lower affinity for H_2O_2 than APX. The peroxisomes are major sites of H_2O_2 production. CAT scavenges H_2O_2 generated in this organelle during photorespiratory oxidation, β-oxidation of fatty acids and other enzyme systems such as XOD coupled to SOD (Scandalios et al., 1997; Corpas et al., 2008). Though there are frequent reports of CAT being present in the cytosol, chloroplast and mitochondria, the presence of significant CAT activity in these is less well established. To date, all angiosperm species studied contain three CAT genes. Willekens et al. (1995) proposed a classification of CAT based on the expression profile of the tobacco genes. Class I CATs are expressed in photosynthetic tissues and are regulated by light. Class II CATs are expressed at high levels in vascular tissues, whereas Class III CATs are highly abundant in seeds and young seedlings. Two of the three catalases in plants typically show day-night rhythms in transcript abundance (Fink et al., 1997; Evans et al., 2004). While the photorespiratory catalase (CAT2 in maize and *Arabidopsis*) shows a peak early in the light period, the CAT3 peak is at the end of the light period. When cells are stressed for energy and are rapidly generating H_2O_2 through catabolic processes, H_2O_2 is degraded by CAT in an energy-efficient manner (Mallick and Mohn, 2000). Environmental stresses cause either enhancement or depletion of CAT activity, depending on the intensity, duration and type of the stress (Moussa and Abdel-Aziz, 2008). In general, stresses that reduce the rate of protein turnover also reduce CAT activity. Stress analysis revealed increased susceptibility of CAT-deficient plants to paraquat, salt and ozone, but not to chilling. In transgenic tobacco plants, having 10% wild-type, CAT activity showed accumulation of glutathione disulfide (GSSG) and a four-fold decrease in AsA, indicating that CAT is critical for maintaining the redox balance during the oxidative stress (Willekens et al., 1997). Overexpression of a CAT gene from *Brassica juncea* introduced into tobacco enhanced its tolerance to Cd-induced oxidative stress (Guan et al., 2009).

13.3.4 GLUTATHIONE REDUCTASE

Glutathione reductase (GR, EC 1.6.4.2), a NADPH-dependent, homodimeric FAD-containing enzyme catalyzes the reduction of GSSG to GSH and, thus, maintains high cellular GSH/GSSG ratio. GR is a thermostable, oxidoreductase which catalyzes the reduction of GSSG to the sulfhydryl form and also acts as a substrate for GSH. The catalytic mechanism involves two

steps: first, the flavin moiety is reduced by NADPH, the flavin is oxidized and a redox active disulfide bridge is reduced to produce a thiolate anion and a cysteine. The second step involves the reduction of GSSG via thiol–disulfide interchange reactions (Edwards et al., 1990). When acting as an antioxidant by participating in enzymic as well as non-enzymic oxidation-reduction cycles, GSH is oxidized to GSSG. In ascorbate–glutathione pathway, also known as Halliwell–Asada pathway, GSH is oxidized in a reaction catalyzed by DHAR. If the reduced enzyme is not reoxidized by GSSG, it can suffer a reversible inactivation. It is universal in occurrence and is found in both prokaryotic as well as eukaryotic organisms. GR is predominantly found in chloroplast. However, a small amount of the enzyme isoforms is also found in mitochondria, cytosol and peroxisomes (Edwards et al., 1990; Jimenez et al., 1997). In leaves, the bulk of GR activity is found in chloroplast, whereas root plastids exhibit a lower proportion of enzyme cellular activity (Foyer and Halliwell, 1976). In higher plants, three types of GR occur in the cytosol, chloroplast and mitochondria, respectively (Creissen et al., 1992; Kubo et al., 1993; Tang and Webb, 1994; Creissen and Mullineaux, 1995; Mullineaux et al., 1996; Kaminaka et al., 1998). Cytosolic isoforms of GRs from rice (RGRC2; Kaminaka et al., 1998) and pea (GOR2; Stevens et al., 2000) have been identified through subcellular fractionation; in addition, chloroplastic GRs have been identified from *Arabidopsis* (AT-2; Kubo et al., 1993) and pea (GOR1; Creissen et al., 1992, 1995). Changes in the GR isoform population between and within subcellular compartments in response to stress were observed in pea (Edwards et al., 1994) and maize (Anderson et al., 1995). Different GR isoforms can be stimulated by different environmental signals and have different functions in the response of plants to stress (Stevens et al., 1997). In chloroplast, GSH and GR are involved in detoxification of H_2O_2 generated by the Mehler reaction. A cDNA encoding GR was cloned by immunoscreening from *Arabidopsis thaliana*. Pastori and Trippi (1992) observed a correlation between the oxidative stress resistance and activity of GR and suggested that oxidative stress caused by paraquat or H_2O_2 could stimulate GR *de novo* synthesis, probably at the level of translation by preexisting mRNA. Antisense-mediated depletion of tomato chloroplast GR has been shown to enhance susceptibility to chilling stress (Shu et al., 2011). Overexpression of GR in *N. tabacum* and Populus plants leads to higher foliar AsA contents and improved tolerance to oxidative stress (Aono et al., 1993). The studies regarding the GR have shown an increased GR activity in various plant species under different types of abiotic stresses. From the studies using transgenic plants, it has been proved that GR plays a prominent role in conferring resistance to oxidative stress caused by drought, ozone, heavy metals, high light, salinity, cold stress, etc. An increased GR activity has been reported in the roots of *C. arientinum* under salt stress, whereas Eyidogan and Oz (2005) found elevated GR activity in the leaf tissue of the same plant under the salt stress conditions. There has also been found an enhanced GR activity in *A. thaliana*, *Vigna mungo*, *Triticum aestivum*, *Capsicum annuum* and *Brassica juncea* following the cadmium treatments. In the same year, Sharma and Dubey identified an increased GR activity in *Oryza sativa* seedlings during drought conditions. Aono et al. (1995) found that decreased expression of GR activity in transgenic tobacco plants increased their susceptibility to MV-induced damage. Overexpression of GR results in abiotic stress tolerance in various crop plants due to efficient ROS-scavenging capacity.

13.3.5 MONODEHYDROASCORBATE REDUCTASE

Monodehydroascorbate reductase (MDHAR, 1.6.5.4) is a FAD enzyme that catalyzes the regeneration of AsA from the MDHA radical using NADPH as the electron donor (Sakihama et al., 2000). It is the only known enzyme to use an organic radical (MDA) as a substrate and is also capable of reducing phenoxyl radicals, which are generated by horseradish peroxidase with H_2O_2 (Hossain et al., 1984). MDHAR activity is widespread in plants. The isoenzymes of MDHAR have been reported to be present in several cellular compartments such as chloroplasts (Dalton et al., 1993), cytosol and mitochondria and peroxisomes (Miyake et al., 1998). In chloroplasts, MDHAR could have two physiological functions: the regeneration of AsA from MDHA and the mediation of the photoreduction of dioxygen to $O_2^{\bullet-}$ when the substrate MDHA is absent (Miyake et al., 1998). The gene expression of MDHAR isoforms was reported to be differentially affected by stress. Some isoforms were induced under salt and osmotic stress while some remained unchanged (Lunde et al., 2006). An increase of MDHAR activity was reported in wheat (Sairam et al., 2002), tomato (Mittova et al., 2004) and rice (Vaidyanathan et al., 2003) under salt stress. Several studies have shown increased activity of MDHAR in plants subjected to environmental stresses such as MDAR is upregulated in tomato by salinity stress (Mittova et al., 2003) and high light intensity (Gechev et al., 2003), in rice by low temperature (Oidaira et al., 2000) and in *Arabidopsis* by UV-C exposure (Kubo et al., 1999). Characterization of membrane polypeptides from pea leaf peroxisomes also revealed MDHAR

to be involved in $O_2^{\bullet-}$ generation (López-Huertas et al., 1999). Overexpression of *Arabidopsis* MDHAR gene in tobacco confers enhanced tolerance to salt and polyethylene glycol stresses (Eltayeb et al., 2007) while the overexpression of a chloroplastic MDHAR from *Avicennia marina* and tomato, respectively, were shown to confer tolerance to salt stress in transgenic tobacco (Kavitha et al., 2010) and to higher temperature and methyl viologen-mediated oxidative stress in transgenic tomato (Li et al., 2010). The overexpression of a cytosolic MDHAR from *Arabidopsis* has been reported to enhance the tolerance to ozone, salt and polyethylene glycol stresses in transgenic tobacco (Eltayeb et al., 2007). However, the overexpression of this gene did not confer tolerance to aluminum stress (Yin et al., 2010).

13.3.6 DEHYDROASCORBATE REDUCTASE

Dehydroascorbate reductase (DHAR, EC 1.8.5.1) catalyzes the reduction of DHA to AsA using GSH as the reducing substrate and, thus, plays an important role in maintaining AsA in its reduced form (Ushimaru et al., 1997). Despite the possibility of enzymic and non-enzymic regeneration of AsA directly from MDHA, some DHA is always produced when AsA is oxidized in leaves and other tissues. DHA, a very short-lived chemical, can either be hydrolyzed irreversibly to 2,3-diketogulonic acid or recycled to AsA by DHAR. Overexpression of DHAR in tobacco leaves, maize and potato is reported to increase AsA content, suggesting that DHAR plays important roles in determining the pool size of AsA (Chen et al., 2003; Qin et al., 2011). DHAR is a monomeric thiol enzyme abundantly found in dry seeds, roots and etiolated as well as green shoots. DHAR has been purified from chloroplast as well as nonchloroplast sources in several plant species, including spinach leaves (Hossain and Asada, 1994) and potato tuber (Dipierro and Borraccino, 1991). Environmental stresses such as drought, metal toxicity and chilling increase the activity of the DHAR in plants. Consistent upregulation of the gene encoding cytosolic DHAR was found in *L. japonicas*, which was found to be more tolerant to salt stress than other legumes. This upregulation of DHAR was correlated to its role in AsA recycling in the apoplast (Rubio et al., 2009). Transgenic potato overexpressing *Arabidopsis* cytosolic AtDHAR1 showed higher tolerance to herbicide, drought and salt stresses (Eltayeb et al., 2011).

Ascorbic acid (ASC), a potent antioxidant, is synthesized in the mitochondria of plants and then transported to other cell compartments to eliminate ROS (Horemans et al., 2000; Wheeler et al., 1998). ASC represents about

10% of the total soluble carbohydrate pool and acts as an electron donor for the detoxification of ROS in plants (Eltayeb et al., 2006; Smirnoff and Wheeler, 2000). ASC content of plants can be increased by enhancing the expression of DHAR (Chen et al., 2003). Overexpressing the wheat DHAR gene increased protection against oxidative damage in tobacco (Chen and Gallie, 2005). Eltayeb et al. in 2006 reported that the elevation of ASC levels via expression of *A. thaliana* DHAR provided a significantly enhanced oxidative stress tolerance to drought and salt. Overexpression of rice DHAR increased the salt tolerance of *Arabidopsis* plants (Ushimaru et al., 2006). The recycling of ASC via wheat DHAR expression can enhance the protection against photooxidative stress (Chen and Gallie, 2008). Overexpression of *A. thaliana* cytosolic DHAR conferred enhanced tolerance to Al stress in tobacco (Eltayeb et al., 2011). Elevating ascorbate contents of potatoes via overexpression of *A. thaliana* DHAR led to higher tolerance to herbicide, drought and salt stresses.

It is very important for plant survival under stress conditions that antioxidants can work in cooperation, thus providing better defense and regeneration of the active reduced forms. The most studied example of the antioxidant network is the ascorbate–glutathione (Halliwell–Asada) pathway in the chloroplasts, where it provides photoprotection (Noctor and Foyer, 1998) by removing H_2O_2. The findings to date indicate that the expression of combinations of antioxidant enzymes in transgenic plants may have synergistic effects on stress tolerance. Due to the complexity of ROS detoxification systems, overexpressing one component of antioxidative defense systems may or may not change the capacity of the pathway as a whole. Aono et al. (1995) reported that co-expression of cytosolic forms of both GR and SOD in tobacco plants provided more substantial protection from MV treatment than either enzyme alone. Several studies have shown that overexpression of combinations of antioxidant enzymes in transgenic plants has a synergistic effect on stress tolerance. Kwon et al. (2002) demonstrated that simultaneous expression of Cu/Zn-SOD and APX genes in tobacco chloroplasts enhanced tolerance to MV stress compared to expression of either of these genes alone. Similarly, enhanced tolerance to multiple environmental stresses has been developed by simultaneous overexpression of the genes of SOD and APX in the chloroplasts (Lim et al., 2007; Kwak et al., 2009), SOD and CAT in cytosol (Tseng et al., 2008) and SOD and GR in cytosol (Lee et al., 2007). Further, simultaneous expression of multiple antioxidant enzymes, such as Cu/Zn-SOD, APX and DHAR, in chloroplasts has shown to be more effective than a single

or double expression for developing transgenic plants with enhanced tolerance to multiple environmental stresses (Lee et al., 2007). Therefore, in order to achieve tolerance to multiple environmental stresses, increased emphasis is now given to produce transgenic plants overexpressing multiple antioxidants.

13.4 ROLE OF GLUTATHIONE IN DETOXIFYING ROS AND MG VIA INVOLVEMENT OF GLYOXALASE SYSTEM

Glutathione participates in detoxification either directly by interacting with ROS and methylglyoxal or by operating as a cofactor for various enzymes. The reduced and oxidized forms of glutathione (GSH and GSSG) act in concert with other redox-active compounds, e.g., NADPH, to regulate and maintain cellular redox status (Jones et al., 2011). MG detoxification is primarily carried out by the ubiquitous glyoxalase system located in cytoplasm, mitochondria and nucleus. The glyoxalase system comprises two enzymes, glyoxalase I and II, both of which act coordinately on MG and convert it into nontoxic lactic acid using GSH as a cofactor (Racker, 1951; Thornalley, 1990). Glyoxalase I used hemiacetal as a substrate, which is formed by spontaneous reaction of MG and GSH and conversion into *S*-lactoylglutathione (SLG), which was further hydrolyzed by glyoxalase II into nontoxic lactic acid and regenerate GSH (Racker, 1951; Crook and Law 1952). Methylglyoxal first spontaneously reacts with glutathione to form a hemithioacetal derivative which is then converted to *S*-lactoylglutathione by the enzyme glyoxalase I:

$$CH_3COCHO + GSH \rightarrow CH_3COCH\,(OH)\text{-}SG$$
$$\rightarrow CH_3CH(OH)\,CO\text{-}SG$$

The *S*-lactoylglutathione is then hydrolyzed by glyoxalase II to release the nontoxic lactic acid and glutathione:

$$CH_3CH\,(OH)\,CO\text{-}SG \rightarrow CH_3CH\,(OH)\,COOH + GSH$$

The deficiency of GSH limits the production of hemiacetal, leading to the accumulation of MG. Therefore, MG detoxification strongly depends on the availability of cellular GSH. The whole process is also known as GSH-dependent MG detoxification. Recent studies have suggested a single-step GSH-independent detoxification of MG to D-lactate by a unique enzyme, glyoxalase III (Ghosh and Islam, 2016; Kwon et al., 2013). Gly III had lower specific activity than the conventional gly I/II, suggesting that gly I protein is the main enzyme to detoxify

MG under normal conditions and gly III performs action when MG accumulation occurs in plants under stress condition (Ghosh and Islam, 2016). The existence of glyoxalase as a multigene family in plants suggests that it has an important role in different biological systems. The relevance of glyoxalase in stress physiology came firstly from the study in tomato where gly I transcript levels showed enhanced induction in the roots, stem and leaves of tomato (Espartero et al., 1995) after stress treatments. Similar results were obtained in *Brassica juncea* (Veena Reddy and Sopory, 1999) when treated with NaCl, mannitol and heavy metals. In addition to this, gly I transcript was also significantly upregulated by white light, salinity, MG and heavy metals in pumpkin seedlings (Hossain et al., 2009) and also through exposure to $ZnCl_2$ and arsenate/arsenite in wheat and rice, respectively (Lin et al., 2010; Chakrabarty et al., 2009). Blomstedt et al, (1998) detected gly I transcript upregulated in dehydration stress in *Sporobolus stapfianus*. Du et al, (2011) detected enhanced accumulation of gly I protein in rice leaves subjected to UV radiation. Like gly I transcript analysis, gly II also reported in different organisms under various stress condition. In *B. juncea*, expression of Bjglyll II was increased under different stresses such as salinity, heavy metals and ABA (Saxena et al., 2005). A similar pattern of gly II transcript was also observed in rice upon exposure of salinity, desiccation, temperature, ABA and salicylic acid (Yadav et al., 2007). Genome-wide expression analysis of glyoxalase gene families in *Arabidopsis*, rice (Mustafiz et al., 2011), *Glycine max* (Ghosh and Islam, 2016) and *Medicago truncatula* (Ghosh, 2017) have revealed their role in different abiotic stress. Recent studies concluded that many stress-responsive elements such as heat shock elements (HSE), ethylene-responsive elements, auxin-responsive elements (AuxxRR-core), abscisic-acid-responsive elements (ABREs) have been identified in the promoter regions of both Gm gly I and Gm gly II family members, suggesting that these genes could be regulated by hormonal and stress response pathways (Devanathan et al., 2014; Ghosh and Islam, 2016). In rice, wheat and tomato, gly I showed induced expression in salt-tolerant varieties rather than in salt-sensitive varieties (Laino et al., 2010; Lee et al., 2009; Sun et al., 2010). Transgenic tobacco plants overexpressing *gly I* of *Brassica juncea* was conferring enhanced tolerance to MG and high salt (Veena Reddy and Sopory, 1999) while overexpression of gly I in *Vigna mungo* had led to salinity tolerance (Bhomkar et al., 2008). *Gly I* of wheat was overexpressed in tobacco, which also provides tolerance to $ZnCl_2$ (Singla-Pareek et al., 2006). Furthermore, overexpression of *gly II* of rice, tobacco and mustard also provides stress tolerance

capacity to MG and salinity (Singla-Pareek et al., 2008; Wani et al., 2011; Saxena et al., 2011). Accordant with these results, transgenic tobacco plant overexpressing both *gly I* and *gly II* from *B. juncea* and *Oryza sativa* respectively were found to be more tolerant to salinity (Singla-Pareek et al., 2003). These double transgenic tobacco plants showed enhanced tolerance capacity to Zn and other heavy metal (Singla-Pareek et al., 2006).

Glutathione also helps in the elimination of reactive oxygen as it is oxidized directly by oxidants such as hydroxyl radical ($\bullet$OH) (Gardner and Aust, 2009; Sagone et al., 1984). Direct oxidation leads to the production of thiyl radicals (GS$\bullet$). The thiyl radicals formed from these reactions can also combine with different molecules, as well as with other thiyl radicals, leading to the formation of oxidized glutathione (glutathione disulfide, GSSG) (Lushchak, 2012). GSSG is also produced in reactions catalyzed by GPx and glutaredoxins (GR). GSH is extensively used as a cosubstrate by GPx, reducing H_2O_2 or organic peroxides (generally abbreviated as ROOH or LOOH in the case of lipid peroxides) with the production of GSSG, water or alcohols. Hydroxyl radical may interact directly with GSH, leading to GSSG formation. Hydrogen peroxide may be removed by catalase or by GPx. The latter requires GSH to reduce peroxide. However, enhanced ROS levels may require not only enhanced GSH action to maintain redox status but also enhanced energy and material consumption to transport consumed GSH to the places where it is needed.

13.5 SIGNALING PATHWAYS INDUCED IN RESPONSE TO OXIDATIVE STRESS

13.5.1 MITOGEN-ACTIVATED PROTEIN KINASES (MAPK) CASCADES

Among the many signaling pathways involved in abiotic stress response in plants, mitogen-activated protein kinase (MAPK) cascade is one of the essential pathways. The MAPK or MPK cascade is highly conserved signaling modules in all eukaryotes. A MAPK cascade is composed of three kinases, i.e., MAP kinase kinase kinases (MAP3Ks/MAPKKKs), MAP kinase kinases (MAP2Ks/MAPKKs) and MAP kinases (MAPKs/MPKs), and these are activated in a sequential phosphorylation-dependent manner (Ichimura et al., 2002; Colcombet and Hirt, 2008). Under stress condition, accelerated plasma membrane activates MAP3Ks. These MAP3Ks are serine/threonine kinases which further activate MAP2K through phosphorylation. Activated MAP2Ks phosphorylated MAPKs on threonine and tyrosine residues. MAPKs are serine/threonine kinases which phosphorylated a broad range of substrate proteins such as other kinases and/or transcription factors (Chang and Karin, 2001; Rodriguez et al., 2010). The MAPK cascade plays a significant role in cell division, plant growth and development, tolerance to pathogens and insect herbivores and to abiotic stresses (Xu and Zhang, 2015; Bigeard et al., 2015; Hettenhausen et al., 2015).

Under oxidative stress, the higher levels of ROS, such as free radicals (hydroxyl, superoxide) or nonradicals (hydrogen peroxide, lipid peroxide), damaged cells or tissue (Mittler, 2002; Gill and Tuteja, 2010). With the help of scavenger enzymes, plants overcome oxidative stress, as catalases (CAT) decompose hydrogen peroxide. Xing et al., (2008) demonstrated that MKK1-MPK6 regulate H_2O_2 metabolism through CAT1. The MEKK1–AtMKK1/AtMKK2–AtMPK4 cascade played an important role in ROS homeostasis. It has been shown that some of the redox-related genes are abnormally expressed in *mekk1* or *atmpk4* mutant plants (Nakagami et al., 2006; Pitzschke and Hirt, 2009). From microarray expression analysis it was revealed that many ROS-related genes' (CSD1, CAT1 and APX1) expression are overlapped in *mekk1*, *atmkk1/atmkk2* and *atmpk4* mutants (Pitzschke et al., 2009). The enzyme analysis in *atmpk4* mutants or *atmpk4/ics1* double mutants showed that Cu/Zn-SOD and APX activities were increased whereas Fe-SOD activity was decreased (Gawronski et al., 2014). Additionally, H_2O_2 activated other MAPKKKs, i.e., ANP1 in *A. thaliana*, which may cause the downstream activation of MPK3 and MPK6 (Kovtun et al., 2000). A novel MAPKKK, oxidative stress-activated MAP triple-kinase1 (OMTK1), was identified in alfalfa, which further activated downstream MAPK, MMK3 (Nakagami et al., 2004). Oxidative stress-induced activation of plant MAPK has also been reported in a variety of plants like maize and pea (Zong et al., 2009). It has been demonstrated that MKP2 positively controls abiotic oxidative stress responses and is a key regulator of MPK3 and MPK6 networks controlling stress responses in plants (Lumbreras et al., 2010).

13.5.2 ABSCISIC ACID

Abscisic acid is an essential phytohormone which has a major role in various environmental cues. It accumulates under stressful environmental conditions such as dehydration, cold temperature, etc. ABA treatment plays a most important role in different physiological processes, for instance, seed dormancy and delay in germination, development of seeds, promotion of stomatal closure, embryo morphogenesis and leaf senescence (Swamy and Smith, 1999). Recently it has been shown that MAPK cascade has an important role in ABA signaling. Treatment of external

ABA induced the activity of several components of MAPK signaling cascade (Jammes et al., 2009; Zong et al., 2009). Application of ABA in *Arabidopsis* induced the transcriptional regulation of MPK3, MPK5, MPK7, MPK18, MPK20, MKK9, MAPKKK1 (ANP1), MAPKKK10 (MEKK3), MAPKKK14, MAPKKK15, MAPKKK16, MAPKKK17, MAPKKK18, MAPKKK19, *Raf6*, *Raf12* and *Raf35*, suggesting a potential role of ABA signaling (Menges et al., 2008; Wang et al., 2011). There is evidence which suggests that H_2O_2 is one of the major oxidative stresses and plays an important role as a second messenger in ABA-induced stomatal closure (Pei et al., 2000; Miao et al., 2006). Zhang et al. (2001) revealed the involvement H_2O_2 in ABA-induced stomatal movement in *V. faba*. The overproduction of ABA induces H_2O_2, which enhances stomatal closure by activating plasma membrane calcium channels (Pei et al., 2000). Abiotic stresses increase ABA/gibberellic acid (GA) ratio, which leads to accumulation of DELLA protein and, hence, reduces H_2O_2 level (Considine and Foyer, 2014). In rice seeds, ABA decreased ROS production, which leads to inhibition of ascorbate and GA accumulation (Ye et al., 2012).

These findings clearly specify that ABA works as a key hormone in inducing abiotic stress responses. Downstream events mediated by MAPK cascade, alterations in Ca^{2+} fluxes and activation of ion channels change the redox state of the cell. Most of the genes such as NADPH oxidases, SOD and extracellular peroxidases expressed in the early signals are involved in the ROS generation, essential for triggering PCD. While other genes are responsible for maintaining ROS levels (CAT and APXs; Gadjev et al., 2008).

13.5.3 TRANSCRIPTION FACTORS

Regulation of gene expression at the transcriptional level controls many of the biological processes in a cell. TFs are proteins that act together with *cis*-elements in the promoter regions of quite a lot of stress-related genes and thus alter the expression of many downstream genes, resulting in imparting abiotic stress tolerance (Agarwal and Jha, 2010). Plant stress responses are regulated by various signaling pathways that activate gene transcription and its downstream machinery. About 1500 TFs reported in *Arabidopsis thaliana* are considered to be involved in stress-responsive gene expression (Riechmann et al., 2000). Transcriptome data in response to ROS in *Arabidopsis* revealed that about 500 TFs (out of 1500) showed either induced or repressed expression (Gadjev et al., 2006). Several TFs from bacteria to multicellular organism are regulated by ROS. Bacterial TFs oxygen-regulated (OxyR) and peroxide regulon repressor (PerR) are directly activated by oxidative stress (Mittler et al., 2004). The tetrameric OxyR TF exists in both reduced and oxidized state, but only the oxidized form is able to initiate transcription (Storz and Imlayt, 1999). Four yeast-encoded TFs, i.e., Yap1, Maf1, Hsf1, and Msn2/4, are regulated by ROS. One of the key TFs in yeast is Yap1, which is involved in oxidative stress response, redox homeostasis and electrophilic response and regulates the transcription of genes such as antioxidant and detoxification enzymes in yeast cells (Marinho et al., 2014). In multicellular organisms, different TFs, specifically AP-1, NRF2, CREB, HSF1, HIF-1, TP53, NF- kB, NOTCH, SP1 and SCREB1, are well-characterized and regulated by ROS (Sewelam et al., 2016). In plants, basic leucine zipper (bZIP) TFs are utilized during oxidative stress to activate protective responses (Assunçáo et al., 2010). Moreover, important plant TFs, namely, ABRE, MYC/MYB, CBF/DREBs and NAC regulate various stress-responsive gene expression. They play a crucial role in providing tolerance to multiple stresses through respective *cis*-elements and DNA-binding domains. These TFs can be genetically engineered to produce transgenics with higher tolerance to drought, salinity, heat, oxidative and cold stress using different promoters. Functional analysis of these TFs will impart more information on regulatory networks involved in abiotic stress responses and in crosstalk between different signaling pathways during stress adaptation (Figure 13.3).

13.6 OMICS TOOL IN ABIOTIC STRESS TOLERANCE

Omics technologies propose improved techniques and strategies for understanding the relationship between the function of specific genes and their phenotypic effects under different environmental conditions. It also helps in the interpretation of the complex molecular regulatory system involved in stress tolerance and adaptation in plants (Soda et al., 2015). Omics refers to the study of large sets of molecules in biology for the detection of genes (genomics), mRNA (transcriptomics), protein (proteomics) and metabolites (metabolomics). It has been employed for a better understanding of the biological processes and molecular/cellular mechanisms involved in plant stress responses.

13.6.1 Transcriptomics

Plants continuously need to adjust their transcriptome profile in response to various abiotic stresses; hence, transcriptomic analysis is used not only to screen

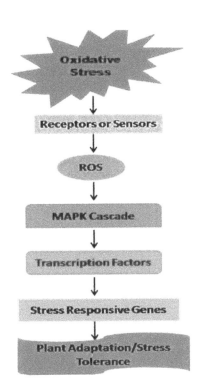

FIGURE 13.3 Schematic representation of MAPK cascade activated by oxidative stress.

candidate genes for abiotic stress management (Jogaiah et al., 2012) but to predict the tentative gene function and their association with the plant phenotype (Francki et al., 2012). It can be performed by various techniques, such as DNA microarrays, serial analysis of gene expression (SAGE) or more recently digital gene expression (DGE) profiling, taking advantage of next-generation sequencing (NGS)-based tools such as RNA sequencing (RNA-Seq) (Cramer et al., 2011; Jogaiah et al., 2012; Lister et al., 2009). The hybridization-based method, such as that used in microarray analyses, together with the availability of completed genome sequences and increasing public repositories of available microarray data and data analysis tools, has opened new avenues to genome-wide analysis of plant stress responses (Mochida and Shinozaki, 2011; Jogaiah et al., 2012). Transcript analysis helps to identify various stress-associated genes such as signal transduction components (e.g., MAP kinase 4), transcription factors (TFs, e.g., RAP2.11 and AP2-EREBP), and active oxygen species scavenging enzymes (e.g., catalase 2), as well as photosynthesis-related genes (e.g., PsaL). These candidate genes with a wide range of biological functions are exploited for genetic improvement of species. SuperSAGE, an improved version of the serial analysis of gene expression technique, has been employed in the analysis of gene expression in chickpea roots in response to drought (Molina et al., 2008). The transcriptome expression profiles of poplar (*Populus simonii* × *Populus nigra*) under salt stress were investigated using Solexa/Illumina digital gene expression technique (Chen et al., 2012). Serine/threonine protein kinases have been reported to confer enhanced multistress tolerance in many plants (Zhao et al., 2009), suggesting that this gene can be a suitable target for biotechnological manipulation with the aim of improving poplar salt tolerance. Transcriptomic technologies have the potential to provide deep coverage and unbiased representation of transcript abundance, which is very important in plants lacking genome sequence information (Trujillo et al., 2008). However, the frequent incongruity between protein levels and the abundance of cognate gene transcripts suggest the need of complementary analysis of the proteome for further validation of candidate genes and pathways (Gygi et al., 1999).

13.6.2 PROTEOMICS

Proteomics deals with determination, identification of proteins, expression profile, post-translational modifications (PTMs) and protein–protein interactions in a comparative manner under stress and non-stress conditions (Hashiguchi et al., 2010; Nam et al., 2012; Mertins et al., 2013; Ghosh and Xu, 2014). It implies two-dimensional electrophoresis and Difference In-Gel Electrophoresis (DIGE) along with protein identification using mass spectrometry. Plant stress proteomics has enabled us to recognize possible candidate genes that can be used for the genetic enhancement and manipulation of plants against stresses (Cushman and Bohnert, 2000; Rodziewicz et al., 2014; Barkla et al., 2009). A large number of proteomic studies are nicely investigated in plants under various abiotic stresses like: salt stress (Nam et al., 2012; Zhu et al., 2012), drought stress (Castillejo et al., 2008; Caruso et al., 2009; Mirzaei et al., 2012; Mohammadi et al., 2012; Cramer et al., 2013), waterlogging (Komatsu et al., 2009, 2010, 2013a,b; Alam et al., 2010a,b) and heat stress (Rollins et al., 2013; Xuan et al., 2013). Evers et al. (2012) have used both transcriptomics and proteomics to study the effects of cold and salt stresses on the leaf transcriptome and proteome of potato (*Solanum tuberosum*). In another example, DIGE has been used to study the effects of high levels of UV radiation on the leaf proteome of artichoke, particularly targeting the levels of inducible antioxidants present in this species (Falvo et al., 2012). The embryo proteome of six rice varieties subjected to water deficit stress predicts the involvement of a total of 28 proteins in stress tolerance (LEA proteins), nutrient reservoir activity, among other proteins implicated in diverse cellular processes potentially related to the stress response (e.g., mitochondrial

import translocase) in this cereal. It is apparent that a variety of proteins involved directly/indirectly in the process of photosynthesis are up/downregulated under saline conditions. A downregulation in RuBisCO activase (Parker et al., 2006) and upregulation of proteins involved in signal transduction, Na^+ compartmentalization, photosynthesis, protein folding and respiration (Wang et al., 2014) was reported under salt stress. Chen et al. (2011) showed that an upregulation of 23 proteins under desiccation were involved in defense against this stress, as well as redox, as observed through proteomic profiling of seeds of recalcitrant tea. The advantages of proteomics are further highlighted by the possibility of studying the PTMs of proteins of key importance in physiological and biochemical responses of plants to stress. Sanchez-Bel et al. (2012) performed comparative proteomic analyses between control and cold-stressed bell pepper fruit to understand the alterations of protein profile associated with redox homeostasis and carbohydrate metabolism. It was observed that the protein abundance in the ascorbate–glutathione cycle was altered and catalase was downregulated. Some of the key proteins involved in glycolysis and the Calvin cycle were also inhibited in chilled fruit. These examples suggest that proteomics not only enables us to understand the responses of plants under various stress but also to procure the candidate genes for generating stress-tolerant transgenic plants using biotechnological approaches.

13.6.3 METABOLOMICS

Plant metabolomics has been defined as the identification and quantitation of all low molecular weight metabolites (primary and secondary) at a given developmental stage of the plants (Arbona et al., 2009; Fiehn, 2001). Metabolomics techniques involve separation of metabolites based on their physical and chemical properties using various analytical tools and technologies. The separation techniques include nuclear magnetic resonance spectroscopy (NMR), mass spectrometry (MS)-based methods including gas chromatography–MS (GC–MS), liquid chromatography–MS (LC–MS), capillary electrophoresis–MS (CE–MS) and Fourier transform ion cyclotron resonance–MS (FTICR–MS), which provide accurate mass measures. Plant metabolomics helps plant breeders to identify the physiological and biochemical effects of different abiotic stress factors, and their association with metabolites, hence, can be used as markers for the selection of cultivars and/or rootstocks with improved yield/abiotic stress tolerance. Although the majority of the metabolic works have been proposed in model species such as *Arabidopsis* (Cook et al., 2004), metabolomic technologies are being used with success in forages (Sanchez et al., 2012), cereals (Sicher and Barnaby, 2012) and other food crops (Hernández et al., 2007). Plants respond to the induced oxidative stress by overproducing antioxidant compounds such as ascorbic acid, glutathione and polyphenols (Urano et al., 2009; Munns and Tester, 2008) while drought stress induces the accumulation of several amino acids such as valine, leucine, isoleucine and agmatine (as a precursor of polyamines) along with carbohydrates and carbohydrate alcohols in combination with proline (Pro). Maize plants, when subjected to water and salinity stress in combination, showed enhanced levels of six metabolites (citrate, fumarate, phenylalanine, valine, leucine, isoleucine) in leaves (Sun et al., 2015). Metabolomic studies of *A. thaliana* plants subjected to drought stress revealed the accumulation of several metabolites, including amino acids such as proline, raffinose family oligosaccharides, gamma aminobutyrate (GABA) and tricarboxylic acid (TCA) cycle metabolites, which are known to respond to drought stress in plants (Urano et al., 2009, 64). Heat-sensitive (Moroberekan) anther has lower levels of sucrose and myo-inositol but higher levels of galactinol and raffinose, while heat-tolerant (N22) anther has lower abundances of glucose-6-P and fructose-6-P (Li et al., 2015). Metabolite profiling of maize and wheat exposed to water stress conditions suggested common changes in the levels of metabolites, including branched-chain amino acids (BCAAs) (Witt et al., 2012; Bowne et al., 2012). Some stress-related metabolites identified include polyols accumulated in phosphorusdeficient roots as well as sugars, providing additional support for the role of these compounds for P stress. The metabolomic data supported the identification of candidate genes involved in common bean root adaptation to P deficiency to be used in improvement programs. A recent study in maize was conducted to understand the combined effects of enhanced atmospheric CO_2 and drought on the stress responses by monitoring foliar metabolites (LC and GC–MS) and transcripts (Sicher and Barnaby, 2012). The concentrations of 28 out of 33 leaf metabolites were altered by drought. Soluble carbohydrates, aconitate, shikimate, serine, glycine, proline and eight other amino acids increased, and leaf starch, malate, fumarate, 2-oxoglutarate and seven amino acids decreased with drought. Isoprene is identified as a metabolite that can act as thermo-protective agent proposed to prevent degradation of photosynthetic enzymes/membrane structures (Sharkey and Singsaas, 1995) and/or as reactive molecule reducing abiotic oxidative stress (Sasaki et al., 2007). The physiological and ecological roles of isoprene in the protection of plants from

environmental stresses were further confirmed by metabolic downregulation of condensed tannins and anthocyanins under high temperature and light intensities in non-isoprene-emitting poplar genotypes. Plant metabolomic technology not only provides information on the numbers of identified metabolites but also their correlations with agronomic important traits; thus, it could link specific metabolites or pathways with yield- or quality-associated traits. The development of metabolic QTL (mQTL) and metabolome-based genome-wide association study (mGWAS) markers will be helpful for crop selection and improvement of abiotic stress tolerance in crops to overcome the problem of varying environmental conditions.

ACKNOWLEDGMENTS

L.C. Rai is thankful to CSIR and ICAR for financial support and DST and DAE for the J.C. Bose National Fellowship and the Raja Ramanna Fellowship, respectively. Ruchi Rai and Shilpi Singh are thankful to DST for WOSA award. Shweta Rai and Antra Chatterjee thank CSIR and UGC respectively for SRF. Alka Shankar thanks DST for N-PDF fellowship. We thank Head and Coordinator, CAS for facilities through FIST and PURSE supports.

REFERENCES

Abogadallah, G. M. (2010) Antioxidative defense under salt stress. *Plant Signaling and Behavior* 5(4): 369–374.

Acquaah, G. (2007)*Principles of Plant Genetics and Breeding*, Blackwell: Oxford, UK, p. 385.

Agarwal, P. K., Jha, B. (2010) Transcription factors in plants and ABA dependent and independent abiotic stress signaling. *Biologia Plantarum* 54(2): 201–212.

Ahmad, P.(Ed) (2014) Oxidative damage to plants *Hydrogen Peroxide (H_2O_2)*, Chapter 19. Academic Press: New York, NY.

Alam, I., Lee, D., Kim, K., Park, C., Sharmin, S. A., Lee, H., Oh, K., Yun, B., Lee, B. (2010a) Proteome analysis of soybean roots under water logging stress at an early vegetative stage. *Journal of Biosciences* 35(1): 49–62.

Alam, I., Sharmin, S. A., Kim, K. H., Yang, J. K., Choi, M. S., Lee, B. H. (2010b) Proteome analysis of soybean roots subjected to short-term drought stress. *Plant and Soil* 333(1–2): 491–505.

Alscher, R. G., Erturk, N., Heath, L. S. (2002) Role of superoxide dismutases (SODs) in controlling oxidative stress in plants. *Journal of Experimental Botany* 53(372): 1331–1341.

Ames, B. N., Shigenaga, M. K., Hagen, T. M. (1993) Oxidants, antioxidants, and the degenerative diseases of aging. *Proceedings of the National Academy of Sciences* 90(17): 7915–7922.

Anderson, M. D., Prasad, T. K., Stewart, C. R. (1995) Changes in isozyme profiles of catalase, peroxidase, and glutathione reductase during acclimation to chilling in mescotyls of maize seedlings. *Plant Physiology* 109(4): 1247–1257.

Aono, M., Kubo, A., Saji, H., Tanaka, K., Kondo, N. (1993) Enhanced tolerance to photooxidative stress of transgenic *Nicotiana tabacum* with high chloroplastic glutathione reductase activity. *Plant and Cell Physiology* 34: 129–135.

Aono, M., Saji, H., Fujiyama, K., Sugita, M., Kondo, N., Tanaka, K. (1995) Decrease in activity of glutathione reductase enhances paraquat sensitivity in transgenic *Nicotiana tabacum. Plant Physiology* 107(2): 645–648.

Apel, K., Hirt, H. (2004) Reactive oxygen species: Metabolism, oxidative stress, and signal transduction. *Annual Review of Plant Biology* 55: 373–399.

Arbona, V., Iglesias, D. J., Talón, M., Gómez-Cadenas, A. (2009) Plant phenotype demarcation using nontargeted LC-MS and GC-MS metabolite profiling. *Journal of Agricultural and Food Chemistry* 57(16): 7338–7347.

Assunção, A. G. L., Herrero, E., Lin, Y. F., Huettel, B., Talukdar, S., Smaczniak, C., Immink, R. G. H., et al. (2010) Arabidopsis thaliana transcription factors bZIP19 and bZIP23 regulate the adaptation to zinc deficiency. *Proceedings of the National Academy of Sciences of the United States of America* 107(22): 10296–10301.

Barkla, B. J., Vera-Estrella, R., Hernandez-Coronado, M., Pantoja, O. (2009) Quantitative proteomics of the tonoplast reveals a role for glycolytic enzymes in salt tolerance. *The Plant Cell* 21(12): 4044–4058.

Bhomkar, P., Upadhyay, C. P., Saxena, M., Muthusamy, A., Shiva Prakash, N., Pooggin, M., Hohn, T., Sarin, N. B. (2008) Salt stress alleviation in transgenic *Vigna mungo* L. Hepper (blackgram) by overexpression of the glyoxalase I gene using a novel *Cestrum* yellow leaf curling virus (CmYLCV) promoter. *Molecular Breeding* 22(2): 169–181.

Bigeard, J., Colcombet, J., Hirt, H. (2015) Signaling mechanisms in pattern-triggered immunity (PTI). *Molecular Plant* 8(4): 521–539.

Blomstedt, C. K., Gianello, R. D., Hamill, J. D., Neale, A. D., Gaff, D. F. (1998) Drought-stimulated genes correlated with desiccation tolerance of the resurrection grass *Sporobolus stapfianus. Plant Growth Regulation* 24(3): 153–161.

Bolouri-Moghaddam, M. R., Le Roy, K., Xiang, L., Rolland, F., Van Den Ende, W. (2010) Sugar signaling and antioxidant network connections in plant cells. *The FEBS Journal* 277(9): 2022–2037.

Bolwell, G. P., Davies, D. R., Gerrish, C., Auh, C. K., Murphy, T. M. (1998) Comparative biochemistry of the oxidative burst produced by rose and French bean cells reveals two distinct mechanisms. *Plant Physiology* 116(4): 1379–1385.

Bolwell, G. P., Bindschedle, L. V., Blee, K. A., Butt, V. S., Davies, D. R., Gardner, S. L., Gerrish, C., Minibayeva, F. (2002) The apoplastic oxidative burst in response to biotic stress in plants: A three-component system. *Journal of Experimental Botany* 53(372): 1367–1376.

Borsetti, F., Tremaroli, V., Michelacci, F., Borghese, R., Winterstein, C., Daldal, F., Zannoni, D. (2005) Tellurite effects on *Rhodobacter capsulatus* cell viability and superoxide dismutase activity under oxidative stress conditions. *Research in Microbiology* 156(7): 807–813.

Bowler, C., Montagu, M. V. Van, Inzé, D. (1992) Superoxide dismutase and stress tolerance. *Annual Review of Plant Physiology and Plant Molecular Biology* 43(1): 83–116.

Bowler, C., Slooten, L., Vandenbranden, S., De Rycke, R., Botterman, J., Sybesma, C., Van Montague, M., Inze´, D. (1991) Manganese superoxide dismutase can reduce cellular damage mediated by oxygen radicals in transgenic plants. *The EMBO Journal* 10(7): 1723–1732.

Bowne, J. B., Erwin, T. A., Juttner, J., Schnurbusch, T., Langridge, P., Bacic, A., Roessner, U. (2012) Drought responses of leaf tissues from wheat cultivars of differing drought tolerance at the metabolite level. *Molecular Plant* 5(2): 418–429.

Caruso, G., Cavaliere, C., Foglia, P., Gubbiotti, R., Guarino, C., Samperi, R. (2009) Analysis of drought responsive proteins in wheat (*Triticum aestivum* durum) by 2D-PAGE and MALDI TOF mass spectrometry. *Plant Science* 177(6): 570–576.

Castillejo, M. A., Maldonado, A. M., Ogueta, S., Jorrín, J. V. (2008) Proteomic analysis of responses to drought stress in sunflower (*Helianthus annuus*) leaves by 2DE gel electrophoresis and mass spectrometry. *The Open Proteomics Journal* 1(1): 59–71.

Chakrabarty, D., Trivedi, P. K., Misra, P., Tiwar, M., Shri, M., Shukla, D., Kumar, S., et al. (2009) Comparative transcriptome analysis of arsenate and arsenite stresses in rice seedlings. *Chemosphere* 74(5): 688–702.

Chang, L., Karin, M. (2001) Mammalian MAP kinase signaling cascades. *Nature* 410(6824): 37–40.

Chaplen, F. W. R. (1998) Incidence and potential implications of the toxic metabolite methylglyoxal in cell culture: A review. *Cytotechnology* 26(3): 173–183.

Chen, G., Asada, K. (1989) Ascorbate peroxidase in tea leaves: Occurrence of two isozymes and the differences in their enzymatic and molecular properties. *Plant and Cell Physiology* 30(7): 987–998.

Chen, Z., Gallie, D. R. (2005) Increasing tolerance to ozone by elevating foliar ascorbic acid confers greater protection against ozone than increasing avoidance. *Plant Physiology* 138(3): 1673–1689.

Chen, Z., Gallie, D. R. (2008) Dehydroascorbate reductase affects non-photochemical quenching and photosynthetic performance. *The Journal of Biological Chemistry* 283(31): 21347–21361.

Chen, F., Zhang, S., Jiang, H., Ma, W., Korpelainen, H., Li, C. (2011) Comparative proteomics analysis of salt response reveals sex-related photosynthetic inhibition by salinity in *Populus cathayana* cuttings. *Journal of Proteome Research* 10(9): 3944–3958.

Chen, S., Jiang, J., Li, H., Liu, G. (2012) The Salt-responsive transcriptome of *Populus simonii* × *Populus nigra* via DGE. *Gene* 504(2): 203–212.

Chen, Z., Young, T. E., Ling, J., Chang, S. C., Gallie, D. R. (2003) Increasing vitamin C content of plants through enhanced ascorbate recycling. *Proceedings of the National Academy of Sciences of the United States of America* 100(6): 3525–3530.

Chen, Z., Young, T. E., Ling, J., Chang, S. C., Gallie, D. R. (2003) Increasing vitamin C content of plants through enhanced ascorbate recycling. *Proceedings of the National Academy of Sciences of the United States of America* 100(6): 3525–3530.

Chew, O., Whelan, J., Millar, A. H. (2003) Molecular definition of the ascorbate-glutathione cycle in *Arabidopsis* mitochondria reveals dual targeting of antioxidant defenses in plants. *The Journal of Biological Chemistry* 278(47): 46869–46877.

Colcombet, J., Hirt, H. (2008) Arabidopsis MAPKs: A complex signaling network involved in multiple biological processes. *The Biochemical Journal* 413(2): 217–226.

Considine, M. J., Foyer, C. H. (2014) Redox regulation of plant development. *Antioxidants and Redox Signaling* 21(9): 1305–1326.

Considine, M. J., Sandalio, L. M., Foyer, C. H. (2015) Unravelling how plants benefit from ROS and NO reactions, while resisting oxidative stress. *Annals of Botany* 116(4): 469–473.

Crook, E. M., Law, K. (1952) Glyoxalase: The role of the components. *The Biochemical Journal* 52(3): 492–499.

Cook, D., Fowler, S., Fiehn, O., Thomashow, M. F. (2004) A prominent role for the CBF cold response pathway in configuring the low-temperature metabolome of *Arabidopsis*. *Proceedings of the National Academy of Sciences* 101(42): 15243–15248.

Corpas, F. J., Palma, J. M., Sandalio, L. M., Valderrama, R., Barroso, J. B., Del Río, L. A. (2008) Peroxisomal xanthine oxidoreductase: Characterization of the enzyme from pea (*Pisum sativum* L.) leaves. *Journal of Plant Physiology* 165(13): 1319–1330.

Cramer, G. R., Urano, K., Delrot, S., Pezzotti, M., Shinozaki, K. (2011) Effects of abiotic stress on plants: A systems biology perspective. *BMC Plant Biology* 11(1): 163.

Cramer, G. R., Van Sluyter, S. C., Hopper, D. W., Pascovici, D., Keighley, T., Haynes, P. A. (2013) Proteomic analysis indicates massive changes in metabolism prior to the inhibition of growth and photosynthesis of grapevine (*Vitis vinifera* L.) in response to water deficit. *BMC Plant Biology* 13(1): 49.

Creissen, G. P., Mullineaux, P. M. (1995) Cloning and characterization of glutathione reductase cDNAs and identification of two genes encoding the tobacco enzyme. *Planta* 197(2): 422–425.

Creissen, G., Edwards, E. A., Enard, C., Wellburn, A., Mullineaux, P. (1992) Molecular characterization of glutathione reductase cDNAs from pea (*Pisum sativum* L.). *The Plant Journal* 2(1): 129–131.

Cushman, J. C., Bohnert, H. J. (2000) Genomic approaches to plant stress tolerance. *Current Opinion in Plant Biology* 3(2): 117–124.

Dalton, D. A., Baird, L. M., Langeberg, L., Taugher, C. Y., Anyan, W. R., Vance, C. P., Sarath, G. (1993) Subcellular localization of oxygen defense enzymes in soybean (*Glycine max* [L.] Merr.) root nodules. *Plant Physiology* 102(2): 481–489.

De Storme, N., Geelen, D. (2014) The impact of environmental stress on male reproductive development in plants: Biological processes and molecular mechanisms. *Plant, Cell and Environment* 37(1): 1–18.

Dean, R. T., Gieseg, S., Davies, M. J. (1993) Reactive species and their accumulation on radical-damaged proteins. *Trends in Biochemical Sciences* 18(11): 437–441.

Demidchik, V., Cuin, T. A., Svistunenko, D., Smith, S. J., Miller, A. J., Shabala, S., Sokolik, A., Yurin, V. (2010) *Arabidopsis* root K+-efflux conductance activated by hydroxyl radicals: Single-channel properties, genetic basis and involvement in stress-induced cell death. *Journal of Cell Science* 123(9): 1468–1479.

Devanathan, S., Erban, A., Perez-Torres, R., Kopka, J., Makaroff, C. A. (2014) *Arabidopsis thaliana* Glyoxalase 2–1 is required during abiotic stress but is not essential under normal plant growth. *PLoS ONE* 9(4): e95971.

Dietz, K. J. (2015) Efficient high light acclimation involves rapid processes at multiple mechanistic levels. *Journal of Experimental Botany* 66(9): 2401–2414.

Dietz, K. J., Turkan, I., Krieger-Liszkay, A. (2016) Redox and reactive oxygen species-dependent signaling in and from the photosynthesizing chloroplast. *Plant Physiology* 171(3): 1541–1550.

Dipierro, S., Borraccino, G. (1991) Dehydroascorbate reductase from potato tubers. *Phytochemistry* 30(2): 427–429.

Doke, N. (1985) NADPH-dependent O_2^--generation in membrane fractions isolated from wounded potato tubers inoculated with *Phytophtora infestans*. *Physiological Plant Pathology* 27(3): 311–322.

Dong, J., Wu, F. B., Zhang, G. P. (2006) Influence of cadmium on antioxidant capacity and four microelement concentrations in tomato (*Lycopersicon esculentum*). *Chemosphere* 64(10): 1659–1666.

Du, H., Liang, Y., Pei, K., Ma, K. (2011) UV radiation-responsive proteins in rice leaves: A proteomic analysis. *Plant and Cell Physiology* 52(2): 306–316.

Duan, J., Zhang, M., Zhang, H., Xiong, H., Liu, P., Ali, J., Li, J., Li, Z. (2012) OsMIOX, a myoinositol oxygenase gene, improves drought tolerance through scavenging of reactive oxygen species in rice (*Oryza sativa* L.). *Plant Science* 196: 143–151.

Edwards, E. A., Enard, C., Creissen, G. P., Mullineaux, P. M. (1994) Synthesis and properties of glutathione reductase in stressed peas. *Planta* 192(1): 137–143.

Edwards, E. A., Rawsthorne, S., Mullineaux, P. M. (1990) Subcellular distribution of multiple forms of glutathione reductase in pea (*Pisum sativum* L.). *Planta* 180(2): 278–284.

Eltayeb, A. E., Kawano, N., Badawi, G. H., Kaminaka, H., Sanekata, T., Morishima, I., Shibahara, T., Inanaga, S., Tanaka, K. (2006) Enhanced tolerance to ozone and drought stresses in transgenic tobacco overexpressing dehydroascorbate reductase in cytosol. *Physiologia Plantarum* 127(1): 57–65.

Eltayeb, A. E., Kawano, N., Badawi, G. H., Kaminaka, H., Sanekata, T., Shibahara, T., Inanaga, S., Tanaka, K. (2007) Overexpression of monodehydroascorbate reductase in transgenic tobacco confers enhanced tolerance to ozone, salt and polyethylene glycol stresses. *Planta* 225(5): 1255–1264.

Eltayeb, A. E., Yamamoto, S., Habora, M. E. E., Yin, L., Tsujimoto, H., Tanaka, K. (2011) Transgenic potato over-expressing *Arabidopsis* cytosolic AtDHAR1 showed higher tolerance to herbicide, drought and salt stresses. *Breeding Science* 61(1): 3–10.

Espartero, J., Sanchez-Aguayo, I., Pardo, J. M. (1995) Molecular characterization of Glyoxalase I from a higher plant: Up regulation by stress. *Plant Molecular Biology* 29(6): 1223–1233.

Evans, M. D., Dizdaroglu, M., Cooke, M. S. (2004) Oxidative DNA damage and disease: Induction, repair and significance. *Mutation Research* 567(1): 1–61.

Evers, D., Legay, S., Lamoureux, D., Hausman, J. F., Hoffmann, L., Renaut, J. (2012) Towards a synthetic view of potato cold and salt stress response by transcriptomic and proteomic analyses. *Plant Molecular Biology* 78(4–5): 503–514.

Eyidogan, F., Oz, M. T. (2005) Effect of salinity on antioxidant responses of chickpea seedlings. *Acta Physiologiae Plantarum* 29(5): 485–493.

Falvo, S., Di Carli, M., Desiderio, A., Benvenuto, E., Moglia, A., America, T., Lanteri, S., Acquadro, A. (2012) 2-D DIGE analysis of UV-C radiation-responsive proteins in globe artichoke leaves. *Proteomics* 12(3): 448–460.

Farrant, J., Brandt, W. F., Lidsey, G. G. (2007) An overview of the mechanisms of desiccation tolerance in selected angiosperm resurrection plants. *Plant Stress* 1(1): 72–84.

Fiehn, O. (2001) Combining genomics, metabolome analysis, and biochemical modelling to understand metabolic networks. *Comparative and Functional Genomics* 2(3): 155–168.

Fink, S. P., Reddy, G. R., Marnett, L. J. (1997) Mutagenicity in *Escherichia coli* of the major DNA adduct derived from the endogenous mutagen malondialdehyde. *Proceedings of the National Academy of Sciences of the United States of America* 94(16): 8652–8657.

Foyer, C. H., Halliwell, B. (1976) The presence of glutathione and glutathione reductase in chloroplasts: A proposed role in ascorbic acid metabolism. *Planta* 133(1): 21–25.

Foyer, C. H., Noctor, G. (2013) Redox signaling in plants. *Antioxidants and Redox Signaling* 18(16): 2087–2090.

Francki, M. G., Crawford, A. C., Oldach, K. (2012) Transcriptomics, proteomics and metabolomics: Integration of latest technologies for improving future wheat productivity. In *Sustainable Agriculture and New Technologies*, Benkeblia N. Ed., CRC Press: Boca Raton, FL, pp. 425–452.

Fu, L. J., Shi, K., Gu, M., Zhou, Y. H., Dong, D. K., Liang, W. S. (2010) Systemic induction and role of mitochondrial alternative oxidase and nitric oxide in a compatible tomato-tobacco mosaic virus interaction. *Molecular Plant–Microbe Interactions Journal* 23(1): 39–48.

Gadjev, I., Stone, J. M., Gechev, T. S. (2008) Programmed cell death in plants: New insights into redox regulation and the role of hydrogen peroxide. *International Review of Cell and Molecular Biology* 270: 87–144.

Gadjev, I., Vanderauwera, S., Gechev, T. S., Laloi, C., Minkov, I. N., Shulaev, V., Apel, K., et al. (2006) Transcriptomic footprints disclose specificity of reactive oxygen species signaling in *Arabidopsis*. *Plant Physiology* 141(2): 436–445.

Gardner, J. M., Aust, S. D. (2009) Quantification of hydroxyl radical produced during phacoemulsification. *Journal of Cataract and Refractive Surgery* 35(12): 2149–2153.

Gawronski, P., Witon, D., Vashutina, K., Bederska, M., Betlinski, B., Rusaczonek, A., Karpinski, S. (2014) Mitogen-activated protein kinase 4 is a salicylic acid-independent regulator of growth but not of photosynthesis in *Arabidopsis*. *Molecular Plant* 7(7): 1151–1166.

Gechev, T. S., Hille, J. (2005) Hydrogen peroxide as a signal controlling plant programmed cell death. *The Journal of Cell Biology* 168(1): 17–20.

Gechev, T., Willekens, H., Van Montagu, M., Inze, D., Van Camp, W., Toneva, V., Minkov, I. (2003) Different responses of tobacco antioxidant enzymes to light and chilling stress. *Journal of Plant Physiology* 160(5): 509–515.

Habibi, G. 2014. Hydrogen Peroxide H_2O_2: Generation, scavenging and signaling in plants. In: *Oxidative Damage to Plants*, Ahmad, P. (ed.). Academic Press, New York, NY: pp. 557–583

Ghosh, A. (2017) Genome-wide identification of glyoxalase genes in *Medicago truncatula* and their expression profiling in response to various developmental and environmental stimuli. *Frontiers in Plant Science* 8: 836.

Ghosh, A., Kushwaha, H. R., Hasan, M. R., Pareek, A., Sopory, S. K., Singla-Pareek, S. L. (2016) Presence of unique glyoxalase III proteins in plants indicates the existence of shorter route for methylglyoxal detoxification. *Scientific Reports* 6: 18358.

Ghosh, D., Xu, J. (2014) Abiotic stress responses in plant roots: A proteomics perspective. *Frontiers in Plant Science* 5: 6.

Gill, S. S., Tuteja, N. (2010) Reactive oxygen species and antioxidant machinery in abiotic stress tolerance in crop plants. *Plant Physiology and Biochemistry: PPB* 48(12): 909–930.

Gilroy, S., Bialasek, M., Suzuki, N., Gorecka, M., Devireddy, A. R., Karpinski, S., Mittler, R. (2016) ROS, calcium and electric signals: Key mediators of rapid systemic signaling in plants. *Plant Physiology* 171(3): 1606–1615.

Godfray, H. C., Beddington, J. R., Crute, I. R., Haddad, L., Lawrence, D., Muir, J. F., Pretty, J., et al. (2010) Food security: The challenge of feeding 9 billion people. *Science* 327(5967): 812–818.

Guan, Z., Chai, T., Zhang, Y., Xu, J., Wei, W. (2009) Enhancement of Cd tolerance in transgenic tobacco plants over expressing a Cd-induced catalase cDNA. *Chemosphere* 76(5): 623–630.

Gupta, A. S., Heinen, J. L., Holaday, A. S., Burke, J. J., Allen, R. D. (1993) Increased resistance to oxidative stress in transgenic plants that overexpress chloroplastic Cu/Zn superoxide dismutase. *Proceedings of the National Academy of Sciences of the United States of America* 90(4): 1629–1633.

Gupta, K. J., Igamberdiev, A. U. (2011) The anoxic plant mitochondrion as a nitrite: NO reductase. *Mitochondrion* 11(4): 537–543.

Gupta, K. J., Igamberdiev, A. U., Mur, L. A. (2012) NO and ROS homeostasis in mitochondria: A central role for alternative oxidase. *The New Phytologist* 195(1): 1–3.

Gygi, S. P., Rochon, Y., Franza, B. R., Aebersold, R. (1999) Correlation between protein and mRNA abundance in yeast. *Molecular and Cellular Biology* 19(3): 1720–1730.

Hasanuzzaman, M., Hossain, M. A., Teixeira da Silva, J. A., Fujita, M. (2012) Plant responses and tolerance to abiotic oxidative stress: Antioxidant defense is a key factor. In *Crop Stress and Its Management: Perspectives and Strategies*, Bandi, V., Shanker, A.K., Shanker, C., Mandapaka, M., Eds. Springer: Berlin, Germany, pp. 261–231.

Hasanuzzaman, M., Nahar, K., Fujita, M. (2013) Extreme temperatures, oxidative stress and antioxidant defense in plants. In *Abiotic Stress—Plant Responses and Applications in Agriculture*, Vahdati, K. Leslie, C., Eds. Rijeka, Yugoslavia. InTech, pp. 169–205.

Hashiguchi, A., Ahsan, N., Komatsu, S. (2010) Proteomics application of crops in the context of climatic changes. *Food Research International* 43(7): 1803–1813.

He, C. J., Finlayson, S. A., Drew, M. C., Jordan, W. R., Morgan, P. W. (1996a) Ethylene biosynthesis during aerenchyma formation in roots of maize subjected to mechanical impedance and hypoxia. *Plant Physiology* 112(4): 1679–1685.

He, C. J., Jin, J. P., Cobb, B. G., Morgan, P. W., Drew, M. C. (1996b) Transduction of an ethylene signal is required for cell death and lysis in the root cortex of maize during aerenchyma formation induced by hypoxia. *Plant Physiology* 112(2): 699–699.

Hernández, G., Ramírez, M., Valdés-López, O., Testaye, M., Graham, M. A., Czechowski, T., Schlereth, A., et al. (2007) Phosphorus stress in common bean: Root transcript and metabolic responses. *Plant Physiology* 144(2): 752–767.

Hettenhausen, C., Schuman, M. C., Wu, J. (2015) MAPK signaling: A key element in plant defense response to insects. *Insect Science* 22(2): 157–164.

Hirayama, T., Shinozaki, K. (2010) Research on plant abiotic stress responses in the post-genome era: Past, present and future. *The Plant Journal* 61(6): 1041–1052.

Hoque, T. S., Hossain, M. A., Mostofa, M. G., Burritt, D. J., Fujita, M., Tran, L. S. P. (2016) Methylglyoxal: An emerging signaling molecule in plant abiotic stress responses and tolerance. *Frontiers in Plant Science* 7: 1341.

Horemans, N., Foyer, C. H., Asard, H. (2000) Transport and action of ascorbate at the plant plasma membrane. *Trends in Plant Science* 5(6): 263–267.

Hossain, M. A., Asada, K. (1984) Purification of dehydroascorbate reductase from spinach and its characterization as a thiol enzyme. *Plant and Cell Physiology* 25(1): 85–92.

Hossain, M. A., Nakano, Y., Asada, K. (1984) Monodehydroascorbate reductase in spinach chloroplasts and its participation in regeneration of ascorbate for scavenging hydrogen peroxide. *Plant and Cell Physiology* 25(3): 385–395.

Hossain, M. A., Teixeira da Silva, J. A., Fujita, M. (2011) Glyoxalase system and reactive oxygen species detoxification system in plant abiotic stress response and tolerance: An intimate relationship. In *Abiotic Stress in Plants—Mechanisms and Adaptations*, Shanker, A. K., Ed. Intech: Rejika, Yugoslavia, pp. 235–266.

Hossain, M., Hossain, A.M., Fujita, Z.M., (2009) Stress-induced changes of methylglyoxal level and Glyoxalase I activity in pumpkin seedlings and cDNA cloning of Glyoxalase I gene. *Australian Journal of Crop Science* 3(2): 53–64.

Huang, S., Van Aken, O., Schwarzlander, M., Belt, K., Millar, A. H. (2016) The roles of mitochondrial reactive oxygen species in cellular signaling and stress responses in plants. *Plant Physiology* 171(3): 1551–1559.

Huang, W., Yang, X., Yao, S., Lwinoo, T., He, H., Wang, A., Li, C., He, L. (2014) Reactive oxygen species burst induced by aluminum stress triggers mitochondria-dependent programmed cell death in peanut root tip cells. *Plant Physiology and Biochemistry* 82: 76–84.

Ichimura, T., Wakamiya-Tsuruta, A., Itagaki, C., Taoka, M., Hayano, T., Natsume, T., Isobe, T. (2002) Phosphorylation-dependent interaction of kinesin light chain 2 and the 14-3-3 protein. *Biochemistry* 41(17): 5566–5572.

Ishikawa, T., Yoshimura, K., Sakai, K., Tamoi, M., Takeda, T., Shigeoka, S. (1998) Molecular characterization and physiological role of a glyoxysome-bound ascorbate peroxidase from spinach. *Plant and Cell Physiology* 39(1): 23–34.

Jammes, F., Song, C., Shin, D., Munemasa, S., Takeda, K., Gu, D., Cho, D., et al. (2009) MAP kinases MPK9 and MPK12 are preferentially expressed in guard cells and positively regulate ROS-mediated ABA signaling. *Proceedings of the National Academy of Sciences of the United States of America* 106(48): 20520–20525.

Jaspers, P., Kangasjarvi, J. (2010) Reactive oxygen species in abiotic stress signaling. *Physiologia Plantarum* 138(4): 405–413.

Jiménez, A., Hernández, J. A., del Río, L. A., Sevilla, F. (1997) Evidence for the presence of the ascorbate-glutathione cycle in mitochondria and peroxisomes of pea leaves. *Plant Physiology* 114(1): 275–284.

Jones, D. P., Park, Y., Gletsu-Miller, N., Liang, Y., Yu, T., Accardi, C. J., Ziegler, T. R. (2011) Dietary sulfur amino acid effects on fasting plasma cysteine/cystine redox potential in humans. *Nutrition* 27(2): 199–205.

Kalapos, M. P. (1999) Methylglyoxal in living organisms: Chemistry, biochemistry, toxicology and biological implications. *Toxicology Letters* 110(3): 145–175.

Kaminaka, H., Morita, S., Nakajima, M., Masumura, T., Tanaka, K. (1998) Gene cloning and expression of cytosolic glutathione reductase in rice (*Oryza sativa* L.). *Plant and Cell Physiology* 39(12): 1269–1280.

Karpinski, S., Szechynska-Hebda, M., Wituszynska, W., Burdiak, P. (2013) Light acclimation, retrograde signalling, cell death and immune defences in plants. *Plant, Cell and Environment* 36(4): 736–744.

Kaur, C., Kushwaha, H. R., Mustafiz, A., Pareek, A., Sopory, S. K., Singla-Pareek, S. L. (2015a) Analysis of global gene expression profile of rice in response to methylglyoxal indicates its possible role as a stress signal molecule. *Frontiers in Plant Science* 6: 682.

Kaur, C., Sharma, S., Singla-Pareek, S. L., Sopory, S. K. (2015b) Methylglyoxal, triose phosphate isomerase and glyoxalase pathway: Implications in abiotic stress and signaling in plants. In *Elucidation of Abiotic Stress Signaling in Plants*, Pandey, G.K., Ed. Springer: New York, NY, pp. 347–366.

Kaur, C., Singla-Pareek, S. L., Sopory, S. K. (2014) Glyoxalase and methylglyoxal as biomarkers for plant stress tolerance. *Critical Reviews in Plant Sciences* 33(6): 429–456.

Kavitha, K., George, S., Venkataraman, G., Parida, A. (2010) A salt-inducible chloroplastic monodehydroascorbate reductase from halophyte *Avicennia marina* confers salt stress tolerance on transgenic plants. *Biochimie* 92(10): 1321–1329.

Kehrer, J. P. (2000) The Haber–Weiss reaction and mechanisms of toxicity. *Toxicology* 149(1): 43–50.

Kerchev, P. I., Waszczak, C., Lewandowska, A., Willems, P., Shapiguzov, A., Li, Z., Alseekh, S., et al. (2016) Lack of glycolate oxidase 1, but not glycolate oxidase 2, attenuates the photorespiratory phenotype of CATALASE2-deficient *Arabidopsis*. *Plant Physiology* 171(3): 1704–1719.

Kim, K., Portis, A. R. (2004) Oxygen-dependent H_2O_2 production by RuBisCO. *FEBS Letters* 571(1–3): 124–128.

Kim, Y., Wang, M., Bai, Y., Zeng, Z., Guo, F., Han, N., Bian, H., et al. (2014) Bcl-2 suppresses activation of VPEs by inhibiting cytosolic Ca^{2+} level with elevated K^+ efflux in NaCl-induced PCD in rice. *Plant Physiology and Biochemistry* 80: 168–175.

Kleine, T., Leister, D. (2016) Retrograde signaling: Organelles go networking. *Biochimica et Biophysica Acta* 1857: 1313–1325.

Komatsu, S., Sugimoto, T., Hoshino, T., Nanjo, Y., Furukawa, K. (2010) Identification of flooding stress responsible cascades in root and hypocotyls of soybean using proteome analysis. *Amino Acids* 38(3): 729–738.

Komatsu, S., Yamada, E., Furukawa, K. (2009) Cold stress changes the concanavalin A-positive glycosylation pattern of proteins expressed in the basal parts of rice leaf sheaths. *Amino Acids* 36(1): 115–123.

Koukalova, B., Kovarik, A., Fajkus, J., Siroky, J. (1997) Chromatin fragmentation associated with apoptotic changes in tobacco cells exposed to cold stress. *FEBS Letters* 414(2): 289–292.

Kovtun, Y., Chiu, W. L., Tena, G., Sheen, J. (2000) Functional analysis of oxidative stress-activated mitogen-activated protein kinase cascade in plants. *Proceedings of the National Academy of Sciences of the United States of America* 97(6): 2940–2945.

Kubo, A., Aono, M., Nakajima, N., Saji, H., Tanaka, K., Kondo, N. (1999) Differential responses in activity of antioxidant enzymes to different environmental stresses in *Arabidopsis thaliana*. *Journal of Plant Research* 112(3): 279–290.

Kubo, A., Sano, T., Saji, H., Tanaka, K., Kondo, N., Tanaka, K. (1993) Primary structure and properties of glutathione reductase from *Arabidopsis thaliana*. *Plant and Cell Physiology* 34(8): 1259–1266.

Kwak, S. S., Lim, S., Tang, L., Kwon, S. Y., Lee, H. S. (2009) Enhanced tolerance of transgenic crops expressing both SOD and APX in chloroplasts to multiple environmental stress. In *Salinity and Water Stress*, Ashraf, M., Ozturk, M. and Athar, H.R., Eds. Springer: Dordrecht, the Netherlands, pp. 197–203.

Kwon, K., Choi, D., Hyun, J. K., Jung, H. S., Baek, K., Park, C. (2013) Novel glyoxalases from *Arabidopsis thaliana*. *The FEBS Journal* 280(14): 3328–3339.

Kwon, S. Y., Jeong, Y. J., Lee, H. S., Kim, J. S., Cho, K. Y., Allen, R. D., Kwak, S. S. (2002) Enhanced tolerances of transgenic tobacco plants expressing both superoxide dismutase and ascorbate peroxidase in chloroplasts against methyl viologen-mediated oxidative stress. *Plant, Cell and Environment* 25(7): 873–882.

Laino, P., Shelton, D., Finnie, C., De Leonardis, A. M., Mastrangelo, A. M., Svensson, B., Lafiandra, D., Masci, S. (2010) Comparative proteome analysis of metabolic proteins from seeds of durum wheat (cv. Svevo) subjected to heat stress. *Proteomics* 10(12): 2359–2368.

Lee, D. G., Ahsan, N., Lee, S. H., Lee, J. J., Bahk, J. D., Kang, K. Y., Lee, B. H. (2009) Chilling stress-induced proteomic changes in rice roots. *Journal of Plant Physiology* 166(1): 1–11.

Lee, K. P., Kim, C., Landgraf, F., Apel, K. (2007a) EXECUTER1-and EXECUTER2-dependent transfer of stress-related signals from the plastid to the nucleus of *Arabidopsis thaliana*. *Proceedings of the National Academy of Sciences of the United States of America* 104(24): 10270–10275.

Lee, Y. P., Kim, S. H., Bang, J. W., Lee, H. S., Kwak, S. S., Kwon, S. Y. (2007b) Enhanced tolerance to oxidative stress in transgenic tobacco plants expressing three antioxidant enzymes in chloroplasts. *Plant and Cell Reports* 26(5): 591–598.

Li, F., Wu, Q. Y., Sun, Y. L., Wang, L. Y., Yang, X. H., Meng, Q. W. (2010) Overexpression of chloroplastic monodehydroascorbate reductase enhanced tolerance to temperature and methyl viologen-mediated oxidative stresses. *Physiologia Plantarum* 139(4): 421–434.

Li, X., Lawas, L. M., Malo, R., Glaubitz, U., Erban, A., Mauleon, R., Heuer, S., et al. (2015) Metabolic and transcriptomic signatures of rice floral organs reveal sugar starvation as a factor in reproductive failure under heat and drought stress. *Plant, Cell and Environment* 38(10): 2171–2192.

Li, Z., Yue, H., Xing, D. (2012). MAP kinase 6-mediated activation of vacuolar processing enzyme modulates heat shock-induced programmed cell death in *Arabidopsis*. *New Phytologist* 195(1): 85–96.

Lim, S., Kim, Y. H., Kim, S. H., Kwon, S., Lee, H., Kim, J., Cho, K., Paek, K., Kwak, S. (2007) Enhanced tolerance of transgenic sweet potato plants that express both CuZnSOD and APX in chloroplasts to methyl viologen-mediated oxidative stress and chilling. *Molecular Breeding* 19(3): 227–239.

Lin, F., Xu, J., Shi, J., Li, H., Li, B. (2010) Molecular cloning and characterization of a novel glyoxalase I gene TaGly I in wheat (*Triticum aestivum* L.). *Molecular Biology Reports* 37(2): 729–735.

Lister, R., Gregory, B. D., Ecker, J. R. (2009) Next is now: New technologies for sequencing of genomes, transcriptomes and beyond. *Current Opinion in Plant Biology* 12(2): 107–118.

Liu, C., Zhao, L., Yu, G. (2011) The dominant glutamic acid metabolic flux to produce gamma-amino butyric acid over proline in *Nicotiana tabacum* leaves under water stress relates to its significant role in antioxidant activity. *Journal of Integrative Plant Biology* 53(8): 608–618.

Liu, J., Li, Z., Wang, Y., Xing, D. (2014) Overexpression of *ALTERNATIVE OXIDASE1a* alleviates mitochondria-dependent programmed cell death induced by aluminium phytotoxicity in *Arabidopsis*. *Journal of Experimental Botany* 65(15): 4465–4478.

Loew (1901) Catalase, new enzyme of general occurrence, with special reference to the tobacco plant. *Science* 68: 47.

López-Huertas, E., Corpas, F. J., Sandalio, L. M., Del Río, L. A. (1999) Characterization of membrane polypeptides from pea leaf peroxisomes involved in superoxide radical generation. *The Biochemical Journal* 337(3): 531–536.

Lumbreras, V., Vilela, B., Irar, S., Solé, M., Capellades, M., Valls, M., Coca, M., Pagès, M. (2010) MAPK phosphatase MKP2 mediates dis-ease responses in Arabidopsis and functionally interacts with MPK3 and MPK6. *The Plant Journal* 63(6): 1017–1030.

Lunde, C., Baumann, U., Shirley, N. J., Drew, D. P., Fincher, G. B. (2006) Gene structure and expression pattern analysis of three monodehydroascorbate reductase (Mdhar) genes in *Physcomitrella patens*: Implications for the evolution of the MDHAR family in plants. *Plant Molecular Biology* 60(2): 259–275.

Lushchak, V. I. (2012) Glutathione homeostasis and functions: Potential targets for medical interventions. *Journal of Amino Acids* 736837.

Lyubushkina, I. V., Grabelnych, O. I., Pobezhimova, T. P., Stepanov, A. V., Fedyaeva, A. V., Fedoseeva, I. V., Voinikov, V. K. (2014) Winter wheat cells subjected to freezing temperature undergo death process with features of programmed cell death. *Protoplasma* 251(3): 615–623.

Madhusudhan, R., Ishiawa, T., Sawa, Y., Shigeoka, S., Shibata, H. (2003) Characterization of an ascorbate peroxidase in plastids of tobacco BY-2 cells. *Physiologia Plantarum* 117(4): 550–557.

Mallick, N., Mohn, F. H. (2000) Reactive oxygen species: Response of algal cells. *Journal of Plant Physiology* 157(2): 183–193.

Mano, J., Miyatake, F., Hiraoka, E., Tamoi, M. (2009) Evaluation of the toxicity of stress-related aldehydes to photosynthesis in chloroplasts. *Planta* 230(4): 639–648.

Marinho, H. S., Real, C., Cyrne, L., Soares, H., Antunes, F. (2014) Hydrogen peroxide sensing, signaling and regulation of transcription factors. *Redox Biology* 2: 535–562.

Marsoni, M., Cantara, C., De Pinto, M. C., Gadaleta, C., DeGara, L., Bracale, M., Vannini, C. (2010) Exploring the soluble proteome of tobacco bright yellow-2 cells at the switch towards different cell fates in response to heat shocks. *Plant, Cell and Environment* 33(7): 1161–1175.

McKersie, B. D., Bowley, S. R., Harjanto, E., Leprince, O. (1996) Water-deficit tolerance and field performance of transgenic alfalfa overexpressing superoxide dismutase. *Plant Physiology* 111(4): 1177–1181.

McKersie, B. D., Chen, Y., De Beus, M., Bowley, S. R., Bowler, C., Inzé, D., D'Halluin, K., Botterman, J. (1993) Superoxide dismutase enhances tolerance of freezing stress in transgenic alfalfa (*Medicago sativa* L.). *Plant Physiology* 103(4): 1155–1163.

Menges, M., Dóczi, R., Okrész, L., Morandini, P., Mizzi, L., Soloviev, M., Murray, J. A., Bögre, L. (2008) Comprehensive gene expression atlas for the *Arabidopsis* MAP kinase signalling pathways. *The New Phytologist* 179(3): 643–662.

Mertins, P., Qiao, J. W., Patel, J., Udeshi, N. D., Clauser, K. R., Mani, D. R., Burgess, M. W., et al. (2013) Integrated proteomic analysis of post-translational modifications by serial enrichment. *Nature Methods* 10(7): 634–637.

Miao, Y., Lv, D., Wang, P., Wang, X. C., Chen, J., Miao, C., Song, C. P. (2006) An *Arabidopsis* glutathione peroxidase functions as both a redox transducer and a scavenger in abscisic acid and drought stress responses. *The Plant Cell* 18(10): 2749–2766.

Mignolet-Spruyt, L., Xu, E., Idanheimo, N., Hoeberichts, F. A., Muhlenbock, P., Brosche, M., Van Breusegem, F., Kangasjarvi, J. (2016) Spreading the news: Subcellular and organellar reactive oxygen species production and signalling. *Journal of Experimental Botany* 67(13): 3831–3844.

Miller, G., Suzuki, N., Ciftci-Yilmaz, S., Mittler, R. (2010) Reactive oxygen species homeostasis and signaling during drought and salinity stresses. *Plant, Cell and Environment* 33(4): 453–467.

Mirzaei, M., Soltani, N., Sarhadi, E., Pascovici, D., Keighley, T., Salekdeh, G. H., Haynes, P. A., Atwell, B. J. (2012) Shotgun proteomic analysis of long-distance drought signaling in rice roots. *Journal of Proteome Research* 11(1): 348–358.

Mittler, R. (2002) Oxidative stress, antioxidants and stress tolerance. *Trends in Plant Science* 7(9): 405–410.

Mittler, R., Blumwald, E. (2010) Genetic engineering for modern agriculture: Challenges and perspectives. *Annual Review of Plant Biology* 61: 443–462.

Mittler, R., Vanderauwera, S., Gollery, M., Van Breusegem, F. (2004) Reactive oxygen gene network of plants. *Trends in Plant Science* 9(10): 490–498.

Mittova, V., Guy, M., Tal, M., Volokita, M. (2004) Salinity upregulates the antioxidative system in root mitochondria and peroxisomes of the wild salt-tolerant tomato species *Lycopersicon pennellii. Journal of Experimental Botany* 55(399): 1105–1113.

Mittova, V., Tal, M., Volokita, M., Guy, M. (2003) Up-regulation of the leaf mitochondrial and peroxisomal antioxidative systems in response to salt-induced oxidative stress in the wild salt tolerant tomato species *Lycopersicon pennellii. Plant, Cell and Environment* 26(6): 845–856.

Miyake, C., Schreiber, U., Hormann, H., Sano, S., Kozi, A. (1998) The FAD-enzyme monodehydroascorbate radical reductase mediates photoproduction of superoxide radicals in spinach thylakoid membranes. *Plant and Cell Physiology* 39(8): 821–829.

Mochida, K., Shinozaki, K. (2011) Advances in omics and Bioinformatics tools for systems analyses of plant functions. *Plant and Cell Physiology* 52(12): 2017–2038.

Mohammadi, P. P., Moieni, A., Hiraga, S., Komatsu, S. (2012) Organ-specific proteomic analysis of drought-stressed soybean seedlings. *Journal of Proteomics* 75(6): 1906–1923.

Molina, C., Rotter, B., Horres, R., Udupa, S. M., Besser, B., Bellarmino, L., Baum, M., et al. (2008) Super SAGE: The Drought Stress, responsive transcriptome of Chickpea Roots. *BMC Genomics* 9: 553.

Moussa, R., Abdel-Aziz, S. M. (2008) Comparative response of drought tolerant and drought sensitive maize genotypes to water stress. *Australian Journal of Crop Sciences* 1(1): 31–36.

Muhlenbock, P., Szechynska-Hebda, M., Plaszczyca, M., Baudo, M., Mateo, A., Mullineaux, P. M., Parker, J. E., Karpinska, B., Karpinski, S. (2008) Chloroplast signaling and LESION SIMULATING DISEASE 1 regulate crosstalk between light acclimation and immunity in *Arabidopsis. The Plant Cell* 20(9): 2339–2356.

Mullineaux, P., Enard, C., Hellens, R., Creissen, G. (1996) Characterisation of a glutathione reductase gene and its genetic locus from pea (*Pisum sativum* L.). *Planta* 200(2): 186–194.

Munns, R., Tester, M. (2008) Mechanisms of salinity tolerance. *Annual Review of Plant Biology* 59: 651–681.

Murata, N., Takahashi, S., Nishiyama, Y., Allakhverdiev, S. I. (2007) Photo inhibition of photosystem II under environmental stress. *Biochimemica et Biophysica Acta* 1767(6): 414–421.

Mustafiz, A., Singh, A. K., Pareek, A., Sopory, S. K., Singla-Pareek, S. L. (2011) Genome-wide analysis of rice and *Arabidopsis* identifies two glyoxalase genes that are highly expressed in abiotic stresses. *Functional and Integrative Genomics* 11(2): 293–305.

Nahar, K., Hasanuzzaman, M., Alam, M. M., Fujita, M. (2015) Exogenous spermidine alleviates low temperature injury in mung bean (*Vigna radiata* L.) seedlings by modulating ascorbate-glutathione and glyoxalase pathway. *International Journal of Molecular Sciences* 16(12): 30117–30132.

Nakagami, H., Kiegerl, S., Hirt, H. (2004) OMTK1, a novel MAPKKK, channels oxidative stress signaling through direct MAPK interaction. *The Journal of Biological Chemistry* 279(26): 26959–26966.

Nakagami, H., Soukupova, H., Schikora, A., Zarsky, V., Hirt, H. (2006) A mitogen-activated protein kinase mediates reactive oxygen species homeostasis in *Arabidopsis. The Journal of Biological Chemistry* 281(50): 38697–38704.

Nam, M. H., Huh, S. M., Kim, K. M., Park, W. J., Seo, J. B., Cho, K., Kim, D. Y., Kim, B. G., Yoon, I. S. (2012) Comparative proteomic analysis of early salt stress-responsive proteins in roots of SnRK2 transgenic rice. *Proteome Science* 10(1): 25.

Nishizawa, A., Yabuta, Y., Shigeoka, S. (2008) Galactinol and raffinose constitute a novel function to protect plants from oxidative damage. *Plant Physiology* 147(3): 1251–1263.

Noctor, G., Foyer, C. H. (1998) Ascorbate and glutathione: Keeping active oxygen under control. *Annual Review of Plant Biology* 49: 249–279.

Noctor, G., Foyer, C. H. (2016) Intracellular redox compartmentation and ROS-related communication in regulation and signaling. *Plant Physiology* 171(3): 1581–1592.

Oidaira, H., Sano, S., Koshiba, T., Ushimaru, T. (2000) Enhancement of oxidative enzyme activities in chilled rice seedlings. *Journal of Plant Physiology* 156(5–6): 811–813.

Parker, R., Flowers, T. J., Moore, A. L., Harpham, N. V. (2006) An accurate and reproducible method for proteome profiling of the effects of salt stress in the rice leaf lamina. *Journal of Experimental Botany* 57(5): 1109–1118.

Pastori, G. M., Trippi, V. S. (1992) Oxidative stress induces high rate of glutathione reductase synthesis in a drought-resistant maize strain. *Plant and Cell Physiology* 33(7): 957–961.

Patterson, W. R., Poulos, T. L. (1995) Crystal structure of recombinant pea cytosolic ascorbate peroxidase. *Biochemistry* 34(13): 4331–4341.

Pei, Z. M., Murata, Y., Benning, G., Thomine, S., Klusener, B., Allen, G. J., Grill, E., Schroeder, J. I. (2000) Calcium channels activated by hydrogen peroxide mediate abscisic acid signaling in guard cell. *Nature* 406(6797): 731–734.

Piterkova, J., Luhova, L., Mieslerova, B., Lebeda, A., Petrivalsky, M. (2013) Nitric oxide and reactive oxygen species regulate the accumulation of heat shock proteins in tomato leaves in response to heat shock and pathogen infection. *Plant Science* 207: 57–65.

Pitzschke, A., Djamei, A., Bitton, F., Hirt, H. (2009) A major role of the MEKK1-MKK1/2MPK4 pathway in ROS signalling. *Molecular Plant* 2(1): 120–137.

Pitzschke, A., Hirt, H. (2009) Disentangling the complexity of mitogen-activated protein kinases and reactive oxygen species signaling. *Plant Physiology* 149(2): 606–615.

Qadir, S., Qureshi, M. I., Javed, S., Abdin, M. Z. (2004) Genotypic variation in phytoremediation potential of *Brassica juncea* cultivars exposed to Cd stress. *Plant Science* 167(5): 1171–1181.

Qin, A., Shi, Q., Yu, X. (2011) Ascorbic acid contents in transgenic potato plants overexpressing two dehydroascorbate reductase genes. *Molecular Biology Reports* 38(3): 1557–1566.

Racker, E. (1951) The mechanism of action of glyoxalase. *The Journal of Biological Chemistry* 190(2): 685–696.

Rajhi, I., Yamauchi, T., Takahashi, H., Nishiuchi, S., Shiono, K., Watanabe, R., Mliki, A., et al. (2011) Identification of genes expressed in maize root cortical cells during lysigenous aerenchyma formation using laser micro dissection and microarray analyses. *The New Phytologist* 190(2): 351–368.

Rajwanshi, R., Kumar, D., Yusuf, M. A., DebRoy, S., Sarin, N. B. (2016) Stress-inducible overexpression of glyoxalase I is preferable to its constitutive overexpression for abiotic stress tolerance in transgenic *Brassica juncea*. *Molecular Breeding* 36(6): 1–15.

Rampino, P., Mita, G., Fasano, P., Borrelli, G. M., Aprile, A., Dalessandro, G., De Bellis, L., Perrotta, C. (2012) Novel durum wheat genes up-regulated in response to a combination of heat and drought stress. *Plant Physiology and Biochemistry* 56: 72–78.

Reginato, M. A., Castagna, A., Furlan, A., Castro, S., Ranieri, A., Luna, V. (2014) Physiological responses of a halophytic shrub to salt stress by Na_2SO_4 and NaCl: Oxidative damage and the role of polyphenols in antioxidant protection. *AoB Plants* 6: plu042.

Riechmann, J. L., Heard, J., Martin, G., Reuber, L., Jiang, C., Keddie, J., Adam, L., et al. (2000) *Arabidopsis* transcription factors: Genome-wide comparative analysis among eukaryotes. *Science* 290(5499): 2105–2110.

Rodriguez, M. C., Petersen, M., Mundy, J. (2010) Mitogen activated protein kinase signaling in plants. *Annual Review of Plant Biology* 61: 621–649.

Rodríguez-Serrano, M., Romero-Puertas, M. C., Sanz-Fernández, M., Hu, J., Sandalio, L. M. (2016) Peroxisomes extend peroxules in a fast response to stress via a reactive oxygen species-mediated induction of the peroxin PEX11a. *Plant Physiology* 171(3). 1665–1674.

Rodziewicz, P., Swarcewicz, B., Chmielewska, K., Wojakowska, A., Stobiecki, M. (2014) Influence of abiotic stresses on plant proteome and metabolome changes. *Acta Physiologiae Plantarum* 36(1): 1–19.

Rollins, J. A., Habte, E., Templer, S. E., Colby, T., Schmidt, J., Korff, M. (2013) Leaf proteome alterations in the context of physiological and morphological responses to drought and heat stress in barley (*Hordeum vulgare* L.). *Journal of Experimental Botany* 64(11): 3201–3212.

Rubio, M. C., Bustos-Sanmamed, P., Clemente, M. R., Becana, M. (2009) Effects of salt stress on the expression of antioxidant genes and proteins in the model legume *Lotus japonicas*. *The New Phytologist* 181(4): 851–859.

Sagone Jr., A. L., Husney, R. M., O'Dorisio, M. S., Metz, E. N. (1984) Mechanisms for the oxidation of reduced glutathione by stimulated granulocytes. *Blood* 63(1): 96–104.

Sairam, R. K., Rao, K. V., Srivastava, G. C. (2002) Differential response of wheat genotypes to long-term salinity stress in relation to oxidative stress, antioxidant activity and osmolyte concentration. *Plant Science* 163(5): 1037–1046.

Saito, R., Yamamoto, H., Makino, A., Sugimoto, T., Miyake, C. (2011) Methylglyoxal functions as hill oxidant and stimulates the photoreduction of O_2 at photosystem I: A symptom of plant diabetes. *Plant, Cell and Environment* 34(9): 1454–1464.

Sakihama, Y., Mano, J., Sano, S., Asada, K., Yamasaki, H. (2000) Reduction of phenoxyl radicals mediated by monodehydroascorbate reductase. *Biochemical and Biophysical Research Communications* 279(3): 949–954.

Sanchez, D. H., Schwabe, F., Erban, A., Udvardi, M. K., Kopka, J. (2012) Comparative metabolomics of drought acclimation in model and forage legumes. *Plant, Cell and Environment* 35(1): 136–149.

Sanchez-Bel, P., Egea, I., Sanchez-Ballesta, M. T., Martinez-Madrid, C., Fernandez-Garcia, N., Romojaro, F., Olmos, E., et al. (2012) Understanding the mechanisms of chilling injury in bell pepper fruits using the proteomic approach. *Journal of Proteomics* 75(17): 5463–5478.

Sasaki, K., Saito, T., Lämsä, M., Oksman-Caldentey, K. M., Suzuki, M., Ohyama, K., Muranaka, T., Ohara, K., Yazaki, K. (2007) Plants utilize isoprene emission as a thermo tolerance mechanism. *Plant and Cell Physiology* 48(9): 1254–1262.

Saxena, M., Bisht, R., Roy, S. D., Sopory, S. K., Bhalla-Sarin, N. (2005) Cloning and characterization of a mitochondrial Glyoxalase II from *Brassica juncea* that is upregulated by NaCl, Zn, and ABA. *Biochemical and Biophysical Research Communications* 336(3): 813–819.

Saxena, M., Roy, S., Singla-Pareek, S., S., Sopory, Bhalla-Sarin, N. (2011) Overexpression of the glyoxalase II gene leads to enhanced salinity tolerance in *Brassica juncea*. *The Open Plant Science Journal* 5(1): 23–28.

Scandalios, G. J., Guan, L., Polidoros, A. N. (1997) Catalases in plants: Gene structure, properties, regulation and expression. In *Oxidative Stress and the Molecular Biology of Antioxidants Defenses*, Scandalios, J.G., Ed. Cold Spring Harbor Laboratory Press: New York, NY, pp. 343–406.

Scandalios, J. G. (2005) Oxidative stress: Molecular perception and transduction of signals triggering antioxidant gene defenses. *Brazilian Journal of Medical and Biological Research* 38(7): 995–1014.

Schopfer, P., Plachy, C., Frahry, G. (2001) Release of reactive oxygen intermediates (superoxide radicals, hydrogen peroxide, and hydroxyl radicals) and peroxidase in germinating radish seeds controlled by light, gibberellin and abscisic acid. *Plant Physiology* 125(4): 1591–1602.

Sewelam, N., Kazan, K., Schenk, M. P. (2016) Global plant stress signaling: Reactive oxygen species at the crossroad. *Frontiers in Plant Science* 7: 187.

Shabala, S. (2009) Salinity and programmed cell death: Unraveling mechanisms for ion specific signalling. *Journal of Experimental Botany* 60(3): 709–712.

Shabala, S., Shabala, L., Barcelo, J., Poschenrieder, C. (2014) Membrane transporters mediating root signaling and adaptive responses to oxygen deprivation and soil flooding. *Plant, Cell and Environment* 37(10): 2216–2233.

Sharkey, T. D., Singsaas, E. L. (1995) Why plants emit isoprene. *Nature* 374(6525): 769–769.

Sharma, P., Dubey, R. S. (2004) Ascorbate peroxidase from rice seedlings: Properties of enzyme isoforms, effects of stresses and protective roles of osmolytes. *Plant Science* 167(3): 541–550.

Sharma, P., Dubey, R. S. (2005) Drought induces oxidative stress and enhances the activities of antioxidant enzymes in growing rice seedlings. *Plant Growth Regulation* 46(3): 209–221.

Shigeoka, S., Ishikawa, T., Tamoi, M., Miyagawa, Y., Takeda, T., Yabuta, Y., Yoshimura, K. (2002) Regulation and function of ascorbate peroxidase isoenzymes. *Journal of Experimental Botany* 53(372): 1305–1319.

Shu, D. F., Wang, L. Y., Duan, M., Deng, Y. S., Meng, Q. W. (2011) Antisense-mediated depletion of tomato chloroplast glutathione reductase enhances susceptibility to chilling stress. *Plant Physiology and Biochemistry* 49(10): 1228–1237.

Sicher, R. C., Barnaby, J. Y. (2012) Impact of carbon dioxide enrichment on the responses of maize leaf transcripts and metabolites to water stress. *Physiologia Plantarum* 144(3): 238–253.

Singla-Pareek, S. L., Reddy, M. K., Sopory, S. K. (2003) Genetic engineering of the glyoxalase pathway in tobacco leads to enhanced salinity tolerance. *Proceedings of the National Academy of Sciences of the United States of America* 100(25): 14672–14677.

Singla-Pareek, S. L., Yadav, S. K., Pareek, A., Reddy, M. K., Sopory, S. K. (2006) Transgenic tobacco overexpressing glyoxalase pathway enzymes grow and set viable seeds in zinc-spiked soils. *Plant Physiology* 140(2): 613–623.

Singla-Pareek, S. L., Yadav, S. K., Pareek, A., Reddy, M. K., Sopory, S. K. (2008) Enhancing salt tolerance in a crop plant by overexpression of glyoxalase II. *Transgenic Research* 17(2): 171–180.

Smirnoff, N., Wheeler, G. L. (2000) Ascorbic acid in plants: Biosynthesis and function. *Critical Reviews in Biochemistry and Molecular Biology* 35(4): 291–314.

Soda, N., Wallace, S., Karan, R. (2015) Omics study for abiotic stress responses in plants. *Advances in Plants and Agricultural Research* 2(1): 00037.

Srivastava, S., Tripathi, R. D., Dwivedi, U. N. (2004) Synthesis of phytochelatins and modulation of antioxidants in response to cadmium stress in *Cuscuta reflexa*—An angiospermic parasite. *Journal of Plant Physiology* 161(6): 665–674.

Stevens, R. G., Creissen, G. P., Mullineaux, P. M. (1997) Cloning and characterisation of a cytosolic glutathione reductase cDNA from pea (*Pisum sativum* L.) and its expression in response to stress. *Plant Molecular Biology* 35(5): 641–654.

Stevens, R. G., Creissen, G. P., Mullineaux, P. M. (2000) Characterization of pea cytosolic glutathione reductase expressed in transgenic tobacco. *Planta* 211(4): 537–545.

Storz, G., Imlayt, J. A. (1999) Oxidative stress. *Current Opinion in Microbiology* 2(2): 188–194.

Sun, C., Gao, X., Fu, J., Zhou, J., Wu, X. (2015) Metabolic response of maize (*Zea mays* L.) plants to combined drought and salt stress. *Plant and Soil* 388(1–2): 99–117.

Sun, W., Xu, X., Zhu, H., Liu, A., Liu, L., Li, J., Hua, X. (2010) Comparative transcriptomic profiling of a salt-tolerant wild tomato species and a salt-sensitive tomato cultivar. *Plant and Cell Physiology* 51(6): 997–1006.

Swamy, P. M., Smith, B. (1999) Role of abscisic acid in plant stress tolerance. *Current Science* 76(9): 1220–1227.

Takagi, D., Takumi, S., Hashiguchi, M., Sejima, T., Miyake, C. (2016) Superoxide and singlet oxygen produced within the thylakoid membranes both cause photosystem I photoinhibition. *Plant Physiology* 171(3): 1626–1634.

Tang, X., Webb, M. A. (1994) Soybean root nodule cDNA encoding glutathione reductase. *Plant Physiology* 104(3): 1081–1082.

Tester, M., Langridge, P. (2010) Breeding technologies to increase crop production in a changing world. *Science* 327(5967): 818–822.

Thornalley, P. J. (1990) The glyoxalase system: New developments towards functional characterization of a metabolic pathway fundamental to biological life. *The Biochemical Journal* 269(1): 1–11.

Thornalley, P. J., Battah, S., Ahmed, N., Karachalias, N., Agalou, S., Babaei-Jadidi, R., Dawnay, A. (2003) Quantitative screening of advanced glycation end products in cellular and extracellular proteins by tandem mass spectrometry. *The Biochemical Journal* 375(3): 581–592.

Trujillo, L. E., Sotolongo, M., Menendez, C., Ochogavia, M. E., Coll, Y., Hernández, I., Borrás-Hidalgo, O., et al. (2008) SodERF3, a novel sugarcane ethylene responsive factor (ERF), enhances salt and drought tolerance when overexpressed in tobacco plants. *Plant and Cell Physiology* 49(4): 512–525.

Tseng, M. J., Liu, C. W., Yiu, J. C. (2008) Tolerance to sulfur dioxide in transgenic Chinese cabbage transformed with both the superoxide dismutase containing manganese and catalase genes of *Escherichia coli*. *Scientia Horticulturae* 115(2): 101–110.

Tsukahara, K., Sawada, H., Kohno, Y., Matsuura, T., Mori, I. C., Terao, T., Ioki, M., Tamaoki, M. (2015) Ozone-induced rice grain yield loss is triggered via a change in panicle morphology that is controlled by aberrant panicle organization 1 Gene. *PLoS ONE* 10(4): e0123308.

Turóczy, Z., Kis, P., Török, K., Cserháti, M., Lendvai, A., Dudits, D., Horváth, G. V. (2011) Overproduction of a rice aldo-keto reductase increases oxidative and heat stress tolerance by malondialdehyde and methylglyoxal detoxification. *Plant Molecular Biology* 75(4–5): 399–412.

Urano, K., Kurihara, Y., Seki, M., Shinozaki, K. (2010) 'Omics' analyses of regulatory networks in plant abiotic stress responses. *Current Opinion in Plant Biology* 13(2): 132–138.

Urano, K., Maruyama, K., Ogata, Y., Morishita, Y., Takeda, M., Sakurai, N., Suzuki, H., et al. (2009) Characterization of the ABA-regulated global responses to dehydration in *Arabidopsis* by metabolomics. *The Plant Journal* 57(6): 1065–1078.

Ushimaru, T., Maki, Y., Sano, S., Koshiba, K., Asada, K., Tsuji, H. (1997) Induction of enzymes involved in the ascorbate-dependent antioxidative system, namely, ascorbate peroxidase, monodehydroascorbate reductase and dehydroascorbate reductase, after exposure to air of rice (*Oryza sativa*) seedlings germinated under water. *Plant and Cell Physiology* 38(5): 541–549.

Ushimaru, T., Nakagawa, T., Fujioka, Y., Daicho, K., Naito, M., Yamauchi, Y., Nonaka, H., et al. (2006) Transgenic *Arabidopsis* plants expressing the rice dehydroascorbate reductase gene are resistant to salt stress. *Journal of Plant Physiology* 163(11): 1179–1184.

Vaahtera, L., Brosche, M., Wrzaczek, M., Kangasjarvi, J. (2014) Specificity in ROS signaling and transcript signatures. *Antioxidants and Redox Signaling* 21(9): 1422–1441.

Vacca, R. A., Valenti, D., Bobba, A., Merafina, R. S., Passarella, S., Marra, E. (2006) Cytochrome c is released in a reactive oxygen species-dependent manner and is degraded via caspase-like proteases in tobacco Bright-Yellow 2cells en route to heat shock-induced cell death. *Plant Physiology* 141(1): 208–219.

Vaidyanathan, H., Sivakumar, P., Chakrabarty, R., Thomas, G. (2003) Scavenging of reactive oxygen species in NaCl-stressed rice (*Oryza sativa* L.) differential response in salt-tolerant and sensitive varieties. *Plant Science* 165(6): 1411–1418.

Veena Reddy, V. S., Sopory, S. K. (1999) Glyoxalase I from *Brassica juncea*: Molecular cloning, regulation and its over-expression confer tolerance in transgenic tobacco under stress. *Plant Journal* 17(4): 385–395.

Wagner, D., Przybyla, D., op den Camp, R., Kim, C., Landgraf, F., Lee, K.P., Würsch, M., et al. (2004) The genetic basis of singlet oxygen-induced stress responses of *Arabidopsis thaliana*. *Science* 306(5699): 1183–1185.

Wang, J., Zhang, H., Allen, R. D. (1999) Overexpression of an *Arabidopsis* peroxisomal ascorbate peroxidase gene in tobacco increases protection against oxidative stress. *Plant and Cell Physiology* 40(7): 725–732.

Wang, L., Liu, X., Liang, M., Tan, F., Liang, W., Chen, Y., Lin, Y., et al. (2014) Proteomic analysis of salt-responsive proteins in the leaves of mangrove *Kandelia candel* during short-term stress. *PLoS ONE* 9(1): e83141.

Wang, R. S., Pandey, S., Li, S., Gookin, T. E., Zhao, Z., Albert, R., Assmann, S. M. (2011) Common and unique elements of the ABA-regulated transcriptome of *Arabidopsis* guard cells. *BMC Genomics* 12: 216.

Wang, Y., Wisniewski, M., Meilan, R., Cui, M., Webb, R., Fuchigami, L. (2005) Overexpression of cytosolic ascorbate peroxidase in tomato confers tolerance to chilling

and salt stress. *Journal of the American Society for Horticultural Science* 130(2): 167–173.

Wani, S. H., Gosal, S. S. S. S. (2011) Introduction of OsGlyII gene into *Oryza sativa* for increasing salinity tolerance. *Biologia Plantarum* 55(3): 536–540.

Welinder, K. G. (1992) Superfamily of plant, fungal and bacterial peroxidases. *Current Opinion in Structural Biology* 2(3): 388–393.

Wheeler, G. L., Jones, M. A., Smirnoff, N. (1998) The biosynthetic pathway of vitamin C in higher plants. *Nature* 393(6683): 365–369.

Willekens, H., Chamnongpol, S., Davey, M., Schraudner, M., Langebartels, C., Van Montagu, M., Inzé, D., Van Camp, W. (1997) Catalase is a sink for H_2O_2 and is indispensable for stress defence in C-3 plants. *The EMBO Journal* 16(16): 4806–4816.

Willekens, H., Inze, D., Van Montagu, M., Van Camp, W. (1995) Catalases in plants. *Molecular Breeding* 1(3): 207–228.

Witt, S., Galicia, L., Lisec, J., Cairns, J., Tiessen, A., Araus, J. L., Palacios-Rojas, N., Fernie, A. R. (2012) Metabolic and phenotypic responses of greenhouse-grown maize hybrids to experimentally controlled drought stress. *Molecular Plant* 5(2): 401–417.

Xing, Y., Jia, W., Zhang, J. (2008) AtMKK1 mediates ABA induced CAT1 expression and H_2O_2 production via AtMPK6-coupled signaling in *Arabidopsis*. *The Plant Journal* 54(3): 440–451.

Xu, H., Xu, W., Xi, H., Ma, W., He, Z., Ma, M. (2013) The ER luminal binding protein (BiP) alleviates Cd^{2+}-induced programmed cell death through endoplasmic reticulum stress-cell death signaling pathway in tobacco cells. *Journal of Plant Physiology* 170(16): 1434–1441.

Xu, J., Yin, H., Li, Y., Liu, X. (2010) Nitric oxide is associated with long- term zinc tolerance in *Solanum nigrum*. *Plant Physiology* 154(3): 1319–1334.

Xu, J., Zhang, S. (2015) Mitogen-activated protein kinase cascades in signaling plant growth and development. *Trends in Plant Science* 20(1): 56–64.

Xuan, J., Song, Y., Zhang, H., Liu, J., Guo, Z., Hua, Y. (2013) Comparative proteomic analysis of the stolon cold stress response between the C4 perennial grass species *Zoysia japonica* and *Zoysia metrella*. *PLoS ONE* 8(9): e75705.

Yabuta, Y., Motoki, T., Yoshimura, K., Takeda, T., Ishikawa, T., Shigeoka, S. (2002) Thylakoid membrane-bound ascorbate peroxidase is a limiting factor of antioxidative systems under photo-oxidative stress. *Plant Journal* 32(6): 915–925.

Yadav, S. K., Singla-Pareek, S. L., Kumar, M., Pareek, A., Saxena, M., Sarin, N. B., Sopory, S. K. (2007) Characterization and functional validation of Glyoxalase II from rice. *Protein Expression and Purification* 51(1): 126–132.

Yadav, S. K., Singla-Pareek, S. L., Ray, M., Reddy, M. K., Sopory, S. K. (2005) Methylglyoxal levels in plants under salinity stress are dependent on glyoxalase I and glutathione. *Biochemical and Biophysical Research Communications* 337(1): 61–67.

Yakimova, E. T., Kapchina-Toteva, V. M., Laarhoven, L. J., Harren, F. M., Woltering, E. J. (2006) Involvement of ethylene and lipid signaling in cadmium-induced programmed cell death in tomato suspension cells. *Plant Physiology and Biochemistry: PPB* 44(10): 581–589.

Ye, N., Zhu, G., Liu, Y., Zhang, A., Li, Y., Liu, R., Shi, L., Jia, L., Zhang, J. (2012) Ascorbic acid and reactive oxygen species are involved in the inhibition of seed germination by abscisic acid in rice seeds. *Journal of Experimental Botany* 63(5): 1809–1822.

Yin, L., Wang, S., Eltayeb, A. E., Uddin, M. I., Yamamoto, Y., Tsuji, W., Takeuchi, Y., Tanaka, K. (2010) Overexpression of dehydroascorbate reductase, but not monodehydroascorbate reductase, confers tolerance to aluminum stress in transgenic tobacco. *Planta* 231(3): 609–621.

Zaefyzadeh, M., Quliyev, R. A., Babayeva, S. M., Abbasov, M. A. (2009) The effect of the interaction between genotypes and drought stress on the superoxide dismutase and chlorophyll content in durum wheat landraces. *Turkish Journal of Biology* 33(1): 1–7.

Zhang, H., Liu, Y., Wen, F., Yao, D., Wang, L., Guo, J., Ni, L., et al. (2014) A novel rice C_2H_2-type zinc finger protein, ZFP36, is a key player involved in abscisic acid-induced antioxidant defense and oxidative stress tolerance in rice. *Journal of Experimental Botany* 65(20): 5795–5809.

Zhang, X., Zhang, L., Dong, F., Gao, J., Galbraith, D. W., Song, C. P. (2001) Hydrogen peroxide is involved in abscisic acid-induced stomatal closure in *Vicia faba Plant Physiology* 126(4): 1438–1448.

Zhao, J., Sun, Z., Zheng, J., Guo, X., Dong, Z., Huai, J., Gou, M., et al. (2009) Cloning and characterization of a novel CBL-interacting protein kinase from maize. *Plant Molecular Biology* 69(6): 661–674.

Zhu, Z., Chenand, J., Zheng, H. L. (2012) Physiological and proteomic characterization of salt tolerance in a mangrove plant, *Bruguiera gymnorrhiza* (L.) *Lam. Tree Physiology* 32(11): 1378–1388.

Zong, X. J., Li, D. P., Gu, L. K., Li, D. Q., Liu, L. X., Hu, X. L. (2009) Abscisic and hydrogen peroxide induce a novel maize group CMAP kinase gene ZmMPK7, which is responsible for the removal of reactive oxygen species. *Planta* 229(3): 485–495.

14 Plant Antioxidant Response During Abiotic Stress
Role of Transcription Factors

Deyvid Novaes Marques, Sávio Pinho dos Reis, Nicolle Louise Ferreira Barros, Liliane de Souza Conceição Tavares, and Cláudia Regina Batista de Souza

CONTENTS

14.1 INTRODUCTION

In natural environments, plants are exposed to a set of unfavorable conditions affecting their survival, growth and productivity. Drought, high salinity, heat and cold are major abiotic stresses that cause severe agricultural yield losses (Mantri et al., 2012; Golldack et al., 2014). It is estimated that more than 50% of worldwide yield losses in major crops are caused by abiotic stresses (Wang et al., 2016a).

Abiotic stress is caused by environmental conditions that reduce growth or yield to levels below the optimum (Cramer et al., 2011). Plants responses to abiotic stress result from integrated events occurring at all levels of the organization, from molecules to anatomy. In addition, the degree and duration of the stress can have a significant effect on the complexity of the response (Cramer et al., 2011).

Several environmental stresses promote the increased production of reactive oxygen species (ROS) in plants. Depending on the imbalance between their production and detoxication rates, ROS could accumulate in cells up to a level beyond which they cause oxidative damage in cells, such as lipid peroxidation, oxidation of proteins, damage of nucleic acids, enzyme inhibition and even the oxidative destruction of the cell, in a process called oxidative stress (Apel and Hirt, 2004; Tripathy and Oelmüller, 2012; Choudhury et al., 2017).

Under optimal growth conditions, ROS are constantly produced at a low level as by-products of metabolism in organelles, particularly chloroplasts, mitochondria and peroxisomes (Miller et al., 2010). The electron transport systems in different cellular compartments and the inevitable reduction of molecular oxygen leads to the formation of ROS, such as superoxide (O_2^-), singlet oxygen ($^1O_2$), hydroxyl radicals ($\cdot OH$) and hydrogen peroxide (H_2O_2). The oxygen has a regulatory function acting as a terminal electron acceptor and alleviating electron pressure in the chain, particularly during stress, when the rate of ROS production is dramatically elevated (Miller et al., 2010; Noctor and Foyer, 2016).

Despite the harmful effects of ROS, their reactive nature also means that they are able to act as signaling molecules (Foyer and Noctor, 2005). ROS function as signaling molecules in plants, being involved in development regulation and in biotic and abiotic stress responses.

Noctor and Foyer (2016) distinguished short-time ROS signaling (seconds to minutes), which would not cause oxidative stress, and longer-term signaling (hours to days), in which sustained ROS production is necessary to reach some threshold perturbation of cell redox state that activates signal pathways. Foyer et al. (2017) defend a view of ROS as universal signaling metabolites. According to this concept, all types of oxidative modification/damage are involved in signaling, including in the induction of repair processes.

In order to keep ROS able to act as signaling molecules, their accumulation must be controlled at different sites. Antioxidants are key players in this process (Noctor and Foyer, 2016). Efficient enzymatic and nonenzymatic antioxidant systems prevent oxidative damage and reestablish the balance between the production and scavenging of ROS. An understanding of the action of antioxidant molecules in response to abiotic stresses can provide strategies to increase plant tolerance.

In this chapter, the action of important components of the antioxidant response of plants against abiotic stresses is presented. First, we present information on transcription factors (TFs) related to abiotic stress tolerance, since which the transcriptional regulation is an important step in the regulation of response to oxidative stress induced by abiotic factors.

14.2 TRANSCRIPTION FACTORS REGULATING THE PLANT RESPONSE TO OXIDATIVE STRESS

When plants suffer oxidative stress induced by abiotic factors, phytohormones (e.g. abscisic acid and ethylene) act as modulators of endogenous response upstream to transcriptional response. Several other molecules, such as protein kinases, receptor-like kinases, phosphatases and ROS (above mentioned) participate in the signal

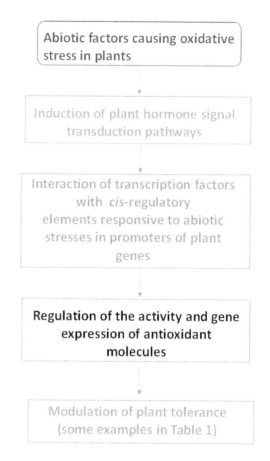

FIGURE 14.1 Regulating of the plant response to oxidative stress induced by abiotic factors.

transduction pathways in response to oxidative stress (Reis et al., 2016) (Figure 14.1).

The direct transcriptional regulation of oxidative stress-responsive genes is performed by transcription factors, whose activity is regulated by signaling molecules. TFs, such as MYC/MYB (Roy, 2016), bZIP (Sornaraj et al., 2016) and NAC (Marques et al., 2017), are transcriptional regulators that interact with conserved DNA sequences (cis-acting regulatory elements) in promoters of oxidative stress-responsive genes and regulate their expression and indirectly modulate the activity of antioxidant molecules and the expression of gene encoding for such molecules.

In this section, information about major TFs involved in tolerance against abiotic stresses mediated by antioxidant molecules is presented. Examples of the importance of these TFs in the modulation of activity and gene expression of antioxidant enzymes are provided.

14.2.1 NAC

Proteins of plant-specific NAC domain family are among the transcriptional regulators related to plant strategies under conditions of abiotic stresses (Marques et al., 2017). NAC is a plant-specific family related to several processes, such as cell division (Kim et al., 2006), senescence (Christiansen et al., 2016), flowering (Ning et al., 2015) and both biotic and abiotic stresses (He et al., 2016a; Wang et al., 2016b).

The NAC acronym is derived from the detection of a conserved domain present in petunia no apical meristem (NAM) and Arabidopsis ATAFI, ATAF2 (Arabidopsis transcription activation factor) and cup-shaped cotyledon (CUC2) proteins (Aida et al., 1997). The conserved NAC domain (presence in the N-terminal region) is composed of approximately 150 amino acids. It is responsible for DNA binding in the cell nucleus and dimer formation. The diverse C-terminal region is related to transcriptional regulation (Olsen et al., 2005).

Some NAC proteins are anchored in the plasma membrane or endoplasmic reticulum membrane (Liang et al., 2015). They are merged into transmembrane NAC protein subfamily named NTL (from NTM1-like). In response to abiotic stresses, their transport to the nucleus occurs after proteolytic cleavage (Seo et al., 2008). For example, MfNACsa (a lipid-anchored NACsa TF in Medicago falcata) is associated with membranes under unstressed conditions. Under drought stress, MfNACsa translocates to the nucleus through de-S-palmitoylation mediated by the thioesterase MtAPT1 (Wang et al., 2017a).

NAC recognition site (NACRS), consensus CGT(G/A) and CDBS (core DNA-binding sequence, CACG) in promoter region of early responsive to dehydration 1 (ERD1) from Arabidopsis are typical regulatory elements found in promoters of responsive genes to abiotic stresses directly regulated by NACs (Marques et al., 2017).

Several studies have reported the relationship of NACs with the regulation of plant responses against oxidative stress induced by several abiotic factors. For example, Mao et al. (2016) verified that drought tolerance generated by overexpression of ZmNAC55 in Arabidopsis is related to transcriptomic changes of genes responsive to H_2O_2 and oxidative stress.

Li et al. (2016a) found that overexpression of SlNAM1 (a NAC gene from tomato) in tobacco reduced H_2O_2 and $O_2^{•-}$ contents and malondialdehyde (MDA, a decomposition product of polyunsaturated fatty acids, which is an indicator of membrane damage and lipid peroxidation) formation in transgenic tobacco under low temperature. Tak et al. (2016) also verified lower MDA content in transgenic lines of banana generated by overexpression of MusaNAC042, related to salinity and drought tolerance.

In the study of Li et al. (2016a), the higher expression levels of NtSOD and NtAPX in transgenic tobacco related to higher antioxidant enzymes superoxide dismutase (SOD) and ascorbate peroxidase (APX) activities to scavenge more H_2O_2 and $O_2^{•-}$ in response to chilling stress was observed. In turn, Zhao et al. (2016a) detected that the transgenic Arabidopsis overexpressing MlNAC9 showed enhanced tolerance to drought and cold stresses, and the activities of antioxidant enzymes SOD, peroxidase (POD) and catalase (CAT) family were significantly increased, enhancing ROS scavenging capacity.

These results are in agreement with a recent evaluation of activities of several enzymatic antioxidants induced by the expression of other NAC proteins, such as JUNGBRUNNEN1 from tomato (Thirumalaikumar et al., 2017), ThNAC13 from Tamarix hispida (Wang et al., 2017b) and PbeNAC1 from Pyrus betulifolia (Jin et al., 2017).

It is important to mention that some NAC proteins have been related to negative effects on tolerance to abiotic stresses. For instance, the modulation of putrescine-associated ROS homeostasis is involved with the negative regulation of drought stress response by PtrNAC72 (Wu et al., 2016a); ONAC095 acts as a positive regulator of cold response, but as a negative regulator of drought response in rice (Huang et al., 2016). Shen et al. (2017) observed that rice OsNAC2 overexpression lines had lower drought and high salinity tolerance compared with wild-type plants.

Such information ratifies the importance of NAC TFs for the regulation of plant antioxidant response system.

14.2.2 MYB

Among TF families, the MYB is one of the largest and most conserved found in all eukaryotes (Roy, 2016), including plants, such as Arabidopsis (Dubos et al., 2010), maize (Du et al., 2012), soybean (Yan et al., 2015) and tomato (Li et al., 2016b). This abbreviation is derived from the first MYB gene identified as the 'oncogene' v-MYB from the avian myeloblastosis virus (Klempnauer et al., 1982). These are involved in regulating various cellular processes, including anthocyanin biosynthesis (Paz-Ares et al., 1987), ABA-dependent pathway of stress signaling (Dubos et al., 2010), cell morphogenesis, biotic and abiotic stress responses (Roy, 2016) and regulation of flavonoid biosynthesis (Yan et al., 2015).

The N-terminal region of these TFs is characterized by the presence of conserved MYB repeats (up to four imperfect amino acid sequence repeats of about 52 amino acids), associated with protein–protein interactions and DNA binding (Ambawat et al., 2013). The C-terminal region is responsible for modulating the regulatory activity of the protein (Dubos et al., 2010; Roy, 2016). The most common type of plant MYB transcription factor is R2R3-MYB (containing two repeats) (Du et al., 2012).

According to Prouse and Campbell (2012), although few of the possible plant R2R3-MYB DNA targets have been characterized, some common elements of plant MYB DNA interactions have emerged (for instance, CNGTT(A/G), ACC(A/T)A(A/C), TTAGGG, AAAATATCT, GATA and TATCCA).

Studies have demonstrated the action of MYBs on the antioxidant response. For example, Su et al. (2014a) observed an enhanced tolerance to drought and cold stress generated by overexpressing *GmMYBJ1* in transgenic *Arabidopsis*. In this study, *GmMYBJ1*-overexpressing plants accumulated less MDA compared with the wild-type (WT) plants after drought treatment. In turn, GbMYB5, from cotton (Chen et al., 2015), and ScMYB2 (a sugarcane R2R3-MYB) (Guo et al., 2017b) play a positive role in plant response to drought, related to modulation of MDA levels.

Other studies have verified that the action of MYBs on the antioxidant response is related to the modulation of antioxidant enzyme activity. Liu et al. (2011) detected that *TaPIMP1*-expressing tobacco plants displayed significantly enhanced tolerances to drought and salt stresses, compared to untransformed tobacco host plants, with the activity of SOD significantly increased.

The transcript levels of the antioxidant genes *SOD*, *CAT* and *GST* and the activities of SOD, POD, CAT and glutathione *S*-transferase (GST) increased markedly by the effect of overexpression of *GbMYB5* in transgenic tobacco as compared to WT under drought stress (Chen et al., 2015).

SoMYB18-expressing plants of tobacco exhibited notably improved tolerances to salt and drought stress, with increased SOD and CAT activities, compared to untransformed tobacco plants (Shingote et al., 2015).

Overexpression of *BplMYB46*, an MYB gene from *Betula platyphylla*, improves salt and osmotic tolerance by affecting the expression of genes (including *SOD*, *POD*) and the SOD and POD activity in transgenic birch plants (Guo et al., 2017a).

14.2.3 bZIP

Basic leucine zipper (bZIP) is a family of dimerizing TFs found in all eukaryotes (Amoutzias et al., 2007). The bZIP gene family codes for proteins containing the conserved bZIP domain (composed of 40–80 amino acids), which consists of a highly charged basic region

(responsible for nuclear localization and DNA binding) and a C-terminal coiled-coil leucine zipper dimerization domain (Hollenbeck et al., 2002; Noman et al., 2017). bZIP was divided into ten different subgroups, according to sequence similarities and functional features. Many bZIPs can form homodimers and others can be combined through heterodimerization to form specific bZIP pairs with distinct functionalities (Llorca et al., 2014).

In plants, bZIP is involved in biological processes (e.g. seed germination, light signaling, hormone and sugar signaling, flower and seed development, organ differentiation and embryogenesis), flavonoid biosynthesis (Malacarne et al., 2016) and responses to stresses, including drought, extreme temperatures and high salinity (abiotic stresses) and defense from pathogens (biotic stresses) (Lindemose et al., 2013; Sornaraj et al., 2016).

Several plant bZIP genes were identified, such as from, maize (Ying et al., 2012), *Arabidopsis* (Inaba et al., 2015), cassava (Hu et al., 2016), soybean (Zhang et al., 2016) and rice (Henriques et al., 2017).

To regulate promoters of most target genes, plant bZIP TFs preferentially bind DNA *cis*-acting elements of palindromic and pseudopalindromic hexamers containing an ACGT core with flanking residues of A-box (TACGTA), C-box (GACGTC), G-box (CACGTG) or abscisic acid-responsive element (ABRE) (ACGTGT/GC) (Lindemose et al., 2013; Sornaraj et al., 2016).

Several studies have highlighted the relationship of bZIPs with the plant antioxidant response. Ectopic expression of *PtrABF* gene (encoding a bZIP transcription factor isolated from *Poncirus trifoliata*) under the control of a CaMV 35S promoter in transgenic tobacco plants enhances dehydration and drought tolerance in tobacco, increasing expression levels of genes encoding SOD, CAT and APX, and activities of three detoxifying enzymes (SOD, CAT and POD) (Huang et al., 2010).

GhABF2, a typical cotton bZIP TF, confers drought and salinity tolerance in *Arabidopsis* and cotton, related to proline contents, and SOD and CAT activities (Liang et al., 2016).

In turn, Wang et al. (2017c) verified that antioxidant enzyme (SOD, POD, CAT and GST) activities were increased by overexpression of *ABP9*, encoding a bZIP transcription factor from maize, in transgenic cotton under NaCl stress conditions.

Transgenic *Arabidopsis* plants overexpressing *TabZIP14-B* from wheat presented salt and freezing tolerance and upregulated expression levels of stress-responsive genes, including *AtGSTF6*, which encodes GST and is involved in oxidative and salt stress (Zhang et al., 2017a).

14.2.4 WRKY

WRKY TFs are named according to the family-defining conserved domain of approximately 60 amino acids which consist of a highly conserved heptapeptide motif WRKYGQK and a zinc-finger structure at the C-terminus, both necessary for the high binding affinity of these TFs to the consensus *cis*-acting elements termed W-box (TTGACT/C) (Bakshi and Oelmüller, 2014). They are categorized into three subgroups, dependent on their number of WRKY domains and the zinc-finger structure (Llorca et al., 2014; Bakshi and Oelmüller, 2014).

WRKYs are found in unicellular eukaryote flagellated protozoan like *Giardia lamblia*, in the soil-living amoeba-like *Dictyostelium discoideum* (Phukan et al., 2016) and in several plant species, such as, carrot (Li et al., 2016c), cassava (Wei et al., 2016), common bean (Wu et al., 2017a) and potato (Zhang et al., 2017b).

In plants, functions of WRKYs are related to several processes such as trichome development, senescence, biosynthesis of alkaloids, responses to biotic and abiotic stresses (Llorca et al., 2014) and regulation of flavonoid biosynthesis (Naoumkina et al., 2008; Amato et al., 2017).

Several studies have revealed the role of WRKYs in modulating the tolerance of various abiotic stresses, including phosphate starvation (Dai et al., 2016), cadmium stress (Yang et al., 2016), drought (He et al., 2016b; Chen et al., 2017a), heat (He et al., 2016b) and salinity (Cai et al., 2017; Hichri et al., 2017; Liang et al., 2017; Wang et al., 2017d).

In addition, recent studies have verified the relationship of WRKY TFs with the antioxidant response. For example, Yang et al. (2016) verified that ThWRKY7 is the upstream regulator of *ThVAc1* and that overexpression of this gene improved plant tolerance of *Arabidopsis* against cadmium stress with increased antioxidant enzymes (SOD, POD, GST, and GPX) activities.

Hichri et al. (2017) verified that salinized *35S::SlWRKY3* tomato plants displayed reduced oxidative stress compared to wild-type, with upregulated genes coding for detoxification enzymes such as GST and POD in transgenic plants.

In turn, the expression of *DgWRKY5* in chrysanthemum increased the activities of SOD, POD and CAT and generated upregulation of the expression of genes coding for antioxidant enzymes (*DgAPX*, *DgCAT*, and *DgCuZnSOD*), whereas the accumulation of H_2O_2, O_2^- and MDA was reduced in transgenic chrysanthemum overexpressing *DgWRKY5* (Liang et al., 2017).

14.2.5 AP2/ERF

APETALA2/Ethylene-Responsive Factor (AP2/ERF) is a large group of TFs conservatively widespread in the plant kingdom and involved in the regulation of fruit ripening, senescence, phytohormone biosynthesis, defense response, metabolism, plant growth and development (Gu et al., 2017). Proteins of this superfamily contain at least one DNA-binding domain (AP2 domain). This domain is composed of 60 amino acid residues that confer a three-dimensional conformation organized into a layer of three antiparallel beta-sheets followed by a parallel alpha helix (Licausi et al., 2013).

Based on copy numbers of this domain, AP2/ERF TFs are classified into AP2, RAV (related to abscisic acid insensitive3/viviparous1), DREB (subgroup A1–A6), ERF (subgroup B1–B6) and others (Phukan et al., 2017). For example, AP2 contain one, or a tandem repeated, AP2 domain, whereas ERF has a single AP2 domain (Gu et al., 2017). GCC-box (AGCCGCC element) and DRE/CRT (dehydration responsive element/C-repeat, RCCGCC element) are the main two *cis*-regulatory elements in promoter of target genes for binding of AP2/ERF proteins, although they have a strong capacity to bind the wide range of *cis*-regulatory elements (Gu et al., 2017).

Several studies have reported differential expression of AP2/ERF proteins in response to salinity (Chen et al., 2007), heat (Mizoi et al., 2013), low temperature (Shu et al., 2016) and drought (Fan et al., 2016).

Studies on transformed plants have confirmed the importance of these proteins in the regulation of plant tolerance against various abiotic stresses, such as salinity (Chen et al., 2007), heat (Mizoi et al., 2013) and drought (Jisha et al., 2015; Kudo et al., 2017). Bouaziz et al. (2015) verified that overexpression of StDREB1 and StDREB2 (members of members of the AP2/ERF family) in transgenic potato plant increased the level of drought tolerance of these transgenic lines, in comparison to wild-type control plants, as measured by H_2O_2 content.

Although this information ratifies the relationship of AP2 TFs with the response of abiotic stresses, which commonly induce oxidative stress, further studies are needed for the identification of genes related to the antioxidant response directly and indirectly regulated by these TFs.

14.3 ANTIOXIDANT MOLECULES RELATED TO ABIOTIC STRESS RESPONSE

Plants have a complex system of antioxidant molecules regulated by TFs. Such complexity includes enzymes and non-enzymatic molecules, which act to prevent oxidative injury.

In this section, information about molecules involved in plant response against oxidative stress induced by abiotic factors is presented. Examples of recent studies on overexpression or silencing of scavenging enzymes in engineered plants, modulating protection against abiotic stresses, are presented in Table 14.1.

14.3.1 ENZYMATIC MOLECULES WITH ANTIOXIDATIVE ACTIVITY

14.3.1.1 Superoxide Dismutase

A crucial group of antioxidant enzymes is the SOD family, which protects cells from the negative effects of O_2^-. SODs have a major role in the degradation of this molecule: they catalyze its dismutation into molecular oxygen and H_2O_2, contributing to ROS detoxification pathways (Miura et al., 2012). As a member of a metalloenzyme family, SOD needs metal ions in its active site. Depending on which metal is present, SODs can be classified into four groups: CuZnSODs, NiSODs,

FeSODs, and MnSODs. Each group displays a different subcellular localization, structural features and expression regulation (Molina-Rueda et al., 2013; Marques et al., 2014).

SODs are some of the first enzymes to be expressed against ROS, being central for the cell initial defense (Xu et al., 2013a; Sheoran et al., 2015), and many studies have confirmed its role in modulating plant tolerance against abiotic stresses (Table 14.1). The H_2O_2 accumulation resultant of SOD action is harmful to cells, so a perfect balance between SODs and several H_2O_2-scavenging enzymes is crucial in determining the steady-state level of H_2O_2 and O_2^-. H_2O_2 can be transformed into water and oxygen by CATs or PODs, which occur mainly in peroxisomes. In addition, the sequestering of metal ions by metal-binding proteins prevents the HO–radical synthesis, which is extremely toxic (Yousuf et al., 2012).

14.3.1.2 Catalase

Another important group of antioxidant enzymes reported in many plant species is the CAT multigene family (H_2O_2 oxidoreductase). They are tetrameric

TABLE 14.1

Examples of Studies Focused on Action of Antioxidant Enzymes in Abiotic Stress Tolerance in Transgenic Plants

Overexpressed or Silenced Genes Coding Antioxidant Enzymes	Abiotic Stress Tolerance Related to Antioxidant Response	Transformed Species	References
OsAPX	Chilling tolerance	Oryza sativa	Sato et al. (2011)
Ss.sAPX	Salt tolerance	Arabidopsis thaliana	Li et al. (2012)
cytAPX; cytCuZnSOD	Salt tolerance	Prunus domestica	Diaz-Vivancos et al. (2013)
GhAPX; GhCAT; GhCuZnSOD	Salt tolerance	Gossypium hirsutum	Luo et al. (2013)
MeCAT; MeCuZnSOD	Cold and drought tolerance	Manihot esculenta	Xu et al. (2013b)
SbpAPX	Drought and salt tolerance	Nicotiana tabacum	Singh et al. (2014)
BoCAT	Heat tolerance	A. thaliana	Chiang et al. (2014)
MeAPX; MeCuZnSOD	Chilling tolerance	M. esculenta	Xu et al. (2014)
SmAPX	Flood tolerance	N. tabacum	Chiang et al. (2015)
cytAPX; cytCuZnSOD	Drought tolerance	N. tabacum	Faize et al. (2015)
PtAPX	Salt tolerance	A. thaliana	Guan et al. (2015)
PgGPX	Drought and salinity tolerance	O. sativa	Islam et al. (2015)
KcCSD (CuZnSOD)	Salinity tolerance	N. tabacum	Jing et al. (2015)
PuCuZnSOD	Salt tolerance	A. thaliana	Wu et al. (2016b)
APX; CuZnSOD	Salt tolerance	Ipomoea batatas	Yan et al. (2016)
PtAPX	Drought and salt tolerance	N. tabacum	Cao et al. (2017)
LcAPX; SmAPX	Flood tolerance	A. thaliana	Chiang et al. (2017)
OsCuZnSOD	Salt tolerance	O. sativa	Guan et al. (2017)
SaCuZnSOD	Cadmium tolerance	A. thaliana	Li et al. (2017b)
PgAPX	Salt tolerance	A. thaliana	Sukweenadhi et al. (2017)
CaAPX	Cold and heat tolerance	N. tabacum	Wang et al. (2017e)
OcMnSOD	Malathion tolerance	Oxya chinensis	Wu et al. (2017b)
SiCSD (CuZnSOD)	Cold and drought tolerance	N. tabacum	Zhang et al. (2017d)

heme-containing enzymes, encoded by three genes and commonly localized in peroxisomes. They also can be localized in mitochondria and glyoxysomes. CATs are crucial in abiotic stresses in order to avoid oxidative damage (Su et al., 2014b). Their main role is to contribute to the removal of excessive H_2O_2 generated during growing metabolism or by abiotic stresses. As a consequence, the levels of CAT activity are inversely correlated with the H_2O_2 amounts in plant cells (Chen et al., 2012). This molecule is transformed into oxygen and water in all aerobic species (Chen et al., 2012; Weisany et al., 2012).

H_2O_2 can be removed by both PODs and CATs. Each of these enzymes uses a specific route for the metabolism in cells. CATs do not require a cofactor because they usually act in a dismutation reaction: one H_2O_2 is converted into water, as an oxidizing molecule; a second H_2O_2 is converted in O_2, as a reducing molecule. On the contrary, peroxidases need a cofactor: a small, reducing molecule in general (Sofo et al., 2015).

14.3.1.3 Peroxidases

PODs are very diverse enzymes. Based on their structures and catalytic properties, they usually are divided into three classes: Class I peroxidases include the intracellular enzymes, such as cytochrome C peroxidase in microorganisms, catalase-peroxidases in bacteria and APX in plants, bacteria and yeast; Class II peroxidases include lignin peroxidase and presents in fungi in an extracellular localization; Class III peroxidases are secreted enzymes with roles into the cell wall or the surrounding medium and the vacuole (Choi and Hwang, 2012).

As mentioned previously, APXs are found in plants: commonly in higher plants, chlorophytes and red algae (Caverzan et al., 2012). There are two common isoforms: APX1 and APX2. APX1 is constitutively expressed in several plant tissues: mainly in roots, leaves and stems, and its expression is significantly upregulated in response to a number of oxidative stresses (Suzuki et al., 2013). APX2 also has a role in the response of plants to oxidative stress. Such as APX1, expression of APX2 is triggered in many tissues but is commonly detected at lower levels. Roots have high activity in response to wounding and general oxidative stress and shoots in response to drought stress (Suzuki et al., 2013).

Together with CATs, they are among the most required key enzymes that scavenge potentially harmful H_2O_2 from the cytosol and chloroplasts of plant cells (Faize et al., 2011). Then, APXs are antioxidant enzymes that play a key role in oxidative stress responses and following recovery from these stresses

(Sofo et al., 2015). For this, APX requires ascorbic acid (AsA) as a specific electron donor to reduce H_2O_2 to water (Caverzan et al., 2012).

14.3.1.4 The Balance Between Antioxidant Enzymes Activity and Reports

In plants, the ROS-scavenging enzymes have been intensely studied in recent years. The main reports have shown that SOD, CAT and APX upregulation are related between each other in response to oxidative stress. In addition, the balance between APX and CAT activities is crucial for the decrease of toxic H_2O_2 levels in plant cells since they are responsible for the main enzymatic H_2O_2-scavenging mechanism in plants. If the balance of ROS-scavenging enzymes changes, compensatory mechanisms are induced (i.e. APX is upregulated when CAT activity is reduced in plants) (Weisany et al., 2012; Sofo et al., 2015).

There are several studies showing the importance of antioxidant enzymes against different oxidative stresses in plants. Mahanty et al. (2012) observed in *Pennisetum glaucum* that different abiotic stresses were able to induce CuZnSOD mRNA expression. Seedlings were exposed to cold (4°C), heat (48°C), drought (withholding water for 8 days), salt (250 mM NaCl) and oxidative (100 uM methyl viologen) stresses, and the mRNA expression occurred gradually during the first 24 hours, with exception to cold stress, when the upregulation occurred up to 12 hours, then decreased. Also, when expressed in *E. coli*, it conferred enhanced tolerance to oxidative stress (induced by methyl viologen) when compared to negative control.

Weisany et al. (2012) reported in soybean root and leaves that oxidative stresses, such as salinity and drought, upregulated both enzymes: CAT and APX. CAT activity increased up to 740% under 66 mM NaCl treatment when compared to control. APX activity increased up to 780% under the same conditions.

Suzuki et al. (2013), studying *Arabidopsis*, obtained results that showed the role of APX2 in plants response to light, heat, salt and oxidative stresses. The deficiency in APX2 resulted in a decreased tolerance to heat stress but an enhanced tolerance to salinity and oxidative stress. Then, the cooperation between APX1 and APX2 was evident during the oxidative stress but not during heat stress. This cooperation is suggested by the activation of redundant mechanisms activated in plants with a deficiency in APX2. In addition, these plants, under heat and drought stresses, produced more seeds. This fact could be possible by the activation of protection pathways of reproductive tissues from the harmful heat and drought stresses.

Choi and Hwang (2012) reported in *Capsicum annuum* leaves that extracellular PODs were induced by: salt, cold, drought and infection by a fungal pathogen. Plants without the peroxidase (by silencing) were less tolerant to these stresses. In *Arabidopsis*, when it was overexpressed, the plants were more tolerant to all stresses. The concentrations of NaCl ranged from 100 to 500 mM, and 4°C was used to cold stress. Mannitol and methyl viologen were used to mimic drought and oxidative stresses, respectively.

On the other hand, ambiguous results were also found. Drought stress in *Populus tremula* caused a downregulation of CuZnSODs and an upregulation of FeSODs in transgenic plants when compared to wild-type. These results show different SODs present differential expression responses to genetic manipulation under abiotic conditions (Molina-Rueda et al., 2013).

There are a number of papers reporting studies with more than one antioxidant enzyme since they are related. Bavita et al. (2012) reported increased activities of SOD, CAT, and APX in both heat-tolerant and heat-sensitive cultivars of wheat under high-temperature stress, with decreased oxidative stress in all pants.

Diaz-Vivancos et al. (2013) reported transgenic plum plants overexpressing CuZnSOD and APX that enhanced the tolerance against salt stress. They were submitted to a 100 mM NaCl treatment, and other enzymatic and non-enzymatic antioxidant molecules were also related. However, under higher salt concentrations (200 mM NaCl), both transformed and wild-type plants were severely damaged. In addition, the shoot length and weight were also compared. Under normal conditions, they behave similarly, but under 100 mM NaCl stress, all transgenic plants were significantly larger than non-transgenic plants, and the weight of some transformed lines was also larger than the weight of wild-type shoots.

Shafi et al. (2015) reported in *Arabidopsis* increased activity of CuZnSOD and APX genes in both wild-type and transformed plants under salt stress, with higher levels in the transgenic plants compared to wild-type because of the overexpression under a constitutive promoter. However, at 150 mM NaCl, the levels decrease in both. Airaki et al. (2012) reported in *Capsicum annuum* an increased activity of CAT in 30-days-old plants under 1–3 days low-temperature stress. In the first two days, the activity improved up to 36%, then decreased in the third day. APX activity increased by 40% after only one day, stabilizing the levels after three days. On the other hand, SOD activity does not change significantly after the three days under low-temperature stress.

There are expression differences between tissues. Asthir et al. (2012), studying wheat, observed significantly elevated activities of APX, CAT and SOD under high-temperature stress in transgenic shoots when compared to control values. APX and SOD were more upregulated in roots while CAT in shoots.

Sometimes there are differences between plants with the same resistance. Another report with wheat observed the leaves' gene expression in two winter cultivars with low-temperature resistance. There were significant differences between them; SOD, APX and CAT were upregulated at the same cultivar compared to the another, with up to two times activity level differences between them (Xu et al., 2012).

Sales et al. (2013) have studied the levels of SOD, APX and CAT activity in two sugarcane cultivars: a drought-resistant and a drought-sensitive. The goal was to test if low-temperature stress influences plants subjected to water deficit. It was observed that SOD was upregulated in both cultivars when plants were subjected to drought and low-temperature stresses alone or simultaneously, after 11 days of treatment. However, at the drought-resistant cultivar, the SOD activity was 4.8 times greater than that observed in drought-sensitive.

In addition, after four days of recovery, the activity levels were maintained only in the first, rising to 30 times the difference between cultivars. The APX activity results were different. In drought-resistant plants, it was increased when plants were under drought and drought/cold stresses. The cold stress alone decreased APX activity. In drought-sensitive plants, the activity was improved only under drought stress, while under drought/cold and only cold there were no significant differences. Between cultivars, drought-resistant plants only not showed higher levels under cold stress (Sales et al., 2013).

Finally, about the CAT activity, only under cold stress the drought-resistant sugarcanes presented improved levels. The drought-sensitive plants did not present increased activity under any treatments. On the contrary, the levels decreased under cold and cold/drought stresses (Sales et al., 2013).

14.3.2 NONENZYMATIC MOLECULES WITH ANTIOXIDATIVE ACTIVITY

14.3.2.1 Glutathione

As a low-weight and water-soluble molecule, the tripeptide glutathione (γ-glutamyl-cysteinyl-glycine) has a thiol group (R-SH) and remarkable reductive activity, due to its central nucleophilic cysteine (Foyer and Noctor, 2003; Das and Roychoudhury, 2014). In addition, the thiol group establishes mercaptide bonds with metals (Nahar et al., 2015), assisting the scavenging process of ROS (Das and Roychoudhury, 2014). Still, while ROS oxidize glutathione (GSH) molecules, they delay the degradation of other cellular constituents (Noctor et al., 2011).

Although they do not perform an enzymatic activity, such molecules function as mediators and, thus, control enzymes related to the antioxidant response, as glutathione peroxidase (GPX) (Foyer and Noctor, 2005). On the other hand, they act as signaling agents during plants' exposure to metals (Jozefczak et al., 2012), therefore indirectly cooperating with the regulation of stress gene expression (Fotopoulos et al., 2010). Likewise, they participate in the regulation of sulfate transportation, synthesis and homeostasis (Mullineaux and Rausch, 2005; Sahoo et al., 2017) since they are responsible for up to 2% of sulfate in plant tissue (Foyer and Noctor, 2003). Among other functions, they contribute to the production of phytochelatins (Roychoudhury et al., 2012a) and the recovery of other antioxidants (Szalai et al., 2009).

As soon as it interacts with those ROS, the reduced GSH is converted to the oxidized version, glutathione disulfide (GSSG) (Hasanuzzaman et al., 2012b), which is essential for the maintenance of the GSH pool. Thereby, it is from GSSG that the levels of reduced glutathione are restored, either using glutathione reductase (GR) (Pang and Wang, 2010) or via *de novo* synthesis (Das and Roychoudhury, 2014). As a result, for the proper functioning of plant metabolism, the relevance of the harmony between GSH and GSSG is higher than the restricted concentrations of each molecule during stressful conditions (Cuypers et al., 2011; Nahar et al., 2015).

The regeneration of GSH can occur in multiple organelles (Christou et al., 2014), considering that it is found in large concentrations in the endoplasmic reticulum, vacuoles, peroxisomes, compartments such as the apoplast and mainly in chloroplast, mitochondria and cytosol, which are places where synthesis happen (Noctor and Foyer, 1998; Mahmood et al., 2010).

It is worth mentioning that the glutathione activity is closely related to AsA, another non-enzymatic antioxidant agent that is also reused after scavenging (Apostolova et al., 2012). Both are part of the AsA-GSH cycle, where occurs the elimination of H_2O_2, whose reduction and conversion into water are catalyzed by APX with the aid of reduced ascorbate (Noshi et al., 2016; Zhao et al., 2016b). Thus, the involvement of GSH ensues when it is utilized

by dehydroascorbate reductase (DHAR) to recycle AsA. Furthermore, GSSG is a product of this reaction (Foyer and Halliwell, 1976; Asada, 1999; Noshi et al., 2016). Despite the cycle, they can also act independently by activating other pathways (Noctor, 2006).

In such studies as in Zhang et al. (2017c), *Echinodorus osiris* Rataj leaves exposed to cadmium (Cd) showed an increase in GSH, since it can attenuate the toxicity of this heavy metal via chelating and transformation into less toxic forms (Clemens, 2006), while in *Arabidopsis*, the application of exogenous GSH favored seed germination and development of mercury-contaminated seedlings, besides root length, which was similar to that under normal circumstances, and effective H_2O_2 and $O_2^{\bullet-}$ scavenging. These parameters aforementioned displayed a worse performance with a GSH synthesis inhibitor (Kim et al., 2017).

In contrast, control tobacco leaves exhibited higher GSH/GSSG ratio in comparison to those submitted to salt stress, whose GSSG concentration was higher than its GSH one, indicating the relationship between the molecules and the stress tolerance response (Da Silva et al., 2016). Similarly, Zhao et al. (2016b) obtained conclusions regarding the oscillations between the reduced and oxidized GSH, but in chloroplasts, in the face of chilling stress on cucumber seedlings.

14.3.2.2 Ascorbic Acid

Ascorbate, a physiologically functional anionic AsA, is produced from the liberation of a H^+ deriving from the hydroxyl in C3 position (Akram et al., 2017) and acts as a cellular redox buffer in favor of the equilibrium between ROS formation and degradation (Hussain et al., 2017). Moreover, it can comprise more than 10% of the plant-soluble carbohydrate portion (Noctor and Foyer, 1998) and integrates the human diet, where plants are the primary source (Smirnoff and Wheeler, 2000). Although it can remove ROS produced by abiotic stresses such as drought, photooxidation and high salinity (Upadhyaya et al., 2010; Kumar et al., 2011), not all plant species have AsA enough concentrations to overcome these harmful environmental conditions (Latif et al., 2016), making exogenous application necessary (Mukhtar et al., 2016).

AsA activity against environmental disturbances include the preservation of lipids and proteins (Khan et al., 2010; Naz et al., 2016), stimulating plant growth, increasing photosynthetic and transpiration rates (Akram et al., 2017), including the restoration of antioxidants as oxidized tocopherol and glutathione (Gallie, 2012; Gest et al., 2013). With regard to regular metabolism, it controls differentiation and cell division, as well as cellular expansion (Gallie, 2013; Akram et al., 2017) given that it is a cofactor for prolyl hydroxylase, essential to the last two processes (Smirnoff, 2000). Similarly, it also acts as a cofactor for enzymes that produce phytohormone (Conklin and Barth, 2004), for violaxanthin de-epoxidase (Müller-Moulé et al., 2002) and for ascorbate peroxidase (Gest et al., 2013), which scavenges H_2O_2 and cooperates to remove $O_2^{\bullet-}$ (Shafiq et al., 2014).

Even if ascorbate is synthesized in mitochondria, it is in chloroplast that it plays an extremely important role in the attenuation of photooxidative stress (Foyer, 2015). As a result of the high light, there is a marked ROS formation and reduction of the photosynthetic electron transport chain (Noshi et al., 2016) starting with the H_2O splitting and electrons flowing from photosystem II (PSII) to I (PSI) (Trubitsin et al., 2014). Accordingly, ascorbate becomes an electron source option for PSII, counteracting this unbalanced scenario by avoiding photoinhibition (Mano et al., 2004; Tóth et al., 2013).

The effects of the exogenous application of AsA have been addressed by studies such as Lukatkin and Anjum (2014), whose results pointed out that under chilling stress at 3°C, cucumber leaf discs (*Cucumis sativus* L.) with AsA evidenced parameters of tolerance as low electrolyte leakage and decrease in $O_2^{\bullet-}$ levels in comparison to the control. The AsA also made it possible to combat the deleterious consequences on NaCl-treated *Citrus aurantium* L. seedlings by means of reducing symptoms such as dehydration, chlorosis, leaf twisting and necrosis, along with it maintaining high levels of chlorophyll (Chl) and carotenoids (Kostopoulou et al., 2015).

The symptoms of exposure to lead (Pb) were attenuated due to the application of AsA in okra plants (*Abelmoschus esculentus* (L.) Moench) cv. Subz-Pari, since fresh and dry weight increased, just as the chlorophyll content and proteins (Hussain et al., 2017). In summary, Chen et al. (2017b) concluded that AsA facilitated electron transfer in the photosynthetic system and promoted a decrease in H_2O_2 and $O_2^{\bullet-}$ concentrations during the heat stress in tall fescue (*Festuca arundinacea* Schreb.).

14.3.2.3 Phenolic Compounds

Phenolic compounds are ubiquitous and abundant in plants because of the variety of functions they assume (Bautista et al., 2016; Mojzer et al., 2016) and are, thus, grouped into flavonoids, tannins, lignins, phenolic acids and coumarins (Gumul et al., 2007). Additionally, the standard composition of such molecules includes one or more aromatic rings having at least one hydroxyl group (Li et al., 2014). Certainly, the performance of these compounds is linked to the structural aspects as substitutions on the aromatic rings, constitution of the lateral

chain (Shahidi et al., 1992), number and distribution of the –OH groups (Król et al., 2014) and free or conjugated configuration (Martins et al., 2016). Even lipophilicity is influenced by the amount of hydroxyl (Hendrich, 2006), although they are more hydrophilic compounds (Tsao, 2010).

When in association with other molecules, phenolic compounds intensify their action and acquire new biological attributes (Mukherjee and Houghton, 2009). On the other hand, they integrate the cell wall and interfere with growth and development of plants through control of auxin transport (Brown et al., 2001). They also participate in plant response to abiotic stresses via ROS scavenging when establishing complexes with metals and minimizing the oxidative enzyme activity (Elavarthi and Martin, 2010). Thereby, the target radicals vary between $O_2^{\bullet-}$, OH•, DPPH• and ABTS•+, the last two ones applied for monitoring the in vitro scavenging process (Mathew et al., 2015; Queiroz et al., 2017). Finally, they combat reactive nitrogen species, lipid peroxidation, parasites, pathogens and herbivores (Mittler, 2002; Mojzer et al., 2016).

Phenolic compounds are found mostly in vacuoles (Niggeweg et al., 2004; Ferreres et al., 2011) and are considered to be secondary antioxidants because they act while the primary enzymatic defense system is suppressed under extreme stress (Fini et al., 2011). In addition, the production of the precursors of these compounds occurs in shikimic and chorismic acid pathways, with the involvement of phenylalanine ammonia lyase, whose activity is stimulated during adverse conditions (Dixon and Paiva, 1995; Schützendübel et al., 2001).

Regarding one of the most studied subgroups, the flavonoids are located in plants frequently connected to glucose or galactose molecules with the configuration of β-glycosides (Lewandowska et al., 2016). Moreover, they are formed by two aromatic rings with three carbons between them, often arranged as an oxygenated heterocyclic compound (Bautista et al., 2016), and retain free radicals by donation of phenolic hydrogen (Hernández et al., 2009).

Results obtained by Becerra-Moreno et al. (2015) illustrate that water stress induced the expression of secondary metabolism genes in carrots (*Daucus carota*), especially those for lignin biosynthesis, which was accumulated throughout this treatment. Whereas wounding stress resulted in positive regulation of genes related to the phenylpropanoid metabolism, as well as those for phenylalanine ammonia lyase, resulting in a high concentration of phenolic compounds during this stress. In synergy, the effects were reinforced.

Likewise, chokeberry fruits (*Aronia melanocarpa*) that received UV-C radiation, microwaves and ultrasound demonstrated, in comparison to the control group, an increment of flavonols, anthocyanins and phenolic acids, varying according to the exposure time and, in some cases, depending on the frequency applied (Cebulak et al., 2017). Furthermore, phenolic acids and flavonoids identified in seeds of *Triplaris gardneriana* possibly preserved cells against oxidative stress by preventing lipid peroxidation and generation of intracellular ROS via scavenging (De Almeida et al., 2017).

14.3.2.4 Carotenoids

Carotenoids are lipid-soluble tetraterpene antioxidants formed by forty carbons and up to thirteen double bonds, also existent in some species of non-photosynthetic bacteria and fungi (Gill and Tuteja, 2010; Ramel et al., 2013; Nisar et al., 2015) and considered the second largest group of pigments in the ecosystem. Additionally, they are categorized as nonoxygenated (β-carotene and lycopene) and oxygenated xanthophylls, which include lutein and zeaxanthin (Domonkos et al., 2013; Ramel et al., 2013), whose proportions in leaf tissue are heterogeneous, predominantly composed by lutein (Jin et al., 2015; Nisar et al., 2015). Likewise, such antioxidants derive from the isoprenoid pathway (Jin et al., 2015), where enzymatic activity is regulated by the amount of ROS (Rosalie et al., 2015).

Considering the location, chloroplasts and chromoplasts do not only produce but also accumulate carotenoid metabolites (Lopez-Juez and Pyke, 2005), because of the lipoprotein substructures of the chromoplasts that sequester these pigments (Vishnevetsky et al., 1999; Li and Yuan, 2013) and because of the thylakoid membrane of chloroplasts (Nisar et al., 2015). Carotenoids have a structural function as they decrease the chloroplast membrane fluidity and propensity to lipid peroxidation (Demmig-Adams and Adams, 2002) and biosynthetic role when they precede the formation of ABA and strigolactones (Jin et al., 2015). Primarily, they capture light energy with wavelengths between 400 and 550 nm for Chl during photosynthesis (Siefermann-Harms, 1987; Taïbi et al., 2016).

Each time that chlorophyll receives a portion of light energy greater than necessary, it establishes the ascorbate-dependent xanthophyll cycle, characterized by the synthesis of zeaxanthin from the precursor violaxanthin and with interposition of the antheraxanthin molecule (Hager, 1980; Müller-Moulé et al., 2002). In order to dissipate excessive light into heat, zeaxanthin retains the energy from the excited singlet chlorophyll (^{1}Chl*) (Frank et al., 2000; Demmig-Adams and Adams, 2006).

Whereas ^{1}Chl* can generate triplet chlorophyll (^{3}Chl*) (Ruban et al., 2012), which gives energy to O_2 resulting in the formation of singlet oxygen ($^1O_2$) (Pospíšil, 2016), this action of carotenoids is vital to neutralize the ROS accumulation and prevent photooxidation. It is noteworthy that these pigments also quench ^{3}Chl* and $^1O_2$ (Taïbi et al., 2016).

However, due to the spatial distancing concerning chlorophyll, β-carotene is not recognized as a ^{3}Chl* quencher (Ramel et al., 2013) but rather as an indicator of stress since, when oxidized, it gives rise to short-chain volatile compounds such as β-ionone and β-cyclocitral (Ramel et al., 2012). In fact, the latterly mentioned derivative is able to regulate the production of $^1O_2$-responsive transcripts in the nucleus (Havaux, 2014), as well as defense and metabolism genes, although repressing those that collaborate with development and biogenesis (Ramel et al., 2013).

As for the correlation between carotenoids and abiotic stresses, Mibei et al. (2017), when applying drought stress to African eggplants, which are traditionally tolerant to numerous adverse conditions, concluded that there was a positive interference on the zeaxanthin content in comparison to the control, the opposite to chlorophylls and carotenes.

On the other hand, Li et al. (2017a) investigated the effects of overexpression of genes that participate in the carotenoid pathway via transformation of sweet potato plants with ζ-carotene desaturase (*IbZDS*) gene and irrigation with NaCl. Thus, the overexpression of *IbZDS* caused an increment in lutein and β-carotene, which may have stimulated ABA synthesis and, subsequently, SOD and APX enzymes, that are able to attenuate the impact of salt stress. Similarly, the overexpression of β-carotene hydroxylase1 (*BCH1*) gene in mulberry, whose mature leaves were exposed to UV stress, generated a rise of β-carotene and lutein and lowered harm to cell membrane stability, besides greater PSII efficiency. When under high light, β-carotene, zeaxanthin and ROS scavenging increased (Saeed et al., 2015).

14.4 CONCLUSION

In this chapter, the prospecting of major antioxidant molecules and transcription factors that regulate the gene expression and activity of such molecules was presented. This approach is essential to the understanding of cellular and molecular mechanisms of plant response and tolerance to oxidative stress imposed by abiotic factors. The identification of new transcription factors, signaling molecules and enzymatic and non-enzymatic antioxidant molecules are also very important to promote better insights into the high complexity of molecular strategies used by plants for their adaptations to abiotic stresses and the generation of tolerant plants at the field level.

REFERENCES

Aida, M., Ishida, T., Fukaki, H., Fujisawa, H., Tasakaet, M. (1997) Genes involved in organ separation in *Arabidopsis*: An analysis of the cup-shaped cotyledon mutant. *Plant Cell* 9: 841–857.

Airaki, M., Leterrier, M., Mateos, R.M., Valderrama, R., Chaki, M., Barroso, J.B., Del Río, L.A., Palma, J.M., Corpas, F.J. (2012) Metabolism of reactive oxygen species and reactive nitrogen species in pepper (*Capsicum annuum*. L.) plants under low temperature. *Plant, Cell and Environment* 35: 281–295.

Akram, N.A., Shafiq, F., Ashraf, M. (2017) Ascorbic acid – A potential oxidant scavenger and its role in plant development and abiotic stress tolerance. *Frontiers in Plant Science* 8: 613.

Amato, A., Cavallini, E., Zenoni, S., Finezzo, L., Begheldo, M., Ruperti, B., Tornielli, G.B. (2017) A grapevine TTG2-Like WRKY transcription factor is involved in regulating vacuolar transport and flavonoid biosynthesis. *Frontiers in Plant Science* 7: 1979.

Ambawat, S., Sharma, P., Yadav, N.R., Yaday, R.C. (2013) MYB transcription factor genes as regulators for plant responses: An overview. *Physiology and Molecular Biology of Plants* 19: 307–321.

Amoutzias, G.D., Veron, A.S., Weiner, J., Robinson-Rechavi, M., Bornberg-Bauer, E., Oliver, S.G., Robertson, D.L. (2007) One billion years of bZIP transcription factor evolution: Conservation and change in dimerization and DNA-binding site specificity. *Molecular Biology and Evolution* 24: 827–835.

Apel, K., Hirt, H. (2004) Reactive oxygen species: Metabolism, oxidation stress, and signal transduction. *Annual Review of Plant Physiology and Plant Molecular Biology* 55: 373–399.

Apostolova, P., Szalai, G., Kocsy, G., Janda, T., Popova, T. (2012) Environmental factors affecting components of ascorbate-glutathione pathway in crop plants. In *Oxidative Stress in Plants: Causes, Consequences and Tolerance*, ed. Anjum, N., Umar, S., Ahmad, A., 52–70. New Delhi, India: I.K. International.

Asada, K. (1999) The water-water cycle in chloroplasts: Scavenging of active oxygens and dissipation of excess photons. *Annual Review of Plant Physiology and Plant Molecular Biology* 50: 601–639.

Asthir, B., Koundal, A., Bains, N.S. (2012) Putrescine modulates antioxidant defense response in wheat under high temperature stress. *Biologia Plantarum* 56: 757–761.

Bakshi, M., Oelmüller, R. (2014) WRKY transcription factors: Jack of many trades in plants. *Plant Signaling and Behavior* 9: e27700.

Bautista, I., Boscaiu, M., Lidón, A., Llinares, J., Lull, C., Donat, M., Mayoral, O., Vincent, O. (2016) Environmentally induced changes in antioxidant phenolic compounds levels in wild plants. *Acta Physiologiae Plantarum* 38: 9.

Bavita, A., Shashi, B., Navtej, S.B. (2012) Nitric oxide alleviates oxidative damage induced by high temperature stress in wheat. *Indian Journal of Experimental Biology* 50: 372–378.

Becerra-Moreno, A., Redondo-Gil, M., Benavides, J., Nair, V., Cisneros-Zevallos, L., Jacobo-Velázquez, D. (2015) Combined effect of water loss and wounding stress on gene activation of metabolic pathways associated with phenolic biosynthesis in carrot. *Frontiers in Plant Science* 6: 837.

Bouaziz, D., Charfeddine, M., Jbir, R., Saidi, M.N., Pirrello, J., Charfeddine, S., Bouzayen, M., Gargouri-Bouzid, R. (2015) Identification and functional characterization of ten AP2/ERF genes in potato. *Plant Cell, Tissue and Organ Culture* 123: 155–172.

Brown, D., Rashotte, A., Murphy, A., Normanly, J., Tague, B., Peer, W., Taiz, L., Muday, G. (2001) Flavonoids act as negative regulators of auxin transport *in vivo* in *Arabidopsis*. *Plant Physiology* 126: 524–535.

Cai, R., Dai, W., Zhang, C., Wang, Y., Wu, M., Zhao, Y., Ma, Q., Xiang, Y., Cheng, B. (2017) The maize WRKY transcription factor ZmWRKY17 negatively regulates salt stress tolerance in transgenic *Arabidopsis* plants. *Planta* 246: 1215–1231.

Cao, S., Du, X., Li, L., Liu, Y., Zhang, L., Pan, X., Li, Y., Li, H., Lu, H. (2017) Overexpression of *Populus tomentosa* cytosolic ascorbate peroxidase enhances abiotic stress tolerance in tobacco plants. *Russian Journal of Plant Physiology* 64: 224–234.

Caverzan, A., Passaia, G., Rosa, S.B. (2012) Plant responses to stresses: Role of ascorbate peroxidase in the antioxidant protection. *Genetics and Molecular Biology* 35: 1011–1019.

Cebulak, T., Oszmiański, J., Kapusta, I., Lachowicz, S. (2017) Effect of UV-C radiation, ultra-sonication electromagnetic field and microwaves on changes in polyphenolic compounds in chokeberry (*Aronia melanocarpa*). *Molecules* 22: 1161.

Chen, J., Nolan, T.M., Ye, H., Zhang, M., Tong, H., Xin, P., Chu, J., Chu, C., Li, Z., Yina, Y. (2017a) *Arabidopsis* WRKY46, WRKY54, and WRKY70 transcription factors are involved in brassinosteroid-regulated plant growth and drought responses. *The Plant Cell* 29: 1425–1439.

Chen, J-H., Jiang, H-W., Hsieh, E-J., Chen, H-Y., Chien, C-T., Hsieh, H-L., Lin, T-P. (2012) Drought and salt stress tolerance of an *Arabidopsis* glutathione S-Transferase U17 knockout mutant are attributed to the combined effect of glutathione and abscisic acid. *Plant Physiology* 158: 340–351.

Chen, K., Zhang, M., Zhu, H., Huang, M., Zhu, Q., Tang, D., Han, X., Li, J., Sun, J., Fu, J. (2017b) Ascorbic acid alleviates damage from heat stress in the photosystem II of tall fescue in both the photochemical and thermal phases. *Frontiers in Plant Science* 8: 1373.

Chen, M., Wang, Q.Y., Cheng, X.G., Xu, S.Z., Li, L.C., Ye, X.G., Xia, L.Q., Ma, Y.Z. (2007) *GmDREB2*, a soybean DRE-binding transcription factor, conferred drought and high-salt tolerance in transgenic plants. *Biochemical and Biophysical Research Communications* 353: 299–305.

Chen, T., Li, W., Hu, X., Guo, J., Liu, A., Zhang, B. (2015) A cotton MYB transcription factor, GbMYB5, is positively involved in plant adaptive response to drought stress. *Plant and Cell Physiology* 56: 917–929.

Chiang, C., Chen, S., Chen, L., Chiang, M., Chien, H., Lin, K. (2014) Expression of the broccoli catalase gene (*BoCAT*) enhances heat tolerance in transgenic *Arabidopsis*. *Journal of Plant Biochemistry and Biotechnology* 23: 266–277.

Chiang, C.M., Chen, C.C., Chen, S.P., Lin, K.H., Chen, L.R., Su, Y.H., Yen, H.C. (2017) Overexpression of the ascorbate peroxidase gene from eggplant and sponge gourd enhances flood tolerance in transgenic *Arabidopsis*. *Journal of Plant Research* 130: 373–386.

Chiang, C.M., Cheng, L.F.O., Shih, S.W., Lin, K.H. (2015) Expression of eggplant ascorbate peroxidase increases the tolerance of transgenic rice plants to flooding stress. *Journal of Plant Biochemistry and Biotechnology* 24: 257–267.

Choi, H.W., Hwang, B.K. (2012) The pepper extracellular peroxidase CaPO2 is required for salt, drought and oxidative stress tolerance as well as resistance to fungal pathogens. *Planta* 235: 1369–1382.

Choudhury, F.K., Rivero, R.M., Blumwald, E., Mittler, R. (2017) Reactive oxygen species, abiotic stress and stress combination. *The Plant Journal* 90: 856–867.

Christiansen, M.W., Matthewman, C., Podzimska-Srok, D., O'Shea, C., Lindemose, S., Møllegaard, N.E., Holme, I.B., Hebelstrup, K., Skriver, K., Gregersen, P.L. (2016) Barley plants over-expressing the NAC transcription factor gene *HvNAC005* show stunting and delay in development combined with early senescence. *Journal of Experimental Botany* 67: 5259–5273.

Christou, A., Manganaris, G., Fotopoulos, V. (2014) Systemic mitigation of salt stress by hydrogen w and sodium nitroprusside in strawberry plants via transcriptional regulation of enzymatic and nonenzymatic antioxidants. *Environmental and Experimental Botany* 107: 46–54.

Clemens, S. (2006) Toxic metal accumulation, responses to exposure and mechanisms of tolerance in plants. *Biochimie* 88: 1707–1719.

Conklin, P., Barth, C. (2004) Ascorbic acid, a familiar small molecule intertwined in the response of plants to ozone, pathogens, and the onset of senescence. *Plant, Cell and Environment* 27: 959–970.

Cramer, G.R., Urano, K., Delrot, S., Pezzotti, M., Shinozaki, K. (2011) Effects of abiotic stress on plants: A systems biology perspective. *BMC Plant Biology* 11: 163.

Cuypers, A., Smeets, K., Ruytinx, J., Opdenakker, K., Keunen, E., Remans, T., Horemans, N., Vanhoudt, N., Van Sanden, S., Van Belleghem, F., Guisez, Y., Colpaert, J., Vangronsveld, J. (2011) The cellular redox state as a modulator in cadmium and copper responses in *Arabidopsis thaliana* seedlings. *Journal of Plant Physiology* 168: 309–316.

Da Silva, C., Fontes, E., Modolo, L. (2016) Salinity-induced accumulation of endogenous H_2S and NO is associated with modulation of the antioxidant and redox defense systems in *Nicotiana tabacum* L. cv. Havana. *Plant Science* 256: 148–159.

Dai, X., Wang, Y., Zhang, W. (2016) *OsWRKY74*, a WRKY transcription factor, modulates tolerance to phosphate starvation in rice. *Journal of Experimental Botany* 67: 947–960.

Das, K., Roychoudhury, A. (2014) Reactive oxygen species (ROS) and response of antioxidants as ROS-scavengers during environmental stress in plants. *Frontiers in Environmental Science* 2: 1–13.

De Almeida, T., Neto, J., De Sousa, N., Pessoa, I., Vieira, L., De Medeiros, J., Boligon, A., Hamers, A., Farias, D., Peijnenburg, A., Carvalho, A. (2017) Phenolic compounds of *Triplaris gardneriana* can protect cells against oxidative stress and restore oxidative balance. *Biomedicine and Pharmacotherapy* 93: 1261–1268.

Demmig-Adams, B., Adams, W. (2002) Antioxidants in photosynthesis and human nutrition. *Science* 298: 2149–2153.

Demmig-Adams, B., Adams, W. (2006) Photoprotection in an ecological context: The remarkable complexity of thermal energy dissipation. *The New Phytologist* 172: 11–21.

Diaz-Vivancos, P., Faize, M., Barba-Espin, G., Faize, L., Petri, C., Hernández, J.A., Burgos, L. (2013) Ectopic expression of cytosolic superoxide dismutase and ascorbate peroxidase leads to salt stress tolerance in transgenic plums. *Plant Biotechnol Journal* 11: 976–985.

Dixon, R., Paiva, N. (1995) Stress-induced phenylpropanoid metabolism. *The Plant Cell* 7: 1085–1097.

Domonkos, I., Kis, M., Gombos, Z., Ughy, B. (2013) Carotenoids, versatile components of oxygenic photosynthesis. *Progress in Lipid Research* 52: 539–561.

Du, H., Feng, B., Yang, S., Huang, S., Tang, Y. (2012) The R2R3-MYB transcription factor gene family in maize. *PLoS One* 7:e37463.

Dubos, C., Stracke, R., Grotewold, E., Weisshaar, B., Martin, C., Lepiniec, L. (2010) MYB transcription factors in *Arabidopsis*. *Trends in Plant Science* 15: 573–581.

Elavarthi, S., Martin, B. (2010) Spectrophotometric assays for antioxidant enzymes in plants. In *Plant Stress Tolerance. Methods in Molecular Biology (Methods and Protocols)*, ed. Sunkar, R. 273–80. Humana Press, New York City, NY.

Faize, M., Burgos, L., Faize, L., Piqueras, A., Nicolas, E., Barba-Espin, G., Clemente-Moreno, M.J., Alcobendas, R., Artlip, T., Hernandez, J.A. (2011) Involvement of cytosolic ascorbate peroxidase and Cu/Zn-superoxide dismutase for improved tolerance against drought stress. *Journal of Experimental Botany* 62: 2599–2613.

Faize, M., Nicolás, E., Faize, L., Díaz-Vivancos, P., Burgos, L., Hernández, J.A. (2015) Cytosolic ascorbate peroxidase and Cu, Zn-superoxide dismutase improve seed germination, plant growth, nutrient uptake and drought tolerance in tobacco. *Theoretical and Experimental Plant Physiology* 27: 215–226.

Fan, W., Hai, M., Guo, Y., Ding, Z., Tie, W., Ding, X., Yan, Y., Wei, Y., Liu, Y., Wu, C., Shi, H., Li, K., Hu, W. (2016) The ERF transcription factor family in cassava: Genome-wide characterization and expression analyses against drought stress. *Scientific Reports* 6: 37379.

Ferreres, F., Figueiredo, R., Bettencourt, S., Carqueijeiro, I., Oliveira, J., Gil-Izquierdo, A., Pereira, D., Valentão, P., Andrade, P., Duarte, P., Barceló, A., Sottomayor, M. (2011) Identification of phenolic compounds in isolated vacuoles of the medicinal plant *Catharanthus roseus* and their interaction with vacuolar class III peroxidase: An H_2O_2 affair? *Journal of Experimental Botany* 62: 2841–2854.

Fini, A., Brunetti, C., Di Ferdinando, M., Ferrini, F., Tattini, M. (2011) Stress-induced flavonoid biosynthesis and the antioxidant machinery of plants. *Plant Signaling and Behavior* 6: 709–711.

Fotopoulos, V., Ziogas, V., Tanou, G., Molassiotis, A. (2010) Involvement of AsA/DHA and GSH/GSSG ratios in gene and protein expression and in the activation of defence mechanisms under abiotic stress conditions. In *Ascorbate-Glutathione Pathway and Stress Tolerance in Plants*, ed. Anjum, N., Chan, M-T., Umar, S., 265–302. Dordrecht, the Netherlands: Springer.

Foyer, C.H. (2015) Redox homeostasis: Opening up ascorbate transport. *Nature Plants* 1: 14012.

Foyer, C.H., Halliwell, B. (1976) The presence of glutathione and glutathione reductase in chloroplasts: A proposed role in ascorbic acid metabolism. *Planta* 133: 21–25.

Foyer, C.H., Noctor, G. (2003) Redox sensing and signalling associated with reactive oxygen in chloroplasts, peroxisomes and mitochondria. *Physiologia Plantarum* 119: 355–364.

Foyer, C.H., Noctor, G. (2005) Oxidant and antioxidant signalling in plants: A re-evaluation of the concept of oxidative stress in a physiological context. *Plant, Cell and Environment* 28: 1056–1071.

Foyer, C.H., Ruban, A.V., Noctor, G. (2017) Viewing oxidative stress through the lens of oxidative signalling rather than damage. *Biochemical Journal* 474: 877–883.

Frank, H., Bautista, J., Josue, J., Young, A. (2000) Mechanisms of nonphotochemical quenching in green plants: Energies of the lowest excited singlet states of violaxanthin and zeaxanthin. *Biochemistry* 39: 2831–2837.

Gallie, D.R. (2012) The role of L-ascorbic acid recycling in responding to environmental stress and in promoting plant growth. *Journal of Experimental Botany* 64: 433–443.

Gallie, D.R. (2013) L-ascorbic acid: A multifunctional molecule supporting plant growth and development. *Scientifica* 2013: 795964.

Gest, N., Gautier, H., Stevens, R. (2013) Ascorbate as seen through plant evolution: The rise of a successful molecule? *Journal of Experimental Botany* 64: 33–53.

Gill, S., Tuteja, N. (2010) Reactive oxygen species and antioxidant machinery in abiotic stress tolerance in crop plants. *Plant Physiology and Biochemistry* 48: 909–930.

Golldack, D., Li, C., Mohan, H., Probst, N. (2014) Tolerance to drought and salt stress in plants: Unraveling the signaling networks. *Frontiers in Plant Science* 5: 151.

Gu, C., Guo, Z., Hao, P., Wang, G., Jin, Z., Zhang, S. (2017) Multiple regulatory roles of AP2/ERF transcription factor in angiosperm. *Botanical Studies* 58: 6.

Guan, Q., Liao, X., He, M., Li, X., Wang, Z., Ma, H., Yu, S., Liu, S. (2017) Tolerance analysis of chloroplast OsCu/Zn-SOD overexpressing rice under NaCl and $NaHCO_3$ stress. *PLoS One* 12: e0186052.

Guan, Q., Wang, Z., Wang, X., Takano, T., Liu, S. (2015) A peroxisomal APX from *Puccinellia tenuiflora* improves the abiotic stress tolerance of transgenic *Arabidopsis thaliana* through decreasing of H_2O_2 accumulation. *Journal of Plant Physiology* 175: 183–191.

Gumul, D., Korus, J., Achremowicz, B. (2007) The influence of extrusion on the content of polyphenols and antioxidant/antiradical activity of rye grains (*Secale cereale* L.). *Acta Scientiarum Polonorum: Technologia Alimentaria* 6: 103–111.

Guo, H., Wang, Y., Wang, L., Hu, P., Wang, Y., Jia, Y., Zhang, C., Zhang, Y., Zhang, Y., Wang, C., Yang, C. (2017a) Expression of the MYB transcription factor gene *BplMYB46* affects abiotic stress tolerance and secondary cell wall deposition in *Betula platyphylla*. *Plant Biotechnology Journal* 15: 107–121.

Guo, J., Ling, H., Ma, J., Chen, Y., Su, Y., Lin, Q., Gao, S., Wang, H., Que, Y., Xu, L. (2017b) A sugarcane R2R3-MYB transcription factor gene is alternatively spliced during drought stress. *Science Reports* 7: 41922.

Hager, A. (1980) The reversible, light-induced conversion of xanthophylls in the chloroplasts. In *Pigments in Plants*, ed. Czygan, F.-C., 57–79. Stuttgart, Germany: Gustav Fischer.

Hasanuzzaman, M., Nahar, K., Alam, M., Fujita, M. (2012b) Exogenous nitric oxide alleviates high temperature induced oxidative stress in wheat (*Triticum aestivum* L.) seedlings by modulating the antioxidant defense and glyoxalase system. *Australian Journal of Crop Science* 6: 1314–1323.

Havaux, M. (2014) Carotenoid oxidation products as stress signals in plants. *The Plant Journal* 79: 597–606.

He, G., Xu, J., Wang, Y., Liu, J., Li, P., Chen, M., Ma, Y., Xu, Z. (2016b) Drought-responsive WRKY transcription factor genes *TaWRKY1* and *TaWRKY33* from wheat confer drought and/or heat resistance in *Arabidopsis*. *BMC Plant Biology* 16: 116.

He, X., Zhu, L., Xu, L., Guo, W., Zhang, X. (2016a) GhATAF1, a NAC transcription factor, confers abiotic and biotic stress responses by regulating phytohormonal signaling networks. *Plant Cell Reports* 35: 2167–2179.

Hendrich, A.B. (2006) Flavonoid-membrane interactions: Possible consequences for biological effects of some polyphenolic compounds. *Acta Pharmacologica Sinica* 27: 27–40.

Henriques, A.R., Farias, D.D.R., Costa de Oliveira, A. (2017) Identification and characterization of the bZIP transcription factor involved in zinc homeostasis in cereals. *Genetics and Molecular Research* 16: 2.

Hernández, I., Alegre, L., Van Breusegem, F., Munné-Bosch, S. (2009) How relevant are flavonoids as antioxidants in plants? *Trends in Plant Science* 14: 125–132.

Hichri, I., Muhovski, Y., Žižková, E., Dobrev, P.I., Gharbi, E., Franco-Zorrilla, J.M., Lopez-Vidriero, I., Solano, R., Clippe, A., Errachid, A., Motyka, V., Lutts, S. (2017) The *Solanum lycopersicum* WRKY3 transcription factor SlWRKY3 is involved in salt stress tolerance in tomato. *Frontiers in Plant Science* 8: 1343.

Hollenbeck, J.J., Mcclain, D.L., Oakley, M.G. (2002) The role of helix stabilizing residues in GCN4 basic region folding and DNA binding. *Protein Science* 11: 2740–2747.

Hu, W., Yang, H., Yan, Y., Wei, Y., Tie, W., Ding, Z., Zuo, J., Peng, M., Li, K. (2016) Genome-wide characterization and analysis of bZIP transcription factor gene family related to abiotic stress in cassava. *Scientific Reports* 6: 22783.

Huang, L., Hong, Y., Zhang, H., Li, D., Song, F. (2016) Rice NAC transcription factor ONAC095 plays opposite roles in drought and cold stress tolerance. *BMC Plant Biology* 16: 203.

Huang, X., Liu, J., Chen, X. (2010) Overexpression of *PtrABF* gene, a bZIP transcription factor isolated from *Poncirus trifoliata*, enhances dehydration and drought tolerance in tobacco via scavenging ROS and modulating expression of stress-responsive genes. *BMC Plant Biology* 10: 230.

Hussain, I., Siddique, A., Ashraf, M., Rasheed, R., Ibrahim, M., Iqbal, M., Akbar, S., Imran, M. (2017) Does exogenous application of ascorbic acid modulate growth, photosynthetic pigments and oxidative defense in okra (*Abelmoschus esculentus* (L.) Moench) under lead stress? *Acta Physiologiae Plantarum* 39: 144.

Inaba, S., Kurata, R., Kobayashi, M., Yamagishi, Y., Mori, I., Ogata, Y., Fukao, Y. (2015) Identification of putative target genes of bZIP19, a transcription factor essential for *Arabidopsis* adaptation to Zn deficiency in roots. *The Plant Journal* 84: 323–334.

Islam, T., Manna, M., Reddy, M.K. (2015) Glutathione peroxidase of *Pennisetum glaucum* (PgGPx) is a functional Cd^{2+} dependent peroxiredoxin that enhances tolerance against salinity and drought stress. *PLoS One* 10: e0143344.

Jin, C., Ji, J., Zhao, Q., Ma, R., Guan, C., Wang, G. (2015) Characterization of lycopene β cyclase gene from *Lycium chinense* conferring salt tolerance by increasing carotenoids synthesis and oxidative stress resistance in tobacco. *Molecular Breeding* 35: 228.

Jin, C., Li, K., Xu, X., Zhang, H., Chen, H., Chen, Y., Hao, J., Wang, Y., Huang, X., Zhang, S. (2017) A novel NAC transcription factor, *PbeNAC1*, of *Pyrus betulifolia* confers cold and drought tolerance via interacting with *PbeDREBs* and activating the expression of stress-responsive genes. *Frontiers in Plant Science* 8: 1049.

Jing, X., Hou, P., Lu, Y., Deng, S., Li, N., Zhao, R., Sun, J., Wang, Y., Han, Y., Lang, T., Ding, M., Shen, X., Chen, S. (2015) Overexpression of copper/zinc superoxide dismutase from mangrove *Kandelia candel* in tobacco enhances salinity tolerance by the reduction of reactive oxygen species in chloroplast. *Frontiers in Plant Science* 6: 23.

Jisha, V., Dampanaboina, L., Vadassery, J., Mithöfer, A., Kappara, S., Ramanan, R. (2015) Overexpression of an AP2/ERF type transcription factor *OsEREBP1* confers biotic and abiotic stress tolerance in rice. *PLoS One* 10: e0127831.

Jozefczak, M., Remans, T., Vangronsveld, J., Cuypers, A. (2012) Glutathione is a key player in metal-induced oxidative stress defenses. *International Journal of Molecular Sciences* 13: 3145–3175.

Khan, A., Iqbal, I., Shah, A., Nawaz, H., Ahmad, F., Ibrahim, M. (2010) Alleviation of adverse effects of salt stress in Brassica (*Brassica campestris*) by pre-sowing seed treatment with ascorbic acid. *American-Eurasian Journal of Agricultural and Environmental Sciences* 7: 557–560.

Kim, Y.-O., Bae, H-J., Cho, E., Kang, H. (2017) Exogenous glutathione enhances mercury tolerance by inhibiting mercury entry into plant cells. *Frontiers in Plant Science* 8: 683.

Kim, Y.S., Kim, S.G., Park, J.E., Park, H.Y., Lim, M.H., Chua, N.H., Park, C.M. (2006) A membrane-bound NAC transcription factor regulates cell division in *Arabidopsis*. *Plant Cell* 18: 3132–3144.

Klempnauer, K.H., Gonda, T.J., Bishop, J.M. (1982) Nucleotide sequence of the retroviral leukemia gene *v-myb* and its cellular progenitor *c-myb*: The architecture of a transduced oncogene. *Cell* 31: 453–463.

Kostopoulou, Z., Therios, I., Roumeliotis, E., Kanellis, A., Molassiotis, A. (2015) Melatonin combined with ascorbic acid provides salt adaptation in *Citrus aurantium* L. seedlings. *Plant Physiology and Biochemistry* 86: 155–165.

Król, A., Amarowicz, R., Weidner, S. (2014) Changes in the composition of phenolic compounds and antioxidant properties of grapevine roots and leaves (*Vitis vinifera* L.) under continuous of long-term drought stress. *Acta Physiologiae Plantarum* 36: 1491–1499.

Kudo, M., Kidokoro, S., Yoshida, T., Mizoi, J., Todaka, D., Fernie, A.R., Shinozaki, K., Yamaguchi-Shinozaki, K. (2017) Double overexpression of DREB and PIF transcription factors improves drought stress tolerance and cell elongation in transgenic plants. *Plant Biotechnology Journal* 15: 458–471.

Kumar, S., Kaur, R., Kaur, N., Bhandhari, K., Kaushal, N., Gupta, K., Bains, T. (2011) Heat-stress induced inhibition in growth and chlorosis in mungbean (*Phaseolus aureus* Roxb.) is partly mitigated by ascorbic acid application and is related to reduction in oxidative stress. *Acta Physiologiae Plantarum* 33: 2091.

Latif, M., Akram, N., Ashraf, M. (2016) Regulation of some biochemical attributes in drought-stressed cauliflower (*Brassica oleracea* L.) by seed pre-treatment with ascorbic acid. *The Journal of Horticultural Science and Biotechnology* 91: 129–137.

Lewandowska, H., Kalinowska, M., Lewandowski, W., Stępkowski, T., Brzóska, K. (2016) The role of natural polyphenols in cell signaling and cytoprotection against cancer development. *The Journal of Nutritional Biochemistry* 32: 1–19.

Li, A-N., Li, S., Zhang, Y-J., Xu, X-R., Chen, Y-M., Li, H-B. (2014) Resources and biological activities of natural polyphenols. *Nutrients* 6: 6020–6047.

Li, K., Pang, C., Ding, F., Sui, N., Feng, Z., Wang, B. (2012) Overexpression of *Suaeda salsa* stroma ascorbate peroxidase in *Arabidopsis* chloroplasts enhances salt tolerance of plants. *South African Journal of Botany* 78: 235–245.

Li, L., Yuan, H. (2013) Chromoplast biogenesis and carotenoid accumulation. *Archives of Biochemistry and Biophysics* 539: 102–109.

Li, M., Xu, Z., Tian, C., Huang, Y., Wang, F., Xionga, A. (2016c) Genomic identification of WRKY transcription factors in carrot (*Daucus carota*) and analysis of evolution and homologous groups for plants. *Scientific Reports* 6: 23101.

Li, R., Kang, C., Song, X., Yu, L., Liu, D., He, S., Zhai, H., Liu, Q. (2017a) A ζ-carotene desaturase gene, *IbZDS*, increases β-carotene and lutein contents and enhances salt tolerance in transgenic sweetpotato. *Plant Science* 26(2): 39–51.

Li, X.D., Zhuang, K.Y., Liu, Z.M., Yang, D.Y., Ma, N.N., Meng, Q.W. (2016a) Overexpression of a novel NAC-type tomato transcription factor, *SlNAM1*, enhances the chilling stress tolerance of transgenic tobacco. *Journal of Plant Physiology* 204: 54–65.

Li, Z., Han, X., Song, X., Zhang, Y., Jiang, J., Han, Q., Liu, M., Qiao, G., Zhuo, R. (2017b) Overexpressing the *Sedum alfredii* Cu/Zn superoxide dismutase increased resistance to oxidative stress in transgenic *Arabidopsis*. *Frontiers in Plant Science* 8: 1010.

Li, Z., Peng, R., Tian, Y., Han, H., Xu, J., Yao, Q. (2016b) Genome-wide identification and analysis of the MYB transcription factor superfamily in *Solanum lycopersicum*. *Plant and Cell Physiology* 57: 1657–1677.

Liang, C., Meng, Z., Meng, Z., Malik, W., Yan, R., Lwin, K.M., Lin, F., Wang, Y., Sun, G., Zhou, T., Zhu, T., Li, J., Jin, S., Guo, S., Zhang, R. (2016) GhABF2, a bZIP transcription factor, confers drought and salinity tolerance in cotton (*Gossypium hirsutum* L.). *Scientific Reports* 6: 35040.

Liang, M., Li, H., Zhou, F., Li, H., Liu, J., Hao, Y., Wang, Y., Zhao, H., Han, S. (2015) Subcellular distribution of NTL transcription factors in *Arabidopsis thaliana*. *Traffic* 10: 1062–1074.

Liang, Q., Wu, Y., Wang, K., Bai, Z., Liu, Q., Pan, Y., Zhang, L., Jiang, B. (2017) Chrysanthemum WRKY gene *DgWRKY5* enhances tolerance to salt stress in transgenic chrysanthemum. *Scientific Reports* 7: 4799.

Licausi, F., Ohme-Takagi, M., Perata, P. (2013) APETALA2/Ethylene Responsive Factor (AP2/ERF) transcription factors: Mediators of stress responses and developmental programs. *New Phytologist* 199: 639–649.

Lindemose, S., O'Shea, C., Jensen, M.K., Skriver, K. (2013) Structure, function and networks of transcription factors involved in abiotic stress responses. *International Journal of Molecular Sciences* 14: 5842–5878.

Liu, H., Zhou, X., Dong, N., Liu, X., Zhang, H., Zhang, Z. (2011) Expression of a wheat MYB gene in transgenic tobacco enhances resistance to *Ralstonia solanacearum*, and to drought and salt stresses. *Functional and Integrative Genomics* 11: 431–443.

Llorca, C.M., Potschin, M., Zentgraf, U. (2014) bZIPs and WRKYs: Two large transcription factor families executing two different functional strategies. *Frontiers in Plant Science* 5: 169.

Lopez-Juez, E., Pyke, K. (2005) Plastids unleashed: Their development and their integration in plant development. *The International Journal of Developmental Biology* 49: 557–577.

Lukatkin, A.S., Anjum, N. (2014) Control of cucumber (*Cucumis sativus* L.) tolerance to chilling stress- evaluating the role of ascorbic acid and glutathione. *Frontiers in Environmental Science* 2: 62.

Luo, X., Wu, J., Li, Y., Nan, Z., Guo, X., Wang, Y., Zhang, A., Wang, Z., Xia, G., Tian, Y. (2013) Synergistic effects of *GhSOD1* and *GhCAT1* overexpression in cotton chloroplasts on enhancing tolerance to methyl viologen and salt stresses. *PLoS One* 8: e54002.

Mahanty, S., Kaul, T., Pandey, P., Reddy, R.A., Mallikarjuna, G., Reddy, C.S., Sopory, S.K., Reddy, M.K. (2012) Biochemical and molecular analyses of copper–zinc superoxide dismutase from a C4 plant *Pennisetum glaucum* reveals an adaptive role in response to oxidative stress. *Gene* 505: 309–317.

Mahmood, Q., Ahmad, R., Kwak, S-S., Rashid, A., Anjum, N. (2010) Ascorbate and glutathione: Protectors of plants in oxidative stress. In *Ascorbate-Glutathione Pathway and Stress Tolerance in Plants*, ed. Anjum, N., Chan, M-T., Umar, S., 209–229. Dordrecht, Netherlands: Springer.

Malacarne, G., Coller, E., Czemmel, S., Vrhovsek, U., Engelen, K., Goremykin, V., Bogs, J., Moser, C. (2016) The grapevine VvibZIPC22 transcription factor is involved in the regulation of flavonoid biosynthesis. *The Journal of Experimental Botany* 67: 3509–3522.

Mano, J., Hideg, E., Asada, K. (2004) Ascorbate in thylakoid lumen functions as an alternative electron donor to photosystem II and photosystem I. *Archives of Biochemistry and Biophysics* 429: 71–80.

Mantri, N., Patade, V., Penna, S., Ford, R., Pang, E. (2012) Abiotic stress responses in plants: Present and future. In *Abiotic Stress Responses in Plants: Metabolism, Productivity an Sustainability*, ed. Ahmad, P., Prasad, M.N.V., 1–19. New York, NY: Springer.

Mao, H., Yu, L., Han, R., Li, Z., Liu, H. (2016) ZmNAC55, a maize stress-responsive NAC transcription factor, confers drought resistance in transgenic *Arabidopsis*. *Plant Physiology and Biochemistry* 105: 55–66.

Marques, A.T., Santos, S.P., Rosa, M.G., Rodrigues, M.A., Abreu, I.A., Frazão, C., Romão, C.V. (2014) Expression, purification and crystallization of MnSOD from *Arabidopsis thaliana*. *Acta Crystallogr Section F: Structural Biology Communications* 70: 669–672.

Marques, D.N., Reis, S.P., Souza, C.R.B. (2017) Plant NAC transcription factors responsive to abiotic stresses. *Plant Gene* 11: 170–179.

Martins, N., Barros, L., Ferreira, I. (2016) *In vivo* antioxidant activity of phenolic compounds: Facts and gaps. *Trends in Food Science and Technology* 48: 1–12.

Mathew, S., Abraham, T., Zakaria, Z. (2015) Reactivity of phenolic compounds towards free radicals under *in vitro* conditions. *Journal of Food Science and Technology* 52: 5790–5798.

Mibei, E., Ambuko, J., Giovannoni, J., Onyango, A., Owino, W. (2017) Carotenoid profiling of the leaves of selected African eggplant accessions subjected to drought stress. *Food Science and Nutrition* 5: 113–122.

Miller, G., Suzuki, N., Ciftci-Yilmaz, S., Mittler, R. (2010) Reactive oxygen species homeostasis and signaling during drought and salinity stresses. *Plant, Cell and Environment* 33: 453–467.

Mittler, R. (2002) Oxidative stress, antioxidants and stress tolerance. *Trends in Plant Science* 7: 405–410.

Miura, C., Sugawara, K., Neriya, Y., Minato, Y. N., Keima, T., Himeno, M., Maejima, K., Komatsu, K., Yamaji, Y., Oshima, K., Namba, S. (2012) Functional characterization and gene expression profiling of superoxide dismutase from plant pathogenic phytoplasma. *Gene* 510: 107–112.

Mizoi, J., Ohori, T., Moriwaki, T., Kidokoro, S., Todaka, D., Maruyama, K., Kusakabe, K., Osakabe, Y., Shinozaki, K., Yamaguchi-Shinozaki, K. (2013) GmDREB2A;2, a canonical DEHYDRATION-RESPONSIVE ELEMENT-BINDING PROTEIN2-type transcription factor in soybean, is posttranslationally regulated and mediates dehydration-responsive element-dependent gene expression. *Plant Physiology* 161: 346–361.

Mojzer, E., Hrnčič, M., Škerget, M., Knez, Z., Bren, U. (2016) Polyphenols: Extraction methods, antioxidative action, bioavailability and anticarcinogenic effects. *Molecules* 21: 901.

Molina-Rueda, J.J., Tsai, C.J., Kirby, E.G. (2013) The populus superoxide dismutase gene family and its responses to drought stress in transgenic poplar overexpressing a pine cytosolic glutamine synthetase (GS1a). *PLoS One* 8: 1–14.

Mukherjee, P.K., Houghton, P.J. (2009). *Evaluation of Herbal Medicinal Products: Perspectives on Quality, Safety and Efficacy*. London, UK: Royal Pharmaceutical Society of Great Britain.

Mukhtar, A., Akram, N., Aisha, R., Shafiq, S., Ashraf, M. (2016) Foliar applied ascorbic acid enhances antioxidative potential and drought tolerance in cauliflower (*Brassica oleracea* L. var. Botrytis). *Agrochimica* 60: 100–113.

Müller-Moulé, P., Conklin, P., Niyogi, K. (2002) Ascorbate deficiency can limit violaxanthin de-epoxidase activity *in vivo*. *Plant Physiology* 128: 970–977.

Mullineaux, P.M., Rausch, T. (2005) Glutathione, photosynthesis and the redox regulation of stress-responsive gene expression. *Photosynthesis Research* 86: 459–474.

Nahar, K., Hasanuzzaman, M., Mahabub Alam, M., Fujita, M. (2015) Exogenous glutathione confers high temperature stress tolerance in mung bean (*Vigna radiata* L.) by modulating antioxidant defense and methylglyoxal detoxification system. *Environmental and Experimental Botany* 112: 44–54.

Naoumkina, M.A., He, X., Dixon, R.A. (2008) Elicitor-induced transcription factors for metabolic reprogramming of secondary metabolism in *Medicago truncatula*. *BMC Plant Biology* 8: 132.

Naz, H., Akram, N., Ashraf, M. (2016) Impact of ascorbic acid on growth and some physiological attributes of cucumber (*Cucumis sativus*) plants under water-deficit conditions. *Pakistan Journal of Botany* 48: 877–883.

Niggeweg, R., Michael, A., Martin, C. (2004) Engineering plants with increased levels of the antioxidant chlorogenic acid. *Nature Biotechnology* 22: 746–754.

Ning, Y.Q., Ma, Z.Y., Huang, H.W., Mo, H., Zhao, T.T., Li, L., Cai, T., Chen, S., Ma, L., He, X.J. (2015) Two novel NAC transcription factors regulate gene expression and flowering time by associating with the histone demethylase JMJ14. *Nucleic Acids Research* 18: 1469–1484.

Nisar, N., Li, L., Lu, S., Khin, N., Pogson, B. (2015) Carotenoid metabolism in plants. *Molecular Plant* 8: 68–82.

Noctor, G. (2006) Metabolic signalling in defence and stress: The central roles of soluble redox couples. *Plant, Cell and Environment* 29: 409–425.

Noctor, G., Foyer, C. (1998) Ascorbate and glutathione: Keeping active oxygen under control. *Annual Review of Plant Physiology and Plant Molecular Biology* 49: 249–279.

Noctor, G., Foyer, C.H. (2016) Intracellular redox compartmentation and ROS-related communication in regulation and signaling. *Plant Physiology* 171: 1581–1592.

Noctor, G., Queval, G., Mhamdi, A., Chaouch, S., Foyer, C. (2011) Glutathione. *The Arabidopsis Book/American Society of Plant Biologists* 9: e0142.

Noman, A., Liu, Z., Aqeel, M., Zainab, M., Khan, M.I., Hussain, A., Ashraf, M.F., Li, X., Weng, Y., He, S. (2017) Basic leucine zipper domain transcription factors: The vanguards in plant immunity. *Biotechnology Letters* 39: 1779–1791.

Noshi, M., Yamada, H., Hatanaka, R., Tanabe, N., Tamoi, M., Shigeoka, S. (2016) *Arabidopsis* dehydroascorbate reductase 1 and 2 modulate redox states of ascorbate-glutathione cycle in the cytosol in response to photooxidative stress. *Bioscience, Biotechnology, and Biochemistry* 81: 523–33.

Olsen, A.N., Ernst, H.A., Leggio, L.L., Skriver, K. (2005) NAC transcription factors: Structurally distinct, functionally diverse. *Trends in Plant Science* 10: 79–87.

Pang, C-H., Wang, B-S. (2010) Role of ascorbate peroxidase and glutathione reductase in ascorbate–glutathione cycle and stress tolerance in plants. In *Ascorbate-Glutathione Pathway and Stress Tolerance in Plants*, ed. Anjum, N., Chan, M-T., Umar, S., 91–113. Dordrecht, the Netherlands: Springer.

Paz-Ares, J., Ghosal, D., Wienand, U., Peterson, P.A., Saedler, H. (1987) The regulatory c1 locus of *Zea mays* encodes a protein with homology to myb proto–oncogene products and with structural similarities to transcriptional activators. *The EMBO Journal* 6: 3553–3558.

Phukan, U.J., Jeena, G.S., Shukla, R.K. (2016) WRKY transcription factors: Molecular regulation and stress responses in plants. *Frontiers in Plant Science* 7: 760.

Phukan, U.J., Jeena, G.S., Tripathi, V., Shukla, R.K. (2017) Regulation of Apetala2/Ethylene response factors in plants. *Frontiers in Plant Science* 8: 150.

Pospíšil, P. (2016) Production of reactive oxygen species by photosystem II as a response to light and temperature stress. *Frontiers in Plant Science* 7: 1950.

Prouse, M.B., Campbell, M.M. (2012) The interaction between MYB proteins and their target DNA binding sites. *Biochimica et Biophysica Acta* 1819: 67–77.

Queiroz, M., Oppolzer, D., Gouvinhas, I., Silva, A., Barros, A., Domínguez-Perles, R. (2017) New grape stems' isolated phenolic compounds modulate reactive oxygen species, glutathione, and lipid peroxidation *in vitro*: Combined formulations with vitamins C and E. *Fitoterapia* 120: 146–157.

Ramel, F., Birtic, S., Ginies, C., Soubigou-Taconnat, L., Triantaphylidès, C., Havaux, M. (2012) Carotenoid oxidation products are stress signals that mediate gene responses to singlet oxygen in plants. *Proceedings of the National Academy of Sciences of the United States of America* 109: 5535–5540.

Ramel, F., Mialoundama, A., Havaux, M. (2013) Nonenzymic carotenoid oxidation and photooxidative stress signalling in plants. *Journal of Experimental Botany* 64: 799–805.

Reis, S.P., Marques, D.N., Lima, A.M, Souza, C.R.B. (2016) Plant molecular adaptations and strategies under drought stress. In *Drought Stress Tolerance in Plants*, ed. Hossain, M.A., Wani, S.H., Bhattacharjee, S., Burritt, D.J., Tran, L-S.P., 91–122. Switzerland: Springer.

Rosalie, R., Joas, J., Deytieux-Belleau, C., Vulcain, E., Payet, B., Dufossé, L., Léchaudel, M. (2015) Antioxidant and enzymatic responses to oxidative stress induced by pre-harvest water supply reduction and ripening on mango (*Mangifera indica* L. cv. 'Cogshall') in relation to carotenoid content. *Journal of Plant Physiology* 184: 68–78.

Roy, S. (2016) Function of MYB domain transcription factors in abiotic stress and epigenetic control of stress response in plant genome. *Plant Signaling and Behavior* 11: e1117723.

Roychoudhury, A., Pradhan, S., Chaudhuri, B., Das, K. (2012a) Phytoremediation of toxic metals and the involvement of *Brassica* species. In *Phytotechnologies: Remediation of Environmental Contaminants*, ed. Anjum, N., Pereira, M., Ahmad, I., Duarte, A., Umar, S., Khan, N., 219–252. Boca Raton, FL: CRC Press.

Ruban, A., Johnson, M., Duffy, C. (2012) The photoprotective molecular switch in the photosystem II antenna. *Biochimica et Biophysica Acta (BBA)—Bioenergetics* 1817: 167–181.

Saeed, B., Das, M., Khurana, P. (2015) Overexpression of β-carotene hydroxylase1 (*BCH1*) in Indian mulberry, *Morus indica* cv. K2, confers tolerance against UV, high temperature and high irradiance stress induced oxidative damage. *Plant Cell, Tissue and Organ Culture* 120: 1003–1014.

Sahoo, S., Awasthi, J., Sunkar, R., Panda, S. (2017) Determining glutathione levels in plants. In *Plant Stress Tolerance. Methods in Molecular Biology*, ed. Sunkar, R., 273–277. New York, NY: Humana Press.

Sales, C.R.G., Ribeiro, R.V., Silveira, J.A., Machado, E.C., Martins, M.O., Lagôa, A.M. (2013) Superoxide dismutase and ascorbate peroxidase improve the recovery of photosynthesis in sugarcane plants subjected to water deficit and low substrate temperature. *Plant Physiol Biochem* 73: 326–336.

Sato, Y., Masuta, Y., Saito, K., Murayama, S., Ozawa, K. (2011) Enhanced chilling tolerance at the booting stage in rice by transgenic overexpression of the ascorbate peroxidase gene, *OsAPXa*. *Plant Cell Reports* 30: 399–406.

Schützendübel, A., Schwanz, P., Teichmann, T., Gross, K., Langenfeld-Heyser, R., Godbold, D., Polle, A. (2001) Cadmium-induced changes in antioxidative systems, hydrogen peroxide content, and differentiation in Scots pine roots. *Plant Physiology* 127: 887–898.

Seo, P.J., Kim, S.G., Park, C. (2008) Membrane-bound transcription factors in plants. *Trends in Plant Science* 13: 550–556.

Shafi, A., Chauhan, R., Gill, T., Swarnkar, M.K., Sreenivasulu, Y., Kumar, S., Kumar, N., Shankar, R., Ahuja, P.S., Singh, A.K. (2015) Expression of SOD and APX genes positively regulates secondary cell wall biosynthesis and promotes plant growth and yield in *Arabidopsis* under salt stress. *Plant Molecular Biology* 87: 615–631.

Shafiq, S., Akram, N., Ashraf, M., Arshad, A. (2014) Synergistic effects of drought and ascorbic acid on growth, mineral nutrients and oxidative defense system in canola (*Brassica napus* L.) plants. *Acta Physiologiae Plantarum* 36: 1539–1553.

Shahidi, F., Janitha, P., Wanasundara, P. (1992) Phenolic antioxidants. *Critical Reviews in Food Science and Nutrition* 32: 67–103.

Shen, J., Lv, B., Luo, L., He, J., Mao, C., Xi, D., Minga, F. (2017) The NAC-type transcription factor *OsNAC2* regulates ABA-dependent genes and abiotic stress tolerance in rice. *Scientific Reports* 7: 40641.

Sheoran, S., Thakur, V., Narwal, S., Turan, R., Mamrutha, H.M., Singh, V., Tiwari, V., Sharma, I. (2015) Differential activity and expression profile of antioxidant enzymes and physiological changes in wheat (*Triticum aestivum* L.) under drought. *Applied Biochemistry and Biotechnology* 177: 1282–1298.

Shingote, P.R, Kawar, P.G., Pagariya, M.C., Kuhikar, R.S., Thorat, A.S., Babu, K.H. (2015) SoMYB18, a sugarcane MYB transcription factor improves salt and dehydration tolerance in tobacco. *Acta Physiologiae Plantarum* 37: 217.

Shu, Y., Liu, Y., Zhang, J., Song, L., Guo, C. (2016) Genome-wide analysis of the AP2/ERF superfamily genes and their responses to abiotic stress in *Medicago truncatula*. *Frontiers in Plant Science* 6: 1247.

Siefermann-Harms, D. (1987) The light-harvesting and protective functions of carotenoids in photosynthetic membranes. *Physiologia Plantarum* 69: 561–68.

Singh, N., Mishra, A., Jha, B. (2014) Over-expression of the peroxisomal ascorbate peroxidase (*Sbp*APX) gene cloned from halophyte *Salicornia brachiata* confers salt and drought stress tolerance in transgenic tobacco. *Marine Biotechnology* 16: 321–332.

Smirnoff, N. (2000) Ascorbic acid: Metabolism and functions of a multi-facetted molecule. *Current Opinion in Plant Biology* 3: 229–235.

Smirnoff, N., Wheeler, G. (2000) Ascorbic acid in plants: Biosynthesis and function. *Critical Reviews in Biochemistry and Molecular Biology* 35: 291–314.

Sofo, A., Scopa, A., Nuzzaci, M., Vitti, A. (2015) Ascorbate peroxidase and catalase activities and their genetic regulation in plants subjected to drought and salinity stresses. *International Journal of Molecular Sciences* 16: 13561–13578.

Sornaraj, P., Luang, S., Lopato, S., Hrmova, M. (2016) Basic leucine zipper (bZIP) transcription factors involved in abiotic stresses: A molecular model of a wheat bZIP factor and implications of its structure in function. *Biochimica et Biophysica Acta* 1860: 46–56.

Su, L.T., Li, J.W., Liu, D.Q., Zhai, Y., Zhang, H.J., Li, X.W., Zhang, Q.L., Wang, Y., Wang, Q.Y. (2014a) A novel MYB transcription factor, GmMYBJ1, from soybean confers drought and cold tolerance in *Arabidopsis thaliana*. *Gene* 538: 46–55.

Su, Y., Guo, J., Ling, H., Chen, S., Wang, S., Xu, L., Allan, A.C., Que, Y. (2014b) Isolation of a novel peroxisomal catalase gene from sugarcane, which is responsive to biotic and abiotic stresses. *PLoS One* 9: 1–11.

Sukweenadhi, J., Kim, Y., Rahimi, S., Silva, J., Myagmarjav, D., Kwon, W.S., Yang, D. (2017) Overexpression of a cytosolic ascorbate peroxidase from *Panax ginseng* enhanced salt tolerance in *Arabidopsis thaliana*. *Plant Cell, Tissue and Organ Culture* 129: 337–350.

Suzuki, Y., Kosaka, M., Shindo, K., Kawasumi, T., Kimoto-Nira, H., Suzuki, C. (2013) Identification of antioxidants produced by *Lactobacillus plantarum*. *Bioscience, Biotechnology, and Biochemistry* 77: 1299–1302.

Szalai, G., Kellős, T., Galiba, G., Kocsy, G. (2009) Glutathione as an antioxidant and regulatory molecule in plants under abiotic stress conditions. *Journal of Plant Growth Regulation* 28: 66–80.

Taïbi, K., Taïbi, F., Abderrahim, L., Ennajah, A., Belkhodja, M., Mulet, J. (2016) Effect of salt stress on growth, chlorophyll content, lipid peroxidation and antioxidant defence systems in *Phaseolus vulgaris* L. *South African Journal of Botany* 105: 306–312.

Tak, H., Negi, S., Ganapathi, T.R. (2016) Banana NAC transcription factor MusaNAC042 is positively associated with drought and salinity tolerance. *Protoplasma* 254: 803–816.

Thirumalaikumar, V.P., Devkar, V., Mehterov, N., Ali, S., Ozgur, R., Turkan, I., Mueller-Roeber, B., Balazadeh, S. (2017) NAC transcription factor JUNGBRUNNEN1 enhances drought tolerance in tomato. *Plant Biotechnology Journal* 16: 354–366.

Tóth, S.Z., Schansker, G., Garab, G. (2013) The physiological roles and metabolism of ascorbate in chloroplasts. *Physiologia Plantarum* 148: 161–175.

Tripathy, B.C., Oelmüller, R. (2012) Reactive oxygen species generation and signaling in plants. *Plant Signaling and Behavior* 7: 1621–1633.

Trubitsin, B.V., Mamedov, M., Semenov, A., Tikhonov, A. (2014) Interaction of ascorbate with photosystem I. *Photosynthesis Research* 122: 215–231.

Tsao, R. (2010) Chemistry and biochemistry of dietary polyphenols. *Nutrients* 2: 1231–1246.

Upadhyaya, C.P., Akula, N., Young, K.E., Chun, S.C., Kim, D.H., Park, S.W. (2010) Enhanced ascorbic acid accumulation in transgenic potato confers tolerance to various abiotic stresses. *Biotechnology Letters* 32: 321–330.

Vishnevetsky, M., Ovadis, M., Vainstein, A. (1999) Carotenoid sequestration in plants: The role of carotenoid-associated proteins. *Trends in Plant Science* 4: 232–235.

Wang, C., Lu, G., Hao, Y., Guo, H., Guo, Y., Zhao, J., Cheng, H. (2017c) ABP9, a maize bZIP transcription factor, enhances tolerance to salt and drought in transgenic cotton. *Planta* 246: 453–469.

Wang, G., Zhang, S., Ma, X., Wang, Y., Kong, F., Meng, Q. (2016b) A stress-associated NAC transcription factor (SlNAC35) from tomato plays a positive role in biotic and abiotic stresses. *Physiologia Plantarum* 158: 45–64.

Wang, H., Wang, H., Shao, H., Tang, X. (2016a) Recent advances in utilizing transcription factors to improve plant abiotic stress tolerance by transgenic technology. *Frontier in Plant Science* 7: 67.

Wang, J., Wu, B., Yin, H., Fan, Z., Li, X., Ni, S., He, L., Li, J. (2017e) Overexpression of *CaAPX* induces orchestrated reactive oxygen scavenging and enhances cold and heat tolerances in tobacco. *BioMed Research International* 4049534.

Wang, K., Wu, Y., Tian, X., Bai, Z., Liang, Q., Liu, Q., Pan, Y., Zhang, L., Jiang, B. (2017d) Overexpression of *DgWRKY4* enhances salt tolerance in chrysanthemum seedlings. *Frontiers in Plant Science* 8: 592.

Wang, L., Li, Z., Lu, M., Wang, Y. (2017b) ThNAC13, a NAC transcription factor from *Tamarix hispida*, confers salt and osmotic stress tolerance to transgenic *Tamarix* and *Arabidopsis*. *Frontiers in Plant Science* 8: 635.

Wang, T., Dong, J., Zhang, R., Zhu, F., Zhang, Z., Gou, L., Wen, J., Duan, M. (2017a) A lipid-anchored NAC transcription factor translocates into nucleus to activate *GlyI* gene expression involved in drought stress. *The Plant Cell* 29: 9.

Wei, Y., Shi, H., Xia, Z., Tie, W., Ding, Z., Yan, Y., Wang, W., Hu, W., Li, K. (2016) Genome-wide identification and expression analysis of the WRKY gene family in cassava. *Frontiers in Plant Science* 7: 25.

Weisany, W., Sohrabi, Y., Heidari, G., Siosemardeh, A., Ghassemi-Golezani, K. (2012) Changes in antioxidant enzymes activity and plant performance by salinity stress and zinc application in soybean (*Glycine max* L.). *Plant Omics* 5: 60–67.

Wu, H., Fu, B., Sun, P., Xiao, C., Liu, J.H. (2016a) A NAC transcription factor represses putrescine biosynthesis and affects drought tolerance. *Plant Physiology* 172: 1532–1547.

Wu, H., Zhang, Y., Shi, X., Zhang, J., Ma, E. (2017b) Overexpression of Mn-superoxide dismutase in *Oxya chinensis* mediates increased malathion tolerance. *Chemosphere* 181: 352–359.

Wu, J., Chen, J., Wang, L., Wang, S. (2017a) Genome-wide investigation of WRKY transcription factors involved in terminal drought stress response in common bean. *Frontiers in Plant Science* 8: 380.

Wu, J., Zhang, J., Li, X., Xu, J., Wang, L. (2016b) Identification and characterization of a *PutCu/Zn-SOD* gene from *Puccinellia tenuiflora* (Turcz.) Scribn. et Merr. *Plant Growth Regulation* 79: 55–64.

Xu, J., Duan, X., Yang, J., Beeching, J.R., Zhang, P. (2013a) Enhanced reactive oxygen species scavenging by overproduction of superoxide dismutase and catalase delays postharvest physiological deterioration of cassava storage roots. *Plant Physiology* 161: 1517–1528.

Xu, J., Duan, X., Yang, J., Beeching, J.R., Zhang, P. (2013b) Coupled expression of Cu/Zn-superoxide dismutase and catalase in cassava improves tolerance against cold and drought stresses. *Plant Signaling and Behavior* 8: e24525.

Xu, J., Li, Y., Sun, J., Du, L., Zhang, Y., Yu, Q., Liu, X. (2012) Comparative physiological and proteomic response to abrupt low temperature stress between two winter wheat cultivars differing in low temperature tolerance. *Plant Biology* 15: 292–303.

Xu, J., Yang, J., Duan, X., Jiang, Y., Zhang, P. (2014) Increased expression of native cytosolic Cu/Zn superoxide dismutase and ascorbate peroxidase improves tolerance to oxidative and chilling stresses in cassava (*Manihot esculenta* Crantz). *BMC Plant Biology* 14: 208.

Yan, H., Li, Q., Park, S.C., Wang, X., Liu, Y.J., Zhang, Y.G., Tang, W., Kou, M., Ma, D.F. (2016) Overexpression of *CuZnSOD* and *APX* enhance salt stress tolerance in sweet potato. *Plant Physiology and Biochemistry* 109: 20–27.

Yan, J., Wang, B., Zhong, Y., Yao, L., Cheng, L., Wu, T. (2015) The soybean R2R3 MYB transcription factor GmMYB100 negatively regulates plant flavonoid biosynthesis. *Plant Molecular Biology* 89:35–48.

Yang, G., Wang, C., Wang, Y., Guo, Y., Zhao, Y., Yang, C., Gao, C. (2016) Overexpression of *ThVHAc1* and its potential upstream regulator, *ThWRKY7*, improved plant tolerance of cadmium stress. *Scientific Reports* 6: 18752.

Ying, S., Zhang, D.F., Fu, J., Shi, Y.S., Song, Y.C., Wang, T.Y., Li, Y. (2012) Cloning and characterization of a maize bZIP transcription factor, ZmbZIP72, confers drought and salt tolerance in transgenic *Arabidopsis*. *Planta* 235: 253–266.

Yousuf, P.Y., Hakeem, U.K., Chandna, R. (2012) Role of glutathione reductase in plan abiotic stress. In *Abiotic Stress Responses in Plants*, ed. Ahmad, P., Prasad, M. N. V., 149–158. New York, NY: Springer-Verlag.

Zhang, L., Sun, L., Zhang, L., Qiu, H., Liu, C., Wang, A., Deng, F., Zhu, J. (2017d) A Cu/Zn superoxide dismutase gene from *Saussurea involucrata* Kar. et Kir., *SiCSD*, enhances drought, cold and oxidative stress in transgenic tobacco. *Canadian Journal of Plant Science* 97: 816–826.

Zhang, C., Wang, D., Yang, C., Kong, N., Shi, Z., Zhao, P., Nan, Y., Nie, T., Wang, R., Ma, H., Chen, Q. (2017b) Genome-wide identification of the potato WRKY transcription factor family. *PLoS One* 12: e0181573.

Zhang, J., Du, H., Chao, M., Yin, Z., Yang, H., Li, Y., Huang, F., Yu, D. (2016) Identification of Two bZIP transcription factors interacting with the promoter of soybean rubisco activase gene (*GmRCAα*). *Frontiers in Plant Science* 7: 628.

Zhang, L., Zhang, L., Xia, C., Gao, L., Hao, C., Zhao, G., Jia, J., Kong, X. (2017a) A novel wheat C-bZIP gene, *TabZIP14-B*, participates in salt and freezing tolerance in transgenic plants. *Frontiers in Plant Science* 8: 710.

Zhang, P., Huang, H., Liu, W., Zhang, C. (2017c) Physiological mechanisms of a wetland plant (*Echinodorus osiris* Rataj) to cadmium detoxification. *Environmental Science and Pollution Research International* 24: 21859–21866.

Zhao, X., Yang, X., Pei, S., He, G., Wang, X., Tang, Q., Jia, C., Lu, Y., Hu, R., Zhou, G. (2016a) The *Miscanthus* NAC transcription factor *MlNAC9* enhances abiotic stress tolerance in transgenic *Arabidopsis*. *Gene* 586: 158–169.

Zhao, H., Ye, L., Wang, Y., Zhou, X., Yang, J., Wang, J., Cao, K., Zou, Z. (2016b) Melatonin increases the chilling tolerance of chloroplast in cucumber seedlings by regulating photosynthetic electron flux and the ascorbate-glutathione cycle. *Frontiers in Plant Science* 7: 1814.

15 Approaches to Enhance Antioxidant Defense in Plants

Hamid Mohammadi, Saeid Hazrati, and Mohsen Janmohammadi

CONTENTS

15.1 INTRODUCTION

Plant growth is highly influenced by the interaction between external and internal signals (Peleg and Blumwald, 2011). On the other hand, growth and development in plants depend on environmental conditions and any factors that might prevent the occurrence of stress (Pandey et al., 2017). Since plants are not able to move, they have to tolerate environmental changes created in their place (Pandey et al., 2017). Plants have different and complex interactions with their environment so they have to adapt their growth and development to environmental factors. Hence, during evolution, they have created special mechanisms that allow them to be consistent under unstable and stressful conditions. It has been stated that under mild stress conditions, plants continue to grow by changing their metabolism

(Claeys and Inze, 2013). Environmental stresses strongly affect the biochemical processes of plants that are sensitive to environmental stresses. Generally speaking, environmental stresses result in the production of reactive oxygen species (ROS), which make the cells more vulnerable, which, in turn, triggers signaling pathways, through which antioxidant systems scavenge these free radicals (Yun et al., 2010). Oxidative stress refers to the conditions in which the production of active oxygen species overrides the existing capacity of maintaining the existing redox (oxidation–regeneration) homeostasis (Halliwell, 2006). Active oxygen species are highly toxic and dangerous and cause disorders in biological systems. Oxidative stress affects a wide range of metabolic activities in plants and leads to damage by generating reactive oxygen species. At this stage, reactive oxygen species can damage cell membrane structure and lead to cell lysis and death (Reczek and Chandel, 2015). In order to maintain a proper balance in plant metabolism processes, plants completely or partially create changes in genetic material, proteins, molecules involved in energy transfer and biochemical compounds (Mittler et al., 2012). Plants have different enzymatic and non-enzymatic mechanisms to protect the cells against the toxicity of active oxygen species and to cope with the damage caused by oxidative stress. Increased antioxidant activity is one of the mechanisms that increase the resistance of plants to adverse environmental conditions (Mittler, 2002). An antioxidant is an agent that can interfere with the processes involved in oxidative stress and prevent oxidative damage. These are substances that slowly increase the oxidation period by reducing the oxidation rate. The overall mechanism is donation of a proton to a free radical to prevent the expansion of oxidation chain reactions. Antioxidants act as a scavenger of free radicals. Increasing or decreasing reactive oxygen species levels in the chloroplast depends on the balance between the generative factors and normal conditions, which is difficult to control by the mechanisms that inhibit their production (Alscher, 1989). The level of some antioxidant enzymes increases in adverse environmental conditions (Heath, 1987). Reactive oxygen species or free radicals are considered as the cells' by-products produced mainly in chloroplast, mitochondria, endoplasmic reticulum, membrane microbodies, cell membrane, and cell wall (Van Breusegem and Dat, 2006). When reactive oxygen species level increases, it results in significant damage to cells' macromolecules such as proteins, fatty acids, and nucleic acid (Pandey et al., 2017; Miller et al., 2010). This chapter focuses on the relationship between antioxidant enzymes and their role in response to stress. Although antioxidant enzymes are highly related to stress response, antioxidants are generally divided into two groups of enzymatic and enzymatic antioxidants. Antioxidant enzymes can be divided into eight groups, including catalase (CAT), peroxidase (POD), superoxide dismutase (SOD), glutathione peroxidase (GPX),

glutathione reductase (GR), monodehydroascorbate reductase (MDHAR), glutathione transferase, and ascorbate peroxidase. Non-enzymatic compounds include ascorbic acid, carotenoids (Car), flavonoids, phenols, and alpha-tocopherols (Miller et al., 2010). This chapter is divided into two section. In the first section, plant responses to stresses with an emphasis on antioxidants will be investigated, and in the second section, the possibility of improving mechanisms by which antioxidant activity can be promoted will be examined.

15.2 THE CONCEPT OF TOLERANCE ENHANCEMENT THROUGH IMPROVEMENT OF ANTIOXIDANTS ACTIVITY

Plants respond to environmental stress in different ways. Short-term changes in environmental factors can lead to cumulative reactions, while gradual changes can lead to adaptation reactions in the plant. The environmental stresses tolerance and compatibility depend on the type and duration of stress. Exposing plants to environmental stresses such as temperature, heavy metals, drought, air pollutants, food shortages, and salinity stresses leads to the generation of reactive oxygen species such as hydroxyl and superoxide radicals (Figure 15.1). Accumulation of ROS can induce defense system and plant adaptation to stress condition, but it sometimes can result in plant death.

Under normal conditions, the ROS level in the cell is very low; however, under stressful conditions, its generation increases. These reactive oxygen species are generated in pathways such as photorespiration, photosynthesis, and mitochondrial respiration (Polle, 2001; Mittler, 2002). In addition, reactive oxygen species also play a useful role in the plants and act as signal molecules in growth processes, cell cycle, cell development, aging, plant death, stomatal conductance, hormonal signal, and gene expression regulation

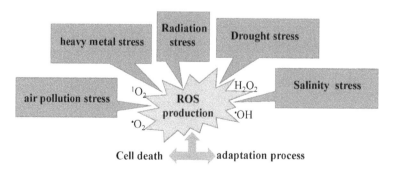

FIGURE 15.1 Environmental stresses causing ROS generation (adapted from Das and Roychoudhury, 2014).

(Kovtun et al., 2000; Neill et al., 2002; Slesak et al., 2007; Inze et al., 2012). As mentioned earlier, under stress conditions, reactive oxygen species level increases in the plant cells, which in turn leads to cell toxicity and death; under such conditions, plants start to develop their antioxidant defense systems to protect mitochondria, chloroplasts, and peroxisomes against these toxic molecules. It is also thought that hydrogen peroxide plays a key role as a secondary signal in signaling pathways under stressful conditions (Mittler, 2002; Khan and Singh; 2008; Gill et al., 2008; Gill and Tuteja, 2010) (Figure 15.1). Many studies have been conducted to increase antioxidant defense capacity under stress conditions, which leads to cellular tolerance and protection. However, some parts of antioxidants' alteration are controlled by plant phonological stages, so that it has been revealed that accumulation of enzymatic antioxidants of winter cereals increases during the vegetative growth and considerably decreases by ignition of reproductive growth (formation double ridge). This status clearly shows the developmental regulation of antioxidants (Janmohammadi et al., 2014).

Requirements for antioxidant accumulation can also vary depending on the type of tolerance. Physiological and genetic evidence suggest that increase in antioxidant defense systems' capacity in plants exposed to oxidative stress is an important defense mechanism so that there is a very close relationship between increasing antioxidant defense systems activity and reducing oxidative damage (Blokhina et al., 2003; Caverzan et al., 2016).

During stress, the cell's internal state, in particular water status, plays a key role in activating defense mechanisms. Typically, the activity of antioxidant enzymes depends on the type of stress. The activity of antioxidant enzymes such as superoxide dismutase and peroxidase are usually higher under stress conditions (Heath, 1987; Shao et al., 2008). In plants, the role of reactive oxygen species in signaling and protecting cells against their toxic effects requires a regulatory network and defense system, where the response rate to each of these factors indicates the degree of tolerance. Since plants are fixed at a place and do not have the ability to escape from adverse environmental conditions, they need a strong and regular signaling network to regulate different processes during growth and development (Figure 15.2).

15.2.1 SIGNALING PATHWAY

The reactive oxygen species play a different signaling role and affect gene expression, so that they show a positive correlation with high levels of tolerance to environmental stresses. The role of hydrogen peroxide in

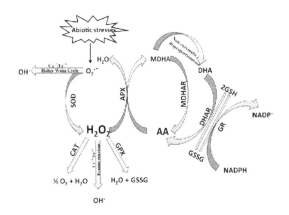

FIGURE 15.2 A part of antioxidant defense system and some involved antioxidants in plants.

signaling pathways is more pronounced due to its higher half-life. It has been reported that there are several compounds involved in ROS generation pathways, such as MAP kinase kinases AtANP1, NtNPK1, MAP kinases AtMPK3/6 and Ntp46MAPK (Samuel et al., 2000; Kovtun et al., 2000). Calmodulin (CAM) is another compound that interferes with hydrogen peroxide. Figure 15.3 shows some of the compounds that are involved in this pathway and alter gene expression through oxidation of some certain proteins in response to environmental stresses. The ROS signaling network is very complex so that in the *Arabidopsis* this network consists of at least 289 genes; some genes have been upregulated and some downregulated.

Hydrogen peroxide is detected by cell receptors which in turn activates mitogen kinase protein (MAPK) and a group of transcription factors that control various pathways in the cell. The hydrogen peroxide production is related to the production of calcium and calmodulin that activates or induces Ca^{2+}/calmodulin kinase, which can also activate or suppress the transcription. Furthermore, it regulates gene expression for activating some defense compounds involved in scavenging, reactive oxygen species, and heat shock proteins (Mittler, 2002).

Fat-soluble antioxidants (vitamin E, carotenoids) and antioxidant enzymes bonded with cell membranes in higher plants are known as the first defensive molecules against reactive oxygen species produced in the cell membrane, while water-soluble antioxidants (vitamin C) eliminate ROS in aquatic phases. The balance between superoxide dismutase and ascorbate peroxidase or catalase in cells is crucial in determining the level of superoxide and hydrogen peroxide radicals. The most important pathways for removing the ROS in plants include superoxide dismutase found in most cell organs, water–water cycle in chloroplasts, glutathione–ascorbate cycles in the cytosol, mitochondria, apoplast, peroxisomes,

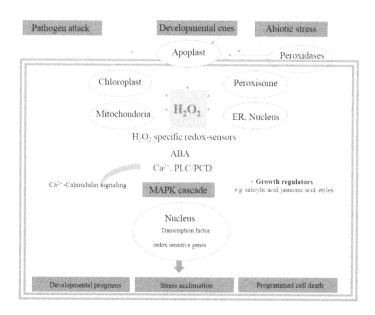

FIGURE 15.3 A schematic diagram or proposed model for activating signal transduction under oxidative stress.

glutathione peroxidase, and catalase in peroxisomes. The glutathione–ascorbate cycle takes place in most cell parts and plays a central role in controlling the level of reactive oxygen species. Catalase is only found in peroxisomes and acts as detoxifier during stress. Therefore, higher amounts of antioxidant compounds play an important role in protecting plants cells against oxidative stresses (Noctor and Foyer, 1998).

The antioxidants performance is classified in one of these two ways:

a. Chain breaker: When a free radical is released, or an electron is abducted, a second radical is formed. This molecule then changes and creates a third molecule. This process goes on to produce unstable products. Therefore, generation of a free radical causes generation of another free radical, creating a cell-destroying chain, so that a free radical can damage millions of other molecules. As long as the generation of free radicals is not stopped, the process won't stop, unless chain-breaker antioxidants, such as beta-carotene, vitamin C, and vitamin E are added.

b. Inhibitors: Antioxidant enzymes such as superoxide dismutase, catalase, and glutathione peroxidase inhibit oxidation by reducing the rate and amount of chain startup. These antioxidants are able to neutralize an oxidation chain. The free radicals have a high degree of chemical reactivity and tend to try and steal an electron from whatever molecule happens to be nearby (Birben et al., 2012; Das and Roychoudhury, 2014).

15.3 SENSING THE ANTIOXIDANT STIMULANTS BY THE PLANTS

Stress recognition and activation of apoplastic and symplastic stress signaling pathways, including ROS, reactive nitrogen species (RNS), hormone, and Ca^{2+} signaling (Zhu, 2016). The initial stress signal is perceived by multiple primary sensors, and these processes are sometimes similar between different abiotic stresses and in some case are very specific. Accordingly, rather than one sensor, there are many sensors that perceive certain stress conditions and control all downstream signals. Different stresses or stimuli could activate overlapping receptors/sensors but produce distinct final outputs which are specific to each stimulus (Sewelam et al., 2016). However, some sensors are very specialized; for example, the main function of two classes of ER stress sensors, membrane-associated basic leucine zipper (bZIP) transcription factors and RNA splicing factors, is a correction of the protein folding (Srivastava et al., 2014). After signal perceiving, cascade of signaling events is initiated by secondary signals such as plant hormones and calcium. In the meantime, all pathways can ultimately induce the activity of ROS scavengers or increase the expression of antioxidant-encoding genes by affecting the genome. However, some reports suggest that internal stress signals can stimulate the production of receptors/sensors. It was reported that H_2O_2 from chloroplasts led to the induced expression of many genes coding for membrane-bound receptor proteins and signaling components (Sewelam et al., 2014).

Now it's well known that abiotic stress can cause increases in the cytosolic free calcium concentration in plants. In this regard, it has been reported that OSCA1, a plasma-membrane calcium-permeable channel, can respond to stress and it increases the concentration of calcium in the cytosol (Yuan et al., 2014). Furthermore, some other protein channels like MscS-like proteins, COLD1, and a large family of cyclic nucleotide-gated channels (CNGCs), as well as a family of glutamate receptor-like (GLR) channels that are potentially very important in generating cytosolic Ca^{2+} signals under stress (Swarbreck et al., 2013; Zhu, 2016). Environmental stress can usually activate these channels by changing the fluidity of the membranes, increasing the cell wall-membrane interaction, and other non-recognized ways. Calcium-dependent protein kinase proteins regulate the downstream components in calcium signaling pathways (You and Chan, 2015). Increase in cytosolic Ca^{2+} requires CAM to deliver its signal to the downstream targets. The content of CAM proteins and the expression of many antioxidant genes in maize were induced by Ca^{2+} elevations (Hu et al., 2007).

The ROS production in plants is mainly localized in the chloroplast, mitochondria, and peroxisomes. There are secondary sites as well, like the endoplasmic reticulum, cell membrane, cell wall and the apoplast (Das and Roychoudhury, 2014). Therefore, most attention has been focused on stimulating antioxidants in these spaces. Although reactive oxygen species (e.g., $O_2^{\bullet-}$, H_2O_2, $OH^{\bullet}$, $^1O_2$) are considered as unavoidable by-products of aerobic metabolism, stress conditions significantly induce ROS generation (Mittler et al., 2011). ROS accumulations affect the reduction–oxidation (redox) processes and seriously challenge the redox homeostasis. Accurate operation of chloroplast and mitochondria and the optimal use of the available light energy is dependent on precise control of redox homeostasis. ROS generation and activity or expression of different antioxidants are two main and influential components of redox homeostasis. It appears that redox signals have a critical role in stimulating the antioxidant under stress conditions (Foyer and Noctor, 2005). ROS and redox cues stimulate acclimation responses through conservative signaling and control whole-plant systemic signaling pathways (Suzuki et al., 2012). The non-enzymatic ROS scavengers (antioxidants), ascorbate (vitamin C), glutathione, carotenoids, and α-tocopherol (vitamin E) are information-rich redox buffers that affect numerous cellular components (Sewelam et al., 2016). On the other hand, the accumulations of antioxidants influence the expression of stress-responsive genes to maximize defense through tuning cellular ROS levels and redox state (Foyer and Noctor, 2005; Janmohammadi et al., 2015).

It seems there is some crosstalking between stress-sensing pathways. One of the well-identified crosstalking is between calcium and reactive oxygen species that originated from NADPH oxidase (Jiang and Zhang, 2003). Plasma membrane-associated NADPH oxidases (Respiratory burst oxidase homologs; RBOH) are the most studied enzymatic source of ROS. Abscisic acid (ABA) and Ca^{2+} also considerably induced the activity of NADPH oxidase. Ca^{2+} can directly bind to NADPH oxidase and activate it or induce it by a Ca^{2+}-dependent phosphorylation pathway that is mediated by calcium-dependent protein kinases (Ogasawara et al., 2008; You and Chan, 2015). Also, it has been suggested that GTP-binding proteins (G proteins) are present in signaling pathways. Extracellular ROS activate the G protein and the signal transmitted to the membrane-bound NADPH oxidases (Sewelam et al., 2016). It has been revealed that MAPK cascades are involved in the activation of NADPH oxidase in response to pathogen signals (Asai et al., 2008).

It has been suggested that each cellular compartment has a specific ROS homeostasis, and this is due to its particular condition; altogether, the different ROS levels in the different compartments can be viewed as generating a specific ROS signature (Choudhury et al., 2017). This suggests that launching the ROS signaling pathway and stimulation of the antioxidants are very different in various situations. This trend can change depending on the type of cell, its developmental stage, or stress level. However, these cellular compartments may crosstalk to relay and further fine-tune the ROS message. Various abiotic stresses and/or different combinations of different stress (abiotic or biotic) may result in the formation of different ROS signatures in plant cells, and decoding these signatures via different ROS sensors can create a stress-specific signal that will tailor the acclimation response to the type of stress/combination affecting the plant (Mignolet-Spruyt et al., 2016; Choudhury et al., 2017).

There is growing evidence that sugars can also play a role in stimulating antioxidants. Disaccharides (sucrose, trehalose), raffinose family oligosaccharides, and fructans are three major types of water-soluble carbohydrates that have a significant effect on redox homeostasis through their close associations with photosynthesis, mitochondrial respiration, and fatty acid β-oxidation (Couée et al., 2006). Intriguingly, both high and low sugar levels can evoke ROS accumulation (Keunen et al., 2013). Besides, it has been suggested that sugars might act as true ROS scavengers in plants, especially when present at higher concentrations (Peshev and Van den Ende, 2013). At low concentrations, however, sugars might still operate as substrate or signal for stress-induced modifications (Van den Ende and Valluru, 2009). Antioxidative functions of sugars may be direct or indirect. The breakdown of cell wall polysaccharides (e.g., oligogalacturonides) under stress condition might generate sugar signals and affect the antioxidants.

High sucrose support fructan biosynthesis by fructosyltransferase and prevent ROS formations during this process; also, sugars can affect the production rate of ascorbate, glutathione, carotenoid, and α-tocopherol (Keunen et al., 2013). Also, accumulation of sugars with the effects on the source-sink relationship can challenge the photosynthetic cycles, and by stimulating the ROS production, stimulate the expression or activity of the antioxidants. Overall, it seems that specific and flexible combinations of sugar and ROS signaling components can trigger adaptation and acclimation responses.

There is accumulating evidence suggesting that a strong relationship between ROS, salicylic acid (SA), ethylene, and jasmonic acid (JA) signaling plays a crucial regulatory role in plant defense responses. The findings suggest that this crosstalking may have a stimulating or inhibitory effect on antioxidants. It was revealed that ethylene signaling is necessary for ROS production (Sewelam et al., 2016). One of the suggested functions for salicylic acid is the inhibition of catalase, a major enzyme scavenging H_2O_2, thereby increasing cellular concentrations of H_2O_2, which acts as a second messenger and activates defense-related genes (Ananieva et al., 2002). Accumulation of ROS can activate the jasmonic acid signaling pathways and induce the scavenging reactions.

Cellular redox state can also affect some chemical reactions that considerably contribute to redox homeostasis and affect antioxidant signaling. Among them, Fenton reaction is very sensitive to a redox imbalance. Fenton reaction can scavenge hydrogen peroxide with ferrous iron as a catalyst. Iron (II) is oxidized by hydrogen peroxide to iron (III), forming a hydroxyl radical and a hydroxide ion in the process (Halliwell and Gutteridge, 2015). Iron (III) is then reduced back to iron (II) by another molecule of hydrogen peroxide, forming a hydroperoxyl radical and a proton. The net effect is a disproportion of hydrogen peroxide to create two different oxygen-radical species, with water ($H^+ + OH^-$) as a by-product. In biological systems, the availability of ferrous ions limits the rate of reaction, but the recycling of iron from the ferric to the ferrous form by a reducing agent can maintain an ongoing Fenton reaction, leading to the generation of hydroxyl radicals (McKersie and Lesheim, 2013).

The findings suggest that plants are equipped with sophisticated and efficient mechanisms to recognize the stress signal and activate the scavenging reactions. The full understanding of the mechanisms involved in redox homeostasis under stress may allow for the engineering of more tolerant plants or the optimization of cultivation practices to improve yield and productivity under stressful conditions.

15.4 THE ACTIVE ANTIOXIDANTS IN RESISTANCE TO ENVIRONMENTAL STRESSES

Biotic and abiotic stresses alter the normal homeostasis of the cell, increase the ROS, and create a secondary stress called oxidative stress. Studies show that there is a very complex network of enzymatic and non-enzymatic scavengers for ROS to maintain the ROS level at normal in different cell components (Foyer and Noctor, 2000; Mittler, 2002). For example, produced superoxide in the photosynthetic electron transfer chain is maintained at the usual level by enzymatic (superoxide dismutase) and non-enzymatic scavengers (ascorbic acid, glutathione, carotenoid, and α-tocopherol) in photosystem I, or produced hydrogen superoxide and hydroxyl radicals in the photosynthetic electron transfer chain is removed by enzymatic scavengers (thioredoxin peroxidase and ascorbate peroxidase) in photosystem II (Edreva, 2005). It should be noted that ROS not only cause oxidative damage to cells under environmental stress conditions but also, as signal transduction molecules, play a key role in the adjustment of reaction to pathogens infections, environmental stresses, and planned cell death in plants (Mittler et al., 2004). Plants have different mechanisms to prevent oxidative damage to cells as follows:

15.4.1 SUPEROXIDE DISMUTASE

This enzyme is a highly active metalloenzyme and acts as the first line of defense against various environmental stresses. Besides, its activity is essentially independent of pH (5–9). In various organisms, the SOD enzyme requires various metal cofactors for its activity, which can be separated from enzymes reversibly. Accordingly, there are three distinct types of SOD isozymes, including Fe-SOD, Mn-SOD, and Cu/Zn-SOD, which place in chloroplasts; mitochondria and peroxisome; chloroplasts and cytosol, respectively (Mittler, 2002). The role of this enzyme is dismutation of the free radical superoxide so that one of the superoxides is reduced to hydrogen peroxide and the other oxidizes to oxygen.

A large number of studies show the positive correlation between SOD activity and tolerance to biotic and abiotic stresses (Gill and Tuteja, 2010; Maksimovic et al., 2013). A significant increase in SOD activity was reported in white clover (Wang and Li, 2008) and rice (Sharma and Dubey, 2005) under drought stress conditions. Moreover, Abedi and Pakniat (2010) indicated that drought stress (60% field capacity [FC]) compared to irrigated condition resulted in a gradual increase in the activity of SOD in rapeseed plants. They also mentioned that this increase in Lincord and Zarfam cultivars was very evident compared to Hyola308 cultivar. On the other hand, in some cultivars, such as Okapy, a decrease in SOD activity was seen in 30% FC. Finally, they concluded that Zarfam and Lincord cultivars' higher yields under drought stress conditions could be related to their conservation by increasing SOD activity. In another study, it was reported that SOD was present in various isoforms and could have different activities during plant adaptation to stress (Jithesh et al., 2006). Ozgur et al. (2013) reported that the defensive strategies among halophytes could be related to the difference in the developmental stage (e.g., identification of new isoenzyme from SOD during growth stage of *Gypsophila oblanceolata* plant compared to SOD isoenzymes during the germination stage), temporal difference (e.g., increase in new isoenzyme from SOD during salinity stress in observed *Centaurea tuzgoluensis*), and the type of tissue (e.g., differences in the activity of antioxidant enzymes in the root and aerial part of reported *Crithmum maritimum*).

There are many reports that indicate the production of stress-tolerant transgenic plants by increasing the expression of various SODs. For example, the results of Wang et al. (2004) showed the activity of various isozymes of SOD in *Arabidopsis* transgenic plants. Increases in SOD activity during leaf aging have been reported in various plant species such as corn (Procházková et al., 2001) and chickpea (Del Rio et al., 2003). It is believed that an increase in SOD activity is a protective mechanism that helps to delay the aging process with the detoxifying of superoxide. In general, it can be said that SOD activity changes in different conditions and it depends on plant species, stress duration, stress intensity, and the used plant organ for measurements.

15.4.2 CATALASE

Catalase is responsible for the detoxification of increased levels of H_2O_2 (in mM) in peroxisomes, and it is assumed that its activity is necessary to maintain the balance of the redox during oxidative stress (Mittler et al., 2004; Willekens et al., 1997). Studies show that under stressful conditions an increase in photorespiratory occurs due to an increase in plant tolerance to environmental stress (Noctor et al., 1999). The produced glycolate in chloroplasts enters the peroxisome through a specific protein carrier present in the wall. In this organelle, initially, glycolate is broken down into glyoxylate and H_2O_2 with the help of a flavin-dependent mononucleotide called glycolate oxidase. Next, H_2O_2 is broken down into water and oxygen by the enzyme catalase (Noctor et al., 1999).

Catalase isoforms are iron porphyrin enzymes that are used as an effective ROS scavenging system to prevent drought and salinity stress-induced oxidative damage (Mittler et al., 2011).

Many reports demonstrated an increase in antioxidant enzyme level including catalase in two levels as follows:

a. Related to the potential of plant species: For example, the *Hordeum marinum* halophyte has higher enzyme levels than the *Hordeum vulgare* one (Seckin et al., 2010). Another report states that *Cakile maritima* has a higher activity compared to *Arabidopsis thaliana* (Ellouzi et al., 2011.)

b. Related to the induction conditions of biotic and abiotic stresses: For example, Mohammadi and Moradi (2016) showed that water stress led to a significant increase in CAT activity of flag leaf of wheat cultivars during 7 to 21 days post-anthesis. Besides, a number of results show that antioxidant enzymes are increased in corn plants (de Azevedo Neto et al., 2006) and sesame (Koca et al., 2007) under salt stress conditions.

15.4.3 PEROXIDASES

Peroxidases are oxidoreductases that can use several organic and inorganic substrates as hydrogen donors in the presence of H_2O_2 (Dawson, 1998). Moreover, PODs are enzymes containing heme that is composed of a single peptide chain. Plant PODs are divided into three groups, including I, II, and III. Class I PODs contain intracellular enzymes in plants, bacteria, and yeast. Class II PODs are extracellular fungal PODs that catalyze the depolymerization of lignin. Class III PODs are secretion PODs and are known as guaiacol oxidases due to their tendency to guaiacol (Hiraga et al., 2001). Studies have shown that class III PODs play an effective role in the ROS detoxification system under stress conditions. For example, it has been determined that there is a good correlation between peroxidase activity and produced H_2O_2 during copper, aluminum, and cadmium-induced stress conditions (Mithofer et al., 2004; Źróbek-Sokolnik et al., 2009).

Furthermore, other reports indicate increased PODs activity in H_2O_2 scavenging under salinity (Caverzan et al., 2014) and drought (Mohammadi and Moradi, 2016).

15.4.4 GLUTATHIONE–ASCORBATE CYCLE

The glutathione–ascorbate cycle is another mechanism that reduces the excess energy by NADPH. This cycle is found in all of the plant cell components, and, in addition, the high affinity of ascorbate peroxidase (APX) for H_2O_2 indicates its crucial role in controlling ROS. Antioxidants, such as ascorbic acid and glutathione, found in high concentrations in chloroplasts and other cellular components, are very important in the plant's defense mechanisms during oxidative stress (Noctor and Foyer, 1998). Therefore, maintaining high ratios of reduced ascorbate and glutathione in cells is essential for appropriate ROS scavenging, and this ratio is retained by GR. MDAR and dehydroascorbate reductase (DHAR) use the NADPH as a reducing agent (Asada, 1999). According to the study of Mohammadi and Moradi (2016) on wheat cultivars under water stress, APX enzyme activity was increased under water stress. Research shows that components of this cycle act either directly through scavenging or indirectly through the activation of defense mechanisms. Glutathione and ascorbate play an important role in the maintenance of cellular redox homeostasis. Besides, the high levels of the reduced form of these two compounds play an important role in activating the defense mechanisms of plants during stress conditions. In addition, glutathione has several roles, including the detoxifying of heavy metals, the transport and storage of sulfur, the regulation of the expression of the gene associated with defense, and protein activity, while ascorbic acid acts as a signal-transfer molecule, the cofactor of some enzymes, and the photoprotection mechanism in the xanthophyll cycle. Increasing the expression of APX genes in several abiotic stresses (Rosa et al., 2010; Caverzan et al., 2014), GPX in wheat plants during salinity stress, H_2O_2, and treatment with ABA (Zhai et al., 2013) were identified.

15.4.5 ASCORBIC ACID

It is an effective chemical quencher of $^1O_2$, which is abundant in plant leaves, and 30–40% of plant ascorbate is in chloroplasts (Triantaphylidès and Havaux, 2009), i.e., ascorbate plays an important role in photosynthesis, and its high concentration in chloroplasts proves this matter. Ascorbate is an important metabolite in plants, acts as an antioxidant, along with other antioxidant

systems, and preserves the plant against oxidative damage caused by various factors, including aerobic metabolism, photosynthesis, and a series of contaminants. Recent research on transgenic and mutated plants shows the key role of ascorbic acid in the ascorbate–glutathione cycle, which protects the plant against oxidative stress. Ascorbate also acts as a cofactor to activate hydroxylase enzymes (for example, hydroxylase propyl) and violaxanthin de-epoxidase. The second enzyme relates ascorbate to photoprotection xanthophyll cycle. Besides, it seems that ascorbate regulates photosynthesis electron-transfer. The ascorbate biosynthesis pathway in plants is still not fully understood and studies are underway to suggest a pathway. Ascorbate is found in the cell wall, which is the first cellular defense line. It has also been shown that ascorbate of the cell wall and ascorbate oxidase in the cell wall are also involved in growth control, and high activity of ascorbate oxidase is associated with rapid development of cells. Ascorbate regulates cell division, which affects progress from stage G1 to S (Noctor and Foyer, 1998). Ascorbate performance is evaluated in three important biochemical aspects as follows (Noctor and Foyer, 1998):

a. Ascorbate acts as an antioxidant by removing hydrogen peroxide (due to lack of catalase in chloroplast). Hydrogen peroxide is formed by the photoreduction of oxygen in the photosystem. The catalyst of this process is ascorbate peroxidase, which traps generated hydrogen peroxide as soon as it is formed in thylakoid.

b. Ascorbate oxidizes to malonyl dehydroascorbate and it acts as a direct electron receptor to photosystem I.

c. Ascorbate acts as a cofactor for violaxanthin de-epoxidase.

It is obvious that by increasing knowledge about this molecule, a wide range of its function as an antioxidant in the mechanism of defense and in photosynthesis as a growth regulator can be studied.

15.4.6 CAROTENOIDS

Carotenoids (Car) are plant pigments and play an important role in protecting photosynthesis components against oxidative damage. Research shows that when singlet oxygen ($^1O_2$) is formed under high light stress condition, the high concentration of Car results in the transfer of $^1O_2$ energy to carotenoids and an excited form of a carotenoid is formed. Finally, it is disappeared in the form of heat. They are also able to directly absorb energy from the excited chlorophyll and lose energy (Taiz and Zeiger, 2006). According to the studies carried out under different stresses, it was found that these compounds increase the ability to cope with stress conditions in the plant (Singh and Dubey, 1995; Mohammadi and Moradi, 2016).

15.4.7 TOCOPHEROL

It is an antioxidant system in the lipid phase of the membrane that acts as a scavenger for singlet oxygen and alkyl peroxyl radical. Tocopherol can form tocopheroxyle by donating an electron to a peroxyl radical of fatty acid. It seems that most of the tocopherols act as singlet oxygen quenching with a physical mechanism and thermal energy distribution, such as carotenoids (Sattler et al., 2004, 2006). Tocopherol, along with other antioxidants, plays a role in reducing the level of ROS (essentially, singlet oxygen and hydroxyl radical) in photosynthesis membranes and limiting the amount of lipid peroxidation by reducing lipid peroxyl radicals to the related hydroperoxides (Munne-Bosch and Alegre, 2002; Munne-Bosch, 2005).

Levels of α-tocopherol are significantly altered during the plant growth, and, in response to environmental stress, they change and turnover because of changing the expression of the genes associated with the pathway. The studies show that an increase in primary levels of α-tocopherol results in a decrease in levels of ROS and consequently prevents oxidative damage. On the other hand, by increasing the stress level and ROS in chloroplasts, levels of α-tocopherol are increased (Hincha, 2008; Espinoza et al., 2013).

15.5 APPLICATION OF ORGANIC COMPOUNDS CONTAINING AMINO ACID AND OTHER COMPOUNDS TO ENHANCE ANTIOXIDANT ACTIVITY

Enzymatic antioxidants in plants include SOD, APX, CAT, GPX, MDHAR, DHAR, GR, glutathione S-transferase (GST), and peroxiredoxin (PRX). These antioxidant enzymes are located in different sites of plant cells and work together to detoxify ROS (You and Chan, 2015). Beside, non-enzymatic antioxidants such as compatible osmolytes (glycinebetaine and proline), ascorbic acid, reduced glutathione, α-tocopherol, amino acids, and polyphenols are very important in redox homeostasis (Janmohammadi et al., 2018). Some evidence suggests that the application of certain organic compound, such as amino acids, amino acid derivatives, and nitrogen-containing compounds on plants, can stimulate ROS scavenging systems.

Amino acids can participate in different processes; they can act as stress-reducing agents, the source of nitrogen, redox buffer, and hormone precursors (Maeda and Dudareva, 2012; Teixeira et al., 2017). In the rhizosphere, amino acids can be found in different forms; however, their half-life is short, and their absorption by plants is only possible due to the presence of transporters in the roots (Jämtgård et al., 2010). Several studies have shown the beneficial effects of foliar application of amino acids (Koukounaras et al., 2013; Sadak et al., 2014; Teixeira et al., 2017). It has been reported that application of the amino acids glutamate, phenylalanine, cysteine, glycine in seed treatment, and foliar application significantly affected the resistance enzymes as polyphenol oxidase and phenylalanine ammonia-lyase in soybean. Furthermore, amino acid application increased the activity of SOD, catalase, and peroxidase (Teixeira et al., 2017). The authors concluded that glutamate, cysteine, phenylalanine, and glycine could act as signaling amino acids in soybean plants since low concentrations are sufficient to improve enzymatic ROS scavenging. In this regard, it seems that amino acids and their derivatives are important components of antioxidant systems in plants. These molecules play crucial roles in the reduction of free radicals and osmoprotection (Gill and Tuteja, 2010).

Among different amino acids, the role of cysteine has been emphasized under oxidative stress (Teixeira et al., 2017). Cysteine functions as a precursor to essential biomolecules, such as vitamins, cofactors, antioxidants such as glutathione, and many defense compounds, such as glucosinolates, thionins, or phytoalexins (Cobbett, 2000). In addition, glutamate can act to diminish oxidative stress indirectly by being the precursor to other amino acids such as arginine and proline, which play an important role in redox homeostasis (Rejeb et al., 2014). Also, glutamate is involved in the biosynthesis of glutathione.

Exogenous application of phenylalanine in the vegetative phase of soybean plants increased the activity of phenylalanine ammonia-lyase (Teixeira et al., 2017), which may increase the biosynthesis of lignin and flavonoids. Phenylpropanoids, particularly flavonoids, have been recently suggested as playing primary antioxidant functions in the responses of plants to a wide range of abiotic stresses (Brunetti et al., 2013). It has been found that foliar application of methionine, ornithine, and arginine on lime plants significantly affected the activity of antioxidant enzymes, catalase, peroxidase, and phenylalanine ammonia-lyase. Methionine spray on plants improved resistance as well as decreased the severity of citrus canker disease by reducing necrotic lesion size (Hasabi et al., 2014). Maxwell and Kieber (2004) indicated the importance of methionine in the biosynthesis of growth regulating substances, e.g., cytokinins, auxins, and brassinosteroids.

Glycine is a precursor of glycinebetaine. Therefore, its foliar application can partially stimulate antioxidant systems of the plant. Glycine is also involved in the route of glyoxylate production, a compound that can reduce hydrogen peroxide content, leading to the reduction of lipid peroxidation (Alhasawi et al., 2015).

Foliar application of ascorbic acid and α-tocopherol on flax under saline conditions increased activities of SOD and CAT while inducing a significant reduction in lipid peroxidation and activities of polyphenol oxidase as well as peroxidase (El-Bassiouny and Sadak, 2015).

Proline, which is usually considered as an osmolyte for osmotic adjustment, also contributes to stabilizing subcellular structures (e.g., membranes and proteins), scavenging free radicals, and buffering cellular redox potential under stress conditions (Ashraf and Foolad, 2007; Hayat et al., 2012). The exogenous application of proline on tobacco-cultured cells' saline conditions significantly reduced malondialdehyde as a by-product of plasma membrane peroxidation (Okuma et al., 2004). Antioxidant enzyme activity such as SOD, CAT, and peroxidase is significantly increased in response to foliar-applied proline in tobacco suspension cultures under salinity stress (Hoque et al., 2007). In these researches, proline antioxidant activity and its inducing effects on scavenging system were evident. Glycinebetaine (GB) is a quaternary ammonium compound that occurs naturally in most biological systems. It has been reported that glycinebetaine can induce expression of fatty acid desaturase and lipoxygenase genes and, therefore, maintain membrane integrity in tomato plants under cold stress (Karabudak et al., 2014). Foliar application of glycinebetaine noticeably increased the ascorbate peroxidase activity and decreased lipid peroxidation under water deficit stress (Cruz et al., 2013). Overall findings suggest that glycinebetaine also can affect the gene expression as signal molecules and alternate the genes expression, accumulation and activity of some ROS scavenger. Altogether, the findings demonstrate a possible interaction between oxidative stress, gene expression and the accumulation of GB under the condition of abiotic stress (Kumar et al., 2017).

Polyamines are group of phytohormone-like aliphatic amine natural compounds with aliphatic nitrogen structure. These compounds widely present in living organisms, are now regarded as a new class of growth substances, and have imperative roles in development and adaptation to environmental stresses. Diamineputrescine, triamine spermidine, and the tetraamine spermine are the main polyamines in plants (Liu et al., 2015). Polyamines have antioxidant functions under environmental stresses. This property is due to a combination of their anion- and cation-binding characteristics involving ROS scavenging and a capacity to restrain both lipid peroxidation and metal-catalyzed oxidative reactions (Groppa and Benavides, 2008). Polyamines hinder the auto-oxidation of metals, which, in turn, inhibits the supply of electrons for the generation of ROS (Shi et al., 2010). A number of studies have demonstrated that exogenous application of polyamines led to increases in endogenous polyamine contents and concomitantly enhanced some antioxidant activity. The application of spermine considerably modified the antioxidants' activity, abscisic acid, and jasmonic acid signals in soybean plants under osmotic stress. Spermine treatment increased activities of CAT, SOD, peroxidase, and polyphenol oxidase activities, and it was associated with a significant decrease in lipid peroxidation (Radhakrishnan and Lee, 2013). These results are consistent with those of Farooq and others (2009), who reported that spermine application alleviated the effect of osmotic stress by enhancing catalase activity in rice plants, which they suggested might result in the removal of H_2O_2. Also, spermidine and spermine application on valerian (*Valeriana officinalis* L.) increased the catalase activity and improved the plant growth (Mustafavi et al., 2016).

In relation to the role of polyamines in the reduction of oxidative–oxidative stress, it has been revealed that foliar application of polyamine biosynthesis inhibitor (O-phospho ethanolamine) considerably reduced the antioxidants' activities and the total contents of phenolics, anthocyanins, flavonols, and hydroxybenzoic acid in roots and different aerial sections of *Echinacea purpurea*, nettle, and dandelion (Hudec et al., 2007). Polyamines' application may also increase the number of non-enzymatic antioxidants such as carotenoids and glutathione (Radhakrishnan and Lee, 2013).

15.6 MOLECULAR AND BIOTECHNOLOGICAL STRATEGIES TO ENHANCE THE ANTIOXIDANT DEFENSE FUNCTION

Environmental stress can be sensed as a stimulus or influence that is outside the normal range of homeostatic control and constantly is associated with ROS production and imbalance in redox homeostasis. ROS signaling cascades can activate ion channels, kinase cascades, accumulation of hormones such as salicylic acid, ethylene, jasmonic acid, and abscisic acid (Pérez-Clemente et al., 2013). These signals ultimately induce expression of antioxidants and some specific subsets of defense genes, which, in, total causes resistance to stress (Jaspers and Kangasjärvi, 2010). Although production of ROS is ubiquitous during metabolism and all plants can cope with them, some subcellular organelles such as chloroplast, mitochondrion, and peroxisome are common sites of ROS production, and more attention is given to these organelles by breeders to reduce the oxidative damages (Ashraf, 2009).

Several studies showed important roles of antioxidative components in ROS homeostasis in crop plants. Plant antioxidative system consists of numerous enzymatic and non-enzymatic antioxidative components that work together with ROS-generating pathway to maintain ROS homeostasis. Enzymatic antioxidants include SOD, CAT, ascorbate peroxidase, MDHAR, DHAR, and GR. The well-known water-soluble non-enzymatic antioxidants are glutathione and ascorbate, and the lipid-soluble non-enzymatic antioxidants are carotenoids and tocopherols (Gomez et al., 2004; Ashraf, 2009). Each of these items can be considered as a target point for improvement of abiotic stress tolerance through gene or protein engineering. SOD acts as the first line of defense converting $O_2^{\bullet-}$ into H_2O_2; CAT, ascorbate peroxidase, and guaiacol peroxidase then detoxify H_2O_2 (You and Chan, 2015). It appears that genetic manipulation of enzymatic antioxidant can considerably improve plants' tolerance to oxidative stress under unfavorable conditions. In rice genome, eight genes encode cytosolic, mitochondrial, or plastidic isoenzymes of SOD (Cu, Zn, Mn). Transgenic rice plants overexpressing SOD gene showed less accumulation of $O_2^{\bullet-}$ under stress conditions (Li et al., 2013). However, it seems that antioxidant functions and scavenging efficiency are highly dependent on the function of antioxidant collection or scavenging system. Therefore, in some cases, the genetic manipulation of an antioxidant is not associated with improvement of stress tolerance. In this regard, Cavalcanti et al. (2004) reported that the survival of cowpea plants under salt stress was not mediated by the underlying antioxidant system entailing SOD, peroxidase, and catalase activities in leaves. A similar result was also reported by Janmohammadi et al. (2012a,b). Likewise, Katsuhara et al. (2005) found that although overexpression of a GST gene reduced the amount of ROS in *Arabidopsis* to some extent, the removal of ROS was not sufficient to influence the overall plant salt tolerance. However, overexpression of genes encoding GST and GPX, as pacemaker enzymes in glutathione biosynthesis pathways, in tobacco plants considerably increased the glutathione accumulation and decreased the oxidative damage (Roxas et al., 2000). Also, down-regulation of homogentisate phytyltransferase (HPT) and γ-tocopherol methyltransferase (γ-TMT) in tobacco plants reduced total tocopherol accumulation by 98%, and transgenic lines were very susceptible to salt stress (Abbasi et al., 2007). Some potential target points for improvement of oxidative stress tolerance are shown in Figure 15.4.

However, it must be remembered that some oxidative signals are essential for growth. The redox balance of cells is controlled by a series of enzymes and

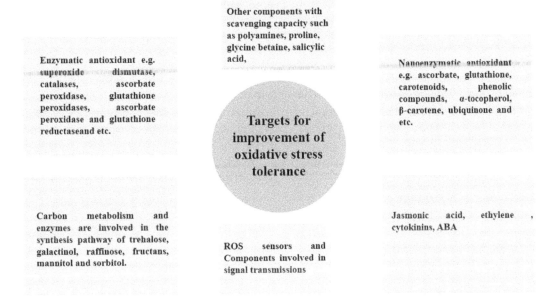

FIGURE 15.4 Molecular and biotechnological target points for improvement of oxidative stress tolerance.

intermediate metabolites, and there is various cross-talking between them; hence, any changes in the activity of antioxidants should be made consciously. In growing cells, the regulated breakdown of cell wall polymers involves OH production (Liszkay et al., 2004) and quenching ROS hinders root growth (Demidchik et al., 2003). Overexpression in *Arabidopsis* of a peroxidase localized mainly in the root elongation zone stimulated root elongation (Passardi et al., 2006).

The imino acid proline is employed by different organisms to compensate cellular imbalances caused by environmental stress. It has been revealed that proline metabolism can affect redox homeostasis by increasing ROS formation in the mitochondria (Liang et al., 2013).

Genes encoding pyrroline-5-carboxylate synthetase (P5CS) and pyrroline-5-carboxylate reductase (P5CR) are the main target points for manipulating the proline biosynthesis, and their overexpression has often been accompanied by an increased tolerance to environmental stress and improvements of redox balance (Verslues and Sharma, 2010). Besides, it has demonstrated that polyamines function in stress tolerance largely by modulating the homeostasis of ROS due to their direct, or indirect, roles in regulating antioxidant systems or suppressing ROS production (Liu et al., 2015). There are three major polyamines in plants, putrescine, spermidine, and spermine. It has been revealed that genes encoding arginine decarboxylase play a critical role in the production of putrescine in *Arabidopsis thaliana*. Putrescine is then converted into spermidine by spermidine synthase, and spermidine is then converted into spermine by spermine synthase. Polyamines synthesis may vary between tissues/organs, and depending on the type of polyamine, the rate-limiting enzyme will be different.

Regarding the interaction between ROS and sugars, it seems that genetic manipulation of sugar metabolism can affect oxidative stress (Nishizawa et al., 2008; You and Chan, 2015). In this context, several studies have been conducted, and trehalose-6-phosphate synthase, trehalose-6-phosphate phosphatase, trehalose phosphorylase, trehalose synthase, galactinol synthase, α-galactosidase, galactinol synthase, UDP-glucose 4-epimerase, levansucrase, sucrose 1-fructosyltransferase, mannitol-1-phosphate dehydrogenase, mannose 6-phosphate reductase, glucitol-6-phosphate dehydrogenase, and sorbitol-6-phosphate dehydrogenase are introduced as the most widely identified candidate enzymes for improving the abiotic stress tolerance and reducing the oxidative damage (Cortina and Culiáñez-Macià, 2005; Han et al., 2005; Liu et al., 2007; Kawakami et al., 2008; Pujni et al., 2007; Tang et al., 2005). The mentioned enzymes are involved in the synthesis pathway of trehalose, galactinol, raffinose, fructans, mannitol, and sorbitol.

15.7 THE ROLE OF PHYTOHORMONES TO ENHANCE THE ANTIOXIDANT ACTIVITY

Studies show that the survival of a plant under stress conditions depends on the plant's potential to receive stimuli, generate and signal transduction, and initiate various physiological and biochemical changes (Shinozaki and Yamaguchi-Shinozaki, 1997). Here, the equilibrium between the interaction of plant growth regulators (including ethylene, salicylic acid, abscisic acid, jasmonates, and cytokinin) and the redox signaling process or signal molecules is very important (Davies and Zhang, 1991; Xiong et al., 2002; Fujita et al., 2006; Bari and Jones, 2009; Verma et al., 2016).

Plant hormones produce ROS as secondary messengers in signaling pathways and ultimately lead to adaptive responses (Bartoli et al., 2012).

Some evidence shows that ABA can increase the superoxide (Jiang and Zhang, 2002) and H_2O_2 production and induce the expression of antioxidant-encoding genes (Kaminaka et al., 1999) and catalase (Anderson et al., 1994). Besides, ABA increases the activity of antioxidant enzymes such as SOD, CAT, APX, and GR in plant tissue (Anderson et al., 1994; Jiang and Zhang, 2002; Mohammadi and Moradi, 2016).

Mohammadi and Moradi (2016) showed that ABA resulted in an increase in the content of H_2O_2 and the activity of CAT, POX, and APX enzymes and subsequent MDA reduction. The contents of non-enzymatic antioxidants such as AsA, Car, and α-tocopherol were increased under ABA treatment. Such an enhancement in the antioxidant defenses is sufficient to scavenge these increased ROS and the oxidative damage expressed as lipid peroxidation.

Research shows that the balance between ROS levels and antioxidants affects auxin levels (Tognetti et al., 2010). For example, there is some evidence that shows antioxidant-encoding genes are one of the primary responsive genes to auxin, and it indicates the important role of auxin in plant defensive responses (George et al., 2010).

Salicylic acid (SA) is a plant hormone, and its role in some defensive responses such as pathogen infections, UV radiation, salinity, and drought was reported (Hayat et al., 2010). The research shows that lower concentration of SA increases the effectiveness of the antioxidant system (Knorzer et al., 1999; Kachroo and Kachroo, 2007;

Khan et al., 2015). SA activates various genes that encode antioxidants, chaperons, and heat shock proteins. These genes are also involved in the biosynthesis of secondary metabolites (Wani et al., 2016). In order to cope with stress conditions, the internal levels of SA should be increased in plants (Munne-Bosch and Penuelas, 2003). Evidence suggests that tolerance to stress conditions, especially drought stress, is associated with an increase in the antioxidant system and reduction of electrolyte leakage (Kang et al., 2013). In other words, the application of exogenous SA may reduce the effects of stress on cell membranes (Bandurska and Stroinski, 2005), which is related to the increased tolerance of plants to stress conditions.

Ethylene is involved in various processes, including seed dormancy, plant growth and aging, adjusting the opening and closing of the stomata, and defense against biotic and abiotic stresses (Achard et al., 2006; Lin et al., 2009). Studies show that ethylene induces ROS production, and H_2O_2 stimulates the expression of reactive proteins to ethylene and involved enzymes in the biosynthesis of ethylene (Vandenabeele et al., 2003).

Jasmonic acid and its related compounds regulate the defensive responses of plants to stress conditions, including infections of pathogens, osmotic stress (Rao et al., 2000; Devoto and Turner, 2005), and heavy metal stress (Maksymiec et al., 2005). Studies have shown that JA is effective in regulating defensive responses by activating the secondary messengers, including H_2O_2 (Orozco-Cárdenas et al., 2001). Also, the expression of responsive genes to JA, including antioxidants and defense-related proteins (such as enzyme-encoding genes involved in the synthesis of ascorbate and glutathione), is increased under stress conditions (Xiang and Oliver, 1998; Wolucka et al., 2005; Shan and Liang, 2010).

Cytokinins can also participate in the removal of ROS from the cell. Cytokinins delay senescence by slowing down macromolecules' decomposition, principally those that are components of photosynthetic apparatus (Wingler et al., 1998). The opposite effects of ROS and cytokinins on the integrity of cell biomolecules and senescence impose the question whether and how cytokinins affect the concentration of ROS. Mohammadi and Moradi (2016) suggest that BAP (benzylaminopurine) application alleviated oxidative stress (H_2O_2 and MDA contents) in wheat cultivars. Also, their results suggested that applying BAP could inhibit the lipid peroxidation of cell membranes in flag leaves, and the tolerance of Pishtaz cultivar (tolerant cultivar) under water stress is associated with low MDA and H_2O_2 contents as well as high Chl, α-tocopherol, and AsA contents which are closely related to its enzymatic antioxidant activity (including CAT, POX, and APX).

15.8 USE OF PLANT METABOLIC ACTIVATORS TO INCREASE ANTIOXIDANT ACTIVITY

Studies have shown that one of the metabolic activators of the plant to enhance antioxidant activity is plant growth-promoting rhizobacteria (PGPR). Reports indicate that inoculation of plants with PGPR can reduce the effects of environmental stresses and increase plant tolerance to environmental stress by producing plant growth hormones such as auxin, increasing the dissolution of low-soluble nutrients such as phosphorus, producing ACC-deaminase, nitrogen fixation, and producing siderophore (Glick, 1995). Besides, PGPR neutralize or modify the harmful effects of plant pathogens by using various antagonistic mechanisms, thereby increasing plant growth. Studies on corn inoculated with *Pseudomonas* spp. under drought stress showed that inoculated plants compared to non-inoculated plants had higher plant biomass, relative moisture content, leaf and root water, proline, sugar, and amino acids. Besides, non-inoculated plants compared to inoculated plants had lower electrolytic leakage and antioxidant enzyme activity (ascorbate peroxidase, catalase, and GPX) under drought stress. Consequently, it showed that inoculated plants compared to non-inoculated plants are less susceptible to stress (Sandhya et al., 2010). In fact, most of the rhizobacteria produce osmotic adjusters (potassium ion, glutamate, trehalose, proline, glycine betaine, proline betaine, ectoine, etc.) to modify their cytoplasmic osmolarity under stress conditions (Talibart et al., 1994). Bacteria like *Pseudomonas* survive under stress conditions through extracellular polymeric substance (EPS) production, which protects microorganisms from water stress and fluctuations in water potential by enhancing water retention and regulating the diffusion of carbon sources in the microbial environment (Sandhya et al., 2009). Exopolysaccharides possess unique water holding and cementing properties, and, thus, they play a vital role in the formation and stabilization of soil aggregates and the regulation of nutrients and water flow cross plant roots through the biofilm formation (Roberson and Firestone, 1992).

Some of the terrestrial bacteria have the ability to produce ACC-deaminase enzymes and thus can modulate the sensitivity of root and leaf growth to soil drying by influencing the signaling of the ethylene hormone. For example, ACC-deaminase activity of *Achromobacter piechaudii* resulted in resistance to drought stress in tomato and pepper. Ethylene production in inoculated plants was reduced compared to inoculated plants, and plants were recovered after restoration (Mayak et al.,

2004). Nadeem et al., (2007) found that inoculation of corn with *P. syringae* and *P. fluorescens* led to an increase in potassium to sodium ratio, relative water content, and chlorophyll and resulted in lower proline content. It is believed that salinity resistance depends on different mechanisms in different plants. Moreover, higher potassium to sodium ratio in *Azospirillum*-inoculated corn was observed under salinity stress (Hamdia et al., 2004).

Mohammadi et al., (2016) results suggest that inoculation of *Satureja hortensis* with certain *Pseudomonas fluorescens* strains can significantly increase the plant biomass and essential oil yields under water stress conditions. However, the inoculation with *P. fluorescens* (PF-135) not only improves the activity of antioxidant enzymes but also induces synthesis of proline, which is one of the osmolytes responsible for maintaining cell turgor under water stress conditions. In addition, the promising strains of *P. fluorescens* improve essential oil yield and carvacrol content under water stress and well-watered conditions. Finally, *P. fluorescens* (PF-135) strain is introduced to minimize the deleterious effects of water stress on plant growth parameters and improve essential oil yield and quality in low input system. Therefore, the use of PGPR could be an encouraging and eco-friendly strategy for increasing antioxidant capacity in plants.

15.9 THE ROLE OF NUTRIENTS IN INCREASING THE ACTIVITY OF ANTIOXIDANTS

Nutrients play a crucial role in regulating interactions between plants and their environment. In addition, micro and macro elements play a key role in controlling plants' growth from germination to full maturity stages. In this regard, nutrients play a significant role in increasing resistance to environmental stresses, which is associated with an increase in the production of antioxidants. Therefore, nutrient management plays an important role in crop growth and production. Macro nutritional elements play a major role in plant structure, and their application under stressful conditions leads to an increase in the production of antioxidants and increase in resistance against unfavorable conditions. However, in some cases, it has been observed that the application of some macro elements under stress conditions can increase plant resistance by increasing antioxidant activity. For instance, nitrogen fertilizer application under stress conditions can increase antioxidant activity and reduce adverse effects of stress. However, the antioxidant response to nitrogen application is different in different species. In *Populus* plants, grown under cadmium stress conditions, nitrogen application increased antioxidant enzyme activity by increasing expression in genes involved in the synthesis of SOD, CAT, etc. (Zhang et al., 2014). Nitrogen also could increase antioxidant activity in *Hordeum vulgare* under water deficit stress conditions (Movludi et al., 2014). By contrast, in a study, nitrogen application decreased flavonoid, glutathione and ascornoc acid content in *Labisiapumila* (Ibrahim et al., 2012). Moreover, an increase in nitrogen application could increase antioxidant activity in *Chrysanthemum morifolium* (Liu et al., 2010). Nitrogen and sulfur are among the most important macro nutrients to support plants growth and also is necessary for synthesis of secondary metabolites. It has been reported that nitrogen and sulfur could change overall antioxidant properties and total phenol content (Li et al., 2008). In addition, an increase in antioxidant activity in tomato plants was observed when P (phosphorous) was sprayed on plants (Ahn et al., 2002; Yogaratnam and Sharples, 1982). Phosphorus application could increase catalase activity in common bean (*Phaseolus vulgaris*). A significant increase in CAT, SOD, and peroxidase activity on account of potassium and calcium application was observed when plants were grown under cadmium stress, which might be due to potassium's key role in enzyme biosynthesis and activation and calcium's role as a secondary signal in relation to calmodulin (Siddiqui et al., 2012; Ahmad, et al., 2015). Microelements play a direct and indirect role in changing the activity of enzymes and the production of certain metabolites in plants in response to environmental factors. In addition, microelements have a potent role in signal transduction in response to environmental stresses (Hajiboland, 2012). There is a strong correlation between the generation and reduction of ROS and the availability of microelements such as iron, zinc, copper, and manganese. Iron is a vital component of many enzymes and plays a very important role in physiological and biochemical processes in plants. Iron as a cofactor performs a crucial role in synthesizing enzymes and hormones, and it has several functions in plants, for example, electron transfer (Kerkeb and Connoly, 2006). Previous studies have shown that iron is a compound that increases antioxidant activity (Molassiotis et al., 2005). Therefore, iron deficiency leads to a marked reduction in CAT and SOD activity. Hence, reducing the activity of these two enzymes is an indicator of iron deficiency in plants (Hajiboland, 2012). In a study, iron application on *Bacops monnieril* could increase peroxidase activity in roots but caused a significant reduction in leaves, whereas ascorbate content increased in both organs (Sinha and Saxena, 2006). Manganese is one of the microelements required by plants. Manganese plays an important role in oxidation and reduction reactions due to the possibility of converting oxidation forms. Manganese deficiency causes inappropriate performance in electron transfer during photosynthesis and leads to oxidative stress. In this situation, the oxygen molecule is converted into reactive oxygen species (Hajiboland, 2012; Papadakis et al., 2007) and acts as a cofactor for SOD or CAT (Humphries et al., 2007). Zinc is a very important element playing a vital role in plants. Zinc plays a critical role in controlling hydrogen peroxide production and increasing antioxidant capacity in different plant species. Zinc has a structural role in more than 300 enzymes. Zinc is the main part of glutamate dehydrogenase (GDH), CAT, and SOD enzymes (Li et al., 2006). Zinc deficiency increases reactive oxygen species generation through increasing activity of NADPH-dependent oxidase (Cakmak, 2000). Zinc exogenous application on wheat led to increased SOD and glutathione reductase activity (Saeidnejad et al., 2016). The boron is also capable of increasing the antioxidant activity. The effects of boron on polyphenols and vitamin C in plants have been previously documented (Camacho et al., 2002). Although some of these elements are essential for plants, the high concentration of these elements can be a limiting factor by enhancing toxic free radicals and inducing oxidative stress. Furthermore, it has been reported that boron can cause cases oxidative stress in plants. For instance, boron contamination in irrigation water reduced catalase and glutathione reductase activity but increased peroxidase and SOD (Keles et al., 2004). In this regard, it has been found that zinc and boron foliar application could increase CAT, SOD, and peroxidase activity in onion (*Allium cepa*) (Denre et al., 2016). Zinc foliar application (at high concentrations) could increase antioxidant activity in barley (Karabal et al., 2003). Considering the wide range of zinc effects, especially in the antioxidant defense systems, it is expected that foliar application would affect plants' resistance against environmental stresses. Copper as a cofactor plays a key role in reducing oxidative stress. This element is considered to be the main producer of CuZnSOD. Copper plays an important role in neutralizing free radicals. CuZn-SOD is located in PSI and converts superoxide into hydrogen peroxide and oxygen molecule; hence, the lack of CuZnSOD increases reactive oxygen species concentration in the cell (Hajiboland, 2012). Molybdenum is one of the elements that alleviate oxidative stress in plants. Molybdenum plays a critical role in synthesizing several enzymes and hormones. It also affects the resistance of plants to environmental stress due to the effect of aldehyde oxidase. Cobalt is effective in nodulation in legumes. Cobalt application increases catalase activity and nodule number in legumes (Ali et al., 2010; Sinha et al., 2012). Selenium is an integral part of the GPX enzyme. According to studies, low levels of selenium is associated with increased resistance against oxidative stress and is required for antioxidant enzymes' gene expression (Pilon-Smits et al., 2009). In a study, selenium application could only increase glutathione reductase and catalase activity (Hajiboland and Amjad, 2007).

15.10 UTILIZATION OF VITAMINS, GROWTH PROMOTERS, AND ASCORBIC ACID TO INCREASE ANTIOXIDANT DEFENSE

Vitamins are the most important antioxidant substances in plants. The main task of these antioxidants is to neutralize the destructive effects of reactive oxygen species. In some studies, exogenous use of vitamins could increase the antioxidants' activity. These compounds are naturally synthesized in plant cells, but its amount changes when stress occurs. Antioxidant vitamins include vitamin E, vitamin C, and beta-carotene, which interact with ROS homeostasis (McDowell et al., 2007). Ascorbic acid (vitamin C) is a non-enzymatic antioxidant (Gill and Tuteja, 2010) and is the most abundant water-soluble antioxidant created to minimize damage caused by ROS in plants. Ascorbic acid can be found in all plants tissues, but meristem cells, photosynthetic tissues, and fruits are the main source of ascorbic acid. The maximum concentration is related to adult leaves with fully developed chloroplasts and high chlorophyll content (Gill and Tuteja, 2010). Also, it has been reported that exogenous application of some growth regulators, such as salicylic acid, could increase the activity of ROS scavenging enzymes (Janmohammadi, 2012). Under normal physiological conditions, ascorbic acid content decreases in leaves' chloroplasts (Smirnoff, 2000). The mitochondria play a central role in ascorbic acid metabolism. L-galactono-1,4-lactone is an ascorbic acid precursor. This pathway is catalyzed by rapid conversion of internal L-galactono-1,4-lactone into ascorbates and mitochondrial enzymes (L-galactono-1,4-lactone in dehydrogenase), which oxidizes L-galactonol,4 catalyzes lactone into ascorbate. The mitochondria play a major role in ascorbic acid synthesis and regeneration. The ascorbic acid regeneration is very important because dehydroascorbate has a short half-life (Foyer and Noctor, 2011; Qian et al., 2014). Ascorbate is an antioxidant and is linked to other antioxidant compounds in plant defense systems against oxidative damage caused by aerobic metabolism, photosynthesis, and pollution. Reducing properties of ascorbate is due to diol groups on carbon numbers 2 and 3. In addition, oxidizing properties of ascorbic acid comes from hydroxyl group on carbon number 3 (Smirnoff, 2000). Recent findings on mutated and transgenic plants suggest a crucial role for ascorbate in the ascorbate–glutathione cycle and in antioxidant defense system against oxidative stress. Ascorbic acid acts as a cofactor in some hydroxylase enzymes and violaxanthin de-epoxidase enzymes. These enzymes couple with ascorbic acid and play a key role in the xanthophyll cycle (Müller-Moulé et al., 2002; Foyer, 2015). The regulatory role of photosynthetic electron transfer is also attributed to ascorbic acid (Ivanov et al., 2005). In different studies, the performance of ascorbic acid has been studied in increasing the antioxidant activity and resistance of plants to environmental stresses.

15.10.1 Results of Exogenous Application of Ascorbic Acid in Plants

The exogenous application of ascorbic acid increases resistance to environmental stress and decreases the lipid peroxidation (Shalata and Neumann, 2001). In this regard, the effect of ascorbic acid on reducing lipid peroxidation in sunflower has been reported under stress conditions (Zhang and Kirkham, 1996). The protective effect of ascorbic acid has been attributed to a reduction in damage caused by reactive oxygen species generated during stress. Reduction in cellular ascorbic acid decreases cell division (mitosis) whereas its exogenous application increases mitosis (Kerk et al., 2000). Ascorbic acid protects plants against oxidative damage through recycling tocopherols (vitamin E) and scavenging reactive oxygen species. It has been reported that the ascorbic acid increases apical meristem growth and induces organogenesis in a pine tree species. Ascorbic acid foliar application could prevent ozone toxicity and increase plants' resistance to ozone (Chen and Gallie, 2005). Ascorbic acid directly suppresses superoxide radical and subsequently oxidizes to monodehydroascorbate and then to dehydroascorbate (Foyer and Noctor, 2011). Ascorbic acid foliar application could reduce H_2O_2 level in durum wheat (*Triticum durum* L.) under salt stress conditions (Fercha et al., 2011; Azzedine et al., 2011). In several studies on wheat (*T. aestivum* L.), ascorbic acid foliar application has been found to increase antioxidant enzyme activity under saline stress conditions (Athar et al., 2009; 2008; Farouk, 2011). Similar results were found in maize (*Zea mays* L.) (Ahmad et al., 2014) and milk thistle (*Silybum marianum* L.) (Ekmekçi and

Karaman, 2012) plants. Generally, antioxidant vitamins increase cellular immunity.

15.10.2 Tocopherols (Vitamin E)

Vitamin E is one of the most important fat-soluble vitamins; it was discovered in 1992 at the University of California, Berkeley. This vitamin is synthesized by all plants, some algae, and cyanobacteria and is abundantly found in the seeds. This vitamin, like vitamin C, has antioxidant properties. Alpha-tocopherol is the main form of vitamin E. This vitamin is placed in the fat layer of the cell walls and also into the cells and prevents cell wall destruction. Vitamin E is a name for a group of molecules that has similar effects to alpha-tocopherol. Vitamin E acts as an antioxidant attached to the membrane. Vitamin E works in trapping free radicals generated from unsaturated fatty acids under oxidative stress conditions. Vitamin E plays a role in breaking fatty acid peroxidation chains. Vitamin E acts as the first protective phosphoric acid under oxidative stress conditions (McDowell, 2000). Vitamin E as an antioxidant is essential for seeding and to prevent lipid peroxidation during germination (Sattler et al., 2004). This vitamin is abundantly found in palm oil, rice bran, soybean, wheat germ, and many oilseeds (Heinonen and Piironen, 1991; Packer et al., 2001), but its amount depends on growth stage, plant species, and growth conditions. Typically, vitamin E content increases under stressful conditions such as high light intensity, drought stress, and cold stress. It has also been observed that vitamin E foliar application could increase resistance to environmental stresses, increase plant growth and antioxidant activity (El Bassiouny et al., 2005; Marzauk et al., 2014). In a study, vitamin E foliar application could increase the activity of antioxidant enzymes such as CAT, SOD, and POD in *Vignaradiata* (Sadiq et al., 2017). For example, wheat seed treatment with vitamin E resulted in increased stress resistance and decreased oxidative stress damage (Kumar et al., 2012). The application of vitamin E on faba bean grown under salt stress conditions increased antioxidant enzyme activity (Orabi and Abdelhamid, 2016). In another study on faba bean, vitamin E reduced adverse effects of salinity stress (Semida et al., 2014). Beta-carotene is a precursor to vitamin A that converts to vitamin A in the human body. Investigations on more than 600 different carotenoids have shown that beta-carotene (one of the most important carotenoids) protects the green, yellow, and orange fruits from sun damage. So it is thought to have such an effect on the body as well. There is no limitation for beta-carotene uptake. Carrot, pumpkin, broccoli, sweet potato, spinach, tomatoes, cantaloupe, peach, and apricot are rich sources of beta-carotene. Beta-carotene is an active antioxidant compound that protects plants against cell damage caused by free radicals (Stahl and Sies, 2003). Due to its unique structure, beta-carotene has specific roles in antioxidant defense mechanisms such as protecting lipophilic compounds or eliminating reactive oxygen species during photooxidation. It is used as a light filter (Stahl and Sies, 2003). Studies have revealed that there is a significant interaction between vitamin C, E, and beta-carotene in eliminating reactive oxygen species (Bestwick and Milne, 1999; Stahl and Sies, 2003).

15.11 CONCLUSIONS AND PERSPECTIVES

Environmental stresses due to cold, heat, salinity, and drought adversely affect plant growth and productivity, which triggers a series of phenomena that result in the production and accumulation of ROS in some cellular organelles. The highest amounts of ROS are produced in the chloroplast and mitochondrial organelles. In addition to organelles, plasma membrane together with apoplast is the main site for ROS generation in response to endogenous signals and exogenous environmental stimuli. Several types of enzymes, such as NADPH oxidases, amine oxidases, polyamine oxidases, oxalate oxidases, and a large family of class III peroxidases, that localized at the cell surface or apoplast are contributed to the production of apoplast ROS. To distinguish how plants sense ROS or respond to oxidative stress, it must be considered that they are subjected to a combination of adverse conditions. This preliminary deliberation is necessary to identify the defense mechanisms against oxidative stress and also to identify strategies to improve stress tolerance. There are many sensors that perceive certain stress conditions and control all downstream signals. However, very few specialized receptors have been identified under oxidative stress conditions. It seems that ROS sensing is carried out by redox-sensitive proteins that can undergo reversible oxidation/reduction and may switch "on" and "off" depending upon the cellular redox state. After perceiving of ROS by the receptors, they can activate MAPK and a group of transcription factors that control various pathways in the cell. This cascade ultimately leads to an increase in the expression of antioxidant genes, and increased activity of antioxidants may lead to redox homeostasis. However, overcoming oxidative stress can be achieved in the presence of ROS scavenging genes; otherwise, the accumulation of ROS can lead to plant death. So far, only plants overexpressing antioxidant enzymes have been engineered, with the aim of increasing stress tolerance by directly modifying the expression of these ROS scavenging enzymes. However, ROS signaling pathways are very complicated and show many interactions with sugar, phytohormones, and calcium signaling pathways, so in breeding, process-related paths should be considered. Some compounds such as proline, sugars, polyamines, carotenoids, and flavonoids can play scavenging roles under stress conditions, and the capacity to accumulate these compounds in the plant can significantly reduce the damage caused by oxidative stress.

REFERENCES

Abbasi, A.R., Hajirezaei, M., Hofius, D., Sonnewald, U., Voll, L.M. (2007) Specific roles of α-and γ-tocopherol in abiotic stress responses of transgenic tobacco. *Plant Physiology* 143(4): 1720–1738.

Abedi, T., Pakniyat, H. (2010) Antioxidant enzyme changes in response to drought stress in ten cultivars of oilseed rape (*Brassica napus* L.). *Czech Journal of Genetics and Plant Breeding* 46(1): 27–34.

Achard, P., Cheng, H., De Grauwe, L., Decat, J., Schoutteten, H., Moritz, T., Van Der Straeten, D., Peng, J., Harberd, N.P. (2006) Integration of plant responses to environmentally activated phytohormonal signals. *Science* 311(5757): 91–94.

Ahmad, I., Basra, S.M.A., Wahid, A. (2014) Exogenous application of ascorbic acid, salicylic acid, hydrogen peroxide improves the productivity of hybrid maize at low temperature stress. *International Journal of Agriculture and Biology* 16: 825–830.

Ahmad, P., Sarwat, M., Bhat, N.A., Wani, M.R., Kazi, A.G., Tran, L.S. (2015) Alleviation of cadmium toxicity in *Brassica juncea* L. (Czern. & Coss.) by calcium application involves various physiological and biochemical strategies. *PLoS One* 10(1): e0114571.

Ahn, T., Schofield, A., Paliyath, G. (2002) Antioxidant enzyme activities during tomato fruit development and in response to phosphorus nutrients. *International Horticultural Congress* (XXVI) S09-0-182.

Alhasawi, A., Castonguay, Z., Appanna, N.D., Auger, C., Appanna, V.D. (2015) Glycine metabolism and antioxidative defense mechanisms in *Pseudomonas fluorescens*. *Microbiological Research* 171: 26–31.

Ali, B., Hayat, S., Hayat, Q., Ahmad, A. (2010) Cobalt stress affects nitrogen metabolism, photosynthesis and antioxidant system in chickpea (*Cicer arietinum* L.). *Journal of Plant Interactions* 5(3): 223–231.

Alscher, R.G. (1989) Biosynthesis and antioxidant function of glutathione in plants. *Physiologia Plantarum* 77(3): 457–464.

Ananieva, E.A., Alexieva, V.S., Popova, L.P. (2002) Treatment with salicylic acid decreases the effects of paraquat on photosynthesis. *Journal of Plant Physiology* 159(7): 685–693.

Anderson, M.D., Prasad, T.K., Martin, B.A., Stewart, C.R. (1994) Differential gene expression in chilling acclimated maize seedlings and evidence for the involvement of abscisic acid in chilling tolerance. *Plant Physiology* 105(1): 331–339.

Asada, K. (1999) The water-water cycle in chloroplasts: Scavenging of active oxygens and dissipation of excess photons. *Annual Review of Plant Physiology and Plant Molecular Biology* 50: 601–639.

Asai, S., Ohta, K., Yoshioka, H. (2008) MAPK signaling regulates nitric oxide and NADPH oxidase-dependent oxidative bursts in *Nicotiana benthamiana*. *The Plant Cell* 20(5): 1390–1406.

Ashraf, M. (2009) Biotechnological approach of improving plant salt tolerance using antioxidants as markers. *Biotechnology Advances* 27(1): 84–93.

Ashraf, M., Foolad, M. (2007) Roles of glycine betaine and proline in improving plant abiotic stress resistance. *Environmental and Experimental Botany* 59(2): 206–216.

Athar, H., Khan, A., Ashraf, M. (2009) Including salt tolerance in wheat by exogenously applied ascorbic acid through different modes. *Journal of Plant Nutrition* 32(11): 1799–1817.

Athar, H., Khan, A., Ashraf, M. (2008) Exogenously applied ascorbic acid alleviates salt induced oxidative stress in wheat. *Environmental and Experimental Botany* 63(1–3): 224–231.

de Azevedo Neto, A.D., Prisco, J.T., Enéas-Filho, J., de Abreu, C.E.B., Gomes-Filho, E. (2006) Effect of salt stress on antioxidative enzymes and lipid peroxidation in leaves and roots of salt-tolerant and salt-sensitive maize genotypes. *Environmental and Experimental Botany* 56(1): 87–94.

Azzedine, F., Gherroucha, H., Baka, M. (2011) Improvement of salt tolerance in durum wheat by ascorbic acid application. *Journal of Stress Physiology and Biochemistry* 7(1): 27–37.

Bandurska, H., Stroinski, A. (2005) The effect of salicylic acid on barley response to water deficit. *Acta Physiologiae Plantarum* 27(3): 379–386.

Bari, R., Jones, J.D. (2009) Role of plant hormones in plant defence responses. *Plant Molecular Biology* 69(4): 473–88.

Bartoli, C.G., Casalongué, C.A., Simontacchi, M., Marquez-Garcia, B., Foyer, C.H. (2012) Interactions between hormone and redox signaling pathways in the control of growth and cross tolerance to stress. *Environmental and Experimental Botany* 94: 73–8.

Birben, E., Sahiner, U.M., Sackesen, C., Erzurum, S., Kalayci, O. (2012) Oxidative Stress and antioxidant defense. *The World Allergy Organization Journal* 5(1): 9–19.

Talibart, R., Jebbar, M., Gouesbet, G., Himdi-Kabbab, S., Wroblewski, H., Blanco, C., Bernard, T. (1994) Osmoadaptation in rhizobia: Ectoine-induced salt tolerance. *Journal of Bacteriology* 176(17): 5210–5217.

Blokhina, O., Virolainen, E., Gagerstedt, K.V. (2003) Antioxidants, oxidative damage and oxygen deprivation stress: A review. *Annals of Botany* 91: 179–194.

Brunetti, C., Di Ferdinando, M., Fini, A., Pollastri, S., Tattini, M. (2013) Flavonoids as antioxidants and developmental regulators: Relative significance in plants and humans.

International Journal of Molecular Sciences 14(2): 3540–3555.

Cakmak, I. (2000) Possible roles of zinc in protecting plant cells from damage by reactive oxygen species. *New Phytologist* 146: 185–205.

Camacho, C.J.J., Anzelotti, D., González, F.A. (2002) Changes in phenolic metabolism of tobacco plants during short-term boron deficiency. *Plant Physiology Biochemistry* 40: 997–1002.

Cavalcanti, F.R., Oliveira, J.T.A., Martins-Miranda, A.S., Viégas, R.A., Silveira, J.A.G. (2004) Superoxide dismutase, catalase and peroxidase activities do not confer protection against oxidative damage in salt-stressed cowpea leaves. *New Phytologist* 163(3): 563–571.

Caverzan, A., Casassola, A., Brammer, S.P. (2016) Antioxidant responses of wheat plants under stress. *Genetics and Molecular Biology* 39(1): 1–6.

Caverzan, C., Bonifacio, A., Carvalho, F.E.L., Andrade, C.M.B., Passaia, G., Schünemann, M., Maraschin, F.S., Martins, M.O., Teixeira, F.K., Rauber, R., et al. (2014) The knockdown of chloroplastic ascorbate peroxidases reveals its regulatory role in the photosynthesis and protection under photo-oxidative stress in rice. *Plant Science* 214: 74–87.

Chen, Z., Gallie, D.R. (2005) Increasing tolerance to ozone by elevating foliar ascorbic acid confers greater protection against ozone than increasing avoidance. *Plant Physiology* 138(3): 1673–1689.

Choudhury, F.K., Rivero, R.M., Blumwald, E., Mittler, R. (2017) Reactive oxygen species, abiotic stress and stress combination. *The Plant Journal* 90(5): 856–867.

Claeys, H., Inze, D. (2013) The agony of choice: How plants balance growth and survival under water limiting conditions. *Plant Physiology* 162(4): 1768–79.

Cobbett, C.S. (2000) Phytochelatin biosynthesis and function in heavy-metal detoxification. *Current Opinion in Plant Biology* 3(3): 211–216.

Cortina, C., Culiáñez-Macià, F.A. (2005) Tomato abiotic stress enhanced tolerance by trehalose biosynthesis. *Plant Science* 169(1): 75–82.

Couée, I., Sulmon, C., Gouesbet, G., El Amrani, A. (2006) Involvement of soluble sugars in reactive oxygen species balance and responses to oxidative stress in plants. *Journal of Experimental Botany* 57(3): 449–459.

Cruz, F.J.R., Castro, G.L.S., Silva Júnior, D.D., Festucci-Buselli, R.A., Pinheiro, H.A. (2013) Exogenous glycine betaine modulates ascorbate peroxidase and catalase activities and prevent lipid peroxidation in mild water-stressed *Carapa guianensis* plants. *Photosynthetica* 51(1): 102–108.

Das, k., Roychoudhury, A. (2014) Reactive oxygen species (ROS) and response of antioxidants as ROS-scavengers during environmental stress in plants. *Frontiers Environmental Science* 2: 1–13.

Davies, W.J., Zhang, J. (1991) Root signals and the regulation of growth and development of plants in drying soil. *Annual Review of Plant Physiology and Plant Molecular Biology* 42(1): 55–76.

Dawson, J.H. (1998) Probing structure-function relations in heme-containing oxygenases and peroxidases. *Science* 240(4851): 430–439.

Del Río, L.A., Luisa, M., Sandalio Deborah, A., Altomare Barbara, A., Zilinskas, B.A. (2003) Mitochondrial and peroxisomal manganese superoxide dismutase: Differential expression during leaf senescence. *Journal of Experimental Botany* 54(384): 923–933.

Denre, M., Bhattacharya, A., Pal, S., Chakravarty, A., Chattopadhyay, A., Mazumdar, D. (2016) Effect of foliar application of micronutrients on antioxidants and pungency in onion. *Notulae Scientia Biologicae* 8(3): 373–379.

Devoto, A., Turner, J.G. (2005) Jasmonate-regulated *Arabidopsis* stress signaling network. *Physiologia Plantarum* 123(2): 161–172.

Edreva, A. (2005) Generation and scavenging of reactive oxygen species in chloroplasts: A submolecular approach. *Agriculture, Ecosystems and Environment* 30: 119–133.

Ekmekçi, B.A., Karaman, M. (2012) Exogenous ascorbic acid increases resistance to salt of *Silybum marianum* (L.). *African Journal of Biotechnology* 11(42): 9932–9940.

El Bassiouny, H.M., Gobarah, M.E., Ramadan, AA. (2005) Effect of antioxidants on growth, yield and favism causative agents in seeds of *Vicia faba* L. plants grown under reclaimed sandy soil. *Journal of Agronomy* 4(4): 281–287.

El-Bassiouny, H., Sadak, M.S. (2015) Impact of foliar application of ascorbic acid and α-tocopherol on antioxidant activity and some biochemical aspects of flax cultivars under salinity stress. *Acta Biológica Colombiana* 20(2): 209–222.

Ellouzi, H., Ben Hamed, K., Cela, J., Munne-Bosch, S., Abdelly, C. (2011) Early effects of salt stress on the physiological and oxidative status of *Cakile maritima* (halophyte) and *Arabidopsis thaliana* (glycophyte). *Physiologia Plantarum* 142(2): 128–143.

Espinoza, A., San Martín, A., López-Climent, M., Ruiz-Lara, S., Gómez-Cadenas, A., Casaretto, J.A. (2013) Engineered drought-induced biosynthesis of α-tocopherol alleviates stress-induced leaf damage in tobacco. *Journal of Plant Physiology* 170(14): 1285–1294.

Farooq, M., Wahid, A., Lee, D.J. (2009) Exogenously applied polyamines increase drought tolerance of rice by improving leaf water status, photosynthesis and membrane properties. *Acta Physiologiae Plantarum* 31(5): 937–945.

Farouk, S. (2011) Ascorbic acid and α-tocopherol minimize salt-induced wheat leaf senescence. *Journal of Stress Physiology and Biochemistry* 7(3): 58–79.

Fercha, A., Hocine, G., Mebarek, B. (2011) Improvement of salt tolerance in durum wheat by ascorbic acid application. *Journal of Stress Physiology and Biochemistry* 7(1): 27–37.

Foyer, C.H., Noctor, G. (2011) Ascorbate and glutathione: The heart of the redox hub. *Plant Physiology* 155(1): 2–18.

Foyer, C.H. (2015) Redox homeostasis: Opening up ascorbate transport. *Nature Plants* 1: 14012.

Foyer, C.H., Noctor, G. (2000) Oxygen processing in photosynthesis: Regulation and signalling. *New Phytologist* 146: 359–388.

Foyer, C.H., Noctor, G. (2005) Redox homeostasis and antioxidant signaling: A metabolic interface between stress perception and physiological responses. *The Plant Cell* 17(7): 1866–1875.

Fujita, M., Fujita, Y., Noutoshi, Y., Takahashi, F., Narusaka, Y., Yamaguchi-Shinozaki, K., Shinozaki, K. (2006) Crosstalk between abiotic and biotic stress responses: A current view from the points of convergence in the stress signaling networks. *Current Opinion in Plant Biology* 9(4): 436–42.

George, S., Venkataraman, G., Parida, A. (2010) A chloroplast-localized and auxin induced glutathione *S*-transferase from phreatophyte Prosopis juliflora confer drought tolerance on tobacco. *Journal of Plant Physiology* 167(4): 311–318.

Gill, S.S., Tuteja, N. (2010) Reactive oxygen species and antioxidant machinery in abiotic stress tolerance in crop plants. *Plant Physiology and Biochemistry* 48(12): 909–930.

Gill, S.S., Anjum, N.A., Khan, N.A., Nazar, R. (2008) Metal-binding peptides and antioxidant defense system in plants: Significance in cadmium tolerance. In *Abiotic Stress and Plant Responses*, ed. Khan, N.A., Singh, S. IK International, New Delhi, India, pp. 159–189.

Glick, B.R. (1995) The enhancement of plant growth by free-living bacteria. *Canadian Journal of Microbiology* 41(2): 109–117.

Gomez, J.M., Jimenez, A., Olmos, E., Sevilla, F. (2004) Location and effects of long-term NaCl stress on superoxide dismutase and ascorbate peroxidase isoenzymes of pea (*Pisum sativum* cv. Puget) chloroplasts. *Journal of Experimental Botany* 55(394): 119–130.

Groppa, M.D., Benavides, M.P. (2008) Polyamines and abiotic stress: Recent advances. *Amino Acids* 34(1): 35–45.

Hajiboland, R., Amjad, L. (2007) Does antioxidant capacity of leaves play a role in growth response to selenium at different sulfur nutritional status? *Plant, Soil and Environment* 53(5): 207–215.

Hajiboland, R. (2012) Effect of micronutrient deficiencies on plants stress responses. In *Abiotic Stress Responses in Plants*, ed. Ahmad, P., Prasad, M. Springer, New York, NY, pp. 215–234.

Halliwell, B. (2006) Oxidative stress and neurodegeneration: Where are we now? *Journal of Neurochemistry* 97(6): 1634–1658.

Halliwell, B., Gutteridge, J.M. (2015) *Free Radicals in Biology and Medicine*. Oxford University Press, New York, NY.

Hamdia, A.B.E., Shaddad, M.A.K., Doaa, M.M. (2004) Mechanisms of salt tolerance and interactive effects of *Azospirillum brasilense* inoculation on maize cultivars grown under salt stress conditions. *Plant Growth Regulation* 44(2): 165–174.

Han, S.E., Park, S.R., Kwon, H.B., Yi, B.Y., Lee, G.B., Byun, M.O. (2005) Genetic engineering of drought-resistant tobacco plants by introducing the trehalose phosphorylase (TP) gene from *Pleurotus sajor-caju*. *Plant Cell, Tissue and Organ Culture* 82(2): 151–158.

Hasabi, V., Askari, H., Alavi, S.M., Zamanizadeh, H. (2014) Effect of amino acid application on induced resistance against citrus canker disease in lime plants. *Journal of Plant Protection Research* 54(2): 144–149.

Hayat, Q., Hayat, S., Irfan, M., Ahmad, A. (2010) Effect of exogenous salicylic acid under changing environment: A review. *Environmental and Experimental Botany* 68: 14–25.

Hayat, S., Hayat, Q., Alyemeni, M.N., Wani, A.S., Pichtel, J., Ahmad, A. (2012) Role of proline under changing environments: A review. *Plant Signaling and Behavior* 7(11): 1456–1466.

Heath, R.L. (1987) The biochemistry of ozone attack on the plasma membrane of plant cells. *Advances in Phytochemistry* 21: 29–54.

Heinonen, M., Piironen, V. (1991) The tocotrienol, and vitamin-E content of the average Finnish diet. *International Journal for Vitamin and Nutrition Research* 61(1): 27–32.

Hincha, D.K. (2008) Effects of alpha-tocopherol (vitamin E) on the stability and lipid dynamics of model membranes mimicking the lipid composition of plant chloroplast membranes. *FEBS Letter* 582(25–26): 3687–3692.

Hiraga, S., Sasaki, K., Ito, H., Ohashi, Y., Matsui, H. (2001) A large family of class III plant peroxidases. *Plant and Cell Physiology* 42(5): 462–468.

Hoque, M.A., Banu, M.N.A., Okuma, E., Amako, K., Nakamura, Y., Shimoishi, Y., Murata, Y. (2007) Exogenous proline and glycinebetaine increase NaCl-induced ascorbate-glutathione cycle enzyme activities, and proline improves salt tolerance more than glycinebetaine in tobacco Bright Yellow-2 suspension-cultured cells. *Journal of Plant Physiology* 164(11): 1457–1468.

Hu, X., Jiang, M., Zhang, J., Zhang, A., Lin, F., Tan, M. (2007) Calcium-calmodulin is required for abscisic acid-induced antioxidant defense and functions both upstream and downstream of H_2O_2 production in leaves of maize (*Zea mays*) plants. *New Phytologist* 173(1): 27–38.

Hudec, J., Burdová, M., Kobida, L.U., Komora, L., Macho, V., Kogan, G., Turianica, I., Kochanová, R., Ložek, O., Hában, M., Chlebo, P. (2007) Antioxidant capacity changes and phenolic profile of *Echinacea purpurea*, nettle (*Urticadioica* L.), and dandelion (*Taraxacum officinale*) after application of polyamine and phenolic biosynthesis regulators. *Journal of Agricultural and Food Chemistry* 55(14): 5689–5696.

Humphries, J.M., Stangoulis, J.C.R., Graham, R.D. (2007) Manganese. In *Handbook of Plant Nutrition*, ed. Barker, A.V., Pilbeam, D.J. CRC Press, Taylor and Francis Group, Boca Raton, FL, pp. 351–374.

Ibrahim, M.H., Jaafar, H.Z., Rahmat, A., Rahman, Z.A. (2012) Involvement of nitrogen on flavonoids, glutathione, anthocyanin, ascorbic acid and antioxidant activities of malaysian medicinal plant *Labisia pumila* Blume (Kacip Fatimah). *International Journal of Molecular Sciences* 13(1): 393–408.

Inze, A., Vanderauwera, S., Hoeberichts, F.A., Vandorpe, M., Van Gaever, T., Van Breusegem, F. (2012) A subcellular localization compendium of hydrogen peroxide-induced proteins. *Plant, Cell and Environment* 35(2): 308–320.

Ivanov, B., Asada, K., Kramer, D.M., Edwards, G. (2005) Characterization of photosynthetic electron transport in bundle sheath cells of maize. I. Ascorbate effectively stimulates cyclic electron flow around PSI. *Planta* 220(4): 572–581.

Jämtgård, S., Näsholm, T., Huss-Danell, K. (2010) Nitrogen compounds in soil solutions of agricultural land. *Soil Biology and Biochemistry* 42(12): 2325–2330.

Janmohammadi, M., Mock, H.P., Matros, A. (2014) Proteomic analysis of cold acclimation in winter wheat under field conditions. *Icelandic Agricultural Sciences* 27(1): 3–15.

Janmohammadi, M. (2012) Alleviation the adverse effect of cadmium on seedling growth of greater burdock (*Aractium lappal* L.) through pre-sowing treatments. *Agriculture and Forestry* 56(1–4): 55–70.

Janmohammadi, M., Abbasi, A., Sabaghnia, N. (2012a) Influence of NaCl treatments on growth and biochemical parameters of castor bean (*Ricinus communis* L.). *Acta Agriculturae Slovenica* 99(1): 31–40.

Janmohammadi, M., Enayati, V., Sabaghnia, N. (2012b) Impact of cold acclimation, de-acclimation and re-acclimation on carbohydrate content and antioxidant enzyme activities in spring and winter wheat. *Icelandic Agricultural Sciences* 25: 3–11.

Janmohammadi, M., Sabaghnia, N., Mahfoozi, S. (2018) Frost tolerance and metabolite changes of rye (*Secale cereale*) during the cold hardening and overwintering. *Acta Physiologiae Plantarum* 40(3): 42–54.

Janmohammadi, M., Zolla, L., Rinalducci, S. (2015) Low temperature tolerance in plants: Changes at the protein level. *Phytochemistry* 117: 76–89.

Jaspers, P., Kangasjärvi, J. (2010) Reactive oxygen species in abiotic stress signaling. *Physiologia Plantarum* 138(4): 405–413.

Jiang, M., Zhang, J. (2002) Role of abscisic acid in water stress-induced antioxidant defense in leaves of maize seedlings. *Free Radical Research* 36(9): 1001–1015.

Jiang, M., Zhang, J. (2003) Cross-talk between calcium and reactive oxygen species originated from NADPH oxidase in abscisic acid-induced antioxidant defence in leaves of maize seedlings. *Plant, Cell and Environment* 26(6): 929–939.

Jithesh, M.N., Prashanth, S.R., Sivaprakash, K.R., Parida, A.K. (2006) Antioxidative response mechanisms in halophytes: Their role in stress defence. *Journal of Genetics* 85(3): 237–254.

Kachroo, A., Kachroo, P. (2007) Salicylic acid, jasmonic acid and ethylene-mediated regulation of plant defense signaling. *Genetic Engineering* 28: 55–83.

Kaminaka, H., Morita, S., Tokumoto, M., Masumura, T., Tanaka, K. (1999) Differential gene expressions of rice superoxide dismutase isoforms to oxidative and environmental stresses. *Free Radical Research* 31: 19–25.

Kang, G., Li, G., Liu, G., Xu, W., Peng, X., Wang, C., Zhu, Y., Guo, T. (2013) Exogenous salicylic acid enhances wheat drought tolerance by influence on the expression of genes related to ascorbate-glutathione cycle. *Biologia Plantarum* 57(4): 718–724.

Karabal, E., Yücel, M., Ökte, H.A. (2003) Antioxidants responses of tolerant and sensitive barley cultivars to boron toxicity. *Plant Science* 164: 925–933.

Karabudak, T., Bor, M., Özdemir, F., Türkan, İ. (2014) Glycine betaine protects tomato (*Solanum lycopersicum*) plants at low temperature by inducing fatty acid desaturase7 and lipoxygenase gene expression. *Molecular Biology Reports* 41(3): 1401–1410.

Katsuhara, M., Otsuka, T., Ezaki, B. (2005) Salt stress-induced lipid peroxidation is reduced by glutathione S-transferase, but this reduction of lipid peroxides is not enough for a recovery of root growth in *Arabidopsis*. *Plant Science* 169(2): 369–373.

Kawakami, A., Sato, Y., Yoshida, M. (2008) Genetic engineering of rice capable of synthesizing fructans and enhancing chilling tolerance. *Journal of Experimental Botany* 59(4): 793–802.

Keles, Y., Oncel, I., Yenice, N. (2004) Relationship between boron content and antioxidant compounds in Citrus leaves taken from fields with different water source. *Plant Soil* 265(1–2): 343–353.

Kerk, N.M., Jiang, K., Feldman, J.L. (2000) Auxin metabolism in the root apical meristem. *Plant Physiology* 122(3): 925–932.

Kerkeb, L., Connoly, E. (2006) Iron transport and metabolism in plants. *Genetic Engineering* 27: 119–140.

Keunen, E.L.S., Peshev, D., Vangronsveld, J., Van Den Ende, W.I.M., Cuypers, A.N.N. (2013) Plant sugars are crucial players in the oxidative challenge during abiotic stress: Extending the traditional concept. *Plant, Cell and Environment* 36(7): 1242–1255.

Khan, M.I.R., Fatma, M., Per, T.S., Anjum, N.A., Khan, N.A. (2015) Salicylic acid-induced abiotic stress tolerance and underlying mechanisms in plants. *Frontiers in Plant Science* 6: 462.

Khan, N.A., Singh, S., Eds. (2008) *Abiotic Stress and Plant Responses*, IK International, New Delhi, India.

Knorzer, O.C., Lederer, B., Durner, J., Boger, P. (1999) Antioxidative defense activation in soybean cells. *Physiologia Plantarum* 107(3): 294–302.

Koca, H., Bor, M., ozdemir, F., Turkan, I. (2007) The effect of salt on lipid peroxidation, antioxidative enzymes and proline content of sesame cultivars. *Environmental and Experimental Botany* 60(3): 344–351.

Koukounaras, A., Tsouvaltzis, P., Siomos, A.S. (2013) Effect of root and foliar application of amino acids on the growth and yield of greenhouse tomato in different fertilization levels. *Journal of Food Agriculture and Environment* 11(2): 644–648.

Kovtun, Y., Chiu, W.L., Tena, G., Sheen, J. (2000) Functional analysis of oxidative stress-activated mitogen-activated protein kinase cascade in plants. *Proceedings of the National Academy of Sciences* 97(6): 2940–2945.

Kumar, S., Singh, R., Nayyar, H. (2012) A-tocopherol application modulates the response of wheat (*Triticum aestivum* L.) Seedlings to elevated temperatures by mitigation of stress injury and enhancement of antioxidants. *Journal of Plant Growth Regulation* 32(2): 307–314.

Kumar, V., Shriram, V., Hoquem, T.S., Hasan, M.M., Burritt, D.J., Hossain, M.A. (2017) Glycinebetaine-mediated abiotic oxidative-stress tolerance in plants: Physiological and biochemical mechanisms. In *Stress Signaling in Plants: Genomics and Proteomics Perspective*, ed. Abdin, M.Z., Sarwat, M., Ahmad, A., Volume 2. Springer International Publishing, Cham, Switzerland, pp. 111–133.

Li, W.Y.F., Wong, F.L., Tsai, S.N., Tsai, S.N., Phang, T.H., Shao, G.H., Lam, H.M. (2006). Tonoplast-located GmCLC1 and GmNHX1 from soybean enhance NaCl tolerance in transgenic bright yellow (by)-2 cells. *Plant Cell Environ* 29: 1122–1137.

Li, J., Zhu, Z., Gerendás, J. (2008) Effects of nitrogen and sulfur on total phenolics and antioxidant activity in two genotypes of leaf mustard. *Journal of Plant Nutrition* 31(9): 1642–1655.

Li, C.R., Liang, D.D., Li, J., Duan, Y.B., Li, H.A.O., Yang, Y.C., Qin, R.Y., Li, L.I., Wei, P.C., Yang, J.B. (2013) Unravelling mitochondrial retrograde regulation in the abiotic stress induction of rice ALTERNATIVE OXIDASE 1 genes. *Plant Cell and Environment* 36(4): 775–788.

Liang, X., Zhang, L., Natarajan, S.K., Becker, D.F. (2013) Proline mechanisms of stress survival. *Antioxidants and Redox Signaling* 19(9): 998–1011.

Lin, Z., Zhong, S., Grierson, D. (2009) Recent advances in ethylene research. *Journal of Experimental Botany* 60(12): 3311–3336.

Liszkay, A., van der Zalm, E., Schopfer, P. (2004) Production of reactive oxygen intermediates ($O_2^{\bullet-}$, H_2O_2, and. OH) by maize roots and their role in wall loosening and elongation growth. *Plant Physiology 136*: 3114–3123.

Liu, D., Liu, W., Zhu, D., Geng, M., Zhou, W., Yang, T. (2010) Nitrogen effects on total flavonoids, chlorogenic acid, and antioxidant activity of the medicinal plant *Chrysanthemum morifolium*. *Journal of Plant Nutrition and Soil Science* 173(2): 268–274.

Liu, H.L., Dai, X.Y., Xu, Y.Y., Chong, K. (2007) Over-expression of OsUGE-1 altered raffinose level and tolerance to abiotic stress but not morphology in *Arabidopsis*. *Journal of Plant Physiology* 164(10): 1384–1390.

Liu, J.H., Wang, W., Wu, H., Gong, X., Moriguchi, T. (2015) Polyamines function in stress tolerance: From synthesis to regulation. *Frontiers in Plant Science* 6: 872.

Maeda, H., Dudareva, N. (2012) The shikimate pathway and aromatic amino acid biosynthesis in plants. *Annual Review of Plant Biology* 63: 73–105.

Maksimovic, J.D., Zhang, J., Zeng, F., Ziivanovic, B.D., Shabala, L., Zhou, M., Shabala, S. (2013) Linking oxidative and salinity stress tolerance in barley: Can root antioxidant enzyme activity be used as a measure of stress tolerance? *Plant and Soil* 365: 141–155.

Maksymiec, W., Wianowska, D., Dawidowicz, A.L., Radkiewicz, S., Ardarowicz, M., Krupa, Z. (2005) The level of jasmonic acid in *Arabidopsis thaliana* and *Phaseolus coccineus* plants under heavy metal stress. *Journal of Plant Physiology* 162(12): 1338–1346.

Marzauk, N.M., Shafeek, M.R., Helmy, Y.I., Ahmed, A.A., Shalaby, M.A.F. (2014) Effect of vitamin E and yeast extract foliar application on growth, pod yield and both green pod and seed yield of broad bean (*Vicia faba* L.). *Middle East Journal of Applied Sciences* 4(1): 61–67.

Maxwell, B., Kieber, J. (2004) Cytokinin signal transduction. In *Plant Hormones. Biosynthesis, Signal Transduction, Action*, ed. Davies, P.J. Kluwer Academic Publishers, Dordrecht, the Netherlands, pp. 321–349.

Mayak, S., Tirosh, T., Glick, B.R. (2004) Plant growth-promoting bacteria confer resistance in tomato plants to salt stress. *Plant Physiology and Biochemistry* 42(6): 565–572.

McDowell, L. R. (2000) Riboflavin. Vitamins in Animal and Human Nutrition. Iowa State University Press, Ames, IA.

McDowell, L. R., Wilkinson, N., Madison, R., & Felix, T. (2007) Vitamins and minerals functioning as antioxidants with supplementation considerations. In Florida Ruminant Nutrition Symposium; Best Western Gateway Grand: Gainesville, FL, USA. pp. 30–31.

McKersie, B.D., Lesheim, Y. (2013) *Stress and Stress Coping in Cultivated Plants*. Springer Science & Business Media.

Mignolet-Spruyt, L., Xu, E., Idänheimo, N., Hoeberichts, F.A., Mühlenbock, P., Brosché, M., Van Breusegem, F., Kangasjärvi, J. (2016) Spreading the news: Subcellular and organellar reactive oxygen species production and signalling. *Journal of Experimental Botany* 67(13): 3831–3844.

Miller, G., Suzuki, N., Ciftci-Yilmaz, S., Mittler, R. (2010) Reactive oxygen species homeostasis and signalling during drought and salinity stresses. *Plant, Cell and Environment* 33(4): 453–467.

Mithofer, A., Schulze, B., Boland, W. (2004) Biotic and heavy metal stress response in plants: Evidence for common signals. *FEBS Letters* 566(1–3): 1–5.

Mittler, R., Finka, A., Goloubinoff, P. (2012) How do plants feel the heat?. Trends in biochemical sciences, 37(3): 118–125.

Mittler, R., Vanderauwera, S., Suzuki, N., Miller, G., Tognetti, V.B., Vandepoele, K., Gollery, M., Shulaev, V., Van Breusegem, F. (2011) ROS signaling: The new wave? *Trends in Plant Science* 16(6): 300–309.

Mittler, R., Vanderauwera, S., Gollery, M., Van Breusegem, F. (2004) Reactive oxygen gene network of plants. *Trends in Plant Science* 9(10): 490–498.

Mittler, R. (2002) Oxidative stress, antioxidants and stress tolerance. *Trends in Plant Science* 7(9): 405–410.

Mohammadi, H., Dashi, R., Farzaneh, M., Hashempour, H., Parviz, L. (2016) Effects of beneficial root pseudomonas on morphological, physiological, and phytochemical characteristics of *Satureja hortensis* (Lamiaceae) under water stress. *Brazilian Journal of Botany* 1(2): 1–8.

Mohammadi, H., Moradi, F. (2016) Effects of growth regulators on enzymatic and non-enzymatic antioxidants in leaves of two contrasting wheat cultivars under water stress. *Brazilian Journal of Botany* 39(2): 495–505.

Molassiotis, A.N., Diamantidis, G.C., Therios, I.N., Tsirakoglou, V., Dimassi, K.N. (2005) Oxidative stress, antioxidant activity and Fe (III)-chelate reductase activity of five *Prunus* rootstocks explants in response to Fe deficiency. *Plant Growth Regulation* 46(1): 69–78.

Movludi, A., Ebadi, A., Jahanbakhsh, S., Davari, M., Parmoon, G. 2014. The effect of water deficit and nitrogen on the antioxidant enzymes' activity and quantum yield of barley (*Hordeum vulgare* L.). *Notulae Botanicae Horti Agrobotanici Cluj-Napoca* 42(2): 398–404.

Müller-Moulé, P., Conklin, P.L., Niyogi, K.K. (2002) Ascorbate deficiency can limit violaxanthin de-epoxidase activity *in vivo*. *Plant Physiology* 128(3): 970–977.

Munne-Bosch, S., Alegre, L. (2002) The function of tocopherols and tocotrienols in plants. *Critical Reviews in Plant Sciences* 21(1): 31–57.

Munne-Bosch, S., Penuelas, J. (2003) Photo- and antioxidative protection, and a role for salicylic acid during drought and recovery in field-grown *Phillyrea angustifolia* plant. *Planta* 217(5): 758–766.

Munne-Bosch, S. (2005) The role of alpha-tocopherol in plant stress tolerance. *Journal of Plant Physiology* 162(7): 743–748.

Mustafavi, S.H., Shekari, F., Maleki, H.H. (2016) Influence of exogenous polyamines on antioxidant defence and essential oil production in valerian (*Valeriana officinalis* L.) plants under drought stress. *Acta agriculturae Slovenica* 107(1): 81–91.

Nadeem, S.M., Zahir, Z.A., Naveed, M., Arshad, M. (2007) Preliminary investigations on inducing salt tolerance in maize through inoculation with rhizobacteria containing ACC deaminase activity. *Canadian Journal of Microbiology* 53(10): 1141–1149.

Neill, S., Desikan, R., Hancock, J. (2002) Hydrogen peroxide signalling. *Current Opinion in Plant Biology* 5(5): 388–395.

Nishizawa, A., Yabuta, Y., Shigeoka, S. (2008) Galactinol and raffinose constitute a novel function to protect plants from oxidative damage. *Plant Physiology* 147(3): 1251–1263.

Noctor, G., Arisi, A.C.M., Jouanin, L., Foyer, C.H. (1999) Photorespiratory glycine enhances glutathione accumulation in both the chloroplastic and cytosolic compartments. *Journal of Experimental Botany* 50(336): 1157–1167.

Noctor, G., Foyer, C.H. (1998) Ascorbate and glutathione: Keeping active oxygen under control. *Annual Review of Plant Physiology and Plant Molecular Biology* 49: 249–279.

Qian, H. F., Peng, X. F., Han, X., Ren, J., Zhan, K. Y., & Zhu, M. (2014) The stress factor, exogenous ascorbic acid, affects plant growth and the antioxidant system in Arabidopsis thaliana. *Russian journal of plant physiology*, 61(4), 467–475.

Ogasawara, Y., Kaya, H., Hiraoka, G., Yumoto, F., Kimura, S., Kadota, Y., Nara, M. (2008) Synergistic activation of the *Arabidopsis* NADPH oxidase AtrbohD by Ca^{2+} and phosphorylation. *Journal of Biological Chemistry* 283(14): 8885–8892.

Okuma, E., Murakami, Y., Shimoishi, Y., Tada, M., Murata, Y. (2004) Effects of exogenous application of proline and betaine on the growth of tobacco cultured cells under saline conditions. *Soil Science and Plant Nutrition* 50(8): 1301–1305.

Orabi, S.A., Abdelhamid, M.T. (2016) Protective role of a-tocopherol on two Viciafaba cultivars against seawater-induced lipid peroxidation by enhancing capacity of anti-oxidative system. *Journal of the Saudi Society of Agricultural Sciences* 15(2): 145–15.

Orozco-Cárdenas, M.L., Narváez-Vásquez, J., Ryan, C.A. (2001) Hydrogen peroxide acts as a second messenger for the induction of defense genes in tomato plants in response to wounding, systemin, and methyl jasmonate. *The Plant Cell* 13(1): 179–191.

Ozgur, R., Uzilday, B., Sekmen, A.H., Turkan, I. (2013) Reactive oxygen species regulation and antioxidant defence in halophytes. *Functional Plant Biology* 40(9): 832–847.

Packer, L., Weber, S.U., Rimbach, G. (2001) Molecular aspects of a-tocotrienol antioxidant action and cell signaling symposium: Molecular mechanisms of protective effects of vitamine in atherosclerosis. *Journal of Nutrition* 131(2): 369–373.

Pandey, S., Fartyal, D., Agarwal, A., Shukla, T., James, D., Kaul, T., Negi, Y.K., Arora, S., Reddy, M.K. (2017) Abiotic stress tolerance in plants: Myriad roles of ascorbate peroxidase. *Frontiers in Plant Science* 8: 581.

Papadakis, I.E., Bosabalidis, A.M., Soiropoulos, T.E., Therios, I.N. (2007) Leaf anatomy and chloroplast ultrastructure of Mn-deficient orange plants. *Acta Physiologiae Plantarum* 29(4): 297–301.

Passardi, F., Tognolli, M., De Meyer, M., Penel, C., Dunand, C. (2006) Two cell wall associated peroxidases from *Arabidopsis* influence root elongation. *Planta* 223(5): 965–974.

Peleg, Z., Blumwald, E. (2011) Hormone balance and abiotic stress tolerance in crop plants. *Current Opinion in Plant Biology* 14(3): 290–295.

Pérez-Clemente, R.M., Vives, V., Zandalinas, S.I., López-Climent, M.F., Muñoz, V., Gómez-Cadenas, A. (2013) Biotechnological approaches to study plant responses to stress. *Biomed Research International* 2013, ID 654120: 1–10. Doi: 10.1155/2013/654120.

Peshev, D., Van den Ende, W. (2013) Sugars as antioxidants in plants. In *Crop Improvement under Adverse Conditions*, ed. Tuteja, N., Gill, S.S. Springer-Verlag, Berlin–Heidelberg, Germany, pp. 285–308.

Pilon-Smits, E.A.H., Quinn, C.F., Tapken, W., Malagoli, M., Schiavon, M. (2009) Physiological functions of beneficial elements. *Current Opinion in Plant Biology* 12(3): 267–274.

Polle, A. (2001) Dissecting the superoxide dismutase-ascorbate peroxidase-glutathione pathway in chloroplasts by metabolic modeling. Computer simulations as a step towards flux analysis. *Plant Physiology* 126(1): 445–462.

Procházková, D., Sairam, R.K., Srivastava, G.C., Singh, D.V. (2001) Oxidative stress and antioxidant activity as the basis of senescence in maize leaves. *Plant Science* 161(4): 765–771.

Pujni, D., Chaudhary, A., Rajam, M.V. (2007) Increased tolerance to salinity and drought in transgenic indica rice by mannitol accumulation. *Journal of Plant Biochemistry and Biotechnology* 16(1): 1–7.

Radhakrishnan, R., Lee, I.J. (2013) Spermine promotes acclimation to osmotic stress by modifying antioxidant, abscisic acid, and jasmonic acid signals in soybean. *Journal of Plant Growth Regulation* 32(1): 22–30.

Rao, M.V., Lee, H.I., Creelman, R.A., Mullet, J.E., Davis, K.R. (2000) Jasmonic acid signaling modulates ozone-induced hypersensitive cell death. *The Plant Cell* 12(9): 1633–1646.

Reczek, C.R., Chandel, N.S. (2015) ROS-dependent signal transduction. *Current Opinion in Cell Biology* 33: 8–13.

Rejeb, K.B., Abdelly, C., Savouré, A. (2014) How reactive oxygen species and proline face stress together. *Plant Physiology and Biochemistry* 80: 278–284.

Roberson, E., Firestone, M. (1992) Relationship between desiccation and exopolysaccharide production in soil *Pseudomonas* sp. *Applied and Environmental Microbiology* 58(4): 1284–1291.

Rosa, S.B., Caverzan, A., Teixeira, F.K., Lazzarotto, F., Silveira, J.A., Ferreira-Silva, S.L., Abreu-Neto, J., Margis, R., Margis-Pinheiro, M. (2010) Cytosolic APx knockdown indicates an ambiguous redox responses in rice. *Phytochemistry* 71(5–6): 548–558.

Roxas, V.P., Lodhi, S.A., Garrett, D.K., Mahan, J.R., Allen, RD. (2000) Stress tolerance in transgenic tobacco seedlings that overexpress glutathione S-transferase/glutathione peroxidase. *Plant and Cell Physiology* 41(11): 1229–1234.

Sadak, M.S.H., Abdelhamid, M.T., Schmidhalter, U. (2014) Effect of foliar application of amino acids on plant yield and some physiological parameters in bean plants irrigated with seawater. *Acta Biológica Colombiana* 20(1): 141–152.

Sadiq, M., Akram, N.A., Ashraf, M. (2017) Foliar application of alpha-tocopherol improves the composition of fresh pods of *Vigna radiate* subjected to water deficiency. *Turkish Journal of Botany* 41(3): 244–252.

Saeidnejad, A.H., Kafi, M., Pessarakli, M. (2016) Interactive effects of salinity stress and Zn availability on physiological properties, antioxidants activity and micronutrients' content of wheat (*Triticum aestivum*) plants. *Communications in Soil Science and Plant Analysis* 47(8): 1048–1057.

Samuel, M.A., Miles, G.P., Ellis, B.E. (2000) Ozone treatment rapidly activates MAP kinase signalling in plants. *The Plant Journal* 22: 367–376.

Sandhya, V., Ali Sk, Z., Grover, M., Reddy, G., Venkateswarlu, B. (2009) Alleviation of drought stress effects in sunflower seedlings by exopolysaccharides producing Pseudomonas putida strain P45. *Biology and Fertility of Soils* 46(1): 17–26.

Sandhya, V, Ali Sk, Z., Grover, M., Reddy, G., Venkateswarlu, B. (2010) Effect of plant growth promoting Pseudomonas spp. on compatible solutes, antioxidant status and plant growth of maize under drought stress. *Plant Growth Regulation* 62(1): 21–30.

Sattler, S.E., Gilliland, L.U., Magallanes-Lundback, M., Pollard, M., DellaPenna, D. (2004) Vitamin E is essential for seed longevity and for preventing lipid peroxidation during germination. *The Plant Cell* 16(6): 1419–1432.

Sattler, S.E., Saffrané, L.M., Farmer, E.E., Krischke, M., Mueller, M.J., Penna, D.D. (2006) Nonenzymatic lipid peroxidation reprograms gene expression and activates defense markers in *Arabidopsis* tocopherol-deficient mutants. *The Plant Cell* 18(12): 3706–3720.

Seckin, B., Turkan, I., Sekmen, A.H., Ozfidan, C. (2010) The role of antioxidant defense systems at differential salt tolerance of *Hordeum marinum* Huds. (sea barley grass) and *Hordeum vulgare* L. (cultivated barley). *Environmental and Experimental Botany* 69(1): 76–85.

Semida, W., Taha, R., Abdelhamid, M., Rady, M. (2014) Foliar-applied α-tocopherol enhances salt-tolerance in *Viciafaba* L. plants grown under saline conditions. *South African Journal of Botany* 95: 24–31.

Sewelam, N., Jaspert, N., Van Der Kelen, K., Tognetti, V.B., Schmitz, J., Frerigmann, H., Stahl, E., Zeier, J., Van Breusegem, F., Maurino, V.G. (2014) Spatial H_2O_2 signaling specificity: H_2O_2 from chloroplasts and peroxisomes modulates the plant transcriptome differentially. *Molecular Plant* 7(7): 1191–1210.

Sewelam, N., Kazan, K., Schenk, P.M. (2016) Global plant stress signaling: Reactive oxygen species at the crossroad. *Frontiers in Plant Science* 7: 187.

Shalata, A., Neumann, P.M. (2001) Exogenouse ascorbic acid (vitamin C) increases Resistance to salt stress and reduced lipid peroxidation. *Experimental Botany* 52(346): 2207–2211.

Shan, C., Liang, Z. (2010) Jasmonic acid regulates ascorbate and glutathione metabolism in Agropyron cristatum leaves under water stress. *Plant Science* 178(2): 130–139.

Shao, H., Chu, L., Shao, M., Abdul Jaleel, C., Hong-Mei, M. (2008) Higher plant antioxidants and redox signaling under environmental stresses. *Comptes Rendus Biologies* 331(6): 433–441.

Sharma, P., Dubey, R.S. (2005) Drought induces oxidative stress and enhances the activities of antioxidant enzymes in growing rice seedlings. *Plant Growth Regulation* 46(3): 209–221.

Shi, J., Fu, X.Z., Peng, T., Huang, X.S., Fan, Q.J., Liu, J.H. (2010) Spermine pretreatment confers dehydration tolerance of citrus in vitro plants via modulation of antioxidative capacity and stomatal response. *Tree Physiology* 30(7): 914–922.

Shinozaki, K., Yamaguchi-Shinozaki, K. (1997) Gene expression and signal transduction in water-stress response. *Plant Physiology* 115(2): 327–334.

Siddiqui, M.H., Al-Whaibi, M.H., Sakran, A.M., Basalah, M.O., Ali, H.M. (2012) Effect of calcium and potassium on antioxidant system of *Vicia faba* L. under cadmium

stress. *International Journal of Molecular Sciences* 13(6): 6604–6619.

Singh, A.K., Dubey, R.S. (1995) Changes in chlorophyll a and b contents and activities of photosystems 1 and 2 in rice seedlings induced by NaCl. *Photosynthetica* 31: 489–499.

Sinha, P., Khurana, N., Nautiyal, N. (2012) Induction of oxidative stress and antioxidant enzymes by excess cobalt in mustard. *Journal of Plant Interactions* 35: 953–960.

Sinha, S., Saxena, R. (2006) Effect of iron on lipid peroxidation, and enzymatic and non-enzymatic antioxidant and bacosideA content in medicinal plant *Bacopa monnieri* L. *Chemosphere* 62(8): 134–135.

Slesak, I., Libik, M., Karpinska, B., Karpinski, S., Miszalski, Z. (2007) The role of hydrogen peroxide in regulation of plant metabolism and cellular signalling in response to environ mental stresses. *Acta Biochimica Polonica* 54(1): 39–50.

Smirnoff, N. (2000) Ascorbic acid: Metabolism and functions of a multifaceted molecule. *Current Opinion in Plant Biology* 3(3): 229–235.

Srivastava, R., Deng, Y., Howell, S.H. (2014) Stress sensing in plants by an ER stress sensor/transducer, bZIP28. *Frontiers in Plant Science* 5: 59.

Stahl, W., Sies, H. (2003) Antioxidant activity of carotenoids. *Molecular Aspects of Medicine* 24(6): 345–351.

Suzuki, N., Koussevitzky, S., Mittler, R.O.N., Miller, G.A.D. (2012) ROS and redox signalling in the response of plants to abiotic stress. *Plant Cell and Environment* 35(2): 259–270.

Swarbreck, S.M., Colaço, R., Davies, J.M. (2013) Plant calcium-permeable channels. *Plant Physiology* 163(2): 514–522.

Taiz, L., Zeiger, E. (2006) *Plant Physiology*, 4th ed. Sinaur Associates, Sunderland, MA.

Tang, W., Peng, X., Newton, R.J. (2005) Enhanced tolerance to salt stress in transgenic loblolly pine simultaneously expressing two genes encoding mannitol-1-phosphate dehydrogenase and glucitol-6-phosphate dehydrogenase. *Plant Physiology and Biochemistry* 43(2): 139–146.

Teixeira, W.F., Fagan, E.B., Soares, L.H., Umburanas, R.C., Reichardt, K., Neto, D.D. (2017) Foliar and seed application of amino acids affects the antioxidant metabolism of the soybean crop. *Frontiers in Plant Science* 8: 327.

Tognetti, V.B., Van Aken, O., Morreel, K., Vandenbroucke, K., Van De Cotte, B., De Clercq, I., Chiwocha, S., Fenske, R., Prinsen, E., Boerjan, W., et al. (2010) Perturbation of indole-3-butyric acid homeostasis by the UDP-glucosyltransferase UGT74E2 modulates *Arabidopsis* architecture and water stress tolerance. *The Plant Cell* 22(8): 2660–2679.

Triantaphylidès, C., Havaux, M. (2009) Singlet oxygen in plants: Production, detoxification and signaling. *Trends in Plant Science* 14(4): 219–228.

Van Breusegem, F., Dat J.F. (2006) Reactive oxygen species in plant cell death. *Plant Physiology* 141(2): 384–390.

Van den Ende W, Valluru R. (2009) Sucrose, sucrosyl oligosaccharides, and oxidative stress: Scavenging and salvaging? *Journal of Experimental Botany* 60(1): 9–18.

Vandenabeele, S., Van Der Kelen, K., Dat, J., Gadjev, I., Boonefaes, T., Morsa, S., Rottiers, P., Slooten, L., Van Montagu, M., Zabeau, M., et al. (2003) A comprehensive analysis of H_2O_2-induced gene expression in tobacco. *Proceedings of the National Academy of Sciences* 100(26): 16113–16118.

Verma, V., Ravindran, P., Kumar, P.P. (2016) Plant hormone-mediated regulation of stress responses. *BMC Plant Biology* 16(1): 86.

Verslues, P.E., Sharma, S. (2010) Proline metabolism and its implications for plant-environment interaction. *The Arabidopsis Book* 8: e0140.

Wang, C.Q., Li, R.C. (2008) Enhancement of superoxide dismutase activity in the leaves of white clover (Trifolium repens L.) in response to polyethylene glycol-induced water stress. *Acta Physiologiae Plantarum* 30(6): 841.

Wang, Y., Ying, Y., Chen, J., Wang, X. (2004) Transgenic *Arabidopsis* overexpressing Mn-SOD enhanced salt-tolerance. *Plant Science* 167(4): 671–677.

Wani, S.H., Kumar, V., Shriram, V., Kumar Sah, S. (2016) Phytohormones and their metabolic engineering for abiotic stress tolerance in crop plants. *The Crop Journal* 4(3): 162–176.

Willekens, H., Chamnongpol, S., Davey, M., Schraudner, M., Langebartels, C., Van Montagu, M., Inzé, D., Van Camp, W. (1997) Catalase is a sink for H_2O_2 and is indispensable for stress defence in C_3 plants. *EMBO Journal* 16(16): 4806–4816.

Wingler, A., Von Schaewen, A., Leegood, R.C., Lea, P.J., Paul Quick, W. (1998) Regulation of leaf senescence by cytokinin, sugars, and light. *Plant Physiology* 116(1): 329.

Wolucka, B.A., Goossens, A., Inze, D. (2005) Methyl jasmonate stimulates the de novo biosynthesis of vitamin C in plant cell suspensions. *Journal of Experimental Botany* 56(419): 2527–2538.

Xiang, C., Oliver, D.J. (1998) Glutathione metabolic genes coordinately respond to heavy metals and jasmonic acid in *Arabidopsis*. *The Plant Cell* 10(9): 1539–1550.

Xiong, L., Schumaker, K.S., Zhu, J.K. (2002) Cell signaling during cold, drought, and salt stress. *Plant Cell* 14: 65–83.

Yogaratnam, N., Sharples, R.O. (1982) Supplementing the nutrition of Bramley's seedling apple with phosphorus sprays II. Effects on fruit composition and storage quality. *Journal of Horticultural Science* 57(1): 53–59.

You, J., Chan, Z. (2015) ROS Regulation during abiotic stress responses in crop plants. *Frontiers in Plant Science* 6: 1092.

Yuan, F., Yang, H.M., Xue, Y., Kong, D.D., Ye, R., Li, C.J., Zhang, J.Y., Theprungsirikul, L., Shrift, T., Krichilsky, B., Johnson, D.M. (2014) OSCA1 mediates osmotic-stress-evoked Ca^{2+} increases vital for osmosensing in *Arabidopsis*. *Nature* 514(7522): 367–71.

Yun, K.Y., Park, M.R., Mohanty, B., Herath, V., Xu, F., Mauleon, R., Wijaya, E., Bajic, V.B., Bruskiewich, R., de los Reyes, B.G. (2010) Transcriptional regulatory network triggered by oxidative signals configures the early response mechanisms of japonica rice to chilling stress. *BMC Plant Biology* 10: 16.

Zhai, C.Z., Zhao, L., Yin, L.J., Chen, M., Wang, Q.Y., Li, L.C., Xu, Z.S., Ma, Y.Z. (2013) Two wheat glutathione peroxidase genes whose products are located in chloroplasts improve salt and H_2O_2 tolerances in *Arabidopsis*. *PLoS ONE* 8(10): e73989.

Zhang, J., & Kirkham, M. B. (1996) Antioxidant responses to drought in sunflower and sorghum seedlings. *New Phytologist*, 132(3), 361–373.

Zhang, F., Wan, X., Zheng, Y., Sun, L., Chen, Q., Zhu, X., Guo, Y., Liu, M. (2014) Effects of nitrogen on the activity of antioxidant enzymes and gene expression in leaves of Populus plants subjected to cadmium stress. *Journal of Plant Interactions* 9(1): 599–609.

Zhu, J.K. (2016) Abiotic stress signaling and responses in plants. *Cell* 167(2): 313–324.

Źróbek-Sokolnik, A., Asard, H., Górska-Koplińska, K., Górecki, R.J. (2009) Cadmium and zinc-mediated oxidative burst in tobacco BY-2 cell suspension cultures. *Acta Physiologiae Plantarum* 31(1): 43–49.

16 Coordination and Auto-Propagation of ROS Signaling in Plants

Suruchi Singh, Abdul Hamid, Madhoolika Agrawal, and S.B. Agrawal

CONTENTS

16.1 INTRODUCTION

To endure different environmental challenges, plants have evolved complicated protective mechanisms. Plants have evolved sophisticated acclimation and defense mechanisms that can be activated in the primary tissue(s) directly exposed to stress, as well as in distal portions not directly challenged. The activation of defense or acclimation responses in systematic or non-challenged tissues is termed systematic acquired resistance (SAR) or systematic acquired acclimation (SAA), respectively, and both play an important role in checking the progression of infection to the entire plant (Shah and Zeier, 2013). Reactive oxygen species (ROS) play a central role in plant defense. Superoxide anion ($O_2^{•-}$), hydrogen peroxide (H_2O_2) and hydroxyl ($•OH$) are the three major forms of ROS. Although ROS represent partially reduced or excited forms of atmospheric O_2 that can react with different cellular components and cause their oxidation (Halliwell and Gutteridge, 2015). ROS are continuously produced in a different compartment of the cells as an unavoidable by-product of aerobic metabolism but are also produced as a result of activation of specific ROS-producing enzymes that mediate signal transduction processes (König et al., 2012). ROS play an intrinsic role as signaling molecules in regulating numerous biological processes and responses to biotic/abiotic stimuli in plants (Baxter et al., 2013), and the signaling network is highly conserved (Mittler et al., 2011). Nevertheless, enhanced levels of ROS overwhelming the cellular antioxidant defense system are implicated in the damage of lipids, proteins, and DNA (Dröge, 2002). ROS were previously thought to lead to cell death as the result of ROS directly killing cells by oxidation via oxidative stress, which is now considered to be the result of ROS triggering a physiological or programmed pathway for cell death. Major ROS producing compartments, related enzymes and their properties in plants are given in Table 16.1. Harnessing the toxic properties of ROS helps in fighting off the stress, and this can be considered as a major evolutionary success. Therefore, the present chapter deals majorly with the role of ROS in regulating various vital biological processes.

16.2 ROS: A DOUBLE-EDGED SWORD

ROS (e.g. $O_2^{•-}$, H_2O_2, $•OH$, $^1O_2$) represent partially reduced or excited forms of atmospheric oxygen that can react with different cellular components and cause their oxidation (Halliwell and Gutteridge, 2015). ROS are continuously formed in different compartments of the cell as an unavoidable by-product of aerobic metabolism but are also found as a result of activation of specific ROS-production enzymes that mediate signal transduction processes (König et al., 2012). The evolutionary analysis of the ROS gene network in plants suggests that ROS-scavenging mechanisms were acquired before ROS-producing mechanisms and that plants first learned to control their intracellular ROS levels and afterwards started using ROS for signaling purposes. An imbalance between formation and detoxification of ROS leads to a metabolic state referred to as oxidative stress. Oxidative stress happens when elevated levels of oxidizing agents are able to abstract an electron from essential organic molecules and disturb cellular function. When ROS production overwhelms antioxidative capacity, cell damage and finally cell death occur. In plants, the major ROS-producing compartments are the chloroplast, which mainly produces $O_2^{\bullet-}$, H_2O_2 and $^1O_2$ as a by-product of photosynthesis; the mitochondrion, which mainly produces H_2O_2 as a by-product of respiration; the peroxisome, which mainly produces H_2O_2 as a by-product of photorespiration; and the apoplast, which mainly produces $O_2^{\bullet-}$ (and subsequently H_2O_2) as a key signaling molecule generated by NADPH oxidases, termed respiratory burst oxidase homolog (RBOH), peroxidases and superoxide dismutases (SODs; Mignolet-Spruyt et al., 2016). Although ROS were originally perceived to be toxic by-products of aerobic metabolism that must be metabolized to check the oxidative destruction of the cell, more recently, it has been discovered that a basal level of ROS is actually required to support life (Mittler, 2017).

The quantity of ROS in cells, therefore, needs to be maintained above a low cytostatic level, but below a high cytotoxic level (Schieber and Chandel, 2014). Thus,

it appears that ROS is indispensable for plant signaling and metabolism, and this gives an indication that life on Earth most likely evolved in the presence of ROS (Mittler, 2017).

16.3 OXIDATIVE STRESS-INDUCED VERSUS DELIBERATE ROS PRODUCTION

Under natural conditions, ROS are unavoidably produced through incomplete reduction of O_2 to H_2O_2 as a consequence of plants' metabolic processes such as respiration and photosynthesis. ROS are potentially toxic, and their overproduction as a result of the responses of the plants to biotic/abiotic stress causes an imbalance in the cell's redox state commonly that may lead to cell death (Møller and Sweetlove, 2010). Plants utilize complex antioxidant machinery to maintain redox homeostasis (Møller et al., 2007; Apel and Hirt, 2004). However, it is now universally accepted that ROS are not exclusively deleterious, but they are also key components in fundamental plant processes, such as physiological responses to environmental changes and the modulation of protein activity or gene expression (Møller et al., 2007; Apel and Hirt, 2004). ROS are important signaling molecules and are generated not only as by-products of other cell processes but also deliberately by enzymatic complexes.

TABLE 16.1
Properties, Localization, and the Mode of Action of Hydrogen Peroxide, Superoxide Anion, and Singlet Oxygen

ROS	$T1/2^a$	Migration Distance	Localization	Mechanism	Mode of Action
Superoxide ($O_2^{\bullet-}$)	1–4 µs	30 nm	Chloroplast Mitochondrion, plasma membrane, peroxisome, cell wall	Photosynthesis ET and PSI or II, respiration ET, excited chlorophyll, NADH oxidase, xanthine oxidase, peroxidase, Mn^{2+}, and NADH	Reacts with Fe-S proteins and dismutates to H_2O_2
Hydrogen peroxide (H_2O_2)	>1 ms	>1 µm	Peroxisome, apoplast, cell wall	Glycolate oxidase, peroxidase, amine oxidase, oxalate oxidase, fatty acid-β-oxidation	Attacks the cysteine and methionine residues of protein. Also, reacts with heme proteins and with DNA
Singlet oxygen ($^1O_2$)	1–4 µs	30 nm	Membrane, chloroplast, nuclei	Excited chlorophyll	Oxidizes lipids and proteins
Hydroxyl ion (•OH)	1 ns	1 nm	Chloroplast	Reacts with iron (Fenton reaction) and H_2O_2	Reacts with almost all biomolecules

[a] half life.

As per previous reports (Apel and Hirt, 2004; Nanda et al., 2010), the complexes involved in ROS generation are:

- Class III peroxidases
- Oxalate oxidases
- Amine oxidases
- Lipoxygenases
- Quinone reductase
- Plant NADPH oxidases (NOXs)

Plant NOXs, RBOHs, play a key role in ROS production (Suzuki et al., 2011). RBOHs are homologs to the catalytic subunit (gp91phox) of mammalian phagocyte NOXs (Nanda et al., 2010). RBOHs constitute a multigene family with ten *AtRboh* genes in the model plant *Arabidopsis*. All RBOH proteins present the same domain structure, with a core C-terminal region that contains six transmembrane domains supporting two haem groups and the functional oxidase domain responsible for the superoxide generation. The haem groups are required for the transfer of electron across the membrane to oxygen, the extracellular (EC) acceptor to generate the superoxide radical (Lambeth, 2004). RBOHs have an N-terminal region; this extension contains regulatory regions such as calcium-binding EF-hands and phosphorylation domains important for the function of the plant oxidases. The structure of the RBOHD, N-terminal is given as Figure 16.1.

RBOH's enzymatic function is to produce ROS for physiological and developmental purposes. The tissue-specific localization of transcript is given in Figure 16.2. Cellular fractionation of plant tissue indicates that RBOH proteins are localized into the plasmalemma membrane (Sagi and Fluhr, 2001; Simon-Plas et al., 2002). The binding sites of different RBOHs are given in Figure 16.3. The homology for different RBOHs from *Arabidopsis* is given in Figure 16.4.

ROS thus formed can function as cellular second messengers that are likely to modulate many different proteins, leading to a variety of responses (Mori and Schroeder, 2004). However, an enzymatic dismutation step must first take place to produce a more stable H_2O_2 derivative from the superoxide that is required for a variable long-range cell-to-cell signal, which can also can easily pass through membranes (Allan and Fluhr, 1997). The superoxide product is membrane impermeable due to its negative charge in the ambient condition of pH. However, under the condition of exceptionally low pH, the superoxide can be protonated and has been shown to functionally cross yeast membrane compartments (Wallace et al., 2004). In plants, the physiological range of EC pH is 5.0, in which 16% of the superoxide would be in the membrane permeable hydroperoxyl (HO_2) form. Thus, the external pH status will moderate the compartmentalization of superoxides produced by RBOH outside the membrane and perhaps enable the participation of cytoplasmic SOD in the formation of H_2O_2. RBOH is involved in regulating various other processes by interacting with many other proteins as is shown in the co-expression network (Figure 16.5). The description of different components, along with their involvement in other biological processes is given in Table 16.2.

16.4 SPECIFICITY OF ROS SIGNALING

There are many diverse opinions on ROS-induced signaling. According to some, the signaling is specific, but others oppose this notion. ROS themselves are the most likely signals of oxidative stress. Different forms of ROS exist, with entirely different molecular properties; for instance, superoxide is a charged molecule under most physiological conditions and cannot passively transfer across the membrane. Favorably, superoxide can be easily converted into H_2O_2, which can readily transfer membranes, passively or through water channels (Miller et al., 2010). Singlet oxygen and the hydroxyl radical are also not very suitable candidates because they are too reactive to carry signals over longer distances (μm) and act as specific signals. This leaves H_2O_2 as the most suitable ROS to act as a messenger (Neill et al., 2002), partly because of its relative stability and partly because it can cross membranes through aquaporins (Bienert et al., 2007). The most acceptable H_2O_2-induced regulation is its involvement in the oxidation of the transcription

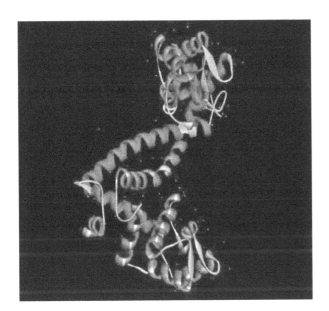

FIGURE 16.1 Structural view of *Arabidopsis* RBOHD N-Terminal.

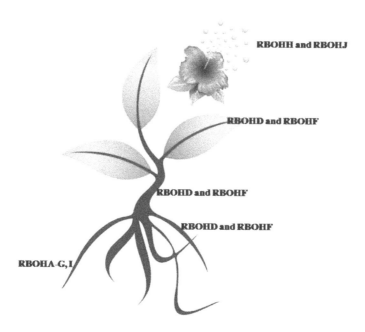

FIGURE 16.2   Localization of different types of RBOH in different parts of plant.

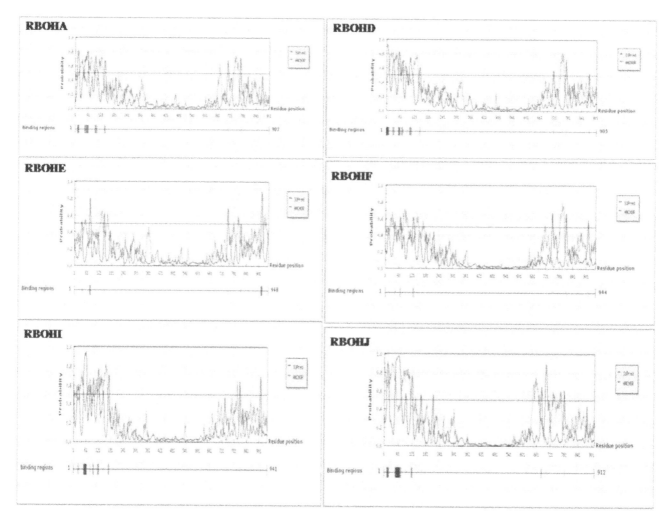

FIGURE 16.3   Binding sites of different types of RBOH proteins.

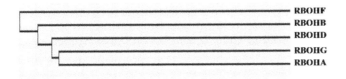

FIGURE 16.4   Homology between different types of RBOH.

factor by affecting its DNA-binding affinity (Imlay, 2008; Lee et al., 2009). In prokaryotic cells with only one compartment, H_2O_2 could be a probable messenger, where stress at any point will affect the entire organism and where all stress-responsive genes need to be activated to deal with the problem. However, in a eukaryotic cell, more complicated organization requires more coordinated signals. H_2O_2 is a simple molecule, which cannot store or transmit complex information (Neill et al., 2002; Kleine et al., 2009).

Also, its interaction partner cannot recognize its point of origin, i.e. an H_2O_2 molecule will have similar properties whether it is derived from a mitochondrion, a chloroplast, a peroxisome, or the apoplast. The information content could be perceived as magnitude and pattern of frequency and amplitude of H_2O_2 waves, similar to that proposed for Ca^{2+} (McAinsh and Hetherington, 1998). Given that many enzymes, both in the organelles and in the cytosol, remove H_2O_2, it is very difficult to see how that could work.

The following points highlight the specific nature of ROS-induced signaling:

- ROS are mainly used as a general signal to facilitate or activate the cellular signaling networks of cells and that other signals function together with ROS to convey specificity. These other signals could be small peptides, hormones, lipids, cell wall fragments and others.
- The ROS signal carries within it a decoded message, which is like calcium signals that have specific oscillation patterns within defined cellular locations.
- The specific features of the signal could be perceived and decoded by specialized mechanisms to trigger specific gene expression patterns.
- Each cellular compartment or individual cell has its own set(s) of ROS receptors to decode ROS signals generated within it, which are then transferred by other networks such as calcium and/or protein phosphorylation.

An interesting example of high specificity in ROS signaling is the analysis of double mutants deficient in APX1 and CAT2 in *Arabidopsis* and APX1 and CAT1 in tobacco. It was later found that the combined lack of APX1 and CAT2 in *Arabidopsis* causes unique ROS signatures in cells that trigger a novel acclimation response involving the activation of DNA repair, cell cycle control and antiprogrammed cell death mechanisms (Vaahtera et al., 2014).

16.5   ROS AND ITS INTERACTING GENE NETWORK

The ROS-interacting gene network of plants includes all the genes that encode not only ROS-detoxifying but also ROS-producing proteins in the cell. Figure 16.6 depicts different processes that contribute to ROS production and their influence on various processes and vice-versa. In *Arabidopsis*, the basic ROS gene network is composed of over 150 genes (Mittler et al., 2004), including

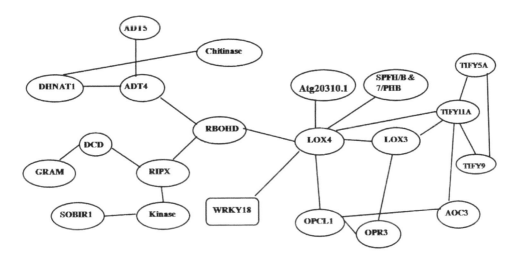

FIGURE 16.5   Correlation network analysis of different RBOH-influenced networks.

TABLE 16.2

Description of the Factors Involved in the Networks Influenced By Respiratory Burst Oxidase Homolog D (RBOHD)

Gene	Description	Biological Process
TIFY11A	Jasmonate ZIM domain protein 5	• Ethylene biosynthetic process • JA biosynthetic process • Hyperosmotic salinity • Abscisic acid and JA-mediated signaling pathway
TIFY5A	Jasmonate ZIM domain protein 8	• Ethylene biosynthetic process • JA biosynthetic process • Response to wounding • Response to JA stimulus
TIFY9	Jasmonate ZIM domain protein 10	• Regulation of JA-mediated signaling pathway • Regulation of systemic acquired resistance • JA biosynthetic process • ABA mediated signaling process
SPFH/B and 7/PHB domain containing membrane associated protein family		• Ethylene biosynthetic process • Response to chitin • Vacuole
Chitinase		• Chitinase activity

ADT4	Arogenate dehydratase 4	• L-phenylalanine biosynthetic process
		• Positive regulation of flavonoid biosynthetic process
		  • Vernalization response
		  • Anthocyanin accumulation in tissue in response to UV-B
		  • Response to karrikin

ADT5	Arogenate dehydratase 5	• L-phenylalanine biosynthetic process • Coumarin biosynthetic process • Vernalization response • Anthocyanin accumulation in tissue in response to UV-B • Carpel development
DHNAT1	1,4-dihydroxy-2-naphthoyl CoA	• Phylloquinone biosynthetic process
OPCL1	OPC-8:0 CoA ligase	• JA biosynthetic process • Phenylpropanoid metabolic process • Response to wounding • Response to chitin • Response to JA stimulus
OPR3	Oxophytodienoate-reductase 3	• Response to ozone • Tryptophan catabolic process • Indole acetic acid biosynthetic process • JA biosynthetic process • JA metabolic process
AOC3	Allene oxide cyclase	• Defense response by callose deposition • JA biosynthetic process • Negative regulation of programmed cell death • ABA mediated signaling pathway • Response to wounding
WRKY18	WRKY-DNA binding protein 18	• Regulation of defense response to virus by host • Response to molecule of bacterial origin • Defense response to fungus • Defense response to bacterium • Response to chitin

(*Continued*)

TABLE 16.2 (CONTINUED)

Description of the Factors Involved in the Networks Influenced By Respiratory Burst Oxidase Homolog D (RBOHD)

Gene	Description	Biological Process
LOX4	Lipoxygenase 4	• Stamen filament development • Anther dehiscence • Response to ozone • Defense response by callose description • Anther development
LOX3	Lipoxygenase 3	• Stamen filament development • Anther dehiscence • Anther development • Ethylene biosynthetic process • JA biosynthetic process
RIPX	Protein kinase superfamily protein	• Cellular response to water deprivation • Response to insects • Galactolipid biosynthetic • Positive regulation of flavonoid biosynthetic process • N-terminal protein myristoylation
Kinase	Protein kinase family	• Response to insects • Response to molecules of bacterial origin • Systemic acquired resistance • Salicyclic acid-mediated signaling pathway • Negative regulation of defense response
DCD	Development and cell death domain protein	• N-terminal protein myristoylation • Negative regulation of programmed cell death • Endoplasmic reticulum unfolded protein response • MAPK cascade • ABA mediated signaling pathway
GRAM domain		• Heat acclimation
SOBIR1	Suppressor of BIR1 protein	• Negative regulation of floral organ abscission • Positive regulation of cell death • Transmembrane receptor protein tyrosine kinase signaling pathway • Protein phosphorylation

genes encoding iron (Fe), manganese (Mn), copper-zinc (CuZn)-SODs that metabolize $O_2^{\bullet-}$; APXs, catalases (CATs), glutathione peroxidases (GPXs), peroxiredoxins (PRXR), and other peroxidases that metabolize H_2O_2; enzymes such as monodehydroascorbate reductase, dehydroascorbate reductase and glutathione reductase, which participate in reducing/recycling the antioxidants, ascorbic acid and glutathione, and are also used in the removal of H_2O_2 and $O_2^{\bullet-}$; and thioredoxins and glutaredoxins, which participate in different redox reactions (Mittler et al., 2004). In addition, the ROS-interacting gene network of plants also includes genes that encode RBOHs and other ROS-producing enzymes such as xanthine, glycolate, and oxalate oxidases. The proteins encoded by the ROS gene network are localized in different cellular

compartments of the cell and mediate ROS production and scavenging. Many of the enzymes encoded by the ROS network function in concert for defined pathways for metabolizing ROS, such as the SOD-Asada-Foyer-Halliwell pathway found in chloroplasts, mitochondria, the cytosol, and peroxisomes. ROS gene networks also include genes that belong to several biosynthetic pathways that mediate the production of different low molecular weight antioxidants, such as ascorbic acid, glutathione and α-tocopherol, carotenoids, flavonols, and other phenolic compounds. Additional enzymes and proteins that play a key role in the network are those that control the level of labile iron in cells. Because iron (Fe^{2+}) can react with H_2O_2 (or dismutated $O_2^{\bullet-}$) to form the highly reactive hydroxyl radical ($\bullet$OH; the

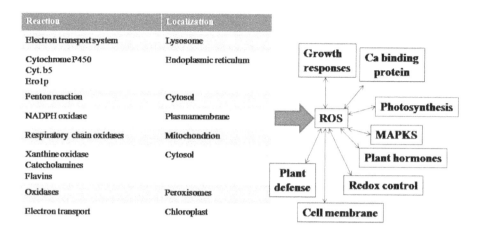

Reaction	Localization
Electron transport system	Lysosome
Cytochrome P450 Cyt. b5 Ero1p	Endoplasmic reticulum
Fenton reaction	Cytosol
NADPH oxidase	Plasmamembrane
Respiratory chain oxidases	Mitochondrion
Xanthine oxidase Catecholamines Flavins	Cytosol
Oxidases	Peroxisomes
Electron transport	Chloroplast

FIGURE 16.6 Different sources of ROS and influence of ROS on different processes and the inverse relationship.

Fenton reaction; Halliwell and Gutteridge, 2015), the level of labile iron in cells must be kept under tight control. This is especially critical under conditions of enhanced $O_2^{\bullet-}$ because $O_2^{\bullet-}$ can react with Fe-S proteins and cause the release of labile Fe from them. Thus, proteins such as ferritin that store Fe, different Fe uptake systems, and different Fe-S biogenesis and repair systems play a key role in a cell under ROS-induced damage (Gammella et al., 2016; López-Millán et al., 2016).

16.6 DELIBERATE ROS PRODUCTION LINKED WITH ROS WAVE PROPAGATION: INNATE IMMUNITY VIA A RELAY OF SIGNALS

The dynamic and rapid nature of ROS signaling in cells is an outcome of two opposite processes, i.e. ROS production and scavenging (Halliwell and Gutteridge, 2007; Foyer and Noctor, 2005). Because these two processes always run in a parallel manner in cells, the balance between scavenging and production rates is disturbed, so it results in rapid alterations in ROS levels, which will eventually generate a signal. In many biological systems, a burst of ROS often occurs as two distinguished peaks, which accompany several different signaling events (Nishimura and Dangl, 2010). Nevertheless, luciferase reporter genes expressed under the control of a rapid ROS-response promoter in plants (Miller et al., 2009) in a H_2O_2/redox-GFP sensor in zebrafish (Niethammer et al., 2009) revealed that the initial burst of ROS production could trigger a cascade of cell-to-cell communication events that results in the formation of a ROS wave that propagates through different tissues and carries the signal over long distances. Thus, the temporal concept of an ROS burst occurring in a cell could be modified into a temporal-spatial concept of an ROS wave. We can,

therefore, envision ROS signaling as a dynamic process that occurs within cells between different organelles, as well as between cells over long distances. Because plants have a high capacity to scavenge ROS, the long-distance aspect of ROS signaling can only be explained by the continuous production of ROS in individual cells along the path of the ROS wave/signal. Such a mechanistic view will imply that the ROS wave is auto-propagating. The initiation of the wave in specific cells must, therefore, be associated with a long-distance signal that causes individual cells along its path to activate ROS production via their own ROS-producing mechanisms.

Initiation of systemic signals upon the perception of high-light waves has been shown to be followed by propagation of plasma membrane electrical signals in a light wavelength-specific manner (Szechynska-Hebda et al., 2010). Membrane potential can be directly affected by ROS, and because the rate of certain electric signals in plants matches the rate of the ROS wave (Miller et al., 2009), it is possible that the generation of an ROS wave affects the formation, amplitude, and/or rate of the electric signal. ROS signaling is integrated with many different signaling networks in plants. These include protein kinase networks, calcium signaling, cellular metabolic networks, and redox responses. In some instances, ROS accumulation was found to precede the activation of signaling through these networks, whereas in other examples, ROS accumulation was found to be a direct result of signaling through these networks. A good example of an ROS-activated signaling network is the mitogen-activated protein kinase (MAPK) cascade. Many different MAPK cascades can be activated after ROS accumulation. These include the ROS-responsive MAPKKK, MEKK1, MPK4, and MPK6 (Teigen et al., 2004) cascades. The MEKK1 pathway is highly active during abiotic and oxidative stress conditions, and MEKK1 is an activator of two highly homologous MAPKKs (MKK1 and MKK2), which function upstream of the MAPKs, MPK4, and MPK6 (Xing et al., 2008). MEKK1 has been suggested to be specifically required for the activation of MPK4 by H_2O_2 (Nakagami et al., 2006).

Direct coupling of ROS signaling with the primary cellular metabolism is a key feature of ROS signaling in cells. ROS accumulation, for example, can cause an inhibition of the tricarboxylic acid cycle in mitochondrion and upregulation of glycolysis and oxidative pentose phosphate pathways (Baxter et al., 2007). In chloroplasts, ROS signaling is coupled to the redox state of the plastoquinone (PQ) pool and plays an important role in the response of plants to changes in environmental conditions (Pfannschmidt et al., 2008). Plants optimize their photosynthetic activity by regulating the association of light-harvesting complexes with thylakoids and by adjusting photosystem stochiometry to rearrange the balance of excitation energy (Pesaresi et al., 2009). A chloroplast sensor kinase has also recently been shown to be required for the regulation of gene expression in chloroplasts in response to changes in the redox state of electron carriers connecting the two photosystems (Puthiyaveetil et al., 2008).

16.7 COORDINATION OF ROS SIGNALING IN PLANTS: FIBERS OF THE FABRIC

ROS signaling is temporally and spatially coordinated in plants, and this leads to rapid changes in the level, composition, and structure of various metabolites, proteins, and RNA molecules. The ROS wave plays a key role in propagating signals from local to systemic tissue and thus in acclimation. The initial perception of abiotic stress-induced release of ROS initiates a cascade of cell-to-cell communication, which propagates through different tissues of the plant over a longer distance (Miller et al., 2009). Early signaling events include ion-fluxes across plasma membrane, increased Ca^{2+} levels in the cytosol, activation of MAPKs, and production of ROS (Finka et al., 2012). There are ample benefits in having ROS as a signaling molecules; for instance, there is tight regulation over the subcellular localization of ROS signals in cells, and they could be used as rapid long-distance auto-propagating signals transferred throughout the plants. Also, ROS induced signaling can be easily integrated with several other signaling pathways, such as calcium and protein phosphorylation networks (Ogasawara et al., 2008). Another key signaling advantage of ROS is their connection with cellular homeostasis and metabolism. Almost any change in cellular homeostasis could lead to a change in the steady-state level of ROS in a particular compartment(s) and vice-versa. Preceding the signal

transduction, there are many physiological and metabolic changes that occurred. For instance, a PM channel can initiate on inward flow of Ca^{2+} after perceiving heat stress, and this Ca^{2+} signal can be related to the regulatory mechanism of ROS-producing enzymes (Mittler et al., 2012).

Phosphoproteomic approaches unveiled the role of MAPKs in inducing phosphorylation of membrane proteins, which include ion channels, calmodulins, protein kinases, protein phosphatases, and proteins associated with auxin signaling as well as RBOHD (Benschop et al., 2007). Under different stresses, different mechanisms were proposed to explain the reason for the development of acclimatory responses in plants. For instance, the early response of plants to high light was studied using transgenic plants expressing a luciferase reporter gene under the control of an APX1, APX2, and ZAT10 promoter. The experiment demonstrated activation of acclimatory responses both in the organ of perception as well as in the distal organs (Karpinski et al., 1999; Rossel et al., 2007; Szechynska-Hebda et al., 2010). These were high-light-induced redox changes in the PQ pool, increased production of ethylene and ROS, reduction of photochemical efficiency and non-photochemical quenching, and changes in extracellular electric potential (Szechynska-Hebda et al., 2010; Rossel et al., 2007; Karpinski et al., 1999).

An instant response to mechanical wounding was also found. In *Arabidopsis*, elevated levels of jasmonic acid accumulated not only in damaged tissue but also in undamaged systemic leaves (Glauser et al., 2009; Koo and Howe, 2009). RBOHD is required for the initiation and amplification of a rapid auto-propagating systemic signal. RBOHD-dependent long-distance signals play an important biological role in the systemic acquired acclimation response of plants to heat or high light (Suzuki et al., 2013).

Rossel et al. (2007) compared the transcriptome in directly challenged and distal organs. About 70% of transcripts upregulated in a directly challenged organ is common to the systemic organ under high-light-mediated SAA. Many studies further confirmed similarities between local and systemic tissue responses to high light (Mühlenbock et al., 2008; Miller et al., 2009; Szechynska-Hebda et al., 2010).

16.8 CONCLUSIONS

Sessile plants can make fine adjustments of physiological and morphological responses to survive. Plants have evolved complicated and integrated cell communication circuits for survival in particular environment conditions. Reactive oxygen species (ROS) are key players in propagating the messages from environment-to-cell,

cell-to-cell, and cell-to-organelles. But the cellular machinery is also well equipped with ROS-scavenging antioxidants, which continually work to bring the ROS level down. ROS below the toxic level are harnessed to fight off invasions from the environment, while ROS at and/or above the toxic level are destructive, damage cells, and lead to cell death.

ACKNOWLEDGMENT

The authors are thankful to the Head, Department of Botany and Coordinator, Centre of Advanced Study in Botany, Banaras Hindu University, Varanasi and DST FIST, Banaras Hindu University, and Council of Scientific and Industrial Research, SERB, APN, and CICERO.

REFERENCES

Allan, A.C., Fluhr, R. (1997) Two distinct sources of elicited reactive oxygen species in tobacco epidermal cells. *The Plant Cell* 9: 1559–1572.

Apel, K., Hirt, H. (2004) Reactive oxygen species: Metabolism, oxidative stress, and signal transduction. *Annual Review of Plant Biology* 55: 373–399.

Baxter, C.J., Redestig,H., Schaufer,N., Repsilbee, D., Patil, K.R., Nielsen, J., Selbig, J., Liu, J., Fernie, A.R., Sweetlove, L.J. (2007) The metabolic response of heterotrophic Arabidopsis cells to oxidative stress. *Plant Physiology* 143, 312–325.

Baxter, A., Mittler, R., Suzuki, N. (2013) ROS as key players in plant stress signaling. *Journal of Experimental Botany* 65: 1229–1240.

Benschop, J.J., Mohammed, S., O'Flaherty, M., Heck, A.J.R., Slijper, M., Menke, F.L.H. (2007) Quantitative phosphoproteomics of early elicitor signaling in *Arabidopsis. Molecular and Cellular Proteomics* 6: 1198–1214.

Bienert, G.P., Møller, A.L., Kristiansen, K.A., Schulz, A., Møller, I.M., Schjoerring, J.K., John, T.P. (2007) Specific aquaporins facilitate the diffusion of hydrogen peroxide across membranes. *Journal of Biological Chemistry* (282): 1183–1192.

Dröge, W. (2002) Free radicals in the physiological control of cell function. *Physiological Reviews* 82: 47–95.

Finka, A., Cuendet, A.F.H., Maathuis, F.J.M., Saidi, Y., Goloubinoff, P. (2012) Plasma membrane cyclic nucleotide gated calcium channels control land plant thermal sensing and acquired thermotolerance. *The Plant Cell* 24: 3333–3348.

Foyer, C.H., Noctor, G. (2005) Redox homeostasis and antioxidant signaling: A metabolic interface between stress perception and physiological responses. *The Plant Cell* 17: 1866–1875.

Gammella, E., Recalcati, S., Cairo, G. (2016) Dual role of ROS as signals and stress agents: Iron tips the balance in favor toxic effects. *Oxidative Medicine and Cellular Longevity* 2016, 8629024.

Glauser, G., Dubugnon, L., Mousavi, S.A.R., Rudaz, S., Wolfender, J.L., Farmer, E.E. (2009) Velocity estimates for signal propagation leading to systemic jasmonic acid accumulation in wounded *Arabidopsis. Journal of Biological Chemistry* 284: 34506–34513.

Halliwell, B., Gutteridge, J.M.C. (2007) Antioxidant defences: Endogenous and diet derived. *Free Radicals in Biology and Medicine* 4: 79–186.

Halliwell, B., Gutteridge, J.M.C. (2015) *Free Radicals in Biology and Medicine.* Oxford University Press, USA.

Imlay, J.A. (2008) Cellular defenses against superoxide and hydrogen peroxide. *Annual Review of Biochemistry* 77: 755–776.

Karpinski, S., Reynolds, H., Karpinska, B., Wingsle, G., Creissen, G., Mullineaux, P. (1999) Systemic signaling and acclimation in response to excess excitation energy in *Arabidopsis. Science* 284: 654–657.

Kleine, T., Voigt, C., Leister, D. (2009) Plastid signaling to the nucleus: Messengers still lost in the mists? *Trends in Genetics* 25 (4): 185–192.

König, J., Muthuramalingam, M., Dietz, K.J. (2012) Mechanisms and dynamics in the thiol/disulfide redox regulatory network: Transmitters, sensors and targets. *Current Opinions in Plant Biology* 15: 261–268.

Koo, A.J.K., Howe, G.A. (2009) The wound hormone jasmonate. *Phytochemistry* 70: 1571–1580.

Lambeth, J.D. (2004) NOX enzymes and the biology of reactive oxygen. *Nature of Reviews Immunology* 4: 181–189.

Lee, S.C., Lan, W., Buchannan, B.B., Luan, S. (2009) A protein kinase-phosphatase pair interacts with ion channels to regulate ABA signaling in plant guard cells. Proceedings of the National Academy of Sciences, USA 102: 4203–4208.

López-Millán, A.F., Duy, D., Philippar, K. (2016) Chloroplast iron transport proteins-function and impact on plant physiology. *Frontiers Plant Science* 7: 1–12.

McAinsh, M.R., Hetherington, A.M. (1998) Encoding specificity in Ca^{2+} signaling systems. *Trends in Plant Science* 3: 32–36.

Mignolet-Spruyt, L., Xu, E., Idänheimo, N., Hoeberichts, F.A., Mühlenbock, P., Brosché, M., Van Breusegem, F., Kangasjärvi, J. (2016) Spreading the news: Subcellular and organellar reactive oxygen species production and signaling. *Journal of Experimental Botany* 67: 3831–3844.

Miller, E.W., Dickinson, B.C., Chang, C.J. (2010) Aquaporin-3 mediates hydrogen peroxide uptake to regulate downstream intracellular signaling. *Proceedings of the National Academy of Sciences USA* 107: 15681–15686.

Miller, G., Schlauch, K., Tam, R., Cortes, D., Torres, M.A., Shulaev, V., Dangl, J.L., Mittler, R. (2009) The plant NADPH oxidase RBOHD mediates rapid systemic signaling in response to diverse stimuli. *Science Signaling* 2: ra45–ra45.

Mittler, R. (2017) ROS are good. *Trends in Plant Science* (22): 11–19.

Mittler, R., Finka, A., Goloubinoff, P. (2012) How do plants feel the heat? *Trends in Biochemical Sciences* 37: 118–125.

Mittler, R., Vanderauwera, S., Gollery, M., Van Breusegem, F. (2004) Reactive oxygen gene network of plants. *Trends in Plant Science* 9: 490–498.

Mittler, R., Vanderauwera, S., Suzuki, N., Miller, G., Tognetti, V.B., Vandepoele, K., Gollery, M., Shulaev, V., Van Breusegem, F. (2011) ROS signaling: The new wave? *Trends in Plant Science* 16: 300–309.

Møller, I.M., Jensen, P.E., Hansson, A. (2007) Oxidative modifications to cellular components in plants. *Annual Review of Plant Biology* 58: 459–481.

Møller, I.M., Sweetlove, L.J. (2010) ROS signaling–specificity is required. *Trends in Plant Science* 15: 370–374.

Mori, I.C., Schroeder, J.I. (2004) Reactive oxygen species activation of plant Ca^{2+} channels. A signaling mechanism in polar growth, hormone transduction, stress signaling, and hypothetically mechanotransduction. *Plant Physiology* 135: 702–708.

Mühlenbock, P., Szechyńska-Hebda, M., Płaszczyca, M., Baudo, M., Mateo, A., Mullineaux, P.M., Parker, J.E., Karpińska, B., Karpiński, S. (2008) Chloroplast signaling and LESION SIMULATING DISEASE1 regulate crosstalk between light acclimation and immunity in *Arabidopsis*. *The Plant Cell* 20: 2339–2356.

Nakagami, H., Soukupová, H., Schikora, A., Zárský, V., Hirt, H. (2006) A mitogen-activated protein kinase kinase kinase mediates reactive oxygen species homeostasis in *Arabidopsis*. *Journal of Biological Chemistry* 281: 38697–38704.

Nanda, A.K., Andrio, E., Marino, D., Pauly, N., Dunand, C. (2010), Reactive oxygen species during plant-microorganism early interactions. *Journal of Integrative Plant Biology* 52: 195–204.

Neill, S.J., Desikan, R., Clarke, A., Hurst, R.D., Hancock, J.T. (2002) Hydrogen peroxide and nitric oxide as signalling molecules in plants. *Journal of Experimental Botany* 53: 1237–1247.

Niethammer, P., Grabher, C., Look, A.T., Mitchison, T.J. (2009) A tissue-scale gradient of hydrogen peroxide mediates rapid wound detection in zebrafish. *Nature* 459: 996–999.

Nishimura, M.T., Dangl, J.L. (2010) *Arabidopsis* and the plant immune system. *The Plant Journal* 61. 1053–1066.

Ogasawara, Y., Kaya, H., Hiraoka, G., Yumoto, F., Kimura, S., Kadota, Y., Hishinuma, H., Senzaki, E., Yamagoe, S., Nagata, K., Nara, M., Suzuki, K., Tanokura, M., Kuchitsu, K. (2008) Synergistic activation of the *Arabidopsis* NADPH oxidase AtrbohD by Ca^{2+} and phosphorylation. *Journal of Biological Chemistry* 283: 8885–8892.

Pesaresi, P., Hertle, Pribil, M., Kleine, T., Wagner, R., Strissel, H., Ihnatowicz, A., Bonardi, V., Scharferberg, M., Schneider, A., Pfannschmidt, T., Leister, D. (2009) *Arabidopsis* STN7 kinase provides a link between short- and long-term photosynthetic acclimation. *The Plant Cell* 21: 2402–2423.

Pfannschmidt, T., Bräutigam, K., Wagner, R., Dietzel, L., Schröter, Y., Steiner, S., Nykytenko, A. (2008) Potential regulation of gene expression in photosynthetic cells by redox and energy state: Approaches towards better understanding. *Annals of Botany* 103: 599–607.

Puthiyaveetil, S., Kavanagh, T.A., Cain, P., Sullivan, J.A., Newell, C.A., Gray, J.C., Robinson, C., Van der Giezen, M., Rogers, M.B., Allen, J.F. (2008) The ancestral symbiont sensor kinase CSK links photosynthesis with gene expression in chloroplasts. *Proceedings of the National Academy of Sciences USA* 105: 10061–10066.

Rossel, J.B., Wilson, P.B., Hussain, D., Woo, N.S., Gordon, M.J., Mewett, O.P., Howell, K.A., Whelan, J., Kazan, K., Pogson, B.J. (2007) Systemic and intracellular responses to photooxidative stress in *Arabidopsis*. *The Plant Cell* 19: 4091–4110.

Sagi, M., Fluhr, R. (2001) Superoxide production by plant homologues of the gp91phox NADPH oxidase. Modulation of activity by calcium and by tobacco mosaic virus infection. *Plant Physiology* 126: 1281–1290.

Schieber, M., Chandel, N.S. (2014) ROS function in redox signaling and oxidative stress. *Current Biology* 24: R453–R462.

Shah, J., Zeier, J. (2013) Long-distance communication and signal amplification in systemic acquired resistance. *Frontiers in Plant Science* 22: 1–16.

Simon-Plas, F., Elmayan, T., Blein, J.P. (2002) The plasma membrane oxidase NtrbohD is responsible for AOS production in elicited tobacco cells. *The Plant Journal* 31: 137–147.

Suzuki, N., Miller, G., Morales, J., Shulaev, V., Torres, M.A., Mittler, R. (2011) Respiratory burst oxidases: The engines of ROS signaling. *Current Opinion in Plant Biology* 14: 691–699.

Suzuki, N., Miller, G., Salazar, C., Mondal, H.A., Shulaev, E., Cortes, D.F., Shuman, J.L., Luo, X., Shah, J., Schlauch, K., Shulaev, V., Mittler, R. (2013) Temporal-spatial interaction between reactive oxygen species and abscisic acid regulates rapid systemic acclimation in plants. *The Plant Cell* 25: 3553–3569.

Szechyńska-Hebda, M., Kruk, J., Górecka, M., Karpińska, B., Karpiński, S. (2010) Evidence for light wavelength-specific photoelectrophysiological signaling and memory of excess light episodes in *Arabidopsis*. *The Plant Cell* 22: 2201–2218.

Teigen, M., Scheikl, E., Eulgem, T., Doczi, R., Ichimura, K., Schinozaki, K., Dangl, J.L., Hert, H. (2004) The MKK2 pathway mediates cold and salt stress signaling in *Arabidopsis*. *Molecular Cell* 15: 141–152.

Vaahtera, L., Brosché, M., Wrzaczek, M., Kangasjärvi, J. (2014) Specificity in ROS signaling and transcript signatures. *Antioxidants and Redox Signaling* 21: 1422–1441.

Wallace, M.A., Liou, L.L., Martins, J., Clement, M.H.S., Barley, S., Longo, V.D., Valentine, J.S., Gralla, E.D. (2004) Superoxide inhibits 4Fe-4S cluster enzymes involved in amino acid biosynthesis. *The Journal of Biological Chemistry* 279: 32055–32062.

Xing, Y., Jia, W., Zhang, J., (2008) AtMKK1 mediates ABA-induced LAT1 expression anf H_2O_2 production via AtMPK6-coupled signaling in *Arabidopsis*. *Plant Journal* 54: 440–451.

17 Regulation of Osmolytes Syntheses and Improvement of Abiotic Stress Tolerance in Plants

Ambuj Bhushan Jha and Pallavi Sharma

CONTENTS

17.1 INTRODUCTION

Plants being sessile organisms are frequently exposed to different kinds of abiotic stresses, such as salinity, drought, chilling, heat and metal toxicity. These stresses are results of either natural processes or various human activities and present major challenges in attaining sustainable food production (Godfray et al., 2010). Abiotic stresses can affect many important physiological and metabolic processes such as carbohydrate and lipid metabolism, carbon partitioning, photosynthesis, protein synthesis and oxidative and osmotic homeostasis, leading to reduced plant growth and productivity (Sharma and Dubey, 2011). They can affect plant metabolism in both general and specific ways (Gueta-Dahan et al., 1997; Beck et al., 2007). Signal transduction cascades are activated in response to slight changes of growth conditions, which in turn activate stress-responsive genes and ultimately lead to changes at the physiological, biochemical and molecular levels (Figure 17.1) (Sharma and Dubey, 2005, 2007, 2011; Kumar et al., 2008; Gupta et al., 2009).

Plants cope with abiotic stresses by employing combinations of stress avoidance and tolerance mechanisms. Stress responses can vary considerably with the type of stress and genotype, but some general reactions, such as accumulation of osmolytes, oxidative stress due to enhanced reactive oxygen species (ROS) production and induction of an antioxidant defense system, occur in all genotypes and appear as the common consequences of exposure to abiotic stresses (Sharma et al., 2012; Gill and Anjum, 2014; Anjum et al., 2016). Osmolytes are highly soluble, naturally accumulated low-molecular-weight organic compounds and are present in a wide range of organisms, including bacteria, algae, plants and animals (Yancey et al., 1982; Schwacke et al., 1999). Osmolytes can be classified into three chemical classes: sugars and polyols, amino acids or quaternary ammonium compounds and tertiary sulfonium compounds (Delauney and Verma, 1993; Rhodes and Hanson, 1993). Besides their major role as osmoprotectants, osmolytes act as antioxidants, signaling molecules, and play an important role in providing redox balance, stabilizing macromolecules, including proteins and membranes, and regulating cellular osmotic adjustment (Yancey, 2005; Slama et al., 2015). They are also known as compatible solutes as they do not interact with macromolecules even at high concentrations and can be safely regulated (up or down) with very little influence on a cell's functions (Brown and Simpson, 1972; Yancey et al., 1982). As a part of plants' strategy for adaptation under adverse environmental conditions, a higher accumulation of compatible solutes during abiotic stress conditions could be

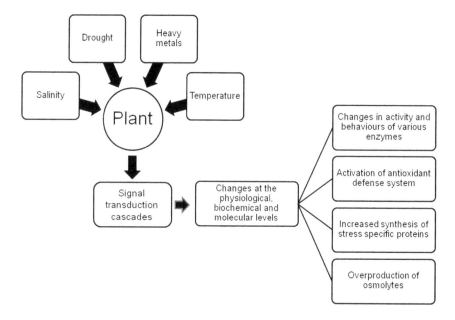

FIGURE 17.1 Abiotic stress induced responses in plants.

attained either by increasing biosynthesis and uptake or by decreasing degradation and export (Yancey et al., 1982; Schwacke et al., 1999; Yancey, 2005). Exogenous applications of osmolytes to plants under abiotic stress conditions have led to amelioration of the adverse effects of abiotic stresses in plants.

Abiotic stresses present major challenges to crop production. Development of abiotic stress-tolerant plants can help satisfy the world's growing food demands. Genetically engineered plants incorporating genes for biosynthesis of osmolytes have shown stress tolerance in some cases. Fine tuning of osmolyte synthesis and degradation in plants can be used to generate plants with higher abiotic stress tolerance. A greater understanding of the mechanisms of metabolic, molecular genetic, transcriptional, post-transcriptional, translational and post-translational regulation of osmolyte biosynthesis and degradation in response to various abiotic stresses will allow a wide application of biotechnology for enhancement of abiotic stress tolerance (Zarattini and Forlani, 2017). In this chapter, we discuss the roles and regulation of key osmolytes in plants during abiotic stresses and advances in engineering abiotic stress tolerant plants using genes involved in their metabolic pathway.

17.2 OSMOLYTES AND THEIR ROLES IN PLANT ABIOTIC STRESS TOLERANCE

Several osmolytes accumulate in plants when exposed to abiotic stresses. These osmolytes can be classified into four chemical classes: (i) sugars (e.g. trehalose and sucrose), (ii) sugar alcohols or polyols (sorbitol,

glycerol, mannitol etc.), (iii) amino acids (proline, glutamate etc.) and (iv) quaternary and tertiary sulfonium compounds (e.g. glycine betaine (GB), alanine betaine, dimethylsulfoniopropionate; Delauney and Verma, 1993). Figure 17.2 shows the various classes of osmolytes. Initially, osmolytes were believed to be involved in the osmotic adjustment and protection of subcellular structure. Later, their absolute concentrations suggested a reassessment of their functional significance. Osmolytes are now known to maintain cell turgor and osmotic balance, protect membranes, proteins, enzymes and other subcellular structures and act as metal chelators, antioxidants and signaling molecules (Murmu et al., 2017). Various roles of osmolytes in plants under abiotic stresses are depicted in Figure 17.3.

17.2.1 SUGARS

Abiotic stress-induced alterations in soluble sugar content have been observed in several plant species (Miller and Langhans, 1990; Guy et al., 1992; Irigoyen et al., 1992; Sánchez et al., 1998; Dubey and Singh, 1999; Kerepesi and Galiba, 2000; Verma and Dubey, 2001; Ranganayakulu et al., 2013). Sugars such as sucrose, trehalose, glucose and fructose are key osmolytes accumulated in plants under abiotic stresses (Briens and Larher, 1982; Yuanyuan et al., 2009). Accumulation of these sugars has been suggested to play a critical role in conferring tolerance to various abiotic stresses by regulating several physiological processes such as seed germination, photosynthesis, flowering, senescence

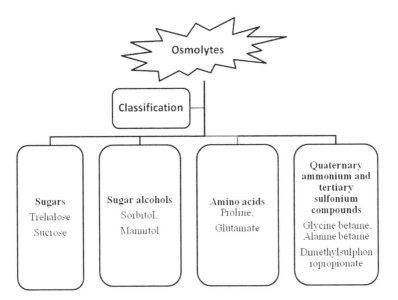

FIGURE 17.2 Classification of osmolytes.

etc. and determining the expression of sugar-regulated genes under unfavorable environmental conditions (Rathinasabapathi, 2000; Sami et al., 2016). Soluble sugars regulate osmotic adjustment, protect membranes and scavenge toxic ROS produced during various abiotic stresses and thus provide protection to plants against abiotic stresses (Williamson et al., 2002; Li et al., 2011; Boriboonkaset et al., 2012; Keunen et al., 2013; Singh et al., 2015). Under abiotic stresses, ROS accumulation is negatively correlated with sugar accumulation (Roitsch, 1999). High concentrations of sugars show the reverse

effect and thus sugar-led protection under abiotic stress is concentration dependent (Sami et al., 2016). In general, accumulation of soluble sugar varies with plant genotype and the stress factor. Cha-Um and Kirdmanee (2009) observed a higher amount of total soluble sugar in a salt-tolerant rice variety compared to the salt-sensitive variety and suggested that accumulation of osmolytes can provide resistance to salt-induced osmotic stress in rice. Increased amount of total sugar content under salt treatment has been suggested to be an important adaptive factor for salt tolerance improvement as it can play

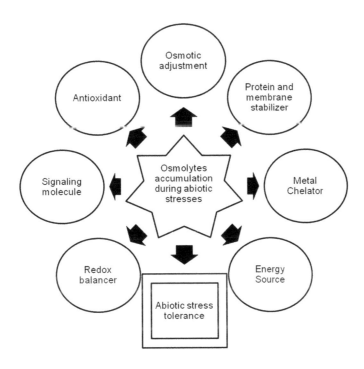

FIGURE 17.3 Roles of osmolytes under abiotic stress conditions.

an important role in translocation and/or compartmentation of Na$^+$ and Cl$^-$ (Liu and Van Staden, 2001). Jha and Dubey (2004) reported enhanced levels of reducing, non-reducing and total sugars in roots as well as shoots of rice cultivars Malviya-36 and Pant-12 under increasing concentration of arsenic in the growth media. Rice seedlings grown under 50 μM As$_2$O$_3$ showed about 46–93% increase in total sugars, 87–156% increase in reducing sugars and 3–56% increase in non-reducing sugars. Glucose and sucrose act as osmoprotectants and play a role in the maintenance of cell homeostasis. Under drought stress, Kerepesi and Galiba (2000) reported the higher accumulation and osmoprotective role of sugars (glucose, sucrose, fructose and fructans), whereas Pilon-Smits et al. (1999) reported the accumulation and protective role of fructan, the reserve carbohydrate in sugar beet. Higher accumulations and osmoprotective roles of glucose, sucrose, fructose and fructans were also reported under salinity by Kerepesi and Galiba (2000) and Murakeözy et al. (2003). Sugar beet plants subjected to osmotic stress accumulated a high concentration of sucrose in their storage root compared to control. It has been suggested that sugar beet adapts to osmotic stress by accumulating more osmolytes, such as Na+ and sucrose (Wu et al., 2016). Rosa et al. (2009) observed an increased amount of sucrose under metal-stressed conditions. *Salvinia natans*, the aquatic weed displayed about a threefold increase in sucrose content upon exposure to Cd, Cr and Zn, whereas exposure to Ni, Pb, Fe, Mn, Co and Cu resulted in about a twofold increase compared to control plants (Dhir et al., 2012).

Trehalose, a non-reducing sugar composed of two glucose units linked in an α,α-1,1-glycosidic linkage, is mainly present in bacteria, fungi and desiccation-tolerant plants (Vinocur and Altman, 2005). It plays a crucial role in abiotic stress tolerance in plants (Garg et al., 2002; Han et al., 2016). Trehalose is found only in trace amounts in angiosperms, and abiotic stresses cause only a moderate increase in its level (Fougere et al., 1991; Garg et al., 2002; Kempa et al., 2008). Its accumulation has mostly been observed in desiccation-tolerant "resurrection plants" (Garg et al., 2002). Trehalose serves as a source of carbon and energy, and as a protective and signaling molecule under abiotic stresses. It regulates osmotic balance, protects membranes as well as proteins and scavenges ROS under various abiotic stresses (Ashraf and Harris, 2004; Koyro et al., 2012; Keunen et al., 2013; Singh et al., 2015). The protective role of trehalose was reported in rice seedlings under various abiotic stresses (Nounjan et al., 2012; Theerakulpisut and Gunnula, 2012). Enhanced synthesis and accumulation of trehalose in root nodules of legumes, *Medicago*

truncatula and *Phaseolus vulgaris* (López et al., 2008), and cassava (Han et al., 2016) was observed under osmotic stress conditions. Trehalose level was highly correlated with dehydration stress tolerance of detached leaves of the cassava varieties. In response to osmotic stress, trehalose concentrations increased 2.3-to-5.5-fold in cassava leaves (Han et al., 2016). Contrary to these findings, the osmotic regulatory role of trehalose was doubtful in alfalfa, a closely related species where low accumulation was observed under salt stress (Fougere et al., 1991). Therefore, it is still uncertain whether a low accumulation of trehalose will provide sufficient protection against stresses.

17.2.2 Sugar Alcohols

Based on their structure, sugar alcohols or polyols are divided into cyclic forms (e.g. myo-inositol and pinitol) and acyclic or linear forms (e.g. sorbitol, mannitol), and accumulation of these polyols facilitates osmotic balance, supports redox regulation and protects membranes, proteins, and enzymes (Murakeözy et al., 2003; Koyro et al., 2012; Singh et al., 2015; Slama et al., 2015). Sugar alcohols are also suggested to act as ROS scavengers and thus inhibit lipid peroxidation and cell damage (Smirnoff and Cumbes, 1989). As sugar alcohols possess water-like hydroxyl groups, they may mimic water structure and therefore maintain an artificial sphere of water molecule around macromolecules (Schobert, 1997).

Stress tolerance in various plants species has been correlated with accumulation of the straight-chain polyols such as mannitol and sorbitol (Stoop et al., 1996). Mannitol, a six-carbon polyol, is found commonly in plants and fungi. An increased amount of mannitol was observed in *Fraxinus excelsior* L. (Guicherd et al., 1997; Patonnier et al., 1999) and olive trees (Dichio et al., 2009) under water deficit conditions. Celery exhibited significant tolerance to salinity, which can be correlated with the presence of mannitol and the capability to accumulate a higher amount under osmotic stress (Kann et al., 1993; Everard et al., 1994; Stoop and Pharr, 1994a). *Salvinia* exposed to heavy metals, Co, Ni, Cd and Cr, showed a two-to-four-times increase in mannitol level compared to control plants (Dhir et al., 2012). Increased mannitol production under stresses has been suggested to play an important role in regulation of coenzymes and cells protection from lipid peroxidation by scavenging free radicals (Prabhavathi and Rajam, 2007; Sickler et al., 2007). Sorbitol, also a six-carbon linear sugar alcohol, has been demonstrated to be associated with tolerance of abiotic stresses in plants. Under salt-stressed conditions, increased synthesis and accumulation of sorbitol was observed in *Bangiopsis subsimplex* (Eggert et al., 2007). Accumulation as well as the osmoprotective role of sorbitol was reported under water deficit in peach (Lo Bianco et al., 2000), cherry (Ranney et al., 1991) and apple (Wang and Stutte, 1992). Ahn et al. (2011) reported increased drought and salt tolerance in *Arabidopsis thaliana* by producing D-ononitol, a seven-carbon sugar alcohol. Methylated cyclitols are key osmolytes in plants. Pinitol, a seven-carbon sugar alcohol, is reported to be accumulated in plants subjected to abiotic stresses. In soybean, a two-to-three-times increase in pinitol accumulation in leaf blades was observed under drought stress (Streeter et al., 2001). Genotypes accumulating high concentrations of pinitol were found to be better adapted to dry areas of China. Although accumulation of all three of pinitol, sugars and proline was observed in leaf blades of soybean plants subjected to drought, the concentration of pinitol in stressed plants was higher than that of proline or sugars. Pinitol has been linked with drought tolerance in soybean (Streeter et al., 2001). In maritime pine populations from Morocco (Tamjoute), water stress led to accumulation of D-pinitol in aerial parts as well as roots of seedlings, whereas myo-inositol accumulation was observed only in roots (Nguyen and Lamant, 1988).

17.2.3 Amino Acids

Amino acids (proline, arginine, histidine, alanine, γ-aminobutyric acid, pipecolic acid, ornithine and citrulline) are important osmolytes that accumulate in plants under various abiotic stresses (Mansour, 2000; Sharma and Dietz, 2006; Hayat et al., 2012; Wu et al., 2016). Amino acids are precursors of proteins and play a key role in the metabolism and development of plants (Hayat et al., 2012). They are potential signaling and regulatory molecules, and precursors for several secondary and energy-associated metabolites, which play an important role in plant growth and adaptive responses to several abiotic stresses. Accumulation of glutamic acid, aspartic acid and glutamine in cotton (Hanower and Brzozowska, 1975); glutamic acid, arginine and ornithine in rice (Yang et al., 2000) and asparagine, serine, glycine and aspartic acid in maize (Slukhai and Shvedova, 1972; Thakur and Rai, 1982) have been reported under various abiotic stress conditions. Amino acids can act as metal chelators, antioxidants and signaling molecules (Kavi Kishor et al., 1995; Sharma and Dietz, 2006; Hayat et al., 2012). They also play an important role in maintaining the NADP+/NADPH ratios required for various parts of metabolism and providing the NAD+ and NADP+ for respiratory and photosynthetic processes, respectively

(Kavi Kishor et al., 1995; Hare et al., 1998). Ajithkumar and Panneerselvam (2013) reported increased concentrations of free amino acids at various stages of growth of *Setaria italica* under drought, whereas Ranganayakulu et al. (2013) reported increased levels of free amino acids in the seedlings of groundnut cultivars, JL-24 and K-134, grown under NaCl stress. Proline is an excellent osmolyte that accumulates in higher plants under abiotic stresses. It is highly soluble in water and possesses no charge at neutral pH. Even at high concentrations, it shows little or no disturbing effect on solvent–macromolecule interactions. It also protects macromolecules and subcellular structures, acts as a protein stabilizer, a metal chelator, a molecular chaperone, an antioxidative defense molecule and a signaling molecule during various stresses (Csonka and Hanson, 1991; Kavi Kishor et al., 2005; Hayat et al., 2012). Proline accumulation enhances cellular osmolarity, increasing water influx and providing the turgor required for cell expansion. It is also known to possess ROS scavenging and singlet oxygen quenching properties (Smirnoff and Cumbes, 1989). Wang et al. (2009) showed alleviation of Hg^{2+} toxicity in rice (*Oryza sativa*) pretreated with proline through scavenging ROS, such as H_2O_2. Under abiotic stress, it reduces the harmful effects of $O_2^{•-}$ and OH• on PSII of isolated thylakoid membranes (Matysik et al., 2002). It stabilizes DNA and membranes and prevents protein aggregation and stabilizes protein during osmotic stress, heavy metal toxicity and extreme temperatures (Sharma and Dubey, 2004). *In vitro*, proline has been shown to reduce enzyme denaturations caused due to heat, NaCl stress, etc. Proline regulates pH and redox balance. It reduces cytoplasmic acidosis and maintains an appropriate ratio of NADP+ and NADPH (Hare and Cress, 1997). Proline synthesis produces NADP+, whereas its degradation generates NADPH. Therefore, synthesis and degradation of proline is required for redox buffering in the plastids and cytosol of cells. Enhanced NAD+ and NADP+ generated during proline synthesis under abiotic stress can maintain respiration and photosynthesis processes. In aqueous media, proline forms hydrophilic colloids, which can interact with a protein backbone. It can also form a hydration layer around phospholipids (Rajendrakumar et al., 1994). Proline can enhance the solubility of proteins that are sparingly soluble depending on the proteins' nature and concentration of proline. It is suggested that hydrophobic portion of proline interacts with the hydrophobic area of proteins and thus increases the hydrophilic area of proteins (Schobert and Tschesche, 1978). The activities of many enzymes such as catalase, peroxidase and polyphenol oxidase were shown to be stimulated by proline. During recovery from various abiotic stresses, proline serves as a reserve source of nitrogen, carbon and energy (Zhang et al., 1997). When the stress period is over, speedy degradation of proline can provide enough reducing agents to support oxidative phosphorylation in mitochondria (Kavi Kishor et al., 2005).

Higher proline accumulation under various abiotic stresses has been reported in several higher plants (Ashraf, 1994; Kavi Kishor et al., 1995; Ali et al., 1999; Öztürk and Demir, 2002; Rhodes et al., 2002; Hsu et al., 2003). For example, higher proline accumulation was reported under drought stress in rice plants (Hsu et al., 2003). Osmotic stress of -1.0 MPa and -1.5 MPa led to increased osmolyte concentrations in the leaf blade, leaf petiole and storage root of sugar beet, suggesting that this plant can adapt to osmotic stress by accumulating more osmolytes, including proline (Wu et al., 2016). Higher transcript levels of the mRNA for P5CS and proline accumulation were observed in the salt-tolerant rice variety Dee-gee-woo-gen (DGWG) compared to salt-sensitive rice variety IR28 (Igarashi et al., 1997) under salinity stress, suggesting a correlation of salt tolerance with expression of the *P5CS* gene and proline accumulation in rice. Increased level of proline content was reported under various metal stresses, such as Cd, in *Lactuca sativa* (Costa and Morel, 1994), Zn and Cu in *Lemna minor* (Bassi and Sharma, 1993), Cu in rice (Chen et al., 2001), and Cu and Zn in *Scenedesmus* (Tripathi and Gaur, 2004). High amounts of proline accumulated in *Silene vulgaris* under metal stresses and accumulation varied under various metals, with Cu producing the highest proline, followed by Zn and Cd (Schat et al., 1997).

17.2.4 QUATERNARY AMMONIUM AND TERTIARY SULFONIUM COMPOUNDS

Quaternary ammonium compounds which are quaternary ammonium cation salts such as proline betaine, β-alanine betaine, GB, pipecolate betaine, hydroxyproline betaine and choline-O-sulfate act as osmolytes and alleviate abiotic stress in plants. Betaine is synthesized from the oxidation of choline, a low-accumulating metabolite, restricting the capability of most plants to produce it. Among many quaternary ammonium compounds, GB (*N,N,N*-trimethyl glycine), which consists of a hydrocarbon moiety with three methyl groups, is one of the most widely found osmolytes under abiotic stresses in many crop plants, such as barley, spinach, sugar beet, wheat and sorghum (Fallon and Phillips, 1989; McCue and Hanson, 1990; Hanson et al., 1991; Rhodes and Hanson, 1993; Venkatesan and Chellappan, 1998; Mansour, 2000;

Guo et al., 2009; Lokhande and Suprasanna, 2012). GB, an amphoteric compound that is highly water soluble, remains electrically neutral over a wide range of physiological pHs and interacts with both hydrophobic and hydrophilic domains of macromolecules including enzymes and protein complexes. GB acts as cellular osmolyte and increases intracellular osmolarity under abiotic stress-induced hyperosmotic conditions. *In vitro*, GB has been shown to maintain membrane integrity and stabilize the activities of enzymes and structures of protein under unfavorable conditions (Gorham, 1995). Due to its 'preferential hydration activity' GB stabilizes the protein's native structure. It maintains a hydration shell around the protein's surface by preferentially excluding from contact with a protein (Arakawa and Timasheff, 1983). In contrast, Schobert (1977) suggested that GB stabilizes proteins by directly interacting with the hydrophobic part of the proteins. GB is found primarily in chloroplasts and maintains the efficiency of photosynthesis by protecting thylakoid membranes (Robinson and Jones, 1986; Genard et al., 1991). In higher plants, GB stabilizes and increases the oxygen-evolving activity of the PSII protein complexes. It inhibits dissociation of extrinsic proteins (the, 18 kD, 23 kD and 33 kD proteins) which are regulatory in nature from the intrinsic components of the PS II. Coordination of the manganese cluster to the protein cleft is stabilized by GB. Further, under salt stress, GB is also proposed to stabilize conformations of many enzymes involved in photosynthesis, such as FBPase, Rubisco, PRKase and FBP aldolase. It maintains these enzymes in a functionally active state (Yang et al., 2008).

GB is found abundantly in the chloroplast and plays a key role in maintaining photosynthetic efficiency by protecting thylakoid membranes (Robinson and Jones, 1986; Genard et al., 1991). The greater capacity of *Salicornia europaea* for osmotic adjustment evident in higher turgor and relative water content was correlated with higher accumulation of GB and Na^+ (Moghaieb et al., 2004). GB provided tolerance to tobacco plants exposed to Cd toxicity through protection of cellular components as well as increasing the activity of antioxidant enzymes (Islam et al., 2014). An increased amount of GB was observed in groundnut cultivars under salinity stress (Ranganayakulu et al., 2013) and in *Setaria italica* under drought stress (Ajithkumar and Panneerselvam, 2014). Similarly, accumulation of GB was reported in aquatic and terrestrial plant species such as *Phragmites* and *Spartina* (Al-Garni, 2006; Islam et al., 2009) under stressful conditions. The concentration of accumulated GB has been correlated with the salt tolerance (Saneoka et al., 1995). The levels of GB increased in *S. natans*, the aquatic weed, when exposed to heavy metals Ni, Co, Cd, Fe, Cu, Zn and Pb and this enhancement could play a significant role in combating heavy metal-induced osmotic stress (Dhir et al., 2012). Tolerant genotypes generally accumulated more GB compared to sensitive genotypes under stress conditions. Instead of accumulating proline, some halophytes, such as *Camphor osmaannua* or *Limonium* spp., accumulated carbohydrate or betaine-derived osmolytes (Gagneul et al., 2007). Accumulation of GB mitigates the adverse effect caused by cold stress by inducing expression of lipoxygenase and fatty acid desaturase 7 gene (Karabudak et al., 2014). GB has been shown to protect some enzymes and proteins against heat denaturation *in vitro* and increase accumulation of heat-shock protein (HSP) and heat tolerance in plants (Li et al., 2014).

Dimethylsulfoniopropionate (DMSP), a tertiary sulfonium compound, is mainly synthesized in algae and some higher plants, including *Spartina*, *Saccharum* spp. and *Wollastonia biflora* (Rhodes and Hanson, 1993; Gage and Rathinasabapathi, 1999; Otte et al., 2004). It plays an important role in detoxification of excess sulfur and ROS and deterrence of herbivores at the normal tissue level (Otte et al., 2004). In the algae *Thalassiosira pseudonana* and *Emiliana huxleyi* (Sunda et al., 2002), DMSP scavenges hydroxyl radicals and other ROS. DMSP is synthesized from methionine in all DMSP-producing plants, but the pathway by which DMSP is synthesized varies between plant groups and species (Stefels, 2000). Methionine is first transaminated to 4-methylthio-2-oxobutyrate in algae, while in higher plants methionine is first methylated to form S-methylmethionine (Trossat et al., 1998). In *W. biflora*, DMSP accumulates under salinity stress in chloroplasts and acts as an osmoprotectant (Trossat et al., 1998).

17.3 REGULATION OF KEY OSMOLYTES DURING ABIOTIC STRESS IN PLANTS

In plant system, the higher amounts of osmolytes accumulated in response to various stressful conditions have been attributed to either an increased rate of synthesis or a decreased rate of degradation (Yancey et al., 1982; Yoshiba et al., 1997; Nanjo et al., 1999). Several investigators have reported variations in sugar accumulation under various abiotic stresses. Regulation of sugars and/or sugar alcohols, including hexoses, sucrose, trehalose, mannitol, pinitol, myo-inositol and sorbitol, can play an important role in their accumulation during various abiotic stresses. In plants, calcineurin B-like protein-interacting protein kinases OsCIPK03 and OsCIPK12

and the MAPK kinase ZmMKK4 stimulate stress-led enhancement of soluble sugar content (Xiang et al., 2007; Kong et al., 2011). In source leaves, sucrose phosphate synthase (SPS) is a key component of carbohydrate metabolism and is involved in control of flux into sucrose. In sucrose synthesis pathway, SPS reversibly catalyzes the first step, i.e. transfer of the glycosyl group from an activated donor sugar to a sugar acceptor forming d-sucrose-6′-phosphate (S6P). S6P is then irreversibly dephosphorylated to sucrose by sucrose phosphate phosphatase (SPP). This is a central regulatory process in the production of sucrose in plants. In sink tissues, invertases cleave sucrose irreversibly and produce glucose and fructose, whereas sucrose synthases form fructose and UDP-glucose capable of undergoing reversible sucrose synthesis (Koch et al., 2000; Vargas and Salerno, 2010). These enzymes are regulated by phosphorylation, Glc-6-P, inorganic phosphate and transcriptional regulators (Harn et al., 1993; Klein et al., 1993; Huber and Huber, 1996; Chávez-Bárcenas et al., 2000; Winter and Huber, 2000). Changes in the activities of these enzymes resulted in increased sucrose content and accumulation of soluble carbohydrates (Daie and Patrick, 1988; Gupta et al., 1993a, b; Kaur et al., 1998). Ser-424 residue present in the SPS of spinach leaves is reported to be involved in osmotic stress-induced activation of this enzyme (Toroser and Huber, 1997). Sucrose synthase and invertase gene expressions have been reported to be regulated under abiotic stresses. In *Arabidopsis* leaves, *Sus1* transcript was differentially regulated when exposed to cold and drought stress, whereas *Sus2* mRNA was induced specifically by oxygen deficiency (Dejardin et al., 1999). An ABA-independent signal transduction pathway has been suggested to regulate *Sus1* expression via perception of a reduction in leaf osmotic potential under stresses. However, *Sus2* expression is independent of sugar osmoticum effects, suggesting a signal transduction mechanism different from *Sus1* expression regulation (Dejardin et al., 1999). Upregulation of extracellular invertase is a general response to a variety of abiotic stresses in plants (Roitsch et al., 2003). In mature leaves of maize, increased accumulation of hexoses was significantly correlated with enhanced activity of vacuolar invertase under drought but activity of cell wall invertase was unaffected (Pelleschi et al., 1999). Under water stress, vegetative sink and source organ-specific increase in acid invertase was significantly correlated with an enhancement in gene transcripts of *Ivr2* and vacuolar invertase proteins (Kim et al., 2000; Roitsch et al., 2003). Water stress leads to increase in ABA concentration and that has been suggested to be involved in regulation of invertase activity in soybeans (Ackerson, 1985).

Mannitol accumulation in plants shows considerable tolerance to stresses, and therefore accumulation of this compound could play a significant role in plants' defense mechanism under abiotic stresses (Stoop et al., 1996). In higher plants, mannitol is synthesized from triose-phosphate via formation of fructose-6-P, mannose-6-phosphate, and mannitol-1-phosphate by enzymatic action of mannose-6-phosphate isomerase (M6PI), NADPH-dependent mannose-6-phosphate reductase (M6PR) and mannose-1-phosphate phosphatase (M1PP), respectively (Rumpho et al., 1983; Loescher et al., 1992). The M6PR enzyme is mainly localized in the cytosol of the green palisade and spongy parenchyma tissues and bundle sheath cells of mature leaves, which are the main sites of mannitol synthesis (Loescher et al., 1992; Everard et al., 1994). Synthesis of mannitol is regulated by developmental stages of plants, with maximum activity observed in mature leaves. Mannitol accumulation under stress is subjected to specific regulation of mannitol synthesizing and catabolizing enzymes. In response to salinity stress, enzymatic adjustments apparently bring about the increased production of mannitol in photosynthetic tissue by way of increased M6PR activity and conservation of mannitol by way of decreased MDH activity in sink tissues (Stoop and Pharr, 1994b). About two-times increase in the activity of M6PR was observed in mature leaves of celery plants irrigated with 300 mM NaCl (Everard et al., 1994). Further, *Arabidopsis*, which normally doesn't produce mannitol, had shown high tolerance to salinity upon overexpressing gene encoding M6PR (Zhifang and Loescher, 2003). In celery plants, NAD-dependent mannitol dehydrogenase (MTD) oxidizes mannitol to mannose and therefore maintains mannitol pool size, which is important for its salt tolerance. It also provides the initial step through which mannitol is committed to central metabolism. In sink tissues, such as expanding leaves and petioles, and in root tips, MTD activity was strongly suppressed (near zero in root tips) when plants were grown at electrical conductivity of 11.9 dS.m^{-1} (Everard et al., 1994). Further, overexpression of the bacterial *mtlD* gene, which encodes mannitol 1-phosphate dehydrogenase and is involved in the conversion of mannitol 1-phosphate to mannitol through non-specific phosphatases led to a high level of tolerance to salinity as well as drought stress through the accumulation of mannitol (Bhauso et al., 2014).

In most of the angiosperms, trehalose occurs in trace amounts, and its level increases only moderately under abiotic stress (Fougere et al., 1991; Garg et al., 2002; Kempa et al., 2008). Trehalose is synthesized in two-step processes in which glucose-6-phosphate is converted to trehalose via trehalose-6-phosphate (T6P) by enzymes

trehalose-6-phosphate synthase (TPS) and trehalose-6-phosphate phosphatase (TPP) (Cabib and Leloir, 1958; Vandesteene et al., 2010; Ponnu et al., 2011). Several investigators have observed the induced expression of genes encoding enzymes involved in trehalose biosynthesis under chilling stress. For example, increased levels of *OsTPP1* and *OsTPP2* in rice (Pramanik and Imai, 2005; Shima et al., 2007), *VvTPPA* in grapevine (Fernandez et al., 2012) and *AtTPPA* in *A. thaliana* (Iordachescu and Imai, 2008) were observed under cold stress resulted in increased accumulation of T6P and trehalose. Similarly, *OsTPP1* was induced in rice under salt stress (Pramanik and Imai, 2005; Shima et al., 2007). A small rise in the amount of trehalose with higher TPS expression was reported in crop plants such as drought-tolerant wheat (El-Bashiti et al., 2005) and cotton varieties (Kosmas et al., 2006) grown under water-stressed conditions. Further, the osmoprotective role of trehalose was reported under salt stress in alfalfa by downregulating trehalase expression (López et al., 2008).

Sorbitol is an important osmolyte that accumulates in plant cells under abiotic stress. In sorbitol biosynthetic process, NADP dependent sorbitol-6-phosphate dehydrogenase (S6PDH) converts glucose-6-phosphate to sorbitol-6-phosphate, which is then converted to sorbitol by sorbitol-6-phosphate phosphatase (Hirai, 1981; Kanayama et al., 1992; Zhou et al., 2003). Increased accumulation of sorbitol and mRNA encoding S6PDH was noticed in apple leaves under osmotic, salt and low-temperature stresses, indicating a correlation between them. S6PDH is regarded as an ABA-inducible gene. ABA application increased S6PDH expression in apple leaves. Kanayama et al. (2006) suggested that ABA-mediated S6PDH is involved in sorbitol biosynthesis in different stress responses and participates in photosynthate translocation (Kanayama et al., 2006). Transformed Japanese persimmon with S6PDH cDNA from apple produced a high amount of sorbitol and showed greater salt tolerance (Gao et al., 2001). Myo-inositol, the most common inositol isomer, is synthesized from glucose-6-phosphate. It serves as a precursor for several metabolites involved in membrane biosynthesis, cell signaling and protection under stressful conditions (Loewus and Murthy, 2000; Sureshan et al., 2009). The enzyme myo-inositol-1-phosphate synthase (InPS) converts glucose-6-phosphate to myo-inositol-1-P, which is then converted to myo-inositol by myo-inositol-1-phosphate phosphatase (InPP) (Rathinasabapathi, 2000). InPS has been shown to be induced by salinity stress (Ishitani et al., 1996). In leaves of ice plant, this enzyme is expressed but in roots it is suppressed (Nelson et al., 1998). At least fivefold enhancement in *Inps1* RNA and a ten-times increase in free myo-inositol amounts was noticed in ice plants under salinity stress. In contrast, no upregulation of *Inps1* and myo-inositol accumulation was observed in *A. thaliana* under salt-stress suggesting differences in glycophytic and halophytic gene expression regulation for this osmolyte. Differential stress-induced upregulation of genes and novel enzymes detected in the ice plants indicates biochemical differences between plants (Ishitani et al., 1996). Various studies conducted in a halophyte and ice plants indicated two-step conversion of myo-inositol to the osmoprotectants D-ononitol and D-pinitol by myo-inositol O-methyltransferase (IMTI) and D-ononitol epimerase, respectively, and these conversions are regulated by stressful conditions (Adams et al., 1992; Vernon and Bohnert, 1992; Nelson et al., 1998). D-ononitol epimerase and myo-inositol O-methyltransferase are exclusive to the ice plant (Rathinasabapathi, 2000). Transgenic plants generated by incorporating genes for enzymes InPS, IMTI and D-ononitol epimerase from the common ice plant had high production of D-pinitol and D-ononitol (Ishitani et al., 1996; Rathinasabapathi, 2000). Further, transgenic tobacco plant accumulating D-ononitol due to expression of IMTl cDNA from ice plant showed higher drought and salt tolerance (Sheveleva et al., 1997).

Higher accumulation of free amino acids, mainly branched-chain amino acids (BCAAs) such as isoleucine, leucine and valine, plays a key role in plant abiotic stress tolerance mechanisms (Planchet and Limami, 2015). Leucine, valine and isoleucine are of equal importance in plants, with the involvement of four common enzymes in their biosynthetic pathways (Joshi et al., 2010). Similarly, isoleucine, methionine and threonine pathways are interconnected in plants. Metabolism of threonine and methionine under changed developmental and environmental conditions affects the availability of isoleucine as those amino acids serve as a substrate for isoleucine biosynthesis (Joshi et al., 2010). Three different synthesis pathways, the glutamate family pathway, the pyruvate family pathway and the aspartate family pathway, operate in plants, and these pathways are under tight regulation during stressful conditions (Planchet et al., 2015). These pathways produce different amino acids that act as compatible osmolytes under adverse environmental conditions. The glutamate family pathway produces proline and γ-aminobutyric acid, whereas the pyruvate family pathway produces alanine, leucine and valine (Planchet and Limami, 2015). Further, the aspartate family pathway produces energy through lysine catabolism.

Proline accumulation and its osmoprotective role under abiotic stresses have been reported in detail;

however, not much has been discussed on the effect of regulatory molecules and signals on the expression of genes involved in proline's biosynthetic pathway. In plants, the plastid and cytosol are the sites for proline biosynthesis, whereas degradation occurs in mitochondria (Ashraf and Foolad, 2007). Higher proline accumulation could be due to either activation of the biosynthetic pathway or inactivation of the degradation. In the proline biosynthetic process, pyrroline-5-carboxylate synthetase (P5CS) catalyzes the conversion of glutamate to glutamate semialdehyde (GSA), and GSA gets converted to pyrroline 5-carboxylate (P5C) by spontaneous cyclization (Kavi Kishor et al., 2005). P5C is reduced to proline by enzyme pyrroline-5-carboxylate reductase (P5CR). Alternatively, in plants, proline is also synthesized from arginine/ornithine (Adams and Frank, 1980; Bryan, 1990). The enzyme arginase converts arginine to ornithine. Ornithine is transaminated to P5C by orinithine-δ-aminotransferase (OAT) (Armengaud et al., 2004; Verbruggen and Hermans, 2008). Degradation of proline in mitochondria is catalyzed by the enzyme proline dehydrogenase or proline oxidase (PDH/PRODH), which converts proline to P5C, and then P5C dehydrogenase (P5CDH) produces glutamate from P5C (Joshi et al., 2010; Szabados and Savouré, 2010; Servet et al., 2012). P5CS is the rate-limiting enzyme for the biosynthesis, whereas PDH is the rate-limiting enzyme for the catabolism of proline (Ashraf and Foolad, 2007; Slama et al., 2015).

Biosynthesis of proline results due to upregulation of biosynthetic genes *P5CS* and *P5CR* and its accumulation under stress conditions could be due to de novo synthesis or decreased degradation or both (Hare et al., 1998). Gene expression of *P5CS* is strongly induced, and *PDH* is inhibited under dehydration conditions, whereas under rehydration conditions, expression of the *PDH* gene is strongly induced and *P5CS* is inhibited (Yoshiba et al., 1997). Thus, P5CS and PDH, which act during the biosynthesis and breakdown of proline are the rate-limiting factors under water stress and these results suggest the regulation of proline accumulation at the transcriptional level. In transgenic tobacco plants, overexpression of *P5CS* resulted in increased concentration of proline and resistance to drought stress (Gubiš et al., 2007). Similarly, Madan et al. (1995) reported increased activities of proline biosynthetic enzymes, P5CR and OAT, in tolerant lines and decreased activity of PDH, a proline-degrading enzyme, in *Brassica juncea* plants grown under stress conditions.

In *Arabidopsis*, two *P5CS* genes, *P5CS1* and *P5CS2*, encode P5CS, and these genes play diverse roles in the regulation of stress and developmental control

(Székely et al., 2008). Adverse conditions such as drought, salinity, and abscisic acid differentially regulate *P5CS* gene transcription. The important role of *P5CS1* in proline accumulation under stress was confirmed by characterization of *p5cs* insertion mutants. Further, knockout mutations of *P5CS1* resulted in decreased proline synthesis along with high susceptibility to salinity as well as ROS accumulation (Székely et al., 2008). Two *P5CS* genes, one housekeeping and one stress-specific *P5CS* isoform, control biosynthesis of proline. Despite sharing a high level of sequence homology in coding regions, transcriptional regulation is different for both *P5CS* genes (Strizhov et al., 1997; Armengaud et al., 2004; Xue et al., 2009). *P5CS1* is activated by light and repressed by brassinosteroids (Hayashi et al., 1997; Ábrahám et al., 2003), whereas *P5CS2* can be activated by avirulent bacteria, salicylic acid and ROS signals (Fabro et al., 2004).

Abscisic acid (ABA) mediated signals regulate the expression of *P5CS* and other genes involved in proline biosynthesis (Thomashow, 1999; Kavi Kishor et al., 2005). In *Arabidopsis*, the causal link between ABA and proline accumulation has been suggested and it has been shown that exogenous application of ABA increases the level of *AtP5CS1* and *AtP5CS2* transcripts (Savouré et al., 1997; Strizhov et al., 1997). In *Arabidopsis*, *P5CS1* expression is triggered by both ABA-dependent and ABA-insensitive as well as H_2O_2-derived signaling pathways (Savouré et al., 1997; Strizhov et al., 1997; Yoshiba et al., 1997; Verslues et al., 2007). Bandurska et al. (2017) observed involvement of ABA in regulating proline synthesis at *P5CS* transcription as well as P5CS enzyme level in drought-stressed barley plants. They also observed that accumulated proline provided resistance to barley genotypes. Different pathways regulate *Arabidopsis P5CS* transcript accumulation under cold and osmotic stress. The signals during salt stress appear to be mediated by ABA that can bring about the expression of stress-related genes and subsequently the synthesis of organic osmolytes. Analysis of the promoter elements in the proline biosynthetic pathway genes such as *AtP5CS1*, *AtP5CS2* and *AtP5CR* indicated that a cis-acting ABA-responsive element sequence is found in the *AtP5CS2* upstream region (Zhang et al., 1997). The involvement of ABA in *P5CS* gene expression was reviewed by Hare et al. (1999) who showed the regulation of *P5CS* gene expression in *Arabidopsis* ABA-insensitive mutants. Further, Knight and coworkers (1997) indicated the role of calcium (Ca^{2+}) in ABA-mediated induction of the *P5CS* gene during drought and salt stress. They suggested that calcium itself is not sufficient and under stressful conditions, additional upstream signaling molecules are required for expression of the *AtP5CS* gene

in proline metabolism besides Ca^{2+}. Yang et al. (2016) reported the involvement of calmodulin (CaM) along with Ca^{2+} in signal transduction events for the accumulation of proline in *J. curcas* seedlings under cold stress. This accumulation is possible by the simultaneous action of enhanced expression of P5CS, an enzyme required for the glutamate pathway in proline biosynthesis, with decreased activity of ProDH, an enzyme involved in the proline degradation pathway. Further, Ca^{2+} treatment enhanced chilling tolerance with decreased malondialdehyde (MDA) content and electrolyte leakage. This action was confirmed by treatment with lanthanum chloride ($LaCl_3$), the plasma membrane Ca^{2+}-channel blocker and calmodulin antagonists, chlorpromazine (CPZ) and trifluoperazine (TFP), as these compounds showed the reverse effects to those of $CaCl_2$ treatment.

Thiery et al. (2004) reported the possible role of phospholipase D besides calcium and ABA for the regulation of proline synthesis. Proline accumulation is negatively controlled by phospholipase D under non-stressed conditions (Knight et al., 1997; Parre et al., 2007). Phospholipase D itself has been suggested to be regulated by calcium. The data also indicated that the application of primary butyl alcohols enhanced the proline responsiveness of seedlings to mild hyperosmotic stress. Higher proline responsiveness was observed to hyperosmotic stress when phospholipase D was abolished (Kavi Kishor et al., 2005). VPS34, class-III phosphatidylinositol 3-kinase (PI3K), which catalyzes synthesis of phosphatidylinositol 3-phosphate (PI3P) from phosphatidylinositol, also controls proline metabolism (Leprince et al., 2015) and this was confirmed by use of the PI3K inhibitor, LY294002, which adversely affects PI3P levels and reduced proline, amino acids and sugars accumulation in salt-stressed *A. thaliana* seedlings. Further, proline accumulation was positively correlated with transcript level of *P5CS1* and negatively correlated with transcript and protein levels of ProDH1. Induced expression of ProDH1 in a pi3k-hemizygous mutant indicates the role of PI3K for ProDH1 regulation and thus proline catabolism (Leprince et al., 2015).

Recently, the role of transcription factors in activating genes coding enzymes involved in the glutamate pathway has been reviewed by Zarattini and Forlani (2017). *In silico* analysis revealed a large number of putative TF binding sites in promoter regions of genes that code for proline synthesis from glutamate, suggesting the role of TFs in regulating expression of proline biosynthetic genes. Several reports have indicated use of mutants in identification of new genes that participate in proline biosynthetic pathway. For example, Park et al. (2017) reported a proline content alterative 21 (*pca21*) mutant, which suppressed the *A. thaliana* ring zinc finger 1 (*atrzf1*) mutant in the osmotic stress response and reduced the expression of proline biosynthetic genes. Further, loss of ribosomal protein L24A (RPL24A), which is involved in proline accumulation under drought stress, also reduces *atrzf1* proline accumulation. Kim et al. (2017) reported a *pca22* mutant that also acts as *atrzf1* suppressor mutant and decreased proline content under drought stress.

β-Alanine betaine is synthesized by the S-adenosylmethionine-dependent N-methylation of β-alanine through intermediates, N-methyl β-alanine and N,N-dimethyl β-alanine (Rathinasabapathi, 2000). Under saline hypoxic conditions, β-alanine betaine is a more appropriate osmoprotectant compared to GB as it involves molecular oxygen in its biosynthetic pathway (Hanson et al., 1991). Since this metabolite is synthesized from β-alanine, a universal primary metabolite, this pathway will avoid using choline, an essential nutrient (Hanson et al., 1994; Duhazé et al., 2003). Proline betaine, synthesized from proline by several methylation steps, accumulates mainly in non-halophytic *Medicago* and *Citrus* species and provides solution to salinity stress (Nolte et al., 1997; Trinchant et al., 2004). GB synthesis occurs in two steps from choline and betaine aldehyde (BAL) in the chloroplast of higher plants (Rhodes and Hanson, 1993). Choline monooxygenase (CMO) catalyzes the oxidation of choline to BAL, which is further oxidized by NAD^+-dependent betaine aldehyde dehydrogenase (BADH) to GB (Chen and Murata, 2002). Salt stress led to upregulation of *OsCMO* in rice and overexpression of this gene in tobacco plants led to enhanced GB content and salt tolerance (Roychoudhury and Banerjee, 2016). Using immunoblot analysis, a functional OsCMO protein was observed in transgenic tobacco. However, no functional OsCMO protein was observed in rice plants. In rice plants, the non-functional OsCMO protein was suggested to block GB synthesis under stress (Luo et al., 2012). BADH1 plays an important role in GB accumulation in rice (Tang et al., 2014). The OsBADH1–RNA interference (RNAi) exhibited much lower cold, salinity and drought tolerance in transgenic rice. The reduced stress tolerance in OsBADH1–RNAi repression lines was correlated with the OsBADH1 downregulation. In addition to BAL, *OsBADH1* also oxidizes other aldehydes and thus participates in alleviation of harmful effects of abiotic stresses without affecting GB content (Tang et al., 2014). Ishitani et al. (1995) reported increased levels of BADH mRNA in leaves and roots of barley plants grown under abiotic stresses such as salinity, drought and ABA treatment. Further, in *Spinacia oleracea*, Zárate-Romero et al. (2016) reported

reversible and partial inactivation of BADH (*SoBADH*) by BAL without using NAD(+). This regulatory mechanism allows synthesis of high amounts of GB without depleting NAD(+) under osmotic stress and prevents accumulation of toxic BAL. Unusual post-transcriptional processing results in many truncated and/or recombinant transcripts of two BADH homologs in rice maize, wheat and barley but not in dicotyledonous species, such as spinach, *Arabidopsis* and tomato (Niu et al., 2007). This explains the differences in GB-synthesizing capacities among different plant species. The site for the deletion/insertion in monocot cereals was found to be altered in response to stress conditions.

17.4 OSMOLYTE-INDUCED IMPROVEMENT IN ABIOTIC STRESS TOLERANCE IN PLANTS

Exogenous application and transgenic plants over=expressing genes involved in the metabolic pathway of osmolytes have revealed the importance of osmolytes under environmental stresses (Tables 17.1 and 17.2). Application of sugar and sugar alcohols such as trehalose, sucrose, myo-inositol, mannitol, D-ononitol and sorbitol improves abiotic stress tolerance in plants. Sucrose application led to increased accumulation of soluble sugar and enhancement of salt tolerance in seedlings of indica rice and *Arabidopsis* (Siringam et al., 2012; Qiu et al., 2014) and chilling tolerance in cucumber (Cao et al., 2014). Enhanced concentrations of soluble sugar in salt-stressed rice seedlings led to alleviation of the degradation of photosynthetic pigments. Reduction in salt led inhibition in shoot growth and salt stress tolerance was observed in maize plants to which mannitol was applied exogenously (Kaya et al., 2013). Trehalose mitigated drought stress-induced damage in maize and fenugreek plants (Ali and Ashraf, 2011; Sadak, 2016). It showed ameliorative effects on biomass production, some important photosynthetic parameters, plant water relation parameters, activities of antioxidant enzymes POD and CAT and content of the non-enzymatic antioxidants phenolics and tocopherols (Ali and Ashraf, 2011; Sadak, 2016). Exogenous application of trehalose and sorbitol led to alleviation of the adverse effect caused by salinity in salt-sensitive rice seedlings KDML105 but not salt-tolerant PK, and the effects were more prominent for trehalose than sorbitol (Theerakulpisut and Gunnula, 2012). Increased photosynthetic pigments, Hill-reaction activity, SOD, CAT, PPO and POD activities, enhanced nucleic acids, K^+, Ca^{2+}, and P contents and decreased Na^+ content were

TABLE 17.1

Enhancement of Abiotic Stress Tolerance in Plants by Exogenous Application of Osmolytes

Osmolyte	Plant Species	Tolerance	References
Sucrose	Rice	Salt	Siringam et al. (2012)
	Arabidopsis	Salt	Qiu et al. (2014)
	Cucumber	Chilling	Cao et al. (2014)
Sorbitol	Rice	Salt	Theerakulpisut and Gunnula (2012)
Mannitol	Maize	Salt	Kaya et al. (2013)
Trehalose	Maize	Drought	Ali and Ashraf (2011)
Proline	Fenugreek	Drought	Sadak (2016)
	Rice	Salt	Theerakulpisut and Gunnula (2012); Nounjan et al. (2012)
	Maize	Salt	Zeid (2009)
	Iodine bush	Drought, salt	Chrominski et al. (1989)
	Maize	Drought	Ali et al. (2007, 2008)
	Faba bean	Salt	Gadallah (1999)
	Broad bean and maize	Salt	Abdel-Samad et al. (2010)
	Soybean	Salt	He et al. (2000); Bai and Wang (2002)
	Tobacco	Salt	Okuma et al. (2004)
	Barley	Salt	Lone et al. (1987)
	Onion	Salt	Mansour (1998)
	Tobacco	Cadmium	Islam et al. (2009)
	Mungbean	Cadmium	Hossain and Fujita (2010)
	Bean	Selenium	Aggarwal et al. (2011)
Glycine betaine	Sun flower	Drought	Hussain et al. (2008); Iqbal et al. (2009)
	Faba bean	Salt stress	Gadallah (1999)
	Rice	Salt stress	Harinasut et al. (1996); Lutts (2000)
	Maize	Salt stress	Yang and Lu (2005)
	Sunflower	Salt stress	Bakhoum and Sadak (2016)
	Tomato plants	Salt stress	Mäkelä et al. (1998a)
	Tomato plants	High temperature	Mäkelä et al. (1998b)
	Potato	Low temperature	Somersalo et al. (1996)
	Strawberry	Low temperature	Rajashekar et al. (1999)
	Arabidopsis	Freezing	Xing and Rajashekar (2001)
	Mungbean	Cadmium	Hossain and Fujita (2010)
	Tobacco	Cadmium	Islam et al. (2009)

observed in rice seedlings pretreated with trehalose (Theerakulpisut and Gunnula, 2012). Trehalose has also been shown to protect thylakoid membranes of winter wheat, reduce Na^+/K^+, strongly decrease endogenous proline upregulated transcription of *P5CS* and *P5CR* and all antioxidant enzyme genes, decrease SOD and POD activities and increase APX activity. Pretreatment of maize grains (Giza 2) with trehalose alleviated the damage induced by salinity on metabolic activity and increased the growth of maize seedlings (Zeid, 2009). Exogenous trehalose showed marked beneficial effect on growth during the recovery period compared with salinity-stressed plants not supplied with exogenous trehalose (Nounjan et al., 2012).

Several studies have indicated increased abiotic stress tolerance in transgenic plants overexpressing genes involved in metabolic pathways of sugar and sugar alcohols. Plants overexpressing the sucrose transporter gene *NtSUT1* and vacuolar sugar carrier *AtSWEET16* gene resulted in improved tolerance of abiotic stresses (Klemens et al., 2013; Kariya et al., 2017). Myo-inositol 1-phosphate synthase, myo-inositol O-methyltransferase and D-ononitol epimerase are involved in myo-inositol metabolism in plants (Ishitani et al., 1996; Sheveleva et al., 1997). Transgenic tobacco plant expressing *IMT1* cDNA from ice plants (encoding D-myo-inositol methyltransferase) showed higher drought and salt tolerance by accumulating D-ononitol (Sheveleva et al., 1997).

TABLE 17.2

List of Transgenic Plant Developed for Abiotic Stress Tolerance Using Genes Involved in Osmolyte Metabolism

Gene	Species	Tolerance	Reference
Escherichia coli mtlD	Peanut	Salinity, drought	Bhauso et al. (2014)
Apium graveolens M6PR	*Arabidopsis*	Salinity	Zhifang and Loescher (2003)
Escherichia coli otsa and *otsb* as a fusion gene	Rice	Drought, salinity and low temperature	Garg et al. (2002)
Escherichia coli bifunctional fusion enzyme *TPSP*	Rice	Drought, salinity and cold	Jang et al. (2003)
Oryza sativa OsTPS1	Rice	Drought, salinity and cold	Li et al. (2011)
Escherichia coli bifunctional fusion of *T6PS* and *T6PP*	Tomato	Drought and salinity	Lyu et al. (2013)
Saccharomyces cerevisiae T6PS and *T6PP*	*Arabidopsis*	Drought, freezing, salinity and heat	Miranda (2017)
Nicotiana tabacum NtSUT1	Tobacco	Aluminum	Kariya et al. (2017)
Arabidopsis Atsweet16	*Arabidopsis*	Freezing	Klemens et al. (2013)
Mesembryanthemum crystallinum plant IMT1	Tobacco	Drought and salinity	Sheveleva et al. (1997)
Porteresia coarctata PcINO1	Mustard	Salinity	Joshi et al. (2010)
Spartina alterniflora SaINO1	*Arabidopsis*	Salinity	Joshi et al. (2010)
Vigna aconitifolia P5CS	Wheat and carrot	Salinity	Sawahel and Hassan (2002); Han and Hwang (2003)
Lilium regale lrp5cs	*Arabidopsis*	Drought and salinity	Wei et al. (2016)
Vigna aconitifolia p5cs	Pigeon pea	Salinity	Surekha et al. (2014)
Arthrobacter codA	*Arabidopsis*	Cold	Hayashi et al. (1997)
Arthrobacter codA	*Arabidopsis*	High temperature	Alia et al. (1998)
Arthrobacter codA	Tomato	Salinity	Wei et al. (2017)
Escherichia coli betA	Tobacco	Salinity	Lilius et al. (1996)
Spinacia oleracea CMO	Rice	Temperature and salinity	Shirasawa et al. (2006)
Spinacia oleracea BADH	Tobacco	Salinity	Yang et al. (2008)
Spinacia oleracea BADH	Tobacco	High temperature	Yang et al. (2005)
Daucus carota BADH	Carrot	Salinity	Kumar et al. (2004)
Atriplex micrantha BADH	Maize	Salinity	Di et al. (2015)
Ammopiptanthus nanus BADH	*Arabidopsis* mutant	Drought and salinity	Yu et al. (2017)
Atriplex hortensis BADH	Wheat	Salinity	Tian et al. (2017)

Inositol and its derivatives influence the overall metabolic pathways and lead to a stress-resistant phenotype (Goswami et al., 2014). *PcINO1* gene from salt-tolerant *Porteresia coarctata* (Roxb.) Tateoka, and *SaINO1* gene from the grass halophyte *Spartina alterniflora* coding for a L-myo-inositol-1-phosphate synthase (MIPS) when introgressed into cultivated mustard, *B. juncea* var B85 and *A. thaliana*, respectively demonstrated increased salinity tolerance. In transgenic *Arabidopsis*, protection of PSII led to higher retention of carotenoids and chlorophylls and lower levels of cellular damage caused by stress (Joshi et al., 2010). Enhancing the level of T6P and/or trehalose by modifying genes involved in their

metabolism has been utilized to increase abiotic stress tolerance in plants (Delorge et al., 2014). Plants overexpressing TPS, TPP and the bifunctional fusion enzyme TPSP, resulted in improved tolerance of abiotic stresses (Garg et al., 2002; Jang et al., 2003; Avonce et al., 2004; Miranda et al., 2007; Lyu et al., 2013). Rice transgenic lines incorporating *Escherichia coli* trehalose biosynthetic genes (*otsA* and *otsB*) showed three-to-ten times the level of trehalose, maintained plant growth, reduced photooxidative damage, and favorable mineral balance under drought, salinity and low-temperature conditions (Garg et al., 2002). A correlation between trehalose accumulation with higher soluble carbohydrate levels and an elevated capacity for photosynthesis under both stress and non-stress conditions was observed, suggesting a role of trehalose in modulating sugar sensing and carbohydrate metabolism. Although expression of yeast trehalose-6-phosphate synthase gene in tomato resulted in improved abiotic stress tolerance, it caused stunted growth and many other changes in appearance. However, transgenic rice generated by the introduction of a gene encoding a bifunctional fusion (TPSP) of the T6PS and T6PP of *E. coli*, led to increased tolerance to drought, salt and cold, as shown by chlorophyll fluorescence and growth inhibition analyses (Jang et al., 2003). Expression of T6PS and T6PP gene fusion from yeast led to an increase in heat, freezing, salt and drought tolerance (Miranda et al., 2007). Expression of a gene encoding a bifunctional fusion of T6PS and T6PP genes from *E. coli* in tomato plants showed increased drought and salt tolerance, enhanced photosynthetic rates and normal growth patterns and appearances (Lyu et al., 2013). However, in certain cases negative effects on growth and development of plant were observed when transgenic plants were engineered to produce increasing levels of sugars, which could be due to participation of sugars in various metabolic pathways.

Proline, a key imino acid, is an excellent osmolyte that is found widely in higher plants and generally accumulates in large amounts under environmental stresses (Kavi Kishor et al., 1995). Proline plays major roles as a signaling molecule, an antioxidant molecule and a metal chelator during various stresses (Kavi Kishor et al., 1995; Hayat et al., 2012). Proline, when applied exogenously, reduced the harmful effects of abiotic stresses on seedling growth (Singh et al., 2015). Its application counteracted the harmful effects of water stress on growth under drought stress in *Allenrolfea occidentalis* (Chrominski et al., 1989) and maize (Ali et al., 2007, 2008). Improved drought tolerance might be associated with proline-led improved nutrient uptake and increased photosynthetic rate, which was positively correlated with stomatal conductance (gs), sub-stomatal CO_2 (Ci) and photosynthetic pigments (Ali et al., 2007, 2008). In rice and *A. occidentalis*, it alleviated the negative effect of salinity on plants (Chrominski et al., 1989; Roy et al., 1993). Proline application decreased membrane injury, increased K^+ uptake and growth and increased chlorophyll contents in *Vicia faba* plants exposed to salinity stress (Gadallah, 1999). Soybean cell cultures, tobacco suspension cells and barley embryo cultures under salt stress showed enhanced tolerance upon application of exogenous proline by increasing the activities of antioxidant enzyme activity and stabilizing membranes (Lone et al., 1987; Mansour, 1998; He et al., 2000; Bai and Wang, 2002;

Okuma et al., 2004). Broad bean and maize plants, when sprayed with proline or phenylalanine, showed enhanced protein and saccharide concentration, decreased endogenous proline level, reduced Na^+ uptake and increased K^+ uptake, K^+/Na^+ ratio, Ca^{++} and P selectivity considerably in comparison to respective controls (Abdel-Samad et al., 2010). Exogenous proline application alleviated the harmful effects of Se and Cd and improved the growth of seedlings by increasing enzymatic and non-enzymatic antioxidants (Islam et al., 2009; Aggarwal et al., 2011). In case of Se, components of the ascorbate-glutathione cycle, and in case of Cd, activities of SOD and CAT increased significantly. Exogenous proline applied 24 h before irradiation reduced lipid peroxidation (Radyukina et al., 2011). Proline has been suggested to protect mungbean against Cd-induced oxidative stress by decreasing H_2O_2 concentration and lipid peroxidation and enhancing the antioxidant defense and methylglyoxal detoxification system (Hossain and Fujita, 2010). Nounjan et al. (2012) observed no improvement in growth inhibition of rice seedlings due to application of proline exogenously during salt stress but remarkable beneficial improvement during the recovery period. Rice seedlings showed higher percentage of growth recovery in proline and salt-treated rice seedlings (162.38%) compared to only salinity-stressed rice seedlings.

Incorporation of the *Vigna P5CS* gene conferred enhanced salt tolerance on transgenic wheat (Sawahel and Hassan, 2002) and carrot (Han and Hwang, 2003). When expressed in *Arabidopsis*, *Lilium regale LrP5CS* gene increased its tolerance to osmotic, drought and salt stress (Wei et al., 2016). P5CS, a rate-limiting enzyme involved in the biosynthesis of praline, is feedback inhibited by proline. In transgenic tobacco plants, incorporation of a mutant form of the P5CS enzyme (P5CSF129A), having feedback inhibition removed by site-directed mutagenesis, accumulated two-times higher proline than tobacco plants incorporating *V. aconitifolia P5CS* (Hong et al., 2000). Transformation of tobacco plants and embryonic structures of pigeonpea with the mutant gene *P5CSF129A* resulted in transgenic plants showing more proline accumulation, increased growth performance, higher relative water and chlorophyll content and lower lipid content under salinity stress (Surekha et al., 2014). Similarly, *N. plumbaginifolia* overexpressing mutant genes having decreased feedback inhibition of proline biosynthesis led to increased salt tolerance (Ahmed et al., 2015).

Quaternary ammonium compounds, which are quaternary ammonium cation salts, act as osmolytes and alleviate abiotic stress in plants. Among several quaternary ammonium compounds, GB is one of the most widely found osmolytes in plants under drought, salinity, extreme temperature and metal stress (Hanson et al., 1991; Guo et al., 2009; Lokhande and Suprasanna, 2012). Exogenously applied GB protects against harmful effects of abiotic stresses. GB counteracted the water stress-induced decrease in head diameter, achene number, weight, yield and oil yield in sunflower plants (Hussain et al., 2008). Its application reduced the inhibitory effects of water stress on net CO_2 assimilation rate, sub-stomatal CO_2 concentration and transpiration rate in sunflower plants (Iqbal et al., 2009). Under salt stress, exogenously supplied GB reduced membrane injury, improved uptake of K^+, growth and enhanced chlorophyll contents in *V. faba* (Gadallah, 1999), reduced Na^+ accumulation, maintained shoot K^+ concentration and improved salt tolerance in rice plants (Harinasut et al., 1996; Lutts, 2000), improved growth, net photosynthesis, leaf water content and the apparent quantum yield of photosynthesis in maize plants (Yang and Lu, 2005), improved growth parameters, free amino acids, photosynthetic pigments, proline, phenolic and total soluble carbohydrate contents, yield and yield components in sunflower (Bakhoum and Sadak, 2016). The fatty acid profile of sunflower oil also exhibited some alterations under the influence of salinity and GB treatments (Bakhoum and Sadak, 2016). GB increased yield (about 40%) of tomato plants grown under salt stress or high temperatures (Mäkelä et al., 1999, 2000). A cryoprotective effect of GB on membranes and enzymes has been reported (Gorham, 1995). Exogenous supply of GB has been shown to improve freezing stress tolerance of plants (Allard et al., 1998; Sakamoto and Murata, 1998; Farooq et al., 2008). GB enhanced low-temperature tolerance in tobacco and strawberry (Rajashekar et al., 1999; Holmström et al., 2000). In *Arabidopsis* plants, GB reduced the freezing temperature from −3.1 to −4.5°C (Xing and Rajashekar, 2001). GB also protected mungbean against Cd-induced oxidative stress by decreasing H_2O_2 concentration and lipid peroxidation and enhancing the antioxidant defense and methylglyoxal detoxification system (Hossain and Fujita, 2010).

Many plant species, such as rice, tobacco, *Arabidopsis* and mustard, do not have the capability to produce GB naturally (Rhodes and Hanson, 1993). Incorporation of genes encoding GB biosynthetic enzymes in these plant species can lead to the production of GB and enhance tolerance to salt, cold, drought or high-temperature stress (Rhodes and Hanson, 1993). Transgenic plants incorporating genes involved in the biosynthesis of GB from different microorganisms and plants showed higher accumulation of GB and enhanced tolerance to abiotic stresses (Lilius et al., 1996; Hayashi et al., 1997;

Alia et al., 1998; Sakamoto and Murata, 1998; Holmström et al., 2000; Waditee et al., 2005; Yang et al., 2005; Chen and Murata, 2008; Yang et al., 2008). Many of these plants were natural GB non-accumulators (Chen and Murata, 2011). GB biosynthesis is catalyzed by choline monooxygenase (CMO) and BADH. Rice contains two BADH genes but no functional CMO gene so GB is not accumulated in rice. Rice seedlings incorporating a copy of the spinach CMO accumulated 0.29–0.43 μmol g^{-1} d. wt and demonstrated enhanced tolerance to temperature and salinity stress (Shirasawa et al., 2006). Low concentration of GB accumulated in the transgenic rice plants suggests different localization of spinach CMO and of endogenous BADHs. The *BADH* gene enhanced photosynthesis under high temperature when expressed in tobacco plants from spinach (Yang et al., 2005) and salt tolerance when expressed in carrot cultured cells, roots and leaves from spinach (Kumar et al., 2004). Transgenic maize plants expressing the *BADH* gene from *Atriplex micrantha* enhanced fresh weight, plant height, chlorophyll content and grain yield, reduced MDA content and enhanced tolerance under salinity stress (Di et al., 2015). Incorporation of the *AnBADH* gene from *Ammopiptanthus nanus* considerably increased the tolerance of the *Arabidopsis* mutant plants to drought and salinity stresses (Yu et al., 2017). Transgenic lines showed higher fresh weight, GB, proline and relative water content, lower MDA content and relative electrolyte leakage in comparison to the untransformed mutant. GB overaccumulation due to incorporation of *BADH* gene can increase salt tolerance by protecting the components and function of thylakoid membranes. Alteration in thylakoid membrane lipids and fatty acid profiles may be responsible for enhanced salt stress tolerance of the transgenic lines (Tian et al., 2017).

Choline is the precursor of GB, so a higher concentration of it can increase osmotic stress tolerance in plants. Overexpression of phosphoethanolamine *N*-methyltransferase, which catalyzes the formation of phosphocholine from phosphoethanolamine by methylation process in transgenic tobacco enhanced the concentrations of phosphocholine by five times and free choline 50 times but phosphatidylcholine content or growth was not affected (McNeil et al., 2001). Transgenic *A. thaliana* expressing *codA* gene for choline oxidase from *Arthrobacter* showed enhanced cold tolerance by synthesizing GB (Hayashi et al., 1997). Tomato plants incorporating choline oxidase gene *codA* from *A. globiformis*, allowed to be targeted to both cytosol and chloroplasts, showed considerably higher photosynthetic rates (Pn), Na^+ efflux and antioxidant enzyme activities and lower K^+ efflux and ROS accumulation under salinity stress in the leaves (Wei et al., 2017). *Arabidopsis* plants transformed with a *codA* gene led to GB accumulation and enhanced tolerance to high temperatures during imbibitions and germination of seeds, as well as during the growth of young seedlings (Alia et al., 1998). Induction of heat-shock proteins, which is a characteristic of heat shock, reduced in these transgenic plants, indicating the alleviation of the harmful effect of heat stress. Transformation of *Arabidopsis* with the *codA* gene encoding the COD enzyme increased viability, retained intracellular ions after freezing treatments and conferred freezing tolerance considerably (Sakamoto and Murata, 2002). Tobacco plants expressing the *bet A* gene from *E. coli*, which encodes for choline dehydrogenase, displayed salinity stress tolerance (Lilius et al., 1996).

17.5 CONCLUSIONS AND FUTURE PROSPECTS

Plants are frequently exposed to different kinds of abiotic stresses, and accumulation of osmolytes during various abiotic stresses is a common response in plants. Higher concentration of osmolytes is attained by increasing biosynthesis and uptake and/or decreasing degradation and export of osmolytes. Osmolytes act as osmoprotectants, metal chelators, antioxidants and signaling molecules and play an important role in balancing redox status, stabilizing and protecting macromolecules, including proteins and membranes, and regulating cellular osmotic adjustment in plants subjected to abiotic stress. Exogenous application of osmolytes under stress conditions has been shown to

confer stress tolerance in plants. Determination of the most effective concentrations, number of applications of osmolytes and most responsive growth stage of different plant species under various abiotic stresses is required for effective utilization of osmolytes in crop production under abiotic stress environments. Higher accumulation of osmolytes in transgenic plants containing genes for osmolyte metabolism has been shown to ameliorate stress effects in some cases. However, in many cases, accumulation of osmolytes in transgenic plants remains low or insufficient and they do not improve the abiotic stress tolerance of plants. Identification of gene regulatory networks and greater understanding of the mechanisms of metabolic, molecular genetic, transcriptional, post-transcriptional, translational and post-translational regulation of osmolyte biosynthesis and degradation in response to various abiotic stresses in the plant cell will be helpful in the generation of stress-tolerant plants with fine-tuning of osmolytes biosynthesis and hydrolysis (Figure 17.4). Other factors, including metabolic fluxes and substrate availability, which influence accumulation of osmolytes during abiotic stresses should be taken into account while developing stress-tolerant plants. Responses of plants challenged with single stress conditions have been subjected to intense studies. However, plants are regularly challenged with a combination of different stresses. These stresses can interact synergistically or antagonistically. Metabolic profiling, expression profiling and forward and reverse genetics approaches can be employed to understand the roles and regulation of osmolytes in providing tolerance to single stresses and combinations of stresses, which will

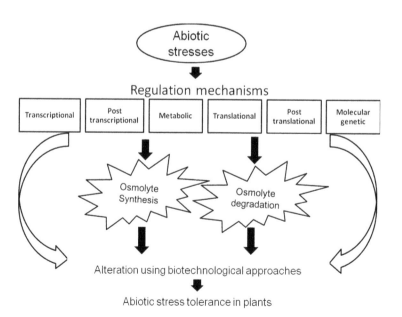

FIGURE 17.4 Osmolyte regulation in plants under abiotic stresses.

be helpful in developing crop plants tolerant to multiple abiotic stresses.

ACKNOWLEDGMENTS

Pallavi Sharma is grateful to the UGC Faculty Recharge program [F.4-5(107-FRP)/2014(BSR)] and DST-SERB Early Carrer Research Award (ECR/2016/000888) for financial support. Financial support from Department of Biotechnology (DBT) Builder project No. BT/PR-9028/INF/22/193/2013 is gratefully acknowledged.

REFERENCES

Abdel-Samad, H.M., Shaddad, M.A.K., Barakat, N. (2010) The role of amino acids in improvement in salt tolerance of crop plants. *Journal of Stress Physiology and Biochemistry* 6: 25–37.

Ábrahám, E., Rigó, G., Székely, G., Nagy, R., Koncz, C., Szabados, L. (2003) Light-dependent induction of proline biosynthesis by abscisic acid and salt stress is inhibited by brassinosteroid in *Arabidopsis*. *Plant Molecular Biology* 51: 363–372.

Ackerson, R.C. (1985) Osmoregulation in cotton in response to water stress III. Effects of phosphorus fertility. *Plant Physiology* 77: 309–312.

Adams, E., Frank, L. (1980) Metabolism of proline and the hydroxyprolines. *Annual Review of Biochemistry* 49: 1005–1061.

Adams, P., Thomas, J.C., Vernon, D.M., Bohnert, H.J., Jensen, R.G. (1992) Distinct cellular and organismic responses to salt stress. *Plant and Cell Physiology* 33: 1215–1223.

Aggarwal, M., Sharma, S., Kaur, N., Pathania, D., Bhandhari, K., Kaushal, N., Kaur, R., Singh, K., Srivastava, A., Nayyar, H. (2011) Exogenous proline application reduces phytotoxic effects of selenium by minimizing oxidative stress and improves growth in bean (*Phaseolus vulgaris* L.) seedlings. *Biological Trace Element Research* 140: 354–367.

Ahmed, A.A.M., Roosens, N., Dewaele, E., Jacobs, M., Angenon, G. (2015) Over-expression of a novel feed-back-desensitized Δ1-pyrroline-5-carboxylate synthetase increases proline accumulation and confers salt tolerance in transgenic *Nicotiana plumbaginifolia*. *Plant Cell, Tissue and Organ Culture* 122: 383–393.

Ahn, C., Park, U., Park, P.B. (2011) Increased salt and drought tolerance by D-ononitol production in transgenic *Arabidopsis thaliana*. *Biochemical and Biophysical Research Communications* 415: 669–674.

Ajithkumar, I.P., Panneerselvam, R. (2013) Osmolyte accumulation, photosynthetic pigment and growth of *Setaria italica* (L.) P. Beauv. under drought stress. *Asian Pacific Journal of Reproduction* 2: 220–224.

Ajithkumar, I.P., Panneerselvam, R. (2014) ROS scavenging system, osmotic maintenance, pigment and growth status of *Panicum sumatrense* roth. under drought stress. *Cell Biochemistry and Biophysics* 68: 587–595.

Al-Garni, S.M.S. (2006) Increasing NaCl-salt tolerance of a halophytic plant *Phragmites australis* by mycorrhizal symbiosis. *American-Eurasian Journal of Agricultural and Environmental Science* 1: 119–126.

Ali, Q., Ashraf, M. (2011) Induction of drought tolerance in maize (*Zea mays* L.) due to exogenous application of trehalose: Growth, photosynthesis, water relations and oxidative defence mechanism. *Journal of Agronomy and Crop Science* 197: 258–271.

Ali, G., Srivastava, P.S., Iqbal, M. (1999) Proline accumulation, protein pattern and photosynthesis in *Bacopa monniera* regenerants grown under NaCl stress. *Biologia Plantarum* 42: 89–95.

Ali, Q., Ashraf, M., Athar, H.U.R. (2007) Exogenously applied proline at different growth stages enhances growth of two maize cultivars grown under water deficit conditions. *Pakistan Journal of Botany* 39: 1133–1144.

Ali, Q., Ashraf, M., Shahbaz, M., Humera, H. (2008) Ameliorating effect of foliar applied proline on nutrient uptake in water stressed maize (*Zea mays* L.) plants. *Pakistan Journal of Botany* 40: 211–219.

Alia, A., Hayashi, H., Sakamoto, A., Murata, N. (1998) Enhancement of the tolerance of *Arabidopsis* to high temperatures by genetic engineering of the synthesis of glycinebetaine. *The Plant Journal* 16: 155–161.

Allard, F., Houde, M., Kröl, M., Ivanov, A., Huner, N.P.A., Sarhan, F. (1998) Betaine improves freezing tolerance in wheat. *Plant and Cell Physiology* 39: 1194–1202.

Anjum, N.A., Sharma, P., Gill, S.S., Hasanuzzaman, M., Khan, E.A., Kachhap, K., Mohamed, A.A., Thangavel, P., Devi, G.D., Vasudhevan, P., et al. (2016) Catalase and ascorbate peroxidase—representative H_2O_2-detoxifying heme enzymes in plants. *Environmental Science and Pollution Research* 23: 19002–19029.

Arakawa, T., Timasheff, S.N. (1983) Preferential interactions of proteins with solvent components in aqueous amino acid solutions. *Archives of Biochemistry and Biophysics* 224: 169–177.

Armengaud, P., Thiery, L., Buhot, N., Grenier-de March, G., Savouré, A. (2004) Transcriptional regulation of proline biosynthesis in *Medicago truncatula* reveals developmental and environmental specific features. *Physiologia Plantarum* 120: 442–450.

Ashraf, M. (1994) Organic substances responsible for salt tolerance in *Eruca sativa*. *Biologia Plantarum* 36: 255–259.

Ashraf, M., Foolad, Mr. (2007) Roles of glycine betaine and proline in improving plant abiotic stress resistance. *Environmental and Experimental Botany* 59: 206–216.

Ashraf, M., Harris, P.J.C. (2004) Potential biochemical indicators of salinity tolerance in plants. *Plant Science* 166: 3–16.

Avonce, N., Leyman, B., Mascorro-Gallardo, J.O., Van Dijck, P., Thevelein, J.M., Iturriaga, G. (2004) The Arabidopsis trehalose-6-P synthase AtTPS1 gene is a regulator of glucose, abscisic acid, and stress signaling. *Plant Physiology* 136: 3649–3659.

Bai, H., Wang, Y. (2002) Effect of exogenous proline on SOD and POD activity for soybean callus under salt stress. *Acta Agriculturae Boreali-Sinica* 17: 37–40.

Bakhoum, G.S.H., Sadak, M.S. (2016) Physiological role of glycine betaine on sunflower (*Helianthus annuus* L.) plants grown under salinity stress. *International Journal of Chemical Technology Research* 9: 158–171.

Bandurska, H., Niedziela, J., Pietrowska-Borek, M., Nuc, K., Chadzinikolau, T., Radzikowska, D. (2017) Regulation of proline biosynthesis and resistance to drought stress in two barley (*Hordeum vulgare* L.) genotypes of different origin. *Plant Physiology and Biochemistry* 118: 427–437.

Bassi, R., Sharma, S.S. (1993) Changes in proline content accompanying the uptake of zinc and copper by *Lemna minor*. *Annals of Botany* 72: 151–154.

Beck, E.H., Fettig, S., Knake, C., Hartig, K., Bhattarai, T. (2007) Specific and unspecific responses of plants to cold and drought stress. *Journal of Biosciences* 32: 501–510.

Bhauso, T.D., Radhakrishnan, T., Kumar, A., Mishra, G.P., Dobaria, J.R., Patel, K., Rajam, M.V. (2014) Overexpression of bacterial mtlD gene in peanut improves drought tolerance through accumulation of mannitol. *The Scientific World Journal* 2014. doi: 10.1155/2014/125967.

Boriboonkaset, T., Theerawitaya, C., Pichakum, A., Cha-um, S., Takabe, T., Kirdmanee, C. (2012) Expression levels of some starch metabolism related genes in flag leaf of two contrasting rice genotypes exposed to salt stress. *Australian Journal of Crop Science* 6: 1579–1586.

Briens, M., Larher, F. (1982) Osmoregulation in halophytic higher plants: A comparative study of soluble carbohydrates, polyols, betaines and free proline. *Plant, Cell and Environment* 5: 287–292.

Brown, A.D., Simpson, J.R. (1972) Water relations of sugar-tolerant yeasts: The role of intracellular polyols. *Microbiology* 72: 589–591.

Bryan, J.K. (1990) Advances in the biochemistry of amino acid biosynthesis. *Intermediary Nitrogen Metabolism* 161–196.

Cabib, E., Leloir, L.F. (1958) The biosynthesis of trehalose phosphate. *Journal of Biological Chemistry* 231: 259–275.

Cao, Y.Y., Yang, M.T., Li, X., Zhou, Z.Q., Wang, X.J., Bai, J.G. (2014) Exogenous sucrose increases chilling tolerance in cucumber seedlings by modulating antioxidant enzyme activity and regulating proline and soluble sugar contents. *Scientia Horticulturae* 179: 67–77.

Cha-Um, S., Kirdmanee, C. (2009) Effect of salt stress on proline accumulation, photosynthetic ability and growth characters in two maize cultivars. *Pakistan Journal of Botany* 41: 87–98.

Chávez-Bárcenas, A.T., Valdez-Alarcón, J.J., Martınez-Trujillo, M., Chen, L., Xoconostle-Cázares, B., Lucas, W.J., Herrera-Estrella, L. (2000) Tissue-specific and developmental pattern of expression of the rice sps1 gene. *Plant Physiology* 124: 641–654.

Chen, T.H.H., Murata, N. (2002) Enhancement of tolerance of abiotic stress by metabolic engineering of betaines and other compatible solutes. *Current Opinion in Plant Biology* 5: 250–257.

Chen, T.H.H., Murata, N. (2008) Glycinebetaine: An effective protectant against abiotic stress in plants. *Trends in Plant Science* 13: 499–505.

Chen, T.H.H., Murata, N. (2011) Glycinebetaine protects plants against abiotic stress: Mechanisms and biotechnological applications. *Plant, Cell and Environment* 34: 1–20.

Chen, C.T., Chen, L.M., Lin, C.C., Kao, C.H. (2001) Regulation of proline accumulation in detached rice leaves exposed to excess copper. *Plant Science* 160: 283–290.

Chrominski, A., Halls, S., Weber, D.J., Smith, B.N. (1989) Proline affects ACC to ethylene conversion under salt and water stresses in the halophyte, *Allenrolfea occidentalis*. *Environmental and Experimental Botany* 29: 359–363.

Costa, G., Morel, J.L. (1994) Water relations, gas exchange and amino acid content in Cd-treated lettuce. *Plant Physiology and Biochemistry* 32: 561–570.

Csonka, L.N., Hanson, A.D. (1991) Prokaryotic osmoregulation: Genetics and physiology. *Annual Reviews in Microbiology* 45: 569–606.

Daie, J., Patrick, J.W. (1988) Mechanism of drought-induced alterations in assimilate partitioning and transport in crops. *Critical Reviews in Plant Sciences* 7: 117–137.

Dejardin, Annabelle, Sokolov, L.N., Kleczkowski, L.A. (1999) Sugar/osmoticum levels modulate differential abscisic acid-independent expression of two stress-responsive sucrose synthase genes in *Arabidopsis*. *Biochemical Journal* 344: 503–509.

Delauney, A.J., Verma, D.P.S. (1993) Proline biosynthesis and osmoregulation in plants. *The Plant Journal* 4: 215–223.

Delorge, I., Janiak, M., Carpentier, S., Van Dijck, P. (2014) Fine tuning of trehalose biosynthesis and hydrolysis as novel tools for the generation of abiotic stress tolerant plants. *Frontiers in Plant Science* 5: 1–9.

Dhir, B., Nasim, S.A., Samantary, S., Srivastava, S. (2012) Assessment of osmolyte accumulation in heavy metal exposed *Salvinia natans*. *International Journal of Botany* 8: 153–158.

Di, H., Tian, Y., Zu, H., Meng, X., Zeng, X., Wang, Z. (2015) Enhanced salinity tolerance in transgenic maize plants expressing a BADH gene from *Atriplex micrantha*. *Euphytica* 206: 775–783.

Dichio, B., Margiotta, G., Xiloyannis, C., Bufo, S., Sofo, A., Cataldi, T. (2009) Changes in water status and osmolyte contents in leaves and roots of olive plants (*Olea europaea* L.) subjected to water deficit. *Trees* 23: 247–256.

Dubey, R.S., Singh, A.K. (1999) Salinity induces accumulation of soluble sugars and alters the activity of sugar metabolizing enzymes in rice plants. *Biologia Plantarum* 42: 233–239.

Duhazé, C., Gagneul, D., Leport, L., Larher, F.R., Bouchereau, A. (2003) Uracil as one of the multiple sources of β-alanine in *Limonium latifolium*, a halotolerant β-alanine betaine accumulating *Plumbaginaceae*. *Plant Physiology and Biochemistry* 41: 993–998.

Eggert, A., Nitschke, U., West, J.A., Michalik, D., Karsten, U. (2007) Acclimation of the intertidal red alga *Bangiopsis subsimplex* (*Stylonematophyceae*) to salinity changes. *Journal of Experimental Marine Biology and Ecology* 343: 176–186.

El-Bashiti, T., Hamamcı, H., Öktem, H.A., Yücel, M. (2005) Biochemical analysis of trehalose and its metabolizing enzymes in wheat under abiotic stress conditions. *Plant Science* 169: 47–54.

Everard, J.D., Gucci, R., Kann, S.C., Flore, J.A., Loescher, W.H. (1994) Gas exchange and carbon partitioning in the leaves of celery (*Apium graveolens* L.) at various levels of root zone salinity. *Plant Physiology* 106: 281–292.

Fabro, G., Kovács, I., Pavet, V., Szabados, L., Alvarez, M.E. (2004) Proline accumulation and *AtP5CS2* gene activation are induced by plant-pathogen incompatible interactions in Arabidopsis. *Molecular Plant-Microbe Interactions* 17: 343–350.

Fallon, K.M., Phillips, R. (1989) Responses to water stress in adapted and unadapted carrot cell suspension cultures. *Journal of Experimental Botany* 40: 681–687.

Farooq, M., Aziz, T., Hussain, M., Rehman, H., Jabran, K., Khan, M.B. (2008) Glycinebetaine improves chilling tolerance in hybrid maize. *Journal of Agronomy and Crop Science* 194: 152–160.

Fernandez, O., Theocharis, A., Bordiec, S., Feil, R., Jacquens, L., Clément, C., Fontaine, F., Barka, E.A. (2012) *Burkholderia phytofirmans* PsJN acclimates grapevine to cold by modulating carbohydrate metabolism. *Molecular Plant-Microbe Interactions* 25: 496–504.

Fougere, F., Le Rudulier, D., Streeter, J.G. (1991) Effects of salt stress on amino acid, organic acid and carbohydrate composition of roots, bacteroids and cytosol of alfalfa (*Medicago sativa* L.). *Plant Physiology* 96: 1228–1236.

Gadallah, M.A.A. (1999) Effects of proline and glycinebetaine on *Vicia faba* responses to salt stress. *Biologia Plantarum* 42: 249–257.

Gage, D.A., Rathinasabapathi, B. (1999) Role of glycine betaine and dimethylsulfoniopropionate in water-stress tolerance. In *Molecular Responses to Cold, Drought, Heat and Salt Stress in Higher Plants*, ed. Shinozaki, K., Yamaguchi-Shinozaki, K., 125–152. Austin: R.G. Landes Co.

Gagneul, D., Aïnouche, A., Duhazé, C., Lugan, R., Larher, F.R., Bouchereau, A. (2007) A reassessment of the function of the so-called compatible solutes in the halophytic Plumbaginaceae *Limonium latifolium*. *Plant Physiology* 144: 1598–1611.

Gao, M., Tao, R., Miura, K., Dandekar, A.M., Sugiura, A. (2001) Transformation of Japanese persimmon (*Diospyros kaki* Thunb.) with apple cDNA encoding NADP-dependent sorbitol-6-phosphate dehydrogenase. *Plant Science* 160: 837–845.

Garg, A.K., Kim, J.K., Owens, T.G., Ranwala, A.P., Choi, Y. Do, Kochian, L. V, and Wu, R.J. (2002) Trehalose accumulation in rice plants confers high tolerance levels to different abiotic stresses.(Abstract). *Proceedings of the National Academy of Sciences of the United States* 99: 15898.

Genard, H., Le Saos, J., Billard, J.P., Tremolieres, A., Boucaud, J. (1991) Effect of salinity on lipid composition, glycine betaine content and photosynthetic activity in chloroplasts of *Suaeda maritima*. *Plant Physiology and Biochemistry* 29: 421–427.

Gill, S.S., Anjum, N.A. (2014) Target osmoprotectants for abiotic stress tolerance in crop plants—glycine betaine and proline. In *Plant Adaptation to Environmental Change: Significance of Amino Acids and Their Derivatives*, ed. Anjum, N.A., Gill, S.S., Gill, R., 97–108. Wallingford: CAB International.

Godfray, H.C.J., Beddington, J.R., Crute, I.R., Haddad, L., Lawrence, D., Muir, J.F., Pretty, J., Robinson, S., Thomas, S.M., Toulmin, C. (2010) Food security: The challenge of feeding 9 billion people. *Science* 327: 812–818.

Gorham, J. (1995) Betaines in higher plants—biosynthesis and role in stress metabolism. In *Amino acids and Their Derivatives in Higher Plants*, ed. Wallsgrove R.M., 173–203. Society for Experimental Biology Seminar Series, Vol 56. Cambridge: Cambridge University Press.

Goswami, L., Sengupta, S., Mukherjee, S., Ray, S., Mukherjee, R., Lahiri Majumder, A. (2014) Targeted expression of L-myo-inositol 1-phosphate synthase from *Porteresia coarctata* (Roxb.) Tateoka confers multiple stress tolerance in transgenic crop plants. *Journal of Plant Biochemistry and Biotechnology* 23: 316–330.

Gubiš, J., Vaňková, R., Červená, V., Dragúňová, M., Hudcovicová, M., Lichtnerová, H., Dokupil, T., Jureková, Z. (2007) Transformed tobacco plants with increased tolerance to drought. *South African Journal of Botany* 73: 505–511.

Gueta-Dahan, Y., Yaniv, Z., Zilinskas, B.A., Ben-Hayyim, G. (1997) Salt and oxidative stress: Similar and specific responses and their relation to salt tolerance in citrus. *Planta* 203: 460–469.

Guicherd, P., Peltier, J.P., Gout, E., Bligny, R., Marigo, G. (1997) Osmotic adjustment in *Fraxinus excelsior* L.: Malate and mannitol accumulation in leaves under drought conditions. *Trees-Structure and Function* 11: 155–161.

Guo, P., Baum, M., Grando, S., Ceccarelli, S., Bai, G., Li, R., Von Korff, M., Varshney, R.K., Graner, A., Valkoun, J. (2009) Differentially expressed genes between drought-tolerant and drought-sensitive barley genotypes in response to drought stress during the reproductive stage. *Journal of Experimental Botany* 60: 3531–3544.

Gupta, A.K., Singh, J., Kaur, N., Singh, R. (1993a) Effect of polyethylene glycol-induced water stress on germination and reserve carbohydrate metabolism in chickpea cultivars differing in tolerance to water deficit. *Plant Physiology and Biochemistry* 31: 369–378.

Gupta, A.K., Singh, J., Kaur, N., Singh, R. (1993b) Effect of polyethylene glycol-induced water stress on uptake, interconversion and transport of sugars in chickpea seedlings. *Plant Physiology and Biochemistry* 31: 743–747.

Gupta, M., Sharma, P., Sarin, N.B., Sinha, A.K. (2009) Differential response of arsenic stress in two varieties of *Brassica juncea* L. *Chemosphere* 74: 1201–1208.

Guy, C.L., Huber, J.L.A., Huber, S.C. (1992) Sucrose phosphate synthase and sucrose accumulation at low temperature. *Plant Physiology* 100: 502–508.

Han, K.H.H., Hwang, C.H. (2003) Salt tolerance enhanced by transformation of a p5cs gene in carrot. *Journal of Plant Biotechnology* 5: 157–161.

Han, B., Fu, L., Zhang, D., He, X., Chen, Q., Peng, M., Zhang, J. (2016) Interspecies and intraspecies analysis of trehalose contents and the biosynthesis pathway gene family reveals crucial roles of trehalose in osmotic-stress tolerance in cassava. *International Journal of Molecular Sciences* 17: 1077.

Hanower, P., Brzozowska, J. (1975) Influence d'un choc osmotique sur la composition des feuilles de cotonnier en acides amines libres. *Phytochemistry* 14: 1691–1694.

Hanson, A.D., Rathinasabapathi, B., Chamberlin, B., Gage, D.A. (1991) Comparative physiological evidence that β-alanine betaine and choline-O-sulfate act as compatible osmolytes in halophytic *Limonium* species. *Plant Physiology* 97: 1199–1205.

Hanson, A.D., Rathinasabapathi, B., Rivoal, J., Burnet, M., Dillon, M.O., Gage, D.A. (1994) Osmoprotective compounds in the *Plumbaginaceae*: A natural experiment in metabolic engineering of stress tolerance. *Proceedings of the National Academy of Sciences* 91: 306–310.

Hare, P.D., Cress, W.A. (1997) Metabolic implications of stress-induced proline accumulation in plants. *Plant Growth Regulation* 21: 79–102.

Hare, P.D., Cress, W.A., Van Staden, J. (1998) Dissecting the roles of osmolyte accumulation during stress. *Plant, Cell and Environment* 21: 535–553.

Hare, P.D., Cress, W.A., Van Staden, J. (1999) Proline synthesis and degradation: A model system for elucidating stress-related signal transduction. *Journal of Experimental Botany* 50: 413–434.

Harinasut, P., Tsutsui, K., Takabe, T., Nomura, M., Takabe, T., Kishitani, S. (1996) Exogenous glycinebetaine accumulation and increased salt-tolerance in rice seedlings. *Bioscience Biotechnology and Biochemistry* 60: 366–368.

Harn, C., Khayat, E., Daie, J. (1993) Expression dynamics of genes encoding key carbon metabolism enzymes during sink to source transition of developing leaves. *Plant and Cell Physiology* 34: 1045–1053.

Hayashi, H., Mustardy, L., Deshnium, P., Ida, M., Murata, N. (1997) Transformation of *Arabidopsis thaliana* with the codA gene for choline oxidase; accumulation of glycinebetaine and enhanced tolerance to salt and cold stress. *The Plant Journal* 12: 133–142.

Hayat, S., Hayat, Q., Alyemeni, M.N., Wani, A.S., Pichtel, J., Ahmad, A. (2012) Role of proline under changing environments: A review. *Plant Signaling and Behavior* 7: 1456–1466.

He, Y., Li, Z., Chen, Y., Wang, Y. (2000) Effects of exogenous proline on the physiology of soyabean plantlets regenerated from embryos *in vitro* and on the ultrastructure of their mitochondria under NaCl stress. *Soybean Science* 19: 314–319.

Hirai, M. (1981) Purification and characteristics of sorbitol-6-phosphate dehydrogenase from loquat leaves. *Plant Physiology* 67: 221–224.

Holmström, K., Somersalo, S., Mandal, A., Palva, T.E., Welin, B. (2000) Improved tolerance to salinity and low temperature in transgenic tobacco producing glycine betaine. *Journal of Experimental Botany* 51: 177–185.

Hong, Z., Lakkineni, K., Zhang, Z., Verma, D.P.S. (2000) Removal of feedback inhibition of Δ-1-pyrroline-5-carboxylate synthetase results in increased proline accumulation and protection of plants from osmotic stress. *Plant Physiology* 122: 1129–1136.

Hossain, M.A., Fujita, M. (2010) Evidence for a role of exogenous glycinebetaine and proline in antioxidant defense and methylglyoxal detoxification systems in mung bean seedlings under salt stress. *Physiology and Molecular Biology of Plants* 16: 19–29.

Hsu, S.Y., Hsu, Y.T., Kao, C.H. (2003) The effect of polyethylene glycol on proline accumulation in rice leaves. *Biologia Plantarum* 46: 73–78.

Huber, S.C., Huber, J.L. (1996) Role and regulation of sucrose-phosphate synthase in higher plants. *Annual Review of Plant Biology* 47: 431–444.

Hussain, M., Malik, M.A., Farooq, M., Ashraf, M.Y., Cheema, M.A. (2008) Improving drought tolerance by exogenous application of glycinebetaine and salicylic acid in sunflower. *Journal of Agronomy and Crop Science* 194: 193–199.

Igarashi, Y., Yoshiba, Y., Sanada, Y., Yamaguchi-Shinozaki, K., Wada, K., Shinozaki, K. (1997) Characterization of the gene for Δ¹-pyrroline-5-carboxylate synthetase and correlation between the expression of the gene and salt tolerance in *Oryza sativa* L. *Plant Molecular Biology* 33: 857–865.

Iordachescu, M., Imai, R. (2008) Trehalose biosynthesis in response to abiotic stresses. *Journal of Integrative Plant Biology* 50: 1223–1229.

Iqbal, N., Ashraf, M., Ashraf, M. (2009) Influence of exogenous glycine betaine on gas exchange and biomass production in sunflower (*Helianthus annuus* L.) under water limited conditions. *Journal of Agronomy and Crop Science* 195: 420–426.

Irigoyen, J.J., Einerich, D.W., Sánchez-Díaz, M. (1992) Water stress induced changes in concentrations of proline and total soluble sugars in nodulated alfalfa (*Medicago sativa*) plants. *Physiologia Plantarum* 84: 55–60.

Ishitani, M., Nakamura, T., Han, S., Takabe, T. (1995) Expression of the betaine aldehyde dehydrogenase gene in barley in response to osmotic stress and abscisic acid. *Plant Molecular Biology* 27: 307–315.

Ishitani, M., Majumder, A.L., Bornhouser, A., Michalowski, C.B., Jensen, R.G., Bohnert, H.J. (1996) Coordinate transcriptional induction of myo-inositol metabolism during environmental stress. *The Plant Journal* 9: 537–548.

Islam, M.M., Hoque, M.A., Okuma, E., Banu, M.N.A., Shimoishi, Y., Nakamura, Y., Murata, Y. (2009) Exogenous proline and glycinebetaine increase antioxidant enzyme activities and confer tolerance to cadmium stress in cultured tobacco cells. *Journal of Plant Physiology* 166: 1587–1597.

Islam, S., Chawdhury, M.H.R., Hossain, I., Sayeed, S.R., Rahman, S., Azam, F.M.S., Rahmatullah, M. (2014) A study on callus induction of *Ipomoea mauritiana*: An Ayurvedic medicinal plant. *American-Eurasian Journal of Sustainable Agriculture* 86–94.

Jang, I.C., Oh, S.J., Seo, J.S., Choi, W.B., Song, S.I., Kim, C.H., Kim, Y.S., Seo, H.S., Do Choi, Y., Nahm, B.H. (2003) Expression of a bifunctional fusion of the *Escherichia coli* genes for trehalose-6-phosphate synthase and trehalose-6-phosphate phosphatase in transgenic rice plants increases trehalose accumulation and abiotic stress tolerance without stunting growth. *Plant Physiology* 131: 516–524.

Jha, A.B., Dubey, R.S. (2004) Carbohydrate metabolism in growing rice seedlings under arsenic toxicity. *Journal of Plant Physiology* 161: 867–872.

Joshi, V., Joung, J.G., Fei, Z., Jander, G. (2010) Interdependence of threonine, methionine and isoleucine metabolism in plants: Accumulation and transcriptional regulation under abiotic stress. *Amino Acids* 39: 933–947.

Kanayama, Y., Mori, H., Imaseki, H., Yamaki, S. (1992) Nucleotide sequence of a cDNA encoding NADP-sorbitol-6-phosphate dehydrogenase from apple. *Plant Physiology* 100: 1607.

Kanayama, Y., Moriguchi, R., Deguchi, M., Yamaki, S., Kanahama, K. (2006) Effects of environmental stresses and abscisic acid on sorbitol-6-phosphate dehydrogenase expression in *Rosaceae* fruit trees. *Acta Horticulturae* 738: 375–381.

Kann, S.C., Everard, J.D., Loescher, W.H. (1993) Growth, salt tolerance and mannitol accumulation in celery. *Horticultural Science* 28: 486.

Karabudak, T., Bor, M., Özdemir, F., Türkan, I. (2014) Glycine betaine protects tomato (*Solanum lycopersicum*) plants at low temperature by inducing fatty acid desaturase7 and lipoxygenase gene expression. *Molecular Biology Reports* 41: 1401–1410.

Kariya, K., Sameeullah, M., Sasaki, T., Yamamoto, Y. (2017) Overexpression of the sucrose transporter gene NtSUT1 alleviates aluminum-induced inhibition of root elongation in tobacco (*Nicotiana tabacum* L.) *Soil Science and Plant Nutrition* 63: 45 54.

Kaur, S., Gupta, A.K., Kaur, N. (1998) Gibberellin A3 reverses the effect of salt stress in chickpea (*Cicer arietinum* L.) seedlings by enhancing amylase activity and mobilization of starch in cotyledons. *Plant Growth Regulation* 26: 85–90.

Kavi Kishor, P., Sangam, S., Amrutha, R.N., Laxmi, P.S., Naidu, K., Rao, K., Rao, S., Reddy, K., Theriappan, P., Sreenivasulu, N. (2005) Regulation of proline biosynthesis, degradation, uptake and transport in higher plants: Its implications in plant growth and abiotic stress tolerance. *Current Science* 88: 424–438.

Kavi Kishor, P.B., Hong, Z., Miao, G.H., Hu, C.A.A., Verma, D.P.S. (1995) Overexpression of Δ-pyrroline-5-carboxylate synthetase increases proline production and confers osmotolerance in transgenic plants. *Plant Physiology* 108: 1387–1394.

Kaya, C., Sonmez, O., Aydemir, S., Ashraf, M., Dikilitas, M. (2013) Exogenous application of mannitol and thiourea regulates plant growth and oxidative stress responses in salt-stressed maize (*Zea mays* L.) *Journal of Plant Interactions* 8: 234–241.

Kempa, S., Krasensky, J., Dal Santo, S., Kopka, J., Jonak, C. (2008) A central role of abscisic acid in stress-regulated carbohydrate metabolism. *PloS One* 3: e3935.

Kerepesi, I., Galiba, G. (2000) Osmotic and salt stress-induced alteration in soluble carbohydrate content in wheat seedlings. *Crop Science* 40: 482–487.

Keunen, E.L.S., Peshev, D., Vangronsveld, J., Van Den Ende, W.I.M., Cuypers, A.N.N. (2013) Plant sugars are crucial players in the oxidative challenge during abiotic stress: Extending the traditional concept. *Plant, Cell and Environment* 36: 1242–1255.

Kim, A.R., Min, J.H., Lee, K.H., Kim, C.S. (2017) PCA22 acts as a suppressor of atrzf1 to mediate proline accumulation in response to abiotic stress in *Arabidopsis. Journal of Experimental Botany* 68: 1797–1809.

Kim, J.Y., Mahé, A., Brangeon, J., Prioul, J.L. (2000) A maize vacuolar invertase, IVR2, is induced by water stress. Organ/tissue specificity and diurnal modulation of expression. *Plant Physiology* 124: 71–84.

Klein, R.R., Crafts-Brandner, S.J., Salvucci, M.E. (1993) Cloning and developmental expression of the sucrose-phosphate-synthase gene from spinach. *Planta* 190: 498–510.

Klemens, P.A.W., Patzke, K., Deitmer, J., Spinner, L., Le Hir, R., Bellini, C., Bedu, M., Chardon, F., Krapp, A., Neuhaus, H.E. (2013) Over-expression of the vacuolar sugar carrier AtSWEET16 modifies germination, growth, and stress tolerance in *Arabidopsis. Plant Physiology* 163: 1338.

Knight, H., Trewavas, A.J., Knight, M.R. (1997) Calcium signalling in *Arabidopsis thaliana* responding to drought and salinity. *The Plant Journal* 12: 1067–1078.

Koch, K.E., Ying, Z., Wu, Y., Avigne, W.T. (2000) Multiple paths of sugar-sensing and a sugar/oxygen overlap for genes of sucrose and ethanol metabolism. *Journal of Experimental Botany* 51: 417–427.

Kong, X., Pan, J., Zhang, M., Xing, X.I.N., Zhou, Y.A.N., Liu, Y., Li, D., Li, D. (2011) ZmMKK4, a novel group C mitogen-activated protein kinase kinase in maize (*Zea mays*), confers salt and cold tolerance in transgenic *Arabidopsis. Plant, Cell and Environment* 34: 1291–1303.

Kosmas, S.A., Argyrokastritis, A., Loukas, M.G., Eliopoulos, E., Tsakas, S., Kaltsikes, P.J. (2006) Isolation and characterization of drought-related trehalose 6-phosphate-synthase gene from cultivated cotton (*Gossypium hirsutum* L.). *Planta* 223: 329–339.

Koyro, H.W., Ahmad, P., Geissler, N. (2012) Abiotic stress responses in plants: An overview. In *Environmental Adaptations and Stress Tolerance of Plants, in the Era of Climate Change*, ed. Ahmad, P., Prasad, M.N.V., 1–28. New York: Springer.

Kumar, K., Rao, K.P., Sharma, P., Sinha, A.K. (2008) Differential regulation of rice mitogen activated protein kinase kinase (MKK) by abiotic stress. *Plant Physiology and Biochemistry* 46: 891–897.

Kumar, S., Dhingra, A., Daniell, H. (2004) Plastid-expressed betaine aldehyde dehydrogenase gene in carrot cultured cells, roots and leaves confers enhanced salt tolerance. *Plant Physiology* 136: 2843–2854.

Leprince, A., Magalhaes, N., De Vos, D., Bordenave, M., Crilat, E., Clement, G., Meyer, C., Munnik, T., Savoure, A. (2015) Involvement of Phosphatidylinositol 3-kinase in the regulation of proline catabolism in *Arabidopsis thaliana*. *Frontiers in Plant Science* 5: 772.

Li, H.W., Zang, B.S., Deng, X.W., Wang, X.P. (2011) Over-expression of the trehalose-6-phosphate synthase gene *OsTPS1* enhances abiotic stress tolerance in rice. *Planta* 234: 1007–1018.

Li, M., Guo, S., Xu, Y., Meng, Q., Li, G., Yang, X. (2014) Glycine betaine-mediated potentiation of *HSP* gene expression involves calcium signaling pathways in tobacco exposed to NaCl stress. *Physiologia Plantarum* 150: 63–75.

Lilius, G., Holmberg, N., Bülow, L. (1996) Enhanced NaCl stress tolerance in transgenic tobacco expressing bacterial choline dehydrogenase. *Nature Biotechnology* 14: 177–180.

Liu, T., Van Staden, J. (2001) Growth rate, water relations and ion accumulation of soybean callus lines differing in salinity tolerance under salinity stress and its subsequent relief. *Plant Growth Regulation* 34: 277–285.

Lo Bianco, R., Rieger, M., Sung, S.S. (2000) Effect of drought on sorbitol and sucrose metabolism in sinks and sources of peach. *Physiologia Plantarum* 108: 71–78.

Loescher, W.H., Tyson, R.H., Everard, J.D., Redgwell, R.J., Bieleski, R.L. (1992) Mannitol synthesis in higher plants evidence for the role and characterization of a nadph-dependent mannose 6-phosphate reductase. *Plant Physiology* 98: 1396–1402.

Loewus, F.A., Murthy, P.P.N. (2000) Myo-inositol metabolism in plants. *Plant Science* 150: 1–19.

Lokhande, V.H., Suprasanna, P. (2012) Prospects of halophytes in understanding and managing abiotic stress tolerance. In *Environmental Adaptations and Stress Tolerance of Plants in the Era of Climate Change*, ed. Ahmad, P., Prasad M.N.V., 29–56. New York: Springer.

Lone, M.I., Kueh, J.S.H., Wyn Jones, R.G., Bright, S.W.J. (1987) Influence of proline and glycinebetaine on salt tolerance of cultured barley embryos. *Journal of Experimental Botany* 38: 479–490.

López, M., Tejera, N.A., Iribarne, C., Lluch, C., Herrera-Cervera, J.A. (2008) Trehalose and trehalase in root nodules of *Medicago truncatula* and *Phaseolus vulgaris* in response to salt stress. *Physiologia Plantarum* 134: 575–582.

Luo, D., Niu, X., Yu, J., Yan, J., Gou, X., Lu, B.R., Liu, Y. (2012) Rice choline monooxygenase (OsCMO) protein functions in enhancing glycine betaine biosynthesis

in transgenic tobacco but does not accumulate in rice (*Oryza sativa* L. ssp. japonica). *Plant Cell Reports* 31: 1625–1635.

Lutts, S. (2000) Exogenous glycinebetaine reduces sodium accumulation in salt-stressed rice plants. *International Rice Research Notes* 25: 39–40.

Lyu, J., Min, S., Lee, J., Lim, Y., Kim, J.K., Bae, C.H., Liu, J. (2013) Over-expression of a trehalose-6-phosphate synthase/phosphatase fusion gene enhances tolerance and photosynthesis during drought and salt stress without growth aberrations in tomato. *Plant Cell, Tissue and Organ Culture* 112: 257–262.

Madan, S., Nainawatee, H.S., Jain, R.K., Chowdhury, J.B. (1995) Proline and proline metabolising enzymes in *in-vitro* selected NaCl-tolerant *Brassica juncea* L. under salt stress. *Annals of Botany* 76: 51–57.

Mäkelä, P., Jokinen, K., Kontturi, M., Peltonen-Sainio, P., Pehu, E., Somersalo, S. (1998a). Foliar application of glycine betaine—a novel product from sugar beet—as an approach to increase tomato yield. *Industrial Crops and Products* 7: 139–148.

Mäkelä, P., Kontturi, M., Pehu, E., Somersalo, S. (1999) Photosynthetic response of drought-and salt-stressed tomato and turnip rape plants to foliar-applied glycine-betaine. *Physiologia Plantarum* 105: 45–50.

Mäkelä, P., Kärkkäinen, J., Somersalo, S. (2000) Effect of glycinebetaine on chloroplast ultrastructure, chlorophyll and protein content, and RuBPCO activities in tomato grown under drought or salinity. *Biologia Plantarum* 43: 471–475.

Mäkelä, P., Munns, R., Colmer, T.D., Condon, A.G., Peltonen-Sainio, P. (1998b) Effect of foliar applications of glycine betaine on stomatal conductance, abscisic acid and solute concentrations in leaves of salt-or drought-stressed tomato. *Functional Plant Biology* 25: 655–663.

Mansour, M.M.F. (1998) Protection of plasma membrane of onion epidermal cells by glycinebetaine and proline against NaCl stress. *Plant Physiology and Biochemistry* 36: 767–772.

Mansour, M.M.F. (2000) Nitrogen containing compounds and adaptation of plants to salinity stress. *Biologia Plantarum* 43: 491–500.

Matysik, J., Alia, Bhalu, B., Mohanty, P. (2002) Molecular mechanisms of quenching of reactive oxygen species by proline under stress in plants. *Current Science* 82: 525–532.

McCue, K.F., Hanson, A.D. (1990) Drought and salt tolerance: Towards understanding and application. *Trends in Biotechnology* 8: 358–362.

McNeil, S.D., Nuccio, M.L., Ziemak, M.J., Hanson, A.D. (2001) Enhanced synthesis of choline and glycine betaine in transgenic tobacco plants that over-express phosphoethanolamine N-methyltransferase. *Proceedings of the National Academy of Sciences* 98: 10001–10005.

Miller, W.B., Langhans, R.W. (1990) Low temperature alters carbohydrate metabolism in Easter lily bulbs. *Horticultural Science* 25: 463–465.

Miranda, J.A., Avonce, N., Suárez, R., Thevelein, J.M., Van Dijck, P., Iturriaga, G. (2007) A bifunctional TPS–TPP enzyme from yeast confers tolerance to multiple and extreme abiotic-stress conditions in transgenic *Arabidopsis*. *Planta* 226: 1411–1421.

Moghaieb, R.E.A., Saneoka, H., Fujita, K. (2004) Effect of salinity on osmotic adjustment, glycinebetaine accumulation and the betaine aldehyde dehydrogenase gene expression in two halophytic plants, *Salicornia europaea* and *Suaeda maritima*. *Plant Science* 166: 1345–1349.

Murakeözy, É.P., Nagy, Z., Duhazé, C., Bouchereau, A., Tuba, Z. (2003) Seasonal changes in the levels of compatible osmolytes in three halophytic species of inland saline vegetation in Hungary. *Journal of Plant Physiology* 160: 395–401.

Murmu, K., Murmu, S., Kandu, C.K., Bera, P.S. (2017) Exogenous proline and glycine betaine in plants under stress tolerance. *International Journal of Current Microbiology and Applied Sciences* 6: 901–913.

Nanjo, T., Kobayashi, M., Yoshiba, Y., Sanada, Y., Wada, K., Tsukaya, H., Kakubari, Y., Yamaguchi-Shinozaki, K., Shinozaki, K. (1999) Biological functions of proline in morphogenesis and osmotolerance revealed in antisense transgenic *Arabidopsis thaliana*. *The Plant Journal* 18: 185–193.

Nelson, D.E., Rammesmayer, G., Bohnert, H.J. (1998) Regulation of cell-specific inositol metabolism and transport in plant salinity tolerance. *The Plant Cell* 10: 753–764.

Nguyen, A., Lamant, A. (1988) Pinitol and myo-inositol accumulation in water-stressed seedlings of maritime pine. *Phytochemistry* 27: 3423–3427.

Niu, X., Zheng, W., Lu, B.R., Ren, G., Huang, W., Wang, S., Liu, J., Tang, Z., Luo, D., Wang, Y. (2007) An unusual posttranscriptional processing in two betaine aldehyde dehydrogenase loci of cereal crops directed by short, direct repeats in response to stress conditions. *Plant Physiology* 143: 1929–1942.

Nolte, K.D., Hanson, A.D., Gage, D.A. (1997) Proline accumulation and methylation to proline betaine in citrus: Implications for genetic engineering of stress resistance. *Journal of the American Society for Horticultural Science* 122: 8–13.

Nounjan, N., Nghia, P.T., Theerakulpisut, P. (2012) Exogenous proline and trehalose promote recovery of rice seedlings from salt-stress and differentially modulate antioxidant enzymes and expression of related genes. *Journal of Plant Physiology* 169: 596–604.

Okuma, E., Murakami, Y., Shimoishi, Y., Tada, M., Murata, Y. (2004) Effects of exogenous application of proline and betaine on the growth of tobacco cultured cells under saline conditions. *Soil Science and Plant Nutrition* 50: 1301–1305.

Otte, M.L., Wilson, G., Morris, J.T., Moran, B.M. (2004) Dimethylsulfoniopropionate (DMSP) and related compounds in higher plants. *Journal of Experimental Botany* 55: 1919–1925.

Öztürk, L., Demir, Y. (2002) *In vivo* and *in vitro* protective role of proline. *Plant Growth Regulation* 38: 259–264.

Park, S.H., Chung, M.S., Lee, S., Lee, K.H., Kim, C.S. (2017) Loss of ribosomal protein L24A (RPL24A) suppresses proline accumulation of *Arabidopsis thaliana* ring zinc finger 1 (atrzf1) mutant in response to osmotic stress. *Biochemical and Biophysical Research Communications* 494: 499–503.

Parre, E., Ghars, M.A., Leprince, A.S., Thiery, L., Lefebvre, D., Bordenave, M., Richard, L., Mazars, C., Abdelly, C., Savouré, A. (2007) Calcium signaling via phospholipase C is essential for proline accumulation upon ionic but not nonionic hyperosmotic stresses in *Arabidopsis*. *Plant Physiology* 144: 503–512.

Patonnier, M.P., Peltier, J.P., Marigo, G. (1999) Drought-induced increase in xylem malate and mannitol concentrations and closure of *Fraxinus excelsior* L. stomata. *Journal of Experimental Botany* 50: 1223–1229.

Pelleschi, S., Guy, S., Kim, J.Y., Pointe, C., Mahé, A., Barthes, L., Leonardi, A., Prioul, J.L. (1999) Ivr2, a candidate gene for a QTL of vacuolar invertase activity in maize leaves. Gene-specific expression under water stress. *Plant Molecular Biology* 39: 373–380.

Pilon-Smits, E.A.H., Terry, N., Sears, T., van Dun, K. (1999) Enhanced drought resistance in fructan-producing sugar beet. *Plant Physiology and Biochemistry* 37: 313–317.

Planchet E., Limami A.M. (2015) Amino acid synthesis under abiotic stress. In *Amino Acids in Higher Plants*, ed. D'Mello J.P.F., 262–276.Wallingford: CAB International.

Ponnu, J., Wahl, V., Schmid, M. (2011) Trehalose-6-phosphate: Connecting plant metabolism and development. *Frontiers in Plant Science* 2: 70.

Prabhavathi, V.R., Rajam, M.V. (2007) Polyamine accumulation in transgenic eggplant enhances tolerance to multiple abiotic stresses and fungal resistance. *Plant Biotechnology* 24: 273–282.

Pramanik, M.H.R., Imai, R. (2005) Functional identification of a trehalose 6-phosphate phosphatase gene that is involved in transient induction of trehalose biosynthesis during chilling stress in rice. *Plant Molecular Biology* 58: 751–762.

Qiu, Z.B., Wang, Y.F., Zhu, A.J., Peng, F.L., Wang, L.S. (2014) Exogenous sucrose can enhance tolerance of *Arabidopsis thaliana* seedlings to salt stress. *Biologia Plantarum* 58: 611–617.

Radyukina, N., Shashukova, A., Makarova, S., Kuznetsov, V. (2011) Exogenous proline modifies differential expression of superoxide dismutase genes in UV-B-irradiated *Salvia officinalis* plants. *Russian Journal of Plant Physiology* 58: 51–59.

Rajashekar, C.B., Zhou, H., Marcum, K.B., Prakash, O. (1999) Glycine betaine accumulation and induction of cold tolerance in strawberry (*Fragaria × ananassa* Duch.) plants. *Plant Science* 148: 175–183.

Rajendrakumar, C.S. V, Reddy, B.V.B., Reddy, A.R. (1994) Proline-protein interactions: Protection of structural and functional integrity of M4 lactate dehydrogenase. *Biochemical and Biophysical Research Communications* 201: 957–963.

Ranganayakulu, G.S., Veeranagamallaiah, G., Sudhakar, C. (2013) Effect of salt stress on osmolyte accumulation in two groundnut cultivars (*Arachis hypogaea* L.) with contrasting salt tolerance. *African Journal of Plant Science* 7: 586–592.

Ranney, T.G., Bassuk, N.L., Whitlow, T.H. (1991) Osmotic adjustment and solute constituents in leaves and roots of water-stressed cherry (*Prunus*) trees. *Journal of the American Society for Horticultural Science* 116: 684–688.

Rathinasabapathi, B. (2000) Metabolic engineering for stress tolerance: Installing osmoprotectant synthesis pathways. *Annals of Botany* 86: 709–716.

Rhodes, D., Hanson, A.D. (1993) Quaternary ammonium and tertiary sulfonium compounds in higher plants. *Annual Review of Plant Biology* 44: 357–384.

Rhodes, D., Nadolska-Orczyk, A., Rich, P.J. (2002) Salinity, osmolytes and compatible solutes. In *Salinity: Environment-Plants-Molecules*, ed. Läuchli A., Lüttge U., 181–204. Dordrecht: Springer.

Robinson, S.P., Jones, G.P. (1986) Accumulation of glycinebetaine in chloroplasts provides osmotic adjustment during salt stress. *Functional Plant Biology* 13: 659–668.

Roitsch, T. (1999) Source-sink regulation by sugar and stress. *Current Opinion in Plant Biology* 2: 198–206.

Roitsch, T., Balibrea, M.E., Hofmann, M., Proels, R., Sinha, A.K. (2003) Extracellular invertase: Key metabolic enzyme and PR protein. *Journal of Experimental Botany* 54: 513–524.

Rosa, M., Prado, C., Podazza, G., Interdonato, R., González, J.A., Hilal, M., Prado, F.E. (2009) Soluble sugars: Metabolism, sensing and abiotic stress: A complex network in the life of plants. *Plant Signaling and Behavior* 4: 388–393.

Roy, D., Basu, N., Bhunia, A., Banerjee, S.K. (1993) Counteraction of exogenous L-proline with NaCl in salt-sensitive cultivar of rice. *Biologia Plantarum* 35: 69–72.

Roychoudhury, A., Banerjee, A. (2016) Endogenous glycine betaine accumulation mediates abiotic stress tolerance in plants. *Tropical Plant Research* 3: 105–111.

Rumpho, M.E., Edwards, G.E., Loescher, W.H. (1983) A pathway for photosynthetic carbon flow to mannitol in celery leaves activity and localization of key enzymes. *Plant Physiology* 73: 869–873.

Sadak, M.S. (2016) Mitigation of drought stress on Fenugreek plant by foliar application of trehalose. *International Journal of Chemical Technology Research* 9: 147–155.

Sakamoto, A., Murata, A.N. (1998) Metabolic engineering of rice leading to biosynthesis of glycinebetaine and tolerance to salt and cold. *Plant Molecular Biology* 38: 1011–1019.

Sakamoto, A., Murata, N. (2002) The role of glycine betaine in the protection of plants from stress: Clues from transgenic plants. *Plant, Cell and Environment* 25: 163–171.

Sami, F., Yusuf, M., Faizan, M., Faraz, A., Hayat, S. (2016) Role of sugars under abiotic stress. *Plant Physiology and Biochemistry* 109: 54–61.

Sánchez, F.J., Manzanares, M., de Andres, E.F., Tenorio, J.L., Ayerbe, L. (1998) Turgor maintenance, osmotic adjustment and soluble sugar and proline accumulation in 49 pea cultivars in response to water stress. *Field Crops Research* 59: 225–235.

Saneoka, H., Nagasaka, C., Hahn, D.T., Yang, W.J., Premachandra, G.S., Joly, R.J., Rhodes, D. (1995) Salt tolerance of glycinebetaine-deficient and -containing maize lines. *Plant Physiology* 107: 631–638.

Savouré, A., Hua, X.J., Bertauche, N., Van Montagu, M., Verbruggen, N. (1997) Abscisic acid-independent and abscisic acid-dependent regulation of proline biosynthesis following cold and osmotic stresses in *Arabidopsis thaliana*. *Molecular and General Genetics* 254: 104–109.

Sawahel, W.A., Hassan, A.H. (2002) Generation of transgenic wheat plants producing high levels of the osmoprotectant proline. *Biotechnology Letters* 24: 721–725.

Schat, H., Sharma, S.S., Vooijs, R. (1997) Heavy metal-induced accumulation of free proline in a metal-tolerant and a nontolerant ecotype of *Silene vulgaris*. *Physiologia Plantarum* 101: 477–482.

Schobert, B. (1977) Is there an osmotic regulatory mechanism in algae and higher plants? *Journal of Theoretical Biology* 68: 17–26.

Schobert, B., Tschesche, H. (1978) Unusual solution properties of proline and its interaction with proteins. *Biochimica et Biophysica Acta -General Subjects* 541: 270–277.

Schwacke, R., Grallath, S., Breitkreuz, K.E., Stransky, E., Stransky, H., Frommer, W.B., Rentsch, D. (1999) LeProT1, a transporter for proline, glycine betaine and γ-amino butyric acid in tomato pollen. *The Plant Cell* 11: 377–391.

Servet, C., Ghelis, T., Richard, L., Zilberstein, A., Savoure, A. (2012) Proline dehydrogenase: A key enzyme in controlling cellular homeostasis. *Frontiers in Bioscience* 17: 607–620.

Sharma, P., Dubey, R. (2005) Drought induces oxidative stress and enhances the activities of antioxidant enzymes in growing rice seedlings. *Plant Growth Regulation* 46: 209–221.

Sharma, P., Dubey, R. (2007) Involvement of oxidative stress and role of antioxidative defense system in growing rice seedlings exposed to toxic concentrations of aluminum. *Plant Cell Reports* 26: 2027–2038.

Sharma, P., Dubey, R.S. (2004) Ascorbate peroxidase from rice seedlings: Properties of enzyme isoforms, effects of stresses and protective roles of osmolytes. *Plant Science* 167: 541–550.

Sharma, P., Dubey, R.S. (2011) Abiotic stress-induced metabolic alterations in crop plants: Strategies for improving stress tolerance. In *Advances in Life Sciences*, ed. Sinha, R.P., Sharma, N.K., Rai, A.K., 1–54. New Delhi: IK International Publishing House Pvt. Ltd.

Sharma, P., Dubey, R.S. (2005) Modulation of nitrate reductase activity in rice seedlings under aluminity toxicity and water stress: Role of osmolytes as enzyme protectant. *Journal of Plant Physiology* 162: 854–864.

Sharma, S.S., Dietz, K.J. (2006) The significance of amino acids and amino acid-derived molecules in plant responses and adaptation to heavy metal stress. *Journal of Experimental Botany* 57: 711–726.

Sharma, P., Jha, A.B., Dubey, R.S., Pessarakli, M. (2012) Reactive oxygen species, oxidative damage and antioxidative defense mechanism in plants under stressful conditions. *Journal of Botany* 2012. doi: 10.1155/2012/217037.

Sheveleva, E., Chmara, W., Bohnert, H.J., Jensen, R.G. (1997) Increased salt and drought tolerance by D-ononitol production in transgenic *Nicotiana tabacum* L. *Plant Physiology* 115: 1211–1219.

Shima, S., Matsui, H., Tahara, S., Imai, R. (2007) Biochemical characterization of rice trehalose-6-phosphate phosphatases supports distinctive functions of these plant enzymes. *The FEBS Journal* 274: 1192–1201.

Shirasawa, K., Takabe, T., Takabe, T., Kishitani, S. (2006) Accumulation of glycinebetaine in rice plants that over-express choline monooxygenase from spinach and evaluation of their tolerance to abiotic stress. *Annals of Botany* 98: 565–571.

Sickler, C.M., Edwards, G.E., Kiirats, O., Gao, Z., Loescher, W. (2007) Response of mannitol-producing *Arabidopsis thaliana* to abiotic stress. *Functional Plant Biology* 34: 382–391.

Singh, M., Kumar, J., Singh, S., Singh, V.P., Prasad, S.M. (2015) Roles of osmoprotectants in improving salinity and drought tolerance in plants: A review. *Reviews in Environmental Science and Biotechnology* 14: 407–426.

Siringam, K., Juntawong, N., Cha-um, S., Boriboonkaset, T., Kirdmanee, C. (2012) Salt tolerance enhancement in indica rice (*Oryza sativa* L. spp. indica) seedlings using exogenous sucrose supplementation. *Plant Omics* 5: 52–59.

Slama, I., Abdelly, C., Bouchereau, A., Flowers, T., Savouré, A. (2015) Diversity, distribution and roles of osmoprotective compounds accumulated in halophytes under abiotic stress. *Annals of Botany* 115: 433–447.

Slukhai, S.I., Shvedova, O.E. (1972) Dynamics of free amino acid content in maize plants in connection with soil water regime. *Fiziol Biokhim Kul'trast* 4: 151–156.

Smirnoff, N., Cumbes, Q.J. (1989) Hydroxyl radical scavenging activity of compatible solutes. *Phytochemistry* 28: 1057–1060.

Somersalo, S. Kyei-Boahen, S., Pehu, E. (1996) Exogenous glycine betaine application as a possibility to increase low temperature tolerance of crop plants. *Nordisk Jordbruksforskning* 78: 10.

Stefels, J. (2000) Physiological aspects of the production and conversion of DMSP in marine algae and higher plants. *Journal of Sea Research* 43: 183–197.

Stoop, J.M.H., Pharr, D.M. (1994a) Mannitol metabolism in celery stressed by excess macronutrients. *Plant Physiology* 106: 503–511.

Stoop, J.M.H., Pharr, D.M. (1994b) Growth substrate and nutrient salt environment alter mannitol-to-hexose partitioning in celery petioles. *Journal of the American Society for Horticultural Science* 119: 237–242.

Stoop, J.M.H., Williamson, J.D., Pharr, D.M. (1996) Mannitol metabolism in plants: A method for coping with stress. *Trends in Plant Science* 1: 139–144.

Streeter, J.G., Lohnes, D.G., Fioritto, R.J. (2001) Patterns of pinitol accumulation in soybean plants and relationships to drought tolerance. *Plant, Cell and Environment* 24: 429–438.

Strizhov, N., Ábrahám, E., Ökrész, L., Blickling, S., Zilberstein, A., Schell, J., Koncz, C., Szabados, L. (1997) Differential expression of two *P5CS* genes controlling proline accumulation during salt-stress requires ABA and is regulated by ABA1, ABI1 and AXR2 in *Arabidopsis*. *The Plant Journal* 12: 557–569.

Sunda, W., Kieber, D.J., Kiene, R.P., Huntsman, S. (2002) An antioxidant function for DMSP and DMS in marine algae. *Nature* 418: 317–320.

Surekha, C.H., Kumari, K.N., Aruna, L. V, Suneetha, G., Arundhati, A., Kishor, P.B.K. (2014) Expression of the *Vigna aconitifolia P5CSF129A* gene in transgenic pigeonpea enhances proline accumulation and salt tolerance. *Plant Cell, Tissue and Organ Culture* 116: 27–36.

Sureshan, K.M., Murakami, T., Watanabe, Y. (2009) Total syntheses of cyclitol based natural products from myo-inositol: Brahol and pinpollitol. *Tetrahedron* 65: 3998–4006.

Szabados, L., Savouré, A. (2010) Proline: A multifunctional amino acid. *Trends in Plant Science* 15: 89–97.

Székely, G., Ábrahám, E., Cséplő, Á., Rigó, G., Zsigmond, L., Csiszár, J., Ayaydin, F., Strizhov, N., Jásik, J., Schmelzer, E., et al. (2008) Duplicated *P5CS* genes of *Arabidopsis* play distinct roles in stress regulation and developmental control of proline biosynthesis. *Plant Journal* 53: 11–28.

Tang, W., Sun, J., Liu, J., Liu, F., Yan, J., Gou, X., Lu, B.R., Liu, Y. (2014) RNAi-directed downregulation of betaine aldehyde dehydrogenase 1 (OsBADH1) results in decreased stress tolerance and increased oxidative markers without affecting glycine betaine biosynthesis in rice (Oryza sativa). *Plant Molecular Biology* 86: 443–454.

Thakur, P.S., Rai, V.K. (1982) Dynamics of amino acid accumulation of two differentially drought resistant *Zea mays* cultivars in response to osmotic stress. *Environmental and Experimental Botany* 22: 221–226.

Theerakulpisut, P., Gunnula, W. (2012) Exogenous sorbitol and trehalose mitigated salt stress damage in salt-sensitive but not salt-tolerant rice seedlings. *Asian Journal of Crop Science* 4: 165–170.

Thiery, L., Leprince, A.S., Lefebvre, D., Ghars, M.A., Debarbieux, E., Savouré, A. (2004) Phospholipase D is a negative regulator of proline biosynthesis in *Arabidopsis thaliana*. *Journal of Biological Chemistry* 279: 14812–14818.

Thomashow, M.F. (1999) Plant cold acclimation: Freezing tolerance genes and regulatory mechanisms. *Annual Review of Plant Biology* 50: 571–599.

Tian, F., Wang, W.W., Liang, C., Wang, X., Wang, G., Wang, W.W. (2017) Overaccumulation of glycine betaine makes the function of the thylakoid membrane better in wheat under salt stress. *The Crop Journal* 5: 73–82.

Toroser, D., Huber, S.C. (1997) Protein phosphorylation as a mechanism for osmotic-stress activation of sucrose-phosphate synthase in spinach leaves. *Plant Physiology* 114: 947–955.

Trinchant, J.C., Boscari, A., Spennato, G., Van de Sype, G., Le Rudulier, D. (2004) Proline betaine accumulation and metabolism in alfalfa plants under sodium chloride stress. Exploring its compartmentalization in nodules. *Plant Physiology* 135: 1583–1594.

Tripathi, B.N., Gaur, J.P. (2004) Relationship between copper- and zinc-induced oxidative stress and proline accumulation in *Scenedesmus* sp. *Planta* 219: 397–404.

Trossat, C., Rathinasabapathi, B., Weretilnyk, E.A., Shen, T.L., Huang, Z.H., Gage, D.A., Hanson, A.D. (1998) Salinity promotes accumulation of 3-dimethylsulfoniopropionate and its precursor S-methylmethionine in chloroplasts. *Plant Physiology* 116: 165–171.

Vandesteene, L., Ramon, M., Le Roy, K., Van Dijck, P., Rolland, F. (2010) A single active trehalose-6-P synthase (TPS) and a family of putative regulatory TPS-like proteins in *Arabidopsis*. *Molecular Plant* 3: 406–419.

Vargas, W.A., Salerno, G.L. (2010) The Cinderella story of sucrose hydrolysis: Alkaline/neutral invertases, from cyanobacteria to unforeseen roles in plant cytosol and organelles. *Plant Science* 178: 1–8.

Venkatesan, A., Chellappan, K.P. (1998) Accumulation of proline and glycine betaine in *Ipomoea pes-caprae* induced by NaCl. *Biologia Plantarum* 41: 271–276.

Verbruggen, N., Hermans, C. (2008) Proline accumulation in plants: A review. *Amino Acids* 35: 753–759.

Verma, S., Dubey, R.S. (2001) Effect of cadmium on soluble sugars and enzymes of their metabolism in rice. *Biologia Plantarum* 44: 117–123.

Vernon, D.M., Bohnert, H.J. (1992) Increased expression of a myo-inositol methyl transferase in *Mesembryanthemum crystallinum* is part of a stress response distinct from Crassulacean acid metabolism induction. *Plant Physiology* 99: 1695–1698.

Verslues, P.E., Kim, Y.S., Zhu, J.K. (2007) Altered ABA, proline and hydrogen peroxide in an *Arabidopsis* glutamate: Glyoxylate aminotransferase mutant. *Plant Molecular Biology* 64: 205–217.

Vinocur, B., Altman, A. (2005) Recent advances in engineering plant tolerance to abiotic stress: Achievements and limitations. *Current Opinion in Biotechnology* 16: 123–132.

Waditee, R., Bhuiyan, M.N.H., Rai, V., Aoki, K., Tanaka, Y., Hibino, T., Suzuki, S., Takano, J., Jagendorf, A.T., Takabe, T. (2005) Genes for direct methylation of glycine provide high levels of glycinebetaine and abiotic-stress tolerance in *Synechococcus* and *Arabidopsis*. *Proceedings of the National Academy of Sciences of the United States of America* 102: 1318–1323.

Wang, Z., Stutte, G.W. (1992) The role of carbohydrates in active osmotic adjustment in apple under water stress. *Journal of the American Society for Horticultural Science* 117: 816–823.

Wang, F., Zeng, B., Sun, Z., Zhu, C. (2009) Relationship between proline and Hg^{2+}-induced oxidative stress in a tolerant rice mutant. *Archives of Environmental Contamination and Toxicology* 56: 723–731.

Wei, C., Cui, Q., Zhang, X.Q., Zhao, Y.Q., Jia, G.X. (2016) Three P5CS genes including a novel one from *Lilium regale* play distinct roles in osmotic, drought and salt stress tolerance. *Journal of Plant Biology* 59: 456–466.

Wei, D., Zhang, W., Wang, C., Meng, Q., Li, G., Chen, T.H.H., Yang, X. (2017) Genetic engineering of the biosynthesis

of glycinebetaine leads to alleviate salt-induced potassium efflux and enhances salt tolerance in tomato plants. *Plant Science* 257: 74–83.

Williamson, J.D., Jennings, D.B., Guo, W.W., Pharr, D.M., Ehrenshaft, M. (2002) Sugar alcohols, salt stress, and fungal resistance: Polyols-multifunctional plant protection? *Journal of the American Society for Horticultural Science* 127: 467–473.

Winter, H., Huber, S.C. (2000) Regulation of sucrose metabolism in higher plants: Localization and regulation of activity of key enzymes. *Critical Reviews in Plant Sciences* 19: 31–67.

Wu, G., Feng, R., Shui, Q. (2016) Effect of osmotic stress on growth and osmolytes accumulation in sugar beet (*Beta vulgaris* L.) plants. *Plant, Soil and Environment* 62: 189–194.

Xiang, Y., Huang, Y., Xiong, L. (2007) Characterization of stress-responsive *CIPK* genes in rice for stress tolerance improvement. *Plant Physiology* 144: 1416–1428.

Xing, W., Rajashekar, C.B. (2001) Glycine betaine involvement in freezing tolerance and water stress in *Arabidopsis thaliana*. *Environmental and Experimental Botany* 46: 21–28.

Xue, X., Liu, A., Hua, X. (2009) Proline accumulation and transcriptional regulation of proline biothesynthesis and degradation in *Brassica napus*. *BMB Reports* 42: 28–34.

Yancey, P.H. (2005) Organic osmolytes as compatible, metabolic and counteracting cytoprotectants in high osmolarity and other stresses. *Journal of Experimental Biology* 208: 2819–2830.

Yancey, P.H., Clark, M.E., Hand, S.C., Bowlus, R.D., Somero, G.N. (1982) Living with water stress: Evolution of osmolyte systems. *Science* 217: 1214–1222.

Yang, X., Lu, C. (2005) Photosynthesis is improved by exogenous glycinebetaine in salt-stressed maize plants. *Physiologia Plantarum* 124: 343–352.

Yang, C.W., Lin, C.C., Kao, C.H. (2000) Proline, ornithine, arginine and glutamic acid contents in detached rice leaves. *Biologia Plantarum* 43: 305–307.

Yang, S.L., Lan, S.S., Deng, F.F., Gong, M. (2016) Effects of calcium and calmodulin antagonists on chilling stress-induced proline accumulation in *Jatropha curcas* L. *Journal of Plant Growth Regulation* 35: 815–826.

Yang, X., Liang, Z., Lu, C. (2005) Genetic engineering of the biosynthesis of glycinebetaine enhances photosynthesis against high temperature stress in transgenic tobacco plants. *Plant Physiology* 138: 2299–2309.

Yang, X. Liang, Z., Wen, X., Lu, C. (2008) Genetic engineering of the biosynthesis of glycinebetaine leads to increased tolerance of photosynthesis to salt stress in transgenic tobacco plants. *Plant Molecular Biology* 66: 73–86..

Yoshiba, Y., Kiyosue, T., Nakashima, K., Yamaguchi-Shinozaki, K., Shinozaki, K. (1997) Regulation of levels of proline as an osmolyte in plants under water stress. *Plant Cell Physiology* 38: 1095–1102.

Yu, H.Q., Zhou, X.Y., Wang, Y.G., Zhou, S.F., Fu, F.L., Li, W.C. (2017) A betaine aldehyde dehydrogenase gene from *Ammopiptanthus nanus* enhances tolerance of *Arabidopsis* to high salt and drought stresses. *Plant Growth Regulation* 83: 265–276.

Yuanyuan, M., Yali, Z., Jiang, L., Hongbo, S. (2009) Roles of plant soluble sugars and their responses to plant cold stress. *African Journal of Biotechnology* 8: 2204–2210.

Zárate-Romero, A., Murillo-Melo, D.S., Mújica-Jiménez, C., Montiel, C., Muñoz-Clares, R.A. (2016) Reversible, partial inactivation of plant betaine aldehyde dehydrogenase by betaine aldehyde: Mechanism and possible physiological implications. *The Biochemical Journal* 473: 873–875.

Zarattini, M., Forlani, G. (2017) Toward unveiling the mechanisms for transcriptional regulation of proline biosynthesis in the plant cell response to biotic and abiotic stress conditions. *Frontiers in Plant Science* 8: 927.

Zeid, I.M. (2009) Trehalose as osmoprotectant for maize under salinity-induced stress. *Research Journal of Agriculture and Biological Sciences* 5: 613–622.

Zhang, C., Lu, Q., Verma, D.P.S. (1997) Characterization of Δ 1-pyrroline-5-carboxylate synthetase gene promoter in transgenic *Arabidopsis thaliana* subjected to water stress. *Plant Science* 129: 81–89.

Zhifang, G., Loescher, W.H. (2003) Expression of a celery mannose 6-phosphate reductase in *Arabidopsis thaliana* enhances salt tolerance and induces biosynthesis of both mannitol and a glucosyl-mannitol dimer. *Plant, Cell and Environment* 26: 275–283.

Zhou, R., Cheng, L., Wayne, R. (2003) Purification and characterization of sorbitol-6-phosphate phosphatase from apple leaves. *Plant Science* 165: 227–232.

18 The Role of Plasma Membrane Proteins in Tolerance of Dehydration in the Plant Cell

Pragya Barua, Dipak Gayen, Nilesh Vikram Lande,
Subhra Chakraborty, and Niranjan Chakraborty

CONTENTS

18.1 INTRODUCTION

The dynamics of plasma membrane (PM) protein responses towards changes in abiotic conditions have been one of the fundamental interests of plant research for decades. However technical limitations, coupled with the difficulty in the extraction of PM and subsequently its proteins, hindered the drive towards comprehensive conclusions. Consequently, we are still far from a consolidated depiction of the role of PM proteins in response to particular abiotic stresses or a combination of them. Nonetheless, intermittent studies conducted so far have depicted overall changes and shifts in the stress-induced PM proteome. Also, targeted studies of PM candidate proteins have shed a fair amount of light on the membrane components of crucial signaling pathways. It is evident that these studies have further underlined the potential importance of PM proteins in performing decisive roles in signaling pathway initiation and modulation. Several studies have been conducted to elucidate the response of PM proteins towards different abiotic stress conditions in an overall, as well as tissue-specific, manner.

Abiotic factors include non-living physical and chemical environmental components, alteration of which affects the living organism and the ecosystem. Since plants spend their entire lifespan in a fixed niche, they have to have a robust defense mechanism that enables them to survive and grow under changing conditions. Deviation towards either extreme (high or low) from the optimum levels of abiotic factors such as light, temperature, and salinity, among others, trigger stress responses at the cellular level.

PM forms the living barrier that defines a cell and its environment. It is much more than just a lipid bilayer punctuated with proteinaceous components. The dynamicity of its lipid and protein composition is reflected in its characteristic functional plasticity. Detergent-resistant microdomains (DRMs) are sporadically distributed across the PM. DRMs are rich in lipid subtypes such as sterols and sphingolipids that can recruit specific proteins and form the site of targeted reactions. In this review, we focus on the PM proteomic studies involving abiotic stress response reported so far. This includes global proteome profiling of the PM, as well as the PM proteins extensively studied for having exemplary roles in stress-related pathways.

18.2 GLOBAL ANALYSES OF PLASMA MEMBRANE PROTEOME UNDER VARIOUS ABIOTIC STRESSES

18.2.1 SALINITY

Salt is among the naturally occurring minerals of soil that affect plant growth and vitality. The salinity of soil takes the shape of stress when the root-water relation is perturbed due to excess accumulation of dissolved mineral solutes. The extent of salinity tolerance varies between plant species and crop types and consequently in yield. The effect of salinity on the PM proteome of rice roots has been studied (Malakshah et al., 2007; Cheng et al., 2009). Both studies applied two-dimensional gel electrophoresis to resolve and identify 24 and 18 differential proteins, respectively. The remorin family of proteins was found to be upregulated in both studies. This protein family is plant specific and is known to be associated with membrane rafts acting as scaffold proteins during early signaling events.

18.2.2 COLD

Cold acclimatization (CA) and ABA-induced changes in the PM proteome were studied in an *Arabidopsis* suspension culture cell line (T87) (Li et al., 2012). The study concluded that the cryostability of the PM in response to freezing-induced dehydration improved freezing tolerance in response to CA and ABA, having both distinct and overlapping components. Proteins known to have dual localizations, such as annexin, which is known to translocate to the PM in response to elevated $[Ca^{+2}]_{cyt}$, showed an altered level of expression. Further, the list of differentially expressed proteins was functionally categorized into a class of transporters, signal transducers,

intracellular traffic and defense-related proteins, among others. Fasciclin domain-containing ABA-responsive cell adhesion proteins were found to be induced by cold in *Arabidopsis*, asserting the role of cell-cell communication in cold tolerance. In yet another cold-induced study of PM proteins in *Arabidopsis* leaf protoplasts (Kawamura et al., 2003), 38 proteins primarily involved in membrane repair, protection, CO_2 fixation, and proteolysis were observed.

18.2.3 FLOOD

Several crop species, including soybean, are susceptible to flood conditions, which can affect plant germination and vegetative as well as reproductive growth. The PM proteome of root and hypocotyl region of soybean seedlings in response to flood condition was studied by Komatsu et al. (2009). They employed the 2-DE approach, which yielded a limited number of identifications, including 14-3-3 and serine-threonine protein kinase. Hydrophobic and less-abundant membrane proteins were not detected. Expansins, involved in cell-wall expansion, were found to be upregulated, which correlates with submergence and subsequently low-oxygen-induced elongation. The number of flood-responsive studies is limited, so this is one of the foremost studies elaborating that PM is directly affected by flooding stress.

18.3 PM PROTEIN FAMILIES IN ABIOTIC STRESS RESPONSE

18.3.1 RECEPTOR-LIKE KINASES (RLKs)

The role of receptor-like kinases (RLKs) in sensing adverse environmental conditions and activating a downstream signaling cascade via activation of its intracellular kinase domain through phosphorylation is well established. A typical RLK structure includes an extracellular receptor, a transmembrane domain, and a cytosolic serine/threonine kinase domain, which facilitates signal perception at the cell surface, membrane anchoring, and relay of signal, respectively. This family of proteins is known to play a precise role in sensing environmental cues and transducing them to its cytosolic target by virtue of its kinase activity (Figure 18.1). Thus, they trigger relay of signals and are also instrumental in the integration of plant hormone signaling with environmental factors (Shiu and Bleecker, 2001b; Dievart and Clark, 2004). They form the largest family of membrane receptors, and among them, the leucine-rich repeat containing LRR-RLKs are the most abundant. In the model species *Arabidopsis* and rice, there

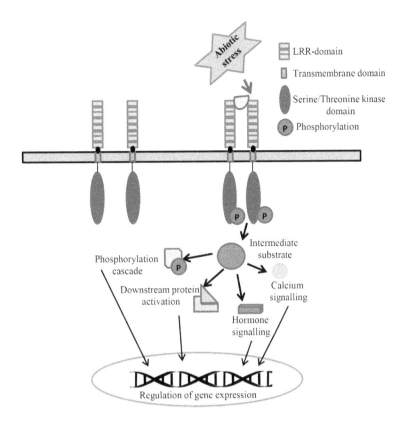

FIGURE 18.1 Schematic representation of abiotic stress induced modulation of PM-resident leucine-rich repeat receptor-like kinase (LRR-RLK) triggering relay of signals and plant defense.

are more than 600 and 1,100 different RLKs, respectively (Hua et al., 2012), implicating their role in multivariate pathways. Although this family of PM-resident receptor kinase proteins is involved in both biotic and abiotic stress response, this review will only focus on the latter. Cold-inducible GsLRPK was found to increase tolerance in *Arabidopsis* and yeast upon overexpression (Yang et al., 2014). Its location in the PM was confirmed by subcellular localization studies. On the other hand, another PM resident LRR-RLK, rice OsGIRL1, was found to elicit a hypersensitive response under salt and light but a hyposensitive response towards gamma irradiation, suggesting its functional plasticity and ability to detect and differentiate between various abiotic stresses. Expressional divergence due to events of tandem duplication has resulted in the functional divergence of duplicated genes (Park et al., 2014). Functional characterization studies of different RLKs conducted in plants have been listed in Table 18.1.

18.3.2 CALCIUM-DEPENDENT PROTEIN KINASES (CDPK)

Calcium-dependent protein kinases (CDPKs), described as positive regulators of abiotic stress response, confer tolerance to plants upon overexpression. They possess both calcium-sensing as well as kinase domains and mediate calcium-binding-induced substrate phosphorylation. This plant-specific family of proteins is characterized by the presence of four domains: a variable N-terminal, kinase, auto-inhibitory junction, and calmodulin-like (CaM) domain. The CaM domain contains EF hand motifs and is activated upon calcium binding. Their substrates primarily consist of ABA-responsive factors (ABFs), ion channels, and transcription factors. Their isoforms appear in both "short-term signaling reactions" and "long-term adaptive processes" (Schulz et al., 2013). They remain associated with the PM by virtue of their N-termini modifications, such as myristoylation or palmitoylation, as demonstrated in *Arabidopsis* (Lu and Hrabak, 2013). They reportedly can undergo N-termini sequence-specific stress-induced phosphorylation, which further relies upon its membrane localization (Witte et al., 2010). There are reportedly as many as 34 CDPKs in *Arabidopsis* (Cheng et al., 2002). A change in the cytosolic calcium concentration $[Ca^{2+}]_{cyt}$ accompanies almost every known abiotic stress (drought, osmotic stress, heat, cold, etc.). The precise stress-specific calcium signatures are then decrypted by downstream signaling networks. $[Ca^{2+}]_{cyt}$ is precisely

TABLE 18.1

List of Different Types of PM Proteins Involved in Multivariate Abiotic Stress-Responsive Pathways

RLK	Organism	Abiotic Stress	Function	Reference
GsLRPK (LRR)	*Glycine soja*	Cold	Positive regulator of stress tolerance.	Yang et al. (2014)
OsGIRL1 (LRR)	*Oryza sativa*	Salinity, heat, gamma-ray irradiation	Hyper- (salinity/heat) and hypo- (irradiation) sensitive response, ABA-induced response.	Park et al. (2014)
RPK1 (LRR)	*Arabidopsis thaliana*	Water deficit	Increases tolerance to water deficit.	Osakabe et al. (2010)
OsRPK1 (LRR)	*Oryza sativa*	Salinity	Negatively regulates root development and polar auxin transport.	Shi et al. (2014) and Zhou et al. (2014)
LP2 (LRR)	*Oryza sativa*	Water deficit	Interacts with water-deficit-responsive aquaporin OsPIP1; 1, OsPIP1; 3, and OsPIP2	Wu et al. (2014)
AtPXL1 (LRR)	*Arabidopsis thaliana*	Cold, heat	Positive regulator of signal transduction pathways under temperature fluctuations.	Jung et al. (2015)
AtRDK1 (LRR)	*Arabidopsis thaliana*	Water deficit	Takes part in ABA signaling by mediating the recruitment of ABI1 onto PM.	Kumar et al. (2017)
PnLRR-RLK27 (LRR)	*Pohlia nutans*	Salinity	Enhances salinity and ABA tolerance.	Wang et al. (2017)
GsRLCK	*Glycine soja*	Salinity, water deficit	Increased tolerance to salinity and water-deficit stresses.	Sun et al. (2013)
AtLPK1	*Arabidopsis thaliana*	Salinity	Enhanced seed germination and cotyledon greening under high salinity conditions.	Huang et al. (2013)
OsSIK2	*Oryza sativa*	Salinity, water deficit	Increased tolerance to salinity and water-deficit stresses.	Chen et al. (2013b)
OsSIK1 (LRR)	*Oryza sativa*	Salinity, water deficit, H_2O_2	Affects stomatal density in the abaxial and adaxial leaf epidermis of rice.	Ouyang et al. (2010)
Srlk (LRR)	*Medicago truncatula*	Salinity	Inhibits root growth under salinity stress.	De Lorenzo et al. (2009)
CRK5 (CRR)	*Arabidopsis thaliana*	Water deficit	Potential positive regulator of ABA signaling.	Lu et al. (2016)
CRK45 (CRR)	*Arabidopsis thaliana*	Water deficit	Positively regulates ABA response, seed germination and seedling development.	Zhang et al. (2013)
GHR1 (LRR)	*Arabidopsis thaliana*	H_2O_2	Instrumental in ABA and H_2O_2 signaling for stomatal movement.	Hua et al. (2012)
PERK4	*Arabidopsis thaliana*	ABA	PERK4 mediates ABA-regulated primary root cell growth.	Bai et al. (2009)
CDPK				
OsCPK10	*Oryza sativa*	Water deficit	Enhances H_2O_2 detoxifying capacity of rice plants during water deficit.	Bundó et al. (2017)
OsCPK17	*Oryza sativa*	Cold	Affects the activity of membrane channels and sugar metabolism.	Almadanim et al. (2017)
CPK3	*Arabidopsis thaliana*	Salinity	Acclimatization to salinity stress.	Mehlmer et al. (2010)
McCPK1	*Mesembryanthemum crystallinum*	Salinity, water-deficit	Undergoes reversible change in subcellular localization.	Chehab et al. (2004)
Ion transporters				
H+-ATPase	*Arabidopsis thaliana*	Cadmium	Key factor of brassinosteroid-induced pathway.	Jakubowska and Janicka (2017)

(Continued)

TABLE 18.1 (CONTINUED)

List of Different Types of PM Proteins Involved in Multivariate Abiotic Stress-Responsive Pathways

RLK	Organism	Abiotic Stress	Function	Reference
AHA2, AHA7	Arabidopsis thaliana	Phosphorous deficiency	AHA2: primary root elongation, AHA7: root hair density	Yuan et al. (2017)
PMH+-ATPase	Vicia faba	Aluminum toxicity	Interacts with 14-3-3 protein regulating citrate exudation.	Chen et al. (2013a)
AtNCL	Arabidopsis thaliana	Salinity	Maintenance of calcium homeostasis.	Wang et al. (2012)
Water channels				
PIP1, PIP2	Triticum turgidum L. subsp. Durum	ABA, salinity	PIP2 increases membrane osmotic water-permeability coefficient.	Ayadi et al. (2011)

regulated by several efflux systems that maintain a very low submicromolar level of Ca^{2+} within the cell. CDPKs are one such class of calcium sensors. In addition to calcium concentration, CDPK activity is also regulated by reversible phosphorylation and 14-3-3 like proteins (Ormancey et al., 2017).

18.3.3 CALCINEURIN B-LIKE PROTEINS (CBLs)

Calcineurin B-like proteins (CBLs) are a family of calcium-binding proteins that undergo PM association by virtue of post-translational modification, such as myristoylation and S-acylation. They act as calcium sensors and regulate CIPKs, a family of protein kinases. They are also found to be present in other cellular locations aside from the PM (CBL1, CBL9), including in the vacuolar membrane. Several such protein isoforms have been identified in *Arabidopsis* and rice (Batistic et al., 2009). Their presence is not restricted to plants only. CBL/CIPK complexes are viewed as fast responders to local bursts of calcium release.

18.3.4 PM H+ATPASES

PM H+ATPases couple ATP hydrolysis with proton transport across the membrane and helps create a potential difference. One of their basic functions in plants is involvement in polar auxin transport and signaling. The several isoforms of PM H+ATPases are encoded by a multigene family. *Arabidopsis* hosts 11 such isoforms, with different members showing ubiquitous as well as tissue or stage-specific expression (Zhang et al., 2017). H+ATPases are also reported to be instrumental in brassinosteroid-mediated signaling under cadmium stress (Jakubowska and Janicka, 2017). Their role in heavy-metal-induced abiotic stress has been highlighted in several studies (Table 18.1).

18.3.5 AQUAPORINS

One of the downstream manifestations of several abiotic stresses is the disturbance in plant-water homeostasis. PM-intrinsic aquaporins (PIPs) function as water channels that also transfer carbon dioxide and small uncharged solutes. The crucial role of these transmembrane proteins in drought-stress tolerance has drawn the attention of plant scientists for development of drought-tolerant crops. The expression of aquaporins is regulated at multiple layers, including transcriptional and post-translational levels, by several different factors including hormones (GA3, ABA, brassinolide) and environmental factors (drought, salinity, temperature, light). Interestingly, the post-translational phosphorylation or heteromerization of aquaporins are influenced by cytosolic pH and Ca^{2+} (Chaumont et al., 2005). In *Arabidopsis*, RLKs have been shown to modulate PIP activity under osmotic stress (Bellati et al., 2016). In spite of extensive work being carried out on aquaporins, their regulation under drought remains a source of debate. The expression pattern of aquaporins under drought shows transcript-level variation depending upon the duration and intensity of stress. While some isoforms undergo upregulation, others remain unaltered or are even downregulated. Therefore, while one school of thought proposes that under stress there is an increase in membrane water permeability, others suggest that plants decrease water loss by downregulation of aquaporins (Zargar et al., 2017).

18.3.6 14-3-3 PROTEINS

This protein family is found ubiquitously in all eukaryotes and is capable of binding with the phosphorylated serine, threonine residues of their interacting partners, influencing their function as well as subcellular localization. They are found in practically all subcellular compartments, including the nucleus, mitochondria, cytoplasm, vacuolar membrane, and PM. They are small and acidic and structurally suitable to function as a scaffold protein, capable of bringing two phosphoproteins in close proximity. Plants harbor the largest family of 14-3-3 paralogs, as compared to other eukaryotes. In PM, the two most prominent binding partners of 14-3-3 are PM H$^+$ ATPase and the K$^+$ channel. It activates the H$^+$ ATPase for stomatal opening induced under blue light and can directly affect membrane potential and the activity of a voltage-gated K$^+$ ion channel. (Wilson et al., 2016). They are a positive regulator of PM ion channels and have been reported to play a crucial role in the regulation of opening of the stomatal aperture.

18.3.7   Ion Channels and Transporters

Abiotic stresses are often accompanied by ionic stresses, as in the case of salinity stress that is manifested with an excess of Na$^+$ ion. Sodium-calcium exchanger (NCX) proteins having 9-11 transmembrane domains play a crucial role in maintaining the calcium homeostasis of the cell. Wang et al. (2012) elaborated the role of an NCX-like protein, AtNCL, in salt stress in *Arabidopsis*. They further reported that *atncl* seedlings showed lower sodium and higher calcium content. NCX is one of the key cellular components for removal of Ca^{2+}. NCX exports single Ca^{2+}, importing three Na$^+$ ions in exchange and can reverse its mode of action when the cellular Na$^+$ concentration goes beyond a critical level. The direction of transfer depends upon the gradient generated by Na$^+$ and Ca^{2+} concentrations across the cell. Although with the availability of massive sequence information it is increasingly evident that NCX proteins are widely distributed in both plants and animals, but so far AtNCL is the only NCX protein characterized in plants (Wang et al., 2012). Functional characterization studies are imperative for comprehensively dissecting these proteins, which are crucial for maintaining Ca^{2+} ion homeostasis.

18.3.8   G-Proteins

Membrane-associated GTP-binding heterotrimeric proteins, or G proteins, classically consist of Gα, Gβ, and Gγ subunits and are vital signal-transducing elements across plants and animals. G-protein-bound membrane receptors relay downstream signals by replacing GDP with GTP upon activation by an external signal. In the model plant *Arabidopsis*, GPA1 has been shown to be involved in the ABA-mediated pathway (Coursol et al., 2003) and blue-light response (Warpeha et al., 2006). Ma et al. (2015) demonstrated the role of the Cold1 receptor — a PM

resident regulator of G-protein interacts with its alpha subunit and activates the calcium channel to sense low temperature and to accelerate GTPase activity. In pea, the Gα subunit was reportedly upregulated by heat and was able to impart heat and salt tolerance when overexpressed in tobacco, whereas the Gβ subunit solely conferred tolerance towards heat (Misra et al., 2007). The role of the Gα subunit in salt stress has also been characterized in *Arabidopsis* (Colaneri et al., 2014), rice, and maize (Urano et al., 2013). G-protein-coupled membrane receptors such as Cold1 play a significant role in various abiotic stress responses.

18.4   PM PROTEINS AND ABIOTIC STRESS-RESPONSIVE PATHWAYS

18.4.1   PM Proteins in ABA-Mediated Abiotic Stress Response

The phytohormone abscisic acid (ABA) regulates plant developmental processes and functions as a major player in abiotic stress response. ABA signaling is fundamental for abiotic stress adaptation and tolerance. The molecular mechanism that connects the core signaling components with those of the PM is indistinct. ABA elicits drought-responsive closure of stomatal pores. Stomatal closure results in a decrease in transpirational water loss. ABA induces an increase in [Ca^{2+}]$_{cyt}$ inside guard cells, which paves the way for a decrease in stomatal aperture. ABA activates PM Ca^{2+} channels in *Arabidopsis* suspension culture cells, indicating that ABA-induced Ca^{2+} influx is a general component of ABA signaling in plants (Murata et al., 2001; Ghelis et al., 2000).

18.4.2   PM Proteins in the Stress-Responsive Calcium-Signaling Pathway

Mediators of signaling such as the plant hormone ABA or secondary messenger calcium have an extensive regulatory hold over many diverse cellular pathways. It becomes increasingly difficult to describe the role of one signaling component without discussing the other. As mentioned earlier, one of the primary functions of ABA in guard cells is activation of PM calcium channels, also known to be activated by ROS. Conversely, one of the most prominent aspects of calcium-signaling pathways is their regulation of calcium-dependent stomatal closure. Regulation of stomatal aperture is pivotal in more than one stress response, including dehydration, cold, light, and osmotic stress. The membranes and transporters controlling calcium influx and efflux across cell membrane as well as different organellar compartments are the sentinels of Ca^{2+} signaling. The two classes of Ca^{2+} transporters residing in the cell membrane are ATP-driven Ca^{2+} pumps, PM Ca^{2+} ATPases (PMCA) and sodium calcium exchangers (Na^+/Ca^{2+} exchangers or NCX), having complementary functions. While PMCA has a very high affinity towards Ca^{2+}, NCX displays low Ca^{2+} affinity but a high transportation speed of 5000 Ca^{2+} ions per second (Carafoli et al., 2002). Also, the NCX PM channel proteins form the gate for selective calcium exchange and in turn are tightly regulated at multiple layers. They work in concert with Ca^{2+}-ATPases and other calcium-binding proteins. Recently, Yang et al. (2017) demonstrated in *Arabidopsis* that PM-localized BON1 protein, together with autoinhibited calcium ATPase 10 (ACA10) and ACA8, controls calcium signaling and regulates stomatal closure. The evolution and enormous effects of different calcium signatures and their control are reviewed elaborately by Edel et al., 2017.

18.4.3 PM PROTEINS IN OTHER HORMONE-MEDIATED ABIOTIC STRESS RESPONSE

Brassinosteroids (BR) are a class of plant steroid hormones that have a role in the development of seeds and modulation of flowering and senescence. However, some components of the BR-signaling pathway can act as multifunctional molecules participating in the crosstalk of several signaling pathways, including stress response. BR are perceived by the BRI1 transmembrane receptor, which again belongs to a class of LRR-RLK. They can dimerize in the PM in a ligand-independent manner (Russinova et al., 2004). The regulation of this receptor is accompanied by phosphorylation events at multiple sites in its cytosolic kinase domains. Brassinosteroid signaling kinase 5 (BSK5), mainly distributed in the PM and nucleus, participates in salt stress response and ABA-mediated drought stress tolerance. This kinase also regulates ABA-mediated stomatal closure under drought and negatively regulates drought stress response through gene expression modulation (Li et al., 2012). BRI1-associated receptor kinase 1 (BAK1) gets activated by transphosphorylation upon binding to BRI1.

18.5 CONCLUSION

The role of PM proteins in response to abiotic stress is diverse and multidimensional. The sporadic distribution of membrane channels and transporters is regulated by lipid rafts and microdomains. The distribution of constituent membrane lipids and carbohydrates also has a regulatory role. However, we have focused only on the proteinaceous components and briefly summarized their role in stress response. Since PM harbors both membranes spanning transmembrane domains and associated proteins, their mode and regulation of action are highly dynamic. Multiple intrinsic membrane channel proteins that also function as enzymes include kinases and ATPases, which not only confer selective permeability to the membrane but also maintain its voltage-gated potential. The PM proteins form the frontier of important signaling cascades such as calcium and hormone signaling. These dynamic, versatile, and highly active proteins will continue to be of interest in plant stress signaling research and to serve as potential candidates for genetic engineering and crop improvement.

ACKNOWLEDGMENTS

This work was supported by grants from the Department of Science and Technology (DST)-SERB (EMR/2015/001870), and PDF/2016/002615, Government of India. The authors thank the Department of Biotechnology (DBT), DST, and Council of Scientific and Industrial Research (CSIR) for providing research fellowships to PB, DG, and NVL, respectively.

REFERENCES

Almadanim, M. C., Alexandre, B. M., Rosa, M. T., Sapeta, H., Leitão, A. E., Ramalho, J. C., ... & Oliveira, M. M. (2017) Rice calcium-dependent protein kinase OsCPK17 targets plasma membrane intrinsic protein and sucrose-phosphate synthase and is required for a proper cold stress response. *Plant, Cell & Environment*, 40(7): 1197–1213.

Ayadi, M., Cavez, D., Miled, N., Chaumont, F., & Masmoudi, K. (2011) Identification and characterization of two plasma membrane aquaporins in durum wheat (Triticum turgidum L. subsp. durum) and their role in abiotic stress tolerance. *Plant Physiology and Biochemistry* 49(9): 1029–1039.

Bai, L., Zhou, Y., & Song, C. P. (2009) Arabidopsis proline-rich extensin-like receptor kinase 4 modulates the early event toward abscisic acid response in root tip growth. *Plant Signaling & Behavior* 4(11): 1075–1077.

Batistič, O., & Kudla, J. (2009) Plant calcineurin B-like proteins and their interacting protein kinases. *Biochimica et Biophysica Acta (BBA)-Molecular Cell Research* 1793(6): 985–992.

Bellati, J., Champeyroux, C., Hem, S., Rofidal, V., Krouk, G., Maurel, C., Santoni, V. (2016) Novel aquaporin regulatory mechanisms revealed by interactomics. *Molecular and Cellular Proteomics* 15: 3473–3487.

Bundó, M., Coca, M. (2017) Calcium-dependent protein kinase OsCPK10 mediates both drought tolerance and blast disease resistance in rice plants. *Journal of Experimental Botany* 68: 2963–2975.

Carafoli, E. (2002) Calcium signaling: a tale for all seasons. *Proceedings of the National Academy of Sciences* 99(3): 1115–1122.

Chaumont, F., Moshelion, M., Daniels, M. J. (2005) Regulation of plant aquaporin activity. *Biology of the Cell* 97: 749–764.

Chehab, E. W., Patharkar, O. R., Hegeman, A. D., Taybi, T., Cushman, J. C. (2004) Auto phosphorylation and subcellular localization dynamics of a salt-and-water deficit-induced calcium-dependent protein kinase from ice plant. *Plant Physiology* 135: 1430–1446.

Chen, Q., Guo, C. L., Wang, P., Chen, X. Q., Wu, K. H., Li, K. Z., Yu, Y. X., Chen, L. M. (2013a) Up-regulation and interaction of the plasma membrane H+-ATPase and the 14-3-3 protein are involved in the regulation of citrate exudation from the broad bean (*Vicia faba* L.) under Al stress. *Plant Physiology and Biochemistry* 70: 504–511.

Chen, L. J., Wuriyanghan, H., Zhang, Y. Q., Duan, K. X., Chen, H. W., Li, Q. T., Lu, X., He, S. J., Ma, B., Zhang, W. K., Lin, Q. (2013b) An S-domain receptor-like kinase, OsSIK2, confers abiotic stress tolerance and delays dark-induced leaf senescence in rice. *Plant Physiology* 163: 1752–1765.

Cheng, S. H., Willmann, M. R., Chen, H. C., Sheen, J. (2002) Calcium signaling through protein kinases. The *Arabidopsis* calcium-dependent protein kinase gene family. *Plant Physiology* 129: 469–485.

Cheng, Y., Qi, Y., Zhu, Q., Chen, X., Wang, N., Zhao, X., … & Zhang, W. (2009) New changes in the plasma-membrane-associated proteome of rice roots under salt stress. *Proteomics* 9(11): 3100–3114.

Colaneri, A. C., Tunc-Ozdemir, M., Huang, J. P., Jones, A. M. (2014) Growth attenuation under saline stress is mediated by the heterotrimeric G protein complex. *BMC Plant Biology* 14: 129.

Coursol, S., Fan, L. M., Le Stunff, H., Spiegel, S., Gilroy, S., Assmann, S. M. (2003) Sphingolipid signaling in *Arabidopsis* guard cells involves heterotrimeric G proteins. *Nature* 423: 651–654.

De Lorenzo, L., Merchan, F., Laporte, P., Thompson, R., Clarke, J., Sousa, C., Crespi, M. (2009) A novel plant leucine-rich repeat receptor kinase regulates the response of *Medicago truncatula* roots to salt stress. *The Plant Cell* 21: 668–680.

Diévart, A., & Clark, S. E. (2004) LRR-containing receptors regulating plant development and defense. Development 131(2): 251–261

Edel, K. H., Marchadier, E., Brownlee, C., Kudla, J., Hetherington, A. M. (2017) The evolution of calcium-based signaling in plants. *Current Biology* 27: R667–R679.

Ghelis, T., Dellis, O., Jeannette, E., Bardat, F., Miginiac, E., Sotta, B. (2000) Abscisic acid plasmalemma perception triggers a calcium influx essential for RAB18 gene expression in *Arabidopsis thaliana* suspension cells. *FEBS Letters* 483: 67–70.

Hua, D., Wang, C., He, J., Liao, H., Duan, Y., Zhu, Z., Guo, Y., Chen, Z., Gong, Z. (2012) A plasma membrane receptor kinase, GHR1, mediates abscisic acid-and hydrogen peroxide-regulated stomatal movement in *Arabidopsis*. The *Plant Cell* 24: 2546–2561.

Huang, P., Ju, H. W., Min, J. H., Zhang, X., Kim, S. H., Yang, K. Y., Kim, C. S. (2013) Overexpression of L-type lectin-like protein kinase 1 confers pathogen resistance and regulates salinity response in *Arabidopsis thaliana*. *Plant Science* 203: 98–106.

Jakubowska, D., Janicka, M. (2017) The role of brassinosteroids in the regulation of the plasma membrane H+-ATPase and NADPH oxidase under cadmium stress. *Plant Science* 264: 37–47.

Jung, C. G., Hwang, S. G., Park, Y. C., Park, H. M., Kim, D. S., Park, D. H., Jang, C. S. (2015) Molecular characterization of the cold-and-heat-induced *Arabidopsis* PXL1 gene and its potential role in transduction pathways under temperature fluctuations. *Journal of Plant Physiology* 176: 138–146.

Kawamura, Y., & Uemura, M. (2003) Mass spectrometric approach for identifying putative plasma membrane proteins of Arabidopsis leaves associated with cold acclimation. *The Plant Journal* 36(2): 141–154.

Kumar, D., Kumar, R., Baek, D., Hyun, T. K., Chung, W. S., Yun, D. J., Kim, J. Y. (2017) *Arabidopsis thaliana* RECEPTOR DEAD KINASE1 functions as a positive regulator in plant responses to ABA. *Molecular Plant* 10: 223–243.

Komatsu, S., Wada, T., Abaléa, Y., Nouri, M. Z., Nanjo, Y., Nakayama, N., … & Furukawa, K. (2009) Analysis of plasma membrane proteome in soybean and application to flooding stress response. *Journal of Proteome Research* 8(10): 4487–4499.

Li, Z. Y., Xu, Z. S., He, G. Y., Yang, G. X., Chen, M., Li, L. C., Ma, Y. Z. (2012) A mutation in *Arabidopsis* BSK5 encoding a brassinosteroid-signaling kinase protein affects responses to salinity and abscisic acid. *Biochemical and Biophysical Research Communications* 426: 522–527.

Lu, S. X., Hrabak, E. M. (2013) The myristoylated amino-terminus of an Arabidopsis calcium-dependent protein kinase mediates plasma membrane localization. *Plant Molecular Biology* 82: 267–278.

Lu, K., Liang, S., Wu, Z., Bi, C., Yu, Y. T., Wang, X. F., Zhang, D. P. (2016) Overexpression of an *Arabidopsis* cysteine-rich receptor-like protein kinase, CRK5, enhances abscisic acid sensitivity and confers drought tolerance. *Journal of Experimental Botany* 67: 5009–5027.

Ma, Y., Dai, X., Xu, Y., Luo, W., Zheng, X., Zeng, D., … & Xiao, J. (2015) COLD1 confers chilling tolerance in rice. *Cell* 160(6): 1209–1221.

Malakshah, S.N., Habibi Rezaei, M., Heidari, M., & Hosseini Salekdeh, G. (2007) Proteomics reveals new salt responsive proteins associated with rice plasma membrane. *Bioscience, Biotechnology, and Biochemistry* 71(9): 2144–2154.

Mehlmer, N., Wurzinger, B., Stael, S., Hofmann-Rodrigues, D., Csaszar, E., Pfister, B., Bayer, R., Teige, M. (2010) The Ca2+-dependent protein kinase CPK3 is required for MAPK-independent salt-stress acclimation in *Arabidopsis*. *The Plant Journal* 63: 484–498.

Misra, S., Wu, Y., Venkataraman, G., Sopory, S. K., Tuteja, N. (2007) Heterotrimeric G-protein complex and G-protein-coupled receptor from a legume (*Pisum sativum*): Role in salinity and heat stress and cross-talk with phospholipase C. *The Plant Journal* 51: 656–669.

Murata, Y., Pei, Z. M., Mori, I. C., Schroeder, J. (2001) Abscisic acid activation of plasma membrane Ca^{2+} channels in guard cells requires cytosolic NAD(P)H and is differentially disrupted upstream and downstream of reactive oxygen species production in abi1-1 and abi2-1 protein phosphatase 2C mutants. *The Plant Cell* 13: 2513–2523.

Ormancey, M., Thuleau, P., Mazars, C., Cotelle, V. (2017) CDPKs and 14-3-3 Proteins: Emerging duo in signaling. *Trends in Plant Science* 22: 263–272.

Osakabe, Y., Mizuno, S., Tanaka, H., Maruyama, K., Osakabe, K., Todaka, D., Fujita, Y., Kobayashi, M., Shinozaki, K.,Yamaguchi-Shinozaki, K. (2010) Overproduction of the membrane-bound receptor-like protein kinase 1, RPK1, enhances abiotic stress tolerance in Arabidopsis. *Journal of Biological Chemistry* 285: 9190–9201.

Ouyang, S. Q., Liu, Y. F., Liu, P., Lei, G., He, S. J., Ma, B., Zhang, W. K., Zhang, J. S., Chen, S. Y. (2010) Receptor-like kinase OsSIK1 improves drought and salt stress tolerance in rice (*Oryza sativa*) plants. *The Plant Journal* 62: 316–329.

Park, S., Moon, J. C., Park, Y. C., Kim, J. H., Kim, D. S., Jang, C. S. (2014) Molecular dissection of the response of a rice leucine-rich repeat receptor-like kinase (LRR-RLK) gene to abiotic stresses. *Journal of Plant Physiology* 171: 1645–1653.

Russinova, E., Borst, J. W., Kwaaitaal, M., Caño-Delgado, A., Yin, Y., Chory, J., de Vries, S. C. (2004) Heterodimerization and endocytosis of *Arabidopsis* brassinosteroid receptors BRI1 and AtSERK3 (BAK1). *The Plant Cell* 16: 3216–3229.

Schulz, P., Herde, M., & Romeis, T. (2013) Calcium-dependent protein kinases: hubs in plant stress signaling and development. *Plant Physiology* 163: 523–530.

Shiu, S. H., & Bleecker, A. B. (2001) Plant receptor-like kinase gene family: diversity, function, and signaling. *Science's STKE*, 2001(113): re22

Shi, C. C., Feng, C. C., Yang, M. M., Li, J. L., Li, X. X., Zhao, B. C., ... & Ge, R. C. (2014) Overexpression of the receptor-like protein kinase genes AtRPK1 and OsRPK1 reduces the salt tolerance of Arabidopsis thaliana. *Plant Science* 217: 63–70

Sun, X., Sun, M., Luo, X., Ding, X., Ji, W., Cai, H., Bai, X., Liu, X., Zhu, Y. (2013) A *Glycine soja* ABA-responsive receptor-like cytoplasmic kinase, GsRLCK, positively controls plant tolerance to salt and drought stresses. *Planta* 237: 1527–1545.

Urano, D., Chen, J. G., Botella, J. R., Jones, A. M. (2013) Heterotrimeric G protein signaling in the plant kingdom. *Open Biology* 3: 120186.

Wang, P., Li, Z., Wei, J., Zhao, Z., Sun, D., Cui, S. (2012) A Na+/Ca2+ exchanger-like protein (AtNCL) involved in salt stress in *Arabidopsis*. *Journal of Biological Chemistry* 287: 44062–44070.

Wang, J., Liu, S., Li, C., Wang, T., Zhang, P., & Chen, K. (2017) PnLRR-RLK27, a novel leucine-rich repeats receptor-like protein kinase from the Antarctic moss Pohlia nutans, positively regulates salinity and oxidation-stress tolerance. *PloS One* 12(2): e0172869.

Warpeha, K. M., Lateef, S. S., Lapik, Y., Anderson, M., Lee, B. S., Kaufman, L. S. (2006) G-protein-coupled receptor 1, G-protein Gα-subunit 1, and prephenate dehydratase 1 are required for blue light-induced production of phenylalanine in etiolated *Arabidopsis*. *Plant Physiology* 140: 844–855.

Wilson, R. S., Swatek, K. N., Thelen, J. J. (2016) Regulation of the regulators: Post-translational modifications, subcellular, and spatiotemporal distribution of plant 14-3-3 proteins. *Frontiers in Plant Science* 7: 611.

Witte, C. P., Keinath, N., Dubiella, U., Demoulière, R., Seal, A., Romeis, T. (2010) Tobacco calcium-dependent protein kinases are differentially phosphorylated in vivo as part of a kinase cascade that regulates stress response. *Journal of Biological Chemistry* 285: 9740–9748.

Wu, F., Sheng, P., Tan, J., Chen, X., Lu, G., Ma, W., Heng, Y., Lin, Q., Zhu, S., Wang, J., Wang, J. (2014) Plasma membrane receptor-like kinase leaf panicle 2 acts downstream of the DROUGHT AND SALT TOLERANCE transcription factor to regulate drought sensitivity in rice. *Journal of Experimental Botany* 66: 271–281.

Yang, D. L., Shi, Z., Bao, Y., Yan, J., Yang, Z., Yu, H., Li, Y., Gou, M., Wang, S., Zou, B., Xu, D. (2017) Calcium pumps and interacting BON1 protein modulate calcium signature, stomatal closure, and plant immunity. *Plant Physiology* 175: 424–437.

Yang, L., Wu, K., Gao, P., Liu, X., Li, G., Wu, Z. (2014) GsLRPK, a novel cold-activated leucine-rich repeat receptor-like protein kinase from *Glycine soja*, is a positive regulator to cold stress tolerance. *Plant Science* 215: 19–28.

Yuan, W., Zhang, D., Song, T., Xu, F., Lin, S., Xu, W., Li, Q., Zhu, Y., Liang, J., Zhang, J. (2017) Arabidopsis plasma membrane H+-ATPase genes *AHA2* and *AHA7* have distinct and overlapping roles in the modulation of root tip H+ efflux in response to low-phosphorus stress. *Journal of Experimental Botany* 68: 1731–1741.

Zargar, S. M., Nagar, P., Deshmukh, R., Nazir, M., Wani, A. A., Masoodi, K. Z., Agrawal, G. K., Rakwal, R. (2017) Aquaporins as potential drought tolerance inducing proteins: Towards instigating stress tolerance. *Journal of Proteomics* 169: 233–238.

Zhang, J., Wei, J., Li, D., Kong, X., Rengel, Z., Chen, L., Yang, Y., Cui, X., Chen, Q. (2017) The role of the plasma membrane H+-ATPase in plant responses to aluminum toxicity. *Frontiers in Plant Science* 8: 1757.

Zhang, X., Yang, G., Shi, R., Han, X., Qi, L., Wang, R., Xiong, L., Li, G. (2013) *Arabidopsis* cysteine-rich receptor-like kinase 45 functions in the responses to abscisic acid and abiotic stresses. *Plant Physiology and Biochemistry* 67: 189–198.

19 Trehalose Metabolism in Plants under Abiotic Stresses

Qasim Ali, Sumreena Shahid, Shafaqat Ali, Muhammad Tariq Javed,
Naeem Iqbal, Noman Habib, Syed Makhdoom Hussain, Shahzad Ali Shahid,
Zahra Noreen, Abdullah Ijaz Hussain, and Muhammad Zulqurnain Haider

CONTENTS

19.1 INTRODUCTION

Altered cellular metabolic activities due to changes in physiological condition under unwanted adverse factors are referred to as stress. Under stress, any change in physical and chemical equilibrium is known as strain (Gaspar et al., 2002). Although the literature gives various meanings of stress, the exact physiological definition in appropriate terms is response to variable environmental conditions. The initiation of responses under a changing environment depends on the flexibility of normal metabolism, which fluctuates regularly daily or in seasonal cycles. So stress is not always any deviation from the optimum value of any environmental factor. Stress is an extremely unpredictable change in regular metabolism up to extent of injury, disease and aberrant physiology. Different biotic or abiotic stresses to which plants are frequently exposed while growing in natural environmental conditions include water deficiency at any stage, extremes of temperature, salinity, flooding, pests and heavy metal toxicity and oxidative stress.

19.2 STRESS REMOVAL MECHANISMS IN PLANTS

Disturbances in plant water relations under stressful conditions have the most effect, leading to variations in further physio-biochemical processes, in turn leading to reduced growth and yield. To counteract adversities due to osmotic imbalances, plants have developed different mechanisms.

A stressful environment adversely affects crop plant biomass production and final yield by causing decreased tissue water status and turgor as a function of inadequate water supply. (Kiani et al., 2007; Hussain et al., 2009). To maintain cellular water content under limited soil water supply, plants have developed mechanisms of osmoregulation to lower cellular osmotic potential by accumulating organic and inorganic solutes without decreasing the actual cellular water content (Serraj and Sinclair, 2002). Even so, even in high concentration, these solutes have no recorded injurious effect on macro-organic molecules, whether metabolic enzymes or membranes. These organic solutes are referred as compatible solutes (Cechin et al., 2006; Kiani et al., 2007). The major and important ones include the sugar alcohols, glycine betaine (GB), soluble sugars, proline, organic acids and trehalose (Cechin et al., 2006; Kiani et al., 2007; Farooq et al., 2008, 2009a, b). In parallel with the maintenance of cellular turgor pressure, they act as molecule chaperones to protect macromolecules, including metabolic enzymes, and cells from the damaging effects of ROS (Farooq et al., 2009a, b). To minimize the adverse effects of stress-induced damage in crop plants, the key adaptation of plants is cellular osmotic adjustment (Blum, 2005). First, it plays a significant role in maintenance of leaf turgor to improve stomatal opening for efficient CO_2 intake (Kiani et al., 2007). Second, it increases the ability of roots to uptake water (Chimenti et al., 2006). Compatible solutes that accumulate under different stresses include free proline, free amino acids, GB and sugars (Manivannan et al., 2007; Farooq et al., 2008). Osmotic adjustment by accumulating a variety of organic and inorganic solutes, along with higher activity from antioxidant enzymes in leaves, is among the most important physiological adaptations of plants grown in

drought-prone environments (Lei et al., 2006). Among others, sugars also play a significant role, not only in non-stress conditions but importantly under varying stress conditions (Huang et al., 2006; Rolland et al., 2006; Grigston et al., 2008).

19.3 ROLE OF SUGARS IN PLANTS

19.3.1 PERCEPTION OF SUGARS AT THE CELLULAR LEVEL

The HEXOKINASE1 (HXK1) protein was the first identified plant sugar sensor of glucose (Rolland et al., 2006). As well as a glucose sensor, it also acts as an enzyme to catalyze the first step of glycolysis but glucose sensing is enzyme dependent, as in yeast. Other than HXK1, the cell membrane also contains a receptor coupled with G-protein. It has been reported that this protein's RGS1 gene was confirmed by the use of mutants that were impaired in glucose sensing (Grigston et al., 2008). G-protein sugar signaling has been confirmed by the study on G-protein mutants that have the GPA1 subunit, showing impaired glucose sensitivity (Huang et al., 2006). Similarly, it was reported that various sensors on cellular membranes for sugar metabolic intermediates and different neutral sugars are present that remain to be identified, such as for disaccharide sucrose. In case of sucrose, its signaling function is difficult to separate from its hydrolysis products, such as fructose and glucose. So sucrose signaling activity is normally considered likely to be a hexose-dependent mechanism. However, this cannot be true because sucrose sensing by a subset of genes cannot be mimicked by hexoses (Rolland et al., 2006) and the nature of the sucrose sensor is still obscure. The intermediates of sugar metabolism can also be sensed; for example, severe growth inhibition of *Arabidopsis* seedlings was recorded when treated with trehalose. Alternatively, the enzyme responsible for trehalose-6-phosphate degradation rescued the seedling from growth inhibition, suggesting that T6P is responsible for inhibition of growth and acts as a sensor (Schluepmann et al., 2003). It was found that T6P has a significant role in embryo development and control of architectural changes and flowering initiation (Chary et al., 2008). More importantly, redox activation of ADP, glucose, pyrophosphorylase controls the starch synthesis (Kolbe et al., 2005).

19.3.2 SUGAR SIGNALING AT THE CELLULAR LEVEL

Based on the underlying signaling pathway, it is confirmed that identifying the sensor for sugars is very difficult. Recently, in the HXK1 pathway, various signaling components were found. It was found that the complex of vacuolar H+-ATPase B1 (VHA-B1) and the 19S regulatory particle of proteasome subunit (RPT5B) activates glucose-regulated genes (Cho et al., 2006) and alternately it depends on the signaling pathway of the abscisic acid (ABA), as ABA-insensitive mutants show insensitivity to glucose (Rolland et al., 2006). There has been recent interest in the AB14 proteins that mediate trehalose-dependent responses (Ramon et al., 2007), also act as important signaling molecules from plastids to the nucleus (Koussevitzky et al., 2007) and also regulate sugar-dependent responses (Bossi et al., 1980). Reports are available showing the role of plant hormones such as auxin, cytokinin, GA, SA and brassinosteroid in sugar signaling in plant developmental processes (Rolland et al., 2006). However, many steps of these signaling pathways remain to be discovered.

A variety of SNF1-related protein kinases (SnRK) that have categorized into various families are present in plants (Polge and Thomas, 2007). Recently, it has been found that in *Arabidopsis*, two kinases KIN10 and KIN11, that similar to SNF1 perform a similar function as that does in yeast and mammals. It has been found that these kinases are involved in stress tolerance because plants with deficits in KIN10 and KIN11 do not performing well under stresses (Baena-González and Sheen, 2008). Moreover, T6P inhibits the activity of KIN10 and KIN11 (Zhang et al., 2009). The KIN10 and KIN11 kinases control the activity of genes through various transcription factors.

19.4 TREHALOSE

19.4.1 GENERAL ACCOUNT

Among the various studied sugars, trehalose is of importance due to its unique properties. It is a non-reducing disaccharide found widely in organisms from unicellular bacteria to multicellular higher plants, as well as in various vertebrates and insects (Elbein et al., 2003). It is also known as mycose or tremalose. It plays a significant role under a variety of abiotic stresses in various organisms. Trehalose was first discovered in 1832 by H.A. Wiggers, while in 1858 Mistcherlich isolated trehalose from mushrooms and named it mycose (Richards et al., 2002). In the same year, Berthelot isolated a new type of sugar from an insect secretion named Trehala manna and named it "trehalic glucose" or trehalose (Richards et al., 2002). At that time, this sugar was categorized as rare and could only be extracted from Trehala manna or the resurrection plant, but later its discovery in yeast made it important for researchers (Koch and Koch, 1925) and they developed protocols for its isolation from yeast (Richards et al., 2002). It had limited commercial application because of its high cost of production. In the 1990s in Japan, the Hayashibara company created a laboratory method for inexpensive trehalose mass production by using starch as a substrate. The enzymes used in this process, maltooligosyl–trehalose trehalohydrolase (MTHase) and maltooligosyl–trehalose synthase (MTSase), were obtained from *Arthrobacter* sp., a non-pathogenic soil bacteria, (Maruta et al., 1995).

19.4.2 TREHALOSE BIOSYNTHESIS PATHWAYS IN LIVING ORGANISMS

Trehalose biosynthesis in living organisms takes places by five known pathways. However, the type of pathway depends on the type of stress affecting the organism (Paul et al., 2008).

19.4.2.1 TPS-TPPP (OtsA-OtsB) Pathway

The common pathway of trehalose biosynthesis in prokaryotes and eukaryotes is the TPS-TPP pathway that completes in two steps (Paul et al., 2008). In the first step, synthesis of intermediate trehalose-6-phosphate takes place from glucose-6-phosphate and uridine diphosphate in a reaction catalyzed by TPP synthase; and in the next step, it converts to trehalose dephosphorylation catalyzed by trehalose-6-phosphate (TPP). However, in bacteria, the enzymes Ots-A and Ots-B directly convert glucose-6-phosphate and UDP to trehalose, a similar pathway in yeast as in plants, but the second step is catalyzed by TPS1 and TPS2 at the place of TPS and TPP, respectively (Bell et al., 1998).

19.4.2.2 TreY-TreZ Pathway

The first discovery of this pathway was in *Arthrobacter* sp. In this pathway, trehalose biosynthesis takes place from maltodextrines in two steps (Maruta et al., 1995). In the first step, maltoligosyntrehalose synthesis takes place from maltopentose by the enzyme maltooligosyl-trehalose synthase (TreY) and then the maltooligosyl-trehalose converts to trehalose in a hydrolysis reaction catalyzed by maltooligosyltrehalose trehalohydrolase (TreZ) (Maruta et al., 1995). This trehalose biosynthesis was also later found in other bacterial species, including *Corynebacterium* (Tzvetkov et al., 2003), *Bradyrhizobium japonicum* (Sugawara et al., 2010) and *Rhizobium* (Maruta et al., 1996a), although it is not found in common bacterial species such as *E. coli* and *Bacillus subtilis*; however, it is frequently found in the archaea *Sulfolobus* (Maruta et al., 1996b).

19.4.2.3 TreS Pathway

In this pathway, trehalose synthesis takes place from maltose by a reversible transglycosylation reaction catalyzed by trehalose synthase (TreS). It was discovered in *Pimelobacter* sp. R48 (Nishimoto et al., 1996) and until now is only found in bacteria (Paul et al., 2008). It activates only under osmotic stress due to the reversible nature of TreS in catabolism of *Pseudomonas syringae* (Freeman et al., 2010). Furthermore, it was also found that TreS also works in a reversible manner to make trehalose in *B. japonicum* (Sugawara et al., 2010).

19.4.2.4 TreT Pathway

This pathway involves trehalose biosynthesis in a reversible manner from ADP-glucose and glucose catalyzed by enzyme trehalose glycosyltransferase (TreT) (Qu et al., 2004). This enzyme is found in archaea as well as bacteria (Paul et al., 2008), and its first discovery is reported in the heat-tolerant archaea *Thermococcus litoralis* (Qu et al., 2004).

19.4.2.5 TreP Pathway

This pathway is found not only in prokaryotes but also in eukaryotes. In this pathway, the biosynthesis of trehalose takes place in a reversible reaction from glucose-1-phosphate (G1P) and glucose catalyzed by trehalose phosphorylase. The pathway was first reported to be discovered in *Euglena gracilis* (Belocopitow and Maréchal, 1970) but later was discovered in bacteria and mushrooms (Paul et al., 2008).

19.5 GENERAL ROLES OF TREHALOSE IN PLANTS AND ANIMALS

Multifarious functions of trehalose have been found but they are species-specific. For example, in microorganisms during germination of spores, it acts as an energy source (Elbein et al., 2003). It accumulates in high concentrations under drought in anhydrobiotic organisms to survive complete dehydration by preserving cellular membranes (Dernnan et al., 1993; Crowe et al., 1984). In mycobacteria, it acts as a membrane structural component, being part of glycolipids (Elbein, 1974). By playing a significant role in membrane stabilization and by preventing protein denaturation, it protects *E. coli* against cold stress, whereas in yeast it enhances tolerance of desiccation, heat and osmotic stresses (Hottiger et al., 1987; Hounsa et al., 1998). In yeast it also performs the function of radical scavenging (Benaroudj, 2001). Insects during flight use trehalose as an energy source (Elbein, 1974). However, in plants trehalose acts as a signaling

molecule affecting the activities of genes responsible for abiotic stress tolerance (Schluepmann et al., 2004; Bae et al., 2005). High accumulation of trehalose has been reported in dormant embryos of brine shrimp and after dormancy it is used as an energy source (Clegg, 1965). During dehydration, high accumulation of trehalose has been reported in nematodes, accounting for up to 20% of total dry weight. It plays a significant role in survival of dehydration (Crowe et al., 1992).

19.5.1 METABOLISM UNDER STRESS

It has been found that trehalose protects proteins and cellular membranes by acting as osmoprotectant (Crowe et al., 1984; Muller et al., 1995). Researchers working in plant sciences who tried to express trehalose genes of microbial origin in various plant species to enhance their stress tolerance gained effect to some extent (Karim et al., 2007; Jang et al., 2003; Cortina and Culiáñez-Macià, 2005; Goddijn et al., 1997; Karim et al., 2007; Miranda et al., 2007) (see Table 19.1). In early trials, overexpression of *E. coli* OtsA or yeast TPS1 were found successful in the induction of stress tolerance in plants even with accumulation at very low levels but an abnormal phenotype was found due to overaccumulation of trehalose-6-phosphate (Goddijn et al., 1997; Romero et al., 1997; Cortina and Culiáñez-Macià, 2005). However, these abnormalities were not found in plants expressing microbial TPS-TPP along with a stress-inducible promoter (Miranda et al., 2007). This problem was overcome by the use of trehalose phosphorylase without the accumulation of trehalose-6-phosphate (Han et al., 2005). Even with these achievements, the trehalose levels were still very low compared with that required for induction of drought tolerance, showing that trehalose does not have a direct role in stress tolerance. Multiple studies have linked trehalose to stress tolerance in plants. In alfalfa (*Medicago sativa* L.), Fougère et al. (1991) reported that trehalose concentration in root and shoot increased 3.5 to 4.4 fold respectively under salt stress. However, this concentration is not enough to declare it as an osmoprotectant. It was found that in rice exposed to salt stress, a small increase in trehalose was found (Garcia et al., 1997). However, exogenous application of trehalose (5 mM) significantly improved salt tolerance, along with reduced Na^+ accumulation, while a dose of trehalose of 10 mM reduced chlorophyll loss as well as maintaining root integrity. Avonce et al. (2004) reported that no phenotypic alteration was found in transgenic plants overexpressing TPS1 but delay in flowering was found. However, drought-stress tolerance was found in plants overexpressing AtPS1 and showing ABA- and glucose-insensitive phenotypes, which

TABLE 19.1

Roles of Trehalose Under Various Abiotic Stresses in Variety of Plants as Its Exogenous Use through Different Modes or by Endogenous Metabolism

Plant Species	Type of Stress (Salt, Drought, Heavy Metal, Temperature etc.)	Source of Trehalose (Endogenous/Exogenous)	Metabolic Response (Attributes Affected)	Effective Concentration (Exogenous Applied or Endogenous Levels)	References
Cicer arietinum	Low temperature stress	Endogenous production	Improvement in morphology, biomass production along with improvement in photosynthesis, water relation and trehalose content	2 mg/g fresh weight	Farooq et al. (2017)
Oryza sativa	Salt stress	Seed priming	The activities of SOD, CAT and POX	25 mM	Abdallah et al. (2016)
Pennisetum glaucum	Downy mildew Disease	Exogenous application (seed priming)	Resistance against Downy Mildew Disease of Pearl Millet	100 and 50 mM	Govind et al. (2016)
Triticum aestivum	Drought	Exogenous application (foliar spray)	Phenolic compounds, flavonoids, Amino acids, reducing sugars, total soluble sugars, protein, proline, POD, APX, catalase, PPO and MDA	10 mM	Ibrahim and Abdellatif (2016)
Manihot esculenta	Drought	Endogenous production	Reletive water content, trehalose, glucose, fructose concentration	05–7 mg/g fresh weight	Han et al. (2016)
Musa acuminate	Drought	Exogenous application (foliar spray)	Improved drought tolerance	20, 60 and 100 mM	Said et al. (2015)
Oryza sativa	Cu stress	Pre-sowing seed treatment	Growth, photosynthesis, improved antioxidant defense mechanism	10 mM trehalose	Mostofa et al. (2015)
Nicotiana tabacum	Cd and Cu stress	AtTPS1 engineered	Improved antioxidative defense mechanism with increased activities of SOD, POD, CAT and APX and also improved sugar metabolism, plant atrer relation, photosynthetic activity and cellular osmotic adjustment	–	Martins et al. (2014)
Brassica species	Drought	Exogenous application (foliar spray)	Improved growth along with water relation, photosynthetic pigments and oxidative defense mechanism	5 mM	Alam et al. (2014)
Zea mays	Heat stress	Exogenous application (foliar spray)	Effect on endogenous hydrogen sulfide and trehalose content	(0,10,15,20 and 25 mM)	Li et al. (2014)
Triticum aestivum	Powdery mildew (fungal infection)	Exogenous application (foliar spray)	Fresh and dry weights, relative expression, LOX activity	44 mM	Tayeh et al. (2014)

(Continued)

TABLE 19.1 (CONTINUED)
Roles of Trehalose Under Various Abiotic Stresses in Variety of Plants as Its Exogenous Use Through Different Modes or by Endogenous Metabolism

Plant Species	Type of Stress (Salt, Drought, Heavy Metal, Temperature etc.)	Source of Trehalose (Endogenous/Exogenous)	Metabolic Response (Attributes Affected)	Effective Concentration (Exogenous Applied or Endogenous Levels)	References
Oryza sativa	Salt stress	Exogenous (foliar spray)	Na$^+$, K$^+$, SOD; peroxidase, POX; ascorbate peroxidase, APX; and upregulation of other antioxidative compounds	10 mM	Nounjan et al. (2012)
Lactococcus lactis	Heat stress	Endogenous production	Trehalose content under stress	170 mM	Carvalho et al. (2011)
Zea mays	Drought	Exogenous application (foliar spray)	Drought tolerance, photosynthesis, water relation, growth	30 mM as foliar spray	Ali and Ashraf (2011)
Lemna giba L.	Cd stress	Exogenous application in Hoagland's nutrient solution in rooting medium	Improved water relation and oxidative defense mechanism	0.5, 1, 2, and 5 mM	Duman et al. (2011)
Zea mays	Salt stress	Seed priming	Growth, physiological and biochemical attributes	2–20 mM	Zeid (2009)
Triticum aestivum	Heat stress	Exogenous (foliar spray)	Improved antioxidative mechanism	50 mmol	Luo et al. (2010)
E. coli	Thermotolerance	Endogenous production	Thermus caldophilus GK24 gene activation	25 percent of total sugar content	Cho et al. (2006)
Zea mays	Salt and drought stress	Endogenous production	Seeds endogenous trehalose	(2.7 mg/g dry weight)	El-Bashiti et al. (2005)
Solanum lycopersicum	Salt and drought stress	Endogenous production	Chlorophyll, starch and trehalose detection	0.2 mg/g fresh weight of leaves	Cortina and Culiáñez-Macià (2005)
Gladiolus communis	Apoptotic cell death	Exogenous application (foliar spray)	Nuclear fragmentation in senescing petals	300 mM as foliar spray	Yamada et al. (2003)
Arabidopsis	Developmental regulation	Endogenous production	Embryo development and photosynthetic Metabolism	1 M	Wingler (2002)
Oryza sativa	Salt and low temperature stress	Endogenous production	Physiological attributes, photosynthetic pigments, trehalose content	17 µg/g fresh weight	Garg et al. (2002)
Hordeum vulgare	Metabolic stress	Exogenous application (foliar spray)	Fructan synthesis in barley leaves	200 and 500 mmol as foliar spray	Müller et al. (2000)
Saccharomyces cerevisiae	Heat stress	Endogenous production	Trehalose synthesis proteins activation (Tps1p, Tps2p, Tps3p)	20% of dry cell trehalose accumulation	Singer and Lindquist (1998)
Nicotiana tobacco	Drought stress	Endogenous production	Photosynthetic pigments, levels of trehalose, glucose, fructose and sucrose	0.20 mg.g^{-1} dry weight	Elizabeth et al. (1998)

demonstrated that trehalose-6-phosphate and/or AtTPS1 play a major role in signaling during the development of seedlings. It was reported that Arabidopsis csp-1 mutant plants, with a point mutation in the synthase domain of another Arabidopsis TPS, AtTPS6, in spite of changing the plant architecture and cell shapes, showed drought tolerance (Chary et al., 2008). Furthermore, it was found that another TPS gene, AtTPS5, showed involvement in plant stress tolerance (Suzuki and Bird, 2008). MBF1c, a transcriptional activator, controls the activity of AtTPS5 in thermotolerance and activates in 20 min of heat-stress treatment, and mutants of tps5 showed no heat tolerance, although the acquired tolerance was similar to that found in wild-type plants. It was found that under heat stress, trehalose was accumulated in wild types with an improvement in stress tolerance. Exogenous application of trehalose improved the tps5 mutant's stress tolerance. Trehalose-applied stress tolerance was also found in mbf1c and mutants that were efficient in salicylic acid (sid2) and ethylene (ein2-1), showing that trehalose biosynthesis is involved in interaction with plant hormones. In rice nine TPSs and TPPs are found. In rice two TPPs (OsTPP1 and OsTPP2) were isolated that were expressed in different stresses such as drought, salt and freezing stresses in a transient manner (Pramanik and Imai, 2005; Shima et al., 2007). This expression was also found in response to ABA application. The expression of OsTPP1 always earlier than OsTPP2, showing a tight synchrony in trehalose biosynthesis under varying abiotic stresses (Shima et al., 2007). Such increments in OsTPP1 in a transient manner has been found under chilling stress (Pramanik and Imai, 2005), and this chilling stress-induced trehalose accumulation coincided with changes in glucose and fructose levels (Pramanik and Imai, 2005). Schluepmann et al. (2004) reported that there was a positive correlation in trehalose-6-phosphate and the activity of several genes involved in abiotic stress tolerance after application of 100 mM trehalose. ATPK19 was induced by cold and salt stress, as well as the protein involved in signaling of calcium and phosphorylation. However, the opposing results were obtained by Bae et al. (2005) in a transcript-level study, where 30 mM trehalose application in combination with 1% sucrose suppressed most of the genes involved in stress tolerance. They reported that this difference might be due to the use of a different trehalose concentration, as well as a different DNA microarray, and the most important were involved in signaling of jasmonate and ethylene. The upregulation of stress-responsive genes due to trehalose application includes major ones for carboxyl methyltransferase, aldo-keto reductase, anthocyanidin synthase, S-adenosyl-L-methionine. However, the repressive ones include 5-adenylsulphate reductase (APR1), cysteine

proteinase (RD21A), plasma membrane intrinsic protein (PIP1C), cytosolic glyceraldehyde-3-phosphate dehydrogenase (GAPDHc), ferritin (At- FER1), vacuolar adenosine triphosphate (ATP) synthase catalytic subunit A, chitinase-like protein 1 (CTL1), lipoxygenase 2 (LOX2), endo-1,3-endo-β-glucanase (BGL1), basic endochitinase (PR) and peroxidase-2 (PRXR2). Trehalose-induced abiotic stress tolerance has been reported in prokaryotes as well as in eukaryotes, with effects found in almost all types of stresses through its endogenous increase or exogenous application. The available information regarding the role of trehalose in abiotic stress tolerance is still incomplete and large gaps are present.

19.5.2 Trehalose Metabolism under Desiccation Stress

Water is a necessary part of every life on Earth and without it no life is possible. However, the requirement depends upon the type of organism. For example, in complete dehydration, anhydrobiotics can survive, and such organisms include some invertebrates, microorganisms and resurrection plants (Crowe et al., 1992). Under dehydration, especially of 99% of water, these organisms can accumulate high trehalose contents (Strom and Kaasen, 1993). Trehalose is the most important of disaccharides in stabilizing cellular membranes under desiccation stress (Crowe et al., 1992), which is the function of vesicle fusion and lipid phase transition (Crowe and Crowe, 1990). It has been reported that the accumulation of trehalose seen in small quantities reduces diffusion of vesicles as well as the transition of dry lipids and maintains their crystalline state under dehydration (Crowe et al., 1992), while trehalose during freezing and desiccation replaces the H_2O_2 molecules. Trehalose has the property of binding of its OH part with polar groups of proteins, as well as with phosphate groups of membranes (Kawai et al., 1992).

Among others, the process of vitrification is another property of trehalose for stress tolerance. It is more stable than other disaccharides due to its non-reducing character. This property leads to holding the molecule firmly and returning bounded molecules to their native structure on rehydration (Crowe and Crowe, 2000). The glassy structure of trehalose is so stable that a small amount of water makes a dehydrate on the outer surface and protects the internal structures (Richards et al., 2002).

19.5.2.1 Trehalose Metabolism in Bacteria under Desiccation

Huge work has been reported about the role of trehalose in desiccation tolerance of bacterial species. Under desiccation stress, *Nostoc commune* accumulated a large amount of trehalose, along with other polysaccharides that have roles in stress tolerance (Cameron, 1962; Sakamoto et al., 2009; Klahn and Hagemann, 2011). Among others that accumulate high amounts of trehalose under stress are *Phormidium autumnale* and *Chroococcidiopsis sp.* (Hershkowitz et al., 1991). It has been found that on rewatering, the trehalose contents decrease (Sakamoto et al., 2009). However, trehalose accumulation under desiccation stress is not same in all studies. For example, in *Anabena* and *Nostoc flagelliforme*, the accumulated content of trehalose is not enough to work as a molecular chaperone for related genes (Higo et al.; 2006; Wu et al., 2010).

In *B. japonicum*, an important rhizobial species inhabiting in the root nodules of soybeans, there are different biosynthetic pathways for trehalose accumulation (Streeter, 2006) and mutants lacking the TreS degradation pathway are found to be sensitive to desiccation stress (Sugawara et al., 2010). The reason behind that is this concentration of trehalose has an effect on the reactivation and refolding of modified proteins and explains that trehalose is quickly degraded after stress was ceased (Singer and Lindquist, 1998). It was found that engineered rhizobacteria with trehalose biosynthetic genes were found effective in improving the drought tolerance of plants inoculated with rhizobia strains having genes for trehalose biosynthesis, as reported in *Phaseolus vulgaris* inoculated with *Rhizobium* overexpressed with trehalose-6-phosphate synthase. A hundred percent recovery was found under drought stress in parallel to plants inoculated with the strain with no trehalose biosynthetic gene, which wilted and died (Suarez et al., 2008). Similarly, maize inoculated with *Azospirillum braziliense* expressing trehalose biosynthetic genes showed an 85% of survival under drought stress as compared with plants inoculated with the untransformed strain (55% survival) (Rodriguez-Salazar et al., 2009).

19.5.2.2 Trehalose Metabolism in Fungi under Desiccation

Accumulation of trehalose is well studied in fungi. In fungi, the accumulation of trehalose under unfavorable conditions occurs especially at vegetative and reproductive stages compared with other growth stages (Thevelein, 1984). In arbuscular mycorrhizal fungi, the accumulation of trehalose is prominent in spores and extra-medical mycelium as compared with other parts (Becard et al., 1991). It was found that log-phase cultures of yeast are susceptible to dehydration due to the low accumulation of trehalose but at the stationary growth phase, a high level of trehalose was found (Elbein et al.,

2003). Hottiger et al. (1987) conducted a study on yeast cells and demonstrated a clear correlation between the cell's trehalose contents and its desiccation tolerance.

19.5.2.3 Trehalose Metabolism in Plants under Desiccation

The discovery of trehalose in higher plants was much later than it was in lower plant species (Bianchi et al., 1993; Drennan et al., 1993; Albini et al., 1994). Studies reveal that there are ten TPP and 11 TPS putative genes in *Arabidopsis*, whereas in rice there are nine each of both TPSs and TPPs. It was found that *Arabidopsis* genetically modified for AtTPS1 exhibited tolerance to shortage of water, as well in sensitivity to ABA and glucose (Avonce et al., 2004) Furthermore, another *Arabidopsis* mutant csp-1 is also found drought resistant that has a point mutation in TPS in the AtTPS6 domain (Chary et al., 2008).

In comparison with other organisms and lower plants, the accumulation of trehalose in higher plants is not sufficient to act as a compatible solute. It was found that genetically modified plants overexpressing bacterial genes for trehalose biosynthesis showed drought tolerance but the accumulation was comparable with that of parent bacteria. Similarly, tomato and tobacco plants with the yeast TPS1 gene for trehalose biosynthesis were found drought tolerant (Romero et al., 1997; Cortina and Culiáñez-Macià, 2005). However, the transgenics that were found drought tolerant did show growth aberrations in morphology, such as studied growth in tobacco transgenics and root aberrations in tomato transgenics and it was found that these growth aberrations are due to the large accumulation of trehalose-6-phosphate, an important regulator of growth in plants. For example, the mutants for homozygous tps1 were lethal embryonically (Eastmond et al., 2002). Furthermore, it was found that trehalose-6-phosphate also inhibits other essential growth regulators such as hexokinase and SnRK1. These also regulate the metabolism and development of plants (Zhang et al., 2009; Paul et al., 2010). Transgenic tobacco plants for trehalose biosynthesis were made drought tolerant without any abnormality by targeting TPS1 expression to the chloroplast as well as by the overexpression of the yeast trehalose synthesis gene (Karim et al., 2007). Elizbeth et al. (1998) reported an endogenous increase (0.20 mg g^{-1} F.Wt.) in trehalose production in transgenic tobacco plants grown under water stress. In a study conducted by Han et al. (2016), an endogenous increase (3 mg g^{-1} F.Wt.) in cassava plants under water-deficit conditions was reported. An endogenous increase (2.7 mg g^{-1} F.Wt. of leaves) in trehalose was reported in wheat plants under drought stress (El-Bashiti at al., 2005).

Furthermore, in tomato plants endogenous production of trehalose (0.2 mg g^{-1} F.Wt.) was reported by Cortina and Culiáñez-Macià (2005). Ibrahim and Abdellatif (2016) reported that exogenous application (10 mM as a foliar spray) of trehalose induced drought-stress tolerance in wheat plants. Furthermore, in banana drought tolerance is associated with exogenous application (foliar application 20, 60 and, 100 mM) of trehalose (Said et al., 2015). Similarly, in another study, exogenous application of trehalose as a foliar spray (5 mM) made brassica plants drought tolerant (Alam et al., 2014). Exogenous application of trehalose as a foliar spray (30 mM) increased water-stress tolerance in maize, which in turn increased photosynthesis and water relation (Ali and Ashraf, 2011).

19.5.3 TREHALOSE METABOLISM UNDER SALT STRESS

High salt content in the growth medium affects plants in two ways. First, it creates a more negative osmotic potential that hinders water uptake. Second, it creates a problem of ion toxicity due to the accumulation of toxic metals. To maintain a proper turgor potential for growth under disputed conditions, the organism has the ability to accumulate high levels of cellular compatible solutes (Ashraf and Foolad, 2007), which alleviate the adverse effects of salinity in two major ways. First, they lower the cellular osmotic potential of the cytoplasm for better uptake of water to maintain the normal turgor pressure of the cells (Kempf and Bremer, 1998). Second, they act as molecular chaperones against toxic ions (Hincha and Hagemann, 2004). Among different compatible solutes, trehalose acts as an anti-stress compound in both prokaryotes and eukaryotes.

19.5.3.1 Trehalose Metabolism in Bacteria under Salt Stress

All organisms from unicellular to multicellular face osmotic stress from salinity. It was found that *E. coli* cells that had accumulated trehalose in high concentration would then degrade it to glucose with the enzyme trehalase after releasing stress, providing an extra source of energy (Strom and Kaasen, 1993). These studies were further confirmed by the findings that mutants of *B. japonicum, E. coli, S. meliloti* showed sensitivity to exposure to salt stress (Strom and Kaasen, 1993; Dominguez-Ferreras et al., 2009; Sugawara et al., 2010).

Studies reveal that cyanobacteria inhibiting fresh water have the ability to withstand a low or moderately high saline environment by accumulating high trehalose contents as an adaptive mechanism (Klahn and Hagemann, 2011). Among species of cyanobacteria, *Rivularia aira* was the first reported to accumulate high concentrations of trehalose under a saline environment (Reed and Stewart, 1983). Up to now, more than 40 different strains of cyanobacteria have been found to tolerate NaCl osmotic stress, and about 20 of these were major accumulators of trehalose as osmotica (Hagemann, 2011), with trehalose performing the function of membrane stabilization. Studies reveal that in most of the bacteria, the transport of compatible solutes for stress tolerance is a potential means but the cyanobacteria rely on the *de novo* synthesis of protectants for stress tolerance and mainly use transporters for their uptake (Hagemann, 2011). *Synechosystis* showed transporters for uptake of glucosylglycerol, sucrose and trehalose (Mikkat et al., 1996).

Furthermore, it was found that *Corynebacterium glutamicum*, a strain of soil bacterium, uses trehalose as a compatible solute to cope with stress. Under N stress, proline was the major compatible solute when N stress was high, but under low N stress, trehalose was the major compatible solute. Furthermore, high contents of trehalose were reported under increased levels of maltose, instead of sucrose (Wolf et al., 2003).

Salt stress severely affects symbiotic relationship and results in reduced seed yield, but the effect was less in strains having the ability to accumulate high trehalose in *B. japonicum*. Furthermore, the strains lacking trehalose accumulation could survive under salt stress (Sugawara et al., 2010). In an earlier study, it was reported that a bacterial strain isolated from nodules of *P. vulgaris* has an efficient ability to accumulate trehalose (Fernandez-Aunion et al., 2010).

19.5.3.2 Trehalose Metabolism in Fungi under Salt Stress

Numerous studies reveal that arbuscular mycorrhizal fungi colonize in the roots of more than 80% of the terrestrial plant species linked with salt tolerance through different modes (Evelin et al., 2009). In relation to trehalose metabolism, the fungal strain *Glomus intraradices* did not show many changes under osmotic stress, but only the activation of neutral trehalose and trehalose-6-phosphate were found (Ocon et al., 2007).

19.5.3.3 Trehalose Metabolism in Plants under Salt Stress

It was found that in rice two genes responsible for trehalose metabolism were activated transiently under abiotic stresses as were to external application of ABA (Pramanik and Imai, 2005; Shima et al., 2007). Garcia et al. (1997) reported that in rice plants exposed to salt stress for three days accumulated a small amount of trehalose in their roots. Not only endogenous but also exogenous application of trehalose was found effective for stress tolerance but was concentration specific. In a study, it was found that exogenous application of a lower dose (5 mM) reduced the Na+ accumulation and as a result improved growth but a 10 mM concentration reduced the leaf chlorophyll loss and maintains root integrity (Garcia et al., 1997). It was found that alfalfa plants grown under salt stress accumulated a small amount of trehalose in roots and bacteroids but the quantity was not enough to integrate it is as osmoprotectant (Fougère et al., 1991). Furthermore, it was found that increased contents of trehalose were found in nodules of *Medicago truncatula* that was associated with decreased activity of trehalose but the accumulation was not enough to integrate it as an osmoprotectant (Lopez et al., 2008). It was found that genetically engineered tomato plants with yeast trehalose synthesis genes showed tolerance against salt, water stress and oxidative stress (Cortina and Culiáñez-Macià, 2005). Furthermore, rice plants genetically engineered for trehalose biosynthesis under the control of tissue-specific promoters showed temperature, drought and salt tolerance without showing any abnormalities in growth (Garg et al., 2002). Along with endogenous biosynthesis of trehalose for salt tolerance, its exogenous application also improved the stress tolerance of crop plants. For example, in rice seeds, priming with 25 mM trehalose was found effective for the induction of salt tolerance (Abdallah et al., 2016). In another study, it was reported that exogenous application of trehalose as a foliar spray (10 mM) increased salt tolerance in rice, which resulted in increasing the activity of SOD, POD, and POX (Nounjan et al., 2012). It was also found that seed priming of maize plants with 2–20 mM of trehalose under salt stress increased growth, and the physiological as well as biochemical attributes of maize (Zeid, 2009). Furthermore, an endogenous increase (0.20 mg g^{-1} D. Wt.) was reported in transgenic tobacco under drought stress, which further increased photosynthetic pigments and levels of trehalose, glucose, fructose and sucrose (Elizabeth et al., 1998).

Studies revealed that trehalose also accumulates under heavy metal stress. Exogenous application of different levels of trehalose was found helpful in inducing heavy metal stress tolerance. For example, in rice 10 mM trehalose application as seed priming was associated with Cu-stress tolerance and helped in improving growth and photosynthetic activity, as well as being an antioxidant defense mechanism (Mostofa et al., 2015). In another study, tobacco plants engineered with the trehalose-6-phosphate gene (AtTPS1) showed resistance against Cd and Cu stress, which helped in improving antioxidative defense mechanism with increased activity of POD, CAT, APX and SOD (Martins et al., 2014).

Furthermore, exogenous trehalose application (0.5, 1, 2 and 3 mM) in duckweed showed resistance against Cd stress and improved plant water relation as well as an antioxidative defense mechanism (Duman et al., 2011).

19.5.4 TREHALOSE METABOLISM UNDER LOW- AND HIGH-TEMPERATURE STRESS

Extremes of temperature significantly disturb plant metabolism under low-temperature stress. The first and foremost effect is on the biological membrane, decreasing fluidity. At present, with global warming, temperature ups and downs pose a major threat to plants, along with drought and salinity. Alterations in extremes of temperature result in the aggregation and denaturation of proteins and trehalose has the ability to prevent protein denaturation under these conditions. Furthermore, trehalose has the ability to stabilize biological membranes by its binding with the membrane phospholipids.

19.5.4.1 Trehalose Metabolism in Bacteria under Temperature Stress

Bacteria under extremes of temperatures have been well studied. In *E.coli* it was found that among various strains, those deficient in trehalose biosynthesis were more sensitive to chilling stress than those that have the ability to accumulate trehalose by biosynthesis (Kandror et al., 2002), but when engineered with the trehalose biosynthesis gene, they gained the ability to tolerate cold. It was reported that tolerance is due to the roles of trehalose in protein protection and cell membrane stabilization, as well as it playing a role against oxidative stress by scavenging ROS. Furthermore, Santos and Costa (2002) reported the bacteria *Thermus thermophilus* and the archaea *Pyrobaculum aerophilum* residing in high temperatures along with salt stress, with biosynthesis of trehalose in high concentration acting as compatible solutes (Santos and Costa, 2002).

It has been reported that under temperature stress, there is increased trehalose content (170 mM) in *Lactococcus lactis* (Carvalho et al., 2011). Endogenous production of trehalose was reported in *E. coli* under heat stress, which increased total sugar contents by 25% (Cho et al., 2006). In another study, it was reported that trehalose production increased (17 µg g^{-1}) in rice under low-temperature stress, which in turn increased photosynthesis and other physiological attributes (Garg et al., 2002).

19.5.4.2 Trehalose Metabolism in Fungi under Temperature Stress

Among those organisms studied regarding trehalose metabolism, fungi are the foremost for detail of study of heat stress. It was found that under heat stress, fungal spores become heat tolerant due to the high accumulation of trehalose and this is due to upgradation of trehalose biosynthesis genes. For example, it was found that under heat shock in *S. cerevisiae*, trehalose accumulation not only stabilized protein structure but also reduced the accumulation of denatured protein, and after heat stress trehalose degraded with denatured protein becoming renatured (Hottiger et al., 1987). Furthermore, it was found that during heat stress, the damage to protein structure is partial and mainly due to ROS accumulation (Davidson et al., 1996). Under high-temperature stress in yeast, trehalose accumulation increased markedly, resulting that resulted in oxidative stress tolerance with decreased membrane damage (Benaroudj et al., 2001).

Trehalose was not only reported under temperature stress but also found under freezing temperatures. In *S. cerevisiae*, tolerance of freezing was associated with a high accumulation of trehalose. This accumulation was due to the activity of the TPS1 and TPS2 genes, while in *Schizosaccharomyces pombe*, the high accumulation was due to the overexpression of TPS1. Due to the overaccumulation of trehalose, they also become tolerant to other stresses besides freezing stress (Mahmud et al., 2010). On relief of stress, the levels of trehalose and its synthesizing enzymes dropped to original levels rapidly (Kandror et al., 2004). Furthermore, in ectomycorrhyzal basidiomycetes and *Hebeloma* spp., high levels of mannitol, trehalose, and arabitol were found that were associated with tolerance of freezing (Tibbett et al., 2002).

19.5.4.3 Trehalose Metabolism in Plants under Temperature Stress

Among higher plants, the best-studied example of the role of trehalose in stress tolerance is the rice plant. In rice exposed to drought, salt and cold, *OsTPP1* and *OsTPP2* were expressed and found to have a role in trehalose biosynthesis. This overexpression was also found under ABA treatment (Pramanik and Imai, 2005). Under chilling stress, the accumulation of trehalose coincided with fructose and glucose levels. In parallel with rice, in *Arabidopsis* the trehalose-synthesizing gene AtTPSS has a role in high-temperature-stress tolerance. This thermal tolerance is furthermore associated with the activation of MBFIc, a transcription activator that has a role in the regulation of thermo tolerance (Suzuki and Bird, 2008). Furthermore, the engineered plants of *Arabidopsis* with yeast TPS1 gene that was under the regulation of 35S promoter or engineered plants with TPS1-TPS2 under the regulation of a stress promoter accumulated trehalose and showed tolerance to different abiotic stresses,

especially freezing stress, but the accumulation of trehalose was very low. However, abnormalities were found in the growth of plants transgenic with only TPS1 but the opposite was found with secondly constructed plants with TPS1 and TPS2 fused genes (Miranda et al., 2007).

It was found in chickpea that under low temperature stress tolerance was associated with high accumulation (2 mg g-1 F.Wt.) of trehalose (Farooq et al., 2017). It has been reported by Li et al. (2014) that exogenous application of trehalose as foliar spay (0, 10, 15, 20 and 25 mM) increased tolerance against heat stress in maize plants. In another study conducted by Luo et al. (2010), it was reported that heat tolerance in wheat plants induced by foliar application (50 mmol) of trehalose improved the oxidative defense mechanism of wheat plants.

19.6 DIFFERENT METABOLISMS OF ENZYMES UNDER TREHALOSE

The first discovery of trehalose was in ergot (a malfunction of fungi) of rye. Later on, trehalose was found in several lower plants such as *Botrychium lunaria* and *Selaginella lepidophylla*. Regarding vascular plants, the discovery of trehalose is later and rare. It is found in ripened fruits of members of the *Apiaceae* family, as well as in leaves of the angiospermic plant *Myrothamnus flabellifolius*. Recently, aiming towards trehalose's role as a stabilizing agent for engineered drought-tolerant food plants/crop, the genes responsible for trehalose biosynthesis were introduced into them from *E. coli* and yeast. However, the accumulation of trehalose is limited due to the presence of enzyme trehalase in plants. It is confirmed by the use of trehalase inhibitors that a high amount of trehalose accumulation was found in transgenic plants. Furthermore, this reported confirmation of trehalose biosynthesis has been found in yeast in a different type of complementation experiment.

19.7 VERSATILE ROLE OF TREHALOSE METABOLISM

It is found that in large number of organisms, trehalose acts as a protectant to cellular membranes. For example, in yeast, protection against different abiotic stresses is associated with the accumulation of trehalose. The significant role of trehalose in plants was found in the cryptobiotic species S. *lepidophylla*, which is well known to be tolerant to desiccation. Accumulation of 12% of total dry weight of the plant was found to play a major role in protection of membranes. After rehydration, the trehalose declined to its original level. However, rare studies were available about the accumulation of trehalose in higher plants. In parallel, in most of plant species, sucrose in the place of trehalose acts as a component of desiccation tolerance. However, reports are available about the accumulation of trehalose in angiosperms. It was found that in *M. flabellifolius*, trehalose accumulated up to 3% of dry weight and slightly increased tolerance of water stress, while increase in sucrose was up to 6%, in parallel with trehalose, which proved sufficient to tolerate the dehydration conditions. Furthermore, trehalose can stabilize cellular membranes and macromolecules such as enzymes more efficiently than other relevant sugars and it has an importance in genetic engineering for efficient tolerance and commercial production that will make a cost-effective way to produce it on a commercial scale. Transgenic tobacco plants overexpressed with yeast TPS1 accumulated 0.08–0.32% of trehalose based on dry weight both in leaves and roots with retarded growth and dehydration loss and had less water as compared with non-transgenics. They found that on rehydration, the recovery of turgor in trehalose accumulation was better than in non-transgenics. Similarly, experiments for drought tolerance in tobacco with *otsA atsB* from *E. coli* found hopeful results. They reported that based on a dry weight basis a clear difference was found in transgenic plants over control ones, although the trehalose accumulation was very low (0.0008–0.0090%) on a fresh weight bases. They also reported improved photosynthesis in transgenics and depicted the role of trehalose as osmoprotectant with a direct role in plant metabolism. Similarly, it was also reported by Nepomuceno et al. (1998) that an EST clone belonging to cotton with a similar TPS to *Arabidopsis* was found upregulated under water stress.

19.8 CONCLUSIONS AND FUTURE PROSPECTIVE

Recent investigations in higher plants on the biosynthesis of trehalose have showed its role in metabolism, physiology and development processes but there are still some gaps in understanding resistance and tolerance mechanisms. Different research results showed that trehalose metabolism could be a vital component in protecting the cell from different abiotic stresses. In particular, trehalose accumulation and mobilization in response to different stress conditions might be of ecological relevance. The knowledge of individual trehalose biosynthesis genes will provide help, not only in understanding the precise role of trehalose in not just abiotic stress but also the signaling mechanism. So more investigation is needed to understand the complex mechanism of trehalose metabolism so that stress-tolerant plants that are better equipped to face environmental adversity may be developed.

REFERENCES

Abdallah, M.S., Abdelgawad, Z.A., El-Bassiouny, H.M.S. (2016) Alleviation of the adverse effects of salinity stress using trehalose in two rice varieties. *South African Journal of Botany* 103: 275–282.

Alam, M.M., Nahar, K., Hasanuzzaman, M., Fujita, M. (2014) Trehalose-induced drought stress tolerance: A comparative study among different *Brassica* species. *Plant Omics* 7(4): 271–283.

Albini, F.C., Murelli, C., Patritti, G., Rovati, M., Zienna, P., Finzi, P.V. (1994) Low-molecular weight substances from the resurrection plant *Sporobolus stapfianus*. *Phytochemistry* 37: 137–142.

Ali, Q., Ashraf, M. (2011) Induction of drought tolerance in maize (*Zea mays* L.) due to exogenous application of trehalose: Growth, photosynthesis, water relations and oxidative defense mechanism. *Journal of Agronomy and Crop Science* 197(4): 258–271.

Avonce, N., Leyman, B., Mascorro-Gallardo, J.O., Van Dijck, P., Thevelein, J.M., Iturriaga, G. (2004) The *Arabidopsis* trehalose-6-P synthase AtTPS1 gene is a regulator of glucose, abscisic acid, and stress signaling. *Plant Physiology* 136(3): 3649–3659.

Bae, H., Herman, E., Bailey, B., Bae, H.J., Sicher, R. (2005) Exogenous trehalose alters *Arabidopsis* transcripts involved in cell wall modification, abiotic stress, nitrogen metabolism, and plant defense. *Physiologia Plantarum* 125(1): 114–126.

Baena-González, E., Sheen, J. (2008) Convergent energy and stress signaling. *Trends in Plant Science* 13(9): 474–482.

Bécard, G., Doner, L.W., Rolin, D.B., Douds, D.D., Pfeffer, P.E. (1991) Identification and quantification of trehalose in vesicular-arbuscular mycorrhizal fungi by in vivo13C NMR and HPLC analyses. *New Phytologist* 118: 547–552.

Bell, W., Sun, W., Hohmann, S., Wera, S., Reinders, A., De Virgilio, C., Thevelein, J.M. (1998) Composition and functional analysis of the *Saccharomyces cerevisiae* trehalose synthase complex. *Journal of Biological Chemistry* 273(50): 33311–33319.

Belocopitow, E., Maréchal, L.R. (1970) Trehalose phosphorylase from *Euglena gracilis*. *Biochimica et Biophysica Acta (BBA)-Enzymology* 198(1): 151–154.

Benaroudj, N., Lee, D.H., Goldberg, A.L. (2001) Trehalose accumulation during cellular stress protects cells and cellular proteins from damage by oxygen radicals. *Journal of Biological Chemistry* 276(26): 24261–24267.

Blum, A. (2005) Drought resistance, water-use efficiency, and yield potential—Are they compatible, dissonant, or mutually exclusive? *Crop and Pasture Science* 56(11): 1159–1168.

Bossi, L., Assael, B.M., Avanzini, G., Batino, D., Caccamo, M.L., Canger, R., Marini, A. (1980) Plasma levels and clinical effects of antiepileptic drugs in pregnant epileptic patients and their newborns. *Obstetrical and Gynecological Survey* 35(9): 561–562.

Carvalho, A.L., Cardoso, F.S., Bohn, A., Neves, A.R., Santos, H. (2011) Engineering trehalose synthesis in *Lactococcus lactis* for improved stress tolerance. *Applied and Environmental Microbiology* 77(12): 4189–4199.

Cechin, I., Rossi, S.C., Oliveira, V.C., Fumis, T.D.F. (2006) Photosynthetic responses and proline content of mature and young leaves of sunflower plants under water deficit. *Photosynthetica* 44(1): 143–146.

Chary, S.N., Hicks, G.R., Choi, Y.G., Carter, D., Raikhel, N.V. (2008b) Trehalose-6-phosphate synthase/phosphatase regulates cell shape and plant architecture in *Arabidopsis*. *Plant Physiology* 146(1): 97–107.

Chimenti, C.A., Marcantonio, M., Hall, A.J. (2006) Divergent selection for osmotic adjustment results in improved drought tolerance in maize (*Zea mays* L.) in both early growth and flowering phases. *Field Crops Research* 95(2): 305–315.

Cho, Y.J., Park, O.J., Shin, H.J. (2006) Immobilization of thermostable trehalose synthase for the production of trehalose. *Enzyme and Microbial Technology* 39: 108–113.

Clegg, J.S. (1965) The origin of trehalose and its significance during the formation of encysted dormant embryos of *Artmia salina*. *Comparative Biochemistry and Physiology* 14(1): 135–143.

Clegg, J.S. (2001) Cryptobiosis-a peculiar state of biological organization. *Comparative Biochemistry and Physiology Part B: Biochemistry and Molecular Biology* 128: 613–624.

Cortina, C., Culiáñez-Macià, F.A. (2005) Tomato abiotic stress enhanced tolerance by trehalose biosynthesis. *Plant Science* 169(1): 75–82.

Crowe, J.H., Crowe, L.M. (1990) Lyotropic effects of water on phospholipids. In: Franks, F. (ed.), *Water Science Reviews*. Cambridge University Press, Cambridge, England, pp. 1–23.

Crowe, J.H., Crowe, L.M., Chapman, D. (1984) Preservation of membranes in anhydrobiotic organisms: The role of trehalose. *Science* 223(4637): 701–703.

Crowe, J.H. Crowe, L.M (2000) Preservation of mammalian cells. learning nature's tricks. *Nature Biotechnology* 18: 145–146.

Crowe, J.H., Hoekstra, F.A., Crowe, L.M. (1992) Anhydrobiosis. *Annual Reviews of Physiology* 54: 579–599.

Drennan, P.M., Smith, M.T., Goldsworthy, D., Van Staden, J. (1993) The occurrence of trehalose in the leaves of the desiccation-tolerant angiosperm *Myrothamnus flabellifolius* welw. *Journal of Plant Physiology* 142(4): 493–496.

Domínguez-Ferreras, A. Munõz, S., Olivares, J., Soto, M.J., Sanjuán, J. (2009) Role of potassium uptake systems in *Sinorhizobium meliloti* osmo-adaptation and symbiotic performance. *Journal of Bacteriology* 191: 2133–2143.

Davidson, J.F., Whyte, B., Bissinger, P.H., Schiestl, R.H. (1996) Oxidative stress is involved in heat-induced cell death in *Saccharomyces cerevisiae*. *Proceedings of National Academy of Sciences of United States of America* 93: 5116–5121.

Duman, F., Aksoy, A., Aydin, Z., Temizgul, R. (2011) Effects of exogenous glycinebetaine and trehalose on cadmium accumulation and biological responses of an aquatic plant (*Lemna gibba* L.). *Water, Air, and Soil Pollution* 217: 545–556.

Eastmond, P.J., van Dijken, A.J.H., Spielman, M., Kerr, A., Tissier, A.F., Dickinson, H.G., Jones, J.D.G., Smeekens, S.C., Graham, I.A. (2002) Trehalose-6-phosphate synthase1, which catalyses the first step in trehalose synthesis, is essential for Arabidopsis embryo maturation. *Plant Journal* 29: 225–235.

Ehlert, U., Erni, K., Hebisch, G., Nater, U. (2006) Salivary α-amylase levels after yohimbine challenge in healthy men. *The Journal of Clinical Endocrinology and Metabolism* 91: 5130–5133.

Elbein, A.D. (1974) The metabolism of α, α-trehalose. *Advances in Carbohydrate Chemistry and Biochemistry* 30: 227–256.

Elbein, A.D., Pan, Y.T., Pastuszak, I., Carroll, D. (2003) New insights on trehalose: A multifunctional molecule. *Glycobiology* 13(4): 17R–27.

El-Bashiti, T., Hamamcı, H., Öktem, H.A., Yücel, M. (2005) Biochemical analysis of trehalose and its metabolizing enzymes in wheat under abiotic stress conditions. *Plant Science* 169(1): 47–54.

Elizabeth, A.H., Pilon-Smits, E.A., Terry, N., Sears, T., Kim, H., Zayed, A., Hwang, S., Goddijn, O.J. (1998) Trehalose-producing transgenic tobacco plants show improved growth performance under drought stress. *Journal of Plant Physiology* 152(4–5): 525–532.

Evelin, H., Kapoor R., Giri, B. (2009) Arbuscular mycorrhizal fungi in alleviation of salt stress: A review. *Annals of Botany* 104: 1263–1280.

Farooq, M., Basra, S.M ., Wahid, A., Cheema, Z.A., Cheema, M.A., Khaliq, A. (2008) Physiological role of exogenously applied glycinebetaine to improve drought tolerance in fine grain aromatic rice (*Oryza sativa* L.). *Journal of Agronomy and Crop Science* 194(5): 325–333.

Farooq, M., Hussain, M., Nawaz, A., Lee, D.J., Alghamdi, S.S., Siddique, K.H. (2017) Seed priming improves chilling tolerance in chickpea by modulating germination metabolism, trehalose accumulation and carbon assimilation. *Plant Physiology and Biochemistry* 111: 274–283.

Farooq, M., Wahid, A., Lee, D.J. (2009) Exogenously applied polyamines increase drought tolerance of rice by improving leaf water status, photosynthesis and membrane properties. *Acta Physiologiae Plantarum* 31(5): 937–945.

Fernandez-Aunion, C., Hamouda, T.B., Iglesias-Guerra, F., Argandona, M., Reina-Bueno, M., Nieto, J.J., Aouani, M.E., Vargas, C. (2010) Biosynthesis of compatible solutes in rhizobial strains isolated from Phaseolus vulgaris nodules in Tunisian fields. *BMS Microbiology* 10(192): 1471–2180.

Fougère, F., Le Rudulier, D., Streeter, J.G. (1991) Effects of salt stress on amino acid, organic acid, and carbohydrate composition of roots, bacteroids, and cytosol of alfalfa (*Medicago sativa* L.). *Plant Physiology* 96(4): 1228–1236.

Freeman, B.C., Chen, C., Beattie, G.A. (2010) Identification of the trehalose biosynthetic loci of *Pseudomonas syringae* and their contribution to fitness in the phyllosphere. *Environmental Microbiology* 12(6): 1486–1497.

Garcia, A.B., Engler, J.D.A., Iyer, S., Gerats, T., Van Montagu, M., Caplan, A.B. (1997) Effects of osmoprotectants upon NaCl stress in rice. *Plant Physiology* 115: 159–169.

Garg, A.K., Kim, J.K., Owens, T.G., Ranwala, A.P., Do Choi, Y., Kochian, L.V., Wu, R.J. (2002) Trehalose accumulation in rice plants confers high tolerance levels to different abiotic stresses. *Proceedings of the National Academy of Sciences* 99: 15898–15903.

Gaspar, T., Franck, T., Bisbis, B., Kevers, C., Jouve, L., Hausman, J.F., Dommes, J. (2002) Concepts in plant stress physiology. Application to plant tissue cultures. *Plant Growth Regulation* 37(3): 263–285.

Goddijn, O.J., Verwoerd, T.C., Voogd, E., Krutwagen, R.W., De Graff, P.T.H.M., Poels, J., Pen, J. (1997) Inhibition of trehalase activity enhances trehalose accumulation in transgenic plants. *Plant Physiology* 113(1): 181–190.

Govind, S.R., Jogaiah, S., Abdelrahman, M., Shetty, H.S., Tran, L.S.P. (2016) Exogenous trehalose treatment enhances the activities of defense-related enzymes and triggers resistance against downy mildew disease of pearl millet. *Frontiers in Plant Science* 7: 1–12.

Grigston, J.C., Osuna, D., Scheible, W.R., Liu, C., Stitt, M., Jones, A.M. (2008) D-glucose sensing by a plasma membrane regulator of G signaling protein, AtRGS1. *FEBS Letters* 582(25–26): 3577–3584.

Hagemann, M. (2011) Molecular biology of cyanobacterial salt acclimation. *FEMS Microbiology Reviews* 35(1): 87–123.

Han, B., Fu, L., Zhang, D., He, X., Chen, Q., Peng, M., Zhang, J. (2016) Interspecies and intraspecies analysis of trehalose contents and the biosynthesis pathway gene family reveals crucial roles of trehalose in osmotic-stress tolerance in cassava. *International Journal of Molecular Sciences* 17: 1–18.

Han, S.E., Park, S.R., Kwon, H.B., Yi, B.Y., Lee, G.B., Byun, M.O. (2005) Genetic engineering of drought-resistant tobacco plants by introducing the trehalose phosphorylase (TP) gene from *Pleurotus sajor-caju*. *Plant Cell, Tissue and Organ Culture* 82(2): 151–158.

Hincha, D.K., Hagemann, M. (2004) Stabilization of model membranes during drying by compatible solutes involved in the stress tolerance of plants and microorganisms. *The Biochemical Journal* 383(2): 277–283.

Hershkovitz, N., Oren, A., Cohen, Y. (1991) Accumulation of trehalose and sucrose in cyanobacteria exposed to matric water stress. *Applied Environmental Microbiology*, 57(3): 645–648.

Hottiger, T., Schmutz, P., Wiemken, A. (1987) Heat-induced accumulation and futile cycling of trehalose in *Saccharomyces cerevisiae*. *Journal of Bacteriology* 169: 5518–5522.

Hounsa, C.G., Brandt, E.V., Thevelein, J., Hohmann, S., Prior, B.A. (1998) Role of trehalose in survival of *Saccharomyces cerevisiae* under osmotic stress. *Microbiology* 144(3): 671–680.

Huang, J., Taylor, J.P., Chen, J.G., Uhrig, J.F., Schnell, D.J., Nakagawa, T., Jones, A.M. (2006) The plastid protein THYLAKOID FORMATION1 and the plasma membrane G-protein GPA1 interact in a novel sugar-signaling mechanism in *Arabidopsis*. *The Plant Cell* 18(5): 1226–1238.

Hussain, M., Malik, M.A., Farooq, M., Khan, M.B., Akram, M., Saleem, M.F. (2009) Exogenous glycinebetaine and salicylic acid application improves water relations, allometry and quality of hybrid sunflower under water deficit conditions. *Journal of Agronomy and Crop Science* 195(2): 98–109.

Ibrahim, H.A., Abdellatif, Y.M. (2016) Effect of maltose and trehalose on growth, yield and some biochemical components of wheat plant under water stress. *Annals of Agricultural Sciences* 61(2): 267–274.

Jang, I.C., Oh, S.J., Seo, J.S., Choi, W.B., Song, S.I., Kim, C.H., Kim, J.K. (2003) Expression of a bifunctional fusion of the Escherichia coli genes for trehalose-6-phosphate synthase and trehalose-6-phosphate phosphatase in transgenic rice plants increases trehalose accumulation and abiotic stress tolerance without stunting growth. *Plant Physiology* 131(2): 516–524.

Kandror, O.K., Bretschneider, N., Kreydin, E., Cavalieri, D., Goldberg, A.L. (2004) Yeast adapt to near-freezing temperatures by STRE/Msn2, 4-dependent induction of trehalose synthesis and certain molecular chaperones. *Molecular Cell* 13(6): 771–781.

Kandror, O., DeLeon, A., Goldberg, A.L. (2002) Trehalose synthesis is induced upon exposure of *Escherichia coli* to cold and is essential for viability at low temperatures. *Proceedings of the National Academy of Sciences of the United States of America* 99(15): 9727–9732.

Karim, S., Aronsson, H., Ericson, H., Pirhonen, M., Leyman, B., Welin, B., Holmström, K.O. (2007) Improved drought tolerance without undesired side effects in transgenic plants producing trehalose. *Plant Molecular Biology* 64(4): 371–386.

Kawai, H., Sakurai, M., Inoue, Y., Chujo, R. Kobayashi, S. (1992) Hydration of oliosaccharides: anomalous hydration ability of trehalose. *Cryobiology* 29(5): 599–606.

Kempf, B. Bremer, E. (1998) Uptake and synthesis of compatible solutes as microbial stress responses to high-osmolarity environments. *Archives of Microbiology* 170(5): 319–330.

Kiani, S.P., Talia, P., Maury, P., Grieu, P., Heinz, R., Perrault, A., Sarrafi, A. (2007) Genetic analysis of plant water status and osmotic adjustment in recombinant inbred lines of sunflower under two water treatments. *Plant Science* 172: 773–787.

Klahn, S., Hagemann, M. (2011). Compatible solute biosynthesis in cyanobacteria. *Environmental Microbiology*, Vol. 13, No. 3, (March 2011), pp. 551-562, ISSN 1462–2920.

Koch, E.M., Koch, F.C. (1925) The presence of trehalose in yeast. *Science* 2961(1587): 570–572.

Kolbe, A., Tiessen, A., Schluepmann, H., Paul, M., Ulrich, S., Geigenberger, P. (2005) Trehalose 6-phosphate regulates starch synthesis via posttranslational redox activation of ADP-glucose pyrophosphorylase. *Proceedings of* the National Academy of Sciences of the United States of America 102(31): 11118–11123.

Koussevitzky, S., Nott, A., Mockler, T.C., Hong, F., Sachetto-Martins, G., Surpin, M., Chory, J. (2007) Signals from chloroplasts converge to regulate nuclear gene expression. *Science* 316(5825): 715–719.

Lei, Y., Yin, C., Li, C. (2006) Differences in some morphological, physiological, and biochemical responses to drought stress in two contrasting populations of *Populus przewalskii*. *Physiologia Plantarum* 127(2): 182–191.

Li, Z.G., Luo, L.J., Zhu, L.P. (2014) Involvement of trehalose in hydrogen sulfide donor sodium hydrosulfide induced the acquisition of heat tolerance in maize (*Zea mays* L.) seedlings. *Botanical Studies* 55(1): 1–9.

Lopez, M., Tejero, N.A., Iribarne, C., Lluch, C., Herrera-Cervera, J.A. (2008) Trehalose and trehalase in root nodules of *Medicago truncatula* and *Phaseolus vulgaris* in response to salt. *Physiologia Plantarum* 134: 575–582.

Luo, Y., Li, F., Wang, X.H., Wang, W. (2010) Exogenously-supplied trehalose protects thylakoid membranes of winter wheat from heat-induced damage. *Biologia Plantarum* 54: 435–501.

Mahmud, S.A., Hirasawa, T., Shimizu, H. (2010) Differential importance of trehalose accumulation in *Saccharomyces cerevisiae* in response to various environmental stresses. *Journal of Biosciences and Bioengineering* 109(3): 262–266.

Manivannan, P., Jaleel, C.A., Sankar, B., Kishorekumar, A., Somasundaram, R., Lakshmanan, G.A., Panneerselvam, R. (2007) Growth, biochemical modifications and proline metabolism in *Helianthus annuus* L. as induced by drought stress. *Colloids and Surfaces B: Biointerfaces* 59: 141–149.

Maruta, K., Mitsuzumi, H., Nakada, T., Kubota, M., Chaen, H., Fukuda, S., Kurimoto, M. (1996) Cloning and sequencing of a cluster of genes encoding novel enzymes of trehalose biosynthesis from thermophilic archaebacterium *Sulfolobus acidocaldarius*. *Biochimica et Biophysica Acta-General Subjects* 1291: 177–181.

Maruta, K., Nakada, T., Kubota, M., Chaen, H., Sugimoto, T., Kurimoto, M., Tsujisaka, Y. (1995) Formation of trehalose from maltooligosaccharides by a novel enzymatic system *Bioscience, Biotechnology and Biochemistry* 59: 1829–1834.

Martins, L.L., Mourato, M.P., Baptista, S., Reis, R., Carvalheiro, F., Almeida, A.M., Cuypers, A. (2014) Response to oxidative stress induced by cadmium and copper in tobacco plants (*Nicotiana tabacum*) engineered with the trehalose-6-phosphate synthase gene (AtTPS1). *Acta Physiologiae Plantarum* 36(3): 755–765.

Miranda, J.A., Avonce, N., Suárez, R., Thevelein, J.M., Van Dijck, P., Iturriaga, G. (2007) A bifunctional TPS–TPP enzyme from yeast confers tolerance to multiple and extreme abiotic-stress conditions in transgenic *Arabidopsis*. *Planta* 226: 1411–1421.

Mikkat, S., Hagemann, M., Schoor, A. (1996) Active transport of glucosylglycerol is involved in salt adaptation of the cyanobacterium Synechocystis sp. strain PCC 6803. *Microbiology* 142(7): 1725–1732.

Mostofa, M.G., Hossain, M.A., Fujita, M. (2015) Trehalose pretreatment induces salt tolerance in rice (*Oryza sativa* L.) seedlings: Oxidative damage and co-induction of antioxidant defense and glyoxalase systems. *Protoplasma* 252: 461–475.

Muller, E.B., Stouthamer, A.H., van Verseveld, H.W.V., Eikelboom, D.H. (1995) Aerobic domestic waste water treatment in a pilot plant with complete sludge retention by cross-flow filtration. *Water Research* 29(4): 1179–1189.

Müller, J., Aeschbacher, R.A., Sprenger, N., Boller, T., Wiemken, A. (2000) Disaccharide-mediated regulation of sucrose: Fructan-6-fructosyltransferase, a key enzyme of fructan synthesis in barley leaves. *Plant Physiology* 123(1): 265–274.

Nepomuceno, A.L., Oosterhuis, D.M., Stewart, J.M. (1998) Physiological responses of cotton leaves and roots to water deficit induced by polyethylene glycol. *Environmental and Experimental Botany* 40: 29–41.

Nishimoto, T., Nakano, M., Nakada, T., Chaen, H., Fukuda, S., Sugimoto, T., Tsujisaka, Y. (1996) Purification and properties of a novel enzyme, trehalose synthase, from *Pimelobacter* sp. R48. *Bioscience, Biotechnology, and Biochemistry* 60: 640–644.

Nounjan, N., Nghia, P.T., Theerakulpisut, P. (2012) Exogenous proline and trehalose promote recovery of rice seedlings from salt-stress and differentially modulate antioxidant enzymes and expression of related genes. *Journal of Plant Physiology* 169: 596–604.

Ocon, A., Hampp, R., Requena, N. (2007) Trehalose turnover during abiotic stress in arbuscular mycorrhizal fungi. *New Phytologist*, 174(4): 879–891.

Paul, M.J., Jhurreea, D., Zhang, Y., Primavesi, L.F., Delatte, T., Schluepmann, H., Wingler, A. (2010) Upregulation of biosynthetic processes associated with growth by trehalose-6-phosphate. *Plant Signaling and Behavior*, 5(4): 368–392.

Paul, M.J., Primavesi, L.F., Jhurreea, D., Zhang, Y. (2008) Trehalose metabolism and signaling. *Annual Review of Plant Biology* 59: 417–441.

Polge, C., Thomas, M. (2007) SNF1/AMPK/SnRK1 kinases, global regulators at the heart of energy control? *Trends in Plant Science* 12(1): 20–28.

Pramanik, H.R.M., Imai, R. (2005) Functional identification of a trehalose 6-phosphate phosphatase gene that is involved in transient induction of trehalose biosynthesis during chilling stress in rice. *Plant Molecular Biology* 58(6): 751–762.

Qu, Z., Thottassery, J.V., Van Ginkel, S., Manuvakhova, M., Westbrook, L., Roland-Lazenby, C., Kern, F.G. (2004) Homogeneity and long-term stability of tetracycline-regulated gene expression with low basal activity by using the rtTA2S-M2 transactivator and insulator-flanked reporter vectors. *Gene* 327(1): 61–73.

Ramon, M., Rolland, F., Thevelein, J.M., Van Dijck, P., Leyman, B. (2007) ABI4 mediates the effects of exogenous trehalose on *Arabidopsis* growth and starch breakdown. *Plant Molecular Biology* 63(2): 195–206.

Reed, R.H., Stewart, W.D.P. (1983) Physiological responses of Rivularia atra to salinity: osmotic adjustment in hyposaline media. *New Phytoogist* 95(4): 595–603.

Richards, A.B., Krakowka, S., Dexter, L.B., Schmid, H., Wolterbeek, A.P.M., Waalkens-Berendsen, D.H., Kurimoto, M. (2002) Trehalose: A review of properties, history of use and human tolerance, and results of multiple safety studies. *Food and Chemical Toxicology* 40(7): 871–898.

Rodriguez-Salazar, J., Suarez, R., Caballero-Mellado, J., Iturriaga, G. (2009). Trehalose accumulation in *Azospirillum brasilense* improves drought tolerance and biomass in maize plants. *FEMS Microbiology Letters*, 296(1): 52–59.

Rolland, F., Baena-Gonzalez, E., Sheen, J. (2006) Sugar sensing and signaling in plants: Conserved and novel mechanisms. *Annual Review of Plant Biology* 57: 675–709.

Romero, C., Belles, J.M., Vaya, J.L., Serrano, R., Culianez-macia, F.A. (1997) Expression of the yeast trehalose-6-phosphate synthase gene in transgenic tobacco plants: pleiotropic phenotypes include drought tolerance. *Planta* 201(3): 293–297.

Said, E.M., Mahmoud, R.A., Al-Akshar, R., Safwat, G. (2015) Drought stress tolerance and enhancement of banana plantlets. *Austin Journal of Biotechnology and Bioengineering* 2(2): 1–7.

Schluepmann, H., Pellny, T., van Dijken, A., Smeekens, S., Paul, M. (2003) Trehalose 6-phosphate is indispensable for carbohydrate utilization and growth in *Arabidopsis thaliana*. *Proceedings of the National Academy of Sciences* 100(11): 6849–6854.

Sakamoto, T., Yoshida, T., Arima, H., Hatanaka, Y., Tkani, Y., Tamaru, Y. (2009) Accumulation of trehalose in response to desiccation and salt stress in the terrestrial cyanobacterium Nostoc commune. *Phycological Research* 57(1): 66–73.

Santos, H., da Costa, M.S. (2002) Compatible solutes of organisms that live in hot saline environments. *Environmental Microbiology* 4(9): 501–509.

Serraj, R., Sinclair, T.R. (2002) Osmolyte accumulation: Can it really help increase crop yield under drought conditions? *Plant, Cell and Environment* 25(2): 333–341.

Schluepmann, H.; van Dijken, A.; Aghdasi, M.; Wobbes, B.; Paul, M. & Smeekens, S. (2004). Trehalose mediated growth inhibition of Arabidopsis seedlings is due to trehalose-6-phosphate accumulation. *Plant Physiology* 135(2): 879–890.

Shima, S., Matsui, H., Tahara, S., Imai, R. (2007) Biochemical characterization of rice trehalose-6-phosphate phosphatases supports distinctive functions of these plant enzymes. *FEBS Journal* 274(5): 1192–1201.

Singer, M.A., Lindquist, S. (1998) Thermotolerance in *Saccharomyces cerevisiae*: The yin and yang of trehalose. *Trends in Biotechnology* 16(11): 460–468.

Streeter, J.G., Gomez, M.L. (2006) Three enzymes for trehalose synthesis in Bradyrhizobium cultured bacteria and in bacteroids from soybean nodules. *Applied and Environmental Microbiology*, 72(6): 4250–4255.

Strøm, AR., Kaasen, I. (1993) Trehalose metabolism in Escherichia coli: stress protection and stress regulation of gene expression. *Molecular Microbiology* 8(2): 205–210.

Sugawara, M., Cytryn, E.J., Sadowsky, M.J. (2010) Functional role of *Bradyrhizobium japonicum* trehalose biosynthesis and metabolism genes during physiological stress and nodulation. *Applied and Environmental Microbiology* 76(4): 1071–1081.

Suarez, R., Wong, A., Ramirez, M., Barraza, A., Orozco, M.D.C., Cevallos, M.S., Lara, M., Hernandez, G., Iturriaga, G. (2008) Improvement of drought tolerance and grain yield in common bean by over expressing trehalose-6-phosphate synthase in rhizobia. *Molecular Plant-Microbe Interactions* 21(7): 958–966.

Suzuki, M.M., Bird, A. (2008) DNA methylation landscapes: Provocative insights from epigenomics. *Nature Reviews Genetics* 9(6): 465–476.

Tayeh, C., Randoux, B., Vincent, D., Bourdon, N., Reignault, P. (2014) Exogenous trehalose induces defenses in wheat before and during a biotic stress caused by powdery mildew. *Phytopathology* 104(3): 293–305.

Tibbett, M., Sanders, F., Cairney, J. (2002) Low-temperature-induced changes in trehalose, mannitol and arabitol associated with enhanced tolerance to freezing in ectomycorrhizal basidiomycetes (*Hebeloma spp.*). *Mycorrhiza* 12(5): 249–255.

Thevelein, J.M. (1984) Regulation of trehalose mobilization in fungi. *Microbiological Reviews* 48(1): 42–59.

Tzvetkov, M., Klopprogge, C., Zelder, O., Liebl, W. (2003) Genetic dissection of trehalose biosynthesis in *Corynebacterium glutamicum*: Inactivation of trehalose production leads to impaired growth and an altered cell wall lipid composition. *Microbiology* 149(7): 1659–1673.

Wingler, A. (2002) The function of trehalose biosynthesis in plants. *Phytochemistry* 60(5): 437–440.

Wolf, A., Kramer, R., Morbach, S. (2003) Three pathways for trehalose metabolism in *Corynebacterium glutamicum* ATCC13032 and their significance in response to osmotic stress. *Molecular Microbiology* 49(4): 1119–1134.

Wu, S., Shen, R., Zhang, X., Wang, Q. (2010). Molecular cloning and characterization of maltooligosyltrehalose synthase gene from Nostoc flagelliforme. *Journal of Microbiology and Biotechnology* 20(3): 579–586.

Yamada, T., Takatsu, Y., Manabe, T., Kasumi, M., Marubashi, W. (2003) Suppressive effect of trehalose on apoptotic cell death leading to petal senescence in ethylene-insensitive flowers of gladiolus. *Plant Science* 164(2): 213–221.

Zeid, I.M. (2009) Trehalose as osmoprotectant for maize under salinity-induced stress. *Research Journal of Agriculture and Biological Sciences* 5: 613–622.

Zhang, J., Yedlapalli, P., Lee, J.W. (2009) Thermodynamic analysis of hydrate-based pre-combustion capture of CO_2. *Chemical Engineering Science* 64: 4732–4736.

20 The Proline Metabolism of Durum Wheat Dehydrin Transgenic Context and Salt Tolerance Acquisition in *Arabidopsis thaliana*

Faical Brini, Hassiba Bouazzi, Kaouthar Feki, and Walid Saibi

CONTENTS

20.1 INTRODUCTION

Proline is one of the 20 DNA-encoded amino acids defined in genetic code, by fore codons such as CCU, CCC, CCA, and CCG. It is classified as a non-essential amino acid, because of the ability of the human body to synthesize it. Proline is unique among the 20 protein-forming amino acids in that the amine nitrogen is bound to two alkyl groups. It is widely recognized as playing a peculiar role in the folding/unfolding transitions of the globular protein. Poverty or near absence of proline in the primary protein sequence may govern the level of disorder in the protein structure Saibi et al. (2015a).

In plants, proline is synthesized mainly from glutamate, which is reduced to glutamate-semialdehyde (GSA) by the pyrroline-5-carboxylate synthetase (P5CS) and spontaneously converted to pyrroline-5-carboxylate (P5C). Then P5C reductase (P5CR) reduces the last to proline. In most plant species, P5CS is encoded by two distinct genes, and one gene encodes P5CR. Proline catabolism occurs in mitochondria via the sequential action of proline dehydrogenase or proline oxidase (PDH or POX) producing P5C from proline, and P5C dehydrogenase (P5CDH), which converts P5C to glutamate. Proline can also be synthesized as an alternative pathway from ornithine, which is transaminated first by ornithine-delta-aminotransferase (OAT), producing GSA and P5C, which is then converted to proline

(Saibi et al., 2015a; Ben Rejeb et al., 2014). The proline metabolism is illustrated in Figure 20.1.

20.2 REGULATION OF PROLINE METABOLISM IN PLANTS

Proline was considered and will remain an important platform, which is implicated in various physiological phenomena, in the microorganism, animal, and vegetal kingdoms Bita and Gerats (2013). While proline metabolism has been studied for more than 60 years in plants, little is known about the signaling pathways involved in its regulation. Proline anabolism is activated and catabolism repressed during dehydration, whereas rehydration triggers the opposite regulation. Proline anabolism is controlled by the activity of two *p5cs* genes in plants, encoding one housekeeping and one stress-specific *p5cs* isoform. Although the duplicated *p5cs* genes share a high level of sequence homology in coding regions, their transcriptional regulation is different. Both *p5cs* genes are active in floral shoot apical meristems and supply proline for flower development Saibi et al. (2015b). It is intriguing that, in *Arabidopsis*, the *p5cs2* gene was identified as one of the targets of CONSTANS (CO), a transcriptional activator that promotes flowering in response to long day length. The *p5cs2* gene can also be activated by virulent bacteria, salicylic acid (SA), and

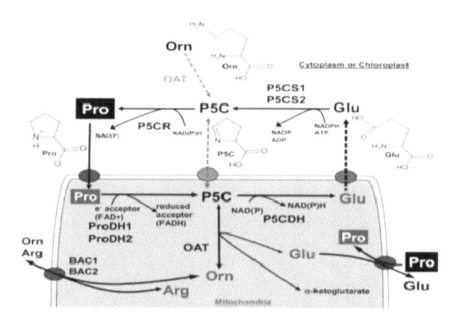

FIGURE 20.1   The Modified Schematic Illustration of the Metabolism of Proline from Glutamic Acid. In Fact, Anabolism of Proline Takes Place in the Chloroplast and the Cytosol and Its Catabolism Takes Place in the Mitochondria.

ROS signals, which trigger a hypersensitive response (HR). *Arabidopsis p5cs1* is induced by osmotic and salt stresses and is activated by abscisic acid (ABA)-dependent and ABA-insensitive 1 (ABI1)-controlled regulatory pathway and H_2O_2-derived signals. Moreover, *p5cs1* activation and proline accumulation are promoted by light and repressed by brassinosteroids (Szabados and Savouré, 2010).

On the other hand, and under non-stressed conditions, various investigations show that phospholipase D (PLD) functions as a negative regulator of proline accumulation, whereas calcium signaling and phospholipase C (PLC) trigger *p5cs* transcription and proline accumulation during salt stress. Indeed, we note that in the halophyte *Thellungiella halophila*, PLD functions as a positive regulator, whereas PLC exerts a negative control on proline accumulation. Calcium signals can be transmitted by a specific CaM4 calmodulin, which interacts with the *myb2* transcription factor and upregulates *p5cs1* transcription (Szabados and Savouré, 2010; Ben Othman et al., 2017).

As well as transcriptional regulation, P5CS activity is submitted for metabolic control, as seen by feedback inhibition of P5CS by proline, representing the end product of the reaction (Hong et al., 2000). Similar allosteric inhibition of bacterial PROB has also been described. Loss of feedback inhibition of P5CS leads to elevated proline accumulation. In mammalian cells, different P5CS isoforms were generated via alternative splicing, which has a function in tissue specificity, organ localization, hormonal regulation, and feedback inhibition.

Conversion of P5C to proline is not a rate-limiting step in proline biosynthesis, yet the control of P5CR activity implies a complex regulation of transcription, which was shown to be under developmental and osmotic regulation. Promoter analysis of *Arabidopsis p5cr* identified a 69-bp promoter region that is responsible for tissue-specific expression. However, trans-acting factors that can bind to this promoter region have not yet been identified (Hong et al., 2000).

20.3   PROLINE ACCUMULATION AND STRESS TOLERANCE

In this case, we noted that salt accumulation in irrigated soils is one of the main factors causing the decrease in crop productivity, since most of the plants are not halophytic (Roy et al., 2014; Shrivastava and Kumar, 2015). Excessive sodium inhibits the growth of many salt-sensitive plants, which include most crop plants. The typical first response of all plants to salt stress is an osmotic adjustment. The accumulation of compatible solutes in the cytoplasm is considered as a primordial mechanism contributing to salt tolerance (Roy et al., 2014; Shrivastava and Kumar, 2015). To counter salt stress, plants increase the osmotic potential of their cells by synthesizing and accumulating compatible osmolytes such as proline and glycine betaine, which participates in osmotic adjustment (Hayat et al., 2012; Slama et al., 2015). Several plants, including halophytes, accumulate high proline levels in response to osmotic stress as a tolerance mechanism to high salinity and water deficit (Szabados and Savouré, 2010).

In plants, this atypical amino acid is anabolized from either glutamate or ornithine. However, the glutamate pathway is the primary route used under osmotic stress or nitrogen limitation conditions, whereas the ornithine pathway is prominent under high nitrogen input. Accumulation of proline in plants under stress is a result of the reciprocal regulation of two pathways: [($\Delta^1\Delta1$-P5CS and P5CR] and repressed activity of proline degradation (Slama et al., 2015). Proline catabolism is catalyzed by pyrroline-5-carboxylate dehydrogenase and PDH, a mitochondrial enzyme, whose activity had been shown to reduce during salt stress. The first two steps of proline biosynthesis are catalyzed by P5CS using its γ-GK and glutamic-γ-semialdehyde dehydrogenase activities. Subsequently, the P5C formed is reduced by P5CR to proline (Szabados and Savouré, 2010).

20.4 SOME OF PROLINE'S PHYSIOLOGICAL ROLES

For a long time, proline was considered to be an inert, compatible osmolyte that protects subcellular structures and macromolecules under osmotic stress. However, proline accumulation can influence stress tolerance in multiple ways. Proline has been shown to be a potential and plausible effector, playing a molecular chaperone role, and also to be able to protect protein integrity and enhance the activities of different enzymes. Examples of such roles include the prevention of protein aggregation and stabilization of M4 lactate dehydrogenase during extreme temperatures, protection of nitrate reductase during heavy metal and osmotic stress, and stabilization of ribonucleases and proteases upon arsenate exposure (Saibi et al., 2012, 2015a).

Several research findings have attributed an antioxidant feature to proline, suggesting ROS-scavenging activity and proline acting as a single oxygen quencher. Proline treatment can decrease ROS levels in fungi, thus preventing programmed cell death, can protect human cells against carcinogenic oxidative stress, and can reduce lipid peroxidation in alga cells exposed to heavy metals. Proline pretreatment also alleviated Hg^{2+} toxicity in rice (*Oryza sativa*) through scavenging ROS, such as H_2O_2. The damaging effects of singlet oxygen and hydroxyl radicals on Photosystem II (PSII) can be reduced by proline in isolated thylakoid membranes (PSII). Free radical levels were reduced in transgenic algae and tobacco plants engineered for a hyperaccumulation of proline by *p5cs* overexpression and acceleration of the proline biosynthetic pathway.

By contrast, compromised proline accumulation in *p5cs1* insertion mutants led to the accumulation of ROS and enhanced oxidative damage. A similar effect was observed in yeast, where low proline levels in PUT1 (PDH)-overexpressing lines led to enhanced ROS, whereas higher proline content in *put1* mutants correlated with increased protection from oxidative damage (Verslues and Sharma, 2010). As an alternative to its direct ROS-scavenging feature, proline can protect and stabilize ROS-scavenging enzymes and activate alternative detoxification pathways. In salt-stressed tobacco cells, proline increased the activities of methylglyoxal detoxification enzymes, enhanced peroxidase, glutathione-S-transferase, superoxide dismutase, and catalase activities, and increased the glutathione redox state. In the desert plant *Pancratium maritimum*, catalase and peroxidase were found to be stabilized by proline during salt stress. The salt-hypersensitive *p5cs1 Arabidopsis* mutant shows reduced activities of key antioxidant enzymes of the glutathione–ascorbate cycle, leading to hyperaccumulation of H_2O_2, enhanced lipid peroxidation, and chlorophyll damage.

As well as having protective or scavenging features, it is feasible that proline metabolism can stabilize cellular homeostasis during stress conditions in a way that is still poorly understood. Accumulation of P5CS1 and P5CR in chloroplasts during salt stress suggests that, under adverse conditions, glutamate-derived proline biosynthesis increases in plastids, where photosynthesis occurs. During stress conditions, the rate of the Calvin cycle is diminished, which prevents oxidation of NADPH and restoration of $NADP^+$. When combined with high light, electron flow in the electron transport chain is suppressed by the insufficient electron acceptor $NADP^+$ pool, leading to singlet oxygen production in the PSI reaction center and accumulation of ROS. Proline biosynthesis is a reductive pathway, it requires NADPH for the reduction of glutamate to P5C and P5C to proline, generating $NADP^+$ that can be used further as an electron acceptor. The phosphorylation of glutamate consumes energy (ATP) and produces ADP, which is a substrate for ATP anabolism during photosynthesis. An enhanced rate of proline biosynthesis in chloroplasts during stress can maintain the low $NADPH/NADP^+$ ratio, contribute to sustaining the electron flow between photosynthetic excitation centers, stabilize the redox balance, and reduce photoinhibition and damage of the photosynthetic apparatus. In transgenic soybean plants, the inhibition of proline biosynthesis and $NADPH–NADP^+$ conversion by antisense *p5cr* led to drought hypersensitivity, whereas overexpression of *p5cr* resulted in moderate drought tolerance, confirming that proline biosynthesis is important for maintaining NADP$^+$ pools during stress. This connection between photosynthesis and proline metabolism is supported by light-dependent proline accumulation, which is regulated by the light-controlled reciprocal *p5cs* and *pdh* gene activation.

In the mitochondria compartment, proline has distinct protective functions. Indeed, after stress, proline pools supply a reducing potential for mitochondria through the oxidation of proline by PDH and P5CDH; they also provide electrons for the respiratory chain and hence contribute to energy supply for resumed growth. The atypical amino acid was shown to protect complex II of the mitochondrial electron transport chain during salt stress and as a consequence the stabilization of the mitochondrial respiration. The recently discovered P5C–proline cycle can deliver electrons to mitochondrial electron transport without producing glutamate and, under particular conditions, can generate more ROS in the mitochondria. In addition, proline catabolism is, therefore, an important regulator of cellular ROS balance and can influence numerous additional regulatory pathways.

20.5 CAUSAL ENZYMOLOGY EXPLAINS THE SALT TOLERANCE PROCESS UNDER DEHYDRIN TRANSGENIC CONTEXT

In the beginning, we note that dehydrins (DHNs) are involved in plant abiotic stress tolerance. The wheat dehydrin DHN-5 was previously isolated and characterized by Brini et al. (2007). To investigate the effect of DHN-5 overexpression in *Arabidopsis thaliana*, the proposed dehydrin transgenic lines expressed *Dhn-5* gene and

TABLE 20.1

Illustration of an Aspect from Causal Enzymology Proving the Role Played by Some Proteases on Salt-Tolerance Acquisition Under Dehydrin Transgenic Lines (Saibi et al., 2016)

Arabidopsis thaliana Protease Contribution Levels on the Salinity-Tolerance Acquisition Process

	Transgenic Context and Salt-Stress Contributions		
	Total Protease Activity	Cys-Protease Activity	Asp-Protease Activity
Wt	+3.36	+4.12	−13.64
DH2	+93.26	+73.08	−50
DH4	+98.26	+88.88	−59.10

the relatively highest level was detected in line 2 and 4, named respectively DH2 and DH4 Brini et al. (2007).

In this section, we reported that salt tolerance of transgenic *Arabidopsis* plants overexpressing durum wheat dehydrin (DHN-5) was closely related to the activation of the proline metabolism enzyme (P5CS) and some antioxidant biocatalysts. Indeed, DHN-5 improved P5CS activity in the transgenic plants generating a significant proline accumulation (Saibi et al., 2015b, 2016). Moreover, salt tolerance of the dehydrin *Arabidopsis* transgenic lines was accompanied by an excellent activation of antioxidant enzymes such as catalase (CAT), superoxide dismutase (SOD), and peroxide dismutase (POD) and generation of a lower level of hydrogen

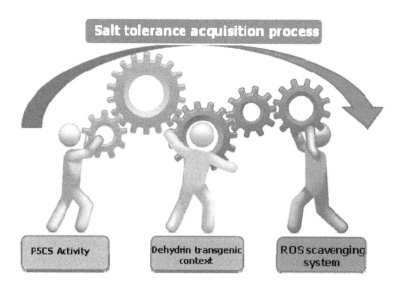

FIGURE 20.2 Schematic Illustration of the Involvement of the Positive Cooperation between the Proline Metabolism, the Dehydrin Transgenic Context and the ROS Scavenging System for the Salt Tolerance Acquisition Process.

peroxide (H_2O_2) in leaves compared to the wild-type plants. The enzyme activities were enhanced in these transgenic plants in the presence of exogenous proline. Nevertheless, proline accumulation was slightly reduced in the transgenic plants promoting chlorophyll levels. All these results suggest the crucial role of DHN-5 in response to salt stress through the activation of enzymes implicated in proline metabolism and in ROS-scavenging enzymes (Saibi et al., 2015b).

In addition, we proved the metal chelation capacity of the wheat dehydrin (DHN-5). This finding will be used to explain salinity tolerance in the dehydrin transgenic lines against wild-type plants. Hence the structural characterization of the wheat dehydrin, over-expressed and purified from *E. coli*, is judged crucial to understand their involvement in the tolerance process (Saibi et al., 2015a). We also report the behavior of protease activities in dehydrin transgenic *Arabidopsis* lines against the wild-type plant under salt stress (100 mM NaCl). Indeed, proteases play key roles in plants, maintaining strict protein quality control and degrading specific sets of proteins in response to diverse environmental and developmental stimuli (Saibi et al., 2016). We proved that durum wheat DHN-5 modulates the activity of some proteases, summarized by the promotion of the cysteinyl protease and the decrease of the aspartyl protease activity. This fact is also upgraded in salt-stress conditions and seems to be illustrated as indicated in Table 20.1.

20.6 CONCLUDING REMARKS

Although proline has long been considered the best-known osmolyte, recent results follow its multifunctionality in stress adaptation, recovery, and signaling. The compartmentalization of proline metabolism adds to its functional diversification complexity and may explain its importance in various fields. The enhanced rate of proline anabolism in chloroplasts can contribute to the stabilization of redox balance and the maintenance of cellular homeostasis by dissipating the excess of reducing potential when electron transport is saturated during various conditions. Proline catabolism in the mitochondria is connected to oxidative respiration and administers energy to resumed growth after stress. The enhancement of proline biosynthesis via increasing the expression of rate-limiting genes in transgenic plants is considered as the best option.

The improvement of drought and/or salt tolerance of crop plants via engineering proline metabolism is a plausible opportunity that should be explored more extensively. Eventually, the fact that proline can act as a signaling effector and influence defense pathways and regulate metabolic and developmental processes offers additional opportunities for plant improvement.

Added to the role of DHN-5 already described during the modulation of proline metabolism and the ROS-scavenging system related to the acquisition of salinity tolerance in transgenic lines, we can add the protein–protein interaction that confers tolerance to salt stress. At this state, the durum wheat dehydrin seems to be a key protein that plays a heat-protective role and interacts with multiple partners involved in different pathways, such as proline metabolism. All those facts may be summarized in Figure 20.2.

REFERENCES

Ben Othman, A., Ellouzi, H., Planchais, S., De Vos, D., Faiyue, B., Carol, P., Abdelly, C., Savouré, A. (2017) Phospholipases Dζ1 and Dζ2 have distinct roles in growth and antioxidant systems in *Arabidopsis thaliana* responding to salt stress. *Planta* 246(4): 721–735.

Ben Rejeb, K., Abdelly, C., Savouré, A. (2014) How reactive oxygen species and proline face stress together. *Plant Physiology and Biochemistry* 80 (Supplement C): 278–284.

Bita, C.E., Gerats, T. (2013) Plant tolerance to high temperature in a changing environment: Scientific fundamentals and production of heat stress-tolerant crops. *Frontiers in Plant Science* 4: 273.

Brini, F., Hanin, M., Lumbreras, V., Amara, I., Khoudi, H., Hassairi, A., Pages, M., Masmoudi, K. (2007) Overexpression of wheat dehydrin DHN-5 enhances tolerance to salt and osmotic stress in *Arabidopsis thaliana*. *Plant Cell Reports* 26(11): 2017–2026.

Hayat, S., Hayat, Q., Alyemeni, M.N., Wani, A.S., Pichtel, J., Ahmad, A. (2012) Role of proline under changing environments: A review. *Plant Signaling and Behavior* 7 (11): 1456–1466.

Hong, Z., Lakkineni, K., Zhang, Z., Verma, D.P. (2000) Removal of feedback inhibition of delta(1)-pyrroline-5-carboxylate synthetase results in increased proline accumulation and protection of plants from osmotic stress. *Plant Physiology* 122(4): 1129–1136.

Roy, S.J., Negrão, S., Tester, M. (2014) Salt resistant crop plants. *Current Opinion in Biotechnology* 26 (Supplement C): 115–124.

Saibi, W., Abdeljalil, S., Masmoudi, K., Gargouri, A. (2012) Biocatalysts: Beautiful creatures. *Biochemical and Biophysical Research Communications* 426(3): 289–293.

Saibi, W., Drira, M., Yacoubi, I., Feki, K., Brini, F. (2015a) Empiric, structural and in silico findings give birth to plausible explanations for the multifunctionality of the wheat dehydrin (DHN-5). *Acta Physiologiae Plantarum* 37(3): 1–8.

Saibi, W., Feki, K., Ben Mahmoud, R., Brini, F. (2015b) Durum wheat dehydrin (DHN-5) confers salinity tolerance to transgenic *Arabidopsis* plants through the regulation of proline metabolism and ROS scavenging system. *Planta* 242(5): 1187–1189.

Saibi, W., Zouari, N., Masmoudi, K., Brini, F. (2016) Role of the durum wheat dehydrin in the function of proteases conferring salinity tolerance in *Arabidopsis thaliana* transgenic lines. *International Journal of Biological Macromolecules* 85: 311–316.

Shrivastava, P., Kumar, R. (2015) Soil salinity: A serious environmental issue and plant growth promoting bacteria as one of the tools for its alleviation. *Saudi Journal of Biological Sciences* 22(2): 123–131.

Slama, I., Abdelly, C., Bouchereau, A., Flowers, T., Savouré, A. (2015) Diversity, distribution and roles of osmoprotective compounds accumulated in halophytes under abiotic stress. *Annals of Botany* 115(3): 433–447.

Szabados, L., Savouré, A. (2010) Proline: A multifunctional amino acid. *Trends in Plant Science* 15(2): 89–97.

Verslues, P.E., Sharma, S. (2010) Proline metabolism and its implications for plant-environment interaction. *The Arabidopsis Book/American Society of Plant Biologists* 8: e0140.

21 Nitric Oxide-Induced Tolerance in Plants under Adverse Environmental Conditions

Neidiquele M. Silveira, Amedea B. Seabra, Eduardo C. Machado, John T. Hancock, and Rafael V. Ribeiro

CONTENTS

21.1 INTRODUCTION

Plants are frequently exposed to environmental stresses in both natural and agricultural systems. Abiotic stresses dramatically impact plant growth and crop yield, with more than 50% of the major crop species being impaired by limiting conditions (Poltronieri et al., 2014). In addition, stressful conditions have been intensified due to global climate changes, which threaten food security (Reddy, 2015; Savvides et al., 2016). Among the stressful conditions plants face, the main abiotic stresses are drought, flooding, heavy metal, salinity, and low or high temperature depending on growing areas.

As plants are sessile, they have a high capacity to respond to abiotic stresses, showing high plasticity under varying conditions. Plants activate a complex network of multiple responses at molecular, cellular, and physiological levels (Larcher, 2000). Such variety of responses occurs because plants have an effective signaling system, which processes and transfers information about environmental changes to other organs and tissues (Roshchina, 2001). These stress-linked signaling networks involve not only classical phytohormones, but also other signaling molecules, including nitric oxide (NO•). Several reports indicate that NO• production and signaling is associated with improvement of plant tolerance to stressful conditions (Silveira et al., 2016, 2017a; Farnese et al., 2016).

In fact, NO• is an important signaling molecule and has a wide range of effects on metabolic and physiological processes (Santisree et al., 2015; Farnese et al., 2016). In general, NO• signaling function is linked to its action as a modulator of the redox state and protein activity. NO• may be a positive or negative regulator of stress responses, sometimes acting as an antioxidant (Silveira et al., 2015, 2017a; Hasanuzzaman et al., 2017) or pro-oxidant (Zaninotto et al., 2006). This double effect of NO•, as reported in the literature, seems to depend predominantly on the type of NO• donor used and its concentration, as well as being genotype-dependent (Murgia et al., 2004; Wieczorek et al., 2006).

This chapter will summarize how NO• modulates physiological and biochemical responses and improves plant acclimation to drought, flooding, heavy metal, salinity, and thermal stresses. The mechanisms of action of NO• donors in plants under those abiotic stress conditions are also discussed. Finally, we will address the most commonly used NO• donors in experiments as well as the potential use of nanoparticles that release this radical.

21.2 NO MODULATES THE BIOCHEMICAL AND PHYSIOLOGICAL RESPONSES OF PLANT SPECIES

The diversity of NO• effects on metabolic and physiological processes is due to its physicochemical properties. NO• has an unpaired electron; it can be reduced or oxidized, forming nitroxyl (NO^-) or nitrosonium (NO^+) ions, respectively. Each of these reactive nitrogen species is capable of interacting differently with biological molecules (Gow and Ischiropoulos, 2001). NO• can also react with atmospheric O_2 producing nitrite (NO_2^-) (Wink et al., 1996) and other reactive nitrogen species, such as nitrogen dioxide (NO_2), or with the superoxide anion ($O_2^{•-}$) forming peroxynitrite ($ONOO^-$) (Radi et al., 2002). Due to its low solubility, NO• is highly mobile in cellular systems, diffusing freely through membranes and can easily vaporize (Mur et al., 2006). Thus, despite the apparent simple structure, its complex chemical characteristics in biological systems allow the formation of multiple secondary and tertiary products. However, Liu et al. (1998) concluded that biological membranes and other hydrophobic compartments of tissues are important sites for NO• disappearance and formation of NO-derived reagents, thus showing that NO• is not as diffusible by membranes.

The modulation of protein activity by NO• may be a consequence of three types of reactions, which represent the main mechanisms by which NO• results in protein modification in plants (Leitner et al., 2009). Firstly, oxidized forms of NO• (*i.e.*, N_2O_3 or HNO_2) can react with thiol groupings (SH) of cysteine residues (Cys) present on proteins, forming an *S*-nitrosothiol (RSNO) in a reaction called *S*-nitrosation (Lindermayr et al., 2005; Aracena-Parks et al., 2006); secondly, the metal ion heme groups may bind to iron-sulfur protein centers that may be characterized using the metal-nitrosylation assay (Besson-Bard et al., 2008); and finally, the NO-derivative peroxynitrite (formed from the reaction of NO• and the superoxide anion) may also cause nitration of tyrosine residues (Astier et al., 2011; Corpas et al., 2013).

Several abiotic stresses often result in an increase in the accumulation of reactive oxygen species (ROS), which can cause cell damage and death, depending on the intensity and duration of stress. The induction of antioxidant systems appears to be a response to various abiotic stresses, and crosstalk between NO• and ROS has been shown to be an important mechanism for plant tolerance under unfavorable conditions. Hormonal signaling cascades represent another primary mechanism activated under abiotic stresses. NO• was revealed as a key mediator in antioxidant responses and hormonal

signaling, modulating gene expression and protein activity through post-translational modifications (Salgado et al., 2017). In this context, we will summarize some recent studies about the role of NO• in plants under abiotic stresses.

21.2.1 DROUGHT

Water deficit is the main abiotic stress affecting plant metabolism, growth, performance, and crop yield worldwide. Drought-induced NO• is found in a wide variety of plant species, suggesting NO• requirement during drought stress signaling (Santisree et al., 2015). Drought-tolerant plants usually have strict control of stomatal movements and a fine balance of cellular metabolism, and both ROS and NO• are important in those processes (Osakabe et al., 2014). One of the first and most important physiological responses induced by drought is the reduction of stomatal opening for preventing excessive dehydration (Garcia-Mata and Lamattina, 2001; Neill et al., 2008). Under water deficit, the dynamics of stomatal movement is directly related to the concentrations of ABA, ROS and NO•. In this process, abscisic acid acts as a regulator inducing the synthesis of NO•, which in turn and together with the ROS, acts synergistically to mediate stomatal closure through the formation of 8-nitro-cyclic guanosine monophosphate (8-nitro-cGMP), a new element in plant signaling under stressful conditions (Joudoi et al., 2013).

The activation of antioxidant mechanisms to maintain ROS homeostasis often involves the participation of NO• (Shi et al., 2014; Hatamzadeh et al., 2015; Silveira et al., 2015). The protective effect of the exogenous application of NO• donor molecules has been attributed to the elimination of superoxide anion ($O_2^{•-}$) and increased activity of antioxidant enzymes (Silveira et al., 2017b; Ren et al., 2017; Siddiqui et al., 2017). Morphological and physiological responses to drought are dependent on genotype and intensity of water deficit (Marchiori et al., 2017). In this sense, Silveira et al. (2017a) demonstrated the NO• metabolism is more active in a drought-tolerant genotype, presenting higher root extracellular NO content, higher nitrate reductase (NR) activity, and lower root *S*-nitrosoglutathione reductase (GSNOR) activity compared to a drought-sensitive genotype.

It is already known that NR can generate NO• from nitrite with NADH as an electron donor and the catalysis site occurs at the site of the enzyme composed of a molybdenum cofactor (Moco) (Harrison, 2002). In addition, *S*-nitrosoglutathione (GSNO), a NO donor, is catabolized by GSNOR resulting in oxidized glutathione (GSSG) and ammonium (NH_4^+) in order to establish intracellular control (Frungillo et al., 2014). Thus, GSNOR regulates the intracellular levels of GSNO, and it is important for maintaining NO• homeostasis.

Recent studies have demonstrated that administration of exogenous NO• donor (GSNO) sprayed on leaves improves plant tolerance to water deficit, with plants presenting a significant improvement in photosynthetic rates under low water availability. Such improvement in leaf CO_2 assimilation was associated, in part, with increases in relative water content, root growth and stomatal conductance under water deficit (Silveira et al., 2016). Besides the diffusive limitation imposed by stomatal closure, plants may face the biochemical limitation of photosynthesis under severe drought. In this sense, Silveira et al. (2017b) also noted that the exogenous GSNO improved CO_2 uptake under water deficit by increasing Rubisco activity. Additionally, NO• induces a slow and continuous increase in the non-photochemical quenching of fluorescence, a well-known photoprotective mechanism when there is excessive light energy at PSII (Ordog et al., 2013). Improvement in plant performance under drought, when supplied with NO• donor, may be a consequence of reduced oxidative damage (Silveira et al., 2017b). Thus, NO-induced tolerance to drought is related to the alleviation of diffusive, biochemical and photochemical limitations of photosynthesis, which benefits plant growth under water deficit (Figure 21.1).

21.2.2 FLOODING

Plants can face excessive irrigation or rainfall when growing in soils with low drainage capacity. In such conditions, oxygen diffusion is reduced, and then root metabolism is affected by flooding. Both total (anoxia) and partial (hypoxia) oxygen deficiencies limit the aerobic metabolism of plants and ATP production by cells (Bailey-Serres and Voesenek, 2008), reducing plant growth and survival. Some species present adaptations to survive under waterlogging conditions, showing anatomical and/or morphological modifications such as increases in shoot growth rate to escape from the excessive water, the formation of aerenchyma for aeration, and formation of adventitious roots, superficial rooting and stem lenticels (Medri et al., 2007).

Before those morphological or anatomical adaptations occur, there is a signaling cascade, and NO• appears to be one of the primary signals in plants during hypoxia. It has been suggested that NO• production under O_2 deficiency occurs in mitochondria, where nitrite (NO_2^-) is reduced to NO• in complexes III and IV of mitochondrial electron transport chain (mETC). NO• produced in mitochondria diffuses to the cytosol where it is oxidized to nitrate (NO_3^-) by non-symbiotic hemoglobin (nsHB) (Pucciariello and Perata, 2016). Thus, NO• accumulation also occurs in plants facing low oxygen availability (Gupta et al., 2011).

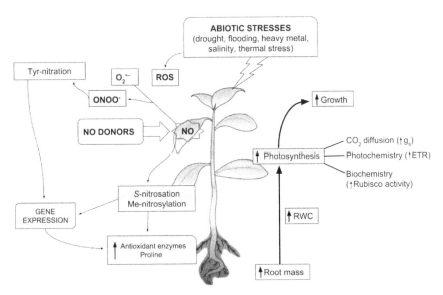

FIGURE 21.1 General schematic representation showing the main effects of supplying exogenous NO donors to plants under environmental stresses. In general, abiotic stresses induce ROS formation, causing oxidative stress. NO production leads to protein modification (tyrosine nitration, *S*-nitrosation, Me-nitrosylation) and regulation of expression of genes involved in antioxidant enzymatic and osmoprotective systems. Such modifications have positive effects on photosynthesis and plant growth under limiting conditions. Abbreviations: NO, nitric oxide; $O_2^{•-}$, superoxide radical; $ONOO^-$, peroxynitrite; ROS, reactive oxygen species; RWC, leaf relative water content; g_s, stomatal conductance; ETR, apparent electron transport rate. Most plant responses of this scheme are based on Silveira et al. (2016, 2017b).

Mugnai et al. (2012) demonstrated that root apex-specific oxygen deprivation dramatically inhibited the oxygen influx peak in the transition zone and simultaneously stimulated a local increase in NO• emission. This local production of NO• was responsible for root acclimation to oxygen deprivation, with maize plants showing a higher survival rate, i.e., the roots were considered to have survived if they resumed elongation after the anoxic treatment (Mugnai et al., 2012). It was also reported that NO• alleviated the hypoxia effects on growth of cucumber seedlings following exogenous NO• application (Fan et al., 2014). NO• treatment induced the activation of antioxidant enzymes, such as superoxide dismutase (SOD), peroxidase (POD), catalase (CAT), and ascorbate peroxidase (APX), reducing lipid peroxidation and membrane damage (Fan et al., 2014). According to Hebelstrup et al. (2012), the regulation of proteins by S-nitrosation might also explain how NO• confers plant tolerance to flooding, besides regulating the production of ethylene.

In salsa seedlings, Chen et al. (2016) verified that application of sodium nitroprusside (SNP), a NO• donor, increased adventitious root formation and enhanced root cell integrity in waterlogging conditions. Besides its role in waterlogging tolerance by enhancing adventitious root formation, NO• is associated with modulation of the NOS pathway as SNP treatment decreased nitrate reductase activity and increased nitric oxide synthase (NOS) activity in adventitious roots. It is worth emphasizing that the mechanism responsible for NO• synthesis in plants is still controversial. Classically, two main pathways for NO• generation are described, where the main substrates are nitrite (NO_2^-) and arginine. In the first case, NO• can be produced by the reduction of nitrite, and several enzymatic and non-enzymatic mechanisms have been proposed (Besson-Bard et al., 2008). In the second case, NO• is synthesized by the oxidation of L-arginine catalyzed by NOS (Alderton et al., 2001). Although there are reports of an L-arginine-dependent NOS in extracts of different plant species (Jasid et al., 2006), its presence in higher plants has not yet been demonstrated. Recently, a homologous gene for this protein was found and characterized in the alga *Ostreococcus tauri* (Foresi et al., 2015), but no equivalent has yet been reported in higher plants. Recently, a bioinformatics study did not identify any sequence for NOS in 1087 transcriptomes sequenced from terrestrial plants (Jeandroz et al., 2016).

Another paper showed that SNP improved the survival rate of maize seedlings subjected to hypoxia through the accumulation of endogenous hydrogen sulfide (H_2S). H_2S is another gaseous molecule, similar to NO•, and is also related to the increase of plant tolerance to stressful conditions (Hancock and Whiteman, 2014). NO-induced

H_2S also enhanced endogenous Ca^{2+} levels as well as the alcohol dehydrogenase activity (ADH), improving the antioxidant capacity as well as the tolerance of maize seedlings to flooding (Peng et al., 2016).

21.2.3 HEAVY METALS

The term heavy metal has been used to group metals and metalloids whose atomic density is greater than 5 g cm^{-3} (Oves et al., 2012). Heavy metals are great concern due to their role in causing environmental contamination. This contamination affects numerous areas worldwide and may continue to have a negative impact for decades or even centuries (Alloway, 2013). The accumulation of these pollutants can occur naturally in the soil or can also be the result of anthropogenic activities such as mining, use of phosphate fertilizers, and disposal of agro-industrial wastes (Gallego et al., 2012). Once released into the environment, they tend to concentrate in the soil and sediments, producing a large reservoir of heavy metals for plant roots. In general, plants are very vulnerable to these elements, presenting physiological damage, structural alterations, as well as reductions in cell area in abaxial epidermis, length and width of stomatal guard cells, size of parenchyma in all seedling organs, and diameter of metaxylem vessels (Kasim, 2005) and reduced growth (John et al., 2009). Oxidative damage also occurs due to the intensification of ROS production in plants exposed to heavy metals (Borges et al., 2018).

An increasing number of studies has been shown the beneficial effects of exogenously applied NO• donors in plants showing heavy metal toxicity symptoms (Mostofa et al., 2014; Silveira et al., 2015). A study investigated the interactive effect of exogenous SNP and GSH on Cu-induced oxidative damage in rice seedlings, and it was observed that both SNP and GSH successfully decreased Cu-induced ROS and MDA levels, also reducing the Cu uptake and showing elevated levels of ascorbate, phytochelatin, and higher activities of catalase (CAT), glutathione peroxidase (GPX), dehydroascorbate reductase (DHAR), and glutathione S-transferase (GST). However, the protective action of GSH + SNP was more efficient than SNP alone (Mostofa et al., 2014). Kopyra and Gwóźdź (2003) verified in *Lupinus luteus* that SNP exerted a considerable effect on the promotion of germination when exposed to lead (Pb) and cadmium (Cd), suggesting that NO• is involved in attenuation of the negative impact of heavy metals on germination. Interestingly, Bethke et al. (2006) demonstrated that volatile cyanide effectively breaks the dormancy of Arabidopsis seeds, and we already know

SNP is a NO• donor that also releases cyanide. Similar responses were found in rice, in which exogenous treatment with NO donors reduced the inhibition of seed germination and seedling growth caused by Cd (He et al., 2014). In a study to investigate the protective role of NO• in plants exposed to Cd, Chen et al. (2013) treated barley seeds with SNP and verified that NO• alleviated the symptoms of Cd toxicity, such as chlorosis and leaf necrosis, and reduced Cd accumulation in shoots and roots. It was also demonstrated that NO• increased the activity of SOD by up to 110% in arsenic-exposed sorghum plants (Saxena and Shekhawat, 2013), as well as increased activities of CAT and APX and activated the ascorbate–glutathione cycle (Hasanuzzaman and Fujita, 2013). Stresses by boron (B) and aluminum (Al) were attenuated by SNP treatment, which favored plant growth and increased the photosynthetic efficiency in stressed plants (Aftab et al., 2012).

Mostofa et al. (2015) revealed that Cu tolerance is induced by supplying exogenous GSNO. As a mechanism, the exogenous application of glutathione (GSH) and SNP led to GSNO production by rice seedlings under aerobic conditions. Then, GSNO can inhibit metal uptake by forming complexes with Cu, or it can be decomposed into GSSG and NH_3 by GSNOR, using NADPH. Alternatively, GSNO can be degraded into NO and GSSG in the presence of reductants such as ascorbic acid and glutathione. Finally, GSNO can transfer NO• to Cys, Tyr, or protein metals using post-translational modifications, thus modifying the function of a broad spectrum of proteins leading to Cu tolerance.

21.2.4 SALINITY

Soil salinity occurs in several regions of the world, being a serious problem in arid and semi-arid areas. Approximately 7% of the world's lands exhibit high salt content (Schroeder et al., 2013). Soil salinity can be increased due to geological events or incorrect practices in agriculture, such as irrigation with salt water and excess of fertilization (Yadav et al., 2011; Yoshii et al., 2013). Salinity is a problem found in 30% of irrigated soils worldwide, and this abiotic stress induces a number of changes in plant metabolism such as ion toxicity, osmotic imbalance, and ROS production, leading to nutritional disturbances, oxidative stress, damage to the photosynthetic apparatus, and consequently limits plant growth and crop yield (Ismail et al., 2014).

The role of NO• in salt tolerance has been demonstrated in many plant species. Gadelha et al. (2017) showed that salinity altered the source-sink relationship, suggesting a severe delay in reserve mobilization of *Jatropha curcas* seeds, restricting initial plant growth. However, NO• seed priming alleviated the harmful effects of salt on reserve mobilization, improving seedling growth under salinity. Salinity also promotes increases in H_2O_2 content and MDA, such effects being alleviated in NO-primed seedlings due to a powerful antioxidant system, including higher levels of GSH and ascorbate (AsA), as well as higher activities of CAT and glutathione reductase (GR). Pretreatment with SNP reduced the deleterious effect of NaCl on both germination and root growth, suggesting the protective role of NO• during osmotic stress in *Lupinus luteus* (Kopyra and Gwóźdź, 2003).

NaCl-induced stress increased H_2S and NO• levels in tobacco plants, these increases correlated with increases in leaf' contents of L-Cys and L-Arg, and with activities of NR, arginase, and other enzymes involved in the biosynthesis of H_2S and NO•. Accordingly, the activity of antioxidant enzymes was considerably decreased upon NO• or H_2S scavenging, showing that the increment in NO• and H_2S levels and their interplay contributed to tobacco survival under extreme salinity conditions (Da Silva et al., 2016).

Pre-exposure to NO• activates a signaling network, with consequent changes in gene and protein expression, and may lead to acclimation to environmental stresses (Tanou et al., 2012). In this sense, some post-translational modifications such as *S*-nitrosation may also contribute to NO• signaling during saline stress (Figure 21.1). In citrus plants, several proteins changed, suggesting that *S*-nitrosation plays a key role in plant response to salinity (Tanou et al., 2009). Proteomic analysis revealed that APX is one of the potential targets of *S*-nitrosation in *Arabidopsis* plants (Fares et al., 2011). In pea plants exposed to 150 mM NaCl, improvements in both APX activity and *S*-nitrosylated APX were found, as well as increases in H_2O_2, NO•, and SNO contents (Begara-Morales et al., 2014).

21.2.5 THERMAL STRESS

Temperature stress—by heating, chilling, or freezing—affects molecular, biochemical, and physiological processes in plants, reducing biomass production and crop yield (Suzuki and Mittler, 2006). In fact, the temperature is one factor determining the distribution and survival of plant species around the world. At the cellular level, heat affects a wide range of structures and functions, altering lipid properties and causing membranes to become more fluid and susceptible to damage. As proteins have an optimum temperature, high temperatures alter their activity, leading to metabolic imbalance and even denaturation. Together, membrane and protein damage will increase ROS production, a well-known consequence of stressful conditions (Hasanuzzaman et al., 2013).

The involvement of NO• in cold and heat tolerance has been demonstrated in the last decade. Wang et al. (2015b) verified that SNP treatment inhibited the development of chilling injury in banana fruit stored at 7°C for up to 20 days. Such cold tolerance of banana fruit induced by supplying NO• was achieved by maintaining high levels of energy status by an increased activity of enzymes involved in energetic metabolism, including H^+-ATPase, Ca^{2+}-ATPase, succinate dehydrogenase, and cytochrome oxidase (Complex IV), during cold storage. In general, plant responses to chilling have been investigated in the context of fruit post-harvest technology, and several studies reported the ability of NO gas fumigation in preventing chilling injury in cold-stored fruits. As a mechanism, NO• acts by improving the antioxidant system (Wu et al., 2014) and reducing ethylene production and respiration (Zaharah and Singh, 2011; Li et al., 2012). Besides post-harvest, low temperature (0.5°C) has an inhibitory effect on the growth of primary roots in *A. thaliana*, with the orientation of actins filaments being modulated by NO (Plohovska et al., 2017). The application of exogenous SNP favored the reorganization of actin filaments network, while either cold or NO• scavenger increased its randomization.

Regarding heat stress, wheat plants exhibited significant increases in the concentration of H_2O_2, $O_2^{•-}$ and lipid peroxidation under 35°C and 40°C, with SNP alleviating oxidative damage through increases in activity of antioxidant enzymes such as SOD, APX, and GR (El-Beltagi et al., 2016). In general, exposure to high temperatures increases ROS production (Bouchardand Yamasaki, 2008; Yu et al., 2014) and supplying exogenous NO• activates the enzymatic and non-enzymatic defense systems (Zhao et al., 2009; Hasanuzzaman et al., 2013).

21.3 NO DONORS

In order to understand how the NO• radical affects metabolic and physiological responses of plants to limiting conditions, exogenous NO• donors have been increasingly used. Major advances recently made for elucidating the role of NO• in plants were possible due to the application of these donors to plants. However, there are several NO• donors that act in different ways, depending on their concentrations, plant species, and even the phenological stage (Floryszak-Wieczorek et al., 2006). Given such variability and specificity, there are some ambiguous and controversial results when considering NO• effects on plants. In order to obtain reliable data,

Neill et al. (2003) suggested the use of NO• synthesis inhibitors or NO-scavengers in parallel to the monitoring of current NO• concentration in plant tissues.

Various NO• donors have been used in plants for releasing NO•, and such donors differ in composition, half-life, NO release rate under light conditions, and presence of reducing agents or in release of other compounds. The most commonly used is SNP probably due to the low cost and ease of handling. However, the mechanism of NO• release from SNP is not clear. Currently, it is known that NO• release from SNP requires either illumination or one-electron reduction and is usually enhanced by thiols. SNP solution is extremely photosensitive and its degradation is promoted by increasing oxygen and temperature (Wang et al., 2002). The reduction and subsequent decomposition of SNP is accompanied by cyanide release, leading to cellular toxicity. The rates of cyanide release increases upon light irradiation. In addition, this cyanide is bioactive and involved in inhibition of nitrate reductase (NR) and cytochrome oxidase, affecting then, NO• production. SNP is also involved in peroxynitrite formation, and also increases the toxicity of H_2O_2 (Wang et al., 2002). Then, released cyanide can impair plant development, leading to misleading interpretation and conclusions about the role of NO• in plants. Alternatively, authors have used positive controls, such as sodium or potassium ferrocyanide (Cui et al., 2010; Liu et al., 2018). Nevertheless, these studies reported the most diverse effects, with NO• being either beneficial or harmful to plants (Zaninotto et al., 2006; Silveira et al., 2015).

S-nitrosothiols (RSNOs) belong to an important group of NO• donors, and the most frequently used RSNOs in plants are *S*-nitrosoglutathione (GSNO) and *S*-nitroso-N-acetylpenicillamine (SNAP). Nonreductive decomposition of RSNOs leads to the formation of disulfides and release of NO•, which are dependent on light, temperature, the presence of metal ions (Cu^{2+}), and pH (Hou et al., 1999). In addition, oxidized forms of endogenous NO• (such as N_2O_3) may react with the thiol group (SH) of cysteine residues (Cys) present in proteins, forming an *S*-nitrosothiol (SNO), a reaction called *S*-nitrosation or *S*-nitrosylation (Lindermayr et al., 2005; Aracena-Parks et al., 2006). Thus, RSNOs, such as GSNO, are the natural reservoir of NO• in biological systems, releasing free NO• during its degradation. The GSNO is an *S*-nitrosated derivative of the most abundant cellular thiol, the glutathione (GSH), which intracellular concentrations may be greater than 10 mM (Hancock and Whiteman, 2018). GSNO itself is not directly absorbed into cells; however, GSNO treatment does cause increases in cellular *S*-nitrosothiol levels under many conditions. It was hypothesized that GSNO decomposes in the extracellular space to release NO•, which is then able to diffuse across the cell membrane to S-nitrosate protein targets (Broniowska et al., 2013).

Decomposition of NO• donors depends on several external factors, such as light, temperature, concentration, the presence of metal ions, and chemical nature of the tissue. As decomposition of NO• donors can be stimulated or inhibited by plant tissue and environmental conditions, it is necessary to monitor the amount of NO• released by each donor type. Floryszak-Wieczorek and colleagues (2006) demonstrated that the highest rate of NO• generation was found with SNAP, followed by GSNO and SNP donors. Although SNAP and GSNO are photodegradated, SNP is the most photosensitive donor. In relation to the stability of donors in solution (expressed as half-life), we have the following order: SNP > GSNO > SNAP (Floryszak-Wieczorek et al., 2006).

Regarding the NO• donor diethylamine nonoate sodium (DETA), there are few studies about its use in plants. Only one paper was found (Planchet et al., 2014), and even then, the effect of DETA was not beneficial in *Medicago truncatula,* decreasing seed germination. In animal cells, NO released from some NO• donors was addressed, and the authors found that DETA decomposed in 3.9 ± 0.2 min, whereas SNAP decomposed around 37 ± 4 h (Ferrero et al., 1999). It appears that the half-life of DETA is very short, which may explain the scarce number of studies with this donor.

Additionally, NO• scavenger substances have been widely explored as a tool for evaluating NO• function in plants. By means of radical sequestration, it is possible to reveal how NO• affects plant species. The most commonly used reagents for NO scavenger are 2-4-carboxyphenyl-4,4,5,5-tetramethylimidazoline-1-oxyl-3-oxide (cPTIO) and, its analogue 2-phenyl-4,4,5,5-tetramethylimidazoline-1-oxyl 3-oxide (PTIO). cPTIO is a stable organic radical and has been used as a control. As a mode of action, cPTIO and PTIO react with NO•, forming carboxy-PTI and nitrogen dioxide (Cao and Reith, 2002). L-NAME [(N(G)-nitro-l-arginine-methyl ester)] is also another known NO inhibitor/scavenger (Zhang et al., 2012) and its supposed mode of action is through inhibition of NOS. However, we have already mentioned that NOS has not yet been found in higher plants.

When selecting NO• donors for agricultural applications, an important point is the understanding of NO• release from the donor, including the potential toxicity and bioactivity of by-products. In such scenarios, RSNOs promise NO• donors as by-products formed are nontoxic (Seabra and Oliveira, 2016).

21.4 NO RELEASE FROM NANOPARTICLES: AN INTERESTING TOOL FOR FUTURE RESEARCH IN PLANTS

In contrast to biomedical applications, the use of nanoparticles in agriculture is a relatively new approach, and general interest about it has increased (Savithramma et al., 2012). This technology has, in fact, a potential for use in agriculture, such as already explored in the efficient and safe delivery of fertilizers, pesticides, and herbicides, and controlled release of the agrochemicals directly to the target site (Grover et al., 2012). In addition, it has been used in packaging to improve the quality of agricultural products (Khan et al., 2017).

Recently, the encapsulation of NO• donors emerged as a strategy for protecting these molecules from rapid degradation and allowing more controlled NO• release, thereby prolonging their effects (Seabra et al., 2010, 2015). A recent study about nanoparticle application in plants demonstrated that the NO• donor S-nitrosomercaptosuccinic acid (S-nitroso-MSA) was more efficient in mitigating the deleterious effects of salt stress in maize plants than the same NO• donor in its free form (Oliveira et al., 2016). In addition, many NO• donors have low molecular weight and are thermally and photochemically unstable, such as RSNOs (Shishido et al., 2003). In order to increase such stability, these donors have been incorporated into polymeric matrices, such as chitosan-based nanoparticles (Cardozo et al., 2014), which changes the NO• release profile. Regarding the matrices, chitosan (CS) is a biodegradable and biocompatible polysaccharide extensively used in drug delivery applications, in particular as a nanocarrier system (Pelegrino et al., 2017a, b). Several different classes of polymeric and metallic nanomaterials can be formulated as NO-releasing systems (Marcato et al., 2013; Seabra et al., 2015; Seabra and Duran, 2010). Although nanomaterials have been successfully and extensively used to carry agrochemicals in plants, and in spite of the extensive uses of NO-releasing nanomaterials in biomedical applications, the use of nanoparticles to carry and to delivery NO• in plants is still in the infancy (Seabra and Duran, 2010; Seabra et al., 2014, 2015). The combination of NO• donors with nanomaterials for agriculture purposes is considered a promising approach that needs to be further investigated.

21.5 CONCLUSION AND PERSPECTIVES

NO• has an important role in primary signaling under abiotic stresses, it causes interesting and positive changes in the physiology of plants, increasing their tolerance

TABLE 21.1

Effects of NO Donors on Plant Performance Under Abiotic Stresses*

Stress	How	Species	NO Donor**	Plant Responses***	Reference
Drought	PEG6000	*Phaseolus vulgaris*	1000 μM SNP (sprayed)	(+) photochemistry (−) ROS production	Zimmer-Prados et al. (2014)
	PEG6000	*Vigna unguiculata*	1000 μM SNP (sprayed)	(+) leaf relative water content (+) stomatal conductance (+) electrolyte leakage	Zimmer-Prados et al. (2014)
	Water withholding	*Agrostis stolonifera* *Festuca aurandinacea*	150 μM SNP (sprayed)	(+) leaf relative water content (+) chlorophyll and proline contents (+) activity of antioxidant enzymes (SOD and APX) (−) ion leakage	Hatamzadeh et al. (2015)
	PEG8000	*Medicago truncatula*	500 μM DETA (dilution in medium)	(−) stomatal conductance (−) seed germination	Planchet et al. (2014)
	PEG8000	*Saccharum* spp.	>100 μM GSNO (sprayed)	(+) leaf gas exchange (+) photochemistry (+) leaf relative water content (+) leaf and root dry matter	Silveira et al. (2016)
	Leaf disc dehydration	*Saccharum* spp.	60 μM NO (fumigation)	(+) photochemistry	Silveira et al. (2016)
	PEG8000	*Saccharum* spp.	100 μM GSNO (sprayed)	(+) photosynthetic rates (+) Rubisco activity (+) activity of antioxidant enzymes in leaves (SOD) and roots (CAT)	Silveira et al. (2017b)
	PEG6000	*Brassica napus*	500 μM SNP (in nutrient solution)	(+) non-enzymatic antioxidant pool (+) activity of antioxidant enzymes (CAT, SOD and POD) (+) glyoxalase system	Hasanuzzaman et al. (2017)
	Water withholding	*Lactuca sativa*	500 μM SNP (sprayed)	(−) mycelia growth within roots. (+) leaf water content	Sánchez-Romera et al. (2018)
	Water withholding	*Cicer arietinum*	100 μM SNP (irrigated)	(+) germination (+) chlorophyll and nitrogen contents	Santisree et al. (2017)
	PEG	*Allium hirtifolium*	40 μM SNP (dilution in medium)	(+) leaf relative water content (+) protecting of photosynthetic pigments (chlorophyll and carotenoids) (+) activity of antioxidant enzymes (SOD and APX) (+) proline and allicin (−) H_2O_2 production and lipid peroxidation	Ghassemi-Golezani et al. (2018)
Flooding	Waterlogged	*Triticum aestivum*	100 μM SNP (seed incubation)	(+) gene expression (+) activity of antioxidant (CAT and GR)	Özçubukçu et al. (2013)
	Root flooding	*Cucumis sativus*	100 μM SNP (sprayed)	(+) activity of antioxidant enzymes (−) lipid peroxidation (+) chlorophyll and protein content during waterlogging	Fan et al. (2014)

(Continued)

TABLE 21.1 (CONTINUED)

Effects of NO Donors on Plant Performance Under Abiotic Stresses*

Stress	How	Species	NO Donor**	Plant Responses***	Reference
	Waterlogged	*Zea mays*	500 µM SNP (nutrient solution)	(+) transpiration rate (+) chlorophyll and nitrogen contents	Jaiswal and Srivastava (2015)
	Waterlogged	*Suaeda salsa*	80 µM SNP (nutrient solution)	(−) growth inhibition (+) adventitious root formation	Chen et al. (2016)
	Waterlogged	*Zea mays*	50 to 700 µM SNP (nutrient solution)	(+) survival rate (+) increased Ca^{2+} levels (+) activity of alcohol dehydrogenase (+) antioxidant defense	Peng et al. (2016)
Heavy metal	Arsenic (1.5 mg L^{-1})	*Pistia stratiotes*	0.1 mg L^{-1} SNP (nutrient solution)	(+) total antioxidant capacity (+) photosynthetic rates (+) pigment content	Farnese et al. (2014)
	Cadmium (5 mg L^{-1})	*Boehmeria nivea*	100 µM SNP (nutrient solution)	(+) S-nitrosothiols content (+) activity of antioxidant enzymes (SOD, APX, and GR) (−) H_2O_2 accumulation	Wang et al. (2015)
	Cooper (100 µM)	*Oryza sativa*	200 µM SNP (nutrient solution)	(−) toxicity symptoms (+) antioxidant capacity	Mostofa et al. (2014)
	Arsenic (50 µM)	*Lactuca sativa*	100 µM SNP (nutrient solution)	(−) concentration of oxidative agents (−) translocation of As to shoots (+) activity of antioxidant enzymes (SOD, CAT, POX, APX, GR, GPX)	Silveira et al. (2015)
	Copper (30 mg kg^{-1})	*Catharanthus roseus*	50 µM SNP (sprayed)	(+) toxicity by regulating mineral absorption, reestablishing ATPase activities, and stimulating secondary metabolites.	Liu et al. (2016)
	Cadmium (10 µM)	*Oryza sativa*	30 µM SNAP (sprayed)	(+) activity of antioxidant enzymes (APX, SOD, GR) (+) glutathione (−) H_2O_2 accumulation.	Yang et al. (2016)
	Arsenic (25 µM)	*Oryza sativa*	30 µM SNP (nutrient solution)	(−) toxicity, modulating regulatory networks involved in As detoxification and jasmonic acid biosynthesis	Singh et al. (2017)
Salinity	120 mM NaCl (nutrient solution)	*Solanum lycopersicum*	100 and 300 µM SNP (nutrient solution)	(+) activity of antioxidant enzymes (SOD, APX, GR, POD) (+) activity of nitrate reductase (NR) and nitrite reductase (NiR) (+) proline and ascorbate contents (−) H_2O_2 accumulation	Manal et al. (2014)
	100 mM NaCl (salt solution)	*Zea mays*	3 and 6 mg L^{-1} SNP (seed soaking and sprayed)	(+) leaf osmotic pressure (+) proline accumulation (+) activity of antioxidant enzymes (POD, SOD and CAT) (+) photosynthetic pigments	Kaya et al. (2015)
	300 mM NaCl (nutrient solution)	*Helianthus annuus*	100 µM SNP (nutrient solution)	(+) indole-3-acetic acid content (+) abscisic acid content	Yurekli and Kirecci (2016)

(Continued)

TABLE 21.1 (CONTINUED)

Effects of NO Donors on Plant Performance Under Abiotic Stresses*

Stress	How	Species	NO Donor**	Plant Responses***	Reference
	100 mM NaCl (nutrient solution)	*Cicer arietinum*	50 μM SNAP (nutrient solution)	(+) growth (+) leaf relative water content (+) photosynthetic pigments (+) production of osmolytes (+) activity of antioxidant enzymes (SOD, CAT, APX)	Ahmad et al. (2016)
	100 mM NaCl (salt solution)	*Jatropha curcas*	75 μM SNP (seed incubation)	(−) accumulation of Na^+ and Cl^- (−) oxidative damage (+) antioxidant system (GSH, CAT, GR, AsA) (+) growth	Gadelha et al. (2017)
	100 mM NaCl (in Petri dishes)	*Solanum lycopersicum*	100 μM SNP (in Petri dishes)	(+) antioxidant capacity (+) chlorophyll a and b (+) proline accumulation (−) H_2O_2, $O_2^{\bullet-}$ and MDA contents (−) electrolyte leakage	Siddiqui et al. (2017)
Thermal stress (High/ Low)	Low temperature ($7 \pm 1°C$)	*Musa* spp.	60 μL L^{-1} NO (injected into sealed bags)	(+) activity of antioxidant enzymes (+) gene expression (−) ROS accumulation	Wu et al. (2014)
	Low temperature (7°C)	*Musa* spp.	50 μM SNP (immersion in solution)	(+) ATP content (+) activity of fructokinase, glucokinase, glucose-6-phosphate dehydrogenase and 6-phosphogluconate dehydrogenase	Wang et al. (2015)
	Low temperature (0.5°C)	*Arabidopsis thaliana*	100 μM SNP (sprayed)	(+) collaborated on actin filaments network reorganization (+) primary roots growth	Plohovska et al. (2017)
	High temperature (45°C)	*Arabidopsis thaliana*	20 μM SNP or SNAP (sprayed)	(+) heat sensitivity	Wang et al. (2014)
	High temperature (35 and 40°C)	*Triticum aestivum*	500 μM SNP (in culture medium)	(−) MDA, H_2O_2 and $O_2^{\bullet-}$ contents (+) activity of antioxidant enzymes (+) non-enzymatic antioxidants contents	El-Beltagi et al. (2016)

*Papers published between 2013 and 2018; **GSNO (*S*-nitrosoglutathione); SNP (sodium nitroprusside); SNAP (*S*-nitroso-N-acetylpA (ennicillamine); DETA (Diethylamine NONOate sodium); ***(+) means increases and (−) means decreases in a given trait or process.

and performance under unfavorable environmental conditions. Plants are often subjected to various types of stress that often act simultaneously. Research focusing on the role of exogenous NO• and its modulation in physiological and biochemical responses is necessary to understand the adjustments that plants need to make to deal with variable and limiting environmental situations, especially in the context of global climate change.

Although supplying exogenous NO• has been used extensively, care must be taken when choosing NO•

donors. First, NO• is either a cytotoxic or a cytoprotective molecule, and plant responses are highly dependent on the donor type used and its concentration. In addition, the amount of NO• released by NO• donors and their release time is often not taken into account. In fact, the stability of these molecules depends on the experimental conditions and their interaction with other molecules. Understanding the releasing characteristics of NO• donors through a systematic evaluation using appropriate controls will allow a critical and better comparison of studies involving NO•

and plants under stressful conditions. Additionally, few studies have explored supplying exogenous GSNO in agricultural systems (Table 21.1), and this is probably due to its high cost. An alternative would be to provide reagents with a lower purity. In future, NO• and nanomaterials might have a great economic impact on agriculture if costs related to NO• supplying are reduced through oriented and multidisciplinary research in field grown crop species, therefore, studies using nanotechnology for encapsulating NO• donors should be stimulated.

ACKNOWLEDGMENTS

Neidiquele M. Silveira acknowledges the São Paulo Research Foundation (Fapesp, Brazil) for their scholarship (Grants no. 2012/19167-0; 2015/21546-7). The authors also acknowledge the fellowships (ABS; ECM; RVR) granted by the National Council for Scientific and Technological Development (CNPq, Brazil).

REFERENCES

Aftab, T., Khan, M.M.A., Naeem, M., Idrees, M., Moinuddin., Da Silva, J.A.T., Ram, M. (2012) Exogenous nitric oxide donor protects *Artemisia annua* from oxidative stress generated by boron and aluminium toxicity. *Ecotoxicology and Environmental Safety* 80: 60–68.

Ahmad, P., Abdel Latef, A. A., Hashem, A., Abd_Allah, E. F., Gucel, S., Tran, L. S. P. (2016). Nitric oxide mitigates salt stress by regulating levels of osmolytes and antioxidant enzymes in chickpea. *Frontiers in Plant Science* 7: 347–358.

Alderton, W.K., Cooper, C.E., Knowles, R.G. (2001) Nitric oxide synthases: Structure, function and inhibition. *Biochemistry Journal* 357: 593–615.

Alloway, B. J. (2013). Sources of heavy metals and metalloids in soils. In Heavy metals in soils (pp. 11–50). *Springer, Dordrecht.*

Aracena-Parks, P., Goonasekera, S.A., Gilman, C.P., Dirksen, R.T., Hidalgo, C., Hamilton, S.L. (2006) Identification of cysteines involved in *S*-nitrosylation, *S*-glutathionylation, and oxidation to disulfides in ryanodine receptor type 1. *The Journal of Biological Chemistry* 281: 40354–40368.

Astier, J., Rasul, S., Koen, E., Manzoor, H., Benson-Bard, A., Lamotte, O. (2011) *S*-nitrosylation: An emerging post-translational protein modification in plants. *Plant Science* 181: 527–533.

Bailey-Serres, J., Voesenek, L.A.C.J. (2008) Flooding stress: Acclimations and genetic diversity. *Annual Review of Plant Biology* 59: 313–339.

Begara-Morales, J.C., Sanchez-Calvo, B., Chaki, M., Valderrama, R., Mata-Perez, C., Lopez-Jaramillo, J., et al. (2014) Dual regulation of cytosolic ascorbate peroxidase (APX) by tyrosine nitration and *S*-nitrosylation. *Journal of Experimental Botany* 65: 527–538.

Besson-Bard, A., Pugin, A., Wendehenne, D. (2008) New insights into nitric oxide signaling in plants. *Annual Review of Plant Biology* 59: 21–39.

Bethke, P.C., Libourel, I.G., Reinöhl, V., Jones, R.L. (2006) Sodium nitroprusside, cyanide, nitrite, and nitrate break *Arabidopsis* seed dormancy in a nitric oxide-dependent manner. *Planta* 223: 805–812.

Borges, K. L. R., Salvato, F., Alcântara, B. K., Nalin, R. S., Piotto, F. Â., & Azevedo, R. A. (2018). Temporal dynamic responses of roots in contrasting tomato genotypes to cadmium tolerance. *Ecotoxicology* 27: 245–258.

Bouchard, J.N., Yamasaki, H. (2008) Heat stress stimulates nitric oxide production in symbiodinium microadriaticum: A possible linkage between nitric oxide and the coral bleaching phenomenon. *Plant and Cell Physiology* 49: 641–652.

Broniowska K.A., Diers A.R., Hogg N. (2013) *S*-Nitrosoglutathione. *Biochimica et Biophysica Acta* 1830: 3173–3181.

Cao, B.J., Reith, M.E. (2002) Nitric oxide scavenger carboxy-PTIO potentiates the inhibition of dopamine uptake by nitric oxide donors. *European Journal of Pharmacology* 448: 27–30.

Cardozo, V.F., Lancheros, C.A.C., Narciso, A.M., Valereto, E.C.S., Kobayashi, R.K.T., Seabra, A.B., Nakazato, G. (2014) Evaluation of antibacterial activity of nitric oxide releasing polymeric particles against *Staphylococcus aureus* and *Escherichia coli* from bovine mastitis. *International Journal of Pharmaceutics* 473: 20–29.

Chen, F., Wang, F., Sun, H., Cai, Y., Mao, W., Zhang, G., et al. (2013) Genotype-dependent effect of exogenous nitric oxide on cd-induced changes in antioxidative metabolism, ultrastructure, and photosynthetic performance in barley seedlings (*Hordeum vulgare*). *Journal of Plant Growth Regulation* 29: 394–408.

Chen, T., Yuan, F., Jie Song, A., Wang, B. (2016) Nitric oxide participates in waterlogging tolerance through enhanced adventitious root formation in the euhalophyte *Suaeda salsa*. *Functional Plant Biology* 43: 244–253.

Corpas, F.J., Palma, J.M., Barroso, J.B. (2013) Protein tyrosine nitration in higher plants grown under natural and stress conditions. *Frontiers in Plant Science* 4: 1–4.

Cui, X.M., Zhang, Y.K., Wu, X.B., Liu, C.S. (2010) The investigation of the alleviated effect of copper toxicity by exogenous nitric oxide in tomato plants. *Plant, Soil and Environment* 6: 274–281.

Da Silva, C.J., Fontes, E.P.B., Modolo, L.V. (2016) Salinity-induced accumulation of endogenous H₂S and NO is associated with modulation of the antioxidant and redox defense systems in *Nicotiana tabacum* L. cv. Havana. *Plant Science* 256: 148–159.

El-Beltagi, H.S., Ahmed, O.K., Hegazy, A.E. (2016) Protective effect of nitric oxide on high temperature induced oxidative stress in wheat (*Triticum aestivum*) callus culture. *Notulae Scientia Biologicae* 8: 192–198.

Fan, H.F., Du, C.X., Ding, L., Xu, Y.L. (2014) Exogenous nitric oxide promotes waterlogging tolerance as related to the activities of antioxidant enzymes in cucumber seedlings. *Russian Journal of Plant Physiology* 61: 366–373.

Fares, A., Rossignol, M., Peltier, J.B. (2011) Proteomics investigation of endogenous *S*-nitrosylation in Arabidopsis. *Biochemical and Biophysical Research Communications* 416: 331–336.

Farnese, F.S., Menezes-Silva, P.E., Gusman, G.S., Oliveira, J.A. (2016) When bad guys become good ones: The key role of reactive oxygen species and nitric oxide in the plant responses to abiotic stress. *Frontiers in Plant Science* 7: 471–486.

Farnese, F.S., Oliveira, J.A., Gusman, G.S., Leão, G.A., Silveira, N.M., Silva, P.M., Ribeiro, C., Cambraia, J. (2014) Effects of adding nitroprusside on arsenic stressed response of *Pistia stratiotes* L. under hydroponic conditions. *International Journal of Phytoremediation* 16: 123–137.

Ferrero, R., Rodríguez-Pascual, F., Miras-Portugal, M.T., Torres, M. (1999) Comparative effects of several nitric oxide donors on intracellular cyclic GMP levels in bovine chromaffin cells: Correlation with nitric oxide production. *British Journal of Pharmacology* 127: 779–787.

Floryszak-Wieczorek, J., Milczarek, G., Arasimowicz, M., Ciszewski, A. (2006) Do nitric oxide donors mimic endogenous NO-related response in plants? *Planta* 224: 1363–1372.

Foresi, N., Mayta, M.L., Lodeyro, A.F., Scuffi, D., Correa-Aragunde, N., García-Mata, C., Casalongué, C., Carrillo, N., Lamattina, L. (2015) Expression of the tetrahydrofolate-dependent nitric oxide synthase from the green alga *Ostreococcus tauri* increases tolerance to abiotic stresses and influences stomatal development in Arabidopsis. *The Plant Journal* 82: 806–821.

Frungillo, L., Skelly, M.J., Loake, G.J., Spoel, S.H., Salgado, I. (2014) *S*-Nitrosothiols regulate nitric oxide production and storage in plants through the nitrogen assimilation pathway. *Nature Communications* 5: 5401–5410.

Gadelha, C.G., Miranda, R.S., Alencar, N.L.M., Costa, J.H., Priscoa, J.T., Gomes-Filho, E. (2017) Exogenous nitric oxide improves salt tolerance during establishment of *Jatropha curcas* seedlings by ameliorating oxidative damage and toxic ion accumulation. *Journal of Plant Physiology* 212: 69–79.

Gallego, S.M., Pena, L.B., Barcia, R.A., Azpilicueta, C.E., Lannone, M.F., Rosales, E.P., Zawoznik, M.S., Groppa, M.D., Benavides, M.P. (2012) Unravelling cadmium toxicity and tolerance in plants: Insight into regulatory mechanisms. *Environmental and Experimental Botany* 83: 33–46.

Garcia-Mata, C., Lamattina, L. (2001) Nitric oxide induces stomatal closure and enhances the adaptive plant responses against drought stress. *Plant Physiology* 126: 1196–1204.

Ghassemi-Golezani, K., Farhadi, N., & Nikpour-Rashidabad, N. (2018). Responses of in vitro-cultured Allium hirtifolium to exogenous sodium nitroprusside under PEG-imposed drought stress. *Plant Cell, Tissue and Organ Culture (PCTOC)* 133: 237–248.

Gow, A.J., Ischiropoulos, H. (2001) Nitric oxide chemistry and cellular signaling. *Journal of Cellular Physiology* 187: 277–282.

Grover, M., Singh, S.R., Venkateswarlu, B. (2012) Nanotechnology: Scope and limitations in agriculture. *International Journal of Nanotechnology* 2: 10–38.

Gupta, K.J., Fernie, A.R., Kaiser, W.M., Van Dongen, J.T. (2011) On the origins of nitric oxide. *Trends in Plant Science* 16: 160–168.

Hancock, J.T., Whiteman, M. (2014) Hydrogen sulfide and cell signaling: Team player or referee? *Plant Physiology and Biochemistry* 78: 37–42.

Hancock, J., Whiteman, M. (2018) Cellular redox environment and its influence on redox signalling molecules. *Reactive Oxygen Species* 5: no pagination.

Harrison, R. (2002) Structure and function of xanthine oxidoreductase: Where are we now? *Free Radical Biology and Medicine* 33: 774–797.

Hasanuzzaman, M., Fujita, M. (2013) Exogenous sodium nitroprusside alleviates arsenic-induced oxidative stress in wheat (*Triticum aestivum* L.) seedlings by enhancing antioxidant defense and glyoxalase system. *Ecotoxicology* 22: 584–596.

Hasanuzzaman, M., Nahar, K., Alam, M., Roychowdhury, R., Fujita, M. (2013) Physiological, biochemical, and molecular mechanisms of heat stress tolerance in plants. *International Journal of Molecular Sciences* 14: 9643–9684.

Hasanuzzaman, M., Nahar K., Hossain S., Anee T.I., Parvin K., Fujita M. (2017) Nitric oxide pretreatment enhances antioxidant defense and glyoxalase systems to confer PEG induced oxidative stress in rapeseed. *Journal of Plant Interactions* 1: 323–331.

Hatamzadeh, A., Nalousi, A.M., Ghasemnezhad, M., Biglouei, M.H. (2015) The potential of nitric oxide for reducing oxidative damage induced by drought stress in two turfgrass species, creeping bentgrass and tall fescue. *Grass and Forage Science* 70: 538–548.

He, J., Y, Ren, Y., Xiulan, C., Chen, H. (2014) Protective roles of nitric oxide on seed germination and seedling growth of rice (*Oryza sativa* L.) under cadmium stress. *Ecotoxicology and Environmental Safety* 108: 114–119.

Hebelstrup, K.H., Van Zanten, M., Mandon, J., Voesenek, L.A., Harren, F.J., Cristescu, S.M., et al. (2012) Haemoglobin modulates NO emission and hyponasty under hypoxia-related stress in *Arabidopsis thaliana*. *Journal of Experimental Botany* 63: 5581–5591.

Hou, Y.C., Janczuk, A., Wang, P.G. (1999) Current trends in the development of nitric oxide donors. *Current Pharmaceutical Design* 5: 417–441.

Ismail, A., Takeda, S., Nick, P. (2014) Life and death under salt stress: Same players, different timing? *Journal of Experimental Botany* 65: 2963–2979.

Jaiswal A., Srivastava, J.P. (2015) Effect of nitric oxide on some morphological and physiological parameters in maize exposed to waterlogging stress. *African Journal of Agricultural Research* 10: 3462–3471.

Jasid, S., Simontacchi, M., Bartoli, C.G., Puntarulo, S. (2006) Chloroplasts as a nitric oxide cellular source. Effect of reactive nitrogen species and chloroplastic lipids and proteins. *Plant Physiology* 142: 1246–1255.

Jeandroz, S., Wipf, D., Stuehr, D.J., Lamattina, L., Melkonian, M., Tian, Z., Zhu, Y., Carpenter, E.J., Wong, G.K-S., Wendehenne, D. (2016) Occurrence, structure, and evolution of nitric oxide synthase-like proteins in the plant kingdom. *Science Signaling* 9: 1–9.

John, R., Ahmad, P., Gadgil, K., Sharma, S. (2009) Heavy metal toxicity: Effect on plant growth, biochemical parameters and metal accumulation by *Brassica juncea* L. *International Journal of Plant Production* 3: 65–76.

Joudoi, T., Shichiri, Y., Kamizono, N., Akaike, T., Sawa, T., Yoshitake, J., Yamada, N., Iwai, S. (2013) Nitrated cyclic GMP modulates guard cell signaling in Arabidopsis. *The Plant Cell* 25: 558–571.

Kasim, W.A. (2005) The correlation between physiological and structural alterations induced by copper and cadmium stress in broad beans (*Vicia faba* L.). *Egyptian Journal of Biology* 7: 20–32.

Kaya, C., Ashraf, M., Sonmez, O., Tuna, A.L., Aydemir, S. (2015) Exogenously applied nitric oxide confers tolerance to salinity-induced oxidative stress in two maize (*Zea mays* L.) cultivars differing in salinity tolerance. *Turkish Journal of Agriculture and Forestry* 39: 909–919.

Khan, M.N., Mobin, M., Abbas, Z.K., AlMutairi, K.A., Siddiqui, Z.H. (2017) Role of nanomaterials in plants under challenging environments. *Plant Physiology and Biochemistry* 110: 194–209.

Kopyra, M., Gwóźdź, E.A. (2003) Nitric oxide stimulates seed germination and counteracts the inhibitory effect of heavy metals and salinity on root growth of *Lupinus luteus. Plant Physiology and Biochemistry* 41: 1011–1017.

Larcher, W. (2000) *Ecofisiologia Vegetal*. São Carlos: Rima Artes e Textos, 531.

Leitner, M., Vandelle, E., Gaupels, F., Bellin, D., Delledonne, M. (2009) NO signals in the haze: Nitric oxide signaling in plant defence. *Current Opinion in Plant Biology* 12: 451–458.

Li, X.P., Wu, B., Guo, Q., Wang, J.D., Zhang, P., Chen, W.X. (2012) Effects of nitric oxide on postharvest quality and soluble sugar content in papaya fruit during ripening. *Journal of Food Processing and Preservation* 38: 591–599.

Lindermayr, C., Saalbach, G., Durner, J. (2005) Proteomic identification of *S*-nitrosylated proteins in Arabidopsis. *Plant Physiology* 137: 921–930.

Liu, S., Yanga, R., Pana, Y., Ren, B., Chen, Q., Li, X., Xiong, X., Tao, J., Cheng, Q., Ma, M. (2016) Beneficial behavior of nitric oxide in copper-treated medicinal plants. *Journal of Hazardous Materials* 314: 140–154.

Liu, W-C., Zheng, S-Q., Yu, Z-D., Gao, X., Shen, R., Lu. (2018) Y-T. WD40-REPEAT 5a represses root meristem growth by suppressing auxin synthesis through changes of nitric oxide accumulation in Arabidopsis. *The Plant Journal* 93: 883–893.

Liu, X., Miller, M.J., Joshi, M.S., Thomas, D.D., Lancaster, J.R. (1998) Accelerated reaction of nitric oxide with O_2 within the hydrophobic interior of biological membranes. *Proceedings of the National Academy of Sciences USA* 95: 2175–2179.

Manai, J., Kalai, T., Gouia, H., Corpas, F.J. (2014) Exogenous nitric oxide (NO) ameliorates salinity-induced oxidative stress in tomato (*Solanum lycopersicum*) plants. *Journal of Soil Science and Plant Nutrition* 14: 433–446.

Marcato, P.D., Adami, L.F., Barbosa, R.M., Melo, P.S., Ferreira, I.R., de Paula, L., Duran, N., Seabra, A.B. (2013) Development of a sustained-release system for nitric oxide delivery using alginate/chitosan nanoparticles. *Current Nanoscience* 9: 1–7.

Marchiori, P.E.R., Machado, E.C., Sales, C.R.G., Espinoza-Núñez, E., Magalhães Filho, J.R., Souza, G.M., Pires, R.C.M., Ribeiro, R.V. (2017) Physiological plasticity is important for maintaining sugarcane growth under water deficit. *Frontiers in Plant Science* 8: 2148.

Medri, M., Ferreira, A., Kolb, R., Bianchini, E., Pimenta, J., Davanso-Fabro, V., Medri, C. (2007) Morpho-anatomical alterations in plants of *Lithraea molleoides* (Vell.) Engl. submitted to flooding. *Acta Scientiarum Biological Sciences* 29: 15–22.

Mostofa, M.G., Seraj, Z.I., Fujita, M. (2014) Exogenous sodium nitroprusside and glutathione alleviate copper toxicity by reducing copper uptake and oxidative damage in rice (*Oryza sativa* L.) seedlings. *Protoplasma* 251: 1373–1386.

Mostofa, M.G., Seraj, Z.L., Fujita, M. (2015) Interactive effects of nitric oxide and glutathione in mitigating copper toxicity of rice (*Oryza sativa* L.) seedlings. *Plant Signaling and Behavior* 10: 1–3.

Mugnai, S., Azzarello, E., Baluska, F., Mancuso, S. (2012) Local root apex hypoxia induces NO-mediated hypoxic acclimation of the entire root. *Plant and Cell Physiology* 53: 912–920.

Mur, L.A., Carver, T.L., Prats, E. (2006) NO way to live; the various roles of nitric oxide in plant-pathogen interactions. *Journal of Experimental Botany* 57: 489–505.

Murgia, I., Pinto, M.C., Delledonne, M., Soavea, C., Gara, L. (2004) Comparative effects of various nitric oxide donors on ferritin regulation, programmed cell death, and cell redox state in plant cells. *Journal of Plant Physiology* 161: 777–783.

Neill, S., Barros, R., Bright, J., Desikan, R., Hancock, J., Harrison, J., Morris, P., Ribeiro, D., Wilson, I. (2008) Nitric oxide stomatal closure, and abiotic stress. *Journal of Experimental Botany* 59: 165–176.

Neill, S.J., Desikan, R., Hancock, J.T. (2003) Nitric oxide signaling in plants. *New Phytologist* 159: 11–35.

Oliveira, H.C., Gomes, B.C.R., Pelegrino, M.T., Seabra, A.B. (2016) Nitric oxide-releasing chitosan nanoparticles alleviate the effects of salt stress in maize plants. *Nitric Oxide* 61: 10–19.

Ordog, A., Wodala, B., Rózsavolgyi, T., Tari, I., Horváth, F. (2013) Regulation of guard cell photosynthetic electron transport by nitric oxide. *Journal of Experimental Botany* 64: 1357–1366.

Osakabe, Y., Osakabe, K., Shinozaki, K., Tran, L.S.P. (2014) Response of plants to water stress. *Frontiers in Plant Science* 5: 1–8.

Oves, M., Khan, M. S., Zaidi, A., & Ahmad, E. (2012). Soil contamination, nutritive value, and human health risk assessment of heavy metals: an overview. In *Toxicity of heavy metals to legumes and bioremediation*, 1–27. Springer, Vienna.

Özçubukçu, S., Ergün, N., İlhan, E. (2013) Waterlogging and nitric oxide induce gene expression and increase antioxidant enzyme activity in wheat (*Triticum aestivum* L.). *Acta Biologica Hungarica* 65: 47–60.

Pelegrino, M.T., Silva, L.C., Watashi, C.M., Haddad, P.S., Rodrigues, T., Seabra, A.B. (2017a) Nitric oxide-releasing nanoparticles: Synthesis, characterization, and cytotoxicity to tumorigenic cells. *Journal of Nanoparticle Research* 19: 57.

Pelegrino, M.T., Weller, R.B., Chen, X., Bernardes, J.S., Seabra, A.B. (2017b) Chitosan nanoparticles for nitric oxide delivery in human skin. *Medicinal Chemistry Communications* 38: 606–615.

Peng, R., Bian, Z., Zhou, L., Cheng, W., Hail, N., Yang, C., Yang, T., Wang, X., Wang, C. (2016) Hydrogen sulfide enhances nitric oxide-induced tolerance of hypoxia in maize (*Zea mays* L.). *Plant Cell Reports* 35: 2325–2340.

Planchet, E., Verdu, I., Delahaie, J., Cukier, C., Girard, C., Paven, M-C., Limami A.M. (2014) Abscisic acid-induced nitric oxide and proline accumulationin independent pathways under water-deficit stress during seedling establishment in *Medicago truncatula*. *Journal of Experimental Botany* 65: 2161–2170.

Plohovska, S.H., Krasylenko, Y.A., Yemets, A.I. (2017) Nitric oxide modulates actin filament organization in *Arabidopsis thaliana* primary root cells at low temperatures. *Cell Biology International*, no pagination.

Poltronieri, P., Taurino, M., Bonsegna, S., Domenico, S.D., Santino, A. (2014) Nitric oxide: Detection methods and possible roles during jasmonate regulated stress response. In *Nitric Oxide in Plants: Metabolism and Role in Stress Physiology*, Khan, M.N., Mohammad, M.M.F., Corpas, F.J. (eds.). Switzerland: Springer, 127–138.

Pucciariello, C., Perata, P. (2016) New insights into reactive oxygen species and nitric oxide signalling under low oxygen in plants. *Plant, Cell and Environment* 40: 473–482.

Radi, R., Cassina, A., Hodara, R. (2002) Nitric oxide and peroxynitrite interactions with mitochondria. *The Journal of Biological Chemistry* 383: 401–409.

Reddy, P.P. (2015) Impacts of climate change on agriculture. In *Climate Resilient Agriculture for Ensuring Food Security*. New Delhi: Springer India, 43–90.

Ren, Y., He, J., Liu, H., Liu, G., Ren, X. (2017) Nitric oxide alleviates deterioration and preserves antioxidant properties in 'Tainong' mango fruit during ripening. *Horticulture, Environment, and Biotechnology* 58: 27–37.

Roshchina, V.V. (2001) *Neurotransmitters in Plant Life*. Enfield: Science Publishers, 283.

Salgado, I., Oliveira, H.C., Gaspar, M. (2017) Plant nitric oxide signaling under environmental stresses. In *Mechanism of Plant Hormone Signaling under Stress*, Pandey, G. (ed.). John Wiley & Sons, 1st edition, 345–370.

Sánchez-Romera, B., Porcel, R., Ruiz-Lozano, J.M., Aroca, R. (2018) Arbuscular mycorrhizal symbiosis modifies the effects of a nitric oxide donor (sodium nitroprusside; SNP) and a nitric oxide synthesis inhibitor (Nω-nitro-L-arginine methyl ester; L-NAME) on lettuce plants under well watered and drought conditions. *Symbiosis* 74: 11–20.

Santisree, P., Bhatnagar-Mathur, P., Sharma, K.K. (2015) NO to drought multifunctional role of nitric oxide in plant drought: Do we have all the answers? *Plant Science* 239: 44–55.

Santisree, P., Bhatnagar-Mathur, P., Sharma, K.K. (2017) Molecular insights into the functional role of nitric oxide (NO) as a signal for plant responses in chickpea. *Functional Plant Biology* 45: 267–283.

Savithramma, N., Ankanna, S., Bhumi, G. (2012) Effect of nanoparticles on seed germination and seedling growth of *Boswellia ovalifoliolata*—An endemic and endangered medicinal. *Tree Taxon Nano Vision* 2: 61–68.

Savvides, A., Ali, S., Tester, M., Fotopoulos, V. (2016) Chemical priming of plants against multiple abiotic stresses: Mission possible? *Trends in Plant Science* 21: 329–340.

Saxena, I., Shekhawat, G.S. (2013) Nitric oxide (NO) in alleviation of heavy metal induced phytotoxicity and its role in protein nitration. *Nitric Oxide* 32: 13–20.

Schroeder, J.I., Delhaize, E., Frommer, W.B., Guerinot, M.L., Harrison, M.J., Herrera-Estrella, L., Sanders, D. (2013) Using membrane transporters to improve crops for sustainable food production. *Nature* 497: 60–66.

Seabra, A.B., Duran, N. (2010) Nitric oxide-releasing vehicles for biomedical applications. *Journal of Materials Chemistry* 20: 1624–1637.

Seabra, A.B., Justo, G.Z., Haddad, P.S. (2015) State of the art, challenges and perspectives in the design of nitric oxide-releasing polymeric nanomaterials for biomedical applications. *Biotechnology Advances* 33: 1370–1379.

Seabra, A.B., Oliveira H.C. (2016) How nitric oxide donors can protect plants in a changing environment: What we know so far and perspectives. *AIMS Molecular Science* 3: 692–718.

Seabra, A.B., Rai, M., Durán, N. (2014) Nano carriers for nitric oxide delivery and its potential applications in plant physiological process: A mini review. *Journal of Plant Biochemistry and Biotechnology* 23: 1–10.

Shi, H., Ye, T., Zhu, J.K., Chan, Z. (2014) Constitutive production of nitric oxide leads to enhanced drought stress resistance and extensive transcriptional reprogramming in Arabidopsis. *Journal of Experimental Botany* 65: 4119–4131.

Shishido, S.M., Seabra, A.B., Loh, W., De Oliveira, M.G. (2003) Thermal and photochemical nitric oxide release from S-nitrosothiols incorporated in Pluronic F127gel:

Potential uses for local and controlled nitric oxide release. *Biomaterials* 24: 3543–3553.

Siddiqui, M.H., Alamri, S.A., Al-Khaishany, M.Y., Al-Qutami, M.A., Ali, H.M., Al-Rabiah, H., Kalaji, H.M. (2017) Exogenous application of nitric oxide and spermidine reduces the negative effects of salt stress on tomato. *Horticulture, Environment, and Biotechnology* 58: 537–547.

Silveira, N.M., Frungillo, L., Marcos, F.C.C., Pelegrino, M.T., Miranda, M.T., Seabra, A.B., Salgado, I., Machado, E.C., Ribeiro, R.V. (2016) Exogenous nitric oxide improves sugarcane growth and photosynthesis under water deficit. *Planta* 244: 181–190.

Silveira, N.M., Hancock, J.T., Frungillo, L., Siasou, E., Marcos, F.C.C., Salgado, I., Machado, E.C., Ribeiro, R.V. (2017a) Evidence towards the involvement of nitric oxide in drought tolerance of sugarcane. *Plant Physiology and Biochemistry* 115: 354–359.

Silveira, N.M., Marcos, F.C.C., Frungillo, L., Moura, B.B., Seabra, A.B., Salgado, I., Machado, E.C., Hancock, J.T., Ribeiro, R.V. (2017b) *S*-nitrosoglutathione spraying improves stomatal conductance, Rubisco activity and antioxidant defense in both leaves and roots of sugarcane plants under water deficit. *Physiologia Plantarum* 160: 383–395.

Silveira, N.M., Oliveira, J.A., Ribeiro, C., Canatto, R.A., Siman, L., Cambraia, J., Farnese, F. (2015) Nitric oxide attenuates oxidative stress induced by arsenic in lettuce (*Lactuca sativa*) leaves. *Water, Air, and Soil Pollution* 226: 379–387.

Singh, P.K., Indoliya, Y., Chauhan, A.S., Singh, S.P., Singh, A.P., Dwivedi, S., Tripathi, R.D., Chakrabarty, D. (2017) Nitric oxide mediated transcriptional modulation enhances plant adaptive responses to arsenic stress. *Scientific Reports* 7: 3592–3605.

Suzuki, N., Mittler, R. (2006) Reactive oxygen species and temperature stress: A delicate balance between signaling and destruction. *Physiologia Plantarum* 126: 45–51.

Tanou, G., Filippou, P., Belghazi, M., Job, D., Diamantidis, G., Fotopoulos, V., Molassiotis, A. (2012) Oxidative and nitrosative-based signaling and associated post-translational modifications orchestrate the acclimation of citrus plants to salinity stress. *The Plant Journal* 72: 585–599.

Tanou, G., Job, C., Rajjou, L., Arc, E., Belghazi, M., Diamantidis, G., et al. (2009) Proteomics reveals the overlapping roles of hydrogen peroxide and nitric oxide in the acclimation of citrus plants to salinity. *The Plant Journal* 60: 795–804.

Wang, D., Liu, Y., Tan, X., Liu, H., Zeng, G., Hu, X., et al. (2015a) Effect of exogenous nitric oxide on antioxidative system and *S*-nitrosylation in leaves of *Boehmeria nivea* (L.) Gaud under cadmium stress. *Environmental Science and Pollution Research* 22: 3489–3497.

Wang, L., Guo, Y., Jia, L., Chu, H., Zhou, S., Chen, K., Wu, D., Zhao, L. (2014) Hydrogen peroxide acts upstream of nitric oxide in the heat shock pathway in Arabidopsis seedlings. *Plant Physiology* 164: 2184–2196.

Wang, P.G., Xian, M., Tang, X., Wu, X., Wen, Z., Cai, T., Janczuk, A.J. (2002) Nitric oxide donors: Chemical activities and biological applications. *Chemical Reviews* 102: 1091–1134.

Wang, Y., Luo, Z., Khan, Z.U., Mao, L., Ying, T. (2015b) Effect of nitric oxide on energy metabolism in postharvest banana fruit in response to chilling stress. *Postharvest Biology and Technology* 108: 21–27.

Wieczorek, J.F., Milczarek, G., Arasimowicz, M., Ciszewski, A. (2006) Do nitric oxide donors mimic endogenous NO-related response in plants? *Planta* 224: 1363–1372.

Wink, D.A., Hanbauer, I., Grisham, M.B., Laval, F., Nims, R.W., Laval, J., et al. (1996) Chemical biology of nitric oxide: Regulation and protective and toxic mechanisms. *Current Topics in Cellular Regulation* 34: 159–187.

Wu, B., Guo, Q., Li, Q., Ha, Y., Li, X., Chen, W. (2014) Impact of postharvest nitric oxide treatment on antioxidant enzymes and related genes in banana fruit in response to chilling tolerance. *Postharvest Biology and Technology* 92: 157–163.

Yadav, S., Irfan, M., Ahmad, A., Hayat, S. (2011) Causes of salinity and plant manifestations to salt stress: A review. *Journal of Environmental Biology* 32: 667–85.

Yang, L., Ji, J., Harris-Shultz, K.R., Wang, H., Wang, H., Abd-Allah, E.F., Luo, Y., Hu, X. (2016) The dynamic changes of the plasma membrane proteins and the protective roles of nitric oxide in rice subjected to heavy metal cadmium stress. *Frontiers in Plant Science* 7: 190.

Yoshii, T., Imamura, M., Matsuyama, M., Koshimura, S., Matsuoka, M., Mas, E., Jimenez, C. (2013) Salinity in soils and tsunami deposits in areas affected by the 2010 Chile and 2011 Japan tsunamis. *Pure and Applied Geophysics* 170: 1047–1066.

Yu, M., Lamattina, L., Spoel, S.H., Loake, G.J. (2014) Nitric oxide function in plant biology: A redox cue in deconvolution. *New Phytologist* 202: 1142–1156.

Yurekli, F., Kirecci, O.A. (2016) The relationship between nitric oxide and plant hormones in SNP administrated sunflower plants under salt stress condition. *Hortorum Cultus - Acta Scientiarum Polonorum* 15: 177–191.

Zaharah, S.S., Singh, Z. (2011) Postharvest nitric oxide fumigation alleviates chilling injury, delays fruit ripening and maintains quality in cold-stored "Kensington Pride" mango. *Postharvest Biology and Technology* 60: 202–210.

Zaninotto, F., La Camera, S., Polverari, A., Delledonne, M. (2006) Cross talk between reactive nitrogen and oxygen species during the hypersensitive disease resistance response. *Plant Physiology* 141: 379–383.

Zhang, L., Zhao, Y.G., Zhai, Y.Y., Gao, M., Zhang, X.F., Wang, K., et al. (2012) Effects of exogenous nitric oxide on glycine betaine metabolism in maize (*Zea mays* L.) seedlings under drought stress. *Pakistan Journal of Botany* 44: 1837–1844.

Zhao, M-G., Chen, L., Zhang, L-L., Zhang, W-H. (2009) Nitric reductase-dependent nitric oxide production is involved in cold acclimation and freezing tolerance in Arabidopsis. *Plant Physiology* 151: 755–767.

Zimmer-Prados, L.M., Moreira, A.S., Magalhaes, J.R., Franca, M.G. (2014) Nitric oxide increases tolerance responses to moderate water deficit in leaves of *Phaseolus vulgaris* and *Vigna unguiculata* bean species. *Physiology and Molecular Biology of Plants* 20: 295–301.

22 Molecular Mechanisms of Polyamines-Induced Abiotic Stress Tolerance in Plants

Ágnes Szepesi

CONTENTS

22.1 INTRODUCTION

Polyamines (PAs), as polycations, can regulate a lot of different developmental processes and have beneficial effects in plant abiotic stress, but sometimes they can be very toxic under certain stress conditions. Modulation of PA contents and related mechanisms can also adjust its potential role to be a good inducer of abiotic stress tolerance, mainly due to its ability to maintain the cellular redox homeostasis, enhance photoprotection, and improve water homeostasis. Nowadays, there are some new challenges facing plants: air pollution and combined stress factors can significantly affect plant stress tolerance responses, and therefore researchers need to think about different modes of action to enhance plant abiotic stress tolerance. PAs can also play an important role in the initiation and regulation of programmed cell death (PCD), a process that is important in plant life. They can activate pro-survival molecules which can alleviate PCD, or they can accelerate depending on the stress conditions and plant species. Transglutaminase (TGase), an enzyme which is responsible for PA conjugation in proteins involved in the process of PCD, is reviewed in Cai et al. (2015). Earlier, it was suggested by some investigations that some PAs exhibit anti-senescence properties in plants.

PAs can act a hub during drought and salt stress as suggested by Sequera-Mutiozabal et al. (2017), who hypothesized in an excellent review that PAs act as hub metabolic molecules. Their novel results highlight the role of PAs in signaling pathways related to central metabolisms, like sugar and lipid homeostasis, and induction of antioxidant defense mechanisms as well. Pál et al. (2015) also argued that PAs play an important role in abiotic stress signaling, which regulates plant stress tolerance. In this chapter, new evidence suggests some novel contributions of PAs for inducing abiotic stress tolerance. Here, I focus on recent investigations of PA-induced stress responses and its relatedness to their tolerance mechanisms. Although the results are impressive, it remains unclear what is the proper mode of action of PAs and how they work under certain combinations of stress.

The following part describes some novel experiments, which emphasize why PAs are good candidates for inducing stress tolerance in plants and why they can act as a double-edged sword in some cases. Sequera-Mutiozabal et al. (2016) provide new evidence using global metabolic profiling by GC-TOF/MS that Arabidopsis polyamine oxidase 4 (AtPAO4) loss-of-function mutations exhibit delayed dark-induced senescence. Constitutively higher levels of important metabolites involved in redox regulation in the central metabolism were detected in pao4 mutant plants, supporting a priming status against oxidative stress. The presence of metabolic interactions between PAs, particularly Spm, with cell oxidative balance and transport/biosynthesis of amino acids can be used as a tool to cope with oxidative damage produced during senescence. This strategy may provide new insight into the role of PA catabolism in modulation of senescence, which is affecting the yield of some important crop plants. Mounting evidence supports the theory that polyamine oxidases (PAOs), FAD-dependent enzymes associated with PA catabolism, play essential roles in abiotic stress responses. Wang and Liu (2015) unraveled six putative PAO genes (CsPAO1-CsPAO6) in sweet orange (Citrus sinensis) using the released citrus genome sequences. The CsPAOs displayed various responses to exogenous treatments with PAs and abscisic acid (ABA) and were differentially altered by abiotic stresses, including cold, salt, and mannitol. Overexpression of CsPAO3 in tobacco demonstrated that Spd and Spm decreased in the transgenic line, while Put was significantly enhanced, implying a potential role of this gene in PA back conversion. This data increases our knowledge for better understanding the pivotal roles of the PAO genes under abiotic stress situations. Sagor et al. (2016) found that a mutant Arabidopsis deficient in polyamine oxidase 4 gene, accumulating about twofold more of Spm than wild-type plants, showed increased sensitivity to cadaverin (Cad), suggesting that endogenous Spm content determines growth responses to Cad in Arabidopsis thaliana. Thermospermine (T-Spm) is a structural isomer of Spm that affects xylem differentiation by limiting auxin signaling (Tong et al. 2014). Alabdallah et al. (2017) suggested that AtPAO5, being involved in the control of T-Spm homeostasis, participates in the tightly controlled interplay between auxin and cytokinins that is necessary for proper xylem differentiation. Gémes et al. (2017) reported that deregulation of apoplastic PAO affects development and salt response of tobacco plants. Apoplastic PAO is a good candidate for further examining a potential approach to breeding plants with enhanced/reduced tolerance to abiotic stress with minimal associated trade-offs. Yoshimoto et al. (2016) applied xylemin, which can inhibit T-Spm biosynthesis and caused excessive xylem differentiation in A. thaliana. Xylemin may be useful not only for the study of T-Spm function in various plant species, but also for the control of xylem induction and woody biomass production. There are some contradictory results in the literature when PAs were added exogenously. For example, Szalai et al. (2017) compared the PA homeostasis of two different crop plants, maize and wheat. They found that maize was more sensitive than wheat and that the adverse effects of PA

treatment were proportional to the accumulation of PA and of the plant hormone salicylic acid in the leaves and roots of both plant species in control conditions. They concluded that changes in the PA pool are important for fine-tuning of PA signaling, which influences the hormonal balance required if putrescine is to exert a protective effect under stress conditions. Further study is needed to decipher the precise mode of action of Put in inducing stress tolerance in maize and wheat.

Our knowledge of plant PA transport is limited. However, there are some novel studies which can help us gain a better understanding of transport mechanism. Two genes, LAT1 and OCT1, are likely to be involved in PA transport in Arabidopsis. Sagor et al. (2016) proposed a model in which the level of Spm content modulates the expression of Organic Cation Transporter 1 (OCT1) and L-Amino acid Transporter 1 (LAT1), and determines Cad sensitivity of Arabidopsis. Recently, Ahmed et al. (2017) revealed that by altering expression of PA transporters Spd can affect the timing of flowering and other developmental response pathways, providing the first genetic evidence of PA transport in the timing of flowering, and indicating the importance of PA transporters in the regulation of flowering and senescence pathways. This role of Spd needs further study in agriculturally important crops and ornamental plants. In a recent study, Patel et al. (2017) showed that plants can use arginine decarboxylase (ADC) and arginase/agmatinase (ARGAH) as a third route for Put synthesis, providing novel insights how plants can adjust their PA homeostasis. Moreover, Dalton et al. (2016) demonstrated that silencing ornithine decarboxylase (ODC) also led to significantly reduced concentrations of PAs with concomitant increases of amino acids ornithine, arginine, aspartate, glutamate, and glutamine, demonstrating that ODC has important roles in facilitating both primary and secondary metabolism in Nicotiana tabacum plants. Sagor et al. (2015) demonstrated that Spm is a novel unfolded protein response (UPR) inducer, mediated by the MKK9-MPK3/MPK6 cascade in Arabidopsis. High night temperature, which is one of the consequences of global climate change, strongly impacts the TCA cycle and amino acid and PA biosynthetic pathways in rice cultivars with different high night temperature sensitivity (Glaubitz et al., 2015). This can raise questions about how detrimental high night temperature is for other agriculturally important crops.

22.2 INTERACTIONS WITH OTHER PHYTOHORMONES

Given the variety of responses reported for all PAs, it is useful to consider how PAs relate to other growth regulator compounds like hormones. One of the most prominent interactions is the PA-nitric oxide (NO) interconnection. A significant example of interactions is the NO-ethylene (ETH) influenced regulatory node in PA biosynthesis, which is linked to drought tolerance/susceptibility in barley. Reduced NO levels during drought stress promote drought tolerance in barley and are associated with elevated PA biosynthesis (Montilla-Bascón et al., 2017).

Nahar et al. (2016a) investigated the PA and NO crosstalk during cadmium (Cd) stress and they found that Put and/or SNP (sodium nitroprusside, a NO donor) reduced Cd uptake, increased phytochelatin (PC) content, reduced oxidative damage enhancing non-enzymatic antioxidants (ascorbate (AsA) and glutathione (GSH)) and activities of enzymes [superoxide dismutase (SOD), catalase (CAT), ascorbate peroxidase (APX), dehydroascorbate reductase (DHAR), glutathione reductase (GR), glutathione S-transferase (GST), and glutathione peroxidase (GPX)]. Exogenous Put and/or SNP treatment could modulate endogenous PA levels, and combined application with NO was more effective indicating that there is a crosstalk between NO and PAs to confer Cd-toxicity tolerance. Khan and Bano (2017) applied growth regulators, salicylic acid, and Put in combination with rhizobacteria and investigated the effect of these treatments on the phytoremediation of chickpea plants. Treating plants with foliar spray containing growth regulators alleviated and improved the growth parameters of the plants; therefore, they are recommended for inducing phytoremediation under stress conditions. Siddiqui et al. (2017) studied the effect of exogenously added NO and Spd on tomatoes during salt stress and they found that when tomato plants were treated with NO and SP simultaneously, NO signaling was further enhanced. Recently, Freitas et al. (2018) found that there is a relationship between metabolisms of ETH and PAs in the processes of salinity acclimation of salt-tolerant and salt-sensitive maize genotypes. Salt-tolerant genotypes did not show stress ETH synthesis because of down-regulation of 1-aminocyclopropane-1-carboxylic acid oxidase (ACO) activity and ZmACO5b gene expression involved in ETH biosynthesis concomitantly increasing PA accumulation. Their results suggested that ETH is intimately involved in salt stress acclimation through activation of a complex pathway of signaling by H_2O_2 that is PA catabolism-dependent. Liu et al. (2017b) found that transgenic expression of LcSAMDC1 could be used to improve the abiotic resistance of crops. LcSAMDC1 was induced in response to cold and could influence the production of PAs involved in stress tolerance of L. chinensis. Overexpression of the main ORF of LcSAMDC1 in transgenic Arabidopsis promoted increased tolerance to cold and salt stress relative to wild-type Arabidopsis. The concentration of PAs, proline (Pro), and chlorophyll, was significantly higher in transgenic Arabidopsis, and Spm of PAs increased more under cold than under salt stress.

Majumdar et al. (2016) suggested that the metabolic interactions between glutamate, ornithine, arginine, Pro, and PAs were regulated at the post-transcriptional level. They used two high-Put producing lines of A. thaliana (containing a transgenic mouse ODC gene). Their main findings suggest that the overall conversion of glutamate to arginine and PAs is enhanced by increased utilization of ornithine for PA biosynthesis by the transgene product; Pro and arginine biosynthesis are regulated independently of PAs and GABA biosynthesis; increased PA biosynthesis results in the increased assimilation of both nitrogen and carbon by the cells. These results focus on the network which can play a critical role in plant development and stress responses.

T-Spm is a structural isomer of Spm, widely distributed in the plant kingdom. It can play a role in repressing xylem differentiation by studies of its deficient mutant, acaulis5 (acl5), in Arabidopsis. Exogenous treatment with T-Spm reduced the expression of MONOPTEROS, an auxin response factor gene, which acts as a master switch for auxin-dependent procambium formation and its target genes. It was also demonstrated that T-Spm had an inhibitory effect on lateral root formation in wild-type seedlings of Arabidopsis plants, suggesting a strong modulatory effect on auxin signaling. Huang et al. (2017) applied Spd soaking treatment on sweet maize seeds and found that Spd application significantly increased endogenous Spd, gibberellins (GA), and ETH contents, and simultaneously reduced ABA concentration in embryos during seed imbibition. Spd contributed to fast seed germination, and high seed vigor of sweet corn might be closely related to the metabolism of hormones including GA, ABA, and ETH, and with the increase of H_2O_2 in the radical produced partly from Spd oxidation, playing an important role in cell membrane integrity and maintaining. To summarize these results, we can conclude that further studies are needed to elucidate the proper mechanism of PA homeostasis in plants in development and stress conditions.

In the following part of the chapter, several important abiotic stress responses are described in order to highlight the main function of exogenous PAs in inducing stress tolerance mechanisms.

22.3 POLYAMINES AND STRESS TOLERANCE

22.3.1 SALINITY STRESS

Salinity is one of the most dangerous abiotic stress factors affecting plant productivity worldwide and it causes drastic yield loss nowadays. Moreover, climate change can enhance this process, so effective progress is needed to induce salt tolerance of plants in a sustainable and environmentally friendly manner. Exogenous PAs are good candidates for treatments to alleviate the salt stress-induced processes. There are numerous studies which demonstrate that PAs can play a significant role in salt stress tolerance in plants (Bouchereau et al., 1999; Szepesi et al., 2009; Szepesi et al., 2011) and these are reviewed by Liu et al. (2015a).

Most of the experiments were done by the application of exogenous Spd. Zhang et al. (2013) reported that exogenous Spd had a beneficial role in nitrogen metabolism in tomato seedlings exposed to saline-alkaline stress. They examined two tomato cultivars with different sensitivities to mixed salinity-alkaline stress. Exogenous Spd alleviated the growth reduction to some extent, especially in the salt-sensitive cultivar. It is suggested that Spd treatment can relieve nitrogen metabolic disturbances caused by salinity-alkalinity stress and eventually plant growth (Zhang et al., 2013). By examining the protective role of Spd against saline-alkaline stress in tomato researchers have investigated by proteomic analysis coupled with bioinformatics that Spd can help tomato seedlings by modulating the defense mechanisms of plants and activating cellular detoxification, thereby protecting plants from oxidative damage induced by saline-alkaline stress (Zhang et al., 2015). Spd can also protect chrysanthemum seedlings against salinity stress damage positively affecting photosynthetic efficiency, reactive oxygen species (ROS) scavenging ability, and the control of ionic balance and osmotic potential (Zhang et al., 2016). Saha and Giri (2017) described in a molecular phylogenomic study by rice cultivars that exogenous Spd, in the presence or absence of salt stress, adjusts the intracellular PA pathways to equilibrate the cellular PAs that would have been attributed to plant salt tolerance. In turfgrass (Kentucky bluegrass (Poa pratensis L.)) cultivars, it was indicated that exogenous Spd might improve turfgrass quality and promote the salinity tolerance through reducing oxidative damages and increasing enzyme activity both at the protein and transcriptional levels (Puyang et al., 2015). The same research group indicated that exogenous Spd pretreatment could enhance salinity tolerance by increasing Pro levels and regulating ion and PA metabolism in Kentucky bluegrass under salinity stress (Puyang et al., 2016). Yin et al. (2016) studied silicon-induced salt tolerance in Sorghum bicolor L. and they suggested that PAs and the ETH precursor, ACC, are involved in Si-induced salt tolerance in sorghum, providing evidence that Si plays a role in mediating salt tolerance. Exogenous Spd could improve the

photosynthetic capacity through regulating the expression of Calvin cycle-related genes and the activity of key enzymes for CO_2 fixation, thus improving cucumber plants' tolerance to salinity, as described by Shu et al. (2014). T-Spm triggers metabolic and transcriptional reprogramming that promotes salt stress tolerance in Arabidopsis, by examining two loss-of-function mutants (atpao5-2 and atpao5-3) that exhibited constitutively higher T-Spm levels, with associated increased salt tolerance. Using global transcriptional and metabolomic analyses, it was revealed that stimulation of ABA and jasmonate (JA) biosynthesis and accumulation of important compatible solutes, such as sugars, polyols, and Pro, as well as TCA cycle intermediates, occurred in atpao5 mutants under salt stress (Zarza et al., 2017).

Tanou et al. (2014) analyzed the impact of exogenous PAs on the oxidative and nitrosative status in citrus plants exposed to salinity, adding new evidence for the interplay among PAs and reactive oxygen and nitrogen species. Expressions of both PA biosynthesis and catabolism-associated genes were systematically upregulated by PAs, oxidative status also was altered by PAs. Moreover, PAs could restore the NaCl-induced upregulation of NO-associated genes, such as nitrate reductase (NR), NADde, NOS-like, and AOX, along with GSNOR and NR activities. While protein carbonylation and tyrosine nitration are depressed by specific PAs, protein S-nitrosylation was elicited by all PAs. In this project, 271 S-nitrosylated proteins were identified that were commonly or preferentially targeted by salinity and individual PAs. Seed pretreatment is also a good candidate for improving salt tolerance of plants during the early developmental phase. Paul et al. (2017) investigated Spd induced salt tolerance mechanisms of three indica rice cultivars with different levels of salt tolerance and they found that seed pretreatment with Spd could alleviate oxidative damages to different extents in the salt-stressed plants by triggering the antioxidants and osmolytes. Another good example for Spd improved antioxidant mechanisms during salt stress is ginseng seedlings (Parvin et al., 2014). Nahar et al. (2016b) reported that in mung bean (Vigna radiata I.) PAs confer salt tolerance by reducing sodium uptake, improving nutrient homeostasis, antioxidant defense, and methylglyoxal (MG) detoxification systems. New evidence showed that overexpression of the SAMS, the key enzyme involved in the biosynthesis of the precursor of PAs, S-adenosylmethionine (SAM), induced salt tolerance in sugar beet M14 (Ma et al., 2017b). It was shown that PAs also can contribute to salinity tolerance in the symbiosis Medicago truncatula-Sinorhizobium meliloti by preventing oxidative damage (López-Gómez et al., 2017). Researchers suggested that PAs induced the expression of genes involved in BRs biosynthesis which support a crosstalk between PAs and BRs in the salt stress response of M. truncatula-S. meliloti symbiosis. Under salt stress, a decrease in polyamine biosynthesis and/or polyamine content has a strong negative effect on leaves as it induced the decline of photosynthetic energy conversion (ΦPSII) and increases salt stress sensitivity in rice seedlings (Yamamoto et al., 2017). Shu et al. (2015) reported investigating cucumber plants that Put regulates protein expression at transcriptional and translational levels by increasing endogenous PAs levels in thylakoid membranes, which may stabilize photosynthetic apparatus under salt stress. Li et al. (2015a) reported that exogenous spermidine also alleviated the stress-induced increases in malondialdehyde content, superoxide radical generation rate, chlorophyllase activity of tomatoes during salinity-alkaline stress, and expression of the chlorophyllase gene and the stress-induced decreases in the activities of antioxidant enzymes, antioxidants, and expression of the porphobilinogen deaminase gene. In addition, exogenous spermidine stabilized the chloroplast ultrastructure in stressed tomato seedlings. Li et al. (2016b) studied the effect of exogenous Spd on polyamine metabolism and salt tolerance in zoyziagrass cultivars exposed to short-term salinity stress. In both cultivars, they found positive correlations between polyamine biosynthetic enzymes (ADC, SAMDC), DAO, and antioxidant enzymes (SOD, POD, CAT), but negative correlations with H_2O_2 and MDA levels, and the Spd + Spm content were observed with an increase in the concentration of exogenous Spd. It can be concluded that exogenous Spd treatment can be an efficient inducer of salinity tolerance against short-term salinity stress. The same research group found that free PAs may be the primary factor influencing the variation of other PA forms during salt stress and applied exogenously Spd. Structural equation modeling (SEM) also indicated that ADC and PAO play a limited role in enhancing zoysiagrass salt tolerance via PA metabolism under salt stress (Li et al., 2017). Hu et al. (2012) investigated the effects of seed pretreatment with exogenous Spd in tomatoes under conditions of mixed salinity-alkalinity stress. They found that Spd application to seeds markedly suppressed the accumulation of free Put, but promoted an increase in free Spd and Spm concentrations, as well as soluble conjugated forms of Spd and insoluble-bound forms of Put in both cultivars. It can be concluded that exogenous Spd promotes the conversion of free Put into free Spd and Spm, and soluble conjugated forms and insoluble-bound forms of PAs under salinity-alkalinity

stress and therefore exogenous Spd treatment can regulate the metabolic status of PAs caused by salinity-alkalinity stress, and eventually enhance tolerance of tomato plants to salinity-alkalinity stress. In addition, Spd treatments have varying effects on different tolerant tomato cultivars. Later, Hu et al. (2014) examined the molecular details underlying PA-mediated photoprotective mechanisms in tomato seedlings under salinity-alkalinity stress. They found that seedlings treated with exogenous Spd had higher zeaxanthin (Z) contents than those without Spd under salinity-alkalinity stress and the chloroplast ultrastructure had a more ordered arrangement in seedlings treated with exogenous Spd than in those without Spd under salinity-alkalinity stress, indicating that exogenous Spd can alleviate the growth inhibition and thylakoid membrane photodamage caused by salinity-alkalinity stress. The Spd-induced accumulation of Z also may have an important role in stabilizing the photosynthetic apparatus but it needs to be further investigated. Sang et al. (2016) also investigated that how exogenous Spd could alleviate harmful effects of salinity. They combined a physiological analysis with iTRAQ-based comparative proteomics of cucumber (Cucumis sativus L.) leaves, treated with 0.1 mM exogenous Spd, 75 mM NaCl, and/or exogenous Spd. They found a total of 221 differentially expressed proteins involved in 30 metabolic pathways, such as photosynthesis, carbohydrate metabolism, amino acid metabolism, stress response, signal transduction, and antioxidant. Cucumber seedlings treated with Spd under salt stress had higher photosynthesis efficiency, upregulated tetrapyrrole synthesis, stronger ROS scavenging ability, and more protein biosynthesis activity than NaCl treatment, suggesting that these pathways may promote salt tolerance under high salinity.

Hu et al. (2016) studied the beneficial role of spermidine in tomato seedlings under salinity-alkalinity stress. They investigated the chlorophyll metabolism and D1 protein content, and they found that expression of chlorophyll biosynthesis gene porphobilinogen deaminase PBGD and psbA, which codes for D1 were enhanced, whereas the expression of Chlase, which codes for chlorophyllase was reduced when Spd was exogenously applied. They concluded that the protective effect of Spd on chlorophyll and D1 protein content during stress might maintain the photosynthetic apparatus, permitting continued photosynthesis and growth of tomato seedlings (Solanum lycopersicum cv. Jinpengchaoguan) under salinity-alkalinity stress.

In order to gain a better understanding of the cucumber responses to NaCl, Yuan et al. (2016) applied exogenously Put to alleviate salt-induced damage. They used comparative proteomic analysis and found that exogenous Put restored the root growth inhibited by NaCl, sixty-two differentially expressed proteins implicated in various biological processes were successfully identified by MALDI-TOF/TOF MS and the four largest categories included proteins involved in defense response (24.2%), protein metabolism (24.2%), carbohydrate metabolism (19.4%), and amino acid metabolism (14.5%). Exogenous Put upregulated most identified proteins involved in carbohydrate metabolism and proteins involved in defense response, and protein metabolism was differently regulated by Put, which indicated the role of Put in stress resistance and proteome rearrangement. Put also improved endogenous PAs contents by regulating the transcription levels of key enzymes in polyamine metabolism. Their results suggest that Put may alleviate NaCl-induced growth inhibition through degradation of misfolded/damaged proteins, activation of stress defense, and the promotion of carbohydrate metabolism to generate more energy. The effects of foliar spray of Put (8 mM) on chlorophyll (Chl) metabolism and the xanthophyll cycle in cucumber seedlings were investigated under saline conditions of 75 mM NaCl by Yuan et al. (2017). They found that Put might improve Chl metabolism and the xanthophyll cycle by regulating enzyme activities and mRNA transcription levels in a way that improved the salt tolerance of cucumber plants because Put treatment alleviated decreases in Chl contents and in a size of the xanthophyll cycle pool under salt stress. Recently, Takács et al. (2017) provided new evidence of the importance of the PAOs in salt stress by applying a PAO specific inhibitor, MDL-72,527, in tomato plants during salt stress.

22.3.2 High Temperature Stress

As the climate change means a stronger threat to the plants, it is necessary to improve the thermotolerance of plants under high temperature stress because it impacts significantly on the yield and growth of plants.

Kumar et al. (2014) studied the effect of exogenous applied Put at pre-anthesis in order to enhance the thermotolerance of wheat (Triticum aestivum L.). They observed that Put before heat stress significantly enhanced the transcript levels of SOD, CAT, cytoplasmic APX (cAPX), and peroxisomal (pAPX) in both the thermotolerant and susceptible cultivars, but the activities of antioxidant enzymes (SOD, CAT, APX and GR), as well as accumulation of antioxidants (AsA and total thiol content), were higher in thermotolerant cultivars than susceptible in response to the combined treatment

of Put and heat stress. Put treatment at pre-anthesis thus modulated the defense mechanism responsible for the thermotolerance capacity of wheat under heat stress. Researchers suggested that elicitors like Put, therefore, need to be further studied for temporarily manipulating the thermotolerance capacity of wheat grown under the field conditions in view of the impending global climate change.

Mellidou et al. (2017) investigated the potential role of apoplastic PAO in the improvement of thermotolerance in Nicotiana tabacum. They used genetically modified N. tabacum plants in this study with altered PA/H_2O_2 homeostasis due to over/underexpression of the ZmPAO gene (S-ZmPAO/AS-ZmPAO, respectively) and these plants were assessed under heat stress (HS). Underexpression of ZmPAO was correlated with increased thermotolerance of the photosynthetic machinery and improved biomass accumulation, accompanied by enhanced levels of the enzymatic and non-enzymatic antioxidants, whereas ZmPAO overexpressors exhibited significant impairment of thermotolerance, but the precise mechanism of how ZmPAO can influence the thermotolerance needs to be further investigated.

Cheng et al. (2012) provided evidence that Spd affects the transcriptome responses to high temperature stress in ripening tomato fruit. They used an Affymetrix tomato microarray to evaluate changes in gene expression in response to exogenous Spd (1 mM) and high temperature (33/27°C) treatments in tomato fruits at the mature green stage. Their results showed that of the 10,101 tomato probe sets represented on the array, 127 loci were differentially expressed in high temperature-treated fruits, compared with those under normal conditions, functionally characterized by their involvement in signal transduction, defense responses, oxidation-reduction, and hormone responses. However, only 34 genes were upregulated in Spd-treated fruits as compared with non-treated fruits, which were involved in primary metabolism, signal transduction, hormone responses, transcription factors, and stress responses. Meanwhile, 55 genes involved in energy metabolism, cell wall metabolism, and photosynthesis were downregulated in Spd-treated fruits. To summarize the findings, Spd can be effective in the regulation of tomato fruit response to high temperatures during the ripening stage.

Sang et al. (2017) investigated the functions of PAs in high-temperature stress responses in tomato plants, and the effects of exogenous Spd were determined in tomato leaves using two-dimensional electrophoresis and MALDI-TOF/TOF MS. A total of 67 differentially expressed proteins were identified in response to high-temperature stress and/or exogenous Spd, which were grouped into different categories according to biological processes. They revealed that exogenous Spd upregulated most identified proteins involved in photosynthesis, implying an enhancement in photosynthetic capacity, and also physiological analysis showed that Spd could improve net photosynthetic rate and the biomass accumulation. They provided evidence that an increased high-temperature stress tolerance by exogenous Spd would contribute to the higher expressions of proteins involved in cell rescue and defense, and Spd regulated the antioxidant enzymes activities and related genes expression in tomato seedlings exposed to high temperatures. This proteomic approach provides valuable insight into how to improve the high-temperature stress tolerance in the global warming epoch indicating that it should be a potent application of Spd to improve the thermotolerance of other crop plants.

Glaubitz et al. (2015) investigated how high night temperature (HNT) impacts the TCA cycle, amino acid, and PA biosynthetic pathways in rice in a sensitivity-dependent manner. They studied twelve cultivars with different HNT sensitivity and compared their metabolic changes in the vegetative stage under HNT. They reported that central metabolism, especially TCA cycle and amino acid biosynthesis, were strongly affected particularly in sensitive cultivars. Levels of several metabolites were correlated with HNT sensitivity. Interestingly, Put, Spd, and Spm showed an increased abundance in sensitive cultivars under HNT conditions. Concomitantly, increased expression levels of ADC2 and ODC1, genes encoding enzymes catalyzing the first committed steps of Put biosynthesis, were restricted to sensitive cultivars under HNT. Additionally, transcript levels of eight PA biosynthesis genes were correlated with HNT sensitivity. It can be concluded that responses to HNT in the vegetative stage result in distinct differences between differently responding cultivars with a dysregulation of central metabolism and an increase of PA biosynthesis restricted to sensitive cultivars under HNT conditions and a pre-adaptation of tolerant cultivars already under control conditions with higher levels of potentially protective compatible solutes.

Nowadays, Shen et al. (2016) identified an important function of the PA transporter lower expression of heat responsive gene 1 (LHR1) in heat-inducible gene expression in A. thaliana. The LHR1 mutant showed reduced induction of the luciferase gene in response to heat stress and was more sensitive to high temperature than the wild-type. By map-based cloning, they identified that the LHR1 gene encodes the PA transporter polyamine uptake transporter 3 (PUT3) localized in the plasma membrane and the LHR1/PUT3 is required for the uptake of extracellular PAs and plays an important role in stabilizing the mRNAs of several crucial heat stress-responsive genes under high temperatures. Their findings revealed an important heat stress response and tolerance mechanism involving PA influx which modulates mRNA stability of heat-inducible genes under heat stress conditions.

22.3.3 Cold Stress

Low temperature stress means a combined stress type for plants, not just a disturbance in the water homeostasis, it also needs a synthesis of stress-responsive compounds. PAs can alleviate the low temperature induced stress injuries. Zhuo et al. (2017) found that a cold responsive ERF from Medicago falcate confers cold tolerance up-regulating PA turnover, antioxidant protection, and Pro accumulation.

Sheteiwy et al. (2017) investigated the effects of seed priming with Spd and 5-aminolevulinic acid (ALA) on seed PA metabolism during chilling stress in rice cultivars. They found that germination percentage, seedling growth, and seedling vigor index was decreased under chilling stress, but these physiological parameters were improved by Spd and ALA priming in both studied cultivars as compared with unprimed seeds. Their results also showed that chilling stress significantly improved SOD, peroxidase (POD), APX and GPX, and further enhancement was observed by Spd and ALA-primed seeds. The enzymes involved in the PAs biosynthesis, ADC, ODC and S-adenosylmethionine decarboxylase (SAMDC), were improved by a priming treatment with Spd and ALA. The relative expressions of genes encoding enzymes involved in PAs biosynthesis increased by Spd and ALA priming. Additionally, priming treatment improved leaf cell and grain structure as compared with the unprimed seeds.

Nahar et al. (2015) reported that exogenous Spd can alleviate low temperature (LT) injury in mung bean (Vigna radiata L.) seedlings by modulating AsA-GSH and glyoxalase pathways. In low temperature affected seedlings, exogenous pretreatment of Spd significantly increased the contents of non-enzymatic antioxidants of the AsA-GSH cycle, which include AsA and GSH. Exogenous Spd decreased dehydroascorbate (DHA), increased AsA/DHA ratio, decreased glutathione disulfide (GSSG), and increased GSH/GSSG ratio under LT stress. Activities of AsA-GSH cycle enzymes such as APX, monodehydroascorbate reductase (MDHAR), DHAR, and GR increased after Spd pretreatment in LT affected seedlings, reducing oxidative stress. Protective effects of Spd also were shown from the reduction of

MG toxicity by improving glyoxalase cycle components, and by maintaining osmoregulation, water status, and improved seedlings growth, suggesting that Spd can alleviate LT injury by affecting AsA-GSH and glyoxalase pathways. Li et al. (2014a) suggested that co-application of SNP and Spd could be an effective approach for the survival of plants under chilling stress through ROS detoxification. They provided evidence that both SNP and Spd increased ginger seedlings' tolerance to chilling stress, by protecting photosystem II (PSII), keeping high level of unsaturated fatty acids, regulating ROS detoxification, and co-application of SNP and Spd was more powerful in improving chilling stress tolerance of ginger, indicating that NO and PAs can stimulate accumulation each other to synergistically enhance chilling tolerance.

Another study which can suggest a relationship between NO and PAs in response to cold stress was investigated in tomato plants by Diao et al. (2016). They investigated the effects of Put and Spd on NO generation and the function of Spd-induced NO in the tolerance of tomato seedling under chilling stress. Their results showed that Spd increased NO release via the NO synthase (NOS)-like and NR enzymatic pathways in the seedlings, whereas Put had no such effect. Compared to chilling treatment alone, Spd enhanced the gene expressions of SOD, POD, CAT, and APX, and their enzyme activities in tomato leaves, but a scavenger or inhibitor of NO abolished Spd-induced chilling tolerance and blocked the increased expression and activity due to Spd of these antioxidant enzymes in tomato leaves under chilling stress. The results proved data that NO-induced by Spd plays a crucial role in tomatoes' response to chilling stress. Diao et al. (2017) examined the interplay between PAs and signal molecules in tomatoes under chilling stress. They elucidated the crosstalk among PAs, ABA, NO, and H_2O_2 under chilling stress conditions using tomato seedlings. During chilling stress, the application of Spd and Spm elevated NO and H_2O_2 levels, enhanced nitrite reductase (NiR), NOS-like, and PAO activities, and upregulated LeNR relative expression, but did not influence LeNOS1 expression. By contrast, Put treatment had no obvious impact. Seedlings pretreated with SNP, showed elevated Put and Spd levels throughout the treatment period, consistent with increased expression in leaves of genes encoding ADC (LeADC, LeADC1), ODC (LeODC), and Spd synthase (LeSPDS) expressions in tomato leaves throughout the treatment period. Exogenously applied SNP did not increase the expression of genes encoding SAMDC (LeSAMDC) and Spm synthase (LeSPMS), consistent with the observation that Spm levels remained constant under chilling stress and during the recovery period.

In contrast, exogenous Put significantly increased the ABA content and the 9-cis-epoxycarotenoid dioxygenase (LeNCED1) transcript level. It is concluded that, under chilling stress, Spd and Spm enhanced the production of NO in tomato seedlings through an H_2O_2-dependent mechanism, via the NR and NOS-like pathways. ABA is involved in Put-induced tolerance to chilling stress, and NO could increase the content of Put and Spd under chilling stress.

22.3.4 DROUGHT AND OSMOTIC STRESS

Drought is also a multifaceted stress condition that inhibits crop growth and reduces yield in agriculture. PAs can defend against drought in plants in different ways, e.g. seed soaking, seed priming, or foliar spray can be optimal treatments against drought conditions (Bala et al., 2016). Ebeed et al. (2017) reported that Put significantly regulates the endogenous PAs by both methods of application such as foliar spraying or seed priming, however, Spm and a mixture of Put and Spm could positively regulate the endogenous PAs and the biosynthetic gene expression by foliar spraying rather than seed priming. This study provided new evidence suggesting that maintenance of water economy through the stabilized cellular structure is an important strategy of drought tolerance by PAs in wheat.

Kotakis et al. (2014) tried to change the bioenergetics of the leaf discs before the exposure to osmotic stress only by exogenously supplied Put, in order to quickly enhance the tolerance against abiotic stress. Treatment with 1 mM exogenous Put 1 h before polyethylene-glycol additions protects the photochemical capacity and inhibits loss of water, confirming the key role of Put in the modulation of plant tolerance against osmotic stress, and indicating that Put is accumulated in lumen during light reactions and may act as a permeable buffer and an osmolyte.

Yin et al. (2014) investigated the effect of exogenous Spd and Spm on drought-induced damage to seedlings of Cerasus humili, as relative water content (RWC), malondialdehyde content, relative electrolyte leakage (EL), superoxide anion generation rate, H_2O_2, endogenous PAs, antioxidant enzymes (SOD and POD) activities, PA-biosynthetic enzymes, (ADC and ODC, SAMDC) activities, as well as photosynthetic parameters, were measured in greenhouse cultured seedlings of C. humili. They found that exogenous Spd or Spm enhanced the activities of ADC, ODC, and SAMDC. The pretreatment with Spd or Spm prevents oxidative damage induced by drought, and the protective effect of Spd was found to be greater than that of Spm. Nowadays, it presents a big problem when crops meet severe water deficits which can limit the photo-assimilate supply during grain filling stages. Yang et al. (2017) investigated the role of ETH and PA in mitigating severe water deficit induced filling inhibition. It was reported that increased ACC and Put concentrations and their biosynthesis-related gene expression reduced Spd biosynthesis under severe water deficit conditions. Peng et al. (2016) provided new evidence that NO has a role in Spd-induced drought tolerance. They studied white clover seedlings under drought stress induced by PEG-6000 solution. Exogenously applied Spd or SNP improved drought tolerance, as indicated by better phenotypic appearance, increased RWC, and decreased EL, and malondialdehyde (MDA) content in leaves as compared to untreated plants. It was also demonstrated that PEG induced an increase in the generation of NO in cells and significantly improved activities of NR and NOS and these responses could be blocked by pretreatment with a Spd biosynthetic inhibitor, dicyclohexyl amine (DCHA), and then reversed by the application of exogenous Spd. It was reported that Spd-induced antioxidant enzyme activities and gene expression also could be effectively inhibited by an NO scavenger as well as inhibitors of NR and NOS, so it can be concluded that Spd was involved in drought stress-activated NR and NOS pathways associated with NO release, which mediated antioxidant defense and thus contributed to drought tolerance in white clover (Peng et al., 2016). Ma et al. (2017a) evaluated the transcriptome changes of creeping bentgrass, which is an important cool-season turfgrass species sensitive to drought. They exposed the plants to drought and exogenous Spd application and used RNA sequencing (RNA-Seq) (see in Table 22.1). This

study is the first which can elucidate that the transcript changes of creeping bentgrass that enriched or differentially expressed transcripts due to drought stress and/or Spd application were primarily associated with energy metabolism, transport, antioxidants, photosynthesis, signaling, stress defense, and cellular response to water deprivation, providing molecular markers in responses to drought and Spd. There is relatively little data about the effect of exogenous PA treatment on seed germination under drought stress. Liu et al. (2016b) studied six wheat genotypes differing in drought tolerance and used exogenous PAs for seed soaking. It was reported that the free Spd accumulation in seeds during the seed germination period favored wheat seed germination under drought stress; however, the free Put accumulation in seeds during the seed germination period may work against wheat seed germination under drought stress. This study clearly revealed that external Spd and Spm significantly increased the endogenous indole-3-acetic acid (IAA), zeatin (Z) + zeatin riboside (ZR), ABA, and GA contents in seeds and accelerated the seed starch degradation and increased the concentration of soluble sugars in seeds during seed germination, suggesting that the effects of Spd and Put on seed germination under drought notably related to hormones and starch metabolism.

Grain filling of cereals is a very sensitive process to drought and PAs are strongly associated with plant resistance against drought. Liu et al. (2016a) investigated how PAs are involved in regulating wheat grain filling under drought stress. They found that the higher activities of SAMDC and Spd synthase in grains promotes the synthetic route from Put to Spd and Spm and notably increases the free Spd and Spm concentrations in grains,

TABLE 22.1

List of Proteomic and Transcriptomic Analyses of Proteomes of Plants Treated with Different Exogenously Applied Polyamines

Plant Species	Stress Factor	Type of PA	Analysis	References
Tomato	High temperature	Spd, 1 mM	MALDI-TOF/TOF MS	Sang et al. (2017)
Cucumber	Salinity stress	Spd, 0.1 mM	iTRAQ-based comparative proteomics	Sang et al. (2016)
Sour orange	Salinity	Put, 1 mM Spd, 1 mM Spm, 1 mM	Nano-LC/MS/MS	Tanou et al. (2014)
Tomato	Salinity-alkaline stress	Spd, 0.25 mM	MALDI-TOF MS	Zhang et al. (2015)
Cucumber	Salinity stress	Put, 0.8 mM	MALDI-TOF/TOF MS	Yuan et al. (2016)
Cucumber	Salinity stress	Spd	MALDI-TOF/TOF MS	Li et al. (2013)
Cucumber	Salinity stress	Put, 8 mM	MALDI-TOF/TOF MS	Shu et al. (2015)
Creeping bentgrass	Drought stress	Spd, 500 µM L-1	RNA-Seq	Ma et al. (2017a)
Tomato	High temperature stress	Spd, 1 mM L-1	Affymetrix tomato microarray	Cheng et al. (2012)

which promotes grain filling and drought resistance in wheat. Also, they provided evidence that the effect of PA on the grain filling of wheat under drought stress was closely related to the endogenous ETH, Z+ZR, and ABA. Interestingly, Spd and Spm significantly increased the Z+ZR and ABA concentrations and decreased the ETH evolution rate in grains, which promoted wheat grain filling under drought. However, Put significantly increased the ETH evolution rate, which led to excessive ABA accumulation in grains, subsequently aggravating the inhibition of drought on wheat grain filling. It can be concluded that the interaction of hormones, rather than the action of a single hormone, was involved in the regulation of wheat grain filling under drought. A challenge for the future will be to investigate this intimate interaction of hormones and PAs in order to improve the grain filling process of cereal crops.

Juzoń et al. (2017) reported that PAs are also involved in yellow lupin tolerance and may play a protective function under soil drought conditions. A drought tolerant and sensitive yellow lupin were exposed to soil drought for 2 weeks and the half of them were additionally sprayed with a solution of PA biosynthesis inhibitor—dl-α-difluoromethylarginine (DFMA). Simultaneously subjecting plants to soil drought and DFMA treatment caused in the tolerant genotype a decline in the number of pods and seeds per plant and seeds dry weight per plant (64, 50 and 54%, respectively), while in the sensitive cultivar a reduction of the number of pods per plant and seeds per pod (32 and 27%, respectively) was observed, confirming that PA biosynthesis is a significant process in drought tolerance. Liu et al. (2017a) studied the responses of PAs and antioxidants in a centipedegrass mutant compared to the wild-type. Centipedegrass (Eremochloa ophiuroides [Munro] Hack.) is an important warm-season turfgrass species, its physiological adaptation is not known precisely to drought stress. Higher levels of antioxidant enzymes activities, non-enzymatic antioxidants, and PAs including Put, Spd, and Spm were more well-maintained in the mutant 22-1 than in WT plants. They revealed that SOD, CAT, APX, and GR activities and ascorbic acid (AsA) content were significantly correlated with both Put and Spd levels, and GSH level was correlated with Put during drought stress. The researchers exogenously applied Put, Spd, and Spm, which could increase drought tolerance and activities of SOD, CAT, APX, and GR in WT plants. The results proved that higher levels of PAs and antioxidant defense systems are associated with the elevated drought tolerance in 22–1, which may improve protection on photosynthesis against drought-induced oxidative damage. Li et al. (2014b) studied the effect

of exogenous Spd on white clover seed germination under water stress induced by PEG 6000. They reported that seed priming with Spd improved seed germination percentage, germination vigor, germination index, root viability and length, and shortened mean germination time under different water stress conditions. They suggested that improved starch metabolism was considered a possible reason for this seed invigoration, since seeds primed with Spd had significantly increased α-amylase/β-amylase activities, reducing sugar, fructose, and glucose content and the transcript level of the β-amylase gene but not the transcript level of the α-amylase gene. Moreover, the physiological effects of exogenous Spd were reflected by lower lipid peroxidation levels, better cell membrane stability and significant higher seed vigor index in seedlings, showing an improvement in seed tolerance to water stress during germination. It can be concluded that exogenous Spd can act a stress-responsive activator during seed germination. Li et al. (2015b) also investigated the effect of the tetramer Spm on drought tolerance in white clover. They found that exogenously applied Spm was effective in amelioration of negative effects induced by drought stress in drought tolerant and also in sensitive cultivars. Accumulation of more water-soluble carbohydrates (WSC), sucrose, fructose, and sorbitol after exogenous Spm treatment in both cultivars under drought stress was reported suggesting that ameliorating drought stress through exogenously applied Spm may be associated with increased carbohydrate accumulation and dehydrins synthesis, but there are some differences between drought-susceptible and resistant white clover cultivars related to Spm regulation of dehydrin expression.

Li et al. (2016c) investigated the effect of exogenous Spd on drought stress tolerance in white clover. After iTRAQ-based proteomic analyses, they found that Spd treatment interacted with GA and ABA signaling pathways and maybe this alteration is in association with improved drought tolerance such as delay of water-deficit development, improved photosynthesis and water use efficiency, and lower oxidative damage. In the future, it would be worthwhile to reveal the exact mechanism of this interaction in order to induce drought tolerance in other crops. Caliskan et al. (2017) investigated the possibility of exogenous application of PAs on maize under PEG-induced drought stress in order to understand the effects of PAs on leaf rolling response and the role of the antioxidant system in the delayed LR process. They reported that exogenous applications of PAs prevented water loss and delayed LR in comparison with the control (seedlings treated only with PEG). Non-enzymatic antioxidant compounds, such as AsA, GSH, and endogenous

PA levels increased as a result of PA applications, while the H_2O_2 content was lower in PA-pretreated plants than in the control in both cultivars. It is suggested that exogenous PAs may have a H_2O_2-mediated regulatory role in the LR process through the induction of antioxidant machinery, and a stimulated antioxidant system decreases oxidative stress damage from excess accumulation of H_2O_2, thus delaying LR, confirming that maize cultivars with late LR by PA applications may be provided as an opportunity for improving yield potential in drought-prone areas.

22.3.5 HEAVY METAL STRESS

Heavy metal stress is one of the most important and dangerous stress types of plants. PAs and related mechanisms can induce some alleviation processes which is recommended in phytoremediation conditions. Yu et al. (2015) reported that elevation of ADC-dependent Put production enhances aluminum (Al) tolerance by decreasing Al retention in root cell walls of wheat suggesting that ADC-dependent Put accumulation plays an important role in providing protection against Al toxicity in wheat plants through decreasing cell wall polysaccharides and increasing the degree of pectin methylation, thus decreasing Al retention in the cell walls. The antioxidative stress response of free-floating aquatic fern (Salvinia natans Linn.) was studied by Mandal et al. (2013) under an increasingly toxic amount of Al and its modulation by exogenous application of PA. In this study, it was revealed that the application of Put of aquatic fern Salvinia natans Linn against Al toxicity, improved the overall antioxidative response, and thus would make it a better candidate to be used as a hyper accumulator of Al and other toxic metals. Another aquatic plant fern (Marsilea minuta Linn.) was investigated under cadmium (Cd) stress and supplied by exogenously Spd (Das et al., 2013). Das et al. (2013) reported that exogenous Spd was effective in alleviating the salt-induced damage by lowering Cd absorption.

Nahar et al. (2017b) investigated the mechanism of PA-induced Al tolerance in mung beans by examining the antioxidant defense and MG detoxification systems. They prove that exogenous Spd positively modulated the endogenous Pas' level and regulated the osmoprotectant molecule (Pro), Spd also improved plant water status under Al stress. Exogenous Spd was potent in preventing the breakdown of Al-induced photosynthetic pigment and in improving growth performances under Al stress. The mechanism by which Spd enhances antioxidant and glyoxalase components might be studied extensively. Spd-induced protection of photosynthetic pigment from

damages and growth enhancement was remarkable and recommended for a further detailed study to understand the mechanism.

Zhu et al. (2016) provided new evidence that Put plays an important role in alleviating Fe deficiency. It can be concluded that Put is involved in the remobilization of Fe from root cell wall hemicellulose in a process dependent on NO accumulation under a Fe-deficient condition in Arabidopsis. Furthermore, Tang et al. (2017) revealed that exogenous Spd enhanced Pb tolerance in Salix matsudana by promoting Pb accumulation in roots and Spd, NO, and antioxidant system levels in leaves. The results suggested that S. matsudana could accumulate a high level of Pb in the roots, and exogenous Spd could enhance S. matsudana tolerance to Pb by the synergistic promotion of Pb accumulation in the roots and the levels of Spd, NO, and antioxidants in the leaves.

Similarly, Legocka et al. (2015) reported that in greening barley leaves exposed to Pb-stress, the changes in free, thylakoid- and chromatin-bound PAs play some role in the functioning of leaves or plants in heavy metal stress conditions. Cd stress causes several negative physiological, biochemical, and structural changes due to the oxidative stress caused through the generation of ROS, leading to a reduction in plant growth.

Nahar et al. (2016c) investigated the physiological and biochemical mechanisms of Spm-induced Cd stress tolerance in mung bean (V. radiata L.) seedlings. They found that the cytotoxicity of MG wasalso reduced by exogenous Spm because it enhanced glyoxalase system enzymes and components. Through osmoregulation, Spm maintained a better water status of Cd-affected mung bean seedlings. Spm prevented the Chl damage and increased its content. Exogenous Spm also modulated the endogenous free Pas' level, which might have arole in improving physiological processes including antioxidant capacity, osmoregulation, and Cd and MG detoxification capacity. The overall Spm-induced tolerance of mung bean seedlings to Cd toxicity was reflected through the improved growth of mung bean seedlings.

Rady and Hemida (2015) reported that presoaking wheat seeds in either PA, increased the seedling growth and the activities of antioxidant enzymes compared to the control, but other attributes were slightly affected. Under Cd stress, presoaking seeds in either PA significantly increased seedling growth, membrane stability index, RWC, concentrations of protein, starch, AsA, total GSH, Spm and Spd, and the activities of SOD and CAT. In contrast, EL, concentrations of Pro, total soluble sugars, MDA, H_2O_2 and Cd2+, and the activities of POD and APX were reduced compared to the control. This study reveals more evidence for the potential of Spd or

Spm to alleviate the harmful effects of Cd stress and offer an opportunity to increase the resistance of wheat seedlings to growth under Cd stress conditions.

Tajti et al. (2018) comparatively studied the effects of Put and Spd pretreatment on Cd stress in wheat as seed soaking or hydroponic conditions. In this study, the increased Put content was also correlated with the highest accumulation of Cd, salicylic acid, and Pro contents in plants treated with a combination of Spd and Cd.

Pál et al. (2017) suggested that PAs may have a substantial influence on PC synthesis at several levels under Cd stress in rice. They found that Put pretreatment enhanced the adverse effect of Cd, while the application of a Put synthesis inhibitor, 2-(difluoromethyl)-ornithine, reduced it to a certain extent and these differences were associated with increased PA content, more intensive PA metabolism, but decreased thiol and PC contents.

Scoccianti et al. (2013) demonstrated that the diamine Put was able to improve pollen tolerance of Actinidia deliciosa to chromium (Cr) metal stress through different mechanisms, mostly depending upon the Cr species, namely via reduced metal uptake or by substituting for calcium.

22.3.6 Combination of Stress Conditions

High temperature and drought stress often occur in combination, and due to global climate change, this kind of phenomenon occurs more frequently and severely, which exerts devastating effects on plants.

Nahar et al. (2017a) investigated how exogenous pretreatment of Spm enhances mung bean seedlings' tolerance to high temperature and drought stress individually and in combination. They found that Spm pretreatment maintained the AsA and GSH levels high, and upregulated the activities of SOD, CAT, GPX, DHAR, and GR which were vital for imparting ROS-induced oxidative stress tolerance under HT and/or drought stress. It was demonstrated that Spm pretreatment modulated endogenous PA levels and osmoregulation and restoration of plant water status were other major contributions of Spm under HT and/or drought stress. They revealed that preventing photosynthetic pigments and improving seedling growth parameters, Spm further confirmed its influential roles in HT and/or drought tolerance.

The roles of Pro and PAs in the drought stress responses of tobacco plants were investigated by Cvikrová et al. (2013) in tobacco plants which overproduce Pro by a modified gene for the Pro biosynthetic enzyme Δ1-pyrroline-5-carboxylate synthetase (P5CSF129A). Drought stress significantly reduced the levels of Spd and Put in the leaves and roots relative to those for controls, and increased the levels of Spm and diaminopropane

(Dap, formed by the oxidative deamination of Spd and Spm). Spd levels may have declined due to its consumption in Spm biosynthesis and/or oxidation by PAO to form Dap, which became more abundant during drought stress. During the rewatering period, the plants' Put and Spd levels recovered quickly, and the activity of the PA biosynthesis enzymes in their leaves and roots increased substantially; this increase was more pronounced in transformants than WT plants. The high levels of Spm observed in drought-stressed plants persisted even after the 24 h recovery and rewatering phase.

22.3.7 Other Stress Types

Sengupta and Raychaudhuri (2017) reported that treatment with Put could partially alleviate oxidative damage caused by gamma rays in Vigna radiata. Treatment with PAs may be a protective compound against gamma rays, e.g. in space. In a study by Li et al. (2016a), the possible relationship was investigated between Put, hydrogen sulfide (H_2S), and H_2O_2 as well as the underlying mechanism of their interaction in reducing UV-B induced damage in hulless barley. They found that H_2S mediated H_2O_2 accumulation, and H_2O_2 induced the accumulation of UV-absorbing compounds and maintained redox homeostasis under UV-B stress, thereby increasing the tolerance of hulless barley seedlings to UV-B stress.

Chen et al. (2014) examined the maize leaf proteome during flooding stress, and they found that the identified proteins were related to energy metabolism and photosynthesis, PCD, phytohormones, and PAs. For better characterization, the role of translationally controlled tumor protein (TCTP) in PCD during a stress response and mRNA expression was examined in different plants: in tobacco, rice, and Arabidopsis by stress-induced PCD. S-adenosylmethionine synthase 2 (SAMS2) and SAMDC mRNA expression were also increased, but ACC synthase (ACS) and ACC oxidase (ACO) mRNA expression were not found in maize leaves following flooding, however, ETH and PA concentrations were increased in response to flooding treatment in maize leaves.

Yiu et al. (2009) studied the effect of exogenously applied Spd and Spm on Welsh onion (Allium fistulosum L) plants during waterlogging. Pretreatment with 2 mM of Spd and Spm was effective to maintain the balance of water content in plant leaves and roots under flooding stress by maintaining antioxidant enzyme activities; an improvement in the protective effect of Spm was discovered. Spd treatment can be a good application to protect rice submerged at the tilling stage because, as Liu et al. (2015b) demonstrated, exogenous Spd alleviates oxidative damage and reduces yield loss. It is worth investigating in other crops the alleviation effects on submerging damage which can negatively affect crop yield.

Ca(NO3)2 stress is one of the most serious constraints to plants production and limits the plant's growth and development. Application of PAs is a convenient and effective approach for enhancing plant salinity tolerance. Du et al. (2016) investigated cucumber plants under Ca(NO3)2 stress and used Spd to mitigate the stress-induced photosynthetic damage. The leaf-applied Spd (1 mM) treatment alleviated the reduction in growth and photosynthesis in cucumber caused by Ca(NO3)2 stress. They found that exogenous Spd could effectively accelerate nitrate transformation into amino acids and improve cucumber plant photosynthesis and C assimilation, thereby enhancing the ability of the plants to maintain their C/N balance, and eventually promote the growth of cucumber plants under Ca(NO3)2 stress.

22.4 FUTURE PERSPECTIVES AND QUESTIONS

The evidence provided in this Chapter clearly shows that PAs have many diverse and important functions in inducing different abiotic stress tolerances. Although much work focused on the application of exogenous PAs, it is only recently that novel, innovative approaches were used to accelerate the investigation of crop improvement adjusting PA homeostasis. In addition, the light-dependent mechanisms of PA biosynthesis and catabolism enzymes have now been questioned and we need to investigate how other signal pathways can influence the PA metabolism. Many questions with regard to the transporters of PAs remain to be answered because up until this point just a few PA transporters were described. Recent evidence suggests that PAs can act as hub molecules (Sequera-Mutiozabal et al., 2017) and this suggests that PAs have a prominent role to play in triggering stress tolerance mechanisms in plants. Using novel molecular and biotechnological approaches can gain a better understanding the precise mode of action of PAs and related signal pathways. In addition, investigating the PA homeostasis of wild crop relatives and tolerant cultivars can help us to decipher those mechanisms which are important to improve stress tolerance by PAs. There is a lack of experiments in field with exogenously applied PAs, which may provide information about how PAs are good growth regulators to induce stress tolerance in agriculture but to achieve this, our knowledge about the different types of treatment techniques (seed soaking, pretreatment, foliar spray, etc.) should be increased by a combination of different stress types (e.g. light and drought or elevated carbon dioxide).

ACKNOWLEDGMENTS

The cited own work was supported by the European Union and the State of Hungary, co-financed by the European Social Fund in the framework of TÁMOP 4.2.4.A/2-11-1-2012-0001 'National Excellence Program' and the Hungarian Scientific Research Fund K101243 project.

REFERENCES

Ahmed, S., Ariyaratne, M., Patel, J., Howard, A.E., Kalinoski, A., Phuntumart, V., Morris, P.F. (2017) Altered expression of polyamine transporters reveals a role for spermidine in the timing of flowering and other developmental response pathways. *Plant Science* 258: 146–155.

Alabdallah, O., Ahou, A., Mancuso, N., Pompili, V., Macone, A., Pashkoulov, D., Stano, P., et al. (2017) The arabidopsis polyamine oxidase/dehydrogenase 5 interferes with cytokinin and auxin signaling pathways to control xylem differentiation. *Journal of Experimental Botany* 68(5): 997–1012.

Bala, S., Asthir, B., Bains, N.S. (2016) Syringaldazine peroxidase stimulates lignification by enhancing polyamine catabolism in wheat during heat and drought stress. *Cereal Research Communications* 44(4), 561–571.

Bouchereau A., Aziz A., Larher F., Martin-Tanguy J. (1999) Polyamines and environmental challenges: recent development. *Plant Science* 140: 103–125.

Cai, G., Sobieszczuk-Nowicka, E., Aloisi, I., Fattorini, L., Serafini-Fracassini, D., Del Luca, S. (2015) Polyamines are common players in different facets of plant programmed cell death. *Amino Acids* 47(1): 27–44.

Caliskan, N.K., Kadioglu, A., Güler, N.S. (2017) Exogenously applied polyamines ameliorate osmotic stress-induced damages and delay leaf rolling by improving the antioxidant system in maize genotypes. *Turkish Journal of Botany* 41: 563–574.

Chen, Y., Chen, X., Wang, H., Bao, Y., Zhang, W. (2014) Examination of the leaf proteome during flooding stress and the induction of programmed cell death in maize. *Proteome Science* 12: 33.

Chen, H., Cao, Y., Li, Y., Xia, Z., Xie, J., Carr, J.P., Wu, B., et al. (2017) Identification of differentially regulated maize proteins conditioning sugarcane mosaic virus systemic infection. *New Phytologist* 215(3): 1156–1172. Doi: 10.1111/nph.14645.

Cheng, L., Sun, R.R., Wang, F.Y., Peng, Z., Kong, F.L., Wu, J., Cao, J.S., et al. (2012) Spermidine affects the transcriptome responses to high temperature stress in ripening tomato fruit. *Journal of Zhejiang University-Science B* 13(4): 283–297.

Cvikrová, M. Gemperlová, L., Martincová, O., Vanková, R. (2013) Effect of drought and combined drought and heat stress on polyamine metabolism in proline-over-producing tobacco plants. *Plant Physiology and Biochemistry* 73: 7–15.

Dalton, H.L., Blomstedt, C.K., Neale, A.D., Gleadow, R., DeBoer, K.D., Hamill, J.D. (2016) Effects of down-regulating ornithine decarboxylase upon putrescine-associated metabolism and growth in *Nicotiana tabacum* L. *Journal of Experimental Botany* 67(11): 3367–3381.

Das, K., Mandal, C., Ghosh, N., Banerjee, S., Dey, N., Adak, M.K. (2013) Effects of spermidine on the physiological activities of *Marsilea minuta* Linn. under cadmium stress. *General and Applied Plant Physiology* 3(3–4): 191–203.

Diao, Q.N., Song, Y.J., Shi, D.M., Qi, H.Y. (2016) Nitric oxide induced by polyamines involves antioxidant systems against chilling stress in tomato (*Lycopersicon esculentum* Mill.) seedling. *Journal of Zhejiang University-Science B* 17(12): 916–930.

Diao, Q.N., Song, Y.J., Shi, D.M., Qi, H.Y. (2017) Interaction of polyamines, abscisic acid, nitric oxide, and hydrogen peroxide under chilling stress in tomato (*Lycopersicon esculentum* Mill.) seedlings. *Frontiers in Plant Science* 8: 203.

Du, J., Shu, S., Shao, Q., An, Y., Zhou, H., Guo, S., Sun, J. (2016) Mitigative effects of spermidine on photosynthesis and carbon-nitrogen balance of cucumber seedlings under Ca(NO3)2 stress. *Journal of Plant Research* 129(1): 79–91. Doi: 10.1007/s10265-015-0762-3.

Ebeed, H.T., Hassan, N.M., Aljarani, A.M. (2017) Exogenous applications of Polyamines modulate drought responses in wheat through osmolytes accumulation, increasing free polyamine levels and regulation of polyamine biosynthetic genes. *Plant Physiology and Biochemistry* 118: 438–448.

Freitas, V.S., Miranda, R.S., Costa, J.H., Oliveira, D.F., Paula, S.O., Miguel, E.C., Freire, R.S., et al. (2018) Ethylene triggers salt tolerance in maize genotypes by modulating polyamine catabolism enzymes associated with H2O2 production. *Environmental and Experimental Botany* 145: 75–86.

Gémes, K., Mellidou, I., Karamanoli, K., Beris, D., Park, K.Y., Matsi, T., Haralampidis, K., et al. (2017) Deregulation of apoplastic polyamine oxidase affects development and salt response of tobacco plants. *Journal of Plant Physiology* 211: 1–12.

Glaubitz, U., Erban, A., Kopka, J., Hincha, D.K., Zuther, E. (2015) High night temperature strongly impacts TCA cycle, amino acid and polyamine biosynthetic pathways in rice in a sensitivity-dependent manner. *Journal of Experimental Botany* 66(20): 6385–6397.

Hu, X., Zhang, Y., Shi, Y., Zhang, Z., Zou, Z., Zhang, H., Zhao, J. (2012) Effect of exogenous spermidine on polyamine content and metabolism in tomato exposed to salinity-alkalinity mixed stress. *Plant Physiology and Biochemistry* 57: 200–209.

Hu, L., Xiang, L., Zhang, L., Zhou, X., Zou, Z., Hu, X. (2014) The photoprotective role of spermidine in tomato seedlings under salinity-alkalinity stress. *PLoS ONE* 9(10): e110855.

Hu, L., Xiang, L., Li, S., Zou, Z., Hu, X.H. (2016) Beneficial role of spermidine in chlorophyll metabolism and D1 protein content in tomato seedlings under salinity-alkalinity stress. *Physiologia Plantarum* 156(4): 468–477.

Huang, Y., Lin, C., He, F., Li, Z., Guan, Y., Hu, Q., Hu, J. (2017) Exogenous spermidine improves seed germination of sweet corn via involvement in phytohormone interactions, H$_2$O$_2$ and relevant gene expression. *BMC Plant Biology* 17(1): 1.

Juzoń, K., Czyczyło-Mysza, I., Marcińska, I., Dziurka, M., Waligórski, P., Skrzypek, E. (2017) Polyamines in yellow lupin (*Lupinus luteus* L.) tolerance to soil drought. *Acta Physiologia Plantarum* 39: 202.

Khan, N., Bano, A. (2017) Effects of exogenously applied salicylic acid and putrescine alone and in combination with rhizobacteria on the phytoremediation of heavy metals and chickpea growth in sandy soil. *International Journal of Phytoremediation* 21: 405–414.

Kotakis, C., Theodoropoulos, E., Tassis, K., Oustamanolakis, C., Ioannidis, N.E., Kotzabasis, K. (2014) Putrescine, a fast-acting switch for tolerance against osmotic stress. *Journal of Plant Physiology* 171(2): 48–51.

Kumar, R.R., Sharma, S.K., Rai, G.K., Singh, K., Choudhury, M., Dhawan, G., Singh, G.P., et al. (2014) Exogenous application of putrescine at pre-anthesis enhances the thermotolerance of wheat (*Triticum aestivum* L.). *Indian Journal of Biochemistry and Biophysics* 51(5): 396–406.

Legocka, J., Sobieszczuk-Nowicka, E., Wojtyla, Ł., Samardakiewicz, S. (2015) Lead-stress induced changes in the content of free, thylakoid- and chromatin-bound polyamines, photosynthetic parameters and ultrastructure in greening barley leaves. *Journal of Plant Physiology* 185–187:15–24.

Li, B., He, L., Guo, S., Li, J., Yang, Y., Yan, B., Sun, J., et al. (2013) Proteomics reveal cucumber Spd-responses under normal condition and salt stress. *Plant Physiology and Biochemistry* 67: 7–14.

Li, X., Gong, B., Xu, K. (2014a) Interaction of nitric oxide and polyamines involves antioxidants and physiological strategies against chilling-induced oxidative damage in *Zingiber officinale* Roscoe. *Scientia Horticulturae* 170: 237–248.

Li, Z., Peng, Y., Zhang, X.Q., Ma, X., Huang, L.K., Yan, Y.H. (2014b) Exogenous spermidine improves seed germination of white clover under water stress via involvement in starch metabolism, antioxidant defenses and relevant gene expression. *Molecules* 19(11): 18003–18024.

Li, J., Hu, L., Zhang, L., Pan, X.B., Hu, X.H. (2015a) Exogenous spermidine is enhancing tomato tolerance to salinity–alkalinity stress by regulating chloroplast antioxidant system and chlorophyll metabolism. *BMC Plant Biology* 15: 303.

Li, Z., Jing, W., Peng, Y., Zhang, X.Q., Ma, X., Huang, L.K., Yan, Y.H. (2015b) Spermine alleviates drought stress in white clover with different resistance by influencing carbohydrate metabolism and dehydrins synthesis. *PLoS ONE* 10(3): e0120708.

Li, Q., Wang, Z., Zhao, Y., Zhang, Y., Zhang, S., Bo, L., Wang, Y., et al. (2016a) Putrescine protects hulless barley from damage due to UV-B stress via H2S- and H2O2-mediated signaling pathways. *Plant Cell Reports* 35(5): 1155–1168.

Li, S., Jin, H., Zhang, Q. (2016b) The effect of exogenous spermidine concentration on polyamine metabolism and salt tolerance in zoysiagrass (*Zoysia japonica* Steud.) subjected to short-term salinity stress. *Frontiers in Plant Science* 7: 1221.

Li, Z., Zhang, Y., Xu, Y., Zhang, X., Peng, Y., Ma, X., Huang, L., et al. (2016c) Physiological and iTRAQ-based proteomic analyses reveal the function of spermidine on improving drought tolerance in white clover. *Journal of Proteome Research* 15(5): 1563–1579.

Li, S., Cui, L., Zhang, Y., Wang, Y., Mao, P. (2017) The variation tendency of polyamines forms and components of polyamine metabolism in zoysiagrass (*Zoysia japonica* Steud.) to salt stress with exogenous spermidine application. *Frontiers in Plant Science* 8: 208.

Liu, J.H., Wang, W., Wu, H., Gong, X., Moriguchi, T. (2015a) Polyamine function in stress tolerance: From synthesis to regulation. *Frontiers in Plant Science* 6: 827.

Liu, M., Chu, M., Ding, Y., Wang, S., Liu, Z., Tang, S., Ding, C., et al. (2015b) Exogenous spermidine alleviates oxidative damage and reduce yield loss in rice submerged at tillering stage. *Frontiers in Plant Science* 6: 919.

Liu, Y., Liang, H., Lv, X., Liu, D., Wen, X., Liao, Y. (2016a) Effect of polyamines on the grain filling of wheat under drought stress. *Plant Physiology and Biochemistry* 100: 113–169.

Liu, Y., Xu, H., Wen, X., Liao, Y. (2016b) Effect of polyamine on seed germination of wheat under drought stress is related to changes in hormones and carbohydrates. *Journal of Integrative Agriculture* 15(12): 2759–2774.

Liu, M., Chen, J., Guo, Z., Lu, S. (2017a) Differential responses of polyamines and antioxidants to drought in a centipedegrass mutant in comparison to its wild type plants. *Frontiers in Plant Science* 8: 792.

Liu, Z., Liu, P., Qi, D., Peng, X., Liu, G. (2017b) Enhancement of cold and salt tolerance of arabidopsis by transgenic expression of the S-adenosylmethionine decarboxylase gene from *Leymus chinensis*. *Journal of Plant Physiology* 211: 90–99.

López-Gómez, M., Hidalgo-Castellanos, J., Munoz-Sanchez, J.R., Lluch, C., Herrera-Cervera, J.A. (2017) Polyamines contribute to salinity tolerance in the symbiosis *Medicago truncatula-Sinorhizobium meliloti* by preventing oxidative damage. *Plant Physiology and Biochemistry* 116: 9–17.

Ma, Y., Shukla, V., Merewitz, E.B. (2017a) Transcriptome analysis of creeping bentgrass exposed to drought stress and polyamine treatment. *PLoS ONE* 12(4): e0175848.

Ma, C., Wang, Y., Gu, D., Nan, J., Chen, S., Li, H. (2017b) Overexpression of S-adenosyl-l-methionine synthetase 2 from sugar beet M14 increased arabidopsis tolerance to salt and oxidative stress. *International Journal of Molecular Sciences* 18(4): pii: E847.

Majumdar, R., Barchi, B., Turlapati, S.A., Gagne, M., Minocha, R., Long, S., Minocha, S.C. (2016) Glutamate, ornithine, arginine, proline, and polyamine metabolic interactions: The pathway is regulated at the post-transcriptional level. *Frontiers in Plant Science* 7: 78.

Mandal, C., Ghosh, N., Maiti, S., Das, K., Gupta, S., Dey, N., Adak, M.K. (2013) Antioxidative responses of Salvinia (*Salvinia natans* Linn.) to aluminium stress and its modulation by polyamine. *Physiology and Molecular Biology of Plants* 19(1): 91–103.

Mellidou, I., Karamanoli, K., Beris, D., Haralampidis, K., Constantinidou, H.A., Roubelakis-Angelakis, K.A. (2017) Underestimation of apoplastic polyamine oxidase improves thermotolerance in *Nicotiana tabacum*. *Journal of Plant Physiology* 218: 171–174.

Montilla-Bascón, G., Rubiales, D., Hebelstrup, K.H., Mandon, J., Harren, F.J.M., Cristescu, S.M., Mur, L.A.J., et al. (2017) Reduced nitric oxide levels during drought stress promote drought tolerance in barley and is associated with elevated polyamine biosynthesis. *Scientific Reports* 7(1): 13311.

Nahar, K., Hasanuzzaman, M., Alam, M.M., Fujita, M. (2015) Exogenous spermidine alleviates low temperature injury in mung bean (*Vigna radiata* L.) seedlings by modulating ascorbate-glutathione and glyoxalase pathway. *International Journal of Molecular Sciences* 16(12): 30117–30132.

Nahar, K., Hasanuzzaman, M., Mahabub Alam, M.D., Rahman, A., Suzuki, T., Fujita, M. (2016a) Polyamine and nitric oxide crosstalk: Antagonistic effects on cadmium toxicity in mung bean plants through upregulating the metal detoxification, antioxidant defense and methylglyoxal detoxification systems. *Ecotoxicology and Environmental Safety* (126): 245–255.

Nahar, K., Hasanuzzaman, M., Rahman, A., Alam, M.M., Mahmud, J.A., Suzuki, T., Fujita, M. (2016b) Polyamines confer salt tolerance in mung bean (*Vigna radiata* l.) by reducing sodium uptake, improving nutrient homeostasis, antioxidant defense, and methylglyoxal detoxification systems. *Frontiers in Plant Science* 7: 1104.

Nahar, K., Rahman, M., Hasanuzzaman, M., Alam, M.M., Rahman, A., Suzuki, T., Fujita, M. (2016c) Physiological and biochemical mechanisms of spermine-induced cadmium stress tolerance in mung bean (*Vigna radiata* L.) seedlings. *Environmental Science and Pollution Research* 23(21): 21206–21218.

Nahar, K., Hasanuzzaman, M., Alam, M.M., Rahman, A., Mahmud, J.A., Suzuki, T., Fujita, M. (2017a) Insights into spermine-induced combined high temperature and drought tolerance in mung bean: Osmoregulation and roles of antioxidant and glyoxalase system. *Protoplasma* 254(1): 445–460.

Nahar, K., Hasanuzzaman, M., Suzuki, T., Fujita, M. (2017b) Polyamines-induced aluminum tolerance in mung bean: A study on antioxidant defense and methylglyoxal detoxification systems. *Ecotoxicology* 26(1): 58–73.

Pál, M., Szalai, G., Janda, T. (2015) Speculation: Polyamines are important in abiotic stress signaling. *Plant Science* 237: 16–23.

Pál, M., Csávás, G., Szalai, G., Oláh, T., Khalil, R., Yordanova, R., Gell, Gy. et al. (2017) Polyamines may influence phytochelatin synthesis during Cd stress in rice. *Journal of Hazardous Materials* 340: 272–280.

Parvin, S., Lee, O.R., Sathiyaraj, G., Khorolragchaa, A., Kim, Y.J., Yang, D.C. (2014) Spermidine alleviates the growth of saline-stressed ginseng seedlings through antioxidative defense system. *Gene* 537(1): 70–78.

Patel, J., Ariyaratne, M., Ahmed, S., Ge, L., Phuntumart, V., Kalinoski, A., Morris, P.F. (2017) Dual functioning of plant arginases provides a third route for putrescine synthesis. *Plant Science* 262: 62–73.

Paul, S. Roychoudhury, A., Banerjee, A., Chaudhuri, N., Ghosh, P. (2017) Seed pre-treatment with spermidine alleviates oxidative damages to different extent in the salt (NaCl)-stressed seedlings of three indica rice cultivars with contrasting level of salt tolerance. *Plant Gene* 11(B): 112–123.

Peng, D., Wang, X., Li, Z., Zhang, Y., Peng, Y., Li, Y., He, X., et al. (2016) NO is involved in spermidine-induced drought tolerance in white clover via activation of antioxidant enzymes and genes. *Protoplasma* 253(5): 1243–1254.

Puyang, X., An, M., Han, L., Zhang, X.. (2015) Protective effect of spermidine on salt stress induced oxidative damage in two Kentucky bluegrass (*Poa pratensis* L.) cultivars. *Ecotoxicology and Environmental Safety* 117: 96–106.

Puyang, X., An, M., Xu, L., Han, L., Zhang, X. (2016) Protective effect of exogenous spermidine on ion and polyamine metabolism in Kentucky bluegrass under salinity stress. *Horticulture, Environment, and Biotechnology* 57(1): 11–19.

Rady, M.M., Hemida, K.A. (2015) Modulation of cadmium toxicity and enhancing cadmium-tolerance in wheat seedlings by exogenous application of polyamines. *Ecotoxicology and Environmental Safety* 119: 178–185.

Saha, J., Giri, K. (2017) Molecular phylogenomic study and the role of exogenous spermidine in the metabolic adjustment of endogenous polyamine in two rice cultivars under salt stress. *Gene* 609: 88–103.

Sagor, G.H., Chawla, P., Kim, D.W., Berberich, T., Kojima, S., Niitsu, M., Kusano, T. (2015) The polyamine spermine induces the unfolded protein response via the MAPK cascade in *Arabidopsis*. *Frontiers in Plant Science* 6:687.

Sagor, G.H., Berberich, T., Kojima, S., Niitsu, M., Kusano, T. (2016) Spermine modulates the expression of two probable polyamine transporter genes and determines growth responses to cadaverine in *Arabidopsis*. *Plant Cell Reports* 35(6): 1247–1257.

Sang, T., Shan, X., Li, B., Shu, S., Sun, J., Guo, S. (2016) Comparative proteomic analysis reveals the positive effect of exogenous spermidine on photosynthesis and salinity tolerance in cucumber seedlings. *Plant Cell Reports* 35(8): 1769–1782.

Sang, Q., Shan, X., An, Y., Shu, S., Sun, J., Guo, S. (2017) Proteomic analysis reveals the positive effect of exogenous spermidine in tomato seedlings' response to high-temperature stress. *Frontiers in Plant Science* 8: 120.

Scoccianti, V., Iacobucci, M., Speranza, A., Antognoni, F. (2013) Over-accumulation of putrescine induced by cyclohexylamine interferes with chromium accumulation and partially restores pollen tube growth in *Actinidia deliciosa*. *Plant Physiology and Biochemistry* 70: 424–432.

Sengupta, M., Raychaudhuri, S.S. (2017) Partial alleviation of oxidative stress induced by gamma irradiation in *Vigna radiata* by polyamine treatment. *International Journal of Radiation Biology* 93(8): 803–817.

Sequera-Mutiozabal, M., Erban, A., Kopka, J., Atanasov, K.E., Bastida, J., Fotopoulos, V., Alcázar, R., et al. (2016) Global metabolic profiling of *Arabidopsis* polyamine oxidase 4 (AtPAO4) loss-of-function mutants exhibiting delayed dark-induced senescence. *Frontiers in Plant Science* 7: 173.

Sequera-Mutiozabal, M., Antoniou, C., Tiburcio, A.F., Alcázar, R., Fotopoulos, V. (2017) Polyamines: Emerging hubs promoting drought and salt stress tolerance in plants. *Current Molecular Biology Reports* 3(1): 28–36.

Shen, Y., Ruan, Q., Chai, H., Yuan, Y., Yang, W., Chen, J., Xin, Z., et al. (2016) The arabidopsis polyamine transporter LHR/PUT3 modulates heat responsive gene expression by enhancing mRNA stability. *Plant Journal* 88(6): 1006–1021.

Sheteiwy, M., Shen, H., Xu, J., Guan, Y., Song, W., Hu, J. (2017) Seed polyamines metabolism induced by seed priming with spermidine and 5-aminolevulinic acid for chilling tolerance improvement in rice (*Oryza sativa* L.) seedlings. *Environmental and Experimental Botany* 137: 58–72.

Shu, S., Chen, L., Lu, W., Sun, J., Guo, S., Yuan, Y., Li, J. (2014) Effects of exogenous spermidine on photosynthetic capacity and expression of Calvin cycle genes in salt-stressed cucumber seedlings. *Journal of Plant Research* 127(6): 763–773.

Shu, S., Yuan, Y., Chen, J., Sun, J., Zhang, W., Tang, Y., Zhong, M., et al. (2015) The role of putrescine in the regulation of proteins and fatty acids of thylakoid membranes under salt stress. *Scientific Reports* 5: 14390.

Siddiqui, M.H., Alamri, S.A., Al-Khaishany, M.Y., Al-Qutami, M.A., Ali, H.M., Al-Rabiah, H., Kalaji, H.M. (2017) Exogenous application of nitric oxide and spermidine reduces the negative effects of salt stress on tomato. *Horticulture, Environment, and Biotechnology* 58(6): 537–547.

Szalai, G., Janda, K., Darkó, É., Janda, T., Peeva, V., Pál, M. (2017) Comparative analysis of polyamine metabolism in wheat and maize plants. *Plant Physiology and Biochemistry* 112: 239–250.

Szepesi, Á., Csiszár, J., Gémes, K., Horváth, E., Horváth, F. Simon, L.M., Tari, I. (2009) Salicylic acid improves acclimation to salt stress by stimulating abscisic aldehyde oxidase activity and abscisic acid accumulation, and increases Na+ content in leaves without toxicity symptoms in *Solanum lycopersicum* L. *Journal of Plant Physiology* 166(9): 914–925.

Szepesi, Á., Gémes, K., Orosz, G., Pető, A., Takács, Z., Vorák, M., Tari, I. (2011) Interaction between salicylic acid and polyamines and their possible roles in tomato hardening processes. *Acta Biologica Szegediensis* 55: 165–166.

Tajti, J., Janda, T., Majláth, I., Szalai, G., Pál, M. (2018) Comparative study on the effects of putrescine and spermidine pre-treatment on cadmium stress in wheat. *Ecotoxicology and Environmental Safety* 148: 546–554.

Takács, Z., Poór, P., Szepesi, Á., Tari, I. (2017) *In vivo* inhibition of polyamine oxidase by a spermine analogue, MDL-72527, in tomato exposed to sublethal and lethal salt stress. *Functional Plant Biology* 44(5): 480–492.

Tang, C., Song, J., Hu, X., Hu, X., Zhao, Y., Li, B., Ou, D., et al. (2017) Exogenous spermidine enhanced Pb tolerance in *Salix matsudana* by promoting Pb accumulation in roots and spermidine, nitric oxide, and antioxidant system levels in leaves. *Ecological Engineering* (107): 41–48.

Tanou, G., Ziogas, V., Belghazi, M., Christou, A., Filippou, P., Job, D., Fotopoulos, V., et al. (2014) Polyamines reprogram oxidative and nitrosative status and the proteome of citrus plants exposed to salinity stress. *Plant, Cell and Environment* 37(4): 864–885.

Tong, W., Yoshimoto, K., Kakehi, J.I., Motose, H., Niitsu, M., Takahashi, T. (2014) Thermospermine modulates expression of auxin-related genes in *Arabidopsis*. *Frontiers in Plant Science* 5: 94.

Wang, W., Liu, J.H. (2015) Genome-wide identification and expression analysis of the polyamine oxidase gene family in sweet orange (*Citrus sinensis*). *Gene* 555(2): 421–429.

Zarza, X., Atanasov, K.E., Marco, F., Arbona, V., Carrasco, P., Kopka, J., Fotopoulos, V., et al. (2017) Polyamine oxidase 5 loss-of-function mutations in *Arabidopsis thaliana* trigger metabolic and transcriptional reprogramming and promote salt stress tolerance. *Plant, Cell and Environment* 40(4): 527–542.

Zhang, Y., Hu, X.H., Shi, Y., Zou, Z.R., Yan, F., Zhao, Y.Y., Zhang, H., et al. (2013) Beneficial role of exogenous spermidine on nitrogen metabolism in tomato seedlings exposed to saline-alkaline stress. *Journal of the American Society for Horticultural Science* 138: 38–39.

Zhang, Y., Zhang, H., Zou, Z.R., Liu, Y., Hu, X.H. (2015) Deciphering the protective role of spermidine against saline–alkaline stress at physiological and proteomic levels in tomato. *Phytochemistry* 110: 13–21.

Zhang, N., Shi, X., Guan, Z., Zhao, S., Zhang, F., Chen, S., Fang, W., et al. (2016) Treatment with spermidine protects chrysanthemum seedlings against salinity stress damage. *Plant Physiology and Biochemistry* 105: 260–270.

Zhu, X.F., Wang, B., Song, W.F., Zheng, S.J., Shen, R.F. (2016) Putrescine alleviates iron deficiency via NO-dependent reutilization of root cell-wall Fe in *Arabidopsis*. *Plant Physiology* 170(1): 558–567.

Zhuo, C., Liang, L., Zhao, Y., Guo, Z., Lu, S. (2017) A cold responsive ERF from *Medicago falcata* confers cold tolerance by up-regulation of polyamine turnover, antioxidant protection and proline accumulation. *Plant, Cell and Environment*. doi: 10.1111/pce.13114.

Yamamoto, A., Shim, I.S., Fujihara, S. (2017) Inhibition of putrescine biosynthesis enhanced salt stress sensitivity and decreased spermidine content in rice seedlings. *Biologia Plantarum* 61: 385.

Yang, W., Li, Y., Yin, Y., Qin, Z., Zheng, M., Chen, J., Luo, Y., et al. (2017) Involvement of ethylene and polyamines biosynthesis and abdominal phloem tissues characters of wheat caryopsis during grain filling under stress conditions. *Scientific Reports* 7. doi: 10.1038/srep46020.

Yin, Z.P., Li, S., Ren, J., Song, X.S. (2014) Role of spermidine and spermine in alleviation of drought-induced oxidative stress and photosynthetic inhibition in Chinese dwarf cherry (*Cerasus humilis*) seedlings. *Plant Growth Regulation* 74(3): 209–218.

Yin, L., Wang, S., Tanaka, K., Fujihara, S., Itai, A., Den, X., Zhang, S. (2016) Silicon-mediated changes in polyamines participate in silicon-induced salt tolerance in *Sorghum bicssolor* L. *Plant, Cell and Environment* 39(2): 245–258.

Yiu, J.C., Liu, C.W., Fang, D.Y., Lai, Y.S. (2009) Waterlogging tolerance of Welsh onion (*Allium fistulosum* L.) enhanced by exogenous spermidine and spermine. *Plant Physiology and Biochemistry* 47(8): 710–716.

Yoshimoto, K., Takamura, H., Kadota, I., Motose, H., Takahashi, T. (2016) Chemical control of xylem differentiation by thermospermine, xylemin, and auxin. *Nature Scientific Reports* 6: 21487.

Yu, Y., Jin, C., Sun, C., Wang, J., Ye, Y., Lu, L., Lin, X. (2015) Elevation of arginine decarboxylase-dependent putrescine production enhances aluminum tolerance by decreasing aluminum retention in root cell walls of wheat. *Journal of Hazardous Materials* 299: 280–288.

Yuan, Y., Zhong, M., Shu, S., Du, N., Sun, J., Guo, S. (2016) Proteomic and physiological analyses reveal putrescine responses in roots of cucumber stressed by NaCl. *Frontiers in Plant Science*. https://doi.org/10.3389/fpls.2016.01035.

Yuan, R.N., Shu, S., Guo, S.R., Sun, J., Wu, J.Q. (2017) The positive roles of exogenous putrescine on chlorophyll metabolism and xanthophyll cycle in salt-stressed cucumber seedlings. *Photosynthetica* 1–10. DOI: 10.1007/s11099-017-0712-5

23 Molecular Approaches for Enhancing Abiotic Stress Tolerance in Plants

Sushma Mishra, Dipinte Gupta, and Rajiv Ranjan

CONTENTS

23.1 INTRODUCTION

Abiotic stress, in simple words, could be referred to as the extreme environmental conditions that affect plant growth and development. These stresses reduce the productivity of major crops by 50 to 70% (Boyer, 1982; Bray et al., 2000; Acquaah, 2007; Mantri et al., 2012). Plants have evolved various mechanisms to counteract abiotic stresses, such as drought, salinity, heat, cold and submergence. Since the world population is expected to reach 11.2 billion by 2100 (Rockville, 2015), world food production needs to be remarkably increased to feed the growing population of the world. Enhancement of plant tolerance to abiotic stresses is one of the primary strategies to increase crop production in the limited land area. A number of approaches have been used in the past to confer abiotic stress tolerance to plants. It is necessary to understand and appreciate the general response of plants to abiotic stresses. Although a number of reviews on plant stress tolerance have been published in the past (Marco et al., 2015; Kumar and Singh, 2016), the main objective of writing this chapter is to present the readers with an overview of major strategies used to confer abiotic stress tolerance to plants, with special emphasis on transgenic approach.

23.2 GENERAL PLANT DEFENSE STRATEGIES AGAINST ABIOTIC STRESS

Various abiotic stresses induce common signaling pathways in plant cells that result in similar defense responses (Ozturk et al., 2002). In fact, almost all the abiotic stresses, ranging from drought to salinity to extreme temperatures, induce secondary stress responses, namely, *osmotic stress* and *oxidative stress* (Wang et al., 2003;

Figure 23.1). Osmotic stress refers to a sudden change in solute concentration in cells, accompanied by movement of water, which mainly results in loss of turgor pressure required for cell growth. Oxidative stress, on the other hand, results in accumulation of harmful reactive oxygen species (ROS) such as superoxides, hydrogen peroxide and hydroxyl radicals that cause oxidative damage to cellular macromolecules like proteins and membrane lipids (Mittler, 2002; Imlay, 2003). At the whole plant level, the effect of any of these stresses is usually perceived as a decrease in photosynthesis and growth.

Plant cells have evolved various mechanisms to counteract different types of abiotic stresses, which could be broadly categorized into *stress avoidance responses* and *stress tolerance responses*. For example, stress avoidance mechanisms of plants under drought conditions include altered leaf angle, leaf rolling, stomatal closure and inhibition of shoot growth, in order to minimize the water loss due to transpiration. This is accompanied by a deeper root system through increased allocation of photosynthates to root tips for increased ability of water uptake. In contrast, the stress tolerance mechanisms, which are the main focus of this chapter, involve

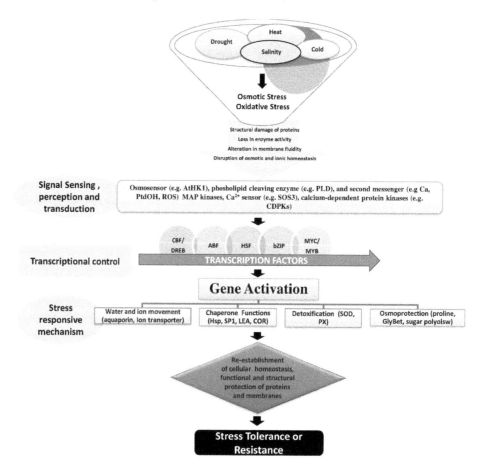

FIGURE 23.1 Flowchart showing the general signal transduction pathway induced in plants in response to abiotic stress. Modified from Wang et al. (2003).

biosynthesis of certain proteins and metabolites that enable plants to counteract stress.

The molecular mechanism underlying stress tolerance in plants have been quite well worked out (Knight and Knight, 2001; Tester and Bacic, 2005; Marco et al., 2015; Kumar and Singh, 2016). The stress signal is generally perceived by a membrane-bound receptor, which undergoes conformational changes and initiates an intracellular signaling cascade. The signal transduction typically involves a series of phosphorylation and dephosphorylation reactions catalyzed by kinases and phasphatases, respectively, which culminates into changes in gene expression (Knight and Knight, 2001; Tester and Bacic, 2005; Marco et al., 2015; Kumar and Singh, 2016). The genes induced by abiotic stresses could be categorized into *functional genes*, which encode for products like osmoprotectants, detoxifying enzymes, chaperones, late embryogenesis abundant (LEA) proteins, etc., and *regulatory genes*, which generally act upstream to several functional genes and mainly encode for transcription factors and protein kinases. Apart from the changes in gene expression, abiotic stress responses in plant cells also include a rapid and transient rise in cytoplasmic Ca^{2+} levels, which function as a secondary messenger in abiotic stress signaling. The influx of Ca^{2+}, either from extracellular sources or from intracellular stores (endoplasmic reticulum or vacuole), lead to a series of phosphorylation and dephosphorylation events, thereby facilitating the transmission of stress signal to the nucleus, where appropriate gene expression changes could take place (Knight, 1999; Chinnusamy et al., 2004). In this chapter, we begin by presenting an overview of the mechanisms involved in counteracting the osmotic and oxidative stresses generated in plant cells, followed by approaches that have been used to engineer crop species to withstand stress tolerance through genetic engineering (Marco et al., 2015).

23.2.1 Counteracting Osmotic Stress

Plants suffer from dehydration not only under drought but also under high salt and freezing conditions, which result in loss of turgor pressure and water loss from the cells. Such osmotic stress created in the plant cell is perceived either by abscisic acid (ABA)-dependent or ABA-independent pathways (Figure 23.2). Although distinct, these are not entirely isolated from each other, and involve points of crosstalks between the pathways via certain transcription factors (TFs) like NAC (Nakashima et al., 2012; Yoshida et al., 2014) and secondary messengers (e.g., Ca^{2+}) (Xiong et al., 2002; Chinnusamy et al., 2004). The NAC acronym is derived from three genes that were initially discovered to contain a particular domain: NAM (for no apical meristem), ATAF1 and -2, and CUC2 (for cup-shaped cotyledon) (Souer et al., 1996; Aida et al., 1997). As a result, some downstream responses, like induction of

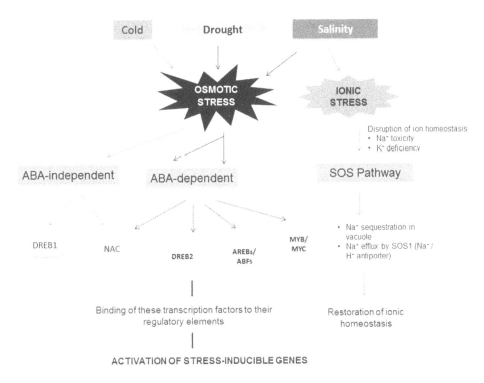

FIGURE 23.2 Schematic representation of generation of osmotic stress as a result of extreme abiotic factors, and different stress tolerance mechanisms to counteract osmotic stress.

RD29A gene transcription and accumulation of proline osmolyte, are mediated by both ABA-dependent and ABA-independent pathways (Yamaguchi-Shinozaki and Shinozaki, 1993).

23.2.1.1 ABA-Dependent Gene Expression

It has been reported that most of the genes induced by drought or salt stress are also induced by abscisic acid treatment (Xiang et al., 2007). In fact, a common and quick response of plant cells to osmotic stress is to increase the concentration of ABA (also known as the stress hormone), which triggers gene expression and also causes the stomatal closure to prevent intracellular water loss (Fujita et al., 2005; Tuteja, 2007). The majority of the stress-induced ABA-regulated genes share the conserved ABA-responsive *cis* element (ABRE; Yamaguchi-Shinozaki and Shinozaki, 2005) in their promoter region. Various transcription factors such as DREB2A/2B, AREB1 and MYC/MYB are known to regulate the ABA-mediated responses through interacting with their corresponding *cis*-acting elements such as DRE/CRT, ABRE and MYCRS/MYBRS, respectively. When ABA binds to its receptor in the cytoplasm, the signal is relayed to AREB TFs to regulate ABA-dependent gene expression.

23.2.1.2 ABA-Independent Gene Expression

ABA-independent gene expression, also known as DREB-mediated pathway, is predominantly mediated by transcription factors like dehydration-responsive element-binding (DREB)/C-repeat-binding (CBF) proteins, which recognize dehydration-responsive element (DRE)/C-repeat (CRT) motifs through a conserved AP2 domain (Yoshida et al., 2014) These TFs are not induced by ABA but are responsive to dehydration, resulting from drought, salinity, heat and freezing conditions. The DREB1 is mainly involved in relaying the cold stress signal. However, the DREB2, which also functions in the ABA-independent pathway, is induced in response to drought, salt and heat stress (Thomashow, 1999; Shinozaki and Yamaguchi-Shinozaki, 2000). Transgenic *Arabidopsis* plants over expressing OsDREB2B (DREB 2 of rice) showed enhanced expression of DREB2A target genes and improved drought and heat-shock stress tolerance (Matsukura et al., 2010). Enhanced drought tolerance through a DREB-mediated pathway is considered to involve LEA protein (Lata and Prasad, 2011; Yoshida et al., 2014).

23.2.1.3 Osmoprotectants

Osmoprotectants are low molecular weight, nontoxic, organic compounds that accumulate in plants cells under various abiotic stresses like drought, salinity and extreme temperatures (Djilianov et al., 2005; Ashraf

and Foolad, 2007). Osmoprotectants exist as chemically diverse biomolecules such as amino acids (proline), amines (glycine betaine and polyamines), sugars (trehalose) and sugar alcohols (mannitol, sorbitol). These compounds primarily function to regulate water uptake and retention (in order to maintain turgor pressure) and to stabilize cellular macromolecules like proteins, nucleic acids and lipids (Chen and Murata, 2002; Djilianov et al., 2005). Some plants are inherently more tolerant to stress in comparison to others, which may be attributed to the accumulation of certain types of osmolytes. Under extreme temperature, salinity and drought stress conditions, the higher concentration of osmotically active solutes helps to maintain the turgor pressure required for cell growth. For example, proline not only acts an excellent osmolyte but as a good metal chelator and an antioxidant too (Kishor et al., 2005; Verbruggen and Hermans, 2008). Proline has been reported to accumulate at very high levels in dehydrated plants, both due to prevention of its degradation and promotion of its biosynthesis. Under stress situations, proline is synthesized from glutamate by the enzymes pyrroline-5-carboxylate synthase (P5CS) and pyrroline-5-carboxylate reductase (P5CR), which have been targeted for engineering drought tolerance (Zhu et al., 1998; Kocsy et al., 2005). Another class of osmoprotectants is polyamines. The polyamines are small nitrogenous compounds, namely, putrescine, spermine and spermidine. These compounds exist as polycations at physiological pH and are involved in neutralizing the polyanionic sites exposed on the macromolecules under stress (Groppa and Benavides, 2008). Another important osmolyte is glycine betaine, which, although it occurs naturally in some crops like spinach, wheat and beet, its levels are too low to carry out osmotic adjustments under stressful conditions (Tian et al., 2017). It is synthesized from choline by the action of two enzymes, choline monooxygenase (CMO) and betaine aldehyde dehydrogenase (BADH). Trehalose is a non-reducing disaccharide which acts as membrane protectant, a compatible solute and a reserve carbohydrate (mobilized during stress). Apart from nitrogenous compounds and quaternary amines, some non-reducing sugars like trehalose also act as effective osmolytes. The trehalose biosynthetic pathway in *E. coli* involves two enzymes: trehalose-6-phosphate synthase and trehalose phosphatase, both of which have been expressed under ABA-inducible promoter to impart drought stress tolerance in *Arabidopsis* (Houtte et al., 2013).

23.2.1.4 Late Embryogenesis Abundant Proteins

Late embryogenesis abundant proteins are low molecular weight (10–30 kDa), hydrophilic proteins that are rich in charged amino acids. These accumulate naturally in embryos during seed desiccation but can also be induced in vegetative tissues during dehydration, cold, salt and ABA treatment (Olvera-Carrillo et al., 2011). The LEA proteins are not plant-specific that have been reported from a wide variety of species ranging from bacteria, fungi, rotifers and nematodes. These proteins are generally classified into six different families on the basis of their amino acid sequence. LEA proteins impart stress tolerance through maintenance of protein structure, membrane integrity by the formation of H-bonds with phosphate groups in membrane phospholipids, sequestration of ions (because of charged residues) and as molecular chaperones and antioxidants (Battaglia et al., 2008). These proteins are localized to different organelles in the cells, i.e., cytoplasm, endoplasmic reticulum, mitochondria and nucleus. Under water deficit conditions, an enzyme undergoes changes in its structural conformation (i.e., through exposure of hydrophobic residues), which leads to loss of enzyme activity. Under such conditions, the LEA proteins prevent such structural changes and help in maintaining the integrity of the enzyme. Overexpression of some LEA genes has been reported to result in enhanced tolerance to dehydration and other abiotic stresses, which has been discussed in Section 23.4.1.3 (Hundertmark & Hincha, 2008; Ling et al., 2016).

23.2.2 COUNTERACTING OXIDATIVE STRESS

23.2.2.1 Role of Antioxidants

Reactive oxygen species include both free radical forms like superoxide radical ($O_2^{\bullet-}$), hydroxyl radical (OH), perhydroxyl radical ($HO_2^{\bullet}$) and alkoxy radicals ($RO^{\bullet}$); and non-radical molecular forms like hydrogen peroxide (H_2O_2) and singlet oxygen ($^1O_2$). These reactive groups cause damage to membrane lipids and other macromolecules (proteins, nucleic acids), resulting in oxidative stress (Figure 23.3). In order to counteract oxidative stress, cells have a defense mechanism involving antioxidants, which neutralize the charge on ROS by donating an electron to free radicals. The antioxidants could be metabolites like ascorbic acid, glutathione, phenolic compounds, alkaloids, non-protein amino acids and α-tocopherols, or enzymes like catalase, ascorbate peroxidase and superoxide dismutase (SOD). ROS are mainly generated in cell organelles such as chloroplasts, mitochondria and peroxisomes. Production of ROS is not necessarily a symptom of cellular dysfunction but might represent a necessary signal in adjusting the cellular machinery to the altered conditions. In higher plants,

SOD isozymes are found in various cellular compartments and provide the first line of defense against ROS. In olive leaves, Mn-SOD is present in mitochondria and peroxisomes. The Fe-SOD has been found mainly in chloroplasts but has also been detected in peroxisomes. However, the CuZnSOD has been localized in the cytosol, chloroplasts, peroxisomes and apoplast (Corpas et al., 2006). SODs have been overexpressed under stress-inducible promoters to raise transgenics with increased tolerance to abiotic stresses (Mckersie et al., 1996; Prashanth et al., 2008). In fact, the transformation of two or more antioxidants (both superoxide dismutase and ascorbate peroxidase in tobacco) has resulted in much-enhanced stress tolerance in double transgenics, in comparison to those that expressed single antioxidant enzymes (Kwon et al., 2002).

Another approach of counteracting oxidative stress and maintaining the cell redox homeostasis is through the glutathione (GSH)/glutathione disulfide (GSSG) redox couple (Figure 23.3). Under optimal growth conditions, the high GSH/GSSG ratio results in maintaining a reducing environment in the cells, which, by preventing the formation of intermolecular disulfide bridges, helps in maintaining appropriate structure and activity of proteins. In response to abiotic stresses, the GSH/GSSG ratio decreases (due to the oxidation of GSH during the detoxification of ROS), which alters the protein structure and, hence, inhibits the enzyme activity (Szalai et al., 2009). Therefore, a high GSH/GSSG ratio, maintained by increased GSH synthesis and/or GSSG reduction, may be necessary for the efficient protection of plants against the abiotic stress-induced accumulation of ROS.

23.2.2.2 Role of MAPK Signaling

In addition to the aforementioned responses, mitogen-activated protein kinase (MAPK) cascade gets activated in response to ROS generation following abiotic stresses. It consists of a series of phosphorylation-activation relay system, mediated by three protein kinases: MAPKKK–MAPKK–MAPK; i.e., a stimulus is perceived by a MAP kinase kinase kinase (MAPKKK), which activates a MAP kinase kinase (MAPKK), and finally the signal is transmitted to their downstream target via a MAP kinase (MAPK) (Mittler, 2002). Generally, the downstream target is a TF whose activity, localization or stability would be affected by the phosphorylation (Sinha et al., 2011; Jaspers and Kangasjärvi, 2010). Overexpression of *Nicotiana* protein kinase 1 (*NPK1*), a MAPKKK, in several crop species resulted in increased tolerance to freezing, heat, salt and drought stress (Banno et al., 1993; Kovtun et al., 2000). Transgenic maize overexpressing *NPK1* under the control of a constitutive

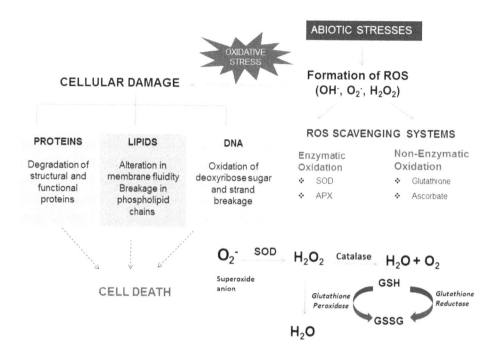

FIGURE 23.3 Schematic representation of generation of reactive oxygen species (ROS) in plants, and ROS scavenging systems to withstand the oxidative stress.

promoter activated an oxidative stress signaling pathway (involving glutathione-S-transferases (GSTs), HSPs, peroxidases and catalases), which led to enhanced tolerance to drought stress, by protecting the photosynthetic machinery of plants from damage. The transgenics also showed increased kernel weight, comparable to that of well-watered plants (Shou et al., 2004). Two *Arabidopsis* MAPKs, ATMPK4 and ATMPK6, are also activated by environmental stresses such as low temperature, low humidity, hyper-osmolarity, touch and wounding (Ichimura et al., 2000). Similar reports involving the role of the MAPK pathway in abiotic stresses have been documented in other species as well (Taj et al., 2010; Zhang et al., 2016).

23.2.3 ROLE OF CALCIUM SIGNALING IN ABIOTIC STRESS RESPONSES

Calcium ion (Ca^{2+}) is a ubiquitous secondary messenger in eukaryotic signal transduction cascades (Cheng et al., 2002). In response to almost all types of abiotic stresses, the cytosolic Ca^{2+} concentration is transiently increased, which is otherwise tightly regulated in the order of 100–200 nM in eukaryotic cells (Sanders et al., 1999; Knight, 1999). This occurs either as a result of the influx of Ca^{2+} from exogenous source or intracellular organelles via the action of Ca^{2+}-selective ion channels. This is followed by an array of phosphorylation and dephosphorylation events, either directly

through calcium-dependent protein kinases (CDPKs) or through Ca^{2+}-binding sensory proteins such as calmodulins and calcineurin B-like proteins (CBLs) or can be independent of calcium by sucrose nonfermenting-related kinase (SnRK2). At the end of the response, Ca^{2+} homeostasis is restored by removal of Ca^{2+} by pumps or antiporters that bring the cytosolic concentration back to basal levels.

In plants, CDPKs are a family of plant serine/threonine protein kinases which are important sensors of environmentally induced Ca^{2+} fluxes. In contrast to calmodulin-dependent protein kinases (CMPKs), CDPKs contain both calcium-sensing domain and kinase effector domain within the single polypeptide. In response to increased levels of Ca^{2+} in the cytosol, CDPKs cause Ca^{2+}-stimulated phosphorylation of target proteins that impart stress tolerance. *OsCDPK7*, one of the rice CDPKs, is a stress-inducible gene whose transcript level increases under cold and salt stress conditions. Expression of *OsCDPK7* driven by the cauliflower mosaic virus 35S promoter confers both cold- and salt/drought-stress tolerance in rice plants (Saijo et al., 2000). The authors have proposed that under normal conditions, OsCDPK7 is kept in an inactive form in the cells; it is activated by Ca^{2+} only under stress conditions (Saijjo et al., 2000).

Calcineurin B-like proteins (CBLs) represent another category of calcium sensors, involved in relaying the abiotic stress signal. These proteins harbor EF-hand motifs as a structural basis for calcium binding and interact specifically with a group of serine/threonine protein kinases designated as CBL-interacting protein kinases (CIPKs) (Shi et al., 1999). In *Arabidopsis*, analyses of loss-of-function and overexpressing transgenics of CBL1 have indicated a crucial function of this calcium sensor protein in drought, salt and cold stress responses (Albrecht et al., 2003). Of the 30 CIPKs reported from rice, at least 20 are induced by one or other type of abiotic stresses; for example, *OsCIPK12*, *OsCIPK3* and *OsCIPK15* are strongly induced under drought, cold and salinity stress, respectively (Xiang et al., 2007). Moreover, these genes, when overexpressed in rice, conferred enhanced tolerance to these stresses. Since the kinase activity of CIPKs depends on their interaction with CBLs, the manipulation of CIPK alone may not always result in conferring stress tolerance.

23.3 APPROACHES FOR CONFERRING ABIOTIC STRESS TOLERANCE

With an overview of stress tolerance mechanisms in plants, it seems to be a straightforward approach to engineer the stress tolerant trait into the sensitive species. However, the fact is that stress tolerance is a quantitative trait (controlled by many genes) which involves a highly complicated molecular mechanism. Hence, the engineering of stress tolerance is difficult, and the outcome is unpredictable. A number of approaches, like marker-assisted breeding, Quantitative Trait Loci (QTL) mapping, genetic engineering, association mapping and the recent omics approaches have been used to incorporate the stress tolerance trait into the desired cultivars. These approaches are briefly discussed in the following sections.

23.3.1 PLANT BREEDING

Plant breeding enables the incorporation of stress resistance genes from a tolerant variety into the desired sensitive cultivar through sexual hybridization. Plant breeders generally rely on genetic variation in wild varieties as a source of stress tolerance genes. For example, *Triticum dicoccoides* and *Hordeum spontaneum*, the wild relatives of wheat and barley, have been extensively used for introgressing salt and drought tolerance traits into durum wheat and barley cultivars (Nevo and Chen, 2010). Major progress by plant breeding has also been made at the Central Soil Salinity Research Institute, Karnal, where salt-tolerant high-yielding rice varieties like CSR10, CSR13 and CSR27 have been developed (Mishra, 1994).

23.3.2 GENETIC ENGINEERING

The second approach of plant improvement is genetic engineering or the *transgenic approach*, which involves the introduction of one or more specific genes into the desired cultivar. Apart from the speed and specificity through which the desired genes could be introduced, it remains the only option when genes of interest originate from cross barrier species, distant relatives or from non-plant sources. The initial attempts of engineering stress tolerance in plants were focused mainly on "single-action genes" or "functional genes"; this was followed by engineering of the "regulatory machinery", like transcription factors that act upstream to and control the expression of several stress-responsive genes. With the advancement in high throughput gene expression analysis and next-generation sequencing (NGS), hundreds of stress-induced genes have been identified as promising targets for use by genetic engineering technology (Bhatnagar-Mathur et al., 2008; Deshmukh et al., 2015; Chunqing et al., 2015). For example: improved stress tolerance was demonstrated by transgenic tomatoes expressing mannitol-1-phosphate dehydrogenase gene (Khare et al., 2010); potato expressing non-specific lipid transfer protein-1 (StnsLTP1) (Gangadhar et al., 2016); chili pepper expressing tobacco osmotin gene (Subramanyam et al., 2011); sweet potato overexpressing betaine aldehyde dehydrogenase gene (Fan et al., 2016) and many more.

23.3.3 QTL MAPPING AND ASSOCIATION MAPPING

Abiotic stress responses, like other complex traits in the plant (flowering time, growth rate and yield), are controlled by several genes and are inherited as quantitative traits. In plants, QTLs were originally mapped in biparental crosses, but they were restricted in allelic diversity and in having a limited genomic resolution (Borevitz and Nordborg, 2003). Moreover, traditional QTL mapping using pedigrees and suitable crosses is clearly not "scalable" to screen a large number of accessions for genetic variations underlying diverse, complex traits.

Several limitations associated with earlier methods of QTL mapping have been overcome by genome-wide association studies (GWAS), which are a method for high-resolution mapping of complex trait loci based on linkage disequilibrium. This method results in higher resolution (often up to gene level) by performing whole-genome scans to identify haplotype blocks that are significantly correlated with quantitative trait variation (Brachi et al., 2011). In addition, GWAS (also known as association mapping) is a quick and effective method that has been used in several crop species to identify candidate genes/QTLs involved in abiotic stress tolerance. This method is based on the screening of single nucleotide polymorphisms (SNPs) across different chromosomes of a tolerant variety and co-relating it with stress tolerance trait. In one of the reports, QTLs for salt tolerance in rice (during the reproductive stage) was identified by screening SNPs in a customized SNP array containing around 6000 stress-responsive genes (Kumar et al., 2015). This genotyping information was co-related with the phenotypic expression of 220 diverse rice accessions for several morphological, biochemical (Na^+ and K^+ content) and agronomic traits. Twelve of the twenty SNPs identified in this study were found in the region of *Saltol*, a major QTL on chromosome 1 in rice, involved in controlling salinity tolerance at seedling stage (Bonilla et al., 2002). Likewise, GWAS in 368 different rapeseed accessions revealed 75 SNPs distributed across 14 chromosomes, which were associated with four salt tolerance-related traits (Wan et al., 2017). The high throughput sequencing, combined with GWAS, has been used to identify potential molecular markers, such as SNPs, insertions and deletions, and copy number variations, which are associated with growth and development and/or stress responses.

23.3.4 Omics Approaches

Since the molecular mechanism of stress tolerance is complex, it requires information at the whole genome level to understand it effectively. The next-generation sequencing technology has greatly facilitated the identification of potential molecular markers, such as single nucleotide polymorphisms, insertions and deletions, and copy number variations, which are associated with growth and development and/or stress responses. With the advent of next-generation genome sequencing, the fields of transcriptomics, proteomics and metabolomics have been developed to cover the genome-wide level of transcripts, proteins and metabolites, respectively. In this regard, these omics technologies are making an enormous contribution to our understanding of abiotic stress signaling at the whole genome level (Deshmukh et al., 2014). For example, NGS technology coupled with high throughput transcriptome profiling has been used to investigate genome-wide changes in transcripts in response to stresses (Molina et al., 2011; Ma et al., 2012; Liu et al., 2015).

Whole genome transcriptome profiling using microarray enables simultaneous analysis of thousands of genes at a time (those present on oligo chip). In fact, the RNA sequence provides an unbiased analysis of all the RNA transcripts of a given cell. Since the proteins represent the products of gene expression, proteomics provides a

better picture of the cellular condition than its transcriptome. It becomes necessary that the analytical platform is equipped to analyze potential post-translational modifications and protein–protein interactions. Among all -*omics* technologies, metabolomics present an even better picture of the phenotype than either transcriptomics or proteomics, as metabolites represent the integration of gene expression, protein interaction and other different regulatory processes (Arbona et al., 2013). An important aspect of metabolome profiling is the combination of extraction and detection techniques to increase the coverage of the study. A concomitant feature of all the omics approaches is the generation of huge amounts of high throughput data, which requires extensive computational resources for storage and analysis.

23.4 CONFERRING ABIOTIC STRESS TOLERANCE BY TRANSGENIC APPROACH

23.4.1 Drought Stress

Drought is the major abiotic factor affecting ~25% arable land (Blum, 1988); this problem is further aggravated due to dwindling supply of irrigation water. As mentioned earlier, water deficit conditions have contrasting effects on root and shoot growth; it enhances root growth while inhibiting shoot growth and leaf area. In fact, the inhibition of shoot growth and reduction in leaf area is often the first visible symptom under drought conditions (Farooq et al., 2012). Water deficit conditions also lead to loss of turgor pressure from plant cells, giving "wilted appearance" to the plant. Perhaps the quickest response of plants to drought is stomatal closure; this occurs to conserve water loss through stomatal pores. A significant disadvantage of this acclimatization response is reduced carbon fixation and, hence, reduced productivity. Drought stress triggers a wide variety of plant responses, like alterations in expression of drought-responsive genes, synthesis of specific proteins like LEAs, chaperones, etc., and rise in the levels of metabolites like ABA, antioxidants and osmoprotectants. The application of these mechanisms in raising drought-tolerant transgenics has been described below.

23.4.1.1 Stress-Responsive Transcription Regulation

As mentioned in Section 23.2.2, plant cells, upon receiving the abiotic stress signal, initiate a signaling cascade that culminates in altering the expression of several stress-responsive genes. Microarray analyses conducted in the model plants *Arabidopsis* and rice have led to the identification of several stress-inducible genes, some of which have been transformed into crop species and have resulted in significant stress tolerance (Zhang et al., 2004; Umezawa et al., 2006). Apart from altering the levels of functional genes, targeting the expression of regulatory genes like transcription factors (which act upstream to a myriad of functional or single effect genes) is another method of imparting stress tolerance to the crop species. An advantage of this strategy is the ease with which the expression of multiple genes (acting downstream to the regulatory gene) could be altered in a single transformation event. The central drought stress-inducible TFs include ABA-responsive element binding protein (AREBs), NAC (an acronym for NAM, *ATAF1-2* and CUC2 myeloblastosis-related proteins [MYBs]) and DREBs. Some of these transcription factors are also involved in salinity response pathway.

AREBs belong to a class of basic leucine zipper (bZIP) domain containing TFs, and they function in ABA signaling during seed maturation and water stress. In response to ABA, AREB binds to conserved *cis*-regulatory element, ABA-responsive element (ABRE), to trigger expression of downstream drought-responsive genes like *rd29B* (Uno et al., 2000; Fujita et al., 2005). Transgenic *Arabidopsis* seedlings overexpressing AREB1 showed ABA hypersensitivity and enhanced drought tolerance, mainly through induction of LEA proteins, which protect the plants from water deficit stress (Fujita et al., 2005). Transgenic tomato plants overexpressing the transcription factor *SlAREB* showed increased tolerance to salt and water stress compared to wild-type, as assessed by physiological parameters such as relative water content, chlorophyll fluorescence and damage by lipoperoxidation (Orellana et al., 2010). Another important class of TFs includes NAC family, which encode plant-specific TFs (Riechmann et al., 2000). Transgenic rice overexpressing *STRESS-RESPONSIVE NAC* (*SNAC1 1*) showed significantly enhanced drought resistance (22–34% higher seed setting than control) under severe drought stress conditions at the reproductive stage while showing no phenotypic changes or yield penalty (Hu et al., 2006). Likewise, *OsNAC6* is highly inducible by both ABA and salinity stress, and its overexpression enhances plant tolerance to drought, salt and blast disease (Nakashima et al., 2007). For a detailed account of the strategies involved in engineering drought tolerance in crops, the readers may refer to other reviews (Zhu, 2002; Umezawa et al., 2006).

23.4.1.2 Stress-Responsive Protein Modifications

After discussing the regulatory role of TFs in controlling the expression of stress-responsive genes, it becomes necessary to mention the role of various post-translational modifications (including phosphorylation, methylation, ubiquitination, sumoylation, etc.) involved in the abiotic stress signal transduction pathway. These protein modifications bring about a change in protein conformation and facilitate in transmitting the stress signal through intermediates from the cytosol to nuclei. For example, the activity of AREB1 TF is regulated by ABA-dependent phosphorylation by SNF1-related protein kinase 2 (SnRK2) (Furihata et al., 2006). The stability of various proteins or intermediates of the signaling pathway is often regulated by ubiquitination. The ubiquitination is a process of ubiquitin-mediated degradation. The covalent attachment of ubiquitin to a lysine residue of a target protein can regulate its stability, activity and trafficking, thereby modulating the protein turnover in response to external stimulus (Lyzenga and Stone, 2011).

23.4.1.3 Stress-Responsive Accumulation of Metabolites

The proline biosynthetic pathway has been targeted to increase the levels of osmoprotectants to confer drought tolerance in various crop plants. Drought-induced overexpression of *P5CS* in rice and *P5CR* in soybean resulted in increased drought tolerance, as revealed by better growth under water deficit conditions (Zhu et al., 1998; Kocsy et al., 2005; Hsieh et al., 2002). Similarly, maize plants transformed with *betA* gene of *E. coli* (encoding choline dehydrogenase) were more tolerant to drought stress than wild-type (Quan et al., 2004). The increased accumulation of glycine betaine provided greater protection of the integrity of cell membrane and greater activity of enzymes and, therefore, results in higher grain yield after drought treatment.

Drought tolerance is often conferred upon by the high levels of LEA proteins. For instance, the *HVA1* gene of barley, which encodes for a group 3 LEA protein, is induced by a variety of abiotic stresses like drought, cold, salinity and heat (Straub et al., 1994). This gene has been overexpressed in many important crop species like rice, wheat and mulberry, where it has been demonstrated to confer tolerance against drought and salinity (Xu et al., 1996; Sivamani et al., 2000; Babu et al., 2004; Lal et al., 2008). In addition, transgenic rice overexpressing *HVA1* exhibited high growth rate and faster recovery under drought stress as compared to wild-type plants. The stress tolerance of transgenics mainly resulted from better cell membrane protection, as indicated by low membrane electrolyte leakage, while there was virtually no difference in the level of osmolytes between wild-type and transgenics (Babu et al., 2004). Like other classes of LEA proteins, group 2 LEA proteins, called dehydrins, are also induced in vegetative tissues following salinity, dehydration, cold and freezing stress and have been used to confer abiotic tolerance in plants (Hanin et al., 2011). Transgenic alfalfa plants expressing manganese superoxide dismutase cDNA also exhibited reduced injury from water deficit stress as determined by chlorophyll fluorescence, electrolyte leakage and regrowth from crowns (Mckersie et al., 1996).

23.4.2 SALINITY STRESS

Approximately 20% of the total cultivated and 33% of the irrigated agricultural land is being affected due to salinity (Jamil et al., 2011). The adverse effects of salt stress on plants mainly results due to osmotic stress and ion toxicity. The high salinity of soil or irrigation water leads to a reduction of soil water potential, ultimately resulting in decreased water uptake ability (Serrano et al., 1999). Also, the inhibition of uptake of K^+ ions (due to high concentrations of Na^+) results in disrupting intracellular Na^+/K^+ homeostasis and inhibition of enzyme activity. Fundamentally, plants cope with salt stress by either avoiding or tolerating salt stress. The mechanisms of salt tolerance in plants are described in the following sections.

23.4.2.1 Targeting Ion Transporters

The characterization of salt overly sensitive (SOS) *Arabidopsis* mutants, which were hypersensitive to salt, revealed the discovery of a novel calcium-regulated salt-stress tolerance pathway in plants (Shi et al., 2000, 2003a). The salt tolerant species manage to maintain a low concentration of Na^+ in the cytosol by regulating the expression and activity of transporters of K^+, Na^+ and H^+. Like other abiotic stresses, salinity stress also induces an increase in cytosolic calcium levels (Ca^{2+}); the role of Ca^{2+} signaling in stress response has been explained in detail in Section 23.2.3. During salinity stress, the increase in the cytosolic concentration of calcium is perceived by a Ca^{2+} sensor, SOS3/CBL4, which interacts physically with the regulatory domain of a calcineurin B-like-interacting protein kinase (CIPK), SOS2/CIPK24. The activated SOS2 kinase regulates the activity of SOS1, a plasma membrane Na^+/H^+antiporter, and NHX1, a tonoplast Na^+/H^+ antiporter (Zhu, 2001a,b). The Na^+/H^+ antiporter protein transports Na^+ into the vacuole, driven by an electrochemical gradient of protons generated by vacuolar H^+-ATPases and H^+-PPases. The SOS pathway functions to maintain cellular homeostasis by promoting Na^+ efflux from the cell (by the action of plasma membrane Na^+/H^+antiporter) and vacuolar compartmentation of excess Na^+ ions (by the action of the

tonoplast antiporter). Plants overexpressing members of SOS pathway have been implicated in enhanced salt tolerance (Zhu, 2001a,b). Other members of SOS pathway include SOS4, a pyridoxal kinase involved in the biosynthesis of pyridoxal-5-phosphate, and SOS5, a putative cell surface adhesion protein that functions during salt stress (Shi et al., 2003b).

23.4.2.2 Targeting Transcription Factors

Some of the TFs involved in drought stress responses have also been reported to regulate salinity responses. For example, transgenic rice overexpressing *OsNAC5* exhibited enhanced salinity stress tolerance with no obvious constraints on plant growth under favorable conditions (Takasaki et al., 2010). In addition, some signaling kinases, like salt-stress-inducible map kinase (SIMK), have been reported to be activated upon hyperosmotic stress treatments (Munnik et al., 1999; Bartels and Sunkar, 2005). Over expression of some drought-responsive genes has been effective for conferring salt tolerance as well. Thus, it re-iterates the fact that multiple abiotic stresses induce common signaling intermediates.

23.4.2.3 Other Strategies for High Salt Tolerance

The enzymes involved in synthesizing osmoprotectants have also been targeted to impart stress tolerance to high salt conditions. For example, the levels of polyamines have been increased in transgenic rice plants by expressing arginine decarboxylase, a key enzyme of polyamine biosynthesis pathway under the control of an ABA-inducible promoter (Roy and Wu, 2001). These transgenics exhibited increased biomass than the wild-type plants under salt stress. These plants were found to accumulate higher levels of spermine and spermidine, as a result of enhanced putrescine levels. On the contrary, wild-type plants failed to accumulate significant polyamine levels and hence were stress sensitive. In another report, the levels of glycine betaine have been increased in transgenic potato plants by overexpressing the *codA* gene of *Arthobacter globiformis* in chloroplast under the control of an oxidative stress-inducible *SWPA2* promoter to impart salt and drought tolerance (Ahmad et al., 2008). The increase in hydrogen peroxide content in these transgenics also contributed to stress tolerance as a ROS scavenger, and also as an important signaling molecule. Similarly, transgenic tobacco plants overexpressing the *P5CS* gene (proline biosynthetic pathway) produced 10- to 18-fold more proline and exhibited better performance under salt stress (Kishor et al., 1995).

Similarly, in another report, a cytosolic copper-zinc SOD from the mangrove plant *Avicennia marina* was overexpressed in *indica* rice. The transgenic plants were more tolerant to methyl viologen-mediated oxidative stress and could withstand salinity stress of 150 Mm NaCl for a period of eight days, while the untransformed control plants wilted at the end of stress treatment (Prashanth et al., 2008). The role of glutathione peroxidase and glutathione-S-transferase during salt stress was demonstrated in transgenic tobacco, where the overexpression of these enzymes increased the GSH content. The transgenics grew faster than the wild-type and showed increased tolerance to salt and chilling stress (Roxas et al., 1997).

23.4.3 Cold Stress

Cold stress is often described by chilling injury or freezing injury. These similar-sounding terms are distinct in the sense that chilling injury occurs at above zero temperatures while freezing injury is the damage caused at sub-zero temperatures. The most harmful effects of chilling stress (that is faced by tropical and subtropical plants) occur on the cell membrane, which undergoes a transition to a more gel-like conformation/less fluid state (Prasad et al., 1994; Thomashow, 1994). In contrast, the effect of freezing is evident in the form of dehydration, which results during freezing and subsequent thawing; the former resulting in the efflux of water, the latter leading to the influx of water into the cells (Thomashow, 1994). This movement of water in and out of the cells during freezing and thawing puts considerable strain on the plasma membrane. Since the effect of cold stress primarily occurs on the cell membrane, cold tolerance involves modifying the lipid composition to regulate membrane permeability. The more gel-like state of the membrane could be regulated to optimum fluidity (maintaining semi-permeable nature) by increasing the level of polyunsaturated fatty acids. The major areas that have been targeted for conferring cold tolerance in plants have been discussed below.

23.4.3.1 Targeting Membrane Fluidity

Under cold stress, membrane fluidity can be brought to optimum levels by overexpression of omega-3 fatty acid desaturase (encoded by *fad7* gene), which increases the proportion of polyunsaturated fatty acids in the cell membrane. Transgenic tobacco plants containing increased levels of trienoic fatty acids (α-linolenic acid [18: 3] and hexadecatrienoic acid [16: 3]) were obtained by introduction of a chloroplast *fad7* gene isolated from *Arabidopsis thaliana* (Kodama et al., 1994). In another strategy, the *fad* gene was expressed under the control of cold-inducible promoter *COR15A*. Survival of *COR15A–FAD7* transgenic lines (40.2–96%) was far superior to the

wild type (6.7–10.2%) when exposed to low temperatures (0.5, 2 or 3.5°C) for up to 44 days (Khodakovskaya et al., 2005).The role of dehydrins, a class II LEA protein, in freezing tolerance has been reported for wheat (Houde et al., 1992), *Vigna ungicuata* (Ismail et al., 1999) and *Arabidopsis* (Puhakainen et al., 2004). Dehydrins, when overexpressed in *Arabidopsis*, were found to impart freezing tolerance to the transgenic plants (Puhakainen et al., 2004). Apart from the above strategies, the general stress responses like a synthesis of osmoprotectants and ROS scavenging systems like SOD and GSH (glutathione S-transferase) have also been overexpressed in crops to impart abiotic stress tolerance.

23.4.3.2 Targeting the Cold-Responsive Transcription Factors

In *Arabidopsis*, CBF/DREB1 TFs induce cold-responsive genes like *COR* (cold-regulated), *KIN* (cold inducible), *RD* (responsive to desiccation) and *ERD* (early dehydration-inducible) to regulate cold responses. Since cold exerts its effects on gene expression largely through an ABA-independent pathway, targeting the cold-responsive genes could be another strategy to confer cold tolerance. Similar to *Arabidopsis*, the DREB TFs of rice (*OsDREB1A* and *OsDREB1B*) are also induced by cold stress, and overexpression of either of these cold-responsive genes has been reported to confer freezing tolerance in transgenic plants (Thomashow, 1998; Ito et al., 2006).

23.4.4 High-Temperature Tolerance

Heat stress is defined as the rise in temperature beyond a threshold level for a period of time sufficient to cause irreversible damage to plant growth and development. Heat tolerance is generally defined as the ability of the plant to grow and produce economic yield under high temperatures (Wahid et al., 2007). Under field conditions, high-temperature stress is frequently associated with reduced water availability, increased membrane permeability and alteration in the structure of proteins (Simoes-Araujo et al., 2003). Various abiotic stresses, especially heat stress, causes dysfunction of structural and functional proteins by denaturing their native conformation, leading to aggregation of denatured/misfolded proteins (Wang et al., 2004).

23.4.4.1 Targeting Chaperones

Heat shock proteins (HSPs), which could be broadly divided into three classes – HSP90, HSP70 and low molecular weight (sHsps, 15–30 kDa), act as molecular chaperones that stabilize protein structures and membranes during high temperature and also assist in re-folding of the misfolded proteins under heat stress

(Wahid et al., 2007). The different classes of HSPs/chaperones cooperate in cellular protection and play complementary and sometimes overlapping roles in the protection of proteins from stress (Wang et al., 2005). It has been reported that the sHsps stabilize the stress-denatured proteins (by preventing their aggregation) for subsequent refolding by members of the Hsp70/Hsp100 chaperone families (Veinger et al., 1998; Lee and Vierling, 2000).

23.4.4.2 Targeting Membrane Fluidity

Contrary to cold stress, heat stress causes increased fluidity of cell membranes, which occurs either due to denaturation of proteins or an increase in the levels of unsaturated fatty acids (Savchenko et al., 2002). As a result, the permeability of cell membranes is increased, often resulting in electrolyte leakage. In heat-tolerant plants like those growing in deserts, the levels of polyunsaturated fatty acids like α-linolenic acid (18: 3) and hexadecatrienoic acid (16: 3) are highly reduced. In order to engineer this trait into heat-sensitive plants, *fad7*, a key target enzyme involved in the biosynthesis of polyunsaturated fatty acids, has been silenced to confer tolerance to heat stress (Murakami et al., 2000). It is interesting to note that the same enzyme could be targeted to regulate membrane fluidity in opposite ways. Its levels could be either overexpressed to make plants more cold stress tolerant, or it could be decreased through gene silencing approach in order to make the plants more heat stress tolerant.

23.4.4.3 Other Strategies for Heat Tolerance

Apart from modifying cellular fluidity, some of the other strategies used to make plants more tolerant to heat stress includes overexpression of heat shock protein genes, alteration of levels of heat shock transcription factor proteins, increasing the levels of osmolytes and increasing levels of cell detoxification enzymes.

23.5 CONCLUSIONS AND FUTURE PERSPECTIVES

The increase in population, along with climate change and a decrease in arable land, necessitates the production of crops that are tolerant to adverse abiotic conditions. Identifying and mining genes involved in stress response represents a key step to unraveling and manipulating stress tolerance in plants. Although the recent omics approach and association mapping have significantly shaped up our understanding of mechanisms of stress tolerance in plants, these technologies are limited to few labs in the world that can afford the expensive, sophisticated infrastructure and computational tools to analyze high throughput data. A feasible approach is genetic engineering, where the use of stress-inducible promoters has revolutionized the field of conferring abiotic stress tolerance to crop species. However, the virtues of genetic engineering could not be fully realized due to certain ethical and biosafety concerns raised by anti-GM organizations. One of the most frequently raised objections is the "transgene escape" through horizontal transfer of antibiotic resistance genes to bacteria (soil or gut bacteria) and weeds. This issue could be solved by raising marker-free transgenics or substituting the antibiotic selection marker genes with alternative marker genes like *isopentenyl transferase, phospho mannose isomerase,* etc. Also, there is a need to relax the regulatory framework that has been plaguing the transgenic crops generated after years of hard work in the laboratory from entering the fields. The story of "Bt brinjal" and "GM mustard" in India is a perfect example, where a lot of hue and cry has been done to stop the commercial release of these transgenic crops in spite of clearing all the biosafety tests and regulations. The outlook and perspective of neighboring countries like Bangladesh are inspiring and need to be appreciated, where the field trials of the former crop and golden rice have been done. Another significant advancement in the field of genetic engineering is the use of genome editing by the CRISPR-Cas9 system. This is a highly specific and precise method of gene targeting; hence, it does not suffer from off-target effects, which were common with the earlier generations of transgenic plants (Haque et al., 2018).

REFERENCES

Acquaah, G. (2007) *Principles of Plant Genetics and Breeding.* Blackwell, Oxford, UK.

Ahmad, R., Kim, M.D., Black, K.H., Kim, H.S., Lee, H.S., Kwon S.Y., Murata, N., et al. (2008) Stress-induced expression of choline oxidase in potato plant chloroplasts confers enhanced tolerance to oxidative, salt, and drought stresses. *Plant Cell Reports* 27:687–698.

Aida, M., Ishida, T., Fukaki, H., Fujisawa, H., Tasaka, M. (1997) Genes involved in organ separation in *Arabidopsis*: An analysis of the *cup-shaped cotyledon* mutant. *Plant Cell* 9:841–857.

Albrecht, V., Weinl, S., Blazevic, D., Angelo, C.D., Blastistic O., Kolukisaoglu, U., Bock, R., et al. (2003) The calcium sensor CBL1 integrates plant responses to abiotic stresses. *Plant Journal* 36:457–470.

Arbona, V., Manzi, M., Ollas, C.D., Gómez-Cadenas, A. (2013) Metabolomics as a tool to investigate abiotic stress tolerance in plants. *International Journal of Molecular Sciences* 14:4885–4911.

Ashraf, M., Foolad, A. (2007) Roles of glycine betaine and proline in improving plant abiotic stress resistance. *Environmental and Experimental Botany* 59:206–216.

Babu, R.C., Zhang, J., Blum, A., Ho, T.H.D., Wu, R., Nguyen, H.T. (2004) HVA1, a LEA gene from barley confers dehydration tolerance in transgenic rice (*Oryza sativa* L.) via cell membrane protection. *Plant Science* 166:855–862.

Banno H, Hirano K, Nakamura T, Irie K, Nomoto S, Matsumoto K, Machida Y (1993) NPK1, a tobacco gene that encodes a protein with a domain homologous to yeast BCK1, STE11, and Byr2 protein kinases. *Molecular Cell Biology* 13:4745–4752.

Bartels, D., Sunkar, R. (2005) Drought and salt tolerance in plants. *Critical Reviews in Plant Sciences* 24:23–58.

Battaglia, M., Olvera-Carrillo, Y., Garciarrubio, A., Campos, F., Covarrubias, A.A. (2008) The enigmatic LEA proteins and other hydrophilins. *Plant Physiology* 148:6–24.

Bhatnagar-Mathur, P., Vadez, V., Sharma, K.K. (2008) Transgenic approaches for abiotic stress tolerance in plants: retrospect and prospects. *Plant Cell Reports* 27:411–424.

Blum, A. (1988) *Plant Breeding for Stress Environments*. CRC Press, Boca Raton, FL.

Bonilla, P., Dvorak, J., Mackill, D., Deal, K., Gregorio, G. (2002) RFLP and SSLP mapping of salinity tolerance genes in chromosome 1 of rice (*Oryza sativa* L.) using recombinant inbred lines. *Philippine Journal of Agricultural Scientist* 85:68–76.

Borevitz, J.O., Nordborg, M. (2003) The impact of genomics on the study of natural variation in *Arabidopsis*. *Plant Physiology* 132:718–725.

Boyer, J.S. (1982) Plant productivity and environment. *Science* 218:443–448.

Brachi, B., Morri, G.P., Borevitz, J.O. (2011) Genome-wide association studies in plants: The missing heritability is in the field. *Genome Biology* 12:232.

Bray, E.A., Bailey-Serres, J.W., Retilnyk, E., Gruissem, W., Buchannan, B., Jones, R. (2000) Responses to abiotic stresses. In *Biochemistry and Molecular Biology of Plants*, American Society of Plant Physiologists:Rockville, MD, 1158–1249.

Chen, T.H., Murata, N. (2002) Enhancement of tolerance of abiotic stress by metabolic engineering of betaines and other compatible solutes. *Current Opinion in Plant Biology* 5:250 257.

Cheng, S., Willmann, M.R., Chen, H.C., Sheen, J. (2002) Calcium signaling through protein kinases. The *Arabidopsis* calcium-dependent protein kinase gene family. *Plant Physiology* 129:469–485.

Chinnusamy, V., Schumaker, K., Zhu, J.K. (2004) Molecular genetic perspectives on cross-talk and specificity in abiotic stress signaling in plants. *Journal of Experimental Botany* 55:225–236.

Chunqing, L., Xuekun, Z., Zhang, K., Hong, A., Hu, K., Wen, J., Shen, J.X., et al. (2015) Comparative analysis of the *Brassica napus* root and leaf transcript profiling in response to drought stress. *International Journal of Molecular Science* 16:18752–18777.

Corpas, F.J., Fernández-Ocaña, A., Carreras, A., Valderrama, R., Luque, F., Esteban, F.J., Rodríguez-Serrano, M., et al. (2006) The expression of different superoxide dismutase

forms is cell-type dependent in olive (*Olea europaea* L.) leaves. *Plant and Cell Physiology* 47:984–994.

Deshmukh, R., Sonah, H., Patil, G., Chen, W., Prince, S., Mutava, R., Vuong, T., Valliyodan, B. and Nguyen, H.T. (2014) Integrating omic approaches for abiotic stress tolerance in soybean. *Frontiers in Plant Science* 5:244.

Djilianov, D., Georgieva, T., Moyankova, D., Atanassov, A., Shinozaki, K., Smeeken, S.C.M., Verma, D.P.S.M., et al. (2005) Improved abiotic stress tolerance in plants by accumulation of osmoprotectants – Gene transfer approach. *Biotechnology and Biotechnological Equipment* 19:63–71.

Fan, Q., Song, A., Jiang Zhang, T., Sun, H., Wang, Y. (2016) CmWRKY1 enhances the dehydration tolerance of *Chrysanthemum* through the regulation of ABA-associated genes. *PLOS One* 11:e0150572.

Farooq, M., Hussain, M., Wahid, A., Siddique, K.H.M. (2012) Drought stress in plants: An overview. In *Plant Responses to Drought Stress*, Springer-Verlag, Berlin, Heidelberg: 2012: 1–33.

Fujita, Y., Fujita M., Satoh, R., Maruyama, K., Mohammad, P.M., Seki, M., Hiratsu, K., et al. (2005) AREB1 is a transcription activator of novel ABRE-dependent ABA signaling that enhances drought stress tolerance in *Arabidopsis*. *Plant Cell* 17:3470–3488.

Furihata, T., Maruyama, K., Fujita, Y., Umezawa, T., Yoshida, R., Shinozaki, K. (2006) Abscisic acid-dependent multisite phosphorylation regulates the activity of a transcription activator AREB1. *Proceeding of the National Academy of Science of the United States of America* 103:1988–1993.

Gangadhar, B.H., Sajeesh, K., Venkatesh, J., Baskar, V., Abhinandan, K., Yu, J.W., Prasad, R., et al. (2016) Enhanced tolerance of transgenic potato plants overexpressing non-specific lipid transfer protein-1 (StnsLTP1) against multiple abiotic stresses. *Frontiers in Plant Science* 7:1228.

Groppa, M.D., Benavides, M.P. (2008) Polyamines and abiotic stress: Recent advances. *Amino Acids* 34:35–45.

Haque, E., Taniguchi, H., Karim, R., Hassan, M., Bhowmik, P., Smiech, M., Zhao, K., et al. (2018). Application of CRISPR/Cas9 genome editing technology for the improvement of crops cultivated in tropical climates: Recent progress, prospects and challenges. *Frontiers in Plant Science*, Vol 9: 617.

Hanin, M., Brini, F., Ebel, C., Toda, Y., Takeda, S., Masmoudi, K. (2011) Plant dehydrins and stress tolerance: Versatile proteins for complex mechanisms. *Plant Signaling and Behavior* 6:1503–1509.

Houde, M., Dhindsa, R.S., Sarhan, F. (1992) A molecular marker to select for freezing tolerance in Gramineae. *Molecular and General Genetics* 234:43–48.

Houtte, V.H., Vandesteene, L., López-Galvis, L., Lemmens, L., Kissel, E., Carpentier, S. (2013) Overexpression of the trehalase gene AtTRE1 leads to increased drought stress tolerance in Arabidopsis and is involved in abscisic acid-induced stomatal closure. *Plant Physiology* 161:1158–1171.

Hsieh, T.H., Lee, J.T., Charng, Y.Y., Chan, M.T. (2002) Tomato plants ectopically expressing *Arabidopsis* CBF1show enhanced resistance to water deficit stress. *Plant Physiology* 130:618–626.

Hu, H., Mingqiu, D., Jialing, Y., Benze, X., Xianghua, L., Zhang, Q., Xiong, L. (2006) Overexpressing a NAM, ATAF, and CUC (NAC) transcription factor enhances drought resistance and salt tolerance in rice. *Proceeding of the National Academy of Science of the United States of America* 103:12987–12992.

Hundertmark, M., Hincha, D.K. (2008) LEA (Late Embryogenesis Abundant) proteins and their encoding genes in *Arabidopsis thaliana. BMC Genomics* 9:118.

Ichimura, K., Mizoguchi, T., Yoshida, R., Yuasa, T., Shinozaki, K. (2000) Various abiotic stresses rapidly activate *Arabidopsis* MAP kinases ATMPK4 and ATMPK6. *The Plant Journal* 24:655–665.

Imlay, J.A. (2003) Pathways of oxidative damage. *Annual Review of Microbiology* 57:395–418.

Ismail, A.M., Hall, A.E., Close, T.J. (1999) Allelic variation of a dehydrin gene co-segregates with chilling tolerance during seedling emergence. *Proceeding of the National Academy of Science of the United States of America* 96:13566–13570.

Ito, K.Y., Kyonoshin, K., Teruaki, M., Masatomo, T., Motoaki, K., Kazuo, S., Kazuko, S. et al. (2006) Functional analysis of rice DREB1/CBF-type transcription factors involved in cold-responsive gene expression in transgenic rice. *Plant and Cell Physiology* 47:141–153.

Jaspers, P., Kangasjärvi, J. (2010) Reactive oxygen species in abiotic stress signaling. *Physiologia Plantarum* 138:405–413.

Jamil, A., Riaz, S., Ashraf, M., Foolad, M.R. (2011) Gene expression profiling of plants under salt stress. *Critical Review in Plant Science* 30:435–458.

Khare, N., Goyary, D., Singh, N.K., Shah, P., Rathore, M., Anandhan, S. (2010) Transgenic tomato cv. Pusa Uphar expressing a bacterial mannitol-1-phosphate dehydrogenase gene confers abiotic stress tolerance. *Plant Cell Tissue Organ Culture* 2:267–277.

Khodakovskaya, M., Li Y., Li J., Vanova, R., Malbeck, J., McAvoy, R. (2005) Effects of cor15a–IPT gene expression on leaf senescence in transgenic petunia and chrysanthemum. *Journal of Experimental Botany* 56:1165–1175.

Kishor, P.B.K., Hong, Z., Miao, C.H., Hu, C.A., and Verma, D.P.S. (1995) Overexpression of Al-Pyrroline-5-Carboxylate Synthetase Increases Proline Production and Confers Osmotolerance in Transgenic Plants. *Plant Physiology* 108:1387–1394.

Kishor, P.B.K., Sangam, S., Amrutha, R.N., Sri Laxmi, P., Naidu, K.R., Rao, K.R.S.S., Rao, S., et al. (2005) Regulation of proline biosynthesis, degradation, uptake and transport in higher plants: Its implication in plant growth and abiotic stress tolerance. *Current Science* 88:424–438.

Knight, H. (1999) Calcium signaling during abiotic stress in plants. *International Review of Cytology* 195:269–324.

Knight, H., Knight, M.R. (2001) Abiotic stress signaling pathways: specificity and cross-talk. *Trends in Plant Science* 6:262–267.

Kocsy, G., Laurie, R., Szalai, G., Szilágyi, V., Simon-Sarkadi, L., Galiba, G., De Ronde, J.A. (2005) Genetic manipulation of proline levels affects antioxidants in soybean subjected to simultaneous drought and heat stresses. *Physiologia Plantarum* 124:227–235.

Kodama, H., Hamada, T., Horiguchi, G., Nishimura, M., Iba, K. (1994) Genetic enhancement of cold tolerance by expression of a gene for chloroplast [omega]-3 fatty acid desaturase in transgenic tobacco. *Plant Physiology* 105:601–605.

Kovtun, Y., Chiu, W.L., Tena, G., Sheen, J. (2000) Functional analysis of oxidative stress-activated mitogen-activated protein kinase cascade in plants. *Proceedings of the National Academy of Sciences* 97:2940–2945.

Kumar, S., Singh, A. (2016) Epigenetic regulation of abiotic stress tolerance in plants. *Advances in Plants and Agriculture Research* 5: 00179.

Kumar, V., Singh, A., Mithra, S.A., Krishnamurthy, S.L., Parida, S.K., Jain, S., Tiwari, K.K., Kumar, P., Rao, A.R., Sharma, S.K., Khurana, J P. (2015) Genome-wide association mapping of salinity tolerance in rice (*Oryza sativa*). *DNA Research* 22:133–145.

Kwon, S., Jeong, Y.J., Lee, H.S., Kim, J.S., Cho, K.Y., Allen, R.D., Kwak, S.S. (2002) Enhanced tolerances of transgenic tobacco plants expressing both superoxide dismutase and ascorbate peroxidase in chloroplasts against methyl viologen-mediated oxidative stress. *Plant, Cell and Environment* 25:873–882.

Lal, S., Gulyani, V.K., Khurana, P. (2008) Overexpression of HVA1 gene from barley generates tolerance to salinity and water stress in transgenic mulberry (*Morus indica*). *Transgenic Research* 17:651.

Lata, C., M. Prasad. (2011) Role of DREBs in regulation of abiotic stress responses in plants. *Journal of Experimental Botany* 62:4731–4748.

Lee, G.J., Vierling, E. (2000) A small heat shock protein cooperates with heat shock protein 70 systems to reactivate a heat-denatured protein. *Plant Physiology* 122: 189–197.

Ling, H., Zeng, X., & Guo, S. (2016) Functional insights into the late embryogenesis abundant (LEA) protein family from Dendrobium officinale (Orchidaceae) using an Escherichia coli system. *Scientific reports* 6:39693.

Liu, C., Zhang, X., Zhang, K., An, H., Hu, K., Wen, J., Shen, J., et al. (2015) Comparative analysis of the *Brassica napus* root and leaf transcript profiling in response to drought stress. *International Journal of Molecular Science* 16:18752–18777.

Lyzenga, W.J., Stone, S.L. (2011) Abiotic stress tolerance mediated by protein ubiquitination. *Journal of Experimental Botany* 63:599–616.

Ma, Y., Qin, F., Tran, L.S.P. (2012) Contribution of genomics to gene discovery in plant abiotic stress responses. *Molecular Plant* 5:1176–1178.

Mantri, N., Patade V., Penna S., Ford R. (2012) Abiotic stress responses in plants: Present and future. *Abiotic Stress Responses in Plants* 1–19.

Marco, F., Bitrián, M., Carrasco, P., Rajam, M.V., Alcázar, R., Tiburcio, A.F. (2015) Genetic engineering strategies for abiotic stress tolerance in plants. In *Plant Biology and Biotechnology*. Springer, New Delhi, India, 579–609.

Matsukura, S., Mizoi, J., Yoshida, T., Todaka, D., Ito, Y., Maruyama, K., Shinozaki, K., et al. (2010) Comprehensive analysis of rice DREB2-type genes that encode transcription factors involved in the expression of abiotic stress-responsive genes. *Molecular Genetics and Genomics* 283:185–196.

Mckersie, B.D., Bowley, S.R., Harjanto, E., Leprince, O. (1996) Water-deficit tolerance and field performance of transgenic alfalfa overexpressing superoxide dismutase. *Plant Physiology* 111:1177–1181.

Mishra, B. (1994) Breeding for salt tolerance in crops. *Salinity Management For Sustainable Agriculture* 226–259.

Mittler, R. (2002) Oxidative stress, antioxidants and stress tolerance. *Trends Plant in Science* 7:405–410.

Molina, C., Zaman Allah M., Khan, M., Fatnassi, N., Homes, R., Rotter, B., Steinhauer, D., et al. (2011) The salt-responsive transcriptome of chickpea root and nodules via deep super SAGE. *BMC Plant Biology* 11:31.

Munnik, T., Ligterink, W., Meskiene, L., Calderini, O., Beyerly, J., Musgrave, A., Hirt, H. (1999) Distinct osmo-sensing protein kinase pathways are involved in signaling moderate and severe hyper-osmotic stress. *The Plant Journal* 20:381–388.

Murakami, Y., Tsuyama, M., Kobayashi, Y., Kodama, H., Iba, K. (2000) Trienoic fatty acids and plant tolerance of high temperature. *Science* 287:476–479.

Nakashima, K., Tran, L.S.P., Van Nguyen, D., Fujita, M., Maruyama, K., Todaka, D., Ito, Y., et al. (2007) Functional analysis of a NAC-type transcription factor OsNAC6 involved in abiotic and biotic stress-responsive gene expression in rice. *The Plant Journal* 51:617–630.

Nakashima, K., Takasaki, H., Mizoi, J., Shinozaki, K., Shinozaki, K.Y. (2012) NAC transcription factors in plant abiotic stress responses. *Biochimica et Biophysica Acta (BBA) Gene Regulatory Mechanisms* 1819:97–103.

Nevo, E., Chen, G. (2010) Drought and salt tolerances in wild relatives for wheat and barley improvement. *Plant, Cell and Environment* 33:670–685.

Olvera-Carrillo, Y., LuisReyes, J.C., Ovarrubias, A.A. (2011) Late embryogenesis abundant proteins: versatile players in the plant adaptation to water limiting environments. *Plant Signaling and Behavior* 6:586–589.

Orellana, S., Yanez, M., Espinoza, A., Verdugo, I., Gonzalez, E., Ruiz-Lara, S., Casaretto, J.A. (2010) The transcription factor SlAREB1 confers drought, salt stress tolerance and regulates biotic and abiotic stress-related genes in tomato. *Plant Cell and Environment* 33:2191–2208.

Ozturk, Z.N., Talame, V., Deyholos, M., Michalowski, C.B., Galbraith, D.W., Gozukirmizi, N., Tuberosa, R., et al. (2002) Monitoring large-scale changes in transcript abundance in drought- and salt stressed barley. *Plant Molecular Biology* 48:551–573.

Prasad, T.K., Anderson, M.D., Stewart C.R. (1994) Acclimation, hydrogen peroxide, and abscisic acid protect mitochondria against irreversible chilling injury in maize seedlings. *Environmental and Stress Physiology* 105:619–627.

Prashanth, S.R., Sadhasivam, V., Parida, A. (2008) Over expression of cytosolic copper/zinc superoxide dismutase from a mangrove plant *Avicennia marina* in indica rice var *Pusa Basmati-1* confers abiotic stress tolerance. *Transgenic Research* 17:281–291.

Puhakainen, T., Hess, M.W., Makela, P., Svensson, J., Heino, P., Palva, E.T. (2004) Overexpression of multiple dehydrin genes enhances tolerance to freezing stress in *Arabidopsis*. *Plant Molecular Biology* 54:843–753.

Quan, R., Shang, M., Zhang, M., Zhao, Y., Zhang, J. (2004) Engineering of enhanced glycine betaine synthesis improves drought tolerance in maize. *Plant Biotechnology Journal* 2:477–486.

Riechmann, J.L., Heard, J., Martin, G., Reuber, L., Jiang, C.Z., Keddie, J., Adam, L., et al. (2000) *Arabidopsis* transcription factors: Genome-wide comparative analysis among eukaryotes. *Science* 290:2105–2110.

Rockville, MD. (2015) World Population Prospects – Population Division – United Nations. Esa.un.org.

Roxas, V.P., Smith, R.K., Allen, E.R., Allen, R.D. (1997) Overexpression of glutathione S-transferase/glutathione peroxidase enhances the growth of transgenic tobacco seedlings during stress. *Nature Biotechnology* 15:988–991.

Roy, M., Wu, R. (2001) Arginine decarboxylase transgene expression and analysis of environmental stress tolerance in transgenic rice. *Plant Science* 160:869–875.

Saijo, Y., Hata, S., Kyozuka, J., Shimamoto, K., Izui, K. (2000) Over-expression of a single protein kinase confers both cold and salt/drought tolerance on rice plants. *The Plant Journal* 23:319–327.

Sanders, D., Brownlee, C., Harper, J.F. (1999) Communicating with calcium. *Plant Cell* 11:691–706.

Savchenko, G.E., Klyuchareva, E.A., Abramchik, L.M., Serdyuchenko, E.V. (2002) Effect of periodic heat shock on the inner membrane system of etioplasts. *Russian Journal of Plant Physiology* 49:349–359.

Serrano, R., Mulet, J.M., Rios, G., Marquez, J.A., de Larrinoa, I.F., Leube, M.P., Mendizabal, I., et al. (1999) A glimpse of the mechanisms of ion homeostasis during salt stress. *Journal of Experimental Botany* 50:1023–1036.

Shi, J., Kim, K.N., Ritz, O., Albrecht, V., Gupta, R., Harter, K., Luan, S., et al. (1999) American Society of Plant Physiologists novel protein kinases associated with calcineurin B–like calcium sensors in *Arabidopsis*. *The Plant Cell* 11:2393–2405.

Shi, H., Ishitani, M., Kim, C., Zhu, J.K. (2000) The *Arabidopsis thaliana* salt tolerance gene SOS1 encodes a putative Na+/H+ antiporter. *Proceedings of the National Academy of Sciences* 97:6896–6901.

Shi, H., Kim, Y., Guo, Y., Stevenson, B., Zhu, J.K. (2003a) The *Arabidopsis* SOS5 locus encodes a putative cell surface adhesion protein and is required for normal cell expansion. *The Plant Cell* 15:19–32.

Shi, H., Lee, B.H., Wu SJ, Z., Hu, J.K. (2003b) Overexpression of a plasma membrane Na⁺/H⁺ antiporter gene improves salt tolerance in *Arabidopsis thaliana*. *Nature Biotechnology* 21:81–85.

Shinozaki, K., Yamaguchi-Shinozaki, K. (2000) Molecular responses to dehydration and low temperature: Differences and cross-talk between two stress signaling pathways. *Current Opinion in Plant Biology* 3:217–23.

Shou, H., Bordallo, P.W., Wang, K. (2004) Expression of the Nicotiana protein kinase (NPK1) enhanced drought tolerance in transgenic maize. *Journal of Experimental Botany* 55:1013–1019.

Simoes-Araujo, J.L., Rumjanek, N.G., Pinheiro, M.M. (2003) Small heat shock proteins genes are differentially expressed in distinct varieties of common bean. *Brazilian Journal of Plant Physiology* 15:1.

Sinha, A.K., Jaggi, M.., Raghuram, B.T., Tuteja, N. (2011) Mitogen-activated protein kinase signaling in plants under abiotic stress. *Plant Signaling and Behavior* 6:196–203.

Sivamani, E., Bahieldin, A., Wraith, J.M., Niemi, T., Dyer, WF, Ho, T.H.D., Qu, R. (2000) Improved biomass productivity and water use efficiency under water deficit conditions in transgenic wheat constitutively expressing the barley HVA1 gene. *Plant Science* 155:1–9.

Souer, E., van Houwelingen, A., Kloos, D., Mol, J., Koes, R. (1996) The no apical meristem gene of Petunia is required for pattern formation in embryos and flowers and is expressed at meristem and primordia boundaries. *Cell* 85:159–170.

Straub, P.F., Shen, Q., Ho, T.H.D. (1994) Structure and promoter analysis of an ABA-and stress-regulated barley gene, HVA1. *Plant Molecular Biology* 26:617–63.

Subramanyam, K., Sailaja, K.V., Subramanyam, K., Rao, D.M., Lakshmidevi, K. (2011) Ectopic expression of an osmotin gene leads to enhanced salt tolerance in transgenic chilli pepper (*Capsicum annum* L.). *Plant Cell, Tissue and Organ Culture* 105:181–192.

Szalai, G., Kellős, T., Galiba, G., Kocsy, G. (2009) Glutathione as an antioxidant and regulatory molecule in plants under abiotic stress conditions. *Journal of Plant Growth Regulation* 28:66–80.

Taj, G., Agarwal, P., Grant, M., Kumar, A. (2010) MAPK machinery in plants: Recognition and response to different stresses through multiple signal transduction pathways. *Plant Signaling and Behavior* 5:1370–1378.

Takasaki, H., Maruyama, K., Kidokoro, S., Ito, Y., Fujita, Y., Shinozaki, K., Yamaguchi-Shinozaki, K., et al. (2010) The abiotic stress-responsive NAC-type transcription factor OsNAC5 regulates stress-inducible genes and stress tolerance in rice. *Molecular Genetics and Genomics* 284:173–183.

Tester, M., Bacic, A. (2005) Abiotic stress tolerance in grasses from model plants to crop plants. *Plant Physiology* 137:791–793.

Thomashow, M.F. (1994) *Arabidopsis thaliana* as a model for studying mechanisms of plant cold tolerance. *Arabidopsis* 807–834.

Thomashow, M.F. (1998) Role of cold-responsive genes in plant freezing tolerance. *Plant Physiology* 118:1–8.

Thomashow, M.F. (1999) Plant cold acclimation: Freezing tolerance genes and regulatory mechanisms. *Annual Review of Plant Physiology and Plant Molecular Biology* 50:571–599.

Tian, F., Wang, W., Liang, C., Wang, X., Wang, G.W. (2017) Over accumulation of glycine betaine makes the function of the thylakoid membrane better in wheat under salt stress. *The Crop Journal* 5:73–82.

Tuteja, N. (2007) Abscisic acid and abiotic stress signaling. *Plant Signaling and Behavior* 2:135–138.

Umezawa, T., Fujita, M., Fujita, Y., Yamaguchi-Shinozaki, K., Shinozaki, K. (2006) Engineering drought tolerance in plants: Discovering and tailoring genes to unlock the future. *Current Opinion in Biotechnology* 17: 113–122.

Uno, Y., Takashi, F., Hiroshi, A., Riichiro, Y., Shinozaki, K., Shinozaki, K.Y. (2000) *Arabidopsis* basic leucine zipper transcription factors involved in an abscisic acid-dependent signal transduction pathway under drought and high-salinity conditions. *Proceeding of the National Academy of Science of the United States of America* 10:11632–11637.

Veinger, L., Diamant, S., Buchner, J., Goloubinoff, P. (1998) The small heat-shock protein ibpb from *Escherichia coli* stabilizes stress-denatured proteins for subsequent refolding by a multi chaperone network. *The Journal of Biological Chemistry* 273:11032–11037.

Verbruggen, N., Hermans, C. (2008) Proline accumulation in plants: a review. *Amino Acids* 35:753–759.

Wahid, A., Gelani, S., Ashraf, M., Foolad, M.R. (2007) Heat tolerance in plants: An overview. *Environmental and Experimental Botany* 61:199–223.

Wan, H., Hen, L., Guo, J., Li, Q., Wen, J., Yi, B., Ma, C., et al. (2017) Genome-wide association study reveals the genetic architecture underlying salt tolerance-related traits in rapeseed (*Brassica napus* L.). *Frontiers in Plant Science* 8:1–15.

Wang, W., Vinocur, B., Altman, A. (2003) Plant responses to drought, salinity and extreme temperatures: towards genetic engineering for stress tolerance. *Planta* 218:1–14.

Wang, W., Vinocur, B., Shoseyov, O., Altman, A. (2004) Role of plant heat-shock proteins and molecular chaperones in the abiotic stress response. *Trends in Plant Science* 5:244–252.

Wang, Y.J., Wisniewski, M., Melian, R., Cui, M.G., Webb, R., Fuchigami, L. (2005) Overexpression of cytosolic ascorbate peroxidase in tomato confers tolerance to chilling and salt stress. *Journal of the American Society for Horticultural Science* 130:167–173.

Xiang, Y., Huang, Y., Xiong, L. (2007) Characterization of stress-responsive CIPK genes in rice for stress tolerance improvement. *Plant Physiology* 144:1416–1428.

Xiong, L., Schumaker, K.S., Zhu, J.K. (2002) Cell signaling during cold, drought, and salt stress. *The Plant Cell* 14:165–183.

Xu, D., Duan, X., Wang, B., Hong, B., Ho, T.H.D., Wu, R. (1996) Expression of a late embryogenesis abundant protein gene, HVA1, from barley confers tolerance to water deficit and salt stress in transgenic rice. *Plant Physiology* 110:249–257.

Yamaguchi-Shinozaki, K., Shinozaki, K. (1993) Characterization of the expression of a desiccation-responsive rd29 gene of *Arabidopsis* thaliana and analysis of its promoter in transgenic plants. *Molecular and General Genetics* 23:331–340.

Yamaguchi-Shinozaki, Y.K., Shinozaki, K. (2005) Organization of cis-acting regulatory elements in osmotic and cold-stress-responsive promoters. *Trends in Plant Science* 10:88–94.

Yoshida, T., Mogami, J., Yamaguchi-Shinozaki, K. (2014) ABA-dependent and ABA-independent signaling in response to osmotic stress in plants. *Current Opinion in Plant Biology* 21:133–139.

Zhang, J.Z., Creelman, R.A., Zhu, J.K. (2004) From laboratory to field. Using information from *Arabidopsis* to engineer salt, cold, and drought tolerance in crops. *Plant Physiology* 135:615–621.

Zhang, X., Xu, X., Yu, Y., Chen, C., Wang, J., Cai, C., Guo, W. (2016) Integration analysis of MKK and MAPK family members highlights potential MAPK signaling modules in cotton. *Scientific Reports* 6:297–281.

Zhu, J.K. (2001a) Plant salt tolerance. *Trends in Plant Science* 6:66–71.

Zhu, J.K. (2001b) Cell signaling under salt, water and cold stresses. *Current Opinion in Plant Biology* 4:401–406.

Zhu, J.K. (2002) Salt and drought stress signal transduction in plants. *Annual Review of Plant Biology* 53:247–273.

Zhu, B., Su, J., Chang, M., Pal, D., Verma, S., Fan, Y., Wu, R. (1998) Overexpression of a Δ^1-pyrroline-5-carboxylate synthetase gene and analysis of tolerance to water- and salt-stress in transgenic rice. *Plant Science* 139:41–48.

24 Genomic Approaches for Understanding Abiotic Stress Tolerance in Plants

Richa Rai, Amit Kumar Rai, and Madhoolika Agrawal

CONTENTS

24.1 INTRODUCTION

Plants are exposed to a varied range of environmental stresses due to their sessile nature. Abiotic stresses that influence plants and crops in the agricultural field include: drought, salinity, heat, cold, chilling temperatures, freezing temperatures, nutrition, high light intensity, pollutants including troposperic ozone (O_3), anaerobic stresses etc. (Cavanagh et al., 2008; Munns and Tester, 2008; Chinnusamy and Zhu, 2009; Mittler and Blumwald, 2010). In natural conditions, combinations of two or more stresses, such as drought and salinity, salinity and heat, and combinations of drought with extreme temperature or high light intensity, are common to many agricultural areas around the world and can influence crop productivity in a different way compared to individual stresses mainly utilized in controlled conditions used in the laboratory. A comparison of all major US weather disasters that exceeded a loss of a billion dollars each between 1980 and 2017 indicated that a combination of drought and heat stress caused extensive agricultural losses of $300 billion (NOAA, 2018 http://www.ncdc.noaa.gov/billions/events). Modeling studies predicted that an enhancement in the frequency and amplitude of heat stress in the near future accompanied by other weather disasters like extended droughts would immensely affect crop production worldwide (Mittler et al., 2012; Li et al., 2013a; IPCC, 2008). So, there is an urgent need to generate crops with an enhanced tolerance to abiotic stress.

The molecular response of plants to a combination of abiotic stresses is unique and cannot be directly extrapolated from the response of plants (Rizhsky et al., 2002, 2004). Many studies have uncovered the responses of plants to different combinations of stresses involving drought, salt, extreme temperature, heavy metals, UV-B, high light intensity, O_3, CO_2, soil compaction and biotic stresses (Mittler, 2006; Mittler and Blumwald, 2010; Alameda et al., 2012; Atkinson and Urwin, 2012; Kasurinen et al., 2012; Srivastava et al., 2012; Perez-Lopez et al., 2013; Rivero et al., 2013). These studies have suggested that, though with a certain degree of overlap, each stress required a unique mechanism of response, tailored to the specific needs of the plant and that each combination of two or more different stresses may also have a specific response. The simultaneous occurrence of different biotic and abiotic stresses was shown to result in a high degree of complexity in plant responses, as the responses to these combined stresses are largely controlled by different signalling pathways that may interact and inhibit one another (Mittler, 2006; Atkinson and Urwin, 2012; Prasch and Sonnewald, 2013; Rasmussen et al., 2013). Metabolic and signaling pathways involved in the response of plants to a combination of stress triggers was found to include transcription factors, photosynthesis, antioxidant mechanisms, pathogen responses, hormone signaling, and osmolyte synthesis (Rizhsky et al., 2004; Atkinson et al., 2013; Iyer et al., 2013; Prasch and Sonnewald, 2013; Rasmussen et al., 2013). However, the majority of the mechanisms underlying the tolerance of plants to stress combinations are still unknown, and studies are required to understand them.

The challenges of abiotic stresses on plant growth and development are evident among the emerging ecological impacts of climate change (Bellard et al., 2012) and the constraints to crop production exacerbated with the increasing human population competing for environmental resources. Climate change is predicted to affect agricultural production, especially in developing countries due to an increase in carbon dioxide and high temperature (Rosenzweig et al., 2014). It is of the utmost importance to develop climate smart crops which are resilient to climate change to overcome constraints to global food supply and a balanced environment (Wheeler and Von Braun, 2013).

For the last few decades, the main goals of agricultural science have been to increase the yield of agronomic crops and some of the main obstacles to attaining these goals are abiotic stresses. So, it is important to develop abiotic stress tolerant crops. Plants have an intrinsic defense mechanism that is highly complex, and it is well known that complex mechanisms generally involve interactions of a number of genes at the molecular level. The identification of candidate genes for stress tolerance and their expression is required to understand the complete metabolic process. Over the last two decades, the application of functional genomics (alteration of genes) have been helpful in understanding the molecular and metabolic pathways involved in the adaptation of plants to environmental challenges. The present chapter deals with different genomic approaches employed to understand complex responses of crops at different levels (biochemical, physiological and phenological) under abiotic stresses.

24.2 GENOMIC APPROACHES

24.2.1 STRUCTURAL GENOMICS (FORWARD GENETICS)

Structural genomics focuses on the physical structure of the genome, aiming to identify, locate, and order genomic features along chromosomes. Together, structural genomics (forward genetics) and functional genomics (reverse genetics) can characterize a genome to its full extent and aid us in understanding causes of different abiotic stresses (Table 24.1).

24.2.1.1 Molecular Markers

Molecular markers are now widely used to track loci and genome regions in several crop-breeding programmes. Marker-Assisted Selection (MAS) refers to the utilization of molecular markers in breeding improved varieties with respect to desired traits, abiotic stress tolerance, and high yield. Molecular markers tightly linked with a large number of agronomic and disease resistance traits are available in major crop species (Phillips and Vasil, 2001; Jain et al., 2002; Gupta and Varshney, 2004). These molecular markers include: (i) hybridization-based markers such as restriction fragment length polymorphism (RFLP), (ii) PCR-based markers: random amplification of polymorphic DNA (RAPD), amplified fragment length polymorphism (AFLP) and microsatellite or simple sequence repeat (SSR), and (iii) sequence-based markers: single nucleotide polymorphism (SNP). The majority of these molecular markers have been developed either from genomic DNA libraries (e.g., RFLPs and SSRs) or random PCR amplification of genomic DNA (e.g., RAPDs) or both (e.g., AFLPs). These DNA markers can be generated in large numbers and are very useful for crop improvement. As these markers can be employed extensively for the preparation of molecular maps (genetical and physical) and for their association with genes/QTLs (Quantitative Trait Loci) controlling the traits of economic importance (Koebner, 2004; Korzun, 2002). SSR markers are mostly recommended for plant breeding applications and are known as markers of choice (Gupta and Varshney, 2004). RFLP is not readily adapted to high sample throughput, and RAPD assays are not sufficiently reproducible or transferable between laboratories. Both SSRs and AFLPs are efficient in identifying polymorphisms. Genomics applications involving molecular markers are largely dominated by single nucleotide polymorphism (SNPs).

The first genome map in plants was reported in maize (Gardiner et al., 1993), followed by rice (McCouch et al., 1988), and then *Arabidopsis* (Nam et al., 1989), using RFLP markers. Maps for several other crops like potato, barley, banana, and members of Brassicaceae have been constructed (Winter and Kahl, 1995). Microsatellite markers are useful in genome mapping. After being completely mapped, these markers may be efficiently used in tagging several individual traits that

TABLE 24.1

Genomic Techniques Employed in Understanding Abiotic Stress in Different Agriculture Crops

Plant Species	Targeted Gene	Abiotic Stress	Genomic Approach	Reference
Zea mays L.	–	Water stress	microarray	Yu and Setter (2003)
Nicotiana tabacum L.	Downregulate activity of putrescence N-methyl transferase (*PMT*)	–	RNAi	Chinatapakorn and Hamill (2003)
Festuca arundinacea L.	Downregulation of caffeic acid O-methyltransferase (*COMT*)-lignin synthesis	–	RNAi	Chen et al. (2004)
Triticum aestivum L.		High temperature stress	Microarray	Gulick et al. (2005)
Capsicum annum L.	–	Cold stress	Microarray	Hwang et al. (2005)
Oryza sativa L.	–	Drought stress	Microarray	Hazen et al. (2005)
O. sativa	486 salt responsive *EST*	Salt stress	Microarray	Chao et al. (2005)
T. aestivum		Salinity	Microarray	Kawaura et al. (2006)
O. sativa and *T. aestivum*	*ERA1, PP2C, AAPK, PKS3* (photosynthesis)	Salinity stress	QTL	Roy et al. (2011)
O. sativa	–	Salt resistance	CRISPR/Cas	Shan et al. (2015)
Z. mays	*ESK1*	Drought	RNAi	Xu et al. (2014)
Z. mays	Yield improvement	–	CRISPR/Cas	Svitaseu et al. (2015)
O. sativa	*OzT8/OzT9*	Ozone	QTLs	Wang et al. (2014)
T. aestivum	Heat responsive transcripts	Heat stress	NGS	Kumar and Jain (2015)

are extremely important for a breeding program like yield, disease resistance, stress tolerance, seed quality, etc. A large number of monogenic and polygenic loci for various traits have been identified in many plants and are currently being exploited by breeders and molecular biologists together to make the dream of marker-assisted selection come true.

The first report on gene tagging was from tomato (Williamson et al., 1994) for the identification of markers which are linked to genes involved in several traits like water-use-efficiency (Martin et al., 1989), resistance to *Fusarium oxysporum* (12 genes) (Sarfatti et al., 1989), leaf rust resistance genes *LR9* and *LR24* (Schachermayr et al., 1995) and root knot nematodes (*Meliodogyne* sp. the *mi* gene) were identified. STMS markers were used as potential diagnostic markers for important traits in plant breeding programs, e.g. (AT) 15 repeat located within a soybean heat shock protein gene, which is about 0.5 cm from (*Rsv*), a gene conferring resistance to soybean mosaic virus (Yu et al., 1994). Similar to RFLPs, STMS, and ASAPs, and arbitrary markers, RAPDs have also played an important role in the saturation of genetic linkage maps and gene tagging. Tripathi et al. (2011) reported mutational and structural alterations in linseed DNA after ozone and UV-B treatment using ten different RAPD primers. Singh et al. (2014) identified reduction

of genome template stability of kidney beans exposed to UVB exposure.

Marker-assisted selections of target QTLs provide powerful support for improving productivity under drought and/or saline conditions which assist selection in the breeding process. One of the major difficulties in abiotic stress QTLs identification in crops is the identification of the key physiological and morphological determinants of stress tolerance. In wheat, QTLs for drought tolerance have been identified through yield and yield measurement under drought conditions (Maccaferri et al., 2008). Ueda et al. (2015) conducted a study using genome-wide association study (GWAS) in rice (*Oryza sativa* L.) exposed to troposheric O_3 to determine candidate loci associated with ozone tolerance in rice (*O. sativa* L.). They also conducted an association mapping study conducted based on more than 30,000 single-nucleotide polymorphism (SNP) markers, which yielded 16 significant markers throughout the genome by applying a significance threshold of $P < 0.0001$ and on the basis of linkage disequilibrium blocks associated with significant SNPs, a total of 195 candidate genes for biomass related traits were identified. Novel polymorphisms in two candidate genes for the formation of visible leaf symptoms, a *RING* and an *EREBP* gene, both of which are involved in cell death and stress defense reactions were also identified. This study demonstrated substantial natural variations in response to O_3 in rice and the possibility of using GWAS in elucidating the genetic factors underlying O_3 tolerance.

Frei et al. (2008) reported quantitative trait loci (QTL) in rice associated with symptom formation (OzT9) and biomass (OzT8) of rice plants exposed to O_3 (100 or 120 ppb 8 h per day) during the vegetative growth stage. In further experiments using chromosome segment substitution lines (SL46) carrying QTLs in the genetic background of their recurrent parent Nipponbare, OzT8 shown to be associated with the ability to maintain the biochemical efficiency of photosynthesis despite O_3 stress (Chen et al., 2011) and SL41 carrying OzT9 was associated with favorable responses of the antioxidant system, including the suppressed expression of an O_3 responsive ascorbate oxidase gene (Frei et al., 2010).

Further studies were undertaken by Wang et al. (2014) to evaluate the combined effects of both QTLs, OzT8 and OzT9, in newly generated crosses containing both loci in the genetic background of the O_3 sensitive variety Nipponbare. Two lines, SL41 (OzT9) and SL46 (OzT8) containing single QTLs were crossed, and four OzT8/OzT9 lines containing both QTLs were selected for evaluation of physiological and yield parameters in a season-long fumigation experiment with all OzT8/OzT9 lines (Table 24.1). The OzT8/OzT9 lines had a constitutively higher level of chlorophyll a and total chlorophyll, total biomass and shoot biomass. Yield was drastically reduced by 55 and 52% in Nipponbare, but for the OzT8/OzT9 lines reduction ranged only between 36–41% and 25–37%, respectively which demonstrates that the pyramiding of QTLs, OzT8, and OzT9 confers enhanced tolerance to ozone as compared to single QTLs, because lines containing tolerance alleles at both loci combine the advantages of oxidative stress tolerance and photosynthetic efficiency. The synergetic effect of both QTLs was effective in limiting O_3 induced yield losses. This represents an important step forward in the molecular breeding of rice varieties adapted to future atmospheric environments.

24.2.1.2 Genome Sequencing Approach

The catalogue pertaining to expressed genes of a particular species is investigated through expressed sequence tags (ESTs). Focusing mainly on functional studies, EST techniques are used which are fast and cost-effective in identifying genes (Iqbal et al., 2013). The availability of EST compilations and cDNA sequences of *Arabidopsis* and rice has encouraged large-scale compilations for other crops as well (Bevan et al., 1999; Tyagi et al.,

2006). However, there are limited studies focusing on ESTs from plants exposed to abiotic stresses. The main importance of the EST technique is to recognize the vital function of the responsible genes. The National Centre for Biotechnological Information (NCBI) database currently has about a million ESTs for crops like maize, rice, soybean, and wheat, along with other plants as well. In order to develop plant EST datasets with respect to stress-responsive genes, it is necessary to build up sequencing programs at different developmental stages based on cDNA libraries from stress-treated plant tissues and organs of different plant species. In comparison to cDNA, ESTs are shorter, and their overlapping provides more information about the organization of parental cDNA. Since para-log genes may result in sequences being misassembled, they must be handled carefully, particularly in species with polyploidy (Rudd, 2003). They are widely employed in crops which comprise of lengthy and repetitive genomes. They have the potential to aid in gene discovery. The latest reports have shown the EST sequencing method as being most suitable for evaluating the range of genotypes under controlled and stressed conditions (AkpJnar et al., 2013; Brenner et al., 2000).

Besides their utility in genome annotation and expression profiling, ESTs are helpful in providing a source of sequences for designing "functional markers." Functional markers refer to polymorphic sites on genes that are attributed to the phenotypic variation of traits among individuals of a species. Functional marker design requires knowledge of the allelic sequences of functionally characterized genes (Gupta and Rustgi, 2004). In contrast to random DNA markers, functional markers are completely linked to the trait of interest, hence these markers are also called "perfect markers." In addition, functional markers may explore natural variations and biodiversity better, particularly compared to random DNA markers with absence/presence of polymorphisms, where allelic variations of a trait exceed that of the linked DNA marker. The importance of functional markers has been highlighted in stress tolerance studies as well (Bagge et al., 2007; Garg et al., 2012).

Innovations in the DNA-sequencing tools have provided us with information about the complete genomic sequences. NGS has offered us the stage with the facility to undergo sequencing economically in contrast to the conventional Sanger sequencing approach (Varshney et al., 2009a, b). NGS has defined a tiling path by making sequencing and re-sequencing of large genomes feasible and helpful in exploiting plant genomes for breeding improved varieties through breeding against abiotic stress. Morrell et al. (2011) complied complete genomic sequences of a number of crops such as maize, rice, sorghum, and soybean along with some model species like *Arabidopsis thaliana* and *Brachypodium distachyon*. Although whole genomic sequences provide a thorough description of coding and non-coding regions, repetitive elements and regulatory sequences in a gene, and manipulation of genes is only possible through functional studies related to abiotic stresses (Mochida et al., 2010). In the course of molecular breeding, genomic sequencing is a fundamental means for crop development.

Further developments in genomics technologies are likely to deliver advanced applications; for example, in wheat crop improvement against abiotic stress using NGS facility. Recent applications include the shotgun sequencing of the *Brassica rapa* genome (Wang et al., 2011), wheat chromosomes 7DS, 7BS, and 4A (Berkman et al., 2011a,b; Hernandez et al., 2011), 5-fold coverage of the wheat cultivar Chinese Spring (http://www.cerealsdb.uk.net/), and D-genome donor *Aegilop stauschii* (http://www.cshl.edu/genome/wheat).

Wheat's large genome size and complex family background has hampered efforts to determine the genetic basis of phenotypic traits. The high proportion of repetitive DNA in the wheat genome complicates genome assembly. DNA sequencing is length-limited, and even the longest sequencing technology was incapable of spanning the long repetitive regions in the wheat genome. The International Wheat Genome Sequencing Consortium (IWGSC, http://www.wheatgenome.org/) was established in 2005 to sequence hexaploid wheat using a physical mapping and a BAC-by-BAC approach (Gill et al., 2004), which consists of BAC library generation, followed by the identification of a minimum tiling path, sequencing and anchoring to a physical map, and finally assembly of the genome. Using this approach genome sequencing of *O. sativa*, *Sorghum bicolor* and *Zea mays* was conducted, and genome size ranged from ~150 Mbp to ~2.3 Gbp (*Arabidopsis* Genome Initiative, 2000; Matsumoto et al., 2005; Paterson et al., 2009; Schnable et al., 2009).

Though this approach is widely accepted as the current standard to produce a "finished" genome sequence it takes substantial time and resources for larger genomes. Wheat genome is large and highly complex compared to other cereal crops like rice with genome size roughly 400 Mbp in size (Goff et al., 2002; Yu et al., 2002) and maize (allotetraploid) with a genome of 2.3 Gbp (Schnable et al., 2009). Wheat genome size is estimated to be 17 Gbp (Paux et al., 2006) because it is an allohexaploid which means it contains three distinct diploid genomes that together function much like

any diploid. They are understood to have combined to produce *Triticum aestivum* in two distinct hybridization events. First, *Triticum urartu* (AA) and an unknown relative of *Aegilops speltoides* (BB) are believed to have produced the tetraploid *Triticum turgidum*, followed by hybridization with *A. stauschii* (DD) to produce the hexaploid (Chantret et al., 2005). Even the wheat genome has a proliferation of repetitive elements, which results in a composition of between 75 and 90% repetitive DNA sequences (Flavell et al., 1974, 1977; Wanjugi et al., 2009).

When assembling a large genome sequence, its complexity needs to be reduced where possible. Shotgun sequencing was applied to individual wheat chromosomes, allowing assembly and the identification of gene-containing contigs, which can then be ordered and oriented based on synteny with related species. The approach was first developed in barley by Mayer et al. (2009) using Roche 454 sequence data for chromosome 1H and since then it has been used for complete barley genome (Mayer et al., 2011). The first application of this approach in wheat applied Illumina sequencing to identify all genes on chromosome arm 7DS (Berkman et al., 2011a), incorporating roughly two-third of the genes into a syntenic build, with the remaining genes included in "additional contigs" (http://www.wheatgenome.info). Subsequent sequencing of arms 7BS and 4AL provided the basis for gene-level delimitation of a previously described7BS/4AL translocation and suggested a total gene-content in wheat of ~77,000 genes (Berkman et al., 2011b). This information may be utilized in understanding complicated pathway related to heat, drought, and O_3 tolerant cultivars. Using the NGS platform, Kumar and Jain (2015) studied whole wheat transcriptome under heat stress, and gene expression profile showed significant differential expression of 1525 transcripts under heat stress out of which 27 transcripts associated with cellular processes such as metabolic processes, protein phosphorylation, and oxidation-reductions were >10 fold upregulated. Such studies enriched with the transcript dataset of wheat are available on public domain and show a de novo approach to discovering the heat-responsive transcripts of wheat, which can accelerate the progress of wheat stress-genomics as well as the course of wheat breeding programs in the era of climate change (Table 24.1).

24.2.2 Functional Genomics (Reverse Genetics)

Functional Genomics led to functional variation by changing the structure or function of proteins. Functional genomics techniques have long been adopted to unravel gene functions and the interactions between genes in regulatory networks, which can be exploited to generate improved varieties. Functional genomics employ use of sequence or hybridization-based technique, which helps in elucidating the role of intermediate components in the cascade of abiotic stress response and to trace out the entire stress response pathway. The basis of functional genomics underlies the evaluation and study of the entire cell or organisms as a whole. The basic outline of functional genomics includes gene identification, expression, and validation. *Arabidopsis* and rice genome sequence outcomes have cemented the means for analyzing the function of genes at genomic scale (Goff et al., 2002; Yu et al., 2002).

24.2.2.1 Hybridization-Based Approach

In addition to sequence-based methodology, hybridization-based gene expression has proved to be a promising technique for analyzing tens of thousands of genes in a single genome. This transcript profile has been encouraged by the development of microarray-based technologies, cDNA microarrays (Ergen et al., 2009), and oligo-nucleotide microarrays (Close et al., 2004). Microarray technologies, based on the selective and differential principle of nucleic acid hybridization can also completely restructure gene expression profiling which is done with probe and target as hybridization partners. Probe is the extension of DNA specific to a particular gene attached to the solid surface and assembles microarray itself together with the labeled DNA or RNA strand in the solution as the "target." The technology has evolved significantly by increasing the number of probes on an array and reducing the surface area of the array. Hence, microarrays have become an indispensable tool in functional genomics and global gene expression analysis.

Globally, cDNA microarray profiling and oligonucleotide base chips are the two major types of microarrays employed to study the gene expression in crops. During stress response studies, microarrays provide a comprehensive evaluation of transcriptional activity by providing new insights into the complicated world of the signaling system governing stress responses and by helping in the recognition of new genes. DNA microarray is merely a PCR amplicon that results from precise amplification of genomic DNA by employing EST-based primers (Jiao et al., 2003). Schena et al. (1995) first reported the use of cDNA microarray to study the differential expression of 45 genes in roots and shoots of *A. thaliana* L. Along with cDNA of *Arabidopsis*, microarrays for rice, strawberry, lima beans, maize, sorghum, and soybean are developed. This technology

is also applicable to plant species which are less known but agriculturally and industrially important such as cassava, tomato, and cotton, but are less emphasized to untangle the stress response (Utsumi et al., 2012; Loukehaich et al., 2012).

Microarrays have been effectively exploited to evaluate the gene regulation at different stages of development (Kehoe et al., 1999). Seki et al. (2001) studied about 7000 full-length cDNA microarrays in *Arabidopsis* to establish the genes associated with stress conditions like cold, salinity, and drought and out of the total identified genes, 53 were cold inducible, 194 were salinity inducible, and 277 were drought inducible. Presently, microarray profiling provides information about prime DNA sequence both in coding and regulatory regions and also their interactions, RNA expression during development, subcellular localization and intermolecular interactions of RNA molecule, polymorphic variations within a species, and physiological response and environmental stresses. In *Arabidopsis*, Genome Gene Chip array was applied to make out the plants' response to a range of biotic and abiotic stresses. As the two stresses were similar, in order to increase the specificity and to differentiate these arrays and full-length cDNA arrays, 50-mer and 70-mer probe arrays were designed (Kane et al., 2000; Ten-Bosch et al., 2001). Zhu et al. (2003) reported that 25-mer oligonucleotide (Gene Chip Rice Genome Array) denoting 21,000 genes of rice cultivar was used to study the gene expression during different stages and to categorize the genes involved in the synthesis and transport of carbohydrates, proteins, and fatty acids. Since the oligonucleotide information is already accessible through databases, there is no need to keep the compilation of cloned DNA molecules. Cooper et al. (2003), using Gene Chip, analyzed the gene expression during seed development and stress response and identified genes that might play a role in development in response to environmental stresses. At the genomic level, microarrays provide an ideal platform to evaluate gene expression and simultaneously create their functional data.

Abiotic stresses are generally complex in nature, and slight differences in the experimental application of stress conditions may produce significant differences in stress responses and further caveat while interpreting microarray studies as many transcripts are known to undergo post-transcriptional and post-translational modifications, resulting in uncorrelated transcriptomic and proteomic data. For species with an available whole genome sequence, a successful expansion of array-based transcript profiling is done using whole genome tiling arrays (Rensink and Buell, 2005). Tiling arrays can identify novel transcriptional units on chromosomes and alternative splice sites and can map transcripts and methylation sites (Yazaki et al., 2007; Mochida et al., 2010). Tiling arrays have already been applied in model species to investigate abiotic stress responses (Zeller et al., 2009; Matsui et al., 2010; Verelst et al., 2013).

Serial analysis of gene expression (SAGE) is an alternative and influential technique exploited for global gene expression. The technique was developed to quantify thousands of transcripts and generates short sequences of 9–17 bp holding adequate information to recognize the transcripts (Tyagi et al., 2006; Saha et al., 2002). These transcripts once translated and sequenced can provide complete information about gene expression (Velculescu et al., 1995; Vega-Sanchez et al., 2007). SAGE may construct an array of tags potentially able to identify >49 genomic sequences which are more than the expected figure of genes in *Arabidopsis* and rice. In a population, these tags offer an explicit update about gene expression at the cellular level, but the scheme is only suited for those organisms whose genome has been fully sequenced.

In plants, SAGE is typically applied with some modification in the actual methodology like SuperSAGE and DeepSAGE (Lorenz et al., 2002; Lee et al., 2002; Matsumura et al., 2003; Nielsen et al., 2006). This technique is mostly applied in humans (Chen et al., 2002), yeast (Velculescu et al., 1995), and mice (Gunnersen et al., 2002). In plants, the approach has been extensively exploited to study gene-related stress-response (Lee and Lee, 2003). In rice seedlings alone, about 10,122 tags from 5921 expressed genes were examined. As revealed by global gene expression in rice leaf and seed, out of 50,519 SAGE tags, 15,131 tags resembled unique transcripts, and 70% occurred only once in both libraries (Gibbings et al., 2003). The first report of SAGE in plants identified novel genes and information about novel functions for known genes in rice seedlings (Matsumura et al., 2003).

A similar tag-based approach, massively parallel signature sequencing (MPSS), involving longer sequence tags that are ligated to microbeads and sequenced in parallel, enabled the study of millions of transcripts simultaneously (Brenner et al., 2000). MPSS identifies genes with greater specificity and sensitivity due to longer tags and high-throughput analysis, and it has the ability to capture rare transcripts especially beneficial for species that lack a whole genome sequence (Reinartz et al., 2002). In plants, apart from mRNA transcripts, MPSS has been utilized in the expression studies of small RNAs (Meyers et al., 2006; Nobuta et al., 2007), which are increasingly used in abiotic stress responses (Sunkar et al., 2007). Currently, plant MPSS expression databases contain expression data for a number of plant species, including important crops such as rice, maize, and soybean (Nakano et al., 2006). This MPSS data can be extracted, compiled, and compared with newly generated MPSS data for functional analysis of gene expression (Jain et al., 2007). MPSS technology further evolved into the current Next Generation Sequence technology, which is now applied as a replacement for previous methods.

24.3 RNA SILENCING (RNAi)

RNA silencing is one of the innovative and efficient molecular biology tools to harness the downregulation of expression of the gene(s) specifically. This family of diverse molecular phenomena has a common exciting feature of gene silencing which is collectively called RNA interference, abbreviated as RNAi. This molecular phenomenon has become a focal point of plant biology and medical research throughout the world. As a result, this technology has turned out to be a powerful tool in understanding the function of the individual gene and has tremendous potential for use in crop improvement.

In Plants, RNAi was first discovered serendipitously by R. Jorgensen in 1990, when he was trying to create petunia flowers with a dark purple hue by introducing numerous copies of a gene with code for a deep purple flower, i.e., Chalcone Synthase (Chs A), but surprisingly some of the final plants yielded white or patchy flowers. Somehow, the transgene silenced the expression of both homologous and endogenous genetic recombination and introduced loci (Napoli et al., 1990); the phenomenon was termed "cosuppression." But it was Fire and Mello's in 1977 who solved this puzzle and explained the phenomena. They observed the presence of dsRNA in *C. elegans* that triggered a marked silencing of genes containing sequence homology to dsRNA (Montgomery and Fire, 1998; Tabara et al., 1998). They termed this unusual phenomenon of gene silencing as RNA interference. Later on, it was observed that both plant cosuppression, as well as plant gene silencing mediated by "Antisense technology," led to the production of cellular dsRNAs. In parallel, several reports showed that plants respond to RNA viruses by targeting their RNAs for destruction using the same gene silencing phenomenon as RNA viruses replicate through dsRNA intermediate. Several such contemporary phenomena of RNAi exist as a result of plant-virus interactions.

Gene silencing mediated by viruses can occur with both types of viruses viz. RNA as well as DNA viruses, which replicate in the cytoplasm and the nucleus, respectively. Viruses may carry exogenous gene sequence in the specific location of its genome, and retain the infectivity of the RNA transcript. When these transcripts are used to infect plants, the foreign sequences also induce and become the target of RNAi response of the host plant. Virus-induced gene silencing (VIGS) is a technology which takes advantage of the ability of viruses to carry and induce RNAi against foreign sequences (Baulcombe, 1999). Several RNA or DNA viruses have been developed as potential VIGS. VIGS is particularly useful in plant species that are recalcitrant for transformation, provided there is a VIGS system available for the species. VIGS may also be useful in analyzing gene function as the infectious transcripts can be applied to mature plants (Waterhouse and Helliwell, 2003). Among VIGS vectors, the TRV-based are more promising because these are capable of inducing meristem gene silencing which has not been possible to achieve with other RNA virus-based vectors.

Tomato golden mosaic geminivirus (TGMV) vectors system has also been reported to be used as meristem gene silencing (Peele et al., 2001). One major limitation with VIGS is the host range, as VIGS have a limited number of hosts, and thus virus-host combination becomes a crucial factor in determining the efficacy of silencing. Another viral silencing system, using satellite RNAs has also been developed. In this satellite virus-induced silencing system (SVISS), the target sequence is inserted into the satellite RNA, which is then co-inoculated with the associated virus. Another RNAi related technology is amplicon mediated RNA silencing, which was first described by Angell and Baulcombe (1999) in potato virus X (PVX). An amplicon is a transgene that comprises target gene sequence and viral genome, but not necessarily all of the genes of the native virus. Plants transfected with amplicon do not show symptoms of being infected with the virus. Amplicon can also overcome the host specificities, which are the major limitation of VIGS. A PVX-GFP amplicon transgene induces GFP-specific RNAi in *Arabidopsis* (Dalmay et al., 2000), even though *Arabidopsis* is not a host of PVX.

Recent findings manifest the RNAi, playing an important role in abiotic stress stimulation in different crops. RNAi technology may be a substitute for complex molecular techniques because it contains several benefits; for example, its specificity, and sequence-based gene silencing. Environmental stress causes plants to over- or under-express certain miRNAs or to synthesize new miRNAs to cope with stress. Several stress-regulated miRNAs have been identified in model plants under various abiotic stress conditions, including nutrient deficiency (Fujii et al., 2005), drought (Liu et al., 2008), cold (Zhao et al., 2007), salinity (Zhou et al., 2010), UV-B

radiation (Navarro et al., 2006), and mechanical stress (Zhou et al., 2007). miRNA-expression profiling under drought stress has now been performed in *Arabidopsis*, rice, and *Populus trichocarpa* and many other plants under drought-stress conditions and some of the miRNAs were shown to be responsive toward this stress in different plants (Jung and Kang, 2007; Jia et al., 2009; Liu et al., 2009).

Zhou et al. (2007) studied miRNAs induced by UV-B radiation in *Arabidopsis*, and of the 21 miRNAs families identified, 11 miRNA were predicted to be upregulated under UV-B stress: miR156/157, miR159/319, miR160, miR165/166, miR167, miR169, miR170/171, miR172, miR393, miR398, and miR401. Studies on ABA or gibberellin (GA) treatment regulated miR159 expression and controlled floral organ development; these studies were conducted by Sunkar and Zhu (2004) and Achard et al. (2004). Zhou et al. (2010), using genome-wide profiling and analysis of miRNAs in drought-challenged rice during different developmental stages from tilling to inflorescence formation, showed that 16 miRNAs (miR156, miR159, miR168, miR170, miR171, miR172, miR319, miR396, miR397, miR408, miR529, miR896, miR1030, miR1035, miR1050, miR1088, and miR1126) were significantly downregulated in response to drought stress and 14 miRNAs (miR159, miR169, miR171, miR319, miR395, miR474, miR845, miR851, miR854, miR896, miR901, miR903, miR1026, and miR1125) were significantly upregulated under drought stress.

Elevated levels of ROS are often associated with plant stress, and superoxide in plants are converted into molecular oxygen and hydrogen peroxide by superoxide dismutases (SODs). CuZn SODs are encoded by *CSD1, CSD2, and CSD3* in *Arabidopsis* and miR398 was predicted to target CSD1 and CSD2 (Jones-Rhoades and Bartel, 2004) Sunkar et al. (2006) confirmed these targets and discovered that miR398 is downregulated under oxidative stress. Downregulation of miR398 is accompanied by an accumulation of CSD1 and CSD2 transcripts. This accumulation did not result from stress-related transcriptional induction of the Cu/Zn SOD genes, but rather resulted from the relaxation of miR398-directed cleavage.

Manavalan et al. (2012) demonstrated that RNAi-mediated disruption of a rice farnesyl transferase/squalene synthase (SQS) by maize SQS improves drought tolerance at both the vegetative and reproductive stages. Twenty-day-old rice seedlings of wild-type (Nipponbare) and seven independent events of transgenic RNAi lines showed no difference in morphology, and at a period of water stress under growth chamber conditions, transgenic positives showed delayed wilting, conserved more soil water, and improved recovery compared with the wild-type, through reduced stomatal conductance and the retention of high leaf relative water content (RWC). After 28 d of slow progressive soil drying, transgenic plants recovered better and flowered earlier than wild-type plants. The yield of water-stressed transgenic positive plants ranged from 14–39% higher than wild-type plants.

24.4 GENOME EDITING

Genome editing (GE) enables us to change the gene expression regulation at predetermined sites and facilitates new insights into the plant functional genomics. GE differs from genetic engineering in which no foreign DNA is made from part of plants, and they may not be modified to be distinguished from parent plants. GE includes a wide variety of tools such as zinc finger nucleases (ZFNs), transcription activator-like effector nucleases (TALENs), clustered regularly interspaced short palindromic repeats (CRISPR), and CRISPR-associated (Cas) proteins (CRISPR/Cas) (Arora and Narula, 2017). GE techniques are modified with engineered nucleases (GEEN) and programmable sequence-specific DNA nuclease to make it more practical and reliable. They have provided more precision to the process of endogenously targeted genomic modifications.

These GE technologies use the cell's endogenous repair system in which specific genomic regions are manipulated using sequence-specific nucleases (SSNs). SSNs can drive double-strand breaks (DSBs) in targeted sites of genomic DNA, and DSBs are repaired by processes known as non-homologous end-joining (NHEJ) (Rouet et al., 1994) and homology-directed repair (HDR) (Bibikova et al., 2002). During the NHEJ repair process, various types of error such as the insertion or deletion of nucleotides by imperfect ligation at the tip of DSB points may be generated while the HDR pathway uses a template for repairing DSBs. The process of HDR is more precise than that of NHEJ.

The first- and second-generation system of GE are ZFNs and TALENs, which are dependent upon proteins' ability to recognize specific DNA sites and nuclease activity of FokI domains to cleave the target sequences.

24.4.1 Zinc Finger Nucleases (ZFN)

ZFNs were one of the oldest gene editing technologies, which were developed in the 1990s and owned by Sangamo BioSciences. ZFNs are premeditated restriction enzymes having sequence-specific DNA binding zinc finger motifs and non-specific cleavage domain of

Fok1endonuclease. An array of 4–6 binding modules combine to form a single zinc finger unit. Each module is recognized as a codon (Pabo et al., 2001). A pair of ZFNs together identifies a unique 18–24 bp DNA sequence and double-stranded breaks (DSBs) are made by Fok1 dimer. FokI nucleases are naturally occurring typeII S restriction enzymes that introduce single-stranded breaks in a double helical DNA. Hence, Fok I functions as a dimer, with each catalytic monomer (nickase) cleaving a single DNA strand to create a staggered DSB with overhangs (Pabo et al., 2001). ZFNs have been successfully employed in genome modification of various plants including tobacco, maize, soybean, etc. (Curtin et al., 2011; Ainley et al., 2013; Baltes et al., 2014). But major drawbacks associated with ZFNs are their time-consuming and expensive construction of target enzymes with low specificity and high off-target mutations that eventually made way for the new technology.

A. thaliana (Osakabe et al., 2010) and *Z. mays* (Shukla et al., 2009) have led to the successful development of herbicide-tolerant genotypes through the insertion of herbicide-resistant genes into targeted sites in the genome by using ZFN (Shukla et al., 2009). ZFN was also used for the targeted modification of the endogenous malate dehydrogenase (MDH) gene in plants; the plants containing modified MDH have shown increased yield (Shukla et al., 2009). Applications include, but are not limited to, the generation of herbicide tolerance, insect resistance, enhanced disease resistance (bacterial and viral), improved nutritional value, and enhanced yield without the introduction of foreign genes, as has been used in the traditional genetic engineering approach for crop development against abiotic stress (Kamburova et al., 2017).

24.4.2 Transcription Activator-Like Effector Nucleases (TALENs)

TALENs are one of the substitutes to ZFNs and are identified as restriction enzymes that could be manipulated for cutting specific DNA sequences. Traditionally, TALENs are considered as long segments of transcription activator-like effector (TALE) sequences that occur naturally and join the Fok1 domain with the carboxylic-terminal end of manipulated TALE repeat arrays (Christian et al., 2010). TALENs contain a customizable DNA-binding domain which is fused with non-specific Fok1 nuclease domain (Christian et al., 2010). As compared to ZFNs, TALENs are involved in the interaction of individual nucleotide repeats of the target site and amino acid sequences of TAL effector proteins. They can generate overhangs by employing Fokl nuclease domain to persuade site-specific DNA cleavage. It has been widely used to generate non-homologous mutations with higher efficiencies in diverse organisms (Joung and Sander, 2012) TALE proteins consist of a central domain responsible for DNA binding, nuclear localization signal, and a domain that activates transcription of the target gene. DNA-binding domain in TALE monomers consists of a central repeat domain (CRD) that confers DNA binding and host specificity. The CRD consists of tandem repeats of 34 amino acid residues and each 34-amino acid long repeat in CRD binds to one nucleotide in the target nucleotide sequence.

Rice is one of the first plant species to undergo gene editing using TALENs (Shan et al., 2015) and Cas9/sgRNA (Feng et al., 2013) due to its ability to obtain biallelic gene modifications in a single generation (Zhou et al., 2014), the opportunity to delete large segments of chromosomes (Zhou et al., 2014), and the availability of gene replacement through homologous recombination (Feng et al., 2013), coupled with positive/negative selection schemes (Shimatani et al., 2015) led to the continued use of rice as a rapid means of generating genetically altered plants for basic understanding of monocot and its ability to target several genes for knockout or modification (Endo et al., 2015) to improve rice varieties against abiotic stresses.

24.4.3 CRISPER Approach

The third generation genome-editing tool is CRISPR, which is a comparatively precise approach to modifying DNA at specific sites. CRISPR has evolved as a principal technique for gene function analysis (Deltcheva et al., 2011; Perez-Pinera et al., 2013; Kanchiswamy et al., 2016) (Figure 24.2). During the last few years, CRISPR-Cas 9 mediated mutagenesis was performed in *Arabidopsis*, sorghum, tobacco, etc. which demonstrated the applicability of this technique in both dicot and monocot plants (Feng et al., 2013; Li et al., 2013b). The functions of CRISPR and Cas genes (CRISPR-associated) are indispensable for adaptive immunity in some bacteria and archaea, and to date, three types of mechanisms have been identified. Type II CRISPR is the highly studied and applied mechanism developed by Bortesi and Fischer (2015). The Type I and III systems involve specific Cas endonucleases which make the pre-crRNAs (Pre-CRISPR RNA), and after attaining maturity, this crRNA assembles into a Cas protein complex with the ability to recognize and cleave nucleic bases complementary to the crRNA (Jinek et al., 2012). The CRISPR-Cas9 Type II is characterized as a small

RNA-based immune system of archaea and bacteria (Haft et al., 2005).

CRISPR-Cas9 system just requires three components, i.e., Cas9, tracer RNA (trRNA), and the function and potential of CRISPR RNA (crRNA) were identified at the start of this decade (Jinek et al., 2012; Hsu et al., 2014; Schaeffer and Nakata, 2016). In Type II CRISPR, attacking viral DNA or plasmids are divided into smaller pieces and integrated into CRISPR locus and then that loci are transcribed, and processed transcripts produce crRNA. These crRNAs directly affect endonuclease's ability to target alien DNA, depending upon the complementarity of the sequence. Cas 9 produce DSBs at the target site, which on the other hand facilitate endogenous DNA repair mechanisms leading to edited DNA (Charpentier and Doudna, 2013; Hsu et al., 2014). The Type II system comprises crRNA and trRNA that combines into one sgRNA (single-guide RNA) (Jinek et al., 2012; Xing et al., 2014). The sgRNA programmed Cas9 is more effective in targeting gene modifications than individual trRNA and crRNA.

Genome-editing protocols have adopted three different types of Cas9 nuclease. The first Cas9 type cut DNA site-specifically, which results in the activation of DSB repair. Cong et al. (2013) introduced advanced Cas 9-D10A, a mutant with very precise nickase activity, because it was target specific it produced less del mutations (Jinek et al., 2012; Cong et al., 2013). D Cas9 may be taken as a tool for either gene silencing or activation by fusion with a variety of effects or domains (Maeder et al., 2013a,b);this technique does not use recombinant DNA, which may help in developing edited plants which can be exempted from current GMO regulations. It has a widespread application of RNA-guided GE in agriculture and plant biotechnology.

Studies using CRISPR-Cas 9 technology for understanding abiotic stress are very limited. In the coming future, agriculture will benefit from better tasting tomatoes, non-browning mushrooms, mildew resistant wheat, and drought resistant corn. The examples listed will be among the first to benefit from the hot new technique CRISPR-Cas 9 technology.

24.5 COMPARATIVE GENOMICS ABOUT UNEXPLORED GENOMES

Comparative genomics help us to understand similarities between model species and their corresponding crop species and further how these model species can be applied in agriculture (Paterson et al., 2005; van de Mortel and Aarts, 2006). The most valuable source for this approach is the

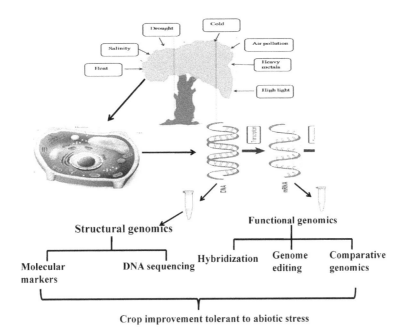

FIGURE 24.1 Development of tolerant plants against abiotic stress using genomic applications.

availability of large-scale plant genomic sequences together with their expression data and the number of stress-related cDNA libraries. Conservation of gene sequences, order, and distribution between the species enable a high quality of putative gene (Brady et al., 2006). The species which are evolved from common ancestors, and their genes that are preserved, have similar gene functions. Stress-related transcription factors (TFs), gene expression patterns, and resemblance in sequences of orthologs in different plant species help to predict genes with similar functions in newly sequenced crop species. Comparative analytical studies of known responsive transcription factors (TFs) in *Arabidopsis* and rice helped to predict the stress-responsive TFs in maize, soybean, barley, sorghum, and wheat (Tran, 2010; Mochida et al., 2011).

Walia et al. (2009) reported that comparative genomics might be useful in analyzing stress-related expressions of plants whose whole genome sequence is not available and to identify stress-related genes by comparing their gene expression profiles with the sequenced ones. This approach will help researchers to predict and understand the function of genes in newly sequenced species and provide appropriate prospective opportunities to detect species-specific stress-responsive genes and regulatory mechanisms within and among model species. Therefore, with the help of this approach, it will be feasible to transfer information from model species to other species, which is important for food security.

The most recent approach to comparative genomic analysis is the Genome Zipper, which is useful in determining the implicit order of genes in a partially

sequenced genome. The approach has been helpful in comparing less studied species with full sequences and marked genomes of crops like *Brachypodium*, rice, and sorghum (Mayer et al., 2009). In the case of Triticeae, genome zipper reveals the evolutionary relationship by

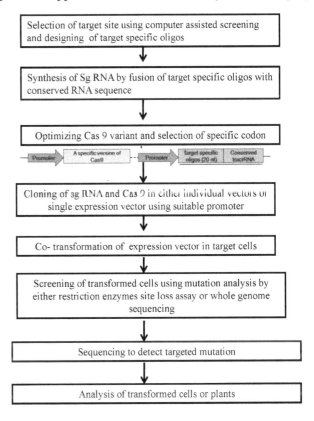

FIGURE 24.2 Strategy for the CRISPR-Cas-mediated plant genome editing (modified from Kumar and Jain, 2015).

providing its close approximation to a reference genome sequence (Mayer et al., 2011). But genome zipper has its limitation due to its dependence on synteny, and it cannot be used to explore newly evolved genes. So, for optimum exploitation of crop genomes, the construction of reference genome sequences by the physical and genetic map is necessary.

24.6 CONCLUSIONS

Abiotic stresses are a threat to global food security and enhancement of agriculture productivity. The knowledge available for understanding the mechanisms of abiotic stress is still insufficient to foster sustainable farming. The information available so far from the genomic and post-genomic approaches will be helpful for the development of crop varieties with different agronomic traits, including abiotic stress tolerance (Figure 24.1). Genomic-based approaches have been helpful in identifying the genomic systems for the improvement of crop plants against various abiotic stresses. Large-scale related resources such as germplasm lines and inbred lines can be developed on the basis of genetic markers by using the genomic techniques which may reveal the function of target genes. The availability of the complete genome sequences of many crops along with the availability of several plant ESTs sequences will be greatly helpful for determining the unknown functions of the majority of plant genes with respect to abiotic stress and mechanisms of stress tolerance. Genome-wide strategies have accelerated the deciphering of complex stress-responsive networks. Genes associated with stress may be manipulated for crop breeding or genetic engineering for development of 'the model stress tolerant crop', which can be considered as the ultimate aim of these efforts in terms of application and outreach.

ACKNOWLEDGMENTS

The authors (Richia Rai and Madhoolika Agrawal) are thankful to the Head of the Department of Botany, DST-FIST, CAS-Botany for all the laboratory and field facilities and to the Science and Enginering Research Board, New Delhi for providing a research grant and fellowship to Richia Rai. Amit Kumar Rai is thankful to DBT, New Delhi for funding and lab facilities.

REFERENCES

Achard, P., Herr, A., Baulcombe, D.C., Harberd, N.P. (2004). Modulation of floral development by a gibberellin-regulated microRNA. *Development* 131: 3357–3365.

Ainley, W.M., Sastry-Dent, L., Welter, M.E., Murray, M.G, Zeitler, B., Amora, R. (2013). Trait stacking via targeted genome editing. *Journal of Plant Biotechnology* 11: 1126–1134.

AkpJnar, B.A., Stuart, J., Budak, L.H. (2013). Genomics approaches for crop improvement against abiotic stress. *The Scientific World Journal.* Doi:10.1155/2013/361921.

Alameda, D., Anten, N.P.R., Villar, R. (2012). Soil compaction effects on growth and root traits of tobacco depend on light, water regime and mechanical stress. *Soil and Tillage Research* 120: 121–129.

Angell SM, Baulcombe DC 1999. Potato virus X amplicon–mediated silencing of nuclear genes. *Plant J.* 20: 357–362. https://www.ncbi.nlm.nih.gov/pubmed/11532181

Arabidopsis Genome Initiative 2000. Analysis of the genome sequence of the flowering plant. *Arabidopsis thaliana.* Nature 408: 796-815.

Arora, L., Narula, A. (2017). Gene editing and crop improvement using CRISPR-Cas 9system. *Frontiers in Plant Science.* Doi: 10.3389/fpls.2017.01932.

Atkinson, N.J, Urwin, P.E. (2012). The interaction of plant biotic and abiotic stresses: From genes to the field. *Journal of Experimental Botany* 63: 3523–3543.

Atkinson, N.J., Lilley, C.J., Urwin, P.E. (2013). Identification of genes involved in the response of Arabidopsis to simultaneous biotic and abiotic stresses. *Plant Physiology* 162: 2028–2041.

Bagge, M., Xia, X., Lubberstedt, T. (2007). Functional markers in wheat. *Current Opinion in Plant Biology* 10: 211–216.

Baltes, N.J., Gil-Humanes, J., Cermak, T., Atkins, P.A., Voytas, D.F. (2014). DNA replicons for plant genome engineering. *Plant Cell* 26: 151–163.

Baulcombe, D.C. (1999). Fast forward genetics based on virus induced gene silencing. *Current Opinion Plant Biology* 2: 109–113.

Bellard, C., Bertelsmeier, C., Leadley, P., Thuiller, W., Courchamp, F. (2012). Impacts of climate change on the future of biodiversity. *Ecology Letters* 15: 365–377.

Berkman, P.J., Skarshewski, A., Lorenc, M., Lai, K., Duran, C., Ling, E.Y.S., Stiller, J. (2011a). Sequencing and assembly of low copy and genic regions of isolated *Triticum aestivum* chromosome arm 7ds. *Plant Biotechnology Journal* 9: 768–775.

Berkman, P.J., Skarshewski, A., Manoli, S., Lorenc, M.T., Stiller, J., Smits, L., Lai, K. (2011b). Sequencing wheat chromosome arm 7bs delimits the 7bs/4al translocation and reveals homoeologous gene conservation. *Theoretical and Applied Genetics.* Doi: 10.1007/S00122-011-1717-2.

Bevan, M., Bancroft, I., Mewes, H.W., Martienssen, R., Mc Combie, R. (1999). Clearing a path through the jungle: Progress in *Arabidopsis* genomics. *Bioassay* 21(30): 110–120.

Bibikova, M., Golic, M., Golic, K.G., Carroll, D. (2002). Targeted chromosomal cleavage and mutagenesis in *Drosophila* cosuppression of homologous genes in trans. *Plant Cell* 2: 279–289.

Bortesi, L., Fischer, R. (2015). The CRISPR system for plant genome editing and beyond. *Biotechnology Advances* 33: 41–52.

Bouchez, D., Hofte, H. (1998). Functional genomics in plants. *Plant Physiology* 118: 725–732.

Brady, S.M., Long, T.A., Benfey, P.N. (2006). Unraveling the dynamic transcriptome. *Plant Cell* 18: 2101–2111.

Brenner, S., Johnson, M., Bridgham, J., Golda, G., Lloyd, D.H., Johnson, D. (2000). Gene expression analysis by massively parallel signature sequencing (MPSS) on micro-bead arrays. *Nature Biotechnology* 18: 630–634.

Cavanagh, C., Morell, M., Mackay, I., Powell, W. (2008). From mutations to MAGIC: Resources for gene discovery, validation and delivery in crop plants. *Current Opinion in Plant Biology* 11: 215–221.

Chantret, N., Salse, J., Sabot, F., Rahman, S., Bellec, A., Laubin, B., Dubois, I. (2005). Molecular basis of evolutionary events that shaped the hardness locus in diploid and polyploid wheat species (*Triticum* and *Aegilops*). *Plant Cell* 17: 1033–1045.

Chao, D.Y., Luo, Y.H., Shi, M., Luo, D., Lin, H.X. (2005). Salt-responsive genes in rice revealed by cDNA microarray analysis. *Cell Research* 15: 796–810.

Charpentier, E., Doudna, J.A. (2013). Rewriting a genome. *Nature* 495: 50–51.

Chen, J., Sun, M., Lee, S., Zhou, G., Rowley, J.D., Wang, S.M. (2002). Identifying novel transcripts and novel genes in the human genome by using novel SAGE tags. *Proceedings of the National Academy of Sciences* 99: 12257–12262.

Chen, C.P., Frei, M., Wissuwa, M. (2011). The *OzT8* locus in rice protects leaf carbon assimilation rate and photosynthetic capacity under ozone stress. *Plant Cell and Environment* 34: 1141–1149.

Chinnusamy, V., Zhu, J.K. (2009). Epigenetic regulation of stress responses in plants. *Current Opinion in Plant Biology* 12: 133–139.

Christian, M., Cermak, T., Doyle, E.L., Schmidt, C., Zhang, F., Hummel, A. (2010). Targeting DNA double-strand breaks with TAL effector nucleases. *Genetics* 186: 757–761.

Chung, C.L., Jamann, T., Longfellow, J., Nelson, R. (2010). Characterization and fine-mapping of a resistance locus for northern leaf blight in maize bin 8.06. *Theoretical Applied Genetics* 121: 205–227.

Close, T.J., Wanamaker, S.I., Caldo, R.A. (2004). A new resource for cereal genomics: 22K Barley Gene Chip comes of age. *Plant Physiology* 134: 960–968.

Cong, L., Ran, F.A., Cox, D., Lin, S., Barretto, R., Habib, N. (2013). Multiplex genome engineering using CRISPR/Cas systems. *Science* 339: 819–823.

Cooper, B., Clarke, J.D., Budworth, P., Kreps, J., Hutchison, D., Park, S. (2003). A network of rice genes associated with stress response and seed development. *Proceedings of the National Academy of Sciences of the United States of America* 100: 4945–4950.

Curtin, S.J., Zhang, F., Sander, J.D., Haun, W.J., Starker, C., Baltes, N.J. (2011). Targeted mutagenesis of duplicated genes in soybean with zinc-finger nucleases. *Plant Physiology* 156: 466–473.

Dalmay, T., Hamilton, A., Mueller, E., Baulcombe, D.C. (2000). Potato virus X amplicons in *Arabidopsis* mediate genetic and epigenetic gene silencing. *Plant Cell* 12: 369–379.

Degenkolbe, T., Do, T.P., Zuther, E., Repsilber, D., Walther, D., Hincha, D.K., Kohl, K.I. (2009). Expression profiling of rice cultivars differing in their tolerance to long term drought stress. *Plant Molecular Biology* 69: 133–153.

Deltcheva, E., Chylinski, K., Sharma, C.M., Gonzales, K., Chao, Y., Pirzada, Z.A. (2011). CRISPR RNA maturation by trans-encoded small RNA and host factor RNase III. *Nature* 471: 602–607.

Endo, M., Mikami, M., Toki, S. (2015). Multigene knockout utilizing offtarget mutations of the CRISPR/Cas9 system in rice. *Plant Cell Physiology* 56: 41–47.

Ergen, N.Z., Thimmapuram, J., Bohnert, H.J., Budak, H. (2009). Transcriptome pathways unique to dehydration tolerant relatives of modern wheat. *Functional and Integrative Genomics* 9: 377–396.

Feng, Z., Zhang, B., Ding, W., Liu, X., Yang, D.L., Wei, P. (2013). Efficient genome editing in plants using a CRISPR/Cas system. *Cell Research* 23: 1229–1232.

Flavell, R.B., Bennett, J., Smith, B. (1974). Genome size and the proportion of repeated nucleotide sequence DNA in plants. *Biochemical Genetics* 12: 257–269.

Flavell RB, Rimpau J, Smith DB 1977. Repeated sequence DNA relationships in four cereal genomes. *Chromosoma* 63:205–222.

Frei, M., Tanaka, J.P., Wissuwa, M. (2008). Genotypic variation in tolerance to elevated ozone in rice: Dissection of distinct genetic factors linked to tolerance mechanisms. *Journal of Experimental Botany* 59: 3741–3752.

Frei, M., Tanaka, J.P., Chen, C.P., Wissuwa, M. (2010). Mechanisms of ozone tolerance in rice: Characterization of two QTLs affecting leaf bronzing by gene expression profiling and biochemical analyses. *Journal of Experimental Botany* 61: 1405–1417.

Fujii, H., Chiou, T.-J., Lin, S.-I., Aung, K., Zhu, J.-K. (2005). A miRNA involved in phosphate-starvation response in *Arabidopsis*. *Current Biology* 15: 2038–2043.

Gardiner, J.M., Coc, F.H., Melia-Hancock, S., Loisington, D.A., Chao, S. (1993). Development of a core RELY map in maize using an immortalized-1 72 population. *Genetics* 134: 917–930.

Garg, B., Lata, C., Prasad, M. (2012). A study of the role of gene *TaMYB2* and an associated SNP in dehydration tolerance in common wheat. *Molecular Biology Reports* 39: 10865–10871.

Gibbings, J.G., Cook, B.P., Dufault, M.R., Madden, S.L., Khuri, S., Turnbull, C.J. (2003). Global transcript analysis of rice leaf and seed using SAGE technology. *Plant Biotechnology Journal* 1: 271–285.

Gill, B.S., Appels, R., Botha-Oberholster, A.M., Buell, C.R., Bennetzen, J.L., Chalhoub, B., Chumley, F. (2004). A workshop report on wheat genome sequencing: International Genome Research on Wheat Consortium. *Genetics* 168: 1087–1096.

Goff, S.A. (1999). Rice as a model for cereal genomics. *Current Opinion in Plant Biology* 2(31): 86–89.

Goff, S.A., Ricke, D., Lan, T.-H., Presting, G., Wang, R., Dunn, M., Glazebrook, J. (2002). A draft sequence of the rice genome (*Oryza sativa* L. Ssp. Japonica). *Science* 296: 92–100.

Gunnersen, J.M., Augustine, C., Spirkoska, V., Kim, M., Brown, M., Tan, S.S. (2002). Global analysis of gene expression patterns in developing mouse neocortex using serial analysis of gene expression. *Molecular and Cellular Neuroscience* 19: 560–573.

Gupta, P.K., Rustgi, S. (2004). Molecular markers from the transcribed/expressed region of the genome in higher plants. *Functional Integrative Genomics* 4: 139–162.

Gupta, P.K., Varshney, R.K. (2004). Cereal genomics: An overview. In: Gupta, P.K., Varshney, R.K. (Eds) *Cereal genomics*. Kluwer Academic Press , Dordrecht, The Netherlands, p. 639.

Haft, D.H., Selengut, J., Mongodin, E.F., Nelson, K.E. (2005). A guild of 45CRISPR-associated (Cas) protein families and multiple CRISPR/Cas sub types existing prokaryotic genomes. *PLoS Computational Biology* 1: 60.

Hernandez, P., Martis, M., Dorado, G., Pfeifer, M., Galvez, S., Schaaf, S., Jouve, N. (2011). Next generation sequencing and syntenic integration of flow-sorted arms of wheat chromosome 4a exposes the chromosome structure and gene content. *The Plant Journal* 69: 377–386.

Hsu, P.D., Lander, E.S., Zhang, F. (2014). Development and applications of CRISPR for genome engineering. *Cell* 157: 1262–1278.

IPCC. 2008. Kundzewicz ZW, Palutikof J, Wu S, eds. *Climate change and water. Technical paper of the Intergovernmental Panel on Climate Change*. Cambridge, UK & New York, NY, USA: Cambridge University Press.

Iqbal, M., Khan, M A., Naeem, M., Aziz, U., Afzal, J., Latif, M. (2013). Inducing drought tolerance in upland cotton (*Gossypium hirsutum* L.), accomplishments and future prospects. *World Applied Sciences Journal* 21: 1062–1069.

Iyer NJ, Tang Y, Mahalingam R. 2013. Physiological, biochemical and molecular responses to a combination of drought and ozone in *Medicago truncatula. Plant, Cell & Environment* 36: 706–720

Jain, S.M., Brar, D.S., Ahloowalia, B.S. (2002). *Molecular techniques in crop improvement*. Kluwer Academic Publishers, The Netherlands.

Jain, M., Nijhawan, A., Arora, R. (2007). F-Box proteins in rice: Genome-wide analysis, classification, temporal and spatial gene expression during panicle and seed development, and regulation by light and abiotic stress. *Plant Physiology* 143: 1467–1483.

Jia, X., Wang, W.X., Ren, L., Chen, Q.-J., Mendu, V., Willcut, B., Dinkins, R., Tang, X., Tang, G. (2009). Differential and dynamic regulation of miR398 in response to ABA and salt stress in *Populus tremula* and *Arabidopsis thaliana. Plant Molecular Biology* 71: 51–59.

Jiao, Y., Yang, H., Ma, L., Sun, N., Yu, H., Liu, T. (2003). A genome-wide analysis of blue-light regulation of *Arabidopsis* transcription factor gene expression during seedling development. *Plant Physiology* 133: 1480–1493.

Jinek, M., Chylinski, K., Fonfara, I., Hauer, M., Doudna, J.A., Charpentier, E. (2012). A programmable dual-RNA-guided DNA endonuclease in adaptive bacterial immunity. *Science* 337: 816–821.

Jinek, M., Jiang, F., Taylor, D.W., Sternberg, S.H., Kaya, E., Ma, E. (2014). Structures of Cas9 endonucleases reveal RNA-mediated conformational activation. *Science* 343: 1247997.

Jones-Rhoades, M.W., Bartel, D.P. (2004). Computational identification of plant microRNAs their targets, including a stress-induced miRNA. *Molecular Cell* 14: 787–799.

Joung, J.K., Sander, J.D. (2012). TALENs: A widely applicable technology for targeted genome editing. *National Review Molecular Cell Biology* 14: 49–55.

Jung, H.J., Kang, H. (2007). Expression and functional analyses of microRNA417 in Arabidopsis thaliana under stress conditions. *Plant Physiology Biochemistry* 45: 805–811.

Kamburova, V.S., Nikitina, E.V., Shermatov, S.E., Buriev, Z.T., Kumpatla, S.P., Emami, C., Abdurakhmonov, I.Y. (2017). Genome editing in plants: An overview of tools and applications. *International Journal of Agronomy* (Article ID: 731531) 1–15.

Kanchiswamy, C.N., Maffei, M., Malnoy, M., Velasco, R., Kim, J.S. (2016). Fine-tuning next-generation genome editing tools. *Trends in Biotechnology* 34: 562–574.

Kane, M.D., Jatkoe, T.A., Stumpf, C.R., Lu, J., Thomas, J.D., Madore, S.J. (2000). Assessment of the sensitivity and specificity of oligonucleotides (50mer) microarrays. *Nucleic Acids Research* 28: 4552–4557.

Kasurinen, A., Biasi, C., Holopainen, T., Rousi, M., Maenpaa, M., Oksanen, E. (2012). Interactive effects of elevated ozone and temperature on carbon allocation of silver birch (*Betula pendula*) genotypes in an open-air field exposure. *Tree Physiology* 32: 737–751.

Kehoe, D.M., Villand, P., Somerville, S. (1999). DNA microarrays for studies of higher plants and other photosynthetic organisms. *Trends in Plant Science* 4: 38–41.

Koebner, R.M.D. (2004). Marker-assisted selection in the cereals: The dream and the reality. In: Gupta, P.K., Varshney, R.K. (Eds) *Cereal genomics*. Kluwer Academic Publishers, Netherlands, p. 199.

Korzun, V. (2002). Use of molecular markers in cereal breeding. *Cell Molecular Biology Letters* 7: 811–820.

Kumar, V., Jain, M. (2015). The CRISPR–Cas system for plant genome editing: Advances and opportunities. *Journal of Experimental Botany* 66: 47–57.

Lee, J.Y., Lee, D.H. (2003). Use of serial analysis of gene expression technology to reveal changes in gene expression in Arabidopsis pollen under-going cold stress. *Plant Physiology* 132: 517–529.

Li, T., Liu, B., Spalding, M.H., Weeks, D.P., Yang, B. (2012). High-efficiency TALEN-based gene editing produces disease-resistant rice. *Nature Biotechnology* 30: 390–392.

Li, J., Lin, X., Chen, A., Peterson, T., Ma, K., Bertzky, M., Ciais, P., et al. (2013a). Global priority conservation areas in the face of 21st century climate change. *PLoS ONE* 8: e54839.

Li, J.F., Norville, J.E., Aach, J., McCormack, M., Zhang, D., Bush, J. (2013b). Multiplex and homologous recombination-mediated genome editing in *Arabidopsis* and *Nicotiana benthamiana* using guide RNAand Cas9. *Nature Biotechnology* 31: 688–691.

Liu, H.-H., Tian, X., Li, Y.-J., Wu, C.-A., Zheng, C.-C. (2008). Microarray-based analysis of stress-regulated microRNAs in *Arabidopsis thaliana*. *RNA* 14: 836–843.

Liu, Q., Zhang, Y.-C., Wang, C.-Y., Luo, Y.-C., Huang, Q.-J., Chen, S.-Y., Zhou, H., et al. (2009). Expression analysis of phytohormone-regulated microRNAs in rice, implying their regulation roles in plant hormone signaling. *FEBS Letters* 583: 723–728.

Liu, L., Chen, Y., Xu, B., Zhang, K., Li, F. (2016). A genome-wide association study reveals new loci for resistance to clubroot disease in *Brassica napus*. *Frontiers in Plant Science* 7: 1483.

Lorenz, W.W., Dean, J.F.D. (2002). SAGE profiling and demonstration of differential gene expression along the axial developmental gradient of lignifying xylem in loblolly pine (*Pinus taeda*). *Tree Physiology* 22: 301–310.

Loukehaich, R., Wang, T., Ouyang, B., Ziaf, K., Li, H., Zhang, J. (2012). SpUSP, an annexin interacting universal stress protein, enhances drought tolerance in tomato. *Journal of Experimental Botany* 63: 5593–5606.

Maccaferri, M., Sanguineti, M.C., Corneti, S. (2008). Quantitative trait loci for grain yield and adaptation of durum wheat (*Triticum durum* Desf.) across a wide range of water availability. *Genetics* 178: 489–511.

Maeder, M.L., Angstman, J.F., Richardson, M.E., Linder, S.J., Cascio, V.M., Tsai, S.Q. (2013a). Targeted DNA demethylation and activation of endogenous genes using programmable TALE-TET1 fusion proteins. *Nature Biotechnology* 31: 1137–1142.

Maeder, M.L., Linder, S.J., Cascio, V.M., Fu, Y., Ho, Q.H., Joung, J.K. (2013b). CRISPR RNA-guided activation of endogenous human genes. *Nature Methods* 10: 977–979.

Manavalan, L.P., Chen, X., Clarke, J., Salmeron, J., Nguyen, H.T. (2012). RNAi mediated disruption of squalene synthase improves drought tolerance and yield in rice. *Journal of Experimental Botany* 63: 163–175.

Martin, B., Nienhuis, S.J., King, G., Schaefer, A. (1989). Restriction fragment length polymorphisms associated with water use efficiency in tomato. *Science* 243: 1725–1728.

Matsui, A., Ishida, J., Morosawa, T. (2010). Arabidopsis tiling array analysis to identify the stress-responsive genes. *Methods in Molecular Biology* 639: 141–55.

Matsumoto, T., Wu, J.Z., Kanamori, H., Katayose, Y., Fujisawa, M., Namiki, N., Mizuno, H. (2005). The map-based sequence of the rice genome. *Nature* 436: 793–800.

Matsumura, H. Reich, S., Ito, A., Saitoh, H., Kamoun, S., Winter, P. (2003). Gene expression analysis of plant host-pathogen interactions by super SAGE. *Proceedings of the National Academy of Sciences of the United States of America* 100: 15718–15723.

Mayer, K.F., Taudien, S., Martis, M., Imkova, H.Š., Suchánková, P., Gundlach, H., Wicker, T. (2009). Gene content and virtual gene order of barley chromosome 1h. *Plant Physiology* 151: 496–505.

Mayer, K.F., Martis, X., Hedley, M., Simková, P.E., Liu, H., Morris, H. (2011). Unlocking the barley genome by chromosomal and comparative genomics. *Plant Cell* 23: 1249–1263.

McCouch, S.R., Kochert, G., Yu, Z.H., Wang, Z.Y., Khush, G.S., Coffman, W. R, Tanksley, S.D. (1988). Molecular mapping of rice chromosomes. *Theoretical Applied Genetics* 76: 815–829.

Meyers, C., Souret, F.F., Lu, C., Green, P.J. (2006). Sweating the small stuff: MicroRNA discovery in plants. *Current Opinion in Biotechnology* 17: 139–146.

Mittler, R. (2006). Abiotic stress, the field environment and stress combination. *Trends in Plant Science* 11: 15–19.

Mittler, R., Blumwald, E. (2010). Genetic engineering for modern agriculture: Challenges and perspectives. *Annual Review of Plant Biology* 61: 443–462.

Mittler, R., Finka, A., Goloubinoff, P. (2012). How do plants feel the heat? *Trends in Biochemical Sciences* 37: 118–125.

Mochida, K., Shinozaki, K. (2010). Genomics and bioinformatics resources for crop improvement. *Plant and Cell Physiology* 51: 497–523.

Mochida, K., Yoshida, T., Sakurai, T., Yamaguchi-Shinozaki, K., Shinozaki, K., Tran, L.S. (2010). Genome-wide analysis of two-component systems and prediction of stress-responsive two-component system members in soybean. *DNA Research* 17: 303–324.

Mochida, K., Yoshida, T., Sakurai, T., Yamaguchi-Shinozaki, K., Shinozaki, K., Tran, L.S. (2011). In silico analysis of transcription factor repertoires and prediction of stress-responsive transcription factors from six major gramineae plants. *DNA Research* 18: 321–332.

Molina, C., Rotter, B., Horres, R., Udupa, S.M., Besser, B., Bellarmino, L., Baum, M., et al. (2008). SuperSAGE: The drought stress responsive transcriptome of chickpea roots. *BMC Genomics* 9: 553.

Montgomery, M.K., Fire, A. (1998). Double-stranded RNA as a mediator in sequence-specific genetic silencing and co-suppression. *Trends Genetics* 14: 255–258.

Morozova, O., Hirst, M., Marra, M.A. (2009). Applications of new sequencing technologies for transcriptome analysis. *Annual Review of Genomics and Human Genetics* 10: 135–151.

Morrell, P.L., Buckler, E.S., Ross-Ibarra, J. (2011). Crop genomics: Advances and applications. *Nature Reviews Genetics* 13(2): 85–96.

Munns, R., Tester, M. (2008). Mechanisms of salinity tolerance. *Annual Review of Plant Biology* 59: 651–681.

Mysore, K.S., Tuori, R.P., Martin, G.B. (2001). Arabidopsis genome sequence as a tool for functional genomics in tomato. *Genome Biology* 2: 1003.

Nakano, M., Nobuta, K., Vemaraju, K., Tej, S.S., Skogen, J.W., Meyers, B.C. (2006). "Plant MPSS databases: Signature-based transcriptional resources for analyses of mRNA and small RNA. *Nucleic Acids Research* 34: D731–D735.

Nam, H.G., Giraudat, J., Den, B.B., Moonan, F., Loos, W.D.B., Hauge, B.M., Goodman, H.M. (1989). Restriction fragment length polymorphism linkage map of *Arabidopsis thaliana*. *Plant Cell* 1: 699–705.

Napoli, C., Lemieux, C., Jorgensen, R. (1990). Introduction of chimeric chalcone synthase gene into Petunia results in reversible using zinc-finger nucleases. *Genetics* 161: 1169–1175.

Navarro, L., Dunoyer, P., Jay, F., Arnold, B., Dharmasiri, N., Estelle, M., Voinnet, O., et al. (2006). A plant miRNA contributes to antibacterial resistance by repressing auxin signaling. *Science* 312: 436–439.

Nielsen, K.L., Hogh, A.L., Emmersen, J. (2006). Deep SAGE-digital transcriptomics with high sensitivity, simple experimental protocol and multiplexing of samples. *Nucleic Acids Research* 34: e133.

Nishitani, C., Hirai, N., Komori, S., Wada, M., Okada, K., Osakabe, K. (2016). Efficient genome editing in apple using a CRISPR/Cas9 system. *Scientific Reports* 6: 314–381.

NOAA (2018). National Centers for Environmental Information (NCEI) U.S. Billion-Dollar Weather and Climate Disasters. https://www.ncdc.noaa.gov/billions/

Nobuta, K., Venu, R.C., Lu, C. (2007). An expression atlas of rice mRNAs and small RNAs. *Nature Biotechnology* 25: 473–477.

Nongpiur, R.C., Pareek, S.L.P., Pareel, A. (2016). Genomics approaches for improving salinity stress tolerance in crop plants. *Current Genomics* 17: 343–357.

Osakabe, K., Osakabe, Y., Toki, S. (2010). Site-directed muta-Genesis in Arabidopsis using custom-designed zinc finger nucleases. *Proceedings of National Academy of Sciences U.S.A.* 107: 12034–12039.

Pabo, C.O., Peisach, E., Grant, R.A. (2001). Design and selection of novel Cys2His2 zinc finger proteins. *Annual Review of Biochemistry* 70: 313–340.

Paterson, A.H., Freeling, M., Sasaki, T. (2005). Grains of knowledge: Genomics of model cereals. *Genome Research* 15: 1643–1650.

Paterson, A.H., Bowers, J.E., Bruggmann, R., Dubchak, I., Grimwood, J., Gundlach, H., Haberer, G. (2009). The *Sorghum bicolor* genome and the diversification of Grasses. *Nature* 457: 551–556.

Paux, E., Roger, D., Badaeva, E., Gay, G., Bernard, M., Sourdille, P., Feuillet, C. (2006). Characterizing the composition and evolution of homoeologous genomes in hexaploid wheat through Bac-End sequencing on chromosome 3b. *Plant Journal* 48: 463–474.

Peele C, Jordan CV, Muangsan N, Turnage M, Egelkrout E, Eagle P, Hanley-Bowdoin L, Robertson D 2001. Silencing of a meristematic gene using Gemini virus-derived vectors. *Plant J.* 27: 357–366. https://www.ncbi.nlm.nih.gov/pubmed/?term=Peele%20C%5BAuthor%5D&cauthor=true&cauthor_uid=11532181

Perez-Lopez, U., Miranda-Apodaca, J., Munoz-Rueda, A., Mena-Petite, A. (2013). Lettuce production and antioxidant capacity are differentially modified by salt stress and light intensity under ambient and elevated CO_2. *Journal of Plant Physiology* 170: 1517–1525.

Perez-Pinera, P., Kocak, D.D., Vockley, C.M., Adler, A.F., Kabadi, A.M., Polstein, L.R. (2013). RNA-guided gene activation by CRISPR-Cas9-based transcription factors. *Nature Methods* 10: 973–976.

Phillips, R.L, Vasil, I.K. (2001). DNA-based markers in plants. In: Phillips R.L., Vasil I.K., (Eds) *DNA-based markers in plants*. Kluwer Academic Publishers, Dordrecht, The Netherlands, p. 497.

Prasch, C.M., Sonnewald, U. (2013). Simultaneous application of heat, drought, and virus to Arabidopsis plants reveals significant shifts in signaling networks. *Plant Physiology* 162: 1849–1866.

Ranjan, A., Pandey, N., Lakhwani, D., Dubey, N.K., Pathre, U.V., Sawant, S.V. (2012). Comparative transcriptomic analysis of roots of contrasting *Gossypium herbaceum* genotypes revealing adaptation to drought. *BMC Genomics* 13: 680.

Rasmussen, S., Barah, P., Suarez-Rodriguez, M.C., Bressendorff, S., Friis, P., Costantino, P., Bones, A.M., et al. (2013). Transcriptome responses to combinations of stresses in *Arabidopsis*. *Plant Physiology* 161: 1783–1794.

Reinartz, J., Bruyns, E., Lin, J.Z. (2002). Massively parallel signature sequencing (MPSS) as a tool for in-depth quantitative gene expression profiling in all organisms. *Briefings in Functional Genomics and Proteomics* 1: 95–104.

Rensink, W.A., Buell, C.R. (2005). Microarray expression profiling resources for plant genomics. *Trends in Plant Science* 10: 603–609.

Rivero, R.M., Mestre, T.C., Mittler, R., Rubio, F., Garcia-Sanchez, F., Martinez, V. (2013). The combined effect of salinity and heat reveals a specific physiological, biochemical and molecular response in tomato plants. *Plant, Cell and Environment* 37(5): 1059–1073.

Rizhsky, L., Liang, H., Mittler, R. (2002). The combined effect of drought stress and heat shock on gene expression in tobacco. *Plant Physiology* 130: 1143–1151.

Rizhsky, L., Liang, H., Shuman, J., Shulaev, V., Davletova, S., Mittler, R. (2004). When defense pathways collide. The response of Arabidopsis to a combination of drought and heat stress. *Plant Physiology* 134: 1683–1696.

Rosenzweig, C., Elliott, J., Deryng, D., Ruane, A.C., Müller, C., Arneth, A. (2014). Assessing agricultural risks of climate change in the 21st century in a global gridded crop model intercomparison. *Proceedings of National Academy of Sciences USA* 111: 3268–3273.

Rouet, P., Smih, F., Jasin, M. (1994). Introduction of doublestrand breaks into the genome of mouse cells by expression of a rare-cutting endonuclease. *Molecular Cell Biology* 14: 8096–8106.

Rudd, S. (2003). Expressed sequence tags: Alternative or complement to whole genome sequences. *Trends in Plant Science* 8: 321–329.

Saha, S., Sparks, A.B., Rago, C., Akmaev, V., Wang, C.J., Vogelstein, B. (2002). Using the transcriptome to annotate the genome. *Nature Biotechnology* 20: 508–512.

Sarfatti, M., Katan, J., Fluhr, R., Zamir, D. (1989). An RFLP marker in tomato linked to the *Fusarium oxysporum* resistance gene I2. *Theoretical and Applied Genetics* 78: 755–758.

Satish, K., Srinivas, G., Madhusudhana, R., Padmaja, P.G., Reddy, N., Murali, S. (2009). Identification of quantitative trait loci for resistance to shoot fly in sorghum (*Sorghum bicolor* (L.) Moench). *Theoretical Applied Genetics* 119: 1425–1439.

Schachermayr, G.M., Messmer, M.M., Feuillet, C., Winzeler, H., Winzeler, M., Keller, B. (1995). Identification of molecular markers linked to the *Agrapyron elongatum* derived leaf resistance gene *Lr24* in wheat. *Theoretical and Applied Genetics* 90: 982–990.

Schaeffer, S.M., Nakata, P.A. (2015). CRISPR-mediated genome editing and gene replacement in plants: Transitioning from lab to field. *Plant Science* 240: 130–142.

Schaeffer, S.M., Nakata, P.A. (2016). The expanding footprint of CRISPR/Cas9 in the plant sciences. *Plant Cell Reproduction* 35: 1451–1468.

Schena, M., Shalon, D., Davis, R.W., Brown, P.O. (1995). Quantitative monitoring of gene expression patterns with a complementary DNA microarray. *Science* 270: 467–470.

Schnable, P.S., Ware, D., Fulton, R.S., Stein, J.C., Wei, F., Pasternak, S., Liang, C.(2009). The B73 maize genome: Complexity, diversity and dynamics. *Science* 326: 1112–1115.

Schupp, J.M., Dick, A.B., Zinnamon, K.N., Keim, P. (2003). Serial analysis of gene expression applied to soybean. In: *Plant and animal genomes XI conference*, January 11–15, 2003. Town and Country Convention, Center, San Diego.

Seki, M., Narusaka, M., Yamaguchi, S.K., Carninci, P., Kawai, J., Hayashizaki, Y. (2001). Arabidopsis encyclopedia using full-length cDNAs and its application. *Plant Physiology Biochemistry* 39: 211–220.

Shan, Q., Zhang, Y., Chen, K., Zhang, K., Gao, C. (2015). Creation of fragrant rice by targeted knockout of the OsBADH2 gene using TALEN technology. *Journal of Plant Biotechnology* 13: 791–800.

Shimatani, Z., Nishizawa, YA., Endo, M., Toki, S., Terada, R. (2015). Positive-negative-selection-mediated gene targeting in rice. *Frontiers in Plant Sciences* 5: 748.

Shukla, V.K., Doyon, Y., Miller, J.C., De-Kelver, R.C., Moehle, E.A., Worden, S.E., Mitchell, J.C., et al. (2009). Precise genome modification in the crop species *Zea mays* using zinc-finger nucleases. *Nature* 459: 437–441.

Singh, S., Sarkar, A., Agrawal, S.B., Agrawal, M. (2014). Impact of ambient and supplemental ultraviolet-B stress on kidney bean plants: An insight to oxidative stress management. *Protoplasma* 251: 1395–1405.

Song, S.Y., Chen, Y., Chen, J., Dai, X.Y., Zhan, W.H. (2011). Physiological mechanisms underlying OsNAC5-dependent tolerance of rice plants to abiotic stress. *Planta* 234: 331–345.

Srivastava, G., Kumar, S., Dubey, G., Mishra, V., Prasad, S.M. (2012). Nickel and ultraviolet-B stresses induce differential growth and photosynthetic responses in *Pisum sativum* L. seedlings. *Biological Trace Element Research* 149: 86–96.

Sternberg, S.H., Redding, S., Jinek, M., Greene, E.C., Doudna, J.A. (2014). DNA interrogation by the CRISPR RNA-guided endonuclease Cas9. *Nature* 507: 62–67.

Sunkar, R., Zhu, J.K. (2004). Novel and stress-regulated microRNAs and other small RNAs from Arabidopsis. *Plant Cell* 16: 2001–2019.

Sunkar, R., Kapoor, A., Zhu, J.K. (2006). Posttranscriptional induction of two Cu/Zn superoxide dismutase genes in *Arabidopsis* is mediated by downregulation of miR398 and important for oxidative stress tolerance. *Plant Cell* 18: 2051–2065.

Sunkar, R., Chinnusamy, V., Zhu, J., Zhu J.K. (2007). Small RNAs as big players in plant abiotic stress responses and nutrient deprivation. *Trends in Plant Science* 12: 301–309.

Sunkar, R., Zhou, X., Zheng, Y., Zhang, W., Zhu, J.-K. (2008). Identification of novel and candidate miRNAs in rice by high throughput sequencing. *BMC Plant Biology* 8: 25.

Tabara, H., Grishok, A., Mello, C.C. (1998). RNAi in *C. elegans*: Soaking in the genome sequence. *Science* 282: 430–431.

Ten-Bosch, J., Seidel, C., Batra, S., Lam, H., Tuason, N., Saljoughi, S. (2001). Operon® array-ready OligoSetsTM provides sequence-optimized 70mers for DNA microarrays. *QIAGEN News* 4: 1.

Tran, L.S., Mochida, K. (2010). Identification and prediction of abiotic stress responsive transcription factors involved in abiotic stress signalling in soybean. *Plant Signalling and Behaviour* 5: 255–257.

Tripathi, R., Sarkar, A., Pandey, S., Agrawal, S.B. (2011). Supplemental ultraviolet-b and ozone: Impact on antioxidants, proteome and genome of linseed (*Linum usitatissimum* L. cv. Padmini). *Plant Biology* 13: 93–104.

Tyagi, A.K., Vij, S., Saini, N. (2006). Functional genomics of stress tolerance. In: Rao, K.V.M., Raghavendra, A.S., Janardhan R.K. (eds.). *Physiology and Molecular Biology of Stress Tolerance in Plants*, 301–334.

Ueda, Y., Frimpong, F., Qi, Y., Matthus, E.,Wu, L.B., Hollei, S., Kiaska, T., Frei, M. (2015). Genetic dissection of ozone tolerance in rice by genome wide association study. *Journal of Experimental Botany* 66: 293–306.

Utsumi, Y., Tanaka, M., Morosawa, T., Kurotani, A., Yoshida, T., Mochida, K. (2012). Transcriptome analysis using a high-density oligo-microarray under drought stress in various genotypes of cassava: An important tropical crop. *DNA Research* 19: 335–345.

van de Mortel, J.E., Aarts, M.G.M. (2006). Comparative transcriptomics – model species lead the way. *New Phytologist* 170: 199–201.

Varshney, R.K., Hiremath, P.J.,Lekha, P., Kashiwagi, J., Balaji, J., Deokar, A.A. (2009a). A comprehensive resource of drought and salinity responsive ESTs for gene discovery and marker development in chickpea (*Cicer arietinum L.*). *BMC Genomics* 10: 523.

Varshney, R.K., Nayak, S.N., May, G.D., Jackson, S.A. (2009b). Next generation sequencing technologies and their implications for crop genetics and breeding. *Trends in Biotechnology* 27: 522–530.

Vega-Sanchez, M.E., Gowda, M., Wang, G.L. (2007). Tag-based approaches for deep transcriptome analysis in plants. *Plant Science* 173: 371–380.

Velculescu, V.E., Zhang, L., Vogelstein, B., Kinzler, K.W. (1995). Serial analysis of gene expression. *Science* 270: 484–487.

Verelst, W., Bertolini, E., de Bodt, S. (2013). Molecular and physiological analysis of growth-limiting drought stress in *Brachypodium distachyon* leaves. *Molecular Plant* 6: 311–322.

Walbot, V. (1999). Genes, genomes, genomics. What can plant biologists expect from the 1998 national science foundation plant genome research program? *Plant Physiology* 119: 1151–1155.

Walia, H., Wilson, C., Ismail, A.M., Close, T.J., Cui, X. (2009). Comparing genomic expression patterns across plant species reveals highly diverged transcriptional dynamics in response to salt stress. *BMC Genomics* 10: 398.

Wang, X., Wang, H., Wang, J., Sun, R., Wu, J., Liu, S., Bai, Y. (2011). The genome of the mesopolyploid crop species *Brassica rapa*. *Nature Genetics* 43: 1035–1039.

Wang, Y., Yang, L., Hoeller, M., Zaisheng, S., Pariasca, T.J., Wissuwa, M., Frei, M. (2014). Pyramiding of ozone tolerance QTLs OzT8 and OzT9 confers improved tolerance to season-long ozone exposure in rice. *Environmental Experimental Botany* 104: 26–33.

Wanjugi, H., Coleman-Derr, D., Huo, N., Kianian, S.F., Luo, M.-C., Wu, J., Anderson, O. (2009). Rapid development of PCR-based genome-specific repetitive DNA junction markers in wheat. *Genome* 52: 576–587.

Waterhouse, P.M., Helliwell, C.A. (2003). Exploring plant genomes by RNA-induced gene silencing. *Nature Reverse Genetics* 4: 29–38.

Wheeler, T., Von Braun, J. (2013). Climate change impacts on global food security. *Science* 341: 508–513.

Williamson, V.M., Ho, J.Y., Wu, F.F., Miller, N., Kaloshian, I. (1994). A PCR-based marker tightly linked to the nematode resistance gene, *Mi*, in tomato. *Theoretical Applied Genetics* 87: 757–763.

Winter, P., Kahl, G. (1995). Molecular marker technologies for plant improvement. *World Journal of Microbiology Biotechnology* 11: 438–448.

Xing, H.L., Li, D., Zhi-Ping, W., Hai-Yan, Z., Chun-Yan, H., Bing, L. (2014). A CRISPR tool kit for multiplex genome editing in plants. *BMC Plant Biology* 14: 327.

Yazaki, J., Gregory, B.D., Ecker, B.D. (2007). Mapping the genome landscape using tiling array technology. *Current Opinion in Plant Biology* 10: 534–542.

Yu, Y.G., Saghai, M.A., Buss, G.R., Maughan, P.J., Tolin, S.A. (1994). RFLP and microsatellite mappng of a gene for soybean mosaic virus resistance. *Phytopathology* 84: 60–64.

Yu, J., Hu, S., Wang, J., Wong, G.K.-S., Li, S., Liu, B., Deng, Y. (2002). A draft sequence of the rice genome (*Oryza Sativa* L. Ssp. Indica). *Science* 296: 79–92.

Zeller, G., Henz, S.R., Widmer, C.K. (2009). Stress-induced changes in the *Arabidopsis thaliana* transcriptome analyzed using whole-genome tiling arrays. *Plant Journal* 58: 1068–1082.

Zhao, B., Liang, R., Ge, L., Li, W., Xiao, H., Lin, H., Ruan, K., et al. (2007). Identification of drought-induced microRNAs in rice. *Biochemistry Biophysics Research Communications* 354: 585–590.

Zhou, H., Liu, B., Weeks, D.P., Spalding, M.H., Yang, B. (2014). Large chromosomal deletions and heritable small genetic changes induced by CRISPR/Cas9 in rice. *Nucleic Acids Research* 42: 10903–10914.

Zhou, X., Wang, G., Zhang, W. (2007). UV-B responsive microRNA genes in *Arabidopsis thaliana*. *Molecular System Biology* 3: 103.

Zhou, L., Liu, Y., Liu, Z., Kong, D., Duan, M., Luo, L. (2010). Genome-wide identification and analysis of drought-responsive microRNAs in *Oryza sativa*. *Journal of Experimental Botany* 61: 4157–4168.

Zhu, J.K. (2003). Regulation of ion homeostasis under salt stress. *Current Opinion in Plant Biology* 6: 441–445.

25 Hallmark Attributes of Plant Transcription Factors and Potentials of *WRKY*, *MYB* and *NAC* in Abiotic Stresses

Sami Ullah Jan, Muhammad Jamil, Muhammad Faraz Bhatti, and Alvina Gul

CONTENTS

25.1 INTRODUCTION

In the context of plant sciences, an increment or decrement in environmental component(s) which affect the development, growth, yield and nutritional measures of plants is termed as stress. All stresses disrupt the plant's processes at the physiological, biochemical and molecular level (Munns, 2002; Ashraf and Harris, 2004). Stresses caused by living organisms, including insects and weeds, are known as biotic stresses while the stresses attained by non-living environmental factors such as chemicals (salts, metals), pH, humidity and temperature are referred to as abiotic stresses.

Plants are immobilized and cannot escape from all the biotic as well as abiotic stresses, resultantly; plants are very susceptible to growth and yield losses (Ahuja et al., 2010; Xu et al., 2010; Aroca et al., 2012), ionic imbalances (Sun et al., 2003; Wang et al., 2012), variable biochemical responses (Hasegawa et al., 2000), altered genome expression (Rasmussen et al., 2013) and reduction of vital nutritional contents of plants (Jamil et al., 2012; Jamil et al., 2013), surviving under numerous stresses. Upon the exposure, plants adapt specialized mechanisms at biochemical-, physiological- and molecular-level processes to combat these stresses (Wang et al., 1998; Yang et al., 2012; Cheng et al., 2013) whereas such adaptations acquired are facilitated through highly organized and complex interlinked signaling mechanisms (Nakashima et al., 2009). Plants possess about 300 genes in their genome dedicated to responding against stresses (Kültz, 2005).

A considerable amount of research and researchers are dedicated globally to investigating these processes; especially, the most primary and life-sustaining issue is to understand the interactions happening among the genome and proteome within a plant cell (Hirayama and Shinozaki, 2010; Shogren-Knaak et al., 2006; Chew and Halliday, 2011). One of these processes is a primary phase of gene expression, the transcription, which is a vital process because the production of the healthy transcriptome is a promising feature for healthy protein yield and stabilized cellular activities. The fundamental elements influencing gene expression are transcription factors (Nakashima et al., 2009).

Transcription factors (TFs) belong to one of the most extensively studied groups of proteins by various research groups across the globe (Riechmann et al., 2000; Reddy, 2001). Despite the efforts, no global consensus has been developed for its name, type and understanding its role due to its highly diverse structural and functional attributes as well as the availability of a repertoire of TFs in a cell. In *Arabidopsis*, 6% genes in nuclear genome codes for about 1600 TFs (Riechmann et al., 2000). Till the beginning of the new century, >600 zinc finger proteins, 21 heat shock-TFs, 133 *MYB*-TFs and 72 *WRKY*-TFs were studied only in *Arabidopsis* (Eulgem et al., 2000; Nover et al., 2001). Discovering new TFs in all plants has been extended at an exponential rate since the 21st century; hence, basic information and a detailed assortment of the TFs is necessary for better understanding and further studies.

This piece of manuscript focuses on many areas of TFs, including a description of the basic knowledge about TFs, reviewing the nomenclature criteria, classification and the generalized role played by TFs under normal condition. Section 25.9 also describes scenarios of various abiotic stress-induced TFs. It will help in producing a much clearer image about TFs as well as contributing recommendations accordingly in terms of plant-based TFs.

25.2 TRANSCRIPTION FACTORS

A transcription factor (TF) is the fundamental component in regulating expression within the genome (He et al., 2016; Liu et al., 2018). As the name indicates, TFs are the factors involved in the process of transcription. The importance of a transcription factor cannot be neglected while referring to a gene's mediated expression specifically due to variable stimuli generated either internally or externally (Atkinson and Urwin, 2012). They also confer stress responses by controlling a variety of interlinked downstream processes as an important part of cell signaling cascade (Xu et al., 2010; Pardo, 2010). TFs are very highly diverse and constitute 5–7% among the total protein-coding genes in each plant genome (Gray et al., 2009).

According to Gray et al. (2009), a transcription factor can be defined as a protein which has the tendency to locate and attach to a conserved sequence of DNA in gene's promoter through employing its DNA-binding domain (DBD) for mediating gene regulation. This role of mediating gene expressivity in a genome performed by TFs is accomplished with high specificity (Nakashima et al., 2009). The cell contains a variety of proteins which possess DNA-binding activity, but TFs are differentiated from these proteins in the way that TFs possess a characteristic structural motif called a DNA-binding domain, which other proteins do not possess, a site for attachment to DNA (Gray et al., 2009; Yamasaki et al., 2005).

25.3 NOMENCLATURE OF TRANSCRIPTION FACTORS

In the past, when few TFs were discovered, these were named without proper criteria. Soon after the completion of *Arabidopsis* genome sequencing, initial steps were taken based on a family-by-family strategy to name a transcription factor (Stracke et al., 2001; Jakoby et al., 2002; Bailey et al., 2003). Although this nomenclature was not a global consensus, still, it was adopted and followed by the scientific community of that era (Guo et al., 2005).

In 2009, Gray and his colleagues (Gray et al., 2009) proposed a system for naming TFs which is commonly followed in current studies. According to them, the name of a transcription factor shall consist of three parts: (i) initials of the name of a species from which the transcription factor is isolated, for example, *Ta* represents *Triticum aestivum, Zm* shows *Zea mays, Os* indicates *Oryza sativa* and *At* means *Arabidopsis thaliana*, (ii) a three to four letter code for the DNA-binding domain family name, for example, *WRKY* for the presence of *WRKYGQR* sequence in DBD, and (iii) a number indicating serial of discovery of TFs in respective species that shows the order in which TFs have been discovered. For example, *ZmMYB3* ("*Zm*": *Zea mays*, "*MYB*": DNA-binding domain, "*3*": Serial of identification).

In addition to the above proposal for TF nomenclature, certain amendments were also proposed: (i) The name of a transcription factor shall be italicized; (ii) in the case of TFs made out of alternateive splicing of mRNAs, there shall be added suffixes (like: .1, .2, .3 and so on) with its name as *ZmMYB3.1* and *ZmMYB3.2;*

(iii) the allelic variants of TFs can be expressed by its identification in superscript, as shown as: *ZmMYB3.1*[W22] and *ZmMYB3.1*[B73] while in case of formatting issues, these allelic variants can also be presented by adding an arrow-head sign as *ZmMYB3.1^W22* and *ZmMYB3.1^B73*; and finally (iv) the number of TFs produced from single gene, if they show variation, may be added with alphabetic letters such as *ZmMYB3.1*[W22]*a* and *ZmMYB3.1*[W22]*b* (Gray et al., 2009).

25.4 STRUCTURE OF TRANSCRIPTION FACTORS

Transcription factors contain highly organized structures of alpha-helices and β-sheets with variable chain lengths and molecular weight (Reid et al., 2010; Hichri et al., 2011). TFs share less structural similarity among themselves because of high numbers and functional variations, but there are certain conserved structural motifs which can be used to access a distinguished type. Three important conserved structural motifs in TFs are "Signal-Sensing Domain (SSD)", "DNA-Binding Domain (DBD)" and "Trans-Activating Domain (TAD)" (Depège-Fargeix et al., 2010). This section includes an introduction to these structures, while their role and complete mechanisms are discussed in later sections (Figure 25.1).

DNA-binding domains are the primary sites for DNA–protein interactions. The part of transcription factor that has a conserved structural motif with capability to bind specific promoter or enhancer-sequence in a gene is called DNA-binding domain while, alternatively, the position/site on DNA where a transcription factor binds refers to Transcription Factor Binding Site (TFBS) or Response Element (RE) (Reid et al., 2010). DBDs show more conserved pattern as compared to the other structural motifs (Atkinson and Urwin, 2012; Guo et al., 2005). DBDs are the key sites in determining the specificity, type and rate of binding a transcription factor to specific loci in DNA (Reid et al., 2010). These are highly organized folded-motifs with a variable number of amino acids, such as 50 in *MYB*-type TFs, 60 in *WRKY*-Type TFs (Liu et al., 2013), while 150–160 in *NAC* (Ooka et al., 2003).

The signal-sensing domain is another conserved structural motif in a transcription factor. In correspondence to its activity of activating the TFs, it is also known as Transcription Factor Activating Domain (TFAD). SSD structure depends upon the nature, type and intensity of signals (Brivanlou and Darnell, 2002). An SSD is usually named accordingly depending upon the type of signal, for example, acidic domain is one that is activated by acidic signals (Triezenberg, 1995). Ligands, ions and hormones also serve as a signal that binds to SSD, resulting in conformational changes in the structure of SSD which entails its activation (Karin and Hunter, 1995).

Trans-activating domains (TADs) are specialized in protein–protein interaction, which serves as a platform for other proteins or enzymes to bind (Wärnmark et al., 2003; Piskacek et al., 2007) to perform their action. After activation and binding to TFBS, TAD of TFs attracts and assembles the whole transcriptional machinery which also includes other proteins like co-activators, corepressors and enzymes like polymerase. The well-characterized study reveals that the size of TADs varies from few to several amino acids depending upon the number of factors required in transcriptional complex assembly; similarly, the position of TADs interacting with other proteins cannot be anticipated accurately (Piskacek et al., 2007). For instance, TATA-Binding Protein (TBP) as well as TBP-Associated Factors (TAFs) constitutively form an initiation complex termed as TFIID (Gangloff et al., 2000).

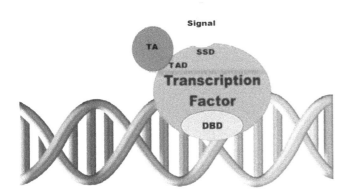

FIGURE 25.1 A generalized structure of a typical transcription factor that binds to major groove in its target site on DNA. A signal binds to Signal-Sensing Domain (SSD) on transcription factor and activates it. Activated transcription factor binds to its target with the help of DBD. Finally, Transcription Activator (TA) approaches and binds to transcription factor at Transcription Activator Domain (TAD) and transcription starts.

25.5 CLASSIFICATION OF TRANSCRIPTION FACTORS

All cellular proteins, especially the TFs, are very diverse in nature, with highly variable structures and functions due to complex cellular process like pre-mRNA alternative splicing (Lorković et al., 2000; Kazan, 2003).

Though it is a difficult job to develop a global consensus for all proteins, including enzymes and TFs, a number of scientists have contributed and succeeded to a level. Up till now, all the TFs have been sorted into 50–60 families on the basis of their DBDs (Gray et al., 2009), yet the complete picture for classification of TFs has not been revealed. To date, TFs are classified categorically on the basis of various parameters like DBD-structure, homology modeling, hidden Markov modeling, function and on the basis of species in which the respective TFs reside (Matus et al., 2008). Pertinent information of each category is given as follows.

25.5.1 CLASSIFICATION OF TRANSCRIPTION FACTORS ON THE BASIS OF ACTIVITY

It was an organized classification based on the role TFs perform in a genome during gene expression and signal transduction (Brivanlou and Darnell, 2002). According to this classification, TFs were placed in two major groups: (i) Constitutively Active TFs; those related with housekeeping genes and are available in a cell all the time, (ii) Conditionally Active TFs, which are activated under certain circumstances as under stress conditions and developmental stages (Brivanlou and Darnell, 2002). Conditionally active TFs are further classified into developmental process-dependent TFs and signal dependent TFs.

25.5.2 CLASSIFICATION OF TRANSCRIPTION FACTORS ON THE BASIS OF STRUCTURE

On the basis of structural attributes, all the eukaryotic TFs are categorized on the basis of their DBDs in an online database entitled TRANSFAC®. Wingender and his colleagues (Wingender et al., 2000) contributed to this database on the basis of nucleotide matrices as a major tool, while Stegmaier and his colleagues (Stegmaier et al., 2004) used hidden Markov models for further description of all the TFs. According to TRANSFAC®, all the TFs lie in four super-classes, namely, (i) Basic Domains: consists of six classes and 21 families, (ii) *Zinc-Coordinating-DNA-Binding Domains*, comprising five classes and 13 families, (iii) Helix-Turn-Helix (HTH), includes six classes and 15 families, (iv) *Beta-Scaffold-Factors*, contacts with minor-groove in DNA and possess 11 classes and 19 families. There are some TFs that are not classified into any of these super-classes, and so until they are placed in a fifth, separate super-class, they are as yet entitled "Other Transcription Factors" (Stegmaier et al., 2004).

25.5.3 CLASSIFICATION OF TRANSCRIPTION FACTORS ON THE BASIS OF SPECIES

An online database PlnTFDB (*Plant Transcription Factor Database*) was established which classified plant TFs with taxonomic ranks (on the basis of species containing TFs) coordinated with the features of DBD of plant TFs (Pérez-Rodríguez et al., 2010). Though this classification is limited to plants, it not only classifies TFs, but it also considers transcription regulators.

A similar classification is also observed maintained in a database named PlantTFDB (*Plant Transcription Factor Database*) which uses the same parameters for classification of TFs, with modifications in defining families of TFs (Zhang et al., 2011).

These databases further provide high integration with the latest tools like BLAST, structures from the Protein Data Bank and sequence along with more information for much better insight into TFs. Due to this reason, these are also termed as "Integrative Plant Transcription Factor Databases". Other databases are also available which classify TFs on various aspects; some are specific to plants while others may also cover the TFs of animals, humans and pathogens. Details of plant-related databases of TFs are listed in Table 25.1.

25.5.4 CLASSIFICATION OF TRANSCRIPTION FACTORS WITH SPECIALIZED PURPOSE

Some databases classify TFs with specific purposes. For example, STIFDB2 categorizes those TFs which respond and act against stress-induced signals (Naika et al., 2013). Similarly, wDBTF is a specialized classification database for wheat, TreeTFDB for trees and SoyDB includes TFs of soybean (Table 25.1).

25.6 STUDY OF TRANSCRIPTION FACTOR IN LABORATORY

Isolation of a specific TF enhances the accuracy of well-characterized study, and to locate the exact TF, binding sites of TF in DNA must be known. Previously, *Electrophoretic-DNA-Mobility-Shift Assay* was the acquired methodology for isolation of transcription factors. In this method, TFBS of DNA is isolated from DNA, followed by passing all cellular contents (Manley et al., 1980; Smith et al., 1999) or only contents from the nucleus (Dignam et al., 1983; Latchman, 2008) upon TFBS. After incubation, the mixture is subjected to gel electrophoresis, and, depending upon the size, exact TF with TFBS is extracted (Latchman, 2008; Fried and Crothers, 1981;

Garner and Revzin, 1981). Initially, these protocols proved to be very useful, but over the course of time, certain discrepancies were observed. The major problem that may arise in this method is that if a certain type of TF exists in very low concentration, then another protein or TF with more or less homology with exact TF may attach to TFBS and thus may lead to false results (Smith et al., 1999).

To overcome the limitations in *Electrophoretic-DNA-Mobility-Shift Assay*, other methods have been developed.

All these other techniques are based on the concept of *Electrophoretic-DNA-Mobility-Shift Assay* with enhanced modifications. Afterward, one of the most commonly used techniques is *DNA-Sepharose Chromatographic Technique* (Kadonaga and Tjian, 1986). In this method, TFBS is multimerized (to increase the number of copies of TFBS) after isolation and fixed on the sepharose column. Then the nuclear or whole cell extract is passed down the column. The resultant mixture is electrophorized,

TABLE 25.1
List of Publically Available Databases Which Classify All the Available Transcription Factors with Different Aspects

Database	Details
TRANSFAC *TRANScription FACtors*	Classifies eukaryotic transcription factors on the basis of the DNA-binding domain they contain and the activity of transcription factors in transcriptional regulation
PlnTFDB *Plant Transcription Factor Database*	Classifies plant transcription factors on the basis of species as well as on the basis of the characteristics of DNA-binding domain. PlnTFDB also includes relevant
PlantTFDB *Plant Transcription Factor Database*	gene expression's regulatory proteins in classification along with transcription factors
PlanTAPDB *Plant Transcription-Associated Proteins Database*	Classifies plant transcription factors and associated proteins on the basis of phylogenetic relationship among transcription factors
DATFAP *Database of Transcription Factors with Alignments and Primers*	Classifies transcription factors of 13 plant species on the basis of sequence using PCR primers which also gives a comparative analysis of relationship among transcription factors
AGRIS *Arabidopsis Gene Regulatory Information Service*	Specialized database for classification of transcription factors found in *Arabidopsis* on the basis of promoter sequence, cDNA sequence and mutant types
RARTF *Riken Arabidopsis Transcription Factor Database*	
GRASSIUS *Grass Regulatory Information Service*	Classifies transcription factors as well as other proteins affecting gene expression in grasses
LegumeTFDB *Legume Transcription Factor Database*	Classifies transcription factors found in legumes
TOBFAC *The Database of Tobacco Transcription Factors*	Classifies transcription factors of tobacco plants on the basis of respective genes of each transcription factor
TreeTFDB *Transcription Factor Database of Tree*	Classifies transcription factors of six sequenced-trees on the basis of their gene annotation (*Populus trichocarpa* [poplar], *Vitis vinifera* [grape vine], *Carica papaya* [papaya], *Ricinus communis* [castor bean], *Manihot esculenta* [Cassava] and *Jatropha curcas* [jatropha])
SoyDB *Database of Soybean Transcription Factors*	Specialized database for classification of soybean transcription factors on the basis of annotated protein sequence by hidden Markov models
wDBTF *Database of Wheat Transcription Factors*	Specialized database for classification of wheat transcription factors on the basis of sequence homology of rice transcription factors
DRTF *Database of Rice Transcription Factors*	Specialized database for classification of rice transcription factors on the basis of gene models
PLACE *PLAnt Cis-acting regulatory DNA Elements*	Classifies plant transcription factors on the basis of their binding to cis-regulatory sequence in DNA
JASPAR *Transcription Factor Binding Profile Database*	Classifies eukaryotic transcription factors on the basis of published literature with similar resources with TRANSFAC
STIFDB2 *Stress-Responsive Transcription Factor Database*	Classifies plant transcription factors which are involved in acting against stress-signals

and TF is isolated. Further developments in this technique also include radioactively labeled TFBS (Jarrett, 2012). Employment of the latest instrumentation such as two-dimensional gel electrophoresis, high-performance liquid chromatography (HPLC) and blotting techniques enhances the certainty of isolating the exact TF.

25.7 TRANSCRIPTION FACTORS UNDER NORMAL CONDITIONS

Transcription factors, the key to gene regulation, can declare to be the worthy objectives to deal with issues like plant growth, yield, quality and environmental concerns at the genomic level (Hymus et al., 2013). TFs serve as the principal factors such that a singular transcription factor can control the expression of a single gene, but also a single TF may confer regulating numerous genes (Nakashima et al., 2009). Besides their broad range of gene-control ability, TFs do not bind to any gene with homologous responsive sites (Vaahtera and Brosche, 2011); rather, TFs form hetero- or homodimers, which restricts its binding activity to certain targets, thus increasing their specificity (Vaahtera and Brosche, 2011; Baudry et al., 2004). The role of TFs is highly signal-dependent, which may be an internal signal arising from the homeostatic condition of the cell (Latchman, 1997) or may be an external signal driven by environmental issues like biotic plus abiotic stresses (Nakashima et al., 2009). However, the current section's focal point is to provide a generalized role of TFs in normal conditions.

Any internal- or external-induced signal in the various forms contacts the interior specialized components or cell membrane, respectively, and stimulates a cascade system of the complex signaling network (Karin and Hunter, 1995). Enzymes like kinases serve as a mediator and transform the signals accordingly, and these transformed signals detect and adhere to respective TFAD, leading to the activation of TFs depending upon the nature of signals and the structure of TFAD (Triezenberg, 1995; Roberts, 2000). The joining of transformed signal and TFAD causes changes in the structural motifs of TF, resulting in activated TF (Piskacek et al., 2007). Once a TF is activated, it leads to regulation (repression and/or activation) of transcription in response to signal received.

25.7.1 TRANSCRIPTION FACTOR IN ACTIVATION OF TRANSCRIPTION

Activated TFs start their activity to perform the distinguished role in various directions and locations within the genome. The first step in their activity is to find out the exact location where it attaches to its relative TFBS using its DBD with high specificity (Roeder, 1996). Diverse roles of activated TFs include: (i) act only for opening the wrapped strings of DNA around histones (Zhang and Wang, 2005; Jin et al., 2009) and (ii) act only for initiation of transcription when TFBS is available on linear DNA (Nikolev and Burley, 1997). It is also possible that TFs first deals with nucleosome particle and release the DNA and then process transcription (Zaret and Carroll, 2011).

Chromatin in condensed form is a complex structure consisting of DNA and histones, collectively termed as nucleosome (Kornberg, 1977), which is further aided with folding that results in a more condensed and higher complex form of chromatin (Schwarz and Hansen, 1994). These highly organized structures influence both the process of transcription and binding of TFs (Shogren-Knaak et al., 2006; Jin and Felsenfeld, 2007; Martino et al., 2009). Previous works about whole-genome-location analysis showed that TFs bind to their targets in DNA with very low rates instead of their high potential (Kaplan et al., 2011) because the chromatin structure and DNA sequence determines when and where TFs have to bind (Zaret and Carroll, 2011). Important parts of a gene necessary for gene expression may be masked in DNA turns around histones. Some TFs, termed as *Pioneer Transcription Factors* (PTFs) by Zaret and Carroll (2011), exist specifically for remodeling the histone particles so that polymerase can bind after opening the loops of DNA and further processes can be accomplished (Kornberg, 1977). Under such circumstances, the target sites for respective TFs are available on DNA loops turning around the histone particles (Luger et al., 1997). PTFs, when binding to these sites, call upon certain enzymes like acetyltransferase p300, which modifies the histone particles and exposes the target sites on DNA (Visel et al., 2009). Exposing the target sites for TFs within chromosome is carried out in two ways: (i) Intrinsic positioning or destabilizing the nucleosome, an ATP-dependent process in which a nucleosome position is changed or destabilized with the cell's own remodeling of DNA process (Shogren-Knaak et al., 2006; Sekinger et al., 2005; Segal et al., 2006). Once a target site is open, polymerases can bind and lead to transcription. (ii) Another strategy followed by TFs occurs in the case of if the target site is not masked by histones and directly leads to initiating the transcription process. In this case, after binding a TF to TFBS in the enhancer or promoter region, a cluster of helper molecules join and form transcription complex (Li et al., 2010). TFBS of certain TFs may lie beyond cis-regulatory elements in enhancer or promoter region linked to a gene with spacer DNA (Nakashima et al., 2009; Van Helden et al., 2000), which can prove the concept that a single TF may affect the expression of more than one gene. These helper molecules may be other TFs and/or proteins (such as coactivators or co-repressors) along enzymes like polymerase (Zaret and Carroll, 2011), while some studies reveal that non-coding long RNAs are also involved (Wang et al., 2011). TFs either serve as activators to initiate, enhancers to enhance or repressors to repress a transcription process (Jin and Felsenfeld, 2007). Though the basic mechanism of regulating transcription process lies in these

mechanisms (Figure 25.2), much more has yet to be characterized about these mechanisms.

25.7.2 Transcription Factor in Repression of Transcription

It was previously considered that TFs are involved in the initiation of transcription only, but soon it was confirmed that TFs might also act as a repressor for transcription (Latchman, 1997). In the normal process, an activator binds to TFBS and activation of gene expression results, but this activation can be prevented by two broad categories of mechanisms which include indirect repression

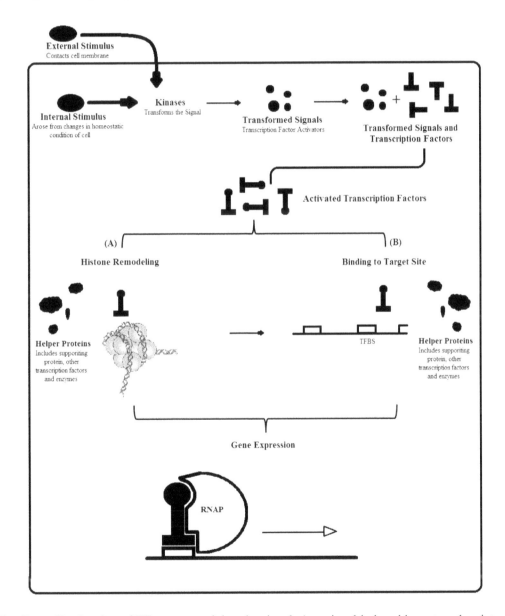

FIGURE 25.2 Generalized action of TFs upon receiving the signal. Any signal induced by external or internal stimulus is transformed into TF-readable signals. These transformed signals bind to TF at SSD. Activated TF proceeds to their targets, which may be (1) histone remodeling for opening DNA loops or (2) direct binding to TFBS in linear DNA. These processes are followed by the formation of transcriptional complex which include RNA-Polymerase (RNAP) and thus transcription starts.

and direct repression (Hanna-Rose and Hansen, 1996). Indirect methods of repression include: (i) a repressor binds to TFBS and, thus, no room exists for activators to bind, (ii) a repressor binds directly to TF and does not allow it to bind TFBS, (iii) a repressor deactivates a TF at its binding site, and this type of mechanism occurs when a repressor possesses structural homology with polymerases or other helper proteins that are required in the formation of transcriptional complex, (iv) a repressor is capable of degrading a TF, thus leaving no TF to bind to TFBS while the direct repression of transcription occurs when a gene is active without TFs and a repressor's molecule gets into the target site and results in the inactivation of gene (Latchman, 1997, 2008; Maldonado et al., 1999).

25.8 TRANSCRIPTION FACTOR REGULATION

Regulation of TFs proves to be the fundamental step toward the integrity of gene regulation in a cell (Wang et al., 2016; Wu et al., 2018). Some regulation processes of TFs are dependent upon the open reading frame and ubiquitination (Hinnebusch, 2005; Galgoczy et al., 2004) while some regulatory processes are complex enough to accomplish at various levels (Edmondson and Olson, 1993). However, the TF regulation happens in two broad categories: (i) at the stage of synthesis and (ii) according to their activity (Roberts, 2000), which are discussed below.

25.8.1 REGULATION OF TRANSCRIPTION FACTORS SYNTHESIS

Regulation of TFs is crucial for maintaining a cell's integrity because errors in the regulation process can lead to various disorders (Latchman, 1997). After completion of the target role, a gene/TF needs to be restricted from their action, and, thus, accurate regulation of TFs prevents a cell from overcrowding and increases control over cellular mechanisms (Vaahtera and Brosche, 2011). Cells possess a variety of strategies to regulate transcription factors, and all of the strategies for regulation of TFs are dependent upon the interacting signals, and molecular interactions at the cellular level play a vital role (Atkinson and Urwin, 2012). These strategies include:

25.8.1.1 Mechanisms of the Regulation of TFs

TFs modify themselves in cysteine residues with redox reactions which make them unavailable for phosphorylation, protein degradation and transportation to the nucleus (Vaahtera and Brosche, 2011; Baudry et al., 2004). However, it is linked with the regulation of the synthesis of TFs because expression of TFs is stopped or enhanced depending upon the cell's requirement under a specific condition (Latchman, 1997).

25.8.1.1.1 Regulation of Activity of TFs

A repertoire of TFs is available in a cell under normal conditions, but upon exposure, the genes responsible for producing TFs start expression efficiently, and more TFs can be produced to combat according to the invading signals (Latchman, 1997, 2008).

25.8.1.1.2 Regulation of Synthesis of TFs

Some TFs are cell/tissue-specific such that a specific type of TF produced by a particular cell may not be required or expressed in another cell (Latchman, 2008). For instance, *MyoD* transcription factor exists in the cells of skeletal muscle. So, overexpression of skeletal-muscle-specific *MyoD* in an undifferentiated fibroblast cell will lead to the formation of skeletal muscle cells (Edmondson and Olson, 1993).

25.9 TRANSCRIPTION FACTORS UNDER ABIOTIC STRESS

Plants are open to several types of abiotic conditions in fields (Mittler and Blumwald, 2010), and a key selection plants can adapt is to acquire the state of tolerance in response to such conditions at molecular stage (Ashraf and Harris, 2004; Chaves et al., 2003; Atkinson et al., 2013). The underlying mechanism in adapting tolerance by plants has much been investigated in various laboratories across the globe. It is understood that plants respond to these stresses through complex alterations in molecular mechanisms, biochemical processes and physiological variations (Atkinson and Urwin, 2012). Transcription factors are utilized by plants in adapting tolerance against stresses at the molecular level (Wang et al., 2016; Wu et al., 2018). Out of the repertoire of TFs in a cell, *WRKY*, *MYB* and *NAC* type are extensively studied TFs which are renowned against abiotic stresses. Hereafter, we will discuss each of them in detail. A list of extensively studied, well-characterized *WRKY-*, *MYB-* and *NAC*-TFs are given in Table 25.2.

25.9.1 WRKY TRANSCRIPTION FACTORS

This is most widely studied and well-characterized among plant TFs. *WRKY*-type transcription factor *SPF1* (*Sweet-Potato-Factor-1*) is the first cloned-TF obtained

TABLE 25.2

List of Well-Studied WRKY-, MYB-, and NAC-Transcription Factors in Various Plants

Transcription Factor	Inducer	Host Plant	Role/Involved In	Reference
TaWRKY2	Multiple abiotic stresses	*Triticum aestivum*	Drought and salt stress tolerance	Niu et al. (2012)
TaWRKY10	Multiple abiotic stresses	*T. aestivum*	Drought and salt stress tolerance	Wang et al. (2013)
TaWRKY19	Multiple abiotic stresses	*T. aestivum*	Drought and salt stress tolerance	Niu et al. (2012)
GmWRKY13	Drought and salt stress	*Glycine max*	Drought and salt stress tolerance	Zhou et al. (2008)
GmWRKY21	Drought, salt and cold stresses	*Glycine max*	Drought, salt and cold stress tolerance	Zhou et al. (2008)
GmWRKY54	Drought and salt stress	*Glycine max*	Drought and salt stress tolerance	Zhou et al. (2008)
AtWRKY2	Salt stress	*Arabidopsis thaliana*	Osmotic stress response and ABA pathway	Jiang and Yu (2009)
AtWRKY3	Pathogen (biotic) stress	*A. thaliana*	Pathogen (biotic) stress response	Dong et al. (2003)
AtWRKY6	Pathogen (biotic) stress	*A. thaliana*	Pathogen (biotic) stress response	Dong et al. (2003)
AtWRKY25	Salt and heat stresses	*A. thaliana*	Salt and heat stress tolerance	Li et al. (2009b)
AtWRKY26	Heat stress	*A. thaliana*	Heat stress tolerance	Li et al. (2009b)
AtWRKY33	Salt and cold stresses	*A. thaliana*	Salt and cold stress tolerance	Li et al. (2009b)
OsWRKY08	Multiple abiotic stresses	*Oryza sativa*	Drought, salt and osmotic stress tolerance	Song et al. (2009)
OsWRKY11	Drought and heat stress	*O. sativa*	Drought and heat stress tolerance	Wu et al. (2009)
OsWRKY23	Salt stress	*O. sativa*	Salt stress tolerance	Jing et al. (2009)
OsWRKY45	Cold stress	*O. sativa*	Cold stress tolerance	Qiu and Yu (2009)
OsWRKY72	Salt and heat stress	*O. sativa*	Salt and heat stress tolerance	Yu et al. (2010)
OsWRKY89	Salt stress	*O. sativa*	Salt stress tolerance	Wang et al. (2007)
OsMYB3R-2	Cold stress	*O. sativa*	Cold stress tolerance	Ma et al. (2009)
AtMYB44	Multiple abiotic stress	*A. thaliana*	Multiple abiotic stress tolerance	Jung et al. (2008)
AtMYB96	Drought stress	*A. thaliana*	Drought stress tolerance	Seo and Park (2010)
GbMYB5	Drought stress	*Cotton*	Drought stress tolerance	Chen et al. (2015)
TaMYB1D	Drought and oxidative stress	*Transgenic tobacco*	Polyprenoid metabolism	Wei et al. (2017)
OsMYB91	Salt stress	*O. sativa*	Salt stress tolerance and plant growth	Zhu et al. (2015)
SbMYB8	Drought tolerance	*Transgenic tobacco*	Flavonoid biosynthesis	Yuan et al. (2015)
GmNAC11	Salt stress	*G. max*	Salt stress tolerance	Hao et al. (2011)
TaNAC69	Poly ethylene glycol	*T. aestivum*	Drought stress tolerance	Xue et al. (2011)
TaNAC6	Drought, salt and cold stress	*T. aestivum*	Drought, salt and cold stress tolerance	Tang et al. (2012)
TaNAC7	Multiple abiotic stresses	*T. aestivum*	Multiple abiotic stress tolerance	Tang et al. (2012)
TaNAC13	Drought, salt and cold stress	*T. aestivum*	Drought, salt and cold stress tolerance	Tang et al. (2012)
OsNAC5	Drought stress	*O. sativa*	Drought stress tolerance	Chung et al. (2018)

from *Ipomoea batatas* (sweet potato) (Ishiguro and Nakamura, 1994; Rushton et al., 2008). *WRKY*-TFs are those TFs which contain the zinc finger motif, positioned at the C-terminus, while one or two domain(s) comprised of 60 highly conserved amino acids (WRKYGQK) to N-terminus (Eulgem et al., 2000; Rushton et al., 2008; Niu et al., 2012). The *WRKY*-domain is comprised of four-stranded β-sheet and single motif comprised of residues of Cys/His for attachment of zinc (Yamasaki et al., 2005). *WRKY*-domain locates and binds to ([T] [T]TGAC[C/T]) sequence, known as W-box, which lies in a gene's promoter region and regulates transcriptional mechanism (Wu et al., 2008). In certain cases, it is also observed that *WRKY* bind to "Sugar Responsive *cis* Element (SURE)" to initiate the process of transcription (Sun et al., 2003). Even though the *WRKY* domain is highly conserved, in rice 19 variants are found, which include WRKYGKK, WRKYGEK, WRMCGQK, WIKYGQK, WKKYGQK, WKRYGQK, WRKYSEK, WSKYEQK and WRICGQK (Xie et al., 2005).

Genes for WRKY have been described in many plants, such as 74 genes in *Arabidopsis thaliana* (de Pater et al., 1996; Deslandes et al., 2002; Dong et al., 2003), 109 in *Oryza sativa*/rice (Kim et al., 2000; Ross et al., 2007), while numerous *WRKY* genes have also been isolated from wheat (Sun et al., 2003), barley (Sun et al., 2003), potato (Dellagi et al., 2000; Beyer et al., 2001) and tobacco (Wang et al., 1998; Yoda et al., 2002).

Initially, WRKY were isolated from plants due to which it was considered as WRKY-TFs are limited to plants only (Eulgem et al., 2000), but later on, these were also found in other eukaryotes (Zhang and Wang, 2005). *WRKY* are not tissue-specific, but their expression rate is high in leaves as compared to other parts. However, it is well understood that in both biotic and abiotic stresses, *WRKY*-TFs are upregulated (Wu et al., 2009). Functions of *WRKY* in numerous plants have also been reported in both abiotic and biotic stresses. In biotic stress, activity of *WRKY* is observed against bacteria (Dong et al., 2003), fungi (Chen and Chen, 2002), viruses (Wang et al., 1998) and oomycetes (Beyer et al., 2001), while in abiotic stresses, *WRKY* has shown its role against single stresses as salinity (Atkinson and Urwin, 2012) and cold (Huang and Duman, 2002) as well as against combined stress like drought and heat (Rizhsky et al., 2002). Much evidence proves the employment of *WRKY*-TFs in developmental mechanisms like growth (Zhang and Wang, 2005; Chen and Chen, 2002), trichome morphogenesis (Johnson et al., 2002), embryo morphogenesis (Alexandrova and Conger, 2002), dormancy of seeds (Pnueli et al., 2002), senescence (Robatzek and Somssich, 2001) and biochemical reactions such as metabolism (Sun et al., 2003).

WRKY-TFs proved very helpful in supporting plants to overcome several abiotic stresses like salt, drought and cold (Chen et al., 2011). Their expression has also been observed at various stages of cell differentiation and different parts of the plant, such as pollen in *Arabidopsis* (Zou et al., 2010). When a plant cell contacts a single abiotic stress, proteases specialized in stress responding cleaves membrane-bounded *WRKY*-TFs due to its proteolytic activity, and the active subunits of TFs are directed toward the nucleus (Atkinson and Urwin, 2012). Histone modification and chromatin remodeling occur, which helps in the exposure/availability of responsive elements where *WRKY*-TFs adhere and results in transcription initiation (Sun et al., 2003; Zhou et al., 2005; Sridha and Wu, 2006). *WRKY*-TFs are also expressed depending upon the concentration of TFs in a cell under certain stimuli so that more transcription factors can be useful in combating abiotic stresses. After a stress is responded to by a plant and *WRKY*-TFs are being synthesized, now, depending upon the concentration, a TF can perform a dual role: activator and/or repressor (Rushlow et al., 2001).

Enhanced production of *WRKY*-TFs with the same role has also been observed in mature pollen of *Arabidopsis* pollen in low temperatures (Zou et al., 2010). Transgenic techniques for overexpression of *WRKY* has also been employed which showed increased tolerance for salt, cold and drought in wheat (Niu et al., 2012). They also respond in plant wounds (Hara et al., 2000) while they react differently to combined stresses as compared to single stress, but, in this case, the response depends upon the type and number combination of stresses such that these combined stresses may produce antagonistic or synergistic approaches (Atkinson and Urwin, 2012).

25.9.2 *MYB* Transcription Factors

MYB-transcription factors (*MYB*-TFs) are a diversely dispersed group of TFs in eukaryotes (Hichri et al., 2011), particularly in the plant kingdom (Li et al., 2016). Although they are not restricted to plants only, *MYB*-genes can be found in high frequency in plants, such as 85, 108, 198 and 127 in *Oryza sativa* (rice), *Vitis vinifera* (grape), *Arabidopsis thaliana* and *Solanum lycopersicum* (tomato), respectively (Matus et al., 2008; Li et al., 2016; Jiang et al., 2004). For the first time, the *MYB*-encoding gene found in plants was *C1* (*Colored-1*) locus in maize plant in 1987 (Paz-Ares et al., 1987). Beyond their diverse function, *MYB*-TFs are identified by their conserved structural motif called *MYB* domain (Yanhui et al., 2006; Dubos et al., 2010). *MYB* domain contains one, two, three or four imperfect repeats (R) of 50 to 52 amino acids sequence in plants, each of which make three-alpha (α) helices that has the tendency to bind the sequence C/TAACG/TG similar to the hexanucleotide E-box on DNA (Dubos et al., 2010; Rosinski and Atchley, 1998). Two of three helices in individual repeat give rise to a three dimensional HTH construction, at the time of attachment to DNA, which possess a central hydrophobic portion (Lipsick, 1996; Ogata et al., 1996). These two helices play a supportive role to the third of three helices, termed "recognition helix", for making direct contact with DNA at its major grooves (Rabinowicz et al., 1999; Jia et al., 2004). Mostly, *MYB*-TFs possess their regulatory (repression or activation) domain at their C-end while their DBD resides on their N-terminus (Yanhui et al., 2006; Dubos et al., 2010).

There are four categories of *MYB*-TFs based on the number of repeats in structural motif, and all of these types can be found in a variety of plants (Ito, 2005). If there existed one, two, three or four repeats, these are categorized as R1-*MYB*, R2-*MYB*, R3-*MYB* and R4-*MYB*, respectively (Martin and Paz-Ares, 1997). R1R2R3-type *MYB*-TFs are common in animals while R2R3-type *MYB*-TFs are endemic in plants (Allan et al., 2008; Zhou et al., 2015; Sun et al., 2015). Two theories exist about the evolution of R2R3 in plants: (i) they might have been evolved by losing R1 from R1R2R3-*MYB* over the course of time, or (ii) a duplication event of R1-*MYB* in the history may have resulted in R2R3-*MYB* in plants (Cominelli and Tonelli, 2009).

The mechanism of involvement of *MYB*-TFs in all plants is shown common pattern; however, here, *Arabidopsis* is taken as an example. It is well-examined that *MYB*-TFs are the key protein in the regulation of cellular proliferation, developmental and physiological processes both in animals and plants (Introna et al., 1994). Many relevant biochemical/biosynthetic and signaling pathways are covered by *MYB*-TFs in plants such as phenylpropanoid pathway yielding flavonoids, phenylpropanoid metabolism, lignins and stilbene compounds (Paz-Ares et al., 1987; Wei et al., 2017; Lu et al., 2017), pathways of proanthocyanidin and anthocyanin to enhance expressing structural genes like those needed for flower coloring and seed coat (Allan et al., 2008; Nakatsuka et al., 2008). All plant-based *MYB*-TFs play a vital role in any of the four broad categories: (i) Identification and destination of cell, (ii) Cell's developmental mechanisms, (ii) Both the primary as well as secondary metabolic reactions, and (iv) Biotic plus abiotic stress-responding network (Yang et al., 2012; Dubos et al., 2010).

During biotic stress in *Arabidopsis*, *MYB*-TFs such as *AtMYB30* may initiate apoptosis or may also induce the process of synthesizing very long chain fatty acids (Li et al., 2009a). Likewise, with response to abiotic stresses, *AtMYB96* and *AtMYB60* are known to regulate pathways like brassinosteroid or regulate structural genes for physiological, structural activities like stomatal cell's opening or closing by participating in the abscisic acid pathway (Raffaele et al., 2008; Seo and Park, 2010). Similarly, other R2R3-type *MYB*-TFs, such as *TaMYB73* in wheat (He et al., 2012) and *OsMYB2* in rice (Yang et al., 2012), also serve in combating major abiotic stresses such as NaCl and drought through vital pathways, mainly both "ABA-independent" and "ABA-dependent" pathways (He et al., 2012). Despite their primary function in developmental processes, *MYB*-TFs have proved to be a weapon to fight both biotic and abiotic stresses, and its enhanced expression is likely to have a positive and supportive role in stresses.

25.9.3 NAC Transcription Factors

NAC-transcription factors (*NAC*-TFs) are collectively referred to as separate group among largest classes of TFs which are found in plants only (Mao et al., 2012). The name "*NAC*" developed as acronym deduced from the initially identified three genes which possessed a unique domain: (i) "N" from "N*AM (No Apical Meristem)*", (ii) "A" from "A*TAF-1 and ATAF-2*", (iii) "C" from "C*UC2 (Cup-shaped Cotyledon)*" (Ooka et al., 2003; Souer et al., 1996; Jensen et al., 2008; Nuruzzaman et al., 2013). Various *NAC*-encoding genes have been discovered from genome-wide sequencing of numerous plants, such as 152 in tobacco (Rushton et al., 2008), 152 in soybean (Thao et al., 2013), 163 in *Populus trichocarpa* (poplar) (Hu et al., 2008), 151 in *Oryza sativa* (Nuruzzaman et al., 2010) and 117 in *Arabidopsis thaliana* (Mao et al., 2012) as well as 79 and 26 in grape and citrus, respectively (Nuruzzaman et al., 2013).

NAC-TFs contain 150–160 amino acids-long *NAC*-domain at their N-terminus (Ooka et al., 2003), which deals systematically, initiating from localization to nucleus, binding to DNA and may also form homo- or heterodimers in association with other *NAC*-TFs (Olsen et al., 2005) while activation domains reside at their C-terminus that serves either as repressor or activator in gene-regulation process (Fan et al., 2018).

The role of *NAC*-TFs covers a wide variety, including developmental processes such as secondary wall formation, root development, flower and embryo development and senescence of leaves (Xie et al., 2005; Zhong et al., 2007). Depending upon externally induced stimuli, *NAC*-TFs support the plant in gaining tolerance to both biotic plus abiotic stresses such as pathogens, salinity, variable temperatures, drought and deficiency of oxygen and growth hormones (Hu et al., 2008; Jeong et al., 2010; Shang et al., 2016; Chung et al., 2018). For instance, *OsNAC10* support the rice plants in enhancing drought tolerance (Jeong et al., 2010). Similarly, many other examples of *NAC*-TFs in rice plants, *OsNAC063*, *OsNAC045* and *OsNAC6*, have been extensively studied, and it is confirmed that these TFs play a vital role as the first line of defense to gain both biotic plus abiotic stress tolerance to individual as well as combined stresses (Nakashima et al., 2007; Zheng et al., 2009). Likewise, *TaNAC8* and *TaNAC4* play the same role in wheat (Xia et al., 2010a,b). The employment of transgenic techniques (Lu et al., 2018) showed that *TaNAC2* from wheat incorporated in *Arabidopsis* has shown an increased tolerance against salt, freezing and drought stresses (Mao et al., 2012). Numerous roles played by *NAC*-TFs identified are: *HvNAC6* in tolerating drought (Delessert et al., 2005) and pathogens (Jensen et al., 2007), *AtNAC102* in germination of seeds even under less-amount of oxygen (Christianson et al., 2009).

25.10 CONCLUSION AND RECOMMENDATIONS

Transcription factors are vital components of plant cell signaling, especially linked with genome for gene expression. The role of TFs in a cell cascade is crucial for the perfect performance of a cell's genome as nature, type, structural attributes, quantity and function determines a gene's expression. Transcription factors are a part of cell cascade system which mainly performs two actions: (i) serves as a mediocre and reads the upcoming signals either from inside or outside of host cell and transforms the signals in terms of gene expression accordingly, and (ii) minimizes the possible direct harmful effects of any type of signals upon host genome.

Along with its purpose in a normal cell, TFs can also lead a plant to the state of tolerance against several biotic and abiotic stresses. These factors can decide the fate of a cell in terms of their genetic response to a wide range of internal as well as external stimuli. However, in normal conditions TFs play as regulators for gene expression depending upon cell's homeostatic condition while in most cases the external stimuli addressing TFs entail the severe condition to which plant must gain tolerance (Wang et al., 1998). Not only single but also combined stresses are administered by plants using TFs.

Despite their importance and vital role, plant TFs have not been fully understood to a level such that a global consensus can be developed for their nomenclature, structure, classification and define a role for a specific class of TFs. The major issue regarding this problem is the fact that TFs are very diverse both in structure and function as well as the organism in which they are found. Nomenclature and classification are interlinked in a biological system, and proper classification is based on a maximum understanding of functional and structural attributes. In the currently available classifications of TFs, some discrepancies exist, which restricts its proper naming and categorization which can be acceptable to all in the scientific community.

The problems with their recommendations include: (i) A need of understanding the structural and functional attributes of TFs at maximum level. Though a tough fact, still this objection can be removed by the employment of the latest and most sophisticated techniques in the study of TFs such as NMR-spectroscopy and electron microscopy for structural attributes while developing knockout models, mutant analysis, transgenic

and cloning approaches for describing the function of TFs. (ii) Global consensus classification of TFs must be developed. For this purpose, one of the options is to review all the classifications available today so that ambiguities of certain categories in classification as "Other Transcription Factors" would be removed. The hierarchy of classification which may be more helpful is, first, to divide the entire transcription factors with respect to the species in which they reside. The second step will is its DNA-binding domain and third is its expression as constitutive or induced, followed by its function as upregulated or downregulated based accordingly. The last but not the least will be the process in which they perform their roles such as embryo development or certain types of stress tolerance. (iii) It may also be sometimes referred to as a problem that specific types of classification have been developed by various scientific groups, usually according to their own needs, like *wDBTF* and *DRTF*. In these cases, phylogenetic analysis remains unsolved and TFs of different origin with the same homology are not clear. This analysis is necessary because it helps in identifying respective genes for further uses as transgenesis. (iv) Though the naming technique previously proposed is quite practical, this system still does not provide information about a TF gene, whether it is constitutive or inducible, as well as we cannot know from a name whether a certain type of transcription factor serves as an activator or as a repressor. However, the solution to this problem is based on detail analysis of the structure and function of TFs. (v) Mostly, plant TFs have been known to act against invading harsh conditions, yet more has to be investigated about the effects of various combinations of stresses (abiotic with abiotic, biotic with abiotic and biotic with biotic). So, more and more experimentation of various combinations of stresses will prove to be very helpful in developing consensus. (vi) The transgenic approach can be utilized in many ways: (a) up- or downregulation of a transcription factor within its host, (b) enhancing the expression of certain type of transcription factor, as well as (c) suppressing a transcription factor depending upon the requirements by sophisticated techniques like RNAi.

Overall, it is much clearer that controlling and directing the transcription factors means directing the gene expression, ultimately, to solve genome-based problems of various issues including abiotic stresses.

REFERENCES

Ahuja, I., de Vos, R.C.H., Bones, A.M., Hall, R.D. (2010) Plant molecular stress responses face climate change. *Trends in Plant Science* 15(12):664–74.

Alexandrova, K.S., Conger, B.V. (2002) Isolation of two somatic embryogenesis-related genes from orchard grass (*Dactylis glomerata*). *Plant Science* 162(2):301–7.

Allan, A.C., Hellens, R.P., Laing, W.A. (2008) *MYB* transcription factors that colour our fruit. *Trends in Plant Science* 13(3):99–102.

Aroca, R., Porcel, R., Ruiz-Lozano, J.M. (2012) Regulation of root water uptake under abiotic stress conditions. *Journal of Experimental Botany* 63(1):43–57.

Ashraf, M., Harris, P.J.C. (2004) Potential biochemical indicators of salinity tolerance in plants. *Plant Science* 166(1):3–16.

Atkinson, N.J., Urwin, P.E. (2012) The interaction of plant biotic and abiotic stresses: From genes to the field. *Journal of Experimental Botany* 63(10):3523–43.

Atkinson, N.J., Lilley, C.J., Urwin, P.E. (2013) Identification of genes involved in the response of *Arabidopsis* to simultaneous biotic and abiotic stresses. *Plant Physiology* 162(4):2028–41.

Bailey, P.C., Martin, C., Toledo-Ortiz, G., et al. (2003) Update on the basic helix-loop-helix transcription factor gene family in *Arabidopsis thaliana*. *The Plant Cell* 15(11):2497–502.

Baudry, A., Heim, M.A., Dubreucq, B., Caboche, M., Weisshaar, B., Lepiniec, L. (2004) TT2, TT8, and TTG1 synergistically specify the expression of *BANYULS* and proanthocyanidin biosynthesis in *Arabidopsis thaliana*. *The Plant Journal* 39(3):366–80.

Beyer, K., Binder, A., Boller, T., Collinge, M. (2001) Identification of potato genes induced during colonization by *Phytophthora infestans*. *Molecular Plant Pathology* 2(3):125–34.

Brivanlou, A.H., Darnell, J.E. Jr. (2002) Signal transduction and the control of gene expression. *Science* 295(5556):813–18.

Chaves, M.M., Maroco. J.P., Pereira, J.S. (2003) Understanding plant responses to drought – From genes to the whole plant. *Functional Plant Biology* 30(3):239–64.

Chen, C., Chen, Z. (2002) Potentiation of developmentally regulated plant defense response by *AtWRKY18*, a pathogen induced *Arabidopsis* transcription factor. *Plant Physiology* 129(2):706–16.

Chen, L., Song, Y., Li, S., Zhang, L., Zou, C., Yu, D. (2011) The role of *WRKY* transcription factors in plant abiotic stresses. *Biochimica et Biophysica Acta (BBA) – Gene Regulatory Mechanisms* 1819(2):120–8.

Chen, T., Li, W., Hu, X., et al. (2015) A cotton *MYB* transcription factor, *GbMYB5*, is positively involved in plant adaptive response to drought stress. *Plant and Cell Physiology* 56(5):917–29.

Cheng, M.C., Liao, P.M., Kuo, W.W., Lin, T.P. (2013) The *Arabidopsis* ETHYLENE RESPONSE FACTOR1 regulates abiotic stress-responsive gene expression by binding to different *cis*-acting elements in response to different stress signals. *Plant Physiology* 162(3):1566–82.

Chew, Y.H., Halliday, K.J. (2011) A stress-free walk from *Arabidopsis* to crops. *Current Opinion in Biotechnology* 22(2):281–6.

Christianson, J.A., Wilson, I.W., Llewellyn, D.J., Dennis, E.S. (2009) The low-oxygen-induced *NAC* domain transcription factor *ANAC102* affects viability of *Arabidopsis* seeds following low-oxygen treatment. *Plant Physiology* 149(4):1724–38.

Chung, P.J., Jung, H., Choi, Y.D., Kim, J.K. (2018) Genome-wide analyses of direct target genes of four rice *NAC*-domain transcription factors involved in drought tolerance. *BMC Genomics* 19(1):40.

Cominelli, E., Tonelli, C. (2009) A new role for plant *R2R3-MYB* transcription factors in cell cycle regulation. *Cell Research* 19:1231–2.

de Pater, S., Greco, V., Pham, K., Memelink, J., Kijne, J. (1996) Characterization of a zinc-dependent transcriptional activator from *Arabidopsis*. *Nucleic Acids Research* 24(23):4624–31.

Delessert, C., Kazan, K., Wilson, I.W., et al. (2005) The transcription factor *ATAF2* represses the expression of pathogenesis-related genes in *Arabidopsis*. *The Plant Journal* 43(5):745–57.

Dellagi, A., Heilbronn, J., Avrova, A.O., et al. (2000) A potato gene encoding a WRKY-like transcription factor is induced in interactions with *Erwinia carotovora* subsp *atroseptica* and *Phytophthora infestans* and is coregulated with class I endochitinase expression. *Molecular Plant-Microbe Interactions* 13(10):1092–101.

Depège-Fargeix, N., Javelle, M., Chambrier, P., et al. (2010) Functional characterization of the HD-ZIP IV transcription factor OCL1 from maize. *Journal of Experimental Botany* 62(1):293–305.

Deslandes, L., Olivier, J., Theuliéres, F., et al. (2002) Resistance to *Ralstonia solanacearum* in *Arabidopsis thaliana* is conferred by the recessive *RRS1-R* gene, a member of a novel family of resistance genes. *Proceedings of the National Academy of Sciences of the United States of America* 99(4):2404–9.

Dignam, J.D., Lebovitz, R.M., Roeder, R.G. (1983) Accurate transcription initiation by RNA polymerase II in a soluble extract from isolated mammalian nuclei. *Nucleic Acids Research* 11(5):1575–89.

Dong, J., Chen, C., Chen, Z. (2003) Expression profiles of the *Arabidopsis WRKY* gene superfamily during plant defense response. *Plant Molecular Biology* 51(1):21–37.

Dubos, C., Stracke, R., Grotewold, E., Weisshaar, B., Martin, C., Lepiniec, L. (2010) *MYB* transcription factors in *Arabidopsis*. *Trends in Plant Science* 15(10):573–81.

Edmondson, D.F., Olson, E.N. (1993) Helix-loop-helix proteins as regulators of muscle-specific transcription. *The Journal of Biological Chemistry* 268(2):755–8.

Eulgem, T., Rushton, P.J., Robatzek, S., Somssich, I.E. (2000) The WRKY superfamily of plant transcription factors. *Trends in Plant Science* 5(5):199–206.

Fan, K., Li, F., Cheng, J., et al. (2018) Asymmetric evolution and expansion of the *NAC* transcription factor in polyploidized cotton. *Frontiers in Plant Science*. 9:47.

Fried, M., Crothers, D.M. (1981) Equilibria and kinetics of lac repressor–operator interactions by polyacrylamide gel electrophoresis. *Nucleic Acids Research* 9(23):6505–25.

Galgoczy, D.J., Cassidy-Stone, A., Llinás, M., et al. (2004) Genomic dissection of the cell-type specification circuit in *Saccharomyces cerevisiae*. *Proceedings of the National Academy of Sciences of the United States of America* 101(52):18069–74.

Gangloff, Y.G., Werten, S., Romier, C., et al. (2000) The human TFIID components TAF$_{II}$135 and TAF$_{II}$20 and the yeast SAGA components ADA1 and TAF$_{II}$68 heterodimerize to form histone-like pairs. *Molecular and Cell Biology* 20(1):340–51.

Garner, M.M., Revzin, A. (1981) A gel electrophoresis method for quantifying the binding of proteins to specific DNA regions: Application to components of the *Escherichia coli* lactose operon regulatory system. *Nucleic Acids Research* 9(13):3047–60.

Gray, J., Bevan, M., Brutnell, T., et al. (2009) A recommendation for naming transcription factor proteins in the grasses. *Plant Physiology* 149(1):4–6.

Guo, A., He, K., Liu, D., et al. (2005) DATF: A database of *Arabidopsis* transcription factors. *Bioinformatics* 21(10):2568–9.

Hanna-Rose, W., Hansen, U. (1996) Active repression mechanisms of eukaryotic transcription repressors. *Trends in Genetics* 12(6):229–34.

Hao, Y.J., Wei, W., Song, Q.X., et al. (2011) Soybean *NAC* transcription factors promote abiotic stress tolerance and lateral root formation in transgenic plants. *The Plant Journal* 68(2):302–13.

Hara, K., Yagi, M., Kusano, T., Sano, H. (2000) Rapid systemic accumulation of transcripts encoding a tobacco *WRKY* transcription factor upon wounding. *Molecular and General Genetics* 263(1):30–7.

Hasegawa, P.M., Bressan, R.A., Zhu, J.K., Bohnert, H.J. (2000) Plant cellular and molecular responses to high salinity. *Annual Review of Plant Physiology and Plant Molecular Biology* 51:463–99.

He, Y., Li, W., Lv, J., Jia, Y., Wang, M., Xia, G. (2012) Ectopic expression of a wheat *MYB* transcription factor gene, *TaMYB73*, improves salinity stress tolerance in *Arabidopsis thaliana*. *Journal of Experimental Botany* 63(3):1511–22.

He, X., Zhu, L., Xu, L., Guo, W., Zhang, X. (2016) GhATAF1, a NAC transcription factor, confers abiotic and biotic stress responses by regulating phytohormonal signaling networks. *Plant Cell Reports* 35(10):2167–79.

Hichri, I., Deluc, L., Barrieu, F., et al. (2011) A single amino acid change within the R2 domain of the *VvMYB5b* transcription factor modulates affinity for protein partners and target promoters selectivity. *BMC Plant Biology* 11(1):117–30.

Hinnebusch, A.G. (2005) Translational regulation of *GCN4* and the general amino acid control of yeast. *Annual Review of Microbiology* 59(1):407–50.

Hirayama, T., Shinozaki, K. (2010) Research on plant abiotic stress responses in the post-genome era: Past, present and future. *The Plant Journal* 61(6):1041–52.

Hu, H., You, J., Fang, Y., Zhu, X., Qi, Z., Xiong, L. (2008) Characterization of transcription factor gene *SNAC2*

conferring cold and salt tolerance in rice. *Plant Molecular Biology* 67(1–2):169–81.

Huang, T., Duman, J.G. (2002) Cloning and characterization of a thermal hysteresis (antifreeze) protein with DNA-binding activity from winter bittersweet nightshade, *Solanum dulcamara*. *Plant Molecular Biology* 48(4):339–50.

Hymus, G.J., Cai, S., Kohl, E.A., et al. (2013) Application of HB17, an *Arabidopsis* class II homeodomain-leucine zipper transcription factor, to regulate chloroplast number and photosynthetic capacity. *Journal of Experimental Botany* 64(14):4479–90.

Introna, M., Luchetti, M., Castellano, M., Arsura, M., Golay, J. (1994) The *MYB* oncogene family of transcription factors: Potent regulators of hematopoietic cell proliferation and differentiation. *Seminars in Cancer Biology* 5(2):113–24.

Ishiguro, S., Nakamura, K. (1994) Characterization of a cDNA encoding a novel DNA-binding protein, SPF1, that recognizes SP8 sequences in the 5' upstream regions of genes coding for sporamin and beta-amylase from sweet potato. *Molecular and General Genetics* 244(6): 563–71.

Ito, M. (2005) Conservation and diversification of three-repeat *MYB* transcription factors in plants. *Journal of Plant Research* 118(1):61–9.

Jakoby, M., Weisshaar, B., Droge-Laser, W., et al. (2002) bZIP transcription factors in *Arabidopsis*. *Trends in Plant Science* 7(3):106–11.

Jamil, M., Bashir, S., Anwar, S., et al. (2012) Effect of salinity on physiological and biochemical characteristics of different varieties of rice. *Pakistan Journal of Botany* 44:7–13.

Jamil, M., Malook, I., Parveen, S., et al. (2013) Smoke priming, a potent protective agent against salinity: Effect on proline accumulation, elemental uptake, pigmental attributes and protein banding patterns of rice (*Oryza Sativa*). *Journal of Stress Physiology and Biochemistry* 9(1):169–83.

Jarrett, H.W. (2012) Proteomic methodologies to study transcription factor function. In *Gene Regulatory Networks: Methods in Molecular Biology (Methods and Protocols)* Deplancke, B., Gheldof, N. (eds). Humana Press: New York, NY, 786:315–34.

Jensen, M.K., Rung, J.H., Gregersen, P.L., et al. (2007) The *HvNAC6* transcription factor: A positive regulator of penetration resistance in barley and *Arabidopsis*. *Plant Molecular Biology* 65(1–2):137–50.

Jensen, M.K., Hagedorn, P.H., de Torres-Zabala, M., et al. (2008) Transcriptional regulation by an *NAC* (*NAM-ATAF1,2-CUC2*) transcription factor attenuates ABA signalling for efficient basal defence towards *Blumeria graminis* f. sp. *hordei* in *Arabidopsis*. *The Plant Journal* 56(6):867–80.

Jeong, J.S., Kim, Y.S., Baek, K.H., et al. (2010) Root-specific expression of *OsNAC10* improves drought tolerance and grain yield in rice under field drought conditions. *Plant Physiology* 153(1):185–97.

Jia, L., Clegg, M.T., Jiang, T. (2004) Evolutionary dynamics of the DNA-binding domains in putative *R2R3-MYB* genes identified from rice subspecies *indica* and *japonica* genomes. *Plant Physiology* 134(2):575–85.

Jiang, W., Yu, D. (2009) *Arabidopsis WKRY2* transcription factor mediates seed germination and postgermination arrest of development by abscisic acid. *BMC Plant Biology* 9:96.

Jiang, C., Gu, X., Peterson, T. (2004) Identification of conserved gene structures and carboxyterminal motifs in the *MYB* gene family of *Arabidopsis* and *Oryza sativa* L. ssp. *indica*. *Genome Biology* 5(7):R46.

Jin, C., Felsenfeld, G. (2007) Nucleosome stability mediated by histone variants H3.3 and H2A.Z. *Genes and Development* 21(12):1519–29.

Jin, C., Zang, C., Wei, G., et al. (2009) H3.3/H2A.Z double variant-containing nucleosomes mark 'nucleosome-free regions' of active promoters and other regulatory regions. *Nature Genetics* 41(8):941–5.

Jing, S., Zhou, X., Song, Y., Yu, D. (2009) Heterologous expression of *OsWRKY23* gene enhances pathogen defense and dark-induced leaf senescence in *Arabidopsis*. *Plant Growth Regulation* 58(2):181–90.

Johnson, C.S., Kolevski, B., Smyth, D.R. (2002) *TRANSPARENT TESTA GLABRA2*, a trichome and seed coat development gene of *Arabidopsis*, encodes a *WRKY* transcription factor. *The Plant Cell* 14(6):1359–75.

Jung, C., Seo, J.S., Han, S.W., et al. (2008) Overexpression of *AtMYB44* enhances stomatal closure to confer abiotic stress tolerance in transgenic *Arabidopsis*. *Plant Physiology* 146(2):623–35.

Kadonaga, J.T., Tjian, R. (1986) Affinity purification of sequence-specific DNA binding proteins. *Proceedings of the National Academy of Sciences of the United States of America* 83(16):5889–93.

Kaplan, T., Li, X.Y., Sabo, P.J., et al. (2011) Quantitative models of the mechanisms that control genome-wide patterns of transcription factor binding during early *Drosophila* development. *PLOS Genetics* 7(2):e1001290.

Karin, M., Hunter, T. (1995) Transcriptional control by protein phosphorylation: Signal transmission from the cell surface to the nucleus. *Current Biology* 5(7):747–57.

Kazan, K. (2003) Alternative splicing and proteome diversity in plants: The tip of the iceberg has just emerged. *Trends in Plant Science* 8(10):468–71.

Kim, C.Y., Lee, S.H., Park, H.C., et al. (2000) Identification of rice blast fungal elicitor-responsive genes by differential display analysis. *Molecular Plant–Microbe Interactions* 13(4):470–4.

Kornberg, R.D. (1977) Structure of chromatin. *Annual Review of Biochemistry* 46:931–54.

Kültz, D. (2005) Molecular and evolutionary basis of the cellular stress response. *Annual Review of Physiology* 67:225–57.

Latchman, D.S. (1997) Transcription factors: An overview. *The International Journal of Biochemistry and Cell Biology* 29(12):1305–12.

Latchman, D.S. (2008) *Eukaryotic Transcription Factors*, 5th Ed. Elsevier/Academic Press: Amsterdam, the Netherlands.

Li, L., Yu, X., Thompson, A., et al. (2009a) *Arabidopsis MYB30* is a direct target of *BES1* and cooperates with *BES1* to regulate brassinosteroid-induced gene expression. *The Plant Journal* 58(2):275–86.

Li, S., Fu, Q., Huang, W., Yu, D. (2009b) Functional analysis of an *Arabidopsis* transcription factor *WRKY25* in heat stress. *Plant Cell Reports* 28(4):683–93.

Li, G., Margueron, R., Hu, G., Stokes, D., Wang, Y.H., Reinberg, D. (2010) Highly compacted chromatin formed in vitro reflects the dynamics of transcription activation in vivo. *Molecular Cell* 38(1):41–53.

Li, Z., Peng, R., Tian, Y., Han, H., Xu, J., Yao, Q. (2016) Genome-wide identification and analysis of the *MYB* transcription factor superfamily in *Solanum lycopersicum*. *Plant Cell Physiology* 57(8):1657–77.

Lipsick, J.S. (1996) One billion years of *MYB*. *Oncogene* 13(2):223–35.

Liu, X., Yang, L., Zhou, X., et al. (2013) Transgenic wheat expressing *Thinopyrum intermedium MYB* transcription factor *TiMYB2R-1* shows enhanced resistance to the take-all disease. *Journal of Experimental Botany* 64(8):2243–53.

Liu, C., Wang, B., Li, Z., Peng, Z., Zhang, J. (2018) *TsNAC1* is a key transcription factor in abiotic stress resistance and growth. *Plant Physiology* 176(2):742–56.

Lorković, Z.J., Kirk, D.A.W., Lambermon, M.H.L., Filipowicz, W. (2000) Pre-mRNA splicing in higher plants. *Trends in Plant Science* 5(4):160–7.

Lu, N., Roldan, M., Dixon, R.A. (2017) Characterization of two *TT2*-type *MYB* transcription factors regulating proanthocyanidin biosynthesis in tetraploid cotton, *Gossypium hirsutum*. *Planta* 246(2):323–35.

Lu, X., Zhang, X., Duan, H., et al. (2018) Three stress-responsive *NAC* transcription factors from *Populus euphratica* differentially regulate salt and drought tolerance in transgenic plants. *Physiologia Plantarum* 162(1):73–97.

Luger, K., Mäder, A.W., Richmond, R.K., Sargent, D.F., Richmond, T.J. (1997) Crystal structure of the nucleosome core particle at 2.8 Å resolution. *Nature* 389(6648):251–60.

Ma, Q., Dai, X., Xu, Y., et al. (2009) Enhanced tolerance to chilling stress in *OsMYB3R-2* transgenic rice is mediated by alteration in cell cycle and ectopic expression of stress genes. *Plant Physiology* 150(1):244–56.

Maldonado, E., Hampsey, M., Reinberg, D. (1999) Repression: Targeting the heart of the matter. *Cell* 99(5):455–8.

Manley, J.L., Fire, A., Cano, A., Sharp, P.A., Gefter, M.L. (1980) DNA-dependent transcription of adenovirus genes in a soluble whole-cell extract. *Proceedings of the National Academy of Sciences of the United States of America* 77(7):3855–9.

Mao, X., Zhang, H., Qian, X., Li, A., Zhao, G., Jing, R. (2012) *TaNAC2*, a NAC-type wheat transcription factor conferring enhanced multiple abiotic stress tolerances in *Arabidopsis*. *Journal of Experimental Botany* 63(8):2933–46.

Martin, C., Paz-Ares, J. (1997) MYB transcription factors in plants. *Trends in Genetics* 13(2):67–73.

Martino, F., Kueng, S., Robinson, P., et al. (2009) Reconstitution of yeast silent chromatin: Multiple contact sites and O-AADPR binding load SIR complexes onto nucleosomes in vitro. *Molecular Cell* 33(3):323–34.

Matus, J.T., Aquea, F., Arce-Johnson, P. (2008) Analysis of the grape *MYB R2R3* subfamily reveals expanded wine quality-related clades and conserved gene structure organization across *Vitis* and *Arabidopsis* genomes. *BMC Plant Biology* 8:83.

Mittler, R., Blumwald, E. (2010) Genetic engineering for modern agriculture: Challenges and perspectives. *Annual Review of Plant Biology* 61:443–62.

Munns, R. (2002) Comparative physiology of salt and water stress. *Plant, Cell and Environment* 25(2):239–50.

Naika, M., Shameer, K., Mathew, O.K., Gowda, R., Sowdhamini, R. (2013) STIFDB2: An updated version of plant stress-responsive transcription factor database with additional stress signals, stress-responsive transcription factor binding sites and stress-responsive genes in *Arabidopsis* and rice. *Plant and Cell Physiology* 54(2):e8(1–15).

Nakashima, K., Tran, L.S.P., Van Nguyen, D., et al. (2007) Functional analysis of a *NAC*-type transcription factor *OsNAC6* involved in abiotic and biotic stress-responsive gene expression in rice. *The Plant Journal* 51(4):617–30.

Nakashima, K., Ito, Y., Yamaguchi-Shinozaki, K. (2009) Transcriptional regulatory networks in response to abiotic stresses in *Arabidopsis* and grasses. *Plant Physiology* 149(1):88–95.

Nakatsuka, T., Haruta, K.S., Pitaksutheepong C., et al. (2008) Identification and characterization of *R2R3-MYB* and *bHLH* transcription factors regulating anthocyanin biosynthesis in gentian flowers. *Plant & Cell Physiology* 49(12):1818–29.

Nikolev, D.B. Burley, S.K. (1997). RNA polymerase II transcription initiation: A structural view. *Proceedings of the National Academy of Sciences of the United States of America* 94(1):15–22.

Niu, C.F., Wei, W., Zhou, Q.Y., et al. (2012) Wheat *WRKY* genes *TaWRKY2* and *TaWRKY19* regulate abiotic stress tolerance in transgenic *Arabidopsis* plants. *Plant, Cell and Environment* 35(6):1156–70.

Nover, L., Bharti, K., Döring P., Mishra, S.K., Ganguli, A., Scharf, K.D. (2001) *Arabidopsis* and the heat stress transcription factor world: How many heat stress transcription factors do we need? *Cell Stress Chaperones* 6(3):177–89.

Nuruzzaman, M., Manimekalai, R., Sharoni, A.M., et al. (2010) Genome-wide analysis of *NAC* transcription factor family in rice. *Gene* 465(1–2):30–44.

Nuruzzaman, M., Sharoni, A.M., Kikuchi, S. (2013) Roles of *NAC* transcription factors in the regulation of biotic and abiotic stress responses in plants. *Frontiers in Microbiology* 4:248.

Ogata, K., Kanei-Ishii, C., Sasaki, M., et al. (1996) The cavity in the hydrophobic core of *MYB* DNA-binding domain is reserved for DNA recognition and *trans*-activation. *Nature Structural and Molecular Biology* 3:178–87.

Olsen, A.N., Ernst, H.A., Lo Leggio, L., Skriver, K. (2005) *NAC* transcription factors: Structurally distinct, functionally diverse. *Trends in Plant Science* 10(2): 1360–85.

Ooka, H., Satoh, K., Doi, K., et al. (2003) Comprehensive analysis of NAC family genes in *Oryza sativa* and *Arabidopsis thaliana*. *DNA Research* 10(6): 239–47.

Pardo, J.M. (2010) Biotechnology of water and salinity stress tolerance. *Current Opinion in Biotechnology* 21(2):185–96.

Paz-Ares, J., Ghosal, D., Wienand, U., Peterson, P.A., Saedler, H. (1987) The regulatory c1 locus of *Zea mays* encodes a protein with homology to *MYB* proto-oncogene products and with structural similarities to transcriptional activators. *The EMBO Journal* 6(12):3553–8.

Pérez-Rodríguez, P., Riaño-Pachón, D.M., Corrêa, L.G.G., Rensing, S.A., Kersten, B., Mueller-Roeber, B. (2010) PlnTFDB: Updated content and new features of the plant transcription factor database. *Nucleic Acids Research* 38:D822–7.

Piskacek, S., Gregor, M., Nemethova, M., Grabner, M., Kovarik, P., Piskacek, M. (2007) Nine-amino-acid transactivation domain: Establishment and prediction utilities. *Genomics* 89(6):756–68.

Pnueli, L., Hallak-Herr, E., Rozenberg, M., et al. (2002) Molecular and biochemical mechanisms associated with dormancy and drought tolerance in the desert legume *Retama raetam*. *The Plant Journal* 31(3):319–30.

Qiu, Y., Yu, D. (2009) Over-expression of the stress-induced *OsWRKY45* enhances disease resistance and drought tolerance in *Arabidopsis*. *Environmental and Experimental Botany* 65(1):35–47.

Rabinowicz, P.D., Braun, E.L., Wolfe, A.D., Grotewold, E. (1999). Maize *R2R3 MYB* genes: Sequence analysis reveals amplification in the higher plants. *Genetics* 153(1):427–44.

Raffaele, S., Vailleau, F., Léger, A., et al. (2008) A *MYB* transcription factor regulates very-long-chain fatty acid biosynthesis for activation of the hypersensitive cell death response in *Arabidopsis*. *The Plant Cell* 20(3): 752–67.

Rasmussen, S., Barah, P., Suarez-Rodriguez, M.C., et al. (2013) Transcriptome responses to combinations of stresses in *Arabidopsis*. *Plant Physiology* 161(4):1783–94.

Reddy, A.S.N. (2001) Nuclear pre-mRNA splicing in plants. *Critical Reviews in Plant Sciences* 20(6):523–71.

Reid, J.E., Evans, K.J., Dyer, N., Wernisch, L., Ott, S. (2010) Variable structure motifs for transcription factor binding sites. *BMC Genomics* 11:30–47.

Riechmann, J.L., Heard, J., Martin, G., et al. (2000) *Arabidopsis* transcription factors: Genome-wide comparative analysis among eukaryotes. *Science* 290(5499):2105–10.

Rizhsky, L., Liang, H., Mittler, R. (2002) The combined effect of drought stress and heat shock on gene expression in tobacco. *Plant Physiology* 130(3):1143–51.

Robatzek, S., Somssich, I.E. (2001) A new member of the *Arabidopsis WRKY* transcription factor family, *AtWRKY6*, is associated with both senescence- and defence-related processes. *The Plant Journal* 28(2):123–33.

Roberts, S.G. (2000) Mechanisms of action of transcription activation and repression domains. *Cellular and Molecular Life Sciences* 57(8–9):1149–60.

Roeder, R.G. (1996). The role of general initiation factors in transcription by RNA polymerase II. *Trends in Biochemical Sciences* 21(9):327–35.

Rosinski, J.A., Atchley, W.R. (1998). Molecular evolution of the *MYB* family of transcription factors: evidence for polyphyletic origin. *Journal of Molecular Evolution* 46(1):74–83.

Ross, C.A., Liu, Y., Shen, Q.J. (2007) The *WRKY* gene family in rice (*Oryza sativa*). *Journal of Integrative Plant Biology* 49(6):827–42.

Rushlow, C. Colosimo, P.F., Lin, M.C., Xu, M., Kirov, N. (2001) Transcriptional regulation of the *Drosophila* gene *zen* by competing Smad and Brinker inputs. *Genes and Development* 15(3):340–51.

Rushton, P.J., Bokowiec, M.T., Han, S., et al. (2008) Tobacco transcription factors: Novel insights into transcriptional regulation in the *Solanaceae*. *Plant Physiology* 147(1):280–95.

Schwarz, P.M., Hansen, J.C. (1994) Formation and stability of higher order chromatin structures. Contributions of the histone octamer. *The Journal of Biological Chemistry* 269(23):16284–9.

Segal, E., Fondufe-Mittendorf, Y., Chen, L., et al. (2006) A genomic code for nucleosome positioning. *Nature* 442(7104):772–8.

Sekinger, E.A., Moqtaderi, Z., Struhl, K. (2005) Intrinsic histone–DNA interactions and low nucleosome density are important for preferential accessibility of promoter regions in yeast. *Molecular Cell* 18(6):735–48.

Seo, P.J., Park, C.M. (2010) *MYB96*-mediated abscisic acid signals induce pathogen resistance response by promoting salicylic acid biosynthesis in *Arabidopsis*. *New Phytologist* 186(2):471–83.

Shang, H., Wang, Z., Zou, C., et al. (2016) Comprehensive analysis of *NAC* transcription factors in diploid *Gossypium*: Sequence conservation and expression analysis uncover their roles during fiber development. *Science China Life Sciences* 59(2):142–53

Shogren-Knaak, M., Ishii, H., Sun, J.M., Pazin, M.J., Davie, J.R., Peterson, C.L. (2006) Histone H4-K16 acetylation controls chromatin structure and protein interactions. *Science* 311(5762):844–47.

Smith, M.D., Dent, C.L., Latchman, D.S. (1999) The DNA mobility shift assay. In *Transcription factors: A practical approach*, 2nd edn., Latchman, D.S. (ed.), Oxford University Press: New York, NY, 1–25.

Song, Y., Jing, S.J., Yu, D.Q. (2009) Overexpression of the stress induced *OsWRKY08* improves the osmotic stress tolerance in *Arabidopsis*. *Chinese Science Bulletin* 54(24):4671–8.

Souer, E., van Houwelingen, A., Kloos, D., Mol, J., Koes, R. (1996) The no apical meristem gene of *Petunia* is required for pattern formation in embryos and flowers and is expressed at meristem and primordial boundaries. *Cell* 85(2):159–70.

Sridha, S., Wu, K. (2006) Identification of *AtHD2C* as a novel regulator of abscisic acid responses in *Arabidopsis*. *The Plant Journal* 46(1):124–33.

Stegmaier, P., Kel, A.E., Wingender, E. (2004) Systematic DNA-binding domain classification of transcription factors. *Genome Informatics* 15(2):276–86.

Stracke, R., Werber, M., Weisshaar, B. (2001) The *R2R3-MYB* gene family in *Arabidopsis thaliana*. *Current Opinion in Plant Biology* 4(5):447–56.

Sun, C., Palmqvist, S., Olsson, H., Borén, M., Ahlandsberg, S., Jansson, C. (2003) A novel WRKY transcription factor, SUSIBA2, participates in sugar signaling in barley by binding to the sugar-responsive elements of the iso1 promoter. *The Plant Cell* 15(9):2076–92.

Sun, X., Gong, S.Y., Nie, X.Y., et al. (2015) A *R2R3-MYB* transcription factor that is specifically expressed in cotton (*Gossypium hirsutum*) fibers affects secondary cell wall biosynthesis and deposition in transgenic *Arabidopsis*. *Physiologia Plantarum* 154(3):420–32.

Tang, Y., Liu, M., Gao, S., et al. (2012) Molecular characterization of novel *TaNAC* genes in wheat and overexpression of *TaNAC2a* confers drought tolerance in tobacco. *Physiologia Plantarum* 144(3):210–24.

Thao, N.P., Thu, N.B.A., Hoang, X.L.T., Ha, C.V., Phan, L.S. (2013) Differential expression analysis of a subset of drought-responsive *GmNAC* genes in two soybean cultivars differing in drought tolerance. *International Journal of Molecular Sciences* 14(12):23828–41.

Triezenberg, S.J. (1995) Structure and function of transcriptional activation domains. *Current Opinion in Genetics and Development* 5(2):190–6.

Vaahtera, L., Brosche, M. (2011) More than the sum of its parts – How to achieve a specific transcriptional response to abiotic stress. *Plant Science* 180(3):421–30.

Van Helden, J., Rios, A.F., Collado-Vides, J. (2000) Discovering regulatory elements in non-coding sequences by analysis of spaced dyads. *Nucleic Acids Research* 28(8):1808–18.

Visel, A., Blow, M.J., Li, Z., et al. (2009) ChIP-seq accurately predicts tissue-specific activity of enhancers. *Nature* 457(7231):854–8.

Wang, Z., Yang, P., Fan, B., Chen, Z. (1998) An oligo selection procedure for identification of sequence- specific DNA-binding activities associated with the plant defence response. *The Plant Journal* 16(4):515–22.

Wang, H., Hao, J., Chen, X., et al. (2007) Overexpression of rice *WRKY89* enhances ultraviolet B tolerance and disease resistance in rice plants. *Plant Molecular Biology* 65(6):799–815.

Wang, K.C., Yang, Y.W., Liu, B., et al. (2011) A long noncoding RNA maintains active chromatin to coordinate homeotic gene expression. *Nature* 472(7341):120–4.

Wang, H., Wu, Z., Zhou, Y., et al. (2012) Effects of salt stress on ion balance and nitrogen metabolism of old and young leaves in rice (*Oryza sativa* L.). *BMC Plant Biology* 12:194.

Wang, C., Deng, P., Chen, L., et al. (2013) A wheat *WRKY* transcription factor *TaWRKY10* confers tolerance to multiple abiotic stresses in transgenic tobacco. *PLOS ONE* 8(6):e65120.

Wang, H., Wang, H., Shao, H., Tang, X. (2016) Recent advances in utilizing transcription factors to improve plant abiotic stress tolerance by transgenic technology. *Frontiers in Plant Science* 7:67.

Wärnmark, A., Treuter, E., Wright, A.P.H., Gustafsson, J.A. (2003) Activation functions 1 and 2 of nuclear receptors: Molecular strategies for transcriptional activation. *Molecular Endocrinology* 17(10):1901–9.

Wei, Q., Zhang, F., Sun, F., et al. (2017) A wheat *MYB* transcriptional repressor *TaMyb1D* regulates phenylpropanoid metabolism and enhances tolerance to drought and oxidative stresses in transgenic tobacco plants. *Plant Science* 265:112–23.

Wingender, E., Chen, X., Hehl, R., et al. (2000) TRANSFAC: An integrated system for gene expression regulation. *Nucleic Acids Research* 28(1):316–9.

Wu, H., Ni, Z., Yao, Y., Guo, G., Sun, Q. (2008) Cloning and expression profiles of 15 genes encoding *WRKY* transcription factor in wheat (*Triticum aestivem* L.). *Progress in Natural Science* 18(6):697–705.

Wu, X., Shiroto, Y., Kishitani, S., Ito, Y., Toriyama, K. (2009) Enhanced heat and drought tolerance in transgenic rice seedlings overexpressing *OsWRKY11* under the control of *HSP101* promoter. *Plant Cell Reports* 28(1):21–30.

Wu, Z., Liang, J., Wang, C., et al. (2018) Overexpression of two novel *HsfA3s* from lily in *Arabidopsis* confer increased thermotolerance and salt sensitivity via alterations in proline catabolism. *Journal of Experimental Botany.* 69(8):2005–21.

Xia, N., Zhang, G., Liu, X.Y., et al. (2010a) Characterization of a novel wheat *NAC* transcription factor gene involved in defense response against stripe rust pathogen infection and abiotic stresses. *Molecular Biology Reports* 37(8):3703–12.

Xia, N., Zhang, G., Sun, Y., et al. (2010b) *TaNAC8*, a novel *NAC* transcription factor gene in wheat, responds to stripe rust pathogen infection and abiotic stresses. *Physiological and Molecular Plant Pathology* 74(5–6):394–402.

Xie, Z., Zhang, Z.L., Zou, X., et al. (2005) Annotations and functional analyses of the rice *WRKY* gene superfamily reveal positive and negative regulators of abscisic acid signaling in aleurone cells. *Plant Physiology* 137(1):176–89.

Xu, H.S., Chen, M., Li, L.C., Ma, Y.Z. (2010) Functions and application of the AP2/ERF transcription factor family in crop improvement. *Journal of Integrative Plant Biology* 53(7):570–85.

Xue, G.P., Way, H.M., Richardson, T., Drenth, J., Joyce, P.A., McIntyre, C.L. (2011) Overexpression of *TaNAC69* leads to enhanced transcript levels of stress up-regulated genes and dehydration tolerance in bread wheat. *Molecular Plant* 4(4):697–712.

Yamasaki, K., Kigawa, T., Inoue, M., et al. (2005) Solution structure of an *Arabidopsis* WRKY DNA binding domain. *Plant Cell* 17(3):944–56.

Yang, A., Dai, X., Zhang, W.H. (2012) A *R2R3*-type *MYB* gene, *OsMYB2*, is involved in salt, cold, and dehydration tolerance in rice. *Journal of Experimental Botany* 63(7):2541–56.

Yanhui, C., Xiaoyuan, Y., Kun, H., et al. (2006) The *MYB* transcription factor superfamily of *Arabidopsis*: expression analysis and phylogenetic comparison with the rice *MYB* family. *Plant Molecular Biology* 60(1):107–24.

Yoda, H., Ogawa, M., Yamaguchi, Y., Koizumi, N., Kusano, T., Sano, H. (2002) Identification of early-responsive genes associated with the hypersensitive response to tobacco mosaic virus and characterization of a WRKY-type transcription factor in tobacco plants. *Molecular Genetics and Genomics* 267(2):154–61.

Yu, S., Ligang, C., Liping, Z., Diqiu, Y. (2010) Overexpression of *OsWRKY72* gene interferes in the abscisic acid signal and auxin transport pathway of *Arabidopsis*. *Journal of Biosciences* 35(3):459–71.

Yuan, Y., Qi, L., Yang, J., Wu, C., Liu, Y., Huang, L. (2015) A *Scutellaria baicalensis* R2R3-MYB gene, *SbMYB8*, regulates flavonoid biosynthesis and improves drought stress tolerance in transgenic tobacco. *Plant Cell, Tissue and Organ Culture* 120(3):961–72.

Zaret, K.S., Carroll, J.S. (2011) Pioneer transcription factors: establishing competence for gene expression. *Genes and Development* 25(21):2227–41.

Zhang, Y., Wang, L. (2005) The *WRKY* transcription factor superfamily: Its origin in eukaryotes and expansion in plants. *BMC Evolutionary Biology* 5:1.

Zhang, H., Jin, J., Tang, L., et al. (2011) PlantTFDB 2.0: Update and improvement of the comprehensive plant transcription factor database. *Nucleic Acids Research* 39:D1114–7.

Zheng, X., Chen, B., Lu, G., Han, B. (2009) Overexpression of a *NAC* transcription factor enhances rice drought and salt tolerance. *Biochemical and Biophysical Research Communications* 379(4):985–9.

Zhong, R., Richardson, E.A., Ye, Z.H. (2007) Two *NAC* domain transcription factors, *SND1* and *NST1*, function redundantly in regulation of secondary wall synthesis in fibres of *Arabidopsis*. *Planta* 225(6):1603–11.

Zhou, C., Zhang, L., Duan, J., Miki, B., Wu, K. (2005) *HISTONE DEACETYLASE19* is involved in jasmonic acid and ethylene signaling of pathogen response in *Arabidopsis*. *The Plant Cell* 17(4):1196–1204.

Zhou, Q.Y., Tian, A.G., Zou, H.F., et al. (2008) Soybean *WRKY*-type transcription factor genes, *GmWRKY13, GmWRKY21*, and *GmWRKY54*, confer differential tolerance to abiotic stresses in transgenic *Arabidopsis* plants. *Plant Biotechnology Journal* 6(5):486–503.

Zhou, M., Sun, Z., Wang, C., et al. (2015) Changing a conserved amino acid in *R2R3-MYB* transcription repressors results in cytoplasmic accumulation and abolishes their repressive activity in *Arabidopsis*. *The Plant Journal* 84(2):395–403.

Zhu, N., Cheng, S., Liu, X., et al. (2015) The *R2R3*-type *MYB* gene *OsMYB91* has a function in coordinating plant growth and salt stress tolerance in rice. *Plant Science* 236:146–56.

Zou, C., Jiang, W., Yu, D. (2010) Male gametophyte-specific *WRKY34* transcription factor mediates cold sensitivity of mature pollen in *Arabidopsis*. *Journal of Experimental Botany* 61(14):3901–14.

26 Application of CRISPR-Cas Genome Editing Tools for the Improvement of Plant Abiotic Stress Tolerance

Pankaj Bhowmik, Md. Mahmudul Hassan, Kutubuddin Molla,
Mahfuzur Rahman, and M. Tofazzal Islam

CONTENTS

26.1 INTRODUCTION

Abiotic stress refers to sub-optimal climatic and/or edaphic conditions that adversely affect cellular homeostasis and ultimately impair the growth and fitness of living organisms including plants (Bray et al., 2000; Schmidt et al., 2018). These stresses, such as drought, flooding, high salt, high or limited oxygen, and too low or too high temperature, adversely affect the growth and productivity of all plants and can reduce the yield of crop species by more than 50% (Bray et al., 2000). Crop species are exposed to increasing occurrences of abiotic stresses due to global warming and climate change (Mahalingam, 2015). Despite some progress during the last decade, proper understanding about how plants survive in a stressed environment, such as in saline soils, drought, cold or high temperature, is still a burning question for the plant scientists. Plant breeders have been using various conventional approaches such as crossing and selection to improve crop performance under abiotic stresses. However, conventional approaches are less effective in directly conferring

abiotic stress tolerance by targeting determinant key genes due to targets because of the complexity of stress tolerance traits. One major problem towards elucidating detailed molecular mechanism is the lack of a robust technique that can be used to analyze functions of responsive genes in plants. In addition, the process of the development of a new variety through conventional breeding takes 7 to 12 years depending on the crop. Alternative approaches are thus required to develop new crop varieties within the shortest possible time that can withstand the harsh environment posed by rapidly changing global climate. Recent studies (Ref) suggest that genome editing could be an economically feasible and consumer acceptable (non-GMO) solution for introducing new plant traits more quickly and precisely. Thus, potentially saving years in delivering the new varieties to farmers.

Genome editing is a technique which generates site-specific insertions, deletions, or substitutions in the genomes of living cellular organisms (Doudna and Charpentier, 2014). It relies on programmable nucleases to cleave DNA, with cellular DNA repair processes inducing desired mutations. Depending on the DNA repair pathway used, mutations can be random or targeted. Genome editing technologies such as clustered regularly interspaced short palindromic repeat (CRISPR)/CRISPR-associated protein (CRISPR/Cas) allow targeted modification of almost any crop genome sequence to generate novel variation and accelerate breeding efforts. This system relies on the ability of short sequences called guide RNA (gRNA) to guide CRISPR-Cas nuclease to cleave target sites and produce site-specific DNA double-strand breaks (DSBs), leading to genome modifications during the repair process (Xing et al., 2014; Adli, 2018). The CRISPR-Cas system has already been successfully used for the improvement of plant traits in multiple plant species (Haque et al., 2018).

The potentials of the CRISPR-Cas system for genome editing of organisms including plants have been described in many earlier reviews (Mickelbart et al., 2015; Sharma et al., 2016; Scheben et al., 2017; Haque et al., 2018; Mushtaq et al., 2018). In this chapter, we describe how the game-changing technology, the CRISPR-Cas system, can be used to understand plant abiotic stress or develop new crop plant varieties that can perform better under abiotic stress environments. We provide the progress of using the CRISPR-Cas system in developing abiotic stress tolerant crop plants or understanding plant abiotic stress tolerance. Some potential traits with the required type of editing for crop improvement under different abiotic stresses are also discussed.

26.2 CRISPR-CAS SYSTEM

This transformative genome engineering technology began with the curiosity about how bacteria fight against viruses. Bacteria have developed many biological systems to defend from phage, and CRISPR is one of the many anti-phage systems. The CRISPR-associated enzyme Cas9 (the blue representation in Figure 26.1) is an RNA-guided protein that uses a small RNA coming from the CRISPR sequence in the genome to detect matching sequences in double-stranded DNA. This is the 20 nucleotide sequences of RNA base paired with DNA, triggering DNA unwinding locally and allowing the enzyme to cut DNA and make a clean double-strand break. This system operates with a second RNA called tracer that creates a structure required for Cas9 binding. The CRISPR and tracer molecule can be linked together to create a single guide form of the RNA. It is a two-component system requiring to generate a single transcript with a guide sequence that can be altered depending on the target sites that we want to modify in a particular genome. The Cas9 undergoes a big conformational change when it binds to the gRNA. Rearrangement of the protein structure occurs when DNA and RNA bind together to accommodate DNA–RNA hybrid.

The DSBs introduced by the CRISPR-Cas9 complex can be repaired by non-homologous end joining (NHEJ) and homologous recombination (HR) (Figure 26.2A). The homologous end joining or recombination is also popularly known as Homology Dependent Repair (HDR). The NHEJ repair can produce two different mutations at each chromosome: heterozygous mutations, biallelic mutations, and two independent identical mutations: homozygous mutations leading to gene insertion or gene deletion (Figure 26.2). In the presence of donor DNA digested with the same endonuclease leaving behind similar overhangs, HEJ can be achieved, leading to gene modification and insertion.

From the discovery to the application of CRISPR-Cas technology for editing the genomes of organisms

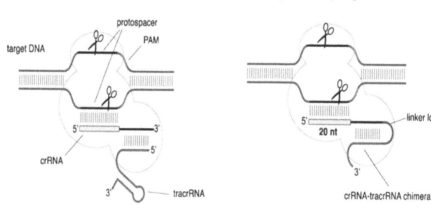

FIGURE 26.1 Cas9 can be programmed using a single engineered RNA molecule combining tracrRNA and crRNA features. (Adapted from Jinek et al., 2012.)

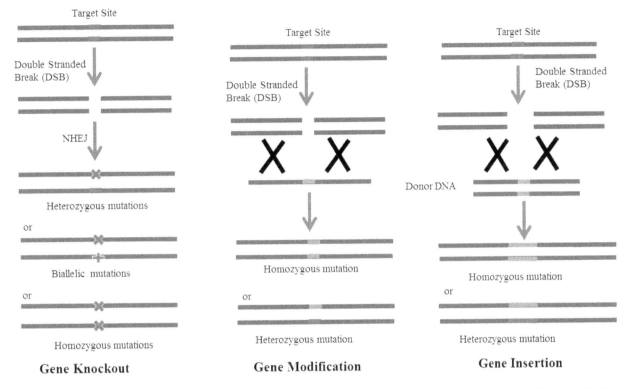

A. Non-homologous end joining (NHEJ)

Target Site

Double Stranded Break (DSB)

NHEJ

Heterozygous mutations

or

Biallelic mutations

or

Homozygous mutations

Gene Knockout

B. Homologous Recombination

Target Site

Double Stranded Break (DSB)

X   X

Homozygous mutation

or

Heterozygous mutation

Gene Modification

Target Site

Double Stranded Break (DSB)

X   X

Donor DNA

Homozygous mutation

or

Heterozygous mutation

Gene Insertion

FIGURE 26.2 Genome editing with site-specific nucleases (SSNs). (Adapted and redrawn from Arora and Narula, 2017.)

for desired improvement and its application, including the improvement of crop plants, is simply phenomenal in the history of science. Chronological advancements of the CRISPR-Cas genome editing tool is presented in Figure 26.3 (adapted from Mushtaq et al., 2018).

26.3 THREE 'DS' FOR EFFICIENT CRISPR

Although there are many recent reports of successful CRISPR-Cas9-mediated plant gene editing, the experimental protocols (Figure 26.3) required to implement these powerful techniques are yet to be accepted by many plant science laboratories as a routine protocol (Doudna and Charpentier, 2014; Haque et al., 2018). There are several factors such as efficient gRNA **designing**, assembling multiple gRNA cassettes, efficient **delivery** of Cas9 and gRNA vectors or Ribonucleoprotein (RNP) complex, selection and regeneration of edited plantlets, efficient **detection** of the gene editing event (Sharma et al., 2016). The frequency of gene editing will depend on whether or not these three 'Ds' or conditions are optimal (Figures 26.4 and 26.5) (Doudna and Charpentier, 2014; Sharma et al., 2016; Haque et al., 2018). The genome editing of plants by CRISPR-Cas-mediated technology is very quick. A tentative timeline required for completion of the four major steps of the

genome editing of a plant by CRISPR-Cas technology is illustrated in Figure 26.6 (Yin et al., 2017).

26.4 HOW CRISPR-CAS GENOME EDITING TOOLS COULD BE UTILIZED IN DEVELOPING ABIOTIC STRESS TOLERANCE

26.4.1 CRISPR-Cytosine Base Editor and HDR

Plant abiotic stress response is considered an extremely complicated network of interactions among members of many gene families, transcription factors, and *cis*-elements. Considering rice as a model crop plant, over the past few decades, about 100 genes were identified to play roles in abiotic stress response (Wang et al., 2016). In order to discuss the applicability of the CRISPR toolbox, the development of submergence tolerance in rice is deliberately discussed here as a case study. Identification of *Sub1A* gene was one of the major breakthroughs in this line as it could be deployed to save an estimated annual yield loss of about US$600 million due to submergence of rice fields in monsoon flash flooding in some South and Southeast Asian countries (Rice Knowledge Bank, IRRI). Rice production in low-lying coastal and deltaic areas of Southeast Asia is projected to face increasing

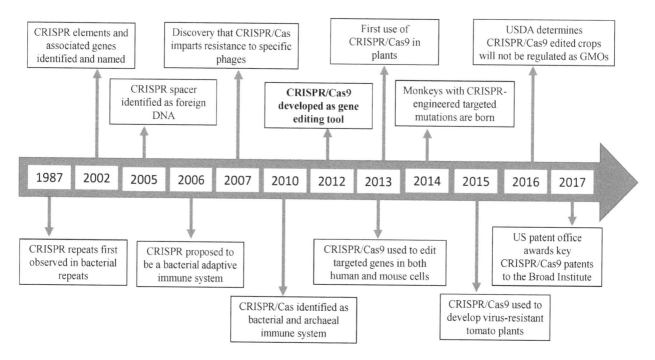

FIGURE 26.3 Timeline of CRISPR-Cas/genome editing key development from initial to present. (Adapted from Mushtaq et al., 2018.)

risk due to sea-level rise, increased temperature, and more intense tropical cyclones (FAOSTAT, 2014). The study revealed that *Sub1A* locus has two allelic forms in submergence tolerant and intolerant indica cultivars. *Sub1A-1* is found only in tolerant accessions and *Sub1A-2* is found in intolerant indica rice accessions. A single nucleotide polymorphism (SNP) is responsible for a substitution from Pro186 (intolerant) (genomic codon-CCG) to Ser186 (tolerant) (genomic codon-TCG) in a mitogen-activated protein kinase (MAPK) site in *Sub1A-1* (Xu et al., 2006). A recent study revealed that mitogen-activated protein kinase 3 (MPK3) specifically phosphorylates the *Sub1A-1* protein but not SUB1A-2 (Singh and Sinha, 2016). Therefore, the intolerant allele could be

switched to the tolerant one by altering the C>T. Here, an obvious question arises, is there any CRISPR tool other than normal CRISPR-mediated gene disruption which could alter specific SNP like the one in *Sub1A* gene? The answer is a big YES! Komor et al. (2016) developed C>T (or G>A) base editor (BE) by fusing a cytidine deaminase with dCas9/nCas9, which specifically alters a target C base to a T base at the protospacer region 4–8. A normal *Streptococcus pyogenes* Cas9 (Cas9) creates a DSB, but dCas9 (dead Cas9) is a mutated form of Cas9, which is catalytically inactive and is unable to cause any DSBs. Similarly, nCas9 (nickase Cas9) is another type of mutated Cas9 which creates a single strand nick instead of a DSB. The Cas9 is targeted to a specific

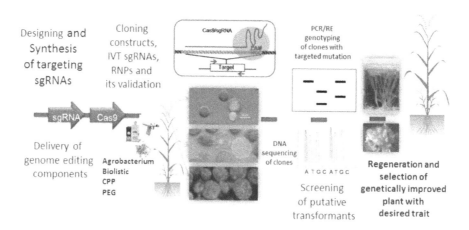

FIGURE 26.4 Basic workflow for the CRISPR-Cas mediated genome editing in plants.

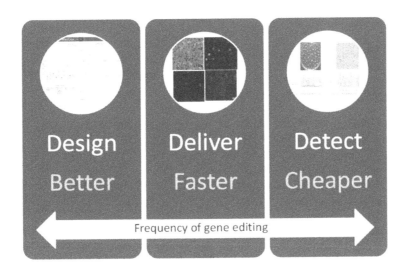

FIGURE 26.5 Three 'Ds' for efficient CRISPR.

genomic region depending on two factors: 1) a 20 bp protospacer sequence in the genome where the sgRNA binds; and 2) an adjacent Protospacer Adjacent Motif (PAM) sequence downstream of the protospacer. For SpCas9, a 3 bp 5'-NGG-3' PAM must be present at target genomic loci. Similarly, the presence of an NGG PAM that places the target C within a 5-nucleotide window near the PAM-distal end of the protospacer is a prerequisite for efficient editing by BE. The absence of NGG PAM 13–17 nucleotides downstream of the target C limits the number of sites that could otherwise be efficiently targeted by BE (Kim et al., 2017). Since a suitable NGG PAM is absent, *Sub1A* target loci fall into this category and the BE developed by Komor et al. (2016) cannot be used. One major development to target AT-rich genomic

regions is the discovery of Cpf1 nuclease, which recognizes 'TTTN' PAM sequence on the 5' side of the protospacer (upstream), unlike the Cas9 (Zetsche et al., 2015). In order to overcome this limitation, different variants of base editors which recognize PAM sequence other than NGG have been developed to target more genomic regions (Table 26.1).

Unfortunately, the target loci of *Sub1A* do not contain any of the alternative PAMs (presented in Table 26.1) at a suitable distance. Hence, a target C in this kind of genomic region cannot be edited to T using the existing BEs. The next question arises here; is there anything left in the CRISPR toolbox that could be utilized for this kind of precise editing? The answer is CRISPR-Cas9-mediated homology-directed repair (HDR). For HDR to

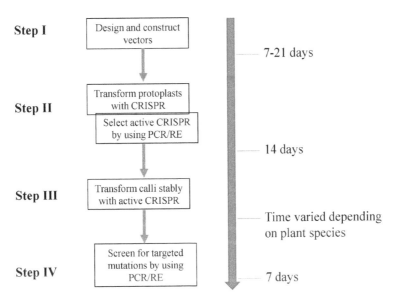

FIGURE 26.6 Time required for the genome editing of a plant. PCR, polymerase chain reaction; RE, restriction enzyme. (Adapted from Yin et al., 2017.)

TABLE 26.1

C>T Base Editors with PAM Requirements Other Than NGG

Base Editors (C>T)	PAM	References
VQR-BE3	NGAN	Kim et al. (2017)
EQR-BE3	NGAG	Kim et al. (2017)
VRER-BE3	NGCG	Kim et al. (2017)
SaBE3	NNGRRT	Kim et al. (2017)
dLbCpf1-eBE	TTTN	Li et al. (2018)
xCas9-BE4	NGG, NG, GAA, and GAT	Hu et al. (2018)

occur at a target site, we need to supply a donor DNA template containing the desired change flanked with homology arm. The Cas9 creates a DSB at a nearby location of the target site, and the donor DNA is utilized by cellular HDR machinery to repair the DSB, and as a result, the desired edit is incorporated at the targeted locus (Jasin and Haber, 2016). The *Sub1A-2* (intolerant) allele could be altered to *Sub1A-1* (tolerant) for developing submergence tolerance using CRISPR-Cas9-HDR-mediated editing. The efficiency of HDR-mediated editing in plant systems is very low because of the inherent low rate of HDR in plant cells and the lack of a suitable method to deliver donor DNA in the vicinity of DSB. In order to make the CRISPR-Cas9-HDR-based precise editing readily usable by any laboratories, we need to further improve the methodology to overcome those limitations.

26.4.2 CRISPR-ADENINE BASE EDITORS

Recently, another group of base editors was developed by David Liu's group at Harvard University (Gaudelli et al., 2017). The base editor could alter precisely an A base to a G base and named as adenine base editor (ABE). The ABE for A>G (/T>C) editing has been developed by fusing an engineered tRNA adenosine deaminase (TAD) with nCas9. As the ABEs are also functionally active in plant systems (Hua et al., 2018, Yan et al., 2018), it could be efficiently utilized for precise editing of functional A>G (/T>C) SNP in any known stress responsive gene in crop plants.

26.4.3 CRISPR/CAS9 GENE KNOCKOUT AND CRISPRI

The revolutionary CRISPR-Cas-mediated genome editing tools are expanding exponentially each day. The

CRISPR-mediated tools have tremendous potential to be utilized in forward genetic approach (genotype to phenotype), which could, in turn, greatly facilitate functional annotation of abiotic stress responsive genes and regulatory genomic elements. One of the simplest ways to enhance plant abiotic stress tolerance by editing is to generate CRISPR-Cas9-mediated knock out/knock down of negative regulatory genes. For example, *Oryza sativa SRFP1* (*OsSRFP1*) is a negative regulator of salinity tolerance (Fang et al., 2015); it could be targeted for disruption to generate novel mutant with enhanced salinity tolerance. Similarly, the *SPINDLY* gene (*SPY-3*) of *Arabidopsis* and its homolog in other crop species is a potential target as it acts as a negative regulator in plant drought and salinity tolerance (Qin et al., 2011). Tang et al. (2016) demonstrated that MODD protein mediates deactivation and degradation of *OsbZIP46* and negatively regulates ABA signaling and drought resistance in rice. Hence, MODD knock out mutant could be generated using CRISPR-Cas9 to increase drought tolerance in rice plants. Knocking out of a gene may have several unknown consequences, including the development of phenotypic cost. Achieving stress tolerance at the expense of phenotypic abnormalities is not desirable. Instead of genetically altering the target loci by knock out, the gene could be silenced using CRISPRi (CRISPR interference) system (Qi et al., 2013). The CRISPRi system requires a nuclease-deactivated Cas9 (dCas9) protein and sgRNA designed with a 20-base-pair complementary region to any gene of interest. The sgRNA directs the binding of dCas9 to the target loci, and this binding inhibits transcription of the gene. The dCas9 does not create any double-strand break, and as a result, the genomic locus is not altered. CRISPRi differs from the traditional RNAi approach on the key point that CRISPRi acts on reducing transcriptional activity of a gene, whereas RNAi acts on the mRNA level. Table 26.2 represents a few potential abiotic stress responsive rice genes that could be targeted for editing using different CRISPR-mediated methods.

26.4.4 CRISPR-MULTIPLEXING

The *cis*-regulatory elements (CREs) are DNA sequences located at the upstream region of a particular coding sequence and bound with transcription factors (TFs) and other regulatory proteins to fine-tune the expression of the gene in response to different stresses. The abscisic acid response element (ABRE), dehydration response element (DRE), ethylene response element (ERE), low temperature responsive elements (LTRE), MYB recognition sequence (MYBRS), MYC

TABLE 26.2

Abiotic Stress Responsive Rice Genes and Their CRISPR Genome Editing Methods

Gene	Trait	Type of Editing Required to Change the Allele	Applicable CRISPR Tool	Reference
OsCTB4a	Booting stage cold tolerance	A few SNPs and 1 indel at promoter region	CRISPR/Cas9-HDR	Zhang et al. (2017)
OsCOLD1	Cold tolerance	A>T SNP alteration in 4th exon	CRISPR/Cas9-HDR	Ma et al. (2015)
OsMODD	Drought tolerance	Disruption	CRISPR/Cas9 knock out/CRISPRi	Tang et al. (2016)
OsTT1	Thermo tolerance	Three SNPs	CRISPR/Cas9-HDR or C>T and A>G editors	Li et al. (2015)
OsSIT1	Salt tolerance	Disruption	CRISPR/Cas9 knock out/CRISPRi	Li et al. (2014)

recognition sequence (MYCRS), and light responsive elements (LRE) are among the well-studied CREs that play important roles in plant abiotic stress response. The identification of novel CREs will boost our basic understanding of gene expression and their regulation in the presence of stress stimuli. To identify the crucial *cis*-region at the upstream of any particular gene, CRISPR-Cas9 multiplex editing tool could be utilized. Multiple guide RNA could be designed to simultaneously target different locations of the upstream region. Screening of the resultant single mutants/double mutants/multiple mutants and subsequent phenotyping would reveal the most important *cis*-regulatory region and their role in altering expression of a particular gene at the onset of stress. Polycistronic tRNA-gRNA (PTG) system, developed by Xie et al. (2015), has been demonstrated to successfully target eight different genomic locations in rice using a single vector. Once an important *cis*-regulatory element is identified, CRISPR-Cas9-HDR-mediated (for a precise change) editing tool or CRISPR-Cas9-mediated disruption tool (for deleting and negative regulatory elements) could be applied to redesign a promoter to generate edited crops with enhanced stress tolerance.

26.5 EVIDENCE OF APPLICATION OF CRISPR-CAS GENOME EDITING SYSTEM FOR IMPROVING PLANT ABIOTIC STRESS TOLERANCE

In recent years, several findings were published where the CRISPR-Cas system was used to discover potential candidate genes as well as to create edited plants with enhanced abiotic tolerance. Lou et al. (2017) targeted osmotic stress/ABA-activated protein kinase 2 (SAPK2)

in rice. They have elucidated the functional properties of SAPK2 using loss of function mutants produced with the CRISPR-Cas9 system and reported very high SAPK2 expression level, which was upregulated by drought, high-salinity, and polyethylene glycol (PEG) treatments. Wang et al. (2017b) targeted mitogen-activated protein kinases (MAPKs) in tomato and used the CRISPR-Cas system to generate *slmapk3* mutants. They used two independent T_1 transgenic lines and wild-type (WT) tomato plants for analysis of drought tolerance. Researchers observed more severe wilting symptoms, higher hydrogen peroxide content, and lower antioxidant enzyme activities in *slmapk3* mutants compared to the WT plants. The *slmapk3* mutants also suffered more membrane damage under drought stress. In the case of maize, the ARGOS8 variants generated by the CRISPR-Cas9 system improved grain yield under field drought stress conditions in the field (Shi et al., 2017). The group researchers inserted the native maize GOS2 promoter into the 5'-untranslated region of the native ARGOS8 gene, which confers a moderate level of constitutive expression. As expected from the manipulation of the GOS2 promoter, the ARGOS8 variants had elevated levels of ARGOS8 transcripts relative to the native allele, and these transcripts were detectable in all the tissues tested. They have also validated their results in a field study and reported increased grain yield by five bushels per acre under stress conditions compared to the WT. In ARGOS8 variants, there was no yield loss under well-watered conditions. These results clearly demonstrate the application of the CRISPR-Cas9 system for improving plant abiotic stress tolerance in multiple plant species. Over the years, the genomes of a good number of crop plants were edited for abiotic stress tolerance. Some of the remarkable successes are summarized in Table 26.3.

TABLE 26.3

Evidence of the Application of CRISPR-Cas Genome Editing Technology in the Development of Abiotic Stress Tolerant Crop Plants

Crop/Plant Species	Target Sequence (TS)	Type of TS	Function of TS	Type of Edit	Result	Reference
Rice	*Oryza sativa* bilateral blade senescence 1 (BBS1)	Gene	Cause early leaf senescence	Deletion	Absence of *OsBBS1* function in rice makes plant sensitive to drought stress	Zeng et al. (2018)
Rice	*Oryza sativa* phosphate transporter 4 (*OsPT4*)	Gene	Help in uptaking arsenic (As) from soil	Deletion	Absence of *OsPT4* activity reduces arsenic uptake from soil in rice plant	Ye et al. (2017)
Rice	*OsNAC14*	Gene	Induces the expression of *OsRAD51A1, Piz-t, DR, OsPAE1*, and *OsFbox341*, which are involved in drought tolerance	Deletion	Absence of *OsNAC14* activity significantly reduces the expression of *OsRAD51A1, Piz-t, DR, OsPAE1*, and *OsFbox341*	Shim et al. (2018)
Rice	*Oryza sativa* annexins 3 (*OsANN3*)	Gene	Presence and expression of *OsANN3* confer cold tolerance	Deletion	Absence of *OsANN3* activity make plant sensitive to cold	Shen et al. (2017)
Rice	Natural Resistance Associated Macrophase Proteins 5 (*OsNRAM5*) of *Oryza sativa*	Gene	*OsNRAMP5* is involved in uptaking cadmium from soil	Deletion	Absence of *OsNRAMP5* reduce the uptake of cadmium from soil	Tang et al. (2017)
Rice	Arsenite-responsive MYB1 of *Oryza sativa* (*OsARM1*)	Gene	Expression of *OsARM1* helps in absorbing arsenic from soil	Deletion	Plant lacking the activity of *OsARM1* shows tolerance to as stress	Wang et al. (2017a)
Rice	*OsHAK1*	Gene	*OsHAK1* helps in uptaking radioactive Cs (^{137}Cs$^+$) from soil in rice plant	Deletion	Inactivation of *OsHAK1* reduces the uptake of ^{137}Cs$^+$ from soil in rice plant	Nieves-Cordones et al. (2017)
Tomato	Mitogen-activated protein kinase 3 (*SlMAPK3*) in *Solanum lycopersicum*	Gene	Presence and expression of *SlMAKP3* helps plant to survive in drought	Deletion	Absence of *SlMAPK3* activity make plant susceptible to drought stress	Wang et al. (2017b)
Rice	ABA-activated protein kinase 2 of *SnRK2* gene family (*SAPK2*)	Gene	Presence and expression of *SAPK2* make plant tolerant to drought and salinity stress	Deletion	Absence of *SAPK2* make rice plant sensitive to drought and salinity stress	Wang et al. (2017b)
Rice	*Oryza sativa* 9-*cis*-epoxycarotenoid dioxygenase (*OsNCED*)	Gene	Expression of *OsNCED* helps in making plant tolerant to drought and salinity stress	Mutation	Absence of *OsNCED* makes plant sensitive to drought and salinity stress	Huang et al. (2018)

(Continued)

TABLE 26.3 (CONTINUED)

Evidence of the Application of CRISPR-Cas Genome Editing Technology in the Development of Abiotic Stress Tolerant Crop Plants

Crop/Plant Species	Target Sequence (TS)	Type of TS	Function of TS	Type of Edit	Result	Reference
Maize	ARGOS8	Promoter	Constitutive expression of ARGOS8 helps in tolerating drought stress in maize. However, it is expressed at low level in different plant parts.	Promoter replacement	High expression of ARGOS8	Shi et al. (2017)
Rice	OsRAV2	Promoter	Expression of OsRAV2 helps in tolerating salinity stress	Mutation in promoter	Functional inactivation of OsRAV2 promoter reduces the plant tolerance to salinity	Duan et al. (2016)

26.6 CHALLENGES AND OPPORTUNITIES OF CRISPR-CAS GENOME EDITING FOR MITIGATION OF ABIOTIC STRESSES IN CROP PRODUCTION

A major challenge for using CRISPR-Cas genome editing technology for improving abiotic stress tolerance is the limited availability of potential candidate genes and relevant transformation and plant regeneration protocols. The discovery of new genes and the optimization of plant tissue culture methods have the potential to address these limitations. Our knowledge of the genetic mechanisms of abiotic stress tolerance that translates to crop yield has vastly advanced in recent years (Mickelbart et al., 2015). The molecular dissection of signaling pathways and characterization of transcription factors associated with abiotic stress responses in plants are becoming more popular for the discovery of new genes (Licausi et al., 2013). A significant number of abiotic stress responsive genes and their loci have been discovered in many major plant species that can be targeted for editing through CRISPR-Cas technology (Tables 26.2 and 26.4). Overexpression studies in *Arabidopsis*, rice, tobacco, and tomato have discovered several heat shock transcription factors (HSFs) that could enhance abiotic stress tolerance (Grover et al., 2013). In a recent study, the functional properties of stress/ABA-activated protein kinase 2 (SAPK2) have been characterized using loss-of-function mutants produced with the CRISPR/Cas9 system (Lou et al., 2017). However, the delivery of CRISPR reagents into plant cells and regeneration of edited plants still remains as the major impediment for mitigation of abiotic stresses in crop species. Over the last several years, a number of approaches, such as the optimization of the promoters to drive and express Cas9 and the utilization of different fluorescent reporters and selection markers (Bhowmik et al., 2018; Čermák et al., 2015; Voytas and Gao, 2014; Wang et al., 2015; Kaur et al., 2018), have been evaluated in different crop species.

A microspore-based gene editing system (Bhowmik et al., 2018) could also provide an innovative platform for functional validation of genes as well as genetic characterization and improvement of abiotic traits. This system, along with mesophyll protoplasts, can be used for high throughput screening of multiple gRNAs simultaneously. Therefore, the microspore-based gene editing system has the potential to accelerate the pace of gene discovery and improvement of abiotic stress tolerance as an alternative to the traditional somatic cell transformation-based gene editing.

Although CRISPR editing is a revolutionary scientific breakthrough, it has some technical limitations. One of the major concerns is the off-target editing. The combination of NGG PAM and specific gRNA make highly specific editing, but there are instances of editing of other unintended genomic sites (Zhang et al., 2015). Although it only has 20% of the binding efficiency of NGG PAM, Cas9 can bind with NAG PAM sequence (Hsu et al., 2013). Another challenge to overcome is

TABLE 26.4

Plant Gene(s) That Can Be Edited by the CRISPR-Cas9 Technology to Improve Tolerance to the Abiotic Stresses

Crop	Abiotic Stress	Target Gene(s) and Locus	References
Banana	Cold and salt	*MaSWEET-4d*, *MaSWEET-1b*, and *MaSWEET-4b*	Miao et al. (2017a)
Banana	Cold and salt	*MaAPS1* and *MaAPL3*	Miao et al. (2017b)
Cassava	Salt, osmosis, cold, drought	*MeKUPs*	Ou et al. (2018)
Cassava	Drought	*MeMAPKKK*	Ye et al. (2017)
Cotton	Drought	*GhPIN1–3* and *GhPIN2*	He et al. (2017)
Cotton	Drought	*GhRDL1*	Dass et al. (2017)
Date palm	Cadmium and chromium toxicity	*Pdpcs* and *Pdmt*	Chaâbene et al. (2018)
Date palm	Heavy metal toxicity	*Pdpcs* and *Pdmt*	Chaâbene et al. (2018)
Papaya	Drought, heat and cold	*CpDreb2*	Arroyo-Herrera et al. (2016)
Papaya	Heat and cold	*CpRap2.4a* and *CpRap2.4b*	Figueroa-Yañez et al. (2016)
Sugarcane	Copper	*ScAPX6*	Liu et al. (2017)
Sugarcane	Drought and chilling	*ScNsLTP*	Chen et al. (2017)
Rice	Flooding	*Sub1A*, *SK1* and *SK2*	Xu et al. (2006); Fukao et al. (2006); Hattori et al. (2009)
Rice	Drought	*DRO1*	Uga et al. (2011, 2013)
Wheat and rice	Saline soil	*TaHKT1;5*	Dubcovsky et al. (1996)
Maize	High Al^{3+}	*MATE1*	Maron et al. (2013)
Wheat	High Al^{3+}	*ALMT*	Zhou et al. (2014)
Rice	High Al^{3+}	*NRAT1*	Li et al. (2014)
Rice	Low Pi	*PSTOL1* at the *Pup1* locus	Gamuyao et al. (2012); Chin et al. (2011)
Wheat and barley	Low temperature	*VRN1* at the *FR1* locus and *CBFs* at the *FR2* locus	Dhillon et al. (2012); Stockinger et al. (2007); Knox et al. (2010); Francia et al. (2007)

that CRISPR cannot cut all sequences of interest due to the restricted PAM requirements by different nucleases. HDR-mediated precise editing is still not efficient enough with CRISPR system; as for the plant system, we do not have a suitable method of supplying abundant donor template at the DSB site.

The global climate change has the potential to reduce the yield of stable food crops up to 70% through increasing occurrence of various abiotic stresses such as soil salinity, drought, flooding, low level of available nutrients in soils and very low and very high temperature. The CRISPR-Cas9 technology is becoming a user-friendly tool for the development of non-transgenic genome-edited crop plants to cope with the changing climate and ensure future food security. We foresee that overcoming the technical and regulatory barriers associated with mass adoption and large-scale application of the genome editing will allow CRISPR-Cas technology to produce a new generation of high-yielding, climate resilient, and abiotic stress tolerant crops.

26.7 CONCLUSION AND FUTURE PERSPECTIVE

Global climate change is anticipated to pose a severe risk to agrarian profitability around the globe and, consequently, challenges food as well as nutritional security. Conventional plant breeding seems unable to meet increasing food and feed demands through the desired improvement of plants to tolerate various abiotic stresses and other environmental challenges. In contrast, CRISPR technology is making genome editing an easily achievable target, and it may revolutionize plant breeding. Recent success stories of the application of CRISPR technology in the improvement of various traits in crop plants are very encouraging. We envision exciting practical applications of CRISPR tools for the improvement of crop plants for abiotic stress tolerance and other desirable traits. However, improvement in technology and necessary changes in the global regulatory environments are needed for its wider application for sustainable food production in the changing climate to feed the ever-increasing population of the world.

ACKNOWLEDGMENTS

This work was partly funded by the Krishi Gobeshona Foundation (KGF), Bangladesh Project No. TF 50-C/17 and by the World Bank through a HEQEP CP # 2071 to TI. TI and KM are thankful to the US State Department for the Fulbright Visiting Fellowship during the preparation of this manuscript.

REFERENCES

Arora, L., Narula, A. (2017) Gene editing and crop improvement using CRISPR-Cas9 system. *Frontiers in Plant Science* 8: 1932. doi:10.3389/fpls.2017.01932

Arroyo-Herrera, A., Figueroa-Yanez, L., Castano, E., Santamari'a, J., Pereira-Santana, A., Espadas-Alcocer, J., et al. (2016) A novel *Dreb2*-type gene from *Carica papaya* confers tolerance under abiotic stress. *Plant Cell, Tissue and Organ Culture* 125(1): 119–133. doi:10.1007/s11240-015-0934-9

Adli, M. (2018) The CRISPR tool kit for genome editing and beyond. *Nature Communications* 9(1): 1911. doi:10.1038/s41467-018-04252-2

Bhowmik, P., Ellison, E., Polley, B., Bollina, V., Kulkarni, M., Ghanbarnia, K., et al. (2018) CRISPR/Cas9-based targeted mutagenesis in wheat microspores. *Scientific Reports* 8: 6502. doi:10.1038/s41598-018-24690-8

Bray, E.A., Bailey-Serres, J., Weretilnyk, E. (2000) Responses to abiotic stresses. In: Buchanan, B.B., Gruissem, W., Jones, R.L., editors. *Biochemistry and Molecular Biology of Plants*. Rockville, MD: American Society of Plant Physiologists, 1158–1203.

Čermák, T., Baltes, N.J., Čegan, R., Zhang, Y., Voytas, D.F. (2015) High-frequency, precise modification of the tomato genome. *Genome Biology* 16(1): 232. doi:10.1186/s13059-015-0796-9

Chaâbene, Z., Rorat, A., Hakim, I., Bernard, F., Douglas, G.C., Elleuch, A., et al. (2018) Insight into the expression variation of metal-responsive genes in the seedling of date palm (*Phoenix dactylifera*). *Chemosphere* 197: 123–134. doi:10.1016/j.chemosphere.2017.12.146

Chen, Y., Ma, J., Zhang, X., et al. (2017) A novel non-specific lipid transfer protein gene from sugarcane (*NsLTPs*), obviously responded to abiotic stresses and signaling molecules of SA and MeJA. *Sugar Technology* 19(1): 17. doi:10.1007/s12355-016-0431-4

Chin, J.H., et al. (2011) Developing rice with high yield under phosphorus deficiency: *Pup1* sequence to application. *Plant Physiology* 156(3): 1202–1216. doi:10.1104/pp.111.175471

Dass, A., Abdin, M.Z., Reddy, V.S., Leelavathi, S. (2017) Isolation and characterization of the dehydration stress-inducible GhRDL1 promoter from the cultivated upland cotton (*Gossypium hirsutum*). *Journal of Plant Biochemistry and Biotechnology* 26(1): 113–119. doi:10.1007/s13562-016-0369-3

Dhillon, T., Pearce, S.P., Stockinger, E.J., Distelfeld, A., Li, C., Knox, A.K., et al. (2012) Regulation of freezing tolerance and flowering in temperate cereals: The *VRN-1* connection. *Plant Physiology* 153(4): 1846–1858. doi.org/10.1104/pp.110.159079

Doudna, J.A., Charpentier, E. (2014) The new frontier of genome engineering with CRISPR-Cas9. *Science* 346(6213): 1258096. doi:10.1126/science.1258096

Duan, Y.B., Li, J., Qin, R.Y., Xu, R.F., Li, H., Yang, Y.C., Ma, H., Li, L., Wei, P.C., Yang, J.B. (2016) Identification of a regulatory element responsible for salt induction of rice *OsRAV2* through ex situ and in situ promoter analysis. *Plant Molecular Biology* 90(1–2): 49–62. doi:10.1007/s11103-015-0393-z

Dubcovsky, J., María, G.S., Epstein, E., Luo, M.C., Dvořák, J. (1996) Mapping of the K⁺/Na⁺ discrimination locus Kna1 in wheat. *Theoretical and Applied Genetics* 92(3–4): 448–454. doi:10.1007/BF00223692

Fang, H., Meng, Q., Xu, J., Tang, H., Tang, S., Zhang, H., Huang, J. (2015) Knock-down of stress inducible *OsSRFP1* encoding an E3 ubiquitin ligase with transcriptional activation activity confers abiotic stress tolerance through enhancing antioxidant protection in rice. *Plant Molecular Biology* 87(4–5): 441–458. doi:10.1007/s11103-015-0294-1

FAOSTAT (2014) http://faostat3.fao.org/

Figueroa-Yañez, L., Pereira-Santana, A., Arroyo-Herrera, A., Rodriguez-Corona, U., Sanchez-Teyer, F., Espadas-Alcocer, J., et al. (2016) RAP2.4a is transported through the phloem to regulate cold and heat tolerance in papaya tree (*Carica papaya* cv. Maradol): implications for protection against abiotic stress. *PLoS ONE* 11(10): 0165030. doi:10.1371/journal.pone.0165030

Francia, E., Barabaschi, D., Tondelli, A., Laidò, G., Rizza, F., Stanca, A.M., Busconi, M., Fogher, C., Stockinger, E.J., Pecchioni, N. (2007) Fine mapping of a *HvCBF* gene cluster at the frost resistance locus *Fr-H2* in barley. *Theoretical and Applied Genetics* 115(8): 1083–1091. doi:10.1007/s00122-007-0634-x

Fukao, T., Xu, K., Ronald, P.C., Bailey-Serres, J. (2006) A variable cluster of ethylene response factor–like genes regulates metabolic and developmental acclimation responses to submergence in rice. *The Plant Cell* 18: 2021–2034. doi:10.1105/tpc.106.043000

Gamuyao, R., Chin, J.H., Pariasca-Tanaka, J., Pesaresi, P., Catausan, S., Dalid, C., et al. (2012) The protein kinase *Pstol1* from traditional rice confers tolerance of phosphorus deficiency. *Nature* 488(7412): 535–539. doi:10.1038/nature11346

Gaudelli, N.M., Komor, A.C., Rees, H.A., Packer, M.S., Badran, A.H., Bryson, D. I., Liu, D.R. (2017) Programmable base editing of A• T to G• C in genomic DNA without DNA cleavage. *Nature* 551(7681): 464. doi:10.1038/nature24644

Grover, A., Mittal, D., Negi, M., Lavania, D. (2013) Generating high temperature tolerant transgenic plants: achievements and challenges. *Plant Science* 205–206:38–47. doi:10.1016/j.plantsci.2013.01.005

Haque, E., Taniguchi, H., Hassan, M.M., Bhowmik, P., Karim, M.R., Smiech, M., Zhao, K., Rahman, M., Islam, T. (2018) Application of CRISPR/Cas9 genome editing technology for the improvement of crops cultivated in tropical climates: recent progress, prospects, and challenges. *Frontiers in Plant Science* 9: 617. doi:10.3389/fpls.2018.00617

Hattori, Y., Nagai, K., Furukawa, S., Song, X.J., Kawano, R., Sakakibara, H., Wu, J., Matsumoto, T., Yoshimura, A., Kitano, H., Matsuoka, M., Mori, H., Ashikari, M. (2009) The ethylene response factors *SNORKEL1* and *SNORKEL2* allow rice to adapt to deep water. *Nature* 460(7258): 1026–1030. doi:10.1038/nature08258

He, P., Zhao, P., Wang, L., Zhang, Y., Wang, X., Xiao, H., et al. (2017) The PIN gene family in cotton (*Gossypium hirsutum*): Genome-wide identification and gene expression analyses during root development and abiotic stress responses. *BMC Genomics* 18: 507. doi:10.1186/s12864-017-3901-5

Hsu, P.D., Scott, D.A., Weinstein, J.A., Ran, F.A., Konermann, S., Agarwala, V. et al. (2013) DNA targeting specificity of RNA-guided Cas9 nucleases. *Nature Biotechnology* 31(9): 827. doi:10.1038/nbt.2647

Hu, J.H., Miller, S.M., Geurts, M.H., Tang, W., Chen, L., Sun, N. et al. (2018) Evolved Cas9 variants with broad PAM compatibility and high DNA specificity. *Nature* 556(7699): 57. doi:10.1038/nature26155

Hua, K., Tao, X., Yuan, F., Wang, D., et al. (2018) Precise A· T to G· C base editing in the rice genome. *Molecular Plant* 11(4): 627–630. doi:10.1016/j.molp.2018.02.007

Huang, Y., Guo, Y., Liu, Y., Zhang, F., Wang, Z., Wang, H., Wang, F., Li, D., Mao, D., Luan S., et al. (2018) 9-*cis*-Epoxycarotenoid dioxygenase 3 regulates plant growth and enhances multi-abiotic stress tolerance in rice. *Frontiers in Plant Science* 9: 162. doi:10.3389/fpls.2018.00162

Jasin, M., Haber, J.E. (2016) The democratization of gene editing: Insights from site-specific cleavage and double-strand break repair. *DNA Repair (Amst)* 44: 6–16. doi:10.1016/j.dnarep.2016.05.001

Jinek, M., Chylinski, K., Fonfara, I., Hauer, M., Doudna, J.A., Charpentier, E. (2012) A programmable dual-RNA-guided DNA endonuclease in adaptive bacterial immunity. *Science* 337(6096): 816–821. doi:10.1126/science.1225829

Kaur, N., Alok, A., Shivani, Kaur, N., Pandey, P., Awasthi, P., et al. (2018) CRISPR/Cas9-mediated efficient editing in phytoene desaturase (PDS) demonstrates precise manipulation in banana cv. rasthali genome. *Functional and Integrative Genomics* 18(1): 89–99. doi:10.1007/s10142-017-0577-5

Kim, Y.B., Komor, A.C., Levy, J.M., Packer, M.S., Zhao, K.T., Liu, D.R. (2017) Increasing the genome-targeting scope and precision of base editing with engineered Cas9-cytidine deaminase fusions. *Nature Biotechnology* 35(4): 371. doi:10.1038/nbt.3803

Komor, A.C., Kim, Y.B., Packer, M.S., Zuris, J.A., Liu, D.R. (2016) Programmable editing of a target base in genomic DNA without double-stranded DNA cleavage. *Nature* 533(7603): 420. doi:10.1038/nature17946

Knox, A.K., Dhillon, T., Cheng, H., Tondelli, A., Pecchioni, N., Stockinger, E.J. (2010) CBF gene copy number variation at Frost Resistance-2 is associated with levels of freezing tolerance in temperate-climate cereals. *Theoretical and Applied Genetics* 121(1): 21–35. doi:10.1007/s00122-010-1288-7

Li, C.H., Wang, G., Zhao, J.L., Zhang, L.Q., Ai, L.F., Han, Y.F., et al. (2014) The receptor-like kinase SIT1 mediates salt sensitivity by activating MAPK3/6 and regulating ethylene homeostasis in rice. *The Plant Cell* 26(6): 2538–2553. doi:10.1105/tpc.114.125187

Li, X.M., Chao, D.Y., Wu, Y., Huang, X., Chen, K., Cui, L.G., Su, L., Ye, W.W., Chen, H., Chen, H.C., Dong, N.Q., Guo, T., Shi, M., Feng, W., Zhang, P., Han, B., Shan, J.X., Gao, J.P., Lin, H.X. (2015) Natural alleles of a proteasome α2 subunit gene contribute to thermotolerance and adaptation of African rice. *Nature Genetics* 47(7): 827–833. doi:10.1038/ng.3305

Li, X., Wang, Y., Liu, Y., Yang, B., Wang, X., Wei, J., et al. (2018) Base editing with a Cpf1–cytidine deaminase fusion. *Nature Biotechnology* 36(4): 324. doi:10.1038/nbt.4102

Licausi, F., Ohme-Takagi, M., Perata, P. (2013) APETALA2/ethylene responsive factor (AP2/ERF) transcription factors: mediators of stress responses and developmental programs. *New Phytologist* 199: 639–649. doi:10.1111/nph.12291

Liu, F., Huang, N., Wang, L., Ling, H., Sun, T., Ahmad, W., et al. (2017) A novel L-ascorbate Peroxidase 6 gene, ScAPX6, plays an important role in the regulation of response to biotic and abiotic stresses in sugarcane. *Frontiers in Plant Science* 8: 2262. doi:10.3389/fpls.2017.02262

Lou, D., Wang, H., Liang, G., Yu, D. (2017) *OsSAPK2* Confers abscisic acid sensitivity and tolerance to drought stress in rice. *Frontiers in Plant Science* 8: 993. doi:10.3389/fpls.2017.00993

Ma, Y., Dai, X., Xu, Y., Luo, W., Zheng, X., Zeng, D., Pan, Y., Lin, X., Liu, H., Zhang, D., et al. (2015) *COLD1* confers chilling tolerance in rice. *Cell* 160(6): 1209–1221. 10.1016/j.cell.2015.01.046

Maron, L.G., Guimarães, C.T., Kirst, M., Albert, P.S., Birchler, J.A., Bradbury, P.J., Buckler, E.S., Coluccio, A.E., Danilova, T.V., Kudrna, D., Magalhaes, J.V., Piñeros, M.A., Schatz, M.C., Wing, R.A., Kochian, L.V. (2013) Aluminum tolerance in maize is associated with higher *MATE1* gene copy number. *Proceedings of the National Academy of Sciences* 110(13): 5241–5246. doi:10.1073/pnas.1220766110

Miao, H., Sun, P., Liu, Q., Miao, Y., Liu, J., Zhang, K., et al. (2017a) Genome-wide analyses of SWEET family proteins reveal involvement in fruit development and abiotic/biotic stress responses in banana. *Scientific Reports* 7: 3536. doi:10.1038/s41598-017-03872-w

Miao, H., Sun, P., Liu, Q., Miao, Y., Liu, J., Xu, B., et al. (2017b) The AGPase family proteins in banana: genome-wide identification, phylogeny, and expression analyses reveal their involvement in the development, ripening, and abiotic/biotic stress responses. *International Journal Molecular Science* 18(8): 1581. doi:10.3390/ijms18081581

Mickelbart, M.V., Hasegawa, P.M., Bailey-Serres, J. (2015) Genetic mechanisms of abiotic stress tolerance that translate to crop yield stability. *Nature Reviews Genetics* 16(4): 237–251. doi:10.1038/nrg390

Mahalingam, R. (ed.) (2015) Consideration of combined stress: a crucial paradigm for improving multiple stress tolerance in plants, In *Combined Stresses in Plants*. Cham, Switzerland: Springer International Publishing, 1–25. doi:10.1007/978-3-319-07899-1_1

Mushtaq, M., Bhat, J.A., Mir, Z.A., Sakina, A., Ali, S., Singh, A. K., Tyagi, A., Salgotra, R.K., Dar, A. A., Bhata, R. (2018) CRISPR/Cas approach: A new way of looking at plant-abiotic interactions. *Journal of Plant Physiology* 224–225:156–162. doi:10.1016/j.jplph.2018.04.001

Nieves-Cordones, M., Mohamed, S., Tanoi, K., Kobayashi, N.I., Takagi, K., Vernet, A., Guiderdoni, E., Périn, C., Sentenac, H., Véry, A.-A. (2017) Production of low-Cs⁺ rice plants by inactivation of the K⁺ transporter *OsHAK1* with the CRISPR-Cas system. *Plant Journal* 92(1): 43–56. doi:10.1111/tpj.13632

Ou, W., Mao, X., Huang, C., Tie, W., Yan, Y., Ding, Z., et al. (2018) Genome-wide identification and expression analysis of the KUP family under abiotic stress in cassava (*Manihot esculenta* Crantz). *Frontiers in Physiology* 9: 17. doi:10.3389/fphys.2018.00017

Qi, L.S., Larson, M.H., Gilbert, L.A., Doudna, J.A., Weissman, J.S., Arkin, A.P., Lim, W.A. (2013) Repurposing CRISPR as an RNA-guided platform for sequence-specific control of gene expression. *Cell* 152(5): 1173–1183. doi:10.1016/j.cell.2013.02.022

Qin, F., Kodaira, K.S., Maruyama, K., Mizoi, J., Tran, L.S.P., Fujita, Y., et al. (2011) *SPINDLY*, a negative regulator of gibberellic acid signaling, is involved in the plant abiotic stress response. *Plant Physiology* 157(4): 1900–1913. doi:10.1104/pp.111.187302

Rice Knowledge Bank, International Rice Research Institute. http://www.knowledgebank.irri.org/

Singh, P., Sinha, A.K. (2016) A positive feedback loop governed by SUB1A1 interaction with MITOGEN-ACTIVATED PROTEIN KINASE3 imparts submergence tolerance in rice. *The Plant Cell* 28(5): 1127–1143. doi:10.1105/tpc.15.01001

Scheben, A., Wolter, F., Batley, J., Puchta, H., Edwards, D. (2017) Towards CRISPR/Cas crops – Bringing together genomics and genome editing. *New Phytologist* 216(3): 682–698. doi:10.1111/nph.14702

Schmidt, R.R., Weits, D.A., Feulner, C.F.J., van Dongen, J.T. (2018) Oxygen sensing and integrative stress signalling in plants. *Plant Physiology* 176(2): 1131–1142. doi:10.1104/pp.17.01394

Sharma, S., Kaur, R., Singh, A. (2016) Recent advances in CRISPR/Cas mediated genome editing for crop improvement. *Plant Biotechnology Reports* 11(4): 193–207. doi:10.1007/s11816-018-0472-0

Shen, C., Que, Z., Xia, Y., Tang, N., Li, D., He, R., Cao, M. (2017) Knock out of the annexin gene OsAnn3 via CRISPR/Cas9-mediated genome editing decreased cold tolerance in rice. *Journal of Plant Biology* 60(6): 539–547. doi:10.1007/s12374-016-0400-1

Shi, J., Gao, H., Wang, H., Lafitte, H.R., Archibald, R.L., Yang, M., Hakimi, S.M., Mo, H., Habben, J.E. (2017) ARGOS8 variants generated by CRISPR-Cas9 improve maize grain yield under field drought stress conditions. *Plant Biotechnology Journal* 15(2): 207–216. doi:10.1111/pbi.12603

Shim, J.S., Oh, N., Chung, P.J., Kim, Y.S., Choi, Y.D., Kim, J.-K. (2018) Overexpression of osnac14 improves drought tolerance in rice. *Frontiers in Plant Science* 9: 310. doi:10.3389/fpls.2018.00310

Stockinger, E.J., Skinner, J.S., Gardner, K.G., Francia, E., Pecchioni, N. (2007) Expression levels of barley Cbf genes at the Frost resistance-H2 locus are dependent upon alleles at Fr-H1 and Fr-H2. *The Plant Journal* 51(2): 308–321. doi:10.1111/j.1365-313X.2007.0141.x

Tang, N., Ma, S., Zong, W., Yang, N., Lv, Y., Yan, C., Guo, Z., Li, J., Li, X., Xiang, Y., et al. (2016) MODD mediates deactivation and degradation of OsbZIP46 to negatively regulate ABA signaling and drought resistance in rice. *The Plant Cell* 28(9): 2161–2177. doi:10.1105/tpc.16.00171

Tang, L., Mao, B., Li, Y., Lv, Q., Zhang, L., Chen, C., He, H., Wang, W., Zeng, X., Shao, Y., et al. (2017) Knockout of OsNramp5 using the CRISPR/Cas9 system produces low Cd-accumulating indica rice without compromising yield. *Scientific Reports* 7: 14438. doi:10.1038/s41598-017-14832-9

Uga, Y., Okuno, K., Yano, M. (2011) *Dro1*, a major QTL involved in deep rooting of rice under upland field conditions. *Journal of Experimental Botany* 62(8): 2485–2494. doi:10.1093/jxb/erq429

Uga, Y., Sugimoto, K., Ogawa, S., Rane, J., Ishitani, M., Hara, N., Kitomi, Y., Inukai, Y., Ono, K., Kanno, N., Inoue, H., Takehisa, H., Motoyama, R., Nagamura, Y., Wu, J., Matsumoto, T., Takai, T., Okuno, K., Yano, M. (2013) Control of root system architecture by DEEPER ROOTING 1 increases rice yield under drought conditions. *Nature Genetics* 45(9): 1097–1102. doi:10.1038/ng.2725

Voytas, D.F., Gao, C. (2014) Precision genome engineering and agriculture: opportunities and regulatory challenges. *PLOS Biology* 12(6): 1–6. doi:10.1371/journal.pbio.1001877

Wang, Z.P., Xing, H.L., Dong, L., Zhang, H.Y., Han, C.Y., Wang, X.C., et al. (2015) Egg cell-specific promoter-controlled CRISPR/Cas9 efficiently generates homozygous mutants for multiple target genes in *Arabidopsis* in a single generation. *Genome Biology* 16: 144. doi:10.1186/s13059-015-0715-0

Wang, H., Wang, H., Shao, H., Tang, X. (2016) Recent advances in utilizing transcription factors to improve plant abiotic stress tolerance by transgenic technology. *Frontiers in Plant Science* 7: 67. doi:10.3389/fpls.2016.00067

Wang, F.-Z., Chen M.-X., Yu, L.-J., Xie, L.-J., Yuan, L.-B., Qi, H., Xiao, M., Guo, W., Chen, Z., Yi, K., et al. (2017a) *Osarm1*, an R2R3 MYB transcription factor, is involved in regulation of the response to arsenic stress in rice. *Frontiers in Plant Science* 8: 1868. doi:10.3389/fpls.2017.01868

Wang, L., Chen, L., Li, R., Zhao, R., Yang, M., Sheng, J., Shen, L. (2017b) Reduced drought tolerance by CRISPR/Cas9-mediated SlMAPK3 mutagenesis in tomato plants. *Journal of Agricultural and Food Chemistry* 65(39): 8674–8682. doi:10.1021/acs.jafc.7b02745

Xie, K., Minkenberg, B., Yang, Y. (2015) Boosting CRISPR/Cas9 multiplex editing capability with the endogenous tRNA-processing system. *Proceedings of the National Academy of Sciences* 112(11): 3570–3575. doi:10.1073/pnas.1420294112

Xing, H.L., Dong, L., Wang, Z.P., Zhang, H.Y., Han, C.Y., Liu, B., Wang, X.C., Chen, Q.J. (2014) A CRISPR/Cas9 toolkit for multiplex genome editing in plants. *BMC Plant Biology* 14: 327. doi:10.1186/s12870-014-0327-y

Xu, K., Xu, X., Fukao, T., Canlas, P., Maghirang-Rodriguez, R., Heuer, S., et al. (2006) Sub1Ais an ethylene-response-factor-like gene that confers submergence tolerance to rice. *Nature* 442(7103): 705–708. doi:10.1038/nature04920

Yan, F., Kuang, Y., Ren, B., Wang, J., Zhang, D., Lin, H., et al. (2018) Highly efficient A· T to G· C base editing by Cas9n-guided tRNA adenosine deaminase in rice. *Molecular Plant* 11(4): 631–634. doi:10.1016/j.molp.2018.02.008

Ye, Y., Li, P., Xu, T., Zeng, L., Cheng, D., Yang, M., Luo, J., Lian, X. (2017) *Ospt4* contributes to arsenate uptake and transport in rice. *Front in Plant Science* 8: 2197. doi:10.3389/fpls.2017.02197

Yin, K., Gao, C., Qiu, J.L. (2017) Progress and prospects in plant genome editing. *Nature Plants* 3: 17107. doi:10.1038/nplants.2017.107

Zeng, D.-D., Yang, C.-C., Qin, R., Alamin, M., Yue, E.-K., Jin, X.-L., Shi, C.-H. (2018) A guanine insert in *OsBBS1* leads to early leaf senescence and salt stress sensitivity in rice (*Oryza sativa* L.). *Plant Cell Reports* 37(6): 933–946. doi:10.1007/s00299-018-2280-y

Zetsche, B., Gootenberg, J.S., Abudayyeh, O.O., Slaymaker, I.M., Makarova, K.S., Essletzbichler, P., et al. (2015) Cpf1 is a single RNA-guided endonuclease of a class 2 CRISPR-Cas system. *Cell* 163(3): 759–771. doi:10.1016/j.cell.2015.09.038.

Zhang, X.H., Tee, L.Y., Wang, X.G., Huang, Q.S., Yang, S.H. (2015) Off-target effects in CRISPR/Cas9-mediated genome engineering. *Molecular Therapy-Nucleic Acids* 40: e264. doi:10.1038/mtna.2015.37

Zhang, Z., Li, J., Pan, Y., Li, J., Zhou, L., Shi, H., Zeng, Y., Guo, H., Yang, S., Zheng, W., et al. (2017) Natural variation in CTB4a enhances rice adaptation to cold habitats. *Nature Communications* 8: 14788. doi:10.1038/ncomms14788

Zhou, G., Pereira, J.F., Delhaize, E., Zhou, M., Magalhaes, J.V., Ryan, P.R. (2014) Enhancing the aluminium tolerance of barley by expressing the citrate transporter genes *SbMATE* and FRD3. *Journal of Experimental Botany* 65(9): 2381–2390. https://doi.org/10.1093/jxb/eru121

27 Beneficial Microorganisms and Abiotic Stress Tolerance in Plants

Antra Chatterjee, Alka Shankar, Shilpi Singh, Vigya Kesari,
Ruchi Rai, Amit Kumar Patel, and L.C. Rai

CONTENTS

27.1 INTRODUCTION

Recurring climate change results in the generation of different abiotic stress factors, such as extreme temperatures (heat or cold), drought, flooding, UV radiation, salinity and heavy metal, around the world, which further become one of the bases for major losses to crop production. Scientific communities across the world are earnestly involved in finding novel solutions of plant stress tolerance strategies without compromising proper growth to fulfill food demands under limited resource availability. Several environment-friendly schemes are in practice for the improvement of crop production under different abiotic stresses. One such application includes the strengthening of plants' natural defense systems by employing beneficial microbes. Research based on predicting plant–microbe interactions at biochemical, physiological and molecular levels revealed that associations with microbes largely influence plant responses towards stress (Farrar et al., 2014). Microbes from rhizosphere of crop plants, degraded soils, sodic and infertile soil and endophytes inhabiting plant tissues are well reported sources for the identification as well as understanding of microbe-mediated stress tolerance in plants (Kaushik and Subhashini, 1985; Belimov et al., 2001; Barka et al., 2006; Rodriguez et al., 2008; Yang et al., 2009; Singh et al., 2013). Beneficial microbes mainly include plant growth promoting bacteria/rhizobacteria (PGPRs) (Glick, 2012; Kaushal and Wani, 2016), mycorrhizal fungi (Azcon et al., 2013; Elhindi et al., 2017), actinomycetes (Grover et al., 2016) and nitrogen-fixing cyanobacteria (Singh et al., 2013). Association of microbes with plants involved not only in ameliorating abiotic stress but also in nutrient mineralization and availability, production of plant growth hormones and protection from plant pests, parasites or diseases. Other microbial machineries' involvement includes the transmission of molecular and genetic information, secondary metabolites production, siderophores generation, quorum-sensing system activation, biofilm formation and cellular transduction signaling. The ultimate unit of response under a definite abiotic stress of a particular organism is the specific gene expression patterns it exhibits, which is further responsible for the production of molecules involved in these interactions. The arrival of next-generation sequencing (NGS) facilities and different omics approaches, viz., genomics, proteomics, metabolomics, etc., supported incredibly in generating a large amount of data for better understanding of the mechanism involved in providing abiotic stress tolerance. The present chapter covers all aspects of beneficial microorganism found in extreme habitats as well as agricultural lands, their stress-specific as well as non-specific mechanism of providing tolerance to associated plants under different abiotic stresses, different omics approaches helpful in revealing microbial stress responsive strategies, identification of important stress-responsive metabolic pathways, their application as biostimulants simultaneously providing stress tolerance and development of a wide range of stress tolerating transgenics utilizing microbial gene mines.

27.2 MICROBE INTERACTIONS UNDER EXTREME ENVIRONMENT

Microorganisms found in different environmental conditions exhibit enormous metabolic capabilities to cope with abiotic stresses. Microbes found in extreme habitats developed a different strategy to adapt to such condition through evolution. Several biomolecules found in microorganism are reported to be involved in providing

stress tolerance. For example, trehalose stabilizes cell membranes and prevents protein denaturation in *E. coli*, hence protecting from cold stress, while in the case of yeast, it provides protection from osmotic (Hounsa et al., 1998), heat and desiccation tolerance (Hottiger et al., 1987) as well as acting as an ROS scavenger (Benaroudj et al., 2001). Symbiotic fungal endophytes are required by native grass species of different coastal and geothermal habitats for salt and heat tolerance, respectively. Such fungal species isolated from habitats devoid of extreme condition do not possess the stress tolerance capacity. Hence, habitat-specific fungal species which provide symbiotically conferred stress tolerance are responsible for the survival of plants in extreme habitats. The endophytic fungus *Curvularia protuberata* protects its host *Dichanthelium lanuginosum*, a tropical panicgrass which grows in geothermal areas of Yellowstone National Park, the United States, from high temperature stress. Marquez et al. (2007) reported that a three-way symbiosis is required for providing thermal tolerance as Curvularia thermal tolerance virus (CThTV)-infected fungus showed resistance against thermal stress whereas noninfected were not able to show the same. *Fusarium culmorum*, an endophytic fungus, confers salt tolerance (300–500 mM NaCl) to coastal dune grass *Leymus mollis* (Rodriguez et al., 2008) in symbiotic association; although, when grown non-symbiotically, the growth of both fungi as well as host were found retarded. *Festuca arundinacea*, commonly known as tall fescue, showed drought tolerance capacity because of the presence of seed transmissible endophytic fungus *Neotyphodium coenophialum* (West et al., 1993). Halotolerant bacteria have been found to be associated with *Psoralea corylifolia* L., an annual weed found in salinity-affected semi-arid regions of western Maharashtra, India. Sorty et al. (2016) evaluated its putative role as plant growth promoting bacteria in wheat under saline condition.

The role of blue-green algae or cyanobacteria in abiotic stress mitigation, more specifically, in salt tolerance, have been widely analyzed and discussed. Many cyanobacterial species such as *Plectonema*, *Nostoc*, *Calothrix*, *Scytonema*, *Hapalosiphon*, *Microchaete*, and *Westiellopsis* have been found to grow naturally in salt-affected soil from India, registered their potential role in the amelioration of saline, saline-sodic and sodic soil and provided tolerance to crop plants (Kaushik, 1990). The colonization of such cyanobacterial species in stressed soil encourages plants to grow in extreme conditions by releasing biologically active metabolites in the rhizosphere, which further induces systemic responses in plants against stress.

27.3 UNDERSTANDING OF PLANT–MICROBE INTERACTION VIA MULTI-OMICS APPROACHES

To understand the mechanism behind plant–microbe interaction in abiotic stress tolerance, omics approaches comprising genomics, metabolomics, phenomics, proteomics and transcriptomics separately as well as in integration can be used. Major outcomes from the interpretation of such massive data might have the noteworthy ability for field implementation. Involvement of computational application in data interpretation proved to be time-saving and have the potential to connect individual genes, proteins, signal molecules and gene cascade with the metabolic pathways for a better vision of understanding. Advancement in technology also assisted in a thorough understanding of RNAi-mediated gene silencing, mutant technology, gene editing systems, post-translational modifications and metabolite profiling, hence adding an extra dimension in resolving microbe-mediated abiotic stress tolerance strategies in plants (Luan et al., 2015). In addition, such technologies assist researchers to find out influential methods to mitigate the adverse effect of abiotic stresses in field grown plants.

To enhance abiotic stress tolerance capacity in plants, manipulation of crop genetics is in vogue, which requires extensive breeding practices. Such breeding practices become more challenging while dealing with low heritability and/or under continuous environmental variations (Manavalan et al., 2009). Marker-assisted breeding programs, which requires a good understanding of genomic loci governing stress tolerant traits, is among the most widely used approaches to develop stress-tolerant cultivars. A recent trend in genetic breeding includes accumulation of silicon (Si) in rice for generating abiotic stress tolerance. PCR-based markers for microsatellite (RM5303) and expressed sequence tag (EST, E60168) have been used in mapping Si transporter gene (Ma et al., 2013). To understand the mechanism of plant–microbe interaction role played by genomics in the area of systemic modulation of gene expression, in the case of Trichoderma atroviride and T. harzianum, interaction with tomato has been studied to demonstrate the effective role in stress alleviation (Tucci et al., 2011). Molecular markers for Sub1 gene, which provide submergence tolerance in sensitive varieties, was utilized by Neeraja et al. (2007) in backcross breeding program with parent variety of the rice Swarna. Sehrawat et al. (2013) had utilized the gene-silencing strategy of Hb1 gene in boosting NO production and upregulation of CBF regulon for enhancing cold tolerance in the crop.

Sheibani-Tezerji et al. (2015) performed whole transcriptome sequencing of Burkholderia phytofirmans PsJN, which colonizes potato (Solanum tuberosum L.) plants, for analyzing the role of strain PsJN in drought tolerance.

Metagenomics is another approach to gather information regarding the habitat-specific distribution of microbial communities with abiotic stress tolerant traits. Metagenomics is one of the best technique to explore novel culturable flora from specific niches (Handelsman et al., 2007). Kapardar et al. (2010) used the metagenomic library to recognize salt tolerance genes in uncultivable bacteria from pond water by growing at inhibitory NaCl concentrations of 750 mM. Metagenomics helped in identifying proteins similar to a putative general stress protein (GspM) having GsiB domain with a enoyl-CoA hydratase (EchM) activity proved its role in salt tolerance. EchM further registered with crotonyl-CoA hydratase activity. Sessitsch et al. (2012) reported that data including metagenomic sequences derived from endophytic cell extracts revealed metabolic functioning of cell-like quorum sensing and detoxification of ROS with a putative role in plant stress resistance.

Comparative transcriptomic is an influential tool to analyze and identify different sets of transcripts responsible for differences between two biologically different expressions in different conditions. Microarray technique analysis has been used in plant–microbe interaction in wild emmer and modern wheat for drought response strategies (Akpinar et al., 2015). Transcriptomics proved as a helpful tool in understanding the role of three Rhizobium strains in inducing drought tolerance in cucumber (Wang et al., 2012). Transcriptome analysis of rapeseed and its symbiont Stenotrophomonas rhizophila resulted in the identification of plant growth regulator spermidine during abiotic stress (Alavi et al., 2013).

Proteomics is an excellent methodology to produce a better understanding of the regulation of biological systems by identifying several proteins as a signal of changes in physiological status due to stress or factors responsible for stress alleviation (Silva-Sanchez et al., 2015). Therefore, a comparative analysis in stressed, non-stressed and microbe-associated plants can help to identify protein targets and networks. The mechanism of cold tolerance in alfalfa for both cold-tolerant (ZD7) and cold-sensitive lines (W5) was elucidated with a proteomics approach (Chen et al., 2015). P. indica involved in ameliorating drought stress by higher antioxidant production in barley was elucidated by analyzing (Ghabooli et al., 2013).

Another omics application is metaproteomics, a novel approach which breaks the limitation of proteomics focus on a single organism. To understand the network generated because of multiple metabolic interactions occurring in an ecosystem, metaproteomics is on trend. Although the analysis of data generated from metaproteome is complex, nevertheless modern techniques in the extraction and analysis of successful environmental metaproteome resulted in significant output and set up a better relationship among the omics data, stress tolerance mechanisms among organisms (Schweder et al., 2008).

The scope of metabolomics involves characterization of all the metabolites elaborated by an organism under the influence of given environmental conditions. Environmental variation in the environment induces an alteration in metabolism, which further results in a change in the secretion pattern of molecules (Martínez-Cortés et al., 2014). Metabolomics approach revealed that under stressful conditions, *Trichoderma* spp. produce auxins which further stimulated plant growth and provide protection against stress (Contreras-Cornejo et al., 2009).

27.4 INDUCED SYSTEMIC TOLERANCE BY DIFFERENT MICROBES

Plants do not comprise mobile immune cells, unlike animals, to cope with the microbial infection mediated by adaptive immunity (Jones and Dangl, 2006; Spoel and Dong, 2012). They strictly depend on the immune system associated with the cell entity to resist microbe and herbivore attacks. Plant immune systems contain a system of multiple layers which may express constitutively or induced by the microbial attack. Induced defense mechanism in plants can be effective locally at the site of attack or may be functional systemically all over the plant body. Plants are able to optimize the effect of inducible defense activity according to the nature of attacking microbes (Pieterse et al., 2009). All plants have the capacity to produce a generalized defense response against a wide range of bacteria and fungi. This defense mechanism is generally activated after the recognition of non-self microbial molecules by transmembrane pattern recognition receptors (PRRs). This phenomenon is commonly known as microbe-associated molecular patterns (MAMPS) or pathogen-associated molecular patterns (PAMPs) (Boller and Felix, 2009). The presence of MAMPS is found to be conserved across the wide range of microbes. Therefore, it is equally produced by harmful microbes as well as beneficial microbes generally found to be present in the rhizosphere (Zamioudis and Pieterse, 2012). Activated

pattern recognition receptors are responsible for inducing gene expression of the MAMP-responsive defense gene (Nicaise et al., 2009). The product of transcription of these genes is responsible for inducing a downstream defense pathway mediated by mitogen-activated protein kinases and WRKY transcription factors. This cascade of plant immunity is commonly known as MAMP/PAMP-triggered immunity and is effective to combat the harmful effect of many microorganisms (Boller and Felix, 2009). Plants are also able to induce programmed cell death of infected cells as well as the production of antibiotic at the site of infection. These local responses in combination trigger strong systemic responses that are responsible to make a plant resistant from a wide range of microorganism (Dong, 2001; Metraux, 2001).

Plant beneficial bacteria share much similar character with harmful bacteria, and they use an almost same mechanism to invade the host plants. There are ample reports supporting the hypothesis that beneficial microbes invading plants are initially considered as foreign particles due to the presence of conserved MAMPS (Zamioudis and Pieterse, 2012). Later on, a complex molecular crosstalk between plant cells and microorganism is required for the establishment of a symbiotic relationship (VanWees et al., 2008; Zamioudis and Pieterse, 2012; Cameron et al., 2013).

Induced systemic resistance is generally used for the enhanced state of immunity of plants against foreign invaders triggered by beneficial microorganism associated with that plant. The root system of plants are mainly responsible for accommodating foreign beneficial microorganisms which are responsible for the maximum storage of plant photosynthetic products in a mini-zone around the root system commonly known as the rhizosphere (Nihorimbere et al., 2011; Metraux, 2001; Westover et al., 1997), the most energy-rich habitat on earth (Bais et al., 2006). Several genera of plant growth promoting bacteria (PGPR) and fungi (PGPF) are present in the rhizosphere which are responsible for enhancing immunity from external invaders along with promoting plant growth (Lugtenberg and Kamilova, 2009; Shoresh et al., 2010). These beneficial microbes are also responsible for providing antibiosis, production of enzymes and phytohormones and activation of the plant's defense system against phytopathogens (Whipps, 2001). PGPR and PGPF are supposed to be responsible for enhancing plant immunity mediated by induced systemic resistance (Meera et al., 1995; Koike et al., 2001; Saldajeno and Hyakumachi, 2011; Murali et al., 2013).

27.5 MICROBIAL STRATEGY TO COPE WITH DIFFERENT ABIOTIC STRESSES

27.5.1 SALT TOLERANCE

Plant-associated microbial communities currently have received increased attention for enhancing crop productivity and providing stress tolerance to plants (Yang et al., 2009). Several reports suggest that inoculation with endophytic bacteria can moderate the effects of salt stress in the plant. For example, *Azospirillum*-inoculated seeds of lettuce (*Lactuca sativa*) exhibited better germination rates and vegetative growth than non-inoculated control plants when exposed to NaCl (Barassi et al., 2006). Moreover, *Arabidopsis thaliana* inoculated with *Paenibacillus yonginensis* DCY84T and *Micrococcus yunnanensis* PGPB7 are more resistant than control plants when challenged with different abiotic stresses, majorly salt stress, drought stress and heavy metal (aluminum) (Sukweenadhi et al., 2015). Bacteria-mediated salt stress tolerance was achieved by the reduced level of endogenous ethylene (Saravanakumar and Samiyappan, 2007). Transgenic plants of canola showed more tolerance to high concentrations of salt than non-transformed control plants because of the expression of a bacterial gene, ACC deaminase (Sergeeva et al., 2006). Under severe salt stress condition, the expression of abscisic acid (ABA) pathway genes got triggered and resulted in higher ABA level, which consecutively closed the stomata and decreased the photosynthesis rate (Zhu, 2002). Suppression of photosynthesis detected in tomato plants exposed to salt stress was less severe in plants inoculated with *Achromobacter* (Mayak et al., 2004a). A higher ratio of K^+/Na^+ was found after the inoculation of salt-stressed maize with ACC deaminase containing *Pseudomonas syringae*, *Enterobacter aerogenes* and *P. fluorescens* (Nadeem et al., 2007). Hamdia et al. (2004) reported that salt-stressed maize inoculated with *Azospirillum* showed higher K^+/Na^+ ratio because of the change of selectivity for Na^+, K^+ and Ca^{2+} ions. Wheat seedlings inoculated with exopolysaccharide-producing bacteria showed a decline in uptake of Na^+ because a higher proportion of seedling root zone was covered in soil sheaths, causing a lower apoplastic flow of Na^+ ions into the stele (Ashraf et al., 2004). Likewise, inoculation of *Bacillus* sp. TW4 to pepper led to relief from osmotic stress (Sziderics et al., 2007). Under salt stress, mycorrhizal maize plants showed a more increased biomass than non-mycorrhizal plants, implying mycorrhizal plants grow better than non-mycorrhizal plants. Similar trends were reported with tomato, cotton and barley (Singh et al., 2011).

27.5.2 DROUGHT TOLERANCE

In recent times, drought-stress is considered as the most limiting factor in plant distribution and productivity (Bahadur et al., 2011; Chatterjee and Solankey, 2015). Inoculation of *Paenibacillus polymyxa* enhanced drought tolerance in *Arabidopsis* (Timmusk et al. 1999). ACC deaminase-producing *Achromobacter piechaudii* ARV8 showed tolerance to drought-stressed pepper and tomato plants (Mayak et al., 2004a,b). In stress condition, plant hormone ethylene regulates plant homeostasis and, thus, reduced root and shoot growth (Glick et al., 2007). Degradation of ACC, which is a precursor of ethylene by ACC deaminase, relieves plants from stress and rescues normal plant development (Figueiredo et al., 2008). Under drought stress conditions, co-inoculation *Rhizobium tropici* and *P. polymyxa* resulted in enhanced bean plant height, shoot and nodule number (Figueiredo et al., 2008). Co-inoculation of *Pseudomonas mendocina* and fungi, *Glomus intraradices* or *G. mosseae* enhanced the proline accumulation in lettuce and thus provides tolerance to plants (Kohler et al., 2008). *Piriformospora indica* fungus, which colonizes both monocot and dicot plants, has the ability to promote plant growth and enhance plant abiotic stress tolerance, with arbuscular mycorrhiza fungi (AMF) (Sun et al., 2010). *Trichoderma spp.*, which colonizes the *Theobroma cacao* tree, initiates a major delay in the onset of many drought-induced physiological changes (Bae et al., 2009). The AM fungus *Glomus fasciculatum* increases the growth of drought-stressed soybean (*Glycine max*) plants by improving their drought tolerance (Busse and Ellis, 1985). *C. protuberata* confers heat and drought tolerance to its host plant *D. lanuginosum* (Rodriguez et al., 2008).

27.5.3 HEAVY METAL TOLERANCE

Ever-increasing concentration of heavy metals due to anthropogenic activities is a serious threat to crop productivity. Plant growth is highly affected by heavy metals, and their contamination in agricultural lands has raised negative impact on human health. Reports suggest that Cd (Mohanpuria et al., 2007), Zn (Tsonev and Lidon, 2012), Pb (Chatterjee et al., 2008), Cu (Dey et al., 2015), Cr (Panda and Choudhury, 2005) and As (Vezzaa et al., 2018) negatively affects plant growth and, hence, productivity. Many bacteria release metal-chelating substances and influence the uptake of various metals such as Fe, Zn and Cu (Dimkpa et al., 2009). Cd-contaminated soil grown barley plants showed higher grain yield and reduced Cd contents when inoculated with the *Klebsiella mobilis* CIAM 880. This happens because free Cd ions can be bound by bacteria into complex forms, thereby limiting their availability to plants (Pishchik et al., 2002). AM fungi have a huge role to ease heavy metal toxicity, as they improve the uptake of nutrient and water by host plants through their mycelia and, thus, protect the host plants from heavy metal toxicity (Hildebrandt et al., 1999). In the presence of PGPR strain, *Brassica juncea* and *Brassica campestris* produce siderophores, which play an important role in the remediation of Ni-, Pb- and Zn-contaminated soil (Burd et al., 1998). *Brassica napus* plant growth is improved through ACC-deaminase activity of PGPR (Belimov et al., 2001), whereas barley plant growth was improved by biological nitrogen fixation and auxin production with PGPR inoculation in Cd-contaminated soil (Belimov and Dietz, 2000). Rhizospheric bacteria enhanced the metal mobility and bioavailability of the plants by releasing chelating agents, acidification, phosphate solubilization and redox changes (Idris et al., 2004). *P. aeruginosa* produced surfactant rhamnolipid, which showed specificity for some of the metals, such as Cd and Pb. *Thiobacillus ferrooxidans* and *Leptospirillum ferrooxidans* are competent enough to oxidize iron and sulfur (Sand et al., 1992). Another bacterium, *Pseudomonas stulzeri* AG259, is able to produce silver-based single crystals, which can decrease the toxicity of metals (Klaus-Joerger et al., 2001). Jiang et al. (2017) revealed that *Boehmeria nivea* get protected from high concentrations of Cd, Pb and Cu with the help of rhizosphere bacteria belonging to genera *Pseudomonas, Cupriavidus, Bacillus* and *Acinetobacter. Pseudomonas fluorescens, Rhizobium leguminosarum bv phaseoli* CPMex46, *Azospirillum lipoferum* UAP40 and UAP154 are siderophore-producing bacteria that protect *M. sativa* seeds from the toxicity of Cu in high concentrations (Carrillo-Castaneda et al., 2003). Inoculation of *P. aeruginosa* in roots and shoots of *Cucurbita pepo* and *Brassica juncea* reduced the Cd uptake (Sinha and Mukherjee, 2008). Tank and Saraf (2009) reported that inoculation of *Pseudomonas* increased plant growth and reduced Ni uptake in chickpea plants. Thus, interactions between plants and useful microbes can improve phytoremediation.

27.5.4 Extreme Temperature Tolerance

Extreme temperature (both high and low) is one of the major causes of crop loss. Inoculation of *Burkholderia phytofirmans* PsJN on potato at two different temperatures revealed that colonization of potato by rhizobacteria might play a role in their adaptation to heat as tuberization was increased by more than 50% in bacteria-treated clones (Bensalim et al., 1998). Additionally, during cold

stress, inoculation of grapevine (*Vitis vinifera*) with *B. phytofirmans* PsJN, reduced the rate of electrolyte leakage and biomass reduction and encouraged post-chilling recovery (Barka et al., 2006). Grapevine plants inoculated with *B. phytofirmans* have higher amounts of carbohydrates than control plants. Also, the rate of photosynthesis and starch deposition was improved. Zhang et al. (1997) investigated the impact of various bacterial strains on soybean growth under low temperature condition and found that bacterial stimulation is temperature dependent. *Curvularia protuberata* provides heat tolerance to its host plant *Dichanthelium lanuginosum*. Above 38°C temperature, neither fungus nor plant can survive (Rodriguez et al., 2008).

27.5.5 Flooding Stress

Flooding stress imposes a number of challenges to normal plant functioning. Flooding affects soil nutrient availability and oxygen depletion. *Pterocarpus officinalis* (Jacq.) seedlings inoculated with the arbuscular mycorrhizal (AM) fungus, *Glomus intraradices*, and the strain of *Bradyrhizobium* under flooded condition made remarkable contributions to the flood tolerance of *P. officinalis* seedlings by improving plant growth and phosphorus (P) acquisition in leaves. Flooding induced nodules both on adventitious roots and submerged parts of the stem (Fougnies et al., 2007). *Casuarina equisetifolia* seedlings might be better adapted to flooding than non-inoculated seedlings because of the increased O_2 diffusion and removal of ethanol through adventitious roots (Rutto et al., 2002). A better adaptation of inoculated *Casuarina equisetifolia* seedlings to flooding than non-inoculated seedlings was due to better development of adventitious roots and lenticels, which increased oxygen availability and consequently AM colonization of plants (Osundina, 1998). Plants double inoculated with *G. intraradices* and *Bradyrhizobium* exhibited a reduction of root nodulation as compared to plants singly inoculated with *Bradyrhizobium* in nonflooded conditions, which suggests a competitive interaction between both endophytes (Ruiz-Lozano et al., 2001).

27.6 GENETIC BASIS OF MICROBE MEDIATED ABIOTIC STRESS TOLERANCE

Due to their sessile nature, plants are comparatively sensitive to high levels of abiotic stresses such as drought, salt, extreme temperatures, etc. As a result, they are less able to survive in extreme environments (Alpert, 2000). Conversely, when an association between plant and microbe is established in an extreme habitat, the plant, together with the microbe, is able to survive that environment (Rodriguez et al., 2008). Without the help of microbes, the same plant is not capable of tolerating habitat-imposed abiotic and biotic stresses (Redman et al., 2011). The genetic basis of abiotic stress resistance in microbes is a unique genetic resource to develop abiotic stress resistance of crops. *Paenibacillus polymyxa–*inoculated *Arabidopsis* showed more resistant than control plants against a bacterial pathogen and drought stress. This effect shows a relationship with an increase in the expression of drought stress responsive genes such *ERD15, RAB18* and biotic stress associated genes like *PR-1, HEL, ATVSP* (Timmusk et al., 1999). One of the key stress-signaling components studied in fungi is high-osmolarity glycerol (HOG1) pathway. The HOG pathway is responsible for the increment of intracellular glycerol which provides protection against osmotic stress (Saito and Posas, 2012). In fungi, cell wall integrity (CWI) pathway is involved in sensing abiotic stresses, for instance, osmotic pressure, cell wall damage and alteration of growth temperature (Levin, 2011). Several novel genes have been reported from cyanobacterial gene pool with potential in abiotic stress tolerance which can be further introduced in the genome of economically important crop plants (Singh et al., 2017; Chaurasia et al., 2017; Agrawal et al., 2015, 2017; Shrivastava et al., 2015b, 2016a,b, 2012; Narayan et al., 2010, 2016; Pandey et al., 2013a,b; Chaurasia and Apte, 2009).

27.7 CATEGORIZATION OF BENEFICIAL MICROBE

Microorganisms that are beneficial to the plants and co-evolved with them in either a symbiotic or free-living association are called plant probiotic microorganisms. This association mainly occurs in the soil, but the presence of microalgae-associated bacteria (Gomez et al., 2012) represents the existence of other types of associations that are beneficial to the plant system. Rhizodeposits and root exudates are major sources of soil organic carbon released by plant roots. By releasing root exudates, plants mediate positive as well as negative interactions (Philippot et al., 2013). Beneficial microbes are involved in positive interaction with associated plants whereas parasitic plants, pathogenic microbes and invertebrate herbivores are involved in negative interactions. Positive interactions are basically involved in stimulating plant growth even under biotic or abiotic plant stresses (Mayak et al., 2004a,b; Gomez et al., 2012; Yang et al., 2009). Categorization of beneficial microbes includes plant growth promoting rhizobacteria, plant

disease-suppressive bacteria and fungi, N_2-fixing cyanobacteria, actinomycetes and soil toxicant-degrading microbes, among others (Singh et al., 2011).

27.7.1 PLANT GROWTH PROMOTING RHIZOBACTERIA

Rhizobacteria inhabiting around/on the root surface promote plant growth and are involved directly in modulating plant hormone levels and assisting in resource acquisition (nitrogen, phosphorus and essential minerals), whereas they are involved indirectly by decreasing the inhibitory effects of various pathogens on plant growth and development in the forms of biocontrol agents. As compared to those from bulk soils, rhizobacteria are more versatile in transforming, mobilizing, solubilizing nutrients (Hayat et al., 2010). On the basis of their existence in the rhizosphere, on the rhizoplane or in the spaces between cells of the root cortex, they can be categorized as extracellular (ePGPR) such as, *Agrobacterium, Arthrobacter, Azotobacter, Azospirillum, Bacillus, Burkholderia, Caulobacter, Chromobacterium, Erwinia, Flavobacterium, Micrococcus, Pseudomonas* and *Serratia* etc. (Bhattacharyya and Jha, 2012), and intracellular (iPGPR), existing inside root cells, generally in specialized nodular structures (Figueiredo et al., 2011), like *Allorhizobium, Azorhizobium, Bradyrhizobium, Mesorhizobium* and *Rhizobium.* The majority of rhizobacteria belonging to this group are Gram-negative rods while some belong to Gram-positive rods, cocci or pleomorphic (Bhattacharya and Jha, 2012). The main PGPB source is soil, but some studies have demonstrated the existence of bacteria associated with microalgae that stimulate *Bacillus okhensis* growth (Gomez et al., 2012). Rhizosphere microbial communities display marvelous plant growth beneficial traits due to the presence of actinomycetes as one of the major components (Bhattacharyya and Jha, 2012). Biocontrol agents against different root fungal pathogens such as *Micromonospora* sp., *Streptomyces* spp., *Streptosporangium* sp., and *Thermobifida* sp., have shown an enormous potential and are worth to mention (Bhattacharyya and Jha, 2012). Scientists from all over the world are trying to explore a wide range of rhizobacteria possessing novel traits like heavy metal detoxifying potentials (Ma et al., 2011a; Wani and Khan, 2010), pesticide degradation/tolerance (Ahemad and Khan, 2012a,b), salinity tolerance (Tank and Saraf, 2010; Mayak et al., 2004a), biological control of phytopathogens and insects (Hynes et al., 2008; Russo et al., 2008; Joo et al., 2005; Murphy et al., 2000) along with the normal plant growth promoting properties such as phytohormone (Ahemad and Khan, 2012c), siderophore (Jahanian et al., 2012), 1-aminocyclopropan e-1-carboxylate, hydrogen cyanate (HCN), and ammonia production, nitrogenase activity (Glick, 2012) phosphate solubilization (Ahemad and Khan, 2012c), etc. Hence, diverse symbiotic (*Rhizobium, Bradyrhizobium, Mesorhizobium*) and non-symbiotic (*Pseudomonas, Bacillus, Klebsiella, Azotobacter, Azospirillum, Azomonas*) rhizobacteria are now being used worldwide as bio-inoculants to promote plant growth and development under various stresses like heavy metals (Ma et al., 2011), herbicides (Ahemad and Khan, 2011), insecticides (Ahemad and Khan, 2011), fungicides (Ahemad and Khan, 2011, 2012b), salinity (Mayak et al., 2004a), etc. It is very important to understand the interactions of plant growth promoting rhizobacteria with biotic and abiotic factors in bioremediation, energy generation processes and in biotechnological industries such as pharmaceuticals, food, chemical and mining (Sagar et al., 2012), which are indispensable. Furthermore, plant growth promoting rhizobacteria plays an important role in reducing chemical fertilizer application, as well as being environmentally beneficial for lower production cost and can be helpful in recognizing the best soil and crop management practices to achieve more sustainable agriculture as well as the fertility of soil (Maheshwari et al., 2012).

27.7.2 Vesicular-Arbuscular Mycorrhiza

Vesicular-arbuscular mycorrhiza (VAM) fungi are often used as a biofertilizer as it is cost-effective, energy saving and environmentally friendly. Mycorrhizae, the root-symbionts, are the product of an association between a fungus and plant root. They provide mineral elements like N, P, K, Ca, S and Zn to the host plant and in return obtain their nutrients from the plant. VAM is formed by the symbiotic association between certain phycomycetous fungi and angiosperm roots. Fossil evidence (Remy et al., 1994) and DNA sequence analysis (Simon et al., 1993) suggest that this mutualism appeared 400–460 million years ago. The significance of VAM in augmenting food production is far and wide; therefore, these can be used in sustainable agriculture. Vesicular-arbuscular mycorrhizal fungi belong to the class Zygomycetes, order Endogonales (Benjamin, 1979) and family Endogonaceae. The fungus colonizes the root cortex, forming a mycelial (aseptate or septate) network and characteristic vesicles (bladder-like structures) and arbuscules (branched finger-like hyphae). The arbuscules, having an absorptive function, are the most characteristic structures, formed intracellularly. The vesicles have a storage function and are terminal swellings of hyphae formed inter and intracellularly. There are six genera of fungi belonging to Endogonaceae which have

been shown to form mycorrhizal associations: *Glomus, Gigaspora, Acaulospora, Entrophospora Sclerocystis* and *Scutellospora*, identified by their characteristic spores and sporocarps formed in the soil surrounding the roots. VAM fungi enjoy a very wide host range, including almost all the families of angiosperm species. Even the roots of some aquatic plants are colonized by VAM fungi, but identification of VAM fungi directly from roots has been difficult. This mutually beneficial partnership between plants and soil fungi has existed as long as there have been plants growing in soil. When reintroduced to the soil, the mycorrhiza colonizes the root system, forming a vast network of filaments. This fungal system retains moisture while producing powerful enzymes that naturally unlock mineral nutrients in the soil for natural root absorption. They extract water and nutrients from a large volume of surrounding soil and bring them to the plant, improving nutrition and growth. By supplying the plant with much more extra water and nutrients, mycorrhizal fungi contribute to superior growth, resistance and health. Mycorrhizal fungi are responsible in improving growth of host plant species due to increased nutrient uptake, production of growth promoting substances, tolerance to drought, salinity and synergistic interactions with other beneficial microorganisms (Sreenivasa and Bagyaraj, 1989). Mycorrhizae have great potential for field application to improve productivity of cereal, fruit and vegetable crops and suppress nematode and fungal infestations. Incorporation of microorganisms, such as arbuscular mycorrhizal fungi, to reduce environmental problems associated with excessive pesticide usage has prompted research on the reduction or elimination of pesticides and increasing consumer demands for organic or sustainably-produced food products.

27.7.3 Others

Soil fungi can colonize diverse microhabitats, and their functional diversity can influence pathogen levels and play a significant role in improving plant health, e.g., *Trichoderma* spp. and mycorrhizal fungi (Smith and Read, 2008). Proteobacteria, Firmicutes and Actinobacteria, along with 33,000 bacterial and archaeal species, were consistently associated with disease suppression. Nonribosomal peptide synthetases govern the disease-suppressive activity of γ-Proteobacteria-associated members. Mechanisms contributing to disease suppression by microbial biocontrol agents include antibiosis, resource competition, parasitism and induced resistance in the host (Thomashow and Weller, 1996; Handelsman and Stabb, 1996). On the other hand, the nitrogen-fixing, photosynthetic cyanobacteria (BGA) that are part of the tropical paddy field ecosystem can be a better alternative to agrochemicals, with significant economic and environmental benefits (Blaak et al., 1993). BGA, along with red and brown algae, have also been utilized as a potential biofertilizer. Algal biofertilizers like the BGA, such as *Nostoc sp.*, *Anabaena sp.*, *Tolypothrix sp.*, *Aulosira* sp., etc., have the potential to fix atmospheric nitrogen and are used in paddy fields. Biofertilizers include symbiotic nitrogen fixers like *Rhizobium* spp. associated with leguminous crop plants. The *Azolla-Anabaena* and *Rhizobium* form the most important group of biofertilizers. Similarly, rhizosphere growth of actinomycetes is stimulated by plant root exudates as they are strongly antagonistic to fungal pathogens, which, in turn, utilize root exudates for growth and synthesis of antimicrobial substances (Crawford et al., 1993). At the same time, actinomycetes synthesize an array of biodegradative enzymes, including chitinases (Blaak et al., 1993; Gupta et al., 1995; Mahadevan and Crawford, 1997), glucanases (Damude et al., 1993; Hopwood, 1990; Trejo-Estrada et al., 1998) and peroxidases along with other enzymes that are needed for mycoparasitic activity. In the order of abundance in soils, the common genera of actinomycetes are *Streptomyces* (nearly 70%), *Nocardia* and *Micromonospora*. Actinomycetous genera which are agriculturally and industrially important have been found in only two families of Actinomycetaceae and Streptomycetaceae.

27.8 IMPROVING MECHANISM OF STRESS TOLERANCE OF PLANT VIA MICROBES

This section focuses on understanding the complex defense response mechanism developed by plants mediated by microbial symbionts against environmental stresses viz., drought, water stress and salinity. These microbial symbionts release signaling molecules which trigger signaling cascade as a defense response in plants, which can cause activation of genes involved in multiple physiological, molecular and biochemical pathways and quantitative traits that control different metabolic processes such as carbohydrate metabolism, protein metabolism and hormone metabolism as well as antioxidant defenses (Singh, 2014). Several scientific research articles explained possible mechanisms developed as defense response by plants that includes change in root morphology and shoot characteristics, increase in relative water content, accumulation of osmoprotectants, biosynthesis of scavenging enzymes, modifying the phytohormone contents and enrichment of nutrient uptake (Singh et al., 2013; Elhindi et al., 2017; Kaushal and Wani, 2016; Van Oosten et al., 2017). Figure 27.1

depicts the strategies (stress-specific as well as common) employed by different microbes in providing stress tolerance to associated plants.

27.8.1 MECHANISM OF BACTERIAL-MEDIATED STRESS TOLERANCE

PGPRs include bacteria of diverse genera such as *Arthrobacter*, *Azotobacter*, *Pseudomonas* and *Serratia* (Gray and Smith, 2005) and *Streptomyces* spp. (Dimkpa et al., 2009; Tokala et al., 2002). PGPRs have an excellent root colonizing ability that helps plants to tolerate environmental stresses by developing adaptive defense mechanisms and, consequently, stimulate crop yield and productivity. Some of the mechanisms that plants have developed for adaptation to abiotic stresses are by inoculation with PGPRs through a process called rhizobacterial-induced drought endurance and resilience (RIDER), which includes physiological and biochemical changes, as explained in 27.8.1.1 with case studies.

27.8.1.1 Modifying the Phytohormone Content

Plant growth and development is linked to several phytohormones (auxins, gibberellins, cytokinins, ethylene and abscisic acid) controlling cell elongation, cell division and re-orientation of growth (Farooq et al., 2009). Auxins, especially indole-3-acetic acid (IAA), has a strong influence on plant development and in particular on lateral root formation and branching. The increment in root length and modifications of the root architecture were correlated with increased IAA concentrations in PGPR-treated plants (Contesto et al., 2010). Increased root and shoot biomass and water content under abiotic stress have been reported in clover plants treated with PGPR, which was linked with increased IAA production (Marulanda et al., 2009). Treatment of *Arabidopsis* plants with PGPR *Phyllobacterium brassicacearum* strain STM196 resulted in increased lateral root length and modifications of the root architecture that led to the observed abiotic stress tolerance (Bresson et al., 2014).

Under stressful conditions (including mechanical wounding, chemicals and metals, flooding, extreme temperatures, salinity, pathogen infection and drought), ethylene and ACC deaminase ethylene are synthesized at higher rates (Johnson and Ecker, 1998). 1-Aminocyclopropane-1-carboxylate (ACC) is the immediate precursor of ethylene in higher plants. Reduced ACC levels lead to a reduction in the levels of endogenous ethylene, thus eliminating its inhibitory effect and plants are able to maintain normal growth (Siddikee et al., 2011). Some PGPR contains the enzyme ACC deaminase that hydrolyzes ACC into ammonia and alpha-Ketobutyrate

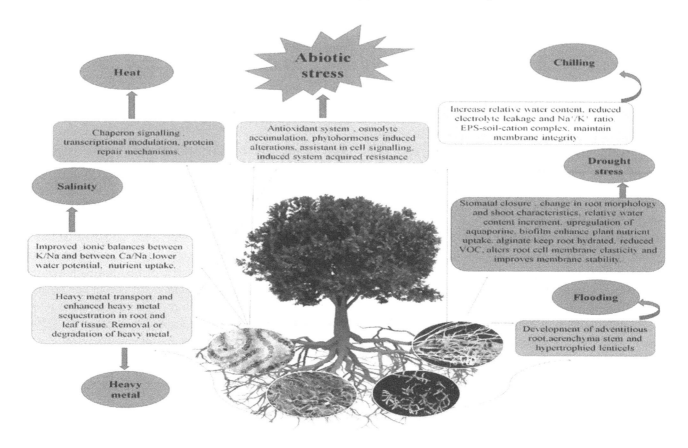

FIGURE 27.1 Different microbes providing different strategies (stress specific as well as common) for providing tolerance against abiotic stress.

(Shaharoona et al., 2006) instead of converting it to ethylene. Studies showed that treatment of pea plants with *Pseudomonas* spp. containing ACC deaminase partially eliminated the effects of drought stress (Arshad et al., 2008). Similarly, treatment of tomato (*Solanum lycopersicum* L.) and pepper (*Capsicum annuum* L.) seedlings with *Achromobacter piechaudii* ARV8 reduced the production of ethylene, which may have contributed to the observed drought tolerance (Mayak et al., 2004b). Lim and Kim (2013) showed that pepper plants treated with PGPR *Bacillus licheniformis* K11 tolerated drought stress and had better survival compared to non-treated plants.

Abscisic acid is crucial for the response to environmental stresses such as drought (Porcel et al., 2014; Cohen et al., 2015). Elevated ABA contents in plant organs under drought stress result in physiological changes that modulate plant growth (Farooq et al., 2009). Studies correlated the observed drought tolerance to the elevated ABA content, for example, lettuce plants treated with *Bacillus* sp. had increased amounts of ABA when compared to non-treated plants (Arkhipova et al., 2007). Similarly, Cohen et al. (2008) reported that *Arabidopsis* plants that were treated with PGPR

Azospirillum brasilense Sp245 had higher ABA content than non-treated plants. Some of the explanations are that ABA enhances drought tolerance via regulation of leaf transpiration and root hydraulic conductivity (Aroca et al., 2006) or via the upregulation of aquaporins (Zhou et al., 2012).

27.8.1.2 Changes in Root Morphology and Shoot Characteristics

Inoculation of various plant species with certain PGPRs enhanced tolerance to abiotic stresses with a positive effect on plant growth as a whole by improved water and nutrient uptake. This is because of the enhancement and alteration of root system topology, increased root growth and/or increased the formation of lateral roots and root hairs. For example, in drought stress, maize plants treated with *Alcaligenes faecalis* (AF3) have increased root length by 10% (Naseem and Bano, 2014), inoculation with PGPR isolate 9 K enhanced root length by 43.3% (Yasmin et al., 2013), and when inoculated with *Burkholderia phytofirmans* strain, PsJN had increased root biomass by 70% and 58% in Mazurka and Kaleo cultivars (Naveed et al., 2014), respectively, as compared to non-inoculated control plants. Wheat plants treated with *Bacillus thuringiensis* AZP2 had two to three times longer root hairs and longer and denser lateral roots; these effects were more pronounced when plants experienced drought stress (Timmusk et al., 2014).

Inhibition of shoot growth is an adaptive response mechanism that helps the plant to tolerate stress, but this will likely decrease plant size and hence limit yield potential (Neumann, 2008; Claeys and Inzé, 2013). However, studies showed that treatment of many crop plants such as sorghum (Grover et al., 2014), sunflower (Castillo et al., 2013), wheat (Arzanesh et al., 2011; Kasim et al., 2013; Timmusk et al., 2014), green gram (Saravanakumar et al., 2011), mung bean (Sarma and Saikia, 2014), maize (Sandhya et al., 2010; Naseem and Bano, 2014; Naveed et al., 2014) and pepper (Lim and Kim, 2013) with selected strains of PGPRs leads to improved shoot growth having higher yield and productivity which help crop plants to tolerate abiotic stress.

27.8.1.3   Increase in Relative Water Content

Several studies have shown that under abiotic stress, PGPR-treated plants maintained relatively higher relative water content (RWC) compared to non-treated plants, leading to the conclusion that PGPR strains that improve survival of plants under drought stress generally increase relative water content in the plants. Casanovas et al. (2002) suggested that the high RWC in maize treated with *Azospirillum brasilense* BR11005 spp. was a result of bacterial abscisic acid that induced stomatal closure and mitigated drought stress. Grover et al. (2014) reported that sorghum plants treated with PGPR *Bacillus* spp. strain KB 129 under drought stress showed 24% increase in RWC over plants that were not treated with PGPR. Similar results have been demonstrated in maize (Sandhya et al., 2010; Vardharajula et al., 2011; Bano et al., 2013; Naveed et al., 2014; Naseem and Bano, 2014). The studies reported above have indicated that higher RWC may help plants counteract the oxidative and osmotic stresses caused by abiotic stress, potentially contributing to greater productivity.

27.8.1.4   Biosynthesis of Scavenging Enzymes

Under stress conditions, the production of reactive oxygen species (ROS) is enhanced, which causes oxidative damage to lipids, proteins and other macromolecules and may ultimately lead to cell death (Farooq et al., 2009; Hasanuzzaman et al., 2014). Major ROS scavenging enzymes include superoxide dismutase (SOD), catalase (CAT), peroxidase (POX), glutathione reductase (GR) and ascorbate peroxidase (APX), which are vital for plant resistance against oxidative damage (Helena and Carvalho, 2008; Simova-Stoilova et al., 2008). Several studies documented that treatment of many plants such as lettuce (Kohler et al., 2010), cucumber (Wang et al., 2012), maize (Vardharajula et al., 2011; Sarma and Saikia, 2014) and wheat (Kasim et al., 2013) with certain strains of PGPR leads to induced/enhanced activity of antioxidant enzymes.

27.8.1.5   Accumulation of Osmoprotectants

Many scientific studies confirmed that plants' resistance mechanism to abiotic stresses (drought and salt) mediated by PGPRs involves accumulation of nitrogen-containing compounds such as glycine, betaine, sugars (e.g., sucrose), polyols (e.g., mannitol), organic acids (e.g., malate), inorganic ions (e.g., calcium) and non-protein amino acids (e.g. proline) (Zhang et al., 2010; Bano et al., 2013). These osmolytes maintain cellular turgor and help plants lower water potential without decreasing actual water content (Serraj and Sinclair, 2002). In many plants, such as maize (Vardharajula et al., 2011; Naseem and Bano, 2014), sorghum (Grover et al., 2014), potato (Gururani et al., 2013), mung bean (Sarma and Saikia, 2014), cucumber (Wang et al., 2012) and *Arabidopsis* (Cohen et al., 2015), it has been demonstrated that treatment with PGPRs leads to an increase in proline levels. Apart from acting as an osmolyte, proline may also function as a scavenger of free radicals, a molecular chaperone to stabilize the structures of protein, enhancing the activity of different enzymes and regulating cellular redox potential (Ashraf and Foolad, 2007; Hayat et al., 2012).

27.8.1.6   Enrichment of Nutrient Uptake

Phosphorous is also a major nutrient for plant growth as it is an integral part of different biochemicals like nucleic acids, nucleotides, phospholipids and phosphoproteins. In most cases, stress, especially salinity, decreased P accumulation in the plant (Rogers et al., 2003). Soil contains a wide range of organic substrates, which can be a source of phosphorus for plant growth. To make this form of P available for plant nutrition, it must be hydrolyzed. The conversion of insoluble phosphate compounds in a form accessible to the plant is an important trait of some of the PGPR genera (such as *Pseudomonas*, *Erwinia*, *Rhizobium* and *Bacillus*) that have phosphatase enzymes (Rodriguez and Fraga, 1999).

27.8.2   Mechanism of Fungal-Mediated Stress Tolerance

The symbiotic relationship between arbuscular mycorrhizal fungi and the roots of higher plants is widespread and plays an important role in nutrient cycling and stress resistance (Al-Karaki and Al-Raddad, 1997). Several studies have demonstrated that AM symbiosis helps plants to cope with adverse abiotic stresses such as water, drought and salinity stress, as demonstrated in a number of host plant and fungal species (Ruiz-Lozano and Aroca, 2010; Sabir et al., 2009; Elhindi et al., 2017). The different mechanisms AMF uses to enhance the drought and salt resistance of host plants include enhancement of nutrient uptake, especially phosphorous (Evelin et al., 2009), which reduces the uptake of sodium and chloride ions and damages their movement to aerial parts (Al-Karaki, 2006; Daei et al., 2009), enhances the uptake of K, Ca, Mg ions, improves ionic balances between K/Na and between Ca/Na (Evelin et al., 2012), enhances water uptake and lowers the leaf water potential to maintain leaf hydration (Ruiz-Lozano and Azcön, 2000; Augé and Moore, 2005), increases the synthesis and activity of scavenging enzymes that leads to enhancement in photosynthetic activity, plant biomass and nutrient status (Baslam and Goicoechea, 2012; Abbaspour et al., 2012), increases photosynthetic efficiency and gas exchange. Other AMF mechanisms help symbiont plants in regulating their osmotic status by increased accumulation of metabolites (proline and sugars), which lowers the osmotic potential and, in turn, leaf water potential and maintains the turgor pressure of leaves in a good state (Ibrahim et al., 2011; Yooyongwech et al., 2013), enhances balance between photosynthesis and transpiration, increases water use efficiency and stomatal conductance (Auge´ et al., 2008; Choe et al., 2006). Moreover, AMF can enhance the physiological processes of the symbiont plant such as water absorption capacity via increasing hydraulic conductivity of roots and regulating the osmotic status, as well as carbohydrate accumulation (Yooyongwech et al., 2013; Baslam and Goicoechea, 2012). Molecular mechanisms activated by AM symbiosis to counteract abiotic stress in plants include gene activation of functional proteins, such as the membrane transporter aquaporins and potentially ion and sugar transporters, in both roots and fungi (Doidy et al., 2012). In addition, AM symbiosis can increase the resistance of plants to drought and salinity through secondary action such as the improvement of soil structural stability.

27.8.3 Mechanism of Algal Mediated Stress Tolerance

Apart from bacterial and fungal species, another group of microorganism is the cyanobacteria that are found symbiotically associated with a range of host plants (Adams et al., 2013), supporting their growth and reproduction even under stress. Cyanobacteria, formerly known as blue-green algae (Cyanophyceae), invade a range of extreme environments of earth and play a significant role in agriculture as they enhance the fertility of the soil by nitrogen fixation (diazotrophs), help in phosphate solubilization and mineral release as well as the synthesis and release of numerous organic substances/metabolites. Various studies on cyanobacteria elucidated their ability to strategically survive in different stressful environments viz., desiccation (Sen et al., 2017; Asthana et al., 2005), salinity (Agrawal et al., 2015; Rai et al., 2013, 2014; Srivastava et al., 2009), UV-B (Shrivastava et al., 2006, 2015; Mishra et al., 2009a,b), heavy metals viz., Cu (Bhargava et al., 2008; Rai and Mallick, 1992), As (Pandey et al., 2012, 2015), Cd (Singh et al., 2015; Atri and Rai, 2013) and herbicides (Agrawal et al., 2014; Kumari et al., 2009). Cyanobacteria produce osmolytes (trehalose, sucrose, glycerol, proline and glycine betaine), antioxidants (glutathione, ascorbic acid and tocopherol), proteins (phytochelatins and exopolysaccharide) and metal-binding proteins. Cyanobacteria provide protection to the plant growth and help them to develop resistance towards biotic and abiotic stress by producing signaling molecules (phytohormones, vitamins, polysaccharides, peptides) that bring changes in genetic expression in their host plants and ultimately synthesis and accumulation of defense response proteins, enzymes, phytochemicals and secondary metabolites (Zaccaro et al., 1991). Cyanobacterial growth promoting effects on other plants such as cotton, several vegetables and herbaceous plants, including *Solanum lycopersicum*, *Cucurbita maxima*, *Cucumis sativus*, *Mentha spicata* and *Satureia hortensis*, has also been reported using cyanobacterial exudates or extracts (Likhitkar and Tarar, 1995; Hashtroudi et al., 2013; Shariatmadari et al., 2013). *Scytonema hofmanni* was found to have a positive effect on the growth of rice seedlings grown in salt stress conditions (Rodriguez et al., 2006).

Cyanobacteria have been extensively used in salt-affected areas for the amelioration of salinity. To cope with salt stress, cyanobacteria employ "salt-out-strategy", that is, to keep the internal ion concentration low and induce the production of compatible solutes to maintain osmotic acclimation. Chatterjee et al. (2017) discussed four strategies of cyanobacteria in ameliorating salinity stress in saline-affected areas. First is extracellular polysaccharides synthesis; second is production and accumulation of osmoregulatory compounds; third is the maintenance of low internal Na$^+$ (either by restricted uptake or efflux); and fourth is induced expression of salt-stress responsive proteins. Sodic soil reclamation required the exchange of sodium ion from sites of clay particles with useful ions such as calcium. *Calothrix braunii*, *Hapalosiphon intricatus*, *Scytonema tolypothricoides*, and *Tolypothrix ceylonica* have been reported to function as gypsum and help in replacement of sodium ion from the root zone of plant and hence protect the plant from the adverse effect of salt stress (Kaushik and Subhashini, 1985).

27.9 ROLE OF BIOSTIMULANTS/ BIOEFFECTORS IN ABIOTIC STRESS TOLERANCE

Microorganisms as biostimulants are involved in stimulating natural processes in plants for better nutrient uptake, abiotic stress tolerance, nutrient efficiency and crop quality. Potential biostimulant bacteria have been isolated from different ecosystems comprising saline, alkaline, acidic and arid soils. Examples of nitrogen-fixing free-living bacterium used as biostimulants include *Azotobacter chroococcum*, which registered its role in salt tolerance (Egamberdiyeva and Höflich, 2003), in *Zea mays* as well as *Triticum aestivum* and temperature tolerance (Egamberdiyeva and Höflich, 2004) in *T. aestivum*. *Azospirillum brasilense*, a plant growth promoting bacteria, provide drought (Pereyra et al., 2006, 2012; Romero et al., 2014) tolerance in *T. aestivum* and salt tolerance in *Cypripedium arietinum* (Hamaoui et al., 2001), *Vicia faba* (Barassi et al., 2006), *Lactuca sativa* (Fasciglione et al., 2015). *Hartmannibacter diazotrophicus* is a phosphate-mobilizing bacterium involved in salt tolerance in *Hordeum vulgare* (Suarez et al., 2015). Symbiotic nitrogen-fixing bacteria such as *Rhizobium leguminosarum* has also registered its role in salt tolerance in V. faba and *Pisum sativum* (Cordovilla et al., 1999). Plant growth promoting proteobacteria *Azospirillum lipoferum* reported providing tolerance against salt stress in *Triticum aestivum* (Bacilio et al., 2004). *Burkholderia phytofirmans* belongs to the ß-Proteobacteria and provides tolerance against cold in *Vitis vinifera* (Fernandez et al., 2012; Theocharis et al., 2012). *Flavobacterium glaciei* psychrophilic obligate aerobic bacterium provided cold tolerance in *Solanum lycopersicum* (Subramanian et al., 2016).

Like bacteria, algae play a crucial role as biostimulants. *Ascophyllum nodosum*, a brown algae involved in cold tolerance in *Kappaphycus alvarezii* (Loureiro et al., 2014), maintain ion homeostasis in *Phyllostachys dulcis* (Saa et al., 2015) and drought tolerance in *Clonorchis sinensis* (Spann and Little, 2011) and is a key player in algal formulations worldwide. One of the most prominently talked about and ubiquitously used biostimulants in the form of commercial formulations are seaweed extracts (SWE) like red, green and brown macroalgae utilized as plant growth promoting factors and involved in providing abiotic stress tolerance to crop plants. Mostly, SWE is derived from *Lithothamnium calcareum*, *Ascophyllum nodosum*, and *Durvillaea potatorum*.

SWE registered its potential role in providing tolerance against different abiotic stresses viz., cold stress in *Arabidopsis thaliana* (Rayirath et al., 2009; Nair et al., 2012), *Z. mays* (Bradáčová et al., 2016), salt stress in *Poa pratensis* (Nabati et al., 1994) and in *L. sativa* (Möller and Smith, 1998), drought stress in *Spinacia oleraceae* (Xu and Leskovar, 2015), *Vitis vinifera* (Mancuso et al., 2006), *Spiraea nipponica* and *Pittosporum eugenioides* (Elansary et al., 2016). Reports suggest that Zn and Mn-rich extracts showed enhanced tolerance under elevated ROS generation. *Azospirillum brasilense* has been reported to provide tolerance against osmotic stress by increasing xylem vessels area and hence higher hydraulic conductance in *T. aestivum* (Pereyra et al., 2012) and *L. esculantum* (Romero et al., 2014). PGPR produced ACC-deaminase to reduce root zone ethylene, thereby maintaining higher root–shoot ratio by plants under water deficit (Nadeem et al., 2010).

Trichoderma spp. and *Glomus* spp. are among the fungal organisms which are used as biostimulants and have registered their role in abiotic stress tolerance vis-a-vis increasing plant productivity and quality (Estrada et al., 2013).

Microorganisms or formulations made from them as biostimulants function in all directions to provide stress tolerance. To maintain ion homeostasis, they cause changes in cell wall composition and accumulate high concentrations of soluble solutes, thereby providing tolerance against osmotic stress. Exopolysaccharides (EPS) and lipopolysaccharides are the known compounds produced by biostimulants which may form a protective biofilm on the root surface and reestablish water potential gradients under water-limiting condition, hence alleviating drought stress (Sandhya et al., 2009) and salinity stress (Paul and Lade, 2014). It has been reported that enhanced levels of IAA is produced by biostimulants to cope with salt stress (Egamberdiyeva, 2009). *Azotobacter* strains used as biostimulants have been shown to facilitate K^+ uptake and Na^+ exclusion together with increasing phosphorous and nitrogen availability in *Z. mays* (Rojas-Tapias et al., 2012). *Azotobacter* strains as biostimulants have also been found to increase biomass, nitrogen content and grain yield under salt stress in *T. aestivum* (Chaudhary et al., 2013).

27.10 TRANSGENIC DEVELOPMENT UTILIZING MICROBIAL GENETIC RESOURCE WITH POTENTIAL IN STRESS TOLERANCE

Microbes are a great source of important genes with potential in abiotic stress tolerance. Such genes have

TABLE 27.1

Transgenics Developed Utilizing Microbial Genetic Resource with Potential in Stress Tolerance

GM Crops	Gene	Microorganism	Stress	Strategy Against Stress	References
Bacteria as gene source					
Nicotiana tabacum	tfdA	Ralstonia eutrophus strain JMP134	2,4 D (herbicide)	detoxification of herbicide	Lyon et al. (1989)
Nicotiana tabacum	gor	Escherichia coli	oxidative Stress	detoxification of ROS	Aono et al. (1991)
Nicotiana tabacum	pcd	Arthrobacter oxydans	phenmedipham (herbicide)	detoxification of herbicide	Streber et al. (1994)
Arabidopsis thaliana	mtlD	Escherichia coli	salt	osmoprotection	Thomas et al. (1995)
Nicotiana tabacum	bet A	Escherichia coli	salt	osmoprotection	Lilius et al. (1996)
Arabidopsis thaliana	merA	Escherichia coli	heavy metal	conversion of toxic Hg^{2+} to relatively inert metallic mercury ($Hg°$).	Rugh et al. (1996)
Arabidopsis thaliana	codA	Arthrobacter globiformis	salt and cold	osmoprotection	Hayashi et al. (1997)
Nicotiana tabacum	otsA and ots B	Escherichia coli	drought	osmoprotection	Pilon-Smits et al. (1998)
Beta vulgaris L.	SacB	Bacillus subtilis	drought	osmoprotection	Pilon-Smits et al. (1999)
Lycopersicon esculentum Mill cv. Heinz 902	acdS	Enterobacter cloacae UW4	heavy metals	metal accumulation, increase metal uptake and root/shoot metal ratio.	Grichko et al. (2000)
Nicotiana tabacum	gdhA	Escherichia coli	herbicide	detoxification of herbicide	Ameziane et al. (2000)
Arabidopsis thaliana, Brassica napus, Nicotiana tabacum	Cox	Arthrobacter pascens	freezing, drought, salt	osmoprotection	Huang et al., (2000)
Nicotiana tabacum	Bar	Stroptomyces hygroscopicus	herbicide tolerance	degrade toxic herbicide into nontoxic form	Lutz et al. (2001)
Nicotiana tabacum	aroA-M1	Escherichia coli	glyphosate (herbicide)	detoxification of herbicide	Wang et al. (2003)
Nicotiana tabacum	pqrA	Ochrobactrum anthropi	herbicide tolerance	increased efflux or decreased uptake of the herbicide	Jo et al. (2004)
Nicotiana tabaccum	ectABC	Halomonas elongata	salt	photosynthesis, osmotic adjustment and nitrogen partitioning	Moghaieb et al. (2006)
Arabidopsis thaliana	proBA	Bacillus subtilis	salt	decreases electrolyte leakage, maintains relative water content.	Chen et al. (2007)
Nicotiana tabacum	katE	Escherichia coli	salt	scavenger of hydrogen peroxide	Al-Taweel et al. (2007)
Maize	CspB	Bacillus subtilis	Water deficit	act as RNA chaperon	Castiglioni et al. (2008)
Petunia hybrida	ACC deaminse and iaaM	Pseudomonas putida and Agrobacterium tumefaciens	heavy metal	heavy metal transport and enhances heavy metal sequestration in root and leaf tissue.	Zhang et al. (2008)
Oryza sativa	katE	Escherichia coli	Salt	modulates the ROS scavenging system	Nagamiya et al. (2007); Motohashi et al. (2010)

(Continued)

TABLE 27.1 (CONTINUED)

Transgenics Developed Utilizing Microbial Genetic Resource with Potential in Stress Tolerance

GM Crops	Gene	Microorganism	Stress	Strategy Against Stress	References
Cucumis sativus	*cAPX, rbcL, rbcS,*	*Bacillus cereus, Bacillus subtilis, Serratia* **spp**	drought	increases antioxidative activity, root morphology change	Wang et al. (2012)
Capsicum annuum	*Cadhn, VA, sHSP* and *CaPR-10*	*Bacillus licheniformis* **K 11**	drought	reduces ethylene production	Lim, and Kim (2013)
Saccharum officinarum	*betA*	*Rhizobium meliloti*	drought	osmoprotection	Waltz (2014)
Fungi as gene source					
Oryza sativa	*Mn-SOD*	*Saccharomyces cerevisiae*	Salt	ROS scavenging	Tanaka et al. (1999)
Nicotiana tabacum	*chit33* and *chit42*	*Trichoderma harzianum*	salinity, and heavy metals	protection against oxidative damage	Dana et al. (2006)
Nicotiana tabacum	*TPS1* and *TPS2*	*Saccharomyces cerevisiae*	drought	osmoprotection	Karim et.al. (2007)
Lycopersicon esculentum Mill.	*SAMDC*	*Saccharomyces cerevisiae*	drought	antioxidant and protects membrane lipid peroxidation	Cheng et al. (2009)
Arabidopsis thaliana	*AgglpF*	*Aspergillus glaucus*	osmotic	transport glycerol along with water and other uncharged solutes that help in osmoregulation	Liu et al. (2015)
Arabidopsis thaliana	*CtHSR1*	*Candida tropicalis* (yeast)	drought	reduced transpiration and increased osmotic potential (proline accumulation)	Martínez et al. (2015)
Arabidopsis thaliana	*hsp70*	*Trichoderma harzianum*	heat, osmotic, salt and oxidative stress	accumulate high amount of heat-shock protein (HSP70) and increase APX1 and SOS1 transcript	Montero-Barrientos et al. (2010)
Nicotiana tabacum	*TvGST*	*Trichoderma virens*	Cd tolerance	lowers level of lipid peroxidation, enhanced antioxidant enzyme activity, lowers Cd accumulation	Dixit et al. (2011)
Nicotiana tabacum	*ThAQGP*	*Trichoderma harzianum*	drought	Increased relative water content, reduced transpiration	Vieira et al. (2017)
Algae as gene source					
Nicotiana tabacum	*desC*	*Anacystis nidulans*	chilling	maintain the fluidity of plasma membrane by increasing polyunsaturated fatty acid	Ishizaki-Nishizawa et al. (1996)
Nicotiana tabacum	*dnaK1*	*Aphanothece halophytica*	salt	maintains growth and photosynthetic activity	Sugino et al. (1999)

(Continued)

TABLE 27.1 (CONTINUED)

Transgenics Developed Utilizing Microbial Genetic Resource with Potential in Stress Tolerance

GM Crops	Gene	Microorganism	Stress	Strategy Against Stress	References
Nicotiana tabacum	*pds*	*Synechococcus PCC 7942 mutant NFZ4*	herbicide	maintain the D1 protein of PSII so less prone to photooxidative damage by herbicide	Wagner et al. (2002)
Nicotiana tabacum	*desC*	*Synechococcus vulcanus*	chilling	increase in amount of polyunsaturated fatty acids in the membrane lipids which maintain the plasma membrane fluidity.	Orlova et al., (2003)
Nicotiana tabacum	*TcGPX* and *TpGPX*	*Chlamydomonas* W80 strain	salt, chilling, oxidative stress	removal of unsaturated fatty acid hydroperoxides from cell membranes, maintenance of membrane integrity	Yoshimura et.al. (2004)
Arabidopsis thaliana	*ApGSMT* and *ApDMT*	*Aphanothece halophytica*	salt	osmoprotection	Waditee et al. (2004)
Nicotiana tabacum	*fld*	*Anabaena PCC 7119*	herbicides, extreme temperatures, high irradiation, water deficit, and UV radiation	preserving the active state of redox sensitive photosynthetic and scavenging enzymes under harmful conditions	Tognetti et al. (2006)
Arabidopsis thaliana	*GPX-2*	*Synechocystis PCC 6803*	oxidative Stress	scavenger of hydrogen peroxide	Gaber et al. (2006)
Solanum tuberosum	*desA*	*Synechocystis sp. PCC 6803*	chilling	maintains fluidity of plasma membrane	Amiri et al. (2010)
Arabidopsis thaliana	*wspa1*	*Nostoc commune*	osmotic stress	Reduction in relative electrolyte leakage, lowers lipid peroxidation, increament in proline accumulation, enhancement of quantum efficiency of Photosystem II	Ai et al. (2014)

been incorporated into higher plants for the improvement of many traits, including abiotic stress tolerance. Stress tolerance genes have been isolated from fungi, bacteria, as well as algae, and incorporated into a large array of higher plants. Transgenics developed following microbial gene incorporation in plants have shown better tolerance compared to wild-type under different stressful conditions. Table 27.1 includes microbes with their putative stress-responsive genes, their strategy to cope

with the stressful condition and the corresponding plant in which the gene has been incorporated.

27.11 CONCLUSION AND FUTURE PERSPECTIVE

After several years of research on plant–microbe interaction, the vast amounts of data generated have shaped a better understanding of beneficial microbes and their mechanism in providing abiotic stress tolerance to economically important crops. Such findings have opened an array of opportunities for applications of beneficial microbes in both agriculture and biotechnology and hence exploiting their ability to alter plant metabolism and provide resistance to abiotic stresses. Different techniques should be used in the identification of novel biostimulants, with a broad and narrow spectrum of specificity regarding the host as well as different abiotic stresses. Furthermore, a good understanding of stress tolerance mechanism should be achieved in terms of generating the next generation of biostimulants with the functionally designed mechanism as per applicability. The present chapter describes in brief novel genes with potential in abiotic stress tolerance and the significant pathways in plant-associated microbes involved in providing stress tolerance. By utilizing different omics approaches, more novel pathways should be revealed from different microbes associated with plants for their proper understanding, utilization and manipulation for a maximum benefit under the adversity of abiotic stresses.

ACKNOWLEDGMENTS

L.C. Rai is thankful to CSIR and ICAR for financial support and DST and DAE for J.C. Bose National Fellowship and Raja Ramanna Fellowship, respectively. Antra Chatterjee thanks UGC for SRF. Alka Shankar thanks DST for N-PDF fellowship. Ruchi Rai and Shilpi Singh are thankful to DST for WOSA award. Vigya Kesari is thankful to NASI for RA fellowship. Amit Kumar Patel is thankful to CSIR-UGC for JRF. We thank Head and Coordinator, CAS for facilities through FIST and PURSE support.

REFERENCES

Abbaspour, H., Saeid-Sar, S., Afshari, H., Abdel-Wahhab, M.A. (2012) Tolerance of mycorrhiza infected pistachio (*Pistacia vera* L.) seedlings to drought stress under glasshouse conditions. *Journal of Plant Physiology* 169(7):704–709.

Adams, D.G., Bergman, B., Nierzwicki-Bauer, S.A., Duggan, P.S., Rai, A.N., Schussler, A. (2013) Cyanobacterial plant symbioses. In *The prokaryotes*, ed Dworkin, M., Falkow, S., Rosenberg, E., Schleifer, K. and Stackebrandt, E. 359–400. Heidelberg–Berlin, Germany: Springer.

Agrawal, C., Sen, S., Chatterjee, A., Rai, S., Yadav, S., Singh, S., Rai, L.C. (2015a) Signal perception and mechanism of salt toxicity/tolerance in photosynthetic organisms: Cyanobacteria to plants. In *Stress responses in plants*, ed Tripathi, B.N. and Muller, M. 79–113, Cham, Switzerland: Springer.

Agrawal, C., Sen, S., Singh, S., Rai, S., Singh, P.K., Singh, V.K., Rai, L.C. (2014) Comparative proteomics reveals association of early accumulated proteins in conferring butachlor tolerance in three N_2-fixing *Anabaena* spp. *Journal of Proteomics* 96:271–290.

Agrawal, C., Sen, S., Yadav, S., Rai, S., Rai, L.C. (2015b) A novel aldo-keto reductase (AKR17A1) of *Anabaena* sp. PCC 7120 degrades the rice field herbicide butachlor and confers abiotic stress tolerance in *E. coli*. *PLOS ONE* 10(9):e0137744.

Agrawal, C., Yadav, S., Rai, S., Chatterjee, A., Sen, S., Rai, R., Rai, L.C. (2017) Identification and functional characterization of four novel aldo/keto reductases in Anabaena sp. PCC 7120 by integrating wet lab with in silico approaches. *Functional and Integrative Genomics* 17(4):413–425.

Ahemad, M., Khan, M.S. (2011) Response of green gram [*Vigna radiata* (L.) Wilczek] grown in herbicide-amended soil to quizalafop-p-ethyl and clodinafop tolerant plant growth promoting *Bradyrhizobium* sp. (vigna) MRM6. *Journal of Agricultural Science and Technology* 13:1209–1222.

Ahemad, M., Khan, M.S. (2012a) Ecological assessment of biotoxicity of pesticides towards plant growth promoting activities of pea (*Pisum sativum*) specific *Rhizobium* sp. strain MRP1. *Emirates Journal of Food Agriculture* 24:334–343.

Ahemad, M., Khan, M.S. (2012b) Effect of fungicides on plant growth promoting activities of phosphate solubilizing *Pseudomonas putida* isolated from mustard (*Brassica compestris*) rhizosphere. *Chemosphere* 86(9): 945–950.

Ahemad, M., Khan, M.S. (2012c) Evaluation of plant growth promoting activities of rhizobacterium *Pseudomonas putida* under herbicide-stress. *Annals of Microbiology* 62(4):1531–1540.

Ai, Y., Yang, Y., Qiu, B., Gao, X. (2014) Unique WSPA protein from terrestrial macroscopic cyanobacteria can confer resistance to osmotic stress in transgenic plants. *World Journal of Microbiology and Biotechnology* 30(9):2361–2369.

Akpinar, B.A., Kantar, M., Budak, H. (2015) Root precursors of microRNAs in wild emmer and modern wheats show major differences in response to drought stress. *Functional and Integrative Genomics* 15(5):587–598.

Alavi, P., Starcher, M.R., Zachow, C., Müller, H., Berg, G. (2013) Root microbe systems: The effect and mode of interaction of stress protecting agent (SPA) *Stenotrophomonas rhizophila* DSM14405T. *Frontier of Plant Science* 4:141.

Al-Karaki, G.N. (2006) Nursery inoculation of tomato with arbuscular mycorrhizal fungi and subsequent performance under irrigation with saline water. *Scientia Horticulturae* 109(1):1–7.

Al-Karaki, G.N., Al-Raddad, A. (1997) Effect of arbuscular mycorrhizal fungi and drought stress on growth and nutrient uptake of two wheat genotypes differing in drought resistance. *Mycorrhiza* 7(2):83–88.

Alpert, P. (2000) The discovery, scope, and puzzle of desiccation tolerance in plants. *Plant Ecology* 151(1):5–17.

Al-Taweel, K., Iwaki, T., Yabuta, Y., Shigeoka, S., Murata, N., Wadano, A. (2007) A bacterial transgene for catalase protects translation of d1 protein during exposure of salt-stressed tobacco leaves to strong light. *Plant Physiology* 145(1):258–265.

Ameziane, R., Bernhard, K., Lightfoot, D. (2000) Expression of the bacterial gdhA gene encoding a NADPH glutamate dehydrogenase in tobacco affects plant growth and development Lightfoot, David, L. *Plant and Soil* 221(1):47–57.

Amiri, R.M., Yur'eva, N.O., Shimshilashvili, K.R., Goldenkova-Pavlova, I.V., Pchelkin, V.P., Kuznitsova, E.I., Tsydendambaev, V.D., et al. (2010) Expression of acyl-lipid Δ12-desaturase gene in prokaryotic and eukaryotic cells and its effect on cold stress tolerance of potato. *Journal of Integrative Plant Biology* 52(3):289–297.

Aono, M., Kubo, A., Saji, H., Natori, T., Tanaka, K., Kondo, N. (1991) Resistance to active oxygen toxicity of transgenic *Nicotiana tabacum* that expresses the gene for glutathione reductase from *Escherichia coli*. *Plant and Cell Physiology* 32(5):691–697.

Arkhipova, T.N., Prinsen, E., Veselov, S.U., Martinenko, E.V., Melentiev, A.I., Kudoyarova, G.R. (2007) Cytokinin producing bacteria enhance plant growth in drying soil. *Plant and Soil* 292(1–2):305–315.

Aroca, R., Ferrante, A., Vernieri, P., Chrispeels, M.J. (2006) Drought, abscisic acid and transpiration rate effects on the regulation of PIP aquaporin gene expression and abundance in *Phaseolus vulgaris* plants. *Annals of Botany* 98(6):1301–1310.

Arshad, M., Shaharoona, B., Mahmood, T. (2008) Inoculation with *Pseudomonas* spp. containing ACC-deaminase partially eliminates the effects of drought stress on growth, yield, and ripening of pea (*Pisum sativum* L.). *Pedosphere* 18(5):611–620.

Arzanesh, M.H., Alikhani, H.A., Khavazi, K., Rahimian, H.A., Miransari, M. (2011) Wheat (*Triticum aestivum* L.) growth enhancement by *Azospirillum* sp. under drought stress. *World Journal of Microbiology and Biotechnology* 27(2):197–205.

Ashraf, M., Foolad, M.R. (2007) Roles of glycine betaine and proline in improving plant abiotic stress resistance. *Environmental and Experimental Botany* 59(2):206–216.

Ashraf, M., Hasnain, S., Berge, O., Mahmood, T. (2004) Inoculating wheat seedlings with exopolysaccharide-producing bacteria restricts sodium uptake and stimulates plant growth under salt stress. *Biology and Fertility of Soils* 40(3):157–162.

Asthana, R.K., Srivastava, S., Singh, A.P., Kayastha, A.M., Singh, S.P. (2005) Identification of maltooligosyltrehalose synthase and maltooligosyltrehalose trehalohydrolase enzymes catalysing trehalose biosynthesis in *Anabaena* 7120 exposed to NaCl stress. *Journal of Plant Physiology* 162(9):1030–1037.

Atri, N., Rai, L.C. (2013) Differential responses of three cyanobacteria to UV-B and Cd. *Journal of Microbiology and Biotechnology* 13(4):544–551.

Augé, R.M., Toler, H.D., Sams, C.E., Nasim, G. (2008) Hydraulic conductance and water potential gradients in squash leaves showing mycorrhiza-induced increases in stomatal conductance. *Mycorrhiza* 18(3):115–121.

Augé, R.M., Moore, J.L. (2005) Arbuscular mycorrhizal symbiosis and plant drought resistance. In *Mycorrhiza: Role and applications*, ed. Mehrotra, V.S. 136–157. New Delhi, India: Allied Publishers Limited.

Azcon, R., Medina, A., Aroca, R., Ruiz-Lozano, J.M. (2013) Abiotic stress remediation by the arbuscular mycorrhizal symbiosis and rhizosphere bacteria/yeast interactions. In *Molecular microbial ecology of the rhizosphere*, Hoboken, ed. de Brujin, F.J. 991–1002. New Jersey: John Wiley & Sons Inc.

Bacilio, M., Rodriguez, H., Moreno, M., Hernandez, J.P., Bashan, Y. (2004) Mitigation of salt stress in wheat seedlings by a gfp-tagged *Azospirillum lipoferum*. *Biology and Fertility of Soils* 40(3):188–193.

Bae, H., Sicher, R.C., Kim, M.S., Kim, S.H., Strem, M.D., Melnick, R.L., Bailey, B.A. (2009) The beneficial endophyte *Trichoderma hamatum* isolate DIS 219b promotes growth and delays the onset of the drought response in Theobroma cacao. *Journal of Experimental Botany* 60(11):3279–3295.

Bahadur, A., Chatterjee, A., Kumar, R., Singh, M., Naik, P.S. (2011) Physiological and biochemical basis of drought tolerance in vegetables. *Vegetable Science* 38(1):1–16.

Bais, H.P., Weir, T.L., Perry, L.G., Gilroy, S., Vivanco, J.M. (2006) The role of root exudates in rhizosphere interactions with plants and other organisms. *Annual Review of Plant Biology* 57:233–266.

Bano, Q., Ilyas, N., Bano, A., Zafar, N., Akram, A., Hassan, F. (2013) Effect of *Azospirillum* inoculation on maize (*Zea mays* L.) under drought stress. *Pakistan Journal of Botany* 45:13–20.

Barassi, C.A., Ayrault, G., Creus, C.M., Sueldo, R.J., Sobrero, M.T. (2006) Seed inoculation with *Azospirillum* mitigates NaCl effects on lettuce. *Scientia Horticulturae* 109(1):8–14.

Barka, E.A., Nowak, J., Clement, C. (2006) Enhancement of chilling resistance of inoculated grapevine plantlets with a plant growth-promoting rhizobacterium, *Burkholderia phytofirmans* strain PsJN. *Applied and Environmental Microbiology* 72(11):7246–7252.

Baslam, M., Goicoechea, N. (2012) Water deficit improved the capacity of arbuscular mycorrhizal fungi (AMF) for inducing the accumulation of antioxidant compounds in lettuce leaves. *Mycorrhiza* 22(5):347–359.

Belimov, A.A., Safronova, V.I., Sergeyeva, T.A., Egorova, T.,Matveyeva, N., Tsyganov, V.A., Borisov, V.E., et al. (2001) Characterization of plant growth promoting rhizobacteria isolated from polluted soils and containing 1- aminocyclopropane-1- carboxylate deaminase. *Canadian Journal of Microbiology* 47(7): 642–652.

Belimov, A.A., Dietz, K.J. (2000) Effect of associative bacteria on element composition of barley seedlings Grown in solution culture at toxic cadmium concentrations. *Microbiological Research* 155(2):113–121.

Benaroudj, N., Lee, D.H., Goldberg, A.L. (2001) Trehalose accumulation during cellular stress protects cells and cellular proteins from damage by oxygen radicals. *Journal of Biology and Chemistry* 276:24261–24267.

Benjamin, R.K. (1979) Zygomycetes and their spores. In *The whole fungus: The sexual asexual synthesis*, ed. Kendrick, B. 573–622. Ottawa, Canada: National Museums of Natural Sciences.

Bensalim, S., Nowak, J., Asiedu, S.K. (1998) A plant growth promoting rhizobacterium and temperature effects on performance of 18 clones of potato. *American Journal of Potato Research* 75(3):145–152.

Bhargava, P., Mishra, Y., Srivastava, A.K., Narayan, O.P., Rai, L.C. (2008) Excess copper induces anoxygenic photosynthesis in *Anabaena doliolum*: A homology based proteomic assessment of its survival strategy. *Photosynthesis Research* 96(1):61–74.

Bhattacharyya, P.N., Jha, D.K. (2012) Plant growth-promoting rhizobacteria (PGPR): Emergence in agriculture. *World Journal of Microbiology and Biotechnology* 28(4):1327–1350.

Blaak, H., Schnellmann, J., Walter, S., Henrissat, B., Schrempf, H. (1993) Characteristics of an exochitinase from *Streptomyces olivaceoviridis*, its corresponding gene, putative protein domains and relationship to other chitinases. *European Journal of Biochemistry* 214(3):659–669.

Boller, T., Felix, G. (2009) A renaissance of elicitors: Perception of microbe-associated molecular patterns and danger signals by pattern-recognition receptors. *Annual Review of Plant Biology* 60(1):379–406.

Bradáčová, K., Weber, N.F., Morad-Talab, N., Asim, M., Imran, M., Weinmann, M., Neumann, G. (2016) Micronutrients (Zn/Mn), seaweed extracts, and plant growth-promoting bacteria as cold-stress protectants in maize. *Chemical and Biological Technologies in Agriculture* 3(1):19.

Bresson, J., Vasseur, F., Dauzat, M., Labadie, M., Varoquax, F., Touraine, B., Vile, D. (2014) Interact to survive: *Phyllobacterium brassicacearum* improves Arabidopsis tolerance to severe water deficit and growth recovery. *PLoS One* 9(9):e107607.

Burd, G.I., Dixonand, D.G., Glick, B.R. (1998) A plant growth promoting bacterium that decreases nickel toxicity in seedlings. *Applied and Environmental Microbiology* 64(10):3663–3668.

Busse, M.D., Ellis, J.R. (1985) Vesicular-arbuscular mycorrhizal (*Glomus fasciculatum*) influence on soybean drought tolerance in high phosphorus soil. *Canadian Journal of Botany* 63(12):2290–2294.

Cameron, D.D., Neal, A.L., VanWees, S.C.M., Ton, J. (2013) Mycorrhiza-induced resistance: More than the sum of its parts? *Trends in Plant Science* 18(10):539–545.

Carrillo-Castaneda, G., Munoz, J.J., Peralta-Videa, J.R., Gomez, E., Gardea-Torresdey, J.L. (2003) Plant growth-promoting bacteria promote copper and iron translocation from root to shoot in alfalfa seedlings. *Journal of Plant Nutrition* 26(9):1801–1814.

Casanovas, E.M., Barassi, C.A., Sueldo, R.J. (2002) *Azospirillum* inoculation mitigates water stress effects in maize seedlings. *Cereal Research Communications* 30(3):343–350.

Castiglioni, P., Warner, D., Bensen, R.J., Anstrom, D.C., Harrison, J., Stoecker, M., Abad, M., et al. (2008) Bacterial RNA chaperones confer abiotic stress tolerance in plants and improved grain yield in maize under water limited conditions. *Plant Physiology* 147(2):446–455.

Castillo, P., Escalante, M., Gallardo, M., Alemano, S., Abdala, G. (2013) Effects of bacterial single inoculation and co-inoculation on growth and phytohormone production of sunflower seedlings under water stress. *Acta Physiologiae Plantarum* 35(7):2299–2309.

Chatterjee, A., Pal, R., Pal, S., Kundu, R. (2008) Morphological and physiological responses of mung bean (*Vigna radiata* var. radiata (L.) Wilczek) seedlings under lead stress. *Journal of Botanical Society of Bengal* 62(1): 55–60.

Chatterjee, A., Singh, S., Agarwal, C., Yadav, S., Rai, R., Rai, L. (2017) Role of algae as biofertilizer. In *Algal green chemistry: Recent progress in biotechnology*, ed. Rastogi, R.P., Madamwar, D., Pandey A. 189–200. Amsterdam, the Netherlands: Elsevier.

Chatterjee, A., Solankey, S.S. (2015) Functional physiology in drought tolerance of vegetable crops-an approach to mitigate climate change impact. In *climate change: The principles and applications in horticultural science. USA: Apple academic press*, ed. Chaudhary, M.L., Patel, V.B., Siddiqui, M.W., Mahdi, S.S. 149–171. Boca Raton, FL: CRC Press.

Chaudhary, D., Narula, N., Sindhu, S.S., Behl, R.K. (2013) Plant growth stimulation of wheat (*Triticum aestivum* L.) by inoculation of salinity tolerant *Azotobacter* strains. *Physiology and Molecular Biology of Plants: an International Journal of Functional Plant Biology* 19(4):515–519.

Chaurasia, A.K., Apte, S.K. (2009) Overexpression of the groESL Operon enhances the heat and salinity stress tolerance of the nitrogen-fixing cyanobacterium *Anabaena* sp. strain PCC7120. *Applied and Environmental Microbiology* 75(18):6008–6012.

Chaurasia, N., Mishra, Y., Chatterjee, A., Rai, R., Yadav, S., Rai, L.C. (2017) Overexpression of phytochelatin synthase (*pcs*) enhances abiotic stress tolerance by altering the proteome of transformed *Anabaena* sp. PCC 7120. *Protoplasma* 254(4):1715–1724.

Chen, J., Han, G., Shang, C., Li, J., Zhang, H., Liu, F., Wang, J., Liu, H., Zhang, Y. (2015) Proteomic analyses reveal differences in cold acclimation mechanisms in freezing-tolerant and freezing-sensitive cultivars of alfalfa. *Frontiers in Plant Science* 6:105.

Chen, M., Wei, H., Cao, J., Liu, R., Wang, Y., Zheng, C. (2007) Expression of *Bacillus subtilis* proAB genes and reduction of feedback inhibition of proline synthesis increases proline production and confers osmotolerance in transgenic *Arabdopsis*. *Journal of Biochemistry and Molecular Biology* 40(3):396–403.

Cheng, L., Zou, Y.J., Ding, S.L., Zhang, J.J., Yu, X.L., Cao, J.S., Lu, G. (2009) Polyamine accumulation in transgenic tomato enhances the tolerance to high temperature stress. *Journal of Integrative Plant Biology* 51(5):489–499.

Choe, K., Toler, H., Lee, J., Ownley, B., Stutz, J.C., Moore, J.L., Auge, R.M. (2006) Mycorrhizal symbiosis and responses of sorghum plants to combined drought and salinity stresses. *Journal of Plant Physiology* 163(5):517–528.

Claeys, H., Inzé, D. (2013) The agony of choice: How plants balance growth and survival under water-limiting conditions. *Plant Physiology* 162(4):1768–1779.

Cohen, A.C., Bottini, R., Piccoli, P.N. (2008) *Azosprillium brasilense* Sp 245 produces ABA in chemically defined culture medium and increases ABA content in *Arabidopsis* plants. *Plant Growth Regulation* 54(2):97–103.

Cohen, A.C., Bottini, R., Pontin, M., Berli, F.J., Moreno, D., Boccanlandro, H., Travaglia, C.N., Picocoli, P.N. (2015) *Azospirillum brasilense* ameliorates the response of *Arabidopsis thaliana* to drought mainly via enhancement of ABA levels. *Physiologia Plantarum* 153(1):79–90.

Contesto, C., Milesi, S., Mantelin, S., Zancarini, A., Desbrosses, G., Varoquaux, F., Bellini, C., Kowalczyk, M., Touraine, B. (2010) The auxin-signaling pathway is required for the lateral root response of *Arabidopsis* to the rhizobacterium *Phyllobacterium brassicacearum*. *Planta* 232(6):1455–1470.

Contreras-Cornejo, H.A., Macías-Rodríguez, L., Cortés-Penagos, C., LópezBucio, J. (2009) *Trichoderma virens*, a plant beneficial fungus, enhances biomass production and promotes lateral root growth through an auxindependent mechanism in *Arabidopsis*. *Plant Physiology* 149(3):1579–1592.

Cordovilla, M.D.P., Berrido, S.I., Ligero, F., Lluch, C. (1999) *Rhizobium* strain effects on the growth and nitrogen assimilation in *Pisum sativum* and *Vicia faba* plant growth under salt stress. *Journal of Plant Physiology* 154(1):127–131.

Crawford, D.L., Lynch, J.M., Whipps, J.M., Ousley, M.A. (1993) Isolation and characterization of actinomycete antagonists of a fungal root pathogen. *Applied and Environmental Microbiology* 59(11):3899–3905.

Daei, G., Ardekani, M.R., Rejali, F., Teimuri, S., Miransari, M. (2009) Alleviation of salinity stress on wheat yield, yield components, and nutrient uptake using arbuscular mycorrhizal fungi under field conditions. *Journal of Plant Physiology* 166(6):617–625.

Damude, H.G., Gilkes, N.R., Kilburn, D.G., Miller, R.C., Warren Jr.,R.A. (1993) Endoglucanase CasA from alkalophilic *Streptomyces* strain KSM-9 is a typical member of family B of beta-1,4-glucanases. *Gene* 123(1):105–107.

Dana, M.M., Pintor-Toro, J.A., Cubero, B. (2006) Transgenic tobacco plants overexpressing chitinases of fungal

origin show enhanced resistance to biotic and abiotic stress agents. *Plant Physiology* 142(2):722–730.

Dimkpa, C.O., Merten, D., Svatoš, A., Büchel, G., Kothe, E. (2009) Siderophores mediate reduced and increased uptake of cadmium by *Streptomyces tendae* F4 and sunflower (*Helianthus annuus*), respectively. *Journal of Applied Microbiology* 107(5):1687–1696.

Dixit, P., Mukherjee, P.K., Ramachandran, V., Eapen, S. (2011) Glutathione transferase from *Trichoderma virens* enhances cadmium tolerance without enhancing its accumulation in transgenic *Nicotiana tabacum*. *PLOS ONE* 6(1):e16360.

Doidy, J., Grace, E., Kühn, C., Simon-Plas, F., Casieri, L., Wipf, D. (2012) Sugar transporters in plants and in their interactions with fungi. *Trends in Plant Science* 17(7):413–422.

Dong, X. (2001) Genetic dissection of systemic acquired resistance. *Current Opinion in Plant Biology* 4(4):309–314.

Egamberdiyeva, D. (2009) Alleviation of salt stress by plant growth regulators and IAA producing bacteria in wheat. *Acta Physiologiae Plantarum* 31(4):861–864.

Egamberdiyeva, D., Höflich, G. (2003) Influence of growth-promoting bacteria on the growth of wheat in different soils and temperatures. *Soil Biology and Biochemistry* 35(7):973–978.

Egamberdiyeva, D., Höflich, G. (2004) Effect of plant growth-promoting bacteria on growth and nutrient uptake of cotton and pea in a semi-arid region of Uzbekistan. *Journal of Arid Environments* 56(2):293–301.

Elansary, H.O., Skalicka-Woźniak, K., King, I.W. (2016) Enhancing stress growth traits as well as phytochemical and antioxidant contents of *Spiraea* and *pittosporum* under seaweed extract treatments. *Plant Physiology and Biochemistry: PPB* 105:310–320.

Elhindi, K.M., El-Din, A.S., Elgorban, A.M. (2017) The impact of arbuscular mycorrhizal fungi in mitigating salt-induced adverse effects in sweet basil (*Ocimum basilicum* L.). *Saudi Journal of Biological Sciences* 24(1):170–179.

Estrada, B., Barea, J.M., Aroca, R., Ruiz-Lozano, J.M. (2013) A native *Glomus intraradices* strain from a Mediterranean saline area exhibits salt tolerance and enhanced symbiotic efficiency with maize plants under salt stress conditions. *Plant and Soil* 366(1–2): 333–349.

Evelin, H., Giri, B., Kapoor, R. (2012) Contribution of *Glomus intraradices* inoculation to nutrient acquisition and mitigation of ionic imbalance in NaCl-stressed *Trigonella graecum*. *Mycorrhiza* 22(3):203–217.

Evelin, H., Kapoor, R., Giri, B. (2009) Arbuscular mycorrhizal fungi in alleviation of salt stress: A review. *Annals of Botany* 104(7):1263–1280.

Farooq, M., Wahid, A., Kobayashi, N., Fujita, D., Basra, S.M.A. (2009) Plant drought stress: Effects, mechanisms and management. *Agronomy for Sustainable Development* 29(1):185–212.

Farrar, K., Bryant, D., Cope-Selby, N. (2014) Understanding and engineering beneficial plant–microbe interactions: Plant growth promotion in energy crops. *Plant Biotechnology Journal* 12(9):1193–1206.

Fasciglione, G., Casanovas, E.M., Quillehauquy, V., Yommi, A.K., Goñi, M.G., Roura, S.I., Barassi, C.A. (2015) *Azospirillum* inoculation effects on growth, product quality and storage life of lettuce plants grown under salt stress. *Scientia Horticulturae* 195:154–162.

Fernandez, O., Theocharis, A., Bordiec, S., Feil, R., Jacquens, L., Clément, C. (2012) *Burkholderia phytofirmans* PsJN acclimates grapevine to cold by modulating carbohydrate metabolism. *Molecular Plant–Microbe Interactions*. 25(4):496–504.

Figueiredo, M.V.B., Burity, H.A., Martínez, C.R., Chanway, C.P. (2008) Alleviation of drought stress in the common bean (*Phaseolus vulgaris* L.) by co-inoculation with *Paenibacillus polymyxa* and *Rhizobium tropici*. *Applied Soil Ecology* 40(1):182–188.

Figueiredo, M.V.B., Seldin, L., Araujo, F.F., Mariano, R.L.R. (2011) Plant growth promoting rhizobacteria: Fundamentals and applications. In *Plant growth and health promoting bacteria*, ed. Maheshwari, D.K. 21–43. Berlin–Heidelberg, Germany: Springer-Verlag.

Fougnies, L., Renciot, S., Mulle, F., Plenchette, C., Prin, Y., de Faria, S.M., Bouvet, J.M., et al. (2007) Arbuscular mycorrhizal colonization and nodulation improve flooding tolerance in *Pterocarpus officinalis* Jacq. seedlings. *Mycorrhiza* 17(3):159–166.

Gaber, A., Yoshimura, K., Yamamoto, T., Yabuta, Y., Takeda, T., Miyasaka, H., Nakano, Y., Shigeoka, S. (2006) Glutathione peroxidase-like protein of *Synechocystis* PCC 6803 confers tolerance to oxidative and environmental stresses in transgenic *Arabidopsis*. *Physiologia Plantarum* 128(2):251–262.

Ghabooli, M., Khatabi, B., Ahmadi, F.S., Sepehri, M., Mirzaei, M., Amirkhani, A., Jorrín-Novo, J.V., Salekdeh, G.H. (2013) Proteomics study reveals the molecular mechanisms underlying water stress tolerance induced by *Piriformospora indica* in barley. *Journal of Proteomics* 94:289–301.

Glick, B.R. (2012) Plant growth-promoting bacteria: Mechanisms and applications. *Scientifica* 2012;963401.

Glick, B.R., Cheng, Z., Czarny, J., Duan, J. (2007) Promotion of plant growth by ACC deaminase-producing soil bacteria. *European Journal of Plant Pathology* 119(3):329–339.

Gomez, C.G., Valero, N.V., Brigard, R.C. (2012) Halotolerant/ alkalophilic bacteria associated with the cyanobacterium *Arthrospira platensis* (Nordstcdt) Gomont that promote early growth in *Sorghum bicolor* (L.). *Agronomía Colombiana* 30(1):111–115.

Gray, E.J., Smith, D.L. (2005) Intracellular and extracellular PGPR: Commonalities and distinctions in the plant-bacterium signalling processes. *Soil Biology and Biochemistry* 37(3):395–412.

Grichko, V.P., Filby, B., Glick, B.R. (2000) Increased ability of transgenic plants expressing the bacterial enzyme ACC deaminase to accumulate Cd, Co, Cu, Ni, Pb and Zn. *Journal of Biotechnology* 81(1):45–53.

Grover, M., Bodhankar, S., Maheswari, M., Srinivasarao, C. (2016) Actinomycetes as mitigators of climate change and abiotic stress. In *Growth promoting actinobacteria*, ed. Subramaniam G., Arumugam S., Rajendran V. 203–212. Singapore: Springer.

Grover, M., Madhubala, R., Ali, S.Z., Yadav, S.K., Venkateswarlu, B. (2014) Influence of *Bacillus* spp. strains on seedling growth and physiological parameters of sorghum under moisture stress conditions. *Journal of Basic Microbiology* 54(9):951–961.

Gupta, R., Saxena, R.K., Chaturvedi, P., Virdi, J.S. (1995) Chitinase production by *Streptomyces viridificans*: Its potential in fungal cell wall lysis. *The Journal of Applied Bacteriology* 78(4):378–383.

Gururani, M.A., Upadhyaya, C.P., Strasser, R.J., Yu, J.W., Park, S.W. (2013) Evaluation of abiotic stress tolerance in transgenic potato plants with reduced expression of PSII manganese stabilizing protein. *Plant Science: an International Journal of Experimental Plant Biology* 198:7–16.

Hamaoui, B., Abbadi, J.M., Burdman, S., Rashid, A., Sarig, S., Okon, Y. (2001) Effects of inoculation with Azospirillum brasilense on chickpeas (*Cicer arietinum*) and faba beans (*Vicia faba*) under different growth conditions. *Agronomie* 21(6–7):553–560.

Hamdia, M.A.E., Shaddad, M.A.K., Doaa, M.M. (2004) Mechanisms of salt tolerance and interactive effects of *Azospirillum brasilense* inoculation on maize cultivars grown under salt stress conditions. *Plant Growth Regulation* 44(2):165–174.

Handelsman, J., Stabb, E.V. (1996) Biocontrol of soilborne plant pathogens. *The Plant Cell* 8(10):1855–1869.

Handelsman, J., Tiedje, J., Alvarez-Cohen, L., Ashburner, M., Cann, I.K.O., Delong, E.F., et al. (2007) In *The new science of metagenomics: Revealing the secrets of our microbial planet*, ed. Washington,DC. National Academy of Science.

Hasanuzzaman, M., Nahar, K., Gill, S.S., Gill, R., Fujita, M. (2014) Drought stress responses in plants, oxidative stress, and antioxidant defense. In *Climate change and plant abiotic stress tolerance*, ed. Tuteja, N., and Gill, S.S. 209–249. Weinheim; Germany: Wiley-VCH Verlag GmbH & Co. KGaA.

Hashtroudi, M.S., Ghassempour, A., Riahi, H., Shariatmadari, Z., Khanjir, M. (2013) Endogenous auxin in plant growth-promoting cyanobacteria *Anabaena vaginicola* and *Nostoc calcicola*. *Journal of Applied Phycology* 25(2):379–386.

Hayashi, H., Alia, Mustardy, L., Deshnium, P., Ida, M., Murata, N. (1997) Transformation of *Arabidopsis thaliana* with the codA gene for choline oxidase: Accumulation of glycinebetaine and enhanced tolerance to salt and cold stress. *The Plant Journal: For Cell and Molecular Biology* 12(1):133–142.

Hayat, R., Ali, S., Amara, U., Khalid, R., Ahmed, I. (2010) Soil beneficial bacteria and their role in plant growth promotion: A review. *Annals of Microbiology* 60(4): 579–598.

Hayat, S., Hayat, Q., Alyemeni, M.N., Wani, A.S., Pichtel, J., Ahmad, A. (2012) Role of proline under changing environments: A review. *Plant Signaling and Behavior* 7(11):1456–1466.

Helena, M.H., Carvalho, C. (2008) Drought stress and reactive oxygen species. *Plant Signaling and Behavior* 3(3):156–165.

Hildebrandt, U., Kaldorf, M., Bothe, H. (1999) The zinc violet and its colonization by arbuscular mycorrhizal fungi. *Journal of Plant Physiology* 154(5–6):709–717.

Hopwood, D. (1990) Antibiotic biosyntheisis in *Streptomyces*. In *Genetics of bacterial diversity*, ed. Hopwood, D.A., and Chater, K. 129–148. London; United Kingdom: Academic Press.

Hottiger, T., Boller, T., Wiemken, A. (1987) Rapid changes of heat and desiccation tolerance correlated with changes of trehalose content in *Saccharomyces cerevisiae* cells subjected to temperature shifts. *FEBS Letters* 220(1):113–115.

Hounsa, C.G., Brandt, E.V., Thevelein, J., Hohmann, S., Prior, B.A. (1998) Role of trehalose in survival of *Saccharomyces cerevisiae* under osmotic stress. *Microbiology* 144(3):671–680.

Huang, J., Hirji, R., Adam, L., Rozwadowski, K.L., Hammerlindl, J.K., Keller, W.A., Selvaraj, G. (2000) Genetic engineering of glycine betaine production toward enhancing stress tolerance in plants: Metabolic limitations. *Plant Physiology* 122(3):747–756.

Hynes, R.K., Leung, G.C., Hirkala, D.L., Nelson, L.M. (2008) Isolation, selection, and characterization of beneficial rhizobacteria from pea, lentil and chickpea grown in Western Canadian. *Canadian Journal of Microbiology* 54(4):248–258.

Ibrahim, A.H., Abdel-Fattah, G.M., Eman, F.M., Abd El-Aziz, M.H., Shohr, A.E. (2011) Arbuscular mycorrhizal fungi and spermine alleviate the adverse effects of salinity stress on electrolyte leakage and productivity of wheat plants. *New Phytology* 51:261–276.

Idris, R., Trifonova, R., Puschenreiter, M., Wenzel, W.W., Sessitsch, A. (2004) Bacterial communities associated with flowering plants of the Ni hyperaccumulator *Thlaspi goesingense*. *Applied and Environmental Microbiology* 70(5):2667–2677.

Ishizaki-Nishizawa, O., Fujii, T., Azuma, M., Sekiguchi, K., Murata, N., Ohtani, T., Toguri, T. (1996) Low-temperature resistance of higher plants is significantly enhanced by a nonspecific cyanobacterial desaturase. *Nature Biotechnology* 14(8):1003–1006.

Jahanian, A., Chaichi, M.R., Rezaei, K., Rezayazdi, K., Khavazi, K. (2012) The effect of plant growth promoting rhizobacteria (pgpr) on germination and primary growth of artichoke (*Cynara scolymus*). *International Journal of Agriculture and Crop Sciences* 4:923–929.

Jang, I.C., Oh, S.J., Seo, J.S., Choi, W.B., Song, S.I., Kim, C.H., Kim, Y.S., et al. (2003) Expression of a bifunctional fusion of the *Escherichia coli* genes for trehalose-6-phosphate synthase and trehalose-6-phosphate phosphatase in transgenic rice plants increases trehalose accumulation and abiotic stress tolerance without stunting growth. *Plant Physiology* 131(2):516–524.

Jiang, J., Pan, C., Xiao, A., Yang, X., Zhang, G. (2017) Isolation, identification, and environmental adaptability of heavy-metal-resistant bacteria from ramie rhizosphere soil around mine refinery. *3 Biotech* 7(1):5.

Jo, J., Won, S.H., Son, D., Lee, B.H. (2004) Paraquat resistance of transgenic tobacco plants over-expressing the *Ochrobactrum anthropi* pqrA gene. *Biotechnology Letters* 26(18):1391–1396.

Johnson, P.R., Ecker, J.R. (1998) The ethylene gas signal transduction pathway: A molecular perspective. *Annual Review of Genetics* 32:227–254.

Jones, J.D.G., Dangl, J.L. (2006) The plant immune system. *Nature* 444(7117):323–329.

Joo, G.J., Kim, Y.M., Kim, J.T., Rhee, I.K., Kim, J.H., Lee, I.J. (2005) Gibberellins-producing rhizobacteria increase endogenous gibberellins content and promote growth of red peppers. *Journal of Microbiology* 43(6):510–515.

Kapardar, R.K., Ranjan, R., Grover, A., Puri, M., Sharma, R. (2010) Identification and characterization of genes conferring salt tolerance to *Escherichia coli* from pond water metagenome. *Bioresource Technology* 101(11):3917–3924.

Karim, S., Aronsson, H., Ericson, H., Pirhonen, M., Leyman, B., Welin, B., Mäntylä, E., et al. (2007) Improved drought tolerance without undesired side effects in transgenic plants producing trehalose. *Plant Molecular Biology* 64(4):371–386.

Kasim, W., Osman, M., Omar, M., Abd El-Daim, I., Bejai, S., Meijer, J. (2013) Control of drought stress in wheat using plant-growth promoting rhizobacteria. *Journal of Plant Growth Regulators* 32(1):122–130.

Kaushal, M., Wani, S.P. (2016) Plant-growth-promoting rhizobacteria: Drought stress alleviators to ameliorate crop production in drylands. *Annals of Microbiology* 66(1):35–42.

Kaushik, B.D. (1990) Cyanobacteria: Some aspects related to their role as ameliorating agent of salt effected soils. In *Prospective of phycology*, ed. Raja Rao, V.N. 405–410. New Delhi, India: Today and Tomorrow Publications.

Kaushik, B.D., Subhashini, D. (1985) Amelioration of salt affected soils with blue green algae: II, Improvement Soil Properties. *Proceedings of the Indian National Science Academy* Section B 51(3):386– 389.

Klaus-Joerger, T.K., Joerger, R., Olsson, E., Granqvist, C.G. (2001) Bacteria as workers in the living factory: Metal accumulating bacteria and their potential for material science. *Trends in Biotechnology* 19(1):15–20.

Kohler, J., Caravaca, F., Roldàn, A. (2010) An AM fungus and a PGPR intensify the adverse effects of salinity on the stability of rhizosphere soil aggregates of *Lactuca sativa*. *Soil Biology and Biochemistry* 42(3):429–434.

Kohler, J., Hernández, J.A., Caravaca, F., Roldán, A. (2008) Plant-growth-promoting rhizobacteria and arbuscular mycorrhizal fungi modify alleviation biochemical mechanisms in water-stressed plants. *Functional Plant Biology* 35(2):141–151.

Koike, N., Hyakumachi, M., Kageyama, K., Tsuyumu, S., Doke, N. (2001) Induction of systemic resistance in cucumber against several diseases by plant growth promoting fungi: Lignifications and superoxide generation. *European Journal of Plant Pathology* 107(5):523–533.

Kumari, N., Narayan, O.P., Rai, L.C. (2009) Understanding butachlor toxicity in *Aulosira fertilissima* using physiological, biochemical and proteomic approaches. *Chemosphere* 77(11):1501–1507.

Levin, D.E. (2011) Regulation of cell wall biogenesis in *Saccharomyces cerevisiae*: The cell wall integrity signaling pathway. *Genetics* 189(4):1145–1175.

Likhitkar, V.S., Tarar, J.L. (1995) Effect of presoaking seed treatment with *Nostoc muscorum* extracts on cotton. *Annual Review of Plant Physiology* 9:113–116.

Lilius, G., Holmberg, N., Bülow, L. (1996) Enhanced NaCl stress tolerance in transgenic tobacco expressing bacterial choline dehydrogenase. *Nature Biotechnology* 14(2):177–180.

Lim, J.H., Kim, S.D. (2013) Induction of drought stress resistance by multi-functional PGPR *Bacillus licheniformis* K11 in pepper. *The Plant Pathology Journal* 29(2):201–208.

Liu, X.D., Wei, Y., Zhou, X.Y., Pei, X., Zhang, S.H. (2015) *Aspergillus glaucus* aquaglyceroporin gene glpF confers high osmosis tolerance in heterologous organisms. *Applied and Environmental Microbiology* 81(19):6926–6937.

Loureiro, R.R., Reis, R.P., Marroig, R.G. (2014) Effect of the commercial extract of the brown alga *Ascophyllum nodosum* Mont. on *Kappaphycus alvarezii* (Doty) Doty ex P.C. Silva *in situ* submitted to lethal temperatures. *Journal of Applied Phycology* 26(1):629–634.

Luan, Y., Cui, J., Zhai, J., Li, J., Han, L., Meng, J. (2015) High-throughput sequencing reveals differential expression of miRNAs in tomato inoculated with *Phytophthora infestans*. *Planta* 241(6):1405–1416.

Lugtenberg, B., Kamilova, F. (2009) Plant-growth-promoting rhizobacteria. *Annual Review of Microbiology* 63:541–556.

Lutz, K.A., Knapp, J.E., Maliga, P. (2001) Expression of *bar* in the plastid genome confers herbicide resistance. *Plant Physiology* 125(4):1585–1590.

Lyon, B.R., Llewellyn, D.J., Huppatz, J.L., Dennis, E.S., Peacock, W.J. (1989) Expression of a bacterial gene in transgenic tobacco plants confers resistance to the herbicide 2,4-dichlorophenoxyacetic acid. *Plant Molecular Biology* 13(5):533–540.

Ma, Y., Rajkumar, M., Luo, Y., Freitas, H. (2011) Inoculation of endophytic bacteria on host and non-host plants-effects on plant growth and Ni uptake. *Journal of Hazardous Materials* 195:230–237.

Ma, Y., Rajkumar, M., Luo, Y., Freitas, H. (2013) Phytoextraction of heavy metal polluted soils using *Sedum plumbizincicola* inoculated with metal mobilizing *Phyllobacterium myrsinacearum* RC6b. *Chemosphere* 93(7):1386–1392.

Mahadevan, B., Crawford, D.L. (1997) Properties of the chitinase of the antifungal biocontrol agent *Streptomyces*

lydicus WYEC108. *Enzyme and Microbial Technology* 20(7):489–493.

Maheshwari, D.K., Dubey, R.C., Aeron, A., Kumar, B., Kumar, S., Tewari, S., Arora, N.K. (2012) Integrated approach for disease management and growth enhancement of *Sesamum indicum* L. utilizing *Azotobacter chroococcum* TRA2 and chemical fertilizer. *World Journal of Microbiology and Biotechnology* 28(10):3015–3024.

Manavalan, L.P., Guttikonda, S.K., Tran, L.S.P., Nguyen, H.T. (2009) Physiological and molecular approaches to improve drought resistance in soybean. *Plant and Cell Physiology* 50(7):1260–1276.

Mancuso, S., Azzarello, E., Mugnai, S., Briand, X. (2006) Marine bioactive substances (IPA extract) improve foliar ion uptake and water stress tolerance in potted *Vitis vinifera* plants. *Advances in Horticultural Science* 20(2):156–161.

Marquez, L.M., Redman, R.S., Rodriguez, R.J., Roossinck, M.J. (2007) A virus in a fungus in a plant: Three-way symbiosis required for thermal tolerance. *Science* 63:545–558.

Martínez, F., Arif, A., Nebauer, S.G., Bueso, E., Ali, R., Montesinos, C., Brunaud, V., Muñoz-Bertomeu, J., Serrano, R. (2015) A fungal transcription factor gene is expressed in plants from its own promoter and improves drought tolerance. *Planta* 242(1):39–52.

Martínez-Cortés, T., Pomar, F., Merino, F., Novo-Uzal, E. (2014) A proteomic approach to *Physcomitrella patens* rhizoid exudates. *Journal of Plant Physiology* 171(17):1671–1678.

Marulanda, A., Barea, J.M., Azcón, R. (2009) Stimulation of plant growth and drought tolerance by native microorganisms (AM Fungi and Bacteria) from dry environments: Mechanisms related to bacterial effectiveness. *Journal of Plant Growth Regulation* 28(2):115–124.

Mayak, S., Tirosh, T., Glick, B.R. (2004a) Plant growth-promoting bacteria confer resistance in tomato plants to salt stress. *Plant Physiology and Biochemistry: PPB* 42(6):565–572.

Mayak, S., Tirosh, T., Glick, B.R. (2004b) Plant growth-promoting bacteria that confer resistance to water stress in tomatoes and peppers. *Plant Science* 166(2):525–530.

Meera, M.S., Shivanna, M.B., Kageyama, K., Hyakumachi, M. (1995) Persistence of induced systemic resistance in cucumber in relation to root colonization by plant growth promoting fungal isolates. *Crop Protection* 14(2):123–130.

Metraux, J.P. (2001) Systemic acquired resistance and salicylic acid: Current state of knowledge. *European Journal of Plant Pathology* 107(1):13–18.

Mishra, Y., Chaurasia, N., Rai, L.C. (2009a) AhpC (alkyl hydroperoxide reductase) from *Anabaena* sp. PCC 7120 protects *Escherichia coli* from multiple abiotic stresses. *Biochemical and Biophysical Research Communications* 381(4):606–611.

Mishra, Y., Chaurasia, N., Rai, L.C. (2009b) Heat pretreatment alleviates UV-B toxicity in the cyanobacterium Anabaena doliolum: A proteomic analysis of cross tolerance. *Photochemistry and Photobiology* 85(3):824–833.

Moghaieb, R.E., Tanaka, N., Saneoka, H., Murooka, Y., Ono, H., Morikawa, H., Nakamura, A., et al. (2006) Characterization of salt tolerance in ectoine-transformed tobacco plants (*Nicotiana tabaccum*): Photosynthesis, osmotic adjustment, and nitrogen partitioning. *Plant, Cell and Environment* 29(2):173–182.

Mohanpuria, P., Rana, N.K., Yadav, S.K. (2007) Cadmium induced oxidative stress influence on glutathione metabolic genes of *Camellia sinensis* (L.) O. Kuntze. *Environmental Toxicology* 22(4):368–374.

Möller, M., Smith, M.L. (1998) The significance of the mineral component of seaweed suspensions on lettuce (*Lactuca sativa* L.) seedling growth. *Journal of Plant Physiology* 153(5–6):658–663.

Montero-Barrientos, M., Hermosa, R., Cardoza, R.E., Gutierrez, S., Nicolás, C., Monte, E. (2010) Transgenic expression of the *Trichoderma harzianum* HSP70 gene increases *Arabidopsis* resistance to heat and other abiotic stresses. *Journal of Plant Physiology* 167(8): 659–665.

Motohashi, T., Nagamiya, K., Prodhan, S.H., Nakao, K., Shishido, T., Yamamoto, Y., Moriwaki, T., et al. (2010) Production of salt stress tolerant rice by overexpression of the catalase gene, *katE*, derived from *Escherichia coli*. *Asia Pacific Journal of Molecular Biology and Biotechnology* 18(1):37–41.

Murali, M., Sudisha, J., Amruthesh, K.N., Ito, S.I., Shetty, H.S. (2013) Rhizosphere fungus *Penicillium chrysogenum* promotes growth and induces defence-related genes and downy mildew disease resistance in pearl millet. *Plant Biology* 15(1):111–118.

Murphy, J.F., Zehnder, G.W., Schuster, D.J., Sikora, E.J., Polston, J.E., Kloepper, J.W. (2000) Plant growth-promoting rhizobacterial mediated protection in tomato against tomato mottle virus. *Plant Disease* 84(7):779–784.

Nabati, D.A., Schmidt, R.E., Parrish, D.J. (1994) Alleviation of salinity stress in Kentucky bluegrass by plant growth regulators and iron. *Crop Science* 34(1):198–202.

Nadeem, S.M., Zahir, Z.A., Naveed, M., Arshad, M. (2007) Preliminary investigations on inducing salt tolerance in maize through inoculation with rhizobacteria containing ACC deaminase activity. *Canadian Journal of Microbiology* 53(10):1141–1149.

Nadeem, S.M., Zahir, Z.A., Naveed, M., Ashraf, M. (2010) Microbial ACC-deaminase: Prospects and applications for inducing salt tolerance in plants. *Critical Reviews in Plant Sciences* 29(6):360–393.

Nagamiya, K., Motohashi, T., Nakao, K., Prodhan, S.H., Hattori, E., Hirose, S., Ozawa, K., et al. (2007) Enhancement of salt tolerance in transgenic rice expressing an *Escherichia coli* catalase gene, katE. *Plant Biotechnology Reports* 1(1):49–55.

Nair, P., Kandasamy, S., Zhang, J., Ji, X., Kirby, C., Benkel, B., Hodges, M.D., et al. (2012) Transcriptional and metabolomic analysis of *Ascophyllum nodosum* mediated freezing tolerance in *Arabidopsis thaliana*. *BMC Genomics* 13(1):643.

Narayan, O.P., Kumari, N., Bhargava, P., Rajaram, H., Rai, L.C. (2016) A single gene *all3940* (Dps) overexpression in *Anabaena* sp. PCC 7120 confers multiple abiotic stress tolerance via proteomic alterations. *Functional and Integrative Genomics* 16(1):67–78.

Narayan, O.P., Kumari, N., Rai, L.C. (2010) Heterologous expression of *Anabaena* PCC 7120 *all*3940 (a Dps family gene) protects *Escherichia coli* from nutrient limitation and abiotic stresses. *Biochemical and Biophysical Research Communications* 394(1):163–169.

Naseem, H., Bano, A. (2014) Role of plant growth-promoting rhizobacteria and their exopolysaccharide in drought tolerance in maize. *Journal of Plant Interactions* 9(1):689–701.

Naveed, M., Mitter, B., Reichenauer, T.G., Wieczorek, K., Sessitsch, A. (2014) Increased drought stress resilience of maize through endophytic colonization by *Burkholderia phytofirmans* PsJN and *Enterobacter* sp. FD 17. *Environmental and Experimental Botany* 97:30–39.

Neeraja, C.N., Maghirang-Rodriguez, R., Pamplona, A., Heuer, S., Collard, B.C.Y., Septiningsih, E.M., Vergara, G., et al. (2007) A marker-assisted backcross approach for developing submergence-tolerant rice cultivars. *TAG. Theoretical and Applied Genetics. Theoretische und Angewandte Genetik* 115(6):767–776.

Neumann, P.M. (2008) Coping mechanisms for crop plants in drought-prone environments. *Annals of Botany* 101(7):901–907.

Nicaise, V., Roux, M., Zipfel, C. (2009) Recent advances in PAMP-triggered immunity against bacteria: Pattern recognition receptors watch over and raise the alarm. *Plant Physiology* 150(4):1638–1647.

Nihorimbere, V., Ongena, M., Smargiassi, M., Thonart, P. (2011) Beneficial effect of the rhizosphere microbial community for plant growth and health. *Biotechnology, Agronomy, and Society and Environment* 15(2): 327–337.

Orlova, I.V., Serebriiskaya, T.S., Popov, V., Merkulova, N., Nosov, A.M., Trunova, T.I., Tsydendambaev, V.D., Los, D.A. (2003) Transformation of tobacco with a gene for the thermophilic acyl-lipid desaturase enhances the chilling tolerance of plants. *Plant and Cell Physiology* 44(4):447–450.

Osundina, M.A. (1998) Nodulation and growth of mycorrhizal *Casuarina equisetifolia* J.R. and G. First in response to flooding. *Biology and Fertility of Soils* 26:95–99.

Panda, S.K., Choudhury, S. (2005) Chromium stress in plants. *Brazilian Journal of Plant Physiology* 17(1):95–102.

Pandey, S., Rai, R., Rai, L.C. (2012) Proteomics combines morphological, physiological and biochemical attribute to unravel the survival strategy of Anabaena sp. PCC7120 under arsenic stress. *Journal of Proteomics* 75(3):921–937.

Pandey, S., Rai, R., Rai, L.C. (2015) Biochemical and molecular basis of arsenic toxicity and tolerance in microbes and plants. In *Handbook of arsenic toxicology*, ed. Flora, S.J.S. 627–674. Amsterdam, Netherlands: Elsevier.

Pandey, S., Shrivastava, A.K., Rai, R., Rai, L.C. (2013a) Molecular characterization of Alr1105 a novel arsenate reductase of the diazotrophic cyanobacterium *Anabaena* sp. PCC7120 and decoding its role in abiotic stress management in *Escherichia coli*. *Plant Molecular Biology* 83(4–5):417–432.

Pandey, S., Shrivastava, A.K., Singh, V.K., Rai, R., Singh, P.K., Rai, S., Rai, L.C. (2013b) A new arsenate reductase involved in arsenic detoxification in *Anabaena* sp. PCC7120. *Functional and Integrative Genomics* 13(1):43–55.

Paul, D., Lade, H. (2014) Plant-growth-promoting rhizobacteria to improve crop growth in saline soils: A review. *Agronomy for Sustainable Development* 34(4):737–752.

Pereyra, M.A., García, P., Colabelli, M.N., Barassi, C.A., Creus, C.M. (2012) A better water status in wheat seedlings induced by *Azospirillum* under osmotic stress is related to morphological changes in xylem vessels of the coleoptile. *Applied Soil Ecology* 53(1):94–97.

Pereyra, M.A., Zalazar, C.A., Barassi, C.A. (2006) Root phospholipids in *Azospirillum* inoculated wheat seedlings exposed to water stress. *Plant Physiology and Biochemistry: PPB* 44(11–12):873–879.

Philippot, L., Raaijmakers, J.M., Lemanceau, P., van der Putten, W.H. (2013) Going back to the roots: The microbial ecology of the rhizosphere. *Nature Reviews. Microbiology* 11(11):789–799.

Pieterse, C.M.J., Leon-Reyes, A., Van der Ent, S., Van Wees, S.C.M. (2009) Networking by small-molecule hormones in plant immunity. *Nature Chemical Biology* 5(5):308–316.

Pilon-Smits, E.A.H., Terry, N., Sears, T., Kim, H., Zayed, A., Hwang, S., van Dun, K., et al. (1998) Trehalose producing transgenic tobacco plants show improved growth performance under drought stress. *Journal of Plant Physiology* 152(4–5):525–532.

Pilon-Smits, E.A.H., Terry, N., Sears Tobin, K.H., Van Dun, K. (1999) Enhanced drought resistance in fructan-producing sugar beet. *Plant Physiology and Biochemistry* 25:313–317.

Pishchik, V.N., Vorobyev, N.I., Chernyaeva, I.I., Timofeeva, S.V., Kozhemyakov, A.P., Alexeev, Y.V., Lukin, S.M. (2002) Experimental and mathematical simulation of plant growth promoting rhizobacteria and plant interaction under cadmium stress. *Plant and Soil* 243(2):173–186.

Porcel, R., Zamarreno, ÁM., Garcia-Mina, J.M., Aroca, R. (2014) Involvement of plant endogenous ABA in *Bacillus megaterium* PGPR activity in tomato plants. *BMC Plant Biology* 14:36.

Rai, L.C., Mallick, N. (1992) Removal and assessment of toxicity of Cu and Fe to *Anabaena doliolum* and *Chloreila vulgaris* using free and immobilized cells. *World Journal of Microbiology and Biotechnology* 8(2):110–114.

Rai, S., Agrawal, C., Shrivastava, A.K., Singh, P.K., Rai, L.C. (2014) Comparative proteomics unveils cross species variations in *Anabaena* under salt stress. *Journal of Proteomics* 98:254–270.

Rai, S., Singh, S., Shrivastava, A.K., Rai, L.C. (2013) Salt and UV-B induced changes in Anabaena PCC 7120: Physiological, proteomic and bioinformatic perspectives. *Photosynthesis Research* 118(1–2):105–114.

Rayirath, P., Benkel, B., Mark Hodges, D., Allan-Wojtas, P., MacKinnon, S., Critchley, A.T., Prithiviraj, B. (2009) Lipophilic components of the brown seaweed, *Ascophyllum nodosum*, enhance freezing tolerance in *Arabidopsis thaliana*. *Planta* 230(1):135–147.

Redman, R.S., Kim, Y.O., Woodward, C.J.D.A., Greer, C., Espino, L., Doty, S.L., Rodriguez, R.J. (2011) Increased fitness of rice plants to abiotic stress via habitat adapted symbiosis: A strategy for mitigating impacts of climate change. *PLOS ONE* 6(7):e14823.

Remy, W., Taylor, T.N., Hass, H., Kerp, H. (1994) Four hundred-million-year-old vesicular arbuscular mycorrhizae. *Proceedings of the National Academy of Sciences of the United States of America* 91(25):11841–11843.

Rodriguez, A.A., Stella, A.M., Storni, M.M., Zulpa, G., Zaccaro, M.C. (2006) Effects of cyanobacterial extracellular products and gibberellic acid on salinity tolerance in *Oryza sativa* L. saline Systems 2:7.

Rodriguez, H., Fraga, R. (1999) Phosphate solubilizing bacteria and their role in plant growth promotion. *Biotechnology Advances* 17(4–5):319–339.

Rodriguez, R.J., Henson, J., Van Volkenburgh, E., Hoy, M., Wright, L., Beckwith, F., Kim, Y.O., Redman, R.S. (2008) Stress tolerance in plants via habitat-adapted symbiosis. *The ISME Journal* 2(4):404–416.

Rogers, M.E., Grieve, C.M., Shannon, M.C. (2003) Plant growth and ion relations in Lucerne (*Medicago sativa* L.) in response to the combined effects of NaCl and P. *Plant and Soil* 253(1):187–194.

Rojas-Tapias, D., Moreno-Galván, A., Pardo-Díaz, S., Obando, M., Rivera, D., Bonilla, R. (2012) Effect of inoculation with plant growth-promoting bacteria (PGPB) on amelioration of saline stress in maize (*Zea mays*). *Applied Soil Ecology* 61:264–272.

Romero, A.M., Vega, D., Correa, O.S. (2014) *Azospirillum brasilense* mitigates water stress imposed by a vascular disease by increasing xylem vessel area and stem hydraulic conductivity in tomato. *Applied Soil Ecology* 82:38–43.

Rugh, C.L., Wilde, H.D., Stack, N.M., Thompson, D.M., Summers, A.O., Meagher, R.B. (1996) Mercuric ion reduction and resistance in transgenic *Arabidopsis thaliana* plants expressing a modified bacterial merA gene. *Proceedings of the National Academy of Sciences of the United States of America* 93(8):3182–3187.

Ruiz-Lozano, J.M., Aroca, R. (2010) Modulation of aquaporin genes by the arbuscular mycorrhizal symbiosis in relation to osmotic stress tolerance. In *Symbioses and stress: Joint ventures in biology, cellular origin, life in extreme habitats and astrobiology*, ed. Seckbach J., Grube M. 359–374. Dordrecht, the Netherlands: Springer Science, Business Media.

Ruiz-Lozano, J.M., Azcön, R. (2000) Symbiotic efficiency and infectivity of an autochthonous arbuscular mycorrhizal

Glomus sp. from saline soils and *Glomus deserticola* under salinity. *Mycorrhiza* 10(3):137–143.

Ruiz-Lozano, J.M., Collados, C., Barea, J.M., Azcon, R. (2001) Arbuscular mycorrhizal symbiosis can alleviate drought-induced nodule senescence in soybean plants. *New Phytologist* 151(2):493–502.

Rus, A.M., Estañ, M.T., Gisbert, C., Garcia-Sogo, B., Serrano, R., Caro, M., Moreno, V., Bolarín, M.C. (2001) Expressing the yeast *HAL1* gene in tomato increases fruit yield and enhances K⁺/Na⁺ selectivity under salt stress. *Plant, Cell and Environment* 24(8):875–880.

Russo, A., Vettori, L., Felici, C., Fiaschi, G., Morini, S., Toffanin, A. (2008) Enhanced micropropagation response and biocontrol effect of *Azospirillum brasilense* Sp245 on *Prunus cerasifera* L. clone Mr.S 2/5 plants. *Journal of Biotechnology* 134(3–4):312–319.

Rutto, K.L., Mizutani, F., Kadoya, K. (2002) Effect of root-zone flooding on mycorrhizal and non-mycorrhiozal peach (*Prunus persica* Batsch) seedlings. *Scientia Horticulturae* 94(3–4):285–295.

Saa, S., Olivos-Del Rio, A., Castro, S., Brown, P.H. (2015) Foliar application of microbial and plant based biostimulants increases growth and potassium uptake in almond (*Prunus dulcis* [Mill.] D.A. Webb). *Frontiers in Plant Science* 6:87.

Sabir, P., Ashraf, M., Hussain, M., Jamil, A. (2009) Relationship of photosynthetic pigments and water relations with salt tolerance of proso millet (*Panicum Miliaceum* L.) accessions. *Pakistan Journal of Botany* 41(6):2957–2964.

Sagar, S., Dwivedi, A., Yadav, S., Tripathi, M., Kaistha, S.D. (2012) Hexavalent chromium reduction and plant growth promotion by *Staphylococcus arlettae* strain Cr11. *Chemosphere* 86(8):847–852.

Saito, H., Posas, F. (2012) Response to hyperosmotic stress. *Genetics* 192(2):289–318.

Saldajeno, M.G.B., Hyakumachi, M. (2011) The plant growth promoting fungus *Fusarium equiseti* and the arbuscular mycorrhizal fungus *Glomus mosseae* stimulate plant growth and reduce severity of anthracnose and damping-off diseases in cucumber (*Cucumis sativus*) seedlings. *Annals of Applied Biology* 159(1):28–40.

Sand, W., Rohde, K., Sabotke, B., Zenneck, C. (1992) Evaluation of *Leptospirillum ferroxidans* for leaching. *Applied and Environmental Microbiology* 58(1):85–92.

Sandhya, V., Ali, A.S., Grover, M., Reddy, G., Venkateswarlu, B. (2009) Alleviation of drought stress effects in sunflower seedlings by the exopolysaccharides producing *Pseudomonas putida* strain GAP-P45. *Biology and Fertility of Soils* 46(1):17–26.

Sandhya, V., Ali, S.Z., Grover, M., Reddy, G., Venkateswarlu, B. (2010) Effect of plant growth promoting *Pseudomonas* spp. on compatible solutes, antioxidant status and plant growth of maize under drought stress. *Plant Growth Regulation* 62(1):21–30.

Saravanakumar, D., Kavino, M., Raguchander, T., Subbian, P., Samiyappan, R. (2011) Plant growth promoting bacteria enhance water stress resistance in green gram plants. *Acta Physiologiae Plantarum* 33(1):203–209.

Saravanakumar, D., Samiyappan, R. (2007) ACC deaminase from *Pseudomonas fluorescens* mediated saline resistance in groundnut (*Arachis hypogaea*) plants. *Journal of Applied Microbiology* 102(5):1283–1292.

Sarma, R.K., Saikia, R. (2014) Alleviation of drought stress in mung bean by strain *Pseudomonas aeruginosa* GGRJ21. *Plant and Soil* 377(1–2):111–126.

Sarmishta, D., P., B.M., S., B.P. (2015) Copper induced changes in growth and antioxidative mechanisms of tea plant (*Camellia sinensis* (L.) O. Kuntze). *African Journal of Biotechnology* 14(7):582–592.

Schweder,T. Markert, S., Hecker, M. (2008) Proteomics of marine bacteria. *Electrophoresis* 29(12):2603–2616.

Sehrawat, A., Gupta, R., Deswal, R. (2013) Nitric oxide-cold stress signalling cross-talk, evolution of a novel regulatory mechanism. *Proteomics* 13(12–13):1816–1835.

Sen, S., Rai, S., Yadav, S., Agrawal, C., Rai, R., Chatterjee, A., Rai, L.C. (2017) Dehydration and rehydration - Induced temporal changes in cytosolic and membrane proteome of the nitrogen fixing cyanobacterium *Anabaena* sp. PCC 7120. *Algal Research* 27:244–258.

Sergeeva, E., Shah, S., Glick, B.R. (2006) Growth of transgenic canola (Brassica napus cv. Westar) expressing a bacterial 1-aminocyclopropane-1-carboxylate (ACC) deaminase gene on high concentrations of salt. *World Journal of Microbiology and Biotechnology* 22(3): 277–282.

Serraj, R., Sinclair, T.R. (2002) Osmolyte accumulation: Can it really help increase crop yield under drought conditions? *Plant, Cell and Environment* 25(2):333–341.

Sessitsch, A., Hardoim, P., Döring, J., Weilharter, A., Krause, A., Woyke, T., et al. (2012) Functional characteristics of an endophyte community colonizing roots as revealed by metagenomic analysis. *Molecular Plant–Microbe Interactions* 25(1):28–36.

Shaharoona, B., Arshad, M., Zahir, Z.A. (2006) Effect of plant growth promoting rhizobacteria containing ACC-deaminase on maize (Zea mays L.) growth under axenic conditions and on nodulation in mung bean (*Vigna radiata* L.). *Letters in Applied Microbiology* 42(2):155–159.

Shariatmadari, Z., Riahi, H., Hastroudi, M.S., Ghassempour, A., Aghashariatmadary, Z. (2013) Plant growth promoting cyanobacteria and their distribution in terrestrial habitats of Iran. *Soil Science and Plant Nutrition* 59(4):535–547.

Sheibani-Tezerji, R., Rattei, T., Sessitsch, A., Trognitz, F., Mitter, B. (2015) Transcriptome profiling of the endophyte *Burkholderia phytofirmans* PsJN indicates sensing of the plant environment and drought stress. *mBio* 6(5):e00621–e00615.

Shoresh, M., Harman, G.E., Mastouri, F. (2010) Induced systemic resistance and plant responses to fungal biocontrol agents. *Annual Review of Phytopathology* 48:21–43.

Shrivastava, A.K., Chatterjee, A., Yadav, S., Singh, P.K., Singh, S., Rai, L.C. (2015a) UV-B stress induced metabolic rearrangements explored with comparative proteomics in three Anabaena species. *Journal of Proteomics* 127(A):122–133.

Shrivastava, A.K., Pandey, S., Dietz, K.J., Singh, P.K., Singh, S., Rai, R., Rai, L.C. (2016a) Overexpression of AhpC enhances stress tolerance and N_2-fixation in *Anabaena* by upregulating stress responsive genes. *Biochimica et Biophysica Acta (BBA)* 1860(11 A):2576–2588.

Shrivastava, A.K., Pandey, S., Singh, P.K., Rai, S., Rai, L.C. (2012) *alr0882* encoding a hypothetical protein of Anabaena PCC7120 protects Escherichia coli from nutrient starvation and abiotic stresses. *Gene* 511(2):248–255.

Shrivastava, A.K., Pandey, S., Yadav, S., Mishra, Y., Singh, P.K., Rai, R., Singh, S., Rai, S., Rai, L.C. (2016b) Comparative proteomics of wild type, An+ahpC and AnΔahpC strains of *Anabaena* sp. PCC 7120 demonstrates AhpC mediated augmentation of photosynthesis, N2-fixation and modulation of regulatory network of antioxidative proteins. *Journal of Proteomics* 140:81–99.

Shrivastava, A.K., Singh, S., Singh, P.K., Pandey, S., Rai, L.C. (2015b) A novel alkyl hydroperoxidase (AhpD) of *Anabaena* PCC7120 confers abiotic stress tolerance in Escherichia coli. *Functional and Integrative Genomics* 15(1):77–92.

Siddikee, M.A., Glick, B.R., Chauhan, P.S., Yim, Wj, Sa, T. (2011) Enhancement of growth and salt tolerance of red pepper seedlings (*Capsicum annuum* L.) by regulating stress ethylene synthesis with halotolerant bacteria containing 1-aminocyclopropane-1-carboxylic acid deaminase activity. *Plant Physiology and Biochemistry: PPB* 49(4):427–434.

Silva-Sanchez, C., Li, H., Chen, S. (2015) Recent advances and challenges in plant phosphoproteomics. *Proteomics* 15(5–6):1127–1141.

Simon, L., Bousquet, J., Levesque, R.C., Lalonde, M. (1993) Origin and diversification of endomycorrhizal fungi and coincidence with vascular land plants. *Nature* 363(6424):67–69.

Simova-Stoilova, L., Demirevska, K., Petrova, T., Tsenov, N., Feller, U. (2008) Antioxidative protection in wheat varieties under severe recoverable drought at seedling stage. *Plant, Soil and Environment* 54(12):529–536.

Singh, L.P., Gill, S.S., Tuteja, N. (2011) Unraveling the role of fungal symbionts in plant abiotic stress tolerance. *Plant Signaling and Behavior* 6(2):175–191.

Singh, P., Tiwari, A., Singh, S.P., Asthana, R.K. (2013) Desiccation induced changes in osmolytes production and the antioxidative defence in the cyanobacterium *Anabaena* sp. PCC 7120. *Physiology and Molecular Biology of Plants: an International Journal of Functional Plant Biology* 19(1):61–68.

Singh, P.K., Shrivastava, A.K., Chatterjee, A., Pandey, S., Rai, S., Singh, S., Rai, L.C. (2015) Cadmium toxicity in diazotrophic *Anabaena* spp. adjudged by hasty up-accumulation of transporter and signaling and severe down-accumulation of nitrogen metabolism proteins. *Journal of Proteomics* 127(A):134–146.

Singh, P.K., Shrivastava, A.K., Singh, S., Rai, R., Chatterjee, A., Rai, L.C. (2017) Alr2954 of *Anabaena* sp. PCC 7120 with ADP-ribose pyrophosphatase activity bestows abiotic stress tolerance in *Escherichia coli*. *Functional and Integrative Genomics* 17(1):39–52.

Singh, S. (2014) A review on possible elicitor molecules of cyanobacteria: Their role in improving plant growth and providing tolerance against biotic or abiotic stress. *Journal of Applied Microbiology* 117(5):1221–1244.

Sinha, S., Mukherjee, S.K. (2008) Cadmium-induced siderophore production by a high Cd-resistant bacterial strain relieved Cd toxicity in plants through root colonization. *Current Microbiology* 56(1):55–60.

Smith, S.E., Read, D.J. (2008) In *Mycorrhizal symbiosis*, ed. Smith, S.E., and Read, D.J. 1–800. London: Academic Press.

Sokhansandzh, A., Neumyvakin, L.V., Moseĭko, N.A., Piruzian, E.S. (1997) Transfer of bacterial genes for proline synthesis in plants and their expression by various plant promotors. *Genetika* 33(7):906–913.

Sorty, A.M., Meena, K.K., Choudhary, K., Bitla, U.M., Minhas, P.S., Krishnani, K.K. (2016) Effect of plant growth promoting bacteria associated with halophytic weed (*Psoralea corylifolia* L) on germination and seedling growth of wheat under saline conditions. *Applied Biochemistry and Biotechnology* 180(5):872–882.

Spann, T.M., Little, H.A. (2011) Applications of a commercial extract of the brown seaweed *Ascophyllum nodosum* increases drought tolerance in container-grown "Hamlin" sweet orange nursery trees. *Horticultural Science* 46(4):577–582.

Spoel, S.H., Dong, X. (2012) How do plants achieve immunity? Defence without specialized immune cells. *Nature Reviews. Immunology* 12(2):89–100.

Sreenivasa, M.N., Bagyaraj, D.J. (1989) Use of pesticides for mass production of vesicular-arbuscular mycorrhizal inoculum. *Plant and Soil* 119(1):127–132.

Srivastava, A.K., Bhargava, P., Kumar, A., Rai, L.C., Neilan, B.A. (2009) Molecular characterization and the effect of salinity on cyanobacterial diversity in the rice fields of Eastern Uttar Pradesh, India. *Saline Systems* 5:4.

Srivastava, A.K., Bhargava, P., Mishra, Y., Shukla, B., Rai, L.C. (2006) Effect of pretreatment of salt, copper and temperature on ultraviolet-B induced antioxidants in diazotrophic cyanobacterium Anabaena doliolum. *Journal of Basic Microbiology* 46(2):135–144.

Streber, W.R., Kutschka, U., Thomas, F., Pohlenz, H.D. (1994) Expression of a bacterial gene in transgenic plants confers resistance to the herbicide phenmedipham. *Plant Molecular Biology* 25(6):977–987.

Suarez, C., Cardinale, M., Ratering, S., Steffens, D., Jung, S., Montoya, A.M.Z., Geissler-Plaum, R., Schnell, S. (2015) Plant growth-promoting effects of Hartmannibacter diazotrophicus on summer barley (Hordeum vulgare L.) under salt stress. *Applied Soil Ecology* 95:23–30.

Subramanian, P., Kim, K., Krishnamoorthy, R., Mageswari, A., Selvakumar, G., Sa, T. (2016) Cold stress tolerance in psychrotolerant soil bacteria and their conferred chilling resistance in tomato (Solanum Lycopersicum Mill.) under low temperatures. *PLOS ONE* 11(8):0161592:.

Sugino, M., Hibino, T., Tanaka, Y.N., Nii, N., Takabe, T., Takabe, T. (1999) Overexpression of DnaK from a halotolerant cyanobacterium Aphanothece halophytice acquires resistance to salt stress in transgenic tobacco plants. *Plant Science* 146(2):81–88.

Sukweenadhi, J., Kim, Y.J., Choi, E.S., Koh, S.C., Lee, S.W., Kim, Y.J., Yang, D.C. (2015) Paenibacillus yonginensis DCY84(T) induces changes in Arabidopsis thaliana gene expression against aluminum, drought, and salt stress. *Microbiological Research* 172:7–15.

Sun, C., Johnson, J.M., Cai, D., Sherameti, I., Oelmüller, R., Lou, B. (2010) Piriformospora indica confers drought tolerance in Chinese cabbage leaves by stimulating antioxidant enzymes, the expression of drought-related genes and the plastid-localized CAS protein. *Journal of Plant Physiology* 167(12):1009–1017.

Sziderics, A.H., Rasche, F., Trognitz, F., Sessitsch, A., Wilhelm, E. (2007) Bacterial endophytes contribute to abiotic stress adaptation in pepper plants (*Capsicum annuum* L.). *Canadian Journal of Microbiology* 53(11):1195–1202.

Tanaka, Y., Hibin, T., Hayashi, Y., Tanaka, A., Kishitani, S., Takabe, T., Yokota, S., Takabe, T. (1999) Salt tolerance of transgenic rice overexpressing yeast mitochondrial *Mn-SOD* in chloroplasts. *Plant Science* 148(2):131–138.

Tank, N., Saraf, M. (2009) Enhancement of plant growth and decontamination of nickel-spiked soil using PGPR. *Journal of Basic Microbiology* 49(2):195–204.

Theocharis, A., Bordiec, S., Fernandez, O., Paquis, S., Dhondt-Cordelier, S., Baillieul, F., Clément, C., Barka, E.A. (2012) *Burkholderia phytofirmans* PsJN primes vitis vinifera L. and confers a better tolerance to low non freezing temperatures. *Molecular Plant-Microbe Interactions* 25(2):241–249.

Thomas, J.C., Sepahi, M., Arendall, B., Bohnert, H.J. (1995) Enhancement of seed germination in high salinity by engineering mannitol expression in *Arabidopsis thaliana*. *Plant, Cell and Environment* 18(7):801–806.

Thomashow, L.S., Weller, D.M. (1996) Current concepts in the use of introduced bacteria for biological disease control: Mechanisms and antifungal metabolites. In *Plant-microbe interactions. N.T.*, ed. Stacey, G., and Keen 187–235. New York, NY: Chapman & Hall.

Timmusk, S., Abd El-Daim, I.A., Copolovici, L., Tanilas, T., Kannaste, A., Behers, L., Nevo, E., et al. (2014) Drought-tolerance of wheat improved by rhizosphere bacteria from harsh environments: Enhanced biomass production and reduced emissions of stress volatiles. *PLOS ONE* 9(5):e96086.

Timmusk, S., Gerhart, E., Wagner, H. (1999) The plant-growth-promoting rhizobacterium *Paenibacillus polymyxa* induces changes in *Arabidopsis thaliana* gene expression: A possible connection between biotic and abiotic stress responses. *Molecular Plant-Microbe Interactions Journal* 12(11):951–959.

Tognetti, V.B., Palatnik, J.F., Fillat, M.F., Melzer, M., Hajirezaei, M.R., Valle, E.M., Carrillo, N. (2006) Functional replacement of ferredoxin by a cyanobacterial flavodoxin in tobacco confers broad-range stress tolerance. *The Plant Cell* 18(8):2035–2050.

Tokala, R.K., Strap, J.L., Jung, C.M., Crawford, D.L., Salove, M.H., Deobald, L.A., Bailey, J.F., Morra, M.J. (2002) Novel plant-microbe rhizosphere interaction involving *S. lydicus* WYEC108 and the pea plant (*Pisum sativum*). *Applied and Environmental Microbiology* 68(5):2161–2171.

Trejo-Estrada, S.R., Paszczynski, A., Crawford, D.L. (1998) Antibiotics and enzymes produced by the biological control agent *Streptomyces violaceusniger* YCED-9. *Journal of Industrial Microbiology and Biotechnology* 21(1–2):81–90.

Tsonev, T., Lidon, F.J.C. (2012) Zinc in plants - An overview. *Emirates Journal of Food and Agriculture* 24(4):322–333.

Tucci, M., Ruocco, M., De Masi, L., De Palma, M., Lorito, M. (2011) The beneficial effect of *Trichoderma* spp. on tomato is modulated by the plant genotype. *Molecular Plant Pathology* 12(4):341–354.

Van Oosten, M.J., Pepe, O., De Pascale, S., Silletti, S., Maggio, A. (2017) The role of biostimulants and bioeffectors as alleviators of abiotic stress in crop plants. *Chemical and Biological Technologies in Agriculture* 4(1):5.

VanWees, S.C.M., Van der Ent, S., Pieterse, C.M.J. (2008) Plant immune responses triggered by beneficial microbes. *Current Opinion in Plant Biology* 11(4):443–448.

Vardharajula, S., Ali, S.Z., Grover, M., Reddy, G., Bandi, V. (2011) Drought-tolerant plant growth promoting *Bacillus* spp: Effect on growth osmolytes, and antioxidant status of maize under drought stress. *Journal of Plant Interactions* 6(1):1–14.

Vezzaa, M.E., Llanesb, A., Travagliab, C., Agostinia, E., Talanoa, M.A. (2018) Arsenic stress effects on root water absorption in soybean plants: Physiological and morphological aspects. *Plant Physiology and Biochemistry: PPB* 123:8–17.

Vieira, P.M., Santos, M.P., Andrade, C.M., Souza-Neto, O.A., Ulhoa, C.J., Aragão, F.J.L. (2017) Overexpression of an aquaglyceroporin gene from *Trichoderma harzianum* improves water-use efficiency and drought tolerance in *Nicotiana tabacum*. *Plant Physiology and Biochemistry: PPB* 121:38–47.

Waditee, R., Bhuiyan, M.N.H., Rai, V., Aoki, K., Tanaka, Y., Hibino, T., Suzuki, S., et al. (2004) Genes for direct methylation of glycine provide high levels of glycinebetaine and abiotic-stress tolerance in in *Synechococcus* and *Arabidopsis*. *PNAS* 102(5):1318–1323.

Wagner, T., Windhövel, U., Römer, S. (2002) Transformation of tobacco with a mutated cyanobacterial phytoene desaturase gene confers resistance to bleaching herbicides. *Zeitschrift für Naturforschung C* 57(7–8):671–679.

Waltz, E. (2014) Beating the heat. *Nature Biotechnology* 32(7):610–613.

Wang, C.J., Yang, W., Wang, C., Gu, C., Niu, D.D., Liu, H.X., Wang, Y.P., Guo, J.H. (2012) Induction of drought tolerance in cucumber plants by a consortium of three plant growth-promoting rhizobacterium strains. *PloS One* 7(12):e52565.

Wang, H.Y., Li, Y.F., Xie, L.X., Xu, P. (2003) Expression of a bacterial aroA mutant, aroA-M1, encoding 5-enolpyruvylshikimate-3-phosphate synthase for the production of glyphosate-resistant tobacco plants. *Journal of Plant Research* 116(6):455–460.

Wani, P.A., Khan, M.S. (2010) *Bacillus* species enhance growth parameters of chickpea (*Cicer arietinum* L.) in chromium stressed soils. *Food and Chemical Toxicology: An International Journal Published for the British Industrial Biological Research Association* 48(11):3262–3267.

West, C.P., Izekor, E., Turner, K.E., Elmi, A.A. (1993) Endophyte effects on growth and persistence of tall fescue along a water-supply gradient. *Agronomy Journal* 85(2):264–270.

Westover, K.M., Kennedy, A.C., Kelly, S.E. (1997) Patterns of rhizosphere microbial community structure associated with co-occuring plant species. *The Journal of Ecology* 85(6):863–873.

Whipps, J.M. (2001) Microbial interactions and biocontrol in the rhizosphere. *Journal of Experimental Botany* 52(Spec Issue):487–511.

Xu, C., Leskovar, D.I. (2015) Effects of A. nodosum seaweed extracts on spinach growth, physiology and nutrition value under drought stress. *Scientia Horticulturae* 183(183):39–47.

Yang, J., Kloepper, J.W., Ryu, C.M. (2009) Rhizosphere bacteria help plants tolerate abiotic stress. *Trends in Plant Science* 14(1):1–4.

Yasmin, H., Bano, A., Samiullah, A. (2013) Screening of PGPR isolates from semi-arid region and their implication to alleviate drought stress. *Pakistan Journal of Botany* 45:51–58.

Yooyongwech, S., Phaukinsang, N., Cha-Um, S., Supaibulwatana, K. (2013) Arbuscular mycorrhiza improved growth performance in *Macadamia tetraphylla* L. grown under water deficit stress involves soluble sugar and proline accumulation. *Plant Growth Regulation* 69(3):285–293.

Yoshimura, K., Miyao, K., Gaber, A., Takeda, T., Kanaboshi, H., Miyasaka, H., Shigeoka, S. (2004) Enhancement of stress tolerance in transgenic tobacco plants overexpressing *Chlamydomonas* glutathione peroxidase in chloroplasts or cytosol. *The Plant Journal: for Cell and Molecular Biology* 37(1):21–33.

Zaccaro, M.C., Caire, G., Cano, M., Halperin, D. (1991) Bioactive compounds from *Nostoc muscorum* (Cyanobacteria). *Cytobios* 66:169–172.

Zamioudis, C., Pieterse, C.M.J. (2012) Modulation of host immunity by beneficial microbes. *Molecular Plant-Microbe Interactions: MPMI* 25(2):139–150.

Zhang, F., Dashti, N., Hynes, R.K., Smith, D.L. (1997) Plant growth promoting rhizobacteria and soybean [*Glycine max* (L.) Merr] growth and physiology at suboptimal root zone temperatures. *Annals of Botany* 79(3): 243–249.

Zhang, H., Murzello, C., Sun, Y., Kim, M.-S., Xie, X., Jeter, R.M., Zak, J.C., Dowd, S.E., Paré, P.W. (2010) Choline and osmotic-stress tolerance induced in *Arabidopsis* by the soil microbe *Bacillus subtilis* (GB03). *Molecular Plant-Microbe Interactions: MPMI* 23(8): 1097–1104.

Zhang, Y., Zhao, L., Wang, Y., Yang, B., Chen, S. (2008) Enhancement of heavy metal accumulation by tissue specific co-expression of iaaM and ACC deaminase genes in plants. *Chemosphere* 72(4):564–571.

Zhou, S., Hu, W., Deng, X., Ma, Z., Chen, L., Huang, C., Wang, J., et al. (2012) Overexpression of wheat aquaporin gene, TaAQP7 enhances drought tolerance in transgenic tobacco. *PLOS ONE* 7(12):e52439.

Zhu, J.K. (2002) Salt and drought stress signal transduction in plants. *Annual Review of Plant Biology* 53:247–273.

Index

Milton Keynes UK
Ingram Content Group UK Ltd.
UKHW050608161024
449569UK00046B/1525